21st Century Astronomy's text, art, and media work together to help students see the universe through the eyes of a scientist.

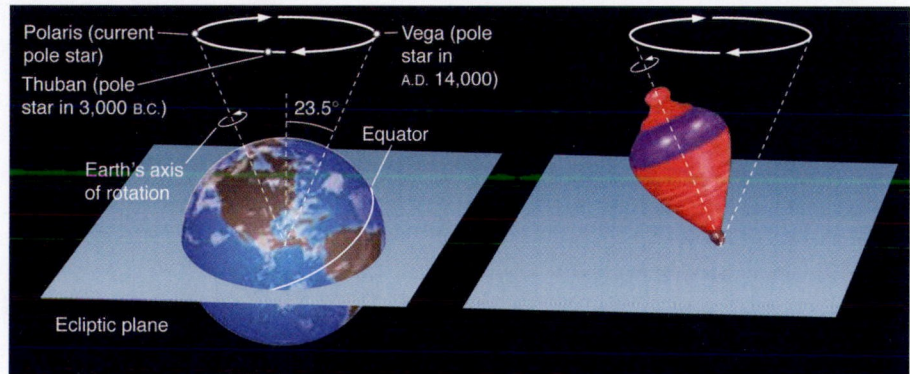

▶❙❙ **AstroTour** animations help students visualize physical processes.

S **StudySpace** is a free and open website that provides study plans that tie together animations, interactive simulations, diagnostic quizzes, vocabulary flash cards, news feeds, and more! Visit **www.wwnorton.com/ studyspace.**

⚙ **SmartWork** online homework provides students with feedback right when they need it, and it offers links to the ebook version of the text. To log in or purchase a registration code, visit **smartwork.wwnorton.com.**

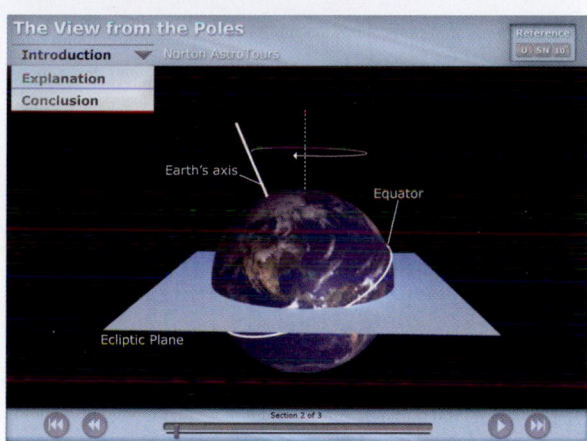

Both text and narrative art help students to understand physical concepts . . .

. . . while AstroTours use animation and interactivity to help students visualize key processes . . .

. . . and SmartWork online homework extends the text's emphasis on reasoning and critical thinking.

21ST CENTURY
ASTRONOMY

THIRD EDITION

THIRD EDITION

21ST CENTURY
ASTRONOMY

Jeff Hester • ARIZONA STATE UNIVERSITY

Brad Smith • UNIVERSITY OF HAWAII—MANOA

George Blumenthal • UNIVERSITY OF CALIFORNIA—SANTA CRUZ

Laura Kay • BARNARD COLLEGE

Howard Voss • ARIZONA STATE UNIVERSITY

 W. W. NORTON & COMPANY • NEW YORK • LONDON

Editor: Rob Bellinger
Senior Editor, Physical Science: Erik Fahlgren
Editorial Assistant: Mary Lynch
Managing Editor, College: Marian Johnson
Associate Managing Editor, College: Kim Yi
Copy Editor: Stephanie Hiebert
Marketing Manager: Kelsey Volker
Developmental Editors: Patricia Longoria, Andrew Sobel
Science Media Editor: Rob Bellinger
Assistant Editor, Electronic Media: Laura Musich
Supplements Editor: Matthew A. Freeman
Senior Production Manager: Jane Searle
Art Director and Designer: Rubina Yeh
Photo Researchers: Junenoire Mitchell, Ramón Rivera Moret
Illustrations: Penumbra Design
Layout: Carole Desnoes
Compositor: Progressive Information Technologies
Manufacturing: R. R. Donnelley & Sons—Willard, Ohio

Library of Congress Cataloging-in-Publication Data
21st century astronomy / Jeff Hester . . . [et al.].—3rd ed.
     p. cm.
   Includes index.
   ISBN: 978-0-393-93198-3 (pbk.)
1. Astronomy.   I. Hester, John Jeffrey.   I I. Title: Twenty-first century astronomy.
QB45.2.A14 2010
520—dc22   2009045134

W. W. Norton & Company, Inc., 500 Fifth Avenue, New York, N.Y. 10110
www.wwnorton.com
W. W. Norton & Company Ltd., Castle House, 75/76 Wells Street, London W1T 3QT

2 3 4 5 6 7 8 9 0

George Blumenthal gratefully thanks his wife, Kelly Weisberg, and his children, Aaron and Sarah Blumenthal, for their support during this project.

Jeff Hester thanks the students over the years who allowed him the pleasure and the privilege of sharing with them something of the extraordinary universe in which we live. He also acknowledges his wife Vicki, his family, and the friends who offered much needed support through trying times.

Laura Kay thanks her partner, M.P.M.

Brad Smith dedicates this Third Edition to his patient and understanding wife, Diane McGregor.

Howard Voss dedicates this book to his wife, Helen Ann, who cheerfully sacrificed much so that he could attend to his teaching and other professional work. Helen Ann passed away at the time of completion of the Second Edition. She is greatly missed by Howard, their family, and her many friends.

Brief Contents

Contents

PART II   The Solar System

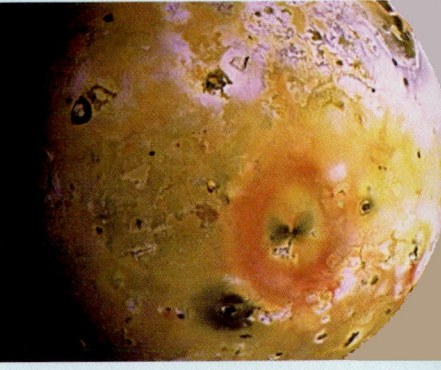

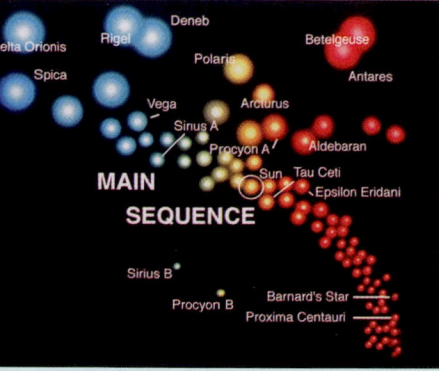

MAIN
SEQUENCE

CHAPTER 14 A Run-of-the-Mill G-Type Star—Our Sun

PART IV   Galaxies, the Universe,
and Cosmology

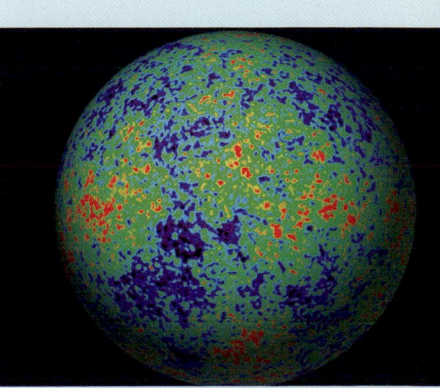

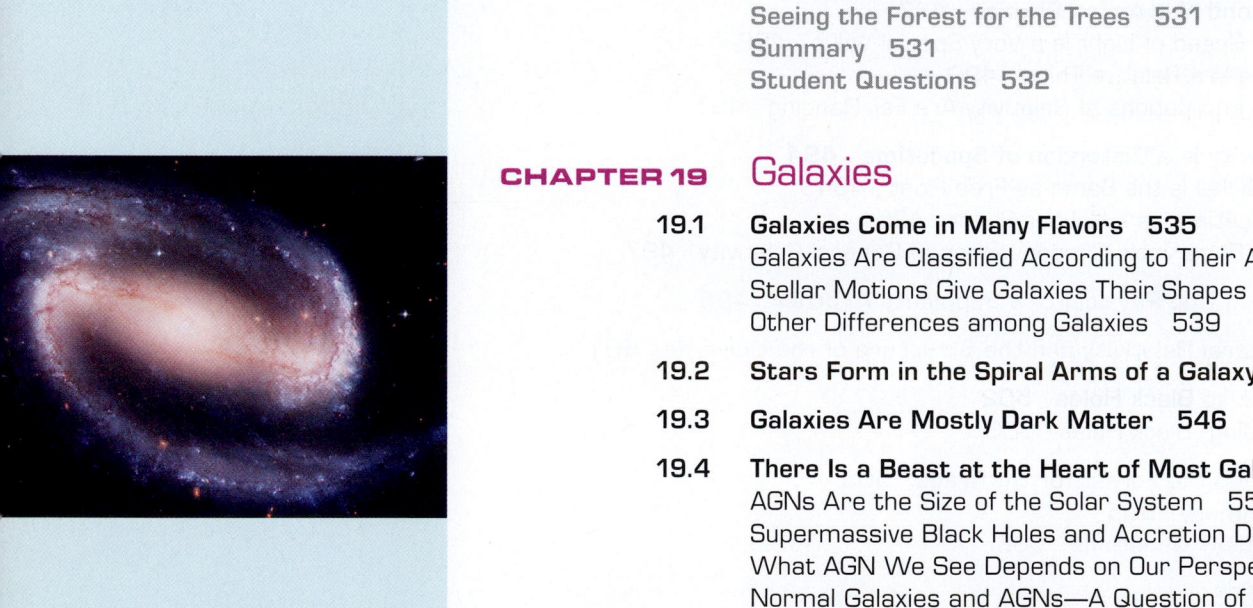

CHAPTER 21   Modern Cosmology

CHAPTER 22   The Origin of Structure

Preface

Astronomy is one of the most rapidly advancing fields in modern science. Its subject matter intrigues many students and simultaneously challenges textbook authors to keep up with the most recent discoveries, interpretations, and definitions. We built this textbook and its supporting media package with one goal in mind: to help students understand the world around them through the eyes of a scientist.

Students who study introductory astronomy typically come from diverse backgrounds; often their introductory astronomy course is the only formal exposure to science they will have in their lives! As instructors, everything we do should turn on the answer to basic questions such as these: What should students carry away from this course that will still be with them 20 years from now? How should this course change the students who take it? And how can we, as scientists and educators, facilitate those changes within our students?

A thoughtful response to these questions does *not* lead to a course that is primarily about memorization of facts. An introduction to science should be an introduction to what it means to think deeply about the universe, finding patterns and relationships within the world that go beyond the specifics of a particular object or setting, and applying those patterns and relationships broadly. If an introductory science course does nothing else, it should encourage students to think of their brains as muscles that grow strong only with exercise and training.

An introductory astronomy course should also show students something of the epistemology of science: knowledge is arrived at as we constantly challenge and test our ideas. How can it be that in science we never really prove things to be true, but instead only fail to show that they are false? It is precisely because science acknowledges lack of certainty that the pinnacle of human knowledge is a well-tested, well-corroborated scientific theory. We will have succeeded if a student comes to recognize the difference between science and pseudoscience, and if 20 years from now a former student still chuckles inwardly at the naïveté of the phrase "it's only a theory."

There is a strong practical side to what we teach about the process of science. Thinking about the world as a scientist means learning to sort information from opinion or happenstance. Physical scientists tend not to buy lottery tickets or call psychic help lines; our students are served well if we pass on a healthy dose of our practical skepticism. A democratic society, which in the long run can fare no better than its citizens can think, is served also.

Yet the real reason it matters that students come to see the world through the eyes of a scientist is the sheer beauty of the view. The majesty of the universe in which we live is staggering. Simply staring at a deep image of distant galaxies or the glow of a newly formed star or a picture of Jupiter's swirling Great Red Spot can be a mind-boggling experience. Add a dose of quantum mechanics or the idea of the tortured fabric of spacetime around a black hole, and we find ourselves in territory that would awe Aristotle. Science—this marvelous expression of human curiosity and passion and reason—has shattered the mental cage in which we have lived since our ancestors first looked to the sky and wondered about what they saw there. It is extraordinary to face these vistas surrounded not by speculation but by the meticulously constructed edifice of science. Finally, to live during that moment in history when we first recognized our own origins in the Big Bang that occurred some 14 billion years ago—*that* is something to write home about!

Science is a vital, exciting, ongoing expression of our humanity. While an introductory astronomy book should teach students about what science is, showing them what science has revealed and giving them powerful tools for thinking about the world, it should also share with them something of the passion scientists feel for the endeavor to which we have devoted our lives.

In order to meet our goals, our book is very different from many textbooks now on the market. Our book tells a story. On a small scale, it tells the story of specific ideas. Understanding comes from thinking carefully and critically about things and how they work. Helping a student understand a concept as a scientist means guiding that student through the concept, making heavy use of examples and analogies, and tying the concept back to everyday phenomena and experiences to which the student can relate. That is what the learning package we have created strives to do. The supporting media, from tutorial animations to online homework, extend the text's emphases to the realm of the interactive.

On a larger scale, *21st Century Astronomy* tells a web of different stories. Some of these stories have to do with what science is and how it is done. Why did Newton choose the form that he did for his universal law of gravitation? What are the fundamental differences between Kepler's empirical "laws" and Newton's theoretical derivation of the same relationships? And if Einstein was "right," why wasn't Newton "wrong"? Other stories explore the human side of science, such as how science has led us to think beyond the box that evolution built for us. Representing science fairly and honestly even means telling those stories that might make some students uncomfortable, like the story of why our understanding of the evolving universe and our place in it is science, while "creation science" is not.

We tell the story of motion, the story of light, the story of matter and energy, the story of Earth and our planetary system, the story of the Sun and stars, the story of life, and the story of the universe as narratives, with one idea flowing into the next and each idea connected to the whole. Knowledge and understanding are interwoven, with each idea and insight given meaning by how it fits into the whole. None of the stories in *21st Century Astronomy* exist in isolation. The stories are threads in a grand intellectual tapestry, woven from the recurring themes of the physics of matter, energy, radiation, and motion.

Our writing style for *21st Century Astronomy* fits this vision of science. The writing you will find here is far less formal than that found in most textbooks. We allow ourselves to be conversational, to pursue the occasional digression, to be irreverent or provocative or idiosyncratic, and to allow the sense of passion and wonder that we feel for the subject to show through from time to time. Our hope is to catch the feel of a pleasant, stimulating conversation with a student during office hours. For example, rather than stating that "the altitude of the north celestial pole above the horizon equals the observer's latitude" and leaving it at that, we talk about what the sky looks like over the course of the day as viewed from the North Pole and why. Then we invite the student to follow along as we head south, seeing how things change along the way.

We hope that *21st Century Astronomy* will appeal to instructors who, like us, have found that a serious nonscience student in today's colleges and universities is capable of thinking more deeply about more conceptually challenging material than is often assumed. Frankly, with the story that *21st Century Astronomy* has to tell—a story that deals head-on with some of the most fundamental, fascinating, and far-reaching questions humans have ever asked themselves—few students can resist being drawn in if approached in the right way. We kept this goal in mind as we created the Third Edition, reconsidering every sentence in the book from a student's perspective. We think the result is, as one peer reviewer put it, "conceptually and analytically enchanting."

One significant tension in the book is between a traditional organization of topics and an organization that more accurately represents the way scientists think and work today. True to our purpose, we chose an approach that reflects the state of our science. For example, we organized the Solar System section around a theme of comparative planetology. We pulled the discussion of tidal interactions, orbital resonances, chaos, and similar phenomena into a separate chapter (Chapter 10) titled "Gravity Is More than Kepler's Laws." We posit the existence of a rotating, collapsing interstellar cloud and a rotating protoplanetary disk and discuss the formation of the Solar System in Chapter 6, *before* discussing the planets themselves. This approach enables us to look at the planets from the outset within the context of how they formed. Enough examples—you get the idea.

Even so, there is flexibility in the text. For example, we feel that the flow of material through the streamlined early chapters discussing physical principles works well. We tried to include enough discussion of the human and social aspects of science to break up the treatment of conceptually difficult material. However, instructors who want to show their students a bit more astronomy and planetary science before hitting the physics might want to start with Chapter 6, on the formation of the Solar System, and then pull in material about orbits and radiation as it is needed in what follows. Another approach is to start with stars in Chapter 13, pull in radiation and orbits as needed to understand the properties of stars, and then insert the Solar System as an extended excursion into the topic of the formation of low-mass stars. Although we wrote the book to tell a complete, well-integrated story, we tried to allow for different paths tailored to the needs and tastes of individual instructors.

Our knowledge of the universe is constantly changing and expanding. For example, the recent and ongoing discoveries by the *Cassini* spacecraft change our perception of Saturn, its moons, and its rings on an almost daily basis. The tireless robots *Spirit* and *Opportunity* continue to roam the surface of Mars, giving us a better understanding of a planet that humans may visit and explore before the end of the 21st century. Another example concerns the dark matter and dark energy that dominate the density of the universe. There is a good chance that the dark matter particle will be discovered at the Large Hadron Collider as this book is going to press or soon after. In the case of dark energy, 20 years ago few scientists would have believed that it even exists. Today we understand that dark energy is *the* major component of our universe.

In closing, we should say a few words about who we are. Among the authors of this text you will find accomplished and well-known scientists who have been at the forefront of many exciting and significant events in astronomy and planetary science in the latter 20th and early 21st centuries. The author team includes members of scientific teams that built three of the cameras that have flown on the Hubble Space Telescope; team members and leaders of many of the major planetary missions; the chancellor of a major research university, who has done fundamental work on dark matter and galaxy formation; an extragalactic astronomer who also teaches about women in science; and a former president of the American Association of Physics Teachers. We have been involved in significant fundamental research on topics ranging from planetary geology, to the origin and evolution of the Solar System, to the formation and evolution of stars, to the structure and dynamics of the interstellar medium, to the nature of galaxies, to the origin of structure in the universe. There are remarkably few fields discussed in this book to which one or more of the authors have not contributed in some way over the years. For us, 21st century astronomy is not a textbook; rather it is the life we live. We hope you find value in our attempt to share with you what *we* see when we look up at the sky at night.

Acknowledgments

There are many at W. W. Norton & Company without whom this project would not have come together. The authors would like to thank our editors: Stephen Mosberg, who first looked across the table at us and said, "We will find a way"; John Byram, who stepped into the First Edition midstream and held it together through difficult times; Leo Wiegman, who spearheaded the Second Edition; and Rob Bellinger, who guided us through the many improvements that characterize this Third Edition and developed its supporting media package. Roby Harrington gave us enough rope to hang ourselves, and then helped us out of trouble when we tried to do just that. We also thank Erik Fahlgren, for providing editorial guidance; Kelsey Volker, for helping bring the Third Edition to the market; Matthew Freeman, for his work in developing the print ancillary program; and Jane Searle and Kim Yi, for their diligent work toward helping us publish this book on schedule. Among those in the trenches were Patricia Longoria and Andrew Sobel, developmental editors; Mary Lynch, editorial assistant; Stephanie Hiebert, copy editor; Laura Musich, assistant editor, electronic media; Trish Marx, manager of photo permissions; Junenoire Mitchell and Ramón Rivera Moret, photo researchers; Stacey Palen, Kevin Marshall, and Trina Van Ausdal, accuracy checkers; Marian Johnson, managing editor, college books; Rubina Yeh, designer; Carole Desnoes, layout artist; Second Avenue Software for our new AstroTour animations and SmartWork online homework; and our ancillary authors, Steven Desch, Gregory Mack, Ann Schmiedekamp, Tammy Smecker-Hane, Ben Sugerman, and Donald Terndrup. Finally, we would like to thank Kelly Paralis Keenan of Penumbra Design, who became a true collaborator on the project rather than simply an artist drawing to spec.

Jeff Hester
Brad Smith
George Blumenthal
Laura Kay
Howard Voss

The Features of *21st Century Astronomy*

The overall themes of *21st Century Astronomy* are emphasizing foundational physical science concepts and stressing scientific literacy for nonmajors by showing the process of scientific discovery. The following features support students as they start chapters, while they read, and as they finish and prepare for the next topic.

Key Concepts boxes at the beginning of each chapter call attention to the main issues and topics that will be discussed in the chapter.

Trailmarker statements—brief, one-sentence summaries—appear throughout each chapter to draw students' attention to fundamental concepts as they read. Redesigned for this edition, these trailmarkers are also a useful scanning tool for review.

Math Tools boxes help show students how mathematics is the language of science. Though we rarely use more than basic algebra, we have listened to our reviewers and separated some of the more difficult concepts from the running text using these boxes. This format allows instructors more flexibility in choosing which mathematical concepts they cover and what reading they assign. We make every effort to supply the equations with real-world values, to make the math and the underlying concepts easier to understand.

Annotated figures throughout the book present clear, accurate science in visually compelling ways.

Seeing the Forest for the Trees essays conclude each chapter with a thematic synthesis and a glimpse ahead to future chapters. These essays also help students make connections between astronomy and the liberal arts.

Summaries provide an outline of the key concepts as they were applied in the chapter.

Double the number of end-of-chapter questions help instructors assess student progress. Many of these questions have been adapted for use in SmartWork, Norton's online homework system. To make creating assignments easier for instructors, we have marked questions that we consider to be of intermediate difficulty with one asterisk (*) and those that are especially challenging with two asterisks (**).

> **Thinking about the Concepts** questions provide an opportunity for review with emphasis on the more important concepts.
>
> **Applying the Concepts** questions involve problem solving and critical thinking.

StudySpace

Each chapter ends with a reminder of additional resources available on the StudySpace website for further chapter review. For more on StudySpace, see page xxix.

In the Third Edition, we have made an effort to reduce the number of boxes in the text, though we retain three types of boxes from the Second Edition. In addition to the new Math Tools boxes mentioned above, these boxes touch on material either because it is of special importance and needs to be highlighted or because it is somewhat out of the main stream of our journey.

> **Foundations** boxes discuss the basic science that is central to our physical understanding of the universe.
>
> **Connections** boxes draw attention to recurring themes—bridges between different parts of our journey.
>
> **Excursions** boxes are short but interesting field trips that highlight how scientists apply what they learn to solving problems.

What's New in the Third Edition?

We were gratified that so many instructors found the First and Second Editions of *21st Century Astronomy* useful teaching tools. For the Third Edition, we called upon the feedback of our adopters and reviewers to build an edition that has evolved with the needs of teachers and students. We have kept the "big story" approach to astronomy that was so well received the first time around, but we place more emphasis on keeping students focused on the big picture. We thoroughly reconsidered every sentence in the text and revised for clarity and style wherever it was necessary. And, as described already, we've introduced a new approach to math and reduced the number of boxes overall.

Changes to Content

We have also introduced important new content—such as the new Chapter 23 on astrobiology—and reorganized our introduction to the process of science. Some of the more important content changes and additions include the following:

> **Chapter 1** brings together all of our introductory coverage on the process of science.
>
> **Chapter 3**, on the work of Newton and Kepler, is more accessible, thanks to our new approach to math.
>
> **Chapter 6** includes a new section on and expanded coverage of extrasolar planets.
>
> **Chapter 11** contains new discoveries about Saturn's rings, its large moon Titan, and the cryovolcanically active moon Enceladus from the *Cassini* mission.
>
> **Chapter 12** includes the properties of the five dwarf planets and new discoveries of the properties of Kuiper Belt objects.
>
> **Chapter 17** provides new material on gravity waves and includes all treatment of the

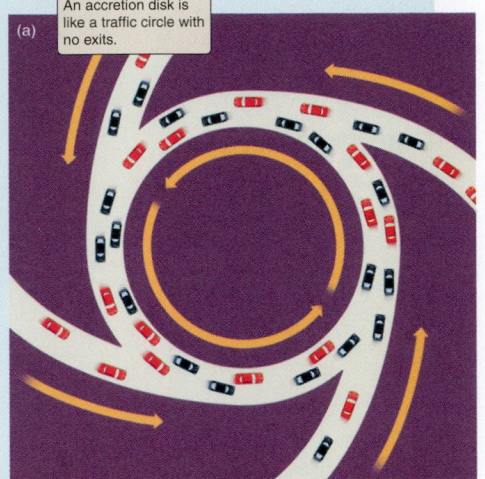

An accretion disk is like a traffic circle with no exits.

(a)

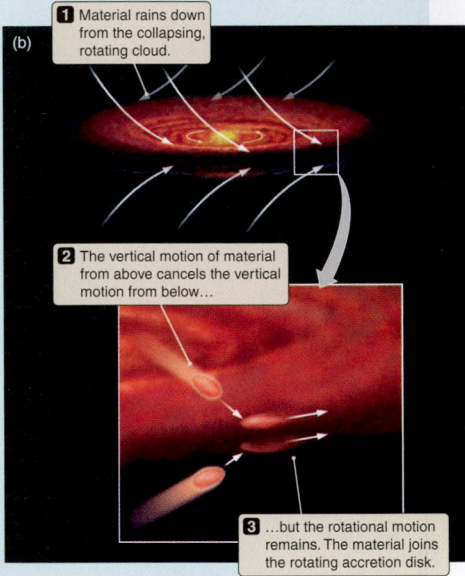

(b)

1 Material rains down from the collapsing, rotating cloud.

2 The vertical motion of material from above cancels the vertical motion from below...

3 ...but the rotational motion remains. The material joins the rotating accretion disk.

VISUAL ANALOGY FIGURE 6.7 (a) Traffic piles up in a traffic circle with entrances but no exits. (b) Similarly, gas from a rotating cloud falls inward from opposite sides, piling up onto a rotating disk.

theories of relativity, including special relativity, which was formerly introduced in Chapter 4.

Chapters 18, 19, and 20 have been reorganized to bring our introduction to galaxies into parallel alignment with our planet and star chapters by starting with the big picture ("Our Expanding Universe") and then going into details of galaxies and the Milky Way. New results from the Wilkinson Microwave Anisotropy Probe on cosmic background radiation are also discussed.

Chapters 21 and 22, on cosmology, have been greatly updated to illustrate the advance of observationally based cosmology and new models and theories. We have also included a discussion of the relative speculative topic of multiverses.

Chapter 23 brings together the discussion about life on Earth and the search for life elsewhere—the science of astrobiology—in a single new chapter.

New Pedagogical Features

We have strived to make the new edition more student-friendly. The following in-text features are designed to make the textbook even easier for students to use.

▶❚❚ **AstroTour** **AstroTour icons** point out when an animation is available to help a student master an important concept. These icons become hyperlinks in the ebook version of the text.

VISUAL ANALOGY Students really like the analogies we make between astronomical concepts and everyday phenomena, such as when we compare a forming accretion disk to cars piling up in a traffic circle with no exits. New **Visual Analogy icons** help students make immediate connections between the written analogies and the images that illustrate them.

Math Tools boxes help communicate to students how mathematics is the language of the scientist. These boxes also separate some of the more difficult mathematical concepts from the running text, allowing instructors flexibility in assigning reading.

MATH TOOLS 7.1

How to Read Cosmic Clocks

As Foundations 7.1 explains, a geologist can calculate the age of a mineral by measuring the relative amounts of a parent radioisotope and its daughter product. The time interval over which a radioactive isotope decays to half its original amount is called its **half-life**. With every half-life that passes, the remaining amount of the radioisotope will decrease by a factor of 2. For example, over three half-lives, the remaining amount of a parent radioisotope will be $\frac{1}{2} \times \frac{1}{2} \times \frac{1}{2} = \frac{1}{8}$ of its original amount. We can make this concept more general with a simple relationship:

$$\frac{P_F}{P_O} = \left(\frac{1}{2}\right)^n,$$

where P_O and P_F are the original and final amounts, respectively, of a parent radioisotope, and n is the number of half-lives that have gone by.

The most abundant isotope of the element uranium (^{238}U, the parent) decays through a series of intermediate daughters to an isotope of the element lead (^{206}Pb, its final daughter). The half-life of ^{238}U is 4.5 billion years. This means that in 4.5 billion years, a sample that originally contained the uranium isotope (the parent) but no lead (its final daughter) would be found instead to contain equal amounts of uranium and lead. If we were to find such a mineral, we would know that half the uranium atoms had turned to lead and that the mineral formed 4.5 billion years ago.

Let's look at another example, this time with a different isotope of uranium (^{235}U) that decays to a different lead isotope (^{207}Pb) with a half-life of 700 million years. Suppose that a lunar mineral brought back by astronauts has 15 times as much ^{207}Pb (the daughter product) as ^{235}U (the parent radioisotope). This means that $^{15}/_{16}$ of the parent radioisotope (^{235}U) has decayed to the daughter product (^{207}Pb), leaving only $^{1}/_{16}$ of the parent remaining in the mineral sample. Noting that $^{1}/_{16}$ is $(\frac{1}{2})^4$, we see that 4 half-lives have elapsed since the mineral was formed, and that this lunar sample is therefore 4×700 million = 2.8 billion years old.

Student Resources

StudySpace: www.wwnorton.com/studyspace

Ann Schmiedekamp, *Pennsylvania State University–Abington*

Norton's free and open student website for the introductory astronomy course has been revised and expanded to support the Third Edition. Study Plans for each chapter tie together all the resources that are available:

Quiz+ diagnostic quizzes

30 AstroTour animations

Interactive simulations

Vocabulary flash cards

News feeds

Links to ebook and SmartWork

StudySpace now includes free Quiz+ diagnostic quizzes. These gradable multiple-choice quizzes generate Custom Study Plans for students, linking them to other site resources that will help students where they need it.

AstroTour animations plus interactive simulations developed at the University of Nebraska–Lincoln reinforce the text's explanation of important physical concepts (see below for details).

From StudySpace, students can also access premium content in the ebook and Smart-Work online homework (see below for details).

AstroTour Animations

We commissioned 12 new animations to complement those we had created for the Second Edition. Many new animations use photorealistic 3-D art to help students visualize important physical concepts. Animations also make use of analogies and art from the printed text.

The Earth Spins and Revolves

The Celestial Sphere and the Ecliptic

The View from the Poles

The Moon's Orbit: Eclipses and Phases

Kepler's Laws

Velocity, Acceleration, Inertia

Newton's Laws and Universal Gravitation

Elliptical Orbits

Light as a Wave, Light as a Photon

Atomic Energy Levels and the Bohr Model

Atomic Energy Levels and Light Emission and Absorption

Doppler Effect

Geometric Optics and Lenses

Solar System Formation

Visual Analogy: Traffic Circle

Processes That Shape the Planets

Continental Drift

Hot Spot Creating a Chain of Islands

Atmospheres: Formation and Escape

Greenhouse Effect

Tides and the Moon

Cometary Orbits

Stellar Spectrum

The H-R Diagram

The Solar Core

Star Formation

Hubble's Law

Big Bang Nucleosynthesis

Active Galactic Nuclei

Galaxy Interactions and Mergers

Dark Matter

AstroTour animations are available from the free StudySpace student website, and they are also integrated into assignable SmartWork exercises. Offline versions of the animations for classroom presentation are available from the Instructor's Resource disc.

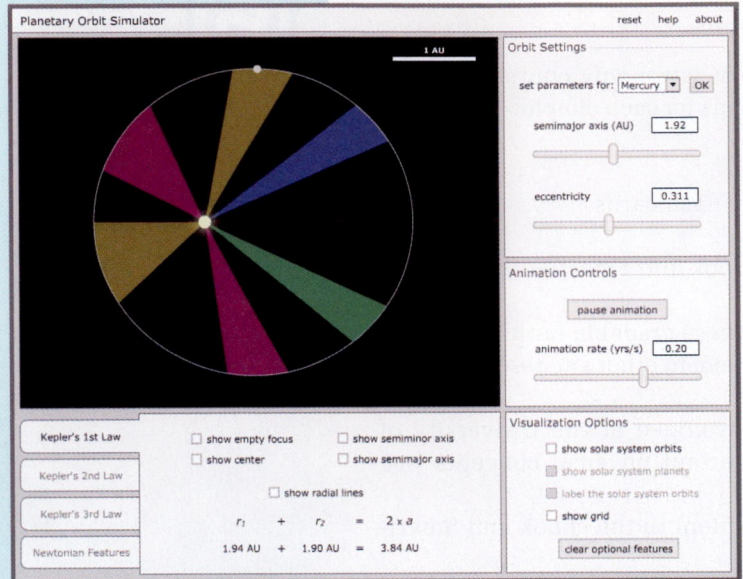

Interactive Simulations

In addition to the AstroTour animations, many of which are interactive, Norton licensed interactive simulations created at the University of Nebraska–Lincoln. These enable students to manipulate variables and work toward understanding physical concepts. All simulations are integrated into Study Plans and Quiz+ feedback on the StudySpace student website. They are also included on the Instructor's Resource disc.

SmartWork Online Homework
www.wwnorton.com/smartwork

SmartWork online homework allows instructors to assign algorithmically generated homework that offers students feedback at just the right teachable moment. Developed by university educators, SmartWork is the most intuitive tutorial and online homework system available.

SmartWork combines the flexibility of traditional paper homework with the administrative ease of computerized grading and the effectiveness of student-focused feedback. Instructors can draw from Norton's bank of high-quality, class-tested, ready-made questions. Or they can easily modify existing questions or write new ones. The system supports a wide variety of question types, including numeric entry, multiple choice, ranking tasks, and short text entry. For the Third Edition, we have also added questions based on the AstroTour animations.

Students may purchase access to Smart-Work bundled with the printed book or ebook, or they may purchase an access code at smartwork.wwnorton.com.

Though most instructors use SmartWork content as-is, a full suite of intuitive authoring tools allows instructors to edit existing problems or create new ones. No programming knowledge is required.

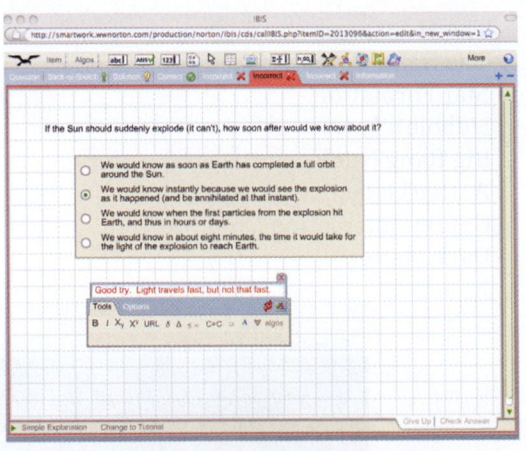

SmartWork's gradebook allows you to view your student's scores in several formats. Here, the At-a-Glance view shows student progress through an assignment.

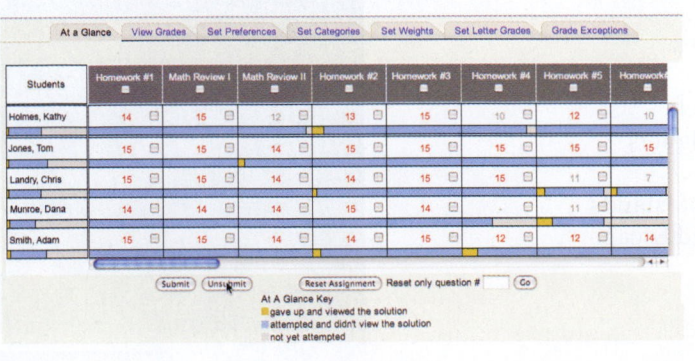

Every SmartWork question includes hints that suggest how to approach the problem, as well as answer-specific feedback that addresses common errors.

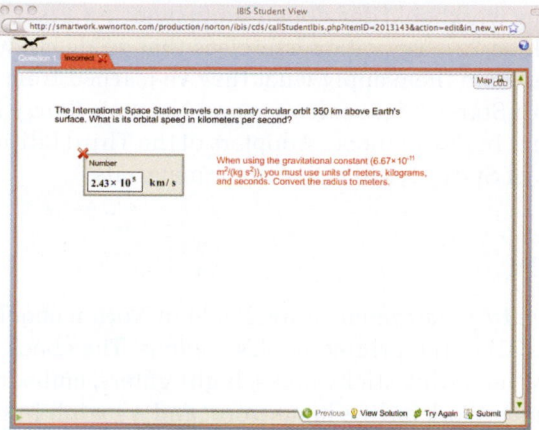

Every question in SmartWork provides a reference link to the complete ebook.

SmartWork offers a wide variety of question types, including new Ranking Task Exercises. These exercises ask students to make comparisons between similar items and rank them in a specific order.

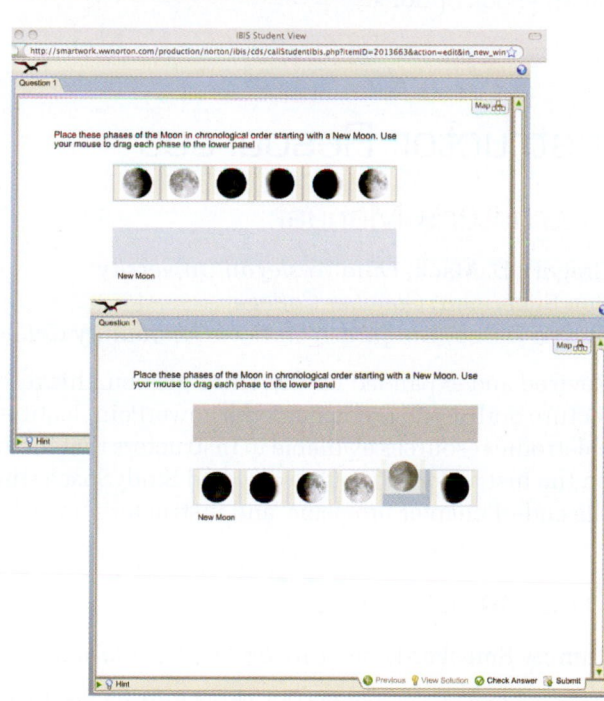

SmartWork and ebook integration   SmartWork is available as a stand-alone purchase or with an integrated ebook version of *21st Century Astronomy*. Links to the ebook make it easy for students to consult the text while completing their homework assignments. Instructors may order any volume of the print text bundled with access to both SmartWork and the ebook for no extra charge. Ask your local Norton representative for details.

Starry Night College CD-ROM and Workbook

Workbook by Steven Desch, *Guilford Technical Community College*, and Donald Terndrup, *The Ohio State University*

The remarkably realistic and user-friendly Starry Night College planetarium software allows students to explore stars and objects in our cosmic neighborhood and beyond. The accompanying workbook, developed exclusively for *21st Century Astronomy*, has been expanded to include 22 observation exercises that guide students' virtual explorations of the night sky and help them apply what they've learned from the text.

Both Starry Night software and Norton's *Starry Night Workbook* contain references to passages in the textbook. Adopters of the Third Edition may also download electronic desk copies of Starry Night's instructor materials.

Ebook

21st Century Astronomy is available in Norton ebook format, a convenient alternative that retains all of the printed book's content. The ebook offers a variety of tools for study and review, including sticky notes, highlighters, embedded links to the AstroTour animations and SmartWork online homework, and a search function.

New copies of the text may be bundled with SmartWork and the ebook version of the text at no extra charge. The SmartWork/ebook package makes it easy for students to consult the text when completing their homework assignments.

For students who prefer a downloadable ebook, a PDF version of the ebook is also available from Powell's Books at powells.com. Go to norton**ebooks**.com for more information on all ebook options.

Instructor Resources

Instructor's Manual

Gregory D. Mack, *Ohio Wesleyan University*
Ben Sugerman, *Goucher College*
Steven Desch, *Guilford Technical Community College*

Revised and expanded for the Third Edition, this resource includes brief chapter overviews, lecture outlines to accompany the PowerPoint lecture slides, a media guide to web and other electronic resources available to instructors that includes notes on the animations contained on the Instructor's Resource disc and StudySpace student website, worked solutions to all of the end-of-chapter problems, and instructor's notes for the *Starry Night Workbook* activities.

Test Bank

Tammy Smecker-Hane, *University of California–Irvine*

Revised and expanded for the Third Edition, each chapter of the Test Bank consists of three question types classified according to Norton's taxonomy of educational objectives:

1. **Factual questions (what?)** test declarative knowledge, including textbook definitions and relationships between two or more pieces of information.

2. **Applied questions (how?)** pose problems in a context different from the one in which a particular concept was learned, requiring students to draw from their declarative and/or procedural understanding of important concepts.

3. **Conceptual questions (why?)** ask students to draw from their prior experience and use critical thinking skills to take part in qualitative reasoning about the real world.

Questions are further classified by section and difficulty, making it easy to construct tests and quizzes that are meaningful and diagnostic. This approach enables instructors to accurately judge students' mastery of the material—what they know, what they don't know, and to what degree—on the basis of the assessment outcomes.

The question types are short answer, multiple choice, and true-false. The Test Bank is available in print, Word RTF, PDF, and ExamView formats.

PowerPoint Lecture Slides

Gregory D. Mack, *Ohio Wesleyan University*

These ready-made lecture slides integrate selected art from the text, "clicker" questions, and offline, lecture-ready versions of the AstroTour animations. Designed to accompany the lecture outlines found in the Instructor's Manual, these lecture slides are fully editable and are available in Microsoft PowerPoint format.

Instructor's Resource Disc

This helpful resource includes the PowerPoint lecture outlines with "clicker" questions, offline versions of the animations, plus selected photographs and all drawn art from the text. The Instructor's Resource disc also contains the interactive simulations used on the StudySpace student website, as well as additional clicker questions developed at the University of Nebraska–Lincoln.

BlackBoard and WebCT Course Cartridges

Course cartridges for BlackBoard and WebCT include access to the animations, a Study Plan for each chapter, multiple-choice tests, and links to ebook and SmartWork premium content.

Reviewers

Production of a book like this is a far larger enterprise than any of us imagined when we jumped on board. We would like to thank the teachers who reviewed portions of the manuscript along the way and helped us build a stronger book:

Timothy Barker, Wheaton College
Edwin Bergin, University of Pittsburgh
William Blass, University of Tennessee–Knoxville
Steve Bloom, Hampden Sydney College
Daniel Boice, University of Texas–San Antonio
Jack Brockway, Radford University
Juan E. Cabanela, Minnesota State University–Moorhead
Amy Campbell, Louisiana State University
Robert Cicerone, Bridgewater State College
Eric M. Collins, California State University–Northridge
John Cowan, University of Oklahoma–Norman
Robert Dick, Carleton University
Tom English, Guilford Technical Community College
John Finley, Purdue University

Matthew Francis, Lambuth University
Kevin Gannon, College of Saint Rose
Bill Gutsch, St. Peter's College
Karl Haisch, Utah Valley University
Barry Hillard, Baldwin Wallace College
Hal Hollingsworth, Florida International University
Olencka Hubickyj-Cabot, San Jose State University
Kevin M. Huffenberger, University of Miami
Steve Kawaler, Iowa State University
Kevin Lee, University of Nebraska–Lincoln
Matt Lister, Purdue University
M. A. K. Lodhi, Texas Tech University
Kevin Marshall, Bucknell University
Chris McCarthy, San Francisco State University
Chris Mihos, Case Western University
Milan Mijic, California State University–Los Angeles

J. Scott Miller, University of Louisville

Scott Miller, Sam Houston State University

Andrew Morrison, Illinois Wesleyan University

Edward M. Murphy, University of Virginia

Kentaro Nagamine, University of Nevada–Las Vegas

Jascha Polet, California State Polytechnic University

Daniel Proga, University of Nevada–Las Vegas

Laurie Reed, Saginaw Valley State University

Masao Sako, University of Pennsylvania

Ann Schmiedekamp, Pennsylvania State University

Jonathan Secaur, Kent State University

Caroline Simpson, Florida International University

Ian Skilling, University of Pittsburgh

Tammy Smecker-Hane, University of California–Irvine

Allyn Smith, Austin Peay State University

Ben Sugerman, Goucher College

Neal Sumerlin, Lynchburg College

Christopher Taylor, California State University–Sacramento

Trina Van Ausdal, Salt Lake Community College

Karen Vanlandingham, West Chester University

Paul Voytas, Wittenberg University

Paul Wiita, Georgia State University

David Wittman, University of California, Davis

We also thank the following colleagues who served as Second Edition reviewers:

Scott Atkins, University of South Dakota

Peter A. Becker, George Mason University

Timothy C. Beers, Michigan State University

Juan Cabanela, Minnesota State University–Moorhead

Judith Cohen, California Institute of Technology

Paul Hintzen, California State University–Long Beach

Paul Hodge, University of Washington

William A. Hollerman, University of Louisiana at Lafayette

Adam Johnston, Weber State University

Laura Kay, Barnard College

Bill Keel, University of Alabama

Leslie Looney, University of Illinois at Urbana–Champaign

Norm Markworth, Stephen F. Austin State University

Scott Miller, Pennsylvania State University

Ata Sarajedini, University of Florida

Paul P. Sipiera, William Rainey Harper College

Donald Terndrup, Ohio State University

Richard Williamon, Emory University

About the Authors

Jeff Hester received his PhD from Rice University and is professor emeritus of physics and astronomy at Arizona State University. His research interests are the interstellar medium in the Milky Way and external galaxies; structure of the diffuse interstellar medium; interstellar shock waves; supernova remnants; pulsar wind interactions; Herbig-Haro objects; and H II region structure.

Brad Smith has served as an associate professor of astronomy at New Mexico State University, a professor of planetary sciences and astronomy at the University of Arizona, and a research astronomer at the University of Hawaii. Through his interest in Solar System astronomy, he has participated as a team member or imaging team leader on several US and international space missions, including *Mars Mariners 6, 7,* and *9; Viking; Voyager;* and the Soviet *Vega* and *Phobos* missions. More recently, Smith's interests have turned to other planetary systems, working as a team member of the Hubble Space Telescope NICMOS experiment. Asteroid 8553 *Bradsmith* was named for Smith. He has four times been awarded the NASA Medal for Exceptional Scientific Achievement. Smith is a member of the IAU Working Group for Planetary System Nomenclature and is chair of the Task Group for Mars Nomenclature. He is now semi-retired but remains affiliated with the Institute for Astronomy at the University of Hawaii, Manoa.

George Blumenthal is chancellor at the University of California–Santa Cruz, where he has been a professor of astronomy and astrophysics since 1972. Chancellor Blumenthal received his BS degree from the University of Wisconsin–Milwaukee and his PhD in physics from the University of California–San Diego. As a theoretical astrophysicist, Chancellor Blumenthal's research encompasses several broad areas, including the nature of the dark matter that

constitutes most of the mass in the universe, the origin of galaxies and other large structures in the universe, the earliest moments in the universe, astrophysical radiation processes, and the structure of active galactic nuclei such as quasars. Besides teaching and conducting research, Chancellor Blumenthal has served as the chair of the UC–Santa Cruz Astronomy and Astrophysics Department, has chaired the Academic Senate for both the UC–Santa Cruz campus and the entire University of California system, and has served as the faculty representative to the UC Board of Regents.

Laura Kay is Ann Whitney Olin professor of physics and astronomy at Barnard College, where she has taught since 1991. She received a BS degree in physics and an AB degree in feminist studies from Stanford University, and MS and PhD degrees in astronomy and astrophysics from the University of California–Santa Cruz. As

a graduate student she spent 13 months at the Amundsen Scott station at the South Pole in Antarctica. She studies active galactic nuclei, using groundbased and X-ray telescopes. She teaches courses on astronomy, astrobiology, women and science, and polar exploration. At Barnard she has served as chair of the Physics & Astronomy Department, chair of the Women's Studies Department, chair of Faculty Governance, and interim associate dean for Curriculum and Governance.

Howard Voss is professor emeritus of physics at Arizona State University, where he taught for over four decades, served as chair of the Department of Physics, and received the Distinguished Faculty Award and the Dean's Teaching Award. He was awarded the Melba Phillips Medal by the American Association of Physics Teach-

ers, which he served as president, secretary, a member of the Executive Board, and in other offices. He also served the American Institute of Physics in several positions, including as chair of the Publishing Policy Committee and as a member of the Governing Board.

THIRD EDITION

21ST CENTURY

ASTRONOMY

The most beautiful thing we can experience is the mysterious.
It is the source of all true art and all science.
He to whom this emotion is a stranger,
who can no longer pause to wonder and stand rapt in awe,
is as good as dead: his eyes are closed.

ALBERT EINSTEIN (1879–1955)

The Coma cluster of galaxies at a distance of more than 300 million light-years.

CHAPTER 1

Why Learn Astronomy?

1.1 Getting a Feel for the Neighborhood

We chose the title of this book—*21st Century Astronomy*—to emphasize that this is the most fascinating time in history to be studying this most ancient of sciences. Loosely translated, the word **astronomy** means "patterns among the stars." But modern astronomy—the astronomy we will talk about in this book—has become far more than looking at the sky and cataloging what is visible there. We are confident in our answers to many of the questions that you might have asked yourself as a child when you looked at the sky. What are the Sun and Moon made of? How far away are they? What are the stars? How do they work? Do they have anything to do with me? In addition, issues as seemingly metaphysical as the origin and fate of the universe and the nature of space and time have become the subjects of rigorous scientific investigation. The answers we are finding to these questions are often far more wondrous than our predecessors could have dreamed. They are changing not only our view of the cosmos, but our view of ourselves as well.

Glimpsing Our Place in the Universe

Most of us have a permanent address—street number, street, city, state, country. It is where the mail carrier delivers our mail. But let's expand our view for a moment. We also live somewhere within an enormously vast universe. What then is our "cosmic address"? It might look something like this: planet, star, galaxy, galaxy group, galaxy cluster.

We all reside on a **planet** called Earth, which is orbiting under the influence of gravity about a **star** called the Sun.

KEY CONCEPTS

Before traveling through unfamiliar terrain, it helps to have some idea of where you are going, what you might see along the way, and what you should pack for the journey. We'll begin this chapter by sketching out a rough map of the universe and our place within it. And we'll present some of the tools that you will need to take along as you look at the wonders of the universe through the eyes of a scientist. In this chapter's overview of what is to come, we will find that

- The universe is vast beyond all human experience, yet it is governed by the same physical laws that shape our daily lives.

- We are a product of the universe; the very atoms of which we are made were formed in stars that died long before the Sun and Earth were formed.

- There is nothing special about our place in the universe; it is indistinguishable from any other place.

- Science—like art, literature, and music—is a creative human activity; it is also a remarkably powerful, successful, and aesthetically beautiful way of viewing the world.

- Understanding comes from thinking carefully and deeply about patterns in the world, not simply from memorization of facts.

- Like climbing a mountain, the journey into astronomy requires effort, but the view from the top is amazing to behold.

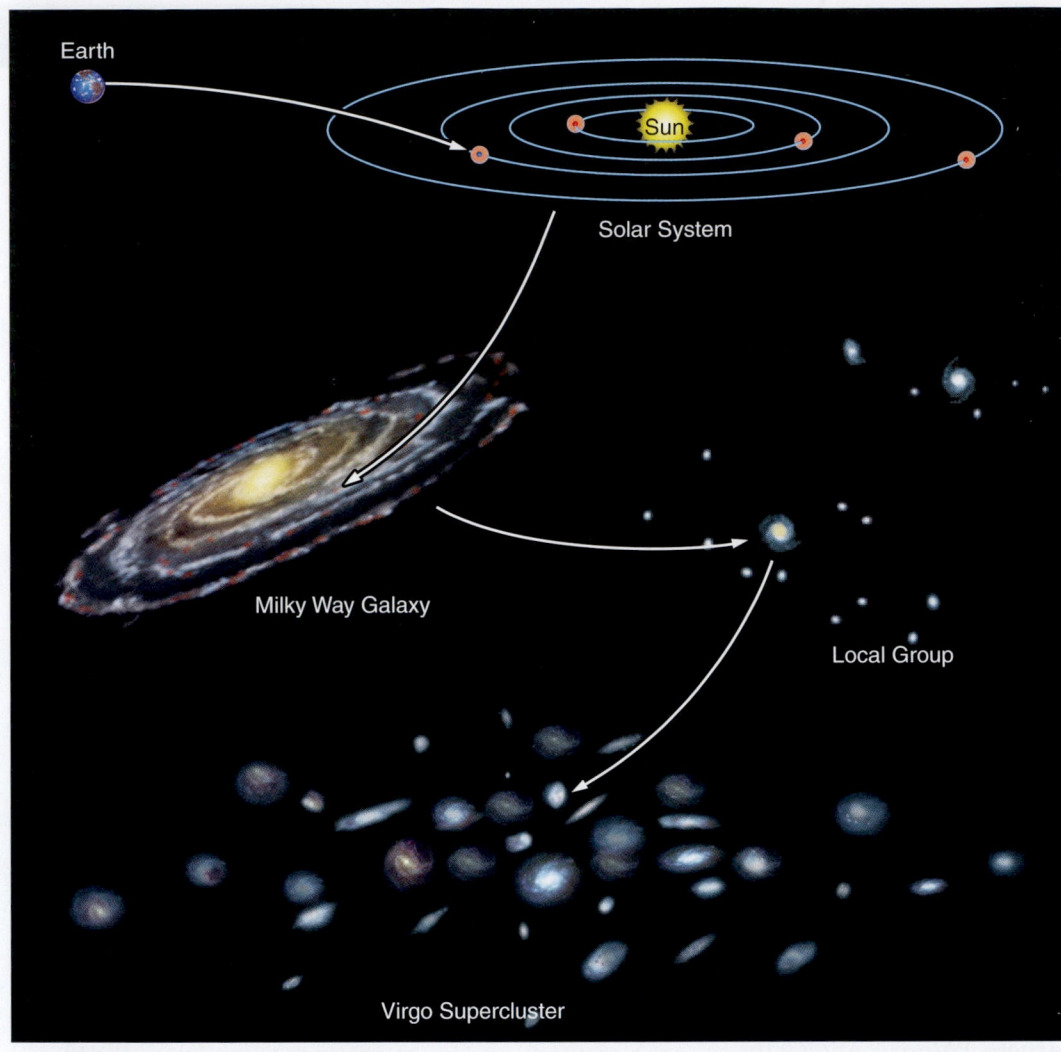

Earth

Solar System

Milky Way Galaxy

Local Group

Virgo Supercluster

FIGURE 1.1 The place where we reside in the universe—our cosmic address: Earth, Solar System, Milky Way Galaxy, Local Group, Virgo Supercluster. We live on Earth, a planet orbiting the Sun in our Solar System, which is a star in the Milky Way Galaxy. The Milky Way is a large galaxy within the Local Group of galaxies, which in turn is located in the Virgo Supercluster.

The Sun is an ordinary, middle-aged star, more massive and luminous than some stars but less massive and luminous than others. The Sun is extraordinary only because of its importance to us within our own **Solar System**. Our Solar System consists of eight **classical planets**—Mercury, Venus, Earth, Mars, Jupiter, Saturn, Uranus, and Neptune. It also contains many smaller bodies, such as **dwarf planets**, **asteroids**, and **comets**. In August 2006 the International Astronomical Union (IAU) redefined what astronomers mean when they call something a planet (see Appendix 8.) Under this new definition, Pluto, known to all of us as the "ninth planet" since its discovery in 1930, was unceremoniously stripped of its planet status and demoted to dwarf planet. We might define the size of our Solar System by its most distant planetary body, a dwarf planet called "Eris," although others could rightfully include the realm of small, icy bodies that extends far beyond the orbit of Eris.

The Sun is located about halfway out from the center of a flattened collection of stars, gas, and dust, referred to as the **Milky Way Galaxy**. Our Sun is just one among approximately 300 billion stars scattered throughout the **galaxy**, and many of these stars are themselves surrounded by planets, suggesting that other planetary systems may be the rule rather than the exception.

The Milky Way, in turn, is a member of a small collection of a few dozen galaxies called the **Local Group**. The Milky Way Galaxy and the Andromeda Galaxy are true giants within the Local Group. Most others are what astronomers call "dwarf galaxies." Looking farther outward, the Local Group is part of a vastly larger collection of thousands of galaxies—a **supercluster**—called the Virgo Supercluster.

We can now define our "cosmic address"—Earth, Solar System, Milky Way Galaxy, Local Group, Virgo Supercluster—as illustrated in **Figure 1.1**. Yet even this address is not complete, because the vast structure we just described is only the *local universe*. The part of the **universe** that we can see extends far beyond—a distance that light takes 13.7 billion years to traverse—and within this volume we estimate that there are *several hundred billion* galaxies, roughly as many galaxies as there are stars in the Milky Way!

We all have a cosmic address.

The Scale of the Universe

One of the first conceptual hurdles that we face as we begin to think about the universe is its sheer size. If a hill is big, then a mountain is really big. If a mountain is really big, then Earth is enormous. But where do we go from there? We quickly run out of superlatives as the scale of what we are talking about comes to dwarf our human experience. One technique that can help us develop a sense for the size of things in the universe is to use a little sleight of hand and move from discussing distance to talking instead about time. If you are driving down the highway at 60 miles per hour (mph), a mile is how far you go in a minute. Sixty miles is how far you go in an hour. Six hundred miles is how far you go in 10 hours. So to get a feeling for the difference in size between 600 miles and 1 mile, you can think about the difference between 10 hours and a single minute.

We can play this same game in astronomy, but the **speed** of a car on the highway is far too slow to be useful. Instead we will use the greatest speed in the universe—the speed of light. Light travels at 300,000 kilometers per second (km/s). At that speed, light can circle Earth (a distance of 40,000 kilometers [km]) in just under $1/7$ of a second—about the time it takes you to snap your fingers. Fix that comparison in your mind. The size of Earth is like—*snap!*—a snap of your fingers. Now follow along in **Figure 1.2** as we move outward into the universe. (This figure is highlighted as a "Visual Analogy." The "Visual Analogy" labels indicate drawings that make analogies between astronomical phenomena and everyday objects more concrete.) We next encounter the Moon, 384,000 km away, or a bit over $1\frac{1}{4}$ seconds when moving at the speed of light. So if the size of Earth is a snap of your fingers, the distance to the Moon is about the time that it takes to turn a page in this book. Continuing on, we find that at this speed the Sun is $8\frac{1}{3}$ minutes away, or the length of a hurried lunch at the student union. Crossing from one side of the orbit of Neptune, the outermost classical planet in our Solar System, to the other takes about 8.3 hours. Think about that for a minute. Let it sink in. Comparing the size of Neptune's orbit to the circumference of Earth is like comparing the time of a good night's sleep to a single snap of your fingers.

In crossing Neptune's orbit, however, we have only just begun our journey. Many steps remain. It takes us a bit over four years to cover the distance from Earth to the nearest star (other than the Sun), or as much time as you spent in high school. At this point, even our analogy using the travel time of light can no longer bring astronomical distance to a human scale. Light takes about 100,000 years to travel across our galaxy—about the time that modern humans (*Homo sapiens*) have walked the surface of Earth. To reach the nearest large galaxies beyond our own takes a few million years—the amount of time that has passed since our australopithecine

> The travel time of light helps in understanding the scale of the universe.

ancestors were on the scene. To reach the limits of the currently observable universe takes 13.7 billion years—the age of the universe, or about three times the age of Earth.

Look at that comparison again. The size of Earth is to the vast expanse of the universe as a single snap of your fingers is to three times the amount of time that has passed since the Sun and Earth were formed! That's something to ponder the next time you look up at a star-filled summer sky.

The Origin and Evolution of the Universe

While seeking knowledge about the universe and how it works, modern astronomy and physics have repeatedly come face-to-face with a number of age-old questions long thought to be solely within the domain of philosophers. **Figure 1.3** depicts a traveler raising the veil of the heavens to see what lies there. Like this fictional traveler, philosophers throughout most of history looked at the universe and saw it as remote and different from Earth—disconnected from our terrestrial existence. When modern astronomers look at the universe, they see instead a network of ongoing processes that we are a part of. Astronomy begins by looking out at the universe, but increasingly that outward gaze turns introspective as we come to appreciate that our very existence is a consequence of those same processes.

The study of the chemical evolution of the universe is such a case. Theory and observation tells us that the universe was created in a "Big Bang" some 13.7 billion years ago. As a result of both observation and theoretical work, we now understand that the only chemical elements found in abundance in the early universe were hydrogen and helium, plus tiny amounts of lithium, beryllium, and boron. Yet we live on a planet with a core consisting mostly of iron and nickel, surrounded by a mantle made up of rocks containing large amounts of silicon and various other elements. Our bodies are built of carbon, nitrogen, oxygen, sodium, phosphorus, and a host of other chemical elements. If these elements that make up Earth and our bodies were not present in the early universe, where did they come from?

The answer to this question lies within the stars. Nuclear fusion reactions occurring deep within the interiors of stars combine atoms such as hydrogen, forming more massive atoms, and thereby accomplishing the alchemist's unattainable dream of transforming one element into another. When a star exhausts its nuclear fuel and nears the end of its life, it often loses much of its mass—including some of the new atoms formed in its interior—by blasting it back into interstellar space. We will talk later about the life and death of stars. For now it is enough to note that our Sun and Solar System formed from a cloud of interstellar gas and dust that had been "polluted" by the chemical effluent from earlier generations of stars. This chemical legacy supplies the building blocks for the inter-

> We are stardust.

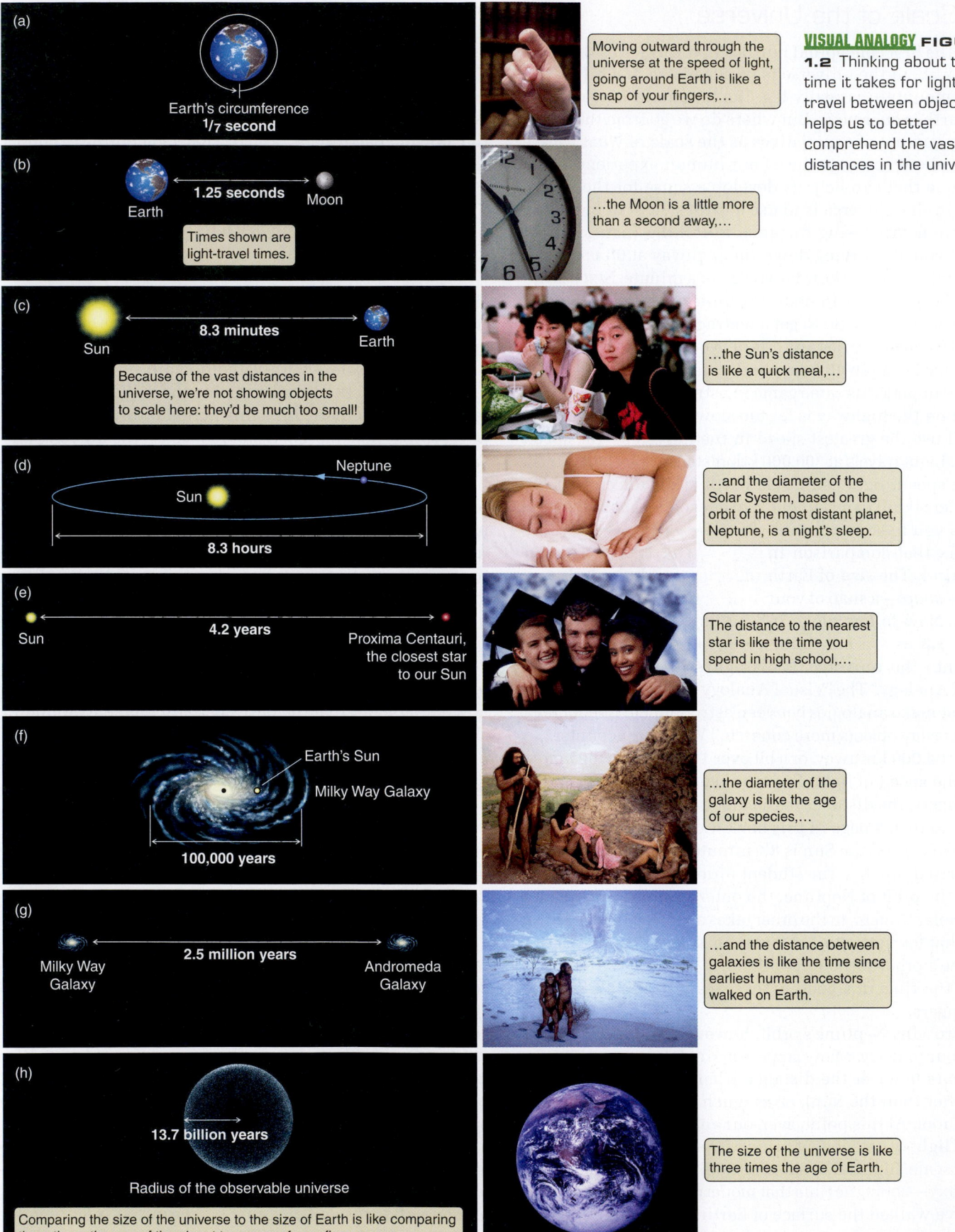

(a) Earth's circumference
1/7 second

Moving outward through the universe at the speed of light, going around Earth is like a snap of your fingers,…

(b) Earth — **1.25 seconds** — Moon

Times shown are light-travel times.

…the Moon is a little more than a second away,…

(c) Sun — **8.3 minutes** — Earth

Because of the vast distances in the universe, we're not showing objects to scale here: they'd be much too small!

…the Sun's distance is like a quick meal,…

(d) Neptune
Sun
8.3 hours

…and the diameter of the Solar System, based on the orbit of the most distant planet, Neptune, is a night's sleep.

(e) Sun — **4.2 years** — Proxima Centauri, the closest star to our Sun

The distance to the nearest star is like the time you spend in high school,…

(f) Earth's Sun
Milky Way Galaxy
100,000 years

…the diameter of the galaxy is like the age of our species,…

(g) Milky Way Galaxy — **2.5 million years** — Andromeda Galaxy

…and the distance between galaxies is like the time since earliest human ancestors walked on Earth.

(h) **13.7 billion years**
Radius of the observable universe

Comparing the size of the universe to the size of Earth is like comparing three times the age of the planet to a snap of your fingers.

The size of the universe is like three times the age of Earth.

VISUAL ANALOGY FIGURE

1.2 Thinking about the time it takes for light to travel between objects helps us to better comprehend the vast distances in the universe.

FIGURE 1.3 Throughout most of history, humans conceived of the rest of the universe as a place apart from us. Here a traveler raises the curtain of the firmament to get a glimpse of what lies beyond Earth.

esting chemical processes that go on around us—chemical processes such as life. **Figure 1.4** symbolizes this intimate relationship between the world around us and our heritage in the stars. Look around you. The atoms that make up everything you see were formed in the hearts of stars. Poets sometimes say that "we are stardust," but this is not just poetry. It is literal truth.

As humans, we have long speculated about our beginnings. Who or what is responsible for our existence? How were Heaven and Earth created? In the modern world, primitive creation myths have largely given way to scien-

FIGURE 1.4 You and everything around you, including beautiful waterfalls, are composed of atoms that were forged in the interior of stars that lived and died before the Sun and Earth were formed. The supermassive star Eta Carinae, shown on the left, is currently ejecting a cloud of chemically enriched material.

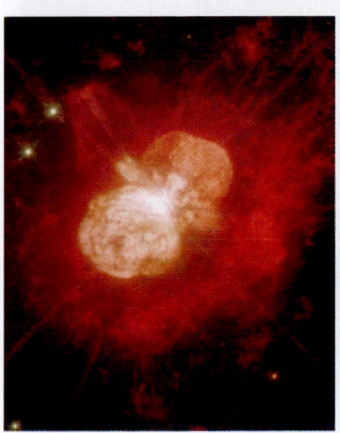 

tific explanations.[1] How and when did the universe begin? What combination of events—some probable, others much less likely—have led to our existence as sentient beings living on a small rocky planet orbiting a typical middle-aged star? Was this a unique happening, or are there others like us scattered throughout the galaxy? We'll look into the genesis of terrestrial life, but we'll also examine the possibilities of life elsewhere in our Solar System and beyond—a subject called **astrobiology**. For life to exist around other stars in the galaxy, there must be life-sustaining planets to support it. We will include the discovery of extrasolar planets and how they compare with the planets of our own Solar System.

1.2 Science Involves Exploration and Discovery

As we look at the universe through the eyes of astronomers, we will also learn something of how science works. It is beyond the scope of this book for us to provide a detailed justification for all that we will say. However, we will try to offer some explanation of where an idea comes from and why we believe it to be valid. We will not present something as fact unless there is a compelling reason to believe it. We will be honest when we are on uncertain, speculative ground, and we will admit it when the truth is that we really do not know. This book is not a compendium of revealed truth or a font of accepted wisdom. Rather, it is an introduction to a body of knowledge and understanding that was painstakingly built (and sometimes torn down and rebuilt) brick by brick.

It is almost impossible to overstate the importance of science in our civilization. One obvious manifestation of science is the technology that has enabled us to explore well beyond our planet. Since the 1957 launch of Sputnik, the first human-made satellite, we have lived in an age of space exploration. Five decades later, we have seen humans walk on the Moon (**Figure 1.5**) and have sent unmanned probes to visit all of the classical planets. Spacecraft have flown past asteroids, comets, and even the Sun. Our inventions have landed on Mars, Venus, Titan (Saturn's largest moon), and an asteroid, and have plunged into the atmosphere of Jupiter. Most of what we know of the Solar System has resulted from these past five decades of exploration.

Satellite observatories in orbit around Earth have also given us many new perspectives on the universe. The very atmosphere that shields us from harmful solar radiation

---

[1]A few cultures still retain their creation myths. And even within "enlightened" society, certain factions have ignored science in favor of nonscientific "creationism" and "intelligent design."

(a)

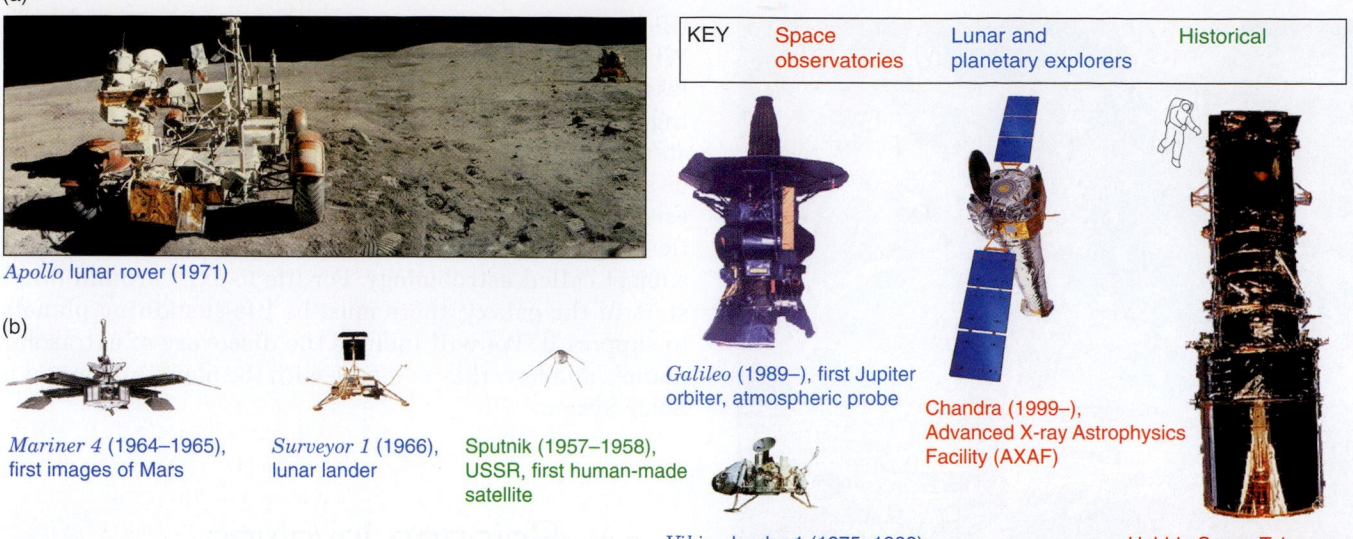

Apollo lunar rover (1971)

(b)

Mariner 4 (1964–1965), first images of Mars

Surveyor 1 (1966), lunar lander

Sputnik (1957–1958), USSR, first human-made satellite

Viking lander 1 (1975–1982), first of two Mars landers

Galileo (1989–), first Jupiter orbiter, atmospheric probe

Chandra (1999–), Advanced X-ray Astrophysics Facility (AXAF)

Hubble Space Telescope (1990–), UV, visible, infrared astronomy

KEY	Space observatories	Lunar and planetary explorers	Historical

FIGURE 1.5 (a) *Apollo 15* astronaut James B. Irwin stands by the lunar rover during an excursion to explore and collect samples from the Moon. (b) Artificial satellites and space probes have grown more complex since the 1957 launch of Sputnik 1. These spacecraft are all shown to the same scale. Some are astronomical observatories that view space from Earth's orbit. Others are interplanetary explorers sent to investigate other worlds within our Solar System.

also blinds us to much of what is going on around us. Space astronomy continues to show us vistas hidden from the gaze of groundbased telescopes by the protective but obscuring blanket of our atmosphere. Satellites capable of detecting the full spectrum of radiation—from the highest-energy gamma rays and X-rays, through ultraviolet and infrared radiation, and to the lowest-energy microwaves—have brought surprising discovery after surprising discovery. Each has forever altered our perception of the universe, further expanding the domain of the human mind. Since the closing years of the 20th century we have witnessed a renewed vigor in astronomical observations from the surface of Earth. The view of the sky seen by radio telescopes, as shown in **Figure 1.6**, illustrates the new perspectives that have been made possible by our growing technological prowess.

> Space exploration has expanded our view of the universe.

When we think of astronomy, telescopes—both on the ground and in space —immediately come to mind. But you may be surprised to learn that a great deal of frontline astronomy is now carried out in large physics facilities like the one shown in **Figure 1.7**. Today astronomers work along with their colleagues in related fields, such as physics, chemistry, geology, and planetary science, to sharpen our understanding of the physical laws that govern the behavior of **matter** and **energy** and to use this understanding to make sense of our observations of the cosmos. Astronomy has also benefited enormously from the computer revolution. The 21st century astronomer spends far more time staring at a computer screen than peering through the eyepiece of a telescope. As astronomers, we use computers to collect and analyze data from telescopes, calculate physical models of astronomical objects, and prepare and disseminate the results of our work.

FIGURE 1.6 In the 20th century, advances in telescope technology opened new windows on the universe. This is the sky as we would see it if our eyes were sensitive to radio waves, shown as a backdrop to the National Radio Astronomy Observatory site in Green Bank, West Virginia.

FIGURE 1.7 The Tevatron high-energy particle collider at Fermilab provides clues about the physical environment during the birth of the universe. Laboratory astrophysics, in which astronomers model important physical processes under controlled conditions, has become an important part of astronomy.

1.3 Science Is a Way of Viewing the World

Given the visible importance of technology in our lives, you might be tempted to say that science *is* technology. It is true that science forms the basis of technology through a mutually supportive relationship in which each enables advances in the other. Yet science is much more than technology. Science can no more be reduced to its practical technological application than the accomplishment of an Olympic athlete can be reduced to utilization of the athlete's Olympic fame to market shoes and other products. Rather, we should understand science as a fundamental way of viewing the world through the *scientific method*, a worldview that has survived constant upheavals and challenges.

The Scientific Method and Scientific Principles

If science is more than technology, you might instead suggest that science is the **scientific method**. When you took science courses in high school, you probably had the scientific method drilled into you. So let's quickly review how the scientific method works. Consider a scientist coming up with an idea that might explain a particular observation or phenomenon. She presents the idea to her colleagues as a **hypothesis**. Her colleagues then look for testable predictions capable of *disproving* her hypothesis. *This is an important property of the scientific method*, because any hypothesis that is not *falsifiable*—in other words, *disprovable*—must be based on faith alone. Faith serves its own purpose in society, but it is not science. As continuing tests fail to disprove a hypothesis, scientists will come to accept it as a **theory**, but *never* as an undisputed fact. Scientific theories are accepted only as long as their predictions are borne out. A classic example is Einstein's theory of relativity, which has withstood more than a century of scientific efforts to disprove its predictions.

> The scientific method involves trying to *falsify* ideas.

Science is sometimes misunderstood because of the special ways that scientists use everyday words. An example is the word *theory*. In everyday language, *theory* may mean something that is little more than a conjecture or a guess: "Do you have a theory about who might have done it?" "My theory is that a third party could win the next election." In everyday parlance a theory is something worthy of little serious regard. "After all," we say, "it is only a theory."

In stark contrast, a *scientific* theory is a carefully constructed proposition that takes into account all the relevant data and all our understanding of how the world works, and it makes testable predictions about the outcome of future observations and experiments. A theory is a well-developed idea that is ready to be confronted by nature. A well-corroborated theory is a theory that has survived many such tests. Rather than being simple speculation, scientific theories represent and summarize bodies of knowledge and understanding that provide our fundamental insights into the world around us. A successful and well-corroborated theory is the pinnacle of human knowledge about the world.

Theories fill a place in a loosely defined hierarchy of scientific knowledge. In science, the word *idea* has its everyday use: an idea is just a notion about how something might be. In science, a *hypothesis* is an idea that leads to testable predictions. A hypothesis may be the forerunner of a scientific theory, or it may be based on an existing theory, or both. When an idea has been thought about carefully enough, has been tied solidly to existing theoretical and experimental knowledge, and makes testable predictions, then that idea has become a theory. Scientists build **theoretical models** that are used to connect theories with the behavior of complex systems. Ultimately, the basis for deciding among competing theories is the success of their predictions. Some theories become so well tested and are of such fundamental importance that we come to refer to them as **physical laws**.

A scientific **principle** is a general idea or sense about how the universe *is* that guides our construction of new theories. **Occam's razor**, for example, is a guiding principle in science stating that when we are faced with two hypotheses that explain a particular phenomenon equally well, we should adopt the simpler of the two. At the heart of modern astronomy is another principle: the **cosmological principle**. Simply put, the cosmological principle states that on a large scale the universe looks the same everywhere. That is, when we look out around us, what we see is representative of what the universe is generally like. In other words, there is nothing special about our particular location. By extension, matter and energy obey the same physical laws throughout space and time as they do today on Earth. This premise means that the same physical laws that we learn about in terrestrial laboratories can be used to understand what goes on in the centers of stars or in the hearts of distant galaxies. Each new success that comes from applying the cosmological principle to observations of the universe around us—each new theory that succeeds in explaining or predicting patterns and relationships among celestial objects—adds to our confidence in the validity of this cornerstone of our worldview. We will discuss the cosmological principle in more detail in Chapter 18.

> There is nothing special about our place in the universe.

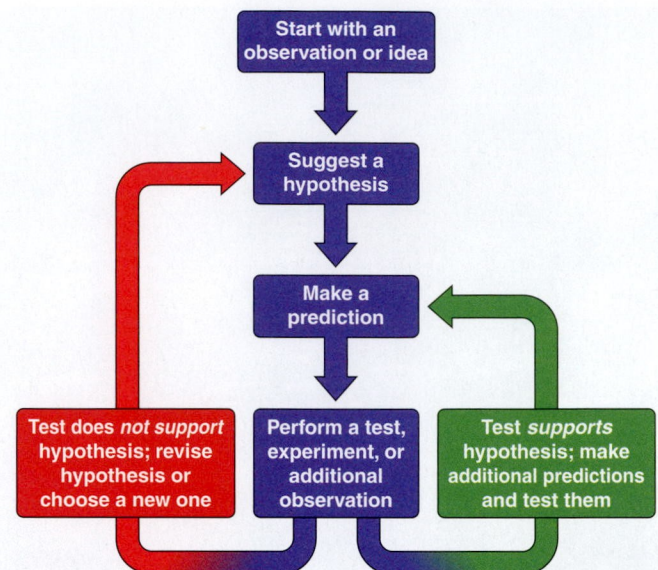

FIGURE 1.8 The scientific method. This is the path by which an idea or observation leads to a falsifiable hypothesis that is either accepted as a tested theory or rejected, on the basis of observational or experimental tests of its predictions. Note that the green loop goes on indefinitely, as scientists continue to test the hypothesis.

Science as a Way of Knowing

The path to scientific knowledge is solidly based on the scientific method. This concept is so important to your understanding of how science works that we should emphasize it once again. The scientific method consists of observation or idea, followed by hypothesis, followed by prediction, followed by further observation or experiments to test the prediction, and ending as a tested theory, as illustrated in **Figure 1.8**. For all practical purposes, it defines what we mean when we use the verb *to know*. It is sometimes said that the scientific method is how scientists prove things to be true, but actually it is a way of proving things to be *false*. Before scientists accept something as true, they work hard to show that it is false. Only after repeated attempts to disprove an idea have failed do scientists begin to accept its likely validity. For a theory to be given serious scientific consideration, it *must be falsifiable*. That is, it must be capable of being shown to be false. Scientific theories are accepted only as long as they are able to be tested and are not shown to be false. In this sense, *all scientific knowledge is provisional*.

> All scientific knowledge is provisional.

Despite being tied so closely to the scientific method, science can no more be said to *be* the scientific method than music can be said to *be* the rules for writing down a musical score. The scientific method provides the rules for asking nature whether an idea is false, but it offers no insight into where the idea came from in the first place or how an experiment was designed. If you were to listen to a group of scientists discussing their work, you might be surprised to hear them using words such as *insight*, *intuition*, and *creativity*. Scientists speak of a beautiful theory in the same way that an artist speaks of a beautiful painting or a musician speaks of a beautiful performance. Yet science is not the same as art or music in one important respect: whereas art and music are judged by a human jury alone, in science it is nature (through application of the scientific method) that provides the final decisions about which theories can be kept and which theories must be discarded. Nature is completely unconcerned about what we *want* to be true. In the history of science, many a beautiful and beloved theory has been abandoned. At the same time, however, there is an aesthetic to science that is as human and as profound as any found in the arts, as **Figure 1.9** illustrates.

Scientific Revolutions

As we have seen, science cannot be defined as either technology or the scientific method, although it is related to both. It is also incorrect to say that science is simply a body of facts. If we limited our definition of science to the body of existing facts, we would fail to convey the dynamic nature of scientific inquiry. Scientists do not pretend to have all the answers, and we constantly have to refine our ideas in response to new data and new insights. (This is, after all, what it means to learn.) The vulnerability of knowledge that is implicit in the scientific method may seem like a weakness at first. "Gee, you really don't know anything," the cynical student might say. But this vulnerability is actually

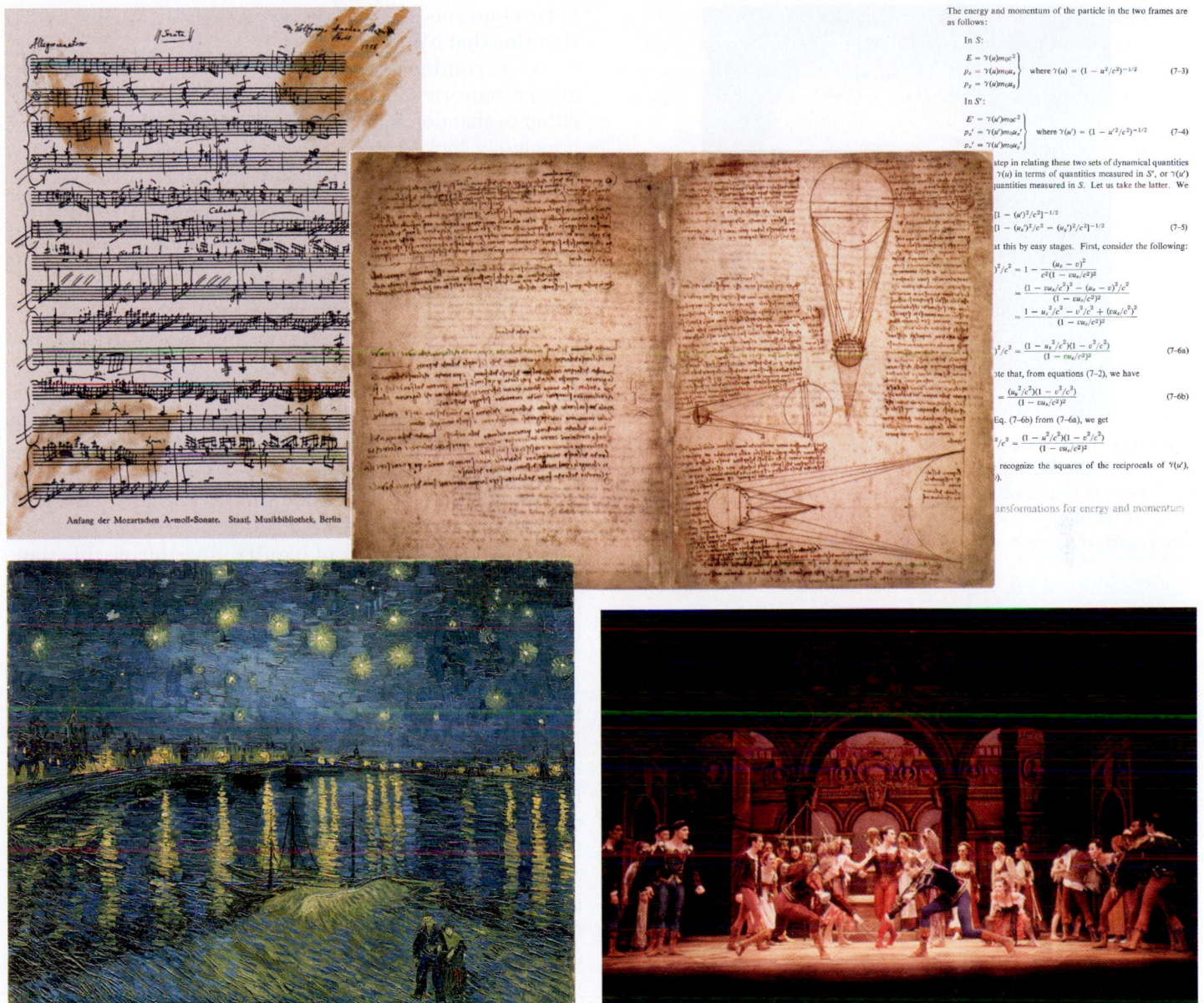

FIGURE 1.9 The scientific worldview is as aesthetically pleasing as that of art, music, or literature. Unlike the arts, however, science relies on nature alone to determine which scientific theories have lasting value.

science's great strength. It is what keeps us honest. Once an idea is declared to be "truth," then all progress stops. In science, even our most cherished ideas about nature remain fair game, subject to challenge by evidence according to the rules of the scientific method. Many of history's best scientists earned their place in the forward march of knowledge by successfully goring a sacred cow.

Scientists spend most of their time working within an established framework of understanding, extending and refining that framework and testing its boundaries. Occasionally, however, major shifts occur in the framework of an entire scientific field. Many books have been written about how science progresses, perhaps the most influential being

The Structure of Scientific Revolutions by Thomas Kuhn. In this work, Kuhn emphasizes the constant tension between the scientist's human need to construct a system of beliefs within which to interpret the world and the occasional (and likely painful) need to drastically overhaul that system of beliefs.

A scientific revolution is not a trivial thing. A new theory or way of viewing the world must be able to explain everything that the previous theory could, while extending this understanding to new territory into which the earlier theory could not go. By the middle of the 19th century, many physicists felt that our fundamental understanding of physical law was more or less complete. For over a cen-

FIGURE 1.10 Albert Einstein, perhaps the most famous scientist of the 20th century, and *Time* magazine's selection as Person of the Century. Einstein helped to usher in two different scientific revolutions, one of which he himself was never able to accept.

tury, the classical physics developed by Sir Isaac Newton had withstood the scrutiny of scientists the world over. It seemed that little remained but cleanup work—filling in the details. Some even went so far as to pronounce this period the "end of science." Yet during the late 19th and early 20th centuries, physics was rocked by a series of scientific revolutions that shook the very foundations of our understanding of the nature of reality.

If one face can be said to represent these scientific revolutions—and modern science—it is that of Albert Einstein (**Figure 1.10**). Einstein's special and general theories of relativity replaced the 300-year-old edifice of Newtonian mechanics, not by proving Newton wrong, but by showing that Newton's mechanics was a special case of a far more general and powerful set of physical laws. Einstein's new ideas unified our concepts of mass and energy and destroyed our conventional notion of space and time as separate things. Yet scientific revolutions are seldom comfortable for those who live through them, and even the greatest of scientists can be left behind. Einstein actually helped to start two scientific revolutions. He saw the first of these—relativity—through and embraced the world that it opened. Yet Einstein was unable to accept the implications of the second revolution he helped start—quantum mechanics—and he went to his grave unwilling to embrace the view of the world it offered.

Quantum mechanics forced us to abandon our everyday understanding of "substance" and even to part with the notion that we live in a universe in which effect follows cause in lockstep. Together, these revolutions led to the birth of what has come to be known as **modern physics**. Although modern physics *contains* Newtonian physics, the understanding of the universe offered by modern physics is far more sublime and powerful than the earlier understanding that it subsumed.

As we continue on our journey, we will encounter many other discoveries and successful ideas that forced scientists either to abandon their treasured notions or be left behind, hopelessly locked into a worldview that had ultimately failed the test of observation and experiment. The point is this: in physical science, we are not just paying lip service to a hollow ideal when we say that the rigorous standards of scientific knowledge respect no authorities. No theory, no matter how central or how strongly held, is immune from the rules.

For those of us who grew up in a world transformed by science, the scientific worldview might seem anything but revolutionary. Throughout much of history, however, knowledge was sought in the pronouncements of "authority" rather than through observation of nature. This authoritarian view slowed the advance of knowledge throughout western Europe for the millennium prior to the European Renaissance, and it was largely the Chinese and Arab cultures that kept the spark of inquiry alive during this time. The greatest scientific revolution of all was the one that overthrew "authority" and replaced it with rational inquiry and the scientific method.

From the perspective of this great scientific revolution, it is clear why science is more than simply a body of facts. Perhaps more than anything, science is a way of thinking about the world. It is a way of relating to nature. It is a search for the relationships that make our world what it is. It is a belief that nature is not capricious, but instead operates by consistent, explicable, inviolate rules. It is a collection of ideas about how the universe works, coupled with an acceptance of the fact that what is known today may be superseded tomorrow. The scientist's faith is that there is an order in the universe and that the human mind is capable of grasping the essence of the rules underlying that order—or at least of inventing ever-better approximations to those rules. The scientist's creed is that nature, through observation and experiment, is the final arbiter of the only thing worthy of the term *objective truth*. Science is an exquisite blend of aesthetics and practicality. And, in the final analysis, science has found such a central place in our civilization because *science works*.

Challenges to Science

Science has not achieved its prominence without criticism and controversy. Philosophers and theologians of the 16th century did not easily give up the long-held belief that Earth is the center of all creation. In the early 20th century, astronomers were arguing bitterly among themselves over the size of the universe. For centuries, conventional wisdom had held that the Milky Way was all that there was—a vast collection of stars and "fuzzy" objects known as "nebulae." That belief collapsed when it was realized that some of

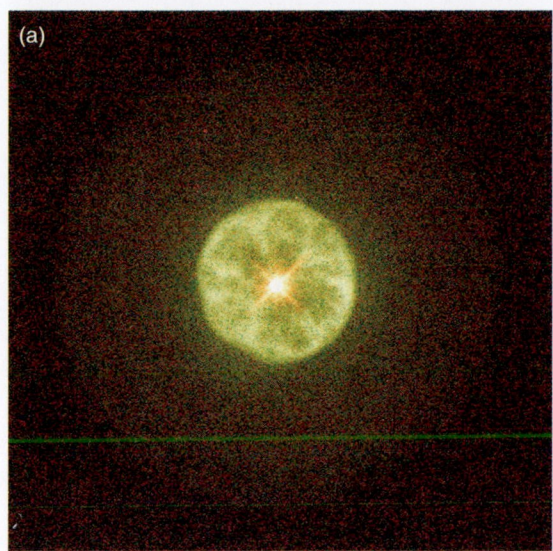 

FIGURE 1.11 Until the early 20th century, astronomers believed that all "fuzzy" celestial objects were located within our Milky Way Galaxy. (a) A planetary nebula approximately 10,000 light-years away within our own galaxy. (b) A spiral galaxy 110 million light-years distant, more than 10,000 times farther than the planetary nebula shown in (a).

these fuzzy objects were actually distant galaxies ("island universes")—similar to but far beyond the confines of our own Milky Way (**Figure 1.11**). For many, it came as a rude awakening that our Milky Way does not, in fact, constitute the entire universe—that it is only an infinitesimal part of a greatly larger universe!

You might feel that science is sometimes arbitrary. For example, the decision as to whether Pluto should be considered a planet or a dwarf planet may seem subjective. But it is important to keep in mind that this controversial decision, unsupported by many astronomers, was really more a matter of semantics than of science. Pluto is still the same scientifically important Solar System body that it always was. It is just that "science" has now put a new label on it. As you will see in discussions throughout this book, the path to knowledge embodied in science does not seem to be any more arbitrary than the logic it is built on.

In recent years, some critics of science have drawn attention to how science is influenced by culture. It is hard to avoid the conclusion that political and cultural considerations strongly influence which scientific research projects are funded. This choice of funding channels the directions in which scientific knowledge advances and can lead to serious ethical issues. For example, moral judgments about nontraditional lifestyles greatly restricted the funding available for AIDS research during the decade or so after its discovery. More recently, progress in stem cell research and solutions to human-caused climate change have been hindered by widely distributed misinformation on the Web and by constant political bickering.

Some critics even carry this view a step further, arguing that scientific knowledge itself is an arbitrary cultural construct. Yet successful scientific theories are *never* arbitrary. Scientific theories must be consistent with all that we know of how nature works, and turning a clever idea into a real theory with testable predictions is a mat-

ter of careful thought and effort. One of the most remarkable aspects of scientific knowledge is its *independence from culture*. Scientists are people, and politics and culture enter into the day-to-day practice of science. But in the end, *scientific theories should be judged not by cultural norms but by whether their predictions are borne out by observation and experiment*. As long as the results of experiments are repeatable and do not depend on the culture of the experimenter—that is, as long as there is such a thing as objective physical reality—scientific knowledge cannot be called a cultural construct. Science is not *just* one of many possible worldviews; science is the most successful worldview in the history of our species because the foundations of the scientific worldview have withstood centuries of fine minds trying to prove them false.

> Nature is the arbiter of science.

It is significant that no philosopher critical of science has ever offered a viable alternative for obtaining reliable knowledge of the workings of nature. Furthermore, no other category of human knowledge is subject to standards as rigorous and unforgiving as those of science. For this reason, scientific knowledge is reliable in a way that no other form of knowledge can claim. Whether you want to design a building that will not fall over, consider the most recent medical treatment for a disease, or calculate the orbit of a spacecraft on its way to the Moon, you had better consult a scientist rather than a psychic—regardless of your cultural heritage.

Finally, we must admit that some disreputable scientists purposefully try to influence results by inventing or ignoring data, often when claiming a "major breakthrough" or challenging a well-established scientific principle. Fortunately, attempts by others to repeat the experiment will eventually expose such scientific misconduct.

1.4 Patterns Make Our Lives and Science Possible

We experience patterns in our everyday life and we find comfort in them. But imagine what life would be like if sometimes when you let go of an object it fell up instead of down. What if one day apples were essential nutrition, but when you bit into an apple the next day you discovered that they were deadly? What if, unpredictably, one day the Sun rose at noon and set at 1:00 P.M., the next day it rose at 6:00 A.M. and set at 10:00 P.M., and the next day the Sun did not rise at all? In fact, objects do fall toward the ground. The biochemistry of the food we eat remains the same. The Sun rises, sets, and then rises again at predictable times. Spring turns into summer, summer turns into autumn, autumn turns into winter, and winter turns into spring. The rhythms of nature produce patterns in our lives, and we count on these patterns for our very survival. If nature did not behave according to regular patterns, then our lives—indeed, life itself—would not be possible.

The patterns that make our lives possible also make science possible. The goal of science is to identify and characterize these patterns and to use them to understand the world around us. Some of the most regular and easily identified patterns in nature are the patterns that we see in the sky. What in the sky will look different or the same a week

FIGURE 1.12 Since ancient times, our ancestors recognized that patterns in the sky change with the seasons. These and other patterns shape our lives.

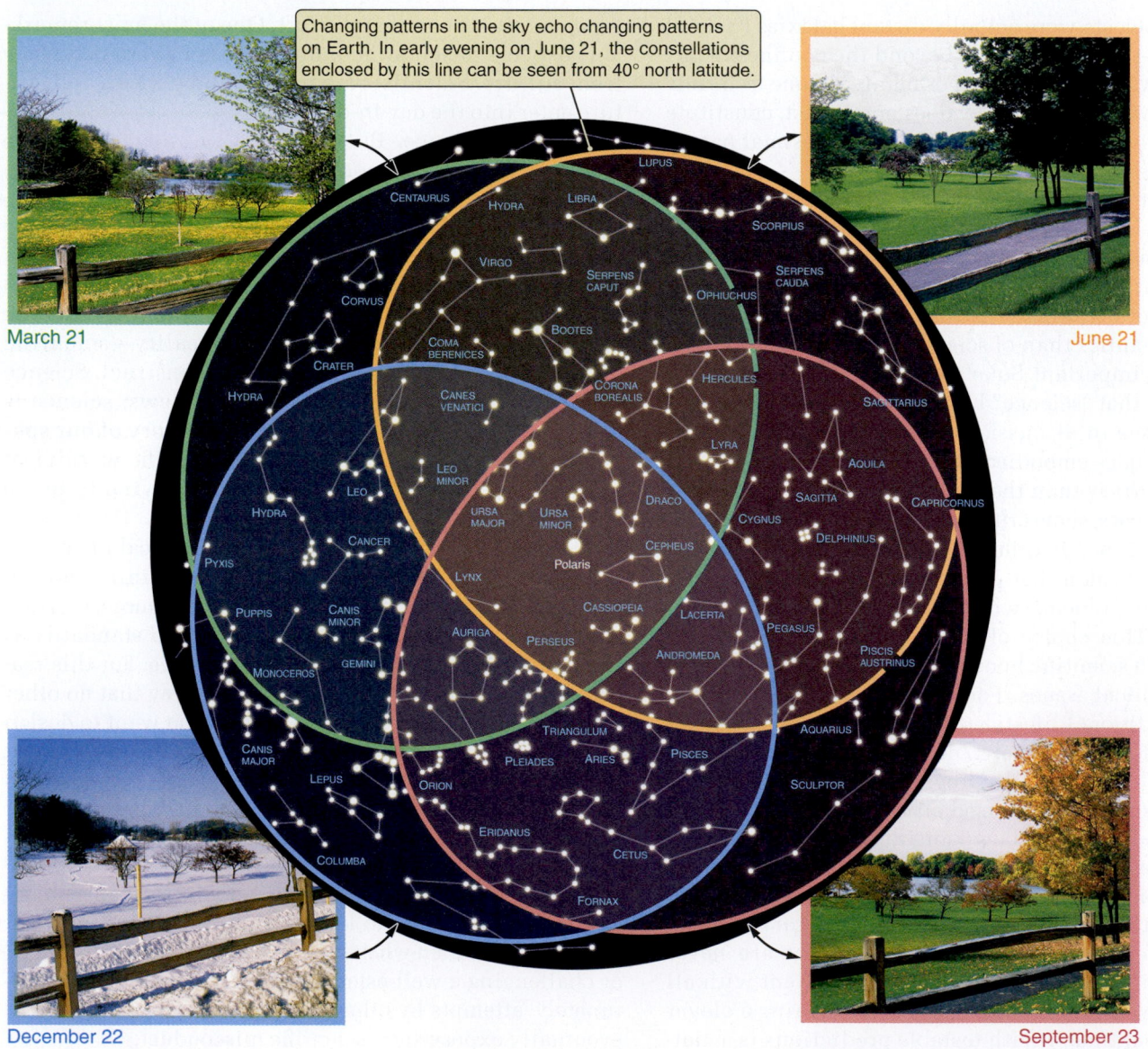

Changing patterns in the sky echo changing patterns on Earth. In early evening on June 21, the constellations enclosed by this line can be seen from 40° north latitude.

March 21

June 21

December 22

September 23

NON SEQUITUR   by WILEY

FIGURE 1.13 Because mathematics helps us identify and analyze patterns in nature, it has become the language of science.

from now? A month from now? A year from now? Most of us probably lead an indoor and in-town existence, removed from an everyday awareness of the patterns in the sky. Away from the smog and glare of our cities, however, the patterns and rhythms of the sky are as easy to see today as they were in ancient times. Patterns in the sky mark the changing of the seasons (**Figure 1.12**) and the planting and harvesting of the crops. Patterns in the sky share the rhythms of our lives. It is no surprise that astronomy, which is the expression of our human need to understand these patterns, is the oldest of all sciences.

One important tool that astronomers use to analyze these patterns is mathematics. There are many branches of mathematics, most of which deal with more than just numbers. Arithmetic is about counting things. Algebra is about manipulating symbols and the relationships between things. Geometry is about shapes. Calculus is about change. Other types of mathematics include topology (the properties of surfaces) and statistics (groups of objects and their relationships). What do all these branches have in common? Why do we consider all of them to be part of a single discipline called "mathematics"? All share one thing: they deal with patterns. The best working definition of mathematics is that "mathematics is the language and science of patterns."

> Mathematics is the science and language of patterns.

If patterns are the heart of science, and mathematics is the language of patterns, it should come as no surprise that *mathematics is the language of science*. Trying to study science while avoiding mathematics is the practical equivalent of trying to study Shakespeare while avoiding the written or spoken word. It quite simply cannot be done, or at least cannot be done meaningfully.

On the other hand, as the authors of this book we understand (and as is humorously depicted in **Figure 1.13**) that for many of you, *math* is not *a* four-letter word—it is *the* four-letter word. Many people decide early in their education that they cannot "do" math, and from that day forward the mere mention of the word causes their eyes to glaze over and their palms to sweat. A distaste for mathematics is one of the most common obstacles standing between a nonscientist and an appreciation of the beauty and elegance of the world as seen through the eyes of a scientist. To move beyond this obstacle, scientist and nonscientist need to find common ground.

Part of the responsibility for moving beyond this obstacle lies with us, the authors. It is our job to take on the role of translators, using words to express as many concepts as possible, even when these concepts are more concisely and accurately expressed mathematically. When we do use mathematics, we will explain in everyday language what the equations mean and try to show you how equations express concepts that you can connect to the world. We will also limit the mathematics to a few basic tools that all college students should have been exposed to.

These basic mathematical tools, as described in **Math Tools 1.1** and Appendixes 1 and 2, enable scientists to convey complex information. Scientific notation, for example, enables astronomers to express in simple terms the vast range of sizes of astronomical objects. Scientists use a basic set of *units* to distinguish among time, distance, mass, and energy. Astronomers also find that a bit of *geometry* is necessary for understanding the sizes, shapes, and volumes of astronomical objects and the distances between them.

Mathematical Tools

Mathematics gives scientists many of the tools that they need to understand the patterns they see and to communicate that understanding to others. As the authors of this text, we are aware that mathematics is not a friend to many of you taking this course, so we have worked to keep the math in this text to a minimum. Even so, there are a few tools that we will need in our study of astronomy:

Scientific notation. Scientific notation is the way that scientists deal with numbers of vastly different sizes. Rather than writing out 7,540,000,000,000,000,000,000 in standard notation, we express the same number in simpler form as 7.54×10^{21}. Rather than writing out 0.000000000005, we write 5×10^{-12}. For example, the distance to the Sun is 149,580,000 km, but astronomers usually express it as 1.4958×10^8 km.

For a more detailed explanation of the use of scientific notation and significant figures, see Appendix 1.

Ratios. Ratios are the most common way that astronomers use to compare things. A star may be "10 times as massive as the Sun" or "10,000 times as luminous as the Sun." These expressions are ratios.

Geometry. To describe and understand objects in astronomy and physics, we use concepts such as distance, shape, area, and volume. Apparent separations between objects in the sky are expressed as *angles*. Earth's orbit is an *ellipse* with the Sun at one *focus*. The planets in the Solar System lie close to a *plane*. Geometry provides the tools for working with these concepts.

Algebra. Algebra provides a way of using and manipulating symbols that represent numbers or quantities. We will use algebra to express relationships that are valid not just for a single case, but for many cases. Algebra lets us conveniently express ideas such as "the distance that you travel is equal to the speed at which you are moving multiplied by the length of time you go at that speed." Written as an algebraic expression, this idea is

$$d = s \times t,$$

where d is distance, s is speed, and t is time.

Algebra also lets us combine these ideas with other ideas to arrive at new relationships.

Proportionality. Often, understanding a concept amounts to understanding the *sense* of the relationships that it predicts or describes. "If you have twice as far to go, it will take you twice as long to get here." "If you have half as much money, you will be able to buy only half as much gas." These are examples of proportionality.

A proportion relevant in astronomy is the relationship among speed, time, and distance. If you are traveling at a constant speed, then time is proportional to distance. We write

$$t \propto d,$$

where $\propto$ means "is proportional to."

Proportionalities often involve quantities raised to a power. For example, a circle of radius r has an area A equal to πr^2, so we say that area is proportional to the square of the radius, and we write

$$A \propto r^2.$$

This means that if you make the radius of a circle three times as large, its area will grow by a factor of 3^2, or 9.

Units. In this book we use the metric system of units, more properly known as the *Système International d'Unités* (*SI*). The United States remains one of very few countries in the world still using the English system of units, but gradual acceptance of the metric system can be seen in everyday life. We buy milk in quart-sized containers, but our soft drinks come in liter-sized bottles. Bank signs often display temperature in both Fahrenheit (°F) and Celsius (°C), also known as Centigrade. In some municipalities road signs show distances in both miles and kilometers.

Conversions to the English units used in the United States can be found in Appendix 2.

Finally, some *algebra*—mostly a few *ratios* and *proportionalities*—will provide a way of expressing the patterns that relate one physical quantity to another. *Basic* does not necessarily mean easy, but it does mean that we will use the most accessible tools that will make our journey of discovery as comfortable and informative as possible.

Your responsibility is to accept the challenge and make an honest effort to think through the mathematical concepts that we use. Do not concede defeat while still in the starting blocks. It is likely that you know what it means to square a number, or to take its square root, or to raise it to the third power. The mathematics in this book is on a par with what it takes to balance a checkbook, build a bookshelf that stands up straight, check your gas mileage, estimate how long it will take to drive to another city, figure your taxes, or buy enough food to feed an extra guest or two at dinner.

1.5 Thinking like an Astronomer

Not everyone is fascinated by mathematics or science, but almost everyone harbors a spark of interest in astronomy. Because you are reading this book, you probably share this spark as well. The spark may have been struck when you were a child looking at the sky and found yourself wondering about what you saw there. The prominence of the Sun, Moon, and stars in cave paintings and rock drawings (such as those in **Figure 1.14**) dating back thousands of years tells us that these questions have long occupied the human imagination. Your initial spark of interest in astronomy may have grown over the years as you saw or read news reports about spectacular discoveries made in your lifetime. Some of these discoveries may have sounded so amazing that it was difficult to draw the line between science fact and science fiction.

If you are like many other people, your conception of astronomy so far may not have gone much beyond learning about the constellations and the names of the stars in them. But if you nurture your spark of interest in astronomy, you may be surprised to find that spark growing into a flame that will fuel your journey through the chapters that follow. This book will take you to places you never imagined going and will lead you to insights and understandings you never imagined having. To those of you who are reading this book for a course in astronomy at your college or university, we have a special note: The authors of this text have taught introductory astronomy many times over the years. We recognize that you may be in this course primarily because you need a science credit to graduate. As you flipped through your course catalog, perhaps you were reminded of your interest in astronomy, and that led you to choose astronomy over other options. (Or perhaps you simply considered astronomy to be the least of the available evils!) Whatever your expectations, the story in *21st Century Astronomy* can fascinate you if you open your mind to it.

Our best suggestion for a successful journey through *21st Century Astronomy* is to nurture your spark of interest in astronomy until it grows enough to draw you in. When, as Einstein phrased it in the opening quote, you "pause to wonder and stand rapt in awe"—*then* you will have learned the secret to thinking like a scientist!

Reading the book and taking the journey with us are two different things. This is only a guidebook. It can lead you to the trailhead and tell you something of what you might find along the way, but *you* have to walk the path! If you become an active participant in this adventure, rather than a passive spectator, then what you gain from the journey will remain a part of you long after the final exam is forgotten. And if along the way you find yourself applying your understanding to new situations and new information, and if you learn to combine your understandings and arrive at new insights that are greater than the sum of their parts, then you will have learned something far more than just astronomy.

This book will likely ask you to think in ways that are different from the ways in which you are accustomed to thinking, and to learn to view the world from new and unfamiliar perspectives. Knowledge and understanding have nothing to do with shoving facts into short-term memory so that they can be regurgitated on an exam, and such a short-term strategy is certainly not how astronomy—or any other field of study, for that matter—should be approached. Changing the way that you think takes more effort. It also means studying in ways that may be different from your normal habits. Above all, remember that building understanding is always an *active* process, never passive! Here are a few practical suggestions for how you might better study this text:

FIGURE 1.14 Ancient petroglyphs often include depictions of the Sun, Moon, and stars.

- **Read the text actively.** Think about each section after you have read it. What major concepts were discussed in the section? How are they related to what you have read so far? Have you run into similar concepts elsewhere? Why are the contents of that section important enough to be included in the book? In your class notes, briefly summarize the section and your thoughts about it.

- **Visualize the concept.** Many physical and mathematical concepts are most easily understood if they are visualized. Review the photos, diagrams, and charts in each chapter to study key concepts. Draw a picture of the concept. If you understand a concept well enough to draw a picture that expresses it, then you probably understand the concept fairly well. Trying to sketch a picture will also help you better identify what things you understand and what things you do not.

- **Ask yourself "What if?"** when trying to understand a concept. What if Earth were more massive? How would that affect Earth's gravity? What if the Sun were hotter? How would that affect the color of the light from the Sun or the amount of energy that the Sun radiates? If you cannot "what if" a concept, then you probably do not really understand it yet.

- **Try to teach it.** It is often said that you do not really understand something until you try to explain it to someone else. At the end of a reading assignment, talk about the material with a friend or family member. Try to find a partner or a group in your class with whom you can meet, and *take turns teaching each other the material*. Each of you should read the assigned text, and then the group should divide up responsibility for who will present which sections. Ask each other lots of questions—the harder the better! (Make a game of "stump the instructor.") Even explaining a concept out loud to yourself can help.

- **Share your ideas and insights** with your discussion group. If a particular concept really "clicks" for you—if you really think it is cool—share both your understanding and your enthusiasm with your group.

- **Be honest with yourself** about what things you understand and what things you do not, and try not to avoid concepts that you find difficult. Getting the most from this journey will come from facing and mastering these challenging concepts. You will find that personal growth comes from this real accomplishment.

- **Focus your discussion on concepts, relationships, and connections.** You have to know the facts, but the facts are the starting point, not the end. Use the key concepts, study questions, and other study aids to identify and concentrate your effort on the most important ideas.

- **Do not let discomfort with math keep you from succeeding.** Mathematical formulas are not magical incantations. Rather, they are expressions of logical ideas. We will always present a plain-English discussion of the idea behind any mathematics that we use. Begin by focusing on this discussion. After you grasp the idea, then look at the math and any equations presented in the separate "Math Tools" boxes. Try to see how the relationships between the quantities in the equation embody the concept and try to follow along with the worked examples, when given. If you need help with basic math skills,

work through the appendixes or explore StudySpace (www.norton.com/studyspace) for other aids available to you. Your instructor is also there to help.

You will occasionally find material set aside in boxes. A topic has been boxed either because it is somewhat out of the mainstream of our journey or because we wish to highlight it. In particular,

- **Foundations** boxes address material that is central to our understanding of the physical nature of the universe.

- **Math Tools** boxes provide additional math information or worked examples of algebraic expressions.

- **Connections** boxes draw attention to recurring themes—bridges between different parts of our journey.

- **Excursions** boxes are short but interesting side trips.

Throughout each chapter you will find **Trailmarkers**, one-sentence summaries of fundamental concepts. They are inserted in the text, set off by light green horizontal bars.

As you study the pages of this book—in concepts and images—you will see that we truly live in a golden age of exploration and discovery. We can be certain that future historians will remember ours as the time when humankind first stepped beyond the world of our birth and began to reach out with our minds and our science to touch the fabric of the universe itself. It is probably safe to say that few things will have a more lasting impact on our culture than this revolution in our understanding of the universe and our place in it. What has yet to be determined is whether future historians will remember us for reaching out to touch the universe and embracing what we found, or whether we will instead be remembered for a loss of spirit—for stepping back and turning away from the frontier of exploration and discovery. The direction that we take from here is a decision in which you—and your willingness to embrace science—will play a part.

The journey of discovery on which we are embarking is not always easy, but few worthwhile journeys are. A hike in the mountains can at times be an easy stroll and at other times a more strenuous climb; but when you arrive, the view from the top is hard to beat. In much the same way, this book will ask you to exercise your mental muscles in different, possibly unaccustomed ways. But, as with the hike in the mountains, we feel certain that you will find the rewards worth the investment.

Seeing the Forest for the Trees

At the end of each chapter of *21st Century Astronomy*, you will find a brief narrative about the content of that chapter. The purpose of "Seeing the Forest for the Trees" is not to rehash the entire contents of the chapter, but rather to pick out a few of the high points and put them into a broader context. Building on the analogy of a hike in the mountains, we will spend a lot of time looking in detail at the rocks and the trees, but every so often we need to step back and look around at the forest as a whole.

We live in a world that has been profoundly shaped by the scientific revolution that took hold of Western thought during the Renaissance. That revolution fundamentally altered our way of thinking about the world, as well as our view of the relationship between ourselves and the universe of which we are a part. A new spirit of rational inquiry was turned on the heavens, dislodging Earth and humankind from the center of the cosmos. Observation, experiment, and rigorously applied reason came to replace dogma and authority as the arbiter of knowledge. The heavens became a realm not of mysticism and magic, but instead of physical law—the same physical law that governs the behavior of matter and energy in laboratories here on Earth.

The closing years of the 20th century saw our knowledge of the universe charge ahead at an ever-accelerating pace. This progress comes courtesy of a great many advances both in our technology and in the sophistication of our physical understanding of matter and energy and of space and time themselves. We have seen many fundamental questions about the origin and fate of the universe—and about the threads that tie our existence to the cosmos—move from the realm of philosophical speculation into the realm of rigorous scientific inquiry. The insights that this age of exploration and discovery have brought are often far more profound and startling than dreamed of even a few decades ago. There can be little doubt that this time will be looked on as one of the more significant moments in the intellectual and cultural history of our species.

Like a hike through the mountains, *21st Century Astronomy* will not be an effortless journey. Many travelers will find that they have to flex a few mental muscles in ways they are not used to, and they may even have to face an old adversary or two on the trail. But muscles that are sore after the first day of a hike grow comfortable and strong with time, and adversaries can become the best of friends.

In Chapter 2 we begin the journey in earnest and, as with most journeys, our starting point is home. What patterns do we see in the skies of our planet Earth, and how do those patterns come to be? This is not an easy or gentle slope on which to begin our trek, but our vistas will change rapidly as we climb.

Summary

- Our Solar System is but a tiny speck in a vast universe.

- There is nothing special about our particular place in the universe.

- We, and everything around us, are composed of atoms that were formed in stars whose lives ended long ago.

- Space exploration has expanded our view of the universe.

- The scientific method is a way of trying to *falsify*, not prove, ideas.

- *All* scientific knowledge is provisional.

- The entire universe is governed by the same physical laws that shape our lives here on Earth.

- Mathematics is the science and language of patterns, and thus it is the language of science.

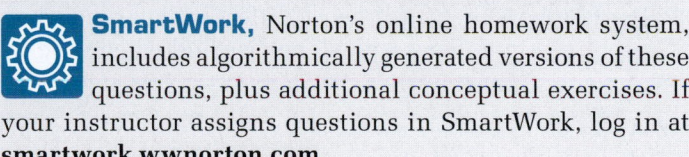 **SmartWork,** Norton's online homework system, includes algorithmically generated versions of these questions, plus additional conceptual exercises. If your instructor assigns questions in SmartWork, log in at **smartwork.wwnorton.com.**

Student Questions

THINKING ABOUT THE CONCEPTS

1. Suppose you lived on imaginary planet Zorg orbiting Alpha Centauri, a nearby star. How would you write your cosmic address?

2. Without looking back in the text,
   a. How many of the eight classical planets can you name?
   b. How many dwarf planets?

3. Assuming that the number of stars in all the dwarf galaxies in our Local Group is negligible compared to the number of stars in our Milky Way Galaxy, estimate the total number of stars in our Local Group.

4. If the Sun suddenly exploded (it can't), how soon after would we know about it?

5. If a star exploded in the Andromeda Galaxy (they sometimes do), how long would it take that information to reach Earth?

*6. It is said that we are made of stardust. Explain why this statement is true.

7. A friend tells you that the reason astronomers put telescopes in space is to get closer to the planets and stars. You know better. How do you explain the real reason to your friend?

8. Your doctor can look for broken bones using X-rays, but an astronomer working with a telescope at your nearby observatory cannot see X-rays coming from a star. Explain why not.

9. List three nonastronomical scientific disciplines that have contributed to our modern knowledge of astronomy.

10. Controversial author Erich von Däniken proposed the theory that Earth was visited by extraterrestrials in the remote past. Would you regard this as scientific or pseudoscientific theory? Is the theory falsifiable? Can you think of any tests that could support or refute the theory?

11. The scientific method requires that scientific theories be falsifiable. List some beliefs or views that you conclude are *not* falsifiable.

*12. Explain how the word *theory* differs in meaning when used in common everyday language and when used by a scientist.

13. What is the difference between the terms *hypothesis* and *theory* as used by scientists?

14. The tabloid newspaper at your local supermarket theorizes that, compared to average children, children born under a full Moon become more intelligent students.
a. Is this theory falsifiable?
b. If so, how could it be tested?

15. A textbook published in 1945 stated that it takes 800,000 years for light to reach us from the Andromeda Galaxy. In this book we say that it takes 2,500,000 years. What does this tell you about a scientific "fact" and how our knowledge evolves with time?

16. Astrology makes testable predictions. For example, it predicts that the horoscope for your star sign on any day should fit you better than do horoscopes for other star signs. Read each of the daily horoscopes in a newspaper or website without regard to your own sign. How many of them might fit the day that you had yesterday? Repeat the experiment every day for a week and keep records. Was your horoscope sign consistently the best description of your experiences?

17. A scientist on television states that it is a known fact that life does not exist beyond Earth. Would you consider this scientist reputable? Explain your answer.

18. Some astrologers use elaborate mathematical formulas and procedures to predict the future. Does this show that astrology is a science? Why or why not?

*19. Imagine yourself living on a planet orbiting a star in a very distant galaxy. What does the cosmological principle tell you about the way you would perceive the universe from this distant location?

20. You run across an old newspaper with the headline "EINSTEIN PROVES NEWTON WRONG!" Did the newspaper get this story right? Explain your answer.

21. List patterns in your own life that repeat regularly. How do these patterns affect you? Which patterns are of your own making, which are set by others, and which are determined by nature?

22. Name some supermarket items that are sold by the liter instead of by the quart.

APPLYING THE CONCEPTS

23. (a) If it takes about 8 minutes for light to travel from the Sun to Earth, and Pluto is 40 times as far from us, how long does it take light to reach Earth from Pluto? (b) Radio waves travel at the speed of light. What does this fact imply about the problems you would have if you tried to conduct a two-way conversation between Earth and a spacecraft orbiting Pluto?

*24. Astronomers find it useful to use the speed of light and the time it takes light to travel a given distance as the basis for discussing astronomical distances. Use the travel time of light and more familiar units, such as miles (you can approximate 1 mile = 1.6 km), to illustrate this point when referring to the diameter of the Milky Way Galaxy and the distance to the Andromeda Galaxy.

*25. Imagine the Sun to be the size of a grain of sand and Earth a speck of dust 83 millimeters (mm) away. (On this scale, each light-minute of distance equals 10 mm.) Look at Figure 1.2. What would be the distance from Earth to the Moon on this scale? From the Sun to Neptune? From Earth to the nearest star? To the nearest large galaxies? (Note that 1 meter = 10^3 mm and that 1 km = 10^3 meters.) At what point do you lose your "feel" for these distances?

26. The average distance from Earth to the Moon is 384,000 km. How many days would it take, traveling at 800 kilometers per hour (km/h)—the typical speed of jet aircraft—to reach the Moon?

27. The surface area of a sphere is proportional to the square of its radius. The radius of the Moon is only about one-quarter that of Earth. How does the surface area of the Moon compare with that of Earth?

28. Write 86,400 (the number of seconds in a day) and 0.0123 (the Moon's mass compared to Earth's) in scientific notation.

29. Write 1.60934×10^3 (the number of meters in a mile) and 9.154×10^{-3} (Earth's diameter compared to the Sun's) in standard notation.

*30. The time (t) it takes for light to reach us from a distant galaxy is equal to the distance (d) of the galaxy divided by the speed of light (c). Use algebra to describe this relationship more simply.

31. A remote Internet Web page may sometimes reach your computer by going through a geostationary satellite orbiting approximately 3.6×10^7 meters above Earth's surface. What is the minimum delay, in seconds, that the Web page takes to reach your computer?

*32. Some people have a theory that a tray of hot water will freeze more quickly than a tray of cold water when both are placed in a freezer.
   a. Does this theory make sense to you?
   b. Is the theory falsifiable?
   c. Do the experiment yourself. Note the results. Was your intuition borne out?

33. If you understand proportionality, then you understand most of the math you need to follow this text. Make a list of five different proportionalities from your daily life. (For example, the price of a bag of apples is proportional to the weight of the bag of apples.) For each proportionality, identify the constant of proportionality (such as the price per pound of apples). How are these constants determined?

34. A pizzeria offers a 9-inch-diameter pizza for $12 and an 18-inch-diameter pizza for $24.
   a. Are both offerings equally economical?
   b. If not, which is the better deal? Show why you think so.

35. The circumference of a circle is given by $C = 2\pi r$, where r is the radius of the circle.
   a. Calculate the approximate circumference of Earth's orbit around the Sun, assuming that the orbit is a circle with a radius of 1.5×10^8 km. You can approximate π as being about equal to 3.
   b. Noting that there are 8,766 hours in a year, how fast, in kilometers per hour, does Earth move in its orbit?
   c. How far along in its orbit does Earth move in one day?

36. Gasoline is sold by the gallon in the United States and by the liter nearly everywhere else. There are approximately 3.8 liters in a gallon. If gas cost $4 per gallon, how much would it cost per liter?

37. There are 4 quarts in a gallon. Using the algebraic expression, 1 quart = $C \times 1$ gallon. What is the constant of proportionality, C?

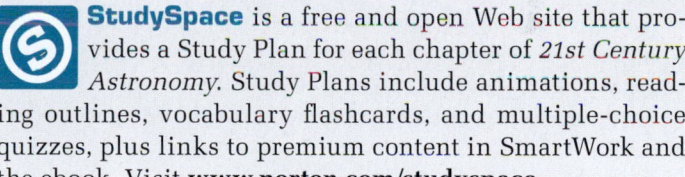 **StudySpace** is a free and open Web site that provides a Study Plan for each chapter of *21st Century Astronomy*. Study Plans include animations, reading outlines, vocabulary flashcards, and multiple-choice quizzes, plus links to premium content in SmartWork and the ebook. Visit **www.norton.com/studyspace.**

. . . marking the conclave of all the night's stars,
those potentates blazing in the heavens
that bring winter and summer to mortal men,
the constellations, when they wane, when they rise.

AESCHYLUS (525–456 B.C.)

The Moon rising over the ruins of the Greek Temple of Poseidon.

CHAPTER 2

Patterns in the Sky— Motions of Earth

2.1 A View from Long Ago

The herds have reached the high meadows where they spend the warm season, and for a time the life of the tribe has settled in as well. The weather is comfortable and the days are long, with none of the hardships that accompany the time of cold and snow and long, dark nights. It is a time of plenty and a time for telling the age-old stories of the tribe around a fire that guards against the chill of the gathering night. You feel a sense of contentment and are thankful to the gods for this time when life is good. As the embers die down, you gaze upward at the familiar canopy of stars overhead and, as you often do in such moments, you wonder about what you see.

To survive, you must learn the subtle patterns of your world. You must know the ways of the herds and recognize the gathering of clouds that heralds a coming storm. When you turn your keen eye toward the heavens, you find subtle and changing patterns there as well—patterns that somehow echo those of your life. The spirits of the great animals dwell in the sky; as a child, you learned to recognize their pictures there. Above you now are the stars that rule the time of the short nights. These are the stars that bring summer and lead the herds to this pleasant place.

Some of the spirits of the sky can be difficult to please. The mischievous planets wander from place to place, using their fearsome powers to sow chaos through the heavens. The Moon sometimes turns blood red, and the Sun is consumed by an ominous beast. But as long as the tribe remembers them, the gods and spirits of the sky will continue to bring the seasons and send the stars to guide the tribe.

KEY CONCEPTS

In this chapter we begin our journey in earnest—starting out as our ancestors did when they first gazed at the Sun, Moon, and stars and tried to understand what they saw. With the benefit of knowledge hard-won over the centuries, we will look at patterns present both in the sky and on Earth, and then look beyond appearances to the underlying motions that cause those patterns. Here we will discover

- How the stars appear to move through the sky as Earth rotates on its axis, and how those motions differ when seen from different latitudes on Earth.

- The fundamental concept of a frame of reference, and how Earth's rotating frame of reference affects our perception of celestial motions.

- How Earth's motion around the Sun and the tilt of Earth's axis relative to the plane of its orbit combine to determine which stars we see at night and which seasons we feel through the year.

- The motion of the Moon in its orbit about Earth, and how that motion, together with the motion of Earth and the Moon around the Sun, shapes the phases of the Moon and the spectacle of eclipses.

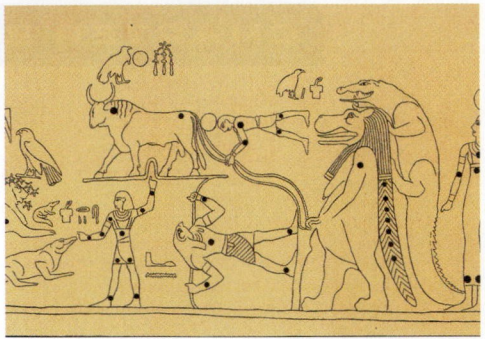
Egyptian—1275 B.C.

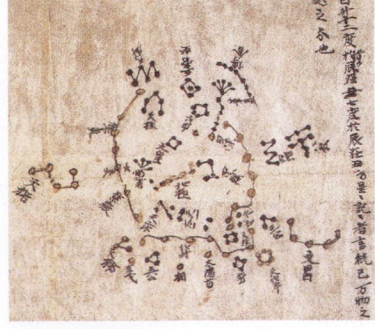
Chinese—940 A.D.

European—1540 A.D.

FIGURE 2.1 The region of the sky around what we now call the Big Dipper (Ursa Major, or the "Great Bear") as viewed by three different civilizations. Constellations are groupings of stars whose images, such as a dipper or bear, exist only in the human mind.

Our ancestors lived their lives attuned to the ebb and flow of nature, and the patterns in the sky are a part of that ebb and flow. The coming of night and day, the changing of the seasons, the rising and falling of the tides, the movement of the herds—all of these march in lockstep with the changes that we see in the sky. The repeating patterns of the Sun, Moon, and stars echo the rhythms that have defined the lives of humans since before the beginning of recorded history. By watching the patterns in the sky, our ancestors found that they could predict when the seasons would change and the rains would come and the herds would move. Knowledge of the sky offered knowledge of the world, and knowledge of the world was power.

> Patterns in the sky have always been important to our species.

It was a small step from here to thinking of the unreachable, untouchable stars not only as a reflection of the patterns in the world, but also as the *cause* of those patterns. The stars found a special place in legend and mythology as the realm of gods and goddesses, holding sway over the lives of humankind. As writing came to replace oral traditions and legends, mythologies of the sky became more elaborate as well. And as humans invented numbers and mathematics to describe and predict and account for things in the world, predictions of the motions of the stars and planets were among their greatest successes. Some of our ancestors came to look upon the orderly and predictable patterns of the sky as the *true* patterns of the world, and our own lives as imperfect reflections of this heavenly reality. They looked for ways to use their knowledge of the sky to find order in the seeming chaos of their everyday lives, and **astrology** was born.

Elements of this same basic history played themselves out many times over and in every part of our globe. From Africa to Asia, from Europe to Central America, from North America to the British Isles, the archaeological record holds evidence of early humans who projected ideas from their own cultures onto what they saw in the sky. As illustrated in **Figure 2.1**, there have been as many different sets of **constellations** and stories to go with them as there have been cultural traditions in human history. But where are these winged horses, dragons, chained maidens, and other imaginary images formed from patterns of stars? The answer may seem obvious: "The constellations are overhead in the sky, for all to see." Yet if you look at the sky, no clear pictures of these images emerge. Instead there is only the random pattern of stars—about 5,000 of them visible to the naked eye—spread out across the sky.[1] Constellations exist only within the imagination of the human mind. They are the ideas and pictures that humans impose on the lights in the sky in an effort to connect our lives on Earth with the workings of the heavens.

Modern constellations visible from the Northern Hemisphere draw heavily from the list compiled 2,000 years ago by the Alexandrian astronomer Ptolemy. Constellations in the southern sky are drawn from the lists put together by European explorers visiting the Southern Hemisphere during the 17th and 18th centuries. Today astronomers use an officially sanctioned set of constellations as a kind of road map of the sky. The entire sky is broken up into 88 different constellations, much as continental landmasses are divided into countries by invisible lines. Every star in the sky lies within the borders of a single constellation, and the names of constellations are used in naming the stars that lie within their boundaries. For example, Sirius, the brightest star in the sky, lies within the boundaries of the constellation Canis Major (meaning the "big dog"). Sirius's official name is therefore Alpha Canis Majoris (*Canis Majoris* is the Latin genitive, or "possessive," form—see Appendix 6), indicating that it is the brightest star in that constellation and earning it its nickname, the "Dog Star." Appendix 5 contains a list

---

[1]Although stars may appear close together in the sky, most are not close together in space. The night sky is filled both with nearby low-luminosity stars and with others that are much more luminous but much farther away.

of the nearest and brightest stars. Appendix 6 provides sky maps showing the constellations.

The connection between the patterns in the sky and the patterns in the lives of the ancients was simply too compelling to be missed. The idea of the sky as a realm of mysticism and magic is deeply rooted in the traditions and beliefs and history of our species. There is no mystery about the currents of mind that led our ancestors to their belief in astrology and other celestial mythologies. At a time when the causes of things were unknown, and humans existed at the seeming whim of forces they could not comprehend, the sky seemed to offer a window into a mystical and powerful world of spirits, gods, devils, and angels.

What an unwelcome shock it must have been when a few remarkable individuals, with names such as Copernicus and Kepler and Galileo and Newton, tugged at the threads of this comfortable and familiar tapestry, *only to discover that it fell apart in their hands*! The stars and other heavenly bodies do not rotate about Earth each day, as humans had thought since they first took notice of the sky. Rather, it is *Earth* that spins on its axis, giving the stars, planets, Sun, and Moon the appearance of following daily paths through the heavens.

Science shattered ancient mystical views of the heavens.

Nor is Earth at the center of all existence, as befits the home of humankind, "the pinnacle of all creation." Earth is just one of eight major planets orbiting the Sun. The subtle complexity of the changing patterns we see in the sky results from the motions of planets and moons as they step through their gravitational dance with the Sun. Even the Sun itself, whose radiant energy makes our world what it is, is only one of countless stars, adrift in a universe whose full extent is unknown even today.

The magic of astrology properly belongs to a time long dead, when in the minds of humans Earth rode on the back of a giant sea turtle, and with each passing month the Sun moved from one stellar "house" to the next. Today it is a matter of experimentally verifiable fact that the imaginary patterns seen in the stars hold no more influence over our lives than do the random patterns of leaves blowing down the street on an autumn day. The astrologers' quest for deep connections between our lives and the patterns in the sky was both understandable and well placed, but knowledge of the true nature of those connections had to wait for the birth of modern science. Today the sky has become a window of knowledge on the *physical* world. This knowledge has proven worth the wait.

Today you had to make a long, hard climb before reaching the meadow by the river where you now camp, and tomorrow's trek promises to be just as demanding. Even so, the evening is pleasant, and you are content. The embers of your campfire have almost died away, when the distant sound of a jetliner interrupts your reverie. Without really thinking, you look up to catch sight of the airplane high overhead and are caught off guard by the blazing spectacle of the summer Milky Way and the thousands of pinpoints of light, which seem so close that you can almost reach out and touch them. For a moment, the thousands of years separating you from a long-dead tribal nomad vanish as you share the same sense of wonder and awe that has always defined humankind's experience of the universe.

It is here that we begin the journey of *21st Century Astronomy*—with the changing patterns in the sky that captured the attention and imagination of that long-ago nomad and that still serve as beacons overhead on dark, cloudless nights. Yet unlike that nomad, we look on those changing patterns with the perspective of centuries of hard-won knowledge. We will find that patterns of change in the sky are often the understandable and even unavoidable consequences of the daily rotation of Earth about its axis and Earth's annual trip around the Sun. The discovery of wonderful variety arising from simple and elegant underlying causes is an example of science at its best. And just as happened over the history of our species, curiosity about the changing patterns in the sky will show us the way outward into a universe far more vast and awesome than our distant ancestor could have imagined.

2.2 Earth Spins on Its Axis

When our remote ancestors first noticed the sky with something approaching human awareness, it was doubtless the daily motion of the Sun in the sky that drew their attention. As civilizations became more aware of the often complex motions of the Sun, Moon, planets, and stars, early astronomers developed models to explain what they saw. The ancient Greeks, for example, devised **geocentric** models of the universe in which heavenly bodies were embedded in a celestial sphere that revolved around a stationary Earth. The most successful was Ptolemy's, which survived for nearly 1,500 years until finally overthrown by Copernicus in the early Renaissance (see Chapter 3.)

Despite the apocryphal stories you may have learned in grade school, Christopher Columbus did not discover that the world is round. Long before his famous (or possibly infamous) journey to the New World, anyone who had read Aristotle or other Greek philosophers (as had Columbus) knew that Earth is a ball. Far more difficult to accept was the idea that the changes occurring in the sky from day to day and month to month are the result of the motions of Earth rather than the motion of the Sun and stars around Earth. The most apparent among these motions is Earth's rotation on its axis, which sets the very rhythm of life on Earth—the passage of day and night.

The Celestial Sphere Is a Useful Fiction

One reason the ancients did not believe that Earth rotates is that they could not perceive the spinning motion of Earth. In fact, as a result of Earth's rotation, the surface of Earth is moving along at a respectable speed—1,674 kilometers per hour (km/h) at the equator (which we calculate by dividing the circumference of Earth by the period of its rotation). Even so, we do not feel that motion any more than we would "feel" the speed of a car with a perfectly smooth ride cruising down a straight highway. Nor do we directly sense the *direction* of Earth's spin, although it is clearly revealed by the hourly motion of the Sun, Moon, and stars across the sky. As viewed from above Earth's **North Pole**, Earth rotates in a counterclockwise direction (**Figure 2.2**), completing one rotation in a 24-hour period. As the rotating Earth carries us from west to east, objects in the sky *appear* to move in the other direction, from east to west. As seen

> Earth's rotation is counterclockwise when viewed from above the North Pole.

FIGURE 2.2 The rotation of Earth and the Moon, the revolution of Earth and the planets about the Sun, and the orbit of the Moon about Earth are counterclockwise as viewed from above Earth's North Pole. (Not drawn to scale.)

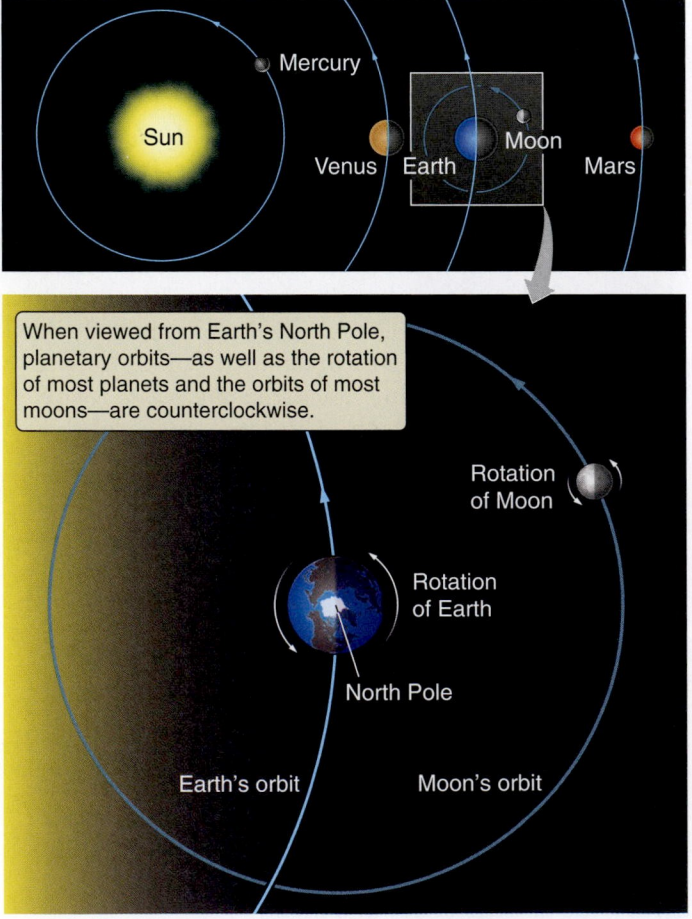

Mercury

Sun

Venus | Earth

Moon

Mars

When viewed from Earth's North Pole, planetary orbits—as well as the rotation of most planets and the orbits of most moons—are counterclockwise.

Rotation of Moon

Rotation of Earth

North Pole

Earth's orbit

Moon's orbit

from Earth's surface, the path each celestial body makes across the sky is called its "apparent daily motion."

▶❙❙ **AstroTour:** The Earth Spins and Revolves

To help visualize the apparent daily motions of the Sun and stars, it is sometimes useful to think of the sky as if it were a huge sphere with the stars on its surface and Earth at its center. (As we have said, from ancient Greek times to the Renaissance, most people believed this to be true.) Astronomers refer to this imaginary sphere as the **celestial sphere** (**Figure 2.3a**).[2] The celestial sphere is a useful concept because it is easy to draw and visualize, but never forget that it is imaginary! Each point on the celestial sphere actually corresponds to a *direction* in space. The direction in which Earth's axis of rotation points, and about which the stars appear to revolve as Earth turns, is called the **north celestial pole** (**NCP**). The direction in space that is at the *zenith* at the South Pole, and about which everything appears to spin as Earth rotates, is the **south celestial pole** (**SCP**).

We divide the celestial sphere into a northern half and a southern half with an imaginary circle called the **celestial equator**. Just as the north celestial pole is the projection of the direction of Earth's North Pole into the sky, the celestial equator is the projection of the plane of Earth's equator into the sky. And just as Earth's North Pole is 90° away from Earth's equator, the north celestial pole is always 90° away from the celestial equator. If you point one arm toward a point on the celestial equator and one arm toward the north celestial pole, your arms will always form a right angle. If you are in the Southern Hemisphere, the same holds true there: the angle between the celestial equator and the south celestial pole is 90° as well.

> The celestial equator is the projection of Earth's equator into space.

The path that the Sun takes along the celestial sphere is called the *ecliptic*. This imaginary circle is inclined 23.5° to the celestial equator (see Section 2.3). To see how the concept of the celestial sphere can be used, let's consider the Sun at noon and at midnight. From our perspective on Earth, and as modeled in the celestial sphere, the Sun appears to move across the sky and reach its highest point in the sky at noon. In more precise terms, astronomers define true "local noon" as the time when the Sun at our location crosses the **meridian**, an imaginary north–south, 360° **great circle** that divides the sky evenly into eastern and western halves. The meridian runs from a point due north on the *horizon*, through a point directly overhead called the **zenith**, to a point due south on the horizon, and then continues below the horizon through the **nadir**, a point directly below and opposite the zenith, and back up to its starting point due north on the horizon (**Figure 2.3b**). True "local midnight" occurs when the Sun again crosses the meridian at its lowest point below the northern horizon. What is really hap-

[2]You can find a description of celestial coordinates used with the celestial sphere in Figure A6.1 in Appendix 6.

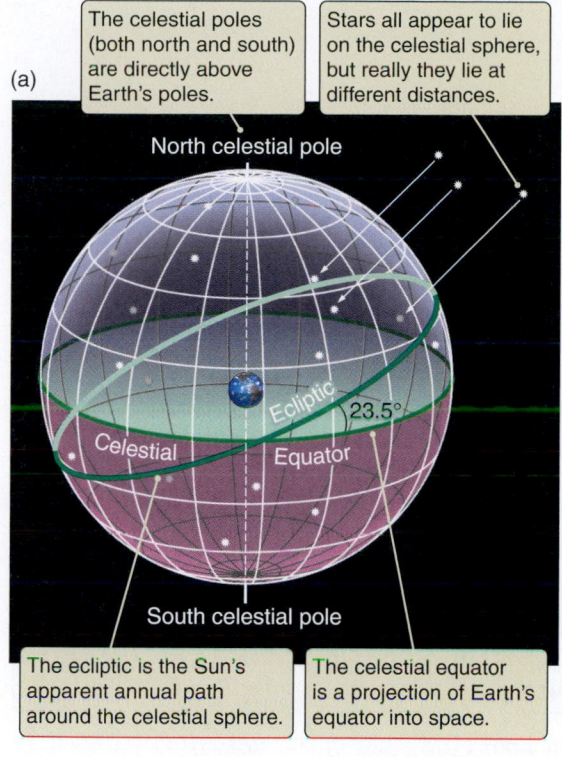

(a)

The celestial poles (both north and south) are directly above Earth's poles.

Stars all appear to lie on the celestial sphere, but really they lie at different distances.

North celestial pole

Ecliptic

Celestial   Equator

23.5°

South celestial pole

The ecliptic is the Sun's apparent annual path around the celestial sphere.

The celestial equator is a projection of Earth's equator into space.

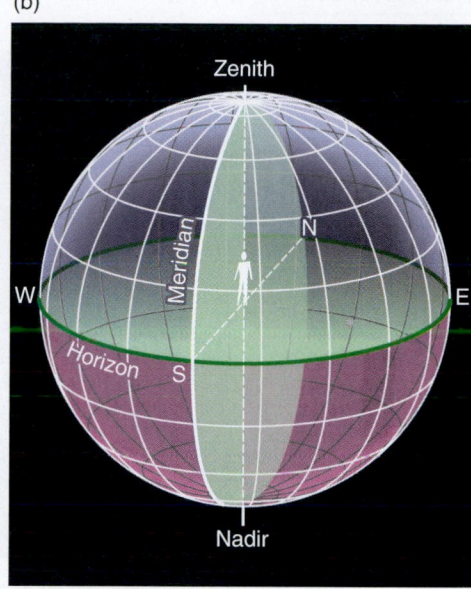

(b)

Zenith

Meridian

W   N   E

Horizon   S

Nadir

FIGURE 2.3 (a) The celestial sphere is a useful fiction for thinking about the appearance and apparent motion of the stars in the sky. (b) Principal features of an observer's coordinate system projected onto the celestial sphere.

pening at noon is that our location on Earth has rotated to face most directly toward the Sun. Half a day later, at midnight, our spot on Earth rotates closest to facing directly away from the Sun.

We will continue to use the celestial sphere model throughout the chapter to show apparent motions of the Sun and stars from different points on the Earth.

The View from the Poles

The apparent daily motions of the stars and the Sun witnessed by ancient nomadic tribes would have depended on where on the surface of the planet they happened to live. The apparent daily motions of celestial objects in northern Europe, for example, are quite different from the apparent daily motions seen from a tropical island. To understand how our location affects our perception of apparent daily motions of celestial bodies, let's look at the daily motions of the stars when viewed from a place where humans did not set foot until the early 20th century—Earth's North Pole. (In science we often start by working out the "easy" or "limiting" cases—the view of the stars from the poles, for example—and then use these to guide our thinking about what happens in more complicated situations.)

Imagine that you are standing on the North Pole watching the sky as shown in **Figure 2.4a**. (Ignore the Sun for the moment and suppose that you can always see stars in the sky.) You are standing where Earth's axis of rotation intersects its surface, which is much the same as standing at the center of a rotating carousel. As Earth rotates, the spot directly above you seems to remain fixed while everything else in the sky appears to revolve in a counterclockwise direction around this spot (**Figure 2.4b**). (If you are having trouble visualizing this, find a globe and, as you spin it, imagine standing at the pole of the globe.) Notice that objects close to the pole appear to follow small circles, while the largest circles are followed by objects nearest to the horizon. ▶‖ **AstroTour:** The View from the Poles

The view from the North Pole is unique because from there we always see the *same* half of the sky (**Figure 2.4c**). Nothing rises or sets as Earth turns beneath you. Of course, regardless of where you are on the surface of our planet, you can never see more than half of the sky at any one time. The other half of the sky is blocked from view by Earth. The boundary between the part of the sky you can see and the part that is blocked by Earth is called the **horizon**. From most locations on Earth, the half of the sky that we can see above the horizon changes constantly as Earth rotates. (The direction in space in which our zenith points right now is different from what it was 12 hours ago, or even 12 seconds ago.) In contrast, Earth's North Pole points in the *same* direction, hour after hour and day after day. For this reason, if you look off toward the horizon, you will see that the objects visible there follow circular paths that keep them always the same distance above the horizon.

> The same half of the sky is always visible from the North Pole.

(a)

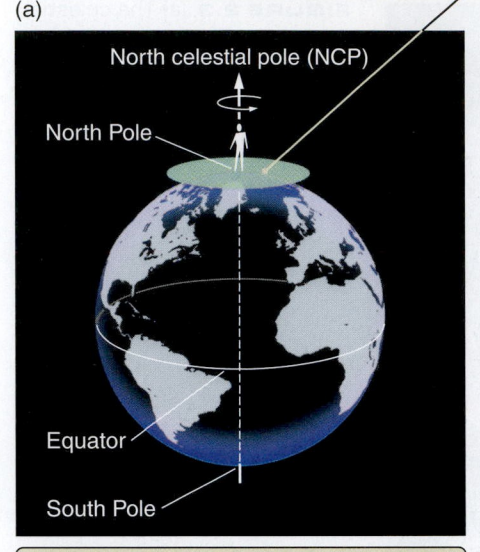

This disk represents the horizon, the boundary between the part of the sky you can see and the part that is blocked from view by Earth.

North celestial pole (NCP)

North Pole

Equator

South Pole

From the North Pole looking directly overhead, the **north celestial pole** is at the zenith.

(b)

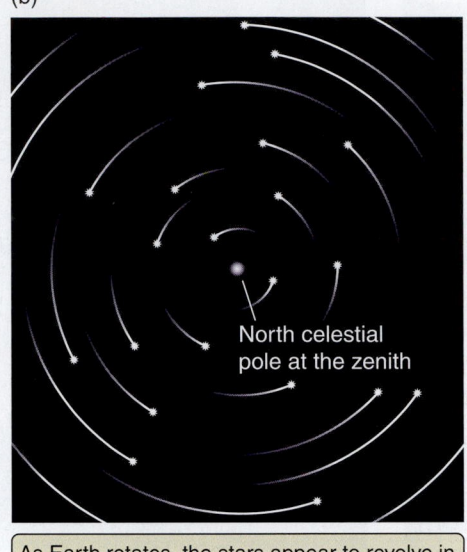

North celestial pole at the zenith

As Earth rotates, the stars appear to revolve in a counterclockwise direction around the **NCP**.

(c)

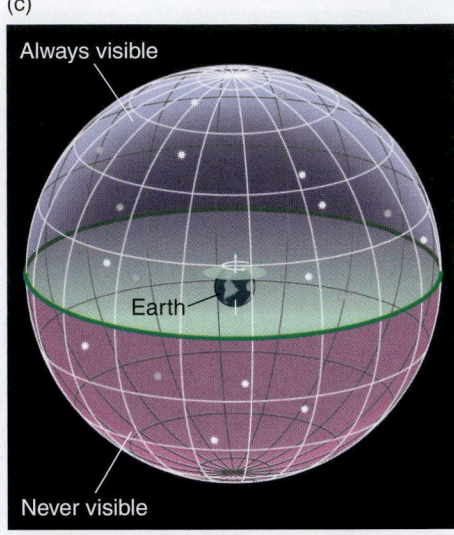

Always visible

Earth

Never visible

From the North Pole, you always see the same half of the sky.

FIGURE 2.4 As viewed from Earth's North Pole (a), stars move throughout the night on counterclockwise, circular paths about the zenith (b). (c) The same half of the sky is always visible from the North Pole.

The view from Earth's **South Pole** is much the same, but with two major differences. First, the South Pole is on the opposite side of Earth from the North Pole, so the half of the sky you see overhead at the South Pole is precisely the half that is hidden from the North Pole. The second difference is that instead of appearing to move counterclockwise around the sky, stars appear to move *clockwise* around the south celestial pole. (To see this, sit in a swivel chair and spin it around from right to left. As you look at the ceiling, things appear to move in a counterclockwise direction; but as you look at the floor, they appear to be moving clockwise.) ▶❙❙ **AstroTour:** The **View from the Poles**

> From the South Pole, the other half of the sky is visible, and stars circle in a clockwise direction.

Away from the Poles, the Part of the Sky We See Is Constantly Changing

Now let's imagine what we see in the sky as we leave the North Pole and travel south to lower latitudes. As you may already know, **latitude**[3] is a measure of how far

north or south we are on the face of Earth. Imagine a line from the center of Earth to your location on the surface of the planet. Now imagine a second line from the center of Earth to the point on the **equator** closest to you. (Refer to **Figure 2.5** for help imagining these lines.) The angle between these two lines is your latitude. At the North Pole, for example, these two imaginary lines form a 90° angle. The latitude of the North Pole is thus 90° north of the equator, which serves as the 0° mark. The South Pole is at latitude 90° south.

Our latitude determines the part of the sky that we can see throughout the year. As we follow the curve of Earth south from the North Pole, our horizon tilts and our zenith moves away from the north celestial pole. At a latitude of 60° north (as shown in **Figure 2.5b**), our horizon is tilted 60° from the north celestial pole. This equality between north latitude and the height of the north celestial pole above the northern horizon holds everywhere. In **Figure 2.5d**, we have reached Earth's equator, at a latitude of 0°. Notice that the north celestial pole is now sitting on the northern horizon. At the same time, we get our first look at the south celestial pole, which is sitting opposite the north celestial pole on the southern horizon. Continuing into the Southern Hemisphere, the south celestial pole is now visible above the southern horizon, while the north celestial pole is hidden from view by the northern horizon. At a latitude of 45° south (**Figure 2.5e**), the south celestial pole lies 45° above the southern horizon. At the South Pole (latitude 90°

---

[3]You can find definitions of the terrestrial coordinates—latitude and longitude—in Appendix 6.

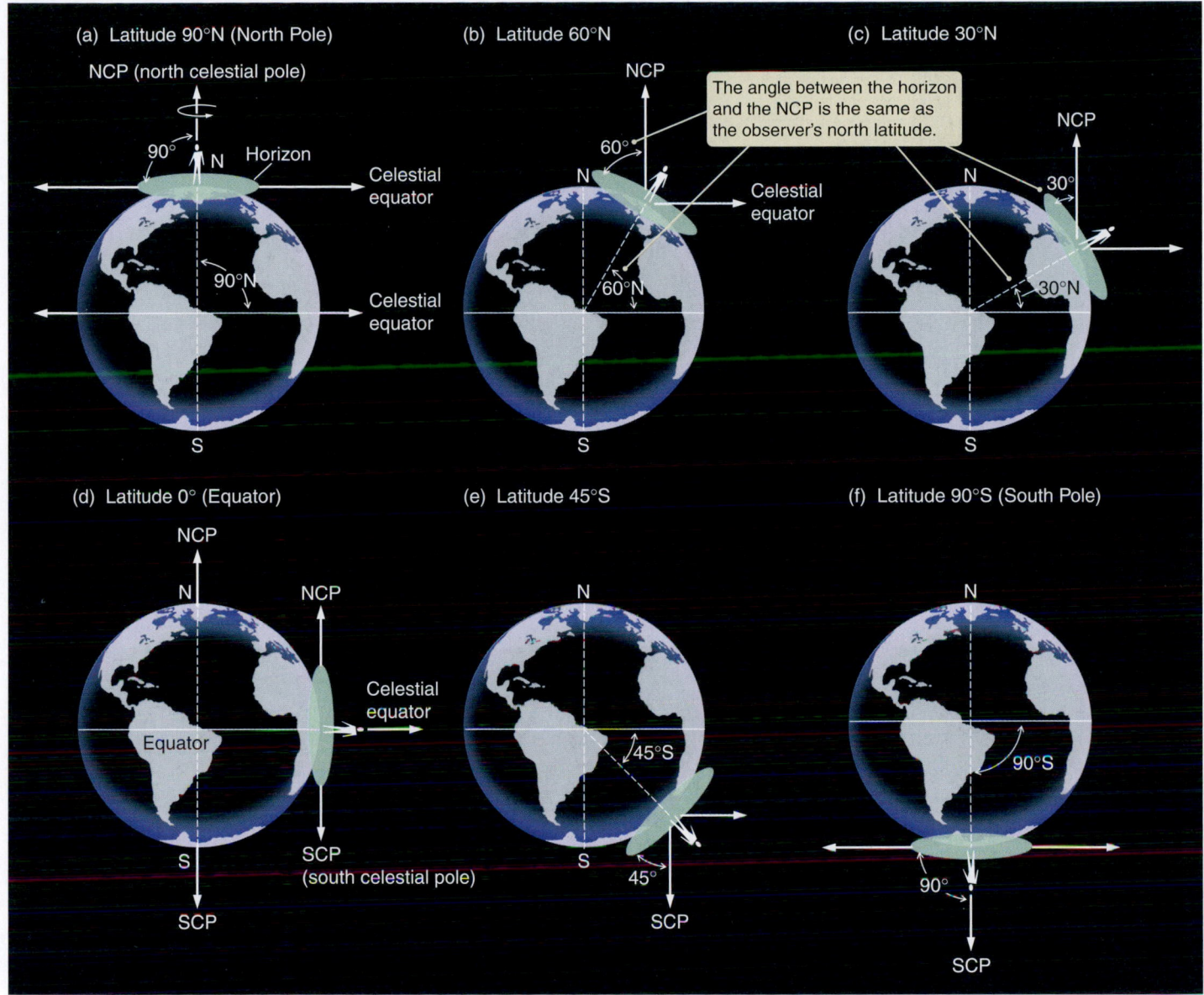

(a) Latitude 90°N (North Pole)

NCP (north celestial pole)

90°

N   Horizon

90°N

Celestial equator

Celestial equator

S

(b) Latitude 60°N

NCP

The angle between the horizon and the NCP is the same as the observer's north latitude.

60°

N

60°N

Celestial equator

S

(c) Latitude 30°N

NCP

N   30°

30°N

S

(d) Latitude 0° (Equator)

NCP

N

Equator

Celestial equator

S

SCP (south celestial pole)

SCP

(e) Latitude 45°S

N

45°S

S

45°

SCP

(f) Latitude 90°S (South Pole)

N

90°S

90°

SCP

FIGURE 2.5 Our perspective on the sky depends on our location on Earth. Here we see how the locations of the celestial poles and celestial equator depend on an observer's latitude.

south—**Figure 2.5f**), the south celestial pole is at the zenith, 90° above the horizon.

Probably the best way to cement your understanding of the view of the sky at different latitudes is to draw pictures like those in Figure 2.5. If you can draw a picture like this for any latitude—filling in the values for each of the angles in the drawing and imagining what the sky looks like from that location—then you will be well on your way to developing a working knowledge of the appearance of the sky. That knowledge will prove useful later, when we discuss a variety of phenomena, such as the changing of the seasons. When practicing your sketches, however, take care not to make the common mistake illustrated in **Figure 2.6**. *The north celes-*

tial pole is not a location in space, hovering over Earth's North Pole. Instead, *it is a direction* in space—the direction parallel to Earth's axis of rotation.

Now that we have shown how the horizon is oriented at different latitudes, let's see how the apparent motions of the stars about the celestial poles differ from latitude to latitude. **Figure 2.7a** shows our view if we are at latitude 30° north. As Earth rotates, the part of the sky visible to us is constantly changing. Of course, from this perspective it is the horizon that seems to remain fixed, while the stars appear to move past overhead. If we focus our attention on the north celestial pole, from this perspective we still see much the same thing we saw from Earth's North Pole. The north celestial pole remains fixed in the sky, and all of

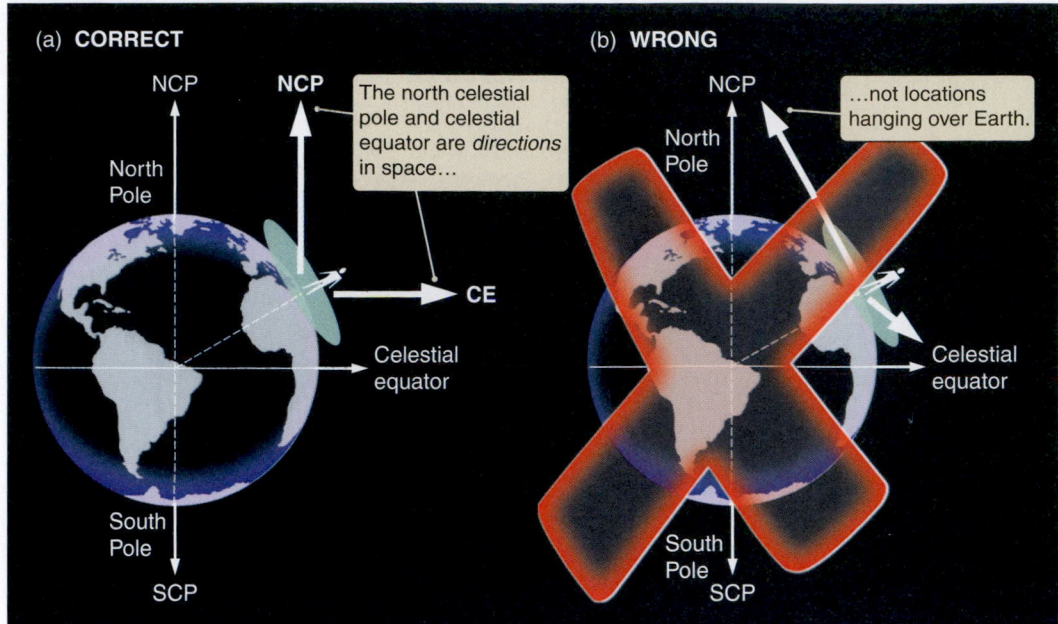

(a) **CORRECT**

NCP

NCP

North Pole

The north celestial pole and celestial equator are *directions* in space…

CE

Celestial equator

South Pole

SCP

(b) **WRONG**

NCP

North Pole

…not locations hanging over Earth.

Celestial equator

South Pole

SCP

FIGURE 2.6 The celestial poles and the celestial equator are directions in space, not fixed locations hanging above Earth. (The red *X* means "avoid this misconception.")

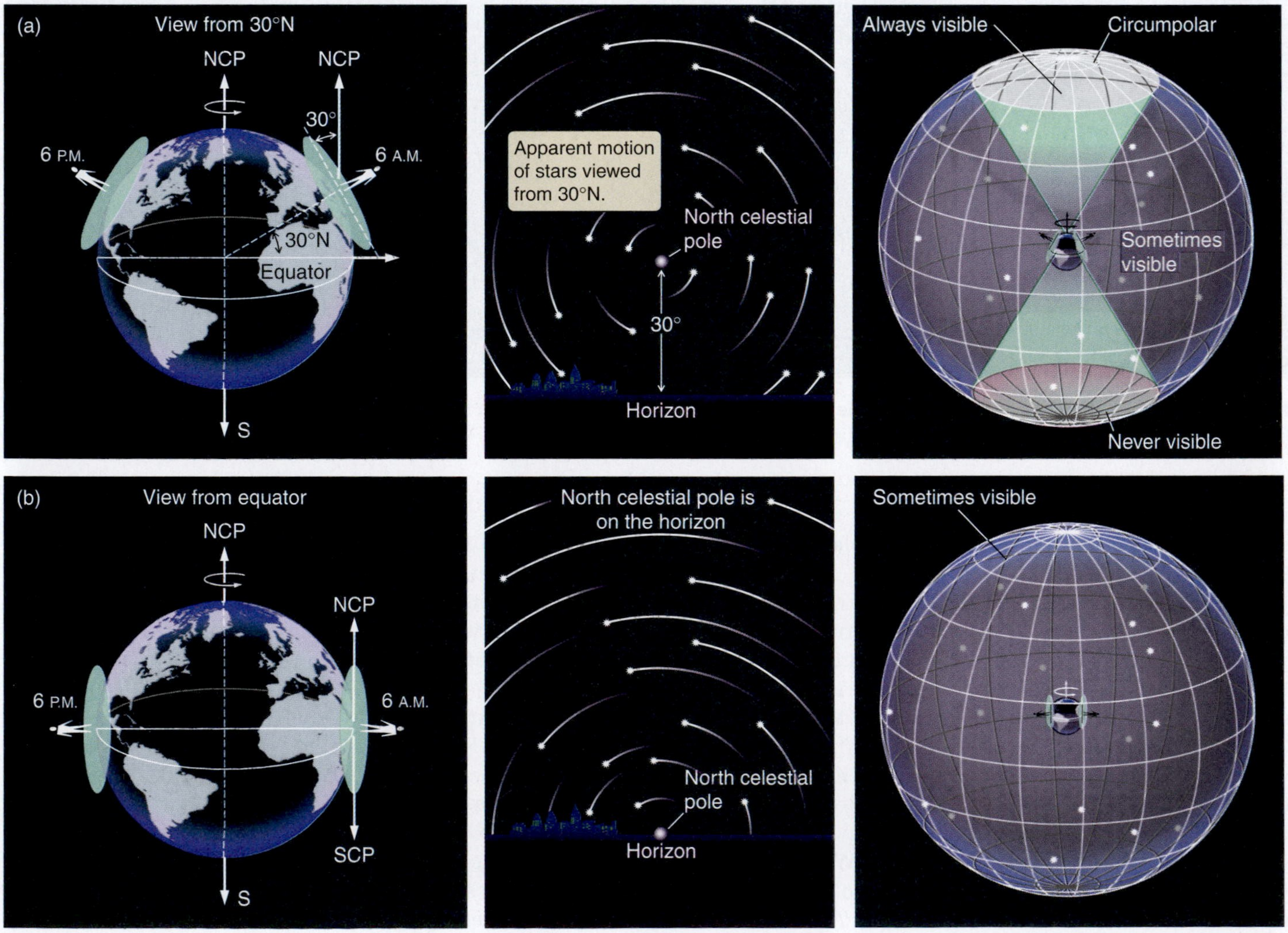

FIGURE 2.7 (a) As viewed from latitude 30° north, the north celestial pole is 30° above the northern horizon. Stars appear to move on counterclockwise paths around this point. At this latitude some parts of the sky are always visible, while others are never visible. (b) From the equator, the north and south celestial poles are seen on the horizon, and the entire sky is visible over a period of 24 hours.

(a) View from 30°N

NCP

NCP

30°

6 P.M.

6 A.M.

30°N

Equator

S

Apparent motion of stars viewed from 30°N.

North celestial pole

30°

Horizon

Always visible

Circumpolar

Sometimes visible

Never visible

(b) View from equator

NCP

NCP

6 P.M.

6 A.M.

SCP

S

North celestial pole is on the horizon

North celestial pole

Horizon

Sometimes visible

From a location in the Canadian woods, the north celestial pole appears high in the sky...

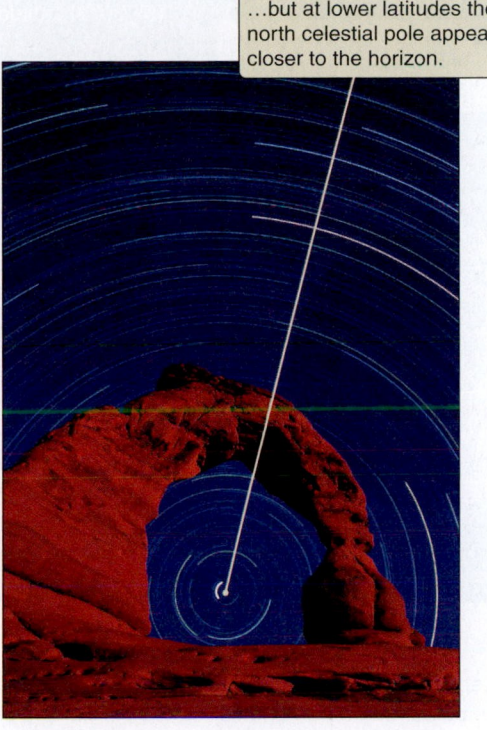

...but at lower latitudes the north celestial pole appears closer to the horizon.

FIGURE 2.8 Time exposures of the sky showing the apparent motions of stars through the night. Note the difference in the circumpolar portion of the sky as seen from the two different latitudes.

the stars appear to move throughout the night in counter-clockwise, circular paths around that point. But because the north celestial pole is no longer directly overhead as it was at the North Pole, the apparent circular paths of the stars are now tipped relative to the horizon. (More correctly, our horizon is now tipped relative to the apparent circular paths of the stars.) ▶❚❚ **AstroTour: The Celestial Sphere and the Ecliptic**

Stars located close enough to the north celestial pole are above the horizon 24 hours a day (even if we can't see them in the daytime) as they complete their apparent paths around the pole (see Figure 2.7a and **Figure 2.8**). This always-visible region of our sky is referred to as being **circumpolar**, which means "around the pole." There is also a part of the sky that can *never* be seen from this latitude. This is the part of the sky near the *south* celestial pole that never rises above your horizon. And between this region and the always-visible circumpolar region lies a portion of the sky that can be seen for *part but not all* of each day. Stars in this intermediate region appear to rise above and set below Earth's shifting horizon as Earth turns. The only place on Earth where you can see the entire sky over the course of 24 hours is the equator. From the equator (**Figure 2.7b**) the north and south celestial poles sit on the northern and southern horizons, respectively, and the whole of the heavens passes through the sky each day.

Look at the location of the celestial equator in **Figure 2.9**.

> Circumpolar stars are always above the horizon.

The points where the celestial equator intersects the horizon are always due east and due west. (The only exception is at the poles, where the celestial equator is coincident with the horizon.) An object on the celestial equator rises due east and sets due west. Objects that are north of the celestial equator rise north of east and set north of west. Objects that are south of the celestial equator rise south of east and set south of west.

Figure 2.9 also shows that regardless of where you are on Earth (again with the exception of the poles), half of the celestial equator is always visible above the horizon. Because half of the celestial equator is always visible, it follows that you can see any object that lies in the direction of the celestial equator half of the time. An object that is in the direction of the celestial equator rises due east, is above the horizon for exactly 12 hours, and sets due west. This is not true for objects that are not on the celestial equator. A look at **Figure 2.9b** shows that from the Northern Hemisphere, you can see more than half of the apparent circular path of any star that is north of the celestial equator. And if you can see more than half of a star's path, then the star is above the horizon for more than half of the time.

As seen from the Northern Hemisphere, stars north of the celestial equator remain above the horizon for more than 12 hours each day. The farther north the star is, the longer it stays up. The circumpolar stars near the north celestial pole that we mentioned already are the extreme example of this phenomenon; they are up 24 hours a day. In contrast, objects south of the celestial equator are above the horizon for less than 12 hours a day, and the farther

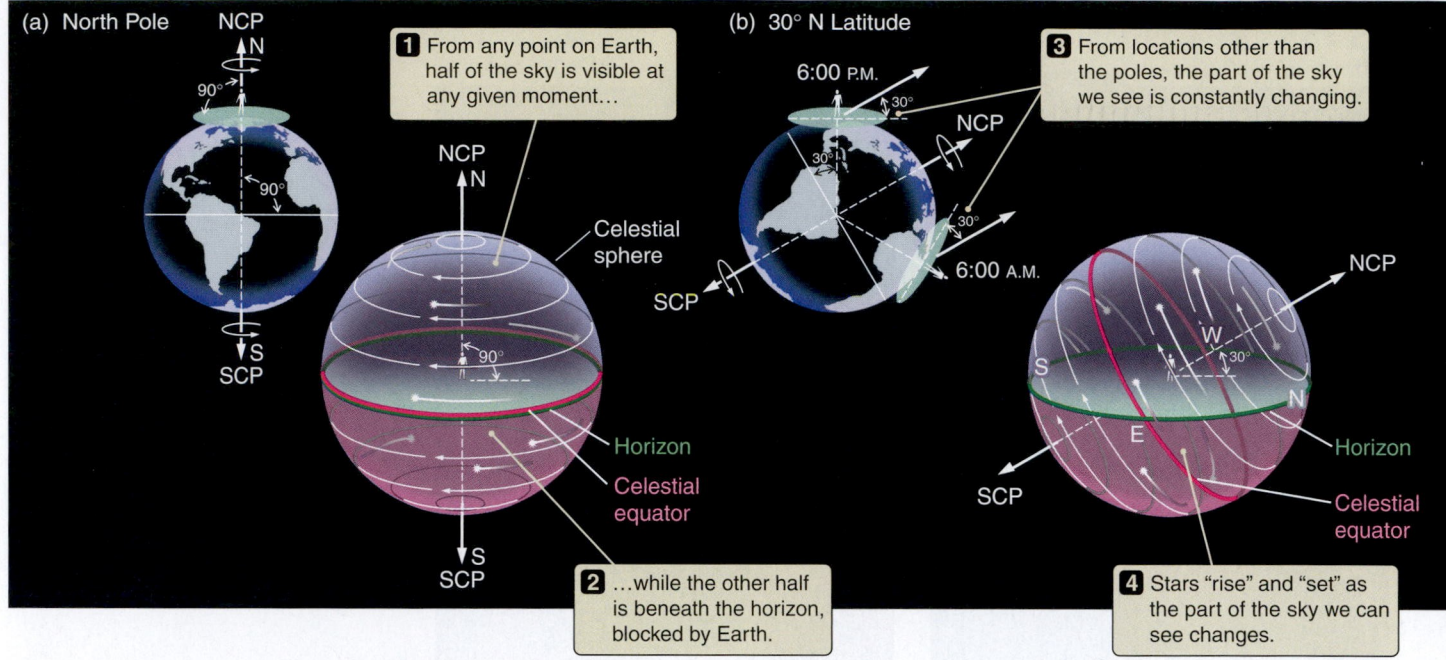

(a) North Pole

1 From any point on Earth, half of the sky is visible at any given moment...

Celestial sphere

Horizon

Celestial equator

2 ...while the other half is beneath the horizon, blocked by Earth.

(b) 30° N Latitude

6:00 P.M.

6:00 A.M.

3 From locations other than the poles, the part of the sky we see is constantly changing.

Horizon

Celestial equator

4 Stars "rise" and "set" as the part of the sky we can see changes.

FIGURE 2.9 The celestial sphere is shown here as viewed by observers at four different latitudes. At all locations other than the poles, stars rise and set as the part of the celestial sphere that we see changes during the day.

south you look, the less time a star is visible. Stars that are located close to the south celestial pole never rise above our horizon.

If you were an observer in the Southern Hemisphere (see **Figure 2.9d**), the reverse of the preceding discussion would be true: Objects on the celestial equator would still be up for 12 hours a day, but now objects south of the celestial equator would be up more than 12 hours, and objects north of the celestial equator would be up less than 12 hours.

For many centuries, travelers, including sailors at sea, have used the stars for navigation. Perhaps the simplest of the navigator's techniques is to use the equality between latitude and the altitude of the north (or south) celestial pole. We can find the north or south celestial poles by recognizing the stars that surround them. In the Northern Hemisphere, a moderately bright star happens by chance to be located within 0.7° of the north celestial pole. This star is called Polaris, or more commonly, the "North Star." If you can find Polaris in the sky and measure the angle between the north celestial pole and the horizon, then you know your latitude. If you are in Phoenix, Arizona, for example (latitude 33.5° north), you will find the north celestial pole 33.5° above your northern horizon. On the other hand, if you are studying astronomy in Fairbanks, Alaska (latitude 64.6° north), the north celestial pole sits much higher overhead, 64.6° above the horizon in the north. The location of the north celestial pole in the sky can also be used to measure the size of Earth (see **Math Tools 2.1**).

> The star Polaris marks the north celestial pole.

2.3 Revolution about the Sun Leads to Changes during the Year

The second motion that we will discuss is the motion of Earth about the Sun. Earth revolves around the Sun in the same direction that Earth spins about its axis—counterclockwise as viewed from above Earth's North Pole. A **year**, by definition, is the time it takes for Earth to complete one revolution around the Sun. The motion of Earth around the Sun is responsible for many of the patterns of change we see in the sky and on Earth, including changes in which stars we see at night. As Earth moves around the Sun, the stars we see overhead at midnight change. Six months from now, Earth will be on the other side of the Sun, and the stars that we see overhead at midnight will be in nearly the opposite direction from the stars we see near overhead at midnight tonight. The stars that were overhead at midnight six months ago are the same stars that are overhead today at noon, but of course we cannot see them today because of the glare of the Sun. ▶❚❚ **AstroTour: The Earth Spins and Revolves**

> Earth's orbital motion is counterclockwise as viewed from above Earth's North Pole.

If you could note the position of the Sun relative to the stars each day for a year, you would find that it traces out a great circle against the background of the stars (**Figure 2.10** on page 34). On September 1, the Sun appears to be in

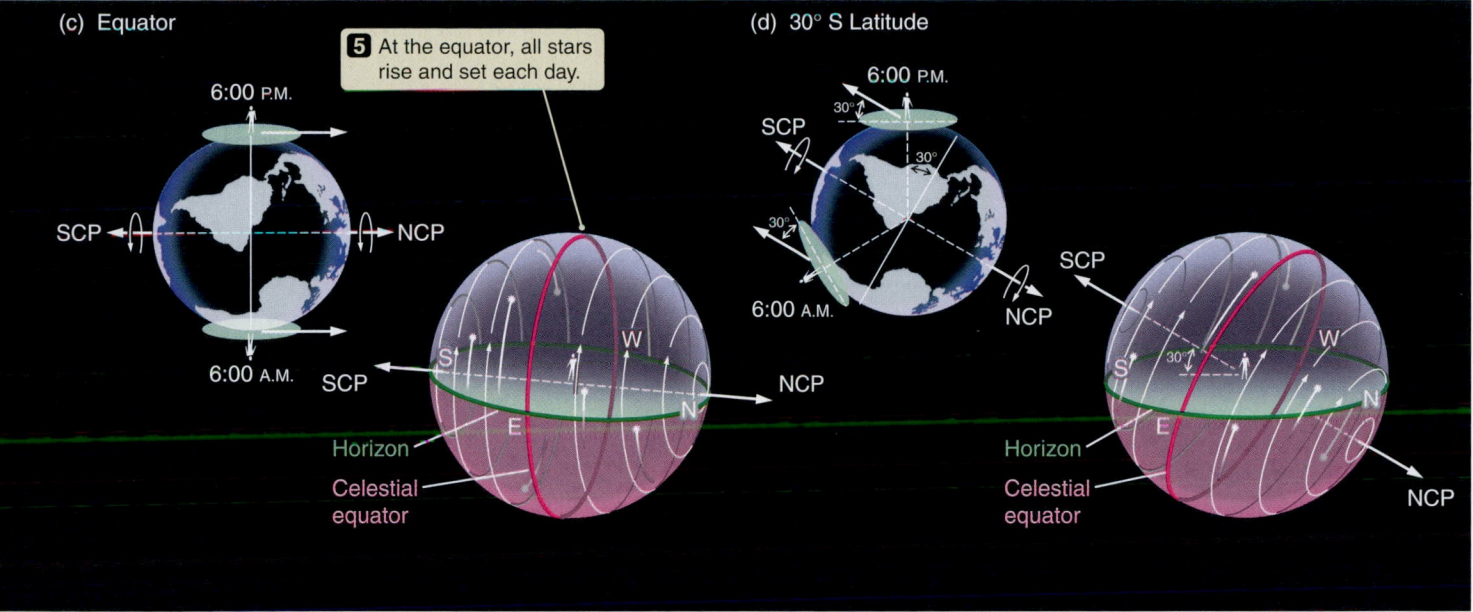

(c) Equator

5 At the equator, all stars rise and set each day.

6:00 P.M.

SCP — NCP

6:00 A.M.   SCP

SCP

Horizon

Celestial equator

(d) 30° S Latitude

6:00 P.M.

SCP

6:00 A.M.

NCP

SCP

NCP

Horizon

Celestial equator

NCP

the direction of the constellation of Leo. Six months later, on March 1, Earth is on the other side of the Sun, and the Sun appears to be in the direction of the constellation of Aquarius. The apparent path that the Sun follows against the background of the stars is called the **ecliptic**. The 12 constellations that lie along the ecliptic and through which the Sun appears to move are called the constellations of the **zodiac**. This is why ancient astrologers assigned special mystical significance to these stars. Actually, the constellations of the zodiac are nothing more than random patterns of distant stars that happen by chance to lie near

> The ecliptic is the Sun's apparent yearly path against the background of stars.

the plane of Earth's orbit about the Sun. ▶❚❚ **AstroTour: The Celestial Sphere and the Ecliptic**

Tiny Deviations in the Direction of Starlight Provide a Measure of Earth's Motion through Space

As difficult as it is to "feel" the effects of Earth's rotation on its axis, it is even harder to sense the motion of Earth around the Sun. As we have seen, through most of the history of our species humans believed that Earth remains stationary while the Sun, the Moon, and the heavens revolve around

MATH TOOLS 2.1

How to Estimate the Size of Earth

You can use the location of the north celestial pole in the sky to estimate the size of Earth. Suppose we start out in Phoenix, Arizona, where we observe the north celestial pole to be 33.5° above the horizon. Now we head north, and by the time we reach the Grand Canyon, about 290 kilometers (km) from Phoenix, we notice that the north celestial pole has risen to about 36° above the horizon. This difference (2.5°) is $1/144$ of the way around a circle. (A circle is 360°, and $2.5°/360° = 1/144$.) This means that we must have traveled $1/144$ of the way around the circumference, C, of Earth. In other words, $1/144 \times C = 290$ km. So, rearranging the expression,

the circumference of Earth must be about 144×290 km, or about 42,000 km. The actual circumference of Earth is just a shade over 40,000 km, so our simple calculation was not too bad, given our sloppy measurements of angles and distances. Recall from geometry class that the circumference of a circle is equal to 2π times its radius. So, the radius of Earth is its circumference (40,000 km) divided by 2π, or about 6,400 km. It was in much this same way that the Greek astronomer Eratosthenes (276–194 B.C.) made the first accurate measurements of the size of Earth, in about 230 B.C. (well before Columbus's time).

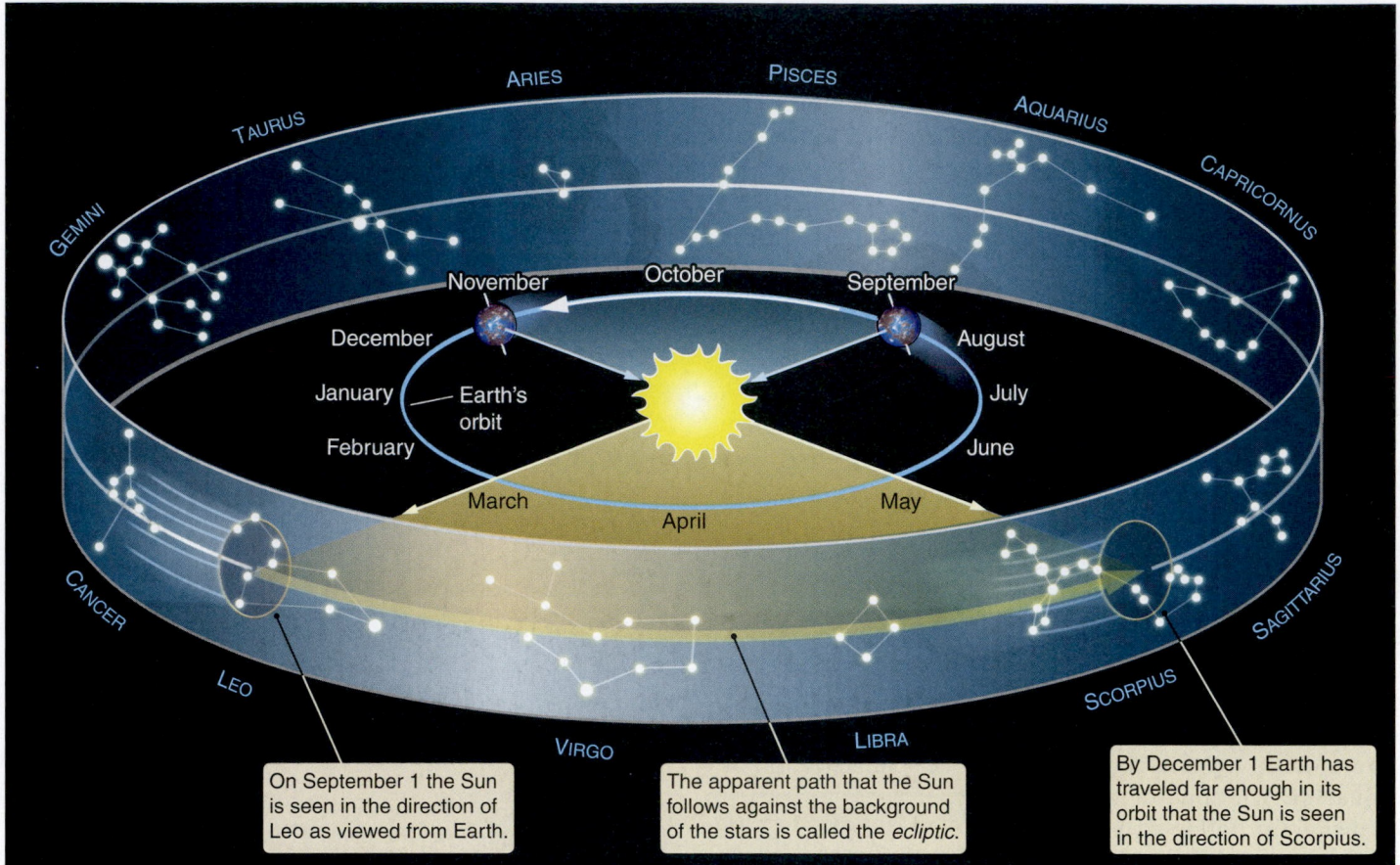

ARIES   PISCES

TAURUS   AQUARIUS

GEMINI   CAPRICORNUS

November   October   September

December   August

January   Earth's   July
orbit

February   June

March   May

April

CANCER   SAGITTARIUS

LEO   SCORPIUS

VIRGO   LIBRA

On September 1 the Sun
is seen in the direction of
Leo as viewed from Earth.

The apparent path that the Sun
follows against the background
of the stars is called the *ecliptic*.

By December 1 Earth has
traveled far enough in its
orbit that the Sun is seen
in the direction of Scorpius.

FIGURE 2.10 As Earth orbits about the Sun, the Sun's apparent position against the background
of stars changes. The imaginary circle traced by the annual path of the Sun is called the ecliptic.
Constellations along the ecliptic form the zodiac.

us. The history of modern astronomy, and to some degree
the story of the rise of modern science, can be told as the
story of how this view was overthrown during the 17th and
18th centuries. However, the first direct measurement of the
effect of Earth's motion was not made until the 18th century.
To understand how this measurement was made, it's helpful first to understand the concept of a **frame of reference**.
Briefly, a frame of reference is a coordinate system within
which an observer measures positions and motions.

Aside from looking out the window or feeling road vibrations, there is no experiment that you could easily do to tell
the difference between riding in a car down a straight section of highway at constant
speed and sitting in the car
while it is parked in your
driveway. Because everything
in the car is moving together, the **relative motions** between
objects in the car are all that count. In fact, the only reason
you can feel the roughness of the road is that it slightly
changes the motion of the car. You feel these brief accelerations as the car's vibration.

The idea that only relative motions count occurs again
and again in astronomy and physics. There are numerous

> Relative motion
> is all that counts.

examples in this chapter alone. For example, even though
Earth is spinning on its axis and flying through space in its
orbit about the Sun, the resulting relative motions between
objects that are near each other on Earth are small. Newton's realization that motions are meaningful only when tied
to the frame of reference of a particular observer is also at
the heart of Einstein's theories of relativity. These theories,
which we will return to later, wound up changing the way
we think about space and time.

We can use a common experience to illustrate what you
might observe within a specific frame of reference. Imagine that you are sitting in a car in a windless rainstorm, as
shown in **Figure 2.11**. If the car is sitting still and the rain
is falling vertically, when you look out your side window
you see raindrops falling straight down. That is, if you hold
a vertical tube out the window, raindrops will fall straight
through the tube. When the car is moving forward, however, the situation is different. Between the time a raindrop
appears at the top of your window and the time it disappears
beneath the bottom of your window, the car has moved forward. The raindrop disappears beneath the window *behind*
the point at which it appeared, which means the raindrop
looks as if it falls at an angle, even though in reality it is

falling straight down. For raindrops to fall directly through the tube you are holding out the window now, you would have to tilt the top of the tube forward. As you go faster, the apparent front-to-back motion of the raindrops increases, and their apparent paths become more tilted. An observer by the side of the road would say the raindrops are coming from directly overhead, but to you in the moving car they are coming from a direction in front of the car. You are observing this apparent motion of raindrops from within your own special frame of reference.

This same phenomenon occurs with starlight, as shown in **Figure 2.12**. The light from a distant star arrives at Earth from the direction to this star. Because Earth moves, however, to an observer on Earth the starlight seems to be coming from a slightly different direction, just as the raindrops appeared to be coming from in front of the car in Figure 2.11.[4] As the direction of Earth's motion around the Sun continuously changes during the year, the apparent position of a star in the sky moves in a small loop. This shift in apparent position is what we refer

> Earth's orbital motion around the Sun was first measured from the aberration of starlight.

[4]Actually, in the frame of reference of Earth, the starlight *is* coming from a different direction; and in the frame of reference of the moving car, the raindrops *are* coming from in front of the car.

(a)

From the vantage point of a person outside, the rain falls vertically, even when the car is moving.

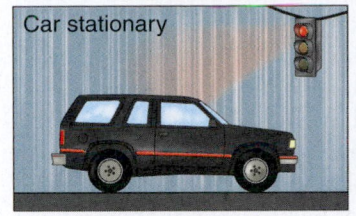 

Car stationary   Car moving

(b)

From the frame of reference of a person inside the car...

...the rain falls vertically if the car is stationary...

...but at an angle if the car is moving.

FIGURE 2.11 On a windless day the direction in which rain falls depends on the frame of reference in which it is viewed. (a) From outside of the car, the rain is seen to fall vertically downward whether the car is stationary or moving. (b) From inside the car, the rain is seen to fall vertically downward if the car is stationary; but if the car is moving, the rain is seen to fall at an angle determined by the speed and direction of the car's motion.

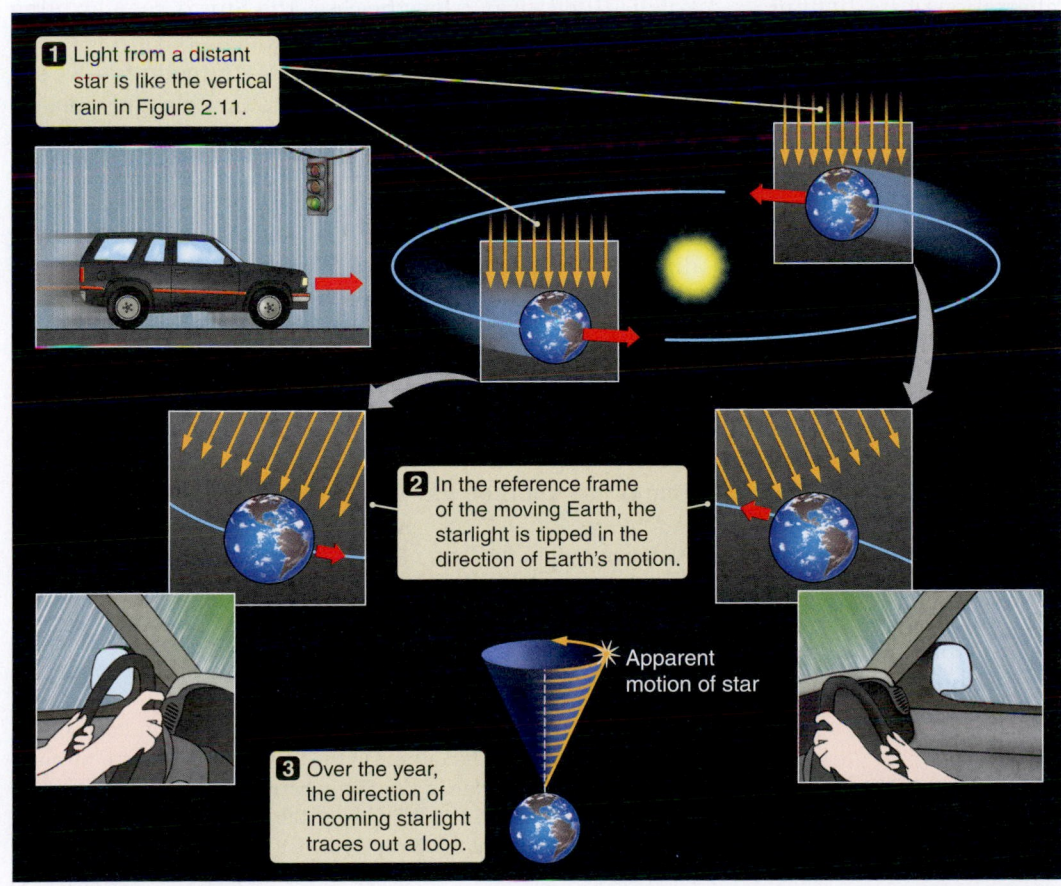

1 Light from a distant star is like the vertical rain in Figure 2.11.

2 In the reference frame of the moving Earth, the starlight is tipped in the direction of Earth's motion.

Apparent motion of star

3 Over the year, the direction of incoming starlight traces out a loop.

VISUAL ANALOGY FIGURE 2.12 The apparent positions of stars are deflected slightly toward the direction in which Earth is moving. As Earth orbits the Sun, stars appear to trace out small loops in the sky. This effect is called *aberration of starlight*.

to as the **aberration of starlight**, and it was first detected in the 1720s by two English astronomers, Samuel Molyneux and James Bradley. Measurement of the aberration of starlight shows that Earth is moving on a roughly (but not exactly) circular path about the Sun with an average speed of just under 30 kilometers per second (km/s). Because distance equals speed multiplied by time, the distance around this near-circle—its circumference—is the speed of Earth (29.8 km/s) multiplied by the length of one year (3.16×10^7 seconds). The circumference of Earth's orbit is then

$$\text{Distance} = \text{Speed} \times \text{Time}$$
$$= \left(29.8 \, \frac{\text{km}}{\text{s}}\right) \times (3.16 \times 10^7 \text{ s})$$
$$= 9.42 \times 10^8 \text{ km}.$$

The radius of Earth's nearly circular orbit is this circumference divided by 2π, or 1.50×10^8 km (150 million kilometers). Astronomers refer to this distance—the average distance between the center of the Sun and the center of Earth[5]—as one **astronomical unit**, abbreviated **AU**. The astronomical unit is a good unit for measuring distances within the Solar System.

> The astronomical unit is the average distance between the Sun and Earth.

Modern measurements of the size of the astronomical unit are made in very different ways, such as bouncing radar signals off Venus. However, the aberration of starlight provided a simple and compelling demonstration that Earth orbits about the Sun—and a pretty good value for the size of Earth's orbit as well.

Seasons Are Due to the Tilt of Earth's Axis

So far we have discussed the rotation of Earth on its axis and the revolution of Earth about the Sun. To understand the changing of the seasons, we need to consider the combined effects of these two motions. If you ask most people why it is cold in the winter and warm in the summer, the reason they are likely to give is that Earth is closer to the Sun in the summer and farther away in the winter. This is a common (and commonsense) idea that does have something to do with the seasons on Mars, but it has virtually *nothing* to do with the seasons on Earth! Earth's orbit around the Sun is an almost perfect circle centered on the Sun,[6] so the distance from the Sun changes little during the year. In fact, Earth is slightly closer to the Sun during the northern winter than it is during the northern summer.

To understand how the combination of Earth's axial tilt and its annual path around the Sun creates seasons, let's look at a special case. If Earth's spin axis were exactly perpendicular to the plane of Earth's orbit (the **ecliptic plane**), then the Sun would always appear to lie on the celestial equator. Because the position of the celestial equator is fixed in our sky, the Sun would follow the same path through the sky day after day, rising due east each morning and setting due west each evening. If the Sun were always on the celestial equator, it would be above the horizon for exactly half the time, and days and nights would always be exactly 12 hours long. In short, if Earth's axis were exactly perpendicular to the plane of Earth's orbit, each day would be just like the last, and there would be no seasons.

However, Earth's axis of rotation is *not* exactly perpendicular to the plane of the ecliptic. Instead it is tilted by 23.5° from the perpendicular.[7] As Earth moves around the Sun, its axis points in almost exactly the same direction throughout the year and from one year to the next.[8] As a result, sometimes Earth's North Pole is tilted more toward the Sun, and at other times it is pointed more away from the Sun. When Earth's North Pole is tilted toward the

> Seasons result from the 23.5° tilt of Earth's axis with respect to a line perpendicular to its orbital plane.

Sun, an observer on Earth sees the Sun as lying *north* of the celestial equator. Six months later, when Earth's North Pole is tilted away from the Sun, the Sun is seen as lying *south* of the celestial equator. If we look at the circle of the Sun's apparent path through the stars—the ecliptic—we see that it is tilted by 23.5° with respect to the celestial equator.

▶‖ **AstroTour:** The Earth Spins and Revolves

To understand the effect that this tilt has on Earth, begin by looking at **Figure 2.13a**. This figure shows the situation on June 21, the day that Earth's North Pole is tilted most directly toward the Sun.[9] Note first that the Sun is north of the celestial equator. We found earlier in the chapter that, from the perspective of an observer in the Northern Hemisphere, an object north of the celestial equator can be seen above the horizon for more than half the time. This is true for the Sun, as well as for any other celestial object. Saying that the Sun is above the horizon for more than half the time is just another way of saying that the days are longer

---

[5] Distances between celestial objects are almost always taken as the distances between their centers.

[6] Planetary orbits are actually *elliptical* in shape, as will be discussed in Chapter 3.

[7] Astronomers use the term *obliquity* to refer to the angle between a planet's equatorial and orbital planes. See, for example, Tables 7.1 and 9.1.

[8] We say *almost* exactly because, as we will soon learn, Earth's axis wobbles slowly, like that of a spinning top, taking about 26,000 years to complete a single wobble.

[9] We are a bit sloppy with language here. Earth's North Pole tilts in the same direction year-round. On the first day of the northern summer, Earth is on the side of the Sun where the tilt of the North Pole is toward the Sun. On the first day of the northern winter, Earth is on the opposite side of the Sun, so the tilt of the North Pole is away from the Sun.

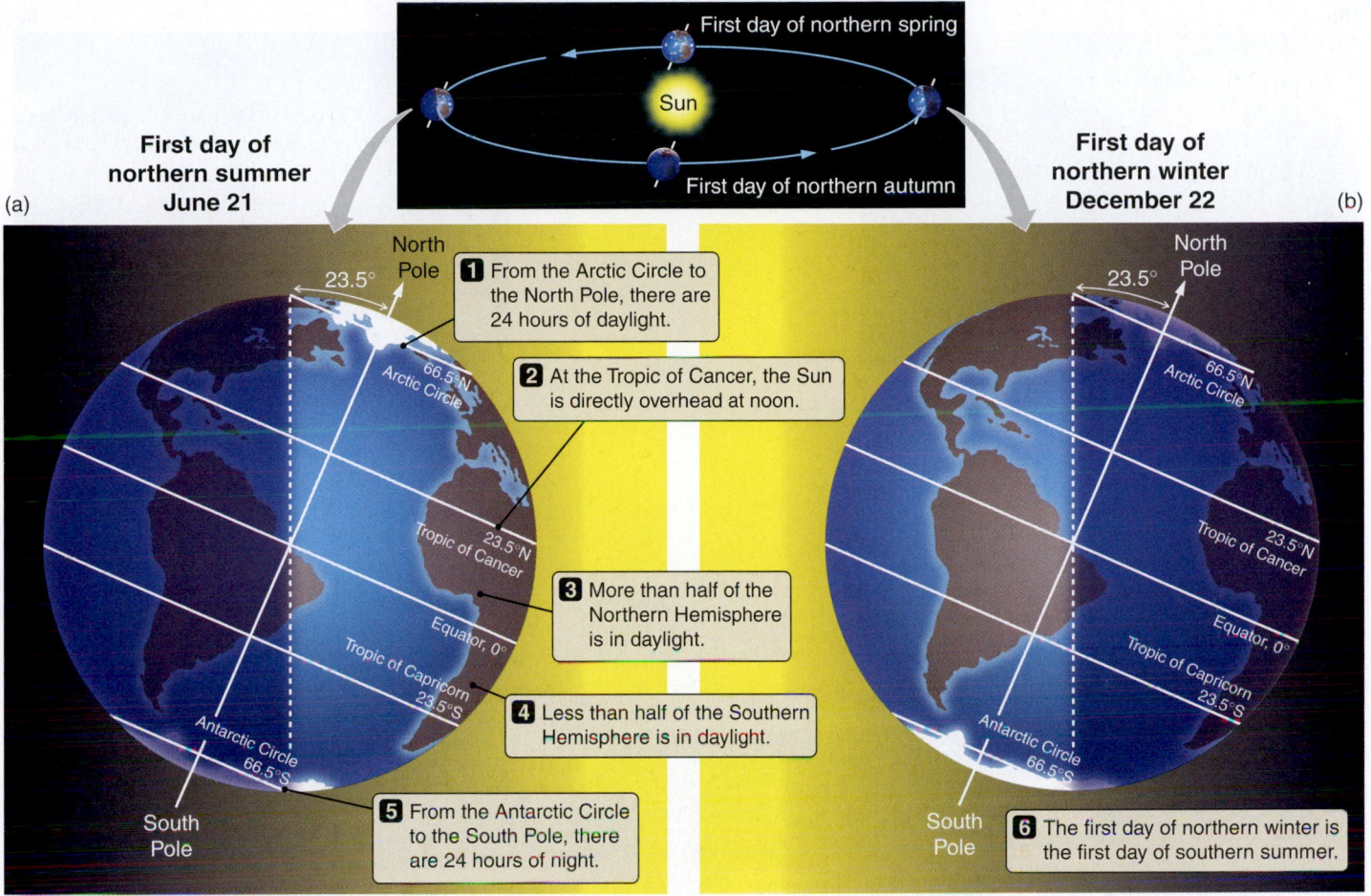

First day of northern spring

Sun

First day of northern autumn

First day of northern summer June 21

(a)

First day of northern winter December 22

(b)

North Pole

23.5°

66.5°N Arctic Circle

1 From the Arctic Circle to the North Pole, there are 24 hours of daylight.

2 At the Tropic of Cancer, the Sun is directly overhead at noon.

23.5°N Tropic of Cancer

Equator, 0°

Tropic of Capricorn 23.5°S

3 More than half of the Northern Hemisphere is in daylight.

4 Less than half of the Southern Hemisphere is in daylight.

Antarctic Circle 66.5°S

South Pole

5 From the Antarctic Circle to the South Pole, there are 24 hours of night.

North Pole

23.5°

66.5°N Arctic Circle

23.5°N Tropic of Cancer

Equator, 0°

Tropic of Capricorn 23.5°S

Antarctic Circle 66.5°S

South Pole

6 The first day of northern winter is the first day of southern summer.

FIGURE 2.13 (a) On the first day of the northern summer (June 21, the summer solstice), the northern end of Earth's axis is tilted most nearly toward the Sun, while the Southern Hemisphere is tipped away. (b) Six months later, on the first day of the northern winter (December 22, the winter solstice), the situation is reversed. Seasons are opposite in the Northern and Southern Hemispheres.

than 12 hours. You can see this directly in Figure 2.13a by noting that when Earth's North Pole is tilted toward the Sun, over half of Earth's Northern Hemisphere is illuminated by sunlight. These are the long days of the northern summer. Six months later the situation is very different. On December 22 (**Figure 2.13b**), Earth's North Pole is tilted away from the Sun, so the Sun appears in the sky south of the celestial equator. Someone in the Northern Hemisphere will see the Sun for less than 12 hours each day. Less than half of the Northern Hemisphere is illuminated by the Sun. It is winter in the north.

Over the course of the year, the length of the day changes, courtesy of the 23.5° tilt of Earth's axis. In the preceding paragraph we were careful to specify the length of the day in the *Northern* Hemisphere because things are very different in the Southern Hemisphere. In fact, things in the Southern Hemisphere are

Seasons in the Southern Hemisphere are the reverse of those in the Northern Hemisphere.

exactly reversed from what is going on in the north. Look again at Figure 2.13. On June 21, while the Northern Hemisphere is enjoying long days and short nights of summer, Earth's South Pole is tilted in the direction away from the Sun. Less than half of the Southern Hemisphere is illuminated by the Sun, and the winter days are shorter than 12 hours. Similarly, on December 22, Earth's South Pole is tilted toward the Sun, and the southern summer days are long.

The differing length of days through the year is part of the explanation for the changing seasons, but we need to consider another important effect: the Sun appears higher in the sky during the summer than it does during the winter, so sunlight strikes the ground *more directly* during the summer than during the winter. To see why this is important, hold a piece of cardboard toward the Sun and look at the size of its shadow. If the cardboard is held so that it is directly face-on to the Sun, then its shadow is large. As you turn the cardboard more edge-on, however, the size of its shadow shrinks. The size of the cardboard's shadow tells you that the cardboard catches less energy from the Sun

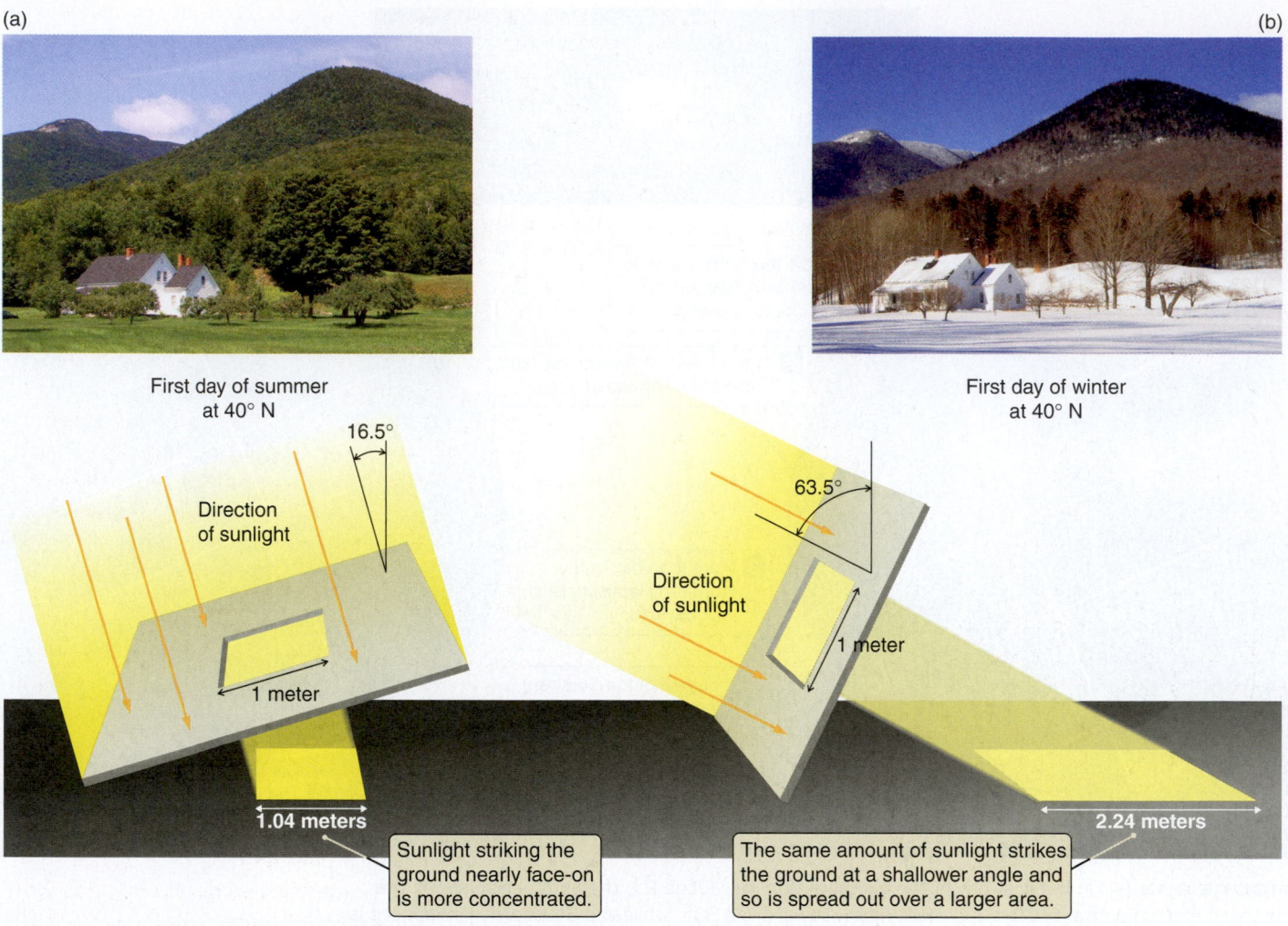

(a)

(b)

First day of summer
at 40° N

16.5°

Direction
of sunlight

1 meter

1.04 meters

Sunlight striking the
ground nearly face-on
is more concentrated.

First day of winter
at 40° N

63.5°

Direction
of sunlight

1 meter

2.24 meters

The same amount of sunlight strikes
the ground at a shallower angle and
so is spread out over a larger area.

FIGURE 2.14 Local noon at latitude 40° north. (a) On the first day of northern summer, sunlight strikes the ground almost face-on. (b) On the first day of northern winter, sunlight strikes the ground more obliquely, and less than half as much sunlight falls on each square meter of ground each second.

each second when it is tilted relative to the Sun than it does when it is face-on to the Sun. This is exactly what happens with the changing seasons. During the summer, Earth's surface is more nearly face-on to the incoming sunlight, so more energy falls on each square meter of ground each second. During the winter, the surface of Earth is more inclined with respect to the sunlight, so less energy falls on each square meter of the ground each second. That is the main reason why it is hotter in the summer and colder in the winter.

> The angle of sunlight to the ground is closer to perpendicular in summer than in winter, so there is more heating per unit area in summer.

To see an example of how this works, look at **Figure 2.14**, which shows the direction of incoming sunlight striking Earth at latitude 40° north, which stretches across middle America from Northern California to New Jersey. At noon on the first day of summer, the Sun is high in the sky—73.5° above

the horizon and only 16.5° away from the zenith. Sunlight strikes the ground almost face-on (**Figure 2.14a**). In contrast, at noon on the first day of winter, the Sun is only 26.5° above the horizon, or 63.5° from the zenith. Sunlight strikes the ground at a rather shallow angle (**Figure 2.14b**). As a result of these differences, more than twice as much solar energy falls on each square meter of ground per second at noon on June 21 as falls there at noon on December 22. Together, these two effects—the directness of sunlight and the differing length of the day—mean that during the summer there is more heating from the Sun and during the winter there is less.

We do not have to wait for the seasons to change to see the effect that the height of the Sun in the sky has on terrestrial climate. We need only compare the climates found at different latitudes on Earth. Near the equator, the Sun passes high overhead every day, regardless of the season. As a result, the climate is warm throughout the year. At high latitudes, however, the Sun is *never* high in the sky, and the climate can be cold and harsh even during the summer.

Four Special Days Mark the Passage of the Seasons

As Earth travels around the Sun over the course of a year, the Sun appears to trace a path along the ecliptic, a great circle that is tilted 23.5° with respect to the celestial equator. Follow along in **Figure 2.15** as we note the four special points on this path that mark the passage of the seasons. Begin with the point in March when Earth's axis is perpendicular to the direction to the Sun (point 1 in **Figure 2.15a**). Here, the Sun's apparent motion along the ecliptic

> The changing seasons are marked by equinoxes and solstices.

crosses the celestial equator moving from the south to the north (point 1 in **Figure 2.15b**). That direction on the celestial sphere, located in the constellation Pisces, is called the **vernal equinox**. The term *vernal equinox* also refers to the day—about March 21—when the Sun appears at this location. When the Sun lies on the celestial equator on the vernal equinox, days are 12 hours long. (The term **equinox** means literally "equal night"; everywhere on Earth, night and day are the same length on the days of the equinoxes.) In the Northern Hemisphere, the vernal equinox is the first day of spring.

As Earth continues its journey around the Sun, the sunward direction moves along toward closer alignment with the tilt of the northern end of Earth's axis, and the Sun climbs higher into the northern sky. The northern end of Earth's axis tilts toward the Sun about 3 months after the vernal equinox. When this happens, the Sun reaches its northernmost point in the sky, located in the constellation Taurus, near its border with Gemini. This day, which marks the longest day of the year and the beginning of summer in the Northern Hemisphere, occurs around June 21. This is the **summer solstice**. (**Solstice** literally means "sun standing still"; the Sun's north–south motion in the sky stops as it reverses its direction.) Note that on this same day, the southern end of Earth's axis is tipped directly away from the Sun. This is the shortest day of the year in the Southern Hemisphere, marking the beginning of the southern winter.

Three months later, around September 23, the Sun is again crossing the celestial equator. This point on the Sun's apparent path, located in the constellation Virgo, is called the **autumnal equinox**. The term refers both to the Sun's location on the celestial sphere and the date when this happens. Because the Sun is again on the celestial equator, days and nights are exactly 12 hours long. It is the first day of autumn in the Northern Hemisphere and the first day of spring in the Southern Hemisphere.

Around December 22, the Sun reaches its southernmost point in the sky as its apparent path takes it through the constellation Sagittarius. This day is called the northern **winter solstice**. Earth's North Pole is tipped most directly away from the Sun on this day. In the Northern Hemisphere this is the shortest day of the year—the first day of winter. As the Sun passes the winter solstice and moves on toward the vernal equinox, the northern days begin growing longer again. Almost all cultural traditions in the Northern Hemisphere include a major celebration of some sort in late December (**Figure 2.16**). Christmas, for example, is celebrated just 3 days after the winter solstice. These winter festivals have many different meanings to their various celebrants, but they all share one thing: they celebrate the return of the source of Earth's light and warmth. The days have stopped growing shorter and are beginning to get longer. It is a "new year," and once again the Sun will hold sway over night. Spring will come again.

Interestingly, there is more to what we feel during the different seasons than the amount of energy we are receiving

FIGURE 2.15 The motion of Earth about the Sun as seen from the frame of reference of (a) the Sun and (b) Earth.

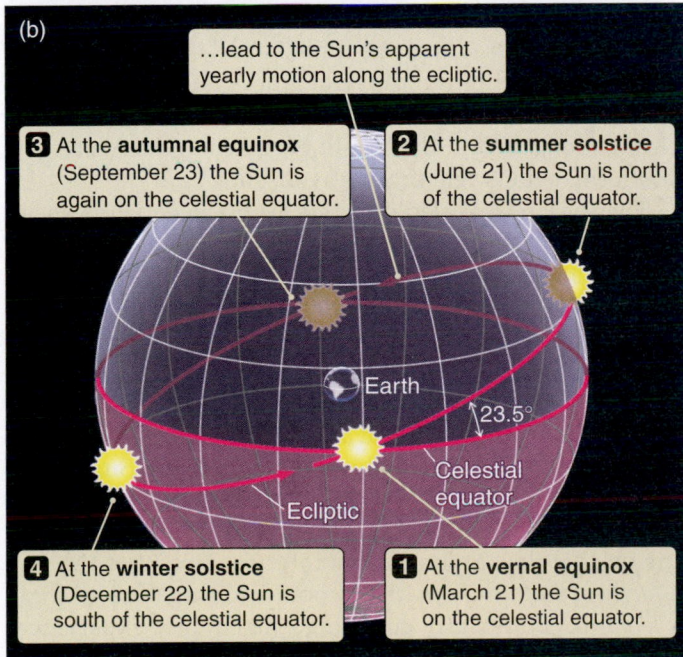

Motion of Earth around the Sun

(a)
1 Vernal equinox
4 Winter solstice
Earth's orbit
Sun
Earth's orbit around the Sun and the tilt of its axis...
3 Autumnal equinox
2 Summer solstice

Apparent motion of the Sun seen from Earth

(b)
...lead to the Sun's apparent yearly motion along the ecliptic.

3 At the **autumnal equinox** (September 23) the Sun is again on the celestial equator.

2 At the **summer solstice** (June 21) the Sun is north of the celestial equator.

Earth
23.5°
Celestial equator
Ecliptic

4 At the **winter solstice** (December 22) the Sun is south of the celestial equator.

1 At the **vernal equinox** (March 21) the Sun is on the celestial equator.

FIGURE 2.16 Most cultural traditions in the Northern Hemisphere include a major celebration in late December, around the time when days begin to grow longer.

from the Sun. Just as it takes time for a pot of water on a stove to heat up when the burner is turned up and time for the pot to cool off when the burner is turned down, it takes time for Earth to respond to changes in heating from the Sun. The hottest months of northern summer are usually July and August, which come *after* the summer solstice, when the days are growing shorter. Similarly, the coldest months of northern winter are usually January and February, which occur *after* the winter solstice, when the days are growing longer. The climatic seasons on Earth lag behind changes in the amount of heating we receive from the Sun.

> Seasonal temperatures lag behind changes in the directness of sunlight.

Our picture of the seasons must be modified somewhat near Earth's poles. At latitudes north of 66.5° north and south of 66.5° south, the Sun is circumpolar for a part of the year surrounding the first day of summer. These lines of latitude are called the **Arctic Circle** and the **Antarctic Circle**, respectively. When the Sun is circumpolar, it is above the horizon 24 hours a day, earning the polar regions the nickname "land of the midnight Sun." The Arctic and

Antarctic regions pay for these long days, however, with an equally long period surrounding the first day of winter when the Sun never rises and the nights are 24 hours long. The Sun never rises high in the Arctic or Antarctic sky, which means that sunlight is never very direct. This is why, even with the long days at the height of summer, the Arctic and Antarctic regions remain relatively cool.

The seasons are also different near the equator. Recall from Figure 2.9c that for an observer on the equator, *all* stars are above the horizon 12 hours a day, and the Sun is no exception. On the equator, days and nights are 12 hours long throughout the year. Changes in the directness of sunlight through the year are also different on the equator. Here the Sun passes directly overhead on the first day of spring and the first day of autumn because these are the days when the Sun is on the celestial equator. Sunlight is most direct at the equator on these days. At the summer solstice, the Sun is at its northernmost point along the ecliptic. It is on this day, and on the winter solstice, that the Sun is *farthest* from the zenith at noon, and therefore sunlight is *least* direct. Strictly speaking, the equator experiences only two seasons: summer, when the Sun passes directly

(a)

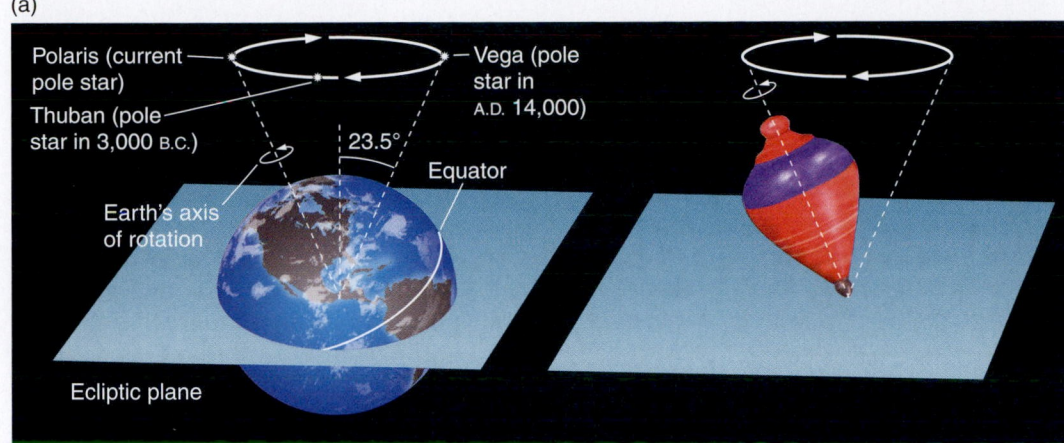

VISUAL ANALOGY FIGURE

2.17 (a) Earth's axis of rotation changes orientation in the same way that the axis of a spinning top changes orientation. (b) This precession causes the projection of the Earth's rotation axis to move in a 47°-diameter circle, centered on the north ecliptic pole (orange cross), with a period of 25,800 years. The red cross shows the projection of Earth's axis on the sky in the early 21st century.

(b)

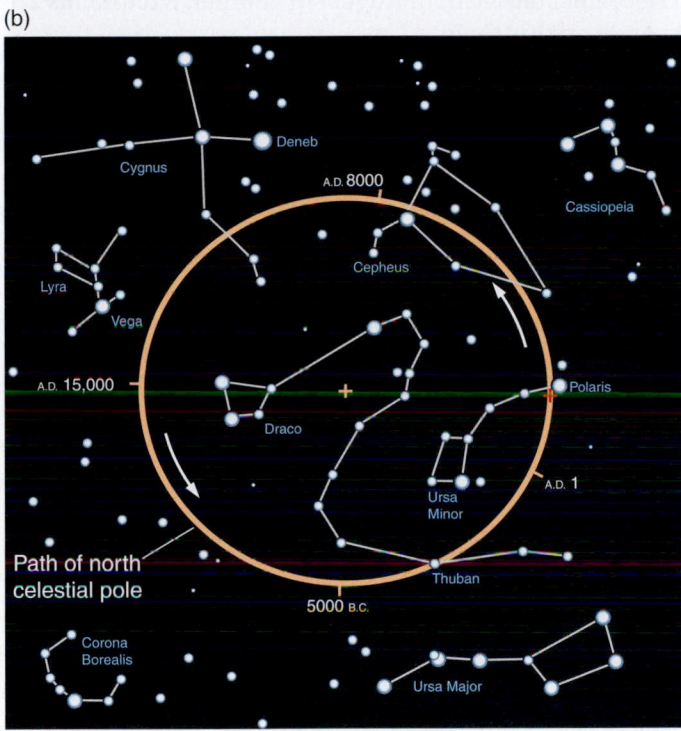

ernmost point and in the constellation of Sagittarius at its southernmost point. Why, then, is the Tropic of Cancer not called the Tropic of Taurus? Why is the Tropic of Capricorn not called the Tropic of Sagittarius? The answer is historical. When these latitude limits were named about 2,000 years ago, the Sun's northernmost point *was* in Cancer. Since then, it has slowly drifted from Cancer to Gemini and is now in Taurus.[10] Likewise, the Sun's southernmost point has drifted from Capricorn to Sagittarius. How could this happen? We are about to find out.

Earth's Axis Wobbles, and the Seasons Shift through the Year

When the Alexandrian astronomer Ptolemy (Claudius Ptolemaeus) and his associates were formalizing their knowledge of the positions and motions of objects in the sky 2,000 years ago, the Sun appeared in the constellation of Cancer on the first day of northern summer and in the constellation of Capricorn on the first day of northern winter—hence, the names of the Tropics. Which leads us to this question: Why have the constellations in which solstices appear changed? The answer has to do with the fact that there are *two* motions associated with Earth and its axis. Earth spins on its axis, but its axis also wobbles like the axis of a spinning top (**Figure 2.17**). The wobble is very slow, taking about 26,000 years to complete one cycle. During this time the north celestial pole makes one trip around a large circle centered on the north ecliptic pole. *Polaris* is a modern name for the star we see near the north celestial pole. If you could travel

> Earth's axis wobbles like the axis of a spinning top.

overhead; and winter, when the Sun is at its northernmost and southernmost points on the ecliptic. However, summer and winter are not very different. The Sun is up for 12 hours a day year-round, and the Sun is always so close to being overhead at noon that the directness of sunlight changes by only 8 percent throughout the year.

If you live between the latitudes of 23.5° south and 23.5° north—in Rio de Janeiro or Honolulu, for example—the Sun will be directly overhead at noon twice during the year. The band between these two latitudes is called the **Tropics**. The northern limit of this region is called the Tropic of Cancer; the southern limit is called the Tropic of Capricorn (see Figure 2.13). (As a challenge, think about what the seasons are like at different locations within the Tropics.)

Wait a minute! We already pointed out that the Sun appears in the constellation of Taurus when at its north-

---

[10] It was only recently, in 1990, that the location of the summer solstice passed from Gemini into Taurus. In about 600 years the vernal equinox will move from Pisces into Aquarius, marking the true beginning of the "Age of Aquarius."

several thousand years into the past or future, you would find that the point about which the northern sky appears to rotate is no longer near Polaris.

Recall that the celestial equator is the set of directions in the sky that are perpendicular to Earth's axis. As Earth's axis wobbles, then, the celestial equator must appear to tilt with it. And as the celestial equator wobbles, the locations where it crosses the ecliptic—the equinoxes—change as well. During each 26,000-year wobble of Earth's axis, the locations of the equinoxes make one complete circuit around the celestial equator. Together, these shifts in position are called the **precession of the equinoxes**.

> A 26,000-year wobble causes the position of the equinoxes to shift gradually.

The tendency of the seasons to shift through the year from century to century has played havoc on human efforts to construct reliable calendars, and history is full of interesting anecdotes related to this difficulty. For example, for much of the 16th, 17th, and 18th centuries, the calendars in Protestant Europe lagged behind the calendars in Catholic Europe by first 10, and later 11, days. Not until 1752 did England and her colonies, including those in America, drop 11 days from their calendars to bring them into line with the Catholic calendar. This conciliatory step was met by riots among the people, who somehow felt that these 11 days had been stolen from them.

Today's calendar (which was not adopted in Russia until the 1917 Bolshevik Revolution) is known as the **Gregorian calendar** and is based on the **tropical year**, which is 365.242199 solar days long. A **solar day** is the 24-hour period of Earth's rotation that brings the Sun back to the same local meridian. The tropical year measures the time from one vernal equinox to the next—from the start of spring to the start of spring. Notice that the tropical year is not an integral number of days long, but is approximately ¼ day longer. An elaborate system of **leap years**—years in which a 29th day is added to the month of February—is used in our calendar to make up for the extra fraction of a day, preventing the seasons from slowly "sliding through" the year (becoming increasingly out of sync with the months): winter in December one year, in August in another. **Excursions 2.1** expands on this topic.

2.4 The Motions and Phases of the Moon

The most prominent object in the sky after the Sun is the Moon. Just as Earth orbits about the Sun, the Moon orbits around Earth. (Actually, Earth and the Moon orbit around each other, and together they orbit the Sun, as we will see

---

EXCURSIONS 2.1

Why Is It Surprising That A.D. 2000 Was a Leap Year?

Almost everyone knows that years that are divisible by 4 are leap years, and A.D. 2000 was no exception to that rule. Yet A.D. 2000 *was* a special case. To understand why, we need to take a look at the way our calendar is constructed, and the purpose that leap years serve.

It takes Earth 365.242199 days to travel from one vernal equinox to the next. This tropical year is *not* exactly equal to the 365 days we normally think of as making up a year. (And thanks to the precession of the equinoxes, it is also not equal to the time—365.256366 days—that it takes for Earth to complete one orbit about the Sun!) For convenience, we count years as starting at midnight on the morning of January 1, and ending 365 days later, at midnight on the night of December 31. But what about that extra fraction of a day (0.24 plus a bit) that we have not accounted for? Here is where leap years come in. A true tropical year is 0.242199 day—or about ¼ day—longer than a 365-day year, which means that after 4 years these extra parts have added up to make about one extra day. So every 4 years we add the extra day back in, and February gets a 29th day. If we did not add this extra

day every four years, our calendar would slip by ¼ day each year, which means that the seasons would shift by not quite a month each century. If we did not correct for leap years, the first day of summer would come in July, then in August, then in September, and so on.

Even after we add a leap year, the calendar still has problems. A 365-day year is short by 0.242199 day, which is a bit *less* than ¼. If we kept adding an extra day *every* 4 years, then over time we would accumulate an average of 0.007801 extra day per year. In 400 years the calendar would have slipped by about 3 days. This difference is what caused the Protestant and Catholic calendars to fall into disagreement by 11 days by the middle of the last millennium. To fix this problem, it is necessary to get rid of 3 days every 400 years. We accomplish this by making century years into common 365-day years, *except* for those century years that are divisible by 400—such as A.D. 2000—which remain leap years. With one slight further revision—making years divisible by 4,000 into common 365-day years—the modern Gregorian calendar now slips by about only 1 day in 20,000 years.

in Chapter 3.) In some respects, the appearance of the Moon is constantly changing, but we begin our discussion of the motion of the Moon by talking about an aspect of the Moon's appearance that does *not* change.

We Always See the Same Face of the Moon

The Moon constantly changes its lighted shape and position in the sky, but one thing that does *not* change is the face of the Moon that we see. If we were to go outside next week or next month, or 20 years from now, or 20,000 *centuries* from now, we would still see

> The Moon rotates on its axis once for each orbit around Earth.

the same side of the Moon that we see tonight. This fact is responsible for the common misconception that the Moon does not rotate. The Moon *does* rotate on its axis—exactly once for each revolution that it makes about Earth.

Imagine walking around the Washington Monument while keeping your face toward the monument at all times (a reasonable thing to do—you want to get a good look at it). By the time you complete one circle around the monument, your head has turned completely around once. (When you were south of the monument, you were facing north; when you were west of the monument, you were facing east; and so on.) But someone looking at you from the monument would never have seen anything other than your face. The

Moon does exactly the same thing, rotating on its axis once per revolution around Earth, always keeping the same face toward Earth, as shown in **Figure 2.18**. This phenomenon is referred to as the Moon's **synchronous rotation**. (The Moon's synchronous rotation is not an accident. In Chapter 10 we will find that its cause is related to why we have tides on Earth.) ▶‖ **AstroTour:** The Moon's Orbit: Eclipses and Phases

The Changing Phases of the Moon

Humans have long been fascinated by the Moon and its changing aspects. We speak of the "man in the Moon," the "harvest Moon," and sometimes a "blue Moon." In mythology, the Moon was the Roman goddess Diana and the Greek goddess Artemis. Among the Aztecs, it was the god Tecciztecatl. The Moon has been the frequent subject of art, literature, and music. Sometimes the Moon appears as a circular disk in the sky. At other times, it is nothing more

> The Moon shines by reflected sunlight.

than a thin sliver. At still others, its face appears dark. Popular culture often refers to the side of the Moon away from Earth as the "dark side of the Moon." This is even the title of one of the most influential rock albums of the 20th century. In fact, however, there is no dark side of the Moon. At any given time, half of the Moon is in sunlight and half is in darkness—just as at any given time, half of Earth is in sunlight and half is in darkness. The side of the Moon that faces away from Earth, the "far side," spends just as much time in sunlight as the side of the Moon that faces toward Earth does. But "I'll see you on the *backside* of the Moon" might not have been as wildly successful as a song lyric.

Our fascination with the Moon can be explained by its unique features and its constantly changing aspect. Unlike the Sun, the Moon has no light source of its own. Like the planets, including Earth, it shines by reflected sunlight. Like Earth, half of the Moon is always in bright daylight, and half is always in darkness. The different **phases** of the Moon result from the fact that the illuminated portion of the Moon that we see is constantly changing. Sometimes (during a new Moon) the side facing away from us is illuminated, and sometimes (during a full Moon) the side facing toward us is illuminated. The rest of the time, only part of the illuminated portion can be seen from Earth.

To help you visualize the changing phases of the Moon, go outside at night and have a friend hold up a soccer ball so that it is illuminated from one side by a nearby streetlight (representing the Sun). Have your friend walk around you in a circle, and watch the changes in the ball's appear-

> The phase of the Moon is determined by how much of its bright side we can see.

ance. When you are between the ball and the streetlight, the face of the ball that is toward you is fully illuminated.

FIGURE 2.18 The Moon rotates once on its axis for each orbit around Earth—an effect called *synchronous rotation*.

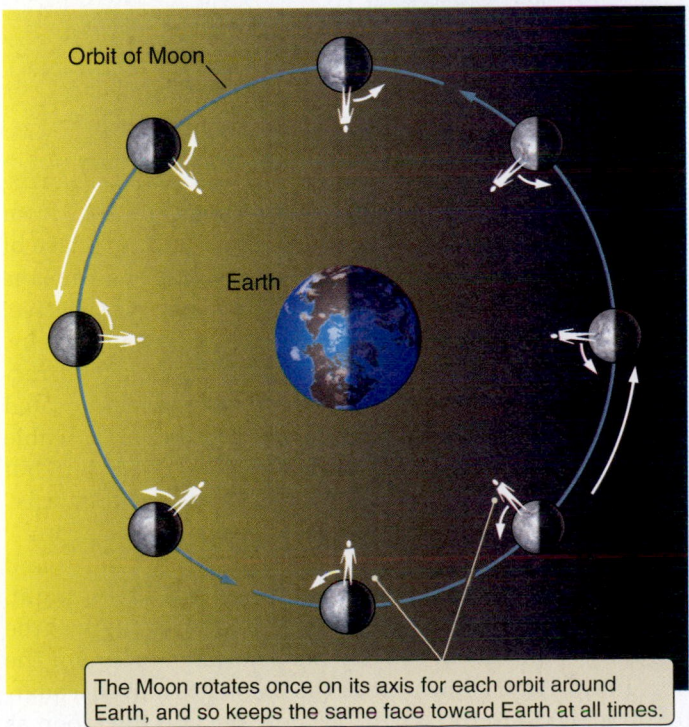

Orbit of Moon

Earth

The Moon rotates once on its axis for each orbit around Earth, and so keeps the same face toward Earth at all times.

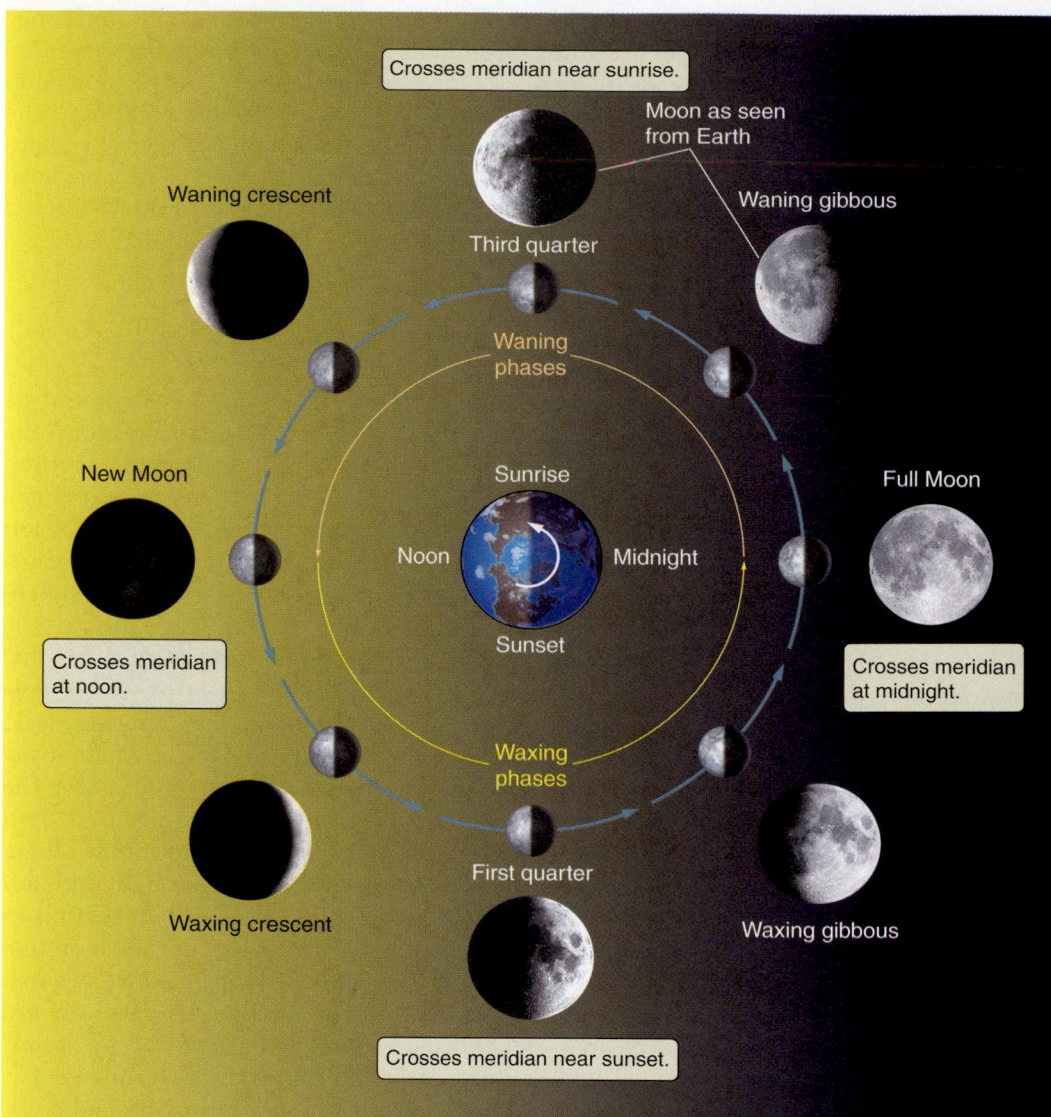

Crosses meridian near sunrise.

Moon as seen from Earth

Waning crescent

Waning gibbous

Third quarter

Waning phases

New Moon

Sunrise

Noon · Midnight

Full Moon

Sunset

Crosses meridian at noon.

Crosses meridian at midnight.

Waxing phases

Waxing crescent

First quarter

Waxing gibbous

Crosses meridian near sunset.

FIGURE 2.19 The inner circle of images (connected by blue arrows) shows the Moon as it orbits Earth, as seen by an observer far above Earth's North Pole. The outer ring of images shows the corresponding phases of the Moon as seen from Earth.

The ball appears to be a bright, circular disk. As the ball moves around its circle, you will see a progression of lighted shapes, depending on how much of the bright side and how much of the dark side of the ball you can see. This progression of shapes exactly mimics the changing phases of the Moon.

Figure 2.19 shows the changing phases of the Moon. When the Moon is between Earth and the Sun, the illuminated side of the Moon faces away from us, and we see only its dark side. This is called a **new Moon**. (Study Figure 2.19 to understand that a new Moon can be "seen" only from the illuminated side of Earth.) It appears close to the Sun in the sky, so it rises in the east at sunrise, crosses the meridian near noon, and sets in the west near sunset. A new Moon is never visible in the nighttime sky. ▶❚❚ **AstroTour: The Moon's Orbit: Eclipses and Phases**

As the Moon continues on its orbit around Earth, a small part of its illuminated hemisphere becomes visible. This

shape is called a **crescent**, from the Latin *crescere*, meaning "to grow." Because the Moon appears to be "filling out" from night to night at this time, the full name for this phase of the Moon is a "waxing crescent Moon." (**Waxing** here means "growing in size and brilliance.") From our perspective, the Moon has also moved away from the Sun in the sky. Because the Moon travels around Earth in the same direction in which Earth rotates, we now see the Moon located to the east of the Sun. A waxing crescent Moon is visible in the western sky in the evening, near the setting Sun but remaining above the horizon after the Sun sets. The "horns" of the crescent always point directly away from the Sun.

As the Moon moves farther along in its orbit, more and more of its illuminated side becomes visible each night, so the crescent continues to fill out. At the same time, the angular separation in the sky between the Moon and the Sun grows. After about a week the Moon has moved a quarter of the way around Earth. We now see half the Moon as

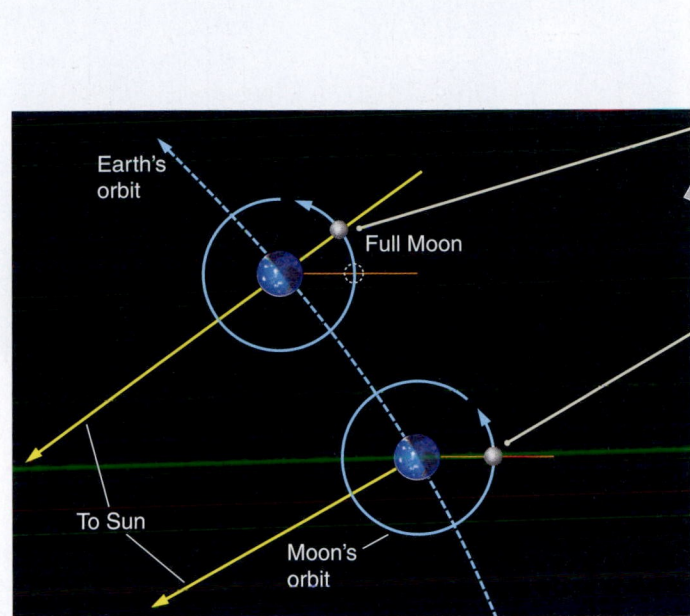

FIGURE 2.20 (a) The Moon completes one sidereal orbit in 27.32 days, but the synodic period (the period between phases seen from Earth) from one full Moon to the next is 29.53 days. (The horizontal orange line to the right of the Moon indicates a fixed direction in space.) (b) The orbits of Earth and the Moon are shown here to scale.

illuminated and half the Moon as dark—a phase that we call **first quarter Moon**. A look at Figure 2.19 shows that the first quarter Moon rises at noon, crosses the meridian at sunset, and sets at midnight. Note that *first quarter* refers not to how much of the face of the Moon that we see illuminated, but rather to the fact that the Moon has completed the first quarter of its cycle from new Moon to new Moon.

As the Moon moves beyond first quarter, we are able to see more than half of its bright side. This phase is called a "waxing gibbous Moon," from the Latin *gibbus*, meaning "hump." The **gibbous** Moon continues nightly to "grow" until finally Earth is between the Sun and the Moon and we see the entire bright side of the Moon—a **full Moon**. The Sun and the Moon now appear opposite each other in the sky. The full Moon rises as the Sun sets, crosses the meridian at midnight, and sets in the morning as the Sun rises.

The second half of the Moon's orbit proceeds just like the first half, but in reverse. The Moon continues in its orbit, again appearing gibbous but now becoming smaller each night. This phase is called a "waning gibbous Moon." (**Waning** means "becoming smaller.") A **third quarter Moon** occurs when we once again see half of the illuminated part of the Moon and half of the dark part of the Moon. A third quarter Moon rises at midnight, crosses the meridian near sunrise, and sets at noon. The Moon continues on its path, visible now as a "waning crescent Moon" in the morning sky, until the Moon again appears as nothing but

a dark circle rising and setting with the Sun, and the cycle begins again.

It takes the Moon 27.32 days to complete one revolution about Earth; this is called its **sidereal period**. However, because of the changing relationship between Earth, the Moon, and the Sun due to Earth's orbital motion, it takes 29.53 days to go from one full Moon to the next; this is called its **synodic period**. **Figure 2.20** shows how this works. You can always tell a waxing Moon from a waning Moon because the side that is illuminated is always the side facing the Sun. When the Moon is waxing, it appears in the evening sky, so its western side is illuminated (the right side as viewed from the Northern Hemisphere). Conversely, when the Moon is waning, the eastern side (the left side as viewed from the Northern Hemisphere) appears bright.

Do not try to memorize all possible combinations of where the Moon is in the sky at what phase and at what time of day. You do not have to. Instead, work on *understanding* the motion and phases of the Moon, and then use your understanding to figure out the specifics of any given case. As a way to study the phases of the Moon, draw a picture like Figure 2.19, and use it to follow the Moon around its orbit. From your drawing, figure out what phase you would see and where it would appear in the sky at a given time of day. You might also enjoy thinking about what phase someone on the Moon would see when looking back at Earth.

FIGURE 2.21 Stonehenge is an ancient artifact in the English countryside, used 4,000 years ago to keep track of celestial events.

2.5 Eclipses: Passing through a Shadow

Put yourself in the place of our nomadic friend from the opening section of this chapter. You are finely attuned to the patterns of the sky, and you view these patterns not as the inexorable consequences of physical law but as visible signs from the gods. Can you imagine any celestial events that would strike more terror in your heart than to look up and see the Sun, giver of light and warmth, being eaten away as if by a giant dragon, or the full Moon turning the ominous color of blood? Archaeological evidence suggests that our ancestors put great effort into trying to figure out the pattern of **eclipses** and thereby bring them into the orderly scheme of the heavens. For example, the ancient and massive stone artifact in the English countryside called Stonehenge, pictured in **Figure 2.21**, may have allowed its builders to predict when eclipses might occur. The lives and motivations of the builders of Stonehenge, 4,000 years dead, may be lost in antiquity. Yet how can

we doubt their desire to exert some control over the terror of eclipses by learning a few of their secrets—and in the process assuring themselves that an eclipse did not mean that all was lost?

Varieties of Eclipses

An eclipse in which Earth moves through the shadow of the Moon is called a **solar eclipse**. Three different types of solar eclipses are possible: *total*, *partial*, and *annular*. To see why, begin by looking at the structure of the shadow of the Sun cast by a round object such as the Moon, as shown in **Figure 2.22**. An observer at point A would be unable to see any part of the surface of the Sun. This darkest, inner part of the shadow is called the **umbra**. If a point on Earth passes through the Moon's umbra, the Sun's light is totally blocked by the Moon. This is called a **total solar eclipse** (**Figure 2.23**). Now look instead at points B and C in Figure

> There are three types of solar eclipses: total, partial, and annular.

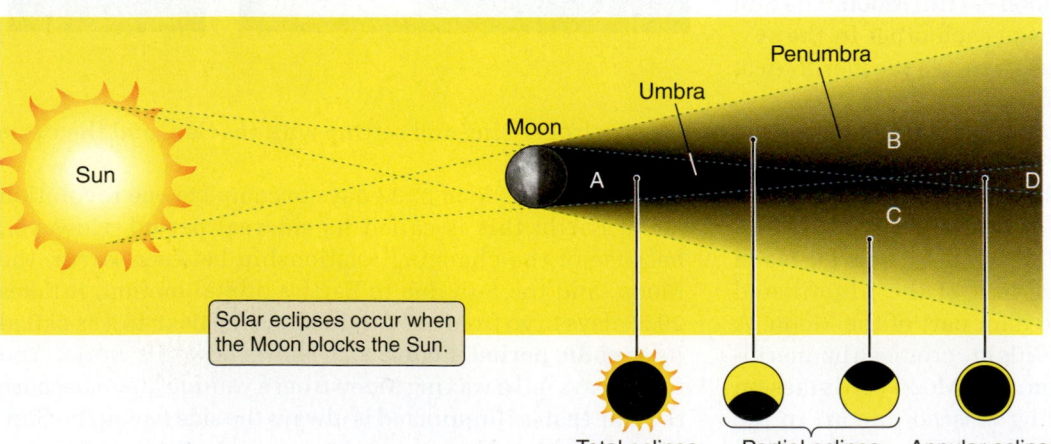

Sun

Moon

Umbra

Penumbra

A

B

C

D

Solar eclipses occur when the Moon blocks the Sun.

Total eclipse   Partial eclipse   Annular eclipse

FIGURE 2.22 Different parts of the Sun are blocked at different places within the Moon's shadow. An observer in the umbra (A) sees a total solar eclipse, observers in the penumbra (B and C) see a partially eclipsed Sun, and observers in region D see an annular solar eclipse.

FIGURE 2.23 The full spectacle of a total eclipse of the Sun.

FIGURE 2.24 An annular eclipse, in which the Moon does not quite cover the Sun. Note that Earth's atmosphere distorts the shape of the Sun when close to the horizon.

2.22. From these points an observer can see one side of the disk of the Sun but not the other. This outer region, which is only partially in shadow, is the **penumbra**. If a point on the surface of Earth passes through the Moon's penumbra, the result is a **partial solar eclipse**, in which the disk of the Moon blocks the light from a portion of the Sun's disk.

In the third type of eclipse, called an **annular solar eclipse**, the Sun appears as a bright ring surrounding the dark disk of the Moon (**Figure 2.24**). An observer at point D is far enough from the Moon that the Moon's angular diameter appears smaller than the Sun's. You may be wondering how one eclipse can be total and another annular. Two things make this possible. One is a fluke of nature: The diameter of the Sun is about 400 times the diameter of the Moon, and the Sun is about 400 times farther away from Earth than the Moon is. As a result, the Moon and Sun have almost exactly the same apparent size in the sky. The other factor is that the Moon's orbit is not a perfect circle. So when the Moon and Earth are a bit closer together than on average, the Moon appears larger in the sky than the Sun. An eclipse occurring at that time will be total. When the Moon and Earth are farther apart than on average, the Moon appears smaller than the Sun, so eclipses occurring during this time will be annular. Sequential pictures of total and annular solar eclipses are shown in **Figure 2.25**. As it turns out, among all solar eclipses, one-third are total, one-third are annular, and one-third are seen only as a partial eclipse.

Figure 2.26a illustrates the geometry of a solar eclipse, with the Moon's shadow falling on the surface of Earth. Note that figures like this (or like Figures 2.18 and 2.19) are seldom drawn to scale. Instead, they show Earth and the Moon much closer together than they are in reality. The rea-

FIGURE 2.25 Time sequences of images of the Sun taken during a total solar eclipse (a) and during an annular solar eclipse (b). The sunspots seen in (b) will be discussed in Chapter 14.

(a)

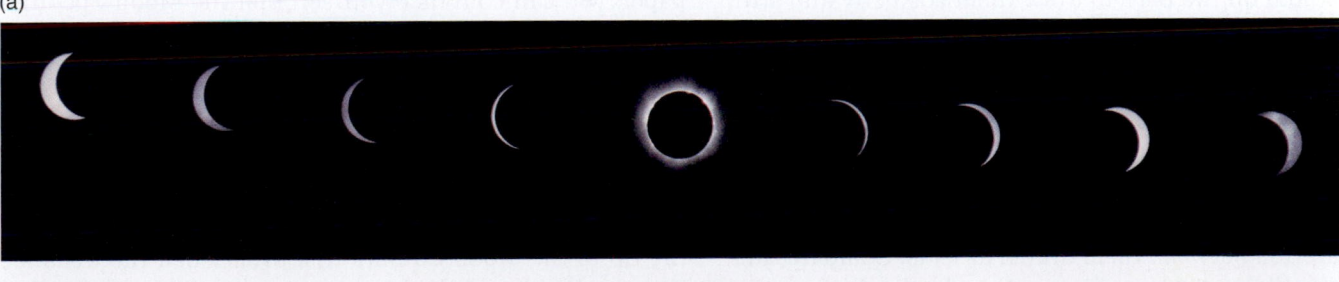

(b)

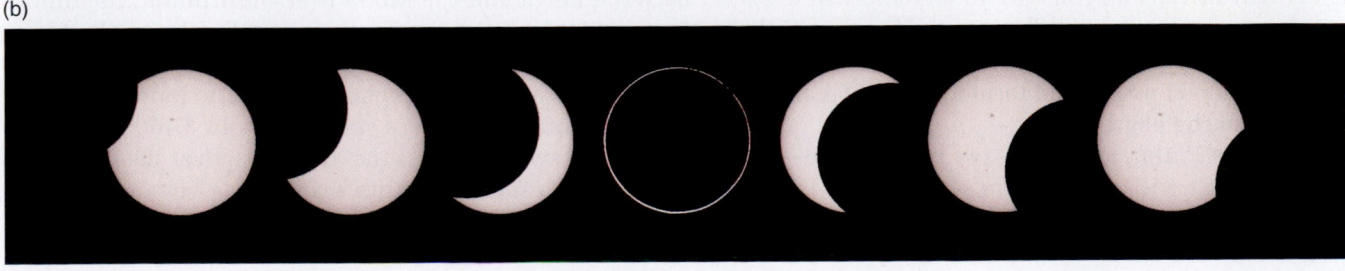

(a) Solar eclipse geometry (not to scale)

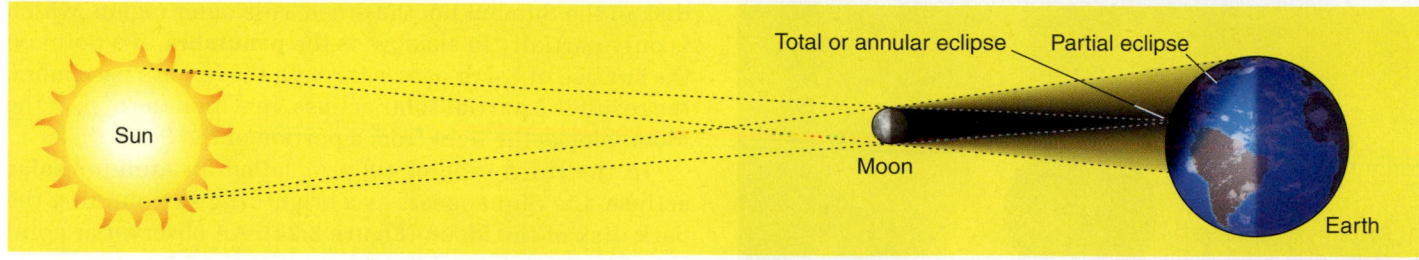

(b) Solar eclipse to scale

(c) Lunar eclipse geometry (not to scale)

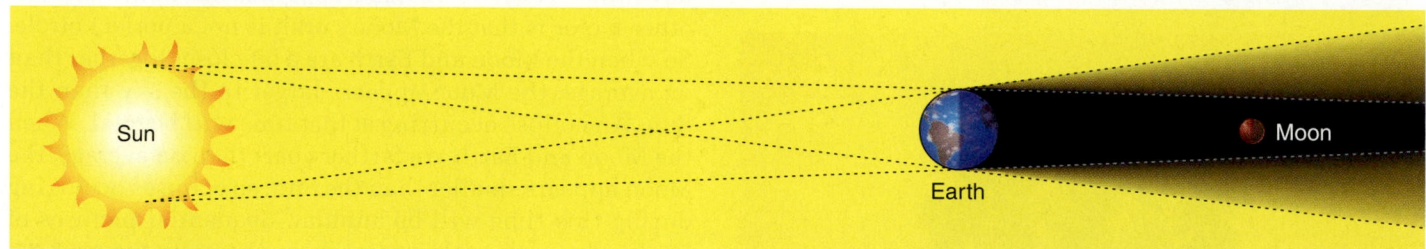

(d) Lunar eclipse to scale

FIGURE 2.26 (a, b) A solar eclipse occurs when the shadow of the Moon falls on the surface of Earth. (c, d) A lunar eclipse occurs when the Moon passes through Earth's shadow. Note that (b) and (d) are drawn to proper scale.

son for distorting figures in this way is simple: there is not enough room on the page to draw them correctly and still keep the smaller details visible. The 384,400-km distance between Earth and the Moon is over 60 times the radius of Earth and over 220 times the radius of the Moon. The relative sizes and distances between Earth and the Moon are roughly equivalent to the difference between a basketball and a baseball placed 7 meters apart. **Figure 2.26b** shows the geometry of a solar eclipse with Earth, the Moon, and the separation between them drawn to scale. Compare this drawing to Figure 2.26a and you will understand why artistic license is normally taken in drawings of Earth and the Moon. If the Sun were drawn to scale in Figure 2.26b, it would be $^{6}/_{10}$ of a meter across and located almost 64 meters off the left side of the page.

From any particular location—say, your home—the probability that you will see a *partial* solar eclipse is very much greater than the likelihood of your seeing a *total* solar eclipse. The reason is that the Moon's penumbra is quite large. In fact, with a bit of thought and a pencil and paper, you can convince yourself that the Moon's penumbra, where it hits Earth, must have a diameter about twice the diameter of the Moon itself, or almost 7,000 km. This part of the shadow is large enough to cover a substantial fraction of Earth, so partial solar eclipses are often seen from much of the planet. In contrast, the path along which a total solar eclipse can be seen (**Figure 2.27**) covers only a tiny fraction of the surface of Earth. Earth is so close to the tip of the Moon's umbra that even when the distance between Earth and the Moon is at a minimum, the umbra is only 269 km wide at the surface of Earth. As the Moon moves along in its orbit, this tiny shadow sweeps across the face of Earth at breakneck speed. The Moon moves in its orbit around Earth at a speed of about 3,400 km/h, and its shadow sweeps across the disk of Earth at the same rate. Earth is also rotating on its axis with a velocity of 1,670 km/h at the equator (and less than that at other latitudes). The situation is further complicated by the fact that the

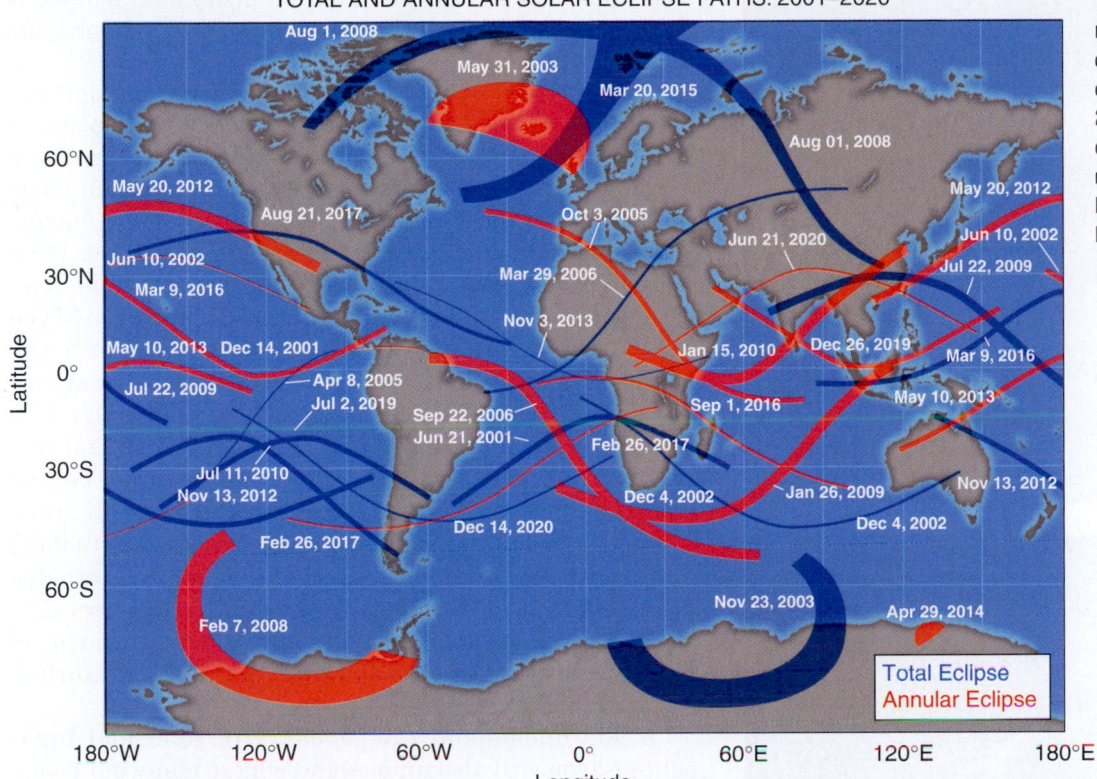

TOTAL AND ANNULAR SOLAR ECLIPSE PATHS: 2001–2020

FIGURE 2.27 The paths of total and annular solar eclipses predicted for the early 21st century. Solar eclipses occurring in Earth's polar regions cover more territory because the Moon's shadow hits the ground obliquely.

ANNULAR SOLAR ECLIPSES: 2001–2025

TOTAL SOLAR ECLIPSES: 2001–2025

Moon's shadow falls on the curved surface of Earth. You may have noticed that the image displayed by an old-style overhead or modern LCD projector is distorted when the beam is not perpendicular to the screen. Similarly, the curvature of Earth often causes the region shaded by the Moon during a solar eclipse to be elongated by differing amounts. The curvature can even cause an eclipse that started out as annular to become total.

When all of these effects are considered, the result is that a total solar eclipse can never last longer than 7½ minutes and is usually significantly shorter. Even so, it is one of the most amazing and awesome sights in nature. People all over the world flock to the most remote corners of Earth to witness the fleeting spectacle of the bright disk of the Sun blotted out of the daytime sky, leaving behind the eerie glow of the Sun's outer atmosphere (see Figure 2.23).

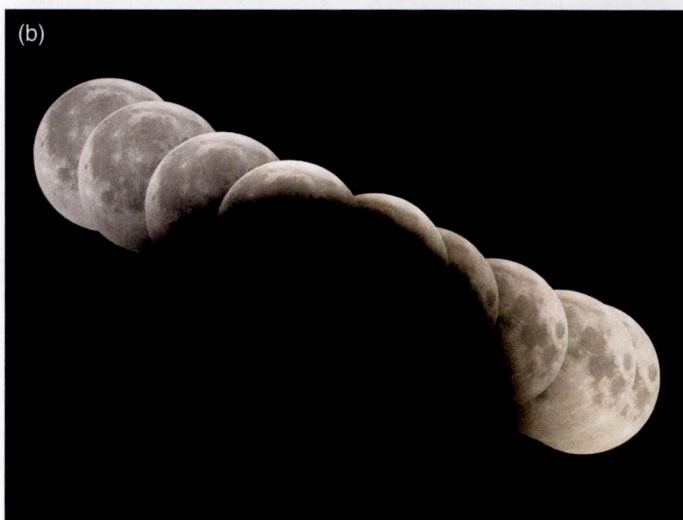

FIGURE 2.28 (a) A total lunar eclipse. (b) The progress of a partial lunar eclipse. Note the size of Earth's shadow compared to the size of the Moon.

Lunar eclipses (Figure 2.28) are very different in character from solar eclipses. The geometry of a lunar eclipse is shown in **Figure 2.26c** (and is drawn to scale in **Figure 2.26d**). Because Earth is much larger than the Moon, the dark umbra of Earth's shadow at the distance of the Moon is about 9,200 km in diameter, or over 2½ times the diameter of the Moon. A **total lunar eclipse** is a much more leisurely affair than a total solar eclipse, with the Moon spending as long as 1 hour and 40 minutes in the umbra of Earth's shadow. A **penumbral lunar eclipse** occurs when the Moon passes through the penumbra of Earth's shadow (to review what the penumbra is, see Figure 2.22). A penumbral eclipse can be unspectacular: its appearance from Earth is nothing more than a fading in the brightness of the full Moon. Although the penumbra of Earth is 16,000 km across at the distance of the Moon—over four

> Lunar eclipses last much longer than solar eclipses.

times the diameter of the Moon—a penumbral eclipse is noticeable only when the Moon passes within about 1,000 km of the umbra.

Total eclipses of the Moon are relatively common, and many of us have seen at least one. As you looked up at that copper-colored disk (**Figure 2.28a**), you may have wondered how it was possible to even see the Moon when completely immersed in Earth's shadow. To visualize how, imagine standing on the Moon during a total lunar eclipse. What would you see? One thing you would *not* see is the Sun, because the Sun would be hiding behind Earth's disk. (If you did see some of the Sun peeking out from behind Earth, you would be standing on a partially eclipsed Moon, as shown in **Figure 2.28b**). What you *would* see is the Earth's dark disk surrounded by a thin, bright, reddish ring. That ring is Earth's atmosphere, with sunlight coming from behind Earth and being scattered by dust and other small atmospheric particles. It is reddish for the same reason that the Sun appears reddened during sunset. (In Chapter 8 we'll discuss scattering and the reddening of sunlight.) Now safely back on Earth, you realize it is that bright, reddish ring of Earth's atmosphere that is lighting up the Moon's surface during a total lunar eclipse.

If you understand the geometry of solar and lunar eclipses, you will also understand why so many more people have experienced a total lunar eclipse than have experienced a total solar eclipse. To see a total solar eclipse, you must be located within that very narrow band of the Moon's shadow as it moves across Earth's surface (see Figure 2.27). On the other hand, when the Moon is immersed in Earth's shadow, anyone located in the hemisphere of Earth that is facing the Moon can see it.

Eclipse Seasons Occur Roughly Twice Every 11 Months

We all know from experience that we don't see a lunar eclipse every time the Moon is full, nor do we observe a solar eclipse every time the Moon is new. So this must be telling us something about how the Moon's orbit around Earth is oriented with respect to Earth's orbit around the Sun. If the Moon's orbit were in exactly the same plane as the orbit of Earth (imagine Earth, the Moon, and the Sun all sitting on the same flat tabletop), then the Moon would pass directly between Earth and the Sun at every new Moon. The Moon's shadow would pass across the face of Earth, and we would see a solar eclipse. Similarly, Earth would pass directly between the Sun and the Moon every synodic month, and each full Moon would be marked by a lunar eclipse.

Solar and lunar eclipses do *not* happen every month, because the Moon's orbit does not lie in exactly the same plane as the orbit of Earth. Look at **Figure 2.29** to see how this works. The plane of the Moon's orbit about Earth is inclined by about 5.2° with respect to the plane of Earth's orbit about the Sun. The line along which the two orbital

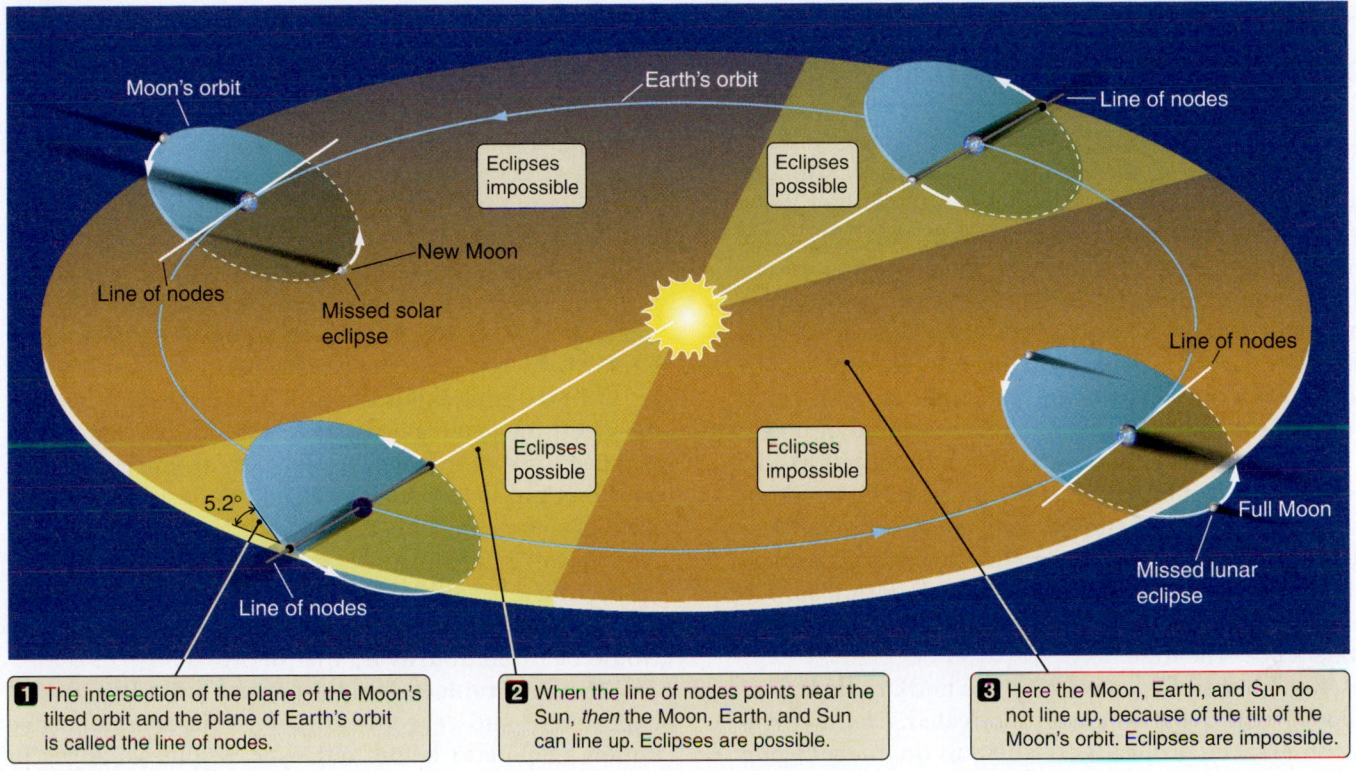

Moon's orbit

Earth's orbit

Line of nodes

Eclipses impossible

Eclipses possible

New Moon

Line of nodes

Missed solar eclipse

Line of nodes

5.2°

Eclipses possible

Eclipses impossible

Full Moon

Line of nodes

Missed lunar eclipse

1 The intersection of the plane of the Moon's tilted orbit and the plane of Earth's orbit is called the line of nodes.

2 When the line of nodes points near the Sun, *then* the Moon, Earth, and Sun can line up. Eclipses are possible.

3 Here the Moon, Earth, and Sun do not line up, because of the tilt of the Moon's orbit. Eclipses are impossible.

FIGURE 2.29 Eclipses are possible only when the Sun, Moon, and Earth lie along (or very close to) an imaginary line known as the *line of nodes*. When the Sun does not lie along the line of nodes, Earth passes under or over the shadow of a new Moon, and a full Moon passes under or over the shadow of Earth.

planes intersect is called the **line of nodes**. For part of the year, the line of nodes passes close to the Sun. During these times, called **eclipse seasons**, a new Moon passes between the Sun and Earth, casting its shadow on Earth's surface and causing a solar eclipse. Similarly, a full Moon occurring during an eclipse season passes through Earth's shadow, and a lunar eclipse results. An eclipse season lasts only 38 days. That is how long the Sun is close enough to the line of nodes for eclipses to occur. Most of the time, the line of nodes points farther away from the Sun, and Earth, Moon, and Sun cannot line up closely enough for an eclipse to occur. A solar eclipse cannot take place, because the shadow of a new Moon passes "above" or "below" Earth. Similarly, no lunar eclipse can occur, because a full Moon passes "above" or "below" the shadow of Earth.

If the plane of the Moon's orbit always had the same orientation, then eclipse seasons would occur twice a year, as suggested by Figure 2.29. In actuality, eclipse seasons occur about every 5 months and 20 days. The roughly 10-day difference is due to the fact that the plane of the Moon's orbit slowly wobbles, much like the wobble of a spinning plate balanced on the end of a circus performer's stick. As it does so, the line of nodes changes direction. This wobble rotates in the direction opposite the direction of the motion of the Moon in its orbit. (That is, the line of nodes moves clockwise as viewed from above Earth's orbital plane.) It takes the Moon's orbit 18.6 years to complete one "wobble" of 360°, so we say that the line of nodes *regresses* at a rate of 360° every 18.6 years, or 19.4° per year. This amounts to about a 20-day regression each year. If January 1 marks the middle of an eclipse season, the next eclipse season will be centered around June 20, and the one after that around December 10.

We have come far since our nomadic ancestor looked at the sky and saw there a mystical reflection of the patterns and events that marked the life of the tribe. Yet when we look at the sky today, our sense of awe and majesty is no less than that experienced by that long-ago tribesman. Our distant ancestors had to look for patterns in the world to survive. This same human impulse to seek patterns led astrologers to look for connections between ourselves and the heavens, and later led people to seek patterns that gave birth to science. We now know that the patterns of the sky are connected to us much more directly than is any mystical link invented by an astrologer. The patterns and changes that we see in the sky are caused by the same forces of nature that bind us to our planet and that cause the wind to blow and the rain to fall. They are the same forces that push the blood through our veins and carry the electric impulses of thought through our brains. So far in our look at the changing patterns of the sky, we have only glimpsed these connections, so it is to these underlying causes that we now turn our attention.

Seeing the Forest for the Trees

Patterns in the sky change in lockstep with the cycles of life on Earth. It is no surprise that our ancestors sought some link between the stars and the events in their lives, but early astrologers were ultimately limited by their misconceptions about the universe. Earth does not reside at the "center of creation," nor does the Sun move from "house" to "house" along the ecliptic. The eerie spectacle of a solar eclipse blotting out the Sun is no more mystical than the shadow you cast on the sidewalk on a sunny afternoon. The heavens are a place not of magic but of physical law, knowable though observation and experiment and subject to test by the scientific method. The signs of the zodiac that grace the entertainment section of your local newspaper and the covers of supermarket tabloids are anachronisms—pictures born from the human imagination and painted onto the random splash of stars across the night sky.

Although astrologers were off the mark in their conclusions, their quest was well motivated. The motions of Earth and the properties of the Sun do, indeed, give rise to the most basic of all the patterns faced by life on Earth. Earth's rotation is responsible for the coming of night and day. Earth's axial tilt and its passage around the Sun bring the changing of the seasons. Changes in the direction in which sunlight falls on Earth cause dramatic differences in climate from the equator to the poles. These patterns set the stage for the evolution of life on Earth, and they remain with us today, buried deep within the genetic code of our species. Even as humans begin to venture beyond Earth and into space, we carry these patterns with us in everything from the length of our cycle of waking and sleeping, to the temperature we prefer, to the amount of light we need in order to see. Much of human culture also has its roots in the apparent motions of celestial objects. Many of our legends and traditions arose in the ancient view of the sky as a place of gods and spirits. At the same time, patterns of objects in the sky spurred the development of mathematics and, as we will see in Chapter 3, the development of a physical understanding of the world around us. Much of who and what we are as a thinking species has its origins in our experience of the larger universe.

The errors in perception that shaped our views of the universe throughout most of human history are both understandable and forgivable. It is remarkably difficult to directly sense the effects of Earth's motion. As you read this, you are probably (depending on your latitude) moving at more than 1,000 km/h on a circular path around Earth's axis, while Earth itself is moving at over 100,000 km/h in its orbit about the Sun. (And the Sun itself is moving through our galaxy at almost 1 million km/h.) Yet you feel none of this motion. Objects around you share your motion, so to you they are motionless. The telltale signs of your motion are far too subtle to perceive directly. As you watch the Sun and stars cross the sky, it certainly seems as if it is you about which the cosmos revolves. In fact, without the benefit of discussions such as ours, it would be difficult to avoid that impression.

When we replace simple perception with careful experiment and reason, however, this commonsense notion evaporates before our eyes. Minute changes in the direction of starlight through the year provide proof of Earth's motion in its orbit, and they enable us to measure our path around the Sun. And as our physical understanding of Earth's motion improves, even the blowing of the wind and the need for a fire on a cold, dark winter night become side effects of Earth's motions through space.

In our journey we will encounter again many of the motions discussed in this chapter, but from a rather different perspective. So far, we have concentrated on *describing* the motion of Earth about the Sun and of the Moon about Earth. From here we will take the step that separates post-Renaissance science from all that came before, by taking up the question of *why* these motions are as they are. The search for understanding will lead us to a discovery that changed our perception of the universe and our place in it—that the same natural forces that dictate the path of a well-hit baseball also govern the clockwork motions of Earth, the Moon, and planets.

Summary

- Earth's rotation on its axis causes the Sun and stars to appear to move through the sky.

- The specific stars we see at night depend on where we are on Earth and where Earth is in its orbit around the Sun.

- A frame of reference is a coordinate system within which an observer measures positions and motions.

- The tilt of Earth's axis determines the seasons.

- The motion of the Moon in its orbit around Earth shapes the phases of the Moon.

- The phase of the Moon is determined by how much of its bright side is visible from Earth.

- Special alignments of the Sun, Earth, and Moon result in solar and lunar eclipses.

 SmartWork, Norton's online homework system, includes algorithmically generated versions of these questions, plus additional conceptual exercises. If your instructor assigns questions in SmartWork, log in at **smartwork.wwnorton.com.**

Student Questions

THINKING ABOUT THE CONCEPTS

1. Some of the mythologies created by early civilizations as they gazed at the sky persist today. Give at least one example.

*2. Explain why the celestial sphere is not really a sphere but is instead a useful fictional feature.

3. Earth has a North Pole, a South Pole, and an equator. What are their equivalents on the celestial sphere?

4. Polaris was used for navigation by seafaring sailors such as Columbus as they sailed from Europe to the New World. When Magellan sailed the South Seas, he could not use Polaris for navigation. Explain why.

5. If you were standing at Earth's North Pole, where would you see the north celestial pole relative to your zenith?

6. If you were standing at Earth's South Pole, which stars would you see rising and setting?

7. Where on Earth can you stand and, over the course of a year, see the entire sky?

8. What do we call the group of constellations through which the Sun appears to move over the course of a year?

9. We tend to associate certain constellations with certain times of year. For example, we see the zodiacal constellation Gemini in the Northern Hemisphere's winter (Southern Hemisphere's summer) and the zodiacal constellation Sagittarius in the Northern Hemisphere's summer. Why do we not see Sagittarius in the Northern Hemisphere's winter (Southern Hemisphere's summer) or Gemini in the Northern Hemisphere's summer?

10. Assume that you are flying along in a jetliner.
    a. Define your frame of reference.
    b. What relative motions take place within your frame of reference?

11. The tilt of Jupiter's rotational axis is 3°. If Earth's axis had this tilt, explain how it would affect our seasons.

*12. Describe the Sun's apparent motion on the celestial sphere
    a. at the time of vernal equinox.
    b. at the time of summer solstice in Earth's Northern Hemisphere.

13. Why is winter solstice not the coldest time of year?

14. There is only a certain region on Earth when at some time of year the Sun can appear precisely at the zenith. Describe this region.

15. Many cities have main streets laid out in east–west and north–south alignments.
    a. Why are there frequent traffic jams on east–west streets during both morning and evening rush hours within a few weeks of the equinoxes?
    b. Considering your conclusion in (a), if you work in the city during the day, would you rather live east or west of the city?

*16. Earth spins on its axis but wobbles like a top.
    a. How long does it take to complete one spin?
    b. How long does it take to complete one wobble?

17. Why do we always see the same face of the Moon?

18. What is the approximate time of day when you see the full Moon near the meridian? At what time is the first quarter (waxing) Moon on the eastern horizon? Use a sketch to help explain your answers.

*19. Assume that the Moon's orbit is circular. Suppose you are standing on the side of the Moon that faces Earth.
    a. How would Earth appear to move in the sky as the Moon made one revolution around Earth?
    b. How would the "phases of Earth" appear to you, as compared to the phases of the Moon as seen from Earth?

*20. Sometimes artists paint the horns of the crescent Moon pointing toward the horizon. Is this depiction realistic? Explain.

21. From your own home, why are you more likely to witness a partial eclipse of the Sun rather than a total eclipse?

22. Why do we not see a lunar eclipse each time the Moon is full or witness a solar eclipse each time the Moon is new?

23. Why does the fully eclipsed Moon appear reddish?

24. The true length of a year is not 365 days, but actually about 365¼ days. How do we handle this extra quarter day to keep our calendars from getting out of sync?

**25. Does the occurrence of solar and lunar eclipses disprove the notion that the Sun and the Moon both orbit around Earth? Explain your reasoning.

APPLYING THE CONCEPTS

26. Earth is spinning along at 1,674 km/h at the equator. Do your own calculation to find Earth's equatorial diameter.

27. Assume that rain is falling at a speed of 5 meters per second (m/s) and you are driving along in the rain at a leisurely 5 m/s (or 18 km/h). Estimate the angle from the vertical at which the rain appears to be falling.

**28. Suppose you are an astronaut on the Moon, which now becomes your inertial frame of reference. Explain the relative motion of Earth as seen from your lunar frame of reference.

29. The Moon's orbit is tilted by about 5° relative to Earth's orbit around the Sun. What is the highest altitude in the sky that the Moon can reach, as seen in Philadelphia (latitude 40° north)?

30. Refer to Appendix 4. On the basis of the tilt of its rotation axis, which planet experiences the most extreme seasons? Explain your answer.

31. Imagine that you are standing on the South Pole at the time of the southern summer solstice.
    a. How far above the horizon will the Sun be at noon?
    b. How far above (or below) the horizon will the Sun be at midnight?

*32. Find out the latitude where you live. Draw and label a diagram showing that your latitude is the same as (a) the altitude of the north celestial pole and (b) the angle (along the meridian) between the celestial equator and your local zenith. What is the noontime altitude of the Sun as seen from your home at the times of winter solstice and summer solstice?

33. The southernmost star in the famous Southern Cross constellation lies approximately 65° south of the celestial equator. Using an atlas, determine from which US states is the entire Southern Cross visible?

34. Let's say you use a protractor to estimate an angle of 40° between your zenith and Polaris. Are you in the continental United States or Canada?

35. Suppose the tilt of Earth's equator relative to its orbit were 10° instead of 23.5°. At what latitudes would the Arctic and Antarctic Circles and the two Tropics be located?

36. Suppose you would like to witness the rarity known as "midnight Sun" (when the Sun appears just above the northern horizon at midnight), but you don't want to travel any farther north than necessary.
    a. How far north (that is, to which latitude) would you have to go?
    b. When would you make this trip?

**37. Assume that Earth is a perfect sphere with a radius of 6,400 km. What distance at Earth's equator corresponds to 1° of longitude? What distance corresponds to 1° of latitude? Estimate the distance corresponding to 1° of longitude at latitudes of 30° and 60°. What would be the distance of 1° of latitude at these same latitudes?

38. Eratosthenes estimated the size of Earth more than 2,000 years ago (see Math Tools 2.1.) Repeat his calculation. The latitude of Alexandria, his hometown, was 30.7° north. The latitude of Syene, 790 km to the south, was 23.5° north. Compare Eratosthenes' calculation with the modern value of 6,371 km for Earth's average radius.

39. The vernal equinox is now in the zodiacal constellation of Pisces. Wobbling of Earth's axis will eventually cause the vernal equinox to move into Aquarius, beginning the legendary, long-awaited "Age of Aquarius." How long, on average, does the vernal equinox spend in each of the 12 zodiacal constellations?

*40. Stonehenge was erected roughly 4,000 years ago. Referring to the zodiacal constellations shown in Figure 2.10, identify the constellation in which these ancient builders saw the vernal equinox.

41. Referring to Figure 2.17a, estimate when Vega, the fifth brightest star in our sky, will once again be the northern pole star.

42. There are 29.53 days between full Moons (the synodic period). On average, how long does it take for the Moon to move from one zodiacal constellation to the next?

43. The apparent diameter of the Moon is approximately ½°. About how long does it take the Moon to move its own diameter through the star field?

44. The Moon has a radius of 1,737 km, with an average distance of 3.780×10^5 km from Earth's surface. The Sun has a radius of 696,000 km, with an average distance of 1.496×10^8 km from Earth. Show why the apparent sizes

of the Moon and Sun in our sky are approximately the same.

45. Earth has an average radius of 6,371 km. If you were standing on the Moon, how much larger would Earth appear in the lunar sky than the Moon appears in our sky?

46. If the Moon were in its same orbital plane, but twice as far from Earth, which of the following would happen?
   a. The phases that the Moon goes through would remain unchanged.
   b. Total eclipses of the Sun would not be possible.
   c. Total eclipses of the Moon would not be possible.

*47. In what way would the length of the "eclipse season" change if the plane of the Moon's orbit were inclined less than its current 5.2° to the plane of Earth's orbit? Explain your answer.

*48. Which, if either, occurs more often: (a) total eclipses of the Moon seen from Earth, or (b) total eclipses of the Sun seen from the Moon? Explain your answer.

49. Suppose you would like to see a solar eclipse without venturing outside the continental United States. Refer to Figure 2.27.
   a. When will be the next opportunity for you to see a total solar eclipse?
   b. When will be the next opportunity for you to see an annular solar eclipse?

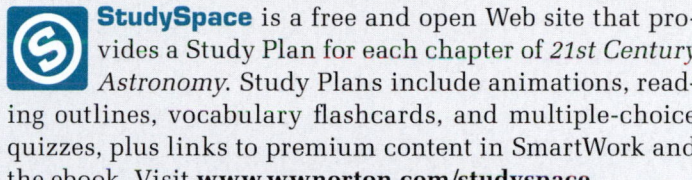

StudySpace is a free and open Web site that provides a Study Plan for each chapter of *21st Century Astronomy*. Study Plans include animations, reading outlines, vocabulary flashcards, and multiple-choice quizzes, plus links to premium content in SmartWork and the ebook. Visit **www.wwnorton.com/studyspace**.

The Newtonian principle of gravitation is now more firmly established, on the basis of reason, than it would be were the government to step in, and to make it an article of necessary faith. Reason and experiment have been indulged, and error has fled before them.

THOMAS JEFFERSON (1743–1826)

The International Space Station in orbit about Earth.

CHAPTER 3

Gravity and Orbits— A Celestial Ballet

3.1 Gravity!

Today, most of us know that the planets, including Earth, *orbit* around the Sun and that **gravity**—one of the fundamental forces of nature—holds them in orbit. Yet this was not always so. Only 500 years or so have passed since a soft-spoken Polish monk named Nicolaus Copernicus (1473–1543) (**Figure 3.1**) started a revolution when he revived the idea, discarded by the Greeks 2,000 years earlier, that the Sun rather than Earth lies at the center of all creation. To those with a 16th century Western philosophical view, this suggestion seemed outlandish. To think that Copernicus wanted us to believe that humankind—the pinnacle of creation—resides anywhere but at the center of all things! To imagine that we occupy but one of several planets circling the Sun's central fire, that we are nothing but a pebble among all the pebbles on the beach—absurd!

It would be easy from our "modern" perspective to chuckle at the naïveté of our ancestors and their **geocentric** (Earth-centered) view of the universe, but to do so would be unfair. After all, a universe with Earth located at its very center was a perfect fit to what our ancestors saw in the world around them. The previous chapter showed that hard evidence of the motions of Earth was remarkably difficult to come by. To a scholar educated in Greek and Roman philosophy, Copernicus's view of the universe simply made no sense. Wishing to avoid the controversy that his theory would certainly cause, Copernicus chose not to publish his ideas until late in his life. In fact, his great work *De revolutionibus orbium coelestium* ("On the Revolutions of the Heavenly Spheres") did not appear until the year of his death. This work pointed the way toward our modern cosmological principle.

KEY CONCEPTS

In this chapter we follow the story of the birth of modern science as astronomers and physicists together discovered regular patterns in the motions of the planets and then went on to explain those patterns with fundamental physical laws. Along the way we will explore

- Empirical rules discovered by Johannes Kepler that describe the elliptical orbits of planets around the Sun.

- Theoretical physical laws discovered by Isaac Newton and Galileo Galilei that govern the motion of all objects.

- How proportionality is used to describe patterns and relationships in nature.

- How Newton's laws of motion and gravitation combine to explain planetary orbits.

- How theories lead to new knowledge, such as the way Newton's derivation of Kepler's third law is used to measure the masses of objects from observations of orbital motions.

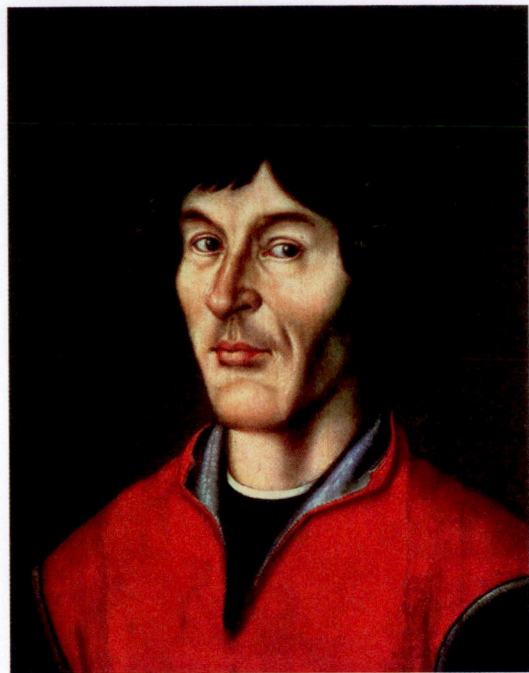

(left) **FIGURE 3.1** Nicolaus Copernicus rejected the ancient Greek belief in an Earth-centered universe and replaced it with one centered on the Sun.

(right) **FIGURE 3.2** Johannes Kepler explained the motions of the planets with three empirically based laws.

Copernicus knew his ideas would be controversial, but he had no way of guessing the long-term consequences of his fresh view of the universe. In retrospect, we see that he knocked the bottom out from under the house of cards representing Western philosophers' view of the world around them, and the following centuries would see that house of cards slowly fall apart. The repercussions of Copernicus's insight not only would shape our understanding of the universe around us, but would change the direction of the progress of human civilization itself.

> Copernicus's heliocentric cosmological model knocked humankind from the center of the universe.

Even today we need to avoid too much complacency about what we think we know. People often confuse knowing the name of something with actually *understanding* something. We happily talk about spacecraft in orbit about Earth, or planets in orbit about the Sun, but remarkably few people understand what those words mean. When asked why astronauts float about the cabin of a spacecraft, most educated adults answer, "The spacecraft has escaped Earth's gravity." Yet the gravitational force acting on a space shuttle orbiting Earth is only slightly weaker than when the shuttle is sitting on the launchpad. In fact, were it not for Earth's gravity, the spacecraft would not orbit Earth at all!

Copernicus himself knew nothing of gravity, but his ideas inevitably led to it. As physicists and astronomers have come to a better understanding of gravity, they have realized that in most respects it is gravity that holds the universe together. Our Solar System is a gravitational symphony. The Sun's gravity shapes the motions of the planets and every other object in its vicinity. These motions range from the almost circular orbits of some planets to the extremely elongated orbits of comets. A comet's orbit may carry it from a location tens of thousands of astronomical units[1] from the Sun to a point somewhere inside the orbit of Mercury and back out again. Within this grand symphony, subthemes arise. A chorus of particles orbiting the giant planets gives rise to majestic systems of rings, which in turn play counterpoint to the gravitational ballet of the planets and their moons. The analogy between the motions of objects in the Solar System and the patterns of music is not new. German mathematician and astronomer Johannes Kepler (1571–1630) (**Figure 3.2**), who will figure prominently in our story, titled his great work of 1619 *Harmonices mundi* ("The Harmony of the Worlds").

As our grasp of the universe has expanded, we have come to realize that our Solar System is but one gravitational opus in a far larger opera. Gravity binds stars into the colossal groups that we call galaxies. It is gravity that holds the planets and stars together and keeps the thin blanket of air we breathe close to the surface of the planet we live on. It is gravity that caused a vast interstellar cloud of gas and dust to collapse 4.5 billion years ago to form our Sun and Solar System. It is gravity that gives space and time their very shape. As we continue our journey outward through the cosmos, we will come to each of these ideas in turn, and time and time again we will find gravity at the center of our growing understanding.

---

[1] Recall from Chapter 2 that an astronomical unit (abbreviated AU) is the average distance between the Sun and Earth.

3.2 An Empirical Beginning: Kepler Describes the Observed Motions of the Planets

Now that we have extolled the wonders of gravity and the role it plays in the universe, you might expect us immediately to discuss what gravity is and how it works. But the great minds that brought us to our modern understanding did not have the benefit of our 20/20 hindsight. All they could do was watch the motions of the planets over the course of months and years and puzzle over what they saw (see **Excursions 3.1**). With no way even to judge the distances to the planets, it is no wonder that it took humans thousands of years to begin to see the reality behind the celestial patterns before their eyes. So, in this chapter we will follow these scientists as each of their discoveries led to our modern understanding of gravity.

In Chapter 1 we painted a picture of science as a worldview in which nature is governed by physical laws and in which mathematical descriptions of these physical laws are used to explain natural phenomena. But how do scientists go about *discovering* these physical laws? When facing phenomena as complex and puzzling as the motions of the planets in the heavens, where can we find a toehold? Just as the wise sailor settles for any port in a storm, the wise scientist knows that when facing a complex and poorly understood phenomenon, there may be little choice but to turn directly to the information our senses provide. We carefully observe the phenomenon under study, systematically recording as much information as we can as accurately as possible. As our observations start piling up, we look for patterns in those observations and start trying to formulate rules that seem to describe what we have seen.

Imagine that you are a scientist from another world, setting foot for the first time on Earth. You notice right away that many interesting structures are sticking out of the ground on this unexplored planet. As you record your observations, you find that some of these structures are large, some are small, some spread out over the ground, some stick up in the air, and so on. But after a time you realize that almost all of these structures are covered with some kind of appendage, and those appendages are green. So you form a descriptive rule about these objects (call them "plants"): most have green appendages. You decide that green appendages must be fundamental to the nature of plants, so you begin to study what makes these appendages green. After a time you discover that the green appearance always comes from the same chemical substance, and when you study that chemical substance you find that it can absorb light and turn water and carbon dioxide into more complex organic molecules. You have discovered chlorophyll and photosynthesis, the process responsible

FIGURE 3.6 Tycho Brahe, known commonly as Tycho, was one of the great astronomical observers of all time.

for powering the majority of life on Earth. *But you did not start out to discover photosynthesis. You started out noting that plants have green leaves.*

The quest to first note and then accurately describe patterns in nature is called **empirical science**. Empirical science often involves a great deal of creativity and not a small amount of pure (but educated) guesswork. Copernicus's theory that Earth and the planets move in circular orbits about the Sun is an example of empirical science. Copernicus did not understand *why* the planets move about the Sun, but he did realize that his **heliocentric** (Sun-centered) picture provided a much *simpler* description of the observed motions of planets than a model with Earth at its center did. Copernicus's work was revolutionary because he was able to see beyond the geocentric prejudice of his time and to think the unthinkable—that perhaps Earth is "merely" one planet among many.

Copernicus's work paved the way for another great empiricist, Johannes Kepler. Science has often benefited from unlikely and fortunate collaborations, and Kepler's is one such story. Kepler, a mathematician who had studied the ideas of Copernicus, worked in 1600 as an assistant to Tycho Brahe (1546–1601) (**Figure 3.6**), a Danish astronomer who was a firm believer in an Earth-centered universe. Although Tycho, as he is commonly known, is described as anything but a pleasant individual, he was also one of the greatest observational astronomers of all time. Toiling away through long nights with primitive equipment, Tycho amassed a wealth of remarkably accurate naked-eye observations of the positions of the planets over the course of decades. Kepler, using Tycho's data, took the next major step

The Motions of the Planets in the Sky

Ancient peoples had long been aware of the five planets ("wandering stars") that moved in a generally eastward direction among the "fixed stars." Ancient astronomers also knew

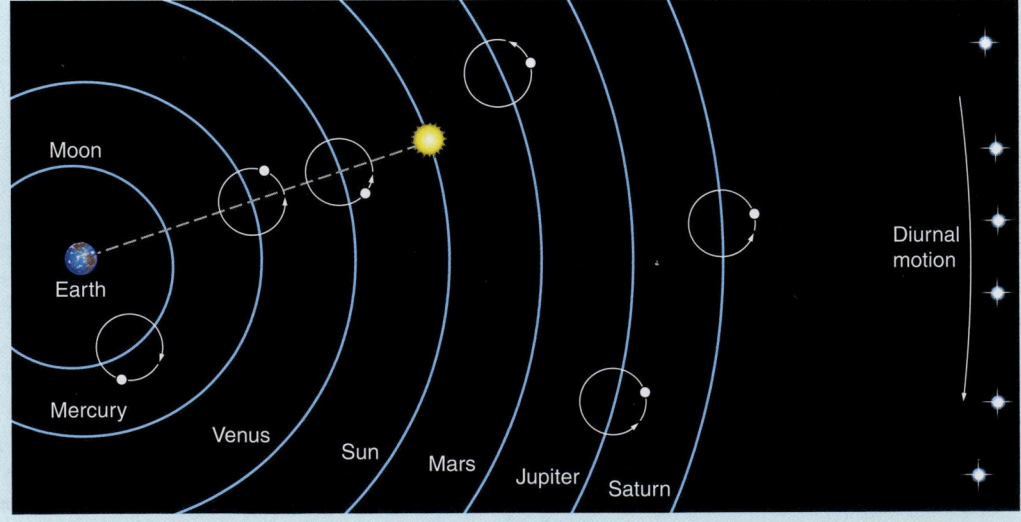

that these moving objects would occasionally turn about in a contrary manner, move westward for a while, and then return to their normal eastward travel. This odd behavior of the five "naked eye" planets—Mercury, Venus, Mars, Jupiter, and Saturn—created a puzzling problem for the geocentric (Earth-centered) model of the Heavens as it was summarized by the Alexandrian astronomer Ptolemy in A.D. 150. Although a few ancient Greek astronomers did consider the possibility that Earth revolved around the Sun, other astronomers of the day were skeptical; "if this were so, we would actually feel Earth's motion," they claimed. Most astronomers therefore preferred the geocentric model, in which the Sun, Moon, and planets all moved in perfect circles around a stationary Earth, as shown in **Figure 3.3a**, with the "fixed stars" being located somewhere way out there beyond the planets. Although a stationary Earth may have "felt right" to astronomers in those early times, the geocentric model in its simplest form failed to explain that curious planetary behavior that we now call **retrograde motion**.

To account for retrograde motion of the planets, Ptolemy had to resort to an embellishment called an "epicycle," a small circle superposed on each planet's larger circle, as seen in **Figure 3.3b**. While traveling along its larger circle, a planet would at the same time be moving along its smaller circle. When its motion along the smaller circle was in a direction oppo-

FIGURE 3.3 (a) The Ptolemaic view of the Heavens. (b) Additional loops called "epicycles" were added to each planet's circle around Earth to explain its retrograde motion. (Diurnal motion refers to daily motion.)

site to the forward motion of the larger circle, its forward motion would be reversed. For nearly 1,500 years, this awkward model of the heavens was the accepted paradigm in the Western world.

In 1543, a *heliocentric* (Sun-centered) model (**Figure 3.4**) proposed by Copernicus provided a simpler explanation of retrograde motion. In this model, the outer of the planets known at that time—Mars, Jupiter, and Saturn—appear to interrupt their eastward **prograde motion** and move backward in a westward direction when Earth overtakes them in their orbits. Likewise, Mercury and Venus appear to reverse direction in their inner orbits when overtaking Earth. With the exception of the Sun, all Solar System objects exhibit retrograde motion, although the magnitude of the effect diminishes with increasing distance from Earth. **Figure 3.5** shows a time-lapse sequence of Mars going through its retrograde "loop."

Retrograde motion is only apparent, not real. We have all experienced how relative motions can fool us. If we are in a car or train and we pass a slower-moving car or train, it can seem to us that the other vehicle is moving backward. Without an external frame of reference, it can be hard to tell which vehicle is moving and in what way. Copernicus provided that frame of reference for the Sun and its planets.

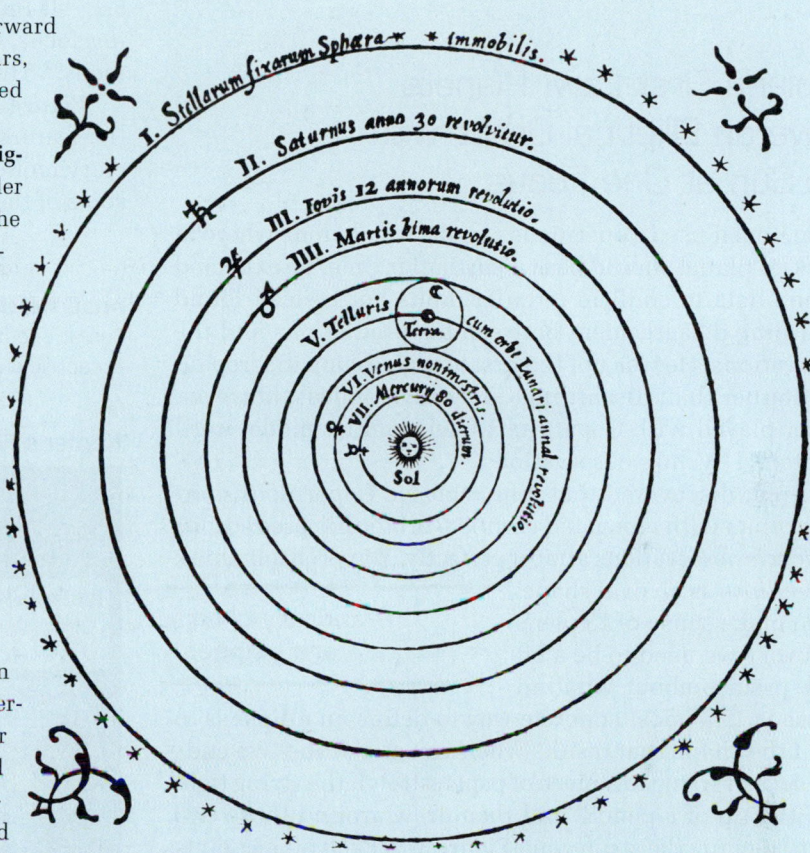

FIGURE 3.4 The Copernican heliocentric view of the Solar System (II through VII) and the fixed stars (I).

FIGURE 3.5 The apparent retrograde motion of Mars through the Taurus-Gemini star field as Earth overtook the red planet in 2007–08.

toward understanding the motions of the planets. Working first primarily with Tycho's observations of Mars, Kepler was able to deduce three empirical rules that elegantly and accurately describe the motions of the planets. These three rules are now generally referred to as **Kepler's laws**.

Kepler's First Law: Planets Move on Elliptical Orbits with the Sun at One Focus

When Kepler used Copernicus's model to calculate where in the sky a planet should be at a particular time, he expected Tycho's data to confirm circular orbits but instead found disturbing disagreement between his predictions and the observations. He was not the first to notice such discrepancies. Rather than discarding Copernicus's ideas, however, Kepler played with Copernicus's heliocentric model until it matched Tycho's observations.

Kepler discovered that if he replaced Copernicus's circular orbits with elongated *elliptical* orbits, his predictions fit Tycho's observations almost perfectly. You probably think of an *ellipse* as an oval shape, but to make sense of Kepler's discovery we need to be a bit more precise about what an ellipse is. The most concrete way to define an **ellipse** is to call it the shape that results when you attach the two ends of a piece of string to a piece of paper, stretch the string tight with the tip of a pencil, and then draw around those two points keeping the string taut (**Figure 3.7**). Each of the points at which the string is attached is called a **focus** (plural: *foci*) of the ellipse. The closer the two foci are to each other, the more nearly circular the ellipse is. In fact, a circle is just an ellipse with the two foci at the same place. (To see this, just think about the shape you would draw if the two ends of the string were attached at the same spot. In this case, each half of the string would become a radius of the circle.) As the two foci are moved farther apart, however, the ellipse

> Planetary orbits are ellipses.

becomes more and more elongated. Kepler found that the orbit of each planet is an ellipse with the Sun located at one focus. This result is now known as **Kepler's first law** of planetary motion. (You might ask, "If the Sun is located at *one* focus, what is at the other focus?" The answer is, nothing but empty space.)

> The Sun is at one focus of a planet's elliptical orbit.

Figure 3.8 illustrates Kepler's first law and shows how the features of an ellipse elegantly match observed planetary motions. Let's look more closely at **Figure 3.8a** to see some of these features. The dashed lines represent the two

FIGURE 3.8 (a) Planets move on elliptical orbits with the Sun at one focus. (b) Ellipses range from circles to elongated eccentric shapes.

(a)
Kepler's First Law

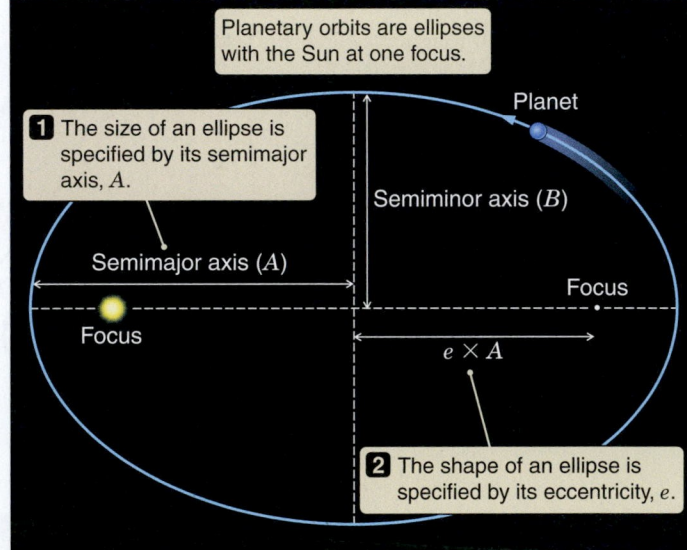

Planetary orbits are ellipses with the Sun at one focus.

Planet

1 The size of an ellipse is specified by its semimajor axis, *A*.

Semiminor axis (*B*)

Semimajor axis (*A*)

Focus

Focus

$e \times A$

2 The shape of an ellipse is specified by its eccentricity, *e*.

(b)

3 The greater the eccentricity, the more elongated the ellipse.

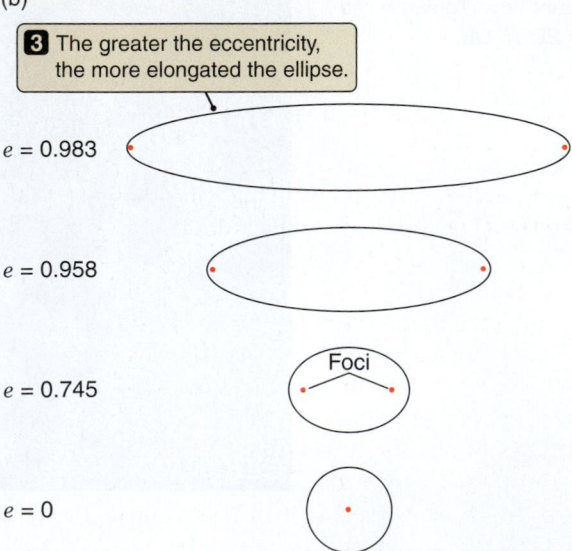

$e = 0.983$

$e = 0.958$

Foci

$e = 0.745$

$e = 0$

FIGURE 3.7 We can draw an ellipse by attaching a length of string to a piece of paper at two points (called *foci*; singular: *focus*) and then pulling the string around as shown.

Focus   Focus

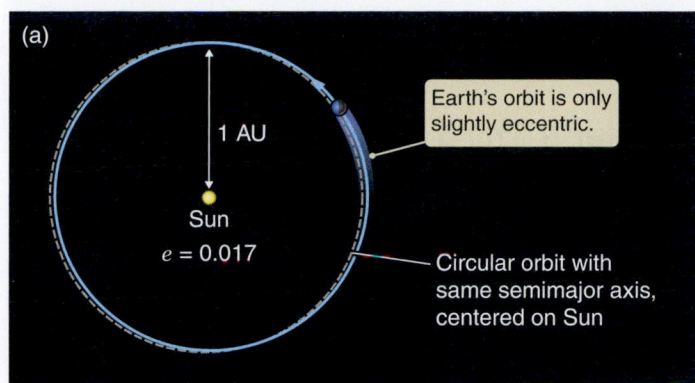

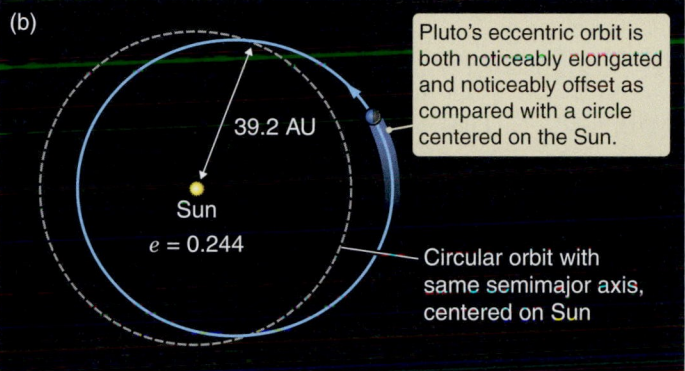

FIGURE 3.9 The shapes of the orbits of Earth (a) and Pluto (b) compared with circles centered on the Sun.

main axes of the ellipse. Half of the length of the long axis of the ellipse is called the **semimajor axis** of the ellipse, often denoted by the letter *A*. The semimajor axis of an orbit turns out to be a handy way to describe the orbit because, apart from being half the longer dimension of the ellipse, it is also the average distance between one focus and the ellipse itself. The average distance between the Sun and Earth, for example, equals the semimajor axis of Earth's orbit. The same is true for the orbits of all the other planets.

▶‖ **AstroTour: Kepler's Laws**

In the case of a circular orbit, the semimajor axis is just the radius of the circle. Some ellipses, on the other hand, are very elongated. When describing the shape of an ellipse, we speak of its **eccentricity**. We define the eccentricity of an ellipse as the separation between the two foci divided by the length of the long axis. A circle has an eccentricity of 0 because the two foci coincide at the center of the circle. The more elongated the ellipse becomes, the closer its eccentricity gets to 1 (**Figure 3.8b**). Most planets have nearly circular orbits with eccentricities close to 0. The eccentricity of Earth's orbit, for example, is 0.017, which means that the distance between the Sun and Earth departs from its average value by only 1.7 percent. You may recall from Chapter 2 that the times when Earth is closest or farthest from the Sun have almost nothing to do with our seasons, and you can perhaps get a better feeling for this when you look at **Figure 3.9a**. It is hard to tell the difference between the orbit

of Earth and a circle centered on the Sun. By contrast, one of the many characteristics that distinguish the dwarf planet Pluto from its classical cousins is its highly eccentric orbit, as seen in **Figure 3.9b**. With an eccentricity of 0.244, the distance between the Sun and Pluto varies by 24.4 percent from its average value. Pluto's orbit is noticeably oblong, and its center is noticeably displaced from the Sun.

Kepler's Second Law: Planets Sweep Out Equal Areas in Equal Times

The next empirical rule that Kepler found has to do with how fast planets move at different places on their orbits. A planet moves most rapidly when it is closest to the Sun and is at its slowest when it is far-thest from the Sun. The aver-age speed of Earth in its orbit about the Sun is 29.8 kilome-ters per second (km/s). When Earth is closest to the Sun, it travels at 30.3 km/s. When it is farthest from the Sun, it travels at 29.3 km/s.

> **Planets move fastest when they are closest to the Sun.**

Kepler found an elegant way to describe the changing speed of a planet in its orbit about the Sun. Look at **Figure 3.10**, which shows a planet at six different points in its orbit. Imagine a straight line connecting the Sun with this planet. We can think of this line as "sweeping out" an area as it moves with the planet from one point to another. Area A (in red) is swept out between times t_1 and t_2, area B (in blue) is swept out between times t_3 and t_4, and area C (in green) is swept out between times t_5 and t_6. When the planet is closest to the Sun

FIGURE 3.10 An imaginary line between a planet and the Sun sweeps out an area as the planet orbits. Kepler's second law states that if the three intervals of time shown are equal, then the three areas A, B, and C will be the same.

Kepler's Second Law

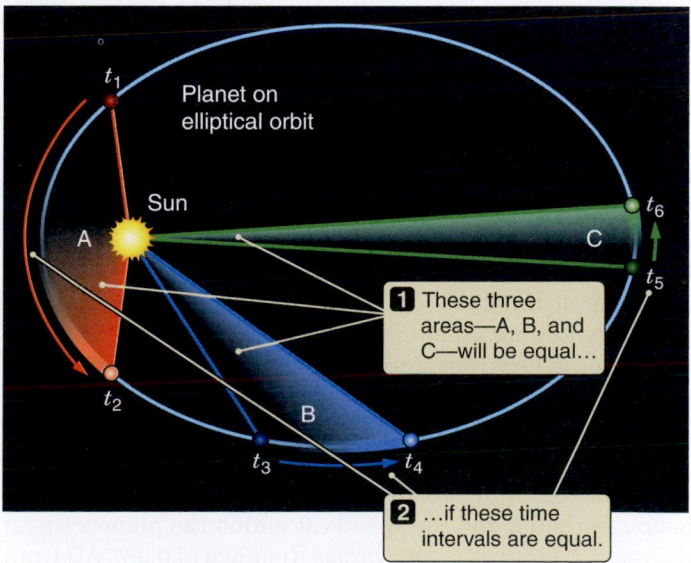

(area A in Figure 3.10) it is moving rapidly, but the distance between the planet and the Sun is small. Kepler realized that changes in the distance between the Sun and a planet and changes in the speed of a planet work together to produce a surprising result: the area swept out by a planet in the same amount of time is always the same, regardless of the location of the planet in its orbit. In Figure 3.10, this means that if the three time intervals are equal (that is, $t_1 \rightarrow t_2 = t_3 \rightarrow t_4 = t_5 \rightarrow t_6$), then the three areas A, B, and C will be equal as well.

This is **Kepler's second law**, which is also referred to as Kepler's **law of equal areas**. It states that the imaginary line connecting a planet to the Sun sweeps out equal areas in equal times, regardless of where the planet is in its orbit. Note that this law applies to only one

> A planet "sweeps out" equal areas in equal times.

planet at a time. The area swept out by Earth in a given time is always the same. Likewise, the area swept out by Mars in a given time is always the same. But the area swept out by Earth and the area swept out by Mars in a given time are *not* the same. ▶❙❙ **AstroTour:** Kepler's Laws

Kepler's Third Law: Planetary Orbits Reveal a Harmony of the Worlds

Kepler's first law describes the shapes of planetary orbits, and Kepler's second law describes how the speed of a planet changes as it travels along its orbit. But neither of these laws tells us how long it takes a planet to complete one orbit about the Sun (referred to as the **period** of the orbit). Nor do these laws tell us how this orbital period depends on the distance between the Sun and a planet. Kepler surely saw this lack of a relationship between orbital period and semimajor axis as a major flaw in his understanding of planetary motions. In such cases one looks for patterns and mathematical relationships.

Planets that are closer to the Sun do not have as far to go to complete one orbit as do planets that are farther from the Sun. Jupiter, for example, has an average distance of 5.2 astronomical units (AU) from the Sun—5.2 times as far from the Sun as Earth is. That means Jupiter must travel 5.2 times farther in its orbit about the Sun than Earth does in its orbit. We might guess, then, that if the two planets were traveling at the same speed,

> Outer planets have farther to go and move more slowly in their orbits around the Sun.

Jupiter would complete one orbit in 5.2 years. But such a guess would be wrong. Jupiter takes almost 12 years to complete one orbit. Clearly, Jupiter not only has farther to go in its orbit but must be *moving more slowly than Earth* as well. This trend holds true for all the planets. As we go farther out from the Sun, the circumferences of planetary orbits become greater, and the speeds at which the planets travel decrease. Mercury, at an average distance of 0.387 AU from

TABLE 3.1

Kepler's Third Law: $P^2 = A^3$

The Orbital Properties of the Classical and Dwarf Planets

Planet	Period P (years)	Semimajor Axis A (AU)	$\frac{P^2}{A^3}$
Mercury	0.241	0.387	$\frac{0.241^2}{0.387^3} = 1.00$
Venus	0.615	0.723	$\frac{0.615^2}{0.723^3} = 1.00$
Earth	1.000	1.000	$\frac{1.000^2}{1.000^3} = 1.00$
Mars	1.881	1.524	$\frac{1.881^2}{1.524^3} = 1.00$
Ceres	4.599	2.765	$\frac{1.559^2}{2.765^3} = 1.00$
Jupiter	11.86	5.204	$\frac{11.86^2}{5.204^3} = 1.00$
Saturn	29.46	9.582	$\frac{29.46^2}{9.582^3} = 0.99*$
Uranus	84.01	19.201	$\frac{84.01^2}{19.201^3} = 1.00$
Neptune	164.79	30.047	$\frac{164.79^2}{30.047^3} = 1.00$
Pluto	247.68	39.236	$\frac{247.68^2}{39.236^3} = 1.02*$
Eris	557.00	67.696	$\frac{557.00^2}{67.696^3} = 1.00$

*These ratios are not exactly 1.00, due to slight perturbations from the gravity of other planets.

the Sun, whizzes along its short orbit at an average speed of 47.9 km/s, completing one revolution in only 88 days. At a distance of 30.1 AU from the Sun, Neptune lumbers along at an average speed of 5.48 km/s, taking 164.8 years to make it once around the Sun.

Kepler discovered a simple mathematical relationship between the period of a planet's orbit and its distance from the Sun. **Kepler's third law** states that the square of the period of

> The square of a planet's orbital period equals the cube of the orbit's semimajor axis.

a planet's orbit, measured in years, is equal to the cube of the semimajor axis of the planet's orbit, measured in astronomical units. Written as an equation, the law says,

$$(P_{\text{years}})^2 = (A_{\text{AU}})^3,$$

where P_{years} represents the period of the orbit expressed in Earth years, and A_{AU} is the semimajor axis of the orbit expressed in astronomical units.

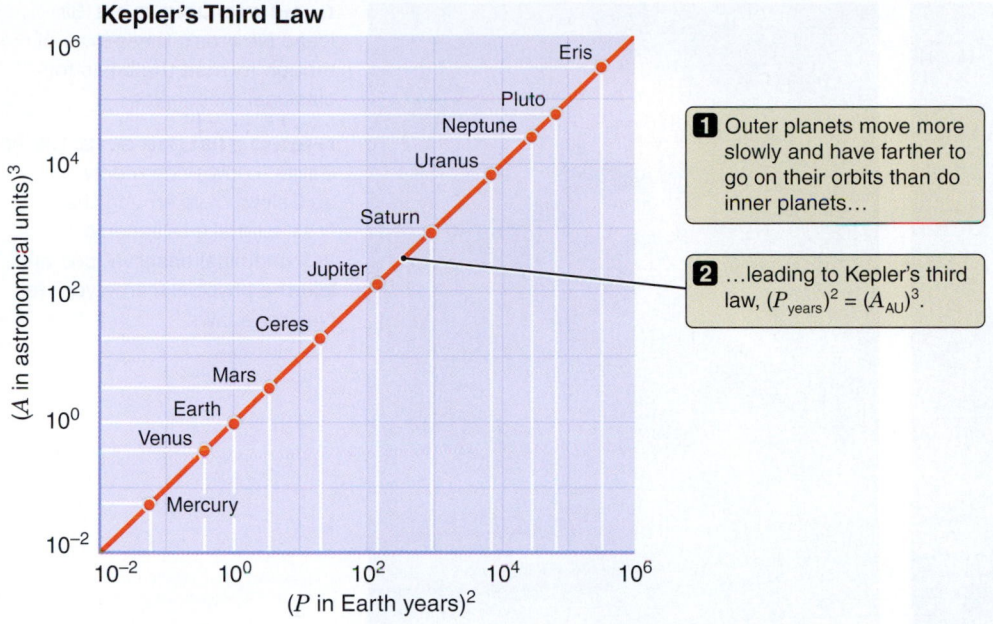

Kepler's Third Law

1 Outer planets move more slowly and have farther to go on their orbits than do inner planets...

2 ...leading to Kepler's third law, $(P_{years})^2 = (A_{AU})^3$.

FIGURE 3.11 A plot of A^3 versus P^2 for the eight classical and three dwarf planets in our Solar System shows that they obey Kepler's third law. (Note that by plotting powers of 10 on each axis, we are able to fit both large and small values on the same plot. We will do this frequently.)

This equation is a case in which astronomers use non-standard units as a matter of convenience. Years are handy units for measuring the periods of orbits, and astronomical units are handy units for measuring the sizes of orbits. When we use years and astronomical units as our units, we get the simple relationship just shown. However, it is important to realize that *our choice of units in no way changes the physical relationship* we are studying. If we instead stayed with standard metric units, this relationship would read $(P_{seconds})^2 = 3 \times 10^{-19} (A_{meters})^3$.

To judge for yourself how well Kepler's third law works, look at **Table 3.1**. Here you can see the periods and semimajor axes of the orbits of the eight classical and three of the dwarf planets, along with the values of the ratio P^2 divided by A^3. These data are also plotted in **Figure 3.11**. This relationship was so beautiful to Kepler that he referred to it as his **harmonic law** or, more poetically, as the "Harmony of the Worlds." ▶❚❚ **AstroTour:** Kepler's Laws

3.3 The Rise of Scientific Theory: Newton's Laws Govern the Motion of All Objects

When investigating a newly discovered or poorly understood phenomenon, an empirical approach is often the only available way to proceed, as we saw in Kepler's case. The development of Kepler's laws of planetary motion was an intellectual accomplishment with few peers. Yet to a modern scientist, the empirical rules that Kepler spent his life

pursuing are only the first step in the study of a phenomenon. Such empirical laws *describe* a particular phenomenon and are even useful in predicting what will happen in the future, but they do little to *explain* that behavior. Taken at face value, empirical rules offer little insight into the more fundamental laws describing nature. Kepler was able to characterize the orbits of planets as ellipses, but he did not understand *why* they should be so.

Once the empirical rules that describe a particular phenomenon have been discovered, a modern scientist will next try to understand those empirical rules in terms of more general physical principles or laws. Beginning with basic physical principles and using the tools of mathematics, the scientist works to *derive* the empirically determined rules. Or sometimes the scientist starts with physical laws and predicts relationships, which are then verified (or falsified) empirically. This technique is sometimes referred to as the "theoretical approach" to science, although in practice, if the relevant physical laws are already understood, the process is often circumvented. A scientist may make a theoretical prediction about the behavior of a system and then compare the prediction with experimental data directly to see how well they fit. In fact, today a great deal of science is done without any attempt to invent an empirical rule, because the relevant physical laws are already known.

When the relevant physical laws are *not* known—as was the case for planetary motion—the empirical rules become a way of *discovering* the physical laws themselves. Can we invent hypothetical physical laws that will enable us to derive the empirical rules? If so, what other predictions might we make on the basis of these hypothetical laws? Are these predictions also borne out by experiment and observation? If so, then we may have discovered something more fundamental about the way the universe works. This is how physical laws are discovered and tested.

(left) FIGURE 3.12 Sir Isaac Newton's three laws of motion formed the basis for classical mechanics.

(right) FIGURE 3.13 Galileo Galilei, known commonly as Galileo, was among the first to make telescopic astronomical observations and laid the physical framework for Newton's laws.

One of the earliest great advances in theoretical science was also arguably one of the greatest intellectual accomplishments in the history of our species. In many ways, the work of Sir Isaac Newton (1642–1727) (**Figure 3.12**) on the nature of motion set the standard for what we now refer to as "scientific theory" and "physical law." Building on the work of Kepler and others, Newton proposed three laws that he believed govern the motions of all objects in the heavens and on Earth. Today **Newton's laws** remain the basis for what is known as **classical mechanics**. Newton's laws themselves are beautifully elegant, and the relationships they describe between such everyday concepts as force, velocity, acceleration, and mass are accessible to all.

> Newton's laws of motion are the basis of classical mechanics.

You might reasonably ask why Newton's laws are an important stop on our journey through *21st Century Astronomy*. "After all," you might say, "this is a book about astronomy, not physics." Yes, but Newton's laws of motion are essential to our understanding of the motions of the planets and all other celestial bodies. In a very real sense, it is with Newton's laws, published in 1687, that truly modern astronomy got its start. It was these laws that enabled Newton to look at the motion of a shot fired from a cannon and see as well the motions of the planets on their orbits around the Sun. It was through Newton's laws that Earth's true place in the universe was finally seen.

The work of great scientists is always built on the foundation of the great scientists who came before,[2] and Galileo's great insight formed the starting point for Newton's tour de force that was to come.

Galileo Describes Objects in Motion

In a strange quirk of history, the physical law almost invariably referred to today as *Newton's first law of motion* did not originate with Newton at all. It was the brainchild of a contemporary of Kepler by the name of Galileo Galilei (1564–1642), known commonly as Galileo (**Figure 3.13**). Galileo is probably best known to the general public as the first person to use a telescope to make significant discoveries about the heavens and to report those discoveries. He was among the first to see craters on the Moon[3] and the first to realize that the "nebulous" Milky Way was actually made up of a myriad of individual stars. In the history of science, however, Galileo's work on the motion of objects is at least as fundamental a contribution as his astronomical observations. For example, by carefully rolling balls down an inclined plane, Galileo demonstrated that gravity accelerates *all* objects at the *same rate*, and he determined the value of Earth's gravitational acceleration, *g*. We will return to *g* later in the chapter. ▶‖ **AstroTour:** Velocity, Acceleration, Inertia

By Galileo's day, Copernicus and others had begun to turn toward the view that knowledge comes from observing nature rather than only from reading the works of clas-

---

[2]Newton himself is credited with the famous quote, "If I have seen further [than you] it is by standing upon the shoulders of giants."

[3]An English astronomer named Thomas Harriot made drawings of the Moon several months before Galileo began his astronomical observations. However, Galileo continued his scientific use of the telescope and published his observations, whereas Harriot did not.

sical Greek and Roman philosophers. Yet even in the 16th and 17th centuries the works of one of the greatest of these philosophers, Aristotle (384–322 B.C.), who had lived almost 2,000 years earlier, still carried the weight of authority. Aristotle believed that the natural state of all objects was to be at rest and that an object in motion would tend toward this natural state. From what Aristotle observed, this seemed to be a good empirical rule about how objects in the world around him behaved. For example, we notice that a cart rolling down the street coasts to a stop when it is no longer being pulled. A bouncing ball eventually settles to the ground. Even an arrow shot from a bow loses much of its speed before striking its target.

It is seldom safe to challenge the entrenched wisdom of your day, and that was especially true in Galileo's time, when intellectual and religious authority and political power resided in the same hands. Within Galileo's lifetime, the Italian priest and philosopher Giordano Bruno had fallen victim to the Inquisition and had been burned at the stake for his beliefs—including Bruno's support of Copernicus, his belief that the universe is infinite, and his suggestion that Earth is but one of many habitable planets. In his writings on motion, Galileo similarly challenged prevailing wisdom by proposing that Aristotle's notion of an object's tendency to come to rest was the result of illusion. On the basis of his own observations and experimental results, Galileo argued that in all of the cases just mentioned—indeed in *every* such case—there are hidden reasons why objects come to rest. There is friction as the axle of the cart rubs against its bearing, resisting the motion and eventually bringing it to a halt. Every time a ball bounces, its shape is distorted, and what we might think of as "internal friction" within the ball causes it to bounce lower each time. The resistance of air, which the arrow must push out of the way and which drags against the arrow's shaft, slows the arrow's progress.

Galileo agreed with Aristotle that an object at rest remains at rest unless something causes it to move. But in disagreement with Aristotle, Galileo asserted that, *left on its own, an object in motion will remain in motion.* Specifically, Galileo said that *an object in motion will continue moving along a straight line*

> Galileo found that an object left in motion remains in motion.

with a constant speed until an unbalanced force acts on it to change its state of motion. When we speak of an **unbalanced force**—as we will frequently throughout this chapter—we mean the *net* force acting on an object.[4] Galileo referred to this resistance to change in an object's state of motion from an unbalanced force as **inertia**.

Galileo developed the idea of inertia relatively early in his career, but much of his later life was taken up by conflict with the Catholic Church over his support of the Copernican system. His astronomical observations of Jupiter's moons

and the phases of Venus had convinced Galileo that the Copernican model was correct and that the church's views were wrong. In 1632, Galileo published his great work, *Dialogo sopra i due massimi sistemi del mondo* ("Dialogue Concerning the Two Chief World Systems"). In the *Dialogo*, the champion of the Copernican (Sun-centered) view of the universe is a brilliant philosopher named Salviati. The *Dialogo*'s defender of Aristotelian authority is named Simplicio and is as much an ignorant buffoon as the name might imply. When Galileo published the *Dialogo*, he actually thought he had the tacit approval of the Vatican, which held to the Aristotelian view. But when he placed a number of the pope's arguments in the unflattering mouth of Simplicio, he found that the Vatican's tolerance had limits. Fortunately for Galileo, he had more friends in high places than Bruno did, so he spent the closing years of his life under house arrest rather than ending up burned at the stake.

To escape a harsher sentence from the Inquisition, Galileo was forced to publicly recant his belief in the Copernican theory that Earth moves around the Sun. In one of the great apocryphal stories of the history of astronomy, it is said that as he left the courtroom following his sentencing, Galileo stamped his foot on the ground and muttered, "And yet it moves!" Galileo's final years were spent compiling his research on inertia and other ideas into a book, *Discorsi e dimostrazioni mathematiche intorno a due nuove scienze attenenti alla meccanica* ("Discourses and Mathematical Demonstrations Relating to Two New Sciences"), which was published in 1638 outside the Inquisition's jurisdiction.

Newton's First Law: Objects at Rest Stay at Rest; Objects in Motion Stay in Motion

It is a tribute to Galileo that his law of inertia became the cornerstone of physics as **Newton's first law of motion**. We can understand inertia in terms of what we discovered during discussions of relative motion and frames of reference in Chapter 2. Recall that within a frame of reference, only the *relative* motions between objects have any meaning. This is the same as saying that there is *no perceptible difference* between an object at rest and an object in uniform motion. What objects are at rest and what objects are in motion anyway? The object at rest beside you on the front seat of your car as you drive down the highway is moving at 60 miles per hour (mph) according to a bystander along the side of the road— but it is moving at 120 mph according to a car in oncoming traffic. These two perspectives are equally valid.

The connection between inertia and the relative nature of motion is so fundamental that a frame of reference moving in a straight line at a constant speed is referred to as an **inertial frame of reference**. Motion is meaningful only when measured relative to an inertial frame of reference,

---

[4]The "net," or "resultant," force is a *single* force that represents a combination of all of the individual forces acting on an object.

and *any* inertial frame of reference is as good as any other. The realization that the laws of physics are the same in any inertial frame of reference is one of the deepest insights ever made into the nature of the universe. When thought of in this way, *of course* an object moving in a straight line at a constant speed remains in motion. As illustrated in **Figure 3.14a**, in the frame of reference of that object, *it is already at rest*.

Newton's Second Law: Motion Is Changed by Unbalanced Forces

Newton often gets credit for Galileo's insight about inertia because it was Newton who took the crucial next step. Newton's first law says that in the absence of an unbalanced force, an object's motion does not change. **Newton's second**

law of motion goes on to say that *if there is an unbalanced force acting on an object, then the object's motion does change*. Even more, Newton's second law tells us *how* the object's motion changes in response to that force.

> Unbalanced forces cause changes in motion.

Before going any further, it would be wise to pause to be sure that we are all together on this. In the preceding paragraphs we spoke of "changes in an object's motion," but what does that phrase really mean? When you are in the driver's seat of a car, a number of controls are at your disposal. On the floor are a gas pedal and a brake pedal. You use these to make the car speed up or slow down. A *change in speed* is one way the motion of an object can change. But also remember the steering wheel in your hands. When you are moving down the road and you turn the wheel,

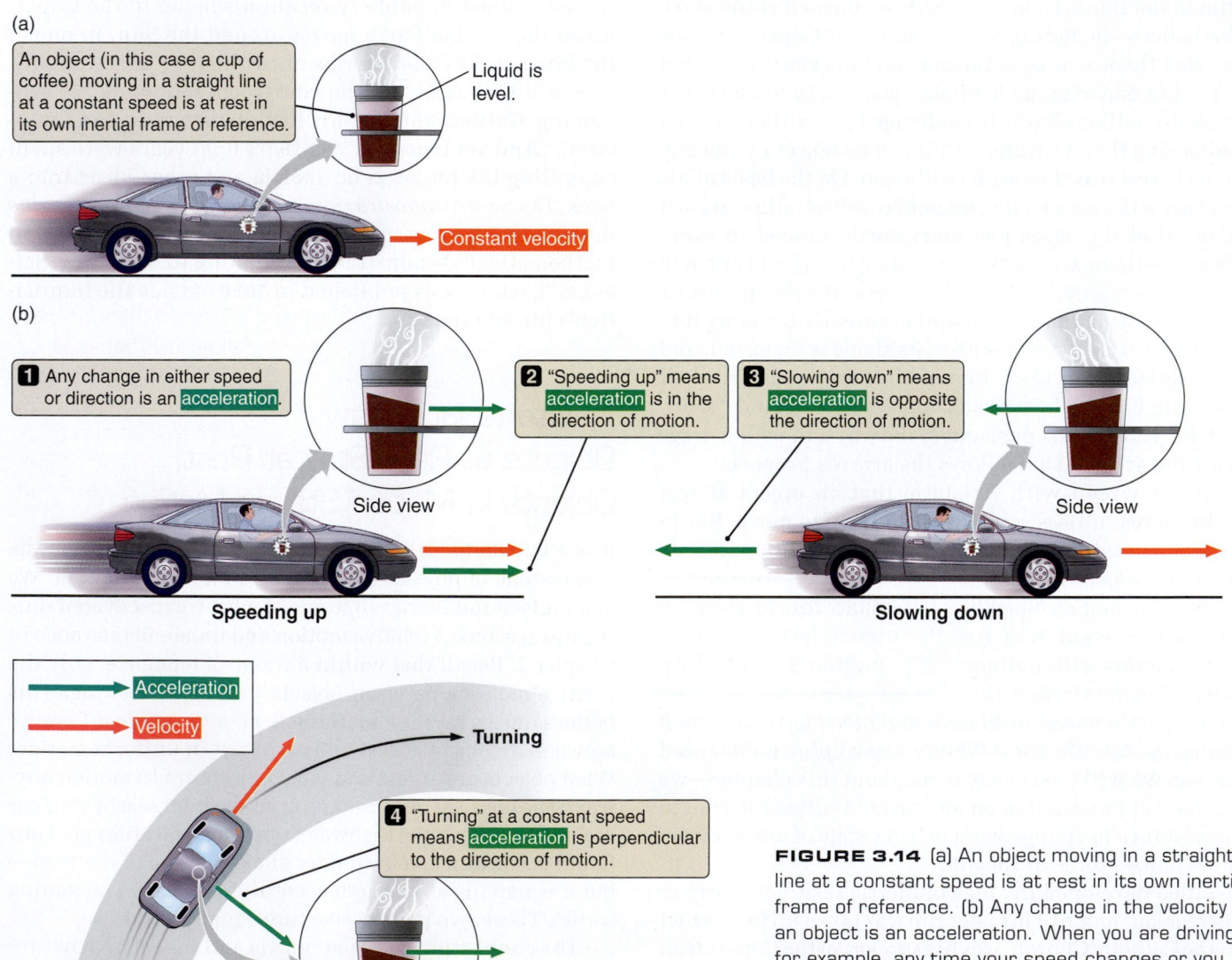

(a)

An object (in this case a cup of coffee) moving in a straight line at a constant speed is at rest in its own inertial frame of reference.

Liquid is level.

Constant velocity

(b)

1 Any change in either speed or direction is an acceleration.

Side view

2 "Speeding up" means acceleration is in the direction of motion.

3 "Slowing down" means acceleration is opposite the direction of motion.

Side view

Speeding up

Slowing down

Acceleration
Velocity

Turning

4 "Turning" at a constant speed means acceleration is perpendicular to the direction of motion.

Rear view

FIGURE 3.14 (a) An object moving in a straight line at a constant speed is at rest in its own inertial frame of reference. (b) Any change in the velocity of an object is an acceleration. When you are driving, for example, any time your speed changes or you follow a curve in the road, you are experiencing an acceleration. (Throughout the text, velocity arrows will be shown as red and acceleration arrows will be shown as green.)

Newton's Second Law:

$$\text{Acceleration } (a) = \frac{\text{Force } (F)}{\text{Mass } (m)}$$

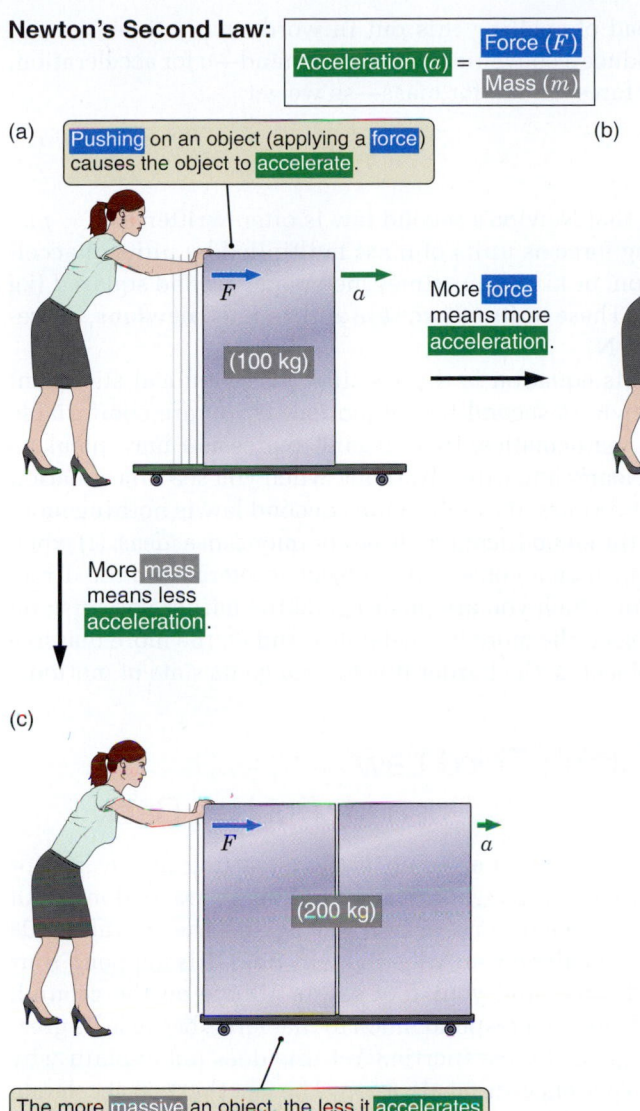

(a) Pushing on an object (applying a force) causes the object to accelerate.

F a

(100 kg)

More force means more acceleration.

More mass means less acceleration.

(c)

F a

(200 kg)

The more massive an object, the less it accelerates in response to a given force, and vice versa.

(b) The harder you push something (more force), the greater the acceleration.

F a

(100 kg)

FIGURE 3.15 Newton's second law of motion says that the acceleration experienced by an object is determined by the force acting on the object, divided by the object's mass. (Throughout the text, force arrows will be shown as blue.)

$$\text{Acceleration} = \frac{\text{How much velocity changes}}{\text{How long the change takes to happen.}}$$

For example, if an object's speed goes from 5 meters per second (m/s) to 15 m/s, then the change in velocity is 10 m/s. If that change happens over the course of 2 seconds, then the acceleration is 10 m/s divided by 2 seconds, which equals 5 meters per second per second. This is the same as saying "5 meters per second squared," which is written as 5 m/s^2 or 5 m s^{-2}. ▶II **AstroTour:** Velocity, Acceleration, Inertia

Because the gas pedal on a car is often called the "accelerator," some people think *acceleration* means that an object is speeding up. But we need to stress that, as used in physics, *any* change in motion is an acceleration. **Figure 3.14b** illustrates the point. Slamming on your brakes and going from 60 to 0 mph in 4 seconds is just as much acceleration as going from 0 to 60 mph in 4 seconds. Similarly, the acceleration you experience as you go through a fast, tight turn at a constant speed is every bit as real as the acceleration you feel when you slam your foot on the gas pedal or the brake pedal. Speeding up, slowing down, turning left, turning right—*if you are not moving in a straight line at a constant speed, you are experiencing an acceleration.*

Newton's second law of motion says that changes in motion—accelerations—are caused by unbalanced forces. The acceleration that an object experiences depends on two things, as shown in **Figure 3.15**. First, it depends on the unbalanced force acting on the object to change its motion. When all the forces acting on an object balance each other—making the total force on the object zero—the object is not accelerating. If the forces acting on the object do *not* add up to zero, then there is an unbalanced force and the object accelerates (**Figure 3.15a**). This is a pretty commonsense idea. The stronger the unbalanced force, the greater the acceleration. In fact, the acceleration of an

your speed does not necessarily change, but the direction of your motion does. A *change in direction* is also a kind of change in motion.

Together, the speed and direction of an object's motion are called the object's **velocity**. The rate at which the velocity of an object changes is called **acceleration**. *Acceleration* actually refers to how *rapidly* the change in velocity happens. For example, if you go from 0 to 60 mph in 4 seconds, you feel yourself being pushed back into the seat, but it is really your seat back that is shoving your body forward, causing you to accelerate along with the car. If you take 2 minutes to go from 0 to 60 mph, on the other hand, the acceleration is so slight that you hardly notice it. To formalize this concept a bit, your acceleration is determined by how much your velocity changes, divided by how long it takes for that change to happen:

> **Acceleration measures how quickly a change in motion takes place.**

> **Greater force means greater acceleration.**

object is *proportional* to the unbalanced force applied (**Figure 3.15b**). Push on something twice as hard and it experiences twice as much acceleration. Push on something three times as hard and its acceleration is three times as great. (The idea of proportionality, discussed in **Foundations 3.1**, will be used over and over again throughout our journey.) The resulting change in motion occurs in the direction in which the unbalanced force is imposed. Push something forward and it speeds up. Push it to the left and it veers in that direction.

The acceleration that an object experiences also depends on the degree to which the object resists changes in motion. Some objects—for example, a baseball—are easily shoved around by humans. A baseball is thrown at great velocity by the pitcher, only to be hit with a bat and have its motion abruptly changed again. The hard-hit line drive comes suddenly to a stop in the glove of the second-base player. A baseball resists changes in its

> Mass is the property of matter that resists changes in motion.

motion; that is, it has inertia—but not *too* much inertia. Other objects are less obliging. A piece of solid iron the size of a baseball would make a poor substitute in the game. A pitcher would be *very* hard-pressed to throw such an iron ball hard enough to get it over home plate (and if he did, the catcher would have an even more dangerous job). A baseball and a ball of iron may be the same size, but they are quite different in the degree to which they resist changes in their motion. The property of an object that determines its resistance to changes in motion—the measure of an object's inertia—is referred to as the object's **mass** (**Figure 3.15c**). In other words, the greater the mass, the greater the resistance.

You probably knew this answer intuitively. The iron ball is "heftier" and therefore harder to throw than a baseball. However, you may not have thought much about what we really mean when we say that a ball of iron "has more mass"—or "is more massive"—than a baseball. You might say that the ball of iron is made up of "more stuff" than the baseball; but again, what is meant by "more stuff"? If you grapple with this question for a time, you may find yourself chasing the question around in circles. When it comes right down to it, *the property of matter that we refer to as "mass" is nothing more and nothing less than the degree*

> Acceleration is force divided by mass.

to which an object resists changes in its motion. So if we want to know how an object's motion is changing, we need to know two things: What unbalanced force is acting on the object, and what is the resistance of the object to that force? We can put this into equation form as follows:

$$\begin{pmatrix} \text{The} \\ \text{acceleration} \\ \text{experienced} \\ \text{by an object} \end{pmatrix} = \frac{\text{The force acting to change the object's motion}}{\text{The object's resistance to that change}} = \frac{\text{Force}}{\text{Mass}}.$$

Instead of spelling this out in words every time, we can introduce a convenient bit of shorthand—*a* for acceleration, *F* for force, and *m* for mass—so we get

$$a = \frac{F}{m}.$$

Note that Newton's second law is often written as $F = ma$, giving force as units of mass multiplied by units of acceleration, or kilograms times meters per second squared (kg m/s^2). These units of force are aptly named **newtons**, abbreviated **N**.

This equation is the succinct mathematical statement of Newton's second law of motion. If you are comfortable with mathematics, this elegant expression may speak to you clearly and directly. If not, when you see this equation remind yourself that Newton's second law is nothing more than the embodiment of three commonsense ideas: (1) when you push on an object, that object accelerates in the direction in which you are pushing; (2) the harder you push on an object, the more it accelerates; and (3) the more massive the object is, the harder it is to change its state of motion.

Newton's Third Law: Whatever Is Pushed, Pushes Back

Imagine that you are a child again, sitting in a wagon or standing on a skateboard and pushing yourself along with your foot. Each shove of your foot against the ground sends you faster along your way. But why does this happen? Your muscles flex and your foot exerts a force on the ground. (Earth does not respond much to that force, because its great mass gives it great inertia.) Yet this does not explain why *you* experience an acceleration. The fact that you accelerate at the same time means that as you push on the ground, *the ground must be pushing back on you.*

Part of Newton's genius was his ability to see sublime patterns in such mundane events. Newton realized that *every* time one object exerts a force on another, a matching force is exerted by the second object on the first. That second force is exactly as strong as the first force but is in exactly the *opposite* direction. The child pushes back on Earth, and Earth pushes the child forward. A canoe paddle pushes backward through the water, and the water pushes forward on the paddle, sending the canoe along its way. A rocket engine pushes hot gases out of its nozzle, and those hot gases push back on the rocket, propelling it into space.

All of these are examples of **Newton's third law of motion**, which says that *forces always come in pairs, and those pairs are always equal in strength but opposite in direction.* The forces in these action-reaction pairs always act on two different objects.

> For every force there is an equal and opposite force.

Your weight pushes down on the floor, and the floor pushes back on you with the same amount of force. For every force

FOUNDATIONS 3.1

Proportionality

Often in this text we will say that one quantity is *proportional* to another. Proportionality is a way of getting the gist of how something works—understanding the relationships between things—without having to actually calculate the details of one case after another.

PROPORTIONALITY

If two quantities are **proportional** to each other, then making one of them larger means making the other quantity larger by the same factor. In other words, the ratio between the two remains constant. For example, think about the weight of a bag of apples and how much the bag costs. Double the weight of the bag of apples and you double the cost. Increase the weight of the bag of apples by a factor of 5, and the cost goes up by a factor of 5 as well. *The cost of a bag of apples is proportional to the weight of the bag of apples.* We write this relationship as

$$\text{Cost} \propto \text{Weight},$$

where the symbol $\propto$ means "is proportional to." This expression captures the essence of the relationship between the cost and the weight of apples. It tells us that the more apples we buy, the more we will pay.

Mass and acceleration provide another example of proportionality. Mass is measured in units of kilograms (kg). An object with a mass of 2 kg is twice as hard to accelerate as an object with a mass of 1 kg. An object with a mass of 9 kg is three times as hard to accelerate as an object with a mass of 3 kg.

INVERSE PROPORTIONALITY

If two quantities are inversely proportional to each other, then making one of them smaller means that the other quantity becomes larger by the same factor. For example, think about a trip to your grandmother's house. If your driving speed is cut in half, the time it takes you to get there is doubled. *The time of travel is inversely proportional to the average travel speed.* We write this relationship as

$$\text{Time for a journey} \propto \frac{1}{\text{Average speed driven}}.$$

This expression tells us that if we drive half as fast, it will take us twice as long to get there. And if we drive twice as fast, our time will be cut in half.

CONSTANTS OF PROPORTIONALITY

Sometimes it is enough to know that two quantities are proportional to each other, but sometimes it is not. What if you need to know how much one of those bags of apples will actually set you back? We know that the full relationship between the cost and weight of a bag of apples is that the cost is equal to the price per pound of apples, multiplied by the weight of the bag. We write

$$\text{Cost} = \text{Price per pound} \times \text{Weight}.$$

Compare this expression with the previous one. When we say that two quantities are proportional to each other, what we mean is that one quantity equals some number *multiplied by* the other quantity. The number by which one quantity is multiplied to get the other number is called the **constant of proportionality**. In our example, the constant of proportionality is just the price per pound of apples.

Look at the difference between the two expressions. The fact that the cost of a bag of apples is proportional to the weight of the apples is a statement about the *relationship* between things. It is a statement about how things work. More apples do not cost *less* than fewer apples. More apples cost *more* than fewer apples. The constant of proportionality—here, the price per pound of apples—means something very different. Hidden within the price per pound of apples is a great deal of information, such as the cost of growing apples, the cost of transporting them from the orchard, and the profit margin that the grocer needs to stay in business. The constant of proportionality carries information about these aspects of the world.

Very often, physical laws work in this same fashion. Proportionalities tell us about *relationships*—how two things vary with one another. They let us get a feel for the "how" in how something works. In this chapter, for example, we will find that gravitational force is proportional to an object's mass. Constants of proportionality more precisely tell us about the way the universe is. The *universal gravitational constant* (G) is a constant of proportionality that tells us about the intrinsic strength of gravitational interactions and enables calculation of the numerical value of this force. Constants of proportionality are needed if we are to turn an understanding of relationships into hard numbers.

Newton's Third Law

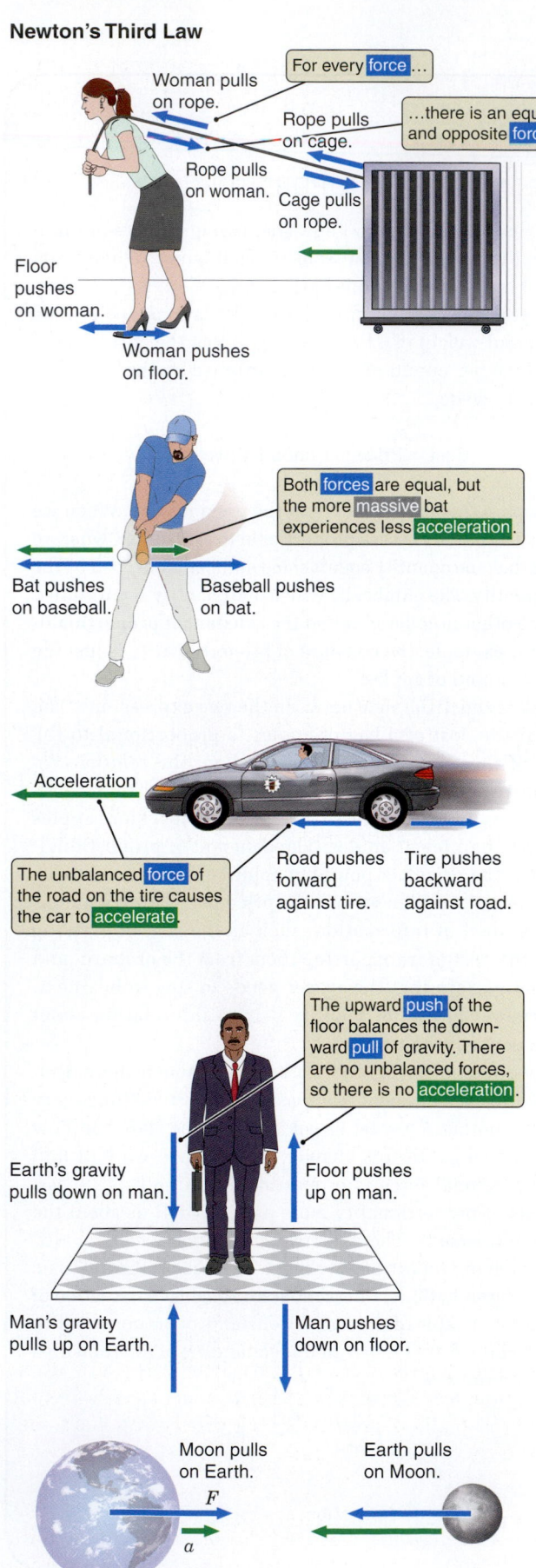

Woman pulls on rope.

For every force ...

Rope pulls on cage.

...there is an equal and opposite force.

Rope pulls on woman.   Cage pulls on rope.

Floor pushes on woman.

Woman pushes on floor.

Both forces are equal, but the more massive bat experiences less acceleration.

Bat pushes on baseball.   Baseball pushes on bat.

Acceleration

The unbalanced force of the road on the tire causes the car to accelerate.

Road pushes forward against tire.   Tire pushes backward against road.

The upward push of the floor balances the downward pull of gravity. There are no unbalanced forces, so there is no acceleration.

Earth's gravity pulls down on man.   Floor pushes up on man.

Man's gravity pulls up on Earth.   Man pushes down on floor.

Moon pulls on Earth.   Earth pulls on Moon.

F

a

there is *always* an equal and opposite force. This is one of the few times when we can say "always" and really mean it. **Figure 3.16** gives a few examples. There is a great game hiding in Newton's third law. It is called "find the force." Look around you at all the forces at work in the world, and for each force find its mate. It will *always* be there!

To see how Newton's three laws of motion work together, think about the situation shown in **Figure 3.17**. An astronaut is adrift in space, motionless with respect to the nearby space shuttle. With no tether to pull on, how can the astronaut get back to the ship? The answer? Throw something. Suppose the 100-kg astronaut throws a 1-kg wrench directly away from the shuttle at a speed of 10 m/s. Newton's second law says that in order to cause the motion of the wrench to change, the astronaut has to apply a force to it in the direction away from the shuttle. Newton's third law says that the wrench must therefore push back on the astronaut with as much force but in the opposite direction. The force of the wrench on the astronaut causes the astronaut to begin drifting toward the shuttle. How fast will the astronaut move? Turn to Newton's second law again. A force that causes the 1-kg wrench to accelerate to 10 m/s will have much less effect on the 100-kg astronaut. Because acceleration equals force divided by mass, the 100-kg astronaut will experience only 1/100 as much acceleration as the 1-kg wrench. The astronaut will drift toward the shuttle, but only at the leisurely rate of 1/100 × 10 m/s, or 0.1 m/s.

Now that we have developed a fundamental understanding of motion, it is time to turn to the concept of gravity.

3.4 Gravity Is a Force between Any Two Objects Due to Their Masses

Drop a ball and the ball falls toward the ground, picking up speed as it falls. It accelerates toward Earth. Newton's second law says that where there is acceleration, there is force. But where is the force that causes the ball to accelerate? Many forces that we see in everyday life involve "direct contact" between objects.[5] The cue ball slams into the eight

---

[5] Actually the "direct contact" between "solid" objects is also force at a distance—electric force acting at a distance between the electrons and protons in the atoms that the objects are made of.

FIGURE 3.16 Newton's third law states that for every force there is always an equal and opposite force. These opposing forces always act on the two different objects in the same pair.

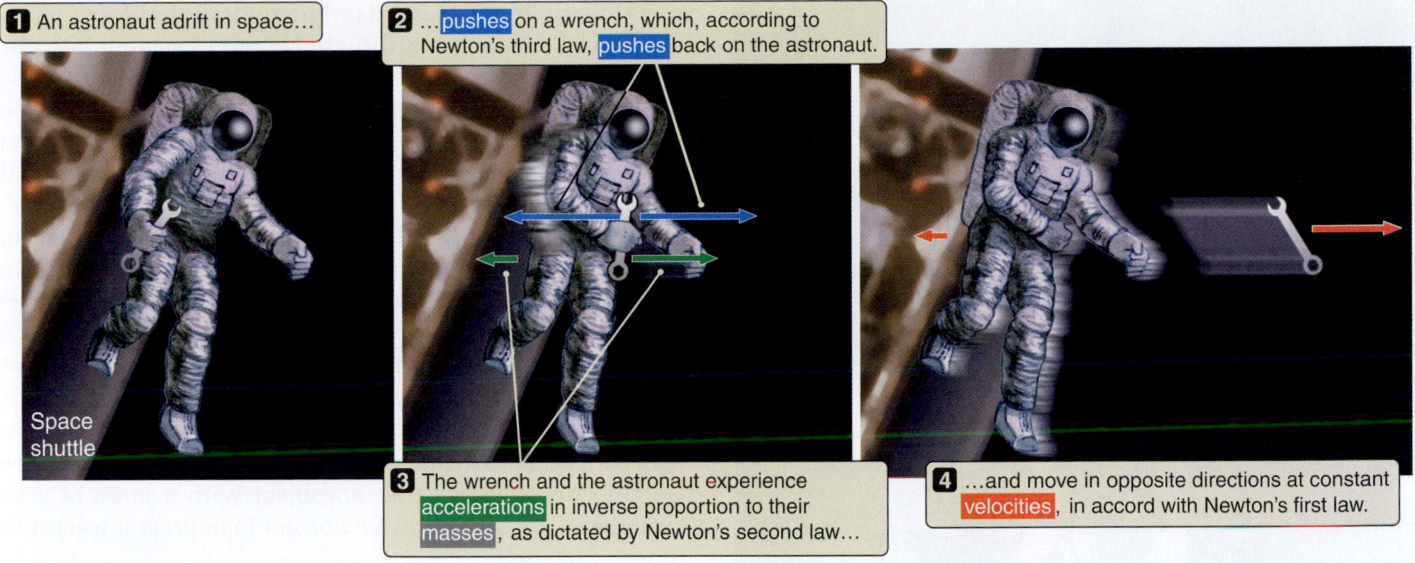

1 An astronaut adrift in space...

2 ...pushes on a wrench, which, according to Newton's third law, pushes back on the astronaut.

Space shuttle

3 The wrench and the astronaut experience accelerations in inverse proportion to their masses, as dictated by Newton's second law...

4 ...and move in opposite directions at constant velocities, in accord with Newton's first law.

FIGURE 3.17 According to Newton's laws, if an astronaut adrift in space throws a wrench, the two will move in opposite directions at speeds that are inversely proportional to their masses. (Acceleration and velocity arrows are not drawn to scale.)

ball, knocking it into the pocket. The shoe of the child pushing along in the wagon presses directly against the surface of the pavement. In cases where there is physical contact between two objects, the source of the forces between them is easy to see. But the ball falling toward Earth is an example of a different kind of force, one that acts at a distance across the intervening void of space. The ball falling toward Earth is accelerating in response to the force of gravity. We began this chapter with a qualitative discussion of the fundamental role that gravity plays in the universe. Having explored both Kepler's empirical description of the motions of planets about the Sun and Newton's laws of motion, we return now to gravity, for it is gravity that unites these two pillars of empirical and theoretical science.

> Gravity is "force at a distance."

You probably will not be surprised to learn that once again it is Newton we turn to for a *universal law of gravitation*. At this point in an introductory textbook it is customary to simply present Newton's law of gravitation as a "done deal" and go straight to its application, but such a leap misses one of the most interesting aspects of this stretch of our journey. A common misconception about how science works is the notion that new theories just spring fully formed into the mind of a scientist as if by magic. This idea is certainly supported by the apocryphal grade-school story of the apple falling on Newton's head, literally knocking the idea of gravity into his brain. One could almost get the idea that scientific theories are arbitrary—that Newton could have invented some *other* law of gravity that would have worked just as well. Although it might seem at first glance that his work was arbitrary,

nothing could be further from the truth. Where did Newton get his ideas about gravity? What guided him in his development of those ideas, and how did he turn them into a theory with testable predictions? How did he confront that theory in the crucible of experiment and observation? By answering these questions, rather than simply stating Newton's law of gravitation, we will gain some insight into how science is done.

Where Do Theories Come From? Newton Reasons His Way to a Law of Gravity

As with inertia, the story of gravity begins with the insight and observation of Galileo. Galileo discovered experimentally that all freely falling objects accelerate toward Earth at the same rate, regardless of their mass. Drop a marble and a cannonball at the same time and from the same height, and they will hit the ground together. (If proof was needed that this is not solely a property of Earth's gravity, it was provided by astronaut David Scott on the lunar surface—**Figure 3.18**.) The gravitational acceleration near the surface of Earth, also measured experimentally by Galileo, is usually written as g and has a value of 9.8 m/s² on average.[6] Whether you drop a marble or a can-

> All objects on Earth fall with the same acceleration, g.

---

[6]Earth's gravitational acceleration, g, is approximately 9.8 m/s², but the value varies with location because Earth's poles are closer to Earth's center than is its equator, and because *centripetal forces* act against

FIGURE 3.18 Astronaut Alan Bean's portrait of fellow astronaut David Scott standing on the Moon and dropping a hammer and a falcon feather together. The two objects reached the lunar surface simultaneously. (Their lunar module was nicknamed "Falcon.")

tional force. In other words, the gravitational force on an object on Earth is, according to Newton's second law, the object's mass multiplied by the acceleration due to gravity, or $F_{grav} = mg$. Make an object twice as massive and you double the gravitational force acting on it. Make an object three times as massive and you triple the gravitational force acting on it.

In precise terms, the gravitational force acting on an object is commonly referred to as the object's **weight**. On the surface of Earth, weight is just mass multiplied by Earth's gravitational constant, g. Because of our everyday use of language, it is easy to see why people confuse mass and weight. We often say that an object with a mass of 2 kg "weighs" 2 kg, but it is more correct to express a weight in terms of newtons:

> Your mass is a property of you, but your weight depends upon where you are.

$$F_{weight} = m \times g,$$

where F_{weight} is an object's weight in newtons, m is the object's mass in kilograms, and g is Earth's gravitational acceleration.

On Earth, an object with a *mass* of 2 kg has a *weight* of 2 kg × 9.8 m/s², or 19.6 N. On the Moon, where the gravitational acceleration is 1.6 m/s², the 2-kg mass would have a weight of 2 kg × 1.6 m/s², or 3.2 N (**Figure 3.19**). Although your mass remains the same wherever you are, your weight depends on where you are. On the Moon your weight is about one-sixth of your weight on Earth.

nonball, after 1 second it will be falling at a speed of 9.8 m/s, after 2 seconds at 19.6 m/s, and after 3 seconds at 29.4 m/s. (These numbers assume that we can neglect air resistance, which is reasonably negligible for relatively dense objects moving at relatively slow speeds.)

After working out the laws governing the motion of objects, Newton saw something deeper in Galileo's findings. Newton realized that if all objects fall with the same acceleration, then the gravitational *force* on an object must be determined by the object's *mass*. To see why, look back at Newton's second law (acceleration equals force divided by mass, or $a = F/m$). The only way gravitational acceleration can be the same for all objects is if the value of the force divided by the mass is the same for all objects. Since force and mass are proportional (see Foundations 3.1), then greater mass *must* be accompanied by a stronger gravita-

and effectively diminish the gravitational force. The value of g ranges from 9.78 m/s² at the equator to 9.83 m/s² at the poles, having a mean value of 9.80 m/s².

FIGURE 3.19 A mass of 2 kg has a different weight (displayed in newtons) on the Moon than it has on Earth.

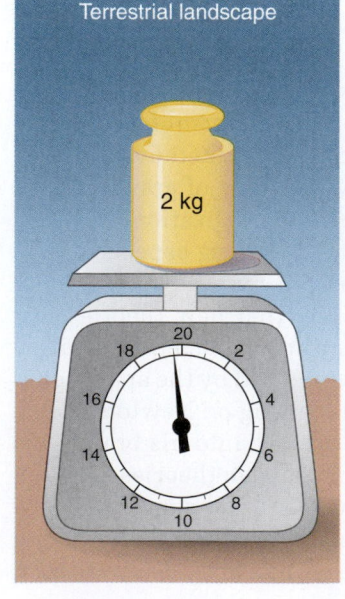
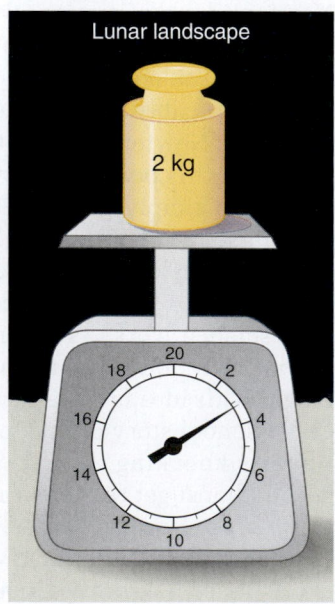

Newton's next great insight came from applying his third law of motion to gravity. Recall that Newton's third law states that for every force there is an equal and opposite force. Therefore, if Earth exerts a force of 19.6 N on a 2-kg mass sitting on its surface, then that 2-kg mass *must* exert a force of 19.6 N on Earth as well. Drop a 20-kg cannonball and it falls toward Earth, but at the same time Earth falls toward the 20-kg cannonball! The reason we do not notice the motion of Earth is that Earth is *very* massive. It has a lot of resistance to a change in its motion. In the time it takes a 20-kg cannonball to fall to the ground from a height of 1 kilometer (km), Earth has "fallen" toward the cannonball by about 3.4×10^{-21} meter, which is only about 1/300 the size of an electron!

Newton reasoned that this relationship should work with either object. If doubling the mass of an object doubles the gravitational force between the object and Earth, then doubling the mass of Earth ought to do the same thing. In short, the gravitational force between Earth and an object must be equal to the product of the two masses multiplied by something:

$$\text{Gravitational force} = \text{Something} \times \text{Mass of Earth} \times \text{Mass of object.}$$

If the mass of the object were three times greater, then the force of gravity would be three times greater. Likewise, if the mass of Earth were three times what it is, the force of gravity would have to be three times greater as well. If *both* the mass of Earth *and* the mass of the object were three times greater, the gravitational force would increase by a factor of 3×3, or 9 times. Because objects fall toward the center of Earth, we know that this force is an attractive force acting along a line between the two masses.

"And by the way," reasoned Newton, "why are we restricting our attention to Earth's gravity?" If gravity is a force that depends on mass, then there should be a gravitational force between *any* two masses. Suppose we have two masses—call them mass 1 and mass 2, or m_1 and m_2 for short. The gravitational force between them is something multiplied by the product of the masses:

The force of gravity is proportional to the product of two masses.

$$\text{Gravitational force between two objects} = \text{Something} \times m_1 \times m_2.$$

Realize that we have gotten this far just by combining Galileo's observations of falling objects with (1) Newton's laws of motion and (2) Newton's belief that Earth is a mass just like any other mass. There has been no wiggle room—there is nothing arbitrary in what we have done. But what about that "something" in the previous expression? Today we have instruments sensitive enough that we can put two masses close to each other in a laboratory, measure the force between them, and determine the value of that something directly. Yet Newton had no such instruments. He had to look elsewhere to continue his exploration of gravity.

It turns out that Kepler had already thought about this question. He reasoned that because the Sun is the focal point for planetary orbits, the Sun must be responsible for exerting an influence over the motions of the planets. Kepler speculated that whatever this influence is, it must grow weaker with distance from the Sun. (After all, it must surely require a stronger influence to keep Mercury whipping around in its tight, fast orbit than it does to keep the outer planets lumbering along their paths around the Sun.) Kepler's speculation went even further. Although he did not know about forces or inertia or gravity, he did know quite a lot about geometry, and geometry alone suggested how this solar "influence" might change for planets progressively farther from the Sun.

As Kepler himself may have speculated, imagine that you have a certain amount of plaster to spread over the surface of a sphere. If the sphere is small, you will get a thick coat of plaster. But if the sphere is larger, the plaster has to spread farther and you get a thinner coat. The surface area of a sphere depends on the square of the sphere's radius. Double the radius of a sphere, and the sphere's surface becomes four times what it was. If you plaster this new, larger sphere, the plaster must cover four times as much area and the thickness of the plaster will be only a fourth of what it was on the smaller sphere. Triple the radius of the sphere and the sphere's surface will be nine times as large, and the thickness of the coat of plaster will be only a ninth as thick.

Kepler thought the influence that the Sun exerts over the planets might be like the plaster in this example. As the influence of the Sun extended farther and farther into space, it would have to spread out to cover the surface of a larger and larger imaginary sphere centered on the Sun. (We will learn later that light works in exactly this way.) If so, then, like the thickness of the plaster, the influence of the Sun should be proportional to 1 divided by the square of the distance between the Sun and a planet. We refer to this relationship as an **inverse square law** (see **Connections 3.1**).

Gravity is governed by an inverse square law.

Kepler had an interesting idea, but not a scientific theory with testable predictions. What he lacked was a good idea of the true source of this influence and the mathematical tools to calculate how an object would move under such an influence. Newton had both. If gravity is a force between *any* two objects, then there should be a gravitational force between the Sun and each of the planets. Might this gravitational force be the same as Kepler's "influence"? If so, then the something in Newton's expression for gravity might be a term that diminishes according to the square of the distance between two objects. In essence, gravity might behave

CONNECTIONS 3.1

Inverse Square Laws

In this chapter we discover that gravity obeys what is called an inverse square law. This means that the force of gravity is "inversely proportional" (see Foundations 3.1) to the square of the distance between two objects, or

$$F_{\text{grav}} \propto \frac{1}{r^2}.$$

If two objects are moved so that they are twice as far apart as they were originally, the force of gravity between them becomes only ¼ of what it was. If two objects are moved three times as far apart, the force of gravity drops to ⅑ of its original value. For example, double the distance between the Sun and a planet, and the force of gravity declines by a factor of $2 \times 2 = 4$, to ¼ of its original strength. Triple the distance and this influence declines by a factor of $3^2 = 9$, becoming ⅑ of its initial strength.

Gravity is only one of several inverse square laws found in nature. The other important one that we will deal with in this book involves radiation. The intensity of radiation from an object is also proportional to 1 divided by the square of the distance between the objects. Our discussion of radiation in Chapter 4 will present a clear picture of why an inverse square law applies in that case.

according to an inverse square law. Newton's expression for gravity now came to look like this:

Gravitational force between two objects =

$$\text{Something} \times \frac{m_1 \times m_2}{(\text{Distance between objects})^2}.$$

There is still a "something" left in this expression, and we realize now that the something is a constant of proportionality (see Foundations 3.1). Newton guessed that the constant was a measure of the intrinsic strength of gravity between objects and that it would turn out to be the same for all objects. He named it the **universal gravitational constant**, written as G. The value of G is 6.673×10^{-11} N m²/kg² (or its equivalent, m³/kg s²).

Putting the Pieces Together: A Universal Law for Gravitation

Newton had good reasons every step of the way in his thinking about gravity—reasons directly tied to observations of how things in the world behave. Newton's chain of logic and reason brought him to what has come to be known as Newton's **universal law of gravitation**. This law, illustrated in **Figure 3.20**, states that gravity is a force between any two objects having mass and has these properties:

1. It is an attractive force acting along a straight line between the two objects.

2. It is proportional to the mass of one object multiplied by the mass of the other object:

$$F_{\text{grav}} \propto m_1 \times m_2.$$

3. It decreases in proportion to 1 divided by the square of the distance between the two objects:

$$F_{\text{grav}} \propto \frac{1}{r^2}.$$

Written as a mathematical formula, the universal law of gravitation states that

$$F_{\text{grav}} = G \times \frac{m_1 \times m_2}{r^2},$$

where F is the force of gravity between two objects, m_1 and m_2 are the masses of objects 1 and 2, r is the distance between the centers of mass of the two objects, and G is the universal gravitational constant.

Anytime you run across a statement like this, get into the habit of pulling it apart to be sure it makes sense. We'll start by analyzing the algebraic expression just presented. First, gravity is an attractive force between two masses that acts along the straight line between the two masses.[7] Regardless of where you stand on the surface of Earth, Earth's gravity pulls you *toward the center of Earth.* Second, the force of gravity depends on the product of the two masses. If you make m_1 twice as large, then the gravitational force between m_1 and m_2 becomes twice as large. Again, this relationship should make sense. Doubling the mass of an object also doubles its weight. A subtle—but in retrospect very important—point lurks in this statement. The mass that appears in the universal law of gravitation is the *same* mass that appears in Newton's laws of motion. *The same property of an object that gives*

---

[7]This is not obvious from the equation alone. More properly, this equation should be written using vector notation, a form of algebra that includes information about both the magnitude and the direction of the force.

Newton's Universal Law of Gravitation:

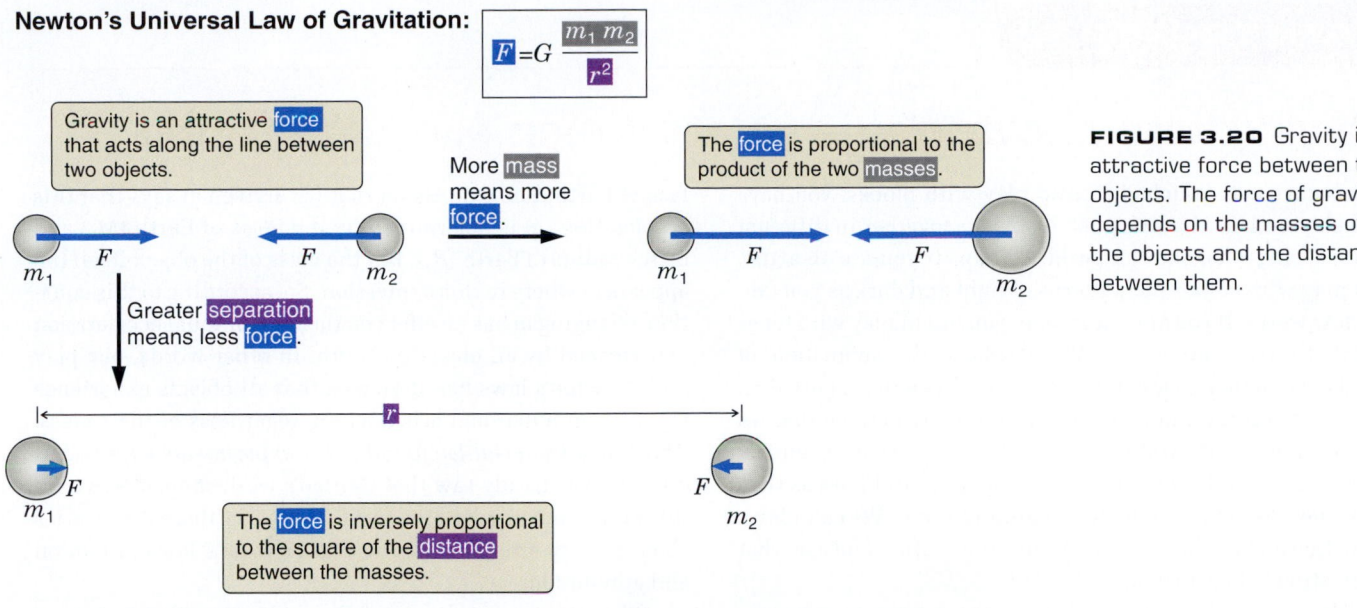

$$F = G \frac{m_1 m_2}{r^2}$$

Gravity is an attractive force that acts along the line between two objects.

More mass means more force.

The force is proportional to the product of the two masses.

Greater separation means less force.

r

The force is inversely proportional to the square of the distance between the masses.

FIGURE 3.20 Gravity is an attractive force between two objects. The force of gravity depends on the masses of the objects and the distance between them.

it inertia is the property of the object that makes it inter-act gravitationally. (This equivalence between the effect of gravitation and the effect of inertia later became the basis for Einstein's general theory of relativity, in which mass literally warps space and time. We will return to this idea later in our journey, when we discuss *space-time* in Chapter 17.)

The third part of Newton's universal law of gravita-tion tells us that the force of gravity is inversely propor-tional to the *square* of the distance between two objects. Doubling the distance between two objects reduces the strength of gravity to $(1/2)^2$, or ¼, of its original value. Tripling the distance between two objects reduces the strength of gravity to $(1/3)^2$, or ⅑, of its original value (**Figure 3.21**). Gravity is only one of several laws we will encounter in which the strength of a particular effect diminishes in proportion to the square of the distance (see **Math Tools 3.1**).

FIGURE 3.21 As two objects move apart, the gravitational force between them decreases by the inverse square of the distance between them.

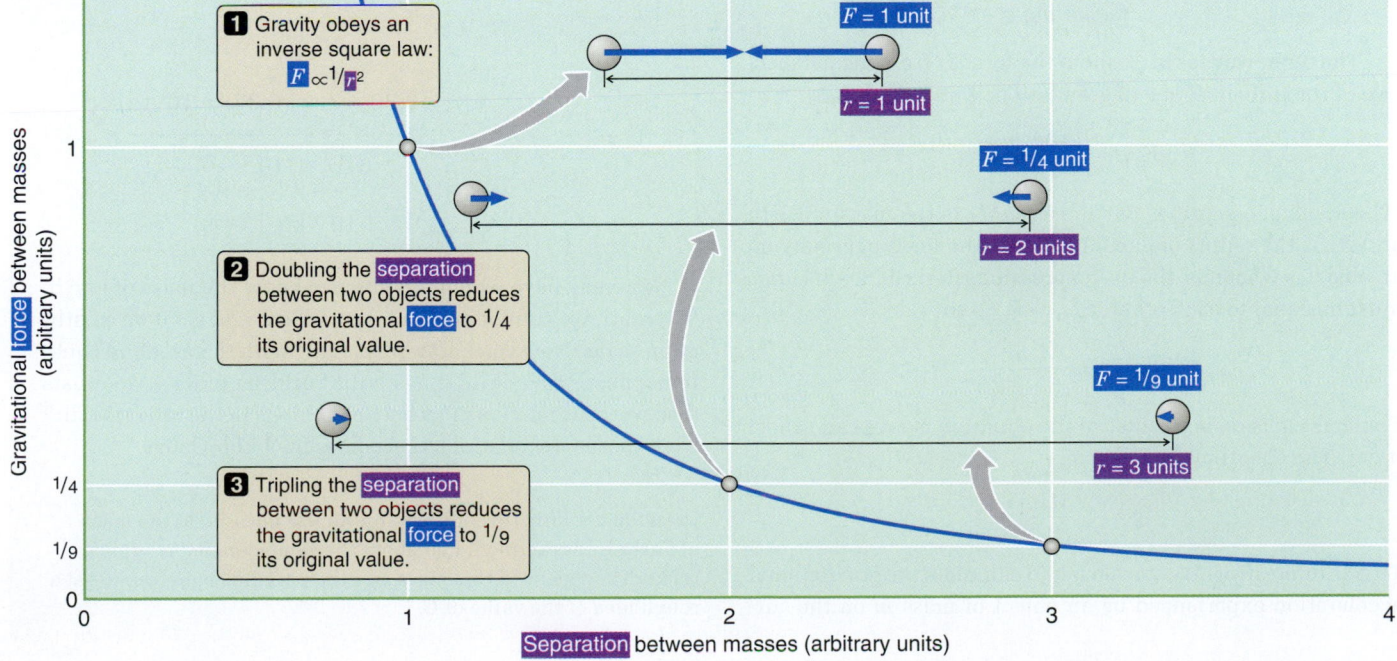

1 Gravity obeys an inverse square law: $F \propto 1/r^2$

F = 1 unit
r = 1 unit

F = ¼ unit
r = 2 units

2 Doubling the separation between two objects reduces the gravitational force to ¼ its original value.

F = ⅑ unit
r = 3 units

3 Tripling the separation between two objects reduces the gravitational force to ⅑ its original value.

Gravitational force between masses (arbitrary units)

Separation between masses (arbitrary units)

Playing with Newton's Laws of Motion and Gravitation

If you have ever watched a child play with blocks, you have probably noticed that she puts the blocks together in different ways, seeing what she can build. Perhaps if you are an artist, you play with colors and patterns of light and dark as you create new works. If you are a musician, you might play with tones and rhythms as you compose. Writers play with combinations of words. All of these uses of the term *play* mean the same thing. Play is very serious business because it is by playing that we explore the world. And in exactly the same sense, scientists often play with the equations describing natural laws as they seek new insights into the world around them. We can play a bit with Newton's laws of gravitation and motion and see what interesting things turn up.

The universal gravitational constant G has a value of $6.67 \times 10^{-11}\,\text{N m}^2/\text{kg}^2$. This indicates that gravity is a *very* weak force. The gravitational force between two 16-pound (7.26-kg) bowling balls sitting 0.3 meter (about a foot) apart is only

$$F_{\text{grav}} = 6.67 \times 10^{-11}\,\frac{\text{Nm}^2}{\text{kg}^2} \times \frac{7.26\,\text{kg} \times 7.26\,\text{kg}}{(0.3\,\text{m})^2}$$
$$= 3.91 \times 10^{-8}\,\text{N},$$

or 0.0000000391 N. This is about equal to the weight on Earth of a single bacterium! Gravity is such an important force in our everyday lives only because Earth is so very massive.

There are two different ways to think about the gravitational force that Earth exerts on an object with mass m. The first is to look at gravitational force from the perspective of Newton's second law of motion: gravitational force equals mass multiplied by gravitational acceleration, or

$$F_{\text{grav}} = mg.$$

The other way to think about the force is from the perspective of the universal law of gravitation, which says that

$$F_{\text{grav}} = G\frac{M_\oplus m}{(R_\oplus)^2}.$$

(The symbol $\oplus$ signifies Earth. Here, $M_\oplus$ is the mass of Earth and $R_\oplus$ is the radius of Earth.) Because the force of gravity on an object is what it is, the two expressions describing this force must be equal to each other. $F_{\text{grav}} = F_{\text{grav}}$, so

$$mg = G\frac{M_\oplus m}{(R_\oplus)^2}.$$

The mass m is on both sides of the equation, so we can cancel it out. The equation then becomes

$$g = G\frac{M_\oplus m}{(R_\oplus)^2}.$$

This is interesting. We started out to calculate the gravitational acceleration experienced by an object of mass m on the sur-face of Earth. The expression that we arrived at says that this acceleration (g) is determined by the mass of Earth ($M_\oplus$) and by the radius of Earth ($R_\oplus$). But the mass of the object itself (m) appears nowhere in this expression. So, according to this equation, changing m has no effect on the gravitational acceleration experienced by an object on Earth. In other words, our play with Newton's laws has shown us that all objects experience the same gravitational acceleration, regardless of their mass. *This is just what Galileo found in his experiments with falling objects!* We already saw that Galileo's work *shaped* Newton's thinking about gravity. Here we find that Galileo's discoveries about gravity are *contained within* Newton's laws of motion and gravitation.

What else can we discover? If we rearrange that last equation a bit, so that the mass of Earth is on the left and everything else is on the right, we get

$$M_\oplus = \frac{g(R_\oplus)^2}{G}.$$

Everything on the right side of this equation is known. Galileo measured a value for g, the acceleration due to gravity on the surface of Earth, almost 400 years ago; and in about 235 B.C. Eratosthenes measured the radius of Earth in the manner described in Chapter 2. The universal gravitational constant G is a bit tougher, but it, too, can be measured in the laboratory. For example, we can determine the value of G by measuring the slight gravitational forces between two large metal spheres. With everything on the right side now known, we can calculate the mass of Earth:

$$M_\oplus = \frac{g(R_\oplus)^2}{G}$$
$$= \frac{\left(9.80\,\frac{\text{m}}{\text{s}^2}\right) \times (6.37 \times 10^6\,\text{m})^2}{6.67 \times 10^{-11}\,\frac{\text{m}^3}{\text{kg s}^2}}$$
$$= 5.97 \times 10^{24}\,\text{kg}$$

You may have wondered how we know the mass of Earth. (After all, we cannot just pick up a planet and set it on a bathroom scale.) Now you know. By playing with theories and equations, much as a child plays with building blocks, scientists discover new relationships between things in the universe, and from those new relationships comes new knowledge.[8]

---

[8]Newton actually turned this around. He *guessed* at the mass of Earth by assuming it had about the same density as typical rocks. Then he used this mass and the previous equation to get a rough idea of the value of G.

3.5 Orbits Are One Body "Falling around" Another

If you have been following our discussion closely, you may be about ready to take us to task. Kepler may have speculated about the dependence of the solar "influence" that holds the planets in their orbits, and Newton may have speculated that this influence is gravity, but physical law is *not* a matter of speculation! Newton could not measure the gravitational force between two objects in the laboratory directly, so how did he test his universal law of gravitation? Again it was Kepler who provided what Newton lacked.

Newton used his laws of motion and his proposed law of gravity to *calculate* the paths that planets should follow as they move around the Sun. When he did so, his calculations predicted that planetary orbits should be ellipses with the Sun at one focus, that equal areas should be swept out during equal times, and that the square of the period of a planet's orbit should vary as the cube of the semimajor axis of that ellipse. In short, Newton's universal law of gravitation *predicted* that planets should orbit the Sun in just the way that Kepler's empirical laws described. This was the moment when it all came together. By *explaining* Kepler's laws, Newton found important corroboration for his law of gravitation. And in the process he moved the cosmological principle out of the realm of interesting ideas and into the realm of testable scientific theories. To see how this happened, we need to look below the surface of how scientists go about connecting their theoretical ideas with events in the real world.

Newton's laws tell us how an object's motion changes in response to forces and how objects interact with each other through gravity. To go from statements about how an object's motion is *changing* to more practical statements about where an object *is*, we have to carefully "add up" the object's motion over time.[9] To see how we can do this, let's begin with a "thought experiment"—the same thought experiment that helped lead Newton to his understanding of planetary motions.

Newton Fires a Shot around the World

Drop a cannonball and it falls directly to the ground, just as any mass does. However, if instead we fire the cannonball out of a cannon that is level with the ground, as shown in **Figure 3.22a**, it behaves differently. The ball still falls to the ground in the same time as before, but while it is falling it is

---

[9]Learning how to make the jump from laws of gravitation and motion to calculations of the paths of the planets about the Sun led Newton to become one of the two co-inventors of the branch of mathematics known as "calculus." Fortunately, we do not need calculus to build a conceptual understanding of such motions.

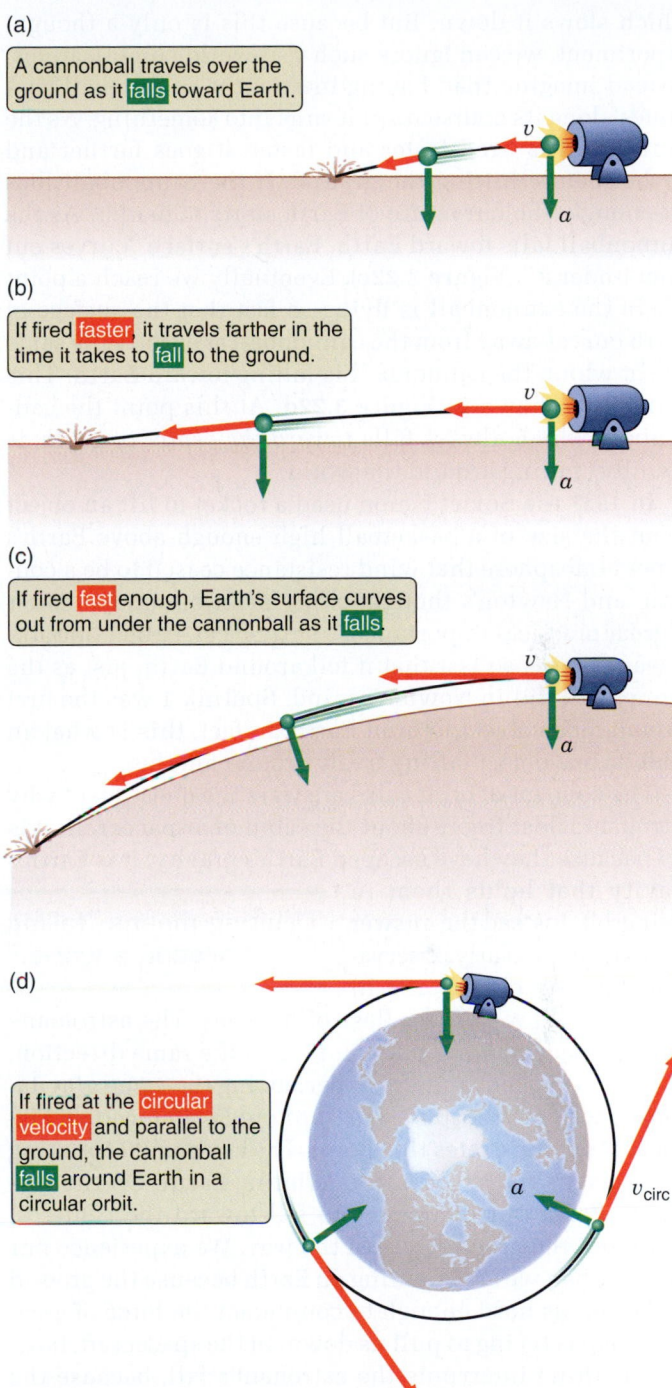

(a) A cannonball travels over the ground as it falls toward Earth.

(b) If fired faster, it travels farther in the time it takes to fall to the ground.

(c) If fired fast enough, Earth's surface curves out from under the cannonball as it falls.

(d) If fired at the circular velocity and parallel to the ground, the cannonball falls around Earth in a circular orbit.

FIGURE 3.22 Newton realized that a cannonball fired at the right speed would fall around Earth in a circle. Velocity (*v*) is indicated by a red arrow and acceleration (*a*) by a green arrow.

also traveling *over* the ground, following a curved path that carries it some horizontal distance before it finally lands. The faster the cannonball is fired from the cannon (**Figure 3.22b**), the farther it will go before hitting the ground.
▶‖ **AstroTour: Newton's Laws and Universal Gravitation**

In the real world this experiment reaches a natural limit. To travel through air the cannonball must push the air out of its way—an effect we normally refer to as "air resistance"—

which slows it down. But because this is only a thought experiment, we can ignore such real-world complications. Instead imagine that, having inertia, the cannonball continues along its course until it runs into something. As the cannonball is fired faster and faster, it goes farther and farther before hitting the ground. If the cannonball flies far enough, the curvature of Earth starts to matter. As the cannonball falls toward Earth, Earth's surface "curves out from under it" (**Figure 3.22c**). Eventually we reach a point where the cannonball is flying so fast that the surface of Earth curves away from the cannonball at exactly the same rate at which the cannonball is falling toward Earth. This is the case shown in **Figure 3.22d**. At this point the cannonball, which always falls *toward the center of Earth*, is literally "falling around the world."

In 1957 the Soviet Union used a rocket to lift an object about the size of a basketball high enough above Earth's upper atmosphere that wind resistance ceased to be a concern, and Newton's thought experiment became a matter of great practical importance.[10] This object, called Sputnik 1, was moving so fast that it fell around Earth, just as the cannonball did in Newton's mind. Sputnik 1 was the first human-made object to orbit Earth. In fact, this is what an **orbit** is: one object falling freely around another.

The concept of orbits also answers the question of why astronauts float freely about the cabin of a spacecraft. It is *not* because they have escaped Earth's gravity; it is Earth's gravity that holds them in their orbit. Instead the answer lies in Galileo's early observation that any object falls in

> Orbiting means "falling around a world."

just the same way, regardless of its mass. The astronauts and the spacecraft are both moving in the same direction, at the same speed, and are experiencing the same gravitational acceleration, *so they fall around Earth together.* **Figure 3.23** demonstrates this point. The astronaut is orbiting Earth just as the spacecraft is orbiting Earth. On the surface of Earth our bodies try to fall toward the center of Earth, but the ground gets in the way. We experience our weight when we are standing on Earth because the ground pushes on us hard enough to counteract the force of gravity, which is trying to pull us down. In the spacecraft, however, nothing interrupts the astronaut's fall, because the spacecraft is falling around Earth in just the same orbit. The astronaut is not truly weightless. Instead the astronaut is in **free fall**.

When one object is falling around another, much more massive object, we say that the less massive object is a satellite of the more massive object. Planets are satellites of the Sun, and moons are natural satellites of planets. Newton's

---

[10]Actually, wind resistance did not totally cease to be a concern. Objects in orbit within a few hundred kilometers of Earth are moving through the thin outer part of Earth's atmosphere. Friction caused by this thin atmosphere will oppose the object's motion and cause its orbit to decay.

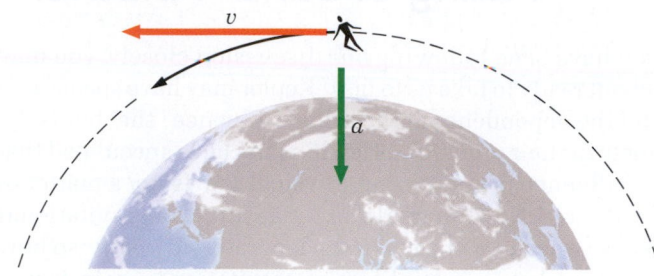

Like Newton's cannonball, an astronaut falls freely around Earth.

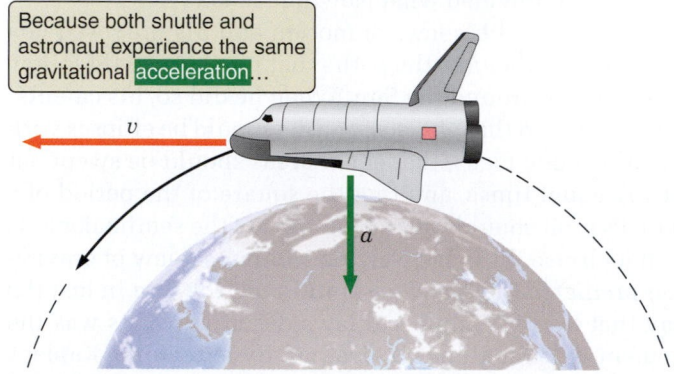

Because both shuttle and astronaut experience the same gravitational acceleration...

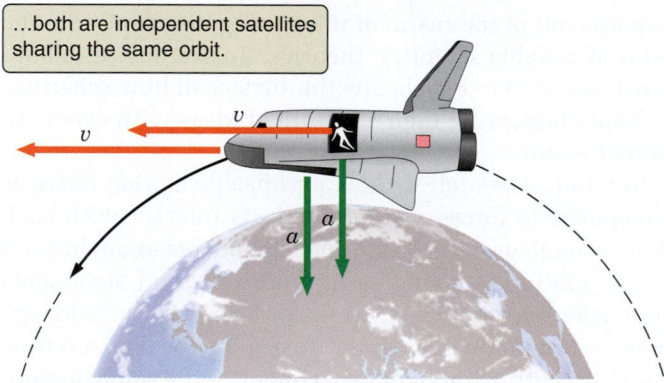

...both are independent satellites sharing the same orbit.

FIGURE 3.23 A "weightless" astronaut has not escaped Earth's gravity. Rather, an astronaut and a spacecraft share the same orbit as they fall around Earth together.

imaginary cannonball is a **satellite**. Sputnik 1, the first artificial satellite (*sputnik* means "satellite" in Russian), was the early forerunner to the spacecraft and the astronauts, which are independent satellites of Earth that conveniently happen to share the same orbit.

How Fast Must Newton's Cannonball Fly?

If fired fast enough, Newton's cannonball falls around the world; but just how fast is "fast enough"? Newton's orbiting cannonball moves along a circular path at constant speed. This type of motion, referred to as **uniform circular motion**, is discussed in more depth in Appendix 7. You are probably familiar with other examples of uniform circular motion. For example, think about a ball whirling around your head on a string, as shown in **Figure 3.24a**. If you were to let go of the string, the ball would fly off in a straight line in whatever direction it was traveling at the time, just as Newton's first law says. It is the string that keeps this from happening. The string exerts a steady force on the ball, causing it constantly to change the direction of its motion, always bending its flight toward the center of the circle. This central force is called a **centripetal force**. Using a more massive ball, speeding up its motion, or making the circle smaller so that the turn is tighter all increase the force needed to keep the ball from being carried off in a straight line by its inertia.

Centripetal forces maintain circular motion.

In the case of Newton's cannonball (or a satellite), there is no string to hold the ball in its circular motion. Instead the centripetal force is provided by gravity, as illustrated in **Figure 3.24b**. For Newton's thought experiment to work, the force of gravity must be just right to keep the cannonball moving on its circular path. In this case we can say that

Gravity provides the centripetal force that holds a satellite in its orbit.

$$\left(\begin{array}{c}\text{The force needed for}\\\text{uniform circular motion}\end{array}\right) = \left(\begin{array}{c}\text{The force provided}\\\text{by gravity}\end{array}\right).$$

In Appendix 7 we derive an expression for the centripetal force needed to keep an object moving in a circle at a steady speed. If we put that expression on the left side of this equation and the universal law of gravitation on the right side (and then do some algebra), we arrive at

$$v_{\text{circ}} = \sqrt{\frac{GM}{r}},$$

where M is the mass of the orbited object and r is the radius of the circular orbit. This value is called the **circular velocity**.

Here is the result we were looking for. If a satellite is in a stable circular orbit, then it *must* be moving at a velocity v_{circ}, where v_{circ} is given by this expression. *If the satellite were moving at any other velocity, it would not be moving in a circular orbit.* Remember the cannonball. If the cannonball were moving too slowly, it would drop below the circular path and hit the ground. Similarly, if the cannonball were moving too fast, its motion would carry it above the circular orbit. Only a cannonball moving at just the right velocity—the circular velocity—will fall around Earth on a circular path. The circular velocity at Earth's surface is about 8 km/s (see **Math Tools 3.2**).

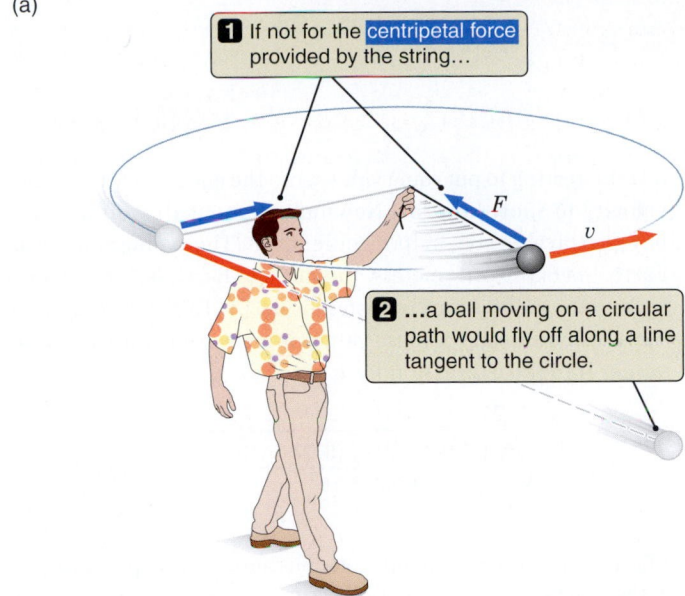

(a)

1 If not for the centripetal force provided by the string...

2 ...a ball moving on a circular path would fly off along a line tangent to the circle.

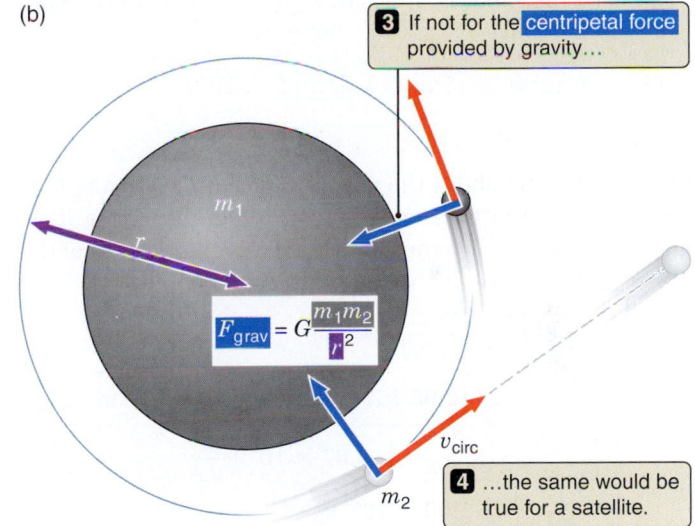

(b)

3 If not for the centripetal force provided by gravity...

$$F_{\text{grav}} = G\frac{m_1 m_2}{r^2}$$

4 ...the same would be true for a satellite.

VISUAL ANALOGY FIGURE 3.24 (a) A string provides the centripetal force that keeps a ball moving in a circle. (We are ignoring the smaller force of gravity that also acts on the ball.) (b) Similarly, gravity provides the centripetal force that holds a satellite in a circular orbit.

Planets Are Just like Newton's Cannonball

We can apply this same idea to the motion of Earth around the Sun. In the case of Earth's orbit, we already know from our discussion of the aberration of starlight that Earth travels at a speed of 2.98×10^4 m/s or (29.8 km/s) on its orbit about the Sun. We also know that the radius of Earth's orbit is 1.50×10^{11} meters. So we

We use Earth's orbit to calculate the Sun's mass.

Achieving Circular Velocity

It is interesting to put some values into the equation for circular velocity to show how fast Newton's cannonball would really have to travel to stay in its circular orbit. The average radius of Earth[11] is 6.37×10^6 meters, the mass of Earth is 5.974×10^{24} kg, and the gravitational constant is 6.67×10^{-11} N m²/kg². (We have seen how each of these values is measured.) Putting these values into the expression for v_{circ} gives

$$v_{circ} = \sqrt{\frac{(6.67 \times 10^{-11}) \times (5.97 \times 10^{24})}{6.37 \times 10^6}} = 7{,}900 \text{ m/s}$$

[11]Earth is not perfectly round. Rotation causes it to bulge somewhat at the equator. Earth's equatorial radius is 6,378 km, its polar radius is 6,357 km, and its mean (average) radius is 6,371 km. When referring to Earth's radius in this book, we will generally use its mean radius, often expressed as 6.37×10^6 meters.

Newton's cannonball would have to be traveling about 8 km/s— over 28,000 kilometers per hour (km/h)—to stay in its circular orbit. That's well beyond the reach of a typical cannon, but just what we routinely accomplish with rockets.

Suppose we wanted to put a satellite into orbit just above the lunar surface. How fast would it have to fly? The radius of the Moon is 1.737×10^6 meters and its mass is 7.35×10^{22} kg. These values give us

$$v_{circ} = \sqrt{\frac{(6.67 \times 10^{-11}) \times (7.35 \times 10^{22})}{1.737 \times 10^6}} = 1{,}680 \text{ m/s}$$

This is actually the speed of the cannonballs (projectiles) fired by Big Bertha, the cannon used by the Germans to bombard Paris during World War I.

know everything about the nearly circular orbit except for the mass of the Sun. If we can skip a step or two, a little algebra applied to the equation for v_{circ} gives the mass of the Sun $(M_\odot)$[12] as

$$M_\odot = \frac{(v_{circ})^2 \times r}{G}$$

$$= \frac{\left(2.98 \times 10^4 \, \frac{m}{s}\right)^2 \times (1.50 \times 10^{11} \text{ m})}{6.67 \times 10^{-11} \, \frac{m^3}{kg^2}}$$

$$= 1.99 \times 10^{30} \text{ kg}.$$

Once again, a bit of play has taken us places we might not have imagined. We began with Newton's thought experiment about a cannonball fired around the world, and ended up knowing the mass of the star that our planet orbits.

As long as we are at it, we can carry our game one step further. Kepler's third law talks about the time for a planet to complete one orbit about the Sun, known as the period, P, of a planet's orbit. The time it takes an object to make one trip around a circle is just the circumference of the circle $(2\pi r)$ divided by the object's speed. (Time equals distance divided by speed.) If the object is a planet in a circular orbit about the Sun, then its speed must be equal to the circular velocity that we calculated. Bringing all this together, we get

$$\text{Period} = \frac{\text{Circumference of orbit}}{\text{Circular velocity}} = \frac{2\pi r}{\sqrt{\frac{GM_\odot}{r}}}.$$

[12]The symbol $\odot$ signifies the Sun.

Some algebra gives us

$$P^2 = \frac{4\pi^2}{GM_\odot} \times r^3.$$

Once again, our play has gotten us somewhere really interesting. The square of the period of an orbit is equal to a constant $(4\pi^2/GM_\odot)$ multiplied by the cube of the radius of the orbit. *This is just Kepler's third law applied to circular orbits.* This is how Newton showed, at least in the special case of a circular orbit, that Kepler's third law—his beautiful "harmony of the worlds"—is a direct consequence of the way objects move under the force of gravity. A more complete treatment of the problem—the problem for which Newton invented calculus—shows that Newton's laws of motion and gravitation predict *all* of Kepler's empirical laws of planetary motion—for elliptical as well as circular orbits.

Most Orbits in the Real World Are Not Circles

Some Earth satellites—like Newton's orbiting cannonball— move along a circular path at constant speed. Just like the ball on a string, satellites traveling at the circular velocity remain the same distance from Earth at all times, neither speeding up nor slowing down in orbit. But now let's change the rules a bit. What if the satellite were in the same place in its orbit and moving in the same direction, but traveling *faster* than the circular velocity? The pull of Earth is as

(a)

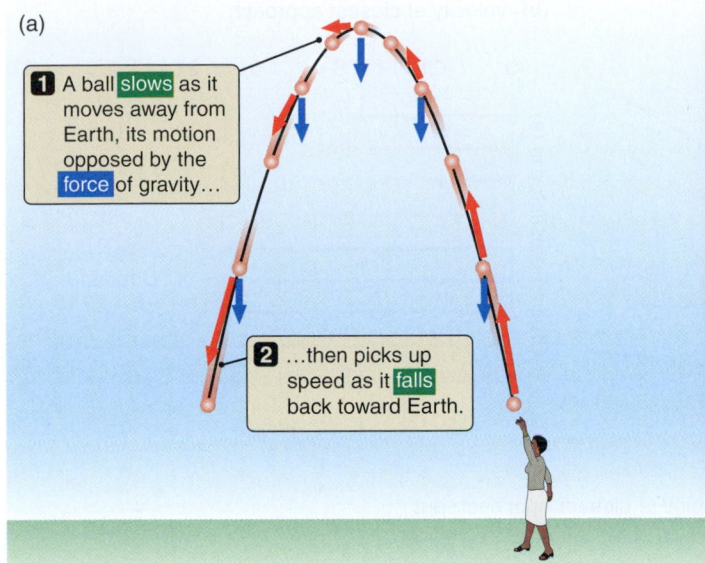

1 A ball slows as it moves away from Earth, its motion opposed by the force of gravity…

2 …then picks up speed as it falls back toward Earth.

(b)

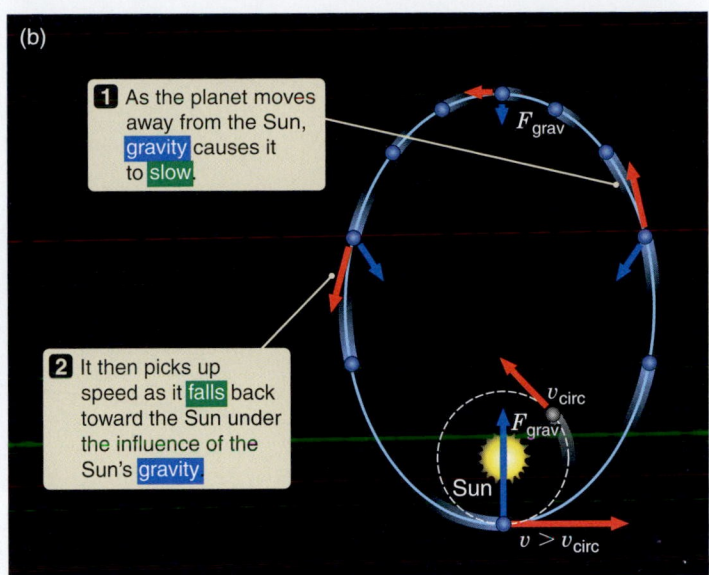

1 As the planet moves away from the Sun, gravity causes it to slow.

2 It then picks up speed as it falls back toward the Sun under the influence of the Sun's gravity

F_{grav}

v_{circ}

F_{grav}

Sun

$v > v_{circ}$

VISUAL ANALOGY FIGURE 3.25 (a) A ball thrown into the air slows as it climbs away from Earth and then speeds up as it heads back toward Earth. (b) A planet on an elliptical orbit around the Sun does the same thing. (Although no planet has an orbit as eccentric as the one shown here, the orbits of comets can be far more eccentric.)

strong as ever, but because the satellite has greater speed, its path is not bent by Earth's gravity sharply enough to hold it in a circle. So the satellite begins to climb above a circular orbit. ▶❚❚ **AstroTour:** Elliptical Orbits

As the distance between Earth and the satellite begins to increase, an interesting thing starts to happen. Think about a ball thrown into the air, as shown in **Figure 3.25a**. As the ball climbs higher, the pull of Earth's gravity opposes its motion, slowing the ball down. The ball climbs more and more slowly until its vertical motion stops for an instant and then is reversed; the ball begins to fall back toward Earth, picking up speed along the way. Our satellite does exactly the same thing as the ball. As the satellite climbs above a circular orbit and begins to move away from Earth, Earth's gravity opposes the satellite's outward motion, slowing the satellite down. The farther the satellite pulls away from Earth, the more slowly the satellite moves—just as happened with the ball thrown into the air. And just like the ball, the satellite reaches a maximum height on its curving path, then begins falling back toward Earth. Now as the satellite falls back toward Earth, Earth's gravity is pulling it along, causing it to pick up more and more speed as it gets closer and closer to Earth. The satellite's orbit has changed from circular to elliptical.

What is true for a satellite orbiting Earth in such an elliptical orbit is also true for any object in an elliptical orbit, including a planet orbiting the Sun. As we saw earlier, Kepler's law of equal areas says that a planet moves fastest when it is closest to the Sun and slowest when it is farthest from the Sun. Now we know why. As shown in **Figure 3.25b**, planets lose speed as they pull away from the Sun and then gain that speed back as they fall inward toward the Sun.

Newton's laws do more than explain Kepler's laws. Newton's laws also predict different types of orbits that are beyond Kepler's empirical experience. **Figure 3.26a** shows a series of satellite orbits, each with the same point of closest approach to Earth but with different velocities at that point, as indicated in **Figure 3.26b**. A look at the figure shows that the greater the speed a satellite has at its closest approach to Earth, the farther the satellite is able to pull away from Earth, and the more eccentric its orbit becomes. Yet no matter how eccentric it becomes, as long as it remains elliptical, an orbit will eventually bring a satellite back to the planet that it orbits or bring a planet back to the Sun. All such orbits are called **bound orbits** because the satellite is gravitationally bound to the object it is orbiting.

> **Ellipses are bound orbits.**

You might imagine, though, that somewhere in this sequence of faster and faster satellites there comes a point of no return—a point when the satellite is moving so fast that gravity is unable to reverse its outward motion, so the satellite coasts away from Earth, never to return. This indeed is possible. The lowest speed at which this happens is called the **escape velocity** (see **Math Tools 3.3**).

> **A satellite moving fast enough will escape a planet's gravity.**

Once an object reaches escape velocity, it is in an **unbound orbit**. That is, the object is no longer gravitationally bound to the primary body that it is orbiting, such as a satellite orbiting a planet or a comet orbiting the Sun. When Newton solved his equations of motion, he found that unbound orbits are shaped as hyperbolas or parabolas. As

(a) Representative orbits

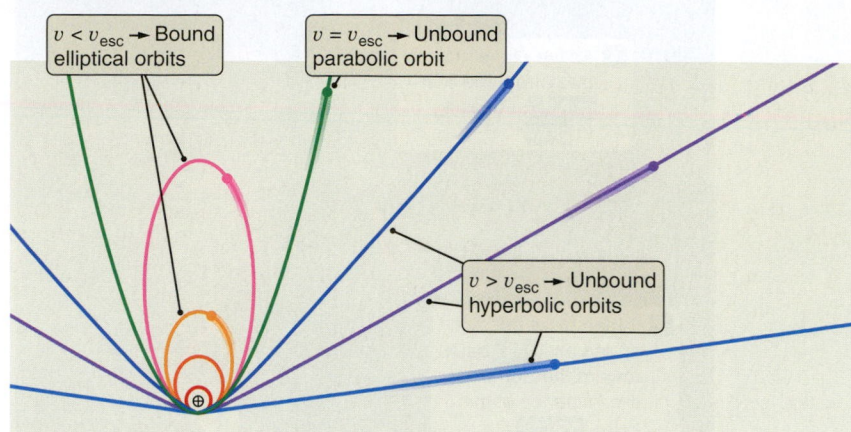

(b) Velocity at closest approach

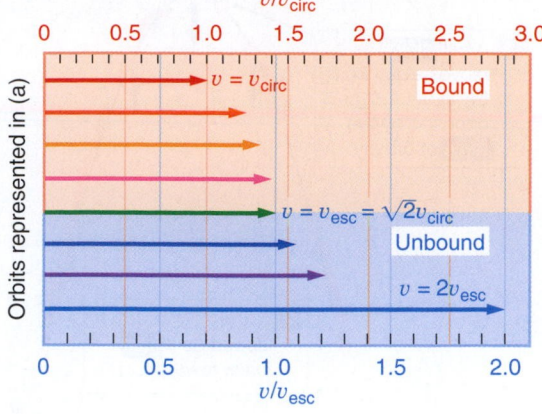

FIGURE 3.26 (a) A range of different orbits that share the same point of closest approach but differ in velocity at that point. (b) Closest-approach velocities for the orbits in (a). An object's velocity at closest approach determines the orbit shape and whether the orbit is bound or unbound.

MATH TOOLS 3.3

Calculating Escape Velocities

The concept of escape velocity occurs frequently in astronomy. For NASA to send a spacecraft to another planet, it must be launched with a velocity greater than Earth's escape velocity. And we will see in Chapter 8 that planets can and do lose their atmospheres when some of their atmospheric atoms are bumped up to escape velocity.

If we were to work through the calculation, we would find that the escape velocity is a factor of $\sqrt{2}$, or approximately 1.414 times the circular velocity. This relation can be expressed as

$$v_{esc} = \sqrt{\frac{2GM}{R}} = \sqrt{2}\, v_{circ},$$

where G is the universal gravitational constant, 6.67×10^{-11} N m^2/kg^2.

Look at this equation for a minute to be sure it makes sense to you. The larger the mass (M) of a planet, the stronger its gravity, so it stands to reason that a more massive planet would be harder to escape from than a less massive planet. Indeed, the equation says that the more massive the planet, the greater the required escape velocity. It also stands to reason that the closer we are to the planet, the harder it will be to escape from its gravitational attraction. Again, the equation confirms our intuition. As the distance R becomes larger (that is, as we get

farther from the planet), v_{esc} becomes smaller (in other words, it is easier to escape from the planet's gravitational pull). We can now calculate the escape velocity from Earth's surface, noting that Earth has an average radius (R) of 6.37×10^6 meters and a mass (M) of 5.97×10^{24} kg.

$$v_{esc} = \sqrt{\frac{2 \times (6.67 \times 10^{-11}) \times (5.97 \times 10^{24})}{6.37 \times 10^6}} =$$

11,180 m/s = 11.18 km/s = 40,250 km/h.

We can also calculate the escape velocity from the surface of Ida, an asteroid that you will meet in Chapter 12. Ida has an average radius (R) of 15,700 meters and a mass (M) of 4.2×10^{16} kg.

$$v_{esc} = \sqrt{\frac{2 \times (6.67 \times 10^{-11}) \times (4.2 \times 10^{16})}{1.57 \times 10^4}} =$$

19 m/s = 0.019 km/s = 68 km/h.

So, a fastball thrown from its surface would easily escape Ida and fly off into interplanetary space.

you can see from Figure 3.26, an object with a velocity *less* than the escape velocity (v_{esc}) will be on an elliptically shaped orbit. Elliptical orbits "close" on themselves. That is, an object traveling in an elliptical orbit will follow the same path over and over again.[13] A hyperbola does not close like an ellipse but instead keeps opening up, as shown in Figure 3.26a. For example, a comet traveling on a "hyperbolic orbit" makes only a single pass around the Sun and then is back off into deep space, never to return. The third type of orbit is the borderline case in which the orbiting object moves at exactly the escape velocity. If it had any less velocity it would be traveling in a bound elliptical orbit; any more, and it would be moving on an unbound hyperbolic orbit. An object moving with a velocity *equal* to the escape velocity follows a "parabolic orbit" and, as in the hyperbolic orbit, passes by the primary body only once. As the object traveling in a parabolic orbit moves away from the primary body, its velocity relative to the primary body gets closer and closer to, but never quite reaches, zero. An object traveling in a hyperbolic orbit, by contrast, always has excess velocity relative to the primary body, even when it has moved infinitely far away.

> Unbound orbits are hyperbolas or parabolas.

Newton's Theory Is a Powerful Tool for Measuring Mass

As mentioned earlier, Kepler's empirical laws describe the motion of the planets but do not explain them. On the basis of Kepler's laws alone, we might imagine that angels carry the planets around in their orbits, just as many people believed during the 16th century! Newton's derivation of Kepler's laws changed all of that. Newton showed that the *same* physical laws that describe the flight of a cannonball on Earth—or the fall of the apocryphal apple on his head—also describe the motions of the planets through the heavens. In this way Newton shattered the prevailing concept of the heavens and Earth, and at the same time opened up an entirely new way of investigating the universe. Copernicus may have dislodged Earth from the center of the universe and started us on the way toward the cosmological principle, but it was Newton who moved the cosmological principle out of the realm of philosophy and into the realm of testable scientific theory. And it was through Newton's work that **astrophysics** was born.

Not only is Newton's method more philosophically satisfying than simple empiricism—it is far more powerful as

well. We have already seen, for example, how Newton's laws can be used to measure the mass of Earth and the Sun. A calculation such as this could never be done with Kepler's empirical rules. This fact is especially important when we remember that Newton's laws apply to *all* objects, not just the Sun and Earth, and it will prove handy as we continue our journey.

Astronomers use Newton's form of Kepler's third law to calculate the masses of planets, stars, and other bodies. In doing so, they often rearrange Newton's form of the law to read

$$M = \frac{4\pi^2}{G} \times \frac{A^3}{P^2}.$$

Everything on the right side of this equation is either a constant (such as 4, π, and G) or a quantity that we can measure (like the semimajor axis A and period P of an orbit). The left side of the equation is the mass of the object at the focus of the ellipse.

It is important to note that we cut a couple of corners to get to this point. For one thing, we arrived at this relationship by thinking about circular orbits and then simply asserting that it holds for elliptical orbits as well. Another corner we cut was assuming that a low-mass object such as a cannonball is orbiting a more massive object such as Earth. Earth's gravity has a strong influence on the cannonball; but as we have seen, the cannonball's gravity has little effect on Earth. For this reason we can imagine that Earth remains motionless while the cannonball follows its elliptical orbit. Similarly, it is a good approximation to say that the Sun remains motionless as the planets orbit about it.

This picture changes, though, when two objects are closer to having the same mass. In this case *both* objects experience significant accelerations in response to their mutual gravitational attraction. The mass M now refers to the *sum* of the masses of the two objects, which are both orbiting about a common **center of mass** located between them. We now must think of the two objects as falling around *each other*, with each mass moving on its own elliptical orbit around their mutual center of mass. So, if we can measure the size and period of an orbit—*any* orbit—then we can use this equation to calculate the mass of the orbiting objects. This is true not only for the masses of Earth and the Sun, but also for the masses of other planets, distant stars, our galaxy and distant galaxies, and vast clusters of galaxies. In fact, it turns out that *almost all of our knowledge about the masses of astronomical objects comes directly from the application of this one equation*. In a sense, this single equation even allows us to tackle the question, What is the mass of the universe itself?

> Objects having similar mass orbit each other.

You may well be wondering why we have taken such a long and sometimes strenuous excursion through the work of Galileo, Kepler, and Newton. Sometimes when we are hiking in the mountains, it is hard to see the summit from the

---

[13]Strictly speaking, this is true only for a single body orbiting a second body. If other bodies are present, the ellipse may not quite close, causing the orientation of the ellipse to swing around slowly in the orbital plane. This effect is referred to as *orbital precession*. Relativistic effects can also cause orbital precession, as we will see in the case of Mercury's orbit in Chapter 17.

perspective of the trail. Now that we have arrived at the top of this particular pass, we can look around and appreciate what we have gained. When Newton carried out his calculations, he found that his laws of motion and gravitation predicted elliptical orbits that agree exactly with Kepler's empirical laws. *This is how Newton tested his theory that the planets obey the same laws of motion as cannonballs and how he confirmed that his law of gravitation is correct.* Had Newton's predictions not been borne out by observation, he would have had to throw them out and go back to the drawing board!

Kepler's empirical rules for planetary motion pointed the way for Newton and provided the crucial observational test for Newton's laws of motion and gravitation. At the same time, Newton's laws of motion and gravitation provided a powerful new understanding of why planets and satellites move as they do. Theory and empirical observation work together hand in hand, and our understanding of the universe strides forward. The work of Copernicus, Galileo, Kepler, and Newton was not just *a* scientific revolution; it was *the* scientific revolution, which changed forever not only our view of the universe but also our very notion of what it means "to know."

> Newton's laws provided a physical explanation for Kepler's empirical results.

We promised we would show you *how* science works. Well, this is how it works.

Seeing the Forest for the Trees

The story of planetary motions is also the story of how we as a species learned to do science. No one taught Copernicus or Galileo or Kepler or Newton about the differences between empiricism and theory, or about how the scientific method could be used to test their ideas. They had to figure these lessons out on their own as they went along. Yet their accomplishments remain near the top of the all-time intellectual feats of humankind, and the trail they blazed pointed the way for all that was to come. Kepler built on Copernicus's revolutionary ideas about planetary motions and uncovered three empirical rules that showed the orbits of all the planets to be reflections of the same underlying patterns. Galileo offered powerful insights into the nature of matter and gravity. Newton combined the two sets of ideas: he built Galileo's insights into a powerful theoretical edifice describing the motions of all objects, and then he tested this edifice against Kepler's empirical reality. The culmination of this intellectual campaign was far greater than the sum of its parts. The walls separating the heavens and Earth came down once and for all, and the science of astrophysics was born.

The model for science that emerged during this era remains the template for how science is done to this day. Careful empiricism uncovers patterns in nature in need of explanation. Scientific theory seeks to discover the fundamental truths underlying all things. And in the meeting of the two—the test of theory against the unforgiving and stalwart challenge of empirical fact—new knowledge and understanding emerge.

Newton's work became the cornerstone of what is often referred to as classical mechanics. All objects have inertia. An object will continue to move in a straight line at a constant speed unless an unbalanced force acts to change its motion. Mass is the property of matter that resists changes in motion. Every force is matched by another force that is equal in magnitude but opposite in direction. Gravity is a force between any two masses, proportional to the product of the two masses and inversely proportional to the square of the distance between them. Putting all of this together, we find that objects "fall around" the Sun and Earth on elliptical, parabolic, or hyperbolic paths. Orbits are ultimately given their shape by the gravitational attraction of the objects involved, which in turn is a reflection of the masses of these objects. We now understand that it is gravity that holds the universe together, giving planets, stars, and galaxies their very shapes, as well as controlling their motions through space. Using Newton's theoretical insight, we look backward along this chain, turning observations of the motions of objects throughout the universe into measurements of the masses of objects that no human has ever or, in most cases, will ever visit.

Thus we arrive at the next stage of our journey. If the ancients could have stepped off Earth and touched the planets, they would never have fallen into the conceptual errors that muddled our thinking for millennia—but they had no such luxury. Today we have sent robotic surrogates to all of the classical planets in the Solar System, and humans have walked on the surface of the Moon. Even so, we have made only the most cursory visits to our immediate neighborhood. Even the nearest stars remain thousands of times more distant than the most far-flung of our robotic planetary explorers. For the most part we, like the ancients, are left with nothing on which to base our knowledge of the universe but the signals reaching us from across space. Far and away the most common of these signals is electromagnetic radiation, which includes light (such as the light by which you are reading this book). Our ability to interpret these signals depends on what we know about light. What is it? How does it originate? How does it interact with matter? What changes does it experience during its journey? In Chapter 4 we turn to these questions.

Summary

- Gravity, as one of the fundamental forces of nature, binds the universe together.

- On the basis of Tycho's observations, Kepler developed empirical solutions to describe the motions of the planets.

- Newton's three laws govern the motion of all objects.

- Unbalanced forces cause changes in motion.

- Proportionality describes patterns and relationships in nature.

- Mass is the property of matter that gives it resistance to changes in motion.

- Gravity is a force between any two objects due to their masses.

- *All* objects near Earth's surface fall with the same acceleration, written as *g*.

- Planets orbit the Sun in bound, elliptical orbits.

- Unbound orbits are shaped as parabolas or hyperbolas.

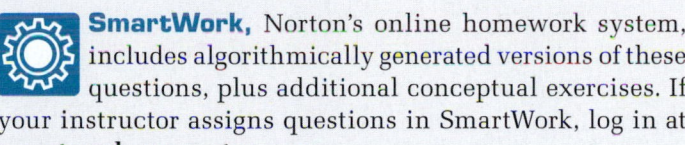 **SmartWork,** Norton's online homework system, includes algorithmically generated versions of these questions, plus additional conceptual exercises. If your instructor assigns questions in SmartWork, log in at **smartwork.wwnorton.com.**

Student Questions

THINKING ABOUT THE CONCEPTS

1. Copernicus and Kepler engaged in what is called empirical science. What do we mean by *empirical*?

*2. The orbits of the planets around the Sun and the orbits of satellites around these planets are always ellipses rather than perfect circles. Why?

3. Each ellipse has two foci. The orbits of the planets have the Sun at one focus. What is at the other focus?

4. Ellipses contain two axes, major and minor. Half the major axis is called the semimajor axis. What is especially important about the semimajor axis of a planetary orbit?

5. What is the eccentricity of a circular orbit?

6. The speed of a planet in its orbit varies in its journey around the Sun.

   a. At what point in its orbit is the planet moving the fastest?
   b. At what point is it moving the slowest?

7. The distance that Neptune has to travel in its orbit around the Sun is approximately 30 times greater than the distance that Earth must travel. Yet it takes nearly 165 years for Neptune to complete one trip around the Sun. Explain why.

*8. Kepler's and Newton's laws all tell us something about the motion of the planets, but there is a fundamental difference between them. What is the difference?

9. Aristotle taught that the natural state of all objects is to be at rest. Even though this idea seems consistent with what we observe around us, explain why Aristotle's reasoning was wrong.

*10. Galileo came up with the concept of inertia. What do we mean by *inertia*?

11. What is the difference between speed and acceleration?

12. Imagine a planet moving in a perfectly circular orbit around the Sun. Because the orbit is circular, the planet is moving at a constant speed. Is this planet experiencing acceleration? Explain your answer.

13. When riding in a car, we can sense changes in speed or direction through the forces that the car applies on us. Do we wear seat belts in cars and airplanes to protect us from speed or from acceleration? Explain your answer.

14. An astronaut standing on Earth could easily lift a wrench having a mass of 1 kg, but not a scientific instrument with a mass of 100 kg. In the International Space Station she is quite capable of manipulating both, although the scientific instrument responds much more slowly than the wrench. Explain why.

*15. In 1920, a *New York Times* editor refused to publish an article based on rocket pioneer Robert Goddard's paper that predicted spaceflight, saying that "rockets could not work in outer space because they have nothing to push against" (that statement the *Times* did not retract until July 20, 1969, the date of the *Apollo 11* Moon landing). You, of course, know better. What was wrong with the editor's logic?

16. Explain the difference between weight and mass.

17. Weight on Earth is proportional to mass. On the Moon, weight is also proportional to mass, but the constant of proportionality is different on the Moon than it is on Earth. Why?

*18. Picture a swinging pendulum. During a single swing, the bob of the pendulum first falls toward Earth and then moves away until the gravitational force between it and Earth finally stops its motion. Would a pendulum swing if it were in orbit? Explain your reasoning.

19. Had Kepler lived on one of a group of planets orbiting a star three times as massive as our Sun, would he have deduced the same empirical laws? Explain your answer.

20. When sending a spacecraft to Mars, we begin by launching it into an Earth orbit with circular velocity.
    a. Describe the shape of this orbit.
    b. What velocity must we give the spacecraft to send it on its way to Mars?

21. Describe the difference between a bound orbit and an unbound orbit.

22. Two comets are leaving the vicinity of the Sun, one traveling in an elliptical orbit and the other in a hyperbolic orbit. What can you say about the future of these two comets? Would you expect either of them to eventually return?

23. Suppose astronomers discovered a comet approaching the Sun in a hyperbolic orbit. What would that say about the origin of the comet?

*24. What is the advantage of launching satellites from space ports located near the equator? Why are satellites never launched in a westerly direction?

**25. We speak of Earth and the other planets all orbiting about the Sun. Under what circumstances do we have to consider bodies orbiting about a "common center of mass"?

APPLYING THE CONCEPTS

26. A sports car accelerates from a dead stop to 100 km/h in 4 seconds.
    a. What is its acceleration?
    b. If it goes from 100 km/h to a dead stop in 5 seconds, what is its acceleration?

27. A train pulls out of a station accelerating at 0.1 m/s². What is its speed, in kilometers per hour, 2½ minutes after leaving the station?

28. Flybynite Airlines takes 3 hours to fly from Baltimore to Denver at a speed of 800 km/h. Wanting to save the cost of fuel, management orders its pilots to reduce their speed to 600 km/h. How long will it now take passengers on this route to reach their destination?

29. You are driving down a straight road at a speed of 90 km/h and see another car approaching you at a speed of 110 km/h along the road.
    a. Relative to your own frame of reference, how fast is the other car approaching you?

    b. Relative to the other driver's frame of reference, how fast are you approaching the other driver's car?

30. You are riding along on your bicycle at 20 km/h and eating an apple. You pass a bystander.
    a. How fast is the apple moving in your frame of reference?
    b. How fast is the apple moving in the bystander's frame of reference?
    c. Whose perspective is more valid?

31. The average distance of Uranus from the Sun is about 19 times Earth's distance from the Sun. How much stronger is the Sun's gravitational grip on Earth than on Uranus?

32. During the latter half of the 19th century, a few astronomers thought there might be a planet circling the Sun inside Mercury's orbit. They even gave it a name: Vulcan. We now know that Vulcan does not exist. If a planet with an orbit a fourth the size of Mercury's actually existed, what would be its orbital period relative to that of Mercury?

33. Earth speeds along at 29.8 km/s in its orbit. Neptune's nearly circular orbit has a radius of 4.5×10^9 km, and the planet takes 164.8 years to make one trip around the Sun. Calculate the speed at which Neptune plods along in its orbit.

34. Assume that a planet just like Earth is orbiting the bright star Vega at a distance of 1 AU. The mass of Vega is twice that of the Sun.
    a. How fast is the Earth-like planet traveling in its orbit around Vega?
    b. How long in Earth years will it take to complete one orbit around Vega?

35. Venus's circular velocity is 35.03 km/s, and its orbital radius is 1.082×10^8 km. Calculate the mass of the Sun.

36. At the surface of Earth, the escape velocity is 11.2 km/s. What would be the escape velocity at the surface of a very small asteroid having a radius 10^{-4} that of Earth's and a mass 10^{-12} that of Earth's? If you were standing on the asteroid and threw a baseball with a strong pitch, what would happen to it?

37. How long does it take Newton's cannonball, moving at 7.9 km/s just above Earth's surface, to complete one orbit around Earth?

38. Earth's mean radius and mass are 6,371 km and 5.97×10^{24} kg, respectively. Show that the acceleration of gravity at the surface of Earth is 9.80 m/s².

39. Two balls, one of gold with a mass of 100 kg and one of wood with a mass of 1 kg, are suspended 1 meter apart. What is the attractive force, in newtons, of
    a. the gold ball acting on the wooden ball?
    b. the wooden ball acting on the gold ball?

*40. Using 6,371 km for Earth's radius, compare the gravitational force acting on NASA's space shuttle when sitting on its launchpad with the gravitational force acting on it when orbiting 350 km above Earth's surface.

41. The International Space Station travels on a nearly circular orbit 350 km above Earth's surface. What is its orbital speed?

42. At some time in the future, scientists might want to launch an interplanetary spacecraft from an orbiting platform traveling at a speed of 7.61 km/s 500 km above Earth. What is the minimum speed at which the spacecraft would have to leave the platform?

43. The radius and mass of the Sun are 696,000 km and 1.99×10^{30} kg, respectively. A comet on a parabolic orbit falls into the Sun. Just before it vaporizes at the Sun's surface, how fast is it traveling?

44. (a) What does an acceleration of 9.8 m/s^2 mean? (b) If you jumped from a stationary balloon, after 1 second you would be falling at a speed of 9.8 m/s. After 2 seconds your speed would be $2 \times 9.8 = 19.6$ m/s, or slightly more than 70 km/h! In the absence of any air resistance, how fast would you be falling after 20 seconds?

45. Weight refers to the force of gravity acting on a mass. We often calculate the weight of an object by multiplying its mass by the local acceleration due to gravity. The value of gravitational acceleration on the surface of Mars is 0.39 times that on Earth. If your mass is 85 kg, your weight on Earth is 830 N ($m \times g = 85$ kg $\times 9.8$ m/s$^2 = 830$ N). What would be your mass and weight on Mars?

46. Suppose a new dwarf planet is discovered orbiting the Sun with a semimajor axis of 50 AU. What would be the orbital period of this new dwarf planet?

*47. Planet Zork takes 7 Earth years to orbit the bright star Achernar at a distance of 7 AU. In terms of solar masses, how massive is Achernar?

48. The asteroid Ida (mass = 4.2×10^{16} kg) is attended by a tiny asteroidal moon, Dactyl, which orbits Ida at an average distance of 108 km. Neglecting the mass of the tiny moon, what is Dactyl's orbital period in hours?

49. A pair of asteroids, each with a mass of 10^{16} kg, orbit about a common center of gravity in 20 hours. What is the separation of these asteroids, in kilometers?

50. The two stars in a binary star system (two stars orbiting about a common center of mass) are separated by 10^9 km and revolve around their common center of mass in 10 years.
    a. What is their combined mass, in kilograms?
    b. The mass of the Sun is 1.99×10^{30} kg. What is their combined mass, in solar masses?

StudySpace is a free and open Web site that provides a Study Plan for each chapter of *21st Century Astronomy*. Study Plans include animations, reading outlines, vocabulary flashcards, and multiple-choice quizzes, plus links to premium content in SmartWork and the ebook. Visit **www.wwnorton.com/studyspace**.

Light brings us the news of the Universe.

WILLIAM HENRY BRAGG (1862–1942)

The setting Sun colors clouds over the Australian bush.

CHAPTER 4

Light

4.1 Let There Be Light

Light is a fundamental part of our experience of the world because it is through light that we most directly perceive the world beyond our physical grasp. (Those who have lost their sight face a challenge greater than most of us can imagine.) Throughout our language and culture, light is a metaphor for knowledge and information. To understand something is to "see it." A "bright idea" is symbolized by a lightbulb going off over our heads. As we leave our ignorance behind, we become "enlightened." A symbol of hope is a "light at the end of the tunnel." At the same time, light plays other, even more essential roles in our lives. Light from the Sun warms Earth, drives the wind and rain, and powers photosynthesis in plants, which lie at the bottom of the terrestrial food chain. The energy you obtain from consuming plants—either directly or indirectly—and expend as you move through your day arrived on the planet in the form of light.

The role of light in astronomy closely parallels the role of light in our everyday existence. Our knowledge of the universe beyond Earth comes overwhelmingly from light given off or reflected by astronomical objects. Fortunately, light is a *very* informative messenger. For any given object, light carries information about that object's *temperature*, composition, *speed*, and even the nature of the material that the light passed through on its way to Earth. Yet light plays a far larger role in astronomy than just being a messenger. Light is one of the main ways in which energy is transported throughout the universe. Light carries energy generated in the heart of a star outward through the star and off into space. Objects as diverse as stars, planets, and vast clouds of gas and dust

KEY CONCEPTS

Unlike the physicist or the chemist, who has control over the conditions in a laboratory, the astronomer must try to glean the secrets of the universe from the light and other particles that reach us from distant objects. On this leg of our journey we turn our attention to light, a most informative messenger, and find that

- Light is not only a messenger but also a means by which energy is transported throughout the universe.

- Light is an electromagnetic wave with a spectrum extending far beyond the colors of the rainbow.

- Light is also a stream of particles called photons.

- Reconciling the two identities of light and matter—its wave and particle characteristics—challenges our everyday ideas about what is "real."

- The wave-particle nature of light and matter gives different types of atoms unique spectral "fingerprints" that we can use to measure the composition and properties of distant objects.

- Temperature measures the thermal energy of an object and determines the amount and spectrum of light that a dense object emits.

filling interstellar space are heated as they absorb light and cooled as they emit light.

Although light allows us to see the world, we cannot actually "see" light as an object. When we view a tree or the stars at night, our eyes and brains are *responding* to light that is reflected from or emitted by these objects. Light is the messenger but is itself invisible. It is hard to study something that cannot be seen, so an understanding of light was a long time coming. The property of light that is easiest to *try* to measure is the speed at which it travels. As we continue our journey, we will find that light sets the standard for what we mean by *when*, *where*, or *how fast* because nothing can travel faster than light.

4.2 How Our Understanding of Light Evolved

Suppose that you are a scientist living in the 17th century. How might you go about measuring the speed of light? One way would be to have a friend stand on a hilltop far away. You uncover a lantern, and the instant your friend sees the light from your lantern, your friend uncovers her own lantern. The time it takes from when you uncover your lantern to when you see your friend's light will be the light's round-trip travel time—plus, of course, your friend's reaction time. Galileo tried this experiment but could not measure any delay. He concluded that the speed of light must be very great indeed, possibly even infinite.

If light travels so rapidly, then to measure its speed we will need either very large distances over which to measure its flight or very good clocks. Galileo had neither at his disposal, but by the end of the 17th century astronomers had both. The great distances were the distances between the planets, whereas the good "clock" was provided by Kepler and Newton. According to Newton's derivation of Kepler's laws, orbital periods are constant, with each orbit taking exactly as much time as the orbit before. This property applies to moons orbiting planets just as it applies to planets orbiting the Sun.

In the 1670s, Danish astronomer Ole Rømer (1644–1710) was studying the moons of Jupiter, measuring the times when each moon disappeared behind the planet. Much to his amazement Rømer found that rather than maintaining a regular schedule, the observed times of these events would slowly drift in comparison with predictions. Sometimes the moons disappeared behind Jupiter too soon, and at other times they went behind Jupiter later than expected. Rømer realized that the difference depended on where Earth was in its orbit. If he began tracking the moons when Earth was closest to Jupiter, then by the time Earth was farthest from Jupiter, the moons were a bit over 16½ minutes "late." But if he waited until Earth was once again closest to Jupiter, the moons "made up"

the lost time and once again passed behind Jupiter at the predicted times.

It is often the case in science that a difference between theoretical predictions and experimental results points the way to new knowledge, and Rømer's work was no exception. Rømer correctly surmised that rather than a failure of Kepler's laws, he was seeing the first clear evidence that light travels at a finite speed. As shown in **Figure 4.1**, the moons appeared "late" when Earth was farther from Jupiter because of the time needed for light to travel the extra distance between the two planets. Over the course of Earth's yearly trip around the Sun, the distance between Earth and Jupiter changes by 2 astronomical units (AU), which is about 3×10^{11} meters. The speed of light equals this distance divided by Rømer's 16.7-minute delay, or about 3×10^8 meters per second (m/s). The value that Rømer actually announced in 1676 was a bit on the low side—2.25×10^8 m/s—because the length of 1 AU was not well known. But Rømer's result was more than adequate to make the point that the speed of light is very great indeed. The International Space Station moves around Earth at a dazzling speed of about 28,000 kilometers per hour (km/h), almost 8,000 m/s. Light travels almost 40,000 times faster than this. It could circle Earth in only ⅐ of a second. No wonder Galileo's attempts to measure the speed of light failed!

A good deal of work has been done to improve on Rømer's original result. Modern measurements of the speed of light made with the benefit of high-speed electronics and lasers give a value of 2.99792458×10^8 m/s in a vacuum. As of October 1983, the length of a meter is now *defined* as the distance traveled by light in a vacuum in 1/299,792,458 of a second.

The speed of light in a vacuum, about 300,000 kilometers per second (km/s), is one of nature's fundamental constants, usually written as c. Keep in mind, however, that this is true *only* in a vacuum. The speed of light through any medium, such as air or glass, is *always* less than c. We refer to the ratio of light's speed in a vacuum to its speed, v, in a medium as the medium's **index of refraction**, n:

> **Rømer used Jupiter's moons to measure the speed of light.**

> **The speed of light is 300,000 km/s in a vacuum.**

$$n = \frac{c}{v}.$$

For typical glass, n is approximately 1.5. Rearranging this equation, we find that the speed of light in glass is

$$v = \frac{c}{n} = \frac{300{,}000 \text{ km/s}}{1.5} = 200{,}000 \text{ km/s}.$$

We will come back to the index of refraction in Chapter 5 when we discuss refraction and refracting telescopes.

Recall that in Chapter 1 we spoke of distances expressed not in kilometers or miles, but in units of time. For example,

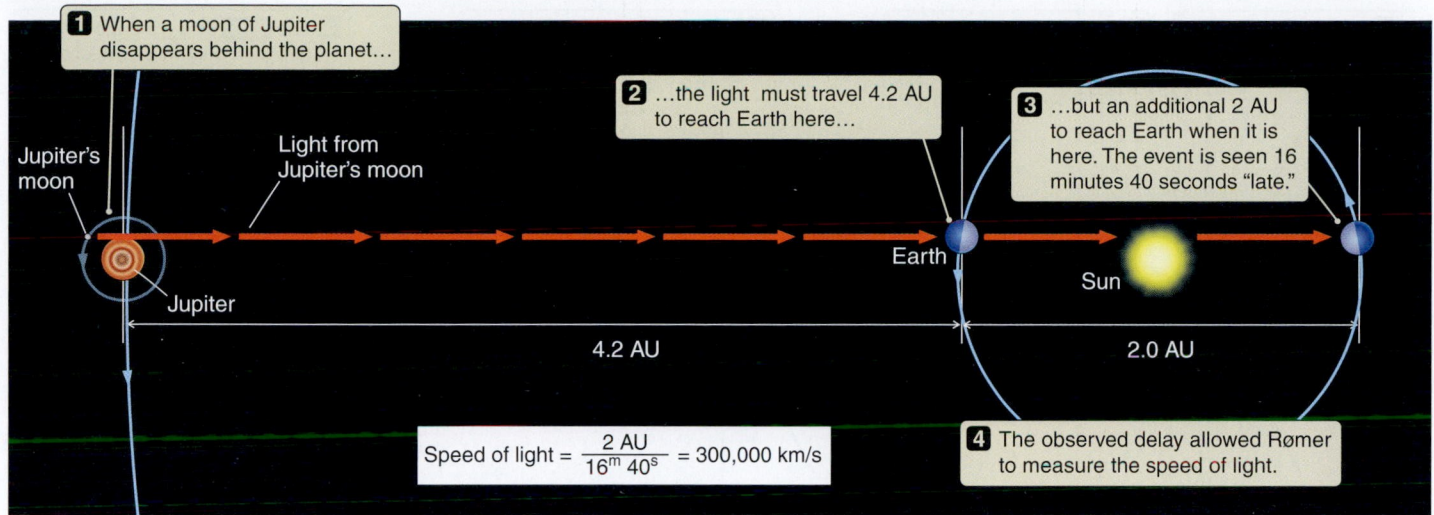

1 When a moon of Jupiter disappears behind the planet...

2 ...the light must travel 4.2 AU to reach Earth here...

3 ...but an additional 2 AU to reach Earth when it is here. The event is seen 16 minutes 40 seconds "late."

Jupiter's moon

Light from Jupiter's moon

Earth

Jupiter

Sun

4.2 AU

2.0 AU

$$\text{Speed of light} = \frac{2 \text{ AU}}{16^m \, 40^s} = 300{,}000 \text{ km/s}$$

4 The observed delay allowed Rømer to measure the speed of light.

FIGURE 4.1 Danish astronomer Ole Rømer realized that apparent differences between the predicted and observed orbital motions of Jupiter's moons depend on the distance between Earth and Jupiter. He used these observations to measure the speed of light. (The superscript letters in the expression "$16^m 40^s$" stand for minutes and seconds of time, respectively.)

the Moon's distance is such that it takes light 1¼ seconds to travel between Earth and Moon. In other words, we can say that the Moon is 1¼ light-*seconds* from Earth. The Sun is 8⅓ light-*minutes* away, and the next-nearest star is 4⅓ light-*years* distant. The travel time of light is a convenient way of expressing cosmic distances, and the basic unit is the **light-year**. (Astronomers frequently use another yardstick to describe stellar and galactic distances, the **parsec**, abbreviated **pc**. One parsec is equal to 3.26 light-years.) A light-year is defined as the distance traveled by light in 1 year, or about 9.5 trillion kilometers (km). Remember, *a light-year is a measure of distance. It is* not *a measure of time.* ("I tell you, it's been simply light-years since we last saw them" is a frequent misuse of the term *light-year*.)

> A light-year is the distance that light travels in one year.

Light Is an Electromagnetic Wave

Since the earliest investigations of light, there has been a good deal of controversy over the question of whether light is composed of particles, as Newton believed, or is instead a wave. (A **wave** is a disturbance that travels from one point to another.) This controversy was seemingly put to rest once and for all in 1873 by the Scottish physicist James Clerk Maxwell (1831–1879). One of Maxwell's many accomplishments was his introduction of the concept that electricity and magnetism are actually two components of the

> A wave is a disturbance that travels away from a source.

same physical phenomenon. An **electric force** is the push or pull between electrically charged particles. Opposite charges attract, and like charges repel. A **magnetic force**, on the other hand, is a force between electrically charged particles arising from their motion.

To describe the electric and magnetic forces, Maxwell introduced the concepts of the **electric field** and the **magnetic field**. A charged particle creates an electric field that points away from the charge if the charge is positive, as shown in **Figure 4.2a**, or toward it if the charge is negative. Because the electric field points directly away from a positively charged particle (or directly toward a negatively charged particle), the force that a second charge feels is either directly toward or directly away from the first charged particle.

The picture gets a bit more interesting if we quickly move the first charged particle (q_1) by some amount, as shown in **Figure 4.2b**. We might expect the force on the second positively charged particle (q_2) to change immediately so that it points away from the new position of the first particle. Yet experiments show that it does not. Immediately after the first charge moves, there is *no* change in the force felt by the second charge. Only later does the second particle feel the change in location of the first (**Figure 4.2c**). The situation is something like what happens if you are holding one end of a long piece of rope and a friend is holding the other end. When you yank your end of the rope up and down, your friend does not feel the result immediately. Instead your yank starts a pulse—a wave—that travels along the rope. Your friend notices the yank only when this wave arrives at his end. Similarly, when you move a charged particle, information about the change travels outward through space as a wave in the electric field. Other charged particles are

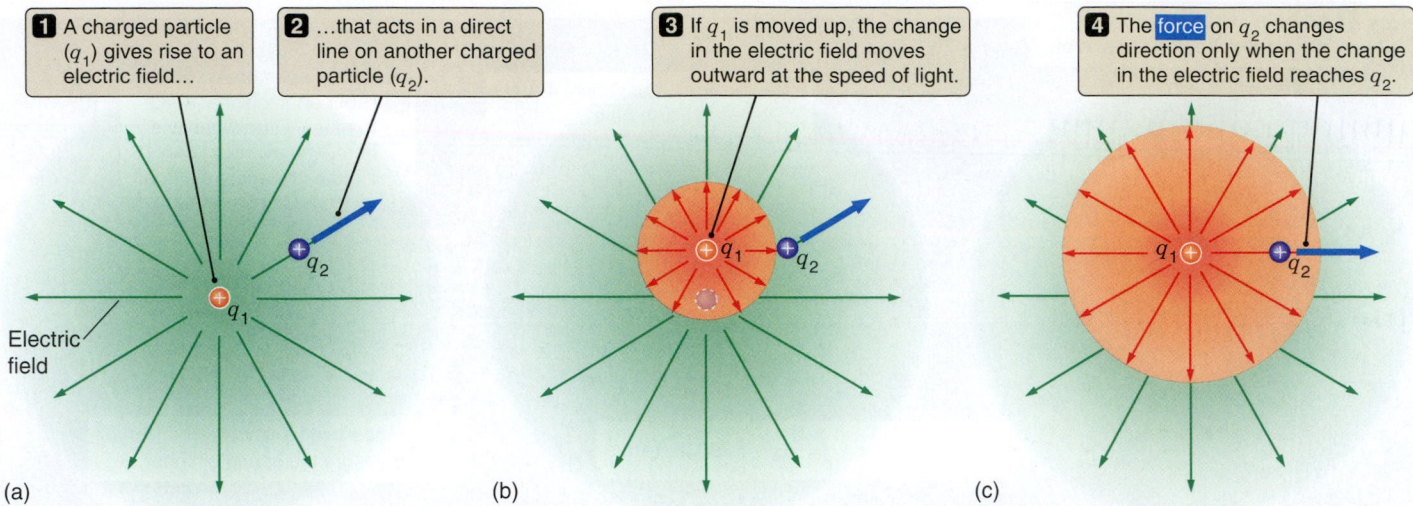

1 A charged particle (q_1) gives rise to an electric field…

2 …that acts in a direct line on another charged particle (q_2).

3 If q_1 is moved up, the change in the electric field moves outward at the speed of light.

4 The force on q_2 changes direction only when the change in the electric field reaches q_2.

Electric field

q_1

q_2

(a)

(b)

(c)

FIGURE 4.2 Maxwell's theory of electromagnetic radiation describes how electrically charged particles move and interact. (a) A positively charged particle, q_1, has an electric field (shown in orange) that acts outward along the field lines. (b) When q_1 accelerates, changes in the electric field move outward at the speed of light. For simplicity, the charge is shown moving instantly from one place to another, but in reality this could not happen. (c) Particle q_2 does not respond until it feels the change in q_1.

not affected by the first particle's movement until the wave reaches them.

Maxwell summarized the behavior of electric and magnetic fields in four elegant equations. Among other things, these equations say that a changing electric field causes a magnetic field, and that a changing magnetic field causes an electric field. These changes "feed" on themselves. A change in the motion of a charged particle causes a changing electric field, which causes a changing magnetic field, which causes a changing electric field, and so on. Once the process starts, a self-sustaining procession of oscillating electric and magnetic fields moves out in all directions through space. Instead of a purely electric wave, an accelerating charged particle gives rise to an **electromagnetic**

> Changing electric and magnetic fields lead to a self-sustaining electromagnetic wave.

wave (**Figure 4.3**). These electromagnetic waves, and the accelerating charges that generate them, are the sources of electromagnetic radiation. By contrast, an electric charge that is moving at a constant velocity is stationary in its inertial frame of reference and so does not radiate.

In addition to predicting that electromagnetic waves should exist, Maxwell's equations predict how rapidly the disturbance in the electric and magnetic fields should move. In short, Maxwell's equations predict the speed at which an electromagnetic wave should travel. When Maxwell carried out this calculation, he discovered that electromagnetic waves should travel at 3×10^8 m/s—which is the speed of light! This agreement could not be simple coincidence. Maxwell had shown that light is an electromagnetic wave.

Maxwell's wave description of light also gives us an idea of how light originates and how it interacts with *mat-*

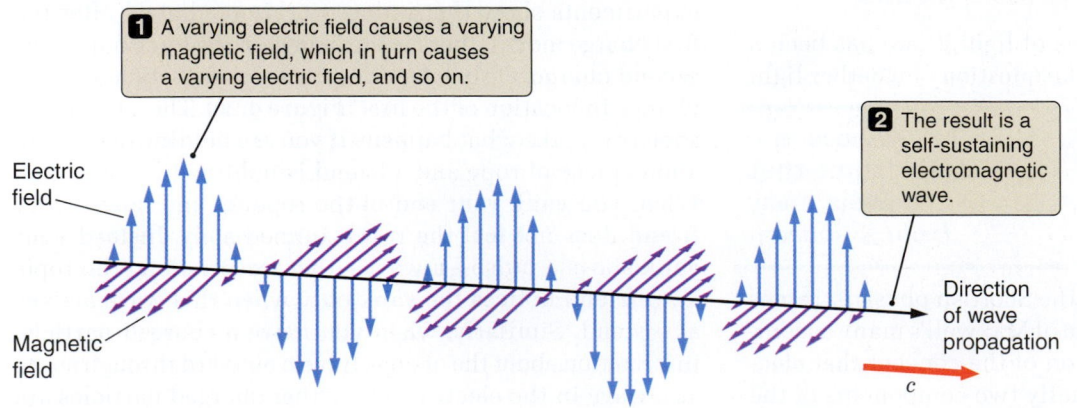

1 A varying electric field causes a varying magnetic field, which in turn causes a varying electric field, and so on.

2 The result is a self-sustaining electromagnetic wave.

Electric field

Magnetic field

Direction of wave propagation

c

FIGURE 4.3 Far from its source, an electromagnetic wave consists of oscillating electric and magnetic fields that are perpendicular both to each other and to the direction in which the wave travels.

(a)

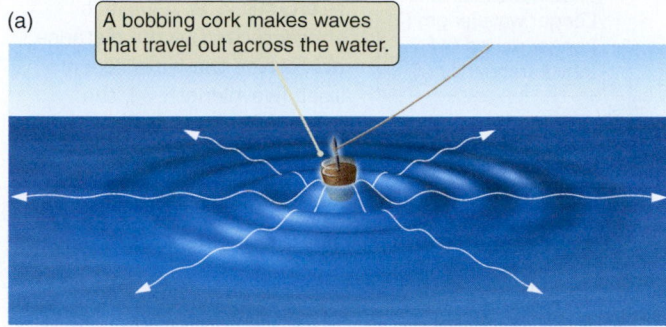

A bobbing cork makes waves that travel out across the water.

(b)

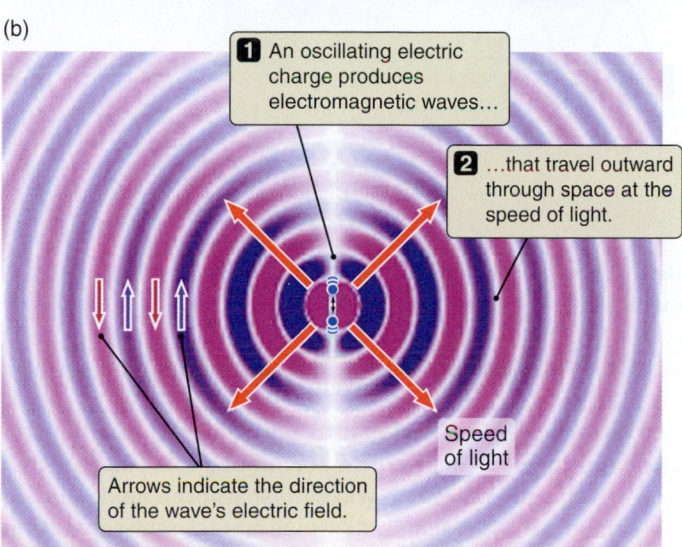

1 An oscillating electric charge produces electromagnetic waves…

2 …that travel outward through space at the speed of light.

Speed of light

Arrows indicate the direction of the wave's electric field.

VISUAL ANALOGY FIGURE 4.4 (a) A fish pulling downward on a hook connected to a fishing cork generates waves that move outward across the water's surface. (b) In similar fashion, an oscillating (accelerated) electric charge generates electromagnetic waves that move away at the speed of light.

(a)

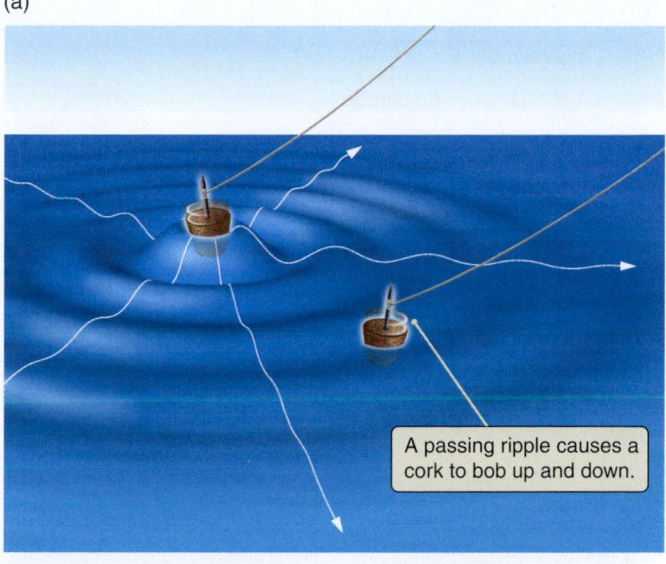

A passing ripple causes a cork to bob up and down.

(b)

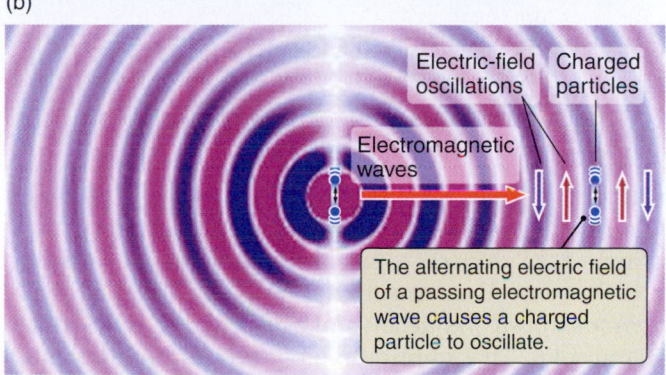

Electric-field oscillations   Charged particles

Electromagnetic waves

The alternating electric field of a passing electromagnetic wave causes a charged particle to oscillate.

VISUAL ANALOGY FIGURE 4.5 (a) When waves moving across the surface of water from one bobbing cork reach a nearby cork, they cause the second cork to bob up and down. (b) Similarly, a passing electromagnetic wave causes an electric charge to wiggle in response to the wave.

ter. Imagine a fishing cork floating on a lake on a perfectly calm day. The surface of the lake is as smooth and flat as a mirror until a fish tugs on the hook and line dangling beneath the cork. The motion of the cork causes a disturbance that moves outward as a ripple on the surface of the lake (**Figure 4.4a**). In much the same way, an oscillating (and therefore accelerating) electric charge (**Figure 4.4b**) causes a disturbance that moves outward through space as an electromagnetic wave. Note, however, that the ripples on the lake result from *mechanical* distortions of the surface of the lake. Mechanical waves such as these involve distortions that are measured as distances and media that have mass. Maxwell showed that light waves are a fundamentally different type of wave. Light waves result not from mechanical distortion of a medium, but from periodic changes in the strength of the electric and magnetic fields.

> Accelerating charges cause electromagnetic waves.

Now imagine that a second cork is afloat on the lake some distance from the first, as in **Figure 4.5a**. The second cork remains stationary until the ripple from the first cork reaches it. As the ripple passes by, the rising and falling of the water causes the second cork to rise and fall as well. Similarly, the oscillating electric field of an electromagnetic wave causes an oscillating force on any charged particle that the wave encounters, and this force causes the particle to move about as well (**Figure 4.5b**). It takes energy to produce an electromagnetic wave, and that energy is carried through space by the wave. Matter far from the source of the wave can absorb this energy. In this way, some of the energy lost by the particles generating the electromagnetic wave is transferred to other charged particles. The *emission* and *absorption* of light by matter are the result of the interaction of electric and magnetic fields with electrically charged particles.

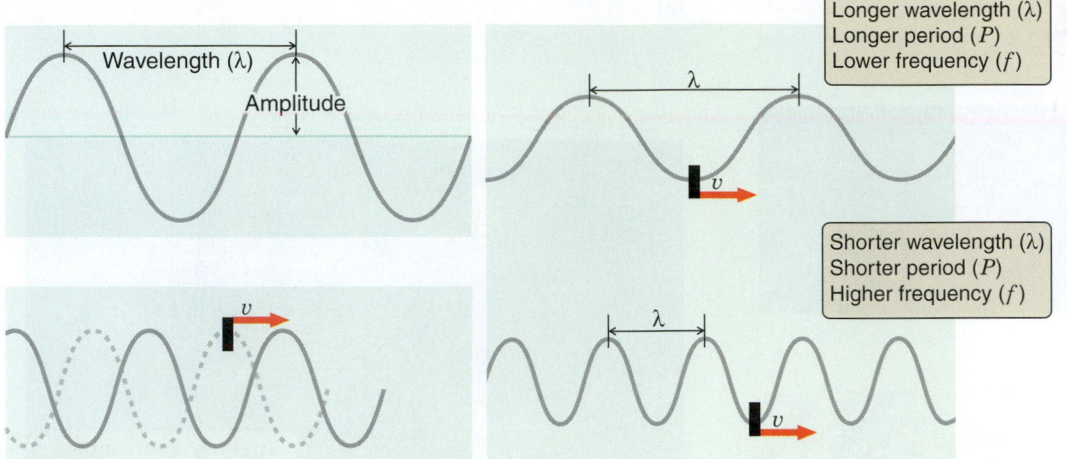

FIGURE 4.6 A wave is characterized by the distance over which the wave repeats itself (wavelength, λ), the maximum excursion from its undisturbed state (amplitude), and the speed (v) at which the wave pattern travels. In an electromagnetic wave, the amplitude is the maximum strength of the electric field, and the speed of light is written as c.

Waves Are Characterized by Wavelength, Speed, Frequency, and Amplitude

Along our journey we will encounter waves of different kinds, ranging from electromagnetic waves crossing the vast expanse of the universe to seismic waves traveling through Earth. We generally characterize waves by four quantities—*amplitude*, *speed*, *frequency*, and *wavelength*—as illustrated in **Figure 4.6**. The **amplitude** of a wave is the maximum deviation from its undisturbed or relaxed position. (In the case of light waves, the amplitude is an indication of the intensity or brightness of the radiation.) Waves travel at a particular speed, which is usually written as v. (In the case of light waves, we have already shown that the speed in a vacuum is 300,000 km/s and is written as c.) The number of wave crests passing a point in space each second is called the wave's **frequency**, denoted by f. The unit of frequency is cycles per second, which is generally referred to as **hertz** (abbreviated **Hz**) after the 19th century physicist Heinrich Hertz (1857–1894), who was the first to experimentally confirm Maxwell's predictions about **electromagnetic radiation**. The time taken for one complete cycle is called the **period**, P, which is measured in seconds.

The distance that a wave travels during one complete oscillation is called the **wavelength**. This is just the distance from one wave crest to the next, or the distance from one wave trough to the next. The wavelength is usually denoted by the Greek letter λ (pronounced "lambda"). There is a clear relationship between the frequency of a wave and its period. If the period of a wave is ½ second—that is, if it takes ½ second for one wave to pass by, crest to crest—then two waves will go by in 1 second. So a wave with a period of ½ second per cycle has a frequency of 2 cycles per second (or 2 Hz). Similarly, if a wave has a period of 1/100 second per cycle, then 100 waves will pass by each second. This wave

has a frequency of 100 Hz. More generally, the frequency of a wave is just 1 divided by its period:

$$\text{Frequency} = \frac{1}{\text{Period}} \quad \text{or} \quad f = \frac{1}{P}.$$

There is also a relationship between the period of a wave and its wavelength. The period of a wave is the time between the arrival of one wave crest and the next. During this time the wave travels a distance equal to the separation between the two wave crests, or one wavelength. So far, so good. Now add to the picture the fact that distance traveled equals speed multiplied by time taken. Change "distance traveled" to one wavelength and change "time taken" to one period, and we find that the wavelength of a wave equals the speed at which the wave is traveling multiplied by the period of the wave:

$$\text{Wavelength} = \text{Speed} \times \text{Period}.$$

Physicists use the letter c to represent the speed of light, so we can say that

$$\lambda = c \times P.$$

Using the relationship between period and frequency just given, we can also write

$$\text{Wavelength} = \frac{\text{Speed}}{\text{Frequency}} \quad \text{or} \quad \lambda = \frac{c}{f}.$$

So if we know the speed of a wave, then knowing one of the three properties—its wavelength, period, or frequency—tells us the other two (see **Math Tools 4.1**.)

Look at this relationship more closely. The longer the length of a wave, the longer you have to wait between wave crests, so the frequency of the wave will be lower. A shorter wavelength means less distance between wave crests, which means a shorter wait until the next wave comes along.

Working with Electromagnetic Radiation

Most of us listen to radios, both AM and FM. When you tune in to a station at, say, "770 on your AM dial," you are receiving a signal that is broadcast at a frequency of 770 kilohertz (kHz), or 7.7×10^5 Hz. We can use the relationship between wavelength and frequency to calculate the wavelength of the AM signal:

$$\lambda = \frac{c}{f} = \left(\frac{3 \times 10^8 \, \text{m/s}}{7.7 \times 10^5/\text{s}} \right) = 390 \, \text{m}.$$

This AM wavelength is about four times the length of a football field!

We can compare this wavelength with that of a typical FM broadcast signal, "99.5 on your FM dial," or 99.5 megahertz (MHz), or 9.95×10^7 Hz:

$$\lambda = \frac{c}{f} = \left(\frac{3 \times 10^8 \, \text{m/s}}{9.95 \times 10^7/\text{s}} \right) = 3 \, \text{m}.$$

This FM wavelength is about 10 feet. As you can see, FM wavelengths are much shorter than AM wavelengths.

The human eye is most sensitive to light in the green to yellow part of the spectrum. This light has a wavelength of about 500–550 *nanometers* (nm). Green light with a wavelength of 520 nm has a frequency of

$$f = \frac{c}{\lambda} = \left(\frac{3.00 \times 10^8 \, \text{m/s}}{5.20 \times 10^{-9}/\text{m}} \right) = \frac{5.80 \times 10^{14}}{\text{s}}.$$

or

$$5.80 \times 10^{14} \, \text{Hz}.$$

This frequency corresponds to 580 *trillion* wave crests passing by each second!

Therefore, a shorter wavelength means a higher frequency. A tremendous amount of information can be carried by waves—intelligible speech, for example, or complex and beautiful music. As we continue our study of the universe, time and again we will find that the information we receive, whether about the interior of Earth or about a distant star or galaxy, rides in on a wave.

> A long wavelength means low frequency, and a short wavelength means high frequency.

A Wide Range of Wavelengths Makes Up the Electromagnetic Spectrum

You have almost certainly seen a rainbow like the one in **Figure 4.7** spread out across the sky, or sunlight split into many different colors by a prism. This sorting of light by colors is really a sorting by wavelength. When we talk about light spread out according to wavelength, we refer to the **spectrum** of the light. On the long-wavelength (and therefore low-frequency) end of the visible spectrum is red light. At the other end of the visible spectrum is violet light, which is the bluest of blue light. A **micrometer**, or **micron**, is the unit often used for measuring the wavelength of electromagnetic radiation. A micrometer is one-millionth (10^{-6}) of a meter. The abbreviation used for a micrometer is **μm** (where

> The spectrum of visible light is seen as the colors of the rainbow.

μ is the Greek letter mu). Another commonly used unit is the **nanometer**, abbreviated **nm**. A nanometer is one-billionth (10^{-9}) of a meter.[1] The wavelengths of the light we perceive as red fall between about 600 and 700 nm. The shortest-wavelength violet light that our eyes can see has a wavelength of about 350 nm. Stretched out between the two, literally in a rainbow, is the rest of the visible spectrum (see Figure 4.7.) The colors in the visible spectrum, in order of decreasing wavelength, can be remembered as a name: "Roy G. Biv," which stands for

Red Orange Yellow Green Blue Indigo Violet.

When we say *visible light*, what we mean is "the light that the light-sensitive cells in our eyes respond to." But this is not the whole range of possible wavelengths for electromagnetic radiation. Radiation can have wavelengths that are much shorter or much longer than our eyes can perceive. The whole range of different wavelengths of light is collectively referred to as the **electromagnetic spectrum**.

> Visible light is only one small segment of the electromagnetic spectrum.

Follow along in **Figure 4.8** as we take a tour of the electromagnetic spectrum, beginning with visible light and work-

---

[1] Astronomers conventionally use nanometers (nm) when referring to wavelengths at visible and shorter wavelengths, micrometers or "microns" (μm) for wavelengths in the infrared, and millimeters (mm), centimeters (cm), and meters (m) for wavelengths in the microwave and radio regions of the electromagnetic spectrum.

FIGURE 4.7 The visible part of the electromagnetic spectrum is laid out in all its glory in the colors of this rainbow.

ing our way next to the part of the spectrum with shorter wavelengths and then to longer wavelengths. We start at the blue end of the visible spectrum. Beyond this short-wavelength, high-frequency side of the visible spectrum, there is light that is "bluer than blue" or actually "more violet than violet." This light, with wavelengths between 40 and 350 nm, is called **ultraviolet (UV) radiation**. You can remember what ultraviolet light is just by looking at the name. The prefix *ultra-* means "extreme," so *ultra*violet light is light that is more "extremely" violet than violet. It is important to remember that ultraviolet light is fundamentally no different from visible light, any more than high C on a piano is fundamentally different from middle C.

As we go to still shorter wavelengths of light (and so to higher frequencies), we pass through the ultraviolet part of the spectrum. At a wavelength shorter than 40 nm, or 4×10^{-8} meter, we stop calling radiation ultraviolet light and instead start calling it **X-rays**. This distinction arose for historical reasons. When X-rays were discovered in the late 19th century, they were given the name "X" by their discoverer, German physicist Wilhelm Conrad Roentgen (1845–1923), to indicate they were "a new kind of ray." As we continue to even shorter wavelengths, we come to another somewhat arbitrary break. Electromagnetic radiation with the very shortest wavelengths (less than about 10^{-10} meter) is referred to as *gamma radiation*, or more commonly as **gamma rays**. Again, the reasons are historical. Gamma rays (or γ-rays) were first discovered as a type of radiation given off by radioactive material. Only later did their true nature become known.

So far, we have been considering ever-shorter wavelengths and higher frequencies. In principle, there is no

FIGURE 4.8 By convention, the electromagnetic spectrum is broken into loosely defined regions ranging from gamma rays to radio waves.

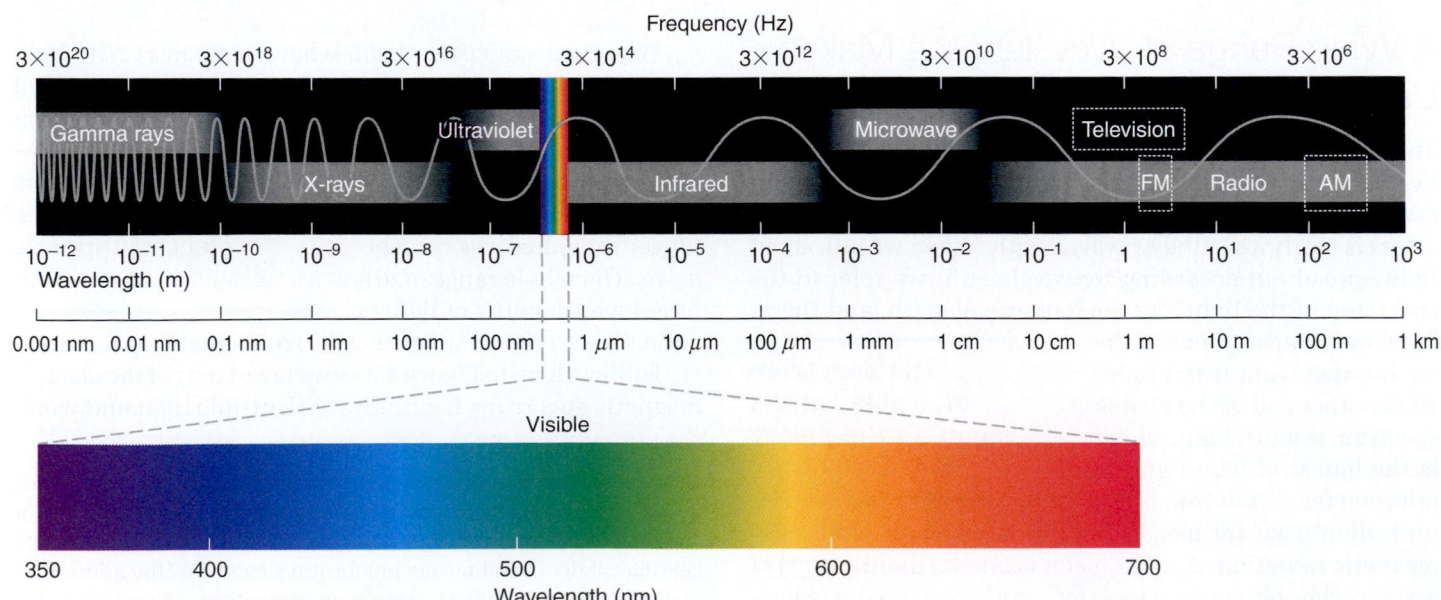

limit to this process. We can conceive of gamma rays of arbitrarily short wavelengths and arbitrarily high frequencies (even though practical considerations eventually come into play). We can also go in the other direction. Just as there is light that is more violet than violet, there is light that is "redder than red." Such light, covering wavelengths longer than about 700 nm and shorter than 500 μm (5×10^{-4} meter), is referred to as **infrared (IR) radiation**. Again, the key to remembering what infrared light is comes from looking at the word itself. *Infra-* is a prefix that means "below." *Infra*red light is light that has a frequency lower than (below) that of red light. When the wavelength of light gets still longer than this, we start calling it **microwave radiation**. The longest-wavelength (and therefore lowest-frequency) electromagnetic radiation, with wavelengths longer than a few centimeters and ranging up to arbitrarily long wavelengths, is called **radio waves**. In Chapter 5 we will discuss the various kinds of telescopes used by astronomers to capture and analyze the wide range of electromagnetic radiation.

Light Is a Wave, but It Is Also a Particle

By showing that light is an electromagnetic wave, Maxwell's work seemed to put to rest the issue of whether light consists of waves or particles. Although the electromagnetic wave theory of light has successfully described many phenomena, there are also many phenomena that it does not describe well. These range from the presence of sharp bright and dark lines, called "spectral lines," at specific wavelengths in the light from some objects, to the shape of the continuous spectrum of light emitted by a lightbulb.

Many of these difficulties with the wave model of light have to do with the way in which light interacts with atoms and molecules. Scientists working in the late 19th and early 20th centuries discovered that many of the puzzling aspects of light could be better understood if light energy came in *discrete packages*. In 1905, Albert Einstein published a paper in which he argued that light consists of particles. He based his argument on the **photoelectric effect**, the emission of electrons from surfaces when the surfaces are illuminated by electromagnetic radiation greater than a certain frequency. Einstein showed that the *rate* at which electrons are ejected depends only on the *intensity* of the incident radiation, and that the electron *velocity* depends only on the *frequency* of the incident radiation.[2] Effectively, scientists were modifying our understanding of light to show that, although it can sometimes be explained as acting like a wave, light can also be explained as acting like a particle. In this model we think about light as being made up of particles called **photons**

(*phot-* means "light," as in *photograph*; and *-on* signifies a particle). As massless particles, photons always travel at the speed of light, and they carry energy.

Recognition of the particle theory of light, however, did not mean that the wave theory had been discarded. The particle description of light is tied to the wave description of light by a relationship between the energy of a photon and the frequency or wavelength of the wave. Specifically, we write

> **The energy of a photon is proportional to its frequency.**

$$E = hf \quad \text{or} \quad E = \frac{hc}{\lambda}.$$

The *h* in this equation is called **Planck's constant** and has the value 6.63×10^{-34} joule-second. (Planck's constant is named after the German physicist Max Planck, 1858–1947.) According to the particle description of light, the electromagnetic spectrum is a spectrum of photon energies. The higher the frequency of the electromagnetic wave, the greater the energy carried by each photon. Photons of shorter wavelength (higher frequency) carry more energy than do photons of longer wavelength (lower frequency). For example, photons of blue light carry more energy than do photons of longer-wavelength red light. Ultraviolet photons carry more energy than do photons of visible light, and X-ray photons carry more energy than do ultraviolet photons. The lowest-energy photons are radio wave photons.

The **intensity** of light measures the *total* amount of energy that a beam of the light carries. A beam of red light can be just as intense as a beam of blue light—that is, it can carry just as much energy—but because the energy of a red photon is less than the energy of a blue photon, it takes more red photons to reach that intensity than it takes blue photons. This relationship is a lot like money. A hundred dollars is a hundred dollars, but it takes a lot more pennies (low-energy photons) than quarters (high-energy photons) to make up a hundred dollars.

When physicists speak of the energy of light as broken into discrete packets called photons, they say that the light energy is **quantized**. The word *quantized*, which has the same root as the word *quantity*, means that something is subdivided into discrete units. A

> **Photons are the *quantum mechanical* description of light.**

photon is referred to as a **quantum of light**. The branch of physics that deals with the quantization of energy and of other properties of matter is called **quantum mechanics**.

Although the predictions of quantum mechanics have been confirmed over and over again by experiment, its fundamental assumptions seem counterintuitive. The wave-particle description of light conflicts with everyday, commonsense ideas about the world. It is hard for us to imagine a single thing sharing the properties of a wave on the ocean *and* a beach ball, yet light does just that. You may find yourself scratching your head in confusion over exactly

---

[2]Interestingly, it was Einstein's work on the photoelectric effect, not special or general relativity, that earned him the Nobel Prize in 1921.

what light really is. If you do, consider yourself in good company. The scientists who developed the seemingly bizarre quantum description of nature had a great deal of trouble thinking about light as well. The wave model of light is clearly the correct description to use in many instances, just as Maxwell showed. At the same time, the particle description of light is also clearly the correct description to use in other cases, as scientists like Planck and Einstein demonstrated. But how can the same thing—light—be described as both a wave *and* a particle?

Light is what light is. The trouble with quantum mechanics lies not with the nature of reality, but with what our brains are used to thinking about. The brains of our ancestors did not have to deal with things moving at nearly the speed of light. Our brains deal well with objects that are sitting still or even moving as fast as a hard-hit fly ball. Basically, our brains cope best with things moving at the speeds of things in nature that we might want to eat or that might want to eat us.

Our trouble with thinking of light as both a wave and a particle only hints at the puzzling and philosophically troublesome world of quantum mechanics. In the next section we will learn that light is not the only thing that shares wave and particle properties. In fact, *all* matter shares wave and particle properties. Sometimes a "particle" such as an electron behaves as if it were a wave, while at other times a "wave" of light clearly exhibits the properties of a discrete particle. Early quantum physicists would sometimes joke that on Monday, Wednesday, and Friday, light and matter were particles, but on Tuesday, Thursday, and Saturday, they were waves. (And on Sunday, it was best just not to think about them at all!)

> Like light, all matter can behave as waves and particles.

It's true that our brains can be bounded by a conceptual box that is defined by the experiences and circumstances that we and our ancestors had to cope with. But they don't have to remain trapped inside that box. One exciting thing about modern physics and astronomy is that they enable us to break down the walls of that conceptual box by giving us the tools for understanding what lies beyond its boundaries. ▶‖ **AstroTour:** Light as a Wave, Light as a Photon

4.3 The Quantum View of Matter

If we want to understand better how light interacts with matter, we need to start by pinning down exactly what we mean by *matter* in the first place. To a physicist, **matter** is anything that occupies space and has mass. Virtually *all* of the matter we have direct experience with is composed of **atoms**. An atom is the smallest piece of any chemical element that retains the properties of that element. This book is made of atoms, and the neurons in your brain that are changing their structure as you read are made of atoms. Atoms are incredibly tiny—so tiny that a single teaspoon of water contains about 10^{23} of them. (There are as many atoms in a single teaspoon of water as there are stars in the observable universe.) When we talk about the interaction of light with matter, what we are really talking about is the interaction of light with atoms and their components.

> Virtually all matter we encounter is composed of atoms.

Atoms are made up of three types of subatomic particles: protons, neutrons, and electrons (**Figure 4.9a**). At the center of the atom is the **nucleus**, which is composed of positively charged **protons** and electrically neutral **neutrons**. An atom may have many protons and neutrons in its nucleus. Surrounding the nucleus of the atom are negatively charged **electrons**. For an atom to be electrically neutral, it must have the same number of electrons as protons. Electrons have much less mass than protons or neutrons have, so almost all the mass of an atom is found in its nucleus. This description naturally leads to a mental picture of an atom as a tiny "solar system," with the massive nucleus sitting in the center and the smaller electrons orbiting about it, much as planets orbit about the Sun (**Figure 4.9b**). We refer to this as the **Bohr model**, after the Danish physicist Niels Bohr (1885–1962), who proposed it in 1913. ▶‖ **AstroTour: Atomic Energy Levels and the Bohr Model**

Unless you have thought about atoms a great deal, the Bohr model probably matches your concept of the structure of an atom. It is much the same picture that scientists in the early 20th century held as well. But it has a fatal problem. In this view, an electron whizzing about in an atom is constantly undergoing an acceleration—the *direction* of its motion is constantly changing. The wave description of electromagnetic radiation says that *any* electrically charged particle that is accelerating must also be giving off electromagnetic radiation. This electromagnetic radiation, in the form of photons, should be carrying away the orbital energy of the electron. (Imagine that electron as the wiggling electric charge in Figure 4.4b.) If you calculate how much energy should be radiated away from the electron, you find that only a tiny fraction of a second should be needed for the electrons in an atom to lose all their energy and fall into the atom's nucleus! Fortunately for us, this does not happen. Atoms exist for very long periods of time, and electrons never "fall into" the nuclei of atoms. So something must be wrong with this concept of an atom.

A way out of this difficulty came when scientists realized that, just as waves of light have particle-like properties, so, too, do particles of matter have wavelike properties. With this realization, the "miniature solar system" model of the atom was modified so that a positively charged nucleus is surrounded *not* by planetlike electrons moving in well-behaved orbits, but by electron "clouds" or electron "waves" (**Figure 4.9c**). This is why we use a featureless cloud to represent electrons in orbit around an atomic nucleus. According to the Bohr model of the atom, the *angular momentum*

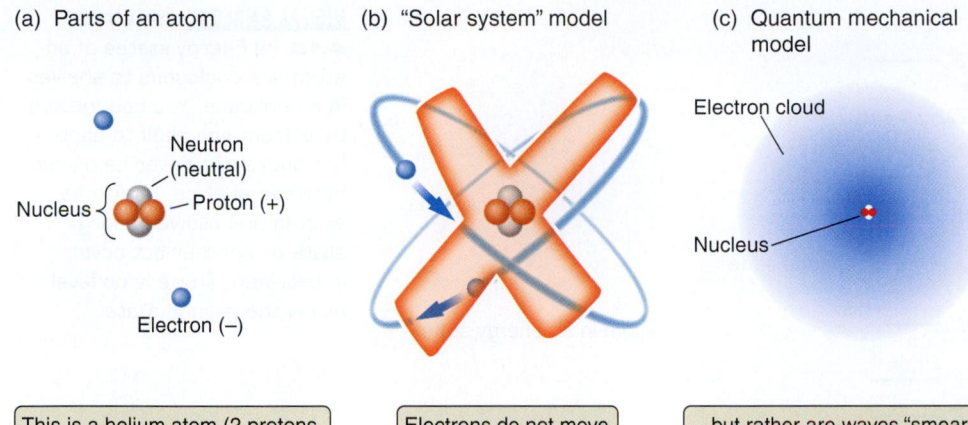

(a) Parts of an atom

Nucleus {
Neutron (neutral)
Proton (+)

Electron (–)

(b) "Solar system" model

(c) Quantum mechanical model

Electron cloud

Nucleus

This is a helium atom (2 protons, 2 neutrons, and 2 electrons).

Electrons do not move in orbits like planets…

…but rather are waves "smeared out" in a cloud of probability held in place by the electric attraction of the nucleus.

FIGURE 4.9 (a) An atom (in this case, helium) is made up of a nucleus consisting of positively charged protons and electrically neutral neutrons and surrounded by less massive negatively charged electrons. (b) Atoms are often drawn as miniature "solar systems," but this model is incorrect, as the red *X* indicates. (c) Electrons are actually smeared out around the nucleus in quantum mechanical clouds of probability.

of the electron in an orbit has to be given *exactly* by an integer multiplied by a constant. There is no room for any uncertainty. In the quantum model, however, we cannot know precisely where the electron is in its orbit. This uncertainty about the electron's location is expressed by the famous **Heisenberg uncertainty principle**, named for the German physicist Werner Heisenberg (1901–1976). Because particles have wave characteristics, we cannot simultaneously pin down both their exact location and exact **momentum** (*p*), defined as the product of mass and velocity ($p = m \times v$). There will *always* be some uncertainty about either the particle's position or its momentum.

> The uncertainty principle tells us we cannot know both the precise momentum and position of a particle.

The bottom line here is that we can be absolutely *certain* that there is *uncertainty* at the root of everything physical. If you are bothered by this concept, you are in good company. It *really* bothered Einstein too. Keep in mind that in the world of the very small, nothing seems intuitive. We have learned that all electromagnetic radiation behaves as both waves and particles. Possibly more surprising was the discovery that things like electrons and protons, which you may have visualized as "solid" particles, also have wave characteristics. There is, of course, a nice symmetry here. Waves have particle-like characteristics and particles have wavelike characteristics. This is not just a curious observation. It has huge implications in both science and technology. For instance, the wave-particle property is the principle by which electron microscopes work, as you will see in the next chapter.

Remember that *velocity* includes both *speed* and *direction*. We can be more quantitative about this. The product of the uncertainty in a particle's position (Δx) and the uncertainty in its momentum (Δp) is always equal to or greater than a particular constant, which is on the order of Planck's constant, *h*. We can express this as a simple equation: $\Delta x \times \Delta p \approx h$. In other words, the more you know about

where something is (Δx approaching zero), the less you can know about *how fast* and *in what direction* it is moving (Δp approaching infinity). Conversely, the better you know the momentum of something, the less you know about its location. This is not a matter of scientists making inferior measurements. You simply *cannot* do better, no matter how precisely you measure!

Another example involves a different form of the Heisenberg uncertainty principle: $\Delta E \times \Delta t \approx h$, where ΔE is the uncertainty of energy and Δt is the time over which the energy is measured. In the next subsection we make the point that atoms can occupy only certain discrete energy states, implying that their electrons are restricted to specific well-defined *excited states*. But can there really be *no* uncertainty whatsoever in these energy states? The answer becomes apparent if we rewrite the previous equation as $\Delta t \approx h/\Delta E$. If we say that $\Delta E = 0$ (no uncertainty in the energy state), then Δt becomes infinite.[3] If an electron were forced to stay in a particular excited state forever, it could never drop to a lower level, and we would never see narrow *spectral emission (bright) lines*. We do, of course, see these bright lines, so there *must* be a certain amount of uncertainty in the electron's energy. Remember that wavelength is related to energy ($E = hc/\lambda$). A mathematical operation using calculus, which is beyond the level of this book, tells us that $\Delta E \propto \Delta \lambda$. A narrow range in energy therefore represents a narrow range in wavelength. So the longer an electron resides in an elevated energy state (Δt is large), the narrower will be the spectral emission line when it finally drops to a lower level (ΔE, and therefore $\Delta \lambda$, is small).

Atoms Have Discrete Energy Levels

Another aspect of the wave-particle nature of electrons is that electron waves in an atom can assume only certain specific forms. (Analogously, the strings on a guitar can

---

[3]If you divide *any* number by zero, the result will be infinite.

(a)

(b)

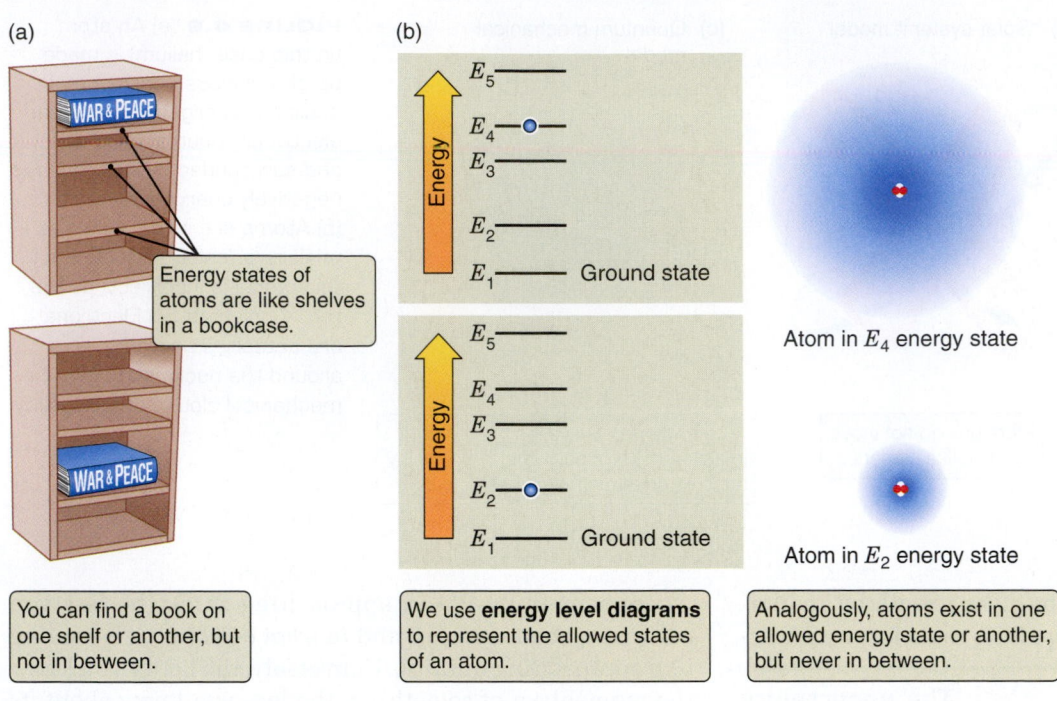

Energy states of atoms are like shelves in a bookcase.

You can find a book on one shelf or another, but not in between.

We use **energy level diagrams** to represent the allowed states of an atom.

Atom in E_4 energy state

Atom in E_2 energy state

Analogously, atoms exist in one allowed energy state or another, but never in between.

VISUAL ANALOGY FIGURE

4.10 (a) Energy states of an atom are analogous to shelves in a bookcase. You can move a book from one shelf to another, but books can never be placed between shelves. (b) Atoms exist in one allowed energy state or another but never in between. There is no level below the ground state.

vibrate only at certain discrete frequencies, giving rise to the distinct notes we hear.) The form that the electron waves take depends on the possible energy states of atoms. We can imagine the energy states of atoms as being a bookcase with a series of shelves, as shown in **Figure 4.10a**. The energy of an atom might correspond to the energy of one shelf or to the energy of the next shelf; but the energy of the atom will *never* be found *between* the two shelves. A given atom may have a tremendous number of different energy states available to it, but these states are *discrete*. An atom might have the energy of one of these allowed states, or it might have the energy of the next allowed state, *but it cannot have an energy somewhere in between.*

> Atoms can have only certain discrete energies, much as guitar strings can play only certain notes.

The lowest possible energy state of an atom—the "floor"—is called the **ground state** of the atom (**Figure 4.10b**.) Allowed energy levels above the ground state are called **excited states** of the atom. When the atom is in its ground state, it has nowhere to go. An electron cannot "fall" into the nucleus because there is no allowed state there with less energy for it to occupy. It cannot move up to a higher-energy state without getting extra energy from somewhere. For this reason an atom will remain in its ground state forever, unless something happens to knock it into an excited state. A book sitting on the floor has nowhere left to fall, and it cannot jump to one of the higher shelves of its own accord.

An atom in an excited state is a very different matter, however. Just as a book on an upper shelf might fall to a lower shelf, an atom in an excited state might **decay** to a lower state by getting rid of some of its extra energy. An important difference between the atom and the book on the shelf, however, is that whereas a snapshot might catch the book falling between the two shelves, the atom will never be caught between two energy states. When the transition from one state to another occurs, the difference in energy between the two states must be carried off all at once. A common way for an atom to do this is to give off a photon. But not just any photon will do. The photon emitted by the atom must carry away exactly the amount of energy lost by that atom as it goes from the higher-energy state to the lower-energy state. In a similar fashion, atoms moving from a lower-energy state to a higher-energy state can absorb only certain specific energies. The specific energies that can be absorbed correspond to the allowed waveforms of their electron clouds.

The Energy Levels of an Atom Determine the Wavelengths of Light That It Can Emit and Absorb

To better understand the relationship between the energy levels of an atom and the radiation it can emit or absorb, imagine a hypothetical atom that has only two available energy states. Call the energy of the lower-energy state (the ground state) E_1 and the energy of the higher-energy state (the excited state) E_2. The energy levels of this atom can be represented in an energy level diagram like those in Figure 4.10, but with only two levels, as shown in **Figure 4.11a**.

To understand the process of *emission*, imagine that the atom begins in the upper state (E_2) and then spontaneously

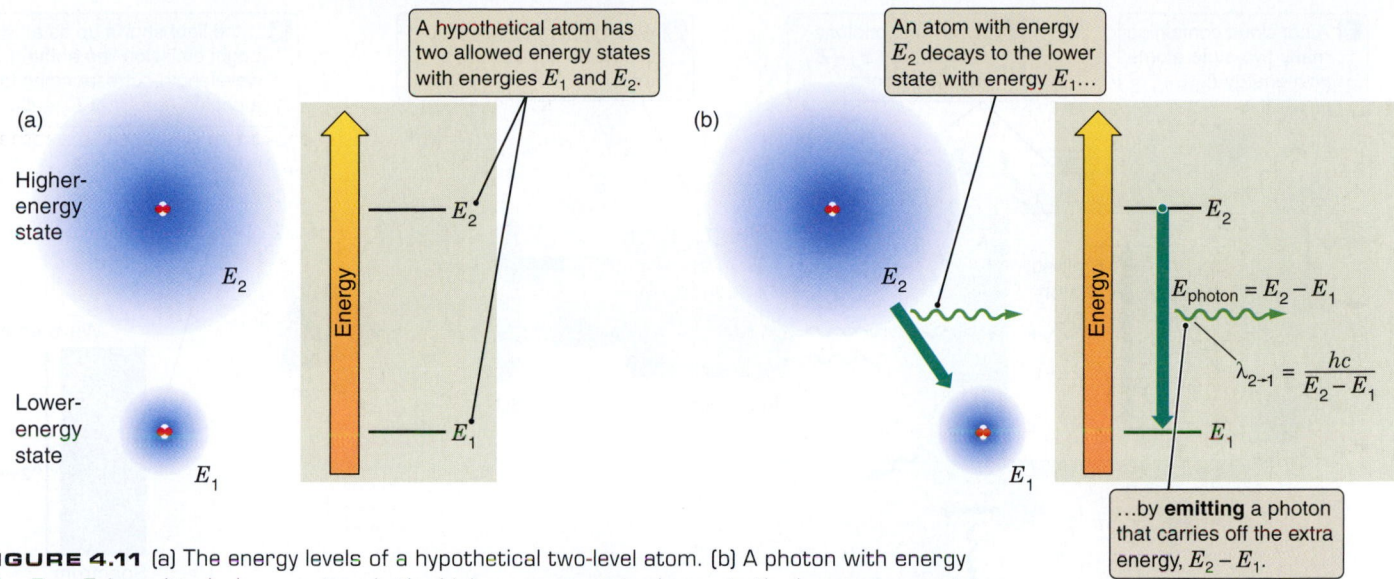

FIGURE 4.11 (a) The energy levels of a hypothetical two-level atom. (b) A photon with energy $hf = E_2 - E_1$ is emitted when an atom in the higher-energy state decays to the lower-energy state.

drops down to the lower-energy state (E_1). This change is illustrated in **Figure 4.11b**, where the downward arrow indicates that the atom went from the upper state to the lower state. The atom just lost an amount of energy equal to the difference between the two states, or $E_2 - E_1$. However, energy is never truly lost or created, so the energy lost by the atom has to show up somewhere. In this case the energy shows up in the form of a photon that is emitted by the atom. The energy of the photon emitted must exactly match the energy lost by the atom, so the energy of the photon must be $E_{\text{photon}} = E_2 - E_1$.

> When an atom drops to a lower energy state, the lost energy is carried away as a photon.

We have already seen how the energy of a photon is related to the frequency or wavelength of electromagnetic radiation. Using this relationship, we can say that the frequency of the photon emitted by a transition from E_2 to E_1, which we will denote as $f_{2\to1}$, is just the energy difference divided by Planck's constant (h):

$$f_{2\to1} = \frac{E_{\text{photon}}}{h} = \frac{E_2 - E_1}{h}.$$

Similarly, the wavelength of the photon is just $\lambda = c/f$, or

$$\lambda_{2\to1} = \frac{c}{f_{2\to1}} = \frac{hc}{E_2 - E_1}.$$

The energy level structure of the atom therefore determines the wavelengths of the photons emitted by an atom—the color of the light that the atom gives off. An atom can emit photons with energies corresponding only to the difference between two of its allowed energy states.

Imagine what the light coming from a cloud of gas consisting of our hypothetical two-state atoms would be like. This case is illustrated in **Figure 4.12**, which shows a cloud of gas made up of our two-state atoms. Any atom that found itself in the upper energy state (E_2) would quickly decay and emit a photon in a random direction, and an enormous number of photons would come pouring out of the cloud of gas. But instead of containing photons of all different energies (that is, light of all different colors, like sunlight), this light would instead contain only photons with the specific energy $E_2 - E_1$ and wavelength $\lambda_{2\to1}$. In other words, all of the light coming from the cloud would be the same color.

We have all seen what happens to sunlight when it passes through a prism. Sunlight contains photons of all different colors, so when it passes through a prism it spreads out into all colors of a rainbow. But if we were to pass the light from our cloud of gas through a slit and a prism, as in Figure 4.12, the results would be very different. This time there would be no rainbow. Instead, all of the light from the cloud of gas

> The spectrum of a cloud of glowing gas contains emission lines.

would show up on the screen as a single bright line. The process just described—the production of a photon when an atom decays to a lower-energy state—is referred to as **emission**. The bright, single-colored feature in the spectrum of the cloud of gas is referred to as an **emission line**.

Until now in this discussion, we have ignored an important question: How did the atom get to be in the excited state E_2 in the first place? An atom sitting in its ground state will remain in the ground state unless it is somehow given just the right amount of energy to kick it up to an excited state. Most of the time this extra energy comes in one of two possible forms: (1) the atom absorbs the energy of a photon (we will talk about this possibility shortly); or (2) the atom collides with another atom, or perhaps an unattached electron, and the collision knocks the atom into an excited state.

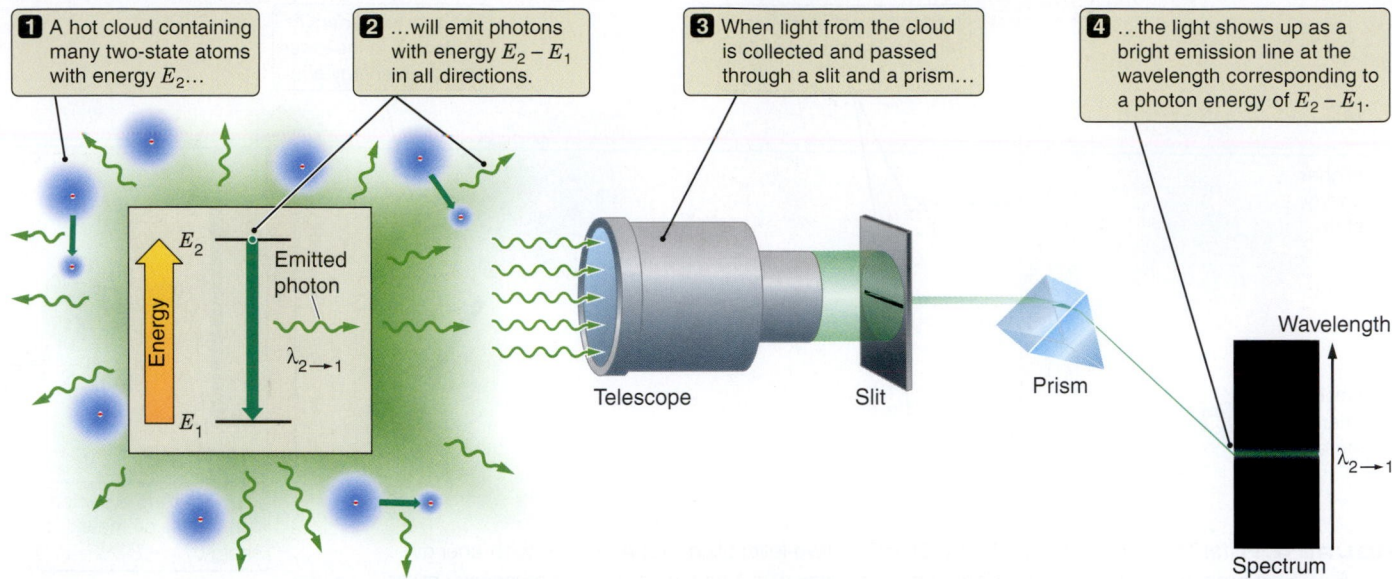

1 A hot cloud containing many two-state atoms with energy E_2...

2 ...will emit photons with energy $E_2 - E_1$ in all directions.

3 When light from the cloud is collected and passed through a slit and a prism...

4 ...the light shows up as a bright emission line at the wavelength corresponding to a photon energy of $E_2 - E_1$.

Telescope   Slit   Prism

Wavelength

Spectrum

$\lambda_{2 \to 1}$

FIGURE 4.12 A cloud of gas containing atoms with two energy states, E_1 and E_2, emits photons with an energy $E = hf = E_2 - E_1$, which appear in the spectrogram (right) as a single *emission line*.

This second possibility is how a neon sign works. When a neon sign is turned on, an alternating electric field is set up inside the glass tube that pushes electrons in the gas back and forth through the neon gas inside the tube. Some of these electrons crash into atoms of the gas, knocking them into excited states. The atoms then drop back down to their ground states by emitting photons, causing the gas inside the tube to glow.

So far, we have focused on the emission of photons by atoms in an excited state, but what about the opposite process? An atom in a low-energy state can absorb the energy of a passing photon and jump up to a higher-energy state as shown in **Figure 4.13**, but not just any photon can be absorbed

FIGURE 4.13 An atom in a lower-energy state may absorb a photon of energy $hf = E_2 - E_1$, leaving the atom in a higher-energy state.

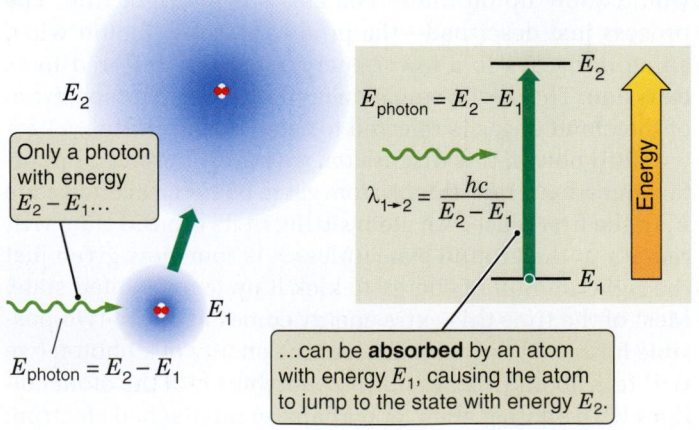

E_2

Only a photon with energy $E_2 - E_1$...

$E_{photon} = E_2 - E_1$

E_1

$E_{photon} = E_2 - E_1$

$\lambda_{1 \to 2} = \dfrac{hc}{E_2 - E_1}$

E_2

Energy

E_1

...can be **absorbed** by an atom with energy E_1, causing the atom to jump to the state with energy E_2.

by the atom. Once again, the energy that it takes to get from E_1 to E_2 is the difference in energy between the two states, or $E_2 - E_1$. For a photon to cause an atom to jump from E_1 to E_2, it must provide just this much energy. Using the relationship that $E_{photon} = hf$, or $f = E_{photon}/h$, we find that the *only* photons capable of exciting atoms from E_1 to E_2 are photons whose frequency and wavelength are, respectively,

$$f_{1 \to 2} = \frac{E_{photon}}{h} = \frac{E_2 - E_1}{h}$$

and

$$\lambda_{1 \to 2} = \frac{c}{f_{1 \to 2}} = \frac{hc}{E_2 - E_1}.$$

These photons have exactly the same energy—the same color of light—that is emitted by the atoms when they decay from E_2 to E_1. This is not a coincidence. The energy difference between the two levels is the same whether the atom is emitting a photon or absorbing one, so the energy of the photon involved will be the same in either case.

> Atoms can absorb photons that have the same energies as the photons they emit.

When we shine light from a lightbulb directly though a glass prism, a rainbow of colors comes out, as shown in **Figure 4.14a**. What might the spectrum of light look like when viewed through a cloud composed of our hypothetical gas of two-state atoms? If we shine photons of all different wavelengths (that is, light of all different colors) through the gas from one side, almost all of these photons will pass through the cloud of gas unscathed. There is only one exception. Rather than passing through the gas, some of the photons with just the right energy ($E_2 - E_1$) might instead be

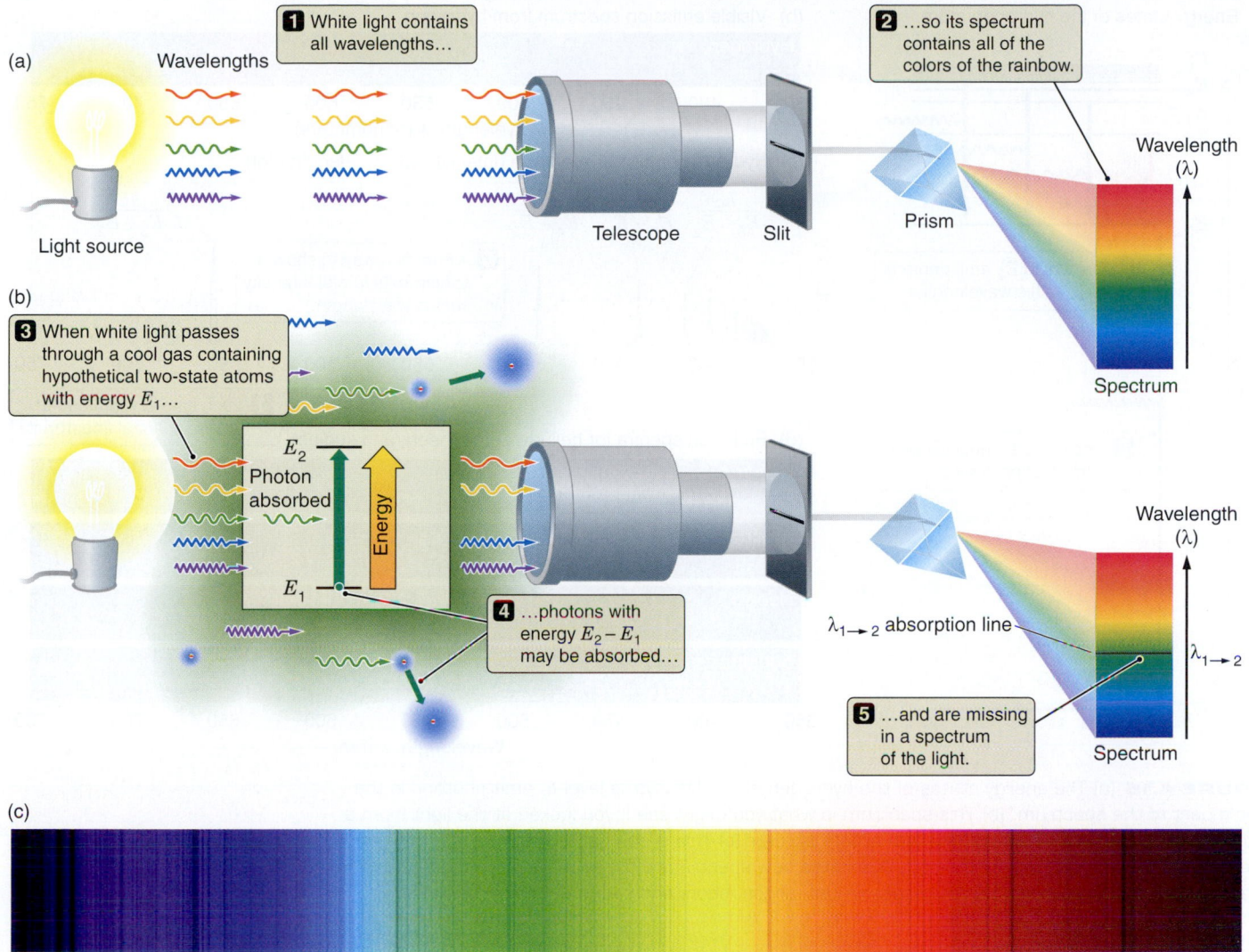

FIGURE 4.14 (a) When passed through a prism, white light produces a spectrum containing all colors. (b) When light of all colors passes through a cloud of hypothetical two-state atoms, photons with energy $hf = E_2 - E_1$ may be absorbed, leading to the dark absorption line in the spectrogram. (c) Absorption lines in the spectrum of a star.

absorbed by atoms. As a result, these photons will be missing in the light passing through the prism. Where the color corresponding to each of these missing photons should be, we will see a sharp, dark line instead (**Figure 4.14b**). The process of atoms capturing the energy of passing photons is referred to as **absorption**, and the dark feature seen in the spectrum is called an **absorption line**. **Figure 4.14c** shows such absorption lines in the spectrum of a star.

One final point is worth making before we leave the subject of emission and absorption of radiation. When an atom absorbs a photon and jumps up to an excited energy state, there is a good chance that the atom will quickly decay back down to the lower-energy state by emitting a photon

> When viewed through a cloud of gas, the spectrum of a lightbulb contains absorption lines.

that has the same energy as the photon it just absorbed. If the atom reemits a photon just like the one it absorbed, you might reasonably ask why the absorption really matters. After all, the photon that was taken out of the passing light was replaced, was it not? The answer is yes and no. The photon was replaced, true enough, but whereas all of the absorbed photons were originally traveling in the *same direction*, the photons that are reemitted travel off in *random directions*. In other words, some of the photons with energies equal to $E_2 - E_1$ are, in effect, diverted from their original paths by their interaction with atoms. If you look at a lightbulb *through* the cloud, you will notice an absorption line at a wavelength of $\lambda_{1\rightarrow2}$; but if you look at the cloud *from the side* (looking perpendicular to the original beam), you will see it as a glowing light with an emission line at this wavelength. ▶❚❚ **AstroTour:** Atomic Energy Levels and **Light Emission and Absorption**

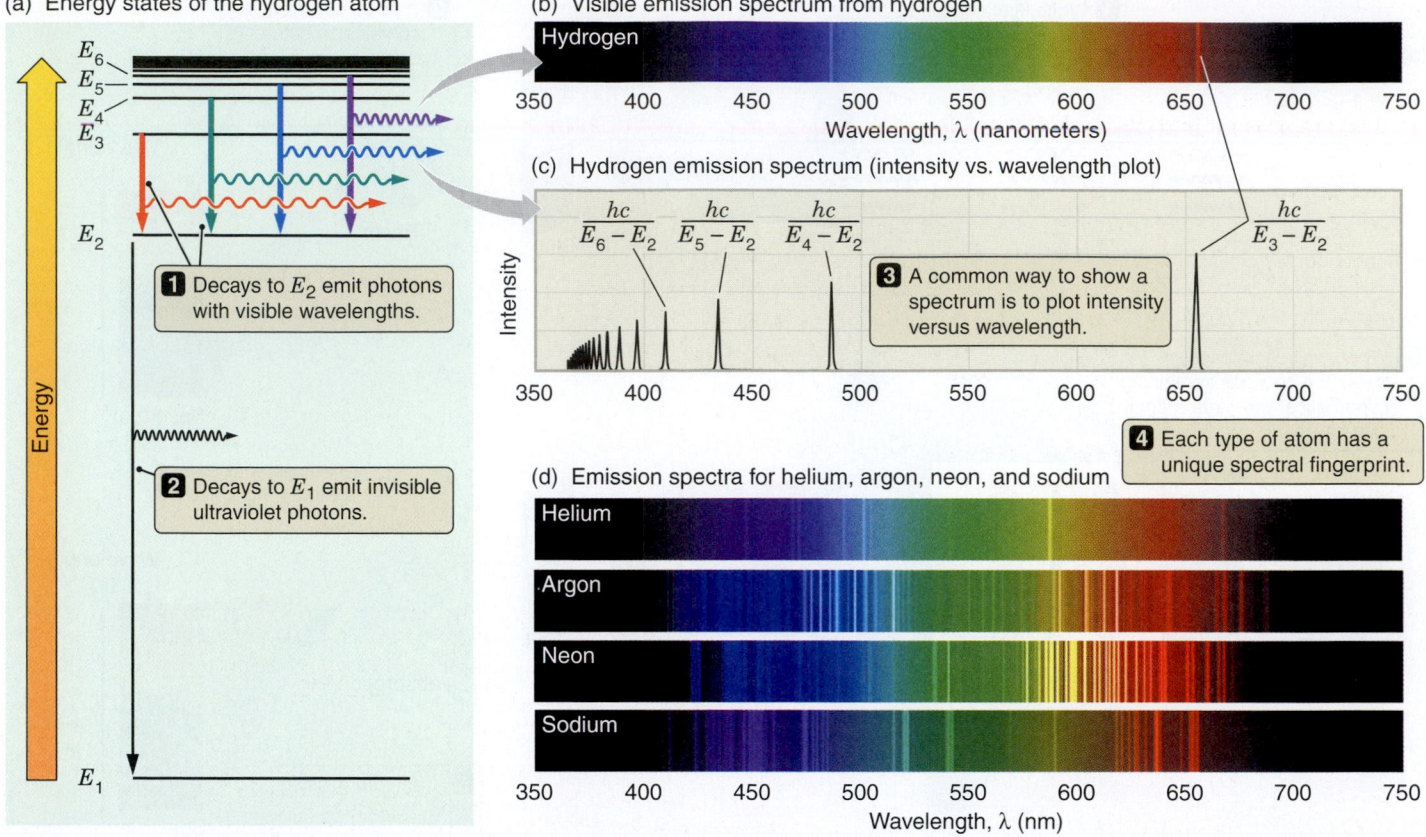

FIGURE 4.15 (a) The energy states of the hydrogen atom. Decays to level E_2 emit photons in the visible part of the spectrum. (b) This spectrum is what you might see if you looked at the light from a hydrogen lamp projected through a prism onto a screen. (c) This graph of the brightness (intensity) of spectral lines versus their wavelength illustrates how spectra are traditionally plotted. (d) Emission spectra from several other types of gases: helium, argon, neon, and sodium.

Emission and Absorption Lines Are the Spectral Fingerprints of Atoms

In the previous subsection we used a hypothetical atom with only two allowed energy states to help us think about the emission and absorption of photons. Real atoms have many more than just two possible energy states that they might occupy, so a given type of atom will be capable of emitting and absorbing photons at many different wavelengths. An atom with three energy states, for example, might jump from state 3 to state 2, or from state 3 to state 1, or from state 2 to state 1. The emission lines from such a gas might have wavelengths of $hc/(E_3 - E_2)$, $hc/(E_3 - E_1)$, and $hc/(E_2 - E_1)$, respectively.

The allowed energy states of an atom are determined by the complex quantum mechanical interactions among the electrons and the nucleus that make up the atom. Every hydrogen atom consists of a nucleus containing one proton, plus a single electron in a cloud surrounding the nucleus. As a result, every hydrogen atom has the same energy states available to it. It follows that all hydrogen atoms are capa-

ble of emitting and absorbing photons that have the same wavelengths. **Figure 4.15a** shows the energy level diagram of hydrogen, along with the spectrum of emission lines for hydrogen in the visible part of the spectrum (**Figures 4.15b and c**).

Every hydrogen atom has the *same* energy states available to it, so all hydrogen atoms are, in principle, capable of producing the same spectral lines. But the energy states of a hydrogen atom are *different* from the energy states available to a helium atom, a lithium atom, or a boron atom, just as the energy states of these kinds of atoms differ from each other. Each different type of atom has a unique set of available energy states and therefore a unique set of wavelengths at which it can emit or absorb radiation. **Figure 4.15d** shows the set of emission lines that are given off by discharge tubes (like those in a neon sign) containing different kinds of atoms. These unique sets of wave-

> The wavelengths at which atoms emit and absorb radiation form unique spectral fingerprints for each type of atom.

lengths serve as unmistakable spectral fingerprints for each type of atom.

Spectral fingerprints are of crucial importance to astronomers. They let us figure out what types of atoms (or molecules) are present in distant objects by doing nothing more than looking at the spectrum of light from those objects. If we see the spectral lines of hydrogen, helium, carbon, oxygen, or any other element in the light from a distant object, then we know that some of that element is present in that object. The strength of a line is determined in part by how many atoms of that type are present in the source. By measuring the strength of the lines from different types of atoms in the spectrum of a distant object, astronomers can often infer the relative amounts of different types of atoms that make up the object. But it gets even better. The fraction of atoms of a given kind that are in a particular energy state (as opposed to a different energy state) is often determined by factors such as the temperature or the **density** of the gas. By looking at the relative strength of different lines from the same kind of atom, it is often possible to determine the temperature, density, and pressure of the material as well.

How Are Atoms Excited, and Why Do They Decay?

Earlier in this discussion we sidestepped an aspect of the emission process that has troubled physicists and philosophers alike since the earliest days of quantum mechanics. To appreciate this question, return to the analogy between the emission of a photon and a book falling off a shelf. If we place a book on a level shelf and do not disturb it, the book will sit there forever. Once the book is resting on the upper shelf, something must *cause* the book to fall off the shelf. So what about the atom? Once an atom is in an excited state, what causes it to jump down to a lower-energy state and emit a photon? What triggers the event? Sometimes an atom in an upper-energy state can be "tickled" into emitting a photon—a process called "stimulated emission"—but under most circumstances the answer is that *nothing causes the atom to jump to the lower-energy state*. There is no trigger. Instead the atom decays *spontaneously*. And while we can say *about* how long the atom is *likely* to remain in the excited state, the rules of quantum mechanics say (and experiment shows) that we cannot know exactly when a given atom will decay until *after* the decay has happened. The atom decays at a *random* time that is not influenced by anything in the universe and cannot be known ahead of time.

You have seen many examples of this rather amazing phenomenon. For example, you may be familiar with toys that "glow in the dark." Photons in sunlight or from a lightbulb are absorbed by the certain "phosphorescent" atoms in the toy, knocking those atoms into excited energy states. Unlike many excited energy states of atoms that tend to decay in a small fraction of a second, the excited states of these atoms in the toy instead tend to live for many seconds before they decay. Suppose, for example, that on average these atoms tend to remain in their excited state for 1 minute before decaying and emitting a photon. In other words, suppose that if we wait 60 seconds, there is a 50-50 chance that any particular atom will have decayed and a 50-50 chance that the atom will remain in its excited state. There are trillions upon trillions of such atoms in the toy. Although it is impossible to say exactly which atoms will decay after a minute, we can say with certainty that *about* half of them *will* decay within 60 seconds. From the standpoint of the glow that we see coming from the toy, it makes little practical difference which half of the atoms decay and which half do not. All we need to know is that if we wait 1 minute, half of the atoms will have decayed, and the brightness of the glow from the toy will have dropped to half of what it was. If we wait another minute, half of the remaining excited atoms will decay, and the brightness of the glow will be cut in half again. Each 60 seconds, half of the remaining excited atoms decay, and the glow from the toy drops to half of what it was 60 seconds earlier. The glow from the toy slowly fades away.

We have now come upon one of the most philosophically troubling aspects of quantum mechanics. In deep space, where atoms can remain undisturbed for long periods of time, there are certain excited states of atoms that live, on average, for tens of million years or even longer. Envision an atom in such a state. It may have been in that excited state for a few seconds, a few hours, or 50 million years when, in an instant, it decays to the lower-level state *without anything causing it to do so*. Newton and virtually every other physicist who lived before the turn of the 20th century envisioned a clockwork universe in which every effect had a cause. They imagined that if we knew the exact properties of every bit of the universe today, it would be just a matter of turning the crank on the laws of physics to predict what the state of the universe will be tomorrow.

Then, quantum mechanics came along and turned this view on its head. Instead of dealing with strict cause-and-effect relationships, physicists found themselves calculating the *probabilities* of certain events taking place and facing fundamental limitations on what can ever be known about the state of the universe. We have mentioned that although Einstein helped start the scientific revolution of quantum mechanics, in the end that revolution left him behind. He could never shake his firm belief in Newton's clockwork, causal universe. "God does not play dice with the universe!" he insisted emphatically. As more of the predictions of quantum mechanics were borne out by experiment, most physicists came to accept the implications of the strange new theory. Einstein, however, went to his grave looking unsuccessfully for a way to save his notion of order in the universe.

> Quantum mechanics undermines the orderly, causal universe of Newtonian physics.

It is interesting to note that although Einstein refused to accept quantum mechanics, our understanding of the quantum mechanical nature of reality owes him a great debt. His was one of the greatest minds of all time, and as he searched tirelessly for flaws in quantum mechanics, he presented challenge after challenge to those who were trying to work out the details of the new theory. It was in responding to Einstein's objections that physicists were forced to confront the full implications of their own work. This struggle continues to this day. A few theoretical physicists are still pursuing Einstein's dream, trying to recast quantum mechanics in a way that recaptures the strict causality that seemed irretrievably lost shortly after the beginning of the 20th century. So far, they have had little success, and most physicists doubt that they ever will. Like Einstein's before them, however, their healthy skepticism has led to ever-deeper understanding of the implications and limitations of the theory of quantum mechanics.

4.4 The Doppler Effect— Is It Moving toward Us or Away from Us?

We have begun to see that to an astronomer, light is far more than just the stuff that bounces off the page and lets you read these words. Light is a tightly packed bundle of information that, when spread into its component wavelengths,

can reveal a wealth of information about the physical state of material located tremendous distances away. The nature of light had forced physicists to abandon many of their most cherished ideas about the nature of matter and energy. Yet we have only begun to explore what light can tell us. It is time to step back from the precipice of the philosophical implications of quantum mechanics and look instead at how light can be used to measure one of the most straightforward questions about a distant astronomical object: Is it moving away from us or toward us, and at what speed?

Have you ever stood outside and listened as a fire truck sped by with sirens blaring? If so, you might have noticed something funny about the way the siren sounded. As the fire truck came toward you, its siren had a certain high pitch; but as it passed by, the pitch of the siren dropped noticeably. If you were to close your eyes and listen, you would have no trouble knowing when the fire truck passed you, just from the change in pitch of its siren. You do not even need a fire engine to hear this effect. The sound of normal traffic behaves in the same way. As a car drives past, the pitch of the sound that it makes suddenly drops.

The pitch of a sound is like the color of light. It is determined by the wavelength or, equivalently, the frequency of the wave. What we perceive as higher pitch corresponds to sound waves with higher frequencies and shorter wavelengths. Sounds that we perceive as lower in pitch are waves with lower frequencies and longer wavelengths. When an object is moving toward us, the waves that it gives off "crowd together" in front of the object. You can see how this works by looking at **Figure 4.16**, which shows the locations of successive wave crests given off by a moving object. If you are standing in front of an object moving toward you, the waves that reach you have a shorter wavelength and therefore a higher frequency than the waves given off by the object when it is not moving. In the case of sound waves, the sound reaching you from the object has a higher pitch than the sound that would be given off by the object if it were stationary. Conversely, if an object is moving away from you, the waves reaching you from the object are spread out. (Again, refer to Figure 4.16.) In the case of sound, this means that the pitch of the sound drops, in line with our experience with the fire truck.

> The motion of a source toward or away from us changes the wavelength of the waves reaching us.

▶❚❚ **AstroTour:** The Doppler Effect

This same phenomenon, which is referred to as the **Doppler effect** and named after Austrian physicist Christian Doppler (1803–1853),[4] occurs with light as well. The wavelength of the light as measured from the source's frame of reference is called the **rest wavelength** (λ_0), as shown in **Figure 4.17a**. If an object is moving toward you, the light

---

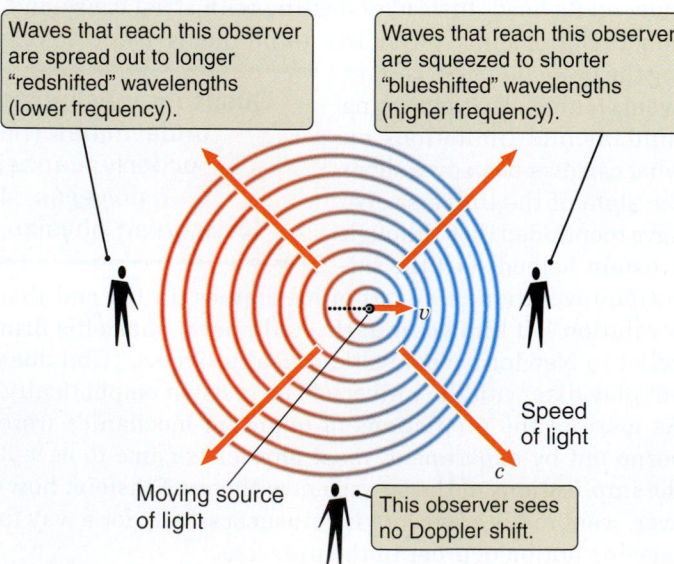

FIGURE 4.16 Motion of a light or sound source relative to an observer may cause waves to be spread out (*redshifted*, or lower in pitch) or squeezed together (*blueshifted*, or higher in pitch). A change in the wavelength of light or the frequency of sound is called a *Doppler shift*.

Waves that reach this observer are spread out to longer "redshifted" wavelengths (lower frequency).

Waves that reach this observer are squeezed to shorter "blueshifted" wavelengths (higher frequency).

Speed of light

c

Moving source of light

This observer sees no Doppler shift.

v

---

[4]The phenomenon is sometimes referred to as the "Doppler-Fizeau effect" to honor the French physicist Hippolyte Fizeau, who discovered the phenomenon independently.

(a)

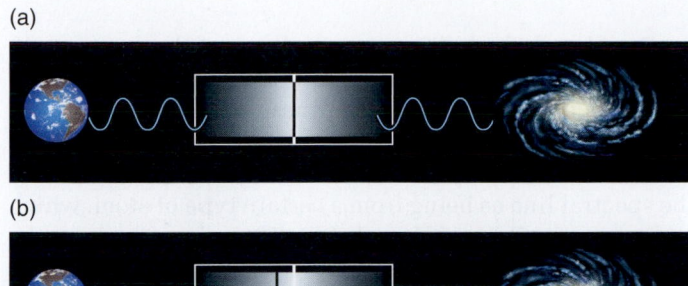

(b)

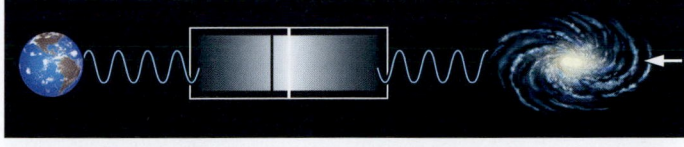

(c)

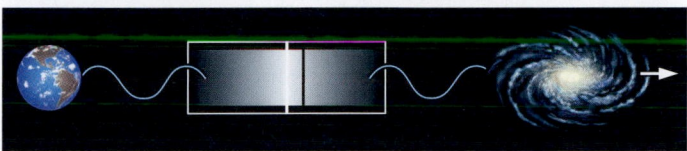

FIGURE 4.17 Spectral lines of astronomical objects will be blueshifted or redshifted from their rest wavelength (a), depending on whether they are moving toward us (b) or away from us (c).

reaching you from the object has a shorter wavelength than its rest wavelength. In other words, the light that you see is bluer than if the source were not moving toward you. We say that the light from an object moving toward us is **blueshifted** (**Figure 4.17b**). In contrast, light from a source that is moving away from you is shifted to longer wavelengths. The light that you see is redder than if the source were not moving away from you, so we say that the light is **redshifted**, as seen in **Figure 4.17c**.

> Light from approaching objects is blueshifted. Light from receding objects is redshifted.

The amount by which the wavelength of light is shifted by the Doppler effect is called the **Doppler shift** of the light.

As long as the speed of an object is much less than the speed of light (which means we can ignore relativistic effects), the observed wavelength of the Doppler-shifted light, λ_{obs}, is given by the equation

$$\lambda_{obs} = \left(1 + \frac{v_r}{c}\right)\lambda_0.$$

In this expression, λ_0 is the rest wavelength of the light. The velocity v_r is the velocity at which the object is moving relative to you. To be more precise, v_r is the rate at which the distance between you and the object is changing. If v_r is positive, the object is getting farther away from you. If v_r is negative, the object is getting closer to you.

Take a moment to be sure this expression makes sense to you. Our argument says that if an object is moving away from you, the wavelength of the light coming to you from that object should be greater than it would be if the object were "at rest" (not moving relative to you). If the object is moving away, v_r will be greater than 0. If v_r is greater than 0, then $1 + v_r/c$ is greater than 1, and therefore $\lambda_{obs} = (1 + v_r/c)\lambda_0$ will be larger than λ_0, so we see a redshift. This is as we expected. What about the opposite case? If an object is moving toward us, v_r is less than 0, so $(1 + v_r/c)$ will be less than 1. Now we find that λ_{obs} is shorter than λ_0. The observed wavelength is shorter than the rest wavelength, so we see a blueshift. Once again, this observation is in line with our expectations (see **Math Tools 4.2**).

Remember that the Doppler shift provides information only about whether an object is moving toward you or away from you. That is what the subscript r in v_r signifies. The variable v_r stands for **radial velocity**, which is the rate at which the distance between you and the object is changing.

MATH TOOLS 4.2

Making Use of the Doppler Effect

A prominent spectral line of hydrogen atoms has a rest wavelength, λ_0, of 656.3 nm (see Figure 4.15b). Suppose that, using a telescope, you measure the wavelength of this line in the spectrum of a distant object and find that instead of seeing the line at 656.3 nm, you see the line at a wavelength, λ_{obs}, of 659.0 nm. You could then infer that the object is moving at a velocity of

$$v_r = \frac{\lambda_{obs} - \lambda_0}{\lambda_0}c = \frac{659.0 \text{ nm} - 656.3 \text{ nm}}{656.3 \text{ nm}} \times (3 \times 10^8 \text{ m/s})$$

$$= 1.2 \times 10^6 \text{ m/s}.$$

The object is moving away from you with a radial velocity of 1.2×10^6 m/s, or 1,200 km/s.

Consider the case of our nearest stellar neighbor, Alpha Centauri, which is moving toward us at a radial velocity of −21.6 km/s (−2.16 × 10⁴ m/s.) What would be the observed wavelength, λ_{obs}, of a magnesium line in Alpha Centauri's spectrum having a rest wavelength, λ_0, of 517.27 nm?

$$\lambda_{obs} = \left(1 + \frac{v_r}{c}\right)\lambda_0 = \left(1 + \frac{-2.16 \times 10^4 \text{ m/s}}{3 \times 10^8/\text{s}}\right) \times 517.27 \text{ nm}$$

$$= 517.23 \text{ nm}.$$

Although the observed Doppler blueshift (517.23 − 517.27) is only −0.04 nm, it is easily measurable with modern instrumentation.

At the moment that the fire truck is passing you, it is neither getting closer to you nor getting farther away from you, so the pitch you hear is the same as the pitch heard by the crew riding on the truck. You can see this directly by looking at Figure 4.16 or by referring to the previous equation. If an object is moving perpendicular to your line of sight, then $v_r = 0$ and $\lambda_{obs} = \lambda_0$. The observed wavelength equals the rest wavelength. The light is neither redshifted nor blueshifted. (In the case of light, we need to stress again the caveat that this result holds true only if the object's velocity is much less than the speed of light. In Chapter 17 you will learn more about how relativity influences an object moving at nearly the speed of light.)

> Radial velocity is measured from Doppler shifts of emission or absorption lines.

Doppler shifts become especially useful when we are looking at an object that has emission or absorption lines in its spectrum. These spectral lines enable astronomers to determine how rapidly the object is moving toward or away from us. To determine the speed, astronomers first identify the spectral line as being from a certain type of atom, which has a unique rest wavelength (λ_0). They then measure the wavelength (λ_{obs}) of the spectrum of the distant object. The difference between the rest wavelength and the observed wavelength enables us to determine the object's velocity. Turning the preceding expression around a bit, we can write

$$v_r = \frac{\lambda_{obs} - \lambda_0}{\lambda_0} c.$$

If you know λ_0, just measure λ_{obs}, plug both values into this expression, and voilà!—you know what v_r is.

FOUNDATIONS 4.1

Equilibrium Means Balance

Equilibrium is the term we use to refer to systems that are in balance. Imagine two well-matched teams struggling in a tug-of-war contest. Each team pulls steadfastly on the rope, but the force of one team's pull is only enough to match but not overcome the force exerted by the other team. Muscles flex, but the scene does not change. A picture taken now and another taken 5 minutes from now would not differ in any significant way. In this book we will frequently encounter this kind of **static equilibrium**, represented by a tug-of-war in which opposing forces just balance each other. The equilibrium between the downward force of gravity and the pressure that opposes it will play a central role in the stories of planetary interiors, planetary atmospheres, and the interiors of stars. Static equilibrium can be stable, unstable, or neutral, as shown in **Figure 4.18**. Consider the image in **Figure 4.18b** or a book standing on its edge, unsupported on either side by other objects or books. If you nudge the book, it will fall over rather than settling back into its original

position. This is an example of an **unstable equilibrium**. When an unstable equilibrium is disturbed, it will move farther away from equilibrium rather than back toward it.

Not all types of equilibrium are static. Equilibrium can also be dynamic, in which the system is constantly changing. In **dynamic equilibrium**, one source of change is exactly balanced by another source of change, so that the configuration of the system remains the same. **Figure 4.19** shows a simple example of dynamic equilibrium. A can with a hole cut in the bottom has been placed under an open water faucet. The depth of the water in the can determines how fast water pours out through the hole in the bottom of the can. Once the water reaches just the right depth, as in **Figure 4.19a**, water continues to pour out of the bottom of the can at exactly the same rate it pours from the faucet into the top of the can. The water leaving the can balances the water entering the can, and equilibrium is established. If you were to take a picture now and another picture a few minutes from now, little of the water in the can would be the same. Even so, the pictures would be indistinguishable.

If a system is not in equilibrium, its configuration will change. Look at **Figure 4.19b**. Here the level of the water in the can is too low, so water will not flow out of the bottom of the can fast enough to balance the water flowing into the can, and thus the water level will rise. A picture taken now and another a short time later will not look the same. The configuration of the system is changing. Conversely, if the water level in the can is too high, as in **Figure 4.19c**, water will flow out of the can faster than it flows into the can, and the water level will fall. Once again, if the system is not in equilibrium, its configuration will change.

Water passing through a can is an example of a **stable equi-**

FIGURE 4.18 Examples of stable (a), unstable (b), and neutral (c) equilibrium. Imagine what happens to the ball if, in each case, you give it a small nudge.

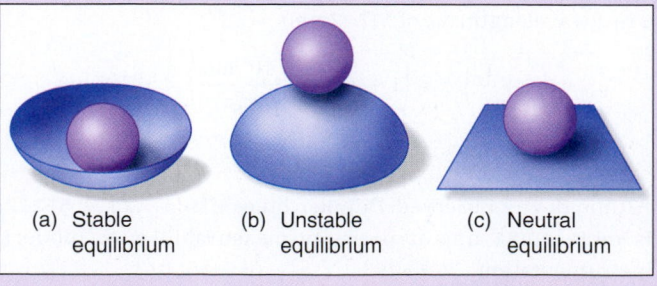

(a) Stable equilibrium   (b) Unstable equilibrium   (c) Neutral equilibrium

4.5 Why Mercury Is Hot and Pluto Is Not

Spectral lines also enable us to determine a distant object's temperature. If you asked an elementary-school student why Mercury is hot and Pluto is cold, he or she would probably look at you like you were hopelessly uninformed and then patiently explain that Mercury is hot because it is close to the Sun, while Pluto is cold because it is much farther away. This explanation is fine as far as it goes, but we should push a bit harder. Why are the planets as hot as they actually are? Why does the surface of Mercury reach temperatures that are hot enough to melt lead while the surface of Pluto remains so cold that even substances such as methane and ammonia remain permanently frozen? And closer to home, why is the surface of Earth hot enough for water to melt over most of the planet, but cold enough for that water to remain as a liquid instead of turning into vapor?

The temperature of any object is determined by two things: what is trying to heat up the object and what is trying to cool it down. If an object's temperature is constant, then these two must be in balance with each other. Your body, for instance, is heated by the release of chemical energy from inside. It is also sometimes heated by energy from your surroundings. If you are standing in sunshine on a hot day, the hot air around you and the sunlight falling on you both are work-

> **If an object's temperature is constant, then heating and cooling must be in balance.**

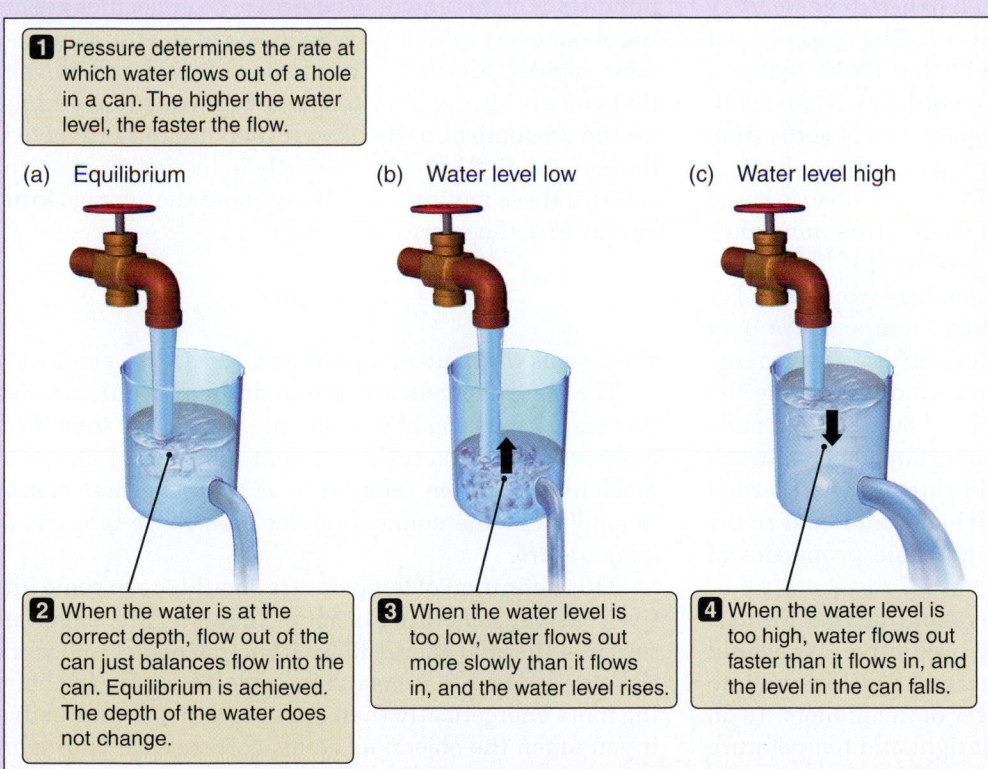

1 Pressure determines the rate at which water flows out of a hole in a can. The higher the water level, the faster the flow.

(a) Equilibrium   (b) Water level low   (c) Water level high

2 When the water is at the correct depth, flow out of the can just balances flow into the can. Equilibrium is achieved. The depth of the water does not change.

3 When the water level is too low, water flows out more slowly than it flows in, and the water level rises.

4 When the water level is too high, water flows out faster than it flows in, and the level in the can falls.

FIGURE 4.19 Water flowing into and out of a can determines the water level in the can. This is an example of dynamic equilibrium.

librium. When a stable equilibrium is *disturbed* (forced away from its equilibrium configuration), it will tend to return to its equilibrium state. If the water level is too high or too low, it will move back toward its equilibrium level. The equilibrium that we discuss in this chapter, between sunlight falling on a planet and thermal energy radiated away into space, is a stable equilibrium. If you take Figure 4.19 and replace "water flows in" with "sunlight is absorbed," "water flows out" with "energy is radiated by the planet," and "water level" with "the planet's temperature," then the stable equilibrium that sets the level of the water in the can becomes the stable equilibrium that sets the temperature of a planet.

ing to heat you up. In response to this heating, your body must have some way to cool itself off. This is where perspiration comes in. When you perspire, water seeps from the pores in your skin and evaporates. It takes energy to evaporate water, and much of this energy comes from your body. Thus, as the perspiration evaporates it carries away your body's **thermal energy**, cooling your body down. For your body temperature to remain stable, the heating must be balanced by the cooling. Such a balance is referred to as **thermal equilibrium**. If your body is out of thermal equilibrium in one direction—if there is more heating than cooling—then your body temperature climbs. If your body is out of thermal equilibrium in the other direction—if there is more cooling than heating—then your body temperature falls.

Planets also have a thermal equilibrium, and electromagnetic radiation plays a crucial role in maintaining that *equilibrium*. The energy from sunlight heats the surface of Earth, driving its temperature up. This is one side of the equilibrium. The other side of Earth's thermal equilibrium

> Earth is heated by sunlight and cooled by the radiation of energy back into space.

is also controlled by light energy. Earth radiates energy back into space, thereby cooling itself. Our eyes are not sensitive to the wavelengths of light that Earth radiates, but that light is there nonetheless. Overall, Earth must radiate away just as much energy into space as it absorbs from the Sun. If there were more heating than cooling, the temperature of Earth would climb. (It would absorb more energy from the Sun than it got rid of, and this imbalance would show up in increasing temperature.) If there were more cooling than heating, the temperature would fall. For Earth to remain at the same average temperature over time—which it has done for many centuries[5]—the energy that Earth radiates into space must exactly balance the energy absorbed from the Sun. Thermal equilibrium must be maintained. **Equilibrium** is an important concept in science. There are many kinds of equilibrium besides thermal equilibrium, some of which we will encounter later in the book. **Foundations 4.1** describes the basic properties of equilibrium (see pp. 110–111).

We now have a qualitative understanding that is interesting but is not yet quantitatively predictive. We would like to turn this intuitive idea of thermal equilibrium into a real prediction for the temperatures of the planets. To do this we need to find out more about light and temperature and the relationship between the two. Before going too far down this path, however, we should start by improving our understanding of what we mean by *temperature*.

---

[5] Over the past century, Earth's average temperature has slightly increased. Many climatologists attribute this rise in temperature to a small decrease in the amount of energy radiated back to space, which is caused by an increase in the amount of "greenhouse" gases in our atmosphere.

Temperature Is a Measure of How Energetically Particles Are Moving About

When we say that something is hot or cold, we know exactly what we mean. In everyday life *hot* and *cold* are defined in terms of our subjective experiences. Something is hot when it *feels* hot or cold when it *feels* cold. But our perceptions of hot and cold are a layer of subjective experience that comes between us and a definable, quantifiable concept called **temperature**. When we talk about temperature, we speak of degrees on a thermometer. The way we define a "degree" is arbitrary—a matter of convention. If you grew up in the United States, for example, you probably think of temperatures in "degrees **Fahrenheit**" (°F), whereas if you grew up anywhere else in the world, you think of temperatures in "degrees **Celsius**" (°C). Both of these are perfectly reasonable scales for measuring temperatures. But what does the thermometer actually measure?

What we refer to as *temperature* is actually a measurement of how energetically the atoms that make up an object are moving about. The air around you is composed of vast numbers of atoms and molecules. Those molecules are moving about every which way. Some move slowly; some move more rapidly. Similarly, the atoms that make up the chair that you are sitting in or the floor that you are walking on (or the anatomical parts of your body that are involved in those two activities) are constantly in motion. We can characterize these motions by talking about the average **kinetic energy** (E_K), the energy of motion:

$$E_K = \frac{1}{2}mv^2,$$

where m is the mass of a particle and v is its velocity.

The more energetically the atoms or molecules in something are bouncing about, the higher is that something's temperature. In fact, the random motions of atoms and molecules are often referred to as their **thermal motions**, to emphasize the connection between these motions and temperature.

This definition of temperature should make some intuitive sense. If something is hotter than you are, experience says that thermal energy flows from that object into you. At the atomic level that means the object's atoms are bouncing more energetically than are the atoms in your body, so if you touch the object, its atoms collide with your atoms, causing the atoms in your body to move faster. Your body gets hotter as thermal energy flows from the object to you. (At the same time, these collisions rob the particles in the object of some of their energy. Their motions slow down. The hotter object becomes cooler.) In general, when we talk about *heating*, we mean processes that increase the average thermal energy of an object's particles, and when we talk about *cooling*, we mean a process that decreases the average thermal energy of those particles. The connection between

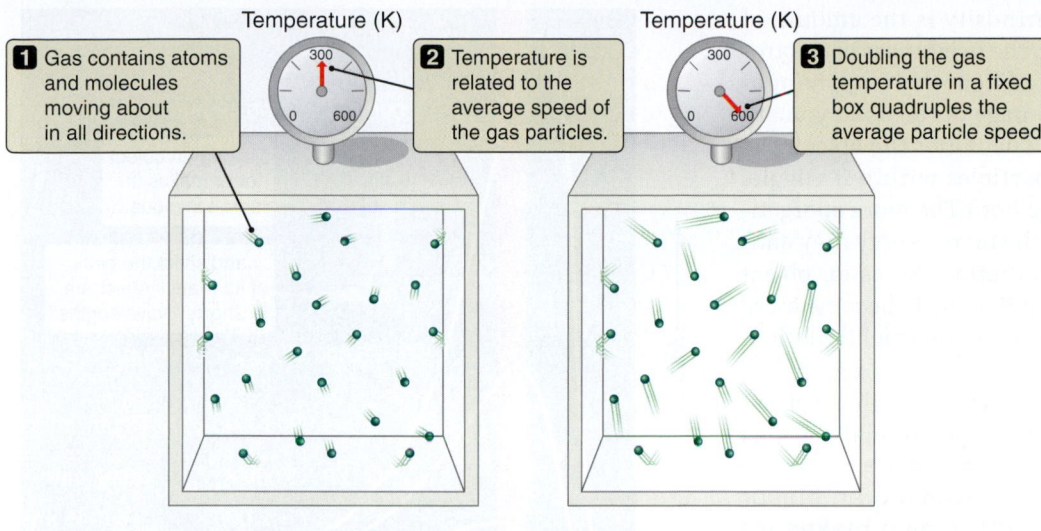

Temperature (K)

1 Gas contains atoms and molecules moving about in all directions.

2 Temperature is related to the average speed of the gas particles.

Temperature (K)

3 Doubling the gas temperature in a fixed box quadruples the average particle speed.

FIGURE 4.20 Hotter gas temperatures correspond to faster motions of the constituent atoms. K stands for *kelvin*, the unit in the *Kelvin scale* of temperature, which is commonly used in scientific investigations.

the temperature and the motion of particles is illustrated in **Figure 4.20**.

The change in thermal energy associated with a change of one unit, or degree, is arbitrary on any temperature scale. On the Fahrenheit scale there are 180 degrees between the melting point (32°F) and the boiling point (212°F) of water at sea level. On the Celsius scale there are 100 degrees between those two temperatures. For these two scales, the temperature corresponding to "zero degrees" is also arbitrary. On the Celsius scale, 0°C is chosen to be the temperature at which water freezes; on the Fahrenheit scale, 0°F is placed at a temperature corresponding to −17.78°C. However, there is a lowest possible physical temperature below which no object can fall. As the motions of the atoms in an object slow down, the temperature drops lower and lower. When the motions of the particles finally stop, things have gotten as cold as they can get. This lowest possible temperature, where thermal motions have come to a standstill, is called **absolute zero**. Absolute zero corresponds to −273.15°C, or −459.57°F.

> Absolute zero, the temperature at which thermal motions stop, is zero on the Kelvin scale.

The preferred temperature scale for most scientists is the **Kelvin scale**. For convenience, the size of one unit on the Kelvin scale, called a **kelvin** (abbreviated **K**) is the same as the Celsius degree. What makes the Kelvin scale special is that 0 kelvins is set equal to that absolute lowest temperature where thermal motions stop—absolute zero. There are no negative temperatures on the Kelvin scale. The importance of the Kelvin scale is that *when temperatures are measured in kelvins, we know that the average thermal energy of particles is proportional to the measured temperature*. The average thermal energy of the atoms in an object with a temperature of 200 K is twice the average thermal energy of the atoms in an object with a temperature of 100 K.

Hotter Means More Luminous and Bluer

So far, we have focused our attention on the way discrete atoms emit and absorb radiation. Our discussion led us to a useful understanding of emission lines and absorption lines and how we might use these lines to learn about the physical state and motion of distant objects. But not all objects have spectra that are dominated by discrete spectral lines. For instance, if you pass the light from an incandescent lightbulb through a prism, as we saw in Figure 4.14a, then instead of discrete bright and dark bands you will see light spread out smoothly from the blue end of the spectrum to the red. Similarly, if you look closely at the spectrum of the Sun, you will see absorption lines, but mostly you will see light smoothly spread out across all colors of the spectrum—red through violet. What is the origin of such **continuous radiation**, and what clues might this kind of radiation carry about the objects that emit it?

> Objects like lightbulbs emit continuous radiation at all wavelengths.

We can think of a dense material as being composed of a collection of charged particles that are being jostled as their thermal motions cause them to run into their neighbors. The hotter the material is, the more violently its particles are being jostled. Anytime a charged particle is subjected to an acceleration, it radiates, so the jostling of particles due to their thermal motions causes them to give off a continuous spectrum of electromagnetic radiation. This is why *any* material that is sufficiently dense for its atoms to be jostled by their neighbors will emit light *simply because of its temperature*. Radiation of this sort is called **thermal radiation**.

We can guess how the radiation from an object changes as the object heats up or cools down. Start with the question of the **power** (energy per second, measured in **watts**) radiated, which is referred to as the object's **luminosity**. To a physicist or an astronomer, luminosity is the amount of light *leaving* a source. By contrast, the **brightness** of electromagnetic radiation is the amount of light that is *arriving* at a particular location, such as the page of the book you are reading or the pupil of your eye. The hotter the object, the more energetically the charged particles within it wiggle. (Again, this is what it *means* to be hot.) The more energetically the charged particles move, the more energy they emit in the form of electromagnetic radiation. So as an object gets hotter, we expect the light that it emits to become more intense. Here is our first intuition about thermal radiation: *hotter means more luminous.*

Now we move to the question of what color light an object emits. Again, as the object gets hotter, the thermal motions of its particles become more energetic. These more energetic motions are capable of producing more energetic photons. So as an object gets hotter, we might expect the average energy of the photons that it emits to increase. In other words, we might expect the average wavelength of the emitted photons to get shorter. The light from the object gets bluer. Here is our second intuition about thermal radiation: *hotter means bluer.*

> Making an object hotter also makes its thermal radiation bluer and more luminous.

Both of these intuitive predictions are borne out by a simple experiment that you can do. An incandescent lightbulb is a good example of an object that emits thermal radiation. The electric current in the lightbulb filament heats the filament. (More precisely, electrons being pushed through the filament by electric fields collide with atoms in the filament, increasing the thermal motions of those atoms.) The hot filament then glows. Somewhere in your home, dorm, or classrooms you can probably find a lightbulb with a rheostat (a dimmer). Turning the knob on the dimmer changes the amount of electric current in the filament in the lightbulb. Turning the dimmer up increases the current, which increases the number and strength of collisions between electrons and atoms, which in turn increases the temperature of the filament. The hotter filament is more luminous. This demonstration confirms the first of our expectations: *hotter means more luminous.*

What about the color of the emitted light? Look again at the lightbulb as you turn up the dimmer. When the bulb first comes on, it glows a dull red; but as the current through the bulb increases, driving up the temperature of the filament, the perceived color of the light changes. When the dimmer is turned all the way up, the light from the bulb has lost its red tint. The hotter the lightbulb gets, the more the highly energetic blue photons become mixed with the less energetic red photons, and the light becomes whiter. The color of the light shifts from red toward blue, confirming our second intuitive expectation: *hotter means bluer.*

These observations offer an intuitive grasp of how the light given off by an object depends on the temperature of the object, but to be really useful they need to be quantified. We need to know *how much* more luminous and *how much*

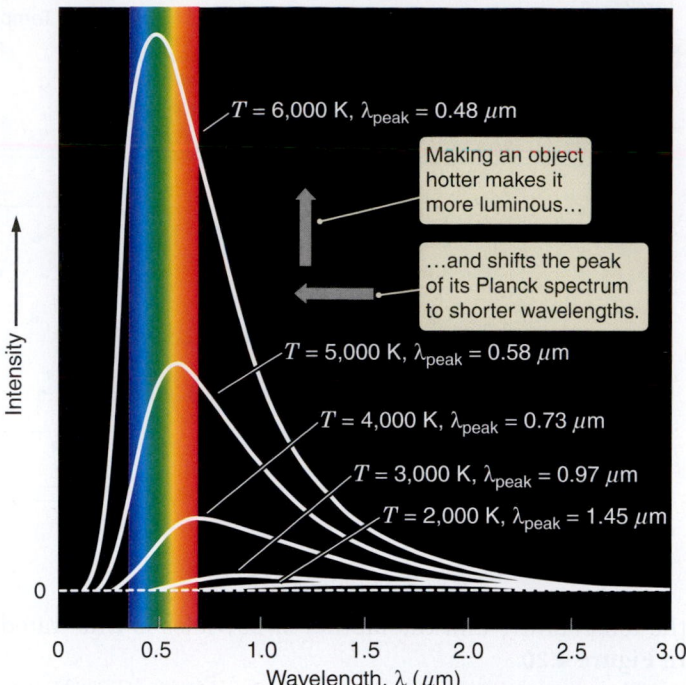

FIGURE 4.21 Planck spectra emitted by sources with temperatures of 2,000 K, 3,000 K, 4,000 K, 5,000 K, and 6,000 K. At higher temperatures the peak of the spectrum shifts toward shorter wavelengths, and the amount of energy radiated per second from each square meter of the source increases.

bluer. The detailed answers to these questions were worked out around 1900 by Max Planck. Planck was thinking about a special situation—a hollow, totally enclosed cavity of material at a specific temperature, T. The crucial point here is that, inside the cavity, all of the radiation emitted by the cavity walls is also absorbed by the walls. In this situation a balance is set up, with each bit of the wall emitting just as much thermal radiation as it is absorbing from its surroundings. Physicists refer to such a special setup as a **blackbody**. Planck used this balance to calculate the spectrum of the light inside such a cavity. The result of his calculation, which beautifully matches the results of experiments, is called a **Planck spectrum** or a **blackbody spectrum**. **Figure 4.21** shows plots of the Planck spectra for objects at several different temperatures.

> A blackbody emits thermal radiation that has a Planck spectrum.

You might reasonably ask what this hypothetical cavity has to do with the light from the filament of a lightbulb. Surely the filament of a lightbulb is not a cavity of this sort! But in a certain sense it is. The light emitted by charged particles within the filament is mostly absorbed by other charged particles within the filament. This is exactly the assumption that Planck made when calculating the shape of the spectrum of a blackbody. As a result, we expect that the radiation existing *inside the filament of the lightbulb* will have a Planck spectrum. This radiation "leaks out"

Working with the Stefan-Boltzmann and Wien's Laws

We can use the Stefan-Boltzmann law to figure out what the surface area of the filament in a lightbulb must be. Suppose the filament in an incandescent bulb operates at a temperature of about 2,500 K. The amount of energy radiated by the bulb is stamped right on the face of the bulb—a 100-W bulb has a luminosity of 100 W, which means that it radiates away 100 joules each second. (A **joule** is a unit of energy, abbreviated J.) If the total amount of light from the filament is equal to the flux ($\mathcal{F}$) multiplied by the surface area of the filament (A), we can write

$$\mathcal{F} = 100\ \text{W} = A \times \sigma T^4.$$

Solving this equation for the area, we get

$$A = \frac{100\ \text{W}}{\sigma T^4}$$

$$= \frac{100\ \text{W}}{\left(5.67 \times 10^{-8}\ \dfrac{\text{W}}{\text{m}^2\text{K}^4}\right) \times (2{,}500\ \text{K})^4} = 4.5 \times 10^{-5}\ \text{m}^2.$$

Note that not much filament surface area is needed to provide the light that turns night into day in our homes and cities.

Wien's law (described in the text) will prove handy as we continue our study of the universe. If we can measure the spectrum of an object emitting thermal radiation and find where the peak in the spectrum is, we can use Wien's law to calculate the temperature of the object. For example, we have no way of dropping a thermometer into the Sun and directly measuring its temperature, but we *can* observe the spectrum of the light coming from the Sun. When we do so, we find that the peak in the spectrum occurs at a wavelength of about 0.5 μm. Wien's law can be written as

$$T = \frac{2{,}900\ \mu\text{m K}}{\lambda_{\text{peak}}}.$$

If we plug the observed peak of the spectrum of the Sun ($\lambda_{\text{peak}} = 0.50$ μm) into this equation, we get

$$T = \frac{2{,}900\ \mu\text{m K}}{0.5\ \mu\text{m}} = 5{,}800\ \text{K}.$$

This is how we know the temperature of the Sun.

Suppose we want to calculate the peak wavelength at which Earth radiates. The average temperature of Earth is 288 K. If we insert this temperature into Wien's law, we get

$$\lambda_{\text{peak}} = \frac{2{,}900\ \mu\text{m K}}{288\ \text{K}} = 10.1\ \mu\text{m}.$$

or slightly more than 10 μm. This result tells us that Earth radiates in the infrared region of the spectrum.

---

of the filament, just as light might leak out of a small hole in the side of Planck's cavity. As a result, the spectrum of radiation from the filament of a lightbulb is very close to a Planck spectrum. The light from stars such as the Sun and the thermal radiation from a planet also often come close to having a blackbody spectrum.

The Stefan-Boltzmann Law Says That Hotter Means Much More Luminous

We now ask the question that scientists must always ask: Does the theoretical prediction agree with observation and experiment? Do the Planck spectra in Figure 4.21 agree with our intuitive ideas and with our experiment with the lightbulb and the dimmer? Begin with luminosity. As the temperature of an object increases, Planck's theory says that the object gives off more radiation at every wavelength, so the luminosity of the object should increase. In fact, it increases in a hurry. Adding up all of the energy in a Planck spectrum shows that the increase in luminosity is proportional to the *fourth power* of the temperature: Luminosity $\propto T^4$. This result is known as the **Stefan-Boltzmann law** because it was discovered in the laboratory by Austrian physicist Josef Stefan (1835–1893) and derived by his student, Ludwig Boltzmann (1844–1906), before Planck's theory came along to explain it.

What the Stefan-Boltzmann law actually says is that the amount of energy radiated *by each square meter* of the surface of an object is given by the equation

$$\mathcal{F} = \sigma T^4.$$

In this equation $\mathcal{F}$ is called the **flux**. It is a measurement of the total amount of energy coming through each square meter of the surface each second. The constant σ (pronounced "sigma") is called the **Stefan-Boltzmann constant**. The value of σ is the same for all cases and is given by $5.67 \times 10^{-8}\ \text{W/(m}^2\text{K}^4)$ (where W stands for watts). To find the total amount of energy emitted by an object in the form of electromagnetic radiation, multiply $\mathcal{F}$ by the surface area of the object (see **Math Tools 4.3**.)

The Stefan-Boltzmann law says that an object rapidly becomes more luminous as the temperature increases. If

> The luminosity of a blackbody is proportional to T^4.

the temperature of an object goes up by a factor of 2, the amount of energy being radiated each second increases by a factor of 2^4, or 16. If the temperature of an object goes up by a factor of 3, then the energy being radiated by the object each second goes up by a factor of 3^4, or 81! A lightbulb with a filament temperature of, say, 3,000 K radiates 16 times as much light as it would if the filament temperature were 1,500 K. Even modest changes in temperature can result in large changes in the amount of power radiated by an object.

> Slight changes in temperature mean large changes in brightness.

Wien's Law Says That Hotter Means Bluer

Look again at Figure 4.21, but this time, instead of paying attention to how high each curve is, notice where the peak of each curve falls along the horizontal axis. As the temperature increases, the *peak* of the Planck spectrum shifts toward shorter wavelengths, which means that the average energy of the photons becomes greater. Just as we surmised, increasing the temperature causes the light from the object to get bluer. The shift in the location of the Planck spectrum's peak wavelength with increasing temperature is given by the equation

> The peak wavelength of a blackbody is inversely proportional to its temperature.

$$\lambda_{peak} = \frac{2,900 \ \mu m \ K}{T}.$$

This result is referred to as **Wien's law**, named for German physicist Wilhelm Wien (1864–1928). In this equation, λ_{peak} (pronounced "lambda peak") is the wavelength where the Planck spectrum is at its peak. It is the wavelength where the electromagnetic radiation from an object is greatest. Wien's law says that the location of the peak in the spectrum is inversely proportional to the temperature of the object. If you increase the temperature by a factor of 2, the peak wavelength becomes half of what it was. If you increase the temperature by a factor of 3, the peak wavelength becomes a third of what it was (see Math Tools 4.3.)

4.6 Twice as Far Means One-Fourth as Bright

You might have noticed that we have consistently spoken of the "luminosity" of objects, where in everyday language we probably would have just said that one object is "brighter" than another. This is a case where everyday language is too sloppy for science. *Luminosity* refers to the amount of light *leaving* a source. The concept of brightness is certainly related to the concept of luminosity. For example, replacing a lightbulb with a luminosity of 50 W with a 100-W bulb succeeds in making a room twice as bright because it doubles the light reaching any point in the room. But brightness also depends on the distance from a source of electromagnetic radiation. If you needed more light to read this book, you could replace the bulb in your lamp with a more luminous one, but it would probably be easier just to move the book closer to the light. Conversely, if a light were too bright for you, you could move away from it. Our everyday experience says that as we move away from a light, its brightness decreases.

The particle description of light provides a convenient way to think about the brightness of radiation and how brightness depends on distance. Suppose you had a piece of cardboard that was 1 meter on an edge. Intuitively you might imagine that making the light that falls on the cardboard twice as bright would mean doubling the number of photons that hit the cardboard each second. Tripling the brightness of the light would mean increasing the number of photons hitting the cardboard each second by a factor of 3, and so on. Here is a beginning point for understanding brightness. Brightness depends on the number of photons falling on each square meter of a surface each second.

> Brightness measures how much light falls per square meter per second.

Working with this idea of brightness, now imagine a lightbulb sitting at the center of a spherical shell, as shown in **Figure 4.22**. Photons from the bulb travel in all directions and land on the inside of the shell. To find the number of photons landing on each square meter of the shell during each second (that is, to find the brightness of the light), take the *total* number of photons given off by the lightbulb each second and divide by the number of square meters over which those photons have to be spread. The surface area of a sphere is given by the formula $A = 4\pi r^2$, where r is the distance between the bulb and the surface of the sphere (thus, r = the radius of the sphere). When this relationship is written as an equation, we find that

$$\begin{pmatrix} \text{Number of photons} \\ \text{striking one square} \\ \text{meter each second} \end{pmatrix} = \frac{\begin{array}{c} \text{Total number of photons} \\ \text{emitted per second} \end{array}}{\begin{array}{c} \text{Number of square meters} \\ \text{the photons are spread over} \end{array}}$$

$$= \frac{\begin{array}{c} \text{Total number of photons} \\ \text{emitted each second} \end{array}}{4\pi r^2}$$

The next step in building an understanding of brightness is to change the size of the spherical shell while keeping the total number of photons given off by the lightbulb each second the same. As the shell becomes larger, the pho-

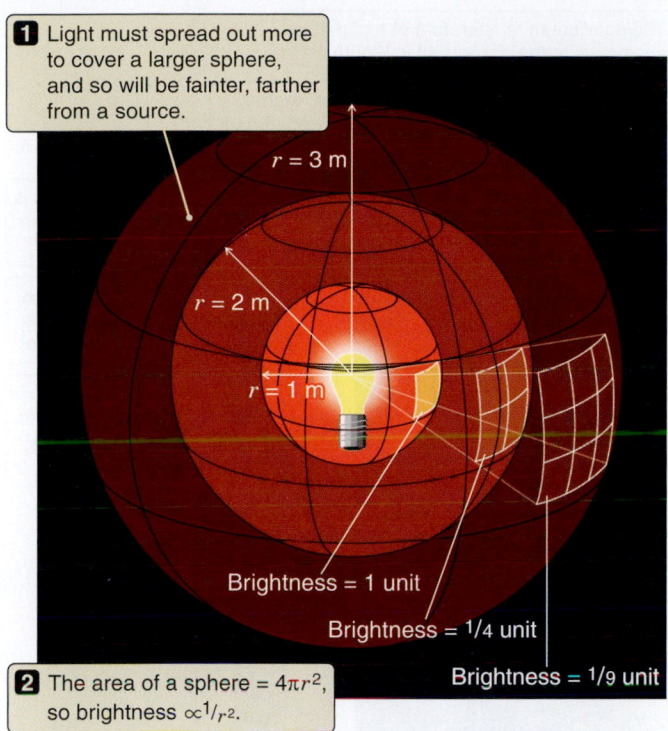

1 Light must spread out more to cover a larger sphere, and so will be fainter, farther from a source.

$r = 3$ m

$r = 2$ m

$r = 1$ m

Brightness = 1 unit

Brightness = 1/4 unit

Brightness = 1/9 unit

2 The area of a sphere = $4\pi r^2$, so brightness $\propto 1/r^2$.

FIGURE 4.22 Light obeys an inverse square law as it spreads away from a source. Twice as far means one-fourth as bright.

The luminosity of an object is the total number of photons given off by the object multiplied by the energy of each photon. Instead of talking about how the number of photons must spread out to cover the surface of a sphere (brightness), we now talk about how the *energy* carried by the photons must spread out to cover the surface of a sphere. When speaking of brightness in this way, we mean the amount of energy falling on a square meter in a second. If L is the luminosity of the bulb, then the brightness of the light at a distance r from the bulb is given by

$$\text{Brightness} = \frac{\text{Energy radiated per second}}{\text{Area over which energy is spread}}$$

$$= \frac{L}{4\pi r^2}.$$

Usually, the only information that astronomers have to work with is the light from a distant object. For this reason we will use our understanding of radiation over and over again throughout our journey. Time spent now thinking carefully about the electromagnetic spectrum, the emission and absorption of photons, Planck radiation, and the inverse square law for brightness will be a *very* good investment for what is to come.

tons from the lightbulb must spread out to cover a larger surface area. Each square meter of the shell receives fewer photons each second, so the brightness of the light decreases. If the shell's surface is moved twice as far from the light, the area over which the light must spread increases by a factor of $2^2 = 2 \times 2 = 4$. The photons from the bulb spread out over four times as much area, so the number of photons falling on each square

> Like gravity, light obeys an inverse square law.

meter each second becomes ¼ of what it was. If the surface of the sphere is three times as far from the light, as illustrated in Figure 4.22, the area over which the light must spread increases by a factor of $3 \times 3 = 3^2 = 9$, and the number of photons per second falling on each square meter becomes ⅑ of what it was originally. We encountered just this kind of relationship when we talked about gravity in Chapter 3. Just like gravity, light obeys an inverse square law. The brightness of the light from an object is inversely proportional to the square of the distance from the object. *Twice as far means one-fourth as bright.*

It is helpful to think of brightness in terms of photons streaming onto a surface from a light because this gives us a good mental picture of the physical nature of brightness and why brightness follows an inverse square law. In practice, however, it is usually more convenient to speak of the *energy* coming to a surface each second, rather than the number of *photons* arriving.

4.7 Radiation Laws Enable Us to Calculate the Equilibrium Temperatures of the Planets

We began our discussion of thermal radiation by asking a straightforward question: Why does a planet have the temperature that it does? In a qualitative way, we said that the temperature of a planet is determined by a balance between the amount of sunlight being absorbed and the amount of energy being radiated back into space. We now have the tools we need to turn this qualitative idea into a real prediction of the temperatures of the planets.

Begin with the amount of sunlight being absorbed. The amount of energy absorbed by a planet is just the area of the planet that is absorbing the energy multiplied by the brightness of sunlight at the planet's distance from the Sun. When we look at a planet, we see a circular disk with a radius equal to the radius of the planet. The area of this circular disk is πR^2, where R is the radius of the planet. The brightness of sunlight at a distance d from the Sun is equal to the luminosity of the Sun ($L_{\odot}$, in watts) divided by $4\pi d^2$, as we saw in Section 4.6 (d here is the same as r was in Section 4.6;

we use d in this case to avoid confusion with the planet's radius, R.) We must consider one additional factor. Not all of the sunlight falling on a planet is absorbed by the planet. The fraction of the sunlight that is reflected from a planet is called the **albedo**, a, of the planet. The corresponding fraction of the sunlight that is absorbed by the planet is 1 minus the albedo. A planet with an albedo of 1 reflects all the light falling on it. A planet that absorbs 100 percent of the sunlight falling on it has an albedo of 0.

Writing this relationship as an equation, we say that

$$\begin{pmatrix} \text{Energy absorbed} \\ \text{by the planet} \\ \text{each second} \end{pmatrix} = \begin{pmatrix} \text{Absorbing} \\ \text{area of} \\ \text{planet} \end{pmatrix} \times \begin{pmatrix} \text{Brightness} \\ \text{of sunlight} \end{pmatrix} \times \begin{pmatrix} \text{Fraction} \\ \text{of sunlight} \\ \text{absorbed} \end{pmatrix}$$

$$= \quad \pi R^2 \quad \times \quad \frac{L_\odot}{4\pi d^2} \quad \times \quad (1-a)$$

Moving to the other piece of the equilibrium, the amount of energy that the planet radiates away into space each second is just the number of square meters of the planet's total surface area multiplied by the power radiated by each square meter. The surface area for the planet is given by $4\pi R^2$. The Stefan-Boltzmann law tells us that the power radiated by each square meter is given by σT^4. So we can say that

$$\begin{pmatrix} \text{Energy radiated by} \\ \text{planet each second} \end{pmatrix} = \begin{pmatrix} \text{Surface area} \\ \text{of planet} \end{pmatrix} \times \begin{pmatrix} \text{Energy radiated by each} \\ \text{square meter each second} \end{pmatrix}$$

$$= \quad 4\pi R^2 \quad \times \quad \sigma T^4.$$

If the planet's temperature is to remain stable—if it is to keep from heating up or cooling down—then it must be radiating away just as much energy into space as it is absorbing in the form of sunlight, as indicated in **Figure 4.23**. Therefore, we can equate these two expressions. We can set the quantity "Energy radiated by planet" equal to the quantity "Energy absorbed by planet." When we do this, we arrive at the expression

$$\begin{pmatrix} \text{Energy radiated by} \\ \text{planet each second} \end{pmatrix} = \begin{pmatrix} \text{Energy absorbed by} \\ \text{planet each second} \end{pmatrix}$$

or

$$4\pi R^2 \sigma T^4 = \pi R^2 \frac{L_\odot}{4\pi d^2}(1-a).$$

Look at this equation for a moment. It may seem rather complex, but when we consider the pieces of this equation separately, it becomes more digestible. On the left side of the equation, $4\pi R^2$ tells how many square meters of the planet's surface are radiating energy back into space, while σT^4 tells how much energy each one of those square meters radiates each second. Put them together and you get the total amount of energy radiated away by the planet each second. On the right side of the equation, πR^2 is the area of the planet as seen from the Sun. That amount multiplied by the brightness of the sunlight reaching the planet,

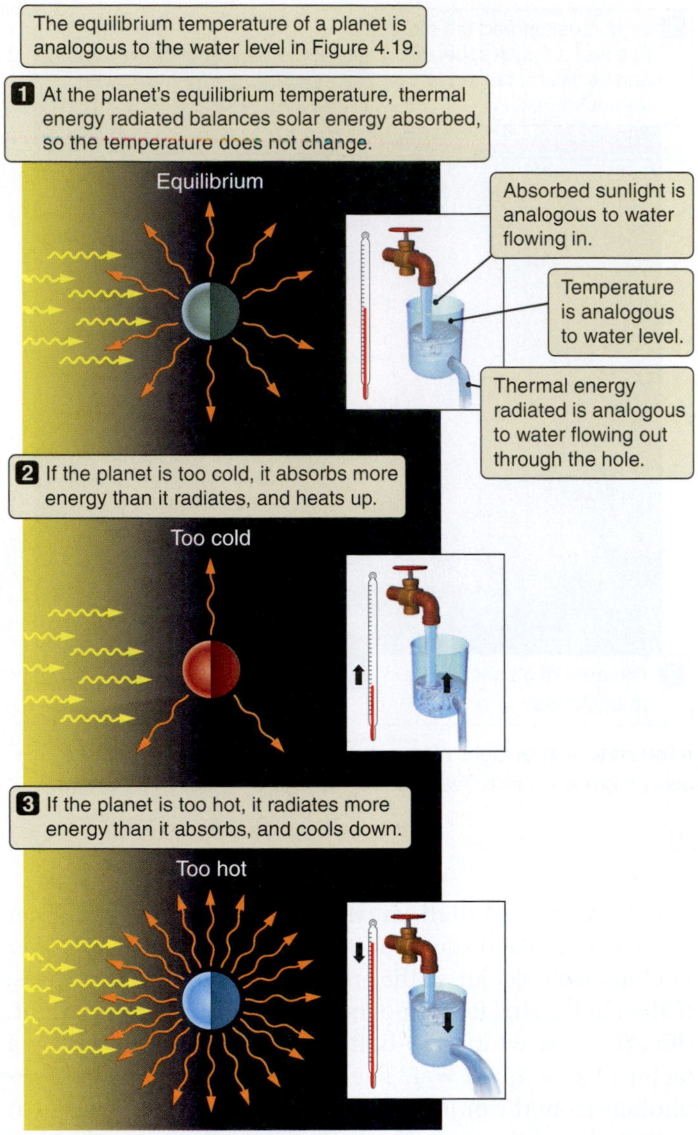

FIGURE 4.23 Planets are heated by absorbing sunlight (and sometimes internal heat sources) and cooled by emitting thermal radiation into space. If there are no other sources of heating or means of cooling, then the equilibrium between these two processes determines the temperature of the planet.

$L_\odot/4\pi d^2$, tells how much energy is falling on the planet each second. The final expression, $(1-a)$, tells how much of that energy the planet actually absorbs. Put everything on the right side of the equation together and you get the amount of energy absorbed by the planet each second. The equal sign says that the energy radiated away needs to balance the sunlight absorbed. There is no magic here. In fact, when broken down, this formidable equation embodies little more than a few straightforward ideas—such as "hotter means more luminous," "twice as far means one-fourth as bright," and "heating and cooling must balance each other." The math just gives us a convenient way to work with these concepts.

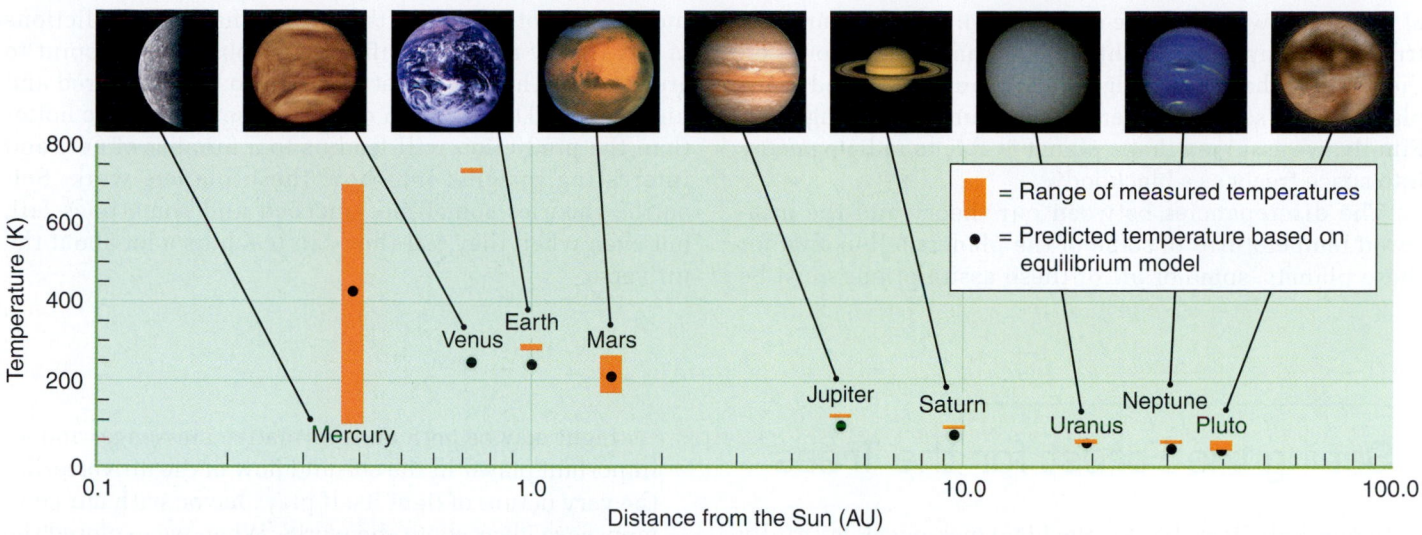

FIGURE 4.24 Predicted temperatures for the classical planets and Pluto, based on the equilibrium between absorbed sunlight and thermal radiation into space, are compared with ranges of observed surface temperatures. Some predictions are correct. Interestingly, others are not.

We started down this path hoping to find a way to predict the temperatures of the planets, and a bit of algebra gets us the rest of the way there. Rearranging the previous equation to put T on one side and everything else on the other gives

$$T^4 = \frac{L_\odot(1-a)}{16\sigma\pi d^2}.$$

If we take the fourth root of each side, we get

$$T = \left[\frac{L_\odot(1-a)}{16\sigma\pi}\right]^{1/4} \times \frac{1}{\sqrt{d}}.$$

Putting in the appropriate numbers, we wind up with

$$T = 279 \times \left(\frac{\sqrt{1-a}}{d_{AU}}\right)^{1/2},$$

where d_{AU} is the distance of the planet from the Sun in astronomical units. For example, we can see that the blackbody ($a = 0$) temperature for a planet at Earth's distance from the Sun (1 AU) is 279 K.

We have now produced a full-fledged physical model for why the temperatures of planets are what they are. Restating the meaning in words, T is the temperature at which the energy radiated by a planet exactly balances the energy absorbed by the planet. If the planet were hotter than this equilibrium temperature, it would radiate energy away faster than it absorbed sunlight, and its temperature would fall. If the planet were cooler than this temperature, it would radiate away less energy than was falling on it in the form

> Balancing cooling and heating sets an equilibrium temperature.

of sunlight, and its temperature would rise. Only at this equilibrium temperature do the two balance.

The equation tells us that as the distance from the Sun increases—in other words, as d gets bigger—the temperature of the planet decreases. No surprise there. But now we know *how much* the temperature should decrease. The temperature should be inversely proportional to the square root of the distance. Using this formula can turn our intuition about why planets that are close to the Sun are hot into a prediction of just how hot they should be.

Figure 4.24 plots the predicted temperatures of the planets. The vertical bars show the range of temperatures found on the surface of each planet (or, in the case of the giant planets, at the tops of their clouds). The black dots show our predictions using the preceding equation. As the figure indicates, overall we are not too far off. That should give us a sense of accomplishment; it says that our basic understanding of *why* planets have the temperatures that they do is probably fairly close to the mark. The data for Mercury, Mars, and the dwarf planet Pluto agree particularly well. (The agreement for Mercury would improve if we took into account the huge difference in temperature between the daytime and nighttime sides of the planet and recomputed our equilibrium accordingly.)

In other cases, however, our predictions are wrong. For Earth and the giant planets the actual temperatures are a bit higher than the predicted temperatures. In the case of Venus, the actual surface temperature is wildly higher than our prediction. Rather than cause for despair, though, these discrepancies between theory and observation are cause for excitement. As we built our physical model for the equilibrium temperatures of planets, we made a number of assumptions. For example, we assumed that the temperature of the planet was the same everywhere. This is clearly not true; for example, we might expect planets to be hotter on the day side than on the night side. We also assumed that a planet's only source of energy is the sunlight falling on it. Finally, we assumed that a planet is able to radiate energy into space freely as a blackbody.

The discrepancies between our theory and the measured temperatures of some of the planets tell us that for these planets, some or all of these assumptions must be incorrect. In other words, the places where the predictions of our theory are not confirmed by observation point to areas where there is something still to be discovered and understood. The question of *why* these planets are hotter than the prediction will lead us to a number of new and interesting insights into how these planets work. Scientific theories sometimes succeed and sometimes fail, but even when they fail they can teach us a lot about the universe.

Seeing the Forest for the Trees

In our daily lives, light is the ideal messenger, faithfully telling us about the world around us. As with any good courier, we usually concentrate on the information that light carries, taking the messenger itself for granted. It is easy to forget that our seemingly immediate visual perception of reality is actually a derived experience—the result of a sophisticated interplay between our eyes, our brains, and the flood of electromagnetic radiation that is emitted, absorbed, transmitted, and reflected by objects in the world around us.

The light that our eyes see is only a tiny portion of the full span of the electromagnetic spectrum. Electromagnetic radiation carries with it a wealth of information about the temperature, density, composition, state of motion, and other physical characteristics of the place of its origin and the material it interacts with en route. In place of the carefully controlled laboratory experiments of many other sciences, the astronomer uses a combination of ingenuity and technology to "slice up" the light reaching us and tease out the information it carries about conditions throughout the universe. The role of electromagnetic radiation in astronomy is far more than that of messenger. Radiation is also a participant in the processes that we study. For example, light carries energy from the Sun outward through the Solar System, heating the planets; and light carries energy away from each planet, allowing it to cool. The balance between these two processes establishes the conditions of our existence.

Light may be both an informative messenger and an important player in the ebb and flow of the universe, but the very nature of light itself plays havoc with our commonsense ideas about the world. When we explored the motions of Earth, the Moon, and the planets, we relied heavily on our intuition about the world. The pull or shove of one object on another and the force of gravity that holds us tightly to the surface of Earth are well within the realm of our everyday experience. But when we consider the properties of light, the boundaries of our experience and intuition are shattered. We are forced beyond the confines of the "box" within which our brains evolved. Answering a simple question like "How fast does light travel?" demands that we abandon our most cherished ideas about the nature of space, time, matter, and energy. We ask, "What is light?" and confront abstract concepts like electric and magnetic fields while running headlong into the seemingly impossible question of how something can be both a wave *and* a particle. We delve into the interaction between light and matter and find that at the scale of atoms and photons, Newton's clockwork universe crumbles. In its place we discover a world of random chance and uncertainty, governed not by the strict march of cause and effect but by laws of probability and statistics. We ask about the nature of light—and we collide squarely with the shortcomings of our intuitive ideas about the nature of reality itself.

If electromagnetic radiation is the messenger, how do we hear and interpret the message? In the next chapter we'll learn about the many and varied tools that astronomers use to detect and decipher the meaning of these communications that come to us from the Solar System and beyond.

Summary

- Light carries both information and energy throughout the universe.

- From gamma rays to visible light to radio waves, all radiation is an electromagnetic wave.

- Light is simultaneously an electromagnetic wave and a stream of particles called photons.

- The speed of light in a vacuum is 300,000 km/s, and nothing can travel faster.

- Nearly all matter is composed of atoms, and light can reveal its identity.

- Atoms absorb and emit radiation at unique wavelengths like spectral fingerprints.

- Light from receding objects is redshifted. Light from approaching objects is blueshifted.

- Temperature is a measure of the thermal energy of an object.

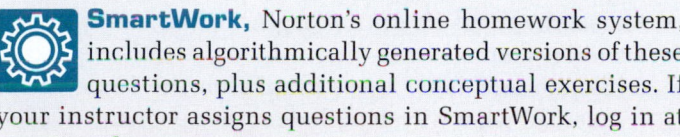 **SmartWork,** Norton's online homework system, includes algorithmically generated versions of these questions, plus additional conceptual exercises. If your instructor assigns questions in SmartWork, log in at **smartwork.wwnorton.com.**

Student Questions

THINKING ABOUT THE CONCEPTS

1. We know that the speed of light in a vacuum is 3×10^8 m/s. Is it possible for light to travel at a lower speed? Explain your answer.

2. Is a light-year a measure of time or distance or both?

3. Would you describe light as a wave or a particle or both? Explain your answer.

*4. In what way can a charged particle create a magnetic field?

5. Name the variables that define a wave.

6. Name the types of electromagnetic radiation that lie at the extreme ends of the electromagnetic spectrum.

7. If photons of blue light have more energy than photons of red light, how can the intensity of a beam of red light carry as much energy as a beam of blue light?

8. What is a continuous spectrum?

*9. Patterns of emission or absorption lines in spectra can uniquely identify individual atomic elements, just as DNA testing uniquely identifies individual human beings. Explain how positive identification of atomic elements can be used as one way of testing the validity of the cosmological principle.

10. Name the three subatomic particles that make up an atom.

*11. Why is it impossible to know the exact position and the exact velocity of an electron simultaneously?

12. Name the two components that together define momentum.

13. Explain the difference between speed and velocity.

14. Describe the ground state of an atom and what this implies for any electron in that state.

15. An atom in an excited state can drop to a lower energy state by emitting a photon. Is it possible to predict exactly how long the atom will remain in the higher energy state? Explain your answer.

16. Spectra of astronomical objects show both bright and dark lines. Describe two properties that each is telling you about the atoms responsible for the spectral lines.

17. Astronomers may describe certain celestial objects as being redshifted or blueshifted. What do these terms tell them about the objects?

18. An object somewhere near you is emitting a pure tone at middle C on the octave scale (262 Hz). You, having perfect pitch, hear the tone as A above middle C (440 Hz). Describe the motion of this object relative to where you are standing.

19. During a popular art exhibition, the museum staff finds it necessary to protect the artwork by limiting the total number of viewers in the museum at any time. New viewers are admitted at the same rate that others leave. Is this an example of static equilibrium or of dynamic equilibrium? Explain.

*20. The temperature of an object has a very specific meaning as it relates to the object's atoms. Explain the relationship.

21. A favorite object for amateur astronomers is the *double star* Albireo, with one of its components a golden yellow and the other a bright blue. What do these colors tell you about the relative temperatures of the two stars?

22. The difference between brightness and luminosity confuses many people. How should you explain the difference to a family member or a friend who is not taking this class?

*23. The stars you see in the night sky cover a large range of brightness. What does that tell you about the distances of the various stars? Explain your answer.

**24. The physical properties of an object are sometimes proportional to its temperature, raised to some power. For example, as we have seen in this chapter, the luminosity of a radiating body is proportional to T^4. Why must the temperature T be expressed in kelvins rather than degrees Celsius or Fahrenheit?

25. Consider two hypothetical planets with no atmosphere. One orbits the Sun at an average distance of 5.0 AU; the other, at an average distance of 10.0 AU. Yet both have the same average surface temperature. Explain how this could be possible.

APPLYING THE CONCEPTS

26. The index of refraction, n, of a diamond is 2.4. What is the speed of light within a diamond?

27. You are tuned to 790 on AM radio. This station is broadcasting at a frequency of 790 kHz (7.90×10^5 Hz). What is the wavelength of the radio signal? You switch to 98.3 on FM radio. This station is broadcasting at a frequency of 98.3 MHz (9.83×10^7 Hz). What is the wavelength of this radio signal?

28. Your microwave oven cooks by agitating water molecules at a frequency of 2.45 gigahertz (GHz), or 2.45×10^9 Hz. What is the wavelength of the microwave's electromagnetic radiation in centimeters?

29. You observe a spectral line of hydrogen at a wavelength of 502.3 nm in a distant galaxy. The rest wavelength of this line is 486.1 nm. What is the radial velocity of this galaxy?

30. Assume that an object emitting a pure tone of 440 Hz is on a vehicle approaching you at a speed of 25 m/s. If the speed of sound at this particular atmospheric temperature and pressure is 340 m/s, what will be the frequency of the sound that you hear? (Hint: Keep in mind that frequency is inversely proportional to wavelength.)

31. If half of the phosphorescent atoms in a "glow-in-the-dark" toy give up a photon every 30 minutes, how bright (relative to its original brightness) will the toy be after 2 hours?

32. Assume that the atoms in a gas remain in an excited state for an average lifetime of 1 minute before decaying to the ground state. How many minutes will pass until only $\frac{1}{32}$ of the atoms remain in the excited state?

33. Imagine that you have been transported to Neptune, 30 AU from the Sun. How bright would the Sun appear, compared to its brightness as seen from Earth?

34. The spacecraft *Voyager 1* is now about 115 AU from the Sun and heading out of the Solar System. Compare the brightness of the Sun seen by *Voyager 1* with that seen from Earth.

35. On a dark night you notice that a distant lightbulb happens to have the same brightness as a firefly that is 5 meters away from you. If the lightbulb is a million times more luminous than the firefly, how far away is the lightbulb?

36. Two stars appear to have the same brightness, but one star is three times more distant than the other. How much more luminous is the more distant star?

37. A panel with an area of 1 square meter (m²) is heated to a temperature of 500 K. How many watts is it radiating into its surroundings?

38. The Sun has a radius of 6.96×10^5 km and a blackbody temperature of 5,780 K. Calculate the Sun's luminosity. (Hint: The area of a sphere is $4\pi R^2$.)

39. You are given a bar of tungsten, a metal with a melting point of 3,640 K and a boiling point of 6,170 K. You heat the bar until it starts to melt and plot a graph of its emitted Planck energy distribution. You then continue to heat the tungsten until it begins to boil, and you plot a similar graph.
   a. Sketch and label these graphs on the same set of axes. At what wavelength does the spectrum of each peak?
   b. How many times more luminous is the boiling tungsten than the melting tungsten?

40. Some of the hottest stars known have a blackbody temperature of 100,000 K.
   a. What is the peak wavelength of their radiation?
   b. Can this wavelength be observed from Earth's surface? Explain your answer.

41. Your body, at a temperature of about 37°C (98.6°F), emits radiation in the infrared region of the spectrum.
   a. What is the peak wavelength, in micrometers, of your emitted radiation?
   b. Assuming an exposed body surface area of 0.25 m², how many watts of power do you radiate?

*42. A planet with no atmosphere at 1 AU from the Sun would have an average blackbody surface temperature of 279 K if it absorbed all the Sun's electromagnetic energy falling on it (albedo = 0).
   a. What would be the average temperature on this planet if its albedo were 0.1, typical of a rock-covered surface?
   b. What would be the average temperature if its albedo were 0.9, typical of a snow-covered surface?

43. Consider a hypothetical planet named Vulcan. If such a planet were in an orbit ¼ the size of Mercury's and had

the same albedo as Mercury, what would be the average temperature on Vulcan's surface? Assume that the average temperature on Mercury's surface is 450 K.

**44. Earth's average albedo is approximately 0.3.
   a. Calculate Earth's average temperature in degrees Celsius.
   b. Does this temperature meet your expectations? Explain why or why not.

45. The orbit of Eris, a dwarf planet, carries it out to a maximum distance of 97.7 AU from the Sun. Assuming an albedo of 0.8, what is the average temperature of Eris when it is farthest from the Sun?

StudySpace is a free and open Web site that provides a Study Plan for each chapter of *21st Century Astronomy*. Study Plans include animations, reading outlines, vocabulary flashcards, and multiple-choice quizzes, plus links to premium content in SmartWork and the ebook. Visit **www.wwnorton.com/studyspace**.

All truths are easy to understand once they are discovered; the point is to discover them.

GALILEO GALILEI (1564–1642)

Robotic rovers *Spirit* and *Opportunity* have roamed the surface of Mars.

CHAPTER 5

The Tools of the Astronomer

5.1 The Optical Telescope— An Extension of Our Eyes

Do you remember the first time you had a close-up view of the Moon's surface? Perhaps your family or a neighbor had a backyard **telescope** like the one shown in **Figure 5.1**. Or the Moon might have been the featured attraction during a visit to your local planetarium. With that first view came

FIGURE 5.1 Amateur astronomers with a modern reflecting telescope. There are several hundred thousand amateur astronomers in the United States alone. Advances in telescope design and electronics have enabled amateur observers to discover new celestial objects and help with long-term research projects of professional astronomers.

KEY CONCEPTS

In the previous chapter we learned how our understanding of the physical and chemical properties of distant planets, stars, and galaxies comes to us in the form of electromagnetic radiation. But this information must first be collected and processed before it can be analyzed and converted to useful knowledge. Here we will learn about the tools that astronomers use to capture and scrutinize that information. We will find that

- Telescopes of various types collect radiation over the entire range of the electromagnetic spectrum—from gamma rays to radio signals.

- Optical telescopes come in two basic types: refractors and reflectors. All of the larger astronomical telescopes are reflectors.

- Telescope resolution increases with aperture; image size is proportional to a telescope's focal length.

- Earth's atmosphere distorts telescopic images and prevents large parts of the electromagnetic spectrum from reaching the ground. Telescopes in orbit overcome this problem.

- The most effective way to study the planets and moons of our Solar System is to go there.

FIGURE 5.2 A replica of Galileo's refracting telescope.

FIGURE 5.3 Newton's reflecting telescope.

recognition of what the Moon really is—a nearby planetary world covered with craters and vast, lava-flooded basins. You might also have been treated to a breathtaking look at the Orion Nebula, a giant assemblage of gas and dust 1,500 light-years away, or the larger and more remote Andromeda Galaxy. These are but a few of many celestial wonders that come alive in the eyepiece of even a small telescope. As it has for so many before you, the telescope can change the way you view the heavens. From such experiences you might understand why the telescope is the astronomer's most important instrument. Yet it is only within the past century and a half that its capabilities have been fully exploited.

> The telescope is the astronomer's most important tool.

As long ago as the late 13th century, craftsmen in Venice were making small disks of glass that could be mounted in frames and worn over the eyes to improve vision. The glass disks were convex on both sides, shaped something like lentils. And so they became known as *lenses*, from *lens*, which

is Latin for "lentil." Looking back, it's rather remarkable that more than 300 years would pass before these lenses would be employed for something other than spectacles!

Hans Lippershey (1570–1619) was a German-born spectacle maker living in the Netherlands around the turn of the 17th century. Legend has it that in 1608, children playing with his lenses put two of them together and saw a distant object magnified. Lippershey looked for himself and mounted the lenses together in a tube to produce a *kijker* ("looker"), as he called it. As you might imagine, news of Lippershey's invention spread rapidly, soon reaching the Italian instrument maker Galileo Galilei. Galileo at once saw the potential of the "looker" for studying the heavens and constructed one of his own (**Figure 5.2**). By 1610, Galileo had become the first to see the phases of Venus and the moons of Jupiter, and among the first to see craters on the Moon.[1] He was also the first to realize that the Milky Way is made

---

[1]As we mentioned in Chapter 3, English astronomer Thomas Harriot made drawings of the Moon several months before Galileo began his

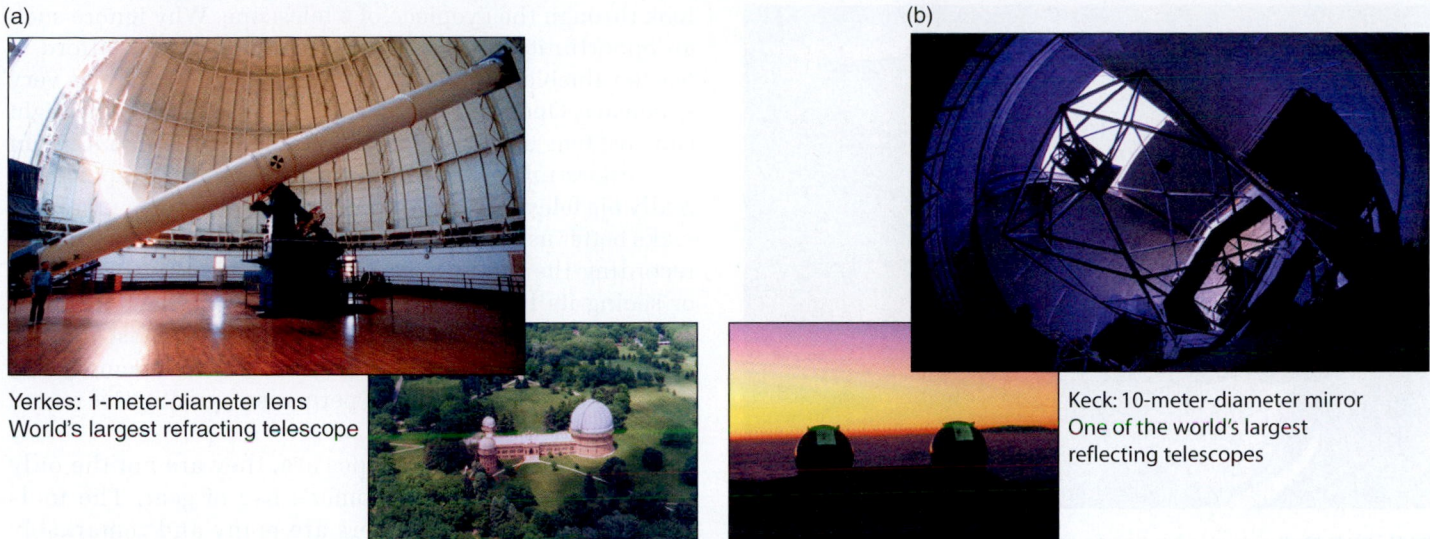

(a)

Yerkes: 1-meter-diameter lens
World's largest refracting telescope

(b)

Keck: 10-meter-diameter mirror
One of the world's largest
reflecting telescopes

FIGURE 5.4 (a) The Yerkes 1-meter refractor uses a lens to collect light. (b) The twin Keck 10-meter reflectors are more compact and use mirrors to collect light.

up of countless numbers of individual stars. As the story goes, Galileo was demonstrating his instrument to guests when one of them christened it the *telescope*, from the Greek meaning "farseeing," and the name stuck. With its ability to see far beyond the range of the human eye, the **refracting telescope**—one that uses lenses instead of mirrors—quickly revolutionized the science of astronomy.

Unfortunately, all simple-lens telescopes (those using a pair of single lenses) suffer from a serious problem called *chromatic aberration*, which forms blue halos around bright objects. As light passes through a simple lens, blue light is brought to a shorter *focus* than the longer visible wavelengths, causing the out-of-focus blue halo effect. Realizing this problem, Sir Isaac Newton in 1668 designed a telescope using mirrors instead of lenses. He cast a 2-inch mirror made of speculum metal (basically copper and tin) and polished it to spherical curvature. He then placed the primary mirror at the bottom of a tube with a secondary flat mirror mounted above it at a 45° angle, which directed the focused light to an eyepiece on the outside of the tube (**Figure 5.3**). As it turned out, others did not share Newton's talent for instrument making, and the **reflecting telescope** remained a curiosity for decades. The spherical mirror surface used by Newton works only for small mirrors. Larger mirrors require a *parabolic* surface to produce a sharply focused image, and parabolic surfaces are much more difficult to

> There are two types of optical telescopes: refractors and reflectors.

fabricate. Not until the latter half of the 18th century did large reflecting telescopes come into their own.

Throughout the 19th century, both refracting and reflecting telescopes continued to grow in size. By 1897, though, refracting telescopes had reached their limit, with the completion of the Yerkes 1-meter (40-inch) refractor (**Figure 5.4a**), destined to become the world's largest operational refracting telescope. Located in Williams Bay, Wisconsin, the Yerkes telescope carries a 450-kilogram (kg) objective lens mounted at the end of a 19.2-meter tube. But herein lies a major problem with refractors: to get the most light-gathering power and produce the largest images, a refractor must suspend a massive piece of glass at the end of a very long tube without sagging unduly under the force of gravity. Reflecting telescopes, on the other hand, seem to have no such size limits. Today the 10-meter twin Keck telescopes located on 4-kilometer (km)-high Mauna Kea in Hawaii (**Figure 5.4b**) are among the world's largest reflecting telescopes. Each of the two Keck telescopes has 4 million times the light-gathering power of the human eye! And even larger reflecting telescopes are in the works. Several organizations are considering telescopes with *apertures* in the 30-meter range, and the European Southern Observatory's (ESO) 42-meter European Extremely Large Telescope (E-ELT) may be in operation by 2017, at an estimated cost of nearly €1 billion (**Figure 5.5**). It seems that the only limitation on the size of reflecting telescopes is the cost of fabricating them. **Table 5.1** lists the world's largest optical telescopes. As you can see, all are reflecting telescopes.

Here is a professional secret we can share with you. Today's working astronomers never—well, hardly ever—

> All of the world's largest telescopes are reflecting telescopes.

---

astronomical observations. However, Galileo published his observations, while Harriot did not.

FIGURE 5.5 Artist's rendering of the European Extremely Large Telescope (E-ELT), a 42-meter reflecting telescope proposed by the European Southern Observatory (ESO).

look through the eyepiece of a telescope. Why ignore such an opportunity? The reason is that they cannot afford to waste valuable observing time. Telescope time can be very expensive: Operating a large telescope for just a single night can cost tens of thousands of dollars! So although it might be exhilarating to glimpse Saturn through the eyepiece of a really big telescope, astronomers can learn much more and make better use of precious observing time by permanently recording the planet's image at a variety of wavelengths or seeing its light spread out into a revealing spectrum. Long after the observing session is over, the wistful peek through the eyepiece would be just a distant memory, but the recorded data remain as a permanent quantitative record for subsequent analysis.

As important as telescopes are, they are not the only instruments in the astronomer's bag of gear. The tools used by modern astronomers are many and remarkably diverse, ranging from powerful supercomputers to high-

TABLE 5.1

The World's Largest Optical Telescopes

Mirror Diameter (meters)	Telescope	Sponsor	Location	Operational Date
11.0	South African Large Telescope (SALT)	South Africa, USA, UK, Germany, Poland, New Zealand	Sutherland, South Africa	2005
10.4	Gran Telescopio CANARIAS (GTC)	Spain, Mexico, University of Florida	Canary Islands	2007
10	Keck I	Caltech, University of California, NASA	Mauna Kea, Hawaii	1993
10	Keck II	Caltech, University of California, NASA	Mauna Kea, Hawaii	1996
9.2	Hobby-Eberly	University of Texas, Penn State, Stanford, Germany	Mount Fowlkes, Texas	1999
2×8.4	Large Binocular Telescope (LBT)	University of Arizona, Ohio State, Italy, Germany, Arizona State, and others	Mt. Graham, Arizona	2008
8.3	Subaru Telescope	Japan	Mauna Kea, Hawaii	1999
4×8.2	Very Large Telescope (VLT)	European Southern Observatory	Cerro Paranal, Chile	2000
8.1	Gemini North	USA, UK, Canada, Chile, Brazil, Argentina, Australia	Mauna Kea, Hawaii	1999
8.1	Gemini South	USA, UK, Canada, Chile, Brazil, Argentina, Australia	Cerro Pachón, Chile	2000
6.5	Magellan I	Carnegie Institute, University of Arizona, Harvard, University of Michigan, MIT	La Serena, Chile	2000
6.5	Magellan II	Carnegie Institute, University of Arizona, Harvard, University of Michigan, MIT	La Serena, Chile	2002

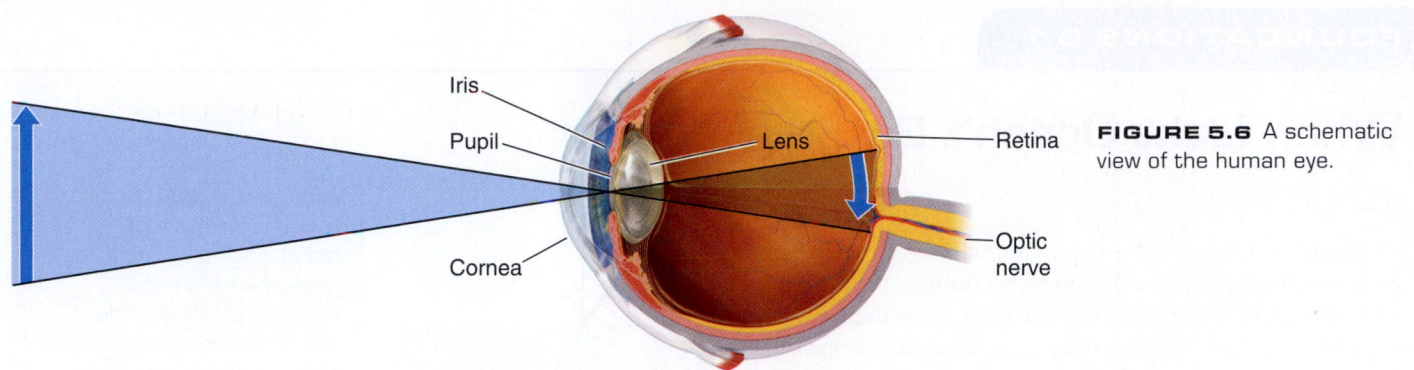

FIGURE 5.6 A schematic view of the human eye.

energy colliders to robotic probes sent to cruise the Solar System. Astronomy's tools are not only diverse; they are also changing rapidly. It seems that every few months we are likely to see the commissioning of a new mountaintop telescope, the launch of a satellite observatory, or the arrival of a spacecraft at a remote planetary destination. In fact, in the time it takes between the final changes we make to this volume and its appearance on your bookshelf, much of what we might say about the latest astronomy tools will already have become yesterday's news. Rather than give in to instant obsolescence, we will instead make use of the technology of the 21st century to bring you a discussion of (what else?) the technology of the 21st century! Keep up to date by visiting the *21st Century Astronomy* Web site.

Refractors and Reflectors

Humanity's use of telescopes dates back to far earlier than the night Galileo first turned his telescope skyward. Our earliest human ancestors had telescopes of their own—their eyes. The human eye essentially operates as a refracting telescope: it uses a simple convex lens to bend the light passing through it, bringing that light to a sharp focus on the retina (**Figure 5.6**). Although the lens forms an upside-down image of an object on the retina, your brain translates the signal from the optic nerve to show you the correct orientation of the object. Evolutionary biologists have concluded that eyes—as crucial sensory organs—evolved *independently* in as many as 65 different species during the history of terrestrial life! Although the human eye is wonderfully evolved to meet our daily needs, it is not so well suited for doing astronomy. For that we need large, powerful telescopes. Now let's look more closely at how telescopes mimic the eye's light-gathering function and how they help us overcome the eye's limitations.

As we have seen, astronomical telescopes come in two basic types: *refracting* and *reflecting*. A refracting telescope has a simple convex lens, called the **objective lens**, whose curved surfaces refract—changes the direction of—the light from a distant object (**Figure 5.7**). (**Foundations 5.1** explores the optics of refracted light.) This refracted light forms an image on the telescope's **focal plane**, which is perpendicular to the optical axis of the telescope. (The "optical axis" is the path that light takes through the center of the lens, or mirror in the case of a reflecting

> A telescope's aperture determines its light-gathering power and resolution.

(a)

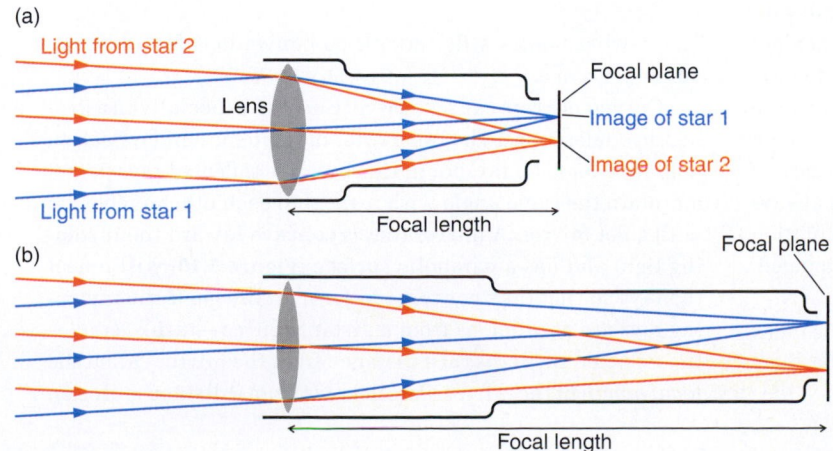

(b)

FIGURE 5.7 (a) A refracting telescope uses a lens to collect and focus light from two stars, forming images of the stars in its focal plane. (b) Telescopes with longer *focal length* produce larger, more widely separated images.

FOUNDATIONS 5.1

When Light Doesn't Go Straight

REFRACTION

As its name suggests, the refracting telescope focuses and enhances the light from a distant object through the process of **refraction**. Refraction is the change in direction of light entering a new medium. When a light wave enters a new medium, its speed changes. Remember from Chapter 4 that the speed of light is always less in any material medium than it is in a vacuum. If the incident light is perpendicular to the surface of the new medium, the speed changes but the direction of travel does not. If the incident light encounters the surface at a different angle, the direction of travel also changes. This change in the direction of the incident light is called refraction. The amount of refraction depends both on the angle of incidence and the relative speeds in the two media, which are defined by their respective indices of refraction (see Chapter 4). In **Figure 5.8a**, light waves are depicted as coming in from the upper left and hitting the surface of the new medium at a certain angle to the perpendicular. The part of each wave that enters the new medium first is slowed before the other part of the wave enters. As a result, the wave bends, taking up a new direction of travel. If the speed of light in the new medium is less than that in the initial medium (that is, if the new medium has a higher refractive index), the light bends toward the perpendicular. If the speed of light in the new medium is greater than that in the initial medium (that is, if the new medium has a lower refractive index), the light bends away from the perpendicular. **Figure 5.8b** shows a green laser beam being refracted by a plastic block.

REFLECTION

Picture a beam of light striking a piece of glass, as illustrated in **Figure 5.9**. When light encounters a different medium—in this case going from air to glass—there will always[2] be a certain amount of **reflection** from the surface of the new medium. In other words, some of the light will change its direction of travel. The most common example occurs when light encounters an ordinary flat mirror. In **Figure 5.9a**, an incoming or incident **ray**, AB, reflects from the surface, becoming the reflected ray BC. The angle between AB and PB, the perpendicular to the surface, is called the "angle of incidence" (i). The angle between BC and PB is called the "angle of reflection" (r). In the case of a flat mirror, the angles of incidence and reflection are always equal. What reflects *from* the mirror is a good representation of what falls *on* it, although left and right are interchanged.

[2]We have to be careful here. In those rare cases in which the index of refraction (see Chapter 4) is *exactly* the same in both media, there will be no reflection or refraction at the surface.

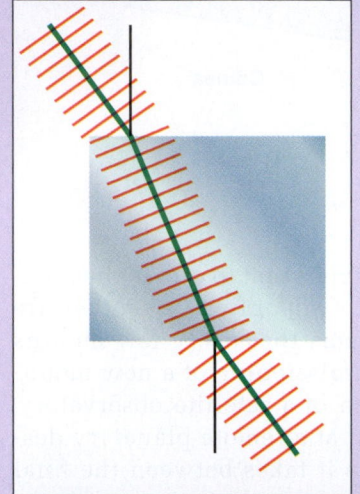

FIGURE 5.8 (a) Light waves are refracted (bent) when entering a medium with a higher index of refraction. They are refracted again as they reenter the medium with the lower index of refraction. (b) Light from a green laser beam is refracted as it enters and exits a plastic block.

(a)

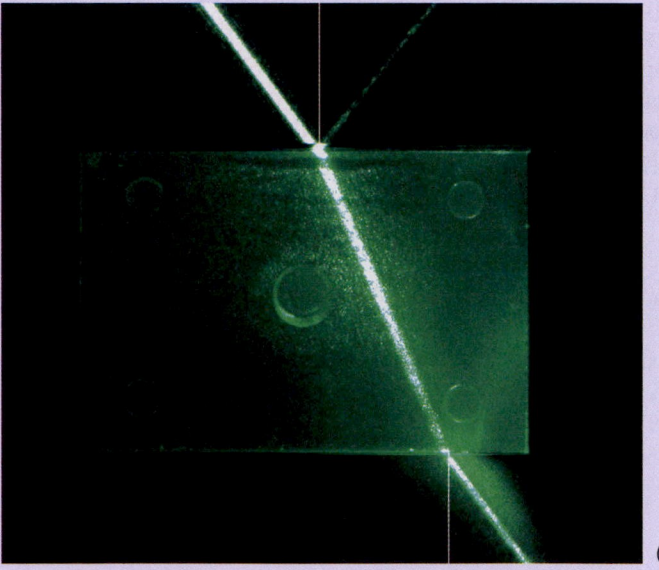

(b)

That's what makes a flat mirror so convenient for admiring our appearance.

Curved mirrors can also be very useful, especially in astronomical telescopes. The same rules of incidence and reflection hold here for each ray, but in this case the reflected rays do not maintain the same angle with respect to each other as they do with a flat mirror. A mirror that is concave toward the incoming light and has a parabolic surface (**Figure 5.10**) will reflect the rays so that they converge to form an image. If the incoming rays are parallel, as from a distant source—think "star"—the reflected rays cross at a distance from the mirror called the *focal length* of the mirror. Parallel rays from a distant source on

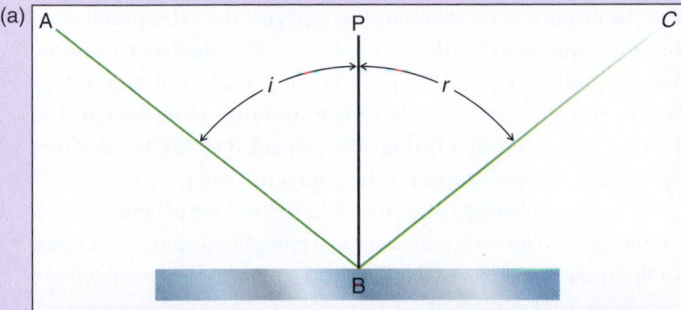

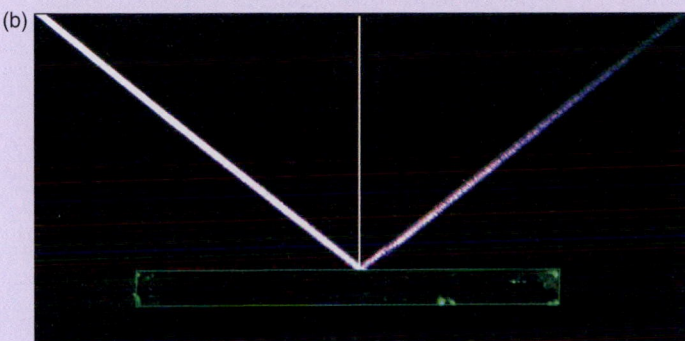

FIGURE 5.9 (a) A ray of incident light (AB) shines on a flat surface. The surface changes the direction of light by refraction and reflection. The angle of incidence (*i*) equals the angle of reflection (*r*). (b) Light from a laser beam is reflected from a flat glass surface.

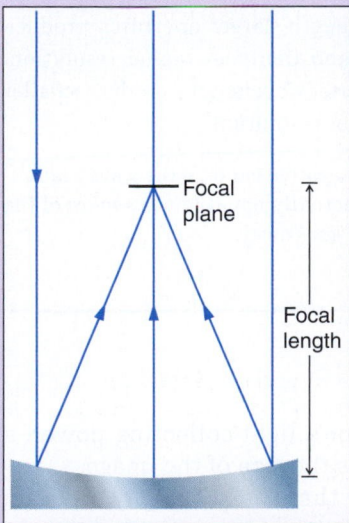

FIGURE 5.10 Parallel rays of light incident on a concave parabolic mirror are brought to a focus in the mirror's *focal plane*.

the axis come together on the axis at a point called the **focus**. The surface at which *all* parallel rays cross is called the focal plane (see Figure 5.10).

DISPERSION

Shine white light through a glass prism, as shown in **Figure 5.11**, and you'll get a rainbowlike spectrum. This effect demonstrates that the refraction, and therefore the speed of light in glass, depends on wavelength. Like most other transparent materials, glass has a refractive index (see Chapter 4) that increases with decreasing wavelength. This means that shorter wavelengths (those toward the blue) are refracted more strongly than longer wavelengths (those toward the red). This wavelength-dependent difference in refraction, which spreads the white light out into its spectral colors, is what we call **dispersion**. Although dispersion is helpful in producing prism spectra, it creates a serious problem called **chromatic aberration** (**Figure 5.12a**). Chromatic aberration causes blue light to come to a shorter focus than the longer visible wavelengths. You can see this effect when you look at a bright object such as a distant streetlight through an inexpensive telescope. (Low-priced telescopes usually have only a simple convex objective lens.) The streetlight will appear to be surrounded by a blue halo, because the blue component of the light is out of focus. Manufacturers of quality cameras and telescopes avoid the use of simple convex lenses in favor of the *compound lens*, like the one shown in **Figure 5.12b**. By using two types of glass, a compound lens corrects for chromatic aberration.

FIGURE 5.11 White light is dispersed into its component colors as it passes through a glass prism.

Continued on next page

FOUNDATIONS 5.1

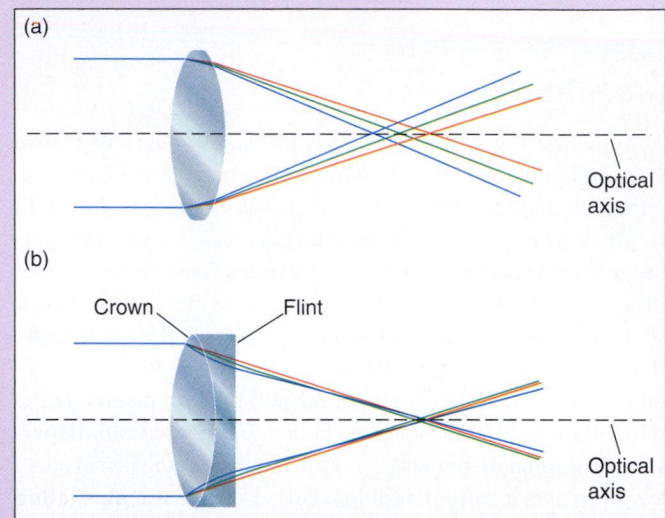

FIGURE 5.12 (a) Light of different wavelengths (different colors) comes to different foci along the optical axis of a simple lens, causing chromatic aberration. (b) A compound lens using two types of glass (crown and flint) with different indices of refraction can compensate for much of the chromatic aberration.

INTERFERENCE AND DIFFRACTION

The intersection of two sets of electromagnetic waves can produce patterns of high and low intensity called **interference** patterns. Say we have a pair of slits and an opaque screen. **Figure 5.13a** illustrates monochromatic light (light having a single wavelength or a very narrow range of wavelengths) going through the slits. Each slit now becomes a source of wavefronts. Notice the regular pattern on the screen where the wavefronts from the two slits intersect. If the intersection point occurs where the amplitudes of both waves are at their maximum positive or maximum negative value, the two add together and the

light will be bright.[3] We call this **constructive interference**. If, on the other hand, one wave is at its maximum positive value and the other is at its maximum negative value, the sum is zero and the result will be darkness. We call this **destructive interference**. When we replace the two slits with a large number of very narrow, very closely spaced parallel slits, we call the result a transmission **grating**. We can achieve the same effect by engraving closely spaced lines on a mirror.

If we now substitute a multiwavelength source of light, we get a similar pattern for each and every wavelength, as shown in **Figure 5.13b** for a reflection grating. For each wavelength, constructive interference takes place at a different point on the screen. In other words, the grating produces a spectrum. Modern spectrographs use a grating to disperse incoming light into its constituent wavelengths. You can see this effect for yourself. Look at light reflected from a CD or DVD. The closely spaced tracks act as a grating and create a respectable spectrum (**Figure 5.13c**).

When light passes through a small opening (or near an opaque edge), the effects of interference come into play. This is called **diffraction**. **Figure 5.14a** shows what happens when monochromatic light from a distant source passes through a lens and is brought to a focus. The image pattern is not a single point as we might expect, but rather is smeared out into a series of bright and dark concentric rings (as seen in **Figures 5.14b and c**), representing constructive and destructive interference from the edge of the aperture. If we make the aperture smaller, the diffracted image grows larger, causing more blurring and limiting how close two images can be and still be resolved. The angular size of the diffraction pattern depends on the ratio between the wavelength of light (λ) and the aperture of the hole (D): λ/D. At any given wavelength, larger apertures produce smaller diffraction patterns and therefore higher resolution. And at any given aperture, shorter wavelengths produce smaller diffraction patterns and higher resolution.

[3]Don't be concerned about the negative value of the wave's amplitude. The intensity of light is actually equal to the *square* of the amplitude, so the negative sign goes away.

---

telescope.) The diameter of the objective lens is known as the telescope's **aperture**, and it determines the light-gathering power of the lens. Note that the light-gathering power of a telescope is proportional to the *area* of its aperture— that is, to the *square* of its diameter. The distance between the telescope lens and the images formed is referred to as the **focal length** of the telescope. Note that longer focal lengths increase the size and separation of objects in the focal plane (see Figure 5.7). Aperture and focal length are the two most important parameters of a telescope. The aper-

ture determines a telescope's light-collecting power, and the focal length establishes the size of the image.

We have already seen that there are physical limits on the size, and thus effectiveness, of refracting telescopes. Early astronomers were also well aware of another serious drawback to simple-lens refracting telescopes: chromatic aberration (see Foundations 5.1). To minimize chromatic aberration, refracting telescopes now use **compound lenses**, which leave only small, residual effects of the blue halo.

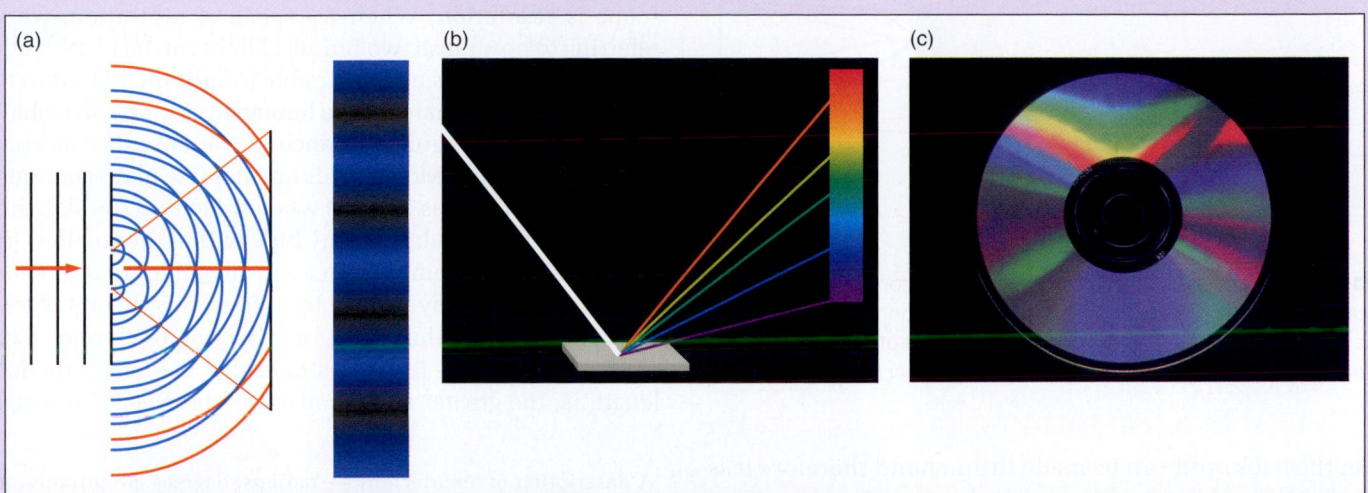

FIGURE 5.13 (a) Constructive and destructive interference patterns are created when monochromatic light passes through a pair of narrow slits. (b) Spectral dispersion is produced by interference when multiwavelength (white) light reflects from a grating. (c) A spectrum is created by the reflection of light from the closely spaced tracks of a CD.

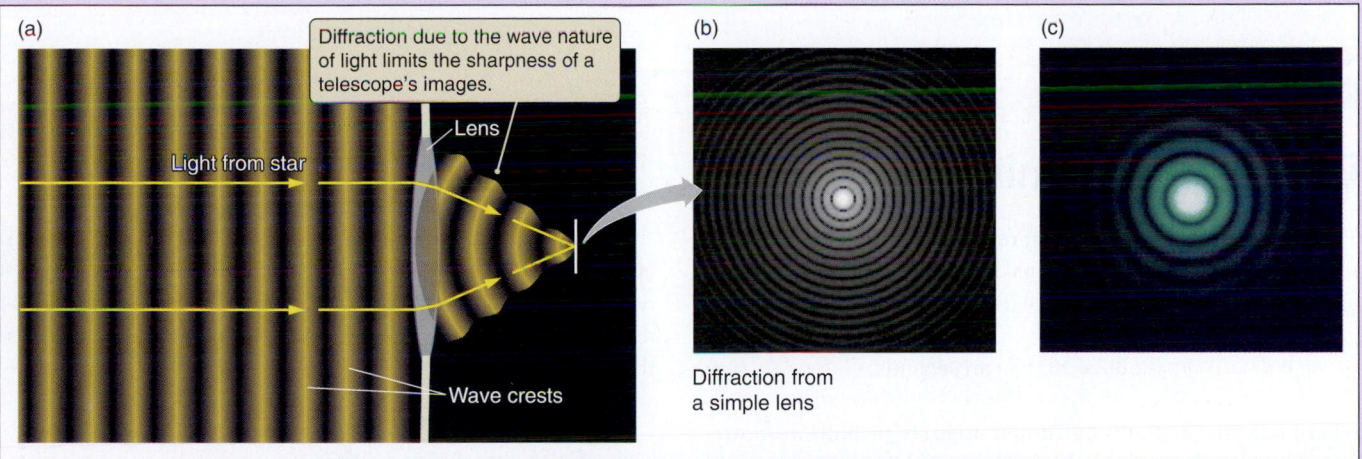

FIGURE 5.14 (a) Light waves from a star are diffracted by the edges of a telescope's lens or mirror. (b) This diffraction causes the stellar image to be blurred, limiting a telescope's ability to resolve objects. (c) Diffraction from a circular aperture illuminated by green laser light.

Reflecting telescopes deal with chromatic aberration and other limitations of refracting in different ways. (See Foundations 5.1 to learn more about the optics of reflected light.) A reflecting telescope forms an image in its focal plane when light is reflected from a specially curved mirror rather than from the refracting telescope's convex lens. We call this mirror the **primary mirror** to distinguish it from the additional mirrors usually employed in modern reflecting telescopes. We can fold the light path from the primary mirror to the focal plane by introducing a **second-**

ary mirror (Figure 5.15). The secondary mirror allows a significant reduction in the length and weight of the telescope. For example, the effective focal length of each of the 10-meter Keck telescopes can be as long as 250 meters, even though the overall length of each telescope is a mere 25 meters.

Reflectors have a number of important advantages over refractors. Because the direction of a reflected ray does not depend on the wavelength of light, chromatic aberration is no longer a problem. Primary mirrors can be supported

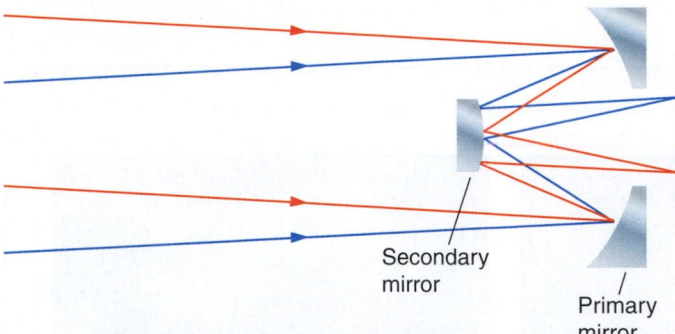

FIGURE 5.15 Reflecting telescopes use mirrors to collect and focus light. Large telescopes typically use a secondary mirror that directs the light back through a hole in the primary mirror to an accessible focal plane behind the primary mirror.

from the back, and can be made thinner and therefore less massive than objective lenses. The lesser mass of a mirror becomes a distinct advantage when we are dealing with astronomers' insatiable appetite for ever-larger telescopes.

▶‖ **AstroTour: Geometric Optics and Lenses**

Optical and Atmospheric Limitations

One of the eye's major limitations as an astronomical telescope is **resolution**. When we speak of resolution, we are referring to how close two points of light can be to each other before a telescope is no longer able to split the light into two separate images. Unaided, the human eye can resolve objects separated by an angular distance of 1 arcminute,[4] or $\frac{1}{30}$ the diameter of the full Moon.[5] This angular size may seem small, and in our daily lives it is; yet when we look at the sky, thousands of stars and galaxies may hide within the smallest area that the unaided human eye can resolve. Figure 5.7a shows the path followed by rays of light from two distant stars as they pass through the lens of a refracting telescope. Comparison with Figure 5.7b illustrates that the longer the focal length is, the greater is the separation between the images.

---

[4]A description of angular units—radians, degrees, arcminutes, and arcseconds—can be found in Chapter 13 (Section 13.2) and in Appendix 1.

[5]Only the sharpest of human eyes achieve a resolution of 1 arcminute. Typical resolution for many of us would be more like 2 arcminutes.

MATH TOOLS 5.1

Diffraction Limit

The ultimate limit on the angular resolution of a telescope, called the diffraction limit, is determined by the ratio of the wavelength of light passing through it to the diameter of the lens:

$$\theta = 2.06 \times 10^5 \left(\frac{\lambda}{D}\right) \text{ arcseconds,}$$

where θ is the diffraction-limited angular resolution in arcseconds, λ is the wavelength of light, and D is the diameter, or aperture, of the telescope.[6] Both λ and D are expressed in the same units, usually meters. As we can see, the smaller the ratio of λ/D, the better will be the resolution of the telescope. We can apply this relationship to the Hubble Space Telescope (HST) operating in the visible part of the spectrum. The HST's primary mirror has a diameter (D) of 2.4 meters. Visible (green) light has a wavelength (λ) of 550 nanometers (nm), or 5.5×10^{-7} meters. Substituting into the previous equation, we have

$$\theta = 2.06 \times 10^5 \left(\frac{5.5 \times 10^{-7}}{2.4}\right) = 0.047 \text{ arcsecond,}$$

or about 1,000 times better than the resolving power of the human eye.

[6]The constant of proportionality here is simply the number of arcseconds in a radian.

Let's look at the human eye. The size of the human pupil (see Figure 5.6) can range from about 2 mm in bright light to 8 mm in the dark. A typical pupil size is about 4 mm, or 0.004 meter. If we substitute this value for D and again assume visible light, we have

$$\theta = 2.06 \times 10^5 \left(\frac{5.5 \times 10^{-7}}{0.004}\right) = 28.3 \text{ arcseconds,}$$

or about 0.5 arcminute. But, as we learned earlier, the best resolution that the human eye can achieve is about 1 arcminute, and more typically it is 2 arcminutes. Why don't we achieve the theoretical resolution with our eyes? The reason is that the optical properties of the lens and the physical properties of the retina (see Figure 5.6) that nature has given us are not perfect.

Electron microscopes take advantage of this property of diffraction to achieve very high resolution by using (what else?) electrons instead of photons to illuminate the target. Recall the dual wave-particle nature of electromagnetic radiation, which we discussed in Chapter 4. Electrons can behave both as particles and as waves, with wavelengths shorter than 0.1 nm. This means that electron microscopes have more than five *thousand* times better resolution than conventional microscopes, which use visible light ($\lambda \approx 550$ nm).

The focal length of a human eye is typically about 20 millimeters (mm). In comparison, telescopes used by professional astronomers often have focal lengths of tens or even hundreds of meters. Such telescopes make images that are far larger than those formed by your eye, and consequently they contain far more detail.

Focal length explains only one difference between the resolution of telescopes and the unaided eye. The other difference results from the wave nature of light. As light waves pass through the lens of a telescope, they spread out from the edges of the lens (see Figure 5.14). The distortion of the wavefront as it passes the edge of an opaque object is called diffraction (see Foundations 5.1). Diffraction "diverts" some of the light from its path, slightly blurring the image made by the telescope. The degree of blurring depends on the wavelength of the light compared with the diameter of the telescope lens. The larger the lens relative to the wavelength of the light it is focusing, the smaller the problem posed by diffraction. The limiting resolution that a given telescope can achieve is known as the **diffraction limit** (see **Math Tools 5.1**).

> Diffraction, or blurring of an image, depends on the ratio of wavelength to telescope aperture.

The diffraction limit tells us that larger telescopes get better resolution. Theoretically, the 10-meter Keck telescopes have a diffraction-limited resolution of 0.0113 arcsecond in visible light, which would allow you to read newspaper headlines 60 km away. But for telescopes with apertures larger than about a meter, Earth's atmosphere stands in the way of better resolution. If you have ever looked out across the desert on a summer day, you have seen the distant horizon shimmer as light from that horizon is constantly bent this way and that by turbulent bubbles of warm air rising off the hot desert floor.

The problem is less pronounced when we look overhead, but the twinkling of stars in the night sky tells us the phenomenon is still there. As telescopes magnify the angular diameter of a planet, they also magnify the shimmering effects of the atmosphere. The limit on the resolution of a telescope on the surface of Earth caused by this atmospheric distortion is called **astronomical seeing**. One advantage of launching telescopes such as the Hubble Space Telescope into orbit around Earth is that from their vantage point above the atmosphere, telescopes get a much clearer view of the universe, unhampered by the astronomical seeing phenomenon. Does that mean that groundbased telescopes are becoming obsolete? Not at all. Modern technology has come to their rescue with computer-controlled **adaptive optics**, which compensate for much of the atmosphere's distortion.

> Earth's atmosphere distorts images.

To better understand how adaptive optics work, we need to look more closely at how Earth's atmosphere smears out an otherwise perfect stellar image. Look again at Figure

5.14a. Light from a distant star arrives at the top of Earth's atmosphere as a series of flat, parallel waves called a **wavefront**. If Earth's atmosphere were perfectly homogeneous, the wavefront would remain flat as it reaches the objective lens or primary mirror of a groundbased telescope. After making its way through the telescope's optical system, the wavefront would produce a tiny diffraction disk in the focal plane, as shown in Figure 5.14b. But Earth's atmosphere is not homogeneous. It is filled with small bubbles of air that have slightly different temperatures than their surroundings. Different temperatures mean different densities, and different densities mean different refractive properties.

These air bubbles act as weak lenses, and by the time the wavefront reaches the telescope it is far from flat, as shown in **Figure 5.16**. Instead of a tiny diffraction disk, the image

FIGURE 5.16 Bubbles of warmer or cooler air in Earth's atmosphere distort the wavefront of light from a distant object.

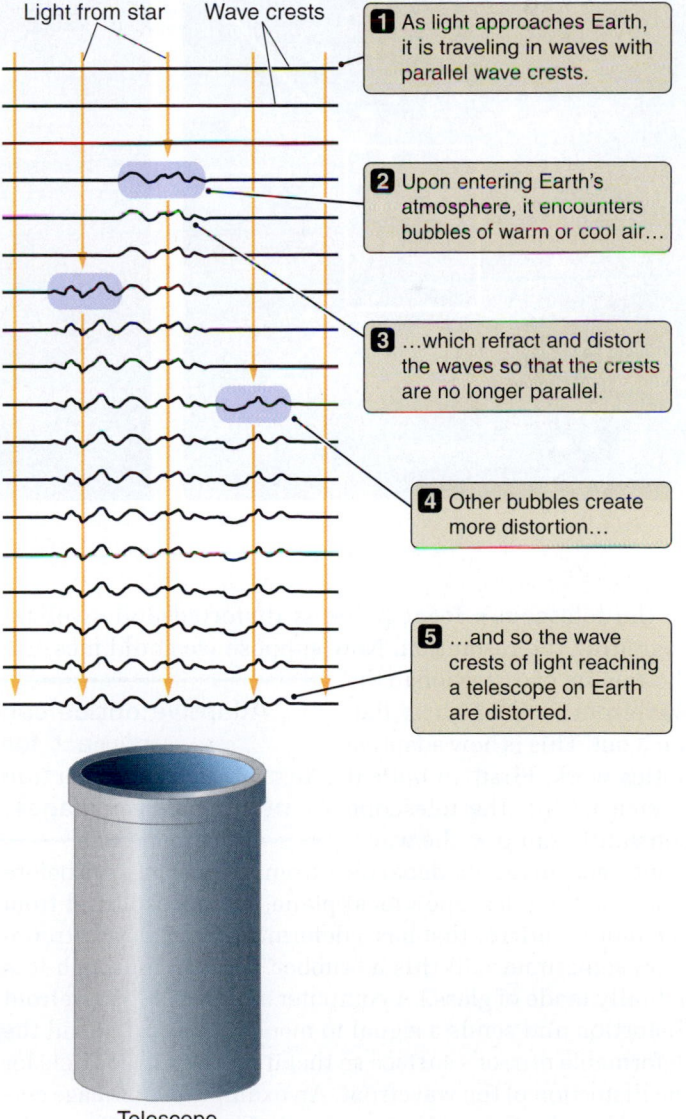

Light from star   Wave crests

1 As light approaches Earth, it is traveling in waves with parallel wave crests.

2 Upon entering Earth's atmosphere, it encounters bubbles of warm or cool air…

3 …which refract and distort the waves so that the crests are no longer parallel.

4 Other bubbles create more distortion…

5 …and so the wave crests of light reaching a telescope on Earth are distorted.

Telescope

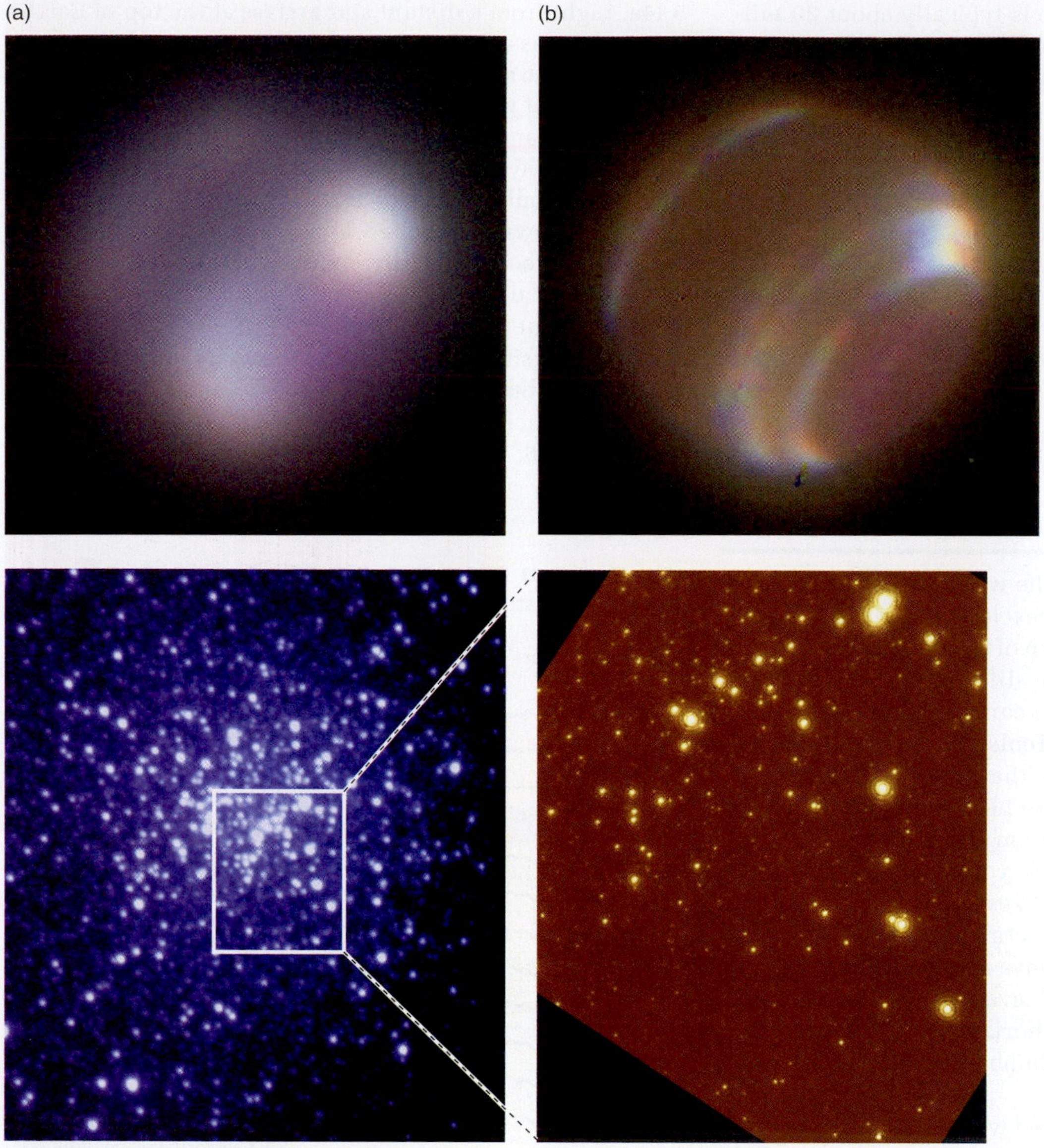

(a)

(b)

FIGURE 5.17 Images of Neptune and a star cluster taken without (a) and with (b) adaptive optics.

in the telescope's focal plane is distorted and swollen, degrading the resolution. Now suppose we could measure the amount of distortion in the wavefront and somehow flatten it out. This is how adaptive optics work. First, an optical device within the telescope constantly samples the wavefront, measuring its departure from flatness. Then, before reaching the telescope's focal plane, light is reflected from yet another mirror that has a deformable surface. (Astronomers sometimes call this a "rubber" mirror, although it is actually made of glass.) A computer analyzes the wavefront distortion and sends a signal to mechanisms that bend the deformable mirror's surface so that it accurately corrects for the distortion of the wavefront. An example of an image corrected by adaptive optics is shown in **Figure 5.17**. The widespread use of adaptive optics has now made the image quality

> Adaptive optics can correct for atmospheric distortion of telescopic images.

of groundbased telescopes competitive with those of the Hubble Space Telescope. But image distortion is not the only problem caused by Earth's atmosphere. Large regions of the electromagnetic spectrum are partially or completely absorbed by various atmospheric molecules.

Another limitation on the human eye is that it is sensitive only to light in the visible part of the electromagnetic spectrum. (That is, after all, why we call it the "visible" part of the spectrum!) Even though visible light is only a small part of the electromagnetic spectrum, it is anything but happenstance that our eyes work in this range of wavelengths. We evolved on a planet with an atmosphere that is transparent to light in the visible part of the spectrum but is opaque to almost all other wavelengths. Nearly all of the X-ray, ultraviolet, and infrared light arriving at Earth is

> Earth's atmosphere blocks much of the electromagnetic spectrum.

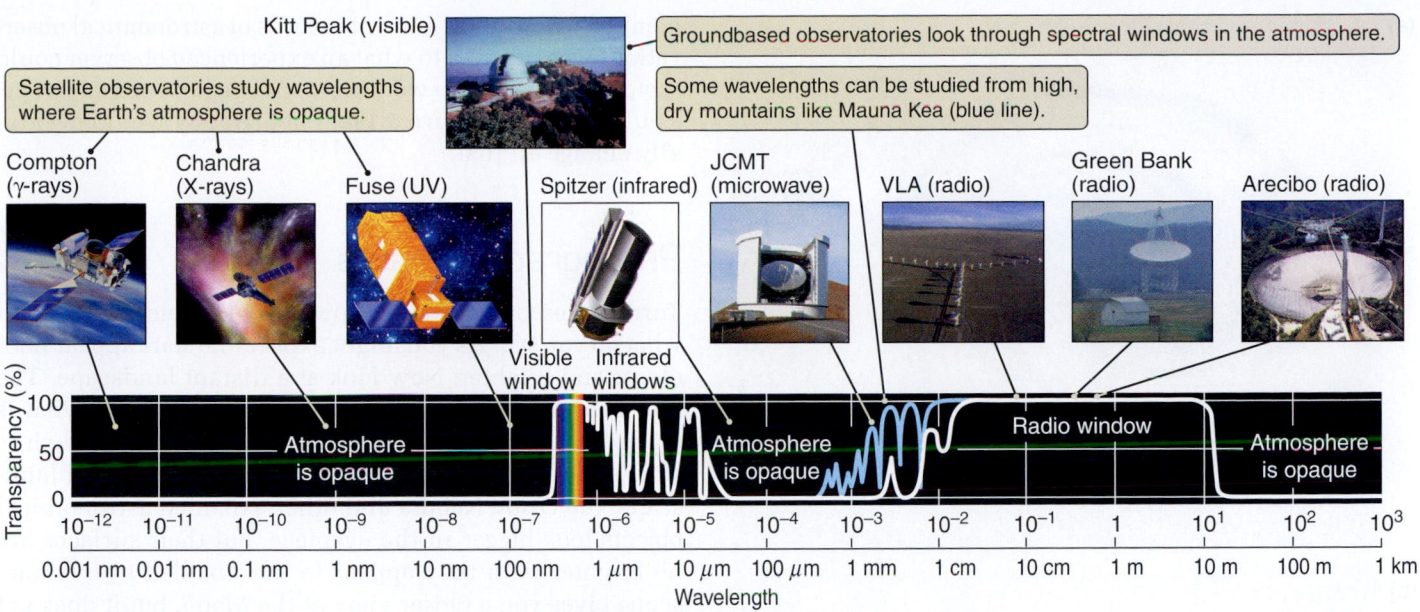

FIGURE 5.18 Earth's atmosphere blocks most electromagnetic radiation.

blocked by Earth's atmosphere before it reaches the ground.[7] The visible part of the spectrum, which is not blocked by our atmosphere, is a fairly narrow range of wavelengths, or window, through which we can view the universe. There are a few other **atmospheric windows** in the spectrum as well, as shown in **Figure 5.18**.

Do not make the mistake of thinking that light that fails to reach the surface of Earth is uninteresting. There are many things that we can learn only by observing the universe outside the visible window. Although radio observations are also possible from the ground, we owe a large fraction of what we know about the universe to a host of ultraviolet, X-ray, gamma-ray, and infrared telescopes that, beginning in the 1960s, were carried above Earth's atmosphere by rockets. We'll discuss space telescopes in more detail later in the chapter.

5.2 Optical Detectors and Instruments

In reality, the sole purpose of an astronomical telescope is simply to collect light waves from the cosmos and bring them to a focus. We turn now to the various ways in which astronomers capture these light waves and convert them to useful information. Detectors are devices placed in a telescope's focal plane to transform light waves into images that

we can see and record. As with our earlier approach, we start with the most basic of telescopes, the human eye.

The detector in the human eye is the retina (see Figure 5.6), and the individual receptor cells that respond to light falling on the retina are called "rods" and "cones." Cones are located near the eye's optical axis at the center of our vision. They provide the highest resolution and enable us to recognize color. The size and spacing of cones—not the 7-mm pupil—determine the 1-arcminute resolution of the human eye. Rods, which are located away from the eye's optical axis and are responsible for our peripheral vision, provide the highest sensitivity to low light levels, but they have poorer resolution and cannot distinguish color. The photons to which the human eye is sensitive have wavelengths ranging from about 400 nm (deep violet) to 700 nm (far red).

As photons from a star enter the aperture or pupil of the eye, they fall on and excite cones at the center of vision. The cones then send a signal to the brain, which interprets this message as "I see a star." So what limits the faintest stars we can see with our unaided eyes, assuming a clear, dark night and good eyesight? This limit is determined in part by two factors that are characteristic of all detectors: *integration time* and *quantum efficiency*.

Integration time is the limited time interval in which the eye can add up photons. For the human eye, the integration time is about 100 milliseconds (ms). If two images on a television or computer screen appear 30 ms apart, you will see them as a single image because your eyes will sum whatever they see over an interval of 100 ms. If the images occur, say, 200 ms apart, you will see them as separate images. So the signals the cones send to your brain include only those photons that arrive within an interval of 100 ms. This relatively brief integration time is the biggest factor limiting our nighttime vision. Stars too faint for us to see

---

[7]Ozone and molecular scattering are the principal atmospheric components that block electromagnetic radiation shorter than visible wavelengths from reaching Earth's surface. Water vapor and carbon dioxide tend to block radiation longer than visible wavelengths.

(a)

(b)

FIGURE 5.19 (a) A drawing of the galaxy M51 made by William Parsons (Lord Rosse, 1800–1867) in 1845. Compare this drawing with the modern photograph of M51 in Figure 5.20. (b) A photograph of the Moon taken by John W. Draper in 1840.

with our unaided eyes are those that produce too few photons for our eyes to process in 100 ms.

Another effect, called **quantum efficiency**, also restricts our nighttime vision. As the name implies, quantum efficiency is the likelihood that a particular photon landing on the retina will, in fact, produce a response. For the human eye, it takes about 10 photons landing on a cone to activate a single response. In other words, the quantum efficiency of our eyes is about 10 percent. Together, integration time and quantum efficiency determine the rate at which photons must arrive on the retina before the brain says, "Aha, I see something."

For more than two centuries after the invention of the telescope, the retina of the human eye was the only astro-

nomical detector. Permanent records of astronomical observations were limited to what an experienced observer could sketch on paper while working at the eyepiece of a telescope, as illustrated in **Figure 5.19a**. Photography would eventually change all that.

Photographic Plates

Turn a telescope or binoculars toward a field of stars and what do you see? As you might expect, the stars appear both closer and brighter. Now look at a distant landscape. The scene seems closer, but its surface is no brighter. What is going on here? It turns out that only "point sources" such as stars appear brighter in a telescope. Like the distant landscape, the Orion Nebula and other extended astronomical objects look bigger in the eyepiece, but their surfaces are no brighter than they appear to the unaided eye. A telescope gives you a closer view of the Moon, but it does not increase the Moon's surface brightness. For more than two centuries after the invention of the telescope, astronomers struggled with this **surface brightness** problem. No matter how big they built their telescopes, nebulae and galaxies might appear larger but their faint details remained elusive. The problem, of course, was not with the telescopes but with the limitations of optics and the human eye. Only with the invention of photography and the later development of electronic cameras were astronomers finally able to discern the faint but intricate fabric of the cosmos.

In 1840, John W. Draper (1811–1882), a New York chemistry professor, created the earliest known astronomical photograph, shown in **Figure 5.19b**. His subject was the Moon. Photography was not quick to catch on among astronomers, though, because the early "daguerreotype" process was slow and very messy. The relatively simple dry emulsion process finally came along in the late 1870s, and with that innovation, astronomical photography took off. Astronomers could now create permanent images of planets, nebulae, and galaxies with ease. Thousands of photographic plates soon filled the "plate vaults" of major observatories. Photography had created its own astronomical revolution.

In the dry emulsion process, a layer of gelatin containing tiny crystals of silver halide is coated onto glass plates or film.[8] During an exposure, photons landing on the emulsion energize the silver halide crystals, creating what is called a "latent" image. "Developing" the emulsion turns these small crystals into black grains of metallic silver, forming a permanent image. The highest density of silver grains occurs where the telescope's image was the brightest. So, bright becomes black, and the photographic image is nega-

> Photography opened the door to modern astronomy.

---

[8]Glass plates, although far more expensive than film, are generally used for imaging and spectroscopy because of their greater geometric stability.

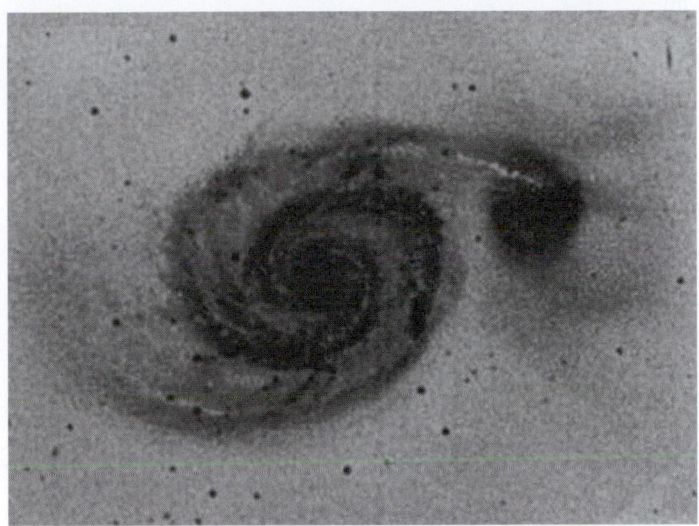

FIGURE 5.20 An image of Galaxy M51 on a photographic plate.

tive, as illustrated in **Figure 5.20**. The quantum efficiency of most photographic emulsions used in astronomy is very low—typically 1–3 percent, which is even poorer than that of the human eye. But unlike the eye, photographic emulsions can overcome poor quantum efficiency by integrating photons over intervals of many hours. Photography made it possible for astronomers to record and study objects much fainter than the human eye can see.

Photography is not without its own problems. Very faint objects often require long exposures that can take up much of an observing night. (Imagine how you would feel if your 10-hour exposure were spoiled because of a mishap.) In addition, the spectral range of photographic emulsions is hardly broader than that of the human eye. In fact, for many years photographic plates were sensitive to only violet and blue light. Another problem is the *nonlinear* response of photographic emulsions to light. This is just another way of saying that the optical density (blackness) of the processed emulsion is *not* proportional to the intensity of light falling on it. Finally, there is the nontrivial matter of economics. Each photographic plate can be used only once, and they are expensive. By the middle of the 20th century, the search was on for electronic detectors that would overcome many of the deficiencies of photographic plates.

Charge-Coupled Devices

Throughout the latter half of the 20th century, astronomers employed various electronic detectors to overcome the sensitivity, spectral range, and nonlinearity problems of photography. Some, such as "photoelectric photometers," are nonimaging devices that receive photons from an object and convert them to an electronic signal that is proportional to the brightness of the object. These devices work extremely well for precision stellar **photometry** because

they have excellent linearity, which means that their electronic output is directly proportional to the intensity of light falling on the photometric detector. But they can measure only one stellar image at a time, so observation is truly labor-intensive. Other detectors, such as "vidicons," are electronic imaging devices with sensitivity far superior to photographic emulsions. Vidicons unfortunately suffer from an electronic instability that causes small geometric distortions of the image. You may have seen pictures taken by early spacecraft that had little plus-shaped marks superimposed on them. These "fiducial" marks were engraved on the faceplates of the vidicons to help with the removal of geometric distortions in the image. The search for a better detector continued.

In 1969, scientists at Bell Laboratories were developing "picture phones"—telephones containing a small camera and viewing screen that could display an image of the person at the other end of the conversation. As it turned out, public opinion declared Bell's picture phones an invasion of personal privacy and they were never commercially produced, but the research led to the invention of a remarkable detector called a **charge-coupled device**, or **CCD**. Astronomers soon realized that this was the detector they had been looking for. Shortly after it was first applied to astronomical imaging in the mid-1970s, the CCD became the detector of choice in almost all astronomical imaging applications. Gone were the problems associated with photographic emulsions, photoelectric photometers, and vidicon-type imagers. The CCD is a photometrically linear imaging device, able to perform precise photometry over large regions of sky. It responds over a wide spectral range from 200 to 1,200 nm and has a high quantum efficiency, typically 80 percent or greater. The output from a CCD is a digital signal that can be sent directly from the telescope to image-processing software or stored electronically for later analysis.

A CCD consists of an ultrathin wafer of silicon, less than the thickness of a human hair, which is divided into a two-dimensional array of picture elements, or **pixels**, as seen in **Figure 5.21a**. When a photon strikes a pixel, it creates a small electric charge within the silicon. As each CCD pixel is "read out," the digital signal that flows to the computer is almost precisely proportional to the accumulated charge. This is what we mean when we say that the CCD is a linear device. Like many electronic detectors, CCDs are subject to thermal noise—a false signal caused by the thermal agitation of charge-carrying atoms within the silicon wafer. But we can minimize thermal noise by cooling the CCDs down to liquid nitrogen temperatures (~80 K). The first astronomical CCDs were small arrays containing no more than a few hundred thousand pixels. The larger CCDs used in astronomy today—like the one seen in **Figure 5.21b**—may contain as many as 100 million pixels. Still-larger arrays are under development, as ever-faster computing power keeps up with image-processing demands.

The impact that CCDs have had on astronomy cannot be overstated! The use of conventional photography by ground-

(a)

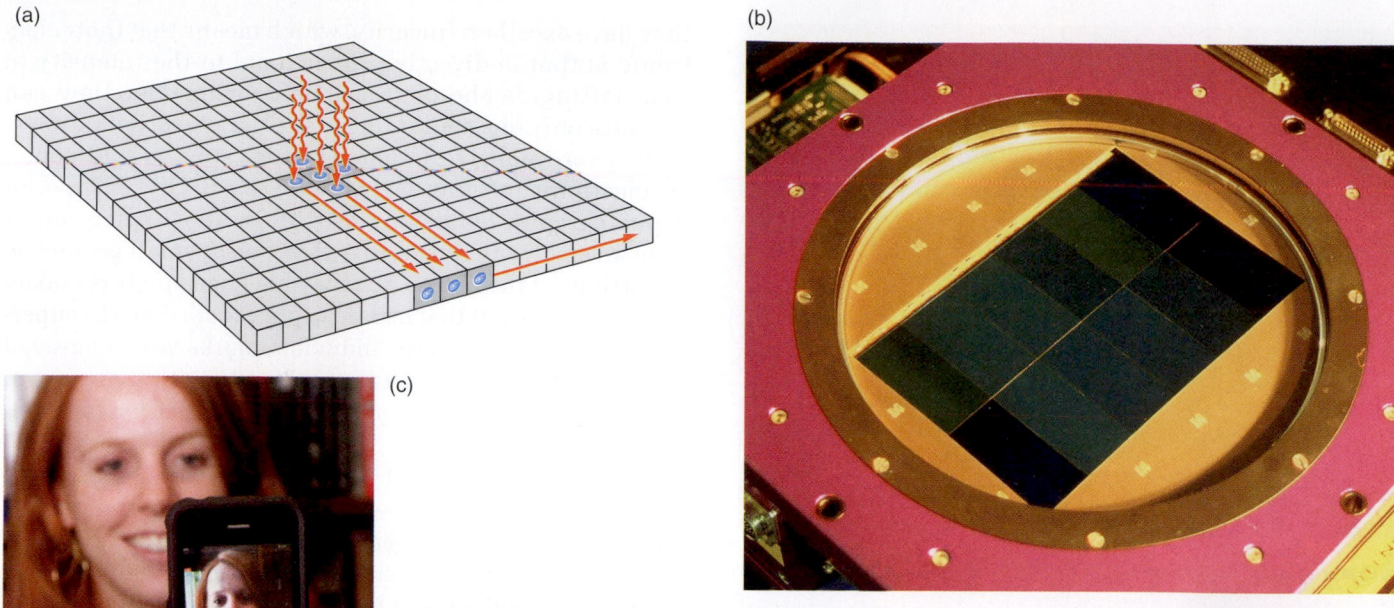

(b)

(c)

FIGURE 5.21 (a) A simplified diagram of a charge-coupled device (CCD). Photons from a star land on pixels (gray squares) and produce free electrons within the silicon. The electron charges are electronically moved sequentially to the collecting register at the bottom. Each row is then moved out to the right to an electronic amplifier, which converts the electric charge of each pixel into a digital signal. (b) This very large CCD contains 12,288 × 8,192 pixels. (c) CCDs have been used in many consumer electronic devices, such as camera phones..

(a)

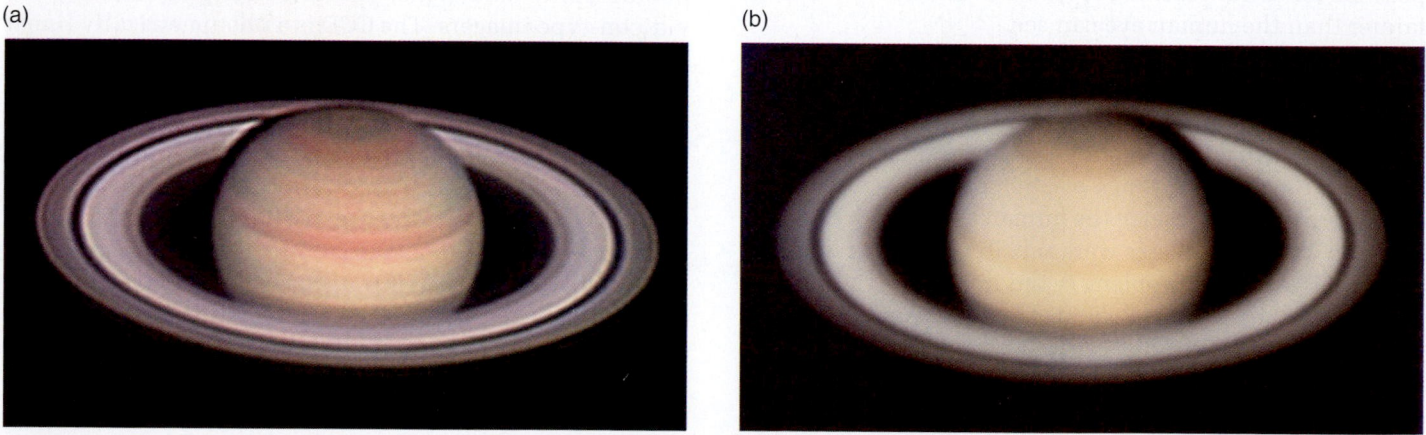

(b)

FIGURE 5.22 (a) A CCD image of Saturn taken in 2005 by an amateur astronomer with a 36-cm telescope. (b) A pre-CCD photograph of Saturn taken in 1974 with a 1.5-meter telescope at the Catalina Observatory of the University of Arizona.

based observatories has steadily declined over the past two decades, and it is seldom employed today. As you surf the Internet, nearly every spectacular astronomical image that pops up on your screen was produced by a CCD, whether from groundbased telescopes or from those in space. And professional astronomers are not the only ones making spectacular photos with CCDs. Amateur astronomers are now using commercially available, ther-

> The CCD is the astronomer's detector of choice.

moelectrically cooled CCD imagers with impressive results. **Figure 5.22a** is an image of Saturn taken by an amateur astronomer with a 36-centimeter (cm) telescope. This image shows more detail than the best professional photographs of Saturn taken before CCDs became available to astronomers, an example of which is shown in **Figure 5.22b**.

You may never have seen a CCD, but in recent years they have found their way into many devices that we now take for granted—digital cameras, digital video cameras, and camera phones, to name just a few (**Figure 5.21c**).

Spectrographs

Spectrographs—or **spectrometers**, as they are often called—are another of the astronomer's essential tools. As we learned in Chapter 4, we can probe the chemical and physical properties of distant objects by studying their spectra. Early spectrographs used glass prisms to disperse the incoming light into its component wavelengths (**Figure 5.23**), creating a spectrum like the ones shown in Figures 4.14 and 5.11. A photographic plate recorded the spectrum for precise measurement of the wavelengths of its spectral lines. One disadvantage of glass prisms is that dispersion (see Foundations 5.1) is not uniform with wavelength. Prism spectrographs produce spectra with more dispersion at the shorter-wavelength (violet) end of the spectrum than at the longer-wavelength (red) end. Another drawback is that glass is opaque to ultraviolet and long-wavelength infrared light. Thus, prism spectrographs are more or less limited to the visible part of the spectrum. Most modern spectrographs use a "diffraction grating" to disperse the light (see Foundations 5.1 and Figure 5.13b) and a CCD to record the spectrum.

Spectrographs may be designed for either low or high dispersion. Astronomers use low-dispersion spectrographs to identify the chemical components of feeble light sources such as nebulae. Low-dispersion spectrographs also measure the reflected spectral energy distribution of faint Solar System objects such as small or distant asteroids. Measurements of temperature and radial velocity, on the other hand, usually require very high dispersion. Today's high-dispersion spectrographs have evolved into substantial scientific instruments—hardly the sort of thing that one would transport from one telescope to another. The high-dispersion spectrographs now associated with most major telescopes tend to be huge and weigh several metric tons—think SUV size. (For obvious reasons, they are not attached directly to the telescope!) A system of mirrors feeds light from the telescope's focal plane into the spectrograph, which is located nearby.

Spectroscopy is the study of an object's electromagnetic radiation in terms of its component wavelengths, and we'll encounter its many applications throughout the chapters to come as we continue in our journey through the Solar System and beyond.

5.3 Radio Telescopes

Photons of visible light are not the only messengers of important information from the cosmos. Karl Jansky (1905–1950) was a young physicist working for Bell Laboratories in the early 1930s when he was assigned the job of identifying sources of static in transatlantic radiotelephone service. Jansky built a pointable antenna and soon identified the major sources of static as nearby and distant thunderstorms. But one source remained a mystery—a faint, steady hiss rose and fell once every 23 hours and 56 minutes. Recall from Chapter 2 that this is the same period as Earth's rotation as seen from the stars. What Jansky had detected was the characteristic time interval of celestial objects far beyond our Solar System. In 1932, Jansky identified the mysterious source. It was located in the Milky Way in the direction of the galactic center, in the constellation of Sagittarius. Excited by his discovery, he submitted a request to build a large dish antenna, or **radio telescope**, to study these signals in more detail. Bell Labs turned down the request. After all, Jansky had already provided the information they needed. Nevertheless, Jansky's discovery marked the birth of radio astronomy. In his honor, the basic unit for the strength of a radio source is called the **jansky (Jy)**.

In 1937, Grote Reber (1911–2002), an American radio engineer and ham radio operator, decided to build his own radio telescope. It consisted of a parabolic sheet of metal, 9 meters in diameter, with a radio receiver mounted at the focus. With this instrument, Reber conducted the first survey of the sky at radio frequencies, and he published the first radio frequency map of the galaxy in 1944. Reber was largely responsible for the rapid advancement in radio astronomy that blossomed in the post–World War II era.

Radio telescopes are yet another of astronomy's indispensable tools. From our Solar System to the most distant galaxies, the penetrating power of radio waves unlocks secrets not possible with shorter-wavelength optical or infrared telescopes. Look back at Figure 5.18 and notice the wide radio window in Earth's atmosphere, covering wavelengths ranging all the way from about a centimeter to 10 meters.[9] This ability of radio waves to pass unattenuated

FIGURE 5.23 A diagram of a prism spectrograph.

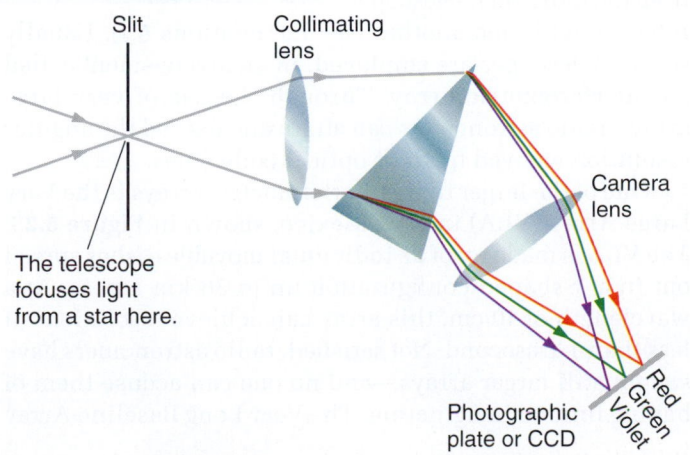

Slit

Collimating lens

The telescope focuses light from a star here.

Camera lens

Photographic plate or CCD

Red
Green
Violet

---

[9]Microwave astronomy is considered a branch of radio astronomy. Microwaves are very high-frequency radio waves with wavelengths ranging from about 1 mm to 10 cm at the short-wavelength end of the radio spectrum. As seen in Figure 5.18, Earth's atmosphere is only partially transparent to microwaves.

(a)

(b)

FIGURE 5.24 (a) A large radio telescope in Australia. (b) The Arecibo radio telescope is the world's largest. The steerable receiver suspended above the dish permits limited pointing toward celestial targets as they pass close to the zenith.

through our atmosphere is also the property that enables us to peer through the vast amounts of gas and dust found in many galaxies. Most radio telescopes are large, steerable parabolic dishes, typically tens of meters in diameter, such as the one shown in **Figure 5.24a**. The world's larg-est radio telescope is the 305-meter Arecibo dish built into a natural bowl-shaped depression in Puerto Rico (**Figure 5.24b**). But there can be a price to pay for size. As you might guess from the picture, this huge structure is too big to steer. Instead, pointing it requires moving the radio receiver suspended in the focal plane above the dish. Arecibo's targets are therefore limited to celestial sources that pass within 20° of the zenith as Earth's rotation carries them overhead.

> Radio telescopes enable astronomers to "see" through obscuring gas and dust.

As large as radio telescopes are, they have relatively poor angular resolution. Recall our earlier discussion about diffraction (see Foundations 5.1 and Math Tools 5.1). A telescope's angular resolution is determined by the ratio λ/D, where λ is the wavelength of electromagnetic radiation and D is the telescope's aperture. (Keep in mind that a larger ratio means poorer resolution.) Radio telescopes have diameters much larger than the apertures of most optical telescopes, and that helps. But the wavelengths of radio waves are typically several hundred thousand times greater than the wavelengths of visible light, and that hurts. Radio telescopes are thus hampered by the very long wavelengths they are designed to receive. Consider the huge Arecibo dish. Its resolution is typically about 1 arcminute, no better than the unaided human eye! So radio astronomers have had to develop their own bag of tricks, and one of the cleverest is the **interferometer**.

> Single radio telescopes have relatively poor resolution . . .

When we combine the signals from two radio telescopes in a certain way, the separation between them—not the diameters of the individual telescopes—determines the angular resolution. For example, if two 10-meter telescopes are located 1,000 meters apart, the D in λ/D is 1,000, not 10. Such an arrangement is called an interferometer because it makes use of the wavelike properties of electromagnetic radiation, in which signals from the individual telescopes *interfere* with one another (see Foundations 5.1). Usually several telescopes are employed, in an arrangement called an **interferometric array**. Through the use of very large arrays, radio astronomers can attain and exceed the angular resolution enjoyed by their optical colleagues.

> . . . but interferometric arrays overcome this problem.

One of the larger radio interferometric arrays is the Very Large Array (VLA) in New Mexico, shown in **Figure 5.25**. The VLA is made up of 27 individual movable dishes spread out in a Y-shaped configuration up to 36 km across. At a wavelength of 10 cm, this array can achieve resolutions of less than 1 arcsecond. Not satisfied, radio astronomers have sought still larger arrays—and no one can accuse them of having limited imagination. The Very Long Baseline Array

FIGURE 5.25 The Very Large Array (VLA) in New Mexico.

FIGURE 5.26 The Very Large Telescope (VLT), operated by the European Southern Observatory in Chile. Movable auxiliary telescopes allow the four large telescopes to operate as an optical interferometer.

(VLBA) employs 10 radio telescopes spread out over more than 8,000 km from the Virgin Islands in the Caribbean to Hawaii in the Pacific. At a wavelength of 10 cm, this array can reach resolutions better than 0.003 arcsecond. It might seem that Earth's diameter would set the ultimate limit on resolution for radio astronomers, but a radio telescope put into near-Earth orbit as part of a Space Very Long Baseline Interferometer (SVLBI) overcomes even this limit. Future SVLBI projects would extend the baseline to as much as 100,000 km, yielding resolutions far exceeding those of any existing optical telescope.

Before leaving our discussion of interferometers, we should point out that radio astronomers are not the only ones using the interferometer's greater resolving power. Optical telescopes can also be arrayed to yield resolutions greater than those of single telescopes, although for technical reasons the individual units cannot be spread as far apart as radio telescopes. The Very Large Telescope Interferometer (VLTI), operated by the European Southern Observatory (ESO) in Chile, makes use of a combination of the four VLT 8-meter telescopes (**Figure 5.26**) and four movable 1.8-meter auxiliary telescopes. It has a baseline of up to 200 meters, yielding angular resolution in the milli-arcsecond range.

5.4 Getting above Earth's Atmosphere: Airborne and Orbiting Observatories

Various components of Earth's atmosphere block infrared photons from reaching astronomical telescopes on the ground, and one of the greatest culprits is water vapor. Imagine strolling through a Hawaiian rain forest and taking in the fragrance of the warm, humid tropical air. In your bliss you might be unaware that 4 km above you on the summit of Mauna Kea, astronomers are at war with the same atmospheric water vapor that has so stimulated your senses. Water vapor is the enemy of the infrared astronomer. We have already seen that Earth's atmosphere distorts telescopic images and that certain molecules in Earth's atmosphere, including water, block large parts of the electromagnetic spectrum from getting through to the ground. It shouldn't surprise us then to find that astronomers have put considerable effort into locating their instruments above as much of the atmosphere as possible. Look for an astronomical observatory and you'll usually encounter it at the summit of a tall mountain. Most of the world's larger astronomical telescopes are located 2,000 meters or more above sea level. Mauna Kea, a dormant volcano and home of the Mauna Kea Observatories (MKO), rises 4,200 meters above the Pacific Ocean. At this altitude the MKO telescopes sit above 40 percent of Earth's atmosphere; but more important, 90 percent of Earth's atmospheric water vapor lies below. Still, for the infrared astronomer the remaining 10 percent is troublesome.

One way to solve the water vapor problem is to make use of high-flying aircraft. NASA's Kuiper Airborne Observatory (KAO), a modified C-141 cargo aircraft, carried a 90-cm telescope and was among the first of these flying observatories. It could cruise at an altitude of 14 km, above 98 percent of Earth's water vapor. NASA retired KAO in 1995 and replaced it with the Stratospheric Observatory for Infrared Astronomy (SOFIA), which is expected to become fully operational in 2011. SOFIA will carry a 2.5-meter telescope and work in the far-infrared region of the spectrum, from 1 to 650 micrometers (μm). It will fly in the stratosphere at an altitude of about 12 km, above 99 percent of the water vapor in Earth's lower atmosphere.

Airborne observatories are able to overcome atmospheric absorption of infrared light by placing telescopes above most of the water vapor in the atmosphere. But gaining full access to the complete electromagnetic spectrum is yet another matter. This means getting completely above Earth's atmosphere.[10] In the late 1940s, scientists put ultraviolet and cosmic-ray instruments in the nose cones of captured German V-2 rockets and launched them from the White Sands Proving Ground in New Mexico to altitudes greater than 100 km. Of course, such observations had to be brief because, thanks to gravity, the rockets and their scientific instruments invariably came back down. The next step was to put astronomical instruments into orbit. The first astronomical satellite was the British Ariel 1, launched in 1962 to study solar ultraviolet and X-ray radiation and the energy spectrum of primary cosmic rays. Today we have a multitude of orbiting astronomical telescopes covering the electromagnetic spectrum from gamma rays to microwaves, with many more in the planning stage (see **Table 5.2**).

Optical telescopes, such as the 2.4-meter Hubble Space Telescope (HST), can operate successfully at modest altitudes in what is called low Earth orbit (LEO), 600 km above Earth's surface. Launched in 1990, HST has been the workhorse for UV, visible, and IR space astronomy for nearly two decades. LEO is also the region where the International Space Station (ISS) and many scientific satellites orbit. For others, 600 km is not nearly high enough. The Chandra X-ray Observatory, NASA's X-ray telescope, cannot tolerate even the tiniest traces of atmosphere and therefore orbits more than 16,000 km above Earth's surface. And even this is not distant enough for some telescopes. NASA's Spitzer Space Telescope, an infrared telescope, is so sensitive that it needs to be completely free from Earth's own infrared radiation. The solution was to put it into a *solar* orbit, trailing tens of millions of kilometers behind Earth. Many future space telescopes, including the James Webb Space Telescope, NASA's replacement for HST, will orbit free of Earth, bound only to the Sun.

> Orbiting observatories explore regions of the spectrum inaccessible from the ground.

5.5 Getting Up Close with Planetary Spacecraft

As we have stressed often in this book, we live in a remarkable time of discovery, when our newfound technological prowess has enabled us to begin the process of exploring our local corner of space. The general strategy for

---

[10]In a sense, there is no definable upper limit to Earth's atmosphere. As we will learn in Chapter 8, our atmosphere simply blends into outer space at an altitude of about 10,000 km.

TABLE 5.2

Selected Present and Future Space Observatories

Telescope	Sponsor(s)	Description	Launch Year
Hubble Space Telescope (HST)	NASA, ESA	Optical, infrared, ultraviolet observations	1990
Chandra X-ray Observatory	NASA	X-ray imaging and spectroscopy	1999
Wilkinson Microwave Anisotropy Probe (WMAP)	NASA	Cosmic background radiation	2001
Spitzer Space Telescope	NASA	Infrared observations	2004
Galaxy Evolution Explorer (GALEX)	NASA	Ultraviolet observations	2003
Swift Gamma-Ray Burst Mission	NASA	Gamma-ray bursts	2004
Fermi Gamma-ray Space Telescope	NASA, European partners	Gamma-ray imaging and gamma-ray bursts	2008
Planck	ESA	Cosmic microwave background	2009
Herschel Space Observatory	ESA	Far-infrared and submillimeter observations	2009
Kepler	NASA	Planet finder	2009
James Webb Space Telescope (JWST)	NASA	Replacement for HST	2014

by, instruments aboard these spacecraft briefly probed the physical and chemical properties of their targets and their environments.

Reconnaissance spacecraft employ **remote sensing** instrumentation much like the remote sensing techniques used by Earth-orbiting satellites to study our own planet. These include tools such as cameras capable of taking images in different wavelength ranges, radar for mapping surfaces hidden beneath obscuring layers of clouds, and spectrometers that spread out the target's light into a diagnosable spectrum. Remote sensing enables planetary scientists to map other worlds, measure the heights of mountains, identify geological features, learn about types of rocks present, watch weather patterns develop, measure the composition of atmospheres, and in general get a feeling for the "lay of the land." Still other instruments make *in situ* measurements of the extended atmospheres and space environment through which they travel.

> Planetary spacecraft take our instruments directly to the planets.

The study of our Solar System from space is a truly international collaboration involving NASA, the European Space Agency (ESA), the Russian Federal Space Agency (Roscosmos), the Japan Aerospace Exploration Agency (JAXA), the China National Space Administration (CNSA), and the Indian Space Research Organisation (ISRO). Other countries may soon join the endeavor.

Flybys and Orbiters

Since the dawn of history, no human had ever seen the far side of the Moon. As we learned in Chapter 2, we always see the same face of the Moon from Earth because its orbital and rotational periods are equal to one another. With one side of the Moon permanently facing Earth, the inappropriately named "dark side" is hidden from our view. Hidden, that is, until October 18, 1959. On that date the Soviet **flyby** probe *Luna 3* sent back humanity's first view of the far side of our nearest celestial neighbor (**Figure 5.27**). The curious features of the Moon's other side, so different from its Earth-facing half, amazed viewers around the world. No matter how powerful we make our groundbased or Earth-orbiting telescopes, *Luna 3* showed us there is nothing quite like going there.

Flyby missions have several distinct advantages in the reconnaissance phase of exploration. First, they are relatively inexpensive and the easiest missions to design and execute. Second, flyby spacecraft such as *Voyager*, shown in **Figure 5.28a**, may be able to visit several different worlds during their travels. The downside of flyby missions is that, thanks to the physics of orbits, these spacecraft must move by very swiftly. They are limited to just a few hours or at most a few days in which to conduct close-up studies of

exploring our Solar System begins with a reconnaissance phase, using spacecraft that fly by or orbit a planet or other body. Now, in the early 21st century, we have conducted preliminary reconnaissance of much of the Solar System. We have sent spacecraft flying by all of the classical planets, giving humanity its first ever close-up views of these distant worlds and their moons. We have even seen comets and asteroids at close range. As they sped

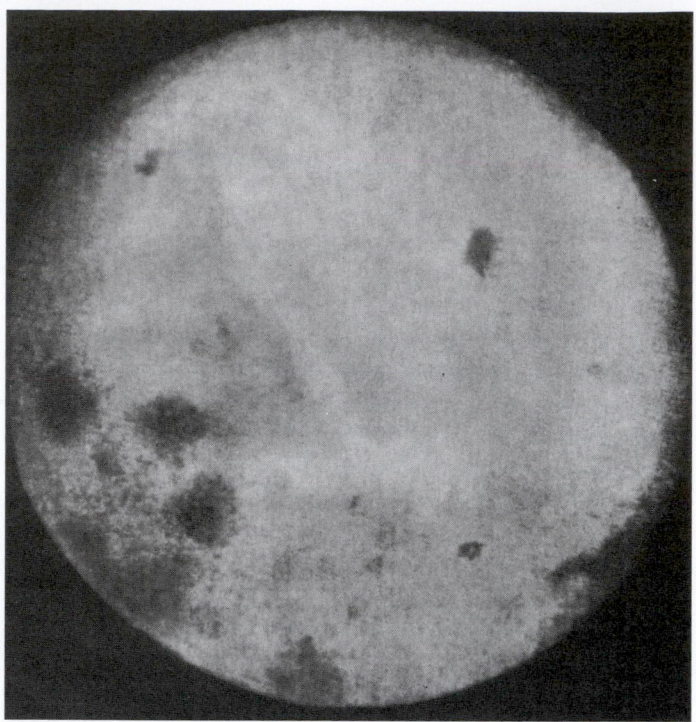

FIGURE 5.27 Humanity's first view of the far side of the Moon, seen in this image sent back by the Soviet probe *Luna 3* in 1959.

their targets. Yet flyby spacecraft give us our first intimate views of our planetary neighbors and provide the details we need to plan follow-up studies.

More detailed reconnaissance work uses spacecraft that orbit around planets. These are intrinsically more diffi-cult missions than flyby missions; but **orbiters** can linger, looking in detail at more of the surface of the object they are orbiting and studying things that change with time, like planetary weather. Spacecraft have orbited the Moon, Venus, Mars, Jupiter, Saturn, and even an asteroid. **Figure 5.28b** shows the *Cassini* spacecraft, which was launched in 1997 and as this book goes to print is still sending us data from its orbit around Saturn.

Landers, Rovers, and Atmospheric Probes

Reconnaissance spacecraft provide a wealth of informa-tion about a planet, but there is no better way of obtaining "ground truth" than to put our instruments where they can get right to the heart of it—within a planet's atmosphere or on solid ground. We have landed spacecraft on the Moon, Mars, Venus, Saturn's large moon Titan, and the asteroid Eros. One spacecraft even shot a massive bullet into a comet nucleus to observe the huge splash that created a tempo-rary cometary atmosphere. These spacecraft have returned pictures of planetary surfaces, measured surface chemis-try, and conducted experiments to determine the physical properties of the surface rocks and soils.

One disadvantage of using landed spacecraft is that only a few landings in limited areas are practical because of the expense, and the results may apply only to the small area around the landing site. Imagine, for example, what a dif-ferent picture of Earth we might get from a spacecraft that landed in Antarctica, as opposed to a spacecraft that landed

FIGURE 5.28 Explorers of the planets. (a) *Voyager* flew past Jupiter, Saturn, Uranus, and Neptune. (b) *Cassini* orbits Saturn.

(a)

b)

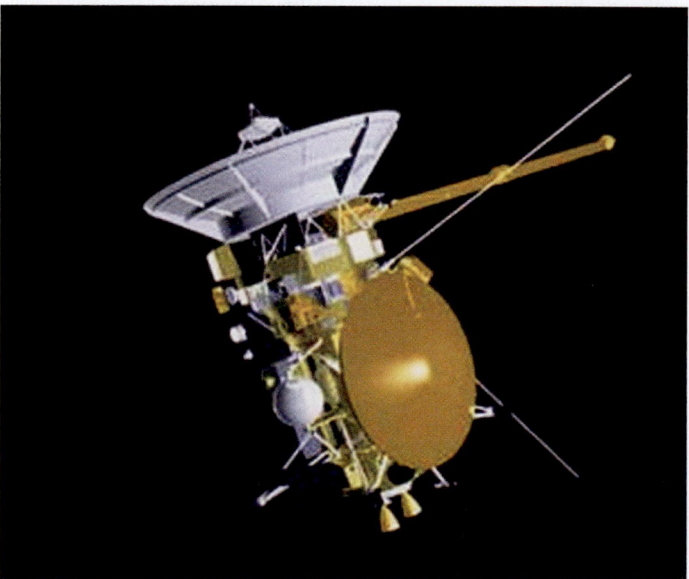

in the caldera (the summit crater) of a volcano or the floor of a dry riverbed. Sites to be explored with landed spacecraft must be very carefully chosen on the basis of reconnaissance data if we are to know what to make of the information they provide. We can mitigate some of the limitations of **landers** by putting their instruments on wheels and sending them from place to place, exploring the vicinity of the landing site. Such remote-controlled vehicles, called **rovers**, were used first by the Soviet Union on the Moon four decades ago, and more recently by the United States on Mars. The opening photo for this chapter shows an artist's view of one of two rovers still roaming the martian landscape.

We have also sent probes into the atmospheres of Venus, Jupiter, and Titan. As they descend, **atmospheric probes** continuously measure and send back data on physical properties such as temperature, pressure, and wind speed, along with other properties, such as chemical composition. Meteorologists take measurements of our terrestrial atmosphere from the surface up by sending their instruments aloft in balloons. By contrast, planetary scientists must work from the top down by suspending their instruments from parachutes. The end result is much the same. Atmospheric probes have survived all the way to the solid surfaces of Venus and Titan, sending back streams of data during their descent. An atmospheric probe sent into Jupiter's atmosphere never reached that planet's surface because, as we will learn later, Jupiter does not have a solid surface in the same sense that terrestrial planets and moons do. After sending back its data, the Jupiter probe eventually melted and vaporized as it dropped into the hotter layers of the planet's atmosphere.

Sample Returns

If you pick up a rock from a road cut, there is a lot you might learn from the rock using tools that you could easily carry in your pocket. But the sophistication of the tools you could carry with you would be limited. It would be much better to pick up a few samples and carry them back to a laboratory equipped with a full range of state-of-the-art instruments capable of measuring chemical compositions, mineral types, radiometric ages (see Chapter 7), and other information needed to reconstruct the story of their origin and evolution. So, too, is the case in Solar System exploration. One of the most powerful methods for investigating remote objects is to collect samples of the objects and bring them back to Earth for detailed study. So far, only samples of the Moon, a comet, and the **solar wind** (a stream of charged particles from the Sun) have been collected and returned to Earth.

As we will learn in Chapter 12, we do know of meteorites that are considered to be part of Mars, but there is a problem with putting them in their proper geological perspective. We just talked about picking up a rock from a road cut. This is quite different from simply grabbing any old rock by the side of the road, because that rock could have come from anywhere. In geological sampling it is important to have samples from a source of known context. Some have claimed that we do not need samples returned from Mars, because we already have the martian meteorites. The problem is that we do not know where on Mars those meteorites came from and, therefore, how they fit into the planet's global geology. Plans are currently under way for unmanned "sample and return" missions to Mars.

> Sample returns provide "ground truth."

Of course, we could not collect specimens in a national park without permission and a scientifically valid reason. Similarly, the return of extraterrestrial samples to Earth is governed by international treaties and standards to ensure that these samples do not contaminate Earth. For example, before the lunar samples brought back by the *Apollo* missions could be studied, they (and the astronauts) had to be placed in quarantine and tested for alien life-forms. The same international standards apply to spacecraft landing on planets. The goal of these standards is to avoid *forward contamination*, or transporting life-forms from Earth to another planet. If there is life on other planets, then not only is there concern about introducing potential harm, but from a scientific perspective we do not want to "discover" life that we, in fact, introduced.

With numerous missions under way and others on the horizon, unmanned exploration of the Solar System is an ongoing, dynamic activity. In our journey we will frequently refer to space missions and the information they return, but today's hot results may be tomorrow's old news in light of other, even more exciting discoveries. We hope that you will make use of the *21st Century Astronomy* Web site as a gateway to the wealth of exciting results that the future holds.

5.6 Other Earth-Based Astronomical Tools

High-profile space missions have sent back stunning images and data from across the electromagnetic spectrum, but there are still other—and perhaps less obvious—tools that astronomers use right here on the ground, including particle accelerators, neutrino and gravitational wave detectors, and high-speed computers. Some are even buried deep underground.

Ever since the early years of the 20th century, physicists have been peering into the structure of the atom by observing what happens when small particles collide. By the 1930s, physicists had developed the means to accelerate charged subatomic particles such as protons to very high

FIGURE 5.29 The high-energy linear particle accelerator at the Fermi National Accelerator Laboratory in Batavia, Illinois.

FIGURE 5.30 The ATLAS particle detector at CERN's Large Hadron Collider near Geneva, Switzerland. The enormous size of this instrument is evident from the figure standing near the bottom center of the picture.

speeds and then observe what happens when they slam into a target. From such experiments (which are continuing even today) physicists have discovered many kinds of subatomic particles[11] and learned about their physical properties. High-energy particle colliders have proven to be an essential tool for physicists studying the basic building blocks of matter.

Why, then, do we regard the high-energy particle collider as a tool of the astronomer? How does the astronomer's interest in the largest objects in the universe relate to what happens on the very smallest scales? As we will see in Chapter 22, to understand the very largest structures we see in the universe—indeed, the large-scale universe itself— we need to understand the

> Colliders provide us with the physics we need to understand the early universe and the formation of structure.

physics that took place during the earliest moments in the universe, when everything was unbelievably hot and dense. Although we have not yet reached that level of comprehension, the high-energy particle colliders that physicists use today are designed to lead us there. As we will find in Chapters 21 and 22, this knowledge may help us understand such issues as the nature of the dominant matter and energy in the universe and whether there really is a beginning or an end to our universe.

Two factors determine the effectiveness of particle accelerators: the energy they can achieve and the number of particles they can accelerate. Whereas the first particle accelerator, the "cyclotron," attained an energy of about 10^{-13} joule, modern particle colliders now reach much higher energies. For example, the accelerator at the Fermi National Accelerator Laboratory (Fermilab—**Figure 5.29**) can accelerate protons up to 1.6×10^{-7} joule, resulting in collision energies of 3.2×10^{-7} joule. This may seem like a small number, but it corresponds to more than a thousand times the rest mass energy (mc^2) of the proton. To put it in perspective, this amount is the energy of a flying mosquito all concentrated into a single tiny proton. Yet this energy is dwarfed by the new Large Hadron Collider near Geneva, Switzerland, built by CERN, the European Organization for Nuclear Research (**Figure 5.30**). This facility can produce collisions with seven times as much energy as the Fermi Tevatron Collider—about 2.2×10^{-6} joule.

In Chapter 14's discussion of the Sun, you will be introduced to the **neutrino**, an elusive elementary particle that plays a major role in the physics of stellar interiors. Of

---

[11] In Chapter 4 we learned about three subatomic particles: the proton, neutron, and electron. In this section we will introduce another, the neutrino; and in Chapter 14 we will meet still another, the positron. However, many other subatomic particles are now known to physicists—in fact, too many to list here.

course, we cannot see beneath the Sun's surface, but observations of neutrinos can give us important insight into what is happening deep within. There's only one problem, and it's a big one: neutrinos are extremely difficult to detect. To study a neutrino you first have to grab one, and they are nearly impossible to catch. In a sense, this is fortunate for us. In less time than it takes you to read this sentence, a thousand trillion (10^{15}) solar neutrinos from the Sun are passing through your body. It doesn't matter a bit if you are reading this at night. Neutrinos are so nonreactive with matter that they can pass right through Earth (and you) as though it (or you) weren't there at all. In fact, half of the neutrinos produced by the Sun would make it through a slab of lead a light-year thick. For us to detect a neutrino, it has to interact with a detector. Several neutrino detectors are in operation today. All are buried deep underground to avoid false detection of cosmic background radiation. Neutrino detectors typically record only one out of every 10^{22} (10 billion trillion) neutrinos passing though them, but that's enough to reveal processes deep within the Sun or witness the violent death of a star 160,000 light-years away.

> Neutrino detectors see the death of stars and measure the Sun's interior.

Another elusive phenomenon is the **gravitational wave**. Gravitational waves are disturbances in a gravitational field, similar to the waves that spread out from the disturbance you create when you toss a pebble into the quiet surface of a pond. There is strong, although indirect, observational evidence for their existence, as we will see in Chapter 17 when we discuss "binary pulsars." However, gravitational waves are so elusive that we have not yet actually detected them. Several facilities, including the Laser Interferometer Gravitational-Wave Observatory (LIGO), have been constructed to detect gravitational waves. Scientists are eager to detect gravitational waves—not so much to confirm their existence but to study the physical phenomena they are likely to reveal, such as the birth and evolution of the universe, stellar evolution, or the very force of gravity itself.

Finally, a survey of the astronomer's tools would not be complete without a discussion of the essential role of computers. Imagine your life without computers. From laptops to the largest servers, we depend on computers to surf the Internet, process our data, and organize our daily lives. Tiny computer processors control every modern convenience from automobiles to washing machines. And, as you might assume, computers are essential in the world of science. Data gathering, analysis, and interpretation are entirely dependent on computers—and the more powerful, the better. Consider, for example, analyzing a night's worth of astronomical images recorded by a very large CCD. A single image may contain as many as 100 million pixels, with each pixel displaying roughly 10,000 levels of brightness. That adds up to a *trillion* pieces of information in each image! And that's only one image. To analyze their data, astronomers typically do calculations on *every single pixel* of an image in order to remove unwanted contributions from Earth's atmosphere or to correct for instrumental effects. From the astronomer's point of view, without high-speed computers the CCD would be just another electronic curiosity.

High-speed computers also play an essential role in generating and testing theoretical models of astronomical objects. Even when we completely understand the underlying physical laws that govern the behavior of a particular object, it is frequently the case that the object is so complex that it would be impossible to calculate its properties and behavior without the assistance of high-speed computers. For example, as we learned in Chapter 3, we can use Newton's laws to easily compute the orbits of two stars that are gravitationally bound to one another because their orbits take the form of simple ellipses. However, it is not so easy to understand the orbits of the hundred billion (10^{11}) stars that make up our Milky Way Galaxy, even though *the underlying physical laws remain the same.*

As if that were not complicated enough, consider the problem involving the collision of *two* such galaxies—and, for good measure, throw in some gas in addition to the stars. We can see the result in **Figure 5.31**. When we apply even the fastest available computers to this problem, the sequence shown in Figure 5.31 is only an approximation of what we believe really happens. Modern computers have enough speed and memory to handle the behavior of a few million stars at best, so we are forced to assume that a single star in the computer simulation really represents hundreds of thousands of real stars.

Similar modeling procedures have worked well in determining the interior properties of stars and planets, including our own Earth. Although we cannot "see" beneath their surfaces, we have a surprisingly good understanding of their interiors, as we will learn in later chapters. We begin a model by assigning well-understood physical properties to tiny volumes within a planet or star. The computer assembles an enormous number of these individual elements into an overall representation of the complete body. When it is all put together, we have a rather good picture of what the interior of the star or planet is like.

We have seen how our view of the sky changes with location and time, discovered the laws that govern the motions of celestial bodies, studied the properties of light, and learned about the tools that astronomers use in their continuing pursuit of scientific knowledge. With this background in the fundamentals now behind us, it's time to step away from Earth and look outward, first to our Solar System and then to the greater universe beyond.

(a)

(b)

(c)

(d)

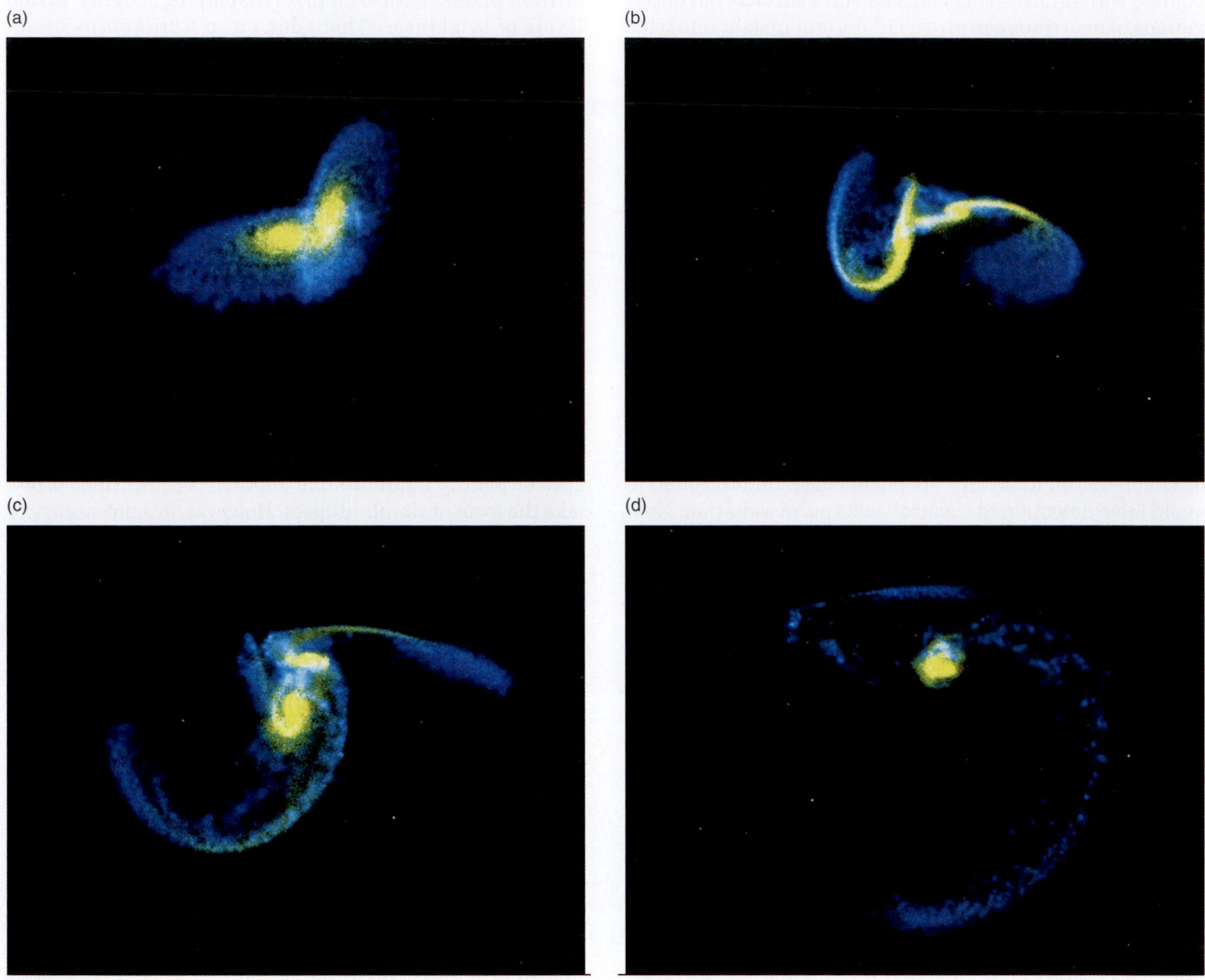

FIGURE 5.31 Numerical simulation of the merging of two galaxies, including the gravitational attraction of all forms of matter. (a) Just before the merge. (b) After the galaxies have passed through one another. (c) As the cores orbit each other and merge. (d) When the central core begins to settle down.

Seeing the Forest for the Trees

From our perspective in the 21st century, we can only imagine what went through the minds of our distant ancestors as they gazed upward toward the heavens. We might guess that they felt completely comfortable with the daily movements of the Sun and the Moon, yet they likely experienced fear when confronted with an eclipse or the appearance of an occasional comet. They probably took for granted the tiny points of light that filled the nighttime sky, but they also must have won-

dered about those mysterious few that seemed to move freely among the others. With nothing more than their eyes to fulfill their curiosity, our early ancestors could only observe and wonder. But wondering alone does not give rise to comprehension. Insight far beyond personal experience was necessary, and it was a long time coming. Somewhat more than two millennia ago, enlightenment of a sort came to the classical world. Greek philosophers concluded that these "wandering stars," or planets, were unlike the distant canopy of fixed stars—that they were much closer—but stopped short of claiming to know *what* they were. And so these and other heavenly mysteries lingered until the turn of the 17th century, when two historic events took place.

It all started when an obscure German-Dutch spectacle maker put a pair of lenses together and saw that distant objects appeared closer. Although Hans Lippershey may have regarded his "looker" as an amusing toy, it was a Tuscan instrument maker who first recognized its potential for studying the heavens. It didn't take Galileo Galilei long to discover that the Moon was a nearby world covered with craters, and that those "mysterious moving points of light" were tiny disks of planets. Galileo's observations marked a turning point. The telescope had changed forever our fundamental perception of the heavens.

Throughout the four centuries since Galileo's historic discoveries, astronomers have continued to perfect the astronomical telescope. Reaching far outside the visible, telescopes now cover the entire electromagnetic spectrum from gamma rays to radio waves. Some telescopes reside on mountaintops; others make their home in Earth orbit or nearby space. Still others make the long and arduous journey through interplanetary space to observe Earth's neighbors up close. But telescopes alone cannot provide all the answers. Tools of a different type have recently joined the astronomer's repertoire of gear. Some are buried deep underground, quietly observing one of nature's most elusive particles. Others sit on desktops crunching numbers. All are devoted to decoding the cryptic messages sent to us from the cosmos.

Having stopped briefly to examine the many tools used by astronomers, we are now ready to resume our outward journey, armed with a new appreciation of how important it is to proceed carefully. Common sense is not enough. We must instead rely on our growing understanding of physical law and on rigorously tested predictions of carefully constructed theories. These also are useful tools that will enable us to look beyond the surface of spectacular vistas that otherwise would be devoid of sense or meaning or connection. Kepler's laws, Newton's laws, Doppler shifts, Wien's law, the Stefan-Boltzmann law, energy states, spectral lines, and the rest—these are the keys we will use to build an understanding of planets, stars, galaxies, and ultimately the universe itself. The first place in which we will bring these tools to bear will be our immediate neighborhood as we consider the nature and origin of our Solar System. And to put our own Solar System in perspective, we'll take a look at some of the many other planetary systems that lie far beyond.

Summary

- Optical telescopes come in two basic types: refractors and reflectors.

- All large astronomical telescopes are reflectors.

- Large telescopes collect more light and have greater resolution.

- The CCD is today's astronomical detector of choice.

- Earth's atmosphere blocks many spectral regions and distorts telescopic images.

- Putting telescopes in space solves problems created by Earth's atmosphere.

- Infrared and radio telescopes can see through vast clouds of cosmic gas and dust.

- Radio and optical telescopes can be arrayed to greatly increase angular resolution.

- Most of what we know about the classical planets comes from spacecraft we have sent there.

- High-speed computers are essential to the acquisition, analysis, and interpretation of astronomical data.

 SmartWork, Norton's online homework system, includes algorithmically generated versions of these questions, plus additional conceptual exercises. If your instructor assigns questions in SmartWork, log in at **smartwork.wwnorton.com.**

Student Questions

THINKING ABOUT THE CONCEPTS

1. Galileo's telescope used simple lenses. What is the primary disadvantage of using a simple lens in a refracting telescope?

2. The largest astronomical refractor has an aperture of 1 meter. List several reasons why it would be impractical to build a still larger refractor with, say, twice the aperture.

3. Name and explain at least two advantages that reflecting telescopes have over refractors.

4. Your camera may have a zoom lens, ranging between wide angle (short focal length) and telephoto (long focal length). How would the size of an object in the camera's focal plane differ between wide angle and telephoto?

5. Optical telescopes reveal much about the nature of astronomical objects. Why do astronomers also need information provided by gamma-ray, X-ray, infrared, and radio telescopes?

6. What optical property is responsible for refraction?

7. When reflecting from a flat surface, the angles of incidence and reflection are the same. Is this also true for the curved surface of a reflecting telescope's primary mirror? Explain with a sketch.

8. What causes a prism to break light into a spectrum?

9. How do manufacturers of refracting telescopes and cameras avoid the problem of chromatic aberration?

**10. Explain constructive and destructive interference.

11. Consider two optically perfect telescopes having different diameters but the same focal length. Is the image of a star larger or smaller in the focal plane of the larger telescope? Explain your answer.

12. Name two ways in which Earth's atmosphere interferes with astronomical observations.

13. "Twinkle, twinkle, little star. How I wonder what you are." Explain *why* stars twinkle. (In a later chapter we will find out *what* the stars are.)

14. Why do we not have groundbased gamma-ray and X-ray telescopes?

*15. Explain adaptive optics and how they improve a telescope's image quality.

*16. Explain integration time and quantum efficiency and how each contributes to the detection of faint astronomical objects.

17. For more than a century, astronomers used photographic plates to record their observations. List at least two shortcomings of the photographic process.

18. Charge-coupled devices are now the most commonly used astronomical detector. List three advantages that CCDs have over photographic emulsions.

19. Explain a major advantage that radio telescopes have over optical telescopes when we are observing the inner parts of galaxies.

20. Why are the world's largest telescopes located on high mountains?

21. Some people believe that we put astronomical telescopes in orbit because doing so gets them closer to the objects they are observing. As an enlightened student taking this class, you know better. Explain what is wrong with this popular misconception.

22. We have now sent various kinds of spacecraft—including flybys, orbiters, and landers—to all of the classical planets. Explain the advantages and disadvantages of each of these types of spacecraft.

*23. If we have meteorites that are pieces of Mars, why is it so important to go to Mars and bring back samples of the martian surface?

24. Humans had our first look at the far side of the Moon only as recently as 1959. Why had we not been able to see it earlier—say, when Galileo first observed the Moon with his telescope in 1610?

25. Where are neutrino detectors located, and why are neutrinos so difficult to detect?

APPLYING THE CONCEPTS

26. Many amateur astronomers start out with a 4-inch (aperture) telescope and then graduate to a 16-inch telescope. How much fainter are the stars that can be seen in the larger telescope?

27. Compare the light-gathering power of a large astronomical telescope (aperture 10 meters) with that of the dark-adapted human eye (aperture 7 mm).

28. Compare the angular resolution of the Hubble Space Telescope (aperture 2.4 meters) with that of a typical amateur telescope (aperture 20 cm).

29. Assume a telescope with an aperture of 1 meter. Compare the telescope's resolution when we are observing in the near-infrared region of the spectrum ($\lambda = 1,000$ nm) with that when we are observing in the violet region of the spectrum ($\lambda = 400$ nm).

*30. Assume that the maximum aperture of the human eye, D, is approximately 8 mm and the average wavelength of visible light, λ, is 5.5×10^{-4} mm.
   a. Calculate the diffraction limit of the human eye in visible light.
   b. How does the diffraction limit compare with the actual resolution of 1–2 arcminutes (60–120 arcseconds)?
   c. To what do you attribute the difference?

31. The diameter of the full Moon's image in the focal plane of an average amateur's telescope (focal length 1.5 meters) is 13.8 mm. How big would the Moon's image be in the focal plane of a very large astronomical telescope (focal length 250 meters)?

*32. The dark-adapted aperture, integration time, and quantum efficiency of the human eye are 7 mm, 100 ms, and 10 percent, respectively. Compare the brightness of the faintest stars that can be seen with the naked eye with those in a 1-hour-exposure photograph using a telescope having a 1-meter aperture and a photographic emulsion with a quantum efficiency of 2 percent.

33. One of the earliest astronomical CCDs had 160,000 pixels, each recording 8 bits (256 levels of brightness). A new generation of astronomical CCDs may contain a billion pixels, each recording 15 bits (32,768 levels of brightness). Compare the number of bits of data that each CCD type produces in a single image.

34. Consider a CCD with a quantum efficiency of 80 percent and a photographic plate with a quantum efficiency of 1 percent. If an exposure time of 1 hour is required to photograph a celestial object with a given telescope, how much observing time would we save by substituting a CCD for the photographic plate?

*35. The VLBA employs an array of radio telescopes ranging across 8,000 km of Earth's surface from the Virgin Islands to Hawaii.
   a. Calculate the angular resolution of the array when radio astronomers are observing interstellar water molecules at a microwave wavelength of 1.35 cm.
   b. How does this resolution compare with the angular resolution of two large optical telescopes separated by 100 meters and operating as an interferometer at a visible wavelength of 550 nm?

36. When operational, the SVLBI may have a baseline of 100,000 km. What will be the angular resolution when we are studying an interstellar molecule emitting at a wavelength of 17 mm from a distant galaxy?

37. The Mars Reconnaissance Orbiter (MRO) flies at an average altitude of 280 km above the martian surface. If its cameras have an angular resolution of 0.2 arcsecond, describe the size of the smallest objects that the MRO can detect on the martian surface?

38. Rovers *Spirit* and *Opportunity* can move across the martian landscape at speeds of up to 5 centimeters per second (cm/s). In contrast, our typical walking speed is about 4 kilometers per hour (km/h).
   a. How long would it take *Spirit* to move the length of a football field (91.44 meters)?
   b. How long would it take you to walk the same distance?

39. *Voyager 1* is now about 115 astronomical units (AU) from Earth, continuing to record its environment as it approaches the limits of our Solar System.
   a. What is the distance of *Voyager 1*, expressed in kilometers?
   b. How long does it take observational data to get back to us from *Voyager 1*?
   c. How does *Voyager 1*'s distance from Earth compare with that of the nearest star (other than the Sun)?

40. As with light, the speed, wavelength, and frequency of gravitational waves are related as $c = \lambda \times f$. If we were to observe a gravitational wave from a distant cosmic event with a frequency of 10 hertz (Hz), what would be the wavelength of the gravitational wave?

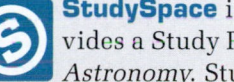 **StudySpace** is a free and open Web site that provides a Study Plan for each chapter of *21st Century Astronomy*. Study Plans include animations, reading outlines, vocabulary flashcards, and multiple-choice quizzes, plus links to premium content in SmartWork and the ebook. Visit **www.wwnorton.com/studyspace**.

PART II The Solar System

You are a child of the universe, no less than the trees
and the stars; you have a right to be here. And whether
or not it is clear to you, no doubt the universe is
unfolding as it should.

MAX EHRMANN (1872–1945), *DESIDERATA*

From clouds of gas and dust, solar systems such as ours are born.

CHAPTER 6

The Birth and Evolution of Planetary Systems

6.1 Stars Form and Planets Are Born

In thinking about how it all began, a good place to start is right here in our own small corner of the universe—a collection of *planets*,[1] *moons*, and other smaller bodies surrounding an ordinary star that we call the Sun. Astronomers refer to such a system of planets surrounding a star as a **planetary system**, and we call our own planetary system the **Solar System** (**Figure 6.1**). But first, a word of caution. Do not confuse our Solar System with "the universe." Planetary systems are infinitesimally small compared to the universe as a whole. For example, light takes about 4 hours to travel here from Neptune, the outermost classical planet in our Solar System. Light from the most distant galaxies takes nearly *14 billion years* to reach us!

> Our Solar System is tiny compared to the universe.

For more than a decade astronomers have realized that our Solar System is but one of an enormous number of other planetary systems scattered throughout the galaxy. It would make sense, then, to begin this chapter by learning how these planetary systems, including our own Solar System, were created. Later, in Part IV, we will learn about a much vaster universe, and how it came to be.

Until the last part of the 20th century, every discussion of the Solar System necessarily started with an accounting of its pieces: "There are *planets* with such and such properties. There are *moons*, there are *asteroids*, and there

[1]The formal meaning of the word *planet*, as defined by the International Astronomical Union, appears in Appendix 8. Curiously, the IAU definition applies only to planets within our own Solar System.

KEY CONCEPTS

By the early years of the 21st century, astronomers had come to see our Solar System as an unmistakable by-product of the birth of the Sun. The discovery of planetary systems surrounding other stars has shown that there is nothing unique about our Solar System. It is this understanding that brings order to what we see around us today. We begin our investigation of the Sun's family by introducing the story of the formation of planetary systems—a story in which we will learn that

- A star forms when a cloud of interstellar gas and dust collapses under its own weight.
- While a star forms it is surrounded by a flat, rotating disk that provides the raw material from which a planetary system might form.
- Dust grains in the disk around a young star stick together to form larger and larger solid objects.
- Temperature differences within the disk determine the kinds of materials from which solid objects can form.
- Planets orbit the Sun in a plane, and they revolve in the same direction that the Sun rotates.
- Giant planets form when solid, planet-sized bodies capture gas from the surrounding disk.
- The atmospheres of smaller terrestrial planets are gases released by volcanism and volatile materials that arrived on comets.
- Planetary systems around other stars are common.

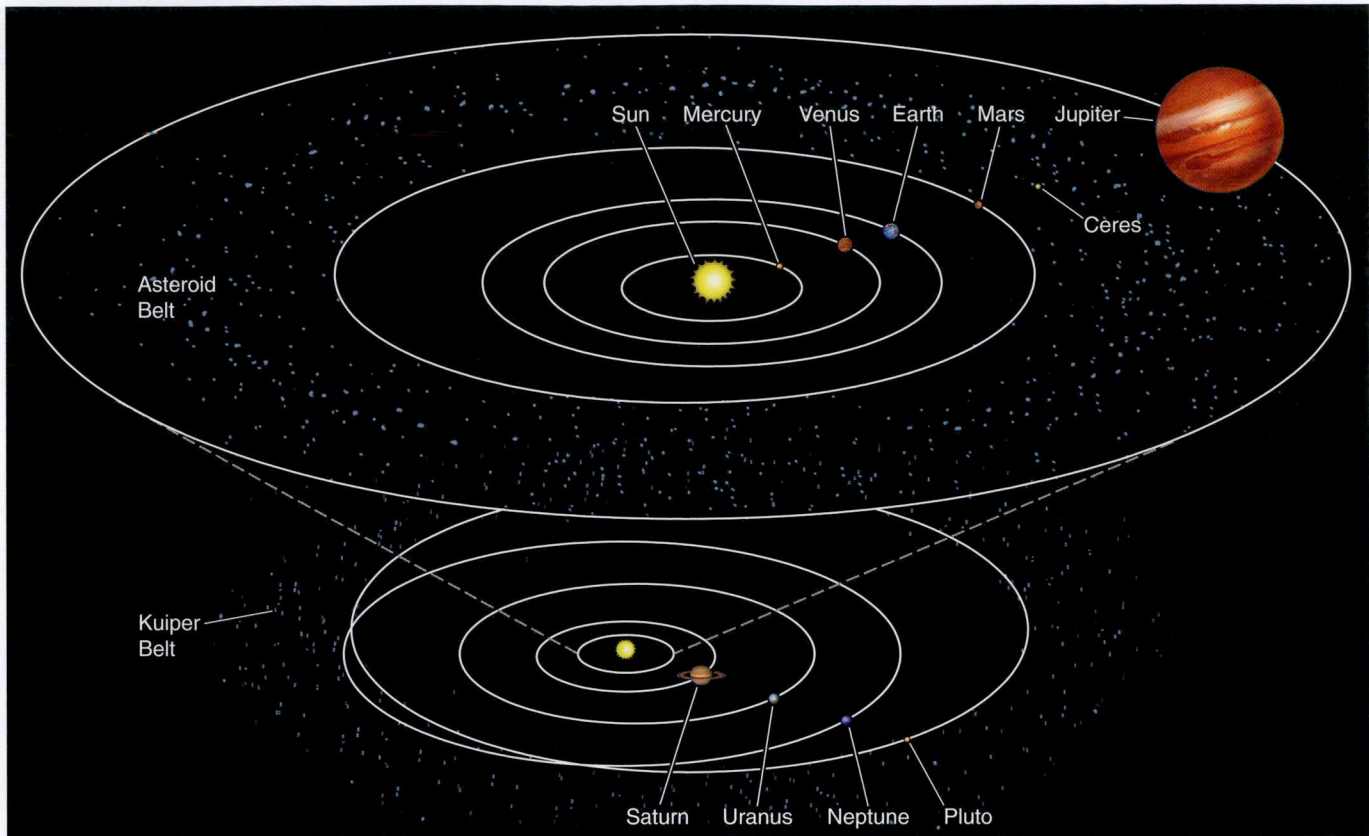

FIGURE 6.1 Our Solar System. There are three other dwarf planets besides Pluto and Ceres that are not shown here.

are *comets*. And so forth." The *origin* of the Solar System, though, remained speculative. Yet, in the last few decades stellar astronomers studying the formation of stars and planetary scientists analyzing clues about the history of the Solar System have found themselves arriving at the *same* picture of our early Solar System—but from two very different directions. This unified understanding provides the foundation for the way we now think about the Sun and the myriad objects that orbit it.

The first plausible theory for the formation of the Solar System, known as the **nebular hypothesis**, was proposed as early as 1734 by the German philosopher Immanuel Kant (1724–1804) and conceived independently a few years later by the French astronomer Pierre-Simon Laplace (1749–1827). Kant and Laplace argued that a rotating cloud of interstellar gas, or **nebula** (Latin for "cloud"), gradually collapsed and flattened to form a disk with the Sun at its center. Surrounding the Sun were rings of material from which the planets formed. Although the nebular hypothesis remained popular throughout the 19th century, it did have serious problems, as we will soon see. Even so, the basic principles of the hypothesis are retained today in our modern theory of planetary system formation.[2]

Modern theory of planetary system formation suggests that, when conditions are right, clouds of interstellar gas collapse under the force of their own *self-gravity* (see Chapter 10) to form stars. In support of the nebular hypothesis, disks of gas and dust have been found surrounding young stellar objects like the ones shown in **Figure 6.2**. From this observational evidence, stellar astronomers have shown that, much like a spinning ball of pizza dough spreads out to form a flat crust, the cloud that produces a star—our Sun, for example—collapses first into a rotating disk. Material in the disk eventually suffers one of three fates: it travels inward onto the forming star at its center, remains in the disk itself to form the planets and other objects, or is thrown back into interstellar space.

Young stars are surrounded by rotating disks.

▶‖ **AstroTour: Solar System Formation**

During the same years that astronomers were beginning to ferret out the secrets of star formation, other groups of scientists with very different backgrounds—mainly geochemists and geologists—were piecing together the *history* of our Solar System. Planetary scientists looking at the current structure of the Solar System inferred what some of its early characteristics must have been. The orbits of all the planets in the Solar System lie very close to a single plane, which says that the early Solar System must have been flat. The fact that all the planets orbit the Sun in the same direc-

---

[2]In the early 20th century a competing hypothesis, known as the "close encounter theory," claimed that planets formed when blobs of material were pulled off the Sun by a close encounter with a passing star.

(a)

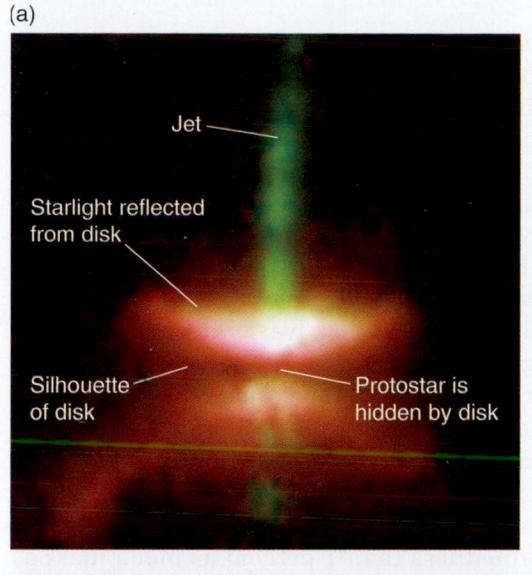

Jet

Starlight reflected from disk

Silhouette of disk

Protostar is hidden by disk

(b)

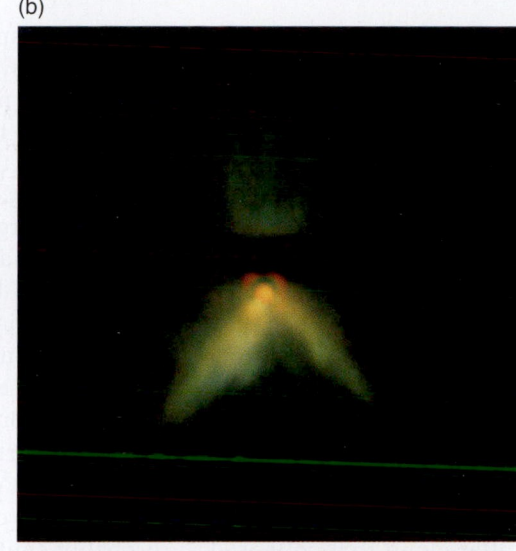

FIGURE 6.2 Hubble Space Telescope images showing disks around newly formed stars. The dark bands are the shadows of the disks seen more or less edge on. Bright regions are dust illuminated by starlight. Some disk material may be expelled in a direction perpendicular to the plane of the disk in the form of violent jets.

(c)

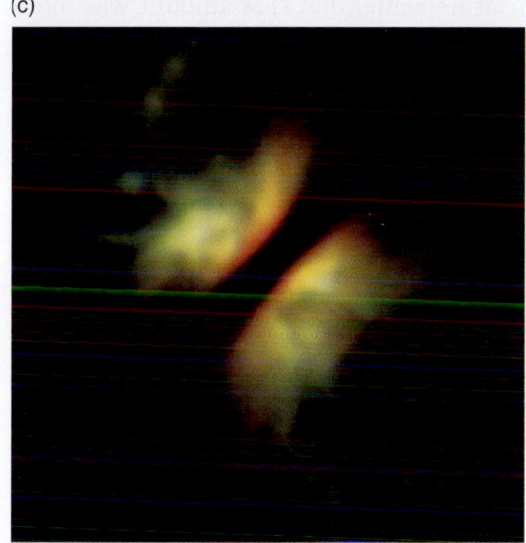

(d)

FIGURE 6.3 Meteorites are the surviving pieces of young Solar System fragments that land on the surfaces of planets. It is clear from this cross section that this meteorite formed from many smaller components that stuck together.

tion says that the material from which the planets formed must have been swirling about the Sun in the same direction as well. Other clues about what the early Solar System was like were harder to puzzle out. **Meteorites**, for example, include bits and pieces of material that are left over from the Solar System's youth. These fragments of the early Solar System can be captured by Earth's gravity and fall to the ground, where they can be picked up and studied. Many meteorites, such as the one in **Figure 6.3**, look something like a piece of concrete in which pebbles and sand are mixed with a much finer filler. This structure is surely telling us *something* about how these pieces of interplanetary debris formed, but what?

Beginning in the 1960s, a flood of information about Earth and other objects in the Solar System poured in from a host of sources, including space probes, groundbased telescopes, laboratory analysis of meteorites, and theo-

The Solar System formed from a rotating disk of gas and dust.

retical calculations. Scientists working with this wealth of information began to see a pattern. What they were learning made sense only if they assumed that the larger bodies in the Solar System had grown from the aggregation of smaller bodies. Following this chain of thought back in time, they came to envision an early Solar System in which the young Sun was surrounded by a flattened disk of both gaseous and solid material. This swirling disk of gas and dust provided the raw material from which the objects in our Solar System would later form.

While reading the previous paragraphs, you may have noticed a remarkable similarity between the disks that stellar astronomers find around young stars and the disk that planetary scientists hypothesize as the cradle of the Solar System. This similarity is not a coincidence. As astronomers and planetary scientists compared notes, they realized they had arrived at the *same* picture of the early Solar System from two completely different directions. The rotating disk from which the planets formed was none other than the remains of the disk that had accompanied the formation of the Sun. The planet we live on, along with all the other orbiting bodies that make up our Solar System, evolved from the remnants of the interstellar cloud that collapsed to form our local star, the Sun. The connection between the formation of stars and the origin and subsequent evolution of the Solar System has become one of the cornerstones of both astronomy and planetary science—a central theme around which a great deal of our understanding of our Solar System revolves.

6.2 In the Beginning Was a Disk

Later in the book we will turn our attention beyond the boundaries of our Solar System to the process of star formation itself. For now it is enough to jump into this story of star formation midstream. Begin by planting the picture shown in **Figure 6.4** firmly in your mind. It presents the newly formed Sun, roughly 5 billion years ago, adrift in

interstellar space. The Sun was not yet a star in the true sense of the word. It was still a **protostar**—a large, hot ball of gas whose nuclear fires had yet to ignite. (The often-used prefix *proto-* means "early form" or "in the process of formation.") As the cloud of interstellar gas collapsed to form the protostar, its gravitational energy was converted into thermal energy and radiation. Surrounding the protostellar Sun was a flat, rotating disk of gas and dust. *Orbiting* is perhaps a better word than *rotating* because each bit of the material in this thin disk was orbiting around the Sun according to the same laws of motion and gravitation that govern the orbits of the planets today. The disk around the Sun was much like the disks that astronomers see today surrounding protostars and newly formed stars elsewhere in our galaxy. This disk is referred to as a **protoplanetary disk**, sometimes abbreviated as *proplyd*. It probably contained less than 1 percent as much mass as the nascent star at its center, but this amount was more than enough to account for the bodies that make up the Solar System today.

Okay, so we know the Solar System formed from a protoplanetary disk and that disks are seen around newly formed stars. But *why* is this the case? What is it about the process of star formation that leads not only to a star itself, but to a flat, orbiting collection of gas and dust as well? The answer to this question lies with a skater spinning on the ice and something called *angular momentum*.

The Collapsing Cloud Rotates

You have likely seen a figure-skater spinning on the ice (**Figure 6.5**). Like any rotating object or isolated group of objects, the spinning ice-skater has some amount of **angular momentum**. The amount of angular momentum that an object possesses depends on three things:

FIGURE 6.4 When you consider the young Sun, think of it as being surrounded by a flat, rotating disk of gas and dust that is flared at its outer edge.

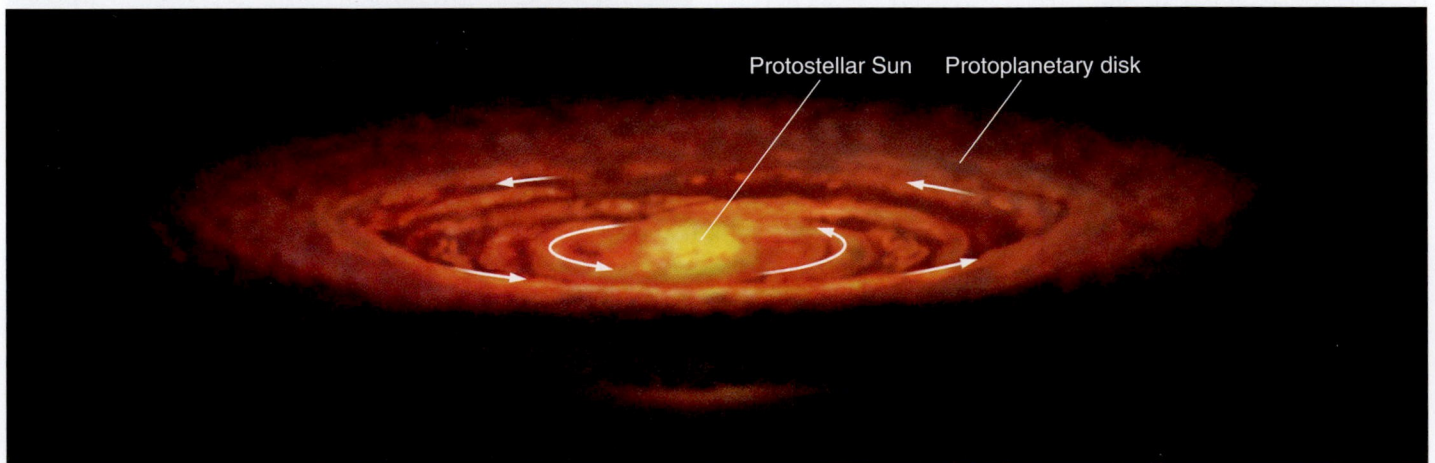

Protostellar Sun   Protoplanetary disk

VISUAL ANALOGY

FIGURE 6.5 A figure-skater relies on the principle of conservation of angular momentum to change the speed with which she spins.

1. How fast the object is rotating. The faster an object is rotating, the more angular momentum it has. A top that is spinning rapidly has more angular momentum than the same top does when it is spinning slowly.

2. The mass of the object. Suppose you compare two spinning tops. Both tops have the same size, shape, and rate of spin. They are the same except that one top is made of metal while the other top is made of wood. The more massive metal top has more angular momentum.

3. How the mass of the object is distributed—how "spread out" the object is. For an object of a given mass and rate of rotation, the more spread out the object is, the more angular momentum it has. An object that is rotating slowly but is very spread out might have more angular momentum than a more rapidly rotating but more compact object has.

It is quite a bit more complicated to determine the *rotational* or *spin* angular momentum of a single object, such as the spinning top just referred to or a rotating planet or interstellar gas cloud. See **Math Tools 6.1** for details.

What makes angular momentum such an important and useful idea in physics and astronomy is that the amount of angular momentum possessed by an object or isolated group of objects does not change unless those objects are affected in just the right way by something other than themselves.

This statement is referred to as the law of **conservation of angular momentum**. In the parlance of physics, if we say that something is "conserved," we mean that the quantity does not change by itself. This idea might remind you of Newton's first law of motion, which says that in the absence of an external force, an object continues to move in a straight line at a constant speed. Indeed, both Newton's first law and the conservation of angular momentum are examples of **conservation laws**. There are many other conservation laws in physics, including the law of conservation of momentum, the law of conservation of energy (discussed later in this chapter), and the law of conservation of electric charge.

> **Angular momentum is conserved.**

Both the ice-skater and the collapsing interstellar cloud are affected by conservation of angular momentum. You have probably noticed that an ice-skater can control how rapidly she spins by doing nothing other than pulling in or extending her arms or legs. A compact object must spin more rapidly to have the same amount of angular momentum as a more extended object with the same mass. As our skater spins, her angular momentum does not change much. (The slow decrease is due to friction, an external force.) When her arms and leg are fully extended, she spins slowly; but as she pulls her arms and leg in, she spins faster and faster. When her arms are held tightly in front of her and one leg is wrapped around the other, the skater's spin becomes a blur. She finishes with a flourish by throwing her arms and leg out—an action that abruptly slows her spin. Despite the dramatic effect, her angular momentum remains the same throughout. This impressive athletic spectacle comes courtesy of the law of conservation of angular momentum and from the difference between an extended object and a compact object.

Now we turn our attention to how conservation of angular momentum affected our protostellar Sun. Remember that its cloud of interstellar gas is collapsing under the force of its own gravity. It might seem most natural for the cloud to collapse directly into a ball—and so it would, but for the cloud's own angular momentum. Interstellar clouds are truly vast objects, light-years in size. (Recall that a light-year is the distance traveled by light in 1 year, or about 9.5 trillion kilometers [km].) As interstellar clouds orbit about the galaxy's center, they are constantly being pushed around by stellar explosions or by collisions with other interstellar clouds. This constant "stirring" guarantees that all interstellar clouds will have *some* amount of rotation. As spread out as an interstellar cloud is, even a tiny amount of rotation corresponds to a huge amount of angular momentum. Imagine our ice-skater now with arms that reach from here to the other side of Earth. Even if she were rotating very slowly at first, think how fast she would be spinning by the time she pulled those long arms to her sides!

> **Interstellar clouds have far more angular momentum than the stars they form have.**

Angular Momentum

We will encounter angular momentum in its various forms many times during our journey. In its simplest form we find that the angular momentum of a system is given by

$$L = m \times v \times r,$$

where m is the mass, v is the speed with which the mass is moving, and r represents how spread out the mass is.

As an example, we can apply this relationship to the *orbital* angular momentum, L_o, of the Moon in its orbit about Earth. Here, the mass (m) of the Moon is 7.35×10^{22} kilograms (kg), the speed of the Moon in orbit (v) is 1.023×10^{3} meters per second (m/s), and the radius of the Moon's orbit (r) is 3.84×10^{8} meters. Putting this together, we have

$$L_o = (7.35 \times 10^{22})(1.023 \times 10^{3})(3.84 \times 10^{8})$$
$$= 2.89 \times 10^{34} \text{ kg m}^2/\text{s}$$

We can compare this with the orbital angular momentum of Earth in its orbit about the Sun. Earth's mass is 5.97×10^{24} kg, its orbital speed is 2.98×10^{4} m/s, and its orbital radius is 1.496×10^{11} meters. This gives us

$$L_o = (5.97 \times 10^{24})(2.98 \times 10^{4})(1.496 \times 10^{11})$$
$$= 2.66 \times 10^{40} \text{ kg m}^2/\text{s}.$$

Earth in its orbit about the Sun has about a million times more orbital angular momentum than the Moon has in its orbit about Earth.

Calculating the *spin* angular momentum of a spinning object, such as a top, a planet, or an interstellar galactic cloud,

is far more complicated. Here we must add up the individual angular momenta of *every tiny mass element* within the body. The mathematical procedure for doing this is beyond the level used in this book. However, we can look at one example. In the case of a uniform sphere, the spin angular momentum is proportional to the square of its radius

$$L_s = \frac{4\pi m R^2}{5P},$$

where R is the radius of the sphere and P is the rotation period of its spin.

Let's compare the Moon's spin angular momentum with its orbital momentum about Earth. (Note that, in this and the following example, we make the unrealistic assumption that the Moon and Earth are uniform spheres.) The radius of the Moon is 1.737×10^{6} meters and its rotation period is 27.3 days $= 2.36 \times 10^{6}$ seconds, the same as its orbital period about Earth. This gives us

$$L_s = \frac{4\pi \times (7.35 \times 10^{22})(1.737 \times 10^{6})^2}{5 \times (2.36 \times 10^{6})} = 2.36 \times 10^{29} \text{ kg m}^2/\text{s}.$$

We can see that the Moon's orbital angular momentum is about 100,000 times greater than its spin angular momentum. We can make the same comparison with Earth, where Earth's average radius is 6.37×10^{6} meters and its rotation period is 86,400 seconds.

$$L_s = \frac{4\pi \times (5.97 \times 10^{24})(6.37 \times 10^{6})^2}{5 \times (8.64 \times 10^{4})} = 7.07 \times 10^{33} \text{ kg m}^2/\text{s}.$$

Earth's orbital angular momentum is nearly 4 million times its spin angular momentum.

Just as the ice-skater speeds up when she pulls in her arms, the cloud of interstellar gas rotates faster and faster as it collapses. However, there is a puzzle here. Suppose, for example, that we tried to form the Sun from a uniform cloud that was about a light-year across—say, 10^{16} meters— and was rotating so slowly that it took a million years to complete one rotation. As we can see in Math Tools 6.1, for a collapsing sphere to conserve spin angular momentum (L_s = constant), its rotation period (P) must be proportional to the square of its radius (R), and its spin rate ($\propto 1/P$) must therefore be *inversely proportional* to the square of its radius. So, by the time such a cloud collapsed to the size of our Sun—a mere 1.4×10^{9} meters across, or only one 10-millionth the size of the original cloud—it would be spinning 50 trillion times faster, completing a rotation in only 0.6 second! This is over 3 million times faster than our Sun actually spins. At this rate of rotation,

the Sun's self-gravity would have to be almost 200 million times stronger to hold the Sun together!

Recall our mentioning in the previous section that there were serious problems with the Kant-Laplace hypothesis. What the conservation of angular momentum seems to be saying is that the nebular hypothesis does not work—that stars cannot form from collapsing interstellar clouds. *Yet there is no other way for stars to form.*

An Accretion Disk Forms

When scientists find what appears to be a contradiction— like the apparent contradiction between the principle of conservation of angular momentum and the fact that a star has far less angular momentum than the cloud that formed it—there is cause for great excitement. Such seeming con-

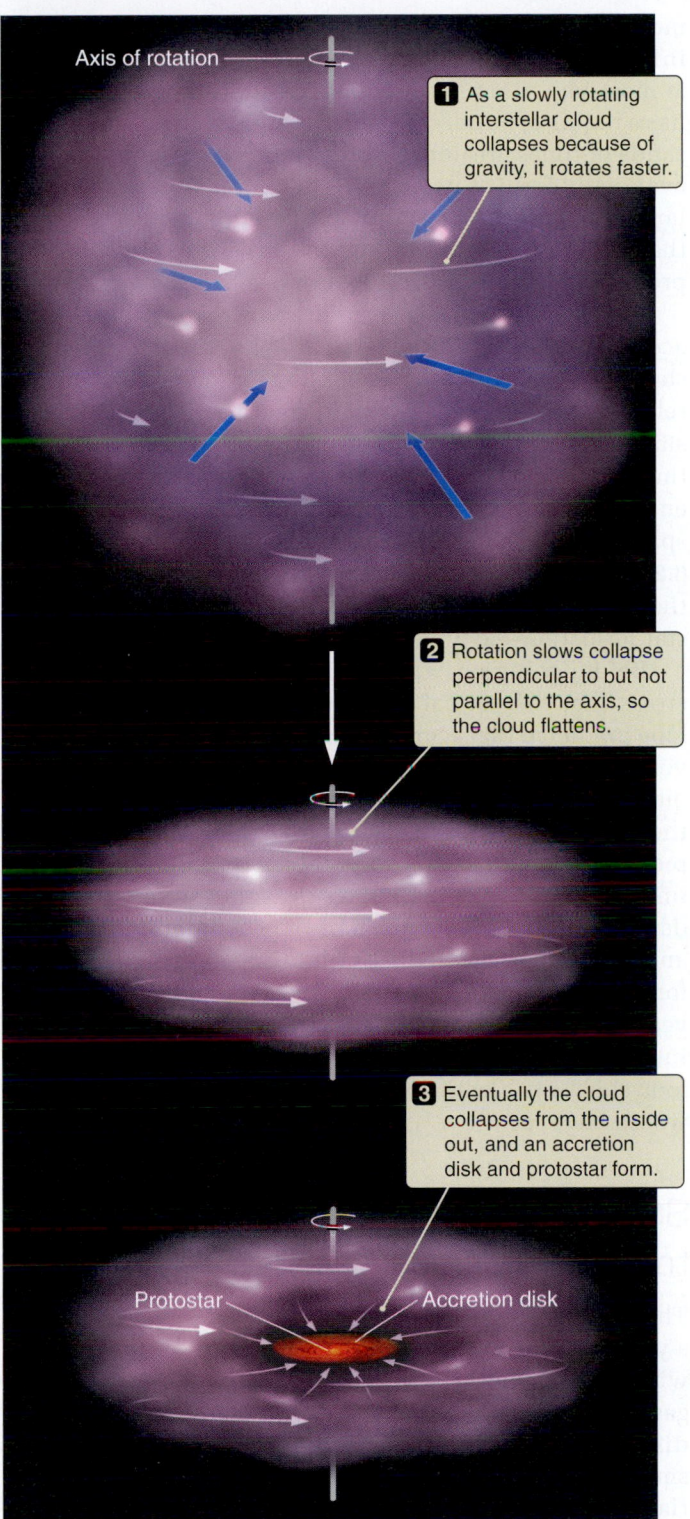

Axis of rotation

1 As a slowly rotating interstellar cloud collapses because of gravity, it rotates faster.

2 Rotation slows collapse perpendicular to but not parallel to the axis, so the cloud flattens.

3 Eventually the cloud collapses from the inside out, and an accretion disk and protostar form.

Protostar   Accretion disk

FIGURE 6.6 A rotating interstellar cloud collapses in a direction *parallel* to its axis of rotation, thus forming an *accretion disk*.

traditions do not mean that nature is breaking the rules. (Nature *never* breaks its own rules!) Instead it means that our understanding of what is going on is incomplete or that we have the wrong rule. It means that we have found a place where there are new things to be learned.

The key to solving the riddle of angular momentum in a collapsing interstellar cloud lies in realizing that the *direction* of the collapse is important. The cloud's rotation may thwart the collapse of the cloud *toward* its axis of rotation, but there is nothing to prevent collapse taking place *parallel to* the axis of rotation. Thus, rotation slows the collapse perpendicular to, but not parallel to, the axis of rotation (**Figure 6.6**). Instead of collapsing into a ball, the interstellar cloud flattens into a disk. As the cloud collapses, the strength of gravity causing the collapse becomes ever greater. Eventually the flattening cloud reaches a point where the inner parts of the cloud begin to fall freely inward, raining down on the growing object at the center. As this happens, the outer portions of the cloud lose the support of the collapsed inner portion, and they start falling inward too. The whole cloud collapses inward, much like a house of cards with the bottom layer knocked out. As this material makes its final inward plunge, it lands on a thin, rotating structure called an **accretion disk**. The accretion disk serves as a way station for material on its way to becoming part of the star that is forming at its center.

> The cloud collapses into a disk rather than directly into a star.

The formation of accretion disks is common in astronomy, so it is worth taking a moment to think carefully about this process. We can use what we learned about orbits in Chapter 3 to better understand what happens during this final stage of the collapse of an interstellar cloud. As the material falls toward the forming star, it travels on curved, almost always elliptical, paths, just as Kepler's laws say it should. These orbits would carry the material around the forming star and back into interstellar space, except for one problem: the path inward toward the forming star is a one-way street. When material nears the center of the cloud, it runs headlong into material that is falling in from the *other* side. Material falling onto the disk from opposite directions comes together at the plane of the accretion disk, which is the plane perpendicular to the cloud's axis of rotation.

To understand how interstellar material collects on the accretion disk, a visual analogy might be helpful. Imagine a traffic engineer's worst nightmare: a huge roundabout, or traffic circle, with multiple entrances but with all exits blocked by incoming traffic (**Figure 6.7**). ▶‖ **AstroTour: Traffic Circle**. As traffic flows into the traffic circle, it has nowhere else to go, resulting in a continuous, growing line of traffic driving around and around in an ever-tightening circle. Eventually, as more and more cars try to pack in, the traffic piles up. This situation is roughly analogous to an accretion disk. Of course, keep in mind that traffic in a roundabout moves on a flat surface, whereas the accretion disk around a protostar forms from material coming in from all directions in three-dimensional space. As material falls onto the disk, its motion *perpendicular* to the disk stops abruptly, but its mass motion *parallel* to the surface of the disk contributes to and adds to the disk's total angular

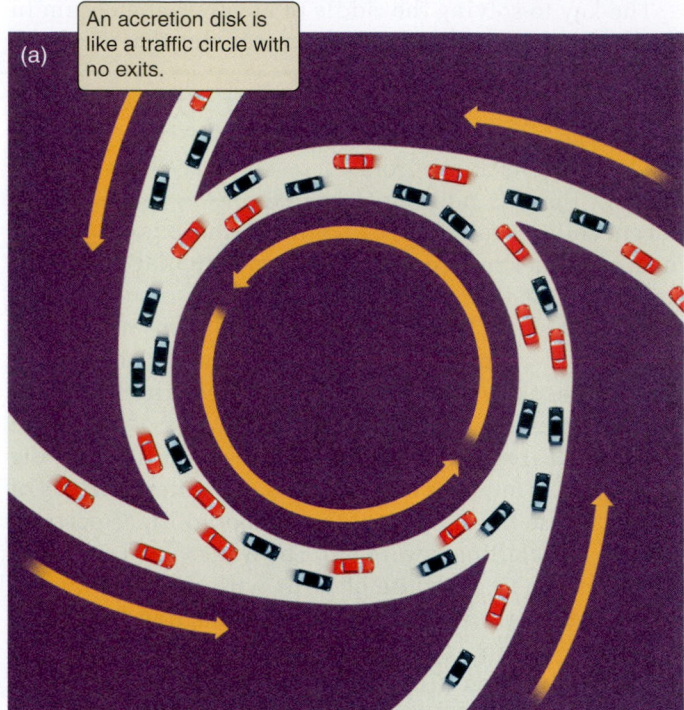

An accretion disk is like a traffic circle with no exits.

(a)

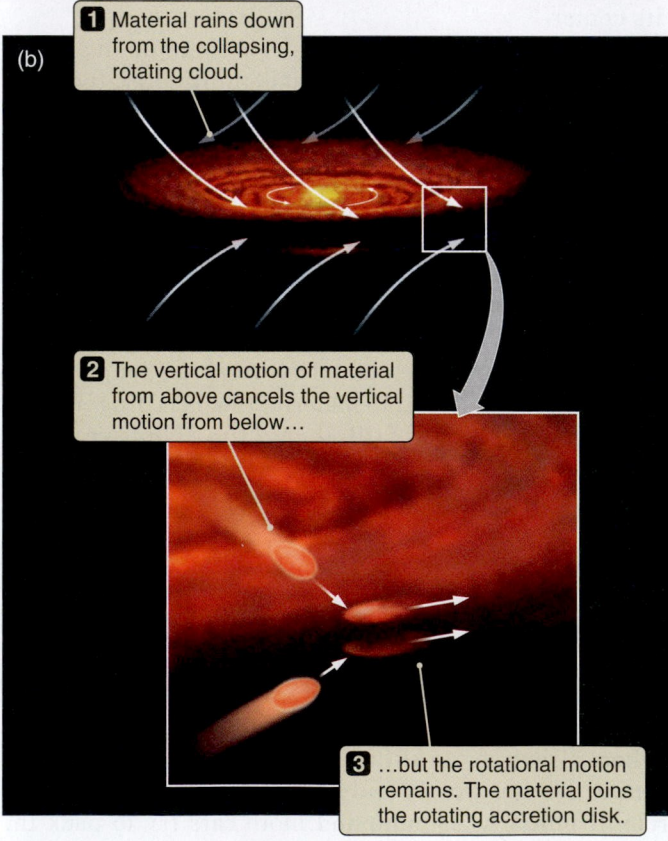

(b)

1 Material rains down from the collapsing, rotating cloud.

2 The vertical motion of material from above cancels the vertical motion from below...

3 ...but the rotational motion remains. The material joins the rotating accretion disk.

VISUAL ANALOGY FIGURE 6.7 (a) Traffic piles up in a traffic circle with entrances but no exits. (b) Similarly, gas from a rotating cloud falls inward from opposite sides, piling up onto a rotating disk.

momentum. In this way, the angular momentum of the infalling material is transferred to the accretion disk. What is most important here is that the rotating accretion disk has a radius of hundreds of astronomical units and is thousands of times greater than the radius of the star that will eventually form at its center. Therefore, most of the angular momentum in the original interstellar cloud ends up in the accretion disk rather than in the central protostar. The problem of the superrotating Sun has been solved.

The next question concerns how material falling onto the accretion disk finds its way inward onto the growing protostar. We will pick up this discussion in Chapter 15, when we return to follow the story of the growth and evolution of the star at the center of the disk. For now it is enough to know that most of the matter that lands on the accretion disk either ends up as part of the star or is ejected back into interstellar space, sometimes in the form of violent jets, as seen in Figure 6.2a. However, a small amount of material is left behind in the disk. It is this leftover disk—the dregs of the process of star formation—to which we next turn our attention.

Before going on with our story, however, we should stress that theoretical calculations by astronomers have long predicted that accretion disks should be found around young stars, and illustrations like Figure 6.4 have been a mainstay in textbooks for years. In the 1990s, however, these cartoon drawings had to make room for images of the real thing. Figure 6.2 shows Hubble Space Telescope (HST) images of edge-on accretion disks around young stars. The dark bands are the shadows of the edge-on disks, the top and the bottom of which are illuminated by light from the forming star. It would be nice if we could go back 5 billion years and watch as our own Sun formed from a cloud of interstellar gas, but we do not have to. All we have to do is look at objects like the ones in Figure 6.2 to know what we would have seen.

Small Objects Stick Together to Become Large Objects

The chain of events that connects the accretion disk around a young star to a planetary system such as our own begins with random motions of the gas within the protoplanetary disk. These motions push the smaller grains of solid material back and forth past larger grains; and as this happens, the smaller grains stick to the larger grains. The "sticking" process among smaller grains is due to the same static electricity that causes dust and hair to cling to plastic surfaces. Starting out at only a few microns (micrometers) across, the slightly larger bits of dust grow to the size of pebbles and then to clumps the size of boulders, which are less susceptible to being pushed around by gas (**Figure 6.8**). Astronomers believe that when clumps grow to about 100 meters across, their rate of growth slows down.

> Gas motions push small particles into larger particles.

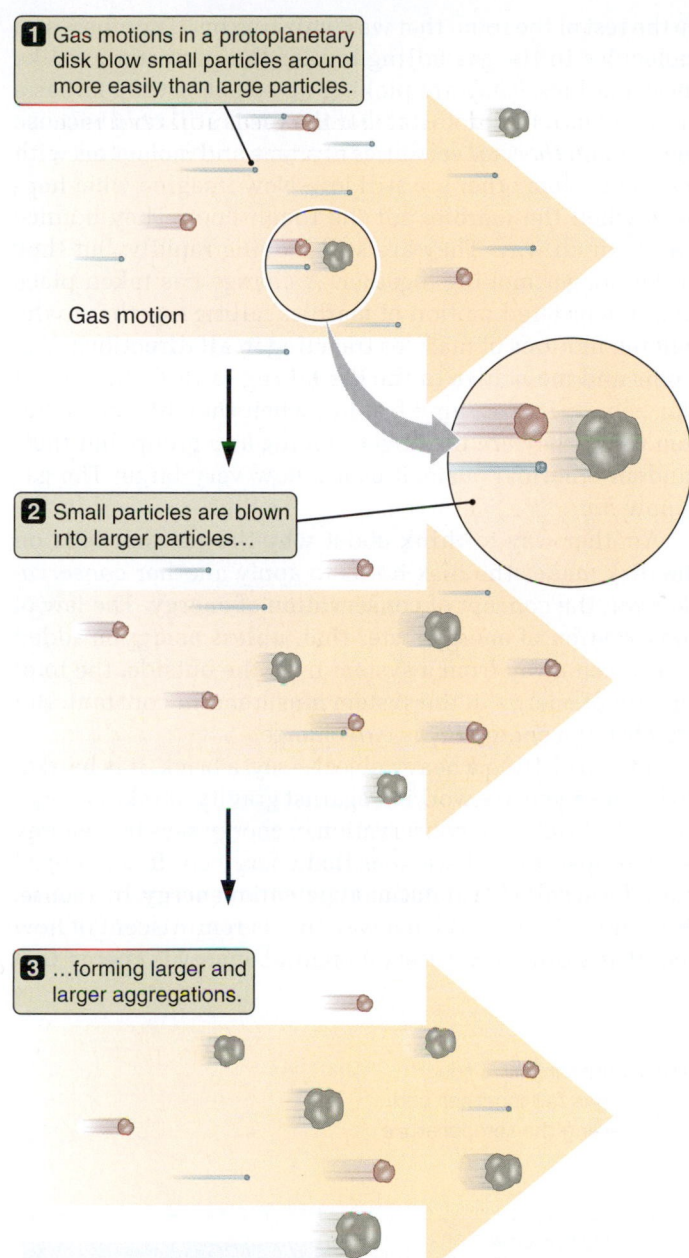

1 Gas motions in a protoplanetary disk blow small particles around more easily than large particles.

Gas motion

2 Small particles are blown into larger particles…

3 …forming larger and larger aggregations.

FIGURE 6.8 Motions of gas in a protoplanetary disk blow smaller particles of dust into larger particles, making the larger particles larger still. This process continues, eventually creating objects many meters in size.

These large objects are so few and far between in the disk that chance collisions become less and less frequent. Even so, the process of growth continues at a slower pace as 100-meter clumps join together to produce still larger bodies. In matters of social dynamics we sometimes hear that "the rich get richer at the expense of the poor." Although we can always debate this social axiom, the principle certainly holds true when it comes to the dynamics within a protoplanetary disk: the larger dust grains get larger at the expense of the smaller grains.

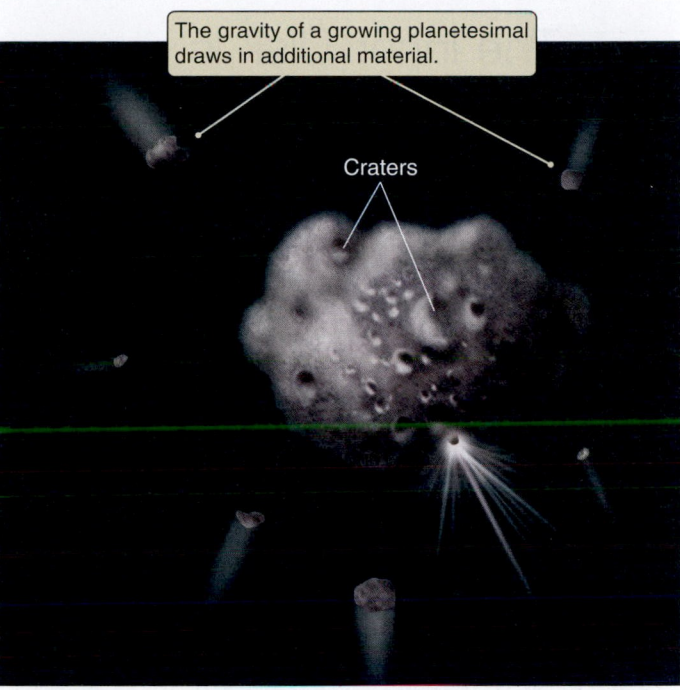

The gravity of a growing planetesimal draws in additional material.

Craters

FIGURE 6.9 The gravity of a planetesimal is strong enough to attract surrounding material, which causes the planetesimal to grow.

In order for two such clumps to stick together, they must bump into each other gently—very gently. Otherwise the energy of collision would cause the two colliding bodies to fragment into many smaller pieces instead of forming a single larger one. Collision speeds cannot be much greater than 0.1 m/s for colliding boulders to stick together. If you were to walk that slowly, it would take you 15 minutes to travel the length of a football field. In a real accretion disk, collisions more violent than this certainly happen on occasion, breaking these clumps back into smaller pieces. Likely there are many reversals in this growth of clumps of boulders.

Up to this point, larger objects have grown mainly by "sweeping up" smaller objects that run into them or that get in their way. As the clumps reach the size of about a kilometer, a different process becomes important. These kilometer-sized objects, now called **planetesimals** (literally "tiny planets"), are massive enough that their gravity begins to dominate, exerting a significant attraction on nearby bodies (**Figure 6.9**). No longer is growth of the planetesimal fed only by chance collisions with other objects; the planetesimal's gravity can now pull in and capture other smaller planetesimals that lie outside its direct path. The growth of planetesimals speeds up, with the larger planetesimals quickly consuming most of the remaining bodies in the vicinity of their orbits. The final survivors of this process are now large enough to be called **planets**. As with the major bodies in orbit about the Sun, some of the planets may be small and others quite large.

Gravity helps planetesimals grow into planets.

6.3 The Inner Disk Is Hot, but the Outer Disk Is Cold

The accretion disks surrounding young stars form from interstellar material that may have a temperature of only a few kelvins, but the disks themselves reach temperatures of hundreds of kelvins or more. What is it that heats up the disk around a forming star? The answer lies with our old friend gravity. Material from the collapsing interstellar cloud falls inward toward the protostar, but because of its angular momentum it "misses" the protostar and instead falls onto the surface of the disk. When this material reaches the surface of the disk, its infalling motion comes to an abrupt halt, and the velocity that the atoms and molecules in the gas had before hitting the disk is suddenly converted into random *thermal* velocities instead. That is to say, the cold gas that was falling toward the disk gets very hot when it lands on the disk.

To help you visualize this, imagine dumping a box of marbles from the top of a tall ladder onto a rough, hard floor below (**Figure 6.10**). The marbles fall, picking up speed as they go. Even though the marbles are speeding up, however, they are all speeding up *together*. As far as one marble is concerned, the other marbles are not moving very fast at all. (If you were riding on one of the marbles, the other marbles would not appear to you to be moving very much; it would

be the rest of the room that was whizzing by.) The atoms and molecules in the gas falling toward the protostar are like these marbles. They are picking up speed as they fall as a group toward the protostar, but the gas is still *cold* because the random *thermal* velocities of atoms and molecules with respect to each other are still low. Now imagine what happens when the marbles hit the rough floor. They bounce every which way. They are still moving rapidly, but they are no longer moving together. A change has taken place from the ordered motion of marbles falling together to the random motions of marbles traveling in all directions. The atoms and molecules in the gas falling toward the central star behave in the same fashion when they hit the accretion disk. They are no longer moving as a group, but their random "thermal" velocities are now very large. The gas is now *hot*.

Another way to think about why the gas that falls on the disk makes the disk hot is to apply another *conservation law*, the concept of **conservation of energy**. The law of conservation of energy states that, unless energy is added to or taken away from a system from the outside, the total amount of energy in the system must remain constant. But the *form* the energy takes *can* change.

Imagine lifting a heavy object—say, a brick. It is hard to do because you are working against gravity. It takes energy to lift the brick, and conservation of energy says that energy is never lost. But where does that energy go? It is changed into a form called **gravitational potential energy**. In a sense, this energy is "stored" in a way that is reminiscent of how energy is stored in a battery. Potential energy is energy that

VISUAL ANALOGY FIGURE 6.10 (a) Marbles dropped as a group fall together until they hit a rough floor, at which point their motions become randomized. (b) Similarly, atoms in a gas fall together until they hit the accretion disk, at which point their motions become randomized, raising the temperature of the gas.

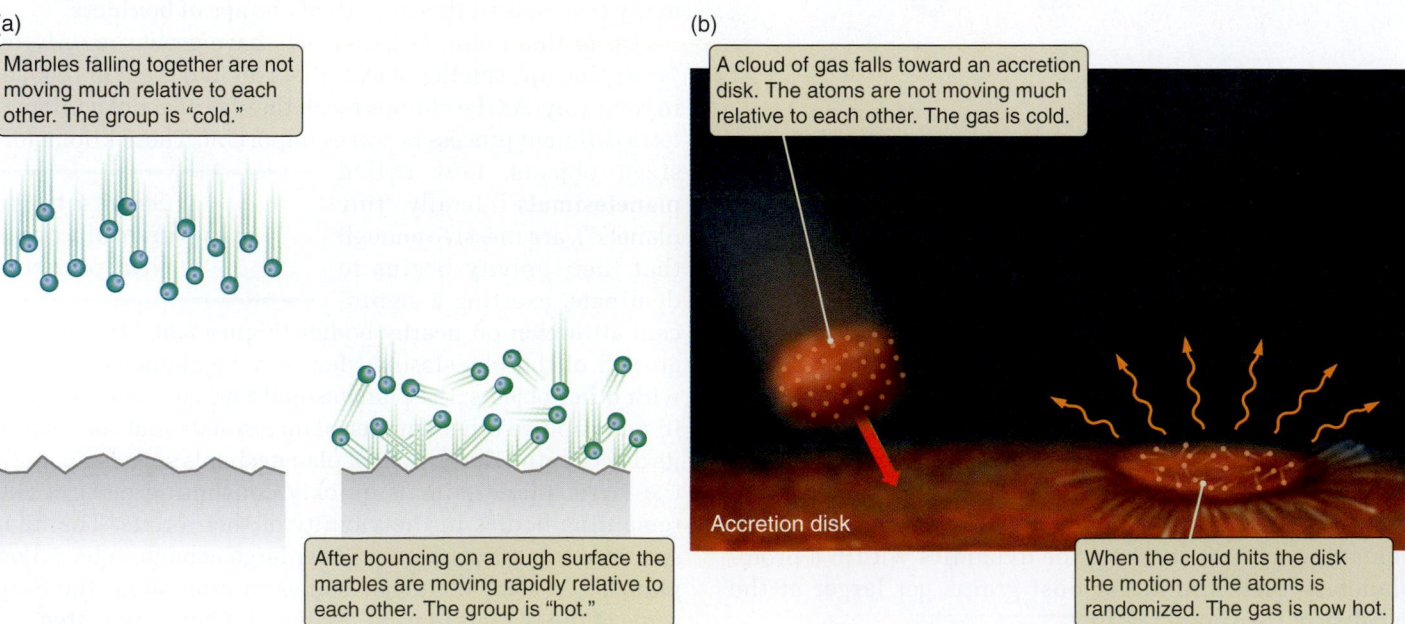

(a)

Marbles falling together are not moving much relative to each other. The group is "cold."

After bouncing on a rough surface the marbles are moving rapidly relative to each other. The group is "hot."

(b)

A cloud of gas falls toward an accretion disk. The atoms are not moving much relative to each other. The gas is cold.

Accretion disk

When the cloud hits the disk the motion of the atoms is randomized. The gas is now hot.

"has potential"—it is waiting to show up in a more obvious form. If you drop the brick it falls, and as it falls it speeds up. The gravitational potential energy that was stored is being converted instead to energy of motion, which, as you may recall from Chapter 4, is called *kinetic energy*. When the brick hits the floor, it stops suddenly. The brick loses its energy of motion, so what form does this energy take now? If the brick cracks, part of the energy goes into breaking the chemical bonds that hold it together. Some of the energy is converted into the sound the brick makes when it hits the floor. Some goes into heating and distorting the floor. But most of the energy is converted into thermal energy. The atoms and molecules that make up the brick are moving about within the brick a bit faster than they were before the brick hit; so the brick and its surroundings, including the floor, grow a tiny bit warmer. Similarly, as

> The gravitational energy of infalling material turns into thermal energy.

gas falls toward the disk surrounding a protostar, gravitational potential energy is converted first to kinetic energy, causing the gas to pick up speed. When the gas hits the disk and stops suddenly, that kinetic energy is turned into thermal energy. (**Foundations 6.1** discusses why it can be useful to think about the same thing—in this case, energy—in different ways.)

In this way, as material falls onto the accretion disk around a forming star, the disk heats up. But the amount of heating depends on *where* the material hits the disk. Material hitting the inner part of the disk (which we will call the "inner disk") has fallen farther and picked up greater speed within the gravitational field of the forming star than has material hitting the disk

> The inner disk is hotter than the outer disk.

farther out. Like a rock dropped from a tall building, material hitting the inner disk is moving quite rapidly when it hits, so it heats the inner disk to high temperatures. In con-

FOUNDATIONS 6.1

Thinking about Energy in Different Ways

In Section 6.3 we discuss how the gas falling onto a protostellar disk heats the disk. First we present the analogy of marbles falling on a floor, noting that the motion of marbles is jumbled up when they hit, like the atoms in the gas hitting the disk. As the correlated motion of gas molecules becomes randomized, the temperature of the gas increases. We then approach the explanation from a different perspective, focusing on how energy is conserved but changes its form from gravitational potential energy to kinetic energy and finally to thermal energy. These may be two different ways of explaining why the disk gets hot, but they are *not* two different *reasons* that the disk gets hot. The disk heats up as a result of the same physical process—the "reason" is the same regardless of the words we may use to describe it. Instead, these two explanations offer two different ways to *think about* why the disk gets hot.

Both ways of thinking about the process are included in the discussion because sometimes students—or scientists—need to look at the same thing from several different angles before they understand it. In this case, however, most scientists would agree that the second way of thinking about the problem is far more *powerful* than the first. The idea of energy changing forms connects how disks are heated with a wide variety of other phenomena. Properly stated, the heating of protostellar disks is an example of one of the most far-reaching patterns in nature: the conservation of energy.

Once you understand the different forms of energy and how energy is conserved, you begin to see this pattern of nature everywhere. For example, when water falls through the tur-

bines in a hydroelectric generator, it turns the generator and produces electric energy that is carried out over power lines and lights up the lights in your home. Where did the energy to light your lights or heat the water that pours from your tap come from? If you have an electric water heater and get your power from a hydroelectric plant, it comes from the gravitational potential energy of the water in a reservoir near you—just as the energy to heat the newly formed Sun came from the gravitational potential energy of the reservoir of gas from which the Sun formed.

Conservation of energy is a powerful way to think about the heating of disks around young stars because it enables astronomers to *calculate* how hot these disks get. If we know the mass of the protostar and the surrounding disk, and how far away from the protostar the gas is falling from, we can calculate how much gravitational energy the gas started out with. According to the law of conservation of energy, this gravitational energy must eventually be converted to thermal energy, allowing us to calculate the expected temperature of the protostar. So without even determining *how* the disk formed, we can say right off the bat how much thermal energy will be deposited onto the disk.

Scientists spend much of their time trying to come up with new ways of thinking about problems, looking for particularly powerful ways that point to new insights and discoveries. The most powerful means of thinking about a problem usually tie the problem to ever-grander patterns in nature. Conservation of energy is one of the grandest and most useful patterns around.

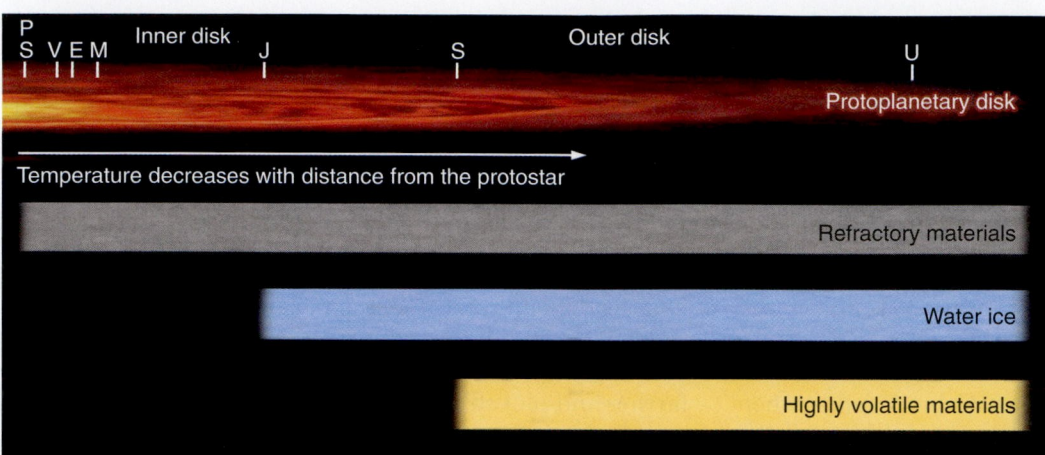

FIGURE 6.11 Differences in temperature within a protoplanetary disk determine the composition of dust grains that then evolve into planetesimals and planets. Shown here are the protostar (PS) and the orbits of Venus (V), Earth (E), Mars (M), Jupiter (J), Saturn (S), and Uranus (U).

trast, material falling onto the outer part of the disk (which we will call the "outer disk") is moving much more slowly (envision a rock dropped from only a foot or so), leaving the outermost parts of the disk at temperatures that may be not much higher than the original interstellar cloud. Stated another way, material falling onto the inner disk converts more gravitational potential energy into thermal energy than does material falling onto the outer disk.

The energy released as material falls onto the disk is not the only source of thermal energy in the disk. Even before the nuclear fires that will one day power the new star have ignited, conversion of gravitational energy into thermal energy drives the temperature at the surface of the protostar to several thousand kelvins, and it also drives the luminosity of the huge ball of glowing gas to many times the luminosity of the present-day Sun. For the same reasons that Mercury is hot while Pluto is not (see Section 4.5), the radiation streaming outward from the protostar at the center of the disk drives the temperature in the inner parts of the disk even higher, increasing the difference in temperature between the inner and outer parts of the disk.

Rock, Metal, and Ice

We have all noticed that temperature affects which materials can and cannot exist in a solid form. On a hot summer day, ice melts and water quickly evaporates, whereas on a cold winter night even the water in our breath freezes into tiny ice crystals before our eyes. Some materials, such as iron, **silicates** (minerals containing silicon and oxygen), and carbon—rocky materials and metals—remain solid even at quite high temperatures. Such materials, which are capable of withstanding high temperatures without melting or being vaporized, are referred to as **refractory materials**. Other materials, such as water, ammonia, and methane, can remain in a solid form

> **Refractory materials remain solid even at high temperatures.**

only if their temperature is quite low. These less refractory substances are called **volatile materials** (or "volatiles" for short). Astronomers generally refer to the solid form of any volatile material as an **ice**.[3]

Differences in temperature from place to place within the disk have a significant effect on the makeup of the dust grains in the disk (**Figure 6.11**). In the hottest parts of the disk (closest to the protostar), only the most refractory substances can exist in solid form. In the inner disk, dust grains are composed of refractory materials only.[4] Somewhat farther out in the disk, some hardier volatiles, such as water ice and certain **organic**[5] substances, can survive in solid form, adding to the materials that make up dust grains. Highly volatile components such as methane, ammonia, and carbon monoxide ices and some organic molecules survive in solid form only in the coldest, outermost parts of the accretion disk, far from the central protostar. The differences in composition of dust grains within the disk are reflected in the composition of the planetesimals formed from that dust. Planets that form closest to the central star tend to be made up mostly of refractory materials such as rock and metals. Those that form farthest from the central star also contain refractory

> **Volatile ices survive in the outer disk, but only refractory solids survive in the inner disk.**

[3] In general astronomical usage, the term *ice* is used to describe the volatile itself, regardless of whether it happens to be in solid, liquid, or gaseous form.

[4] Some geochemists believe that certain volatiles, such as water, did survive in the hot inner disk because they were bound chemically to refractory materials and were thus able to withstand the high temperatures.

[5] The term *organic* does not mean "life" but instead refers to a large class of chemical compounds containing the element carbon. All terrestrial life is organic, but not all organic compounds come from living organisms.

materials, but in addition they contain large quantities of ices and organic materials.

As we turn to a study of our own Solar System, we will find that the trend in composition expected in a protoplanetary disk is closely echoed in the makeup of the solid bodies orbiting the Sun. The inner planets are composed of rocky material surrounding metallic cores of iron and nickel. In contrast, objects in the outer Solar System, including moons, giant planets, and comets, are composed largely of ices of various types. In the years to come, as astronomers learn more about planetary systems around other stars, it seems likely that this trend will prove to be quite common. In fact, our understanding of the way stars and planetary systems form suggests that this change in composition with distance from the central star would be almost unavoidable.

Even so, chaotic encounters like those we will discuss in Chapter 10 can shuffle the deck, adding diversity to the organization of planetary systems. In a process called **planet migration**, gravitational scattering can force some planets to end up far from the place of their birth. For example, many planetary scientists believe that Uranus and Neptune originally formed near the orbits of Jupiter and Saturn, but were then driven outward to their current locations by gravitational encounters with Jupiter and Saturn. A planet can also migrate when it gives up some of its orbital angular momentum to the disk material that surrounds it. Such a loss of angular momentum causes the planet to slowly spiral inward toward the central star. We will see examples when we discuss *hot Jupiters* in Section 6.5.

Solid Planets Gather Atmospheres

Once a solid planet has formed, it may continue growing by capturing gas from the protoplanetary disk. To do so, though, it must act quickly. Young stars and protostars are known to be sources of strong *stellar winds* (see Chapters 19 and 20) and intense radiation that can quickly disperse the gaseous remains of the accretion disk. Gaseous planets such as Jupiter probably have only about 10 million years or so to form and to grab whatever gas they can. Tremendous mass is a great advantage in a planet's ability to accumulate and hold on to the hydrogen and helium gas that makes up the bulk of the disk. Because of their strong gravitational fields, more massive young planets are thought to create their own mini accretion disks as gas from their surroundings falls toward them. What follows is much like the formation of a star and protoplanetary disk, but on a smaller scale. Just as happened in the accretion disk around the star, gas from a mini accretion disk moves inward and falls onto the solid planet.

We refer to the gas that is captured by a planet at the time of its formation as the planet's **primary atmosphere**. The primary atmosphere of a large planet can become massive enough to dominate the mass of the planet, as in the case of giant planets such as Jupiter. Some of the solid material in the mini accretion disk might stay behind to coalesce into larger bodies in much the same way that particles of dust in the protoplanetary disk came together to form planets. The result is a mini "solar system"—a group of moons that orbit about the planet.

A less massive planet may also capture some gas from the protoplanetary disk, only to lose its prize. Here again, more massive planets have the advantage. As we will learn in Chapter 8, the gravity of small planets may not be strong enough to prevent less massive atoms and molecules such as hydrogen or helium from escaping back into space. Even if a small planet is able to gather some hydrogen and helium from its surroundings, this temporary primary atmosphere will be short-lived. The atmosphere that remains around a small planet like our Earth is a **secondary atmosphere**. A secondary atmosphere forms later in the life of a planet. Carbon dioxide and other gases released from the planet's interior by widespread volcanism are probably one important source of a planet's secondary atmosphere. In addition, as we will see later, volatile-rich comets that formed in the outer parts of the disk continue to fall inward toward the new star long after its planets have formed, and will sometimes collide with planets. Comets possibly provide a significant source of water, organic compounds, and other volatile materials on planets close to the central star.

> Less massive planets lose their primary atmospheres and then form secondary atmospheres.

6.4 A Tale of Eight Planets

We are now at a point in our discussion where we can take our general ideas about the evolution of the material in a protoplanetary disk and apply them to our own Solar System. Only in the closing years of the 20th century did our knowledge progress to the point where this story could be told. It is still sketchy in places and doubtless wrong in some ways. But error and uncertainty are part of the advance of science. In this section we bring together a wealth of information taken both from what we know of our local star and planetary neighbors and from what we have learned about stars forming around us today. It is a synthesis of the painstaking efforts of hundreds of astronomers and planetary scientists over the course of decades. Many more scientists will devote their lives to unraveling this story before its details are fully known. There are good reasons to believe that the explanations we give in this section are basically correct, although we know they are not yet complete. Returning to our early discussion of what science is and how science works, the story we tell here is one that many people have tried to prove wrong but that has withstood all such tests . . . so far.

FIGURE 6.12 Large impact craters on Mercury (and on solid bodies throughout the Solar System) record the final days of the Solar System's youth, when planets and planetesimals grew as smaller planetesimals rained down on their surfaces.

Nearly 5 billion years ago, our Sun was still a protostar surrounded by a protoplanetary disk of gas and dust. Over the course of a few hundred thousand years, much of the dust in the disk had collected into planetesimals—clumps of rock and metal near the emerging Sun and aggregates of rock, metal, ice, and organic materials in the parts of the disk that were more distant from the Sun. Within the inner few astronomical units of the disk, several rock and metal planetesimals, probably fewer than a half dozen, quickly grew in size to become the dominant masses at their respective distances from the Sun. With their ever-strengthening gravitational fields, they either captured most of the remaining planetesimals or ejected them from the inner part of the disk. These dominant planetesimals had now become planet-sized bodies with masses ranging between that of Earth and about ¹⁄₂₀ of that value. They were to become the **terrestrial planets**. Mercury, Venus, Earth, and Mars are the surviving terrestrial planets. Planetary scientists think that one or two others may have formed in the young Solar System but were later destroyed. (As we will see later in this section, one of them may have been responsible for the creation of our Moon.)

For several hundred million years following the formation of the four surviving terrestrial planets, leftover pieces of debris still in orbit around the Sun continued to

> Rocky terrestrial planets formed in the inner Solar System.

rain down on their surfaces. Much of this barrage may have originated in the outer Solar System, where the gravitational tug of the massive, newly formed outer planets acted like slingshots shooting debris inward. Today we can still see the scars of these postformation impacts on the cratered surfaces of all the terrestrial planets, such as the surface of Mercury shown in **Figure 6.12**. This rain of debris continues today, albeit at a much lower rate.

Before the proto-Sun emerged as a true star, gas in the inner part of the protoplanetary disk was still plentiful. During this early period the two larger terrestrial planets—Earth and Venus—may have held on to weak primary atmospheres of hydrogen and helium. If so, these thin atmospheres were soon lost to space. For the most part the terrestrial planets were all born devoid of thick atmospheres and remained so until the formation of the secondary atmospheres that now surround Venus, Earth, and Mars. Mercury's proximity to the Sun and the Moon's small mass must have prevented these bodies from retaining significant secondary atmospheres. They remain nearly airless today.

Farther out in the nascent Solar System, 5 astronomical units (AU) from the Sun and beyond, planetesimals coalesced to form a number of bodies with masses about 5–10 times that of Earth. Why such large bodies formed in the region beyond the terrestrial planets remains an unanswered question. Located in a much colder part of the accretion disk, these planet-sized objects formed from planetesimals containing volatile ices and organic compounds in addition to rock and metal. Four such massive bodies would later become the cores of the **giant planets**, which we know today as Jupiter, Saturn, Uranus, and Neptune. In a process astronomers call **core accretion**, mini accretion disks formed around these planetary cores, capturing massive amounts of hydrogen and helium and funneling this material onto the planets.

> The giant planets formed cores from planetesimals and then captured gaseous hydrogen and helium onto their cores.

Jupiter's massive solid core was able to capture and retain the most gas—a quantity roughly 300 times the mass of Earth, or 300 $M_\oplus$. (The symbol $\oplus$ signifies Earth.) The other planetary cores captured lesser amounts of hydrogen and helium, perhaps because their cores were less massive or because there was less gas available to them. Saturn ended up with less than 100 $M_\oplus$ of gas, whereas Uranus and Neptune were able to grab only a few Earth masses' worth of gas. As we pointed out in the previous section, some planetologists believe that all of the giant planets formed closer to where Jupiter is now, and that their mutual gravitational interactions caused them to migrate to their present orbits.

Before moving on, we should mention that some planetary scientists do not believe that our protoplanetary disk could have survived long enough to form gas giants such as Jupiter through the general process of core accretion. The core accretion model indicates that it could take up to 10 million years for a Jupiter-like planet to accumulate. However, all the gas in the protoplanetary disk likely dispersed in a little more than half that time, which would have cut off Jupiter's supply of hydrogen and helium.

This is not just a Solar System predicament. The apparent conflict between the time needed for a Jupiter-type planet to form and the availability of gases within that time period applies as well to other protoplanetary disks and to the formation of their massive planets (see Section 6.5). To get around this time dilemma, some scientists have proposed a process called "disk instability," in which the protoplanetary disk suddenly and quickly fragments into massive clumps equivalent to that of a large planet. Although core accretion and disk instability appear to be competing processes, they may not be mutually exclusive. It is possible that both played a role in the formation of our own and other planetary systems.

For the same reasons that a forming protostar is hot—namely, conversion of gravitational energy into thermal energy—the gas surrounding the cores of the giant planets compressed under the force of gravity and became hotter. Proto-Jupiter and proto-Saturn probably became so hot that they actually glowed a deep red color, similar to the heating element on an electric stove. Their internal temperatures may have reached as high as 50,000 kelvins (K). In a sense, the more massive protoplanets were trying to become stars. But for them, this was an unreachable goal. In Chapter 15 we will find that a ball of gas must have a mass at least 0.08 times that of the Sun (0.08 $M_\odot$) for it to become a star. This minimum mass is some 80 times the mass of Jupiter. Science fiction films notwithstanding, Jupiter never had a chance.

Some of the material remaining in the mini accretion disks surrounding the giant planets coalesced into small bodies, which became moons. (A **moon** is any natural satellite in orbit about a planet or asteroid.) The composition of the moons of the giant planets followed the same trend as

> Moons formed from the mini accretion disks around the giant planets.

the planets that formed around the Sun: the innermost moons formed under the hottest conditions and therefore contained the smallest amounts of volatile material. As very young moons, Io and Europa may have experienced nearby Jupiter glowing so intensely that it rivaled the distant Sun. The high temperatures created by the glowing planet would have evaporated most of the volatile substances in the inner part of its mini accretion disk. Io today contains no water at all. However, water is relatively plentiful on Europa, Ganymede, and Callisto because these moons formed farther from hot Jupiter.

Not all planetesimals in the protoplanetary disk went on to become planets. Jupiter is a true giant of a planet. Its gravity kept the region of space between it and Mars so stirred up that most planetesimals there never coalesced into a single planet. (The one exception is Ceres, once considered to be the largest asteroid but now redefined, along with Pluto, as a **dwarf planet** under the new IAU planet definition; see

Appendix 8.) This region, now referred to as the **asteroid belt**, contains many planetesimals that remain from this early time. In the outermost part of the Solar System as well, planetesimals persist to this day. Born in a "deep freeze," these objects retained most of the highly volatile materials found in the grains present at the formation of the proto-

> Asteroids and comet nuclei are planetesimals that survive to this day.

planetary disk. Unlike conditions in the crowded inner part of the disk, planetesimals in the outermost parts of the disk were too sparsely distributed for large planets to grow. Icy planetesimals in the outer Solar System remain today as **comet nuclei**—relatively pristine samples of the material from which our planetary system formed. Frozen Pluto and the distant dwarf planet Eris are especially large examples of these denizens of the outer Solar System.

The early Solar System must have been a remarkably violent and chaotic place. Many objects in the Solar System show evidence of cataclysmic impacts that reshaped worlds. A dramatic difference in the terrain of the northern and southern hemispheres on Mars, for example, has been interpreted as the result of one or more colossal collisions. Mercury has a crater on its surface from an impact so devastating that it caused the crust to buckle on the opposite side of the planet. In the outer Solar System, one of Saturn's moons, Mimas, sports a crater roughly one-third the diameter of the moon itself. Uranus suffered a collision that was violent enough to literally knock the planet on its side. Today its axis of rotation is tilted at an almost right angle to its orbital plane.

Not even our own Earth escaped devastation by these cataclysmic events. In addition to the four terrestrial planets that remain, there was at least one other terrestrial planet in the early Solar System—a planet about the same size and mass as Mars. As the newly formed planets were settling into their present-day orbits, this fifth planet suffered a grazing collision with Earth and was completely destroyed. The remains of the planet, together with material knocked from Earth's outer layers, formed a huge cloud of debris encircling Earth. For a brief period Earth may have displayed a magnificent group of rings like those of Saturn. In time, this debris coalesced into the single body we know as our Moon.

6.5 There Is Nothing Special about Our Solar System

We began this chapter by discussing the formation of a generic planetary system around a generic star. Only later did we turn to the specifics of our own Solar System. We did this to make a very important point: according to our current understanding, *there is absolutely nothing special about the conditions that led to the formation of our Sun*

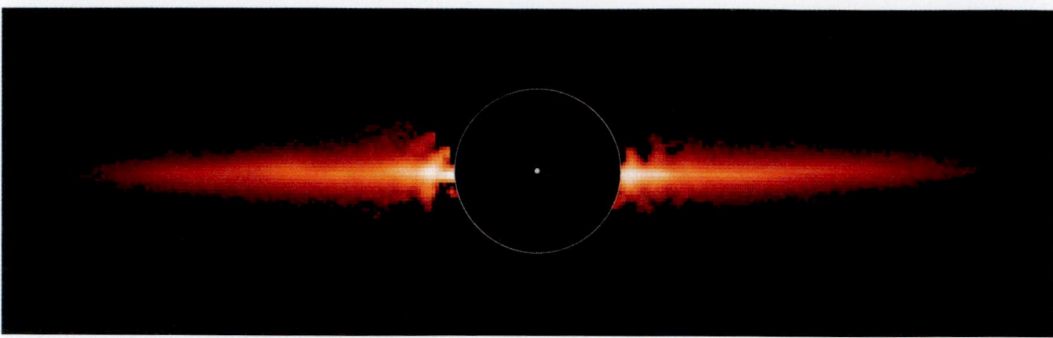

FIGURE 6.13 An edge-on circumstellar dust disk is seen extending outward to 60 AU from the young (12-million-year-old) star AU Microscopii. The star itself, whose brilliance would otherwise overpower the disk, is hidden behind an occulting disk placed in the telescope's focal plane. Its position is represented by the dot.

and our Solar System. When astronomers turn their telescopes to young stars around us, they see just the sorts of disks from which our own Solar System formed (**Figure 6.13**). From what we know, the physical processes that led to the formation of our Solar System should be commonplace wherever new stars are being born.

In 1994, astronomers found their first confirmed **extrasolar planet**,[6] a Jupiter-sized body orbiting surprisingly close to 51 Pegasi, a solar-type star. Today the number of known extrasolar planets, sometimes called "exoplanets," has grown to more than 400, with new discoveries occurring almost daily.

The discovery of extrasolar planets raises the question of what we mean by the term *planet*. Within our own Solar System, we feel reasonably comfortable with our definition of a planet.[7] But what about those extrasolar bodies? The International Astronomical Union defines an extrasolar planet as a body that orbits a star and has a mass less than 13 Jupiters (13 M_J). As we will see in Chapter 15, objects more massive than this, but less than 0.08 $M_\odot$, are known as *brown dwarfs*.

The Search for Extrasolar Planets

How do astronomers search for extrasolar planets, and why were they not found earlier? The "why" is easy to answer. Until the closing years of the 20th century, astronomers lacked the sophisticated instruments needed to detect these elusive objects. We turn now to "how."

The most successful technique employed so far has been the "spectroscopic radial velocity method." More than 370 extrasolar planets have been detected with this technique. As a planet orbits about a star, the planet's gravity causes

the position of the star to wobble back and forth ever so slightly. By observing the Doppler shift of the star, we can sometimes detect this wobble and infer the properties of the planet—its *mass* and its *distance* from the star. (We will use an almost identical method in Chapter 13 to measure the masses of pairs of stars that orbit each other.)

> **Several techniques are being used to find extrasolar planets.**

We can see how this would work by using our own Solar System as an example. We'll start with the simplest of cases by assuming that the Solar System consists only of the Sun and Jupiter. In fact, this makes sense because Jupiter's mass is greater than the mass of all the other planets, asteroids, and comets combined! Imagine an alien astronomer pointing her spectrograph toward the Sun. Both the Sun and Jupiter orbit a common center of gravity that lies just outside the surface of the Sun, as shown in **Figure 6.14**. Our alien astronomer would find that the Sun's radial velocity varies by ±12 m/s, with a period equal to Jupiter's orbital period of 11.86 years. From this information, she would rightly conclude that the Sun has at least one planet with a mass comparable to Jupiter's. Without greater precision, she would be unaware of the other less massive major planets. But, spurred on by the excitement of her discovery, she would improve the sensitivity of her instruments. If she could measure radial velocities as small as 2.7 m/s, she would be able to detect Saturn, and if the precision of her spectrograph extended to radial velocities as small as 0.09 m/s, she would be able to detect Earth.

The technology of today limits the precision of our own radial velocity instruments to about 1 m/s. This technique enables astronomers to detect giant planets around solar-type stars, but we are still far from being able to reveal bodies with masses similar to those of our own terrestrial planets.

Another technique for finding extrasolar planets is the "transit method," in which we observe the effect of a planet passing in front of its parent star. Try picturing the geometry required to observe a transit, however, and you may see a major limitation. For a planet to pass in front of a star, Earth must necessarily lie nearly in the orbital plane of

---

[6]In 1988, astronomers announced the first discovery of an extrasolar planet. However, the announcement was questioned because of the quality of the observational data. That object was finally confirmed in 2002.

[7]Not entirely comfortable, as it turns out. The argument among astronomers continues to rage over whether dwarf planets should be defined as a kind of planet or put in an entirely different category.

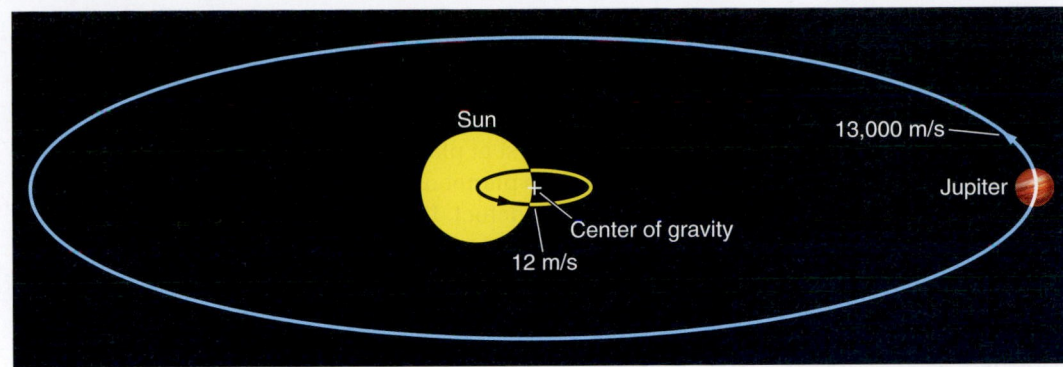

FIGURE 6.14 Both the Sun and Jupiter orbit around a common center of gravity, which lies just outside the Sun's surface. Spectroscopic measurements made by an extrasolar astronomer would reveal the Sun's radial velocity varying by ±12 m/s over an interval of 11.86 years, Jupiter's orbital period.

the planet. When an extrasolar planet passes in front of its parent star, the light from the star will diminish by a tiny amount, as illustrated in **Figure 6.15**. Whereas the radial velocity method yields the mass of the planet and its orbital distance from a star, the transit method provides a measure of the *size* of a planet. Current groundbased technology limits the sensitivity of the transit method to about 0.1 percent of a star's brightness, but that has been good enough for this method to detect more than 60 extrasolar planets. (In fact, amateur astronomers have confirmed the existence of several extrasolar planets by observing transits using CCD cameras mounted on telescopes with apertures as small as 20 centimeters [cm]!)

Getting back to our alien astronomer, she could infer the existence of Earth if she were located somewhere in the plane of Earth's orbit (that is the only way she would be able to see Earth pass in front of the Sun) and could detect a 0.009 percent drop in the Sun's brightness. Our own astronomers now have space instruments with almost that capability. Using the transit method, the French CoRoT satellite has already discovered a planet with a diameter only 1.7 times that of Earth. In 2009, NASA launched a solar-orbiting[8] telescope called Kepler with even greater capability. Its photometric instruments should be able to detect transits of Earth-sized planets.

Still another technique is the "microlensing method," which involves a relativistic effect called *gravitational lensing*[9] (see Chapter 17). The gravitational field of an unseen planet bends light from a distant star in such a way that it causes the star to brighten temporarily while the planet is passing in front of it. Microlensing, like the radial velocity method, provides an estimate of the mass of the planet. So far, nine extrasolar planets have been found with this technique. Like the transit method used in space, lensing is also capable of detecting Earth-sized planets.

A fourth method is "direct imaging." This is the most difficult technique because it involves searching for a relatively faint planet in the overpowering glare of a bright star—a challenge far more difficult than looking for a firefly in the dazzling brilliance of a searchlight. Even when an object is detected by direct imaging, there remains the major problem of determining whether or not the observed object is indeed a planet. Suppose we detect a faint object

FIGURE 6.15 As a planet passes in front of a star, it blocks some of the light coming from the star's surface, causing the brightness of the star to decrease slightly. (The brightness decrease has been exaggerated in this illustration.)

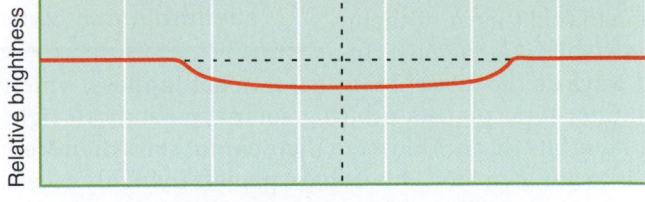

---

[8] To keep stray light from our own planet out of its field of view, Kepler was placed in an Earth-trailing heliocentric orbit.

[9] *Gravitational lensing* generally refers to the relativistic lensing of distant quasars by intervening galaxies. When referring to the lensing of stars by small bodies such as planets, astronomers use the term *microlensing*.

near a bright star. Might it be a distant star that just happens to be in the line of sight? Future observations could tell if it shares the bright star's motion through space. Could it be a brown dwarf (see the next subsection) rather than a true planet? Again, further observations to determine the object's mass would be required.

As of this writing, astronomers have identified 11 possible planets by direct imaging. Some recent discoveries appear to be promising as actual planets. Four of the recent candidates (three around a single star) were discovered by large, groundbased telescopes operating in the infrared region of the spectrum (**Figure 6.16**). The first visible-light discovery was made from space while HST was observing Fomalhaut, a bright naked-eye star only 25 light-years away. The planet, temporarily named Fomalhaut b, has a mass about three times that of Jupiter and orbits within a dusty debris ring some 17 billion km from the central star (**Figure 6.17**).

The most exciting discoveries, however, could come in about a decade, when NASA plans to launch its Terrestrial Planet Finder (TPF) and ESA puts its hopes on Darwin. As designed, both observatories will not only detect Earth-like planets around nearby stars; they will also measure the planets' physical and chemical characteristics. As we will learn later, in Excursions 8.1, certain telltale gases in a terrestrial planet's atmosphere would be a strong indicator of life.

Planetary Systems Seem to Be Commonplace

Our searches for extrasolar planets have been remarkably successful. Since the first was confirmed in 1994, astronomers have added more than 400 to their inventory, bringing the total number of planetary systems to greater than 340, including over 40 with multiple planets. A representative few are shown in **Figure 6.18**. Keep in mind that few are terrestrial-type planets. The least massive found so far is estimated to have a mass equal to 1.9 $M_\oplus$, and this object orbits much too close to its parent star to support an atmosphere.

> More than 400 extrasolar planets have been discovered so far.

As the number of known planetary systems grows, we are provided with an opportunity to see how our own Solar System compares with other planetary systems. One of the first things astronomers noticed about these planetary systems is that most of them *do not look like our own*. In what ways do the differ? Many contain **hot Jupiters**, which are Jupiter-type planets orbiting solar-type stars in tight circular orbits that are closer to their parent stars than Mercury is to our own Sun. Others have planets in highly eccen-

> Most planetary systems found to date do *not* resemble our own.

tric orbits, unlike the relatively circular planetary orbits in our Solar System. Still others contain planets far more massive than Jupiter. Does this mean that our Solar System is an oddball? Not necessarily—but maybe.

Let's look first at planetary systems containing hot Jupiters. A massive planet orbiting close to its parent star is the easiest kind to detect. If you understand from the previous subsection how extrasolar planets are detected, you may have already guessed why. A very massive planet orbiting very close to its parent star will create large radial velocity variations in the star. This means that hot Jupiters are relatively easy targets when we are using the spectroscopic radial velocity method. In addition, large planets orbiting close to their parent stars are more likely to move in front of the star periodically and reveal themselves with the transit method. Therefore, those hot Jupiter systems that we see may *not* be representative of *most* planetary systems. Scientists refer to this bias as the "selection effect."

Knowing why we find so many hot Jupiters is one thing. Knowing how they got there is another. Many astronomers were surprised by the hot Jupiters because these giant, volatile-rich planets should not be able to form so close to their parent stars. Recall from Section 6.3 that Jupiter-type planets form in the more distant, cooler regions of the protoplanetary disk, where the volatiles that make up much of their composition are able to survive. It seems clear, then, that hot Jupiters must have formed much farther away from their parent star and subsequently migrated inward. The mechanism by which a planet migrates so far inward is not well understood, but it must involve an interaction with gas or planetesimals in which orbital angular momentum is somehow transferred from the planet to its surroundings. Recalling from Section 6.2 that decreased angular momentum means decreased radius, we can see that a loss of angular momentum would cause the planet to spiral inward.

The selection effect does *not* explain the large number of extrasolar planets having highly eccentric orbits. Perhaps chaotic gravitational scattering takes place within most planetary systems, knocking their planets out of the circular orbits in which they formed. Why, then, should *our* Solar System seem so stable? This is a question we cannot yet answer.

The discovery of planetary systems containing supermassive planets raises an interesting question. When is a "supermassive planet" too large to be labeled a planet? In Chapter 15 we will introduce you to a curious object called a brown dwarf, an in-between body not massive enough to be considered a star, yet too massive to be called a planet. In their searches for planets, astronomers have come across many brown dwarfs—some orbiting stars, some alone in space. Although the distinction between the most massive planets and the least massive brown dwarfs is somewhat arbitrary, the International Astronomical Union has placed the boundary at 13 Jupiter masses (13 M_J).

Our studies of planetary systems, many unlike are own, challenge some aspects of our understanding of planet

(a)

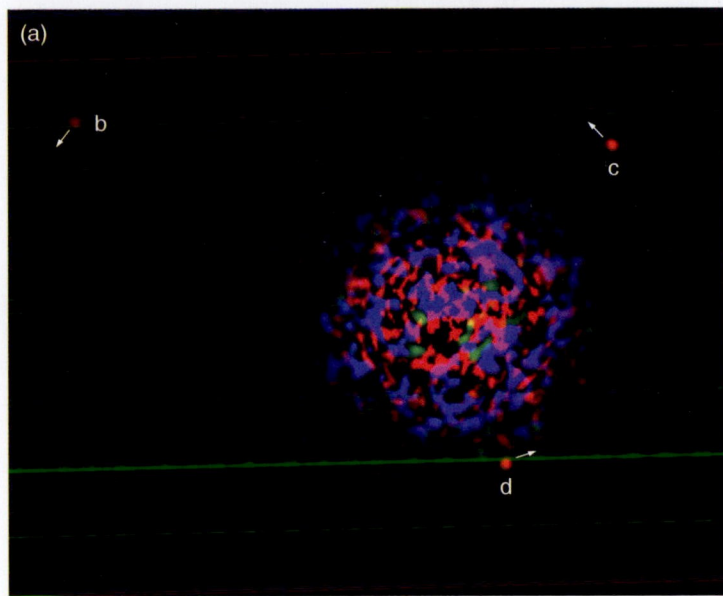

(b)

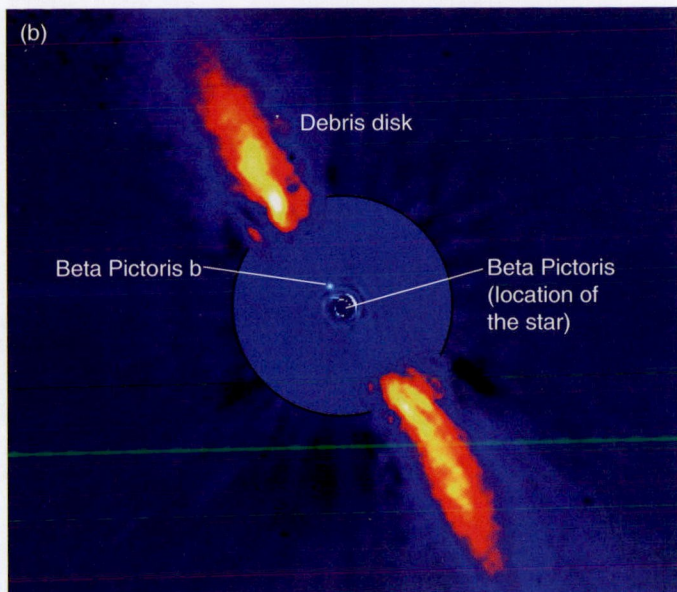

FIGURE 6.16 Direct images of probable planets around two nearby stars. (a) An infrared image shows three planets (labeled "b," "c," and "d"), each with a mass several times that of Jupiter, orbiting the star HR 8799 (hidden behind a mask.) (b) Beta Pictoris b is seen orbiting within a dusty debris disk that surrounds the bright naked-eye star Beta Pictoris. The planet's estimated mass is eight times that of Jupiter. The star is hidden behind an opaque mask, and the planet appears through a semitransparent mask used to subdue the brightness of the dusty disk.

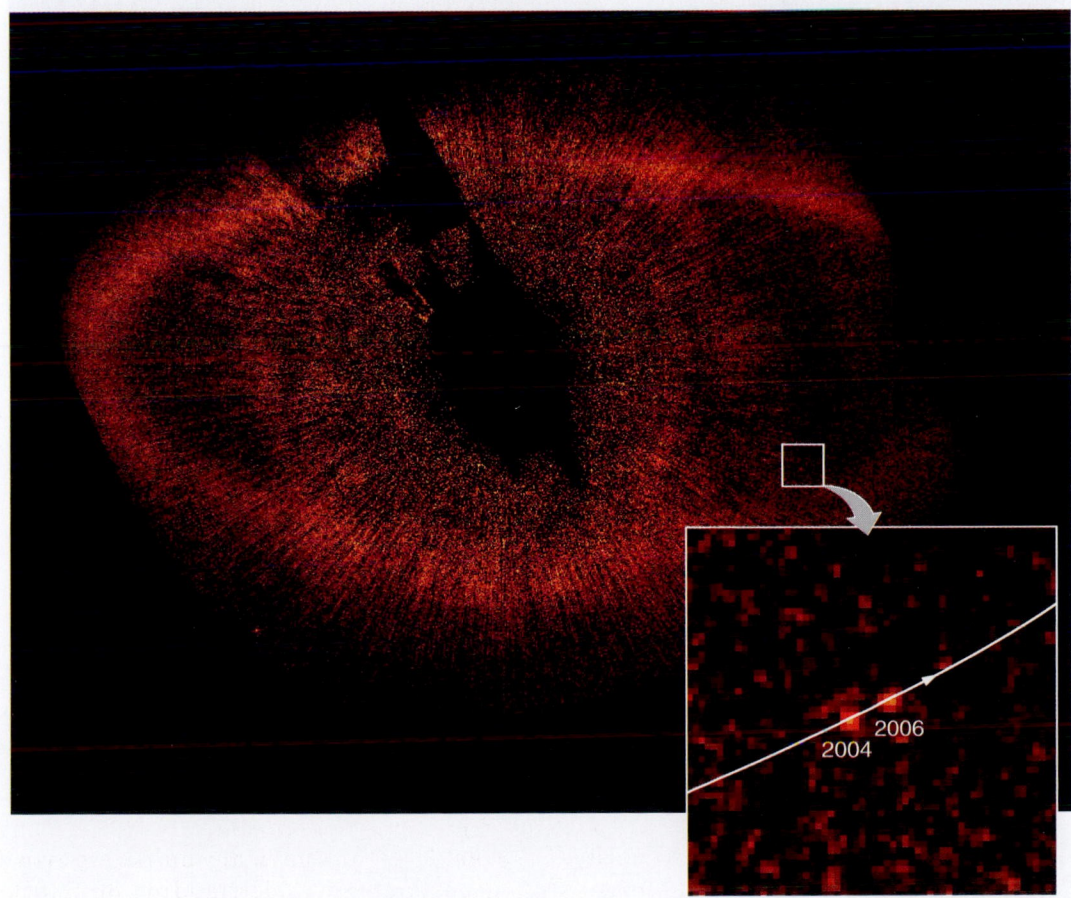

FIGURE 6.17 Fomalhaut b is seen here moving in its orbit around Fomalhaut, a nearby star easily visible to the naked eye. The parent star, hidden by an obscuring mask, is about a billion times brighter than the planet, which is located within a dusty debris ring that surrounds the star.

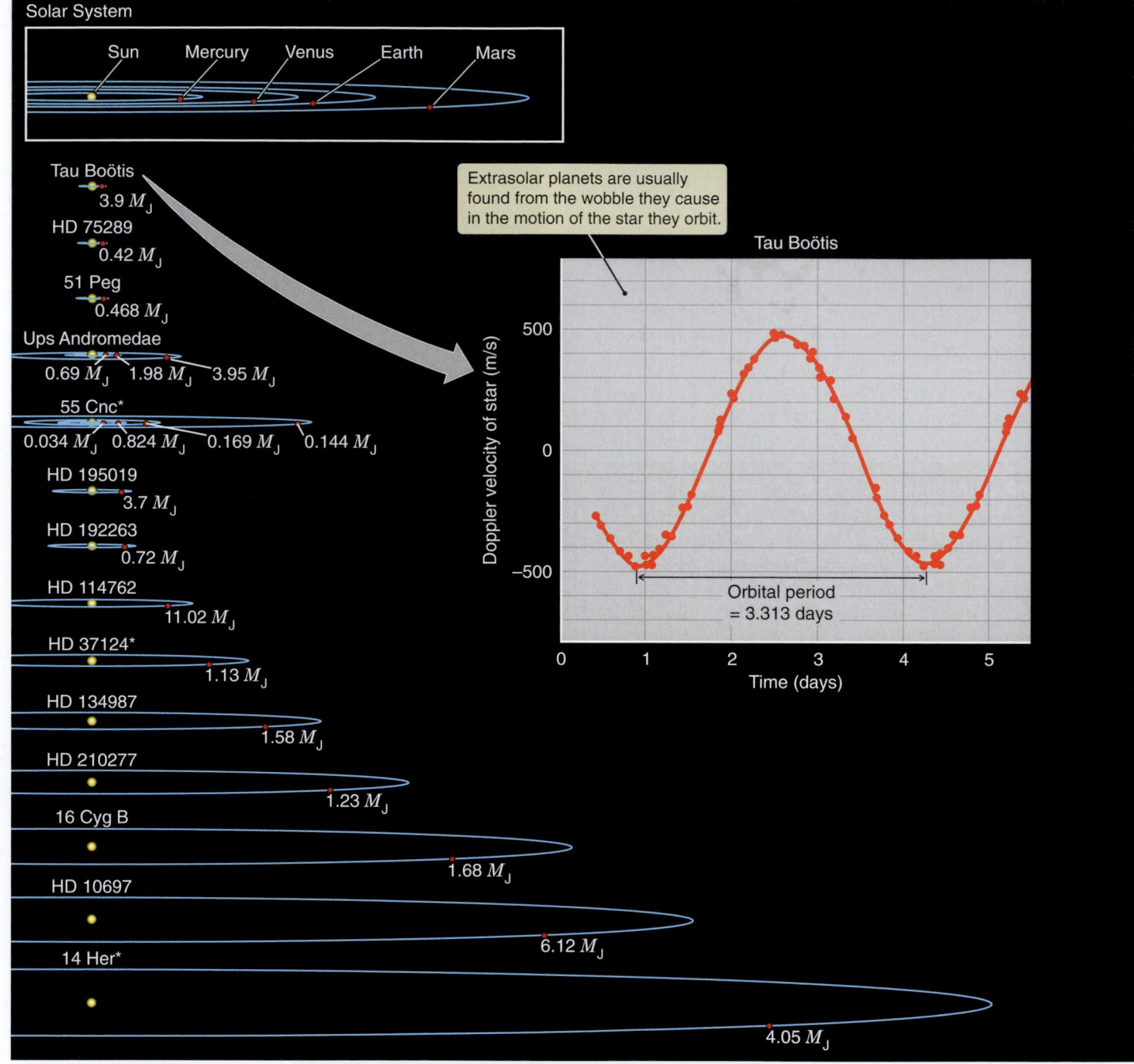

Solar System

Sun   Mercury   Venus   Earth   Mars

Extrasolar planets are usually found from the wobble they cause in the motion of the star they orbit.

Tau Boötis
  3.9 M_J
HD 75289
  0.42 M_J
51 Peg
  0.468 M_J
Ups Andromedae
  0.69 M_J 1.98 M_J 3.95 M_J
55 Cnc*
  0.034 M_J 0.824 M_J 0.169 M_J 0.144 M_J
HD 195019
  3.7 M_J
HD 192263
  0.72 M_J
HD 114762
  11.02 M_J
HD 37124*
  1.13 M_J
HD 134987
  1.58 M_J
HD 210277
  1.23 M_J
16 Cyg B
  1.68 M_J
HD 10697
  6.12 M_J
14 Her*
  4.05 M_J

Tau Boötis

Doppler velocity of star (m/s)

500

0

−500

Orbital period = 3.313 days

0   1   2   3   4   5
Time (days)

Multiple planet system with other planets not shown due to size constraints.
55 Cnc: 3.835 M_J at 5.8 AU
HD 37124: 0.61 M_J at 0.53 AU, 0.68 M_J at 3.2 AU, and 0.6 M_J at 1.6 AU
14 Her: 2.1 M_J at 6.9 AU

FIGURE 6.18 Planetary systems have been discovered around hundreds of stars other than the Sun, confirming what astronomers have long suspected—that planets are a natural and common by-product of star formation. A few of these planetary systems are represented here, along with an example of the radial velocity method. (Masses of planets are given in units of Jupiter masses, M_J.) The inner Solar System is shown to scale.

formation. Yet the message conveyed by our discoveries is clear: the formation of planets frequently, and perhaps always, accompanies the formation of stars. The implications of this conclusion are profound. Planets are a common by-product of star formation. So, in a galaxy of a 100 billion stars, and a universe of hundreds of billions of galaxies, how many planets (or even moons) with Earth-like conditions might exist? And with all of these Earth-like worlds in the universe, how many might play host to the particular category of chemical reactions that we refer to as "life"? In Chapter 23 we will explore the idea of what kind of planet life might choose for its home.

In learning about the formation of planetary systems, then about our Solar System, and finally about extrasolar planets, our journey has come full circle. We began by looking outward in wonder at the lights in the night sky. Now that outward exploration becomes instead a look inward as we discover the processes and events that set the stage for our own existence. In a sense, the insights we have gained in this chapter complete the revolution that Copernicus started long ago when he had the audacity to challenge authority and suggest that Earth was not the center of all things. Not that long ago, our

> Modern discoveries about the formation of the Sun and Solar System "complete" the Copernican revolution.

ancestors viewed Earth and the heavens as fundamentally different regimes—forever apart, each with its own rules and reality. Today we know there is nothing unusual about the conditions of our own existence. All we have to do to see our own history is to look at the stars that continue to form around us today.

It is ironic that at a time when science has turned such a dazzling spotlight on our place in the universe, many otherwise educated individuals continue to cling to outdated and fanciful notions about the heavens and Earth.[10] For someone interested in truly mind-boggling insights, the speculations and "revelations" of the mystic pale beside the discoveries of the scientist. Through the sometimes stodgy, often painstaking, and always uncompromising standards of science, we have come to appreciate that we are not *apart from* the universe. Rather we are *a part of* the universe. And the processes and events that link us to this larger universe are fascinating and wondrous indeed.

---

[10]We cannot resist the temptation to quote the late, famous science fiction author Arthur C. Clarke at this juncture: "In one sense, of course, every age renews itself, as indeed it should. But the nitwits currently parroting this slogan ["New Age"] seem unable to understand that their 'New Age' is exactly the opposite, being about a thousand years past its sale date."

Seeing the Forest for the Trees

In this chapter we have seen two great lines of investigation merge into a single picture that shapes our understanding of the context of our own existence. Working first from the perspective of the stellar astronomer, basic physical principles—in particular, the conservation of angular momentum—demand that when a star like our Sun forms, it will be surrounded by a thin, orbiting disk of gas and dust. This conclusion went from hypothesis to fact with the discovery of such disks around numerous young stars. In the meantime, as stellar astronomers were trying to understand star formation, planetary scientists were scrutinizing the worlds that make up our Solar System.

The pieces of the planetary scientists' puzzle range from specimens from space—meteorites collected in Antarctica and lunar samples brought back by *Apollo* astronauts—to data sent back by spacecraft visiting remote worlds. Those pieces fit together to form a clear picture of a flat, swirling cloud of gas and dust from which Earth and its neighbors coalesced about 5 billion years ago. When the stellar astronomers and the planetary scientists compared notes, they realized that they had arrived at exactly the same description. The disks from which

our Solar System and the many other planetary systems formed were none other than the stellar accretion disks that surrounded the young suns.

The joining of these two great rivers of thought, observation, and exploration—the link between the accretion disks that surround young stars and the local collection of planets that we call home—is the starting point for a modern study of the Solar System. It also ties our journey of understanding within the Solar System to the course we will later chart outward into the larger universe. Stated more bluntly, all that we know about the Sun, planets, and the world of our birth makes sense only when viewed within the context of the evolving universe as a whole. We have had a first glimpse of the power of this insight in the current chapter. As we move on to look at the planets, moons, asteroids, and comets that orbit the Sun, we will time and again see the fingerprints left behind by our birth among the stars.

The journey that we are taking in *21st Century Astronomy* is one of scientific discovery, but the philosophical implications of the connection between star formation and the formation of our Solar System and other planetary systems should not be overlooked. Our very existence is set within the context of the larger universe. We are the legacy of the processes we see at work all around us, even today.

Summary

- Stars and their planetary systems form from collapsing interstellar clouds of gas and dust.

- Planets are a common by-product of star formation, and many stars are surrounded by planetary systems.

- Planets grew from a protoplanetary disk of gas and dust that surrounded the forming Sun.

- Our Solar System formed about 5 billion years ago, nearly 9 billion years after the birth of the universe.

- Solid terrestrial planets formed in the inner disk, where temperatures were high.

- Giant gaseous planets formed in the outer disk, where temperatures were low.

- Dwarf planets formed in the asteroid belt and in the region beyond the orbit of Neptune.

- Asteroids and comet nuclei remain today as leftover debris.

- Hundreds of extrasolar planets have been found orbiting other stars within our galaxy.

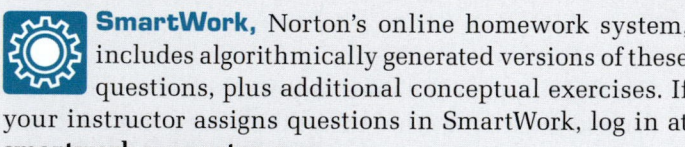 **SmartWork,** Norton's online homework system, includes algorithmically generated versions of these questions, plus additional conceptual exercises. If your instructor assigns questions in SmartWork, log in at **smartwork.wwnorton.com.**

Student Questions

THINKING ABOUT THE CONCEPTS

1. Compare the size of our Solar System with the size of the universe.

2. What is the source of the material that now makes up the Sun and the rest of the Solar System?

3. In a poetic sense, planets are sometimes referred to as "children of the Sun." A more accurate description, though, would be "siblings of the Sun." Explain why.

*4. Describe the different ways that stellar astronomers and planetary scientists each came to the same conclusion about how planetary systems form.

*5 Explain the nebular hypothesis.

6. What is a protoplanetary disk?

7. Physicists describe certain properties, such as angular momentum and energy, as being "conserved." What does this mean? Do these conservation laws imply that an individual object can never lose or gain angular momentum or energy? Explain your reasoning.

8. How does the law of conservation of angular momentum control a figure-skater's rate of spin?

*9. Explain why the law of conservation of angular momentum created problems for early versions of the nebular hypothesis.

10. What is an accretion disk?

11. Describe the process by which tiny grains of dust grow to become massive planets.

12. There are two reasons that the inner part of a protoplanetary disk is hotter than the outer disk. What are they?

13. During the process in which an interstellar cloud collapses to form an accretion disk, energy exists in three forms. Name and describe them.

14. Why do we find rocky material everywhere in the Solar System, but large amounts of volatile material only in the outer regions? Would you expect the same to be true of other solar systems? Explain your answer.

15. When we think of ice, we tend to think of water ice, but among scientists this is a very restrictive definition. Explain why.

16. Why were the four giant planets able to collect massive gaseous atmospheres, whereas the terrestrial planets could not?

17. What is the meaning of the term *organic*?

18. Explain the source of the atmospheres now surrounding three of the terrestrial planets.

19. What happened to all the leftover Solar System debris after the last of the planets formed?

20. Name the terrestrial and the giant planets in our Solar System.

21. Describe four methods that astronomers use to search for extrasolar planets.

22. Why is the transit method so restrictive as a method for finding extrasolar planets?

23. Why has it been so difficult for astronomers to take a picture of an extrasolar planet?

24. Many of the planets that astronomers have found orbiting other stars have been giant planets with masses more like that of Jupiter than of Earth, and with orbits located very close to their parent stars. Does this prove that our Solar System is unusual? Explain your answer.

25. Step outside and look at the nighttime sky. Depending on the darkness of the sky, you may see dozens or hundreds of stars. Would you expect many or very few of those stars to be orbited by planets? Explain your answer.

APPLYING THE CONCEPTS

26. Use information about the planets given in Appendix 4 to answer the following:
    a. What is the total mass of all the planets in our Solar System, expressed in Earth masses ($M_\oplus$)?
    b. What fraction of this total planetary mass does Jupiter represent?
    c. What fraction does Earth represent?

*27. Compare Jupiter's orbital angular momentum with its spin angular momentum using the following values: $m = 1.9 \times 10^{27}$ kg, $v = 13.1$ kilometers per second (km/s), $r = 5.2$ AU, $R = 7.15 \times 10^4$ km, and $P = 9$ hours 56 minutes. Assume Jupiter to be a uniform body. What fraction does each component (orbital and spin) contribute to Jupiter's total angular momentum? Refer to Math Tools 6.1 for help.

28. Assume that the Sun is a uniform sphere with a radius of 700,000 km and a rotation period of 27 days. Near the end of its life, the remnant of the Sun will be a white dwarf with a radius of only 5,000 km. Assuming that the mass remains unchanged, what will be the Sun's new rotation period as a white dwarf?

**29. The mass of the Sun ($M_\odot$) is 2×10^{30} kg. Using information provided in the previous two questions, compare Jupiter's orbital angular momentum with the Sun's spin angular momentum. What does this comparison tell you about the distribution of angular momentum in the Solar System?

30. Venus has a radius 0.949 times that of Earth and a mass 0.815 times that of Earth. Its rotation period is 243 days. What is the ratio of Venus's spin angular momentum to that of Earth? Assume that Venus and Earth are uniform spheres.

31. Jupiter has a mass equal to 318 times Earth's mass, an orbital radius of 5.2 AU, and an orbital velocity of 13.1 km/s. Earth's orbital velocity is 29.8 km/s. What is the ratio of Jupiter's orbital angular momentum to that of Earth?

*32. In the text we give an example of an interstellar cloud having a diameter of 10^{16} meters and a rotation period of 10^6 years collapsing to a sphere the size of the Sun (1.4×10^9 meters in diameter). We point out that if all the cloud's angular momentum went into that sphere, the sphere would have a rotation period of only 0.6 second. Do the calculation to confirm this result.

33. The asteroid Vesta has a diameter of 530 km and a mass of 2.7×10^{20} kg.
    a. Calculate the density of Vesta.
    b. The density of water is 1,000 kilograms per cubic meter (kg/m³), and that of rock is about 2,500 kg/m³. What does this difference tell you about the composition of this primitive body?

34. If an alien astronomer observed a plot of the light curve as Jupiter passed in front of the Sun, by how much would the Sun's brightness drop during the transit?

35. A hot Jupiter nicknamed "Osiris" has been found around a solar-mass star, HD 209458. It orbits the star in only 3.525 days.
    a. What is the orbital radius of this extrasolar planet?
    b. Compare its orbit with that of Mercury around our own Sun. What environmental conditions must Osiris experience?

*36. The extrasolar planet Osiris passes directly in front of its solar-type parent star, HD 209458 (diameter = 1.7×10^6 km), every 3.525 days, decreasing the brightness of the star by about 1.7 percent (0.017).
    a. What is the diameter of Osiris?
    b. Compare the diameter of this extrasolar planet with that of Jupiter (mean diameter = 139,800 km).

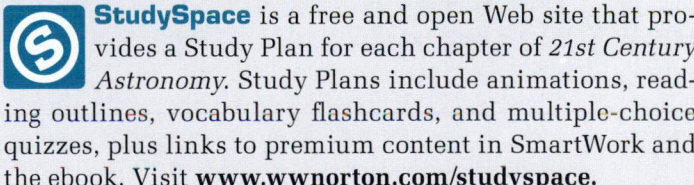 **StudySpace** is a free and open Web site that provides a Study Plan for each chapter of *21st Century Astronomy*. Study Plans include animations, reading outlines, vocabulary flashcards, and multiple-choice quizzes, plus links to premium content in SmartWork and the ebook. Visit **www.wwnorton.com/studyspace**.

We've sent a man to the Moon, and that's [240,000] miles away. The center of the Earth is only 4,000 miles away. You could drive that in a week, but for some reason nobody's ever done it.

ANDY ROONEY (1919–)

Mercury, Venus, and the Moon rising over the Australia Telescope Compact Array in New South Wales.

CHAPTER 7

The Terrestrial Planets and Earth's Moon

7.1 How Are Planets the Same, and How Are They Different?

Most of what we know about our planetary neighbors we have learned only in the most recent decades. The second half of the 20th century was a wonderfully exciting time of exploration and discovery about Earth and its sibling worlds. It was a time that saw unmanned probes visit every classical planet and watched astronauts walk on the surface of the Moon. (See Chapter 5 for a discussion of the variety of techniques used to explore the Solar System.) The information returned from these missions has revolutionized our understanding of our planetary system, offering us insights into the current state of each of our neighbors and clues about their histories.

The four innermost planets in our Solar System are Mercury, Venus, Earth, and Mars, collectively known as the **terrestrial planets** (**Table 7.1**). Although technically the Moon is known as Earth's lone natural satellite, we include it here in this chapter because of its close similarity to the terrestrial planets.

The vast quantity of information about the planets returned by space probes can be hard to digest. What information is truly fundamental and exciting, and what information is flashy but less important? The key to sorting through this information has

> We learn about planets by comparing them with each other.

proven to be comparison among the different planets. The ways the planets are alike and how they differ draw our attention to the most fundamental issues, helping us ask and

KEY CONCEPTS

The objects that formed in the inner part of the protoplanetary disk around the Sun were relatively small rocky worlds, one of which we call home. Comparing those worlds with one another teaches us lessons about what shapes a planet's fate. Among the lessons we will learn in this chapter are

- That each terrestrial planet is shaped by impacts, tectonism, volcanism, and erosion.
- How impacts scarred planets early in the history of the Solar System and still occur on occasion today.
- Why the concentration of craters on a planetary surface tells us how old the surface is.
- How radiometric dating tells us the ages of rocks.
- How radiometric dating of lunar rocks is used to calibrate the cratering clock.
- How we test predictions about Earth's interior by using seismic waves from earthquakes.
- That larger worlds remain geologically active longer because smaller planets cool off sooner.
- That tectonism takes different forms on the terrestrial planets, and plate tectonics is unique to Earth.
- That among the volcanoes found on Mercury, Venus, Earth, and Mars, the most colossal are on Mars.
- The many ways in which erosion modifies and wears down the surface features that other processes form.

179

TABLE 7.1

Comparison of Physical Properties of the Terrestrial Planets and the Moon

	Mercury	Venus	Earth	Mars	Moon*
Orbital radius (AU)	0.387	0.723	1.000	1.524	384,000 km
Orbital period (years)[†]	0.241	0.615	1.000	1.881	27.32^d
Orbital velocity (km/s)	47.9	35.0	29.8	24.1	1.02
Mass ($M_\oplus = 1$)	0.055	0.815	1.000	0.107	0.012
Equatorial diameter (km)	4,880	12,104	12,756	6,794	3,476
Equatorial diameter ($D_\oplus = 1$)	0.383	0.949	1.000	0.533	0.272
Density (water = 1)	5.43	5.24	5.52	3.93	3.34
Sidereal rotation period[†]	58.65^d	243.02^d	$23^h\ 56^m$	$24^h\ 37^m$	27.32^d
Obliquity (degrees)[‡]	0.04	177.36	23.45	25.19	6.68
Surface gravity (m/s^2)	3.70	8.87	9.78	3.71	1.62
Escape speed (km/s)	4.25	10.36	11.18	5.03	2.38

*The Moon's orbital radius and orbital period are given in kilometers and days, respectively.
[†]The superscript letters ^{d}, ^{h}, and ^{m} stand for days, hours, and minutes of time, respectively.
[‡]An obliquity greater than 90° indicates that the planet rotates in a retrograde, or backward, direction.

answer the right questions. The correct explanation for a particular aspect of one planet must be consistent with what we know about the other planets. For example, when we explain why the Moon is covered with craters, our reasoning must also allow for the fact that preserved craters on Earth are rare. An explanation for why Venus has such a massive atmosphere should point to reasons that Earth and Mars do not. Such comparisons—an approach called **comparative planetology**—have provided the guideposts to planetary scientists as they sift through mountains of data. When making comparisons, we need a place to start. Earth is the planet we know best, so it is here at home that we begin our appraisal of the worlds of the inner Solar System.

7.2 Four Main Processes Shape Our Planet

For most of human history we have looked upon Earth as a vast, almost limitless expanse. This view of our planet changed forever with a single snapshot taken by *Apollo 8*

astronauts looking back at Earth from space (**Figure 7.1**). For many, this was a philosophical turning point, compelling us to view the world for what it truly is: "a small blue marble," a tiny and fragile lifeboat adrift in the vastness of space—our home. As seen in a closer view from space (**Figure 7.2**), Earth is awash with color. White clouds drift in our atmosphere, and white snow and ice cover the planet's frozen poles. The blue of oceans, seas, lakes, and rivers of liquid water—Earth's **hydrosphere**—covers most of the planet. Brown shows us the outer rocky shell of Earth, referred to as Earth's **lithosphere**. And green is the telltale sign of vegetation, part of Earth's **biosphere**—the most extraordinary among Earth's many distinctions.

The psychological impact of these images is rivaled only by the change in our scientific perspective. We see Earth in the context of the larger universe, shaped by processes that happen throughout its vast expanse. Earth is a place of change: geological processes constantly work to reshape our planet. Some geological processes originate in the interior of Earth, powered by the thermal energy generated there. Earthquakes are sudden reminders of the ongoing deformation of Earth's lithosphere, referred to as **tectonism**. Tectonism folds and breaks Earth's crust, forming mountain ranges, valleys,

FIGURE 7.1 Our first view of Earth seen from deep space. In December 1968, *Apollo 8* astronauts photographed our planet rising above the Moon's limb.

FIGURE 7.2 Seen from space, the colors of Earth tell of the diversity of our planet's features.

FIGURE 7.3 Volcanism, such as this eruption in Hawaii, spills molten rock and other materials onto planetary surfaces.

and deep ocean trenches. **Volcanism**—another geological process affecting Earth and the other terrestrial planets—is a form of **igneous activity** in which gas and molten rock, or **magma**, erupt at the surface. Volcanic eruptions, like the one shown in **Figure 7.3**, can spill sheets of lava and ash over vast areas, forming mountains or plains in the process.

▶❚❚ **AstroTour: Processes That Shape the Planets**

Another process that affects Earth is external. Collisions involving planetary objects—a process referred to as **impact cratering**—are extremely important in our planet's history. Most of us have seen **meteors**,[1] the bright streaks

that flash across the sky when **meteoroids**, or chunks of material from outer space, hit our atmosphere; but for at least one woman, the experience with celestial debris is more intimate. In 1954 a **meteorite** (a meteoroid that reaches a planet's surface) crashed through the roof of the woman's home in Alabama, striking her hand and hip. Fortunately, the roof of the house slowed the meteorite, so she was left with only a bad bruise.[2] When a very large object hits a planet, as still happens occasionally in the Solar System, the resulting devastation can be global. Distinctive scars in Earth's crust tell of impacts in our past. Such catastrophic events have altered Earth's climate and have changed the history of life itself (see Chapter 23).

These three processes—tectonism, volcanism, and impact cratering—affect Earth's surface in their own characteristic ways. Together, they work to form **topographic relief**—mountains, valleys, and ocean basins. Meanwhile, a fourth process, called **erosion**, works slowly but persistently to level Earth's surface. Erosion by running water, wind, and other agents wears down hills, mountains, and continents. The eroded debris collects downslope, filling in valleys, lakes, and ocean basins. Left on its own, ero-

---

[1] The terms *meteor, meteoroid, meteorite,* and even *comet* are confusing to many people. These terms are all explained more completely in Chapter 12.

[2] You should not lose much sleep worrying about meteorites landing in your lap. Unlike being struck by lightning, being hit by a meteorite is even less likely than winning the state lottery.

sion would eventually leave the surface of our planet smooth and featureless. Because Earth is a geologically and biologically active world, however, its surface is an ever-changing battleground between processes that build up topography and those that tear it down.

> Tectonism, volcanism, and impact cratering rough up planetary surfaces; erosion smooths them.

Each of the four types of geological processes at work shaping the surface of Earth leaves its own distinctive signature. Planetary scientists have learned to read these signatures on the surfaces of other planets as well.

FIGURE 7.4 (a) Stages in the formation of an impact crater. (b) A lunar crater photographed by *Apollo* astronauts, showing the crater wall and central peak, surrounded by ejected material (*ejecta*), rays, and secondary craters—all typical features associated with impact craters. (c) A rayed crater on Mercury.

(a)

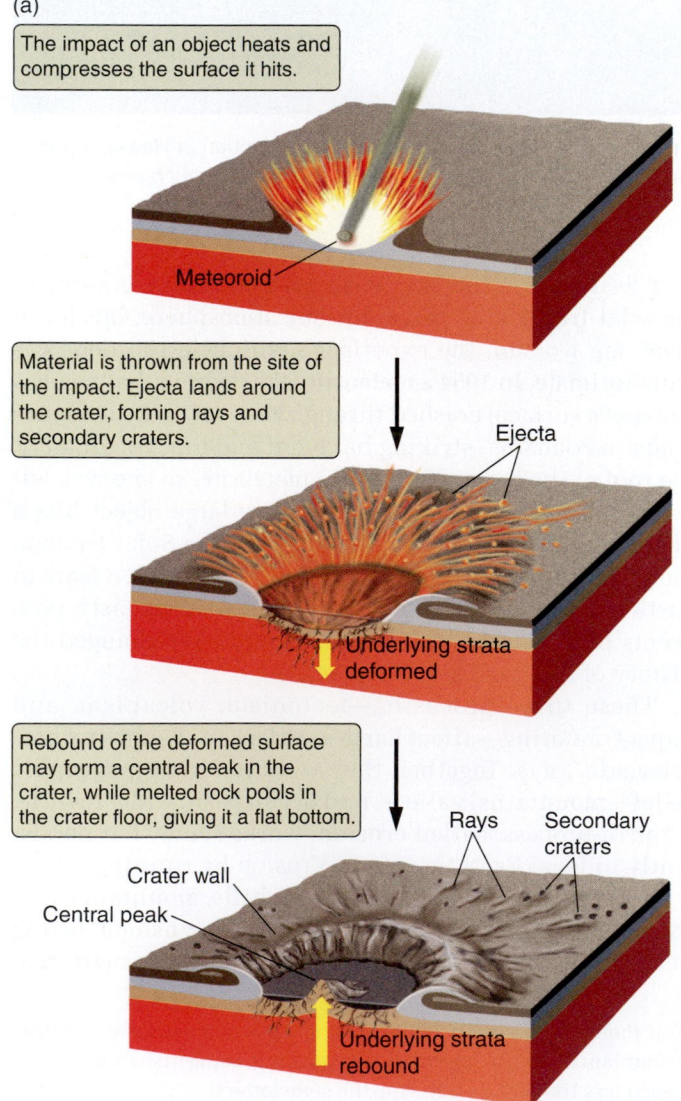

(b)

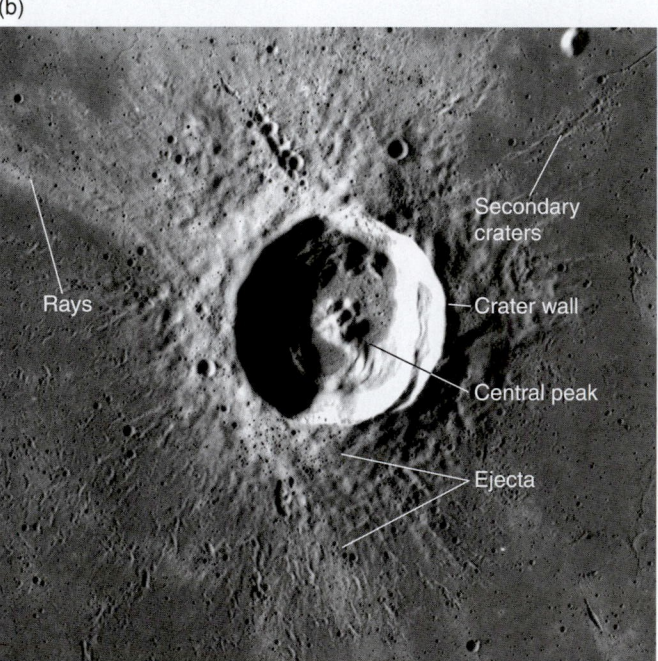

(c)

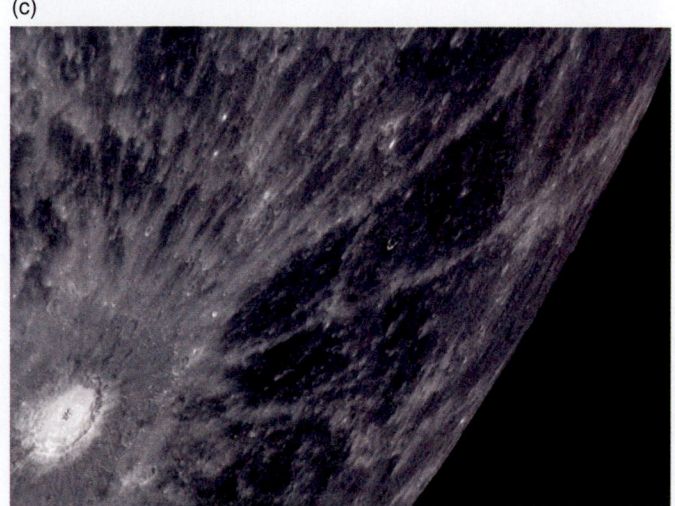

7.3 Impacts Help Shape the Evolution of the Planets

Although all terrestrial planets are subject to impact cratering, tectonism, volcanism, and erosion, the relative intensity of these processes varies among the planets. Of these four geological processes, however, large impacts account for by far the most concentrated and sudden release of energy. Planets and other objects orbiting the Sun move at very high speeds. Earth moves at about 30 kilometers per second (km/s) in its orbit around the Sun, and meteoroids can enter Earth's atmosphere at relative speeds in excess of 70 km/s. Because the kinetic energy of an object is proportional to the square of its speed ($E_K = \frac{1}{2}mv^2$), collisions between such objects release huge amounts of energy. When an object hits

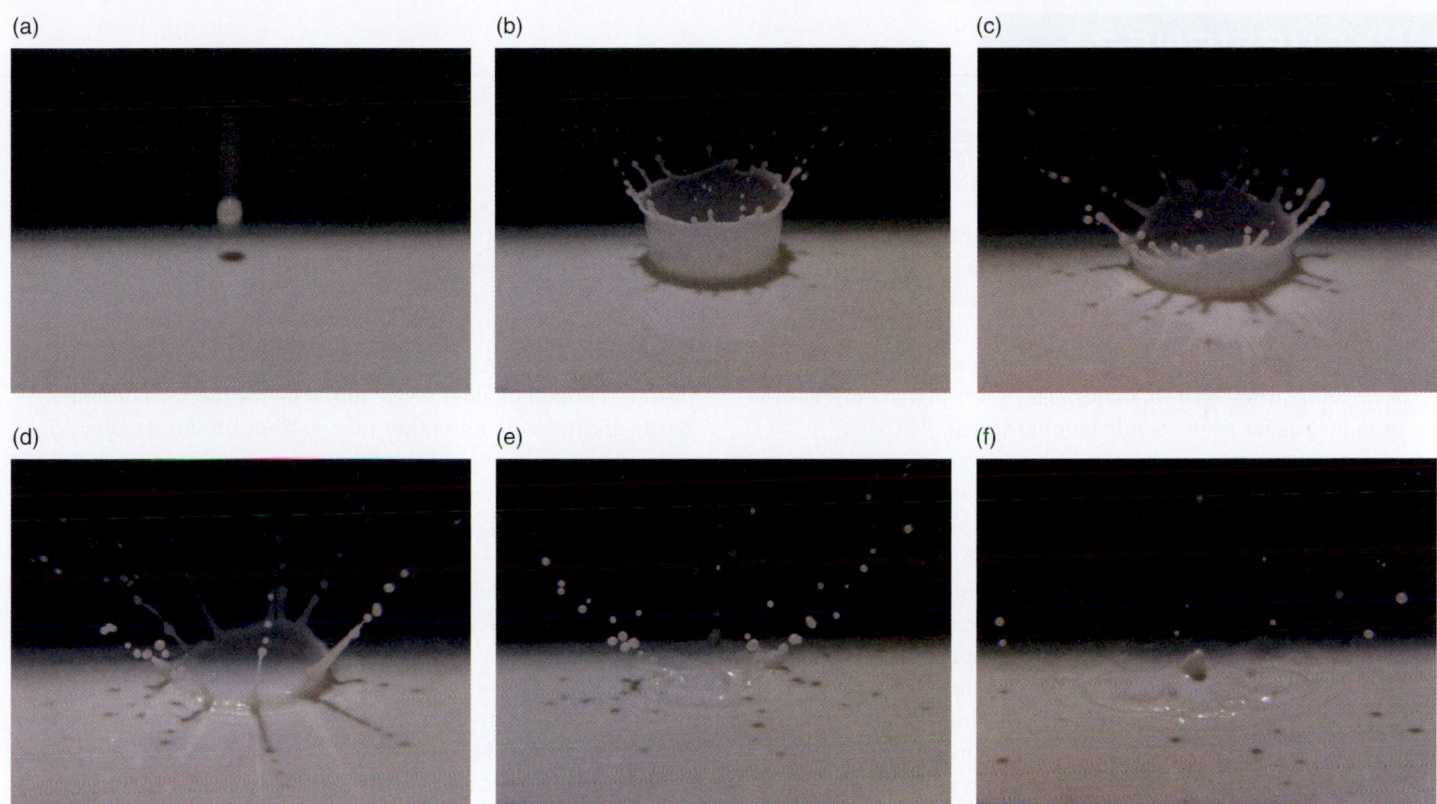

(a)   (b)   (c)

(d)   (e)   (f)

VISUAL ANALOGY FIGURE 7.5 A drop hitting a pool of milk illustrates the formation of features in an impact crater, including crater walls (b and c), secondary craters (d and e), and a central peak (f).

a planet, its kinetic energy goes into heating and compressing the surface that it strikes and throwing material far from the resulting **impact crater** (**Figure 7.4**). Sometimes material thrown from the crater, called **ejecta**, falls back to the surface of the planet with enough energy to cause **secondary craters**. The rebound of heated and compressed material can also lead to the formation of a central peak or a ring of mountains on the crater floor. These processes are similar, on a much smaller scale, to what happens when a drop lands in a glass of milk, as shown in **Figure 7.5**.

The energy of an impact can be great enough to melt or even vaporize rock. The floors of some craters are the cooled surfaces of pools of rock melted by the impact. The energy released in an impact can also lead to the formation of new minerals. In fact, some minerals, such as shock-modified quartz, are known to form *only* during the momentary fury of an impact. Geologists look for these distinctive minerals as evidence of ancient impacts on Earth's surface.

Meteor Crater in Arizona is one of the best-preserved impact structures on our planet (**Figure 7.6**). It is thought to be the result of the impact of a nickel-iron meteorite about 50 meters in diameter with a mass of about 300 million kilo-

> Impacts can melt and vaporize rock.

FIGURE 7.6 Meteor Crater (also known as Barringer Crater), located in northern Arizona, is an impact crater 1.2 kilometers (km) in diameter that formed some 50,000 years ago by a nickel-iron meteoroid's collision with Earth.

grams (kg) that hit Earth traveling at 13 km/s about 50,000 years ago. Planetary scientists believe that approximately half of the original mass was vaporized in the atmosphere before hitting the ground. Such a collision would have released about 300 times as much total energy as the first atomic bomb, which was detonated in New Mexico in 1945.

Where Have All the Dinosaurs Gone?

In 1994 a host of astronomical instruments watched as about two dozen pieces of Comet Shoemaker-Levy 9 slammed one after another into the clouds of Jupiter, leaving temporary scars that were visible through backyard telescopes. The Jupiter comet crash provided graphic evidence that although impacts are not as frequent today as they were when the Solar System was young, they still do occur. The giant planets are not the only targets for such cosmic bombardment.

In 2004 a very small asteroid, whose rocky body measured about 30 meters across, whizzed past Earth at nearly 25,000 kilometers per hour (km/h), missing our planet by only 43,000 km. A close call to be sure, but it was after all a relatively small body that probably would have broken up or exploded in our atmosphere, causing only local damage had we been hit. That same year, University of Hawaii astronomers discovered an asteroid that was given the rather forgettable name 2004 MN_4. Forgettable, that is, until further observations and refined calculations showed that 2004 MN_4 (now named Apophis), a mass of rock 270 meters in diameter, will come within 30,000 km of Earth's surface on April 13, 2029 (Friday, the 13th). Although the chance of a collision with Earth is very unlikely, the energy of such an impact would be 880 megatons—15 times more powerful than the most powerful hydrogen bomb ever detonated. When large impacts happen on Earth, they can have far-reaching consequences for Earth's climate and for terrestrial life.

Earth's fossil record shows that, on occasion, large numbers of species vanish from the face of the planet in a geological blink of an eye. The most famous of these extinctions occurred 65 million years ago, when more than 70 percent of *all living species*, including the dinosaurs, became extinct. This mass extinction is marked in Earth's fossil record by the **Cretaceous-Tertiary boundary**, or simply the "K-T boundary."[3] (The Cretaceous Period lasted from 146 million years ago to 65 million years ago. The Tertiary Period started when the Cretaceous Period ended and lasted until 1.8 million years ago. We live in the Quaternary Period of Earth's history, which started at the end of the Tertiary Period.) In older layers found below the K-T boundary, fossils of dinosaurs and other now-extinct life-forms abound. In the newer rocks above the K-T boundary, well more than half of all previous species are absent, and in their place are found fossils of newly evolving species. Big winners in the new order were the mammals—our own distant ancestors—which moved into ecological niches vacated by extinct species.

The K-T boundary is marked in the fossil record in many areas by a layer of clay. Studies at more than 100 locations around the world have found that this layer contains large amounts of the element iridium, as well as traces of soot. Iridium is very rare in Earth's crust but is common in meteorites. The soot at the K-T boundary tells of a time when widespread fires burned the world over. The thickness of the layer of clay at the K-T boundary and the concentration of iridium increase as we move toward what is today the Yucatán Peninsula in Mexico. Although the original crater has been erased by erosion, geophysical surveys and rocks from drill holes in this area show a highly deformed subsurface rock structure, similar to that seen at known impact sites. Together, these results provide compelling evidence that 65 million years ago an asteroid 10 km in diameter struck the area, throwing great clouds of red-hot dust and other debris into the

[3]The *K* comes from *Kreide*, German for "Cretaceous."

Yet, at only 1,200 meters in diameter, Meteor Crater is tiny compared with impact craters seen on the Moon or ancient impact scars on Earth. (**Excursions 7.1** describes the consequences of an especially violent impact that took place 65 million years ago.)

One of the most obvious differences among the terrestrial planets is the visible extent of impact cratering. On some planets the surfaces are covered by impact craters; on others (especially Earth and Venus), impact craters seem to be rare. For example, the Moon has millions of craters of all different sizes, one on top of another (**Figure 7.8**). Nearly all of these craters are the result of impacts. By comparison, fewer than 200 impact craters, or scars from impacts, have been identified on Earth; and about 1,000 on Venus. The primary culprits behind Earth's crater shortage are plate tectonics in Earth's ocean basins and erosion on land. Most craters on land have been obliterated by wind and water.

There is another reason for the shortage of small craters on Earth. Whereas the surface of the Moon is directly exposed to this cosmic bombardment, the surface of Earth is partly protected by the blanket of Earth's atmosphere. For example, rock samples from the Moon show craters smaller than a pinhead, formed by micrometeoroids. In contrast, most meteoroids smaller than 100 meters in diameter are either burned up or broken up by friction in Earth's atmosphere before they reach the surface. With an atmosphere far more massive than that of Earth, Venus is even better protected. Objects even larger than 100 meters across may be burned up or broken apart by Venus's atmosphere before reaching the surface, creating clusters of craters in a pattern similar to a shotgun blast, or sometimes leaving only a dark "splotch" on the surface with no crater at all.

The characteristics of a crater also depend on the properties of the planetary surface. An impact in a deep ocean

FIGURE 7.7 This artist's rendition depicts an asteroid or comet, perhaps 10 km across, striking Earth 65 million years ago in what is now the Yucatán Peninsula in Mexico. The lasting effects of the impact killed off most forms of terrestrial life, including the dinosaurs.

atmosphere (**Figure 7.7**) and igniting a worldwide conflagration. The energy of the impact is estimated to have been more than that released by 5 *billion* nuclear bombs.

An impact of this magnitude clearly would have had a devastating effect on terrestrial life. Could this cosmic impact have been responsible for the sudden disappearance of forms of life that had ruled Earth for 150 million years? Many scientists believe so. In addition to the nearly global firestorm ignited by the impact, dust thrown into Earth's upper atmosphere would have remained there for years, blocking out sunlight and plunging Earth into decades of cold and darkness. The firestorms,

temperature changes, and decreased food supplies could have led to mass starvation that would have been especially hard on large animals such as the dinosaurs.

Not all paleontologists believe that this mass extinction was the result of an impact. They point out that the evolution of species is a complex process and that simple answers are seldom complete. However, the evidence is compelling that a great impact did occur at the end of the Cretaceous Period. The rock record also shows other instances of mass extinctions associated with colossal impacts, suggesting that impacts have played a central role in the saga of life on Earth.

on Earth might create an impressive wave but leave no lasting crater. In contrast, an impact scar formed in granite can be preserved for billions of years. Planetary scientists can tell a lot about the surface of a planet by studying its craters. For example, craters on the Moon's pristine surface are often surrounded by strings of smaller secondary

> The forms of impact scars give clues about the surfaces they are on.

craters formed from material thrown out by the impact, like those shown in Figure 7.4b. Some craters on Mars have a very different appearance. These craters are surrounded by what appears to be flows of material, much like the pattern you might see if you threw a rock into mud (**Figure 7.9**). The apparent flows seem to indicate that, unlike the lunar case, the martian surface rocks contained water or ice at the time of the impact. Not all martian crater ejecta deposits look

like this, so the water or ice must have been concentrated in only some areas, and these icy locations might have changed with time.

It is possible that these craters were formed at a time in the past when there was liquid water on the surface of Mars. Canyons and dry riverbeds seen on Mars today attest to this possibility. But there is another intriguing possibility. Today the surface of Mars is dry and mostly frozen, suggesting that water that might once have been on the surface has soaked into the ground, much like the water frozen in the ground in Earth's polar regions. The energy released

> Mars was once wetter than it is today.

by the impact of a meteoroid would melt this ice, possibly turning the surface material into a slurry with a consistency much like wet concrete. When thrown from the crater by the force of the impact, this slurry would have hit the sur-

FIGURE 7.8 The Moon's terrain, shown here in a photograph taken during the *Apollo 17* mission, has been heavily cratered by impacts.

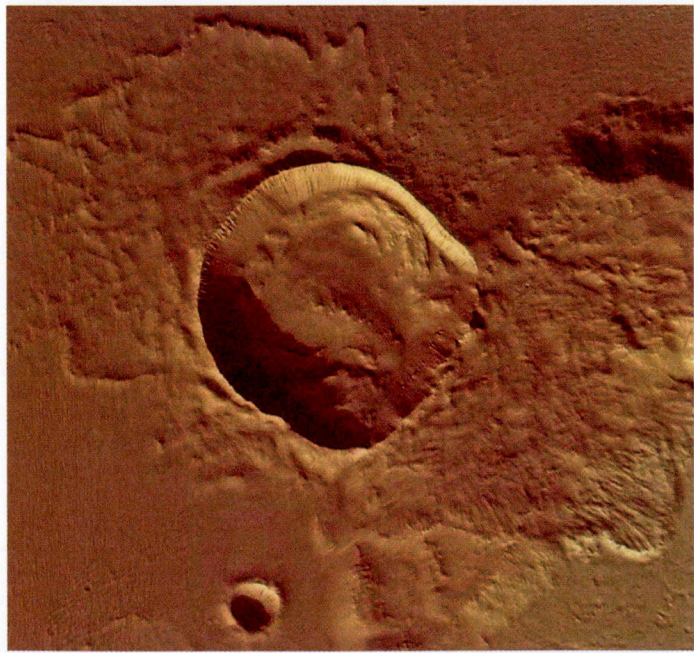

FIGURE 7.9 Some craters on Mars look like those formed by rocks thrown into mud, suggesting that material ejected from the crater contained large amounts of water. This crater is about 20 km across.

rounding terrain and slid out across the surface, forming the mudlike craters we see today.

Calibrating a Cosmic Clock

Every terrestrial planet experienced a similar bombardment early in its history. Although many factors affect the formation of craters, the biggest difference among planets is the rate at which craters are destroyed. Earth experienced an impact history similar to that of the Moon and the other terrestrial planets, and yet preserved impact craters are rare on Earth, as previously noted. The combined action of erosion, tectonic processes, and volcanism throughout geological time has obliterated most of Earth's impact scars. Geologically active planets such as Earth, Mars, and Venus, whose interior thermal energy fuels volcanism and tectonic processes, bear the scars of only the more recent impacts. By contrast, the Moon's surface still preserves the scars of craters dating from the early years of the Solar System. With no atmosphere or surface water for erosion and a cold, geologically dead interior, the lunar surface has remained much the same for more than a billion years. Mercury's well-preserved craters suggest that it, like the Moon, has been geologically inactive for a long time. Planetary scientists use this cratering record to estimate the ages and geological histories of planetary surfaces—extensive cratering means an older planetary surface and minimal geological activity.

> The number of craters on a surface indicates the age of the surface.

We can use the amount of cratering as a clock to measure the ages of surfaces, but first we need to know how fast that clock runs. We need to be able to say that a surface with *this many* craters is *this* old, but a surface with *that many* craters is *that* old. In other words, we need a way of "calibrating the cratering clock."

The key to calibrating the cratering clock came mostly from our exploration of the Moon. The surface of the Moon is not uniform. Some parts of the Moon are heavily cratered, with craters overlapping their neighbors. Other parts are much smoother, telling of more recent geological activity. Between 1969 and 1976, *Apollo* astronauts and Soviet unmanned probes visited the Moon and brought back samples taken from nine different locations on the lunar surface. By measuring relative amounts of various radioactive elements and the elements into which they decay, scientists were able to assign ages to these different lunar regions (see **Foundations 7.1** and **Math Tools 7.1**). The results of that work were surprising. Although smooth areas on the Moon were indeed found to be younger than heavily cratered areas, the differences in age were not great. The oldest, most heavily cratered regions on the Moon

> Most of the surface of the Moon is older than 3.4 billion years.

Determining the Ages of Rocks

Look at a picture of the Grand Canyon (**Figure 7.10**). The rock layers tell the story of the canyon's geological history. Most of the layers were laid down by a process called "sedimentation," in which material carried by water or wind buries what lies below. Volcanism also contributes to the layering as lava flows over Earth's surface. At the top of the stack are the latest deposits, such as those found on the rim of the Grand Canyon. Moving down through these layers takes us progressively further and further back into Earth's past.

To assign real dates to these different layers—or to rocks from any location, including the Moon—scientists use a technique called **radiometric dating**. Radiometric dating makes use of the steady decay of radioactive **parent elements** into more stable **daughter products**. Some minerals can contain radioactive isotopes as part of their chemical structure. (Different **isotopes** of a particular element have the same number of protons in their nuclei but different numbers of neutrons.) Chemical analysis of such a mineral immediately after its formation would reveal the presence of the radioactive parents, but the daughter products of the radioactive decay would be absent. As radioactive atoms decay over time, however, the amount of parent elements decreases and the amount of trapped daughter products builds up. Chemical analysis of an old sample of such a mineral would reveal both remaining radioactive parent atoms and daughter products trapped within the structure of the mineral. By comparing the relative amounts of radioactive parent and daughter products, scientists can determine when the mineral was formed and hence the age of the rock (see Math Tools 7.1).

FIGURE 7.10 Layers of rock within the Grand Canyon reveal its geological history.

How to Read Cosmic Clocks

As Foundations 7.1 explains, a geologist can calculate the age of a mineral by measuring the relative amounts of a parent radioisotope and its daughter product. The time interval over which a radioactive isotope decays to half its original amount is called its **half-life**. With every half-life that passes, the remaining amount of the radioisotope will decrease by a factor of 2. For example, over three half-lives, the remaining amount of a parent radioisotope will be $\frac{1}{2} \times \frac{1}{2} \times \frac{1}{2} = \frac{1}{8}$ of its original amount. We can make this concept more general with a simple relationship:

$$\frac{P_F}{P_O} = \left(\frac{1}{2}\right)^n,$$

where P_O and P_F are the original and final amounts, respectively, of a parent radioisotope, and n is the number of half-lives that have gone by.

The most abundant isotope of the element uranium (^{238}U, the parent) decays through a series of intermediate daughters to an isotope of the element lead (^{206}Pb, its final daughter). The half-life of ^{238}U is 4.5 billion years. This means that in 4.5 billion years, a sample that originally contained the uranium isotope (the parent) but no lead (its final daughter) would be found instead to contain equal amounts of uranium and lead. If we were to find such a mineral, we would know that half the uranium atoms had turned to lead and that the mineral formed 4.5 billion years ago.

Let's look at another example, this time with a different isotope of uranium (^{235}U) that decays to a different lead isotope (^{207}Pb) with a half-life of 700 million years. Suppose that a lunar mineral brought back by astronauts has 15 times as much ^{207}Pb (the daughter product) as ^{235}U (the parent radioisotope). This means that $^{15}/_{16}$ of the parent radioisotope (^{235}U) has decayed to the daughter product (^{207}Pb), leaving only $\frac{1}{16}$ of the parent remaining in the mineral sample. Noting that $\frac{1}{16}$ is $(\frac{1}{2})^4$, we see that 4 half-lives have elapsed since the mineral was formed, and that this lunar sample is therefore 4×700 million = 2.8 billion years old.

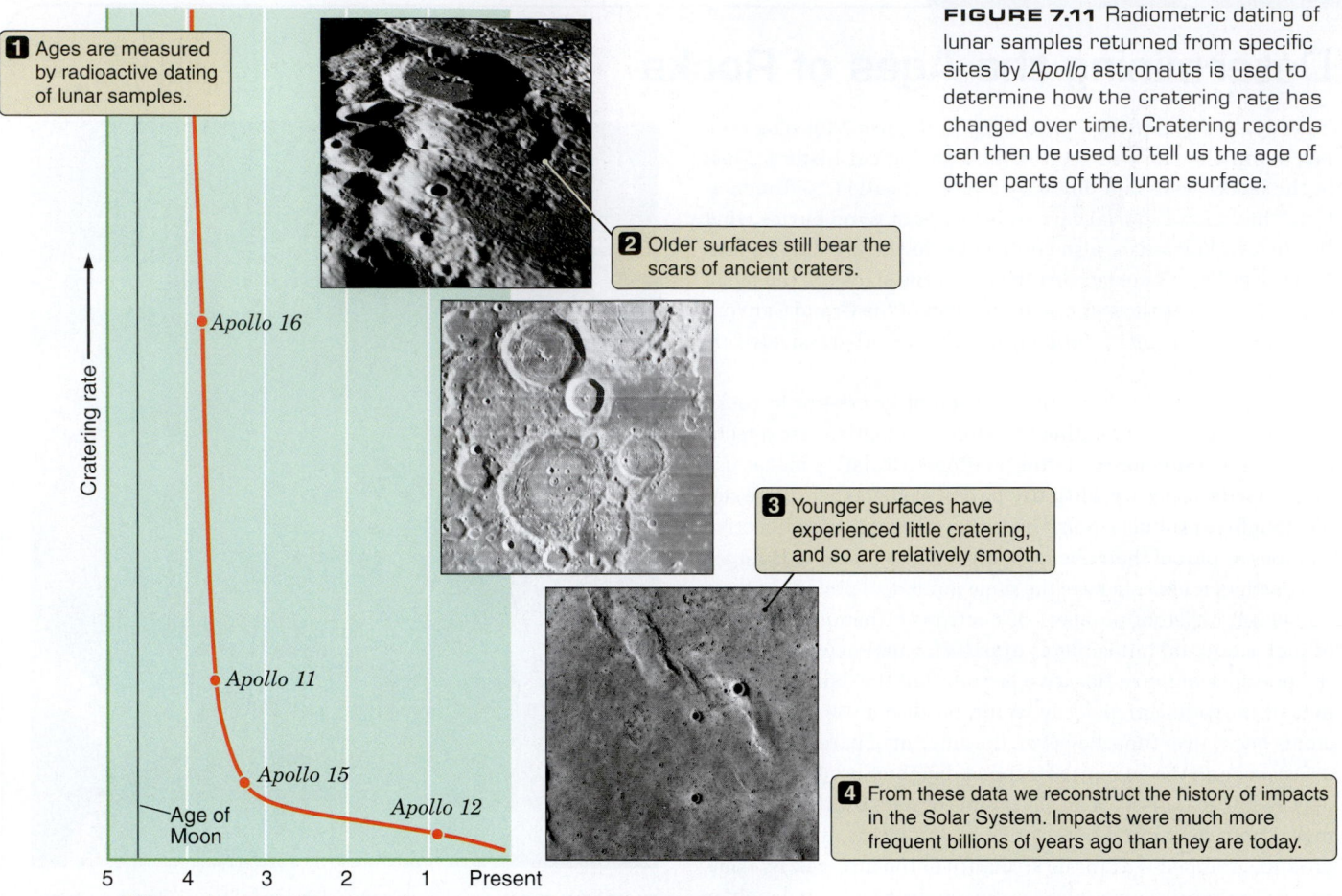

1 Ages are measured by radioactive dating of lunar samples.

2 Older surfaces still bear the scars of ancient craters.

3 Younger surfaces have experienced little cratering, and so are relatively smooth.

4 From these data we reconstruct the history of impacts in the Solar System. Impacts were much more frequent billions of years ago than they are today.

Apollo 16

Apollo 11

Apollo 15

Apollo 12

Age of Moon

Cratering rate

5   4   3   2   1   Present
Billions of years before present

FIGURE 7.11 Radiometric dating of lunar samples returned from specific sites by *Apollo* astronauts is used to determine how the cratering rate has changed over time. Cratering records can then be used to tell us the age of other parts of the lunar surface.

date back to about 4.4 billion years ago, whereas most of the smoother parts of the lunar surface are typically 3.1 billion to 3.9 billion years old.[4] The clear implication is that the vast preponderance of the cratering in the Solar System took place within the first billion years of the Solar System's formation (**Figure 7.11**). Heavily cratered surfaces such as those of Mercury and the Moon are ancient indeed.

7.4 The Interiors of the Terrestrial Planets Tell Their Own Tales

To understand the processes responsible for all but wiping away the visible record of impacts in Earth's past and for continually remaking the surface of our planet, we begin by looking below its surface. What lies hundreds and thou-

[4]Although they have not been radiometrically dated, some of the youngest flows show very few impact craters and are therefore believed to be no more than 1 billion to 2 billion years old.

sands of kilometers below our feet, and how do we know? These are the questions we turn to now.

We Can Probe the Interior of Earth

For all the effect that human activity has had on Earth, we have literally only scratched the surface of our planet. The deepest holes ever drilled have gone only about 12 km below the surface, and Earth's center is 6,350 km deeper! Even so, scientists are confident about what the interior of Earth is like. Information about Earth's interior comes to us in many ways. For example, the size of Earth and the strength of Earth's gravity, together with Newton's universal law of gravitation, tell us the mass and hence the density of Earth. From this information we know that the average density of Earth is 5,500 kilograms per cubic meter (kg/m^3), or five and a half times the density of water (which is 1 gram per cubic centimeter [g/cm^3], or 1,000 kg/m^3). Even though we have no direct samples of material from deep within Earth's interior, this number alone tells us that the compo-

> The density of matter inside terrestrial planets increases with depth.

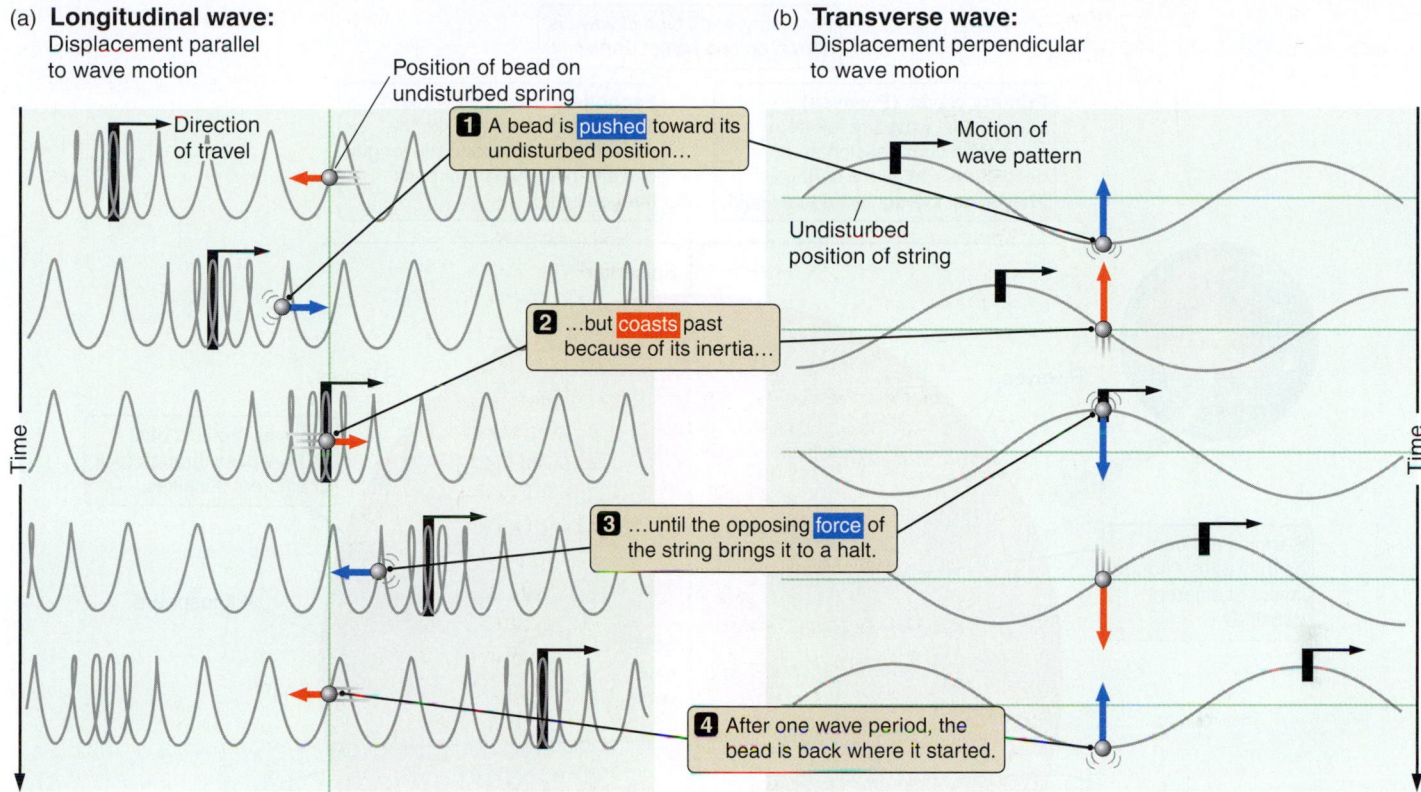

FIGURE 7.12 Mechanical waves result from forces that try to even out disturbances. (a) A longitudinal wave involves oscillations along the direction of travel of the wave. (b) A transverse wave involves oscillations that are perpendicular to the direction in which the wave travels. Primary seismic waves are longitudinal and secondary seismic waves are transverse.

sition of the interior of our planet has a density higher than that of the surface, which averages about 2,900 kg/m³. Other clues about Earth's interior come from studies of meteorites. Because these fragments are left over from a time when the Solar System was young and Earth was forming from similar materials, the overall composition of Earth should resemble the composition of meteorite material, which includes minerals with abundant amounts of iron, whose density is nearly 8,000 kg/m³.

By far the most important source of information about Earth's interior comes from monitoring the vibrations from earthquakes. When an earthquake occurs, vibrations spread out through and across the planet as **seismic waves**. There are two different classes of seismic waves. As their name implies, **surface waves** travel across the surface of a planet, much like waves on the ocean. If conditions are right, surface waves from earthquakes can be seen rolling across the countryside like ripples on water. These waves are responsible for much of the heaving of Earth's surface during an earthquake, causing damage such as the buckling of roadways.

> Seismic waves provide information about the interior configuration of Earth.

The other type of wave travels *through* Earth, rather than along its surface, probing the interior of our planet. Waves in this category include primary waves and secondary waves. **Primary waves** (P waves) result from alternating compression and decompression of a material. P waves are **longitudinal waves** (**Figure 7.12a**) that distort the material they travel through, much as sound waves travel through air or water or compression waves move along the length of a spring. **Secondary waves** (S waves) are more like the motion that occurs when we pluck a guitar string. S waves are **transverse waves** (**Figure 7.12b**) that result from the sideways motion of material. Unlike primary waves, which travel through rock because rock rebounds after being compressed, secondary waves travel through rock because rock springs back after being bent.

The progress of seismic waves through Earth's interior depends on the characteristics of the material they are moving through (**Figure 7.13**). Primary waves can travel through either solids or liquids, but secondary waves cannot travel through liquids, because liquids do not "spring back" when they are "bent." The speed of seismic waves provides additional information about Earth's interior. Seismic waves travel at different speeds depending on the density and composition of rock. As a result, seismic waves

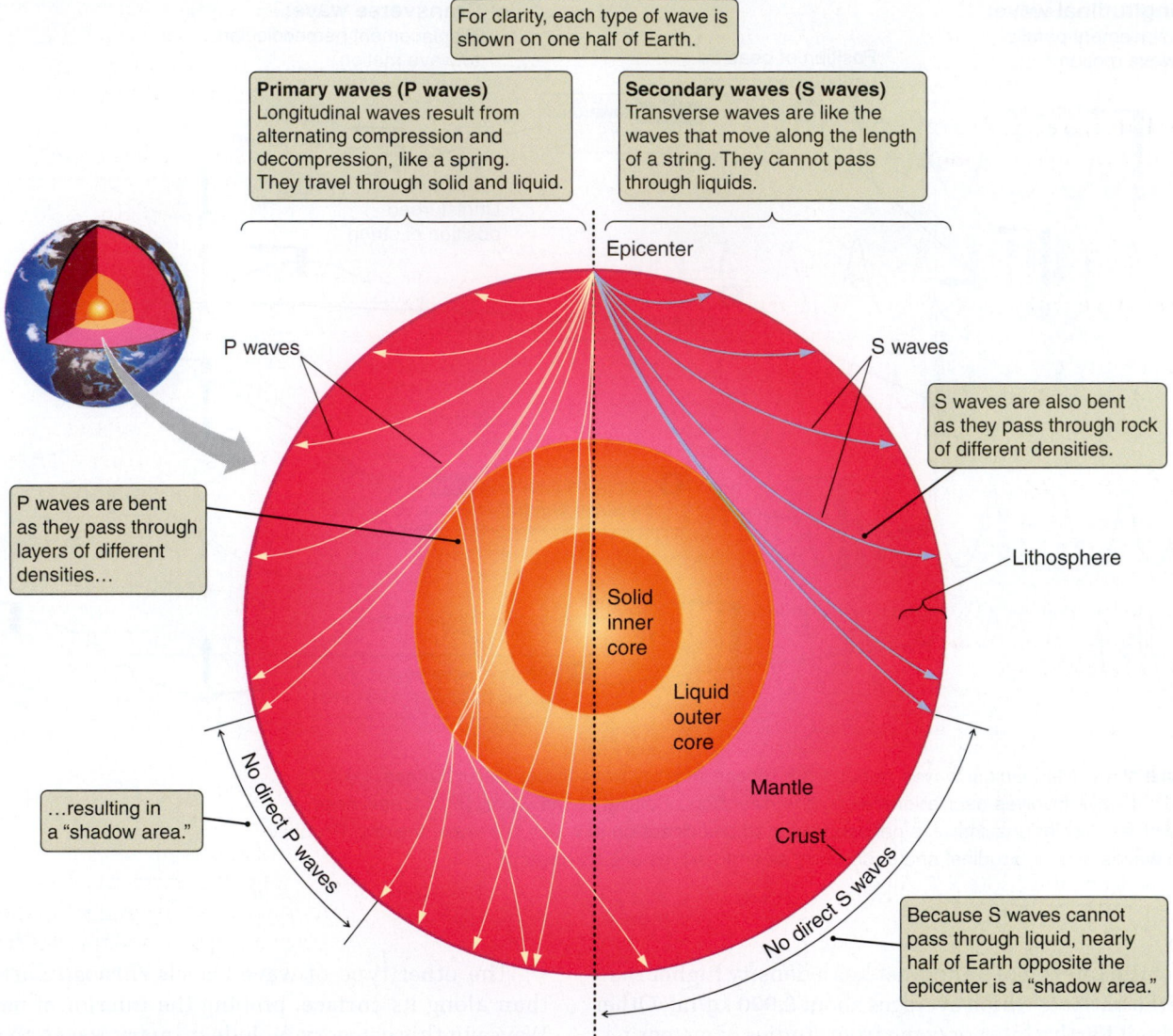

For clarity, each type of wave is shown on one half of Earth.

Primary waves (P waves)
Longitudinal waves result from alternating compression and decompression, like a spring. They travel through solid and liquid.

Secondary waves (S waves)
Transverse waves are like the waves that move along the length of a string. They cannot pass through liquids.

Epicenter

P waves

S waves

S waves are also bent as they pass through rock of different densities.

P waves are bent as they pass through layers of different densities…

Lithosphere

Solid inner core

Liquid outer core

No direct P waves

…resulting in a "shadow area."

Mantle

Crust

No direct S waves

Because S waves cannot pass through liquid, nearly half of Earth opposite the epicenter is a "shadow area."

FIGURE 7.13 Primary and secondary seismic waves move through the interior of Earth in distinctive ways. Measurements of when and where different types of seismic waves arrive after an earthquake enable us to test predictions of detailed models of Earth's interior. Note the "shadow areas" caused by the refraction of primary waves (yellow) at the outer boundary of the liquid outer *core* and the inability of secondary waves (blue) to pass through the liquid outer core.

moving through rocks of varying densities or composition are bent in much the same way that waves of light are bent when they enter or leave glass.[5] In fact, when a wave comes to a place where the density of rock layers varies abruptly, the wave can be refracted (bent) or even reflected just as light is refracted or reflected by a pane of glass (see Foundations 5.1).

The refraction of primary waves at the outer edge of Earth's liquid outer core and the inability of secondary

waves to penetrate the liquid outer core create "shadows" of the liquid core on the side of Earth opposite an earthquake's epicenter, as shown in Figure 7.13. Much of our understanding of the properties of Earth's liquid outer core is due to these shadow areas and the individual characteristics of P and S waves.

Scientists use instruments called **seismometers** to measure the distinctive patterns of seismic waves. For nearly a hundred years, thousands of seismometers scattered around the globe have measured the vibrations from countless earthquakes and other seismic events, such as volcanic eruptions and nuclear explosions. When seismic waves arrive at a seismometer station, geologists ask many questions, including these: What types of waves were measured?

---

[5]The reason light does not take a curved path through glass is that the index of refraction remains constant throughout. Seismic waves take a curved path through Earth's interior because its density is continuously changing.

How strong were they? When were they received at the station? Alone, a single seismometer can record ground motion at only one place on Earth. But when we combine such measurements with those of many other seismometers placed all over Earth, we can use the data to get a comprehensive picture of the interior of our planet.

Building a Model of Earth's Interior

How geologists go from raw seismic data to an understanding of Earth's interior is a very good example of the interplay between theory and observation in modern science. To construct a model of Earth's interior, geologists begin with some obvious clues from the seismic data. Are there liquid regions where secondary waves cannot penetrate? Are there jumps in the density of the rock from which waves are reflected? How do waves bend, and what does that bending say about the density profile of Earth? The model must also be consistent with Earth's average density of 5,500 kg/m³.

To go beyond this basic sketch, geologists turn to the laws of physics, combined with knowledge of the properties of materials and how they behave at different temperatures and pressures. For example, scientists take into account the fact that the pressure at any point in Earth's interior must be just high enough to balance the weight of all the material above it. To see why this is so, think about what would happen if it were not. If the outward pressure at some point within a planet were *less* than the weight per unit area of the overlying material, then that material would fall inward, crushing what was underneath it. If the pressure at some point within a planet were *greater* than the weight per unit area of the overlying material, then the material would be able to expand and push outward, lifting the overlying material. The situation is stable only when the weight of matter above is just balanced by the pressure within the whole interior of the planet. This balance between pressure and weight is known as **hydrostatic equilibrium**. We will see this balance repeatedly as we talk about the structure of planetary interiors, planetary atmospheres, and the structure and evolution of stars.

With such considerations in mind, scientists then construct a model of Earth's interior that follows the rough outline of their basic sketch but that is also physically consistent. They next "set off" earthquakes in their model, calculate how seismic waves would propagate through a model Earth with that structure, and predict what those seismic waves would look like at seismometer stations around the globe. They then test their model by comparing these predictions with actual observations of seismic waves from real earthquakes. The extent to which the predictions agree with observations points out both strengths and weaknesses of the model. The structure of the model is adjusted (always remaining consistent with the known physical properties of materials) until a good match is found between prediction and observation.

This method is how geologists arrived at our current picture of the interior of Earth, as shown in Figure 7.13. The major subdivisions of Earth's interior include a two-component **core**, a thick **mantle**, and the lithosphere. At the center, Earth's core consists primarily of iron, nickel, and other dense metals. In contrast, the outer parts of Earth are made of lower-density materials. The mantle, which surrounds the core, is made up of medium-density materials. The lithosphere, which includes the uppermost part of the mantle and the **crust**, consists of the lowest-density materials. The crust, which is the outermost part of the lithosphere, comes in two components: the low-density, silica-rich crust that forms the continents; and the higher-density crust of the ocean floor. Common continental rocks include **granite**, a coarse-grained volcanic rock rich in silicon and oxygen; most oceanic crust, however, consists of **basalt**, a heavy, dark volcanic rock that is rich in iron and magnesium.

As Figure 7.13 illustrates, the interior structure of Earth is far from uniform in composition. Geologists speak of Earth's interior as being "differentiated" and the process of separating materials by density as **differentiation**. Differentiation of the interiors of Earth, other terrestrial planets, and the Moon results from the fact that these planetary interiors were once molten. When rocks of different types are mixed together, they tend to stay mixed. Once this rock melts, however, the denser materials sink to the bottom and the less dense materials float to the top (just as less dense oil floats on denser vinegar in a bottle of salad dressing). Today, little of Earth's interior is molten, but the differentiated structure of the planet tells of a time when Earth was much hotter and its interior was liquid throughout. The cross sections in **Figure 7.14** show the differentiated structure of each of the terrestrial planets and Earth's Moon. As we continue our study of the composition of objects in the Solar System, differentiation will be an important concept. For example, in Chapter 12 we will find that by analyzing the chemical composition of meteorite material, we can often determine whether it was once part of a body that was chemically differentiated.

> Differentiation of planets shows that they once were molten.

The Moon Was Born from Earth

As Figure 7.14 shows, the Moon has only a tiny core, and its core is composed mostly of material that is similar to that found in Earth's mantle. The best model explaining the Moon's composition involves a catastrophic impact on Earth. Planetary scientists believe that when Earth was very young, a Mars-sized pro-

> The Moon was probably formed by a collision between a protoplanet and the young Earth.

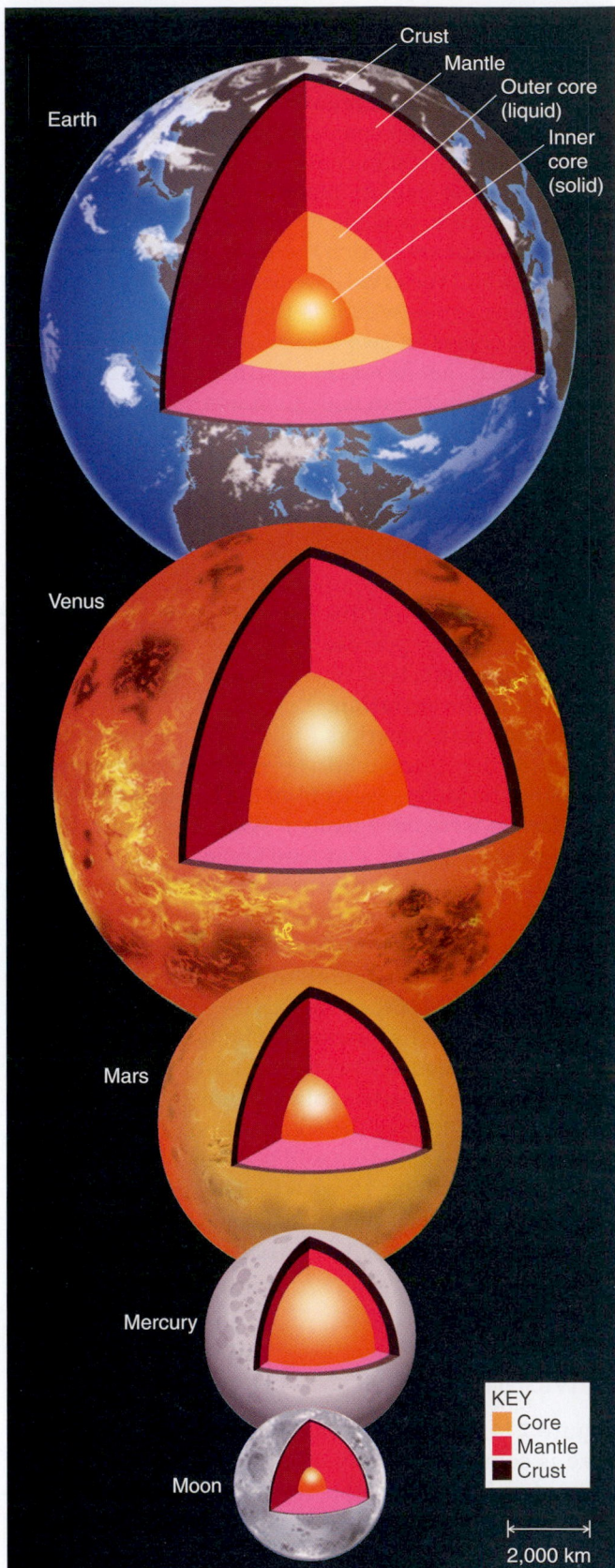

FIGURE 7.14 A comparison of the interiors of the terrestrial planets and Earth's Moon. Some fractions of the cores of Mercury, Venus, and Mars are probably liquid.

Earth

Crust
Mantle
Outer core (liquid)
Inner core (solid)

Venus

Mars

Mercury

Moon

KEY
Core
Mantle
Crust

2,000 km

FIGURE 7.15 An artist's rendition of a Mars-size protoplanet colliding with young Earth. Debris from the impact coalesced around Earth and formed our Moon.

toplanet collided with Earth, blasting off and vaporizing parts of Earth's partly differentiated crust and mantle (**Figure 7.15**). This debris condensed into orbit around Earth, evolving into our Moon. This model explains the similarities in composition between the Moon and Earth's mantle, and the model can account for the Moon's lack of water and other volatiles (see Chapter 6) while Earth and its closest neighbors—Mars and Venus—are volatile-rich. According to this model, during the vaporization stage of the collision, most gases were lost to space, leaving only the nonvolatiles to condense as the Moon. Earth, however, was large enough to keep its volatiles, which continued to be released from the interior. Because of the stronger gravity of Earth, these gases were retained as part of our atmosphere.

The Evolution of Planetary Interiors Depends on Heating and Cooling

A general feature of planetary interiors is that the deeper we go within the planet, the higher the temperature climbs. When we look below a planet's surface, we find that the thermal energy in its interior drives the planet's geological activity. Thermal equilibrium, which we introduced in Chapter 4, governs the complex interplay of heating and cooling within a planet.

Part of the thermal energy in the interior of Earth is left over from when Earth formed. We know that the tremendous energy liberated by the collisions during the accretion process responsible for the formation of Earth, together with energy from short-lived radioactive elements, was enough to melt the planet. The differentiated structure of Earth is evidence of this fact. The surface of Earth then cooled rather rapidly by radiating energy away into space, forming a solid crust on top of a molten interior. Because a solid crust does not conduct thermal energy well, it served as an insulator—a "wool sweater," if you will. (Anyone who has watched a lava flow "crust over" and then walked across its surface has experienced this fact. Molten rock is still there, possibly only centimeters beneath the adventurous walker's feet.) But the crust is not a perfect insulator. Over the eons, energy from the interior of the planet continued to leak through the crust and radiate into space. The interior of the planet slowly cooled, and the mantle and the inner core solidified.

However, there must be more to the story than leftover energy from the time of Earth's formation. If this were the only source of heating in Earth's interior, calculations show that the interior of Earth would be much cooler than it is today, and Earth would have long ago solidified completely. There must be additional sources of energy continuing to heat the interior of Earth if we are to account for the high temperatures that persist there today.

One source of heating Earth's interior is friction generated by tidal effects of the Moon and Sun. When we discuss tides in Chapter 10, we will find that tidal heating is responsible for keeping the interior of Jupiter's moon Io molten. However, tidal heating fails by a wide margin to explain the elevated temperature of Earth's interior. Instead, most of the extra energy in Earth's interior comes from long-lived radioactive elements. As these radioactive elements trapped in the interior of Earth decay, they liberate energy, heating the planet's interior. Today the temperature of Earth's interior is determined by dynamic equilibrium (see Foundations 4.1) between the heating of the interior and the loss of energy that is radiated away into space. As radioactive "fuel" in Earth's interior is consumed by decay, the amount of thermal energy generated declines, and Earth's interior becomes cooler as it ages.

Temperature plays an important role in a planet's interior structure, but it's not the only influence at work. The structure of Earth's core results from an interesting interplay between increasing temperature and pressure. In our everyday experience we notice that temperature alone determines whether a material is solid or liquid. When something gets hot enough, it melts and becomes a liquid. When something gets cold enough, it freezes and becomes a solid. The center of Earth is the hottest location in Earth's interior. With

Collisions and radioactive heating made the forming Earth molten. As it ages, however, Earth is cooling off.

a temperature of perhaps 6,000 kelvins (K), it is hotter than the surface of the Sun. Yet the center of Earth is solid! It is the *outer* core of Earth that is molten, even though it is many hundreds of kelvins cooler than the inner core. How can we explain this solid, superhot inner core coexisting with a cooler, molten outer core?

It turns out that whether a material is solid or liquid depends on pressure as well as on temperature. With most materials (water ice being a rare exception), the solid form of the material is more compact than the liquid form. Putting the material under higher pressure forces atoms and molecules closer together and makes the material more likely to become a solid. Moving toward the center of Earth, the effects of temperature and pressure oppose each other: The higher temperature makes it more likely that material will melt, but the higher pressure favors a solid form. Only in the outer core of Earth does the high temperature win, allowing the material to exist in a molten state. At the center of Earth, even though the temperature is higher, the pressure is so great that the inner core of Earth is solid. We will find even stranger physical properties of the cores of the giant planets when we turn to them in Chapter 9.

Like stars, planets lose their internal thermal energy through a combination of convection, conduction, and radiation. We already discussed *radiation* of energy by planetary bodies in Chapter 4. We will describe the mechanism of *convection* as it relates to plate tectonics in Section 7.5. Finally, we will cover *thermal conduction* when we talk about energy transfer within the Sun's interior in Chapter 14. For now, we'll simply say that liquids transfer heat by convection and solids transfer heat by conduction.

The internal temperature of a planet depends on the planet's size. A planet's *volume*—and thus the amount of radioactive material ("fuel") it contains—determines the amount of interior heating produced. On the other hand, the planet's ability to get rid of the thermal energy in its interior depends on the planet's *surface area* because thermal energy has to escape through the planet's surface. (By analogy, when you want to keep warm, you curl up into a ball, reducing the amount of exposed skin through which thermal energy can escape. But when you want to cool off, you spread out your arms and legs, exposing as much skin as possible and allowing it to get rid of thermal energy.)

Generally, smaller terrestrial planets cool faster than larger terrestrial planets.

A simple mathematical relationship describes the connection between a planet's size and its temperature (see **Math Tools 7.2**). Smaller planets have less energy to lose in relation to surface area and are thus cooler. Larger planets have more energy to lose per square meter of surface and so remain hotter. It should not be surprising to learn that the smaller objects (Mercury and the Moon) are thus geologically inactive in comparison to the larger terrestrial planets (Venus, Earth, and Mars).

How Planets Cool Off

Let's assume that all terrestrial planets formed with the same percentage of radioactive materials in their bulk composition, and that these radioactive materials are their sole source of internal thermal energy. A planet's volume therefore would determine the total amount of the thermal energy–producing material it contains. A planet loses its internal energy by radiating it away at its surface. A planet's cooling surface area therefore determines the rate at which it can get rid of its thermal energy. The energy-producing volume of a planet is proportional to the cube of the planet's radius (volume $\propto R^3$), whereas the cooling surface area of the planet is proportional to only the square of the radius (surface area $\propto R^2$). The ratio of the two—the amount of energy there is to lose, divided by

the surface area through which thermal energy can escape—is proportional to R^3/R^2, or R.

Of course, we have to remember that a planet's ability to transfer internal energy from its hot core to its cooling surface depends on factors such as its own internal convection and conduction properties. Nevertheless, all things being equal, larger planets retain their internal energy longer than smaller planets do. Or, to put in another way, smaller planets lose their internal thermal energy more quickly. For example, Mars has a radius about half that of Earth, so it has been losing its internal thermal energy to space about twice as fast as our own planet has. This is one reason why Mars is less geologically active than Earth.

Most Planets Generate Their Own Magnetic Fields

Earth's magnetic field is not difficult to detect, but what causes it? As most schoolchildren know, you can detect Earth's magnetic field with a compass consisting of an arrow-shaped needle attached to a small bar magnet that is allowed to swing about freely. The compass needle lines up with Earth's magnetic field and points "north" and "south." But if we were to map the orientation of compass needles at every place on Earth, the north-pointing arrows would all converge not at the geographic North Pole, but at a location in northern Canada. This is Earth's north magnetic pole. An opposite magnetic pole (Earth's south magnetic pole) is located just off the coast of Antarctica, 2,800 km from Earth's geographic South Pole. How do compasses work the way they do? Bar magnets attract or repel one another, and Earth behaves as if it contained a giant bar magnet that was slightly tilted with respect to Earth's rotation axis and had its two endpoints near the two magnetic poles (**Figure 7.16**).

Wait a minute! Did you look *carefully* at Figure 7.16? Did you notice that Earth's hypothetical bar magnet has its *south* pole in Earth's *Northern* Hemisphere? Can you guess why? The answer is quite simple when you think about it. Bar magnets are attracted to each other's *opposite* poles. Traditionally, we have defined the north-pointing end of a compass's bar magnet as its *north* pole. Having made this decision, we must in turn define Earth's magnetic field as

> Earth's magnetic field behaves like a giant bar magnet, but the orientation of the poles changes with time.

we show it in Figure 7.16, with the north magnetic pole of a compass attracted to Earth's south magnetic pole.

Earth's magnetic field actually originates deep within the interior of the planet, and the processes responsible for generating Earth's magnetism are not understood in detail. But there's one thing we know for certain: Earth's magnetic field is *not* due to "permanent magnets" (naturally occurring magnetic materials whose individual atoms are magnetically aligned) buried within the planet. Even though naturally occurring magnetic materials do exist, permanent magnets cannot explain the fact that Earth's magnetic field is constantly changing. Some changes in Earth's magnetic field, such as a shift in the exact location of the magnetic poles, can occur over periods that are much shorter than a human lifetime.[6] In addition, the geological record shows that much more dramatic changes in the magnetic field have occurred over the history of our planet.

The study of **paleomagnetism**—the fossil record of Earth's changing magnetic field—is an important part of geology. When a magnetic material such as iron gets hot enough, it loses its magnetization. (This is another reason why permanent magnets cannot be responsible for Earth's magnetic field. At the high temperatures found in Earth's interior, permanent magnets lose their magnetization.) As the material cools, it again becomes magnetized by any magnetic field in which it is immersed. Iron-bearing minerals thus record the direction of the Earth's magnetic field *at the very instant in time that they cooled*. In this way a memory

---

[6]At the moment, the north magnetic pole is on the move, traveling several tens of kilometers per year toward the northwest. If this rate were to continue, the north magnetic pole could leave Canada and be in Siberia by the end of the century. The north and south magnetic poles wander independently of each other.

(a)

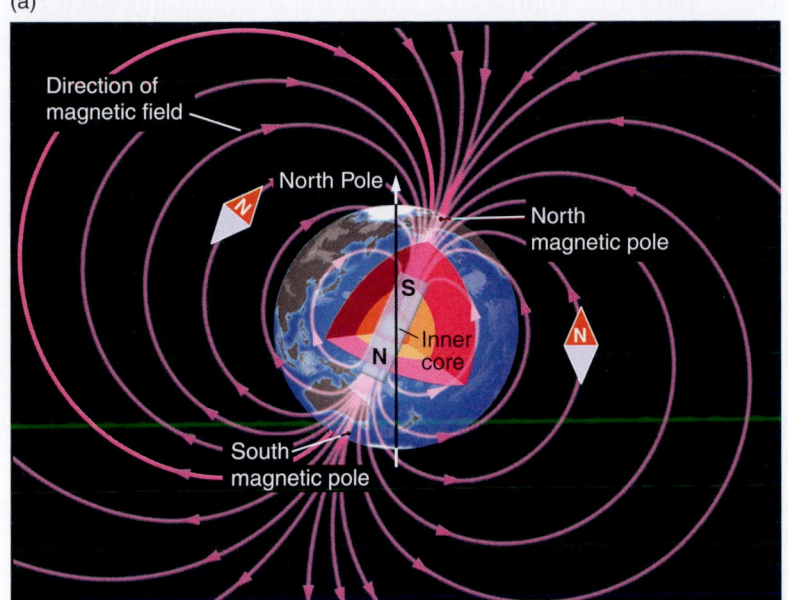

(b)

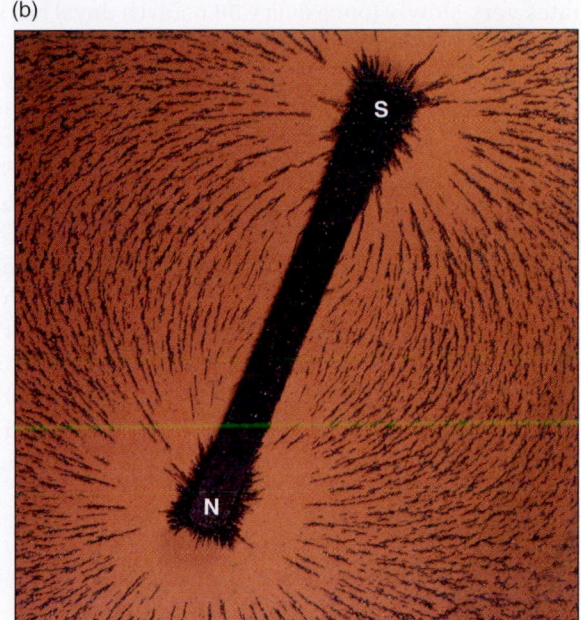

VISUAL ANALOGY **FIGURE 7.16** (a) Earth's magnetic field can be visualized as though it were a giant bar magnet tilted relative to Earth's axis of rotation. Compass needles line up along magnetic field lines and point toward Earth's north magnetic pole. Note that because bar magnets are attracted to each other's opposite poles, Earth's hypothetical bar magnet has its *south* pole coincident with Earth's *north* magnetic pole. (b) Iron filings sprinkled around a bar magnet help us visualize such a magnetic field.

of that magnetic field becomes "frozen" into the material. For example, lava extruded from a volcano carries a record of Earth's magnetic field at the moment when it cooled. By using the radiometric techniques discussed in Foundations 7.1 to date these materials, geologists obtain a record of how Earth's magnetic field has changed over time. Although Earth's magnetic field has probably existed for billions of years, the north-south *polarity* reverses from time to time. On average, these reversals in Earth's magnetic field take place about every 500,000 years.

Although the details are not known, we have a general idea of how Earth's magnetic field originates in the motions of material in Earth's interior. Magnetic fields result from electric currents, which are moving electric charges. Earth's magnetic field is thought to be a side effect of three factors: Earth's rotation about its axis; an electrically conducting, liquid outer core; and fluid motions including convection within the outer core. The interior of Earth acts as if it were a giant **dynamo**,[7] converting mechanical energy into magnetic energy. (Other objects in the Solar System, including the Sun and the giant planets, are also thought to act like dynamos.)

The Moon is the most magnetically surveyed astronomical object other than Earth. During the *Apollo* program,

astronauts used surface "magnetometers" (devices for measuring magnetic fields) to make local measurements; and on two missions small satellites were placed into orbit to search for global magnetism. Results show that the Moon either lacks a magnetic field today or, at the most, has a very weak field. The lack of a lunar magnetic field can probably be understood because of the small size of the Moon and its correspondingly cooler interior. It also has a very small core. Overall, it would be difficult to make a lunar dynamo work today. However, remnant magnetism is preserved in lunar rocks from an earlier time when the Moon likely had a molten core and a magnetic field.

> The Moon had a magnetic field long ago, but it lacks one today.

When we study the other terrestrial planets, we encounter a few surprises. Other than Earth, Mercury is the only terrestrial planet with a significant magnetic field today. The existence of Mercury's magnetic field is understandable. The planet has the ingredients to form a dynamo: rotation and a large iron core, parts of which seem to be molten and circulating. But the lack of a detected magnetic field from Venus presents a puzzle. Venus should, like Earth, have an iron-rich core and partly molten interior. Its lack of a magnetic field might be attributed to its extremely slow rotation, which could make its dynamo ineffective. On the other hand, Mercury also

> The lack of magnetic fields on Venus and Mars is a puzzle.

---

[7]Although the term *dynamo* is frequently used to describe the creation of magnetic fields in the interiors of planets and stars, it is not technically correct. A true dynamo generates an electric current from a moving magnetic field, not the other way around.

rotates very slowly (once every 58.6 Earth days) but still has a planetary magnetic field.

Like the Moon, Mars has a weak magnetic field, presumably frozen in place when its dynamo stopped working many billions of years ago. This pronounced remnant magnetic field was discovered by the *Mars Global Surveyor* orbiter in the late 1990s. The magnetic signature occurs only in the ancient crustal rocks, showing that early in the history of Mars, some sort of an internally generated magnetic field must have existed. Geologically younger rocks lack this residual magnetism, so the planet's original magnetic field has long since disappeared. The lack of a strong magnetic field today on Mars might be the result of its small core. However, given that it is expected to have a partly molten interior and rotates rapidly, the lack of a field is still surprising.

7.5 Tectonism— How Planetary Surfaces Evolve

With a better understanding of planetary interiors, we now turn to their surfaces. As you drive through mountainous or hilly terrain, take a look at places like that

FIGURE 7.17 Tectonic processes fold and warp Earth's crust, as seen in these rocks along a roadside in Israel.

shown in **Figure 7.17**, where the roadway has been cut through rock. These cuts show layers of rock that have been bent, broken, or fractured into pieces. Sometimes the force responsible for tectonism—the deformation of Earth's crust—is just gravity. For example, for more than 60 million years, rivers have dumped trillions of metric tons of rock and sediment into the Gulf of Mexico. Layers of sediment over 25 km thick have built up, and some have solidified into rock. This enormous mass has been pulled downward by gravity, causing the rock layers to bend or to break along **faults** (fractures in Earth's crust). Faults and folds in these rocks form traps for the accumulation of petroleum, creating some of the richest oil fields in the world.

Continents Drift Apart and Come Back Together

Although the weight of the crust is responsible for some of the deformation of Earth's crust, most of the faulting and buckling that we see at Earth's surface originates instead from forces deep within Earth's interior. Early in the 20th century, some scientists recognized that Earth's continents could be fit together like pieces of a giant jigsaw puzzle. The fit was particularly striking between the Americas and Africa-plus-Europe. Other evidence also suggested that this fit was more than coincidence. For example, the layers in the rock on the east coast of South America and the fossil records they hold match those on the west coast of Africa. On the basis of evidence such as this, in the 1920s the German scientist Alfred Wegener (1880–1930) proposed that over millions of years the continents had shifted their positions. This theory is popularly referred to as **continental drift**. Wegener proposed that the continents were originally joined in one large landmass that subsequently broke apart as the continents began to "drift" away from each other. ▶ll **AstroTour:** Continental Drift

Originally, the idea of continental drift was met with great skepticism among geologists because they could not conceive of a mechanism that could move such huge landmasses. In the 1960s, however, paleomagnetic studies of the ocean floor provided compelling evidence for continental drift (**Figure 7.18**). These surveys showed surprising characteristics in bands of basalt found on both sides of the ocean rifts. Ocean floor rifts such as the Mid-Atlantic Ridge are **spreading centers**. These are locations where convection causes hot material to rise toward Earth's surface and fill the gap between tectonic plates, thereby becoming new ocean floor.

Remember that when hot material cools, it becomes magnetized along the direction of the local magnetic field. So as new ocean floor is created, it cools and is carried away from the spreading center, "remembering" the direction of Earth's magnetic field at the time it was formed. In this way the spreading ocean floor carries with it a record of the

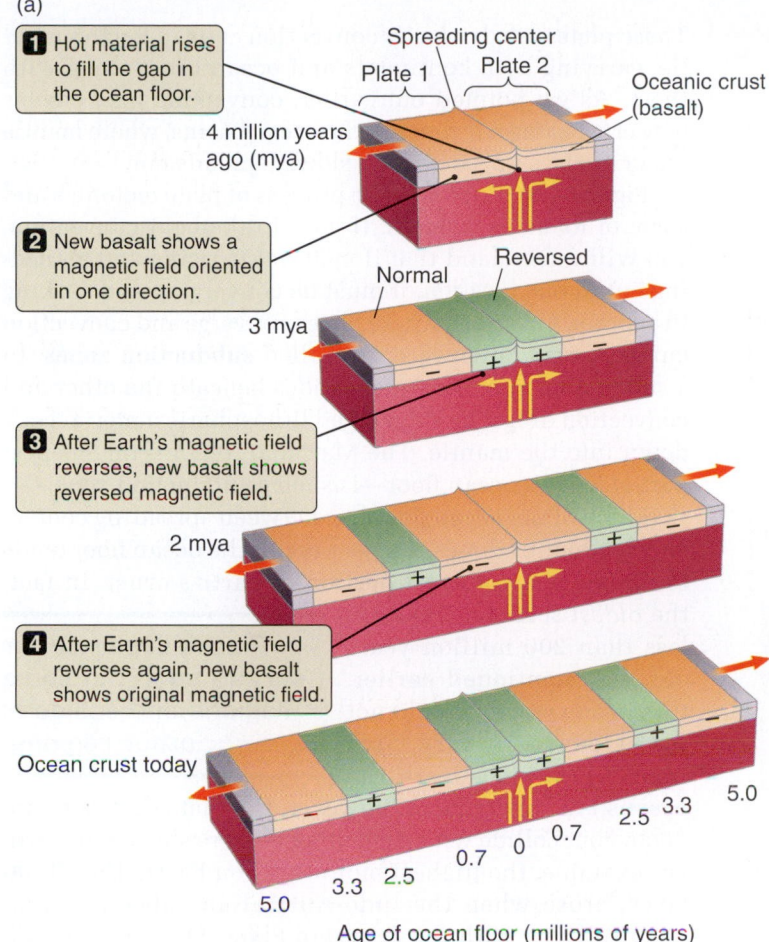

(a)

1 Hot material rises to fill the gap in the ocean floor.

Spreading center

Plate 1   Plate 2

Oceanic crust (basalt)

4 million years ago (mya)

2 New basalt shows a magnetic field oriented in one direction.

Normal   Reversed

3 mya

3 After Earth's magnetic field reverses, new basalt shows reversed magnetic field.

2 mya

4 After Earth's magnetic field reverses again, new basalt shows original magnetic field.

Ocean crust today

5.0   3.3   2.5   0.7   0   0.7   2.5   3.3   5.0

Age of ocean floor (millions of years)

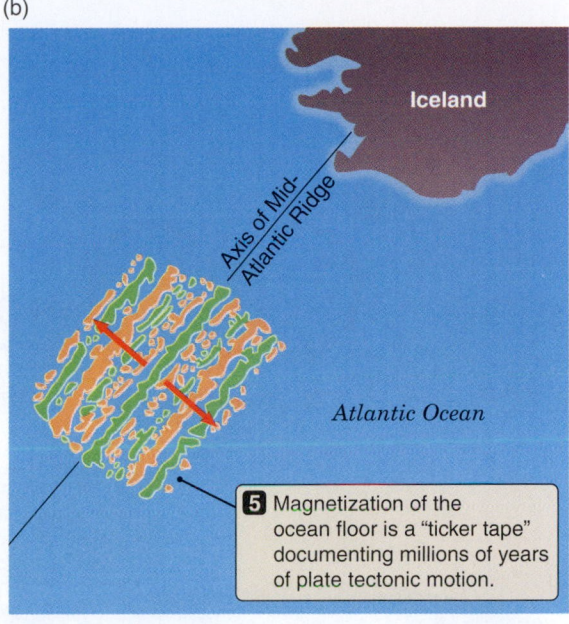

(b)

Iceland

Axis of Mid-Atlantic Ridge

Atlantic Ocean

5 Magnetization of the ocean floor is a "ticker tape" documenting millions of years of plate tectonic motion.

FIGURE 7.18 (a) As new seafloor is formed at a spreading center, the cooling rock becomes magnetized. The magnetized rock is then carried away by tectonic motions. (b) Maps like this one of banded magnetic structure in the seafloor near Iceland provide support for the theory of plate tectonics.

changes in Earth's magnetic field over time. Greater distance from a spreading center indicates an older ocean floor and an earlier time, as reflected by its magnetization. The ocean floor acts much like an old ticker-tape recorder for Earth's magnetic history. The ocean surveys also revealed that this banded magnetic structure is often symmetrical about rifts. If a change in the magnetization of the ocean floor is seen 100 km on one side of a rift, then the same change will usually be seen about 100 km on the other side of the rift. Combined with radiometric dates for the rocks, this magnetic record proved that the spreading of the seafloor and the motions of the plates have continued over long geological time spans.

Precise surveying techniques and satellite positioning methods (such as the global positioning system, or GPS) now allow locations on Earth to be determined to within a few centimeters. These measurements confirm that Earth's lithosphere is indeed moving. Some areas are being pulled apart by more than 15 centimeters (cm)—about the length of a pencil—each year. This rate is slower than a snail's pace, but over millions of years of geological time, such motions add up. Over 10 million years—not a long time by geological standards—15 cm/year becomes 1,500 km, by which time the maps definitely need to be redrawn!

Today, geologists recognize that Earth's outer shell is composed of a number of relatively brittle segments, or **lithospheric plates**, and that motion of these plates is constantly changing the surface of Earth. This theory, which is perhaps the greatest advance in 20th century geology, is referred to as **plate tectonics**. Plate tectonics is ultimately responsible for a wide variety of geological features on our planet.

> The theory explaining motions of Earth's lithosphere is called *plate tectonics.*

Plate Tectonics Is Driven by Convection

The forces required to set lithospheric plates into motion are immense. We now understand that these forces are the result of thermal energy escaping from the interior of Earth through the process of **convection**. If you have ever noticed how water moves about in a heated pot on a stovetop, then you have seen convection at work (**Figure 7.19a**). Thermal energy from the stove warms water at the bottom of the pot. The warm water expands slightly, becoming less dense than the cooler water above it, and the cooler water with higher density sinks, displacing the warmer water upward. When

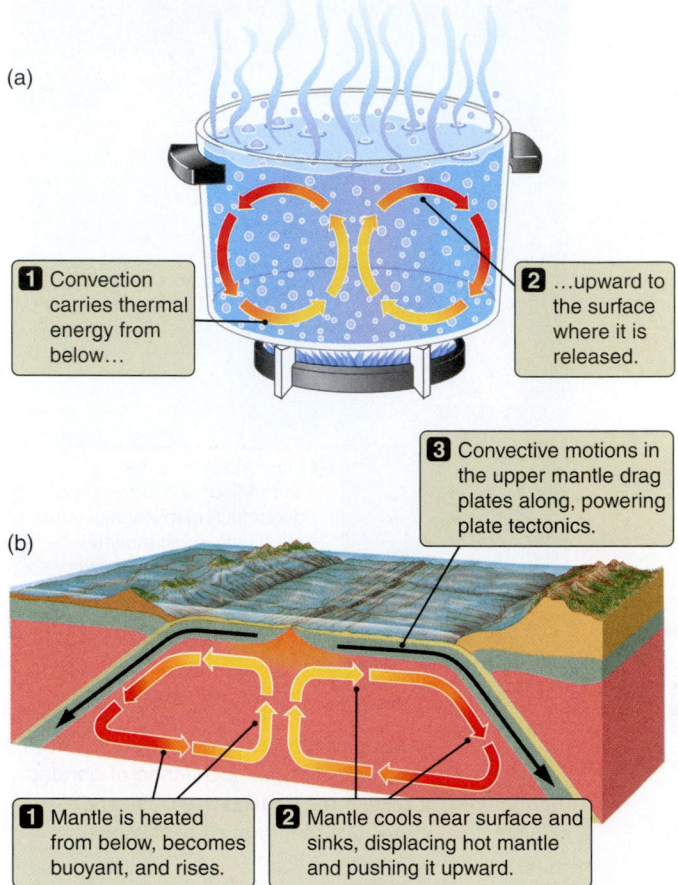

(a)

1 Convection carries thermal energy from below...

2 ...upward to the surface where it is released.

3 Convective motions in the upper mantle drag plates along, powering plate tectonics.

(b)

1 Mantle is heated from below, becomes buoyant, and rises.

2 Mantle cools near surface and sinks, displacing hot mantle and pushing it upward.

VISUAL ANALOGY **FIGURE 7.19** (a) Convection occurs when a fluid is heated from below. (b) Convection in Earth's mantle drives plate tectonics.

the lower-density water reaches the surface, it gives up part of its energy to the air, and in so doing cools, becomes denser, and sinks back toward the bottom of the pot. Water cannot rise and fall in the same place, so convection creates an organized circulating pattern in which warm water rises in some locations and cool water sinks in others. As we will see in later chapters, convection plays an important role in planetary atmospheres and in the structure of the Sun and stars.

In the case of Earth, thermal energy generated by radioactive decay in the interior of the planet causes convection to take place in the mantle (**Figure 7.19b**). Earth's mantle is not molten (if it were, secondary seismic waves could not travel through it), but it is somewhat mobile. You can think of the mantle as having the plastic consistency of hot glass. This consistency allows convection to take place, albeit very slowly. Careful mapping shows that Earth's lithosphere is divided into about seven major plates and about a half-dozen smaller ones.

> Earth's outer rocky shell is divided into seven major plates and about six smaller ones.

These plates are driven by convection cells in Earth's mantle, carrying both continents and ocean crust along with them. As we pointed out earlier, convection also creates new crust along rift zones in the ocean basins, where mantle material rises up, cools, and slowly spreads out.

Figure 7.20 illustrates the process of plate tectonics and some of its consequences. If you think about convection, you will understand that if material is rising and spreading out in one location, it must be converging and sinking in another. Locations where plates converge and convection currents turn downward are called **subduction zones**. In a subduction zone, one plate slides beneath the other and convection drags the submerged lithospheric material back down into the mantle. The Mariana Trench—the deepest part of Earth's ocean floor—is such a subduction zone.

Much of the ocean floor lies between spreading centers and subduction zones, and as a result the ocean floor tends to remain the youngest portion of Earth's crust. In fact, the *oldest* seafloor rocks are less than 200 million years old. (As mentioned earlier, this is one reason we do not see evidence for very large impact craters beneath the oceans.) In some places, however, the plates are not subducted but collide with each other and are shoved upward. For example, the highest mountains on Earth, the Himalayas, arose when the Indo-Australian subcontinental plate collided with the Eurasian Plate. The Indo-Australian Plate is continuing to move northward, causing the Himalayas to rise at a rate of 0.5 meter per century. In still other places, lithospheric plates meet at oblique angles and slide along past each other. A type of fault called a **transform fault** marks the actively slipping fracture zone between plate boundaries. One such zone is the San Andreas Fault in California, where the Pacific Plate shears past the North American Plate.

> Plates separate, or spread apart, in some regions and collide in other regions.

Locations where plates meet tend to be very active geologically. In fact, one of the best ways to see the outline of Earth's plates is to look at a map of where earthquakes and volcanism occur, like that shown in **Figure 7.21**. At locations where plates run into each other, enormous stresses build up. Earthquakes result as the friction between the two plates finally gives way and the plates slip past each other, relieving the stress. Volcanoes are created when friction between plates melts rock, which is then pushed up through cracks to the surface. Lithospheric plates can be thousands of kilometers across and range in thickness from about 5 to 100 km. As they shift, some parts move more rapidly than others, causing the plates to stretch, buckle, or fracture. These effects are readily seen on the surface as folded and faulted rocks. Mountain chains also are common near converging plate boundaries, where plates buckle and break.

> Most volcanoes and earthquakes occur along plate boundaries.

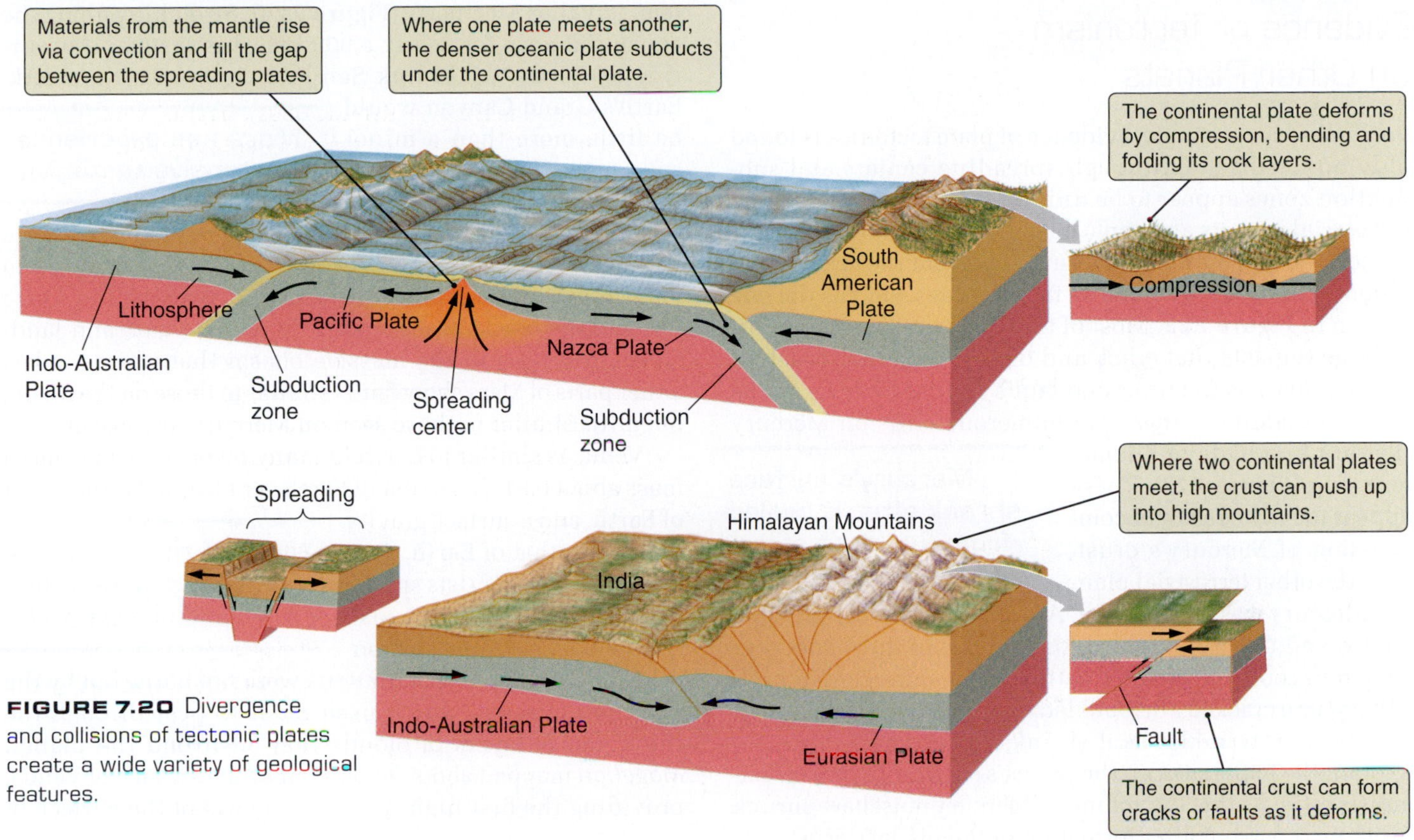

Materials from the mantle rise via convection and fill the gap between the spreading plates.

Where one plate meets another, the denser oceanic plate subducts under the continental plate.

The continental plate deforms by compression, bending and folding its rock layers.

Lithosphere

Pacific Plate

Nazca Plate

South American Plate

Compression

Indo-Australian Plate

Subduction zone

Spreading center

Subduction zone

Spreading

Himalayan Mountains

Where two continental plates meet, the crust can push up into high mountains.

India

Indo-Australian Plate

Eurasian Plate

Fault

The continental crust can form cracks or faults as it deforms.

FIGURE 7.20 Divergence and collisions of tectonic plates create a wide variety of geological features.

FIGURE 7.21 Major earthquakes and volcanic activity are often concentrated along the boundaries of Earth's principal tectonic plates.

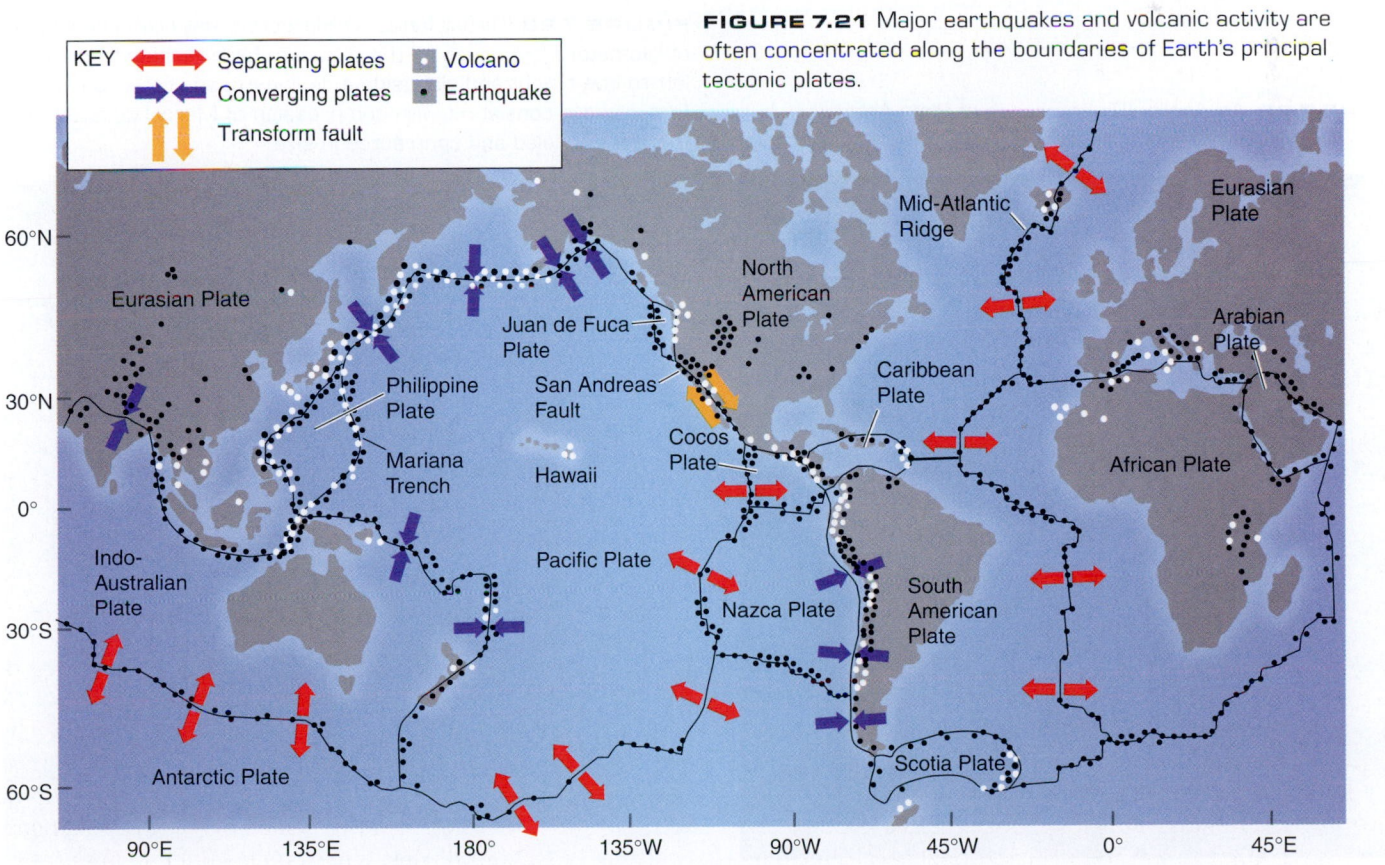

KEY
Separating plates — Volcano
Converging plates — Earthquake
Transform fault

Eurasian Plate

Mid-Atlantic Ridge

Eurasian Plate

Arabian Plate

Juan de Fuca Plate

North American Plate

Caribbean Plate

African Plate

Philippine Plate

San Andreas Fault

Mariana Trench

Hawaii

Cocos Plate

Indo-Australian Plate

Pacific Plate

Nazca Plate

South American Plate

Antarctic Plate

Scotia Plate

60°N
30°N
0°
30°S
60°S

90°E   135°E   180°   135°W   90°W   45°W   0°   45°E

Evidence of Tectonism on Other Planets

Somewhat surprisingly, evidence of plate tectonics is found only on Earth. But although spreading centers and subduction zones appear to be unique to our planet, *all* of the terrestrial planets and some of their moons show evidence of tectonic disruptions. Fractures have cut the crust of the Moon in many areas, leaving fault valleys such as the one shown in **Figure 7.22**. Most of these features are the result of large impacts that crack and distort the lunar crust.

Mercury has fractures and faults similar to those on the Moon. In addition, there are numerous cliffs on Mercury that are hundreds of kilometers long (**Figure 7.23**). These appear to be the result of compression of Mercury's crust. Like the other terrestrial planets, Mercury was once molten. After the surface of the planet cooled and the crust formed, the interior of the planet continued to cool and shrink. As the planet shrank, Mercury's lithosphere cracked and buckled in much the same way that a grape skin wrinkles as it shrinks to become a raisin. To explain the faults seen on the planet's surface, planetary scientists estimate that the volume of Mercury must have shrunk by about 5 percent after formation of the planet's crust.

> Mercury's surface shrank after it cooled from a molten state.

On Mars the most impressive tectonic feature, and possibly the most impressive tectonic feature in the Solar System, is Valles Marineris (**Figure 7.24**). Stretching along the equatorial zone for nearly 4,000 km, this system, if occurring on Earth, would link San Francisco with New York. Earth's Grand Canyon would be little more than a minor spur on the side of this chasm. Valles Marineris includes a series of massive cracks in the lithosphere of Mars that are thought to have formed as local forces, perhaps related to mantle convection, pushed it upward from below. Once formed, the cracks were eroded by wind, water, and landslides, resulting in the massive chasm that we see today. Other parts of Mars show faults similar to those on the Moon, but cliffs similar to those seen on Mercury are absent.

> Mars has experienced extensive tectonism.

Venus is similar to Earth in many respects. Venus has a mass about 0.81 times that of Earth, a radius 0.95 times that of Earth, and a surface gravity 0.91 times that of Earth. As a result, many scientists speculated that Venus might also show evidence of plate tectonics. However, these speculations were not borne out by the *Magellan* mission, which used radar to peer through the thick, dense layers of clouds that enshroud the planet. *Magellan* mapped about 98 percent of the surface of Venus, providing the first high-resolution views of the surface of

> Most of the surface of Venus is less than 1 billion years old.

FIGURE 7.22 An *Apollo 10* photograph of Rima Ariadaeus, a 2-km-wide valley between two tectonic faults on the Moon.

FIGURE 7.23 Thrust faults on Mercury create cliffs hundreds of kilometers long, such as the one seen here running from upper left to lower right and alongside a double-ringed crater. Such features are consistent with compression of Mercury's crust as the planet cooled and contracted in size.

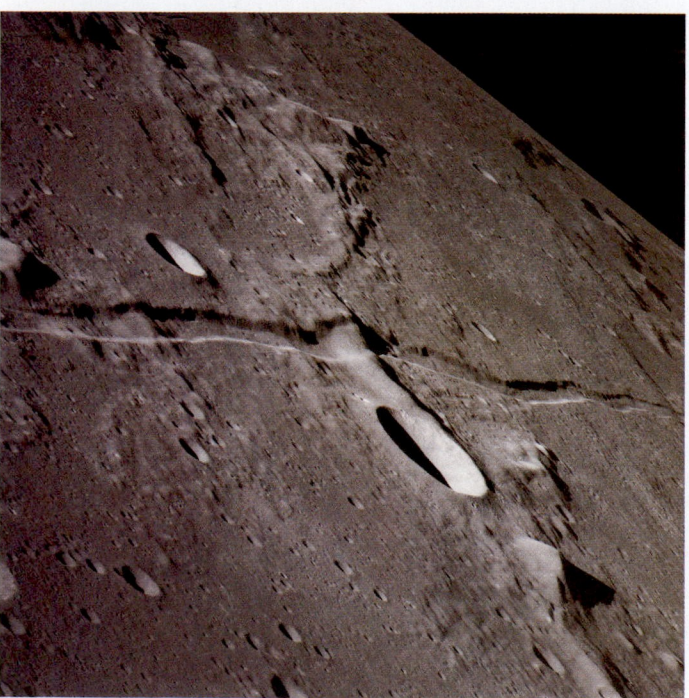

(a)

(b)

FIGURE 7.24 (a) A mosaic of *Viking Orbiter* images shows Valles Marineris, the major tectonic feature on Mars, stretching across the center of the image from left to right. This canyon system is more than 4,000 km long. The dark spots on the left are huge shield volcanoes in the Tharsis region. (b) This close-up perspective view of the canyon wall was photographed by the European Space Agency's *Mars Express* spacecraft.

FIGURE 7.25 The atmosphere of Venus blocks our view of the surface in visible light. This false-color view of Venus is a radar image made by the *Magellan* spacecraft. Bright yellow and white areas are mostly fractures and ridges in the crust. Some circular features seen in the image may be regions of mantle upwelling, or *hot spots*. Most of the surface is formed by lava flows, shown in orange.

the planet. *Magellan*'s view of one face of Venus is shown in **Figure 7.25**. Although Venus has a wealth of volcanic features and tectonic fractures, there is no evidence of lithospheric plates or plate motion of the sort seen on Earth. Yet the relative scarcity of impact craters on Venus suggests that most of its surface is less than 1 billion years old.

The absence of plate tectonics on Venus is a puzzle. The interior of Venus should be very much like the interior of Earth, and convection should be occurring in its mantle. However, this model presents a problem: how does the thermal energy from the interior of Venus escape from the planet? On Earth, mantle convection and plate tectonism provide a means for thermal energy to escape from the interior of the planet. Earth also has a few **hot spots**, where upwellings of hot mantle material rise, releasing thermal energy. The Hawaiian Islands are the result of one such hot spot. On Venus, hot spots may be the principal way that thermal energy escapes from the planet's interior. Circular fractures (called "coronae") on the surface of Venus, ranging from a few hundred kilometers to more than 1,000 km across, may be the result of upwelling plumes of hot mantle that have fractured Venus's lithosphere.

Some planetary scientists have suggested that a radically different form of tectonism may be at work on Venus. They believe that hot spots are not adequate to allow the thermal energy generated within the planet to escape, and that as a result energy may continue to build up in the interior until large chunks of the lithosphere melt and overturn. This melting and overturning would suddenly release an enormous amount of energy, after which the surface of the planet would cool and resolidify. This idea remains highly controversial, but it could help explain the relatively young surface of Venus. It also drives home the point of how different the geological histories of the various planets appear to be.

Why Venus and Earth should have such different styles of tectonism remains an unsolved puzzle. The segmentation of Earth's lithosphere into moving plates seems to be unique among the planets and moons of the Solar System.

> **Earth's tectonic plates are unique in the Solar System.**

7.6 Igneous Activity: A Sign of a Geologically Active Planet

Like tectonism, igneous activity is an important process that shapes planetary surfaces. On Earth, violent igneous events—including volcanic eruptions and some earthquakes—can also result in widespread human casualties. In December 2004 a strong earthquake off the coast of Sumatra triggered an enormous tsunami, which swept across the Indian Ocean. More than 225,000 people died in this disaster, one of the deadliest in modern history. Like earthquakes, volcanic eruptions throughout history have created natural disasters on a grand scale. The eruption of Mount Vesuvius in A.D. 79 that buried Pompeii; the 1883 explosion of Krakatoa in the western Pacific that led to the loss of 36,000 lives; the hot ash flows from Mount Pelée that demolished the Caribbean port of Saint-Pierre in 1902, killing all but two of its 30,000 inhabitants—these are only a few of a long list of examples of death and destruction brought on by volcanic eruptions.

Terrestrial Volcanism Is Related to Tectonism

How do volcanoes form, and why are they found in some regions and not in others? Is there evidence of volcanoes on other planets? To answer these questions we must first understand how and where magma—the main component of volcanism—originates.

Magma does not come to Earth's surface from its molten core, as is sometimes believed. We know from seismic signals that magma originates in the lower crust and upper mantle, where sources of thermal energy combine. These thermal-energy sources include rising convection cells in the mantle, frictional heating generated by movement in

the lithosphere and between tectonic plates, and concentrations of radioactive elements that produce energy from radioactive decay.

Because the thermal-energy sources are not uniformly distributed inside our planet, volcanoes tend to be located only in specific areas, most notably (but not exclusively) over hot spots and along plate boundaries. Maps of geological activity, such as the one shown in Figure 7.21, leave little doubt that most terrestrial volcanism is ultimately linked to the same forces responsible for plate motions. There is a tremendous amount of friction as plates slide under each other at a subduction zone. This friction generates a great deal of thermal energy, raising the temperature and pushing rock toward its melting point.

> **Friction between moving plates generates thermal energy and leads to volcanism.**

When we studied Earth's core, we found the counterintuitive result that even though the inner core of the planet is hotter than its surroundings, it is *solid* while its surroundings are liquid. The reason is that the inner core is under more pressure than its surroundings, and this higher pressure forces the material to stay solid. A similar effect occurs near Earth's surface. Material at the base of a lithospheric plate is under a great deal of pressure because of the weight of the plate pushing down on it. This pressure drives up the melting point of the material, forcing it to remain solid even though its temperature is above its normal melting point on Earth's surface. But as this material is forced up through the crust, its pressure drops; and as its pressure drops, so does its melting point. Because of this declining pressure, material that started out solid at the base of a plate may become molten as it nears the surface.

An obvious place to look for volcanic activity is along spreading centers, where convection carries hot mantle material toward the surface. As mentioned in the previous paragraph, the decrease in pressure as the material nears the surface may allow it to become molten. Spreading centers are indeed found to be frequent sites of eruptions. Iceland, which is one of the most volcanically active regions in the world, sits astride one such spreading center—the Mid-Atlantic Ridge (see Figures 7.18 and 7.21).

Once lava reaches the surface of Earth, it can form many types of structures (**Figure 7.26**). Flows from spreading centers often form vast sheets, especially if the eruptions come from long fractures called **fissures**. If very fluid lava flows from a single "point source," it can spread out over the surrounding terrain or ocean floor, forming what is known as a **shield volcano** (**Figure 7.26a**), so named because it resembles an upside-down warrior's shield. Viscous, pasty lava flows, alternating with pyroclastic (explosively generated) rock deposits, can form a steep-sided structure called a **composite volcano** (**Figure 7.26b**) or smaller circular mounds called "volcanic domes."

A third setting for terrestrial volcanism is found where convective plumes rise toward the surface in the interi-

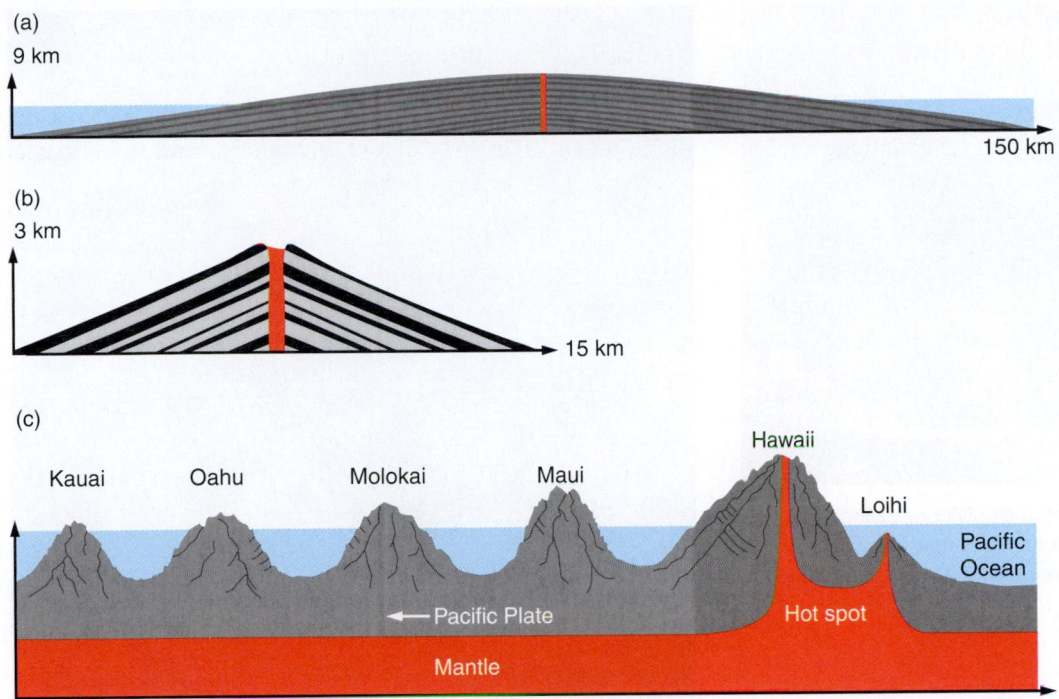

FIGURE 7.26 Magma reaching Earth's surface commonly forms (a) shield volcanoes, such as Mauna Loa, which have gently sloped sides built up by fluid lava flows; and (b) composite volcanoes, such as Vesuvius, which have steeply symmetrical sides built up by viscous lava flows. (c) Hot spots are convective plumes of lava that form a successive series of volcanoes as the plate above them slides by.

ors of lithospheric plates, creating local hot spots (**Figure 7.26c**). Volcanism over hot spots works much like volcanism at a spreading center, except that the convective upwelling occurs at a single spot rather than along the length of a spreading center (see Figure 7.20). These hot spots can melt mantle and lithospheric material and force it toward the surface of Earth. ▶❚❚ **AstroTour: Hot Spot Creating a Chain of Islands**

There are numerous hot spots on Earth, including the region around Yellowstone Park and the Hawaiian Islands. The Hawaiian Islands are a chain of shield volcanoes that formed as the lithospheric plate on which they ride was dragged across a relatively stationary hot spot. Volcanoes erupt over a hot spot, building an island. The island ceases to grow as the plate motion carries the island away from the hot spot, which is the source for the volcanic activity. The slower process of erosion, going on since the island's inception, continues to wear the island away. In the meantime, a new island grows over the hot spot. Today the Hawaiian hot spot is located off the southeast coast of the big island of Hawaii, where it continues to power the active volcanoes. On top of the hot spot, the newest Hawaiian volcano is already forming. This volcano, called Loihi, remains submerged under the surface of the Pacific Ocean. Viewed another way, however, Loihi is already a massive shield volcano, rising more than 3 km above the ocean floor. Geologists expect that it will eventually break the surface of the ocean and merge with the big island of Hawaii—but not anytime soon. Loihi is not expected to show itself above sea level for perhaps another 100,000 years!

Volcanism Also Occurs Elsewhere in the Solar System

Although Earth is the only planet on which plate tectonics is an important process, evidence of volcanism is found throughout the Solar System, including several moons of the outer planets. Even before the *Apollo* astronauts set foot on the lunar surface, photographs showed it to have what appeared to be flowlike features in the dark regions. Some of the first observers to use telescopes thought that these dark areas looked like seas—thus the name **maria** (singular: *mare*), Latin for "seas." Their appearance suggested to planetary scientists that these are vast lava flows similar to basalts on Earth. Because the maria contain relatively few craters, we know that these volcanic flows occurred after the period of heavy bombardment ceased.

> The dark areas on the Moon's surface are ancient lava flows.

When the *Apollo* astronauts returned rock samples from the lunar maria, the rocks were indeed found to be basalts. Many of these Moon rocks contained gas bubbles typical of volcanic materials (**Figure 7.27**). Experiments show that when this lava flowed across the lunar surface, it must have been extremely fluid—something like the consistency of motor oil at room temperature. The fluidity of the lava, due partly to its iron- and titanium-rich chemical composition, explains why lunar basalts form vast sheets that fill low-lying areas such as impact basins. It also explains the Moon's lack of classic volcanoes like Mount Rainier: motor oil poured from a can does not pile up; it spreads out (**Figure 7.28**).

FIGURE 7.27 This rock sample from the Moon, collected by the *Apollo 15* astronauts from a lunar lava flow, shows gas bubbles typical of gas-rich volcanic materials. This rock is about 6 by 12 cm.

FIGURE 7.28 The lava flowing across the surface of Mare Imbrium on the Moon must have been extremely fluid to spread out for hundreds of kilometers in sheets that are only tens of meters thick.

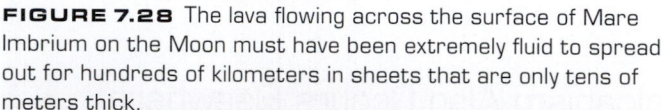

Edges of lava flow

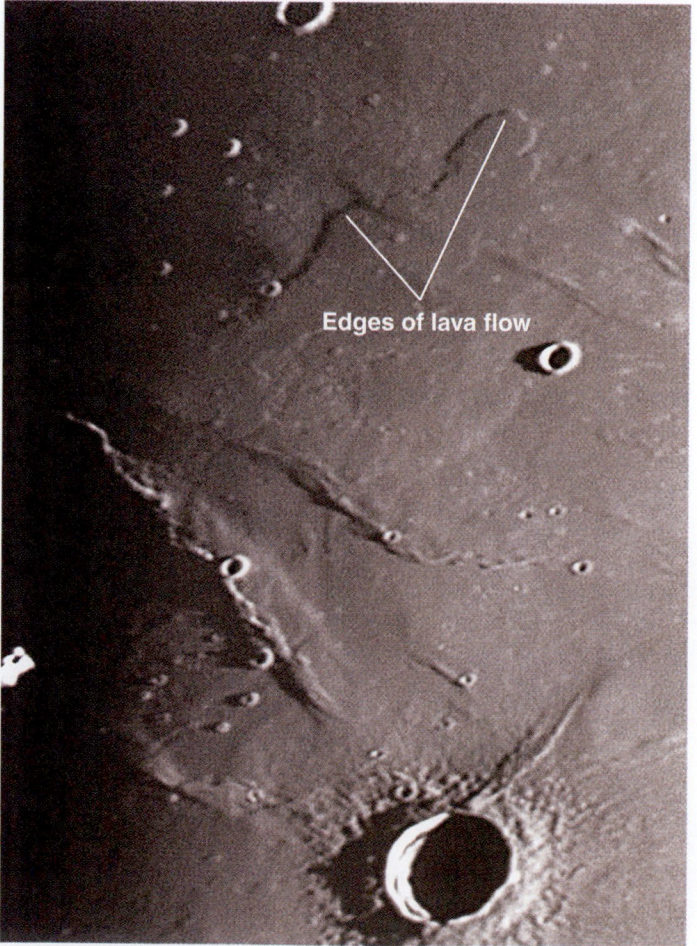

FIGURE 7.29 The largest volcano yet seen on Mercury (right of center at the top) lies on the edge of the Caloris Basin. The volcano is about 50 km across.

The samples also showed that most of the lunar lava flows are older than 3 billion years! Only in a few limited areas of the Moon are younger lavas thought to exist; most of these have not been sampled directly. Samples from the heavily cratered terrain of the Moon also originated from magma, which shows that the young Moon went through a molten stage. These rocks cooled from a "magma ocean" and are more than 4 billion years old, preserving the early history of the Solar System. Thus, most of the sources of heating and volcanic activity on the Moon must have shut down some 3 billion years ago—unlike on Earth, where volcanism continues. This conclusion is certainly consistent with our earlier argument that smaller planets should cool more efficiently and thus be less active than larger planets.

Like the Moon, Mercury also shows evidence for past volcanism. *Mariner 10* and *Messenger* missions have revealed smooth plains similar in appearance to the lunar maria. These sparsely cratered plains are the youngest areas on Mercury and, like those on the Moon, are almost certainly volcanic in origin, where fluid lavas flowed into and filled huge impact basins. High-resolution imaging by *Messenger* has also identified a number of volcanoes (**Figure 7.29**).

More than half the surface of Mars is covered with volcanic rocks. Plain-forming lavas covered huge regions of Mars, flooding the older, cratered terrain. Few of the vents or fissures for these flows are visible, suggesting

FIGURE 7.30 A Mars crater flooded by lava is seen in this perspective view taken by ESA's *Mars Express* spacecraft.

FIGURE 7.31 The largest known volcano in the Solar System, Mars's Olympus Mons is a 27-km-high shield-type volcano, similar to but much larger than Hawaii's Mauna Loa. (a) A partial view of Olympus Mons taken by the *Mars Global Surveyor*. (b) This oblique view was created from an overhead *Viking* image and topographic data provided by the *Mars Orbiter* laser altimeter.

(a)

(b)

that most were buried under the lava that poured forth from them (**Figure 7.30**). Among the most impressive features on Mars are the enormous shield volcanoes. These volcanoes are the largest mountains in the Solar System. Olympus Mons, standing 27 km high at its peak and 550 km wide at its base (**Figure 7.31**), would tower over Mount Everest and dwarf Hawaii's Mauna Loa. Standing a "mere" 9 km above the floor of the Pacific Ocean and spreading out to cover an area 120 km across, Mauna Loa is the largest mountain on Earth.

Despite the difference in size, most of the very large volcanoes of Mars are shield volcanoes, just like their Hawaiian cousins. Olympus Mons and its neighbors grew as the result of hundreds of thousands of individual eruptions that sent lava flows running down their flanks. The difference in size between Olympus Mons and Mauna Loa could be a result of the absence of plate tectonics on Mars. As discussed earlier, the motion of the plate that Hawaii rides on carries the Hawaiian volcanoes away from their hot spot after only a few million years. The martian volcanoes, on the other hand, have remained over their respective hot spots for billions of years, growing ever taller and broader with each successive eruption.

Although no samples have been returned directly from Mars, analysis by instruments on landed spacecraft, remote sensing data, and the shapes of the lava flows and volcanoes all suggest that martian lavas are basalts much like those found on Earth and the Moon. However, chemical analyses at the *Pathfinder* lander site suggest that the rocks contain slightly more silica than typical basalts contain, which could mean that the magma was partly differentiated chemically before it erupted. On the other hand, chemical analyses obtained in Gusev Crater by the Mars exploration rover *Spirit* reveal basalts that are rich in iron, more like those of the Moon. In Chapter 12 we will discuss evidence that certain meteorites found on Earth were probably blasted from the surface of Mars by impacts. Chemical analysis of these meteorites supports the view that volcanism on Mars involved basaltic lavas.

Of all the terrestrial planets, Venus has the largest population of volcanoes. Radar images reveal a wide variety of volcanic landforms. These include highly fluid flood lavas covering thousands of square kilometers, shield volcanoes approaching those of Mars in size and complexity, dome volcanoes, and lava channels thousands of kilometers long. These lavas must have been extremely hot and fluid to flow for such long distances. Some of the volcanic eruptions on Venus are thought to have been associated with deformation of Venus's lithosphere above hot spots such as the circular fractures mentioned earlier.

> Venus has the largest population of volcanoes among the terrestrial planets.

Although we know little about the composition of the volcanic rocks on Venus, the Soviet *Venera* landers did measure some surface compositions; for the most part, the results suggest that lavas on Venus are basalts, much like the lavas on Earth, the Moon, and Mars. It is presumed that the lavas on Mercury are basalts as well.

In summary, what can we say about the volcanic histories of the terrestrial planets and the Moon? After going through a molten state—a sort of "magma ocean" phase—shortly after its birth, the Moon developed an ever-thickening lithosphere overlying a mantle. Pockets of radioactive materials in the Moon's interior generated local reservoirs of magma. Some large impacts penetrated these reservoirs or otherwise triggered the release of magma to the surface through fractures. Most of this volcanism ceased about 3 billion years ago, although some minor eruptions could have continued sporadically for another billion years. In all, less than 18 percent of the lunar surface is covered with volcanic rocks (excluding those cooled from the "magma ocean"). Volcanism on Mercury probably mimicked that of the Moon. Many of the volcanic plains on Mercury are also associated with impact scars. The ages of these plains are not known, but from superimposed impact craters we can conclude that the plains are probably billions of years old.

Lava flows and other volcanic landforms span nearly the entire history of Mars, estimated to extend from the formation of crust some 4.4 billion years ago to geologically recent times, and to cover more than half of the red planet's surface. But remember, "recent" in this sense could still be more than 100 million years ago, or back to our own dinosaur age. Although some "fresh-appearing" lava flows were identified on Mars, until rock samples are radiometrically dated, we will not know the age of these latest eruptions. Mars could, in principle, experience eruptions today.

Most of Venus is covered with volcanic materials or tectonically disrupted rocks of presumed volcanic origin. A geological timescale for Venus has not yet been devised, but from its relative lack of impact craters, most of the surface is considered to be less than 1 billion years old. When volcanism began on Venus and whether volcanoes are still active today remain unanswered questions.

Earth remains the champion for the diversity of volcanism. Volcanic rocks are found throughout the rock record, while compositions of magma span the spectrum of silica-rich to super-iron-rich materials erupted directly from the mantle.

7.7 Erosion: Wearing Down the High Spots and Filling In the Low

Erosion is the great leveler of planetary surfaces. The term *erosion* covers a wide variety of processes that together serve to smooth out planetary terrain, wearing down the high spots and filling in the low. The first step in the process of erosion is called "weathering," in which rocks are broken into smaller pieces and may be chemically altered. For example, rocks on Earth are physically weathered along shorelines, where they are broken into beach sand by pounding waves; and along streambeds, where they are slammed together. Other weathering processes include chemical reactions, such as the combining of oxygen in the air with iron in rocks to form a type of rust. One of the most efficient forms of weathering involves freeze-thaw cycles, during which liquid water runs into crevices and then freezes, expanding and shattering the rock.

After weathering occurs, the resulting debris can be carried away by flowing water, glacial ice, or blowing wind and deposited in other areas as sediment. Where material is eroded, we can see features such as river valleys, wind-sculpted hills, or mountains carved by glaciers. Where eroded material is deposited, we see features such as river deltas, sand dunes, or piles of rock at the bases of mountains and cliffs. It is not surprising to find that erosion is most

> The actions of water and wind produce the greatest amount of erosion on Earth.

FIGURE 7.32 A view of the surface of Mars taken by the *Phoenix* lander showing the rock-littered surface of the northern plains. Fine-grained material between the rocks and pebbles is wind-transported sand and dust.

FIGURE 7.33 This high-resolution image taken by a camera aboard the *Mars Global Surveyor* shows windblown, frost-covered sand dunes in the north polar region of Mars.

efficient on planets where water and wind are present. On Earth, where water and wind are so dominant, most impact craters are worn down and filled in even before they are turned under by tectonic activity. If other processes were not at work to form mountains, valleys, and other topographic relief, erosion would eventually wear planets like Earth as smooth as billiard balls.

Even on the Moon and Mercury, which have no atmospheres or running water, a type of erosion (albeit *very* slow) is still at work. Radiation from the Sun and from deep space slowly decomposes some types of minerals, effectively weathering the rock. Such effects are only "skin deep"—usually a few millimeters at most—and could be considered a kind of rock "sunburn." Impacts of micrometeoroids also chip away at rocks. In addition, landslides can occur wherever gravity and differences in elevation are present. Although landslide activity is enhanced by the lubricating effects of water, landslides are also seen on dry planets like Mercury and the Moon. Debris from landslides has even been seen on the tiny moons of Mars and on asteroids (see Chapter 12).

Earth, Mars, and Venus, on the other hand, do have atmospheres, and all three planets show the effects of windstorms. Images of Mars (**Figure 7.32**) and Venus returned by spacecraft landers show surfaces that have clearly been subjected to the forces of wind. Likewise, orbiting spacecraft have returned pictures showing sand dunes, wind-eroded hills, and surface patterns called "wind streaks." Planet-encompassing dust

> The surfaces of Mars and Venus are also modified by winds.

storms have been seen on Mars. These storms have been known to blot out the visibility of the surface of the planet for months on end.

Sand dunes are common on Earth and Mars, and some have been identified on Venus. They occur frequently wherever moderately strong winds blow and there is a supply of loose grains. The largest field of windblown sand on Mars (**Figure 7.33**) is in the north polar area. Covering more than 700,000 square kilometers (km^2), it is comparable in size to the largest fields of sand dunes on Earth. The most common wind-related features on Mars and Venus are wind streaks (**Figure 7.34**). These surface patterns appear, disappear, and change in response to winds blowing sediments around hills, craters, and cliffs. They serve as local "wind vanes," telling planetary scientists about the direction of local prevailing surface winds.

Today Earth is the only planet where the temperature and atmospheric conditions allow extensive liquid surface water to exist. Water is an extremely powerful agent of erosion and dominates erosion on Earth. Every year, rivers and streams on Earth deliver about 10 billion metric tons of sediment into the oceans. Even though today there is no liquid water on the surface of Mars, at one time water flowed across its surface in such vast quantities as to make the Amazon River look like a backyard irrigation ditch. Huge, dry riverbeds such as those shown in **Figure 7.35** attest to tremendous floods that poured across the martian surface. In addition, many regions on Mars show small networks of valleys that are thought to have been carved by flowing water. Some parts of Mars may even have once contained oceans and glaciers.

The search for water is a primary quest in the exploration of Mars. In 2004, NASA sent two instrument-equipped roving vehicles, *Opportunity* and *Spirit*, to search for evidence

(a)

(b)

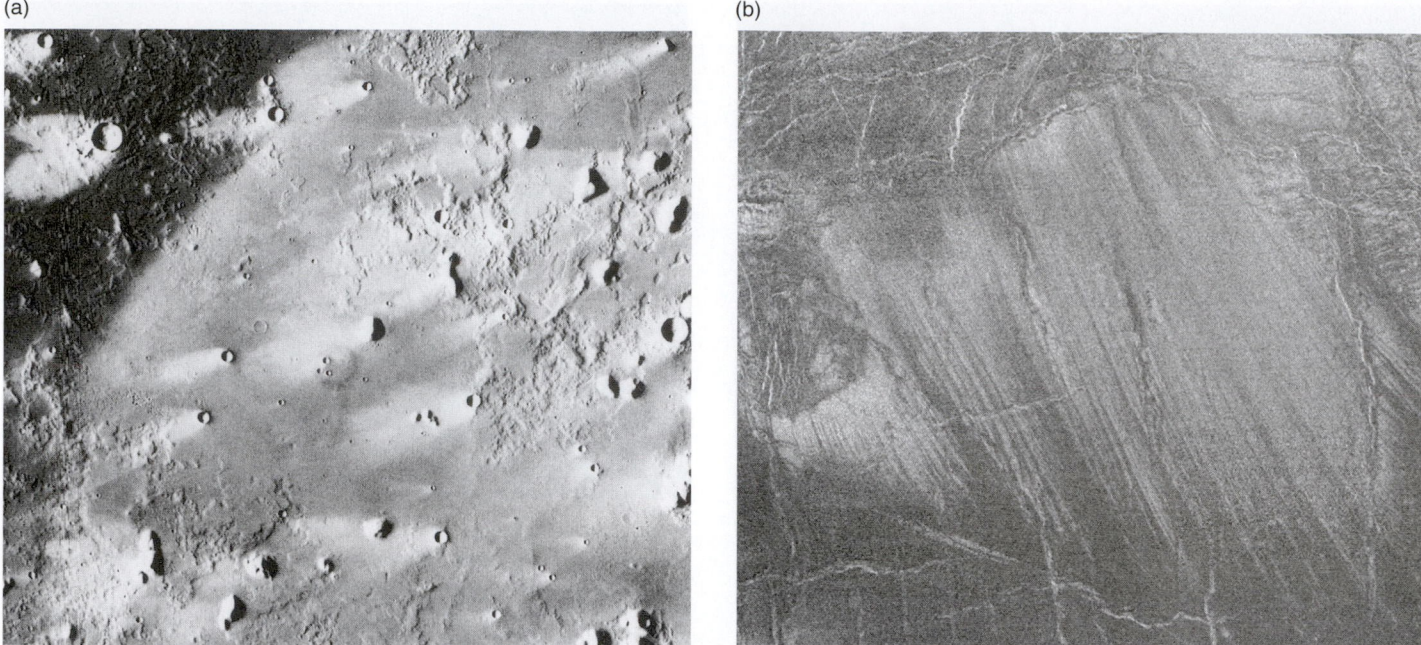

FIGURE 7.34 (a) The bright patterns associated with these craters and hills on Mars resulted from winds that redistribute sand and dust on the surface. The winds forming these streaks blew from right to left. The area shown is about 160 by 185 km. (b) Bright patches of windblown dust streak across the surface of Venus.

FIGURE 7.35 *Viking Orbiter* images showing channels on Mars carved long ago by flowing water. Liquid water cannot exist on the surface of Mars today. The area shown is about 100 km wide. Inset: High-resolution *Mars Global Surveyor* image showing geologically recent "gullies" that may also have been carved by water.

Water channels

of water on Mars (see the opening photograph of Chapter 5). The choice for *Opportunity*'s landing site was based on orbital remote sensing data that suggested the presence of hematite in an area near the martian equator. "Hematite" is an iron-rich mineral that commonly forms in the presence of water, and the goal for this mission was to test the validity of the remote sensing information. Both Mars rovers were surrounded by inflatable airbags to cushion their impact on the martian surface. Upon landing, the whole *Opportunity* package bounced across the surface, finally coming to rest inside a small crater like a well-driven golf ball (**Figure 7.36a**). The real payoff came when the walls of the crater revealed a treasure trove of information. For the first time, outcrops of martian rocks were observed and available for study. "Outcrops" are places were exposed rocks remain in the original order in which they were laid down. Previously, the only rocks that landers and rovers had come across were those that had been dislodged from their original settings by either impacts or river floods.

The layered rocks at the *Opportunity* site (**Figure 7.36b**) revealed several lines of evidence indicating that they had once been soaked in or transported by water. First, the form of the layers was typical of what we see here on Earth in sandy sediments that have been laid down by gentle currents of water.[8] Then, rover instruments detected the presence of a mineral commonly found in acid lakes and thermal springs, with so much sulfur content that it had almost certainly formed by precipitation from water. Finally, magnified images of the rocks showed "blueberries," small spherical grains a few millimeters across that appeared to have formed in place among the layered rocks in a manner similar to terrestrial features that form by the percolation of water through sediments. Analysis of the "blueberries" revealed abundant hematite, confirming the interpretations of the remote sensing data. Observations by the European Space Agency's *Mars Express* orbiter and NASA's *Mars Odyssey* show the hematite signature and the presence of sulfur-rich compounds in a vast area surrounding the *Opportunity* landing site. These observations suggest the existence of an ancient martian sea larger than the combined area of the Great Lakes and as much as 500 meters deep.

The floor of Gusev, a 170-km impact crater, was chosen for *Spirit* because it showed signs of ancient flooding by a now dry river, and it was hoped that surface deposits would provide evidence for that past liquid water. Surprise, and perhaps some disappointment, followed when *Spirit* revealed that the flat floor of Gusev consisted primarily of basaltic rock. Only when the rover ventured cross-country to some low hills located 2.5 km from the landing site did it find evidence of water in the martian past. Here the basaltic rocks showed clear signs of chemical alteration by liquid

[8] Some geologists believe that this layering was caused by deposits of volcanic ash rather than aqueous sediments. We find here an example of honest disagreement among scientists, all of whom are looking at the same data.

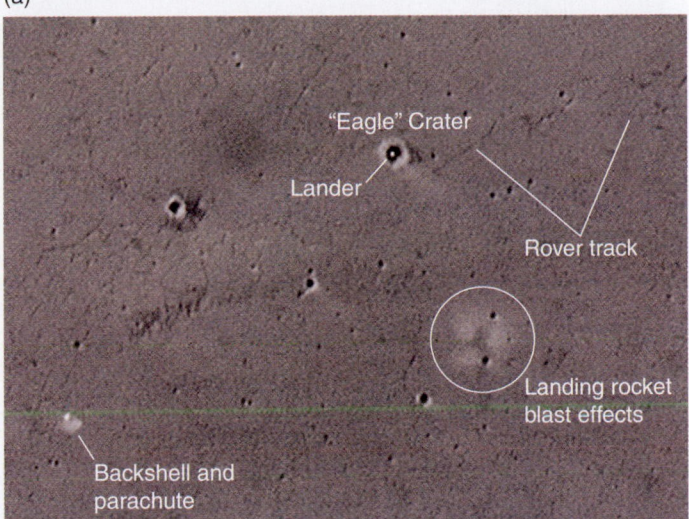

(a)

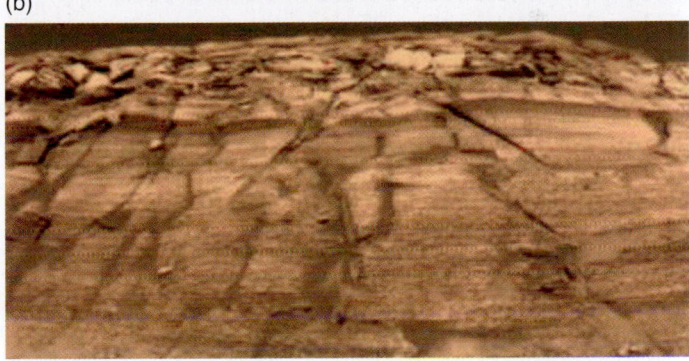

(b)

FIGURE 7.36 (a) An orbital view of *Opportunity*'s landing site. After landing, the protected package bounced along the surface and ended up in the "Eagle" crater—a landing that mission scientists called a "hole in one." The white dot in the middle of the crater is the lander. Tracks leading out of the crater were made by the Mars rover *Opportunity* as it began its exploration of Mars. (b) Layered rock seen near the *Opportunity* landing site may represent sediments deposited by water.

water. As of this writing, *Opportunity* and *Spirit* continue to explore the martian surface.

Where is the water and ice on Mars today? At least some ice is locked in the polar regions, just as the ice caps on Earth hold much of our planet's water. But, unlike our own polar caps, those on Mars are a mixture of frozen carbon dioxide and frozen water, so water must be hiding elsewhere on Mars. Small amounts of water can be found on the surface, as seen in **Figure 7.37**, and NASA's *Phoenix* lander has found water ice just a centimeter or so beneath surface soils at high northern latitudes (**Figure 7.38**). However, much if not most of the water on Mars appears to

> **Water ice could exist on the Moon and Mercury as it does on Mars.**

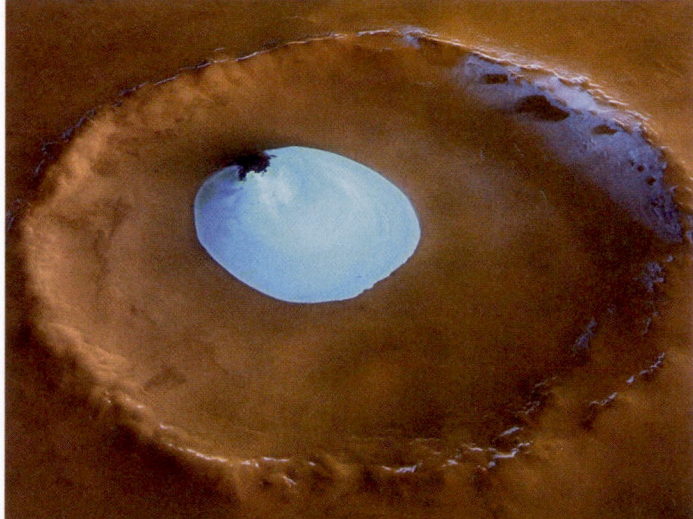

FIGURE 7.37 This image, taken by ESA's *Mars Express* mission, shows water ice in a crater in the north polar region of Mars. The crater is about 35 km in diameter, and the ice is estimated to be about 200 meters thick. The 300-meter-high crater wall blocks much of the sunlight at this high latitude, keeping the ice from vaporizing.

FIGURE 7.38 Water ice appears a few centimeters below the surface in this trench dug by a robotic arm on the *Phoenix* lander. The trench measures about 20 by 30 cm.

be trapped well below the surface of the planet. Recent radar imaging by NASA's *Mars Reconnaissance Orbiter* indicates huge quantities of subsurface water ice, not only in the polar areas as expected but also at lower latitudes (**Figure 7.39**).

It is also intriguing that although Earth and Mars are the only terrestrial planets that show evidence for liquid water at any time in their histories, water ice could exist on Mercury and the Moon today. Some deep craters in the polar regions of both Mercury and the Moon have floors that are in perpetual shadow. Because these planets lack atmospheres, temperatures in these permanently shadowed areas remain below 180 K. For many years, planetary scientists speculated that ice—perhaps implanted by impact comets—could be found in these craters. In the early 1990s, Earth-based radar measurements of Mercury's north pole and infrared measurements of the Moon's polar areas by the joint Department of Defense/NASA *Clementine* mission returned information that seemed to support this possibility. In 1998, NASA sent another spacecraft, *Lunar Prospector*, to take a new look. Again the results suggested subsurface water ice in the polar regions. When its primary mission was completed, NASA crashed *Lunar Prospector* into a crater near the Moon's south pole while groundbased telescopes searched for evidence of water vapor above the impact site. None was seen. Perhaps it was the wrong place to look or the amount of water was too small to be detected. In 2009, NASA's *Lunar Reconnaissance Orbiter* was put into lunar orbit to continue the search for possible sources of subsurface water ice.

FIGURE 7.39 Radar observations by NASA's *Mars Reconnaissance Orbiter* indicate huge quantities of water ice underlying these mid-latitude martian craters.

Although these observations do not prove one way or another whether ice exists in these regions, the possibility is exciting and important. The existence of ice on the Moon could make any future lunar colonization much more practical than if all water had to be either brought up from Earth or synthesized from hydrogen and oxygen extracted from the lunar soil.

Seeing the Forest for the Trees

In Chapter 6 we placed the formation of the Solar System and our Earth into the broader context of the ongoing process of star formation that we see occurring around us in the universe today. Earth is but one of several rocky planets that coalesced from the dust surviving in the hot inner parts of the disk that surrounded the young Sun. Mercury, Venus, Earth, and Mars share this common heritage and have been further shaped by the same fundamental processes of impact cratering, tectonism, volcanism, and erosion over the ensuing 4.5 billion years.

Earth is our benchmark for understanding the nature and effects of these processes. The cosmological principle not only suggests that we can apply physical laws discovered in terrestrial laboratories throughout the universe; it also guides us in our exploration of the Solar System. We live in a remarkable age of exploration and discovery, when our species has launched probes to visit almost all of the major bodies in the Solar System and many of the minor bodies as well. The resulting flood of information is a testament to 20th and 21st century technology; but it is our understanding of Earth that has guided scientists as they have pieced together the story told by these data.

The impression we have gleaned from the results of the last five decades of planetary exploration is one of remarkable diversity. The surfaces of the four terrestrial planets and Earth's moon have all been stressed and fractured over the eons. One of the most startling discoveries of the last century was the fact that the outer shell of Earth itself consists of multiple plates that are in constant motion as they ride slow but inexorable convection currents within Earth's mantle. Exploration of the inner Solar System has provided new perspectives on the drifting motion of our planet's lithosphere by showing us that other possibilities exist. Whereas Mercury's surface is cracked and wrinkled as a result of compression, tectonic activity on Mars fractured the planet, forming a vast canyon system. Even our sister planet, Venus, shows hints of a particularly violent form of tectonism in which the entire surface of the planet may have overturned and released bursts of pent-up thermal energy from the planet's interior. Likewise, igneous activity of several kinds has played an important role in the history of our planet, building such monuments to the power of volcanism as the Hawaiian Islands and the Cascade Range. Yet when looking at our neighbors, we find that Earth's volcanoes shrink into relative insignificance beside the towering heights of Olympus Mons on Mars.

Along the way we have again seen the process of science at work. The realm of our species encompasses only the tiniest fraction of our planet. As alluded to in Andy Rooney's tongue-in-cheek quote that opened the chapter, we are creatures of the *surface* of Earth, as far from direct exploration of the inner reaches of our own world as we are from direct exploration of the heart of the Sun. Even so, on this leg of our journey we have seen how the methods of science have enabled us to construct a detailed picture of the interior of Earth. We have built up that picture by applying physical laws to construct models of Earth and then tested the predictions of those models by analyzing the echoes from earthquakes. At first glance this might seem an insecure foundation for knowledge, but that is far from the case. We go through our lives building knowledge of the world around us through light waves that reach our eyes and sound waves that reach our ears, both interpreted by brains shaped by evolution to accomplish the task. Is it so different to say that we know about the interior of Earth from the seismic waves that reach our instruments, interpreted through computer models that are shaped for the job by our understanding of physical law? So it will be throughout the rest of our journey as we use the techniques and tools of science (and a carefully thought-out notion of what it means "to know") to "sense" the nature of the universe.

We have seen several ways in which the study of terrestrial planets has forced us to change the way we view our planet. Earth is but one of several rocky planets, and our world is only one small corner of the range of the possible. Yet no result of Solar System exploration has more dramatically changed our perspective on our own history—and perhaps our own future—than our understanding of the role of impact cratering in the inner Solar System. The active Earth is very effective at erasing its own memory. Plate tectonics, igneous activity, and erosion are continually at work, wiping the record of our history clean. It took the ancient surfaces of Mercury and the Moon, and even the less ancient surfaces of Mars and Venus, to show us that impacts of objects from space have played a significant role in shaping Earth. With this realization, the cosmological principle truly comes full circle.

Earth may provide many of the clues that we need to understand the Solar System, but it is the Solar System that provides the immediate context for the existence of Earth. Life on our planet has had its course altered time and again by sudden and cataclysmic events when asteroids and comets have slammed into Earth. It seems very likely that we owe our existence to the luck of our remote ancestors—small rodentlike mammals—that could live amid the destruction following such an impact 65 million years ago. The time since that event is but a brief moment in the history of our planet. The time until the next such event will almost certainly be no more than another short span of geological time.

Summary

- Comparative planetology is the key to understanding the planets.

- Impacts, volcanism, and tectonism create topography on the terrestrial planets.

- Relative concentrations of impact craters divulge the ages of planetary surfaces.

- Radiometric dating tells us the ages of rocks.

- Seismic waves reveal Earth's interior structure.

- The Moon was created when a Mars-sized protoplanet collided with Earth.

- Earth has a strong magnetic field, but Venus and Mars do not.

- Smooth areas on the Moon and Mercury are ancient lava flows.

- The largest mountains in the Solar System are volcanoes on Mars.

- Erosion wears down the surfaces of Venus and Mars by wind and the surface of Earth by wind and water.

- Mars today has large amounts of subsurface water ice and once had liquid water on its surface.

SmartWork, Norton's online homework system, includes algorithmically generated versions of these questions, plus additional conceptual exercises. If your instructor assigns questions in SmartWork, log in at **smartwork.wwnorton.com.**

Student Questions

THINKING ABOUT THE CONCEPTS

1. Name the four terrestrial planets.

2. In discussing the terrestrial planets, why do we include our Moon?

*3. What do we mean by *comparative planetology*, and why is it important?

4. Name the four forces that have shaped Earth.

5. One region on the Moon is covered with craters, while another is a smooth volcanic plain. Which is older? How do we know?

*6. Can all rocks be dated with radiometric methods? Explain.

7. Explain how we know that rocks at the bottom of Arizona's Grand Canyon are older than those found on the rim.

8. A current theory suggests that a mass extinction occurred as a consequence of an enormous impact on Earth 65 million years ago. What is the evidence for this theory?

*9. Describe the sources of heating that are responsible for the generation of Earth's magma.

10. Explain why the Moon's core is cooler than Earth's.

11. Name the three components that make up Earth's interior.

12. What is meant by *differentiation* of a planet, and what causes it?

*13. Explain the difference between longitudinal waves and transverse waves.

14. How do we know that Earth's core includes a liquid zone?

15. Explain what is meant by *hydrostatic equilibrium*.

16. Compare and contrast tectonism on Venus, Earth, and Mercury.

17. Explain plate tectonics and identify the only planet on which this process has been observed.

18. Volcanoes have been found on all of the terrestrial planets. Where are the largest volcanoes in the inner Solar System?

*19. Describe and explain the criteria you would apply to images with adequate resolution in order to distinguish between a crater formed by an impact and one formed by a volcanic eruption.

20. What are the differences between a spreading center and a subduction zone?

21. Describe and explain the evidence for reversals in the polarity of Earth's magnetic field.

22. What is meant by *erosion*? What processes contribute to erosion?

23. What are the primary reasons that the surfaces of Venus, Earth, and Mars are younger than those of Mercury and the Moon?

24. Explain some of the evidence that Mars once had liquid water on its surface.

25. Why do we not find liquid water on Mars today?

APPLYING THE CONCEPTS

26. Assume that Earth and Mars are perfect spheres with radii of 6,371 km and 3,390 km, respectively.
    a. Calculate the surface area of Earth.
    b. Calculate the surface area of Mars.
    c. If 0.72 (72 percent) of Earth's surface is covered with water, compare the amount of Earth's land area with the total surface area of Mars.

27. Compare the kinetic energy of a 1-gram piece of ice (about half the mass of a dime) entering Earth's atmosphere at a speed of 50 km/s with that of a 2–metric ton SUV (mass = 2×10^3 kg) speeding down the highway at 90 km/h.

28. The object that created Arizona's Meteor Crater was estimated to have a diameter of 50 meters and a mass of 300 million kg. Calculate the density of the impacting object, and explain what that may tell you about its composition.

29. Using the information in Table 7.1, determine the relative rates of internal energy loss experienced by Earth and the Moon.

30. Although oceans cover 72 percent of Earth's surface, they represent but a tiny fraction of our planet's mass. Earth's mass is 6.0×10^{24} kg, and its oceans have a total mass of 1.5×10^{21} kg.
    a. What fraction of Earth's total mass do our oceans represent?
    b. Does surface and atmospheric water represent Earth's total aqueous inventory? Explain.

31. Earth's mean radius is 6,371 km, and its mass is 6.0×10^{24} kg.
    a. Calculate Earth's average density. Show your work. Do not look this value up.
    b. The average density of Earth's crust is 2,600 kg/m³. What does this tell you about Earth's interior?

*32. Suppose you find a piece of ancient pottery and take it to the laboratory of a physicist friend. He finds that the glaze contains radium, a radioactive element that decays to radon and has a half-life of 1,620 years. He tells you that there could not be any radon in the glaze when the pottery was being fired, but that it now contains three atoms of radon for each atom of radium. How old is the pottery?

33. Assume that the east coast of South America and the west coast of Africa are separated by an average distance of 4,500 km. Assume also that GPS measurements indicate that these continents are now moving apart at a rate of 3.75 cm/year. If this rate has been constant over geological time, how long ago were these two continents joined together as part of a supercontinent?

*34. Shield volcanoes are shaped something like flattened cones. The volume of a cone is equal to the area of its base multiplied by one-third of its height. The largest volcano on Mars, Olympus Mons, is 27 km high and has a base diameter of 550 km. Compare its volume with that of Earth's largest volcano, Mauna Loa, which is 9 km high and has a base diameter of 120 km.

*35. Earth's core represents about one-third of Earth's total mass. Assume that the diameter of Earth's core is 20 percent of Earth's diameter. What fraction of Earth's total volume does the core represent, and what does that tell you about the composition of Earth's core?

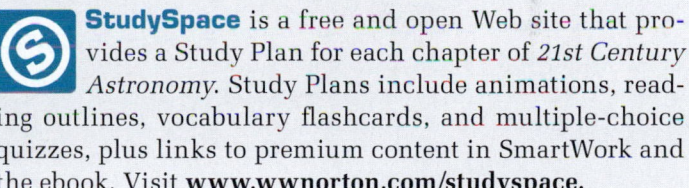 **StudySpace** is a free and open Web site that provides a Study Plan for each chapter of *21st Century Astronomy*. Study Plans include animations, reading outlines, vocabulary flashcards, and multiple-choice quizzes, plus links to premium content in SmartWork and the ebook. Visit **www.wwnorton.com/studyspace.**

Woe to you wicked spirits! hope not
Ever to see the sky again. I come
To take you to the other shore across,
Into eternal darkness, there to dwell
In fierce heat and in ice.

DANTE ALIGHIERI (1265–1321), *THE DIVINE COMEDY: INFERNO*, CANTO III

Lightning strikes the ground near Kitt Peak National Observatory in Arizona.

CHAPTER 8

Atmospheres of the Terrestrial Planets

8.1 Atmospheres Are Oceans of Air

Earth's atmosphere surrounds us like an ocean of air. We see it in the blueness of the sky and feel it in the breezes that tousle our hair. People who live in large cities can sometimes even smell it. It can bring joy into our lives in the form of a spectacular sunset, or apprehension with the approach of a *hurricane* or *tornado*. It is responsible for all of our weather, be it pleasant or stormy. Without our atmosphere there would be neither clouds nor rain; nor streams, lakes, or oceans. There would be no living creatures. Without an atmosphere, Earth would look somewhat like the Moon. And we, quite simply, would not exist.

Among the five terrestrial bodies that we discussed in Chapter 7, only Venus and Earth have dense atmospheres (**Figure 8.1**). Mars has a very low-density atmosphere, and the atmospheres of Mercury and the Moon are so sparse that they can hardly be detected. Why should some of the terrestrial planets have dense atmospheres while others have little or essentially none? Are atmospheres created right along with the planets they envelop, or do they appear later? For answers to these questions we need to look back nearly 5 billion years to the story told in Chapter 6—to a time when the planets were just completing their growth.

Some Atmospheres Appeared Very Early

Planetary atmospheres form in a series of phases, as illustrated in **Figure 8.2**. The young planets at that time were still enveloped by the remaining hydrogen and helium that

KEY CONCEPTS

A thick blanket of atmosphere warms and sustains Earth's temperate climate. On Venus, by contrast, a thick, sulfurous atmosphere pushes the planet's surface beyond medieval poet Dante Alighieri's worst descriptions of hell. A thin atmosphere leaves Mars frozen. Just as we did with their surfaces in Chapter 7, we now compare the atmospheres of the terrestrial planets. Among many interesting insights, on this leg of our journey we will discover

- That terrestrial planets owe their atmospheres to volcanism and to volatiles captured from comets.

- Why some planets hold on to their atmospheres while others do not.

- That differences among Earth, Venus, and Mars are due largely to the atmospheric greenhouse effect.

- How Earth's atmosphere has been reshaped by life.

- That atmospheres are layered by convection and differences in how they are heated, while atmospheric pressure steadily falls at higher and higher altitudes.

- That Earth's magnetic field interrupts the flow of the solar wind and traps charged particles in a huge magnetosphere responsible for auroras.

- That Venus has a hot, dense atmosphere and Mars has a cold, thin atmosphere.

- The unsettling fact that we are living through an uncontrolled experiment in climate modification.

215

(a)

(b)

(c)

FIGURE 8.1 Global views of the atmospheres of Venus (a), Earth (b), and Mars (c). Mercury and the Moon have no atmospheres to speak of.

filled the protoplanetary disk surrounding the Sun, and they were able to capture some of this surrounding gas. Gas capture must have continued until the gaseous disk ultimately dissipated (soon after formation of the planets) and the supply of gas ran out. The gaseous envelope collected by a newly formed planet is called its **primary atmosphere**. Although the giant planets still retain most of their original primary atmospheres, the terrestrial planets probably lost theirs soon after the protoplanetary disk was blown away by the emerging Sun. Why did only the terrestrial planets lose their primary atmospheres? The answer lies in their relatively small masses.

Primary atmospheres consist of captured gas.

The terrestrial planets, with their weak gravity, lack the ability to hold light gases such as hydrogen and helium. When the supply of gas in the disk ran out, their primary atmospheres began leaking back into space.[1] How can gas molecules escape from a planet? All it takes for any object—from a molecule to a spacecraft—to escape a planet is a speed greater than the escape velocity (see Chapter 3) and pointed in the right direction. Intense radiation from the Sun—which is the primary source of thermal, or kinetic, energy in the atmospheres of the terrestrial planets—raises the temperature and thus the kinetic energy of atmospheric atoms and molecules enough that some may escape. The kinetic energy of any object is determined by its mass and its speed. Hotter molecules have higher kinetic energies than do cooler molecules and therefore move faster.

A planet's atmosphere can sometimes escape into space.

Let's look more closely at how molecules move within a planetary atmosphere. When a volume of air contains molecules with different masses, the average kinetic energy tends to be distributed equally among the different types. In other words, each type of molecule, from light to massive, will have the same average kinetic energy. But if each

type has the same average energy, then the less massive molecules must be moving faster than the more massive ones. For example, in a mixture of hydrogen and oxygen at room temperature, hydrogen molecules will be rushing around at about 2,000 meters per second (m/s) on average, while the much more massive oxygen molecules will be poking along at a sluggish 500 m/s. Remember, though, that these are the *average* speeds. A small fraction of the molecules will always be moving much slower than average, while a few will be moving much faster than average.

Deep within a planet's atmosphere, these high-speed molecules will almost certainly collide with other molecules before they have a chance to escape. During the collision process, there is an exchange of energy. The high-speed molecule usually emerges with a lower speed, and the slower one tends to move faster. After a collision, then, both are likely to move with speeds closer to the average. There are fewer surrounding molecules near the top of the atmosphere, and a high-speed molecule has a good chance of escaping before colliding with another molecule if it is heading more or less upward. Less massive molecules and atoms, such as hydrogen and helium, move faster and are more quickly lost to space than more massive molecules such as nitrogen or carbon dioxide.

Heated by the Sun and lacking a strong gravitational grasp, the terrestrial planets soon lost the hydrogen and helium they had temporarily acquired from the protoplanetary disk. Thus, the primary atmospheres were lost early in the evolutionary development of the terrestrial planets. Born naked, they were naked once more. The giant planets, on the other hand, are far more massive than their terrestrial cousins and are situated in the cooler environment of the outer Solar System. Stronger gravity and lower temperatures have enabled them to retain nearly all of their massive primary atmospheres. ▶‖ **AstroTour:** Atmospheres: **Formation and Escape**

Some Atmospheres Developed Later

Given that Earth's primary atmosphere was lost early in its history, what is the source of the air we breathe today? Three

---

[1] Not all atmospheric loss comes from slow leakage. Impacts by large planetesimals may have blasted away substantial amounts of the terrestrial planets' primary atmospheres.

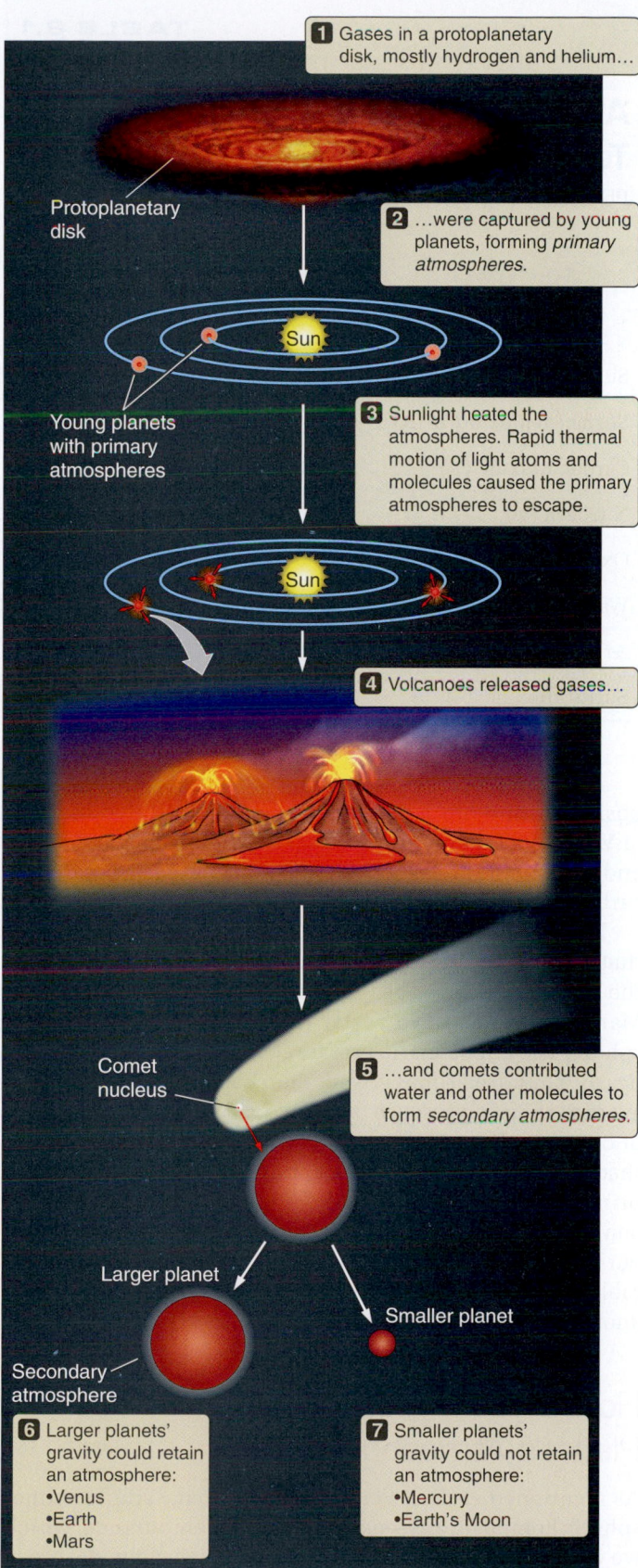

1. Gases in a protoplanetary disk, mostly hydrogen and helium...

Protoplanetary disk

2. ...were captured by young planets, forming *primary atmospheres*.

Sun

Young planets with primary atmospheres

3. Sunlight heated the atmospheres. Rapid thermal motion of light atoms and molecules caused the primary atmospheres to escape.

Sun

4. Volcanoes released gases...

Comet nucleus

5. ...and comets contributed water and other molecules to form *secondary atmospheres*.

Larger planet

Smaller planet

Secondary atmosphere

6. Larger planets' gravity could retain an atmosphere:
   • Venus
   • Earth
   • Mars

7. Smaller planets' gravity could not retain an atmosphere:
   • Mercury
   • Earth's Moon

FIGURE 8.2 Phases in the formation of the atmosphere of a terrestrial planet.

sources likely contributed to what is known as a **secondary atmosphere**: accretion, volcanism, and impacts. During the planetary accretion process, minerals containing water, carbon dioxide, and other volatile matter collected in the interiors of the terrestrial planets. Later, as the interiors heated up, the higher temperatures released these gases from the minerals that had held them. Volcanism then brought the various gases to the surface, where they accumulated and created our secondary atmosphere. Many planetary scientists now believe that there was another important source of gas that formed the secondary atmospheres of the terrestrial planets: impacts by huge numbers of comets, which had formed in the outer parts of the Solar System and were therefore rich in volatiles (see Chapter 6).

> Secondary atmospheres are a result of accretion, volcanism, and comet impacts.

Why did these icy bodies come into the inner Solar System? Their orbits were disrupted by the growth of the giant planets. As the giant planets of the outer Solar System grew to maturity, their gravitational perturbations stirred up the entire population of icy planetesimals (comet nuclei) that orbited within their domain. Many of these icy bodies were flung outward by the giant planets to form the parts of our Solar System that we call the Kuiper Belt and the Oort Cloud. (We will discuss these remote regions in Chapter 12.) Other comets were scattered into the inner parts of the Solar System, where they could rain down on the surfaces of the terrestrial planets. These comet nuclei brought with them ices such as water, carbon monoxide, methane, and ammonia. Cometary water mixed with the water that had been released into the atmospheres by volcanism. On Earth, and perhaps Mars as well, most of the water vapor then condensed as rain and flowed into the lower areas to form the earliest oceans.

Other cometary gases were not able to survive in their original form. Ultraviolet (UV) light from the Sun easily fragments cometary molecules such as ammonia and methane. Ammonia, for example, is broken down into hydrogen and nitrogen. When this happens, the lighter hydrogen atoms quickly escape to space, leaving behind the much heavier nitrogen atoms. The nitrogen atoms then combine to form more massive nitrogen molecules, making it even less likely that these molecules will escape into space. Decomposition of ammonia by sunlight became the primary source of molecular nitrogen in the atmospheres of the terrestrial planets and on one of Saturn's moons, Titan. Molecular nitrogen makes up the bulk of Earth's and Titan's atmospheres.

Among the terrestrial planets, today only Venus, Earth, and Mars have significant secondary atmospheres. What happened in the case of

> Venus, Earth, and Mars have significant secondary atmospheres.

Mercury and the Moon? Even if these two bodies had experienced less volcanism than the other terrestrial planets, they could hardly have escaped the early bombardment of comet nuclei from the outer Solar System. Some carbon dioxide and water must have accumulated during volcanic eruptions and comet impacts. Where are these gases now?

It appears that Mercury's relatively small mass and its proximity to the Sun caused it to lose all of its secondary atmosphere to space, just as it had earlier lost its primary atmosphere. More massive molecules, such as carbon dioxide, can escape from a small planet if the temperature is high enough. Furthermore, intense ultraviolet radiation from the Sun can break molecules into less massive fragments, which are lost to space even more quickly. Because the Moon is much farther from the Sun than Mercury is, the Moon is much cooler than Mercury, but its mass is so small that even at relatively low temperatures molecules can easily escape. From the beginning, a combination of their small mass and relative proximity to the Sun doomed both Mercury and the Moon to remain almost totally "airless."

> Mercury and the Moon have basically no atmosphere.

8.2 A Tale of Three Planets—The Evolution of Secondary Atmospheres

So far, we've discussed how the terrestrial planets gained and lost their primary atmospheres; how their secondary atmospheres formed; and how only Earth, Venus, and Mars retained their secondary atmospheres. We'll now compare and contrast these secondary atmospheres, which are quite different from one another. Would we have expected such differences? To answer this question, we need to examine the processes that affected the evolution of these atmospheres.

Similarities and Differences

Let's look at the similarities first. All three of these planets either are volcanically active today or have been volcanically active in their geological past, and all must have shared the intense cometary showers of the distant past. Their similar geological histories suggest that the early secondary atmospheres of these planets might also have been quite similar. In addition, two of the terrestrial planets—Venus and Earth—are similar in both mass and composition, and they have adjacent orbits that are less than 0.3 astronomical unit (AU) apart. The third—Mars—is also similar in com-

TABLE 8.1

Atmospheres of the Terrestrial Planets
Physical Properties and Composition

	Planet		
	Venus	**Earth**	**Mars**
Surface pressure (bars)	92	1.0	0.006
Surface temperature (K)	737	288	210
Carbon dioxide (%)	96.5	0.039	95.3
Nitrogen (%)	3.5	78.1	2.7
Oxygen (%)	0.00	20.9	0.13
Water (%)	0.002	0.1 to 3	0.02
Argon (%)	0.007	0.93	1.6
Sulfur dioxide (%)	0.015	0.02	0.00

position, but it has a mass only about a tenth that of Earth or Venus. In **Table 8.1** we see that the atmospheres of Venus and Mars today are nearly identical in composition: mostly carbon dioxide, with much smaller amounts of nitrogen.

Now let's explore their differences. Table 8.1 shows us that despite the similarities in atmospheric composition, the surface pressures of Venus and Mars are very different. Mars and Venus have vastly different *amounts* of atmosphere. The atmospheric pressure on the surface of Venus is nearly a hundred times greater than Earth's. By contrast, the average surface pressure on Mars is less than a hundredth of our own. Earth differs in another important respect in that, alone among the planets, its atmosphere is made up primarily of nitrogen and oxygen, with only a trace of carbon dioxide. Although all of these planets must have started out with atmospheres of similar composition and comparable quantity, they ended up being very different from one another. Why did they evolve so differently?

How Mass Affects a Planet's Atmosphere

As mentioned earlier, Venus and Mars have similar atmospheric compositions—except for water abundance, which we will discuss later. Both planets experienced widespread volcanism at some time during their history. Evidence suggests that Venus might still be volcanically active, and Mars has certainly been volcanically active in the recent past. Carbon dioxide and water vapor must have poured out into the emerging secondary atmospheres of both Venus and Mars in the form of volcanic gases. Decomposed cometary ammonia was the likely source of nitrogen on both planets.

The major distinction between the two planets cannot be explained in terms of planetary mass alone. Venus has nearly eight times as much mass as Mars, so we can assume that it probably had about eight times as much carbonate within its interior to produce carbon dioxide, the principal secondary-atmosphere component of both planets. Even allowing for the differences in mass, however, Venus today has more than 2,500 times more atmospheric mass than Mars. Why such a large difference? We can find the answer by considering the relative strengths of their surface gravity, which involves both the mass and radius of a planet. Venus has the gravitational pull necessary to hang on to its atmosphere; Mars has less gravitational attraction to keep its atmosphere. Furthermore, when a planet such as Mars loses so much of its atmosphere to space, the process begins to take on a *runaway* behavior. With less atmosphere, there are fewer intervening molecules to keep breakaway molecules from escaping, and the rate of escape increases. This process in turn leads to even less atmosphere and even greater escape rates.

> Venus retained its atmosphere more effectively than Mars did.

The Atmospheric Greenhouse Effect

Differences in the present-day masses of the atmospheres of Venus, Earth, and Mars have a large effect on their surface temperatures, and thus on the evolution of their atmospheres. Solar radiation can be trapped by the **atmospheric greenhouse effect**. In Chapter 4, we calculated the expected temperatures of the planets, balancing absorbed solar radiation against emitted thermal radiation, and found that Earth is somewhat warmer than expected, while Venus is very much hotter than our simple model predicts. When the predictions of the model fail, the implication is that we are leaving something out of the model. In this case the "something" is the atmospheric greenhouse effect.

The atmospheric greenhouse effect in planetary atmospheres and the conventional *greenhouse effect* operate in somewhat different ways, although the end results are much the same. Planetary atmospheres and the interiors of greenhouses are both heated by trapping the Sun's energy, but here the similarities end. A good example of the conventional **greenhouse effect** is what happens in a car on a sunny day when you leave the windows closed. Sunlight pours through the car's windows, heating the interior and raising the internal air temperature. With the windows closed, convection is unable to carry the hot air away, and temperatures can climb to as high as 180°F. Heating by solar radiation is most efficient if the enclosure is transparent, which is why the walls and roofs of real greenhouses are made mostly of glass.

In the case of planetary atmospheres, convection operates freely, and it is not hot air that is trapped, but rather the electromagnetic energy received from the Sun. The atmospheric greenhouse effect is illustrated in **Figure 8.3**. Atmospheric gases such as nitrogen, oxygen, carbon dioxide, and water vapor freely transmit visible solar energy, allowing the Sun to warm the planet's surface. The warmed surface now tries to radiate the excess energy back into space according to the temperature of that surface (see Chapter 4), which is much lower than that of the Sun. At the typical temperatures of planetary surfaces, this energy is reradiated as *infrared* (IR) radiation. But carbon dioxide and water vapor—among other kinds of atmospheric molecules—strongly absorb IR radiation, converting it to thermal energy. These same molecules subsequently reradiate this thermal energy in all directions—some of the thermal energy continues into

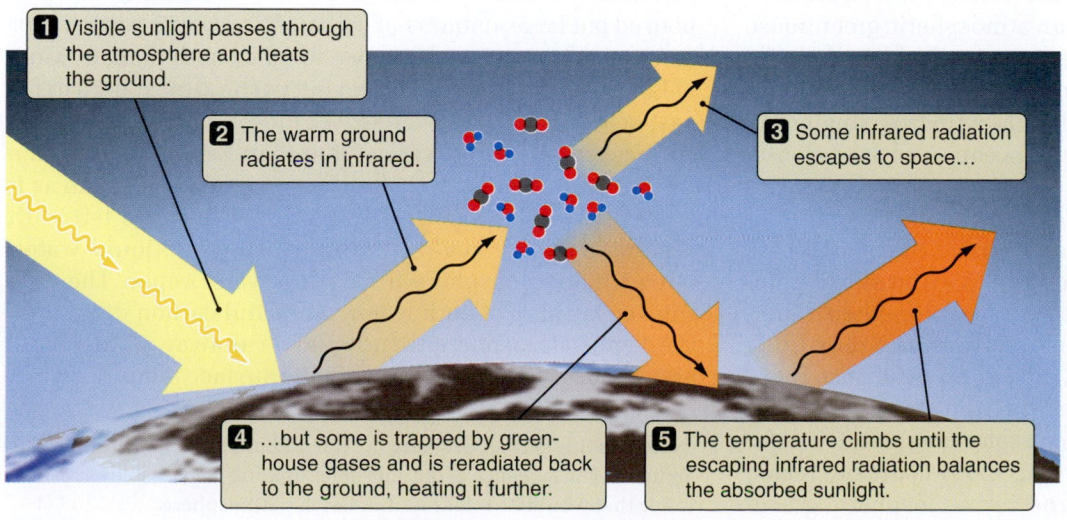

1 Visible sunlight passes through the atmosphere and heats the ground.

2 The warm ground radiates in infrared.

3 Some infrared radiation escapes to space…

4 …but some is trapped by greenhouse gases and is reradiated back to the ground, heating it further.

5 The temperature climbs until the escaping infrared radiation balances the absorbed sunlight.

FIGURE 8.3 Greenhouse gases such as water and carbon dioxide trap infrared radiation, increasing a planet's temperature.

space, but much of it goes back to the ground. The surface is now receiving thermal energy both from the Sun and from the atmosphere.

Molecules such as water vapor and carbon dioxide that transmit visible radiation but absorb IR radiation are called **greenhouse molecules**. (Methane, nitrous oxide, and chlorofluorocarbons [CFCs] are other greenhouse molecules found in Earth's atmosphere.) Thus, the presence of greenhouse molecules in a planet's atmosphere will cause its surface temperature to rise. This rise in temperature continues until the surface becomes sufficiently hot—and therefore radiates enough energy—that the fraction of infrared radiation leaking out through the atmosphere is just enough to balance the absorbed sunlight. Convection also helps, by transporting thermal energy to the top of the atmosphere, where it can be more easily radiated to space. In short, the temperature rises until equilibrium between absorbed sunlight and thermal energy radiated away by the planet is reached, just as our earlier discussion in Chapter 4 said it must be. Even though the mechanisms are somewhat different, the conventional greenhouse effect and the atmospheric greenhouse effect produce the same net result: the local environment is heated by trapped solar radiation.

▶❚❚ **AstroTour:** Greenhouse Effect

Let's look more closely at how the atmospheric greenhouse effect operates on Mars, Earth, and Venus. We already know that solar radiation can be trapped by greenhouse molecules, but what really matters is the actual *number* of greenhouse molecules in a planet's atmosphere, not the *fraction* they represent. For example, even though the atmosphere of Mars is composed almost entirely of carbon dioxide—an effective greenhouse molecule—its tenuous atmosphere contains relatively few greenhouse molecules compared to the atmospheres of Venus or Earth. As a result, the atmospheric greenhouse effect on Mars raises the mean surface temperature by only about 5 kelvins (K). Earth's more massive atmosphere is more efficient. Temperatures on Earth are 35 K warmer than they would be in the absence of an atmospheric greenhouse effect, produced mainly by water vapor and carbon dioxide. Without this greenhouse warming, the mean global temperature of Earth would be well below the freezing point, leaving us with a world of frozen oceans and ice-covered continents!

Nowhere in the Solar System, however, is the atmospheric greenhouse effect more dramatic than on Venus. Its massive atmosphere of carbon dioxide and sulfur compounds raises its surface temperature by more than 400 K, to about 737 K. At such high temperatures, any remaining water and most carbon dioxide locked up in surface rocks would long ago have been driven into the atmosphere, further enhancing the atmospheric greenhouse effect.

The conditions existing on Venus today could be created on an Earth-like planet by a runaway atmospheric green-

> Without the greenhouse effect, Earth would freeze.

house effect. Imagine a hypothetical situation in which large quantities of carbon dioxide or another greenhouse gas suddenly appeared in Earth's atmosphere. The increased warming would raise surface temperatures, driving more carbon dioxide and water vapor into the atmosphere, which would in turn increase the strength of the atmospheric greenhouse effect. The result would be even greater warming and still larger amounts of atmospheric water vapor, ultimately creating a surface pressure at least 300 times as great as it is now[2] and a surface temperature that might exceed 800 K. Long before reaching this stage, our planet would have become devoid of all life.

In reality, the process is more complicated. Increased cloud cover caused by increased water in the atmosphere might decrease the amount of sunlight reaching Earth's surface to a point where the runaway effect would be turned off. There are other complicating factors. Ocean currents are important in transporting energy from one part of Earth to another, but how they would be affected by increased warming is something we really do not know. In fact, the process is so complicated, wherein tiny changes lead to very large results, that we still cannot predict the long-term outcome of the small changes that humans are now making in the composition of Earth's atmosphere. In a real sense, we are *experimenting* with the atmosphere of Earth. We are asking the question (whether we know it or not), "What happens to Earth's climate if we steadily increase the number of greenhouse molecules in its atmosphere?" We do not yet know the answer to this question, but we will eventually know by seeing the results. Whether we will be happy with the results is another matter. We will explore this issue again in Section 8.3.

It is relatively easy to understand the present-day differences between Venus and Mars, but why does the atmospheric greenhouse effect make the composition of Earth's atmosphere so different from that of the other two? We may find the answer in Earth's special location in the Solar System. Consider early Earth and early Venus, each having about the same mass, but with Venus orbiting somewhat closer to the Sun than Earth does. Volcanism must have poured out large amounts of carbon dioxide and water vapor to form early secondary atmospheres on both planets. Most of Earth's water quickly rained out of the atmosphere to fill vast ocean basins. But Venus was closer to the Sun, and its surface temperatures were higher than Earth's. Most of the rainwater on Venus immediately re-evaporated, much as it does today in Earth's desert regions. Venus was left with a planetwide surface containing very little liquid water, but with an atmosphere filled with water vapor. The continuing buildup of both water vapor and carbon dioxide in the Venus atmosphere then led to a runaway atmospheric greenhouse effect that drove up the surface temperature of

---

[2]Note that this surface pressure is even higher than that of present-day Venus—a consequence of the enormous amount of available terrestrial water that would be released into Earth's atmosphere.

the planet. The surface of the planet became so hot that no liquid water could survive there.

This early difference between a watery Earth and an arid Venus forever changed the ways that their atmospheres and surfaces would evolve. On Earth, water erosion caused by rain and rivers continually exposed fresh minerals, which then reacted chemically with

Earth's atmosphere evolved very differently from that of Venus.

atmospheric carbon dioxide to form solid carbonates. This reaction removed some of the atmospheric carbon dioxide, burying it within Earth's crust as a component of a rock called "limestone." Later, the development of life in Earth's oceans accelerated the removal of atmospheric carbon dioxide. Tiny sea creatures built their protective shells of carbonates, and as they died they built up massive beds of limestone on the ocean floors. As a result of water erosion and the chemistry of life, all but a trace of Earth's total inventory of carbon dioxide is now tied up in limestone beds. Earth's particular location in the Solar System seems to have spared it from the runaway atmospheric greenhouse effect. But what if Earth had formed a bit closer to the Sun? Look now at **Table 8.2**. If all the carbon dioxide now in limestone beds had not been locked up by these reactions, Earth's atmospheric composition would resemble that of Venus or Mars.

Differences in the amount of water on Venus, Earth, and Mars are not so well understood. Geological evidence tells us that liquid water was once plentiful on the surface of Mars (see Chapter 7), and the *Mars Odyssey* spacecraft has found evidence that significant amounts of water still exist in the form of subsurface ice—far more than the atmospheric abundance indicated in Table 8.1. Earth's liquid and solid water supply is even greater: about 10^{21} kilograms (kg), or 0.03 percent of its total mass. More than 97 percent of Earth's water is in the oceans, which have an average depth of about 4 kilometers (km). Earth today has 100,000 times more water than Venus. What happened to all the water on

Venus? One possibility is that water molecules high in its atmosphere were broken apart into hydrogen and oxygen by solar ultraviolet radiation. Hydrogen atoms, being of very low mass, were quickly lost to space. Oxygen, however, would eventually have migrated downward to the planet's surface, where it would have been removed from the atmosphere by oxidizing surface minerals.

8.3 Earth's Atmosphere—The One We Know Best

Now that we understand some of the overall processes that have influenced the evolution of the terrestrial planet atmospheres, we will look in depth at each of them. We begin with Earth's atmosphere, not only because we know it best, but also because it provides an introduction to atmospheric structure and weather phenomena that will help us better understand the atmospheres of Venus, Mars, and Titan, and even the atmospheres of the giant planets.

In simplest terms, we can describe Earth's atmosphere as a blanket of gas that is several hundred kilometers deep and has a total mass of approximately 5×10^{18} kg (about 5,000 trillion metric tons.) As enormous as this amount may seem, it represents less than one-millionth of Earth's total mass. Yet the weight of Earth's

Earth's atmosphere has a surface pressure equal to a 10-meter layer of water.

atmosphere creates a force of approximately 100,000 newtons (N) acting on each square meter of our planet's surface. We express this amount of *pressure* as a unit called a **bar** (from the Greek *baros*, meaning "weight" or "heavy"). Earth's average atmospheric pressure at sea level is approximately 1 bar. (Meteorologists frequently quote atmospheric pressures in millibars [mb], which are thousandths of a bar.) One bar of pressure is equivalent to what we might experience underwater at a depth of 10 meters, or 33 feet. Yet we seem to be largely unaware of Earth's atmospheric pressure. How can this be? The reason is that the very same pressure exists both within and outside of our bodies. The two precisely balance one another—another example of the concept of *hydrostatic equilibrium* that we introduced in the previous chapter.

As we learned in Chapter 7, hydrostatic equilibrium tells us that the pressure at any point within a planet must be great enough to balance the weight of the overlying layers. The same principle holds true in a planetary atmosphere. The atmospheric pressure on a planet's surface must be great enough to hold up the weight of the overlying atmosphere. Of course, different *forms* of matter provide the pressure within a planet's interior and in its atmosphere. In the interior of a solid planet, solid materials exert pressure as they resist being compressed. In a planetary atmosphere, the motions of gas molecules exert sufficient pressure to support the

TABLE 8.2

Terrestrial Planet Atmospheres
If All Available Carbon Dioxide Were Included

	Planet		
	Venus	Earth	Mars
Carbon dioxide (%)	96.5	98.0	96.0
Nitrogen (%)	3.5	1.6	2.7
Oxygen (%)	0.0	0.4	0.1
All other constituents (%)	0.0	0.0	1.2

What Is a Gas?

As we continue our journey of discovery through the universe, we will find that the gaseous state is the most common form of matter. More than those of any other form of matter, it is the properties of gases that we will turn to as we seek to understand the workings of planets, stars, and galaxies.

Matter is composed of atoms and molecules, and different forms of matter result from dissimilarities in how those atoms and molecules interact. In a *solid*, molecules are packed tightly, held in place by adjacent molecules like bricks in a wall. In a *liquid*, molecules are free to move about and are constantly jostling one another. You can picture molecules in a liquid as people in a crowded subway station. The crowd is free to flow, but the individuals that make up the crowd are still limited in their movement by the people around them. The molecules in a *gas*, on the other hand, go their own way, traveling relatively long distances without interacting with other molecules. When you think of a gas, picture a swarm of tiny atoms and molecules flying about, each with its own speed and direction.

As we have already seen in our discussion of escape velocity, the temperature of a gas is a measure of the average kinetic energy of the individual molecules as they fly about. Two things determine a molecule's kinetic energy: the mass of the molecule and the speed at which it is moving. The average speed of a molecule in a gas is inversely proportional to the square root of its mass. Oxygen molecules, for example, are 16 times more massive than hydrogen molecules. In a gas containing both hydrogen and oxygen, the hydrogen molecules will therefore be moving four times ($\sqrt{16}$) as fast as the oxygen molecules. As we have seen, this difference explains why Earth can hold onto the oxygen in its atmosphere but loses the hydrogen to space.

These motions of gas molecules create the pressure needed to hold up a planet's atmosphere. As we study gases in different astronomical settings, we will find that one of the most important properties of a gas is how hard it pushes on its surroundings. Imagine a box containing gas, as shown in **Figure 8.4a**. Molecules are constantly bouncing off the walls of the box, pushing outward. This outward push, measured in units of force per square meter of the surface of the box, is referred to as **pressure**. (Anytime you think about the pressure of a gas, this is the mental picture that you should bring to mind: atoms and molecules slamming against the walls of a box, pushing outward.)

Armed with nothing more than this mental picture, we can draw some interesting conclusions about the pressure of a gas. First, if we increase the number of molecules in the box, more molecules will hit the walls of the box each second and the pressure will increase. Doubling the density of the gas but keeping the temperature the same (**Figure 8.4b**) doubles its pressure.

Increasing the temperature of the gas increases the average speed at which molecules move, which also increases the pressure of the gas. There are two reasons for this effect. First, if the molecules in the box are moving faster, they will hit the walls *more frequently* (**Figure 8.4c**). More molecules hitting the walls of the box each second means that the pressure is higher. Second, faster-moving molecules hit the wall *harder*, exerting more force. Together, these two effects mean that doubling the temperature of a gas doubles the pressure of the gas.

Pressure is proportional to density, and also to temperature. We can combine these two relationships to get

$$\text{Pressure} \propto \text{Density} \times \text{Temperature}.$$

This relationship of density, temperature, and pressure is called the **ideal gas law**. The empirically discovered ideal gas law has a special place in the history of science. It provided strong support for the atomic theory of matter, in much the same way that Kepler's empirical laws of planetary orbits provided support for Newton's theories of motion and gravitation. The fact that labora-

---

atmosphere. (The relationship among density, temperature, and pressure of a gas is discussed in **Foundations 8.1**.)

The composition of Earth's atmosphere is relatively constant on a global scale, but temperatures can vary widely. Two principal gases make up our atmosphere. About four-fifths is nitrogen and one-fifth is oxygen, although there are many important minor constituents, such as water vapor and carbon dioxide. The amounts of atmospheric water vapor and carbon dioxide are somewhat variable, depending on global location and season. Atmospheric temperatures near Earth's surface can range from as high as 60°C in the deserts to as low as −90°C in the polar regions. The mean global temperature is about 15°C.

Life Controls the Composition of Earth's Atmosphere

As Table 8.1 shows, the composition of Earth's atmosphere is very different from that of Venus and Mars. We have already discussed the differences in carbon dioxide content, but what truly sets Earth apart from all other known planets is its oxygen. Earth's atmosphere contains abundant amounts of oxygen, while other planets' atmospheres do not. Why should this be so?

> Only Earth's atmosphere contains abundant oxygen.

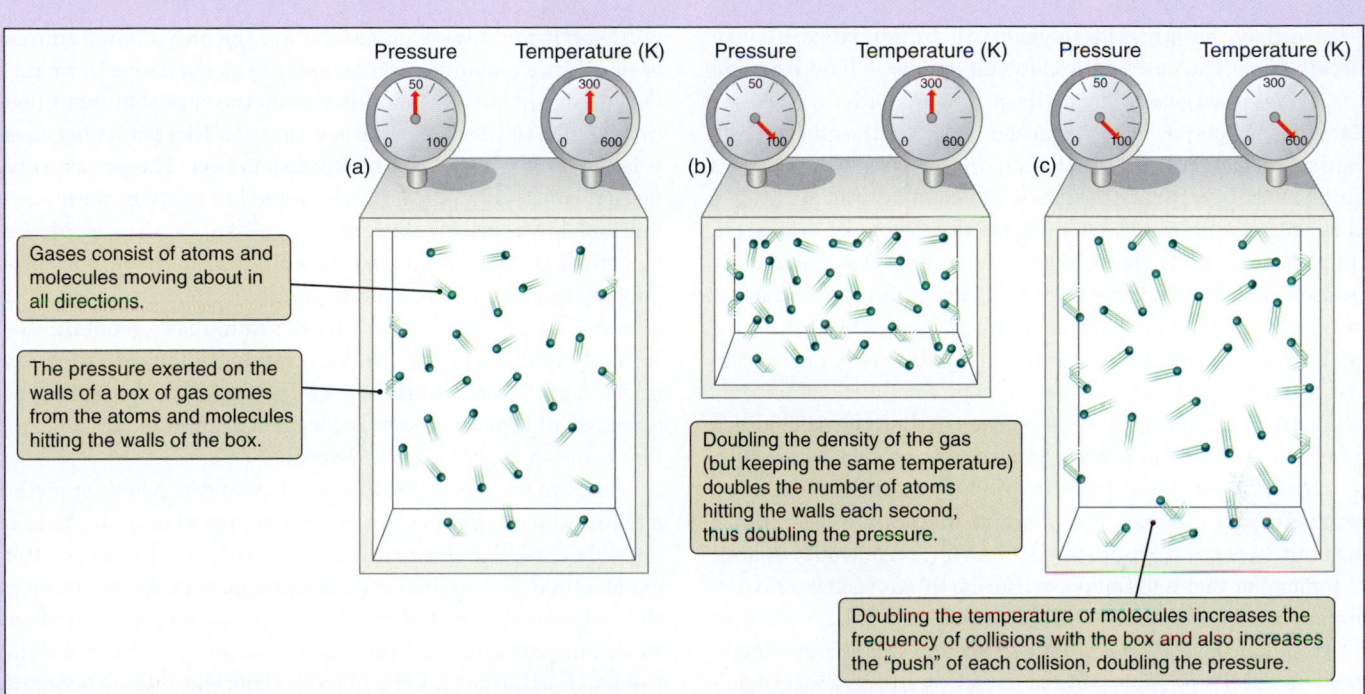

Pressure Temperature (K)   Pressure Temperature (K)   Pressure Temperature (K)

(a)   (b)   (c)

Gases consist of atoms and molecules moving about in all directions.

The pressure exerted on the walls of a box of gas comes from the atoms and molecules hitting the walls of the box.

Doubling the density of the gas (but keeping the same temperature) doubles the number of atoms hitting the walls each second, thus doubling the pressure.

Doubling the temperature of molecules increases the frequency of collisions with the box and also increases the "push" of each collision, doubling the pressure.

FIGURE 8.4 (a) The pressure of a gas comes from the motions of atoms and molecules. Making a gas denser (b) or hotter (c)—that is, increasing the speed of atoms or molecules—increases the pressure of the gas.

tory gases come very close to obeying the ideal gas law provided compelling early evidence that atoms and molecules are real.

We need to know about one other property of gases as our journey continues. If you compress a gas, you increase not only its density, but also its temperature. Conversely, when a gas is allowed to expand, it cools off. There are many every-day examples of this behavior. Pump up a bicycle tire, and the pump becomes hot because the air is being compressed. Hold down the nozzle on an aerosol can, and it feels cold because the propellant gas is expanding. An air conditioner works by alternately compressing a gas to make it hot, letting that gas cool, and then allowing the gas to expand and get really cold.

Over and over again on our journey, our understanding of gases will be our guide—equally as valid when applied to the hot cores of stars as to a gentle breeze on a summer day. Every time we talk about an object made of gas, think back to the picture of molecules bouncing around in a box, and remember the commonsense ideas that explain the behavior of the gas.

Oxygen, it turns out, is a highly reactive gas. It chemically combines with, or "oxidizes," almost any material it touches. Witness the rust (iron oxide) that forms on steel, for example. For a planet to retain significant amounts of this reactive gas in its atmosphere, there would need to be a means for replacing what is lost through oxidation. Such a means exists on Earth—we call it "plant life" (see **Excursions 8.1**).

The oxygen concentration in Earth's atmosphere has changed over the history of the planet, as shown in **Figure 8.5**. When Earth's secondary atmosphere first appeared about 4 billion years ago, it was almost totally free of oxygen. About 2.8 billion years ago, an ancestral form of cyanobacteria (single-celled organisms that contain chlorophyll) began releasing oxygen into Earth's atmosphere as a waste product of their metabolism. At first, this biologically generated oxygen combined readily with exposed metals and minerals in surface rocks and soils and so was removed from the atmosphere as quickly as it

Life is responsible for the oxygen in Earth's atmosphere.

formed. In this way, emerging life dramatically changed the very composition and appearance of Earth's surface—the first of many such widespread modifications imposed on our planet by living organisms. Ultimately, the explosive growth of plant life accelerated the production of oxygen, building up atmospheric concentrations that approached

On Atmospheres and Life

Take a deep breath. The oxygen you inhale sustains you. Breathe out. The carbon dioxide you exhale is food for plant life. If these two gases were to disappear suddenly, most life on Earth would perish.[3] As we learned earlier in the chapter, volcanism is a major source of carbon dioxide, so this gas occurs quite naturally in our atmosphere. Molecular oxygen, however, is an entirely different matter. We would not expect to find it in the atmosphere of a planet like Earth. Oxygen is so chemically reactive that it quickly destroys itself by combining with surface minerals to form oxides. The reddish surface of Mars is coated with oxidized iron-bearing minerals, and this is one reason the martian atmosphere is almost completely free of oxygen. What, then, about Earth? As we learned earlier, the metabolism of ancient, tiny organisms produced oxygen as a by-product, much as present-day animal life produces carbon dioxide as a respiratory by-product. Today, plants maintain oxygen levels at about 20 percent of our atmospheric total. We would do well to remember that without plant life, Earth's oxygen would disappear and all animal life would vanish with it.

You may have already jumped ahead to an obvious conclusion: one of the possible ways of detecting life on a terrestrial-type extrasolar planet is by finding atmospheric gases that, in the absence of life, should not be there. We have already identified molecular oxygen and ozone as among the telltale gases

[3]Certain microorganisms, called "anaerobes," live in the total absence of oxygen. In fact, most anaerobes are poisoned by even small amounts of oxygen.

of life on Earth. Living organisms are the only known source of significant amounts of molecular oxygen or ozone in a planetary atmosphere. Methane is yet another gas that would be unexpected in the atmosphere of an Earth-like planet because it is readily destroyed by *photodissociation*. The presence of methane means that a source is needed to keep up with photodissociative loss. As with oxygen, a likely source would be biological. Certain microorganisms produce methane as a metabolic by-product. Those bubbles you see rising to the surface of a stagnant pond—sometimes called "swamp gas"—contain biologically produced methane. Yet, certain geological processes can also produce methane, so its presence in the atmosphere of a terrestrial-type exoplanet is a weaker argument for life than is the presence of oxygen and ozone.

So there we have it. Find molecular oxygen, ozone, or methane in the atmosphere of an Earth-like planet and you have a possible candidate for extraterrestrial life. Is this a realistic expectation? Yes, and it could happen as soon as we develop the instrumentation to detect Earth-like planets around other stars. Suppose a future orbiting observatory isolates the light from an Earth-like planet and feeds that light into its onboard spectrograph. Suppose further that the spectra reveal substantial amounts of molecular oxygen in the planet's atmosphere. The most likely conclusion would be that the planet harbors life. Of course, the observation would not tell us what kind of life, or even provide positive proof of its existence, but it would certainly mark that planet as an object of considerable interest for further study.

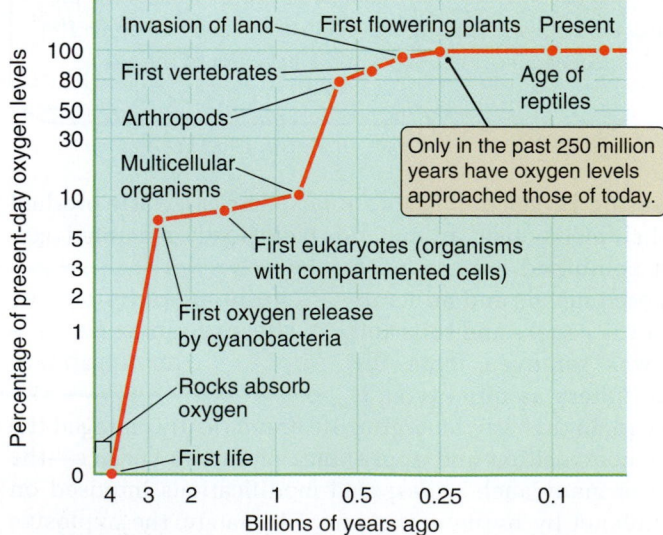

FIGURE 8.5 The amount of oxygen in Earth's atmosphere has built up over time as a result of plant life on the planet.

today's levels only about 250 million years ago. All true plants, from tiny green algae to giant redwoods, use the energy of sunlight to build carbon compounds out of carbon dioxide and produce oxygen as a metabolic waste product in a process called "photosynthesis." Earth's atmospheric oxygen content is held in a delicate balance primarily by plants. If plant life on our planet were to disappear, so, too, would nearly all of Earth's atmospheric oxygen.

Several minor gases in Earth's atmosphere also affect our daily lives. Over the current range of terrestrial temperatures, water is a volatile substance, so its atmospheric abundance varies from time to time and from place to place. In warm, moist climates, water vapor may account for as much as 3 percent of the total atmospheric composition. In cold, arid climates, it may be less than 0.1 percent. The continuous process of condensation and evaporation of water involves the exchange of thermal and other forms of energy, making water vapor a major contributor to Earth's weather.

Carbon dioxide is another variable component of Earth's atmosphere, and a complex pattern of "sources" (places where it originates) and "sinks" (places where it goes) determines how much of it will be present at any one time. Plants consume carbon dioxide in great quantities as part of their metabolic process. Coral reefs are colonies of tiny ocean organisms that build their protective shells with carbonates produced from dissolved carbon dioxide. Fires, decaying vegetation, and human burning of fossil fuels all release carbon dioxide back into the atmosphere. This balance between sources and sinks can and does change with time. Meteorological records show that the amount of carbon dioxide in our atmosphere has been increasing for almost two centuries—since the Industrial Revolution. As noted earlier, this increase in carbon dioxide in turn has had a direct impact on global temperature because carbon dioxide is also a powerful greenhouse gas. It should come as no surprise to learn that Earth's mean global temperature seems to be increasing with the growing abundance of this gas.

Another minor constituent in our atmosphere is "ozone" (O_3). This important molecule is formed when ultraviolet light from the Sun breaks molecular oxygen (O_2) into its individual atoms (O) in a process called **photodissociation**. These atoms can then recombine with other oxygen molecules to form ozone ($O_2 + O \rightarrow O_3$)—a process called **recombination**. Most of Earth's natural ozone is concentrated in the upper atmosphere at altitudes between 20 and 50 km. There it acts as a very strong absorber of ultraviolet sunlight. Without the ozone layer, this lethal radiation would reach all the way to Earth's surface, making it completely uninhabitable for nearly all forms of life.

> High-altitude ozone protects life. Our continuing survival depends on it.

In the 1980s, scientists began noticing that the measured amount of ozone in our upper atmosphere had been decreasing since the 1970s, primarily over the polar latitudes in both the Northern and Southern hemispheres. They referred to these depleted regions as "ozone holes,"[4] although they are more like depressions than real "holes" in the ozone layer. Ozone depletion appears to be caused by a seasonal buildup of atmospheric "halogens"—mostly chlorine, fluorine, and bromine, such as those found in industrial refrigerants. Halogens readily diffuse upward into the stratosphere, where they destroy ozone without themselves being consumed. Such agents are called **catalysts**—materials that participate in and accelerate chemical reactions but are not themselves modified in the process. Because they are not used up, halogens may remain in Earth's upper atmosphere for decades or even centuries.

Why should we be so concerned about the loss of a minor constituent from so high in our atmosphere? Even in tiny amounts, ozone filters out harmful solar ultraviolet radiation, preventing it from reaching Earth's surface. Scientists have predicted that the continuing removal of ozone from the high atmosphere could spell trouble for terrestrial life as more and more UV radiation reaches the ground. Measured increases in the levels of UV radiation appear to be related to increases in skin cancer in humans, and we do not yet understand the effects it may have on other life-forms to which we are inexorably linked. By the late 1980s, the production of CFCs and other ozone-depleting chemicals was being phased out by international agreements; and by 2007, satellite measurements showed some recovery of ozone in the polar latitudes. In 2008, however, the ozone hole deepened again. Full recovery is not expected until the late 21st century at the earliest.

Although the ozone in our upper atmosphere protects life, it may not be so beneficial when found elsewhere. Ozone in the lower atmosphere occurs primarily in urban and suburban environments. It is a human-made pollutant and a health hazard, raising the risk of asthma and heart problems.

Earth's Atmosphere Is Layered like an Onion

Our atmosphere is made up of several distinct layers. Earth's lowermost atmospheric layer, the one in which we live and breathe, is called the **troposphere (Figure 8.6)**. It contains 90 percent of Earth's atmospheric mass and is the source of all of our weather. At Earth's surface, usually referred to as

> Pressure and temperature decrease with altitude in the troposphere.

"sea level," the troposphere has an average temperature of 15°C (288 K) and an average pressure of 1.013 bars. Within the troposphere, atmospheric pressure, density, and temperature all decrease with increasing altitude. For example, at an altitude of 5.5 km, a few thousand feet below the summit of Denali in Alaska, the atmospheric pressure and density are only 50 percent of their sea-level values and the average temperature has dropped to −20°C. Still higher, at an altitude of 12 km, where commercial jets cruise, the temperature is a frigid −60°C and the density and pressure are less than one-fifth what they are at sea level. Mountain climbers and astronomers are very much aware of this behavior of Earth's troposphere. At the Mauna Kea Observatories in Hawaii, even the most dedicated astronomers, surrounded by thin air and subfreezing temperatures, have been known to gaze longingly at the sunny beaches some 4 km below, where it is a pleasant 30°C warmer.

Why, though, does the atmosphere get colder as we climb to higher elevations? From what we have already learned, we might guess correctly that it is warmer near

---

[4] A common misconception confuses the greenhouse effect with the so-called ozone hole. Both are caused by the buildup of certain atmospheric gases and both are a cause for concern, but the individual causes and effects are very different.

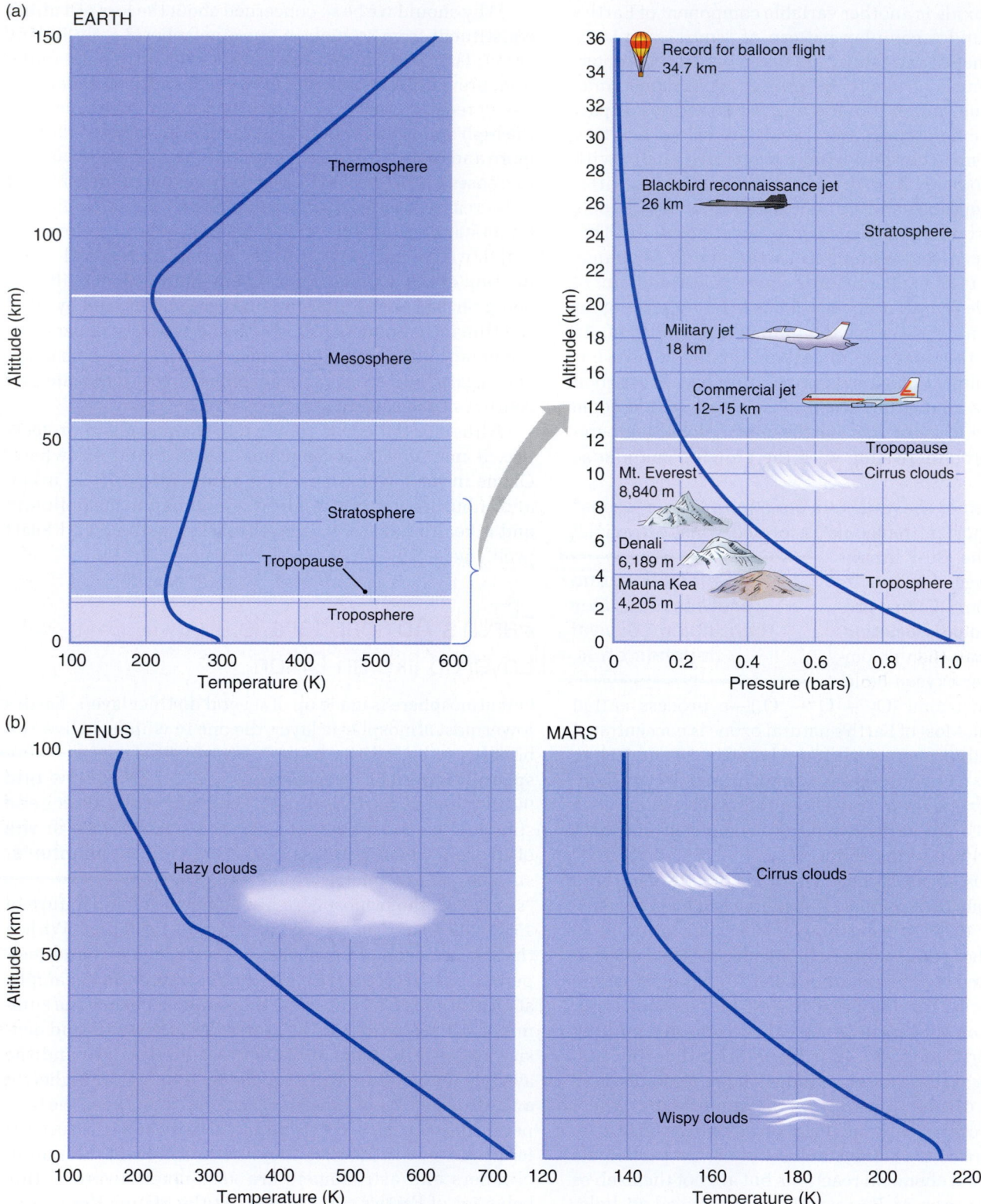

FIGURE 8.6 (a) Temperature and pressure are plotted for Earth's atmospheric layers as a function of height. (b) Atmospheric temperatures for Venus and Mars are shown for comparison.

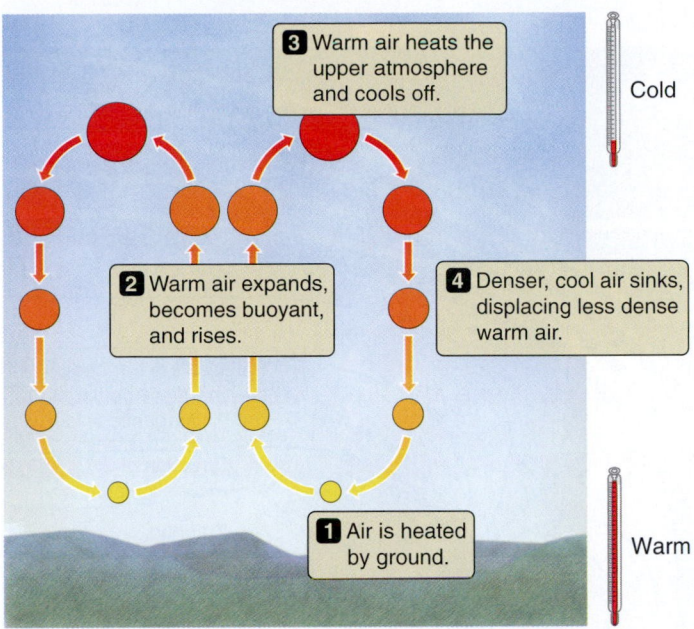

3 Warm air heats the upper atmosphere and cools off.

Cold

2 Warm air expands, becomes buoyant, and rises.

4 Denser, cool air sinks, displacing less dense warm air.

1 Air is heated by ground.

Warm

FIGURE 8.7 Atmospheric convection carries thermal energy from the Sun-heated surface of Earth upward through the atmosphere.

Earth's surface because the air is in contact with the sunlight-heated ground, which also warms the air by its infrared radiation. It also makes sense that the atmosphere would be cooler at very high altitudes because there the atmosphere can freely radiate its thermal energy into space. In fact, it would get colder with increasing altitude even faster if not for **convection**. We first encountered convection in our Chapter 7 discussion of the motions of material in Earth's mantle that drive plate tectonics. **Figure 8.7** illustrates how convection carries thermal energy upward through Earth's atmosphere. At a given pressure, cold air is more dense than warm air. So when cold air encounters warm air, the denser cold air slips under the less dense warm air, pushing the warm air upward. (Rather than "warm air rises," we should perhaps say "cold air sinks.") This convection sets up air circulation between the lower and upper levels of the atmosphere, and such circulation tends to diminish the extremes caused by heating at the bottom and cooling at the top.

Convection also affects the vertical distribution of atmospheric water vapor. The ability of air to hold water in the form of vapor depends very strongly on the air temperature. The warmer the air, the more water vapor it can hold. We refer to the amount of water vapor in the air relative to what the air could hold at a particular temperature as the **relative humidity**. Air that is saturated with water vapor has a relative humidity of 100 percent. As air is convected upward, it cools, limiting its capacity to hold water vapor.

> Convection carries thermal energy upward through the troposphere.

When the air temperature decreases to the point at which the air can no longer hold all its water vapor, it becomes saturated. Water begins to condense out in the form of tiny droplets, which in large numbers become visible to us as clouds. When these droplets coalesce to form large drops, convective updrafts can no longer support them, and they fall as rain. For this reason most of the water vapor in Earth's atmosphere stays within 2 km of the surface. At an altitude of 4 km, the Mauna Kea Observatories are higher than approximately one-third of Earth's atmosphere, but they lie above nine-tenths of the atmospheric water vapor. This is important for astronomers who observe the heavens in the infrared region of the spectrum, because water vapor strongly absorbs IR light (see Figure 5.18).

> Most atmospheric water stays close to Earth's surface.

The top of the troposphere, called the **tropopause**, is defined as the height at which temperature no longer decreases with increasing altitude (see Figure 8.6a). This change in atmospheric behavior is caused by heating from absorbed sunlight within the atmospheric layers that lie above the tropopause. And because temperature no longer decreases with altitude above the tropopause, atmospheric convection also dies out. The tropopause varies between 10 and 15 km above sea level, depending on latitude, and is highest at the equator.

Above the tropopause and extending upward to an altitude of 50 km above sea level is the **stratosphere**. This is a region in which little convection takes place, because the temperature-altitude relationship actually reverses at the tropopause and the temperature begins to *increase* with altitude. The reason for this reversal is the presence of ozone, which warms the stratosphere by absorbing sunlight.

> The stratosphere, mesosphere, and thermosphere lie above the troposphere.

The region above the stratosphere is called the "mesosphere." It extends from 50 km to an altitude of about 90 km. In the mesosphere there is no ozone to absorb sunlight, so temperatures once again decrease with altitude. The base of the stratosphere and the upper boundary of the mesosphere are two of the coldest levels in Earth's atmosphere (see Figure 8.6a).

At altitudes above 90 km, solar ultraviolet radiation and high-energy particles from the solar wind "ionize" (strip electrons from) atmospheric atoms and molecules, causing the temperature once again to increase with altitude. This region is called the "thermosphere," and it is the hottest part of the atmosphere. The temperature can reach 1,000 K near the top of the thermosphere, at an altitude of 600 km. The gases within and beyond the thermosphere are ionized by ultraviolet photons and high-energy particles from the Sun to form a **plasma**. (A plasma is any gas that is made up largely of electrically charged particles rather

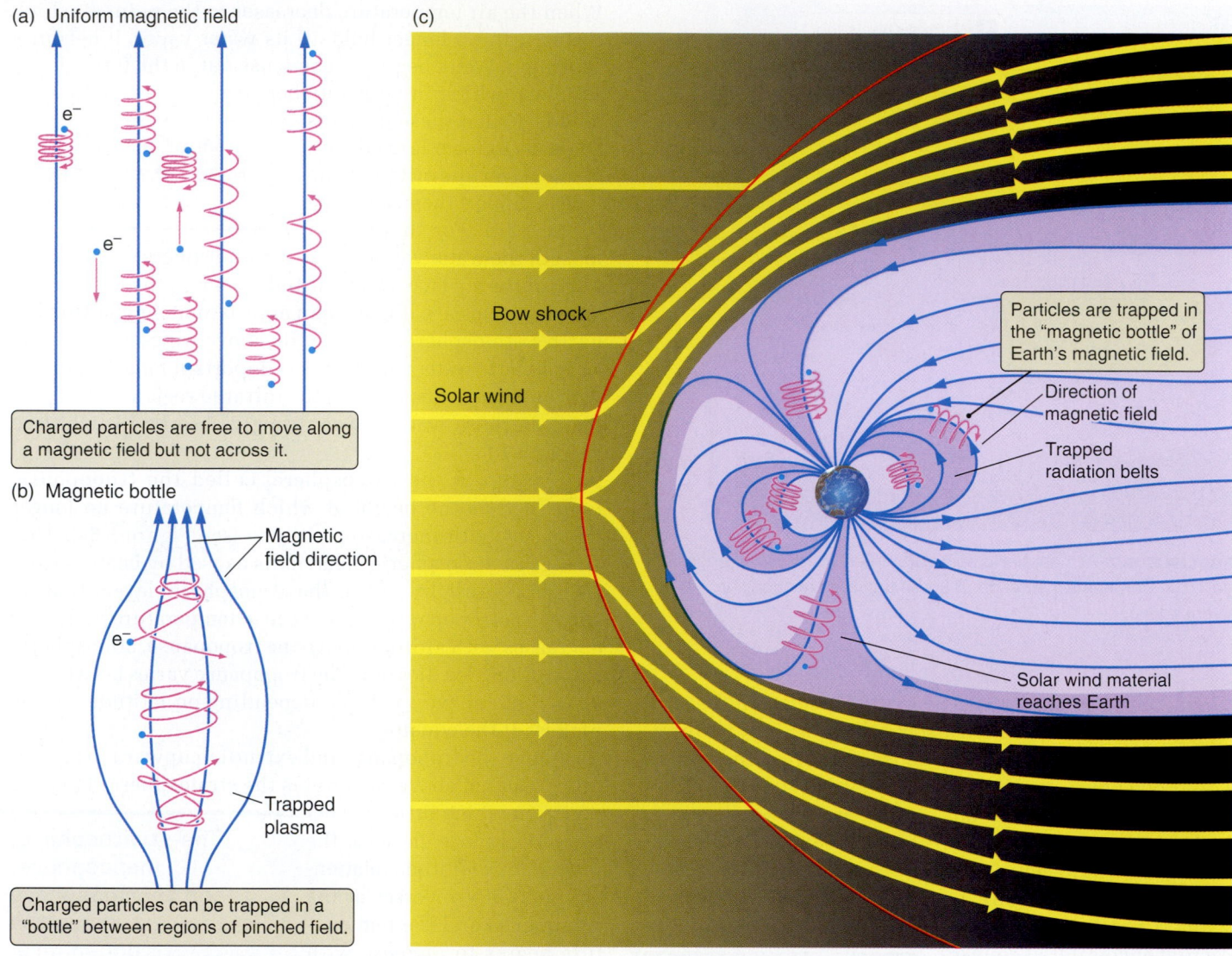

(a) Uniform magnetic field

e⁻

e⁻

Charged particles are free to move along a magnetic field but not across it.

(b) Magnetic bottle

Magnetic field direction

e⁻

Trapped plasma

Charged particles can be trapped in a "bottle" between regions of pinched field.

(c)

Bow shock

Solar wind

Particles are trapped in the "magnetic bottle" of Earth's magnetic field.

Direction of magnetic field

Trapped radiation belts

Solar wind material reaches Earth

FIGURE 8.8 (a) The motion of charged particles, in this case electrons, in a uniform magnetic field. (b) When the field is pinched, charged particles can be trapped in a "magnetic bottle." (c) Earth's magnetic field acts like a bundle of magnetic bottles, trapping particles in Earth's magnetosphere.

than only neutral atoms and molecules.) This region of ionized atmosphere is called the **ionosphere**. The ionosphere is important to us in part because it reflects certain frequencies of radio waves back to the ground. For example, the frequencies used by AM radio bounce back and forth between the ionosphere and the surface, enabling radio receivers to pick up stations at great distances from the transmitters. Amateur radio operators are able to communicate with each other around the world by bouncing their signals off the ionosphere.

Earth and its atmosphere are surrounded by a large region filled with electrons, protons, and other charged particles from the Sun that have been captured by Earth's magnetic field. This region, called Earth's **magnetosphere**, has a radius approximately 10 times the radius of Earth, filling a volume over 1,000 times the volume of the planet itself. To appreciate Earth's magnetosphere, we need to begin by looking more carefully at magnetic fields and the force they apply to charged particles. Magnetic fields have no effect on charged particles unless the particles are moving. Charged particles are free to move *along* the direction of the magnetic field, but if they try to move *across* the direction of the field they experience a force that is perpendicular both to their motion and to the magnetic-field direction. This force causes them to loop around the direction of the magnetic field, as illustrated in **Figure 8.8a**. It is almost as if charged particles were beads on a string, free to slide along the direction of the magnetic field but unable to cross it.

The picture of how charged particles move in a magnetic field gets even more interesting if the field is pinched together at some point. As particles move into the pinch, they feel a magnetic force that (if conditions are right) pushes them back along the direction from which they came. If charged particles are located in a region in which the field is pinched on both ends, as shown in **Figure 8.8b**, then they may bounce back and forth many times. This magnetic-field configuration is called a "magnetic bottle." Earth's magnetic field is pinched together at the two magnetic poles and spreads out around the planet. This configuration is like taking many magnetic bottles and bending them over, attaching them to Earth at either end.

Earth and its magnetic field are immersed in the **solar wind**, a constant stream of charged particles from the Sun. When these particles first encounter Earth's magnetic field, the smooth flow is interrupted and they drop suddenly from **supersonic** to **subsonic** speed at a point

> Earth's magnetic field traps charged particles from the Sun.

called the **bow shock**. As they stream by, they are diverted by Earth's magnetic field as a river is diverted around a boulder. As they flow past, some of these charged particles become trapped by Earth's magnetic field, where they bounce back and forth between Earth's magnetic poles as illustrated in **Figure 8.8c**.

An understanding of Earth's magnetosphere is of great practical importance for space travel. Regions in the magnetosphere that contain especially strong concentrations of energetic charged particles, called **radiation belts**, can be very damaging to both electronic equipment and astronauts. Yet we need not leave the surface of the planet to witness beautiful and dramatic effects of the magnetosphere. Disturbances in Earth's magnetosphere can lead to changes in Earth's magnetic field that are large enough to trip power grids, causing blackouts, and to wreak havoc with communications. Earth's magnetic field also funnels energetic charged particles down into the ionosphere in two rings located around the magnetic poles. These charged particles (mostly electrons) collide with atoms such as oxygen, nitrogen, and hydrogen in the upper atmosphere, causing them to glow like the gas in a neon sign. These glowing rings, called **auroras**, can be seen from space (**Figure 8.9a**). When viewed from the ground (**Figure 8.9b**), they appear as eerie, shifting curtains of multicolored light. People living far from the equator are often treated to spectacular displays of the aurora borealis ("northern lights") in the Northern Hemisphere or the aurora australis in the Southern Hemisphere. Auroras have also been seen on Venus, Mars, the giant planets, and some moons.

Although our discussion has concentrated on the atmosphere of our own planet, it is important to know that the structure we have described here is not limited to Earth's atmosphere. The major vertical structural components—troposphere, tropopause, stratosphere, and ionosphere—also exist in the atmospheres of Venus and Mars, as well as in the atmospheres of Titan and the giant planets. And, as we will see in the following chapter, the magnetospheres of the giant planets are among the largest structures in the Solar System.

FIGURE 8.9 Auroras result when particles trapped in Earth's magnetosphere collide with molecules in the upper atmosphere. (a) An auroral ring around Earth's north magnetic pole, as seen from space. (b) Aurora borealis—the "northern lights"—viewed from the ground.

(a)

(b)

Why the Winds Blow

Winds are the natural movement of air, both locally and on a global scale. Variation in solar heating is the chief reason for differences in the ground-level temperature of Earth's atmosphere from place to place at similar altitudes and throughout the year. It is

> Nonuniform solar heating creates our weather.

usually warmer in the daytime than at night, warmer in the summer than in winter, and warmer at the equator than in the polar regions. Large bodies of water, such as oceans, also affect atmospheric surface temperatures. As we discovered in Foundations 8.1, heating a gas increases its pressure, which in turn causes it to push into its surroundings. The horizontal component of these pressure differences is the reason we have winds, and the strength of the winds is closely related to the magnitude of the difference in temperature and pressure from place to place.

Think about what happens as air in Earth's equatorial regions, heated by the warm surface, begins to rise because of convection. The warmed surface air displaces the air above it, which then has no place to go but toward

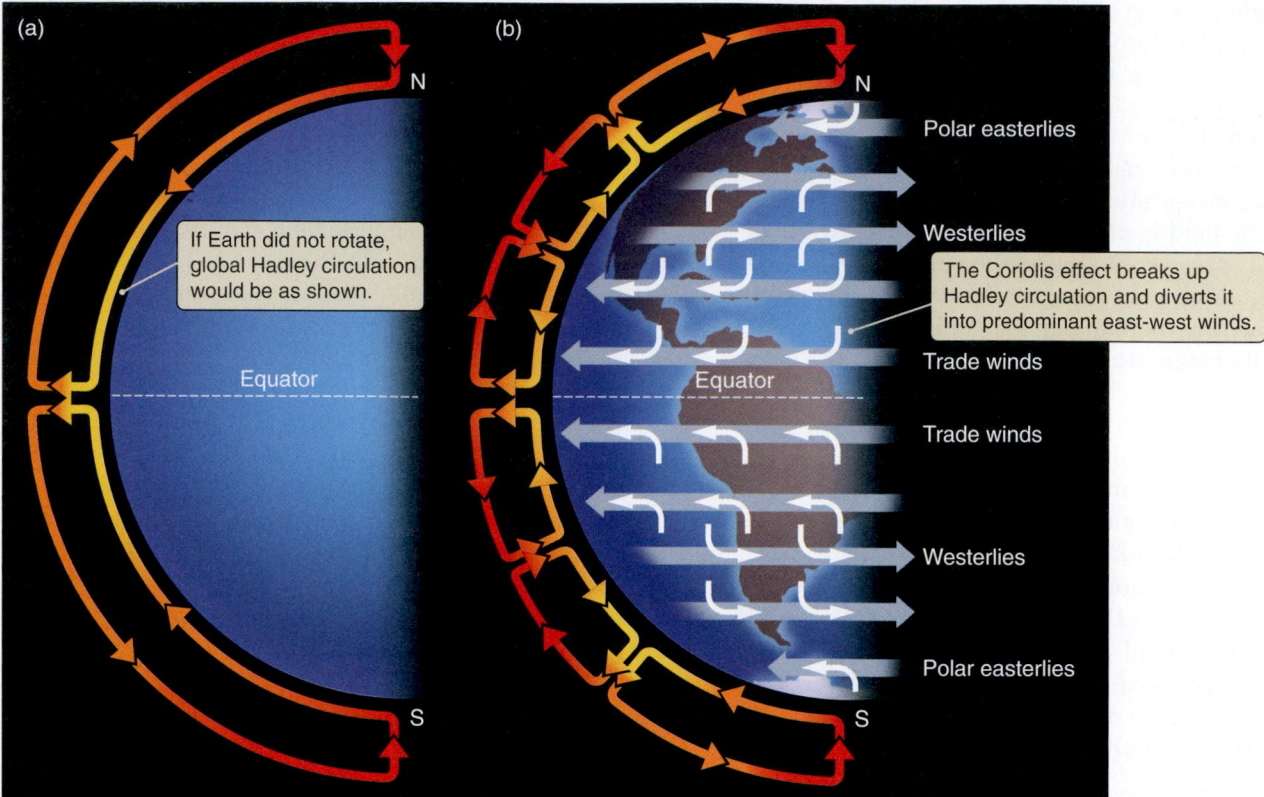

FIGURE 8.10 (a) The classic Hadley circulation. (b) Hadley flow often breaks up into smaller circulation cells. The poleward-equatorward flow is diverted into zonal flow by the Coriolis effect.

the poles. Cooling and becoming denser as it moves poleward, the displaced air now descends in the polar regions. There it displaces the surface polar air, which is forced back toward the equator, completing the circulation. As a result, the equatorial regions remain cooler and the polar regions remain warmer than they otherwise would be. Such planetwide flow of air between equator and poles is called **Hadley circulation** (**Figure 8.10a**). Global Hadley circulation, it turns out, seldom occurs in planetary atmospheres because other factors break up the planetwide flow into a series of smaller "Hadley cells." Planet rotation is a major factor here. Most planets and their atmospheres are rotating rapidly, and the effects produced by this rotation strongly interfere with Hadley circulation by redirecting the horizontal flow (**Figure 8.10b**).

The effect of Earth's rotation on winds—and on the motion of any object—is called the **Coriolis effect** (see **Foundations 8.2**). On a rapidly rotating planet, air is not free to flow in just any direction. When a volume of air starts to move directly toward or away from the poles, the Coriolis effect diverts it into relative motion that is more or less parallel to the planet's equator. This change in motion creates winds that blow predominantly in an east-west direction (see Figure 8.10b). Meteorologists call these **zonal winds**. In general, the more rapid a planet's rotation is, the stronger the Coriolis effect and the stronger its zonal winds

will be. We will encounter very strong zonal winds when we discuss the atmospheres of the giant planets in Chapter 9. Zonal winds are often confined to relatively narrow bands of latitude. Between the equator and the poles in most planetary atmospheres, the zonal winds alternate between "easterly" (those blowing *from the east* and toward the west) and "westerly" (those blowing *from the west* and toward the east). Confusing? Very! This unfortunate terminology is a historical carryover from early terrestrial meteorology, in which winds are labeled not by the direction toward which they are blowing but by the direction from which they come.

> Nonuniform heating, together with the Coriolis effect, causes east-west zonal winds.

In Earth's atmosphere, several bands of alternating zonal winds lie between the equator and the poles of both hemispheres. This zonal pattern is called Earth's **global circulation** because its extent is planetwide. The best-known zonal currents are the subtropical trade winds—more or less easterly winds that once carried sailing ships from Europe westward to the Americas—and the midlatitude prevailing westerlies that carried them home again (see Figure 8.10b).

Embedded within Earth's global circulation pattern are systems of winds associated with large high- and

FOUNDATIONS 8.2

The Coriolis Effect

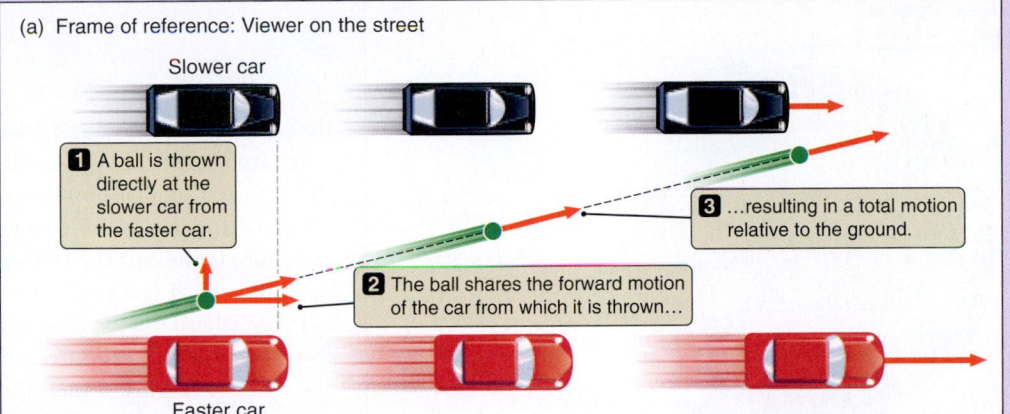

(a) Frame of reference: Viewer on the street

Slower car

1 A ball is thrown directly at the slower car from the faster car.

2 The ball shares the forward motion of the car from which it is thrown…

3 …resulting in a total motion relative to the ground.

Faster car

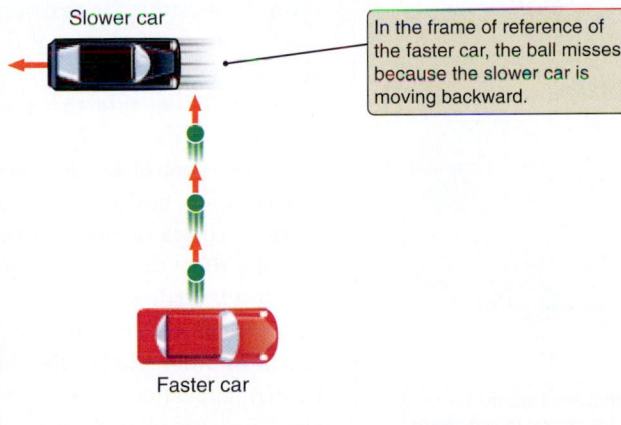

(b) Frame of reference: Viewer in faster car

Slower car

In the frame of reference of the faster car, the ball misses because the slower car is moving backward.

Faster car

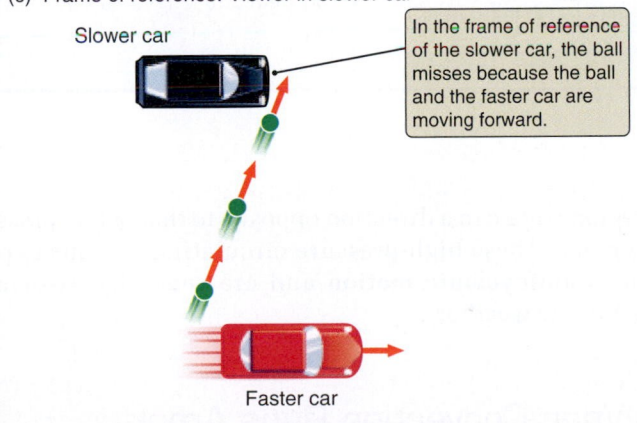

(c) Frame of reference: Viewer in slower car

Slower car

In the frame of reference of the slower car, the ball misses because the ball and the faster car are moving forward.

Faster car

FIGURE 8.11 The motion of an object depends on the frame of reference of the observer.

Earth's rotation influences things as diverse as the motion of weather patterns on Earth and how an artillery gunner must aim at a distant target. Any object sitting on the surface of Earth follows a circle each day as Earth rotates on its axis. This circle is larger for objects near Earth's equator and smaller for objects closer to one of Earth's poles. But because Earth is a solid body, all objects must complete their circular motion in exactly one day. Because an object closer to the equator has farther to go each day than an object nearer a pole, the object nearer the equator must be moving *faster* than the object at a higher latitude. If an object starts out at one latitude and then moves to another, its apparent motion over the surface of Earth is influenced by this difference in speed.

Now recall our discussion of frame of reference in Chapter 2 to understand this apparent motion. Imagine that two cars are driving down the road at different speeds, as shown in **Figure 8.11**. Ignoring for the moment any real-world complications like wind resistance, if you were to throw a ball from the faster-moving car directly at the slower-moving car as the two cars passed, you would miss. The ball shares the forward motion of the faster car, so the ball outruns the forward motion of the slower car. From your perspective in the faster car (**Figure 8.11b**), the slower car lagged behind the ball. From the slower car's perspective (**Figure 8.11c**), your car and the ball sped on ahead.

Now do the same experiment, but instead of two cars think about two locations at different latitudes. Suppose you fire a cannon directly north from a point in the Northern Hemisphere, as shown in **Figure 8.12b**. Because the cannon is located nearer to the equator than its target is, the cannon itself is moving toward the east faster than its target. Even though the cannonball is fired toward the north, it shares the eastward velocity of the cannon itself. This means that the cannonball is *also* moving toward the east faster than its target! Recall how the ball thrown from the faster car outpaced the slower-moving car in Figure 8.11. Similarly, as the cannonball flies north, it finds itself moving toward the east faster than the ground underneath it is. To an observer on the ground, the cannonball appears to curve toward the east as it outruns the eastward motion of the ground it is crossing. The farther north the cannonball flies, the greater is the difference between

Continued on next page

FOUNDATIONS 8.2

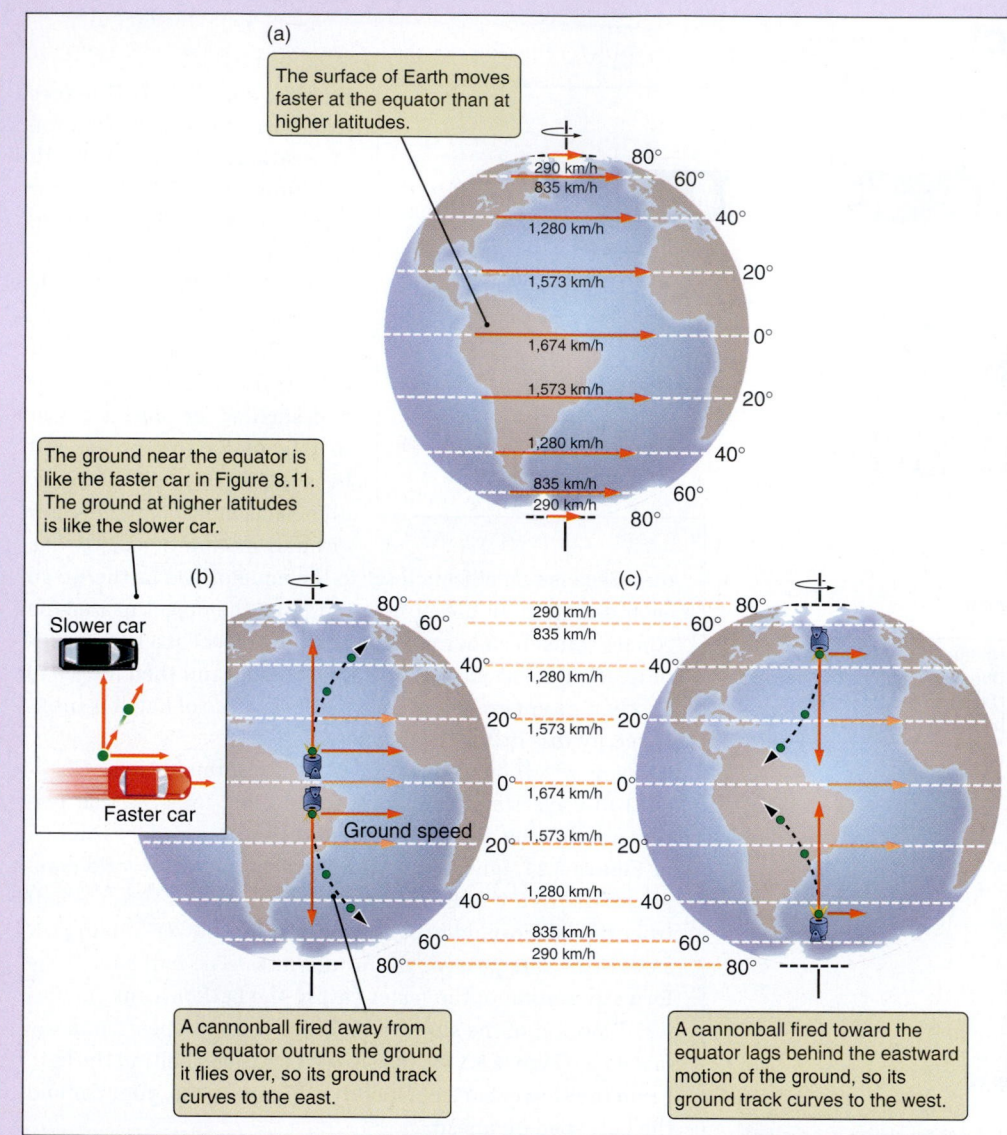

(a)

The surface of Earth moves faster at the equator than at higher latitudes.

290 km/h
835 km/h
1,280 km/h
1,573 km/h
1,674 km/h
1,573 km/h
1,280 km/h
835 km/h
290 km/h

80°
60°
40°
20°
0°
20°
40°
60°
80°

The ground near the equator is like the faster car in Figure 8.11. The ground at higher latitudes is like the slower car.

(b)

Slower car

Faster car

Ground speed

A cannonball fired away from the equator outruns the ground it flies over, so its ground track curves to the east.

(c)

290 km/h
835 km/h
1,280 km/h
1,573 km/h
1,674 km/h
1,573 km/h
1,280 km/h
835 km/h
290 km/h

80°
60°
40°
20°
0°
20°
40°
60°
80°

A cannonball fired toward the equator lags behind the eastward motion of the ground, so its ground track curves to the west.

FIGURE 8.12 The Coriolis effect causes objects to appear to be deflected as they move across the surface of Earth.

its eastward velocity and the eastward velocity of the ground. Thus, the cannonball follows a path that appears to curve more and more to the east the farther north it goes. If you are located in the Northern Hemisphere and fire a cannonball *south* toward the equator (**Figure 8.12c**), the opposite effect will occur. Now the cannon is moving toward the east more slowly than its target. As the cannonball flies toward the south, its eastward motion lags behind that of the ground underneath it, and the cannonball appears to curve toward the west.

This effect of Earth's rotation is called the Coriolis effect. In the Northern Hemisphere the Coriolis effect causes a cannonball fired north to drift to the east as seen from the surface of Earth. In other words, the cannonball appears to curve to the right. A cannonball fired south appears to curve to the west,

low-pressure regions. A combination of a low-pressure region and the Coriolis effect produces a circulating pattern called **cyclonic motion**. Cyclonic motion is associated with stormy weather, including hurricanes.[5] Similarly, high-pressure systems are localized regions where the air pressure is higher than average. We think of these regions of greater-than-average air concentration as "mountains" of air. Owing to the Coriolis effect, high-pressure

regions rotate in a direction opposite to that of low-pressure regions. These high-pressure circulating systems experience **anticyclonic motion** and are generally associated with fair weather.

When Convection Runs Amok

It takes the absorption of thermal energy to turn liquid water into vapor. Water in Earth's oceans, lakes, and rivers is evaporated by thermal energy acquired from the absorption of sunlight. The water vapor then carries this thermal energy along with it as it circulates throughout the atmo-

[5]Cyclones are regions characterized by rising moist air and are therefore associated with stormy weather. Conversely, anticyclones are vast regions of sinking dry air and thus tend to produce fair weather.

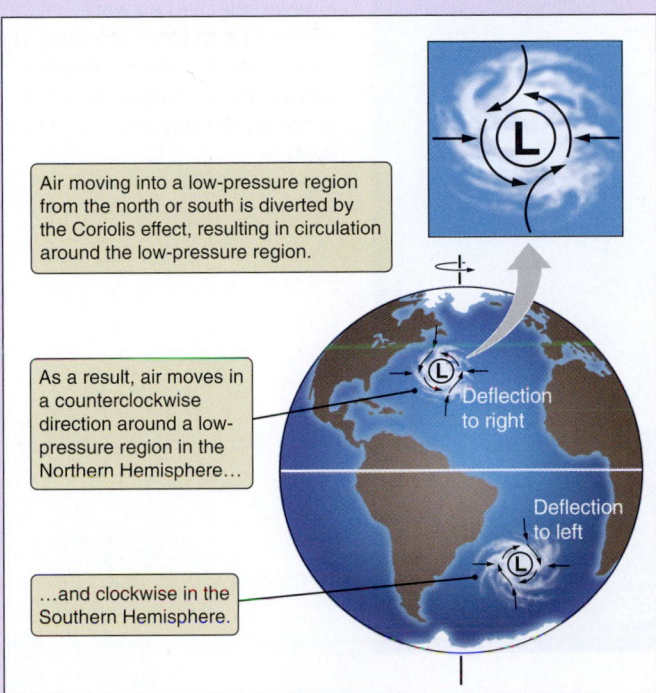

Air moving into a low-pressure region from the north or south is diverted by the Coriolis effect, resulting in circulation around the low-pressure region.

As a result, air moves in a counterclockwise direction around a low-pressure region in the Northern Hemisphere…

…and clockwise in the Southern Hemisphere.

Deflection to right

Deflection to left

FIGURE 8.13 As a result of the Coriolis effect, air circulates around regions of low pressure on the rotating Earth.

weather systems. As air is pushed from regions of higher pressure toward regions of lower pressure, these motions are influenced by the Coriolis effect. Think about a low-pressure region in the Northern Hemisphere. When air is pushed toward this region of low pressure from the south, the Coriolis effect deflects this flow of air toward the east. Similarly, air moving toward the region of low pressure from the north is deflected to the west by the Coriolis effect. The net effect is that as air moves toward a region of low pressure in the Northern Hemisphere, the Coriolis effect deflects it into a counterclockwise circulation (**Figure 8.13**). (Think about this carefully. The Coriolis effect deflects objects toward the right as you face north in the Northern Hemisphere, causing weather patterns that rotate toward the *left*.) The next time you see a television weather map with a low-pressure region, look at the direction of the winds around the region and you will see this counterclockwise flow. The most spectacular example of this cyclonic motion is the swirl of wind and clouds around the deep low-pressure area at the eye of a hurricane or typhoon. As the air moves in closer and closer to the central region of low pressure, it rotates faster and faster, giving rise to the winds of hundreds of kilometers per hour that make hurricanes so destructive.

The Coriolis effect is only one example of the type of effect associated with relative motions. However, the Coriolis effect has nothing to do with the direction that water swirls in a toilet bowl, as many people imagine. The difference in the speed of Earth's motion between the two sides of a toilet bowl is not enough to matter much. Other effects, such as the direction in which the water flows into the bowl, are much more important. However, the Coriolis effect is enough to deflect a fly ball hit north or south into deep left field in a stadium in the northern United States by about a half a centimeter. At one time or another the Coriolis effect has probably determined the outcome of a ball game.

which also gives it the appearance of curving to the right. In the Northern Hemisphere the Coriolis effect seems to deflect things to the *right*. If you think through this example for the Southern Hemisphere, you will see that south of the equator the Coriolis effect seems to deflect things to the *left*. In between, at the equator itself, the Coriolis effect vanishes.

These differences in the Coriolis effect between the Northern and Southern hemispheres are seen in the rotation of

sphere. When the water vapor recondenses, it gives up its thermal energy to its surroundings. This is the process that powers rainstorms, thunderstorms, hurricanes, and a host of other dramatic weather phenomena.

Rainstorms and Thunderstorms A rainstorm begins when Earth's surface, heated by the Sun, warms moist air close to the ground (see Figure 8.15). The moist air is convected upward, cooling as it gains altitude. Those puffy white clouds that you see on a summer day are created when this warm, moist air is convected upward to cooler atmospheric levels a few kilometers above the surface. Here the moisture condenses out of the air to form the myriad

tiny water droplets that we call cumulus clouds. Cooling then causes the water vapor in the moist air to condense as rain. As it condenses, the water vapor gives up its thermal energy to the surrounding air, warming it and thus increasing the strength of the convection. For the most part this is a gentle process. Water, falling back to the surface as rain, eventually returns to the lakes and oceans, wearing down mountains, eroding the soil, and nourishing life as it flows. From the oceans to the air and back again—this is the "water cycle," or hydrological cycle. (In Chapter 11 we will see that methane plays the same role on Titan that water plays on Earth in what some planetary scientists call a "methanological cycle.")

FIGURE 8.14
Thunderstorms are powered by convection and by thermal energy released as water vapor condenses to form droplets. The "anvil" top of a thunderhead is caused by stratospheric winds shearing the top of the convective system.

Convection can also be violent. With strong solar heating and an adequate supply of moist air, this self-feeding process can grow within minutes to become a violent thunderstorm. Summer is the peak time for thunderstorms and lightning. The process that creates thunderstorms is the same as that which forms cumulus clouds, but the amount of energy involved is far greater. Thunderstorms tend to form in the afternoon when solar heating of the ground reaches its maximum. They begin as familiar cumulus clouds; but if the supply of warm, moist air at the surface is great enough, the clouds will continue to grow vertically, and we have a case of convection run amok. As more and more moist air rises and the moisture condenses within the cloud, the heat released by condensation continues to warm the surrounding air, forcing convection ever higher.

> Runaway atmospheric convection can lead to violent weather.

We can easily recognize these cumulonimbus clouds—known popularly as "thunderheads"—by their flat, anvil-shaped tops, as seen in **Figure 8.14**. This upper surface visibly marks the tropopause, the level in the atmosphere where convection finally ceases (see Figure 8.6a). In the midlatitudes of the continental United States, the tropopause occurs at an altitude of about 12 km above sea level. Convection in the more violent storms can be so strong that the tops of the thunderheads punch right through the tropopause, carrying cloud-forming moisture up into the stratosphere to heights of 20 km or more. The anvil shape occurs when strong stratospheric winds pull ice crystals from the top of the cloud and spread them out horizontally.

For every parcel of air that rises within a thunderstorm, another must come back down (**Figure 8.15**). Downdrafts, including the rain and hail that descend with them, produce ferocious winds that can exceed 100 kilometers per hour (km/h) in the immediate vicinity of the storm. The winds and turbulence associated with thunderstorms become so great that even commercial jet aircraft prefer to keep their distance. At its base, a thunderstorm may be several kilometers across and travel for tens of kilometers across the landscape. Thunderstorms cause billions of dollars in crop damage every year and are responsible for hundreds of deaths.

Lightning Lightning is essentially a gigantic electric spark that results from billions of volts of potential difference—a huge example of the "static electricity" you sometimes generate when walking on a carpet—and is usually associated with thunderstorms and rain. However, lightning can also be created by snow, sand and dust storms, volcanic eruptions, earthquakes, and nuclear explosions. Ice formation in cumulonimbus clouds is a key factor for starting the "electric generator" that produces most lightning. Falling small ice pellets in cumulonimbus clouds become negatively charged by friction as they move through the surrounding air, while small supercooled cloud droplets that strike them bounce off the ice pellets and become positively charged. The supercooled cloud droplets rise on updrafts to the top of the storm cloud while the ice pellets fall and melt in the lower regions of the cloud or, as often is the case, fall all the way to the ground.

As a result, a difference in electric potential is created between the top and bottom of the cloud that can exceed a

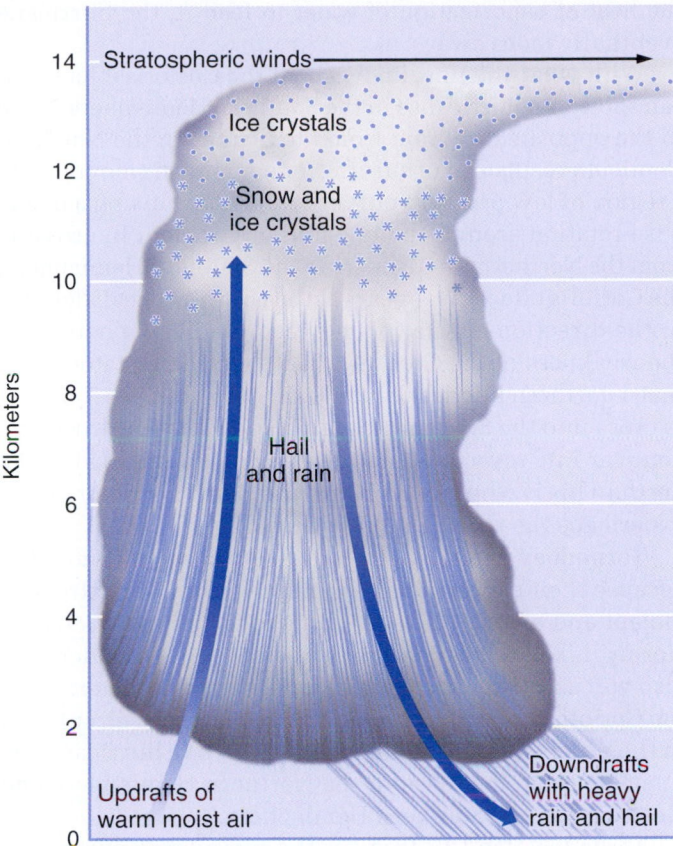

FIGURE 8.15 Convection within a thunderstorm.

FIGURE 8.16 Lightning bolts carry enormous electric currents, killing or injuring hundreds of people each year in the United States alone.

billion volts. At some point the potential difference between parts of the cloud becomes so great that the electrical resistance of the air breaks down[6] (the molecules become ionized) and a series of lightning bolts flashes between the positively and negatively charged regions of the cloud. This is called "cloud-to-cloud lightning." About a third of the time, lightning travels between the cloud and the ground. It is here that lightning can be so dangerous. In the United States lightning kills an average of 100 people every year and injures more than 300. Worldwide, lightning strikes occur about 100 times every second.

Lightning bolts are typically 3–5 km long and can carry tens of thousands of amperes of current at speeds of 200 kilometers per second (km/s) (**Figure 8.16**). On average, the energy of a single lightning bolt could keep a 100-watt (W) lightbulb burning for several months. Although the brilliance of lightning bolts makes them appear huge, they are really no larger in diameter than a quarter or a half-dollar. As it travels through the air, lightning heats the air to temperatures as high as the surface of the Sun, causing the volume of air along the path of the bolt to expand rapidly.

[6]Some atmospheric physicists now believe that all lightning is initiated by secondary cosmic-ray particles (see Chapter 20), which trigger the electrical breakdown (ionization) of the air along the path of the lightning bolt.

The thermal energy causing this rapid expansion of the air is converted into sonic energy, resulting in the familiar boom of thunder.

Lightning also creates powerful electromagnetic waves, some of which can be heard on AM radio receivers as "lightning static." When the *Pioneer Venus* spacecraft was orbiting our sister planet during the 1980s, its radio receiver picked up many bursts of lightning static—so many that Venus appears to have more lightning activity than our own planet. On Venus, as on Earth, lightning is created in the clouds; but Venus's clouds are so high—typically 55 km above the surface of the planet—that the lightning bolts never hit the ground. All of this lightning is of the cloud-to-cloud type. The *Pioneer Venus* results supported observations made by the Soviet lander *Venera 9*, whose optical spectrometer had detected flashes of lightning on Venus's dark side 5 years earlier. As we will learn in the next chapter, lightning has also been detected in the atmospheres of Jupiter, Saturn, Uranus, and Neptune.

> Lightning has been detected in the atmospheres of Venus and all four giant planets.

Hurricanes, Tornadoes, and Dust Devils Hurricanes are examples of Coriolis forces acting on air rushing into regions of low atmospheric pressure, creating a huge **vortex** (see Chapter 9.) But these most powerful of storms are much more complicated than the simple systems described in the preceding discussion. Hurricanes are huge heat engines. They derive their enormous energy from a very

> Coriolis forces can create powerful, and often deadly, atmospheric vortices.

common physical phenomenon: the heat of vaporization of water. The conditions for formation must be just right—warm tropical seawater, light winds, and a region of low pressure in which air spirals inward. As warm seawater evaporates, moisture-laden air rises and releases its heat of vaporization as it condenses at cooler levels. (Remember, this is the same process that leads to cumulonimbus thunderstorms.) When the supply of warm seawater is sufficient, a complex of thunderstorms develops. Then, if the winds aloft are light, the complex remains intact and grows in size and strength. Convection ceases at the tropopause, located about 15 km above sea level at tropical latitudes; but the number of individual storm cells in the complex continues to increase. The stage is now set for the birth of a hurricane.

As surface winds driven by the Coriolis effect rush inward to replace the air rising upward in the cumulonimbus complex, the hurricane grows in size and strength. Sustained winds near the center of the storm can reach speeds greater than 300 km/h, causing widespread damage and fatalities when a hurricane moves ashore. In 1900, a hurricane in Galveston, Texas, took 8,000 human lives—more than any other natural disaster in US history. Hurricane Andrew caused $12 billion in damage when it hit Florida in 1992. In 2005, Hurricane Katrina (pictured in **Figure 8.17**) devastated much of the Gulf Coast and the city of New Orleans, taking more than 1,000 lives and costing more than $200 billion in damage. The eye of a hurricane, however, is relatively calm and free of clouds. The eye is typically 40–50 km wide, whereas the hurricane itself may extend outward for more than 600 km. A hurricane may last weeks and travel thousands of kilometers, as long as it remains over open ocean. But what happens if the hurricane moves ashore? Over land it loses its principal supply of warm, moist air and therefore its source of energy. Without

the heat of vaporization of water to feed it, the hurricane eventually fades away.

With what you now know about the Coriolis effect, you can show that hurricanes in the Southern Hemisphere rotate in the opposite direction from hurricanes in the Northern Hemisphere. Instead of curving to the right, air moving into a region of low pressure curves to the left, causing a clockwise rotation around the low-pressure region. In crossing from the Northern Hemisphere to the Southern Hemisphere, the Coriolis effect disappears at the equator. The difference in the direction of rotation between the hemispheres and the weakness of the Coriolis effect near the equator mean that a northern hurricane would literally collapse if it tried to cross into the Southern Hemisphere. (Imagine throwing your car into reverse while traveling down the road at 150 km/h.) This is why countries around Earth's equator do not experience the ravages of hurricanes.

Tornadoes generally last only a few dozen minutes, but because their energy is so concentrated they are extremely violent and among the most dangerous and destructive of storms. Like their larger hurricane cousins, tornadoes are also vortices. They are usually too small to be governed by the Coriolis effect; but the general atmospheric circulation in their vicinity causes many tornadoes, like hurricanes, to rotate counterclockwise in the Northern Hemisphere and clockwise in the Southern Hemisphere.

Tornadoes tend to form in the vicinity of hurricanes and violent thunderstorms where strong, thermally generated updrafts are present. When surrounding air rushes in to replace the rising air in the updraft, it may start spinning, creating a vortex. As the column of rising air extends upward, its diameter shrinks. And because the circulating air must obey the law of conservation of angular momentum, the vortex spins ever faster as it continues to shrink. (Recall the skater in Chapter 6.) Wind speeds in the most severe tornadoes have been estimated to reach 800 km/h! The base of an average tornado is about 400 meters in diameter. Atmospheric pressure at the base can be extremely low—more than 200 mb lower than the surrounding air. This low pressure causes the tornado to act like a gigantic vacuum cleaner, picking up dust, cars, and even buildings. Over the past 50 years, tornado-related deaths in the United States have averaged about 90 per year. The debris swept up by tornadoes makes them highly visible and reveals their characteristic funnel shape, as in **Figure 8.18**. When a tornado passes over a lake or the ocean, it picks up water and is called a "waterspout."

Dust devils are similar in structure to tornadoes, but they are generally smaller and less intense and they usually occur in fair weather. Diameters range from a few meters to a few dozen meters, with average heights of several

> **Dust devils occur on both Earth and Mars.**

hundred meters. As with tornadoes, the lifetime of a typical dust devil is brief, limited to a dozen or so minutes. Dust devils form in areas of strong surface heating, usually at

FIGURE 8.17 A satellite view of Hurricane Katrina in the Gulf of Mexico approaching New Orleans and the Gulf Coast in 2005.

FIGURE 8.18 A large tornado made visible by the debris it is sweeping up. For scale, note the power line structures at lower right.

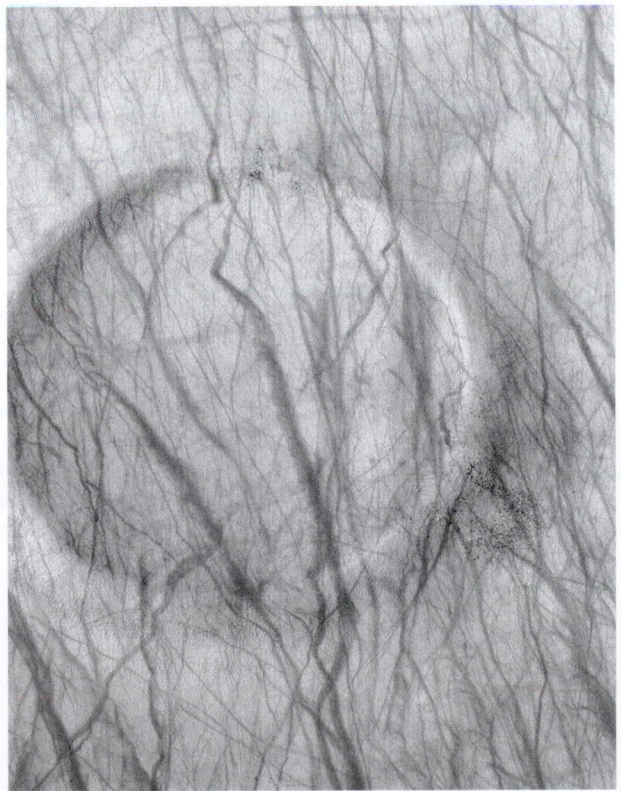

FIGURE 8.19 Meandering dark trails cross a martian crater where numerous dust devils have removed bright surface dust.

the interface between different surface types, such as asphalt and dirt or even irrigated fields and dirt roads. Typically they occur under clear skies and light winds, when air close to the ground is heated to temperatures much hotter than the cooler air above it—a very unstable condition. Though their wind speeds seldom exceed 25 m/s, dust devils can still be destructive as they lift dust and other debris into the air. Small structures can be damaged and even destroyed if they have the misfortune of being located in the path of a strong dust devil. Dust devils are a common sight in the deserts of the American Southwest.

Dust devils have also been seen on the arid surface of Mars, first by the *Viking* orbiters in 1976. In 1997, *Mars Pathfinder* sensed one passing right over it. More recently, *Mars Global Surveyor* spotted a large number of dust devils, made easily visible by the shadows they cast on the martian surface. Most dust devils leave dark meandering trails behind them where they have vacuumed up bright surface dust, revealing the dark surface rock that lies beneath the dust (**Figure 8.19**). Dust devils on Mars—typically higher, wider, and stronger than their Earth counterparts—reach heights of up to 8 km and have diameters ranging from a few dozen to a few hundred meters.

Is Earth Getting Warmer?

The state of Earth's atmosphere at any given time and place is what we commonly call "weather." **Climate** is the term we use to define the *average* state of Earth's atmosphere, including temperature, humidity, winds, and so on. Earth's climate appears to go through lengthy temperature cycles, usually lasting hundreds of thousands of years and occasionally producing shorter cold periods called "ice ages." These oscillations in the mean global temperature are far smaller than typical geographic or seasonal temperature variations. But Earth's

> Climate is the average state of Earth's atmosphere.

atmosphere is so sensitive to even small temperature changes that it takes a drop of only a few degrees in the mean global temperature to plunge our climate into an ice age. We still do not understand all of the mechanisms for these climate-changing temperature swings. An external influence, such as small changes in the Sun's energy output, may be the cause. Changes in Earth's orbit or the inclination of its rotation axis also have been suggested. Or these temperature changes may be triggered internally by volcanic eruptions (which can produce global sunlight-blocking clouds or hazes) or long-term interactions between Earth's oceans and its atmosphere.

Nearly all climatologists now believe that changes in Earth's climate have been accelerating recently and that Earth is becoming warmer. Temperature measurements over the past century show a steady increase in the mean global temperature. Does this trend represent the beginning of a long-term change caused by the buildup of human-made greenhouse gases (**Figure 8.20**)—as most computer models suggest—or merely a temporary short-term cycle? This is a question that some still debate. However, we know that our atmosphere is a delicately bal-anced mechanism. Tiny changes can produce enor-mous and often unexpected results. Earth's climate is an example of the sort of com-plex, chaotic system we will discuss in Chapter 10. To add to the complexity, Earth's climate is intimately tied to ocean temperatures and currents. We see examples of this connec-tion in the periodic El Niño and La Niña conditions, when small shifts in ocean temperature cause much larger global changes in air temperature and rainfall. Recent studies sug-gest that changes in the flow of the Gulf Stream in the North Atlantic can have very large and unpredictable effects on the climates of North America and northern Europe—and that these changes may take place not over centuries but within a matter of decades! Are we unknowingly jeopardiz-ing our future by meddling with our atmosphere? In a sense, we are much like children playing with matches.

> Increases in Earth's greenhouse gases can lead to disastrous climate change.

8.4 Venus Has a Hot, Dense Atmosphere

Venus and Earth are similar in many ways—so similar that they might be thought of as sister planets. Indeed, when we used the laws of radiation in Chapter 4 to predict tempera-tures for the two planets, we concluded that they should be very close. But that was before we considered the green-house effect and the role of carbon dioxide in blocking the infrared radiation that a planetary surface typically emits. The atmospheric pressure at the surface of Venus is a crush-ing 92 times greater than that at the surface of our own planet—equal to the water pressure at an ocean depth of 900 meters. (The pressure at this ocean depth would crush the hull of a World War II–era submarine.) Most of this mas-sive atmosphere (96 percent) is carbon dioxide, with a small amount (3.5 percent) of nitro-gen and still lesser amounts of other gases. This thick blan-ket of carbon dioxide effectively traps the infrared radiation from Venus, driving the temperature at the surface of the planet to a sizzling 737 K, which is hot enough to melt lead. Whereas Earth is a lush paradise, the runaway greenhouse effect has turned Venus into a convincing likeness of hell—an analogy made complete by the presence of choking amounts of sulfurous gases. Venus may be our "sister" planet in many respects, but it will be a *very* long time before humans visit its surface, if ever.

> Venus has a massive carbon dioxide atmosphere, making it a "poster child" for the greenhouse effect.

As on Earth, the atmospheric temperature of Venus decreases continuously throughout the planet's troposphere, dropping to a low of about 160 K at the tropopause. At an altitude of approximately 50 km, the Venus atmosphere is similar to our own atmosphere at sea level in both pres-sure and temperature. At altitudes between 50 and 80 km (see Figure 8.6b), the atmosphere is cool enough for sulfu-rous oxide vapors to react with water vapor to form dense clouds of concentrated sulfuric acid droplets (H_2SO_4). These

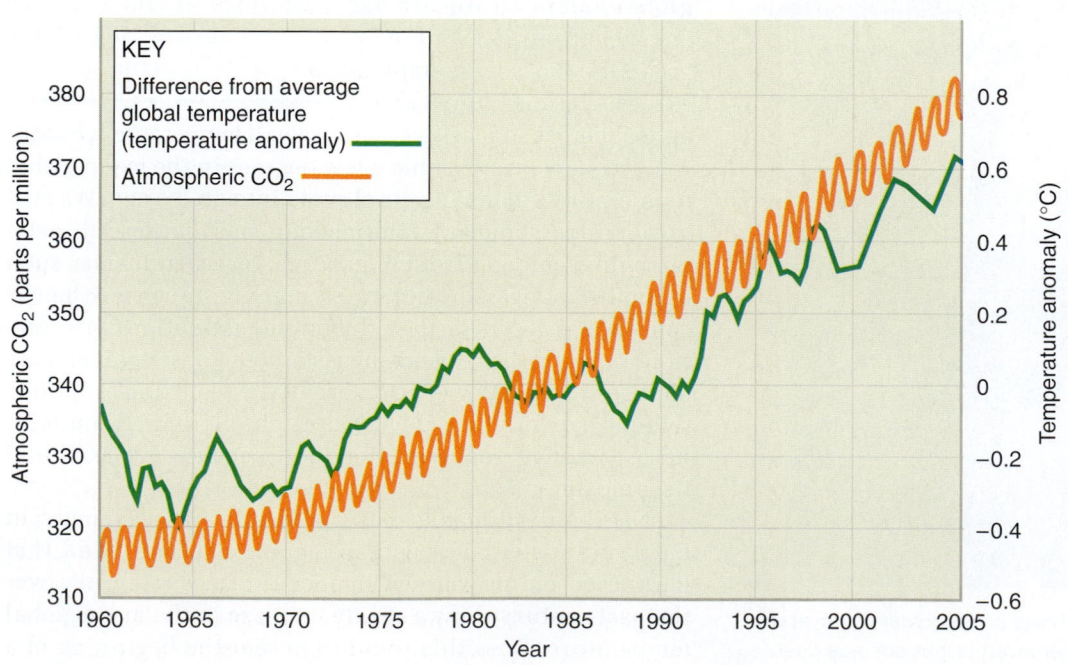

FIGURE 8.20 Our climate is delicately balanced, and carbon dioxide plays a vital role in that balance. This graph shows that global temperatures are climbing along with concentrations of CO_2. Annual variations in atmospheric CO_2 can be attributed to seasonal variations in plant life and fossil fuel use, while the steady climb is due to human activities.

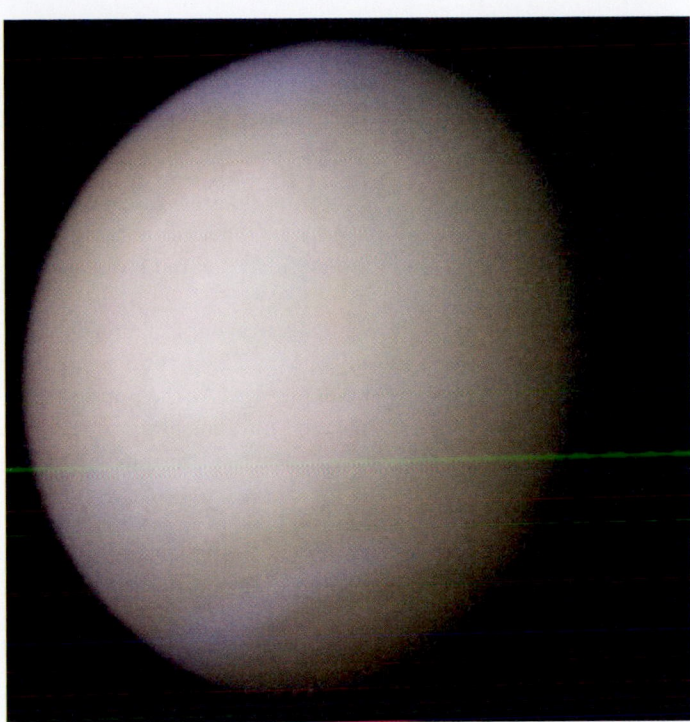

FIGURE 8.21 Thick clouds obscure our view of the surface of Venus.

dense clouds completely block our view of the surface of Venus, as **Figure 8.21** shows. In the 1960s, spacecraft with cloud-penetrating radar provided low-resolution views of the surface of Venus. But it was not until 1975, when the Soviet Union succeeded in landing cameras there, that we got a clear picture of the surface. Radar images taken by the *Magellan* spacecraft in the early 1990s (see Figure 7.24) produced a global map of the surface of Venus.

Imagine yourself standing (and surviving) on the surface of Venus. Since sunlight cannot easily penetrate the dense clouds above you, noontime on the surface of Venus is no brighter than a very cloudy day on Earth. High temperatures and very light winds keep the lower atmosphere of Venus free of clouds and hazes. The local horizon can be seen clearly, but distant mountains are not as clear. Molecules in even a pure gas will scatter light, and the *scattering* efficiency increases sharply with decreasing wavelength. Strong scattering by molecules in the dense atmosphere of Venus would greatly soften any view you might have of distant scenes. Molecular scattering, always stronger at the shorter wavelengths, as discussed in **Excursions 8.2**, causes a loss of contrast and adds a bluish cast to distant terrain. (We see the same effect, but to a lesser extent, in our own atmosphere.) The high atmospheric temperatures on Venus also mean that neither liquid water nor liquid sulfurous compounds can exist on its surface, leaving an extremely dry lower atmosphere with only 0.01 percent water and sulfur dioxide vapor.

Unlike most other planets, Venus rotates on its axis in the opposite sense of its motion around the Sun. Relative to the stars, Venus spins once every 243 Earth days, but a solar day on Venus—the time it takes for the Sun to return to the same place in the sky (see Chapter 2)—is only 117 Earth days. Regardless, this slow rotation means that Coriolis effects on the atmosphere are small, resulting in a global circulation that is quite close to a classic Hadley pattern (see Figure 8.10a). Venus is the only planet known to behave in this way. Its massive atmosphere is highly efficient in transporting thermal energy around the planet, so the polar regions are only a few degrees cooler than equatorial regions, and there is almost no temperature difference between day and night. Because the Venus equator is nearly in the plane of its orbit, seasonal effects are quite small, producing only negligible changes in surface temperature. (Recall the discussion in Chapter 2 about how the seasons change because of the tilt of Earth's equator relative to the plane of its orbit around the Sun.) Such small temperature variations also mean that wind speeds near the surface of Venus are quite low, typically about a meter per second. High in the rarefied atmosphere, though, where temperature differences can be larger, winds reach speeds of 110 m/s, circling the planet in only 4 days. The variation of this high-altitude wind speed with latitude can be seen in the chevron, or V-shaped, cloud patterns in **Figure 8.23**.

> Surface temperatures on Venus vary little from pole to equator or from day to night.

Large variations in the observed amounts of sulfurous compounds in the high atmosphere of Venus suggest to planetary scientists that the source of sulfur may be sporadic episodes of volcanic activity. This evidence strengthens the possibility that Venus remains a volcanically active planet.

8.5 Mars Has a Cold, Thin Atmosphere

Compared to Venus, the surface of Mars is almost hospitable. For this reason we can confidently expect that humans will eventually set foot on the red planet, quite likely before the end of the 21st century and possibly much sooner. What they will see is a stark, waterless landscape, colored reddish by the oxidation of iron-bearing surface minerals. The sky will sometimes be a dark blue but more often a pinkish color caused by windblown dust (**Figure 8.24**). The lower density of the Mars atmosphere makes it more responsive than Earth's to heating and cooling, so temperature extremes are greater. Near the equator at noontime, future astronauts may experience a comfortable 20°C—about the same as a cool room temperature. Nighttime temperatures typically drop to a frigid −100°C, and during the polar night the air temperature can reach

> The surface of Mars is cold and the air is thin.

EXCURSIONS 8.2

Blue Skies, White Clouds, and Red Sunsets

Have you ever noticed the beam from a flashlight or the headlights of a car on a foggy night? As the beam of light shines through the cloud bank, some of the light bounces off tiny water droplets, and you see the light that happens to bounce in your direction. In this way, you "see the beam of light."

The bouncing of light off small particles in its path is called **scattering**. If the particles that are scattering light are much larger than the wavelength of that light, as is the case for water droplets in a fog bank, then photons of all colors are equally likely to be scattered. As a result, the light scattered by a cloud is the same color as the light shining on the cloud. Sunlight is white, so clouds that are illuminated by direct sunlight are also white.

Things get more interesting if the scattering particles are smaller than the wavelength of the light they are interacting with. In such instances, shorter-wavelength photons are more likely to be scattered than longer-wavelength photons. Molecules in Earth's atmosphere, for example, scatter blue light ($\lambda = 400$ nanometers, or nm) about seven times more effectively than they scatter red light ($\lambda = 650$ nm). When the Sun is high overhead, sunlight follows a short path through the atmosphere and so is relatively unaffected by scattering. Even so, a small

fraction of the blue photons in the sunlight are scattered, and when you look at the sky your eyes detect the blue photons that were scattered in your direction. As illustrated in **Figure 8.22**, that is why the sky is blue.

This situation changes dramatically as evening approaches. As the Sun drops lower in the sky, the light from the Sun must pass through more air before it reaches you. As the Sun nears the horizon, sunlight passes through hundreds of times more air than it did when the Sun was high overhead. So much of the blue light is scattered away that by the time the light reaches you, the Sun looks orange. Tiny particles of dust and other materials in the atmosphere (which are similar in size to the wavelength of light) scatter away additional blue light, leaving only a glorious red.

The next time you are captivated by the beauty of a sunset, remember also to look up at the deepening blue of the sky overhead. The red sunset is white sunlight minus the blue light that was scattered away. The blue sky is scattered blue light only. The red of sunset, the blue of the sky, and the white of a billowing cloud are three facets of the same gem—a gem called scattering.

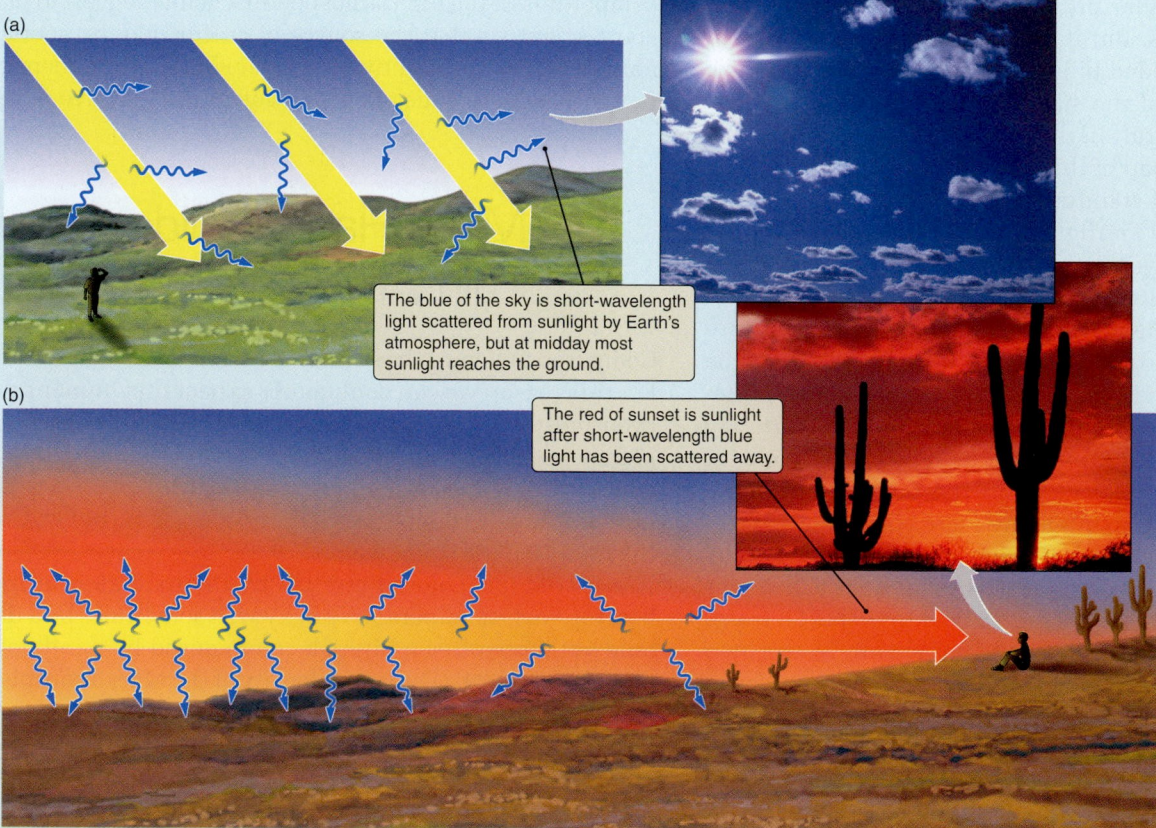

(a)

The blue of the sky is short-wavelength light scattered from sunlight by Earth's atmosphere, but at midday most sunlight reaches the ground.

The red of sunset is sunlight after short-wavelength blue light has been scattered away.

(b)

FIGURE 8.22 The blue of a daytime sky (a) and the red of a sunset (b) are both due to differences in the way Earth's atmosphere scatters light of different wavelengths.

FIGURE 8.23 This *Pioneer Venus* image, taken in ultraviolet light, shows clouds of sulfur compounds in the atmosphere of Venus. Variations in wind speed with latitude cause the clouds to streak out into a "chevron," or V-shaped, pattern.

FIGURE 8.24 A view of the dust-filled pink sky of Mars as seen from the *Viking* lander. In the absence of dust, the sky's thin atmosphere would appear deep blue.

−150°C—cold enough to freeze carbon dioxide out of the air in the form of a dry-ice frost. For human visitors, the low surface pressure will certainly be uncomfortable. The average atmospheric surface pressure of Mars is equivalent to the pressure at an altitude of 35 km on Earth, far higher than our highest mountain. There is, of course, no "sea level" on Mars. Surface pressure varies from 11.5 mb (1.1 percent of Earth's pressure at sea level) in the lowest impact basins of Mars, to 0.3 mb at the summit of Olympus Mons.

Like Earth, Mars does have some water vapor in its atmosphere, but the low temperatures condense much of it out as clouds of ice crystals (see Figure 8.6b). Early-morning ice fog in the lowlands (**Figure 8.25a**) and clouds hanging over the mountains will give Earth visitors some reminders of their home planet. Although the cold, thin air might be endured, what is seriously lacking and what future astronauts must carry with them is oxygen. Without plants, Mars has only a tiny trace of this life-sustaining gas. Like Venus, the atmosphere of Mars is composed almost entirely of carbon dioxide (95 percent), with a lesser amount of nitrogen (2.7 percent). The near absence of oxygen also means that Mars can have very little ozone, and this lack of ozone allows solar ultraviolet radiation to reach all the way to the surface. To survive on Mars, any surface life-forms would have to develop protective layers that could shield against the lethal ultraviolet rays.

The recent discovery of significant amounts of methane in the martian atmosphere has refueled arguments among scientists over the possibility of extant life on Mars (see Excursions 8.1). Atmospheric methane is quickly destroyed by solar ultraviolet radiation, typically lasting only a few hundred years, so its very survival in the martian atmosphere implies the existence of a continuous source to replace the losses. Life is certainly one possibility. Methane in Earth's atmosphere is largely a product of microorganisms, and similar life-forms might exist in protected environments far below the martian surface. On the other hand, many scientists argue that geological processes remain as reasonable alternative possibilities. The conversion of iron oxide into certain complex minerals releases methane as a by-product. Ancient methane trapped within ice crystals is another possible source. The likelihood of life existing on Mars today continues to be one of the great unanswered questions of our time.

The inclination of the Mars equator to its orbital plane is similar to Earth's, so it has similar seasons (see Chapter 2). But the effects are larger for two reasons: Mars varies more in its annual orbital distance from the Sun than does Earth, and the low density of the Mars atmosphere makes it more responsive to seasonal change. The large diurnal, seasonal, and latitudinal surface temperature differences on Mars often create locally strong winds, some estimated to be higher than 100 m/s. High winds can stir up huge quantities of dust (**Figure 8.25b**) and dis-

> Seasonal changes affect climate more on Mars than on Earth.

(a)   (b)

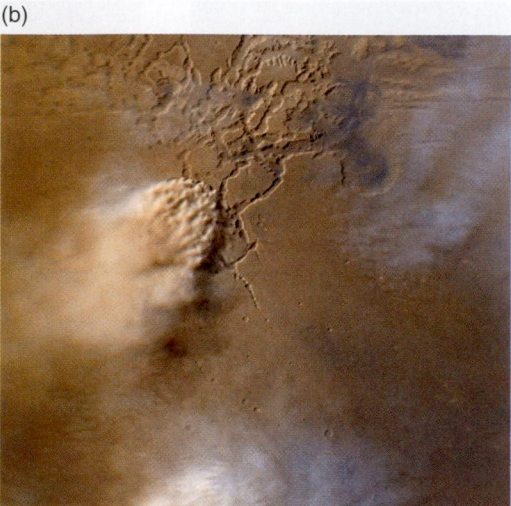

FIGURE 8.25 (a) Patches of early-morning water vapor fog forming in canyons on Mars. (b) A dust storm raging in the canyon lands of Mars.

tribute it around the planet's surface. For more than a century, astronomers have watched the seasonal development of springtime dust storms on Mars. The stronger ones spread quickly and within a few weeks can envelop the entire planet in a shroud of dust (**Figure 8.26**). Such large amounts of windblown dust can take many months to settle out of the atmosphere. Seasonal movement of dust from one area to another alternately exposes and covers large areas of dark, rocky surface. This phenomenon led some astronomers of the late 19th and early 20th centuries to believe that they were witnessing the seasonal growth and decay of vegetation on Mars. Public imagination carried the astronomers' interpretations a step further—to stories about advanced civilizations on Mars and invasions of Earth by warlike Martians (a theme found in some movies).

Mars likely had a more massive atmosphere in the distant past, and geological evidence suggests that liquid water once flowed across its surface, as we saw in Chapter 7, but its low gravity was responsible for the loss of much of this earlier atmosphere. This is a good example of a runaway atmosphere, as we discussed in Section 8.2.

8.6 Mercury and the Moon Have Almost No Atmosphere

Our story would not be complete if we did not include the kinds of "atmosphere" that exist today around the remaining two terrestrial objects, Mercury and the Moon. There is actually little to say. The atmospheres of Mercury and the Moon would hardly be noticed by a visitor from Earth. They are less than a million-billionth (10^{-15}) as dense as our own, and they probably vary somewhat with the strength of the solar wind and the atoms of hydrogen and helium they capture from it. Other atoms, such as sodium, calcium, and even water-related ions, were seen in Mercury's atmosphere by the *Messenger* spacecraft, and may have been blasted loose from Mercury's surface by the solar wind or micrometeoroids. Such ultrathin atmospheres can have no effect whatsoever on local surface temperatures. Unless you are one of the few astronomers interested in the interaction between solar wind and solid bodies, you may comfortably ignore the atmospheres of Mercury and the Moon.

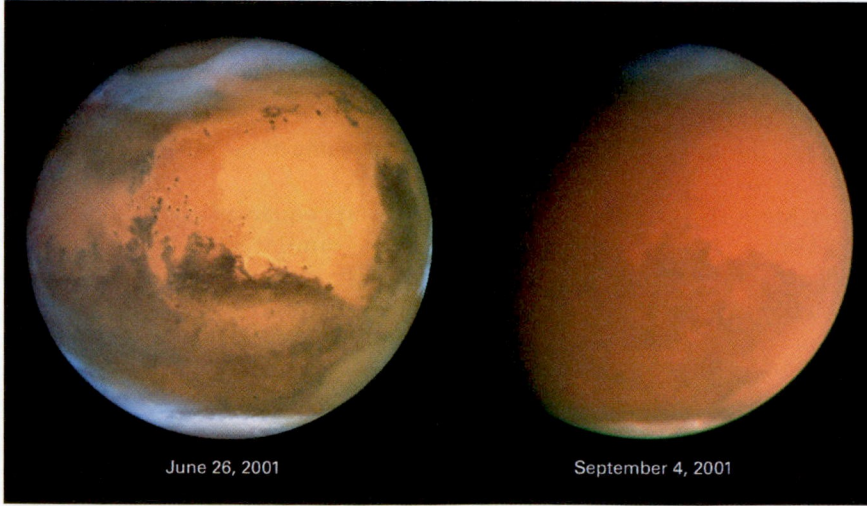

June 26, 2001   September 4, 2001

FIGURE 8.26 Hubble Space Telescope images showing the development of a global dust storm that enshrouded Mars in September 2001

Seeing the Forest for the Trees

While writing these words, one of the authors is sitting in his backyard on a pleasant summer morning. As birds fly overhead, he is warmed by the rays of the Sun and cooled by a gentle morning breeze. Such an idyllic setting might seem far from the path of our journey through astronomy, but nothing could be further from the truth. Some of the atoms drifting past arrived at Earth as it formed, later to be expelled by volcanism into Earth's forming secondary atmosphere. The rest, especially volatile materials, arrived in the rain of ice-laden comets that fell on the young Earth. During the ensuing billions of years, a delicate chemical and physical dance has taken place. The temperature of the young Earth was such that water could condense and fall on the surface as rain. The presence of liquid water served to scrub our atmosphere, absorbing carbon dioxide and locking it up in carbon-bearing rock such as limestone.

Liquid water also provided the bath in which organic chemistry could take place, leading first to molecules that had the useful property of being able to reproduce themselves at the expense of other molecules, and later to life itself. Early life on Earth consumed molecules such as carbon dioxide. The waste products formed by this life included what, at the time, was a deadly poison—oxygen. Over the eons the amount of oxygen in our planet's atmosphere increased, and new forms of life evolved to take advantage of its presence. We are among these later life-forms. To us, the reactive, corrosive gas called oxygen is the breath of life.

In contrast to Earth, Mercury and the Moon were simply too small to hold on to either their primary or their secondary atmospheres, leaving them as airless rocks. Looking inward from Earth toward the Sun, we find a planet that would seem more like our own home. Venus has much the same size and mass as Earth and it is a similar distance from the Sun. We might have imagined Venus to be the ideal world for future human colonization. Yet its history has been different from that of Earth. Slightly warmer than the young Earth, the young Venus was too hot for liquid water to pool on its surface, allowing the same greenhouse effect responsible for maintaining Earth's balmy climate to run away with itself. The result is an environment that is hellish beyond even the fevered imagination of Dante Alighieri.

If we step outward from the Sun, we find Mars. Like Venus, Mars is a planet that is not so different from Earth. Although smaller and less massive, it still has strong enough gravity to hold on to its atmosphere. But whereas Venus was too hot, Mars when it was young was too cold. It had a thick atmosphere, and rain fell in torrents. Images of the surface of the planet show flood basins into which the Amazon would have been but a minor tributary. Liquid water on the surface of Mars was too effective at scrubbing the planet's atmosphere of carbon dioxide. The process that prevented Earth from becoming a Venus-like hothouse got out of hand on Mars, and the temperature fell until the planet's atmosphere nearly froze out.

We owe our lives to the thin blanket of atmosphere that covers our planet. At one time we may have felt justified in taking this atmosphere for granted, but after looking around us in the Solar System we have come to appreciate that our world is maintained by the most delicate of balances. Over billions of years, life has shaped the atmosphere of our planet, and today through the activities of humans, life is reshaping our planet's atmosphere once again.

Great political debate surrounds issues such as the release of ozone-destroying chlorofluorocarbons and other halogen compounds into Earth's atmosphere, as well as the destruction of oxygen-producing plant life and the release of huge amounts of carbon dioxide through the burning of fossil fuels. In the rush for profit and convenience, there are those who say that it is too early to worry about the effects that humans are having on the planet. "Earth is too complex," they say, "and our models too primitive to accurately predict the effect that human activities will have on our atmosphere and climate."

This burden of proof is incorrectly placed, however. We know that factors such as the greenhouse effect made our planet what it is today. In the absence of the greenhouse effect, Earth would have an average temperature below the freezing point of water. We also know that human activities are measurably changing the chemical balance of our atmosphere. We are undeniably playing with the knobs that regulate our climate.

The only intellectually honest way to look at the situation is to start with the hypothesis that unchecked human activity such as the destruction of rain forests and the burning of fossil fuels will have a significant, adverse effect on the future of our planet. We point out that this hypothesis has *not* been disproved.

Summary

- Primary atmospheres consist mainly of hydrogen and helium captured from the protoplanetary disk.

- The terrestrial planets lost their primary atmospheres to space soon after the planets formed.

- Secondary atmospheres are created by volcanic gases, and from volatiles brought in by impacting comets.

- The atmospheric greenhouse effect keeps Earth from freezing, but it turns the atmosphere of Venus into an inferno.

- Plant life is responsible for the oxygen in Earth's atmosphere.

- Temperature and pressure decrease with altitude in the troposphere of Venus, Earth, and Mars.

- Earth's magnetosphere shields us from the solar wind.

- Nonuniform absorption of solar energy is the cause of all our weather.

- The Coriolis effect causes hurricanes to rotate.

- Venus has a massive, hot carbon dioxide atmosphere.

- Mars has a thin, cold carbon dioxide atmosphere.

- The atmospheres of the Moon and Mercury are almost nonexistent.

- Human modification of Earth's atmosphere may lead to unintended consequences.

SmartWork, Norton's online homework system, includes algorithmically generated versions of these questions, plus additional conceptual exercises. If your instructor assigns questions in SmartWork, log in at **smartwork.wwnorton.com.**

Student Questions

THINKING ABOUT THE CONCEPTS

1. Describe the origin and fate of primary atmospheres.

2. Primary atmospheres of the terrestrial planets were composed almost entirely of hydrogen and helium. Explain why they contained only these gases and not others.

3. How were the secondary atmospheres of the terrestrial planets created?

4. Nitrogen, the principal gas in Earth's atmosphere, was not a significant component of the protostellar disk from which the Sun and planets formed. Where did Earth's nitrogen come from?

5. Some of Earth's water was released aboveground by volcanism. What is another likely source of Earth's water?

*6. The force of gravity holds objects tightly to the surfaces of the terrestrial planets. Yet atmospheric molecules are constantly escaping into space. Explain how these molecules are able to overcome gravity's grip. How does the mass of a molecule affect its ability to break free?

*7. Explain what is meant by a runaway atmosphere and how it works.

8. We attribute the warming of a planet's surface to the atmospheric greenhouse effect. How does this mechanism differ from the warming of a conventional greenhouse?

9. Name at least two greenhouse molecules other than carbon dioxide.

10. In what way is the atmospheric greenhouse effect beneficial to terrestrial life?

*11. Why is Venus very hot and Mars very cold if both of their atmospheres are dominated by carbon dioxide, an effective "greenhouse" molecule?

12. In what ways does plant life affect the composition of Earth's atmosphere?

13. Ozone is often vilified as an urban pollutant, yet our very lives depend on it. Explain how ozone protects terrestrial life.

14. Eventually astronomers will have the technical capability to detect Earth-like planets around other stars. When we find such objects, what observations would indicate with near certainty that they harbor some form of life as we know it?

*15. You check the barometric pressure and find that it is reading only 920 mb. Two possible effects could be responsible for this lower-than-average reading. What are they?

16. What mechanism produces the aurora borealis (northern lights)?

17. What is the principal cause of winds in the atmospheres of the terrestrial planets?

18. How does the solar wind affect Earth's upper atmosphere, and what effects can it have on society?

19. Global warming appears to be responsible for increased melting of the ice in Earth's polar regions.
   a. Why does the melting of arctic ice, which floats on the surface of the Arctic Ocean, *not* affect the level of the oceans?

b. How is the melting of glaciers in Greenland and Antarctica affecting the level of the oceans?

20. Why are we unable to get a clear view of the surface of Venus, as we have so successfully done with Mars?

21. Assume you are somehow able to survive on the surface of Venus. Describe your environment.

22. Explain why surface temperatures on Venus hardly vary between day and night and between the equator and the poles.

23. In 1975 the Soviet Union landed two camera-equipped spacecraft on Venus, giving planetary scientists their first (and only) close-up views of the planet's surface. Both cameras ceased to function after only an hour. What environmental conditions most likely led to their demise?

24. Humans may eventually travel to the surface of Mars. Describe the environment they will experience.

*25. Mars has seasons similar to those of Earth, but more extreme. Explain why.

26. Mercury and the Moon have extremely tenuous atmospheres. What are two possible sources of these atmospheres?

APPLYING THE CONCEPTS

27. The total mass of Earth's atmosphere is 5×10^{18} kg. Carbon dioxide (CO_2) makes up about 0.06 percent of Earth's atmospheric mass.
    a. What is the mass of CO_2 (in kilograms) in Earth's atmosphere?
    b. The annual global production of CO_2 is now estimated to be 2.6×10^{13} kg. What annual fractional increase does this represent?

28. The ability of wind to erode the surface of a planet is related in part to the wind's kinetic energy.
    a. Compare the kinetic energy of a cubic meter of air at sea level on Earth (mass 1.23 kg) moving at a speed of 10 m/s with a cubic meter of air at the surface of Venus (mass 64.8 kg) moving at 1 m/s.
    b. Compare the terrestrial case with a cubic meter of air at the surface of Mars (mass 0.015 kg) moving at a speed of 50 m/s.

c. What would you guess is the reason we do not see more evidence of wind erosion on Earth?

*29. Suppose you seal a rigid container that has been open to air at sea level when the temperature is 0°C (273 K). The pressure inside the sealed container is now exactly equal to the outside air pressure: 10^5 newtons per square meter (N/m^2).
    a. What would be the pressure inside the container if it were left sitting in the desert shade where the surrounding air temperature was 50°C (323 K)?
    b. What would be the pressure inside the container if it were left sitting out in an Antarctic night where the surrounding air temperature was –70°C (203 K)?
    c. What would you observe in each case if the walls of the container were not rigid?

30. Oxygen molecules (O_2) are 16 times more massive than hydrogen molecules (H_2). Carbon dioxide molecules (CO_2) are 22 times more massive than H_2.
    a. Compare the average speed of O_2 and CO_2 molecules in a volume of air.
    b. Does this ratio depend on air temperature?

*31. Using the average density of air at sea level (1.225 kilograms per cubic meter, or kg/m^3) and the average mass of Earth's atmosphere above sea level per square meter (1.033×10^4 kg/m^2), what would be the total depth of Earth's atmosphere (in kilometers) if its density were the same at all altitudes? (This is called a scale height [H], a useful quantity for comparing Earth's atmosphere with the atmospheres of other planets.)

32. The average surface pressure on Mars is 6.4 mb. Using Figure 8.6a, estimate how high you would have to go in Earth's atmosphere to experience the same atmospheric pressure that you would experience if you were standing on Mars.

33. Water pressure in Earth's oceans increases by 1 bar for every 10 meters of depth. Compute how deep you would have to go to experience pressure equal to the atmospheric surface pressure on Venus.

 StudySpace is a free and open Web site that provides a Study Plan for each chapter of *21st Century Astronomy*. Study Plans include animations, reading outlines, vocabulary flashcards, and multiple-choice quizzes, plus links to premium content in SmartWork and the ebook. Visit **www.wwnorton.com/studyspace**.

Before the starry threshold of Jove's Court
My mansion is ...
Above the smoke and stir of this dim spot
Which men call earth.

JOHN MILTON (1608–1674)

Magnificent Saturn looms before the cameras of the *Cassini* spacecraft.

CHAPTER 9

Worlds of Gas and Liquid— The Giant Planets

9.1 The Giant Planets— Distant Worlds, Different Worlds

The four largest planets in our Solar System are Jupiter, Saturn, Uranus, and Neptune, and all have characteristics that clearly distinguish them from the terrestrial, or Earth-like, planets. The most obvious difference is that they are all enormous compared with their terrestrial cousins. Even the smaller of them, Uranus and Neptune, are nearly four times the size of Earth. Another distinguishing feature is their very low density. All are composed almost entirely of light materials such as hydrogen, helium, and water, rather than the rock and metal that make up the terrestrial planets. Collectively, we call Jupiter, Saturn, Uranus, and Neptune the **giant planets** (**Table 9.1**), although you may sometimes hear them referred to as the "Jovian planets," after Jupiter, the largest ("Jove" is another name for Jupiter, the highest-ranking Roman deity).

Even though the giant planets share many characteristics, there are significant differences among them. Jupiter and Saturn are similar to one another in size and are composed primarily of hydrogen and helium. Uranus and Neptune are also similar to one another in size, and both contain much larger amounts of water and other ices than do Jupiter or Saturn. These differences are sufficiently large that many astronomers feel it makes sense to divide the giant planets into two classes: Jupiter and Saturn as

> Jupiter and Saturn are gas giants. Uranus and Neptune are ice giants.

KEY CONCEPTS

Unlike the solid planets of the inner Solar System, four worlds in the outer Solar System were able to capture and retain gases and volatile materials from the Sun's protoplanetary disk and swell to enormous size and mass. Examining these giant planets, we will discover

- Two gas giants, composed primarily of hydrogen and helium; and two ice giants, composed primarily of water and other volatile materials.

- Atmospheres and oceans, but no solid surfaces.

- How changes in temperature and pressure with increasing depth lead to changes in the chemical composition of clouds in giant planet atmospheres.

- Gravitational energy being converted into thermal energy in the interiors of three giant planets, driving strong convection in their atmospheres.

- How the Coriolis effect on these rapidly rotating worlds turns convective motions into powerful winds, huge storms, and planet-spanning bands of multi-hued clouds.

- Extreme conditions deep within the interiors of the giant planets.

- Brilliant auroras and glowing doughnuts of gas associated with the huge magnetospheres and strong magnetic fields of the giant planets.

TABLE 9.1

Physical Properties of the Giant Planets

	Jupiter	Saturn	Uranus	Neptune
Orbital radius (AU)	5.2	9.6	19.2	30.0
Orbital period (years)	11.9	29.5	84.0	164.8
Orbital velocity (km/s)	13.1	9.7	6.8	5.4
Mass ($M_\oplus = 1$)	318	95	14.5	17.1
Equatorial diameter (1,000 km)	143	120.5	51.1	49.5
Equatorial diameter ($D_\oplus = 1$)	11.2	9.45	4.0	3.9
Oblateness	0.065	0.098	0.023	0.017
Density (water = 1)	1.3	0.7	1.3	1.6
Rotation period (hours)	9.9	10.7	17.2	16.1
Obliquity (degrees)	3.1	26.7	97.8	29.3
Surface gravity (m/s^2)	25.1	10.5	9.0	11.2
Escape speed (km/s)	60.2	35.5	21.3	23.5

(a)

(b)

FIGURE 9.1 The gas giants. (a) Jupiter imaged by *Cassini*. (b) Saturn seen via the Hubble Space Telescope.

gas giants (**Figure 9.1**) and Uranus and Neptune as **ice giants** (**Figure 9.2**). Other differences between the gas giants and the ice giants[1] will become evident as we more closely examine their physical and chemical characteristics.

The giant planets orbit the Sun far beyond the orbits of Earth and Mars. The closest to the Sun, and to us as well, is Jupiter, and even Jupiter is more than five times as far from the Sun as Earth is. Neptune, the most distant, is 4½ billion kilometers (km) away, or some 30 times farther from the Sun than we are. To put this distance into perspective, if you were traveling at the speed of a commercial jetliner it would take you more than 500 years to reach Neptune.

> Giant planets orbit much farther from the Sun than do the terrestrial planets.

---

[1]Remember from Chapter 6 that planetary scientists often refer to volatile substances as *ices* even when they are in their liquid or gaseous form. Uranus and Neptune may be called ice giants, but that does not mean that they are solid, frozen worlds.

(a)

(b)

FIGURE 9.2 The ice giants. Uranus (a) and Neptune (b) imaged in visible light by *Voyager*.

Two Were Known to Antiquity, and Two Were Telescopic Discoveries

Although very far away, the giant planets are so large that all but Neptune can be seen with the unaided eye. Jupiter and Saturn are a familiar sight in the nighttime sky, comparable to the brightest stars. Along with Mercury, Venus, and Mars, they were among the five *planets* (a Greek word meaning "wandering stars") known to ancient cultures. In contrast, Uranus appears only slightly brighter than the faintest stars visible on a dark night, and Neptune cannot be seen without the aid of binoculars.

Uranus was the first planet to be "discovered," but it was not found until late in the 18th century, more than 170 years after Galileo made his first astronomical observations with a telescope. In 1781 William Herschel, a German-English professional musician and amateur astronomer, came upon it quite by accident. Herschel was producing a catalog of the sky at his home in Bath, England, when he noticed a tiny disk in the eyepiece of his six-inch telescope. At first he thought he had found a comet, but the object's slow nightly motion soon convinced him that it was a new planet beyond the orbit of Saturn. Politically astute, Herschel proposed calling his new planet Georgium Sidus ("George's Star") after King George III of England. Obviously pleased, the monarch rewarded Herschel with a handsome lifetime pension. The astronomical community, however, later rejected Herschel's suggestion, preferring the name Uranus instead. For 65 years Uranus would remain the most distant known planet.

Over the decades that followed Herschel's discovery, astronomers found to their dismay that Uranus was straying from its predicted path in the sky. Mathematicians viewed this aberrant behavior with grave concern. By then, Newton's laws of motion had been firmly established and were the basis for predicting the motion of the newly discovered planet. Could something be wrong with the theory? As a reasonable explanation, a few astronomers suggested that the gravitational pull of an unknown planet might be responsible for this "unacceptable" behavior of Uranus. Using measured positions of Uranus provided by astronomers, two young mathematicians, Urbain-Jean-Joseph Le Verrier (1811–1877) in France and John Couch Adams (1819–1892) in England, independently predicted where the hypothetical planet should be. Although Adams was first to compute his predictions, he was unable to gain the attention of England's Astronomer Royal, so the opportunity for England to gain credit for a second planetary discovery was lost. Meanwhile, Le Verrier was having similar problems convincing French astronomers. It was a German who would finally triumph. Armed with Le Verrier's predictions, Johann Gottfried Galle (1812–1910) began a search at the Berlin Observatory. He found the planet on his first observing night, just where Le Verrier and Adams had predicted it would be. Galle's "discovery" of Neptune in 1846 became a triumph for math-

The Sun shines only dimly and provides very little warmth in these remote parts of the Solar System. If we were to journey to Jupiter, we would see the Sun as but a tiny disk, $\frac{1}{27}$ as bright as it appears from Earth. At Neptune, the Sun would no longer show a disk but would appear as a brilliant starlike point of light about 500 times brighter than the full Moon in our own sky. Daytime on Neptune is equivalent to a perpetual twilight here on our own planet. With so little sunlight available for warmth, daytime temperatures hover around 123 kelvins (K) at the cloud tops on Jupiter, and they can dip to just 37 K on Neptune's moon Triton.

ematical prediction based on physical law—and for the subsequent confirmation of theory by observation.

Neptune became the eighth known planet, and it remains the outermost classical planet in our Solar System. As we will learn in Chapter 12, erroneous discrepancies between the predicted and observed positions of Uranus and Neptune during the 19th century led to the fortuitous discovery of Pluto in 1930. Pluto was immediately declared to be the ninth planet. In 2006, however, the International Astronomical Union officially reclassified Pluto as a dwarf planet. Indeed, Pluto's characteristics are much closer to those of a huge comet nucleus than to either a giant or a terrestrial planet. For this reason we will discuss Pluto in Chapter 12 when we talk about the smaller bodies of the Solar System.

9.2 How Giant Planets Differ from Terrestrial Planets

Historically, the vast distances to the giant planets have made them difficult objects for scientific study. All of this changed with the arrival of the space age and with the simultaneous development of more powerful optical and electronic groundbased instruments during the latter decades of the 20th century. Modern instruments on both groundbased telescopes and the Hubble Space Telescope (HST) have provided new insight into the composition and physical structure of the giant planets; but our greatest leaps of knowledge have come from close-up observations made possible by the planetary probes: *Pioneer*, *Voyager*, *Galileo*, and *Cassini*. Over the past several decades, one or another of these spacecraft has visited each of the four giant planets, sending back a wealth of scientific data with a level of detail that cannot be obtained from the vicinity of Earth. Although groundbased telescopes and the HST have made significant contributions to our knowledge of the giant planets, much of what follows is based on what we have learned from planetary probes.

The Giant Planets Are Large and Massive

Figure 9.3 compares the true relative sizes of the giant planets with their relative sizes as seen from Earth. Jupiter is not only the largest of the eight planets; it is more than one-tenth the size of the Sun itself and more than 11 times the size of Earth. Saturn is only slightly smaller than Jupiter, with a diameter of 9.5 Earths. Uranus and Neptune are each about 4 Earth diameters across, with Neptune being the slightly smaller of the two. The most accurate measurements of planet sizes have come from observing the length of time it takes a planet to eclipse, or "occult," a star. We call these

FIGURE 9.3 (a) Images of the giant planets, shown to the same physical scale. (b) The same images, scaled according to how the planets would appear as seen from Earth.

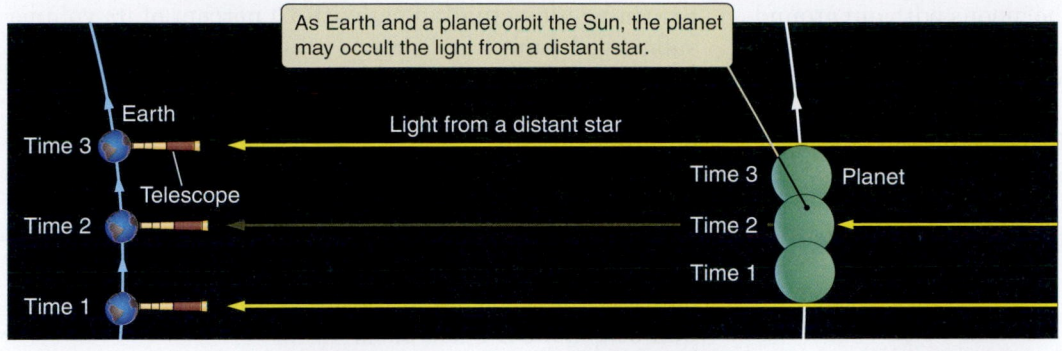

As Earth and a planet orbit the Sun, the planet may occult the light from a distant star.

Earth

Time 3

Telescope

Time 2

Time 1

Light from a distant star

Time 3   Planet

Time 2

Time 1

FIGURE 9.4 Occultations occur when a planet, moon, or ring passes in front of a star. Careful measurements of changes in the starlight's brightness and the duration of these changes give information about the size and properties of the occulting object.

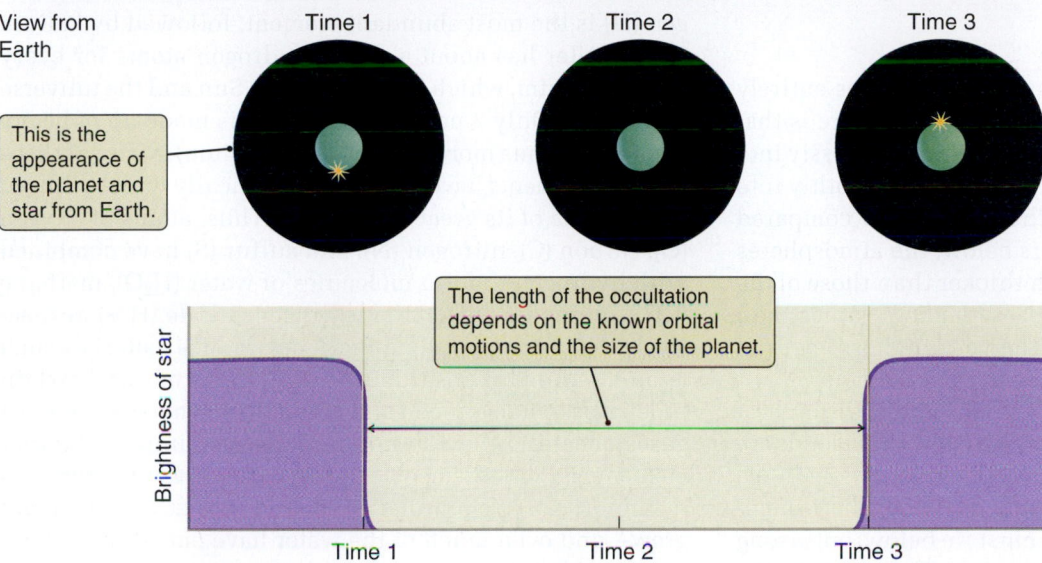

View from Earth

Time 1   Time 2   Time 3

This is the appearance of the planet and star from Earth.

The length of the occultation depends on the known orbital motions and the size of the planet.

Brightness of star

Time 1   Time 2   Time 3

events **stellar occultations** (**Figure 9.4**). For example, Newton's laws may tell us that a particular planet is moving along at a speed of precisely 25 kilometers per second (km/s) relative to Earth's own motion, and we observe that this planet takes exactly 2,000 seconds to pass directly in front of a star. The planet then must have a diameter equal to the distance it traveled during the 2,000 seconds, or 50,000 km. The center of the planet rarely passes directly in front of a star, of course; but observations of occultations from several widely separated observatories provide the geometry necessary to calculate both the planet's size and its shape. Occultations of the radio signals transmitted from orbiting spacecraft and images taken by the spacecraft cameras have also provided accurate measures of the sizes and shapes of planets and their moons. You may be surprised to learn that the giant planets are not perfectly round. Later in the chapter we will see why.

> Stellar occultations let us determine a planet's diameter.

The giant planets represent an overwhelming 99.5 percent of all the nonsolar mass in our Solar System. The vast multitude of other Solar System objects—terrestrial planets, dwarf planets, moons, asteroids, and comets—are all included in the remaining 0.5 percent. Jupiter alone is 318 times as massive as Earth and some 3½ times as massive as Saturn, its closest rival. Jupiter, in fact, weighs in at more than twice the mass of all the other planets combined. Even so, its mass is only about a thousandth that of the Sun. Uranus and Neptune, the ice giants, are the lightweights among the giant planets, but each still has the equivalent of approximately 15 Earth masses ($M_\oplus$).

How do we measure the mass of a planet? In Chapter 3 we learned how a planet's gravitational attraction can affect the motion of a nearby small body—say, one of its moons or a passing spacecraft. We found that the motion of the small body can be accurately predicted if we know the planet's mass and apply Newton's law of gravitation and Kepler's third law. Having seen how this works, we should now be able to invert the procedure and calculate the planet's mass by observing the motion of a small body. Prior to the space age, we measured a planet's mass by observing the motions of its moons. This technique, of course, worked only with planets that have moons. (We had to estimate the masses of Mercury and Venus from their size.) The accuracy of those early calculations was limited by how precisely we

> The motions of a planet's moons allow us to determine the planet's mass.

could measure the positions of the moons with our ground-based telescopes. Planetary spacecraft have now made it possible to measure the masses of planets with unparalleled accuracy. As a spacecraft flies by, the planet's gravity tugs on it and deflects its trajectory. By tracking and comparing the spacecraft's radio signals using several antennae here on Earth, we can detect minute changes in the spacecraft's trajectory, thereby providing a highly accurate measure of the planet's mass.

We See Only Atmospheres, Not Surfaces

The giant planets are made up mostly or perhaps entirely of gases and liquids. Their characteristic structure is that of a relatively shallow atmosphere merging seamlessly into a deep liquid "ocean," which in turn merges smoothly into a denser liquid or solid core. Although shallow compared with the depth of the liquid layers below, the atmospheres of the giant planets are still much thicker than those of the terrestrial planets—thousands of kilometers rather than hundreds. As with Venus, only the very highest levels of their atmospheres are visible to us. In the case of Jupiter or Saturn, we see the tops of a layer of thick clouds, the highest of many others that lie below. Although a few thin clouds are visible on Uranus, we find ourselves looking mostly into a clear, seemingly bottomless atmosphere. Atmospheric models tell us that thick cloud layers must lie below, but strong scattering of sunlight by molecules (see Chapter 8) in the clear part of the atmosphere prevents us from seeing these lower cloud layers. Neptune displays a few high clouds with a deep, clear atmosphere showing between them.

Jupiter's Chemical Composition Is More like the Sun's Than Earth's Is

In Chapter 7 we learned that the terrestrial planets are composed mostly of rocky minerals, such as silicates, along with various amounts of iron and other metals. It is true that the atmospheres of the terrestrial planets contain lighter materials, but the masses of these atmospheres—and even of Earth's oceans—are insignificant compared with the total planetary mass. The terrestrial planets are thus very compact, with densities ranging from 3.9 (Mars) to 5.5 (Earth) times that of water. They are the densest objects in the Solar System.

In contrast, the giant planets are composed almost entirely of lighter materials, such as hydrogen, helium, and water. Consequently, their densities are much lower than the those of the rock-and-metal terrestrial planets. Neptune is the most compact among the giant planets, having a density about 1.6 times that of water. Saturn is the least compact—only 0.7 times the density of water. This means that, placed in a sufficiently large and deep body of water,

Saturn would actually float with 70 percent of its volume submerged. Jupiter and Uranus have densities intermediate between those of Neptune and Saturn. As we will see later, however, the size of the giant planets is not determined by density alone.

The chemical compositions of the giant planets are not all the same. Jupiter's chemical composition is quite similar to that of the Sun. Astronomers take the relative abundance of the elements in the Sun as a standard reference, termed **solar abundance**. As illustrated in **Figure 9.5**, hydrogen (H) is the most abundant element, followed by helium (He). Jupiter has about a dozen hydrogen atoms for every atom of helium, which is typical of the Sun and the universe as a whole. Only 2 percent of its mass is made up of **heavy elements** (atoms more massive than helium). Many of these massive elements have combined chemically with hydrogen (H) because of its great abundance. Thus, atoms of oxygen (O), carbon (C), nitrogen (N), and sulfur (S) have combined with hydrogen to form molecules of water (H_2O), methane (CH_4), ammonia (NH_3), and hydrogen sulfide (H_2S), respectively. More complex combinations produce materials such as ammonium hydrosulfide (NH_4HS). Helium—and certain other gases, such as neon and argon—are what we call **inert gases**, meaning they do not normally combine with other elements or with themselves. Most of the iron—the remains of the original rocky planet around which the gas giant grew—and even much of the water have ended up in Jupiter's liquid core.

Proportionally, Saturn has a somewhat larger inventory of massive elements than Jupiter. In Uranus and Neptune, massive elements are so abundant that they are major compositional components of these two planets. The principal compositional differences among the four giant planets lie in the amounts of hydrogen and helium that each of them contains. This turns out to be an important clue to the process by which the giant planets formed—a subject we will return to in Section 9.7.

> **Jupiter's chemistry is similar to the Sun's.**

Short Days and Nights on the Giant Planets

Still another distinguishing characteristic of the giant planets is their rapid rotation, meaning that their days are short. A day on Jupiter is just under 10 hours long, and Saturn's is only a little longer. Neptune and Uranus have rotation periods of 16 and 17 hours, respectively, giving them days that are intermediate in length between those of Jupiter and Earth.

The rapid rotation of the giant planets distorts their shapes. If they did not rotate at all, these fluid bodies would be perfectly spherical. In Chapter 6 we used the analogy of a spinning disk of pizza dough to see why a collapsing, rotating cloud must settle into a disk. The same principle

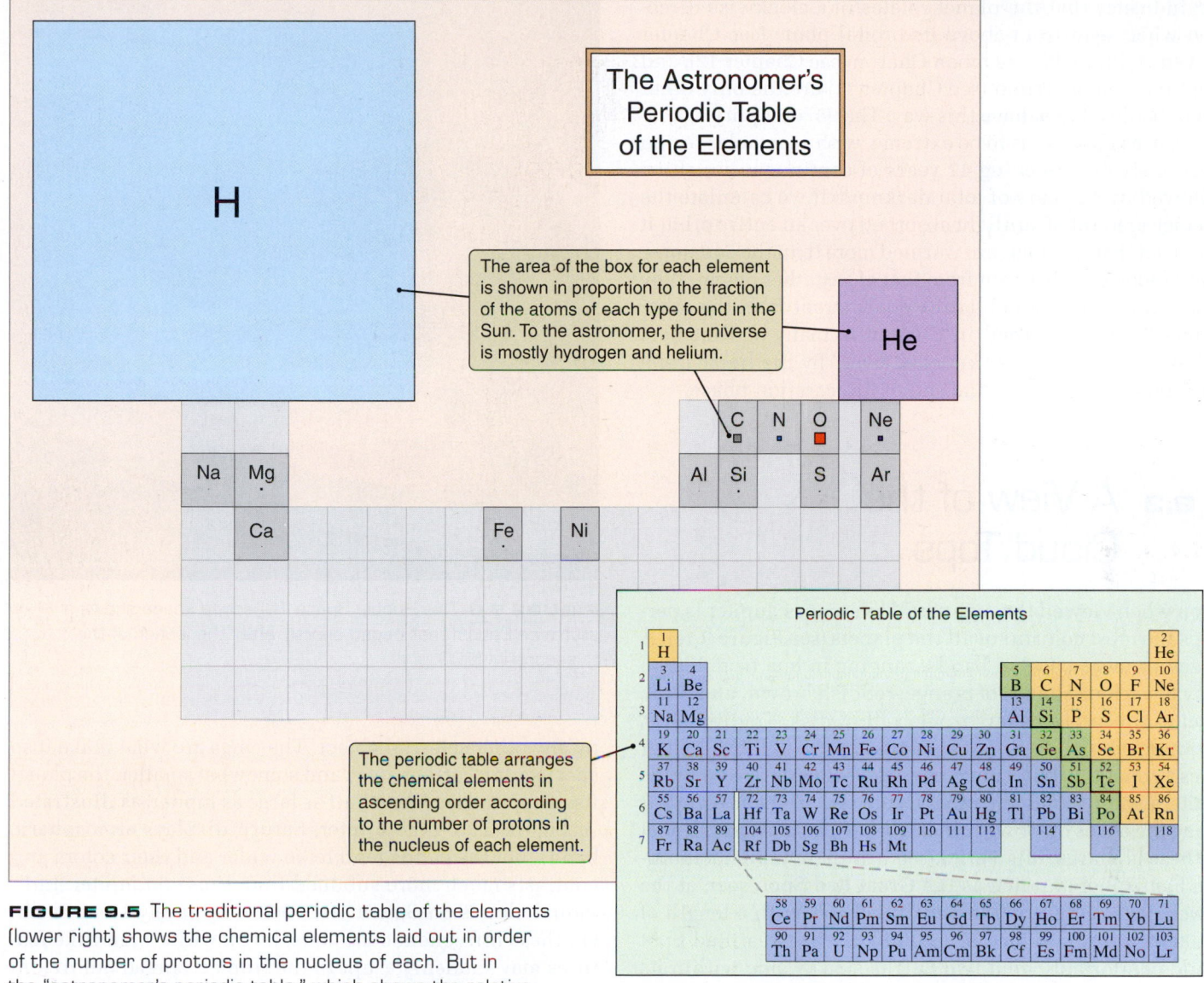

The Astronomer's Periodic Table of the Elements

The area of the box for each element is shown in proportion to the fraction of the atoms of each type found in the Sun. To the astronomer, the universe is mostly hydrogen and helium.

The periodic table arranges the chemical elements in ascending order according to the number of protons in the nucleus of each element.

FIGURE 9.5 The traditional periodic table of the elements (lower right) shows the chemical elements laid out in order of the number of protons in the nucleus of each. But in the "astronomer's periodic table," which shows the relative abundances of the Sun's elements in terms of size, hydrogen and helium appear as the most abundant. See Appendix 3 for a full Periodic Table of the Elements.

is at work in the rapidly rotating giant planets as well, causing the planets to bulge at their equators and giving them an overall flattened appearance. We call this flattening **oblateness**, and it is defined as the difference between a planet's equatorial radius (R_E) and its polar radius (R_P), divided by its equatorial radius:

$$\text{Oblateness} = \frac{R_E - R_P}{R_E}$$

Saturn's oblateness is especially noticeable because its equatorial diameter is almost 10 percent greater than its polar diameter (see Figure 9.3 and Table 9.1). In comparison, the oblateness of Earth is only 0.3 percent.

> Rapid rotation flattens the shapes of the giant planets.

A planet's **obliquity**—the inclination of its equatorial plane to its orbital plane—is a major factor in determining the prominence of its seasons. (Recall from Chapter 2 that Earth's obliquity of 23.5° is responsible for our distinct seasons.) The obliquities of the giant planets are shown in Table 9.1. With an obliquity of only 3°, Jupiter has almost no seasons at all. The obliquities of Saturn and Neptune are slightly larger than those of Earth or Mars, giving these planets moderate but well-defined seasons. Curiously, Uranus spins on an axis that lies nearly in the plane of its orbit. Viewed from Earth, then, the planet appears to be either spinning face-on to us or rolling along on its side (or something in between), depending on where Uranus happens to be in its orbit.

Uranus is one of five major Solar System bodies with very high obliquities. Its obliquity is 98°, and a value greater than

90° indicates that the planet rotates in a clockwise direction when seen from above its orbital plane (see Chapter 2). Venus, Pluto, Pluto's moon Charon (see Chapter 12), and Neptune's moon Triton (see Chapter 11) are the only other major bodies that behave this way. The 98° obliquity of Uranus causes its seasons to be extreme, with each polar region alternately experiencing 42 years of continuous sunshine followed by 42 years of total darkness. If we calculate the average amount of sunlight absorbed over an entire orbit, it turns out that the poles are warmed more than the equator—a situation quite different from that of any other planet. Why does Uranus have an obliquity so different from the other planets? As we learned in Chapter 6, many astronomers believe the planet was "knocked over" by the impact of a huge planetesimal near the end of its accretion phase.

9.3 A View of the Cloud Tops

Even when viewed through small telescopes, Jupiter is perhaps the most colorful of all the planets (see Figure 9.1a). A dozen or more parallel bands, ranging in hue from bluish gray to various shades of orange, reddish brown, and pink, stretch out across its large, pale yellow disk. Traditionally, astronomers call the darker bands "belts" and the lighter ones "zones." Many small clouds—some dark and some white, some circular and others more oval in shape—appear along the edges of, or within, the belts. The most prominent of these is a large, often brick-red feature in Jupiter's southern hemisphere known as the **Great Red Spot**, seen at the lower right in Figure 9.1a. Oval in shape, with a length of 25,000 km and a width of 12,000 km, the Great Red Spot could comfortably hold two Earths side by side within its boundaries (see Figure 9.12b).

No one really knows how long this huge feature has been circulating in Jupiter's atmosphere, but it was first spotted more than three centuries ago, shortly after the invention of the telescope. Since then, the Great Red Spot has varied

> The Great Red Spot is a giant anticyclone.

unpredictably in size, shape, color, and motion as it drifts among Jupiter's clouds. Small clouds seen moving around the periphery of the Great Red Spot show that it circulates like a giant vortex. Its cloud pattern looks a lot like that of a terrestrial hurricane, but it rotates in the opposite direction, exhibiting *anticyclonic* rather than cyclonic flow. (You may recall from Chapter 8 that anticyclonic flow is indicative of a high-pressure system.) Because of its colorful appearance and unpredictable changes, the Great Red Spot has long been a favorite among amateur astronomers.

Our first look at Saturn through a telescope is a moment not quickly forgotten (see Figure 9.1b). Adorned by its magnificent system of rings, Saturn provides a sight unmatched

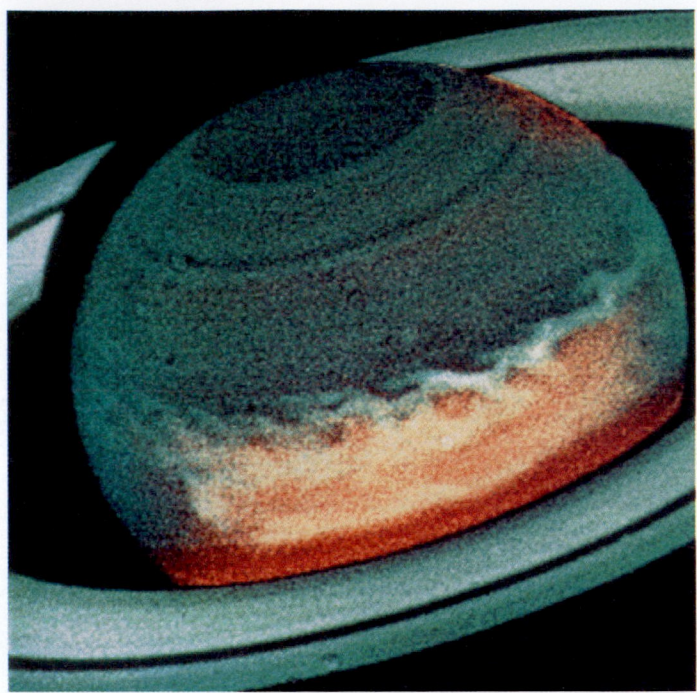

FIGURE 9.6 This Hubble Space Telescope image shows a storm on Saturn that began shortly after the launch of the telescope in 1990.

by any other celestial object. The rings are what make it so spectacular. Farther away and somewhat smaller, the planet itself appears less than half as large as Jupiter, as illustrated in Figure 9.3b. Like Jupiter, Saturn displays atmospheric bands, but the bands tend to be wider and their colors and contrasts much more subdued than those on Jupiter. Individual clouds on Saturn are seen only rarely from Earth. On these infrequent occasions, large, white, cloudlike features may suddenly erupt in the tropics, spread out in longitude, and then fade away over a period of a few months (**Figure 9.6**).

In even the largest telescopes, Uranus and Neptune appear visually as tiny, featureless disks with a pale bluish green color (similar to the *Voyager* view in Figure 9.2a). As we will see, however, infrared imaging reveals a number of individual clouds and belts, giving these distant planets considerably more character.

Observed from close up, the giant planets suddenly appear as real and tangible worlds. The clouds of Jupiter are a landscape of variegated color and intricate formations (see Figure 9.1a). Time-lapse imaging shows a roiling, swirling giant with atmospheric currents and vortices so complex that we still do not fully understand the details of how they interact with one another, even after decades of analysis. The Great Red Spot alone displays more structure than was visible over all of Jupiter prior to the space age. Dynamically, it also reveals some rather bizarre behavior, such as "cloud cannibalism." In a series of time-lapse images, *Voyager* observed a number of Alaska-sized clouds

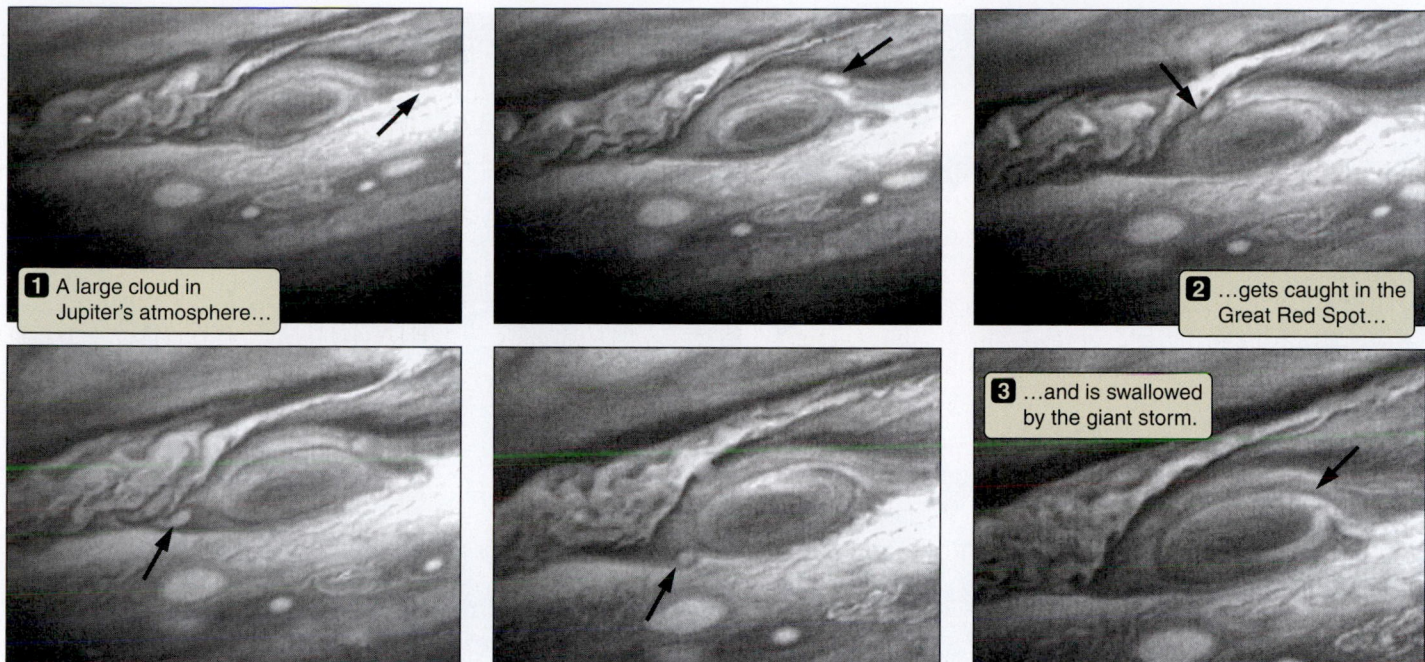

1 A large cloud in Jupiter's atmosphere...

2 ...gets caught in the Great Red Spot...

3 ...and is swallowed by the giant storm.

FIGURE 9.7 This sequence of images, obtained by the *Voyager* spacecraft during its encounter with Jupiter, shows the swirling, anticyclonic motion of Jupiter's Great Red Spot.

being swept into the Great Red Spot vortex. Some of these clouds were carried around the vortex a few times and then ejected, while others were swallowed up and never seen again (**Figure 9.7**). Other smaller clouds with structure and behavior similar to that of the Great Red Spot are seen in Jupiter's middle latitudes.

Broad bands and individual clouds show prominently in *Voyager* and *Cassini* images of Saturn's atmosphere. A relatively narrow, meandering band in the midnorthern latitudes encircles the planet in a manner similar to our own terrestrial jet stream (**Figure 9.8**). The largest features are about the size of the continental United States, but many that we see are smaller than terrestrial hurricanes. Close-up views from the *Cassini* spacecraft (**Figure 9.9**) show immense lightning-producing storms in a region of Saturn's southern hemisphere known to mission scientists as "storm alley."

Uranus and Neptune present a rather bland appearance at visible wavelengths. Yet in the near infrared, beyond the spectral range of our eyes, these planets take on a very different appearance, showing atmospheric bands and small clouds suggestive of those seen on Jupiter and Saturn, but more subdued (**Figure 9.10**). The atmospheres of Uranus and Neptune are rich in methane, and methane is a strong absorber of infrared light. The strong absorption of reflected sunlight by methane causes the atmospheres of Uranus and Neptune to appear dark in the near infrared, allowing the highest clouds and bands to stand out in contrast against the dark background. Here we find an illustration of how

> **Saturn has jet streams similar to Earth's.**

(a)

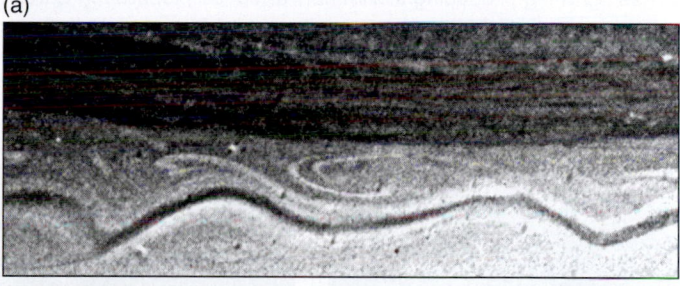

(b)

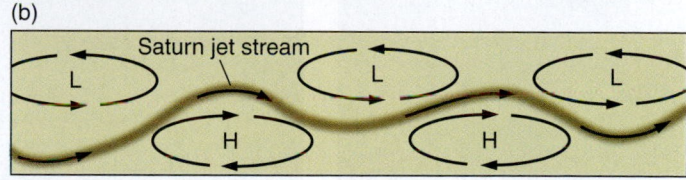

Saturn jet stream

FIGURE 9.8 (a) *Voyager* image of a jet stream in Saturn's northern hemisphere, similar to jet streams in our terrestrial atmosphere. (b) There is a dynamic relationship between the jet stream and vortices of low and high pressure. The jet stream dips equatorward below regions of low pressure and is forced poleward above regions of high pressure.

observations made in different spectral regions can add significantly to our understanding of astronomical objects.

A number of bright cloud bands appear in HST images of Neptune's atmosphere (**Figure 9.10b**). Located near the planet's **tropopause** (see Chapter 8), these cloud bands cast their shadows downward through the clear upper atmosphere onto a dense cloud layer 50 km below (**Figure 9.11**). Some of Neptune's clouds are dark. A large, dark, oval

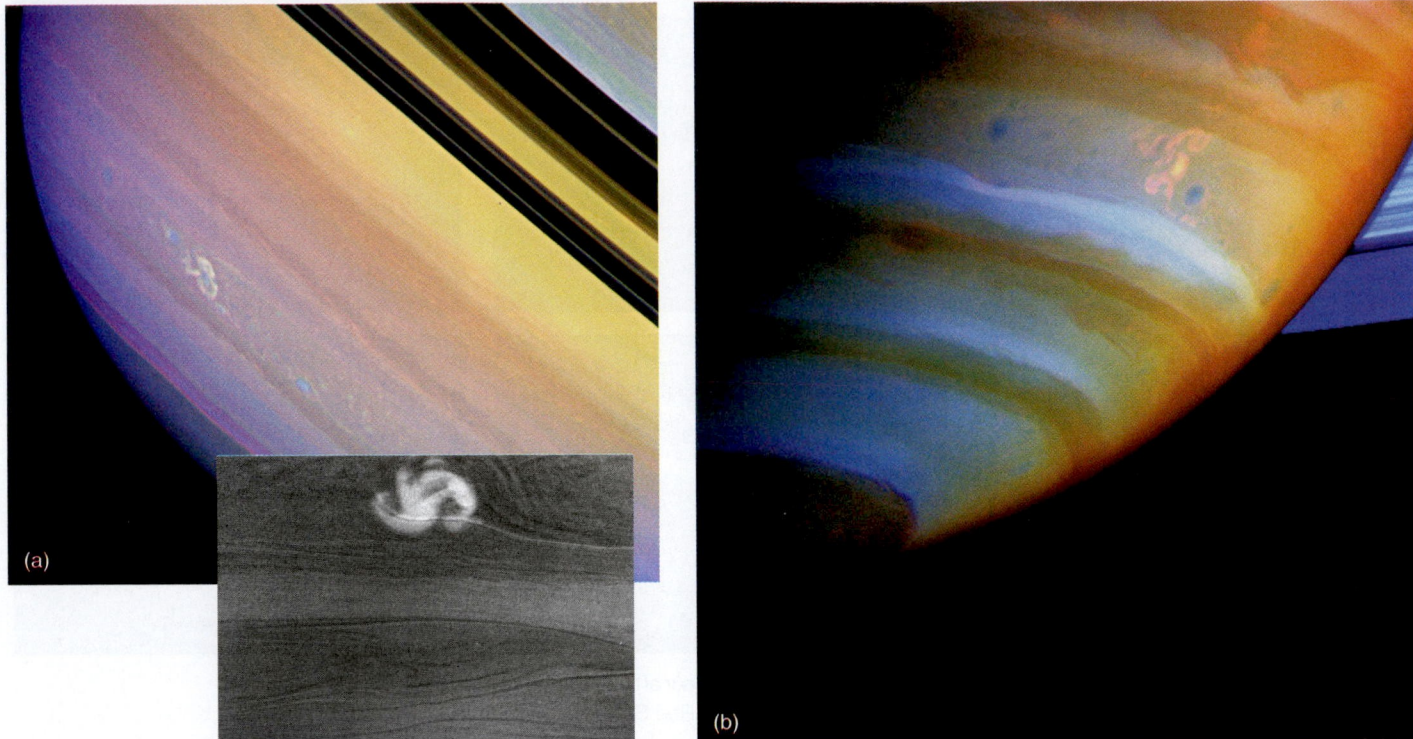

FIGURE 9.9 Violent storms are known to erupt on Saturn. (a) An enhanced *Cassini* image of an intense lightning-producing storm (left of center) located in Saturn's "storm alley." The inset shows a similar storm on Saturn's night side illuminated by sunlight reflecting off Saturn's rings. (b) A near-infrared *Cassini* image showing a storm nicknamed "the dragon."

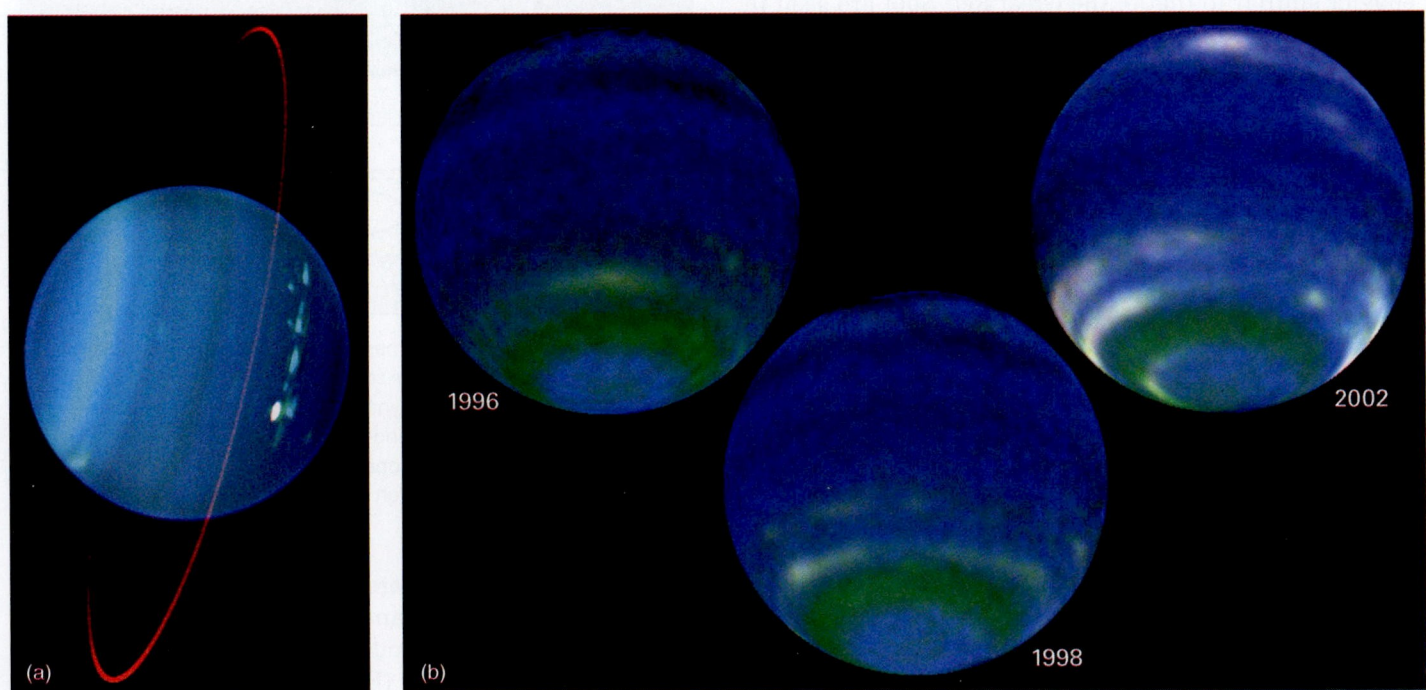

FIGURE 9.10 Hubble Space Telescope images of Uranus (a) and Neptune (b), taken at a wavelength of light that is strongly absorbed by methane. The visible clouds are high in the atmosphere. The rings of Uranus show prominently because the brightness of the planet has been subdued by methane absorption. Seasonal changes in cloud formation on Neptune are evident over a 6-year interval.

FIGURE 9.11 Neptune's clouds cast shadows through a clear atmosphere onto the cloud deck 50 km below in this *Voyager 2* image.

9.4 A Journey into the Clouds

Our visual impression of the giant planets is based on our two-dimensional view of their cloud tops. Atmospheres, though, have depth. They are three-dimensional structures whose temperature, density, pressure, and even chemical composition vary with height and over horizontal distances. As a rule, atmospheric temperature, density, and pressure all decrease with increasing altitude, although temperature will sometimes reverse itself at very high altitudes. The **stratospheres** (see Chapter 8) above the cloud tops appear relatively clear, but closer inspection shows that they contain layers of thin haze that show up best when seen in profile above the limbs of the planets. The composition of the haze particles remains unknown, but we believe that they are **photochemical**—smoglike—products created when ultraviolet sunlight acts on hydrocarbon gases such as methane.

What lies beneath the cloud tops of the giant planets? Imagine riding along in the *Galileo* probe as it descends through Jupiter's atmosphere. At first the only change we note is the expected increase in the outside pressure and temperature as our altitude rapidly decreases. Suddenly

> **Different volatiles produce different clouds at different heights.**

we find ourselves passing through dense layers of cloud, separated by regions of relatively clear atmosphere. Each of these cloud layers is composed of a different chemical substance. In Earth's atmosphere water is the only substance that can condense into clouds, but the atmospheres of Jupiter and the other giant planets contain a variety of volatiles that condense at different temperatures and atmospheric

feature in the southern hemisphere (**Figure 9.12a**) is reminiscent of Jupiter's Great Red Spot (**Figure 9.12b**). Predictably, astronomers called it the Great Dark Spot. However, the Neptune feature was gray rather than red, changed its length and shape more rapidly than the Great Red Spot does, and it lacked the permanence of the Great Red Spot. When the HST trained its corrected optics on Neptune in 1994, the Great Dark Spot had disappeared. It has not been seen since. However, a different dark spot of comparable size had appeared in Neptune's northern hemisphere.

FIGURE 9.12 The Great Dark Spot on Neptune (a) and the Great Red Spot on Jupiter (b) reproduced approximately to scale. Earth is shown for comparison. The Great Red Spot has persisted for centuries, but the Great Dark Spot disappeared between the time *Voyager* flew by Neptune in 1989 and the time HST images were obtained in 1994.

(a) Neptune's Great Dark Spot   Earth   (b) Jupiter's Great Red Spot

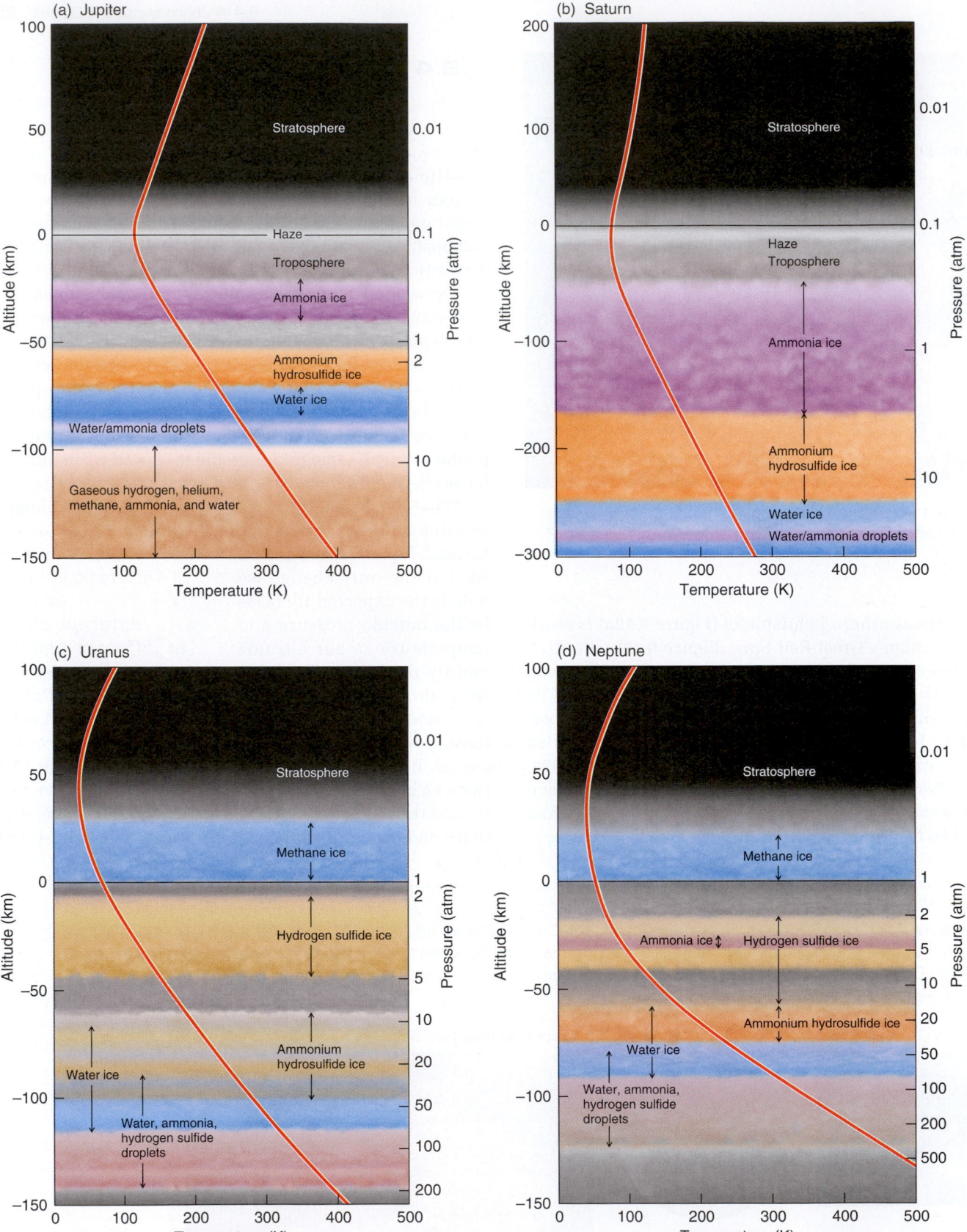

FIGURE 9.13 Volatile materials condense at different levels in the atmospheres of the giant planets, leading to chemically different types of clouds at different depths in the atmosphere. The red line in each diagram shows how atmospheric temperature changes with height. The arbitrary zero points of altitude are at 0.1 atmosphere (atm) for Jupiter (a) and Saturn (b) and 1.0 atm for Uranus (c) and Neptune (d). The value of 1.0 atm corresponds to the atmospheric pressure at sea level on Earth (approximately 1 bar). Note that Saturn's vertical scale is compressed to show the layered structure better.

pressures (**Figure 9.13**). A descent through Jupiter's cloud layers thus becomes a journey that explores the many minor constituents of its atmosphere. Each kind of volatile, such as water, condenses at a particular temperature and pressure, and each therefore forms clouds at a different altitude. Below the condensation cloud layer, each volatile is freely mixed as a gaseous atmospheric constituent. Above, it is highly depleted. We can see the reason for this difference: As the planet's atmosphere convects (a process explained in Chapter 7 and discussed further in Chapter 8), volatile materials are carried upward along with all other atmospheric gases. When a particular volatile reaches an altitude where the temperature is low enough, the condensation process removes most of it from the other gases, leaving it depleted in the air above.

During our descent we find that ammonia has condensed near the top of Jupiter's **troposphere** (see Chapter 8) at a temperature of about −140°C (133 K). Next we pass through a layer of ammonium hydrosulfide clouds at a temperature of about −80°C (193 K). Not long after this, information from the probe ends. (While descending slowly via parachute to an atmospheric pressure of 22 bars and a temperature of about 100°C [373 K], the *Galileo* probe failed, presumably because its transmitter got too hot.) What lies below this level in Jupiter's atmosphere must for now be left to our theories and atmospheric models.

The distance of each planet from the Sun partly determines its tropospheric temperature. The farther the planet is from the Sun, the colder its troposphere will be. Distance from the Sun thus determines the altitude at which a particular volatile, such as ammonia or water, will condense to form a cloud layer on each of the planets (see Figure 9.13). If temperatures are too high, some volatiles may not condense at all. The highest clouds in the frigid atmospheres of Uranus and Neptune are crystals of methane ice. The highest clouds on Jupiter and Saturn are made up of ammonia ice. Methane exists only in gaseous form throughout the warmer atmospheres of Jupiter and Saturn.

In their purest form, the ices that make up the clouds of the giant planets are all white, similar to snow on our own planet. Why, then, are some clouds so colorful, especially Jupiter's? These tints and hues must come from impurities in the ice crystals, similar to the way syrups color snow cones. Although the identities of these impurities remain unknown, our prime suspects include elemental sulfur and phosphorus and various organic materials produced by the photochemical action of sunlight on atmospheric hydrocarbons. Ultraviolet photons from the Sun, absorbed by molecules of hydrocarbons such as methane, acetylene, and ethane, among others, have enough energy to break these molecules apart. The molecular fragments can then recombine to form complex organic compounds that condense into solid particles, many of which are quite colorful. Photochemical reactions also occur in our own terrestrial atmosphere. Some of the pho-

> **Clouds on Jupiter and Saturn are colored by impurities.**

tochemical products produced close to the ground are rather obvious; we call them "smog."

The upper tropospheres of Uranus and Neptune, unlike those of Jupiter and Saturn, are relatively clear, with only a few white clouds—probably methane ice—appearing here and there. Uranus and Neptune are not bluish green because of clouds. Instead they take on a bluish color for much the same reason that Earth's oceans are blue. Methane gas is much more abundant in the atmospheres of Uranus and Neptune than in Jupiter and Saturn. Like water, methane gas tends to selectively absorb the longer wavelengths of light—yellow, orange, and red. Absorption of the longer wavelengths leaves only the shorter wavelengths—green and blue—to be scattered from the relatively cloud-free atmospheres of Uranus and Neptune, giving them a characteristic bluish green color. Earlier we described the atmosphere of Uranus as appearing nearly "bottomless." Molecular scattering also contributes to the bluish color and is so strong in the clear Uranus atmosphere that it completely hides the thick ammonia and water cloud layers that lie far below.

9.5 Winds and Storms— Violent Weather on the Giant Planets

The rapid rotation and resulting strong Coriolis effects (see Chapter 8) in the atmospheres of the giant planets create much stronger zonal winds than we see in the atmospheres of the terrestrial planets, even though less thermal energy is available. **Figure 9.14** shows the zonal wind patterns on the giant planets. On Jupiter the strongest winds are equatorial westerlies, which have been clocked at speeds of 550 kilometers per hour (km/h), as seen in **Figure 9.14a**. (Remember from Chapter 8 that westerly winds are those that blow *from*, not toward, the west.) At higher latitudes the winds alternate between easterly and westerly in a pattern that seems to be related to Jupiter's banded structure, but scientists are not sure. Near a latitude of 20° south, the Great Red Spot appears to be caught between a pair of easterly and westerly currents with opposing speeds of more than 200 km/h. If you think this fact might imply something about the relationship between zonal flow and vortices, you are right.

The equatorial winds on Saturn are also westerly, but stronger than those on Jupiter. In the early 1980s, *Voyager* measured speeds as high as 1,690 km/h. Later, HST recorded maximum speeds of 990 km/h, and more recently *Cassini* has found them to be intermediate between the *Voyager* and HST measurements. What can be happening here? As **Math Tools 9.1** explains, winds on other planets are measured by the motions of their clouds. Saturn's winds appear

> **Equatorial winds are fast on Jupiter and even faster on Saturn and Neptune.**

How Wind Speeds on Distant Planets Are Measured

How do we measure wind speeds on planets that are so far away? It turns out to be surprisingly easy. If we can see individual clouds in their atmospheres, we can measure their winds. As on Earth, clouds are typically carried right along with the local winds. By measuring the positions of individual clouds and noting how much they move during an interval of a day or so, we are able to calculate the local wind speed. We need to know one important additional piece of information, though: how fast the planet itself is rotating. The reason is that we want to measure the speed of the winds with respect to the planet's rotating surface.

In the case of the giant planets, of course, there is no solid surface against which to measure the winds. We must instead assume a hypothetical surface—one that rotates as though it were somehow "connected" to the planet's deep interior. How can we know how rapidly the invisible interior of a planet is rotating? Periodic bursts of radio energy caused by the rotation of a planet's magnetic field tell us how fast the interior of the planet is rotating.

Let's see an example of how this works using a small white cloud in Neptune's atmosphere. The cloud, on Neptune's equator, is observed to be at longitude 73.0° west on a given day. (This longitude system is anchored in the planet's deep interior.) The spot is then seen at longitude 153.0° west exactly 24 hours later. Neptune's equatorial winds have carried the white spot 80.0° in longitude in 24 hours (86,400 seconds).

The equatorial radius of Neptune is 24,760 km. Multiplying the radius by 2π gives the planet's circumference (360° of longitude) as 155,570 km. Thus, each degree of longitude on Neptune's equator is equal to 155,570/360 (approximately 432) km. So, Neptune's equatorial winds have carried the spot $432 \times 80 = 34,560$ km (3.456×10^7 meters) in one day (86,400 seconds). Therefore, the wind speed is 400 meters per second (m/s), the fastest observed zonal wind speed in the Solar System. So we know the equatorial winds are very strong, but in what direction are they blowing? To an external observer, Neptune's west longitude is seen to increase with time. In other words, west longitude increases in a direction *opposite* the direction in which the planet is rotating. The fact that the white spot's west longitude is increasing with time means that Neptune's equatorial winds are blowing in a direction *opposite* the planet's rotation. As we learned in Chapter 8, such winds are called "easterly winds." On Earth, their much slower equivalents are called "trade winds."

to decrease with height, so the *apparent* time variability of Saturn's equatorial winds may be nothing more than changes in the height of the equatorial cloud tops. Alternating easterly and westerly winds also occur at higher latitudes; but unlike Jupiter's case, this alternation seems to bear no clear association with Saturn's atmospheric bands (**Figure 9.14b**). This behavior is but one example of unexplained differences among the giant planets.

As mentioned earlier, Saturn's jet stream is a narrow meandering river of air with alternating crests and troughs (see Figure 9.8a). This feature, located near latitude 45° north, is similar to our own terrestrial jet streams, where high-speed winds blow generally from west to east but with alternate wanderings toward and away from the pole. Nested within the crests and troughs of Saturn's jet stream are anticyclonic and cyclonic vortices. They appear remarkably similar in both form and size to terrestrial high- and low-pressure systems, which bring us alternating periods of fair and stormy weather as they are carried along by our terrestrial jet streams. This similarity in jet streams on Saturn and Earth is a good illustration of the many reasons we study other planets—for the ability to compare them to each other and to similar phenomena on Earth. From observing and analyzing similar atmospheric systems on other planets, we often learn more about the way our own weather works.

Our knowledge of global winds on Uranus (**Figure 9.14c**) is poorer than that of the other giant planets because of the relatively few clouds we have been able to see and track (see Math Tools 9.1). When *Voyager 2* flew by Uranus in 1986, the few clouds we saw were all in the southern hemisphere because the northern hemisphere was in complete darkness at the time. The strongest winds observed were 650-km/h westerlies in the middle to high southern latitudes, and no easterly winds were detected on the part of the planet that could be seen. More recently, as Uranus approached *equinox* season (see Chapter 2), terrestrial observers have reported wind speeds of up to 900 km/h.

Because its peculiar orientation gives Uranus a "reversed" temperature pattern wherein the poles are warmer than the equator, some astronomers had thought earlier that Uranus might have a global wind system very different from that of the other giant planets. Even though the Sun was shining almost straight down on the south pole of Uranus when *Voyager 2* observed it, the dominant winds on Uranus turned out to be zonal, just as they are on the other giant planets. The Coriolis effect is thus more influential than individual atmospheric temperature patterns in determining the fundamental structure of the global winds on all the giant planets.

As Uranus has moved along in its orbit over the course of our recent observations, regions previously unseen by modern telescopes have become visible (**Figure 9.15**). In December

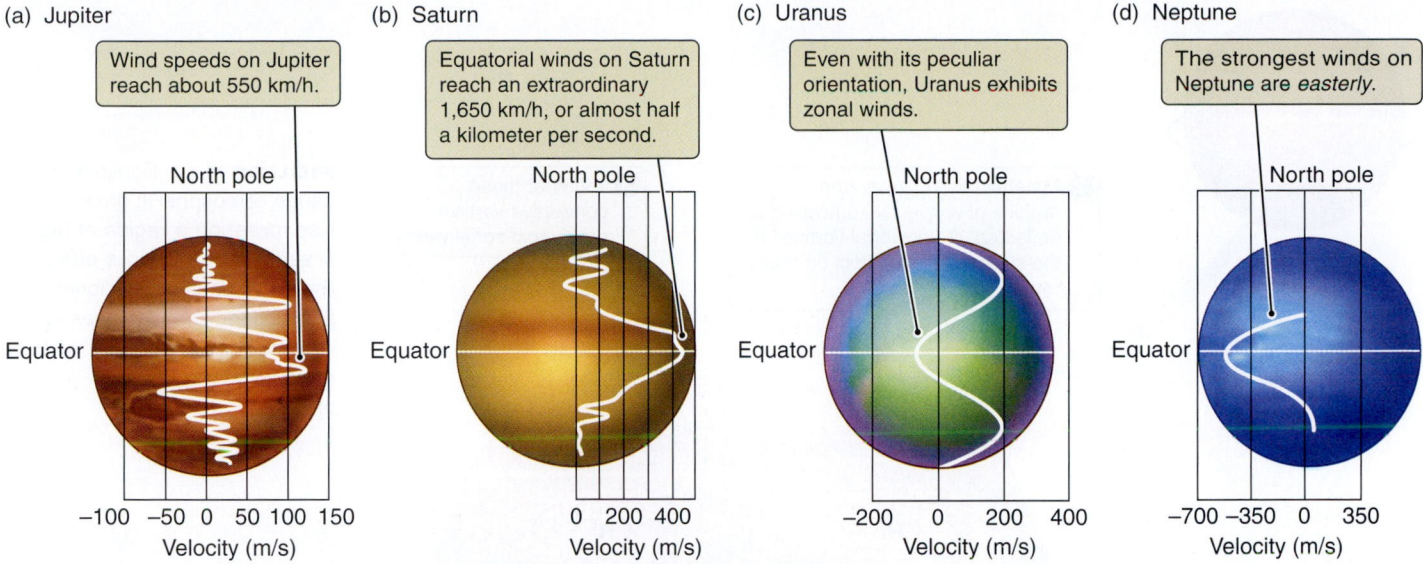

(a) Jupiter

Wind speeds on Jupiter reach about 550 km/h.

North pole

Equator

−100 −50 0 50 100 150
Velocity (m/s)

(b) Saturn

Equatorial winds on Saturn reach an extraordinary 1,650 km/h, or almost half a kilometer per second.

North pole

Equator

0   200   400
Velocity (m/s)

(c) Uranus

Even with its peculiar orientation, Uranus exhibits zonal winds.

North pole

Equator

−200   0   200   400
Velocity (m/s)

(d) Neptune

The strongest winds on Neptune are *easterly*.

North pole

Equator

−700 −350   0   350
Velocity (m/s)

FIGURE 9.14 Strong winds blow the atmospheres of the outer planets, driven by powerful convection and the Coriolis effect on these rapidly rotating worlds.

2007, Uranus arrived at equinox (see Chapter 2) and spring had come to its northern hemisphere. Observations by HST and groundbased telescopes showed bright cloud bands in the far north extending over 18,000 km in length and revealed wind speeds of up to 900 km/h. As we approach northern summer solstice in the year 2027, we can expect to learn much more about the "unseen northern hemisphere" of Uranus.

As expected, the strongest winds on Neptune occur in the tropics (**Figure 9.14d**). The surprise was that they are easterly rather than westerly, with speeds in excess of 2,000 km/h. Westerly winds with speeds of 900 km/h and higher were seen in the south polar regions. With wind speeds five times greater than those of the fiercest hurricanes on Earth, Neptune and Saturn are the windiest planets we know of. Summer solstice arrived in Neptune's southern hemisphere in 2005, so much of the north is still in darkness. We'll have to wait a while before we can get a good look at its northern hemisphere. Each season on Neptune lasts 40 Earth years!

Chapter 8 taught us that vertical temperature differences can create the localized type of atmospheric motion called **convection**. On the giant planets the thermal energy that drives convection comes both from the Sun and from the hot interiors of the planets themselves. As heating drives air up and down, the Coriolis effect shapes that convection into atmospheric vortices, examples of which are familiar to us on Earth as high-pressure systems, hurricanes, and thunderstorms. On the giant planets, convective vortices are visible as isolated circular or oval cloud structures. The Great Red Spot on Jupiter and the Great Dark Spot on Neptune are classic examples. Observations of small clouds distributed within the Great Red Spot show that it is an enormous atmospheric whirlpool or eddy, swirling around in a counterclockwise direction with a period

Thermal energy drives powerful convection on the giant planets.

of about 1 week. On a rapidly rotating planet, winds generated by Coriolis effects can be very strong. Clouds circulating around the circumference of the Great Red Spot have been clocked at speeds of up to 1,000 km/h. Such weather systems dwarf our terrestrial storms in both size and intensity. Comparable whirlpool-like behavior is observed in many of the smaller oval-shaped clouds found elsewhere in Jupiter's atmosphere, as well as in similar clouds observed in the atmospheres of Saturn and Neptune.

FIGURE 9.15 Uranus is approaching equinox in this 2006 HST image. Much of its northern hemisphere is becoming visible. The dark spot in the northern hemisphere (to the right), is similar to but smaller than the Great Dark Spot seen on Neptune in 1989.

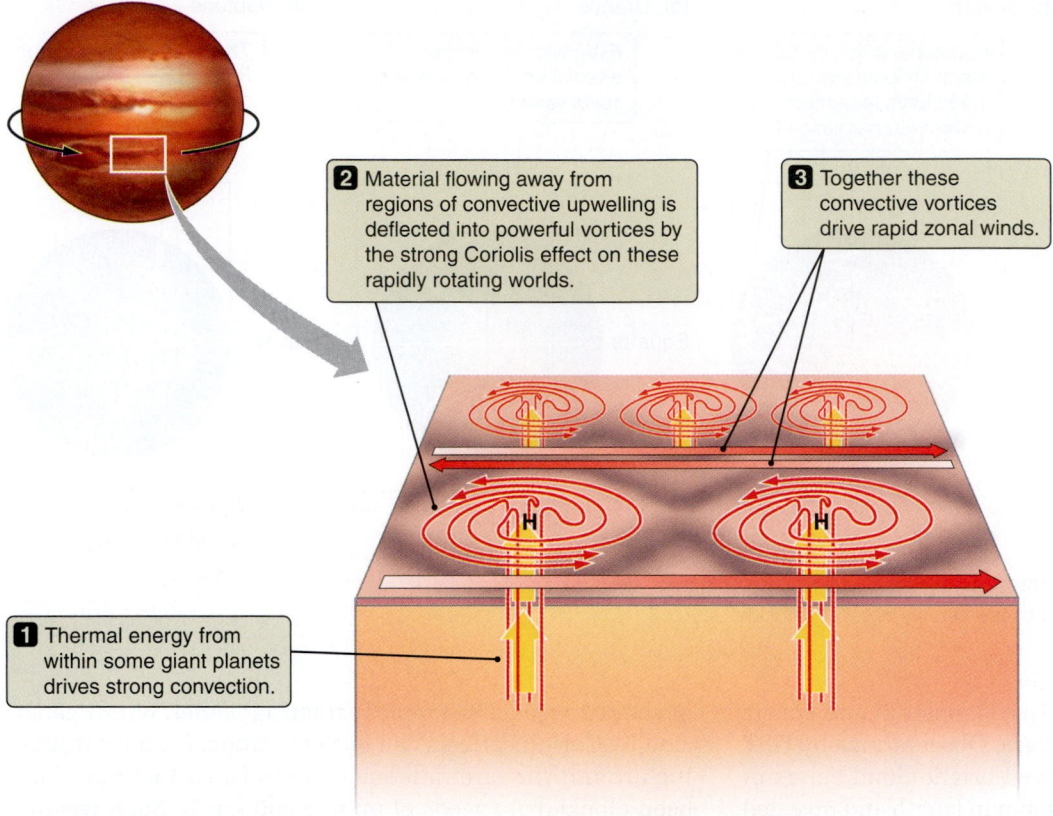

2 Material flowing away from regions of convective upwelling is deflected into powerful vortices by the strong Coriolis effect on these rapidly rotating worlds.

3 Together these convective vortices drive rapid zonal winds.

1 Thermal energy from within some giant planets drives strong convection.

FIGURE 9.16 Convection can cause atmospheric gases to rise, creating a region of high pressure. The Coriolis effect then produces anticyclonic flow around the high pressure region. Such convective vortices drive zonal winds on the giant planets.

As the atmosphere ascends near the centers of the vortices, it expands and cools. As we learned in Chapter 8, cooling condenses certain volatile materials into liquid droplets, which then fall as rain. As they fall, the raindrops collide with surrounding air molecules, stripping electrons from the molecules and thereby developing tiny electric charges in the air. The cumulative effect of countless falling raindrops can generate an electric charge and resulting electric field so great that they break down a conductive path through the atmosphere, creating a surge of current and a flash of lightning. A single observation of Jupiter's night side by *Voyager 1* revealed several dozen lightning bolts within an interval of 3 minutes. We estimate the strength of these bolts to be equal to or greater than the "superbolts" that occur in the tops of high convective clouds in the terrestrial tropics. Although various constraints have prevented an imaging search for lightning on the other giant planets, radio receivers on *Voyager* picked up lightning static on all of them.

How are these atmospheric vortices connected to the global circulation of the giant planets? From a careful study of the interaction between vortices and zonal winds, scientists believe that the massive vortices actually drive zonal wind currents. We have noted that the narrow zonal jet on Saturn moves with a wavelike motion (see Figure 9.8a). Nestled in each of its crests and troughs are clockwise (anticyclonic) and counterclockwise (cyclonic) features, strongly suggesting a dynamic relationship between these systems and the zonal jet. Jupiter's Great Red Spot is situated between

> **Vortices created by convection drive the strong zonal winds of the giant planets.**

pairs of strong zonal winds flowing in opposite directions, as was Neptune's Great Dark Spot. Is this by chance? Similar relationships observed between other isolated vortices and the zonal wind flow suggest that it is not. The enormous wind energy developed within numerous vortices seems to drive the alternating easterly and westerly zonal winds that characterize the global circulation of the giant planets (**Figure 9.16**).

9.6 Internal Thermal Energy Drives Weather Patterns

We saw in Chapter 8 how the uneven heating of a planet, along with the planet's rotation, drives global atmospheric circulation, and how temperature-related differences in pressure from place to place drive winds. In fact, virtually all weather on every planet is driven by the interplays of thermal energy within the planet's atmosphere. On Earth and the other terrestrial planets, the source of this thermal energy is as clear as the Sun shining on a summer sky. Sunlight powers our climate. The Sun, however is not the primary source of heat for the giant planets.

This is not a new insight. In Chapter 4 we learned about the equilibrium between the absorption of sunlight and the radiation of infrared light into space, and in Chapter 8 we saw how the resulting equilibrium temperature is modified by the greenhouse effect on Venus, Earth, and Mars. Yet when we calculate this equilibrium for the giant planets, as we did in Chapter 4, we find that something seems amiss. According to these calculations, the equilibrium temperature for Jupiter, for example, should be 109 K, but when it is measured we find instead an average temperature of about 124 K. A difference of 15 K might not seem like much, but remember that according to the Stefan-Boltzmann law (see Chapter 4), the energy radiated by an object depends on its temperature raised to the fourth power. Applying this relationship to Jupiter, we get

> **Jupiter has a large internal heat source, as do Saturn and Neptune.**

$$\left(\frac{T_{\text{true}}}{T_{\text{expected}}}\right)^4 = \left(\frac{124\ \text{K}}{109\ \text{K}}\right)^4 = 1.67.$$

The implications of this result are somewhat startling: Jupiter is radiating roughly two-thirds more energy into space than it absorbs in the form of sunlight! Thus, almost half of the thermal energy powering Jupiter's weather comes from somewhere other than the Sun. Similarly, the internal energy escaping from Saturn is observed to be about 1.8 times greater than the sunlight that it absorbs. Neptune emits 2.6 times as much energy as it absorbs from the Sun. Strangely, whatever internal energy may be escaping from Uranus is small compared with the absorbed solar energy.

With energy continually escaping from the interiors of the giant planets, we can easily wonder how they have maintained their high internal temperatures over the past 4½ billion years. The short answer is that they are still shrinking in size, converting gravitational potential energy into thermal energy. This continual production of thermal energy is sufficient to replace the energy that is escaping from their interiors.

We can think of the internal energy that lies deep within the giant planets as being primordial. In other words, it is left over from their creation. Though they have much smaller masses than the Sun, the giant planets formed much like collapsing protostars (see Chapter 6). Recall that as a mass of gas collapses under the force of its own self-gravity (see Chapter 10), it converts its gravitational potential energy to thermal energy. The gaseous planets continue to contract indefinitely, releasing their gravitational potential energy as they shrink. This is the primary energy source for replacing the internal energy that leaks out of the interior of Jupiter (see **Excursions 9.1**) and is probably an important source for the other giant planets as well.

The giant planets are still contracting and converting their gravitational potential energy into thermal energy today as they did when they first formed, but they are doing it more slowly. The annual amount of contraction necessary to sustain their internal temperature is only a tiny fraction of their radius. For Jupiter, this is only 1 millimeter (mm) or so—a hundred-billionth of its radius—per year. If this same rate were to continue for the next billion years, Jupiter would shrink by only 1,000 km, a little more than 1 percent of its radius. In Saturn's case, and perhaps Jupiter's as well, there is an additional source of internal energy. Under the right conditions, liquid helium separates from a hydrogen-helium mixture and "rains" downward toward the core. As the droplets of liquid helium sink, they release their gravitational potential energy as thermal energy. Planetary physicists believe that most of Saturn's internal energy and perhaps some of Jupiter's come from this separation of liquid helium.

9.7 The Interiors of the Giant Planets Are Hot and Dense

At depths of a few thousand kilometers, the atmospheric gases of Jupiter and Saturn are so compressed by the weight of the overlying atmosphere that they liquefy. When the interior pressure climbs to 2 *megabars* and the temperature reaches 10,000 K, hydrogen molecules are battered so violently that their electrons are stripped free, and the hydrogen becomes an electrical conductor like a liquid metal. This

---

EXCURSIONS 9.1

Jupiter—A Star That Failed?

In the popular literature Jupiter is frequently and erroneously labeled as "a star that failed"—probably because it is gaseous, has a composition similar to the Sun, has a high internal temperature, and releases a large amount of primordial energy. Although these statements are true, to imply that it is "almost a star" is most certainly an exaggeration. Jupiter's central temperature is probably not much greater than 35,000 K, whereas the temperatures necessary to initiate the self-sustaining thermonuclear reactions that occur in stars are in the tens of millions of degrees. To achieve such temperatures, Jupiter would require a total mass 80 times greater than it actually has. In other words, the least massive stars must still have masses that are approximately 80 times greater than Jupiter's. Jupiter is merely a large planet and never came close to being a star.

happens at a depth of about 20,000 km in Jupiter's atmosphere and 30,000 km in Saturn's. Uranus and Neptune are less massive than Jupiter and Saturn, have lower interior pressures, and contain a smaller fraction of hydrogen—conditions that do not favor the formation of metallic hydrogen. Thus, their interiors probably contain only a small amount of liquid hydrogen,

> Hydrogen-helium oceans lie at depths of a few thousand kilometers on Jupiter and Saturn.

with little or none of it in a metallic state. This transition from a gas to a liquid is so subtle as to be hardly noticeable. To put it another way, the *physical* difference between a liquid and a highly compressed, very dense gas is something that could be appreciated only by—well—a physicist. Thus, unlike the well-defined surface between Earth's atmosphere and its oceans, on Jupiter and Saturn there is no clear boundary between the atmosphere and the "ocean" of liquid hydrogen and helium that lies below. The depths of these hydrogen-helium oceans are measured in tens of thousands of kilometers, making them the largest structures within the interiors of any of the giant planets. Uranus and Neptune do not have these oceans of liquid hydrogen and helium.

Figure 9.17 shows the interior structure of the giant planets. At the center of each giant planet is a dense, liquid core consisting of a very hot mixture of heavier materials such as water, rock, and metals. In Chapter 7 we learned that denser materials, such as iron and other metals, sank to the centers of the terrestrial planets when they were in an earlier, more molten state. This process, which we call **differentiation**, deposited most of the metals in what became the cores of the terrestrial planets, leaving their mantles and crusts relatively depleted of metals.

The cores of the giant planets did not form in the same manner, however. Much of the material now in their cores was already there in the original bodies that captured hydrogen and helium from the Sun's protoplanetary disk to form what ultimately became giant planets. Differentiation, however, has occurred and is still occurring in Saturn, and perhaps Jupiter too. On Saturn, helium condenses out of the hydrogen-helium oceans. (Helium can also be compressed to a metal, but it does not reach this metallic state under the physical conditions existing in the interiors of the giant planets.) Because these droplets of helium are denser than the hydrogen-helium liquid in which they condense, they sink toward the center of the planet. This process tends to enrich helium concentration in the core while depleting it in the upper layers. At the same time the process heats the planet by converting gravitational potential energy to thermal energy. In Jupiter's interior, by contrast, the liquid helium is mostly dissolved together with the liquid hydrogen. Within Saturn's interior, temperatures are lower than on Jupiter, making the helium less soluble. Those who cook know that you can easily dissolve large amounts of sugar in hot water but relatively little when the water is cold. So it is with helium and hydrogen.

In the cores of the gas giants, temperatures may be in the tens of thousands of degrees, with pressures of tens of **megabars**. As we learned in Chapter 8, a pressure of 1 bar corresponds closely to Earth's atmospheric pressure at sea level. A megabar is a million times as great as 1 bar. For

> Jupiter's core is liquid water and rock at a temperature of about 35,000 K.

comparison, when the submersible research vessel *Alvin* cruises Earth's ocean bottoms 10,000 meters below the surface, it experiences a surrounding pressure of about 1,000 bars, or 1/1,000 of a megabar. It may seem strange to be talking about water that is still liquid at temperatures of tens of thousands of degrees, but there is really nothing peculiar

FIGURE 9.17 The central cores and outer liquid shells of the interiors of the giant planets. Only Jupiter and Saturn have significant amounts of the molecular and metallic forms of liquid hydrogen surrounding their cores

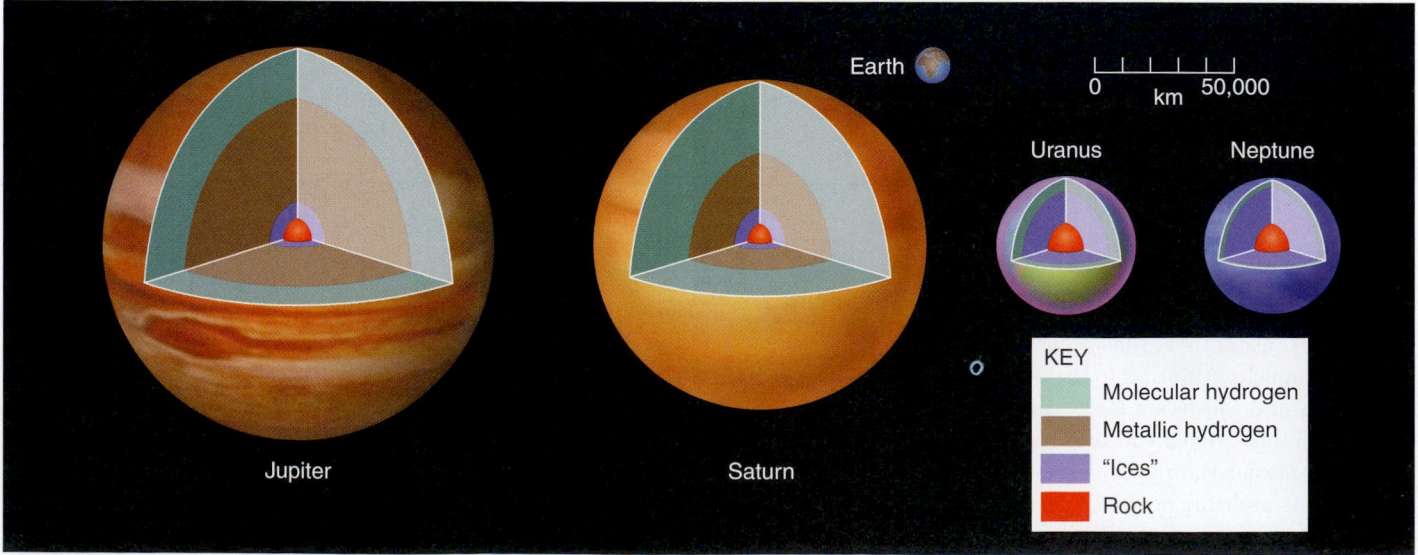

EXCURSIONS 9.2

Strange Behavior in the Realm of Ultrahigh Pressure

You might naturally assume that gaseous, Jovian-type planets with mass greater than Jupiter's should also be larger, but to do so would be wrong. Jupiter's mass is well more than three times as great as Saturn's, even though both have approximately the same composition and are nearly the same size. This apparent anomaly occurs because Jupiter's massive self-gravity—caused by its greater mass—compresses its interior more than Saturn's can. For planets the size of Jupiter and Saturn, adding more hydrogen does not make them much bigger; it makes them denser instead. We now find ourselves with a curious situation that goes against our intuition: if we were to add much more hydrogen to Jupiter, its diameter would actually get *smaller* rather than larger. A planet with a mass, say, 10 times greater than Jupiter would be about 10 percent smaller. The additional overlying mass creates higher internal pressures, which in turn compress the interior further. The decreased volume caused by increased pressure more than makes up for the increased volume of the additional hydrogen.

It turns out that, by chance, Jupiter is almost the largest that any planet can be in this or any other solar system. Planets that are either less massive or more massive will be smaller! Does this mean that stars should also be smaller than Jupiter? They would be except for one important difference. Stars have nuclear reactions going on within their cores and thus have *very* much higher internal temperatures than any planet has. These ultra-high temperatures create enormous internal pressures that better resist gravitational compression, keeping even the smallest "normal" stars much larger than the largest planets. We say "normal" stars because later, in Chapters 16 and 17, we will introduce you to dying stars that are even smaller than Earth.

---

about this. Like a super pressure cooker, the extremely high pressures at the centers of the giant planets prevent the water from turning to steam. The temperature at Jupiter's center is thought to be as high as 35,000 K, and the pressure may reach 45 megabars. Central temperatures and pressures of the other, less massive giant planets are correspondingly lower than those of Jupiter.

Uranus and Neptune Are Different from Jupiter and Saturn

You might suppose that the average densities of the giant planets would tell us how much heavy material they contain. In practice, it is not so simple. Although the actual numbers are very uncertain, the heavy-element components of the cores of Jupiter and Saturn have masses of about 5–10 $M_\oplus$. As Table 9.1 shows, Jupiter and Saturn have total masses of 318 and 95 $M_\oplus$, respectively. The heavy materials in their cores, then, contribute little to their average chemical composition. This means we can think of both Jupiter and Saturn as having approximately the same composition as the Sun and the rest of the universe: about 98 percent hydrogen and helium and only 2 percent of everything else. Why, then, with nearly identical compositions, should Jupiter's density be nearly twice as great as Saturn's? In **Excursions 9.2** we find that the internal pressure created by Jupiter's much greater mass compresses its hydrogen and helium and its core to a higher average density than in the core of Saturn. What does all of this imply about Uranus and Neptune?

If Uranus and Neptune were also of solar composition—that is, if they were made primarily of hydrogen and helium—their average densities would be less than half the density of water, even less than that of Saturn. Their lower mass would not be as effective as Saturn's in compressing their hydrogen and helium. But such low average densities are not what we observe. Instead we find Uranus and Neptune to be about twice as dense as Saturn (see Table 9.1). We now have a clear indication that, unlike Jupiter and Saturn, Uranus and Neptune must have denser material dominating their chemical composition. Is this denser material water or rock? Our observations should be able to tell us. Neptune, the densest of the giant planets, is about 1½ times as dense as uncompressed water and only about half as dense as uncompressed rock. Uranus is even less dense than Neptune. These observations tell us that both Uranus and Neptune must contain more water than rock. But we must also keep in mind that the high pressures within the interiors of Uranus and Neptune cause both water and rock to be more dense than in their uncompressed states. Thus, water and other low-density ices, such as ammonia and methane, must be the major compositional components of Uranus and Neptune, with lesser amounts of silicates and metals. The total amount of hydrogen and helium in these planets is probably limited to no more than 1 or 2 $M_\oplus$, with most of these gases residing in their relatively shallow atmospheres. On the basis of density alone, as we suggested at the beginning of this chapter, neither Uranus nor Neptune very well fit the description of a gas giant. It is more appropriate to refer to them as ice giants.

The water that makes up so much of Uranus and Neptune is probably in the form of deep oceans. Dissolved gases and salts would make these oceans electrically conducting. All of the giant planets have

Uranus and Neptune are ice giants with deep, salty oceans.

deep layers of a conducting fluid—metallic hydrogen in the case of Jupiter and Saturn, and a saltwater brine in the case of Uranus and Neptune. Currents in these conducting layers are likely the source of the giant planets' intense magnetic fields, which we will discuss in Section 9.8.

Differences Are Clues to Origins

That each of the giant planets formed around rocky-metal cores but turned out so differently is an important clue to their origins. Why do Jupiter and Saturn have so much hydrogen and helium compared with Uranus and Neptune? Why is hydrogen-rich Jupiter so much more massive than hydrogen-rich Saturn? The answers may lie both in the time that it took for these planets to form and in the distribution of material from which they formed. We think that all of the hydrogen and helium in the giant planets was captured from the protoplanetary disk by the strong gravitational attraction of their massive cores.[2] The much lower hydrogen-helium content of Uranus and Neptune suggests that their cores were smaller and formed much later than those of Jupiter and Saturn, at a time when most of the gas in the protoplanetary disk had been blown away by the emerging Sun. Why should the cores of Uranus and Neptune have formed so late? Probably because the icy planetesimals from which they formed were more widely dispersed at their greater distances from the Sun. With more space between planetesimals, it would take longer to build up their cores. Saturn may have captured less gas than Jupiter both because its core formed somewhat later and because less gas was available at its greater distance from the Sun.

9.8 The Giant Planets Are Magnetic Powerhouses

All of the giant planets have magnetic fields that are much stronger than Earth's. Planetary magnetic fields are produced by the motions of electrically conducting liquids deep within planet interiors. In Jupiter and Saturn, magnetic fields are generated in deeply buried layers of metallic hydrogen. In Uranus and Neptune, magnetic fields arise in salt brine oceans. Although their origins are complex, we can schematically illustrate the geometry of the magnetic fields of the giant planets as if they came from bar magnets (see Chapter 7) in the interiors of the planets, as shown in **Figure 9.18**.

Jupiter's magnetic-field axis is inclined 10° to its rotation axis—an orientation similar to Earth's—but it is displaced about a tenth of a radius from the planet's center (**Figure 9.18a**). Note that the direction of Jupiter's magnetic field is

[2]Although many planetary scientists believe that the giant planets formed by core accretion, others have proposed a disk instability process, as we learned in Chapter 6.

opposite to that of Earth as defined by where the north end of a compass would point. The total strength of Jupiter's magnetic field is nearly 20,000 times that of Earth's. On the other hand, Jupiter is huge compared to Earth. By the time Jupiter's magnetic field

> **Jupiter's magnetic field is 20,000 times as strong as Earth's.**

emerges from the cloud tops, the field has dropped to about 4.3 *gauss*—only 15 times Earth's surface field. The strength of a magnetic field is often measured in **gauss**, named for the German mathematician and scientist Carl Friedrich Gauss (1777–1855).

The bar magnet used to simulate Saturn's magnetic field is located almost precisely at the center of Saturn and is almost perfectly aligned with the planet's rotation axis (**Figure 9.18b**). Saturn's magnetic field is much weaker than Jupiter's, but overall it is still more than 500 times stronger than Earth's. Because Saturn's diameter is much greater than Earth's, the magnetic-field strength at its cloud tops is about 0.5 gauss—similar to the strength at Earth's surface. As on Jupiter, a compass would point south on Saturn.

Voyager 2 found that the magnetic-field axis of Uranus is inclined nearly 60° to its rotation axis, and its center is displaced by a third of a radius from the planet's center (**Figure 9.18c**). Considering the strange spin orientation of Uranus, this observation did not come as any great surprise to the *Voyager* scientists. The total strength of the field averages 50 times Earth's field, but the large displacement of Uranus's field from the planet's center causes the field at the cloud tops to vary between 0.1 and 1.1 gauss.

The really big surprise came when *Voyager 2* reached Neptune. The orientation of Neptune's rotation axis is similar to that of Earth, Mars, and Saturn. But Neptune's magnetic-field axis is inclined 47° to its rotation axis, and the center of this magnetic field is displaced from the planet's center by more than half the radius—an offset even greater than that of Uranus (**Figure 9.18d**). The displacement of the field is primarily toward Neptune's southern hemisphere, thereby creating a large asymmetry in the field strength at the cloud tops, with 1.2 gauss in the southern hemisphere and only 0.06 gauss in the north. The overall strength of Neptune's magnetic field is only half as great as that of Uranus.

The reason for the unusual geometry of the magnetic fields of Uranus and Neptune remains unknown, but it is clearly not related to the orientations of their rotation axes.

Giant Planets Have Giant Magnetospheres

Just as Earth's magnetic field traps energetic charged particles to form Earth's **magnetosphere**, the magnetic fields of the giant planets also trap energetic particles to form magnetospheres of their own. Although Earth's magnetosphere may have a radius over 10 times that of our planet itself, our magnetosphere is tiny in comparison with the vast clouds

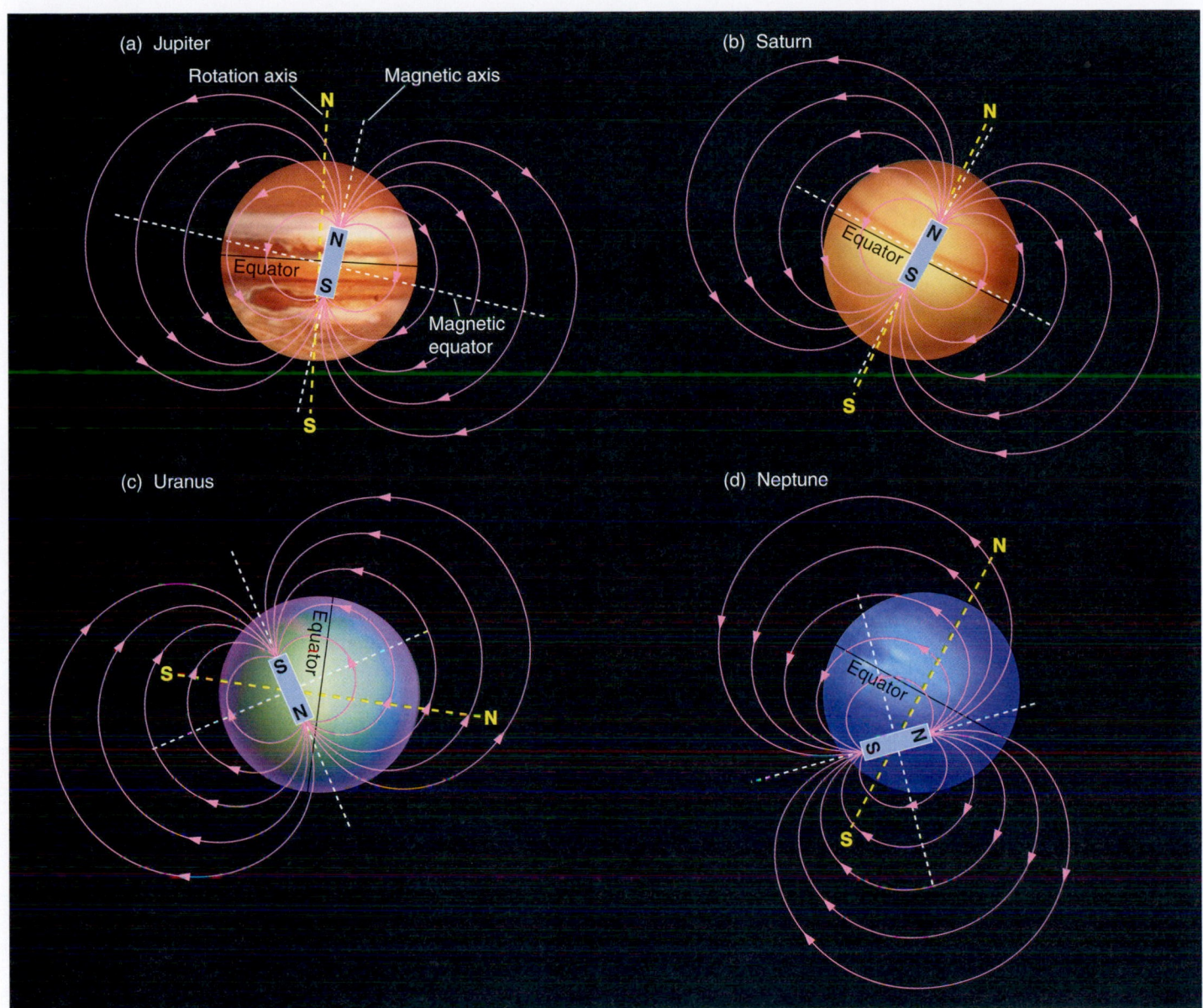

VISUAL ANALOGY FIGURE 9.18 The magnetic fields of the giant planets can be approximated by the fields from bar magnets offset and tilted with respect to the planets' axes. Compare these with Earth's magnetic field, shown in Figure 7.16a.

of plasma held together by the much more powerful magnetic fields of the giant planets. By far the most colossal of these is Jupiter's magnetosphere. Its radius is as much as 100 times that of the planet itself, or about 7 million km. That is roughly 10 times the radius of the Sun! Although the magnetospheres of the other giant planets are much smaller, even the relatively weak magnetic fields of Uranus and Neptune form magnetospheres that are comparable in size to the Sun.

> **The magnetospheres of the giant planets are enormous.**

The **solar wind** (see Chapter 8) does more than supply some of the particles for a magnetosphere. The pressure of the solar wind also pushes on and compresses a magnetosphere.

The size and shape of a planet's magnetosphere can change a great deal depending on how the solar wind is blowing at any particular time. Planetary magnetic fields also divert the solar wind, which flows around magnetospheres the way a stream flows around boulders. Just as a rock in a river creates a wake that extends downstream, the magnetosphere of a planet produces a wake that can extend downstream for great distances. The wake of Jupiter's magnetosphere (**Figure 9.19**) extends over 6 astronomical units (AU) outward from the planet—well past the orbit of Saturn. Jupiter's magnetosphere is the largest permanent "object" in the Solar System, surpassed in size only by the tail of an occasional comet. The magnetic wakes of Uranus and Neptune have a curious structure. Because of the tilt and the large displacements of their magnetic fields from the centers of these planets, their magnetospheres wobble as the planets rotate. This wobble causes the wakes of their magnetospheres to twist like corkscrews as they stretch away from the planets.

Magnetospheres Produce Synchrotron Radiation

We need not send spacecraft to the outer Solar System, or even call on telescopes orbiting Earth, to see evidence of the giant planets' magnetospheres. Rapidly moving electrons in planetary magnetospheres spiral around the direction of the magnetic field, and as they do so they emit *synchrotron radiation*, as discussed in **Foundations 9.1**. If our eyes were sensitive to radio waves, then the second brightest object in the sky would be Jupiter's magnetosphere. The Sun would still be brighter, but it would not appear larger; even at a distance from Earth of 4.2–6.2 AU, Jupiter's magnetosphere would appear roughly twice the size of the Sun in the sky. Saturn's magnetosphere would also be large enough to see, but it would be much fainter than Jupiter's. Even though Saturn has a strong magnetic field, pieces of rock, ice, and dust in Saturn's spectacular rings act like sponges, soaking up magnetospheric particles. Magnetospheric particles typically collide with ring material soon after those particles enter the magnetosphere. With far fewer magnetospheric electrons, there is much less radio emission from Saturn.

We can learn a great deal about planets by studying the synchrotron radiation from their magnetospheres. For example, precise measurement of periodic variations in the radio signals "broadcast" by the giant planets tells us the planets' true rotation periods. The magnetic field of each planet is locked to the conducting liquid layers deep within the planet's interior; the magnetic field thus rotates with exactly the same period as the deep interior of the planet. Given the fast and highly variable winds that push around clouds in the atmospheres of the giant planets, measurement of radio emission is the only way we have to determine the true rotation periods of the giant planets.

> Radio signals broadcast a planet's true rotation period.

Radiation Belts and Auroras

As a planet rotates on its axis, it drags its magnetosphere around with it. In the enormous magnetosphere of a rapidly rotating planet like Jupiter, charged particles are swept around at high speeds. These fast-moving charged particles slam into neutral atoms (which do not share the motion of material in the magnetosphere), and the energy released in the resulting high-speed collisions heats the plasma to extreme temperatures. In 1979, while passing through Jupiter's magnetosphere, *Voyager 1* encountered a region of tenuous plasma with a temperature of over 300 million K; this is 20 times the temperature at the center of the Sun! The density of the plasma (about 10,000 atoms per cubic meter) was much lower than that of the best vacuum we can produce on Earth, however, so the spacecraft was in no danger.

Charged particles trapped in planetary magnetospheres are concentrated in certain regions called **radiation belts**. Although Earth's radiation belts are so severe as to worry astronauts, the radiation belts that surround Jupiter are searing in comparison. In 1973 the *Pioneer 11* spacecraft passed through the radiation belts of Jupiter. During its brief

FIGURE 9.19 The solar wind compresses Jupiter's (or any other) magnetosphere in the direction toward the Sun and draws it out into a magnetic wake in the direction away from the Sun. Jupiter's wake stretches beyond the orbit of Saturn.

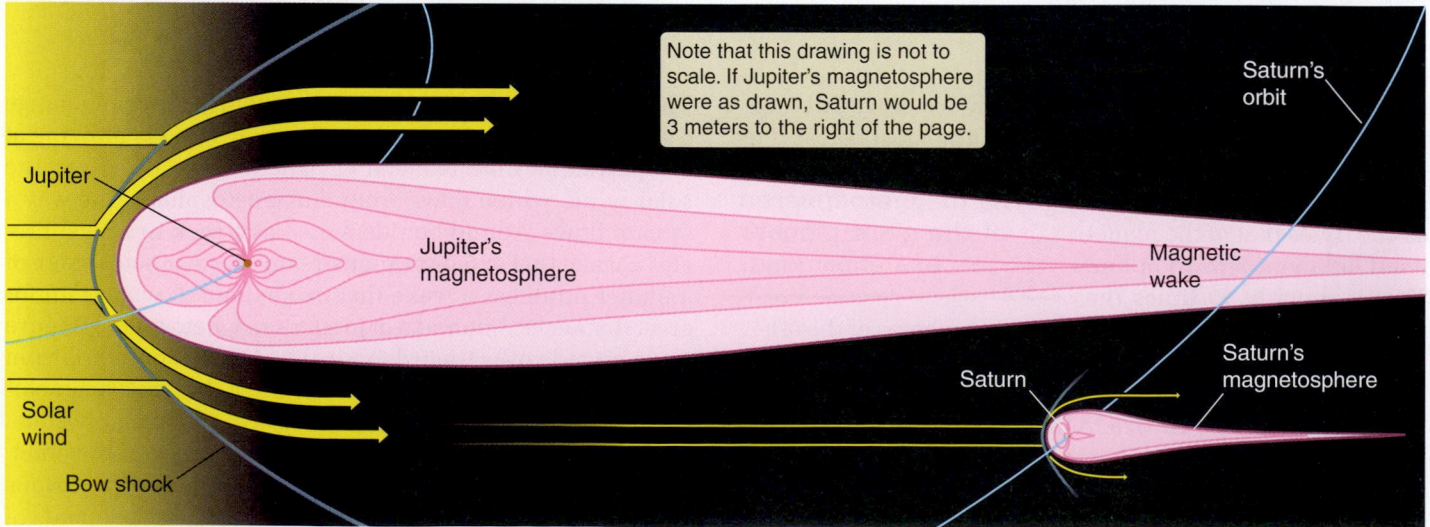

FOUNDATIONS 9.1

Synchrotron Radiation—From Planets to Quasars

In Chapter 4 we found that anytime a charged particle experiences an acceleration, the particle emits electromagnetic radiation. The accelerations resulting from forces exerted by magnetic fields are no exception. As we have seen, charged particles moving in a magnetic field experience a force that is perpendicular both to the motion of the particle and to the direction of the magnetic field. This force produces an acceleration, causing the particles to follow helical paths around the direction of the magnetic field. The accelerations also cause the particles to radiate.

If the particles are traveling at a significant fraction of the speed of light, then relativistic effects cause the radiation they emit to be beamed in the direction in which they are traveling. The situation is illustrated in **Figure 9.20**. The resulting radiation is called **synchrotron radiation**, so named because it was first discovered in a type of particle accelerator called a "synchrotron."

The amount of radiation from a particle depends on the amount of acceleration the particle experiences. For a given amount of force, a less massive particle experiences more acceleration. Because electrons are much less massive than protons or any **ion**, it is the electrons in a magnetized plasma that experience the greatest acceleration. Combining these ideas, we see that it must be the electrons in a magnetized plasma that produce the overwhelming majority of its synchrotron radiation.

The magnetospheres of the giant planets contain energetic electrons moving in strong magnetic fields, so they satisfy the requirements for synchrotron radiation. Synchrotron radiation is unlike thermal (Planck) radiation because instead of being strongly peaked in one part of the electromagnetic spectrum, synchrotron radiation from a single source can range from radio waves to X-rays. The spectrum of synchrotron radiation is determined by the strength of the magnetic field and how energetic the radiating particles are. Synchrotron radiation

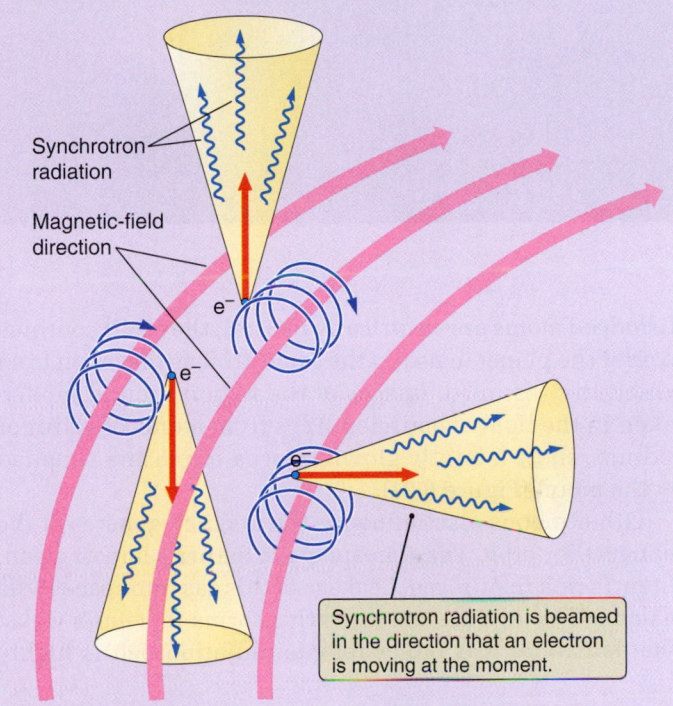

Synchrotron radiation is beamed in the direction that an electron is moving at the moment.

FIGURE 9.20 Rapidly moving charged particles loop around the direction of a magnetic field, giving off electromagnetic radiation as a result of the acceleration they experience. The radiation, called synchrotron radiation, is beamed by relativistic effects in the direction of particle motion.

from planetary magnetospheres is concentrated in the low-energy radio part of the spectrum.

This is our first encounter with synchrotron radiation, but it will not be our last. As we move outward into the universe we will find many objects, from quasars to the remnants of supernovae, that emit synchrotron radiation throughout the electromagnetic spectrum.

---

encounter, *Pioneer 11* picked up a radiation dose of 400,000 rads, or about 1,000 times the lethal dose for humans. Several of the instruments on board were permanently damaged as a result, and the spacecraft itself barely survived to continue its journey to Saturn.

Jupiter has intense radiation belts.

In addition to protons and electrons from the solar wind, the magnetospheres of the giant planets contain large amounts of various elements, including sodium, sulfur, oxygen, nitrogen, and carbon. These elements come from several sources, including the planets' extended atmospheres and the moons that orbit within them. The most

intense radiation belt in the Solar System is a toroidal (that is, torus- or doughnut-shaped) ring of plasma associated with Io, the innermost of Jupiter's four Galilean moons. As you will learn in Chapter 11, Io is the most volcanically active body in the Solar System. Because of its low surface gravity and the violence of its volcanism, some of the gases erupting from its interior can escape the moon and become part of Jupiter's radiation belt. As charged particles are slammed into the moon by the rotation of Jupiter's magnetosphere, even more material is knocked free of the surface and ejected into space. If the

Io is a source of magnetospheric particles.

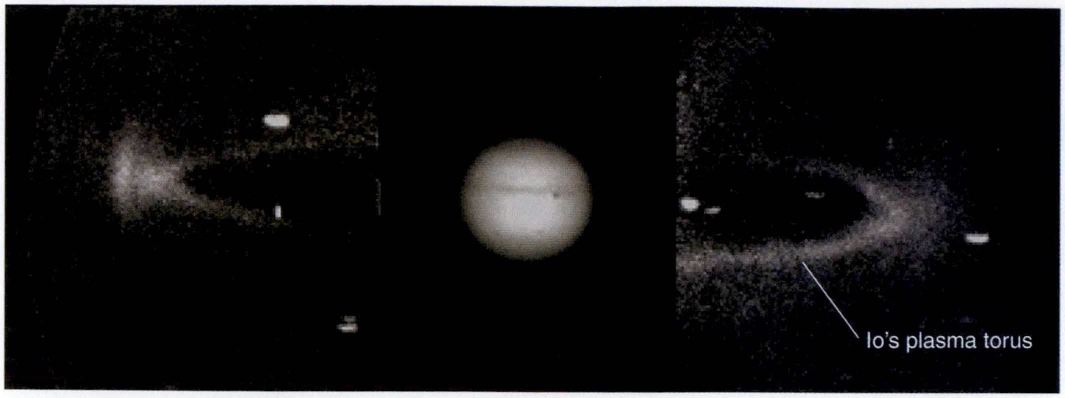

FIGURE 9.21 A faintly glowing torus of plasma surrounds Jupiter (center). The torus is made up of atoms knocked free from the surface of Io by charged particles. Io is the innermost of Jupiter's Galilean moons. A semitransparent mask was placed over the disk of Jupiter to prevent it from overwhelming the much fainter torus.

Io's plasma torus

dislodged atoms are electrically neutral, they will continue to orbit the planet in nearly the same orbit as the moon from which they escaped. Images of the region around Jupiter, taken in the light of emission lines from atoms of sulfur or sodium, show a faintly glowing torus of plasma supplied by the moon (**Figure 9.21**).

Other moons also influence the magnetospheres of the planets they orbit. The atmosphere of Saturn's largest moon, Titan, is rich in nitrogen. Leakage of this gas into space is the major source of a plasma torus that forms in Titan's wake. The density of this rather remote radiation belt is highly variable because the orbit of Titan is sometimes within and sometimes outside Saturn's magnetosphere, depending on the strength of the solar wind. When Titan is outside Saturn's magnetosphere, any nitrogen molecules lost from the moon's atmosphere are carried away by the solar wind.

Charged particles spiral along the magnetic-field lines of the giant planets, bouncing back and forth between the two magnetic poles, just as they do around Earth. As with Earth, these energetic particles collide with atoms and molecules in a planet's atmosphere, knocking them into excited energy states that decay and emit radiation. The results are bright

FIGURE 9.22 Hubble Space Telescope images of auroral rings around the poles of Jupiter (a) and Saturn (b). The auroral images were taken in ultraviolet light and then superimposed on visible-light images. (High-level haze obscures the ultraviolet views of the underlying cloud layers, as the insets show.) The bright spot and trail outside the main ring of Jupiter's auroras are the footprint and wake of Io's *flux tube*.

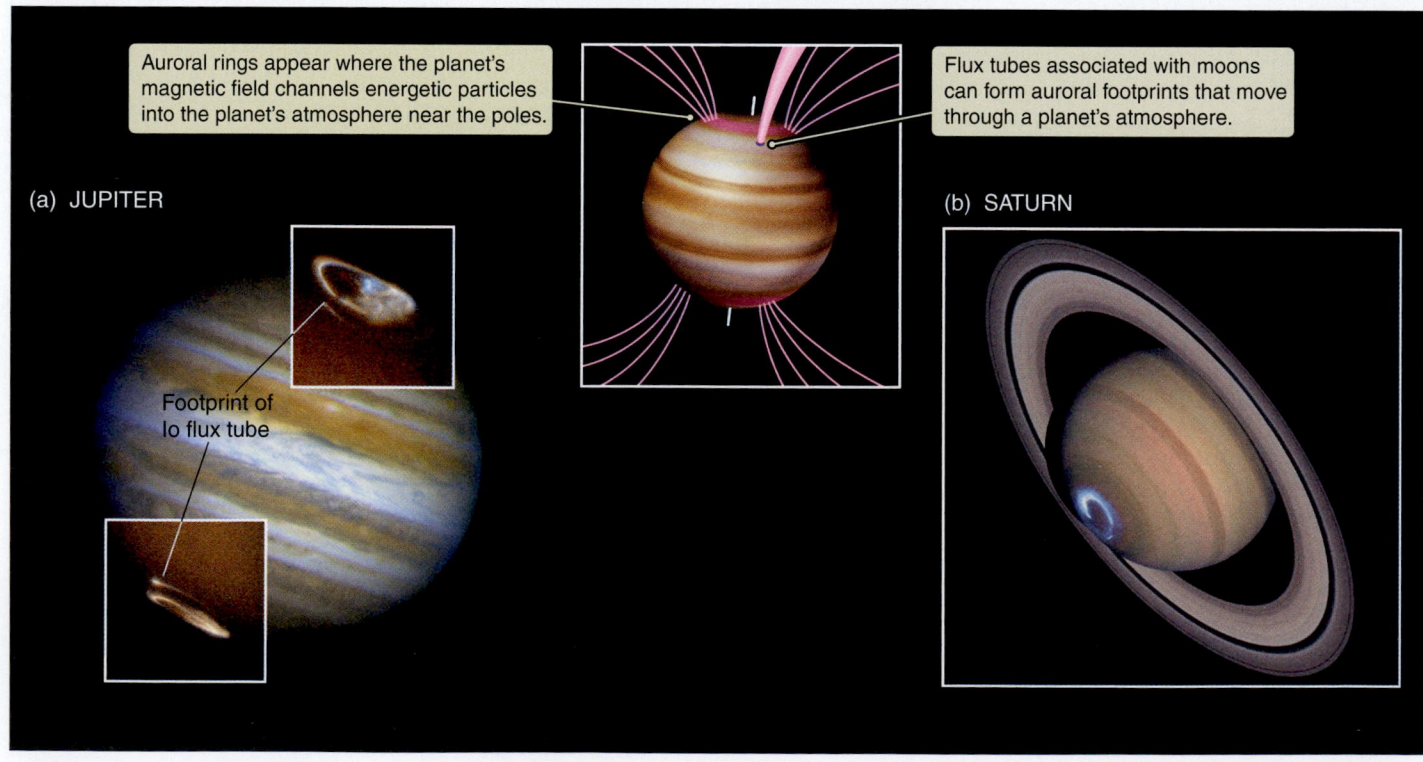

Auroral rings appear where the planet's magnetic field channels energetic particles into the planet's atmosphere near the poles.

Flux tubes associated with moons can form auroral footprints that move through a planet's atmosphere.

(a) JUPITER

(b) SATURN

Footprint of Io flux tube

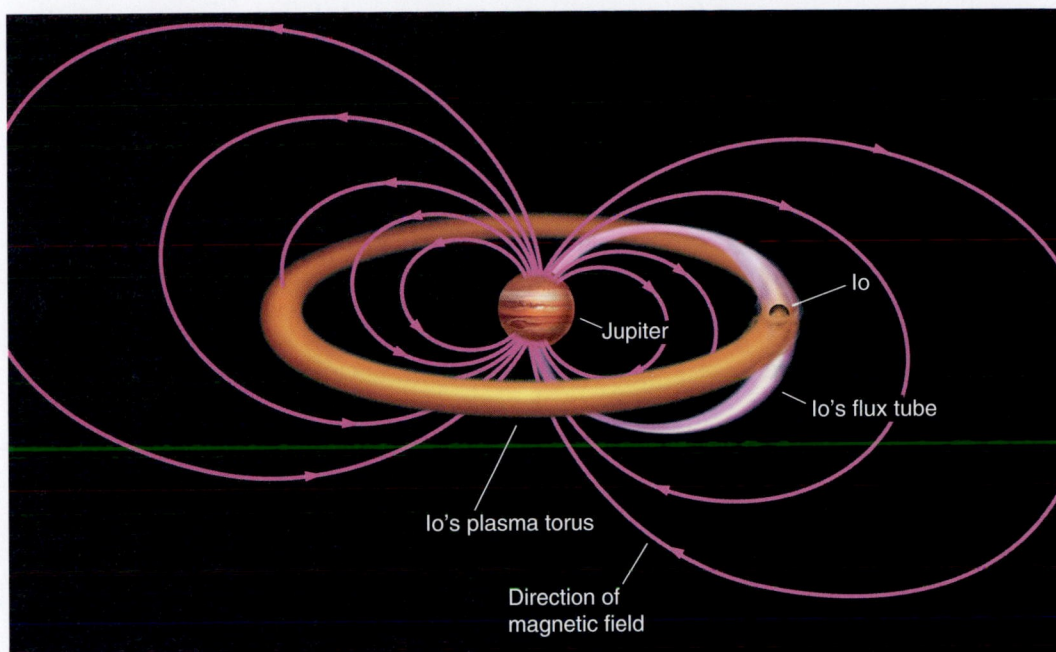

FIGURE 9.23 The geometry of Io's plasma torus and flux tube.

auroral rings (**Figure 9.22**) that surround the poles of the giant planets—just as the aurora borealis and aurora australis ring the north and south magnetic poles of Earth.

Jupiter's auroras have an added twist that we do not see on Earth, however. As Jupiter's magnetic field sweeps past Io, it behaves like a dynamo, generating an electric potential of 400,000 volts (V). Electrons accelerated by this enormous electric field spiral along the direction of Jupiter's magnetic field. The result is a magnetic channel, called a **flux tube**, that connects Io with Jupiter's atmosphere near the planet's magnetic poles (**Figure 9.23**). Io's flux tube carries an electric current of 5 million amperes (amps), amounting to 2 trillion watts (W) of power or about ⅙ of the total electric power produced by generating stations on Earth. Much of the power generated within the flux tube is radiated away as radio energy. These radio signals are received at Earth as intense bursts of synchrotron radiation. However, a substantial fraction of the energy of particles in the flux tube is deposited into Jupiter's atmosphere. At the very location where Io's flux tube intercepts Jupiter's atmosphere, there is a spot of intense auroral activity. As Jupiter rotates, this spot leaves behind an auroral trail in Jupiter's atmosphere. The footprint of Io's plasma torus, along with its wake, can be seen outside the main auroral ring in Figure 9.22a.

Seeing the Forest for the Trees

In our discussion of the terrestrial planets, a major theme was diversity. We used Earth as the basis for understanding our planetary neighbors, and at the same time we learned from those neighbors what it is about Earth that sets it apart. On this leg of our journey we have come to understand that even the range of conditions from the frozen poles of Mars to the hellish surface of Venus is modest in comparison with what is possible. As the Solar System formed, planets in the cold outer reaches grew more massive than the planets closer to the Sun. They became massive enough to capture and hold on to the hydrogen and helium gas that was so abundant in the disk surrounding the young Sun.

In exploring the giant planets, we encounter worlds with scales beyond human experience. In place of Earth's gentle trade winds, we find vast atmospheric bands streaming around the giant planets at speeds greater than 2,000 km/h. The most powerful terrestrial hurricanes would be but inconsequential eddies within Jupiter's Great Red Spot, a system so enormous that it could swallow two Earths whole. On Earth we marvel at the sight of a terrestrial thunderstorm towering overhead, lit up by powerful flashes of lightning. Can we begin to imagine the appearance of multi-hued clouds of water, ammonia, methane, and other compounds, tinted by sulfur, phosphorus, and photochemical smog, as they roil up from within the depths of a giant planet? Can we envision the bolts of lightning that sear through those clouds?

Of course, these analogs of our terrestrial experience only scratch the surface of the differences between the giant planets and Earth. In the case of the terrestrial planets, the distinction between planet and atmosphere is clear-cut. On our Earth, clouds billow in a blue sky, adrift over the palpable surface of land and ocean. As we descend into the giant planets, there is no sudden transition to solid or liquid. Instead there is only the steady and inexorable increase in pressure, a smooth transition from the tops of the clouds to a core that in the case of Jupiter may have an absolute temperature about six times that at the surface of the Sun, and a crushing pressure 45 million times that at the surface of Earth. The higher-than-expected temperatures of the giant planets that we found in Chapter 4 are signposts of these internal differences. Whereas Earth's climate is defined by the interplay between energy from the Sun and the physics and chemistry of our atmosphere and oceans, the fierce weather systems on the giant planets are powered largely by the conversion of gravitational energy to thermal energy within their interiors.

Although the bulk of the giant planets dwarfs that of their terrestrial cousins, their influence does not end here. Powerful magnetic fields trap energetic charged particles streaming outward from the Sun, leading to the formation of giant magnetospheres—the largest permanent structures in the Solar System.

In the following chapter, the time has come for a brief digression. The systems of moons and rings that surround the giant planets are gravitational playgrounds, rich with phenomena that go far beyond Kepler's laws. Before we take the next step outward, we will pause and return our attention to our old friend gravity.

Summary

- The giant planets are large and massive, and less dense than the terrestrial planets.

- Jupiter and Saturn are made up mostly of hydrogen and helium—a composition similar to that of the Sun.

- Uranus and Neptune contain much larger amounts of "ices" such as water, ammonia, and methane.

- We see only atmospheres, not solid surfaces, on the giant planets.

- Clouds on Jupiter and Saturn are composed of various kinds of ice crystals colored by impurities.

- Powerful convection and the Coriolis effect drive high-speed winds in the upper atmospheres of the giant planets.

- Jupiter, Saturn, and Neptune have large internal heat sources. Uranus seems to have little or none.

- The interiors of the giant planets are very hot and very dense.

- The giant planets have enormous magnetospheres that emit synchrotron radiation.

SmartWork, Norton's online homework system, includes algorithmically generated versions of these questions, plus additional conceptual exercises. If your instructor assigns questions in SmartWork, log in at **smartwork.wwnorton.com.**

Student Questions

THINKING ABOUT THE CONCEPTS

1. Identify and describe the two subclasses of giant planets, and indicate which planets fall into each subclass.

2. Describe the ways in which the giant planets differ from the terrestrial planets.

3. What observations led Adams and Le Verrier to predict the location of Neptune?

4. Jupiter's chemical composition is more like the Sun's than Earth's is. Yet both planets formed from the same protoplanetary disk. Explain the difference.

5. What can we learn about a Solar System object when it "occults" a star?

6. Explain two methods used to determine the mass of a giant planet.

7. Why can we not see the surfaces of the giant planets?

8. Jupiter is sometimes described as "a star that failed." Explain why this is a gross exaggeration.

9. Astronomers take the unusual position of lumping together all atomic elements other than hydrogen and helium into a single category, which they call "heavy elements." Why is this a reasonable thing for astronomers to do?

10. None of the giant planets is truly round. Explain why they have a flattened appearance.

*11. Why do the individual cloud layers in the atmospheres of the giant planets have different chemical compositions?

12. What is the source of color in Jupiter's clouds?

13. Uranus and Neptune, when viewed through a telescope, appear distinctly bluish green in color. What are the two reasons for their striking appearance?

14. Which of the giant planets have seasons similar to Earth's, and which one experiences extreme seasons?

15. What creates metallic hydrogen in the interiors of Jupiter and Saturn, and why do we call it "metallic?"

*16. Jupiter's core is thought to consist of rocky material and ices, all in a liquid state at a temperature of 35,000 K. How can materials such as water be liquid at such high temperatures?

17. Explain how astronomers measure wind speeds in the atmospheres of the giant planets.

18. The Great Red Spot is a long-lasting atmospheric vortex in Jupiter's southern hemisphere. Winds rotate counterclockwise around its center. Is the Great Red Spot cyclonic or anticyclonic? Is it a region of high or low pressure? Explain.

*19. What drives the zonal winds in the atmospheres of the giant planets?

20. Lightning has been detected in the atmospheres of all of the giant planets. How is it detected?

*21. Jupiter, Saturn, and Neptune all radiate more energy into space than they receive from the Sun. Does this violate the law of conservation of energy? What is the source of the additional energy?

*22. Saturn has a source of internal thermal energy that Jupiter may not have. What is it and how does it work?

23. Explain how we know that Uranus and Neptune are not gas giants like Jupiter and Saturn.

24. When viewed by radio telescopes, Jupiter is the second brightest object in the sky. What is the source of this radiation?

25. What creates auroras in the polar regions of Jupiter and Saturn?

APPLYING THE CONCEPTS

26. The Sun appears 400,000 times brighter than the full Moon in our sky. How far from the Sun (in astronomical units) would you have to go for the Sun to appear only as bright as the full Moon appears in our nighttime sky? Compare your answer with the semi-major axis of Neptune's orbit.

27. Uranus occults a star at a time when the relative motion between Uranus and Earth is 23.0 km/s. An observer on Earth sees the star disappear for 37 minutes and 2 seconds and notes that the center of Uranus passed directly in front of the star.
    a. On the basis of these observations, what value would the observer calculate for the diameter of Uranus?
    b. What could you conclude about the planet's diameter if its center did not pass directly in front of the star?

28. Jupiter's equatorial radius is 71,500 km, and its oblateness is 0.065. What is Jupiter's polar radius?

29. Ammonium hydrosulfide (NH_4HS) is a molecule in Jupiter's atmosphere responsible for many of its clouds. Using Figure 9.5, calculate the molecular weight of an ammonium hydrosulfide molecule, where the atomic weight of a hydrogen atom is 1. (Hint: The weight of a molecule is equal to the sum of the weights of its component atoms.)

30. Jupiter is an oblate planet with an average radius of 69,900 km, compared to Earth's average radius of 6,370 km.
    a. Remembering that volume is proportional to the cube of the radius, how many Earth volumes could fit inside Jupiter?
    b. Jupiter is 318 times as massive as Earth. Show that Jupiter's average density is about ¼ that of Earth's.

31. The obliquity of Uranus is 98°. If you were located at one of the planet's poles, how far from the zenith would the Sun appear at the time of summer solstice?

**32. A small cloud in Jupiter's equatorial region is observed to be at a longitude of 122.0° west in a coordinate system that rotates at the same rate as the deep interior of the planet. (West longitude is measured along a planet's equator toward the west.) Another observation made exactly 10 Earth hours later finds the cloud at a longitude of 118.0° west. Jupiter's equatorial radius is 71,500 km. What is the observed equatorial wind speed in kilometers per hour? Is this an easterly or a westerly wind?

33. The equilibrium temperature for Saturn should be 82 K, but instead we find an average temperature of 95 K. How much more energy is Saturn radiating into space than it absorbs from the Sun?

*34. Neptune radiates into space 2.6 times as much energy as it absorbs from the Sun. Its equilibrium temperature (see Chapter 4) is 47 K. What is its true temperature?

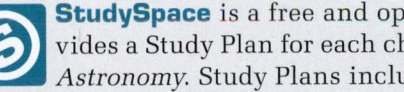 **StudySpace** is a free and open Web site that provides a Study Plan for each chapter of *21st Century Astronomy*. Study Plans include animations, reading outlines, vocabulary flashcards, and multiple-choice quizzes, plus links to premium content in SmartWork and the ebook. Visit **www.wwnorton.com/studyspace**.

Then there are the Tides, so useful to man . . .
We must be grateful for the Moon's existence on that
account alone.

JAMES NASMYTH (1808–1890)

Jupiter's moon Io, showing reddish fallout from an enormous erupting volcano.

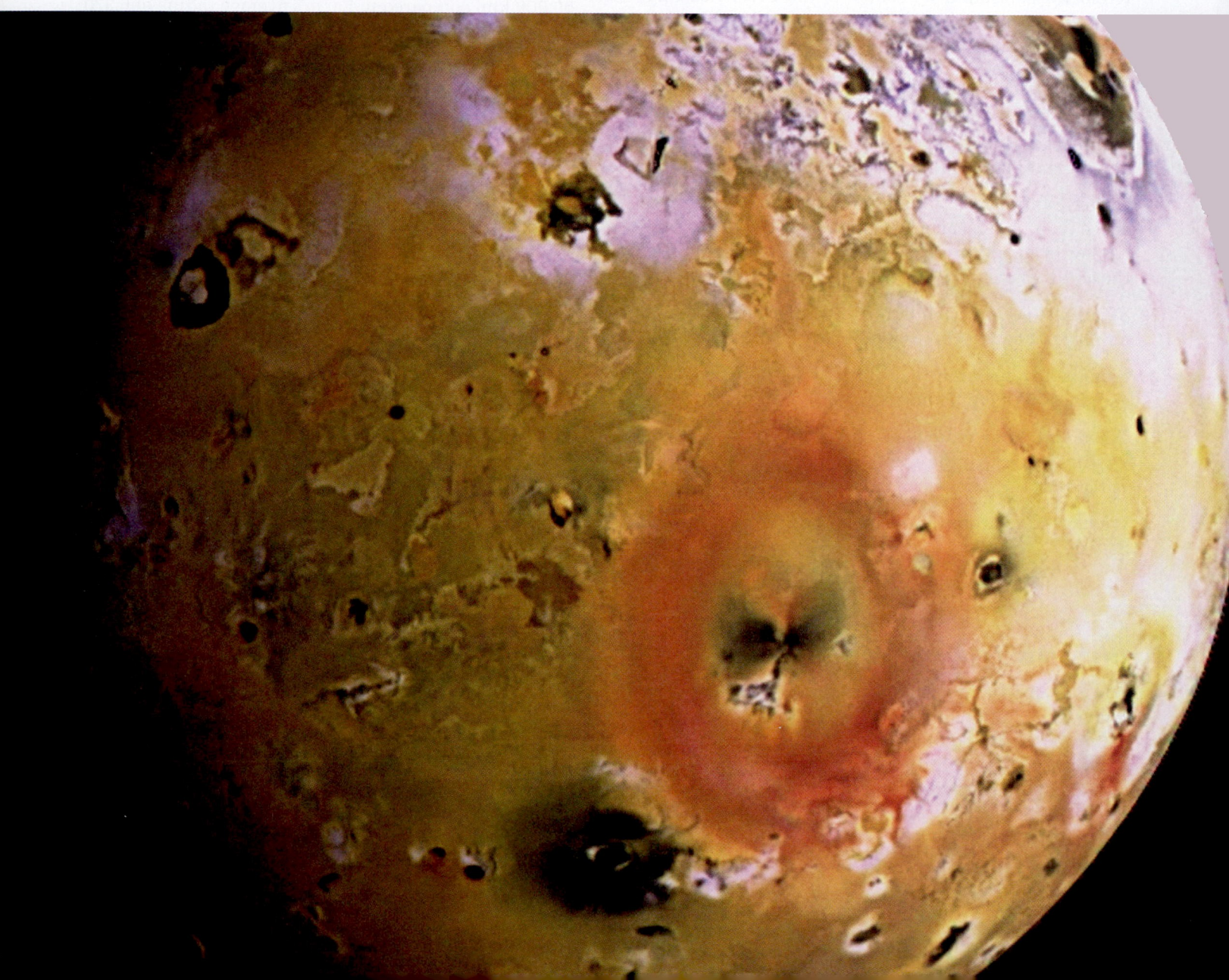

CHAPTER 10

Gravity Is More than Kepler's Laws

10.1 Gravity Once Again

Long before Kepler wrote down the laws describing the motions of the planets about the Sun, or Newton explained and interpreted these motions in terms of the effects of gravity, our ancestors knew that the Moon and the Sun have a direct influence on Earth. Those living near the ocean were especially attuned to these effects. Twice each day—in harmony with the changing position of the Moon in the sky—they saw the seawater rise and then recede once again. They would have noted that this effect, Earth's **tides**, is greatest when the Sun and the Moon are either together in the sky or at opposite extremes (that is, during a full Moon or a new Moon) and is more subdued when the Sun and the Moon are separated in the sky by 90° (first quarter Moon or third quarter Moon). Doubtless the unarguable association between the Moon and tides had a great deal to do with the development of superstitions about the power of the Moon over our lives.

Today we understand that tides are the result of the gravitational pull of the Moon and the Sun on Earth. More to the point, tides are the result of *differences* between how hard the Moon and Sun pull on one part of our planet in comparison with their pull on other parts of the planet. So far in our discussion of gravity we have focused on the gravitational interaction between two whole objects. Such interactions explain the motion of the planets about the Sun, the Moon about Earth, and a space shuttle about our globe. Yet the concept of gravity can explain far more than Newton's derivation of Kepler's laws. While concentrating on the elliptical, parabolic, and hyperbolic orbits of one mass about another, we have overlooked many other fascinating and important manifestations of gravity. We now turn our attention to some of these effects.

KEY CONCEPTS

Newton's law of gravitation is simplicity itself. As we saw in Chapter 3, it helps explain Kepler's laws of planetary motion. Yet when applied to extended rather than pointlike masses or when acting among three or more objects, this simple law gives rise to a surprising diversity of phenomena. In this chapter we apply Newton's law beyond Kepler's elegant laws of planetary motion and discover

- How symmetry enables us to say a great deal about the gravity of an object without actually calculating anything.

- That the gravity within a spherical object is determined only by the mass within a given radius.

- Tides on Earth result from the fact that gravity from the Moon and the Sun pull harder on one side of Earth than the other.

- Tidal interactions between planets and moons (including Earth's Moon) that lock a moon's rotation to its orbit.

- Comets that have been shattered by tidal stresses, and tortured moons alive with tide-powered volcanism.

- Orbital interactions that nudge asteroids from their orbits and sweep out gaps in planetary rings.

10.2 Gravity Differs from Place to Place within an Object

A good place to begin a discussion of the additional effects of gravity is with the way an object interacts gravitationally with *itself*. You can think of Earth, for example, as a collection of small masses, each of which feels a gravitational attraction toward every other small part of Earth. This gravitational attraction between different parts of Earth is what holds our planet together. We call the mutual gravitational attraction that occurs between all parts of the same object self-gravity. ▶‖ **AstroTour:** Tides and the Moon

> Self-gravity holds Earth together.

In a certain sense you are one of these small pieces of Earth. You are perhaps slightly more mobile than the average lump of clay; but as far as the gravitational makeup of our planet goes, you serve more or less the same purpose. As you sit reading this book, you are exerting a gravitational attraction on every other fragment of Earth, and every other fragment of Earth is exerting a gravitational attraction on you, as illustrated in **Figure 10.1a**. Your gravitational interaction is strongest with the parts of Earth closest to you. The parts of Earth that are on the other side of our planet are much farther from you, so their pull on you is correspondingly less.

As you know from experience, the net effect of all these forces is to pull you toward the center of Earth. If you drop a hammer, it falls directly down toward the ground. We can understand why this is so just by thinking about the shape of Earth. Because Earth is nearly spherical, for every piece of Earth pulling you toward your right a corresponding piece of Earth is pulling you toward your left with just as much force. For every piece of Earth pulling you forward a cor-

responding piece of Earth is pulling you backward. And, because Earth is almost **spherically symmetric**, all of these "sideways" forces cancel out, leaving behind an overall force that points toward Earth's center.

Understanding the size of the force is a bit trickier. Some parts of Earth are closer to you while others are farther away, but there must be an "average" or "characteristic" distance between you and each of the small fragments of Earth that is pulling on you. Not surprisingly, this average distance turns out to be the distance between you and the center of Earth. So, as illustrated in **Figure 10.1b**, *the overall pull that you feel is the same as if all the mass of Earth were concentrated at a single point located at the very center of the planet.* This relationship is true for any spherically symmetric object. As far as the rest of the universe is concerned, the gravity from such an object behaves as if all the mass of that object were concentrated at a point at its center. We have already made extensive use of this result. For example, it was an implicit assumption in our discussion in Chapter 3 of orbits and Kepler's laws that the force acting on an object in orbit about Earth is equal to $GM_\oplus m_{object}/r^2$, where $M_\oplus$ is the mass of Earth and r is the distance between the center of Earth and the orbiting object.

> Gravity from a sphere acts as though all of its mass were concentrated at the sphere's center.

We now have a way of accounting for the net gravitational attraction of Earth on an object on or above the surface of Earth, but what about the effect of gravity *within* Earth? What is the overall gravitational attraction felt by an object buried deep inside our planet? To answer this question, let's begin by imagining a physically impossible but theoretically plausible scenario: you are in a room at the center of Earth, as shown in **Figure 10.2**. (Ignore for a

> Net gravity is zero at the center of Earth.

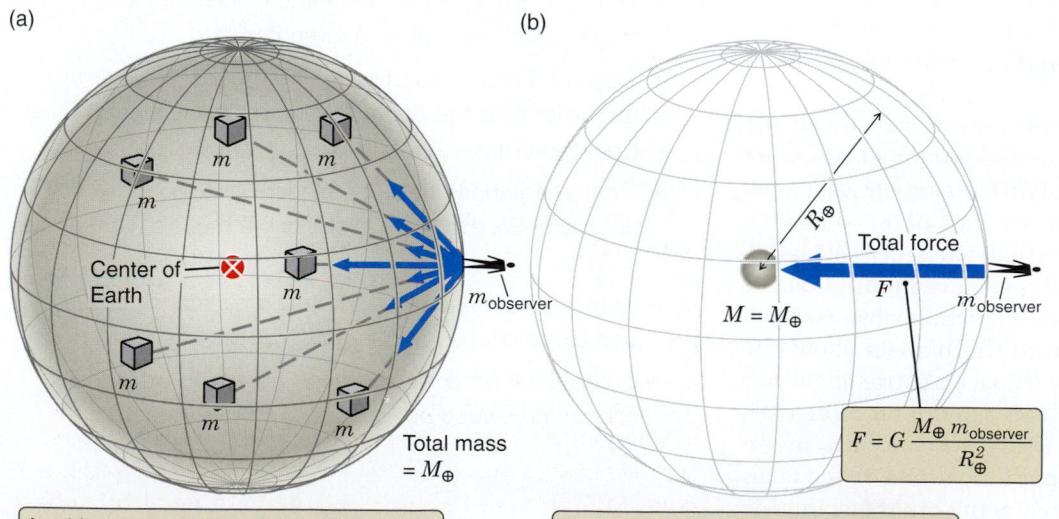

(a)

Center of Earth

Total mass = $M_\oplus$

An object on the surface of a spherical mass (such as Earth) feels a gravitational attraction toward each small part of the sphere.

(b)

$R_\oplus$

Total force

$M = M_\oplus$

F

$m_{observer}$

$$F = G\frac{M_\oplus\, m_{observer}}{R_\oplus^2}$$

The net force is the same as if we scooped up the mass of the entire sphere and concentrated it at a point at the center.

FIGURE 10.1 The net gravitational force due to a spherical mass is the same as the gravitational force from the same mass concentrated at a point at the center of the sphere.

(a)

In an imaginary (and unreachable) room at the center of Earth the gravity from each small piece of Earth would pull you outward.

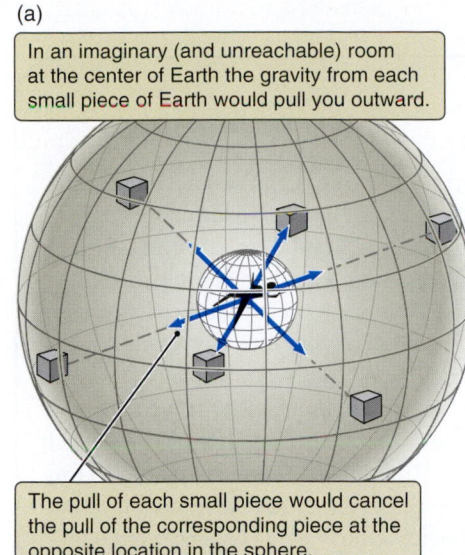

The pull of each small piece would cancel the pull of the corresponding piece at the opposite location in the sphere.

(b)

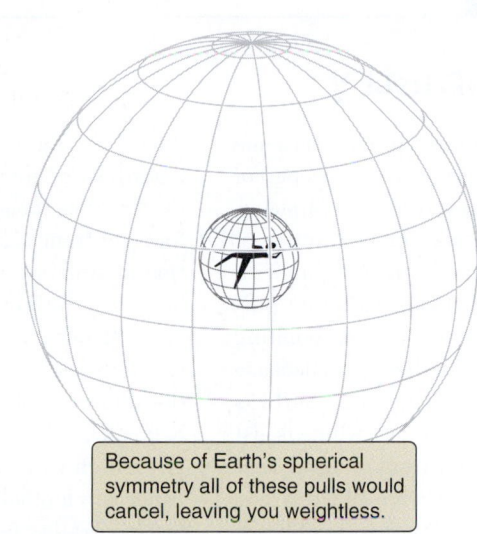

Because of Earth's spherical symmetry all of these pulls would cancel, leaving you weightless.

FIGURE 10.2 In an imaginary room at the center of Earth, the gravitational pull from each small piece of Earth would cancel, leaving you weightless.

moment the crushing pressure or immense temperatures that would surround you!) Every small part of Earth would be exerting a slight gravitational force on you, each of which would be *outward*, away from the center of Earth. Again, Earth's spherical *symmetry* tells us that, overall, these gravitational forces must sum to zero. For each part of Earth pulling you outward in one direction, a corresponding piece of Earth on the other side would pull you outward in exactly the opposite direction, as shown in **Figure 10.2a**. These two forces would cancel each other so that the net force you would feel from these two small parts of the planet would be zero (**Figure 10.2b**). The same holds true for *each and every* small part of Earth. The net effect of the sum of all the gravitational forces acting on an object at the center of Earth is zero! If you were in a room at the center of Earth, you would hang there, surrounded by the enormous mass of a planet but truly weightless. (See **Foundations 10.1** for more discussion of symmetry.)

Now imagine that your room is located partway out from the center of the planet. What gravitational forces would you feel at this new location? To tackle this question we need to think of Earth as being made up of two different pieces, as shown in **Figure 10.3**. The first piece is just a spherical ball containing all the parts of Earth that are closer to the center of Earth than your imaginary room is. You could think of this inner ball as a "planet within a planet." Your room is sitting on the surface of this imaginary sphere. From the discussion in the preceding paragraphs, we know the net effect of the gravity of this inner sphere. It is equivalent to taking all the mass contained within that imaginary ball and placing it at the center of Earth. The force that you would feel from this inner planet is

$$F = \frac{G \times M_{\text{inner}} \times m}{r^2},$$

where M_{inner} is the mass of this inner ball (which depends on where within the planet your room is located), m is

your mass, and r is the distance from the center of Earth to your room.

So far so good; but what about the gravitational attraction that you would feel from the *rest* of Earth—the shell of material that surrounds this "inner planet"? The parts of the shell that are closest to you, and so individually pull on you most strongly, are above you. The force from this part of the planet would pull you away from the center of Earth. However, *most* of the mass in the shell is on the side where its gravitational attraction pulls you *toward* the center of Earth. Although this material is farther away from you—which means that each small part of it pulls on you less strongly—there is a good bit more of it than the material immediately over your head. If we were to work it out in detail, we would find that the gravitational attraction from the part of the shell pulling you away from the center of Earth is exactly canceled by the gravitational attraction of the part of the shell that is pulling you toward the center of Earth. The net effect—the overall force—is zero. If you were inside a huge spherical shell, the overall gravitational attraction from the material in that shell would be zero. (This example is just a more general case of the "room at the center of Earth" problem illustrated in Figure 10.3.)

Let's pull all of this information together. Regardless of where we are within Earth, we feel a net gravitational force only from the part of the planet that lies deeper within the planet than we are, and that mass acts as if it were all concentrated at the center of Earth. If we want to know what the net gravitational attraction is at any point within a spherical object, all we need to do is determine how much mass is closer to the center of the sphere than is our point of interest. This mass then acts as if it were concentrated at the center of the sphere. This general result is extremely important. We have already seen something of this effect in our discussion of the interiors of planets. It will become even more important as we consider the forces responsible for the structure of stars.

FOUNDATIONS 10.1

The Power of Symmetry

In our discussion of gravity we have managed to arrive at some very interesting results based on nothing but the way one part of an object matches up with another—a property called **symmetry**. We were able to say that the overall gravitational attraction of all parts of Earth must point along a line connecting us with the center of Earth, simply by noting that each sideways pull from a different part of Earth is balanced by a corresponding pull in the other direction. We could make this claim because the distribution of mass to our right as we stand on Earth is just the same as the distribution of mass to our left, which is just the same as the distribution of mass to our front or back. That is to say, Earth is *symmetric*. Similarly, we were able to argue that the net effect of gravity on an object at the center of Earth must be zero because every small force from one part of Earth is balanced by a corresponding force from the corresponding part of Earth on the opposite side of us. Again, no calculation was needed—only an appeal to the symmetry of the distribution of mass within Earth.

If you look around, you will find symmetry everywhere; and when you find it, you will often find a clue to why something works the way it does. The tire on a bicycle rolls because

it has "circular symmetry," meaning that its shape is the same regardless of how it rotates about the axis running through its center. Your image in a mirror is flipped right for left and left for right from your actual appearance—an example of "reflection symmetry." You hardly notice the difference, however, because your body itself is nearly symmetric between its right and left sides. A child's cubical blocks can be stacked any of six ways because when you rotate them by 90° they still look the same—a special type of "rotational symmetry." A soccer ball can be kicked in any direction at any time, and it will roll any which way—both results of its "spherical symmetry." An American football makes for a lousy game of soccer but a fantastic spiraling forward pass because of its rotational symmetry about a single axis.

Science progresses by finding and making use of patterns that exist in the universe. Symmetry turns out to be an especially elegant and powerful type of pattern. Physicists and astronomers can often learn a great deal about the properties and behavior of an object or system without making a single calculation, simply by understanding the symmetry of the object or system that they are studying.

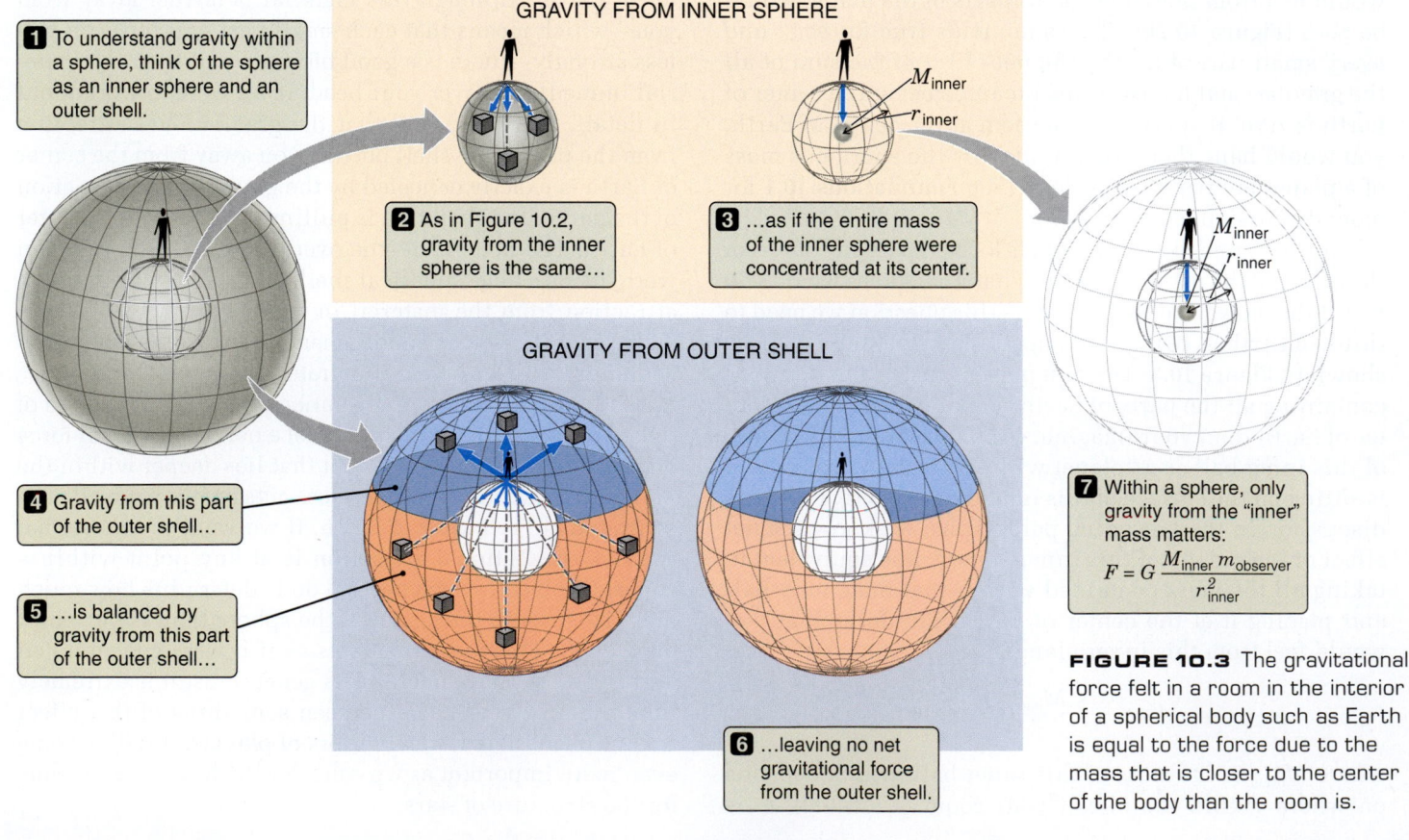

1 To understand gravity within a sphere, think of the sphere as an inner sphere and an outer shell.

GRAVITY FROM INNER SPHERE

M_{inner}
r_{inner}

2 As in Figure 10.2, gravity from the inner sphere is the same…

3 …as if the entire mass of the inner sphere were concentrated at its center.

M_{inner}
r_{inner}

GRAVITY FROM OUTER SHELL

4 Gravity from this part of the outer shell…

5 …is balanced by gravity from this part of the outer shell…

6 …leaving no net gravitational force from the outer shell.

7 Within a sphere, only gravity from the "inner" mass matters:
$$F = G \frac{M_{inner}\, m_{observer}}{r_{inner}^2}$$

FIGURE 10.3 The gravitational force felt in a room in the interior of a spherical body such as Earth is equal to the force due to the mass that is closer to the center of the body than the room is.

10.3 Tides Are Due to Differences in the Gravitational Pull from External Objects

We have seen that each small part of an object feels a gravitational attraction toward every other small part of the object, and that this self-gravity differs from place to place. In addition, each small part of an object feels a gravitational attraction toward every other mass in the universe, and these *external* forces differ from place to place within the object as well.

The most notable local example of an external force involves the effect of the Moon's gravity on Earth. Overall, the Moon's gravity pulls on Earth as if the mass of Earth were concentrated at the planet's center, as we just saw. When astronomers calculate the orbits of Earth and the Moon assuming that the mass of each is concentrated at the center point of each body, their calculations perfectly match the observed orbits. Yet there is more going on, because of the finite size of Earth. The side of Earth that faces the Moon is *closer* to the Moon than is the rest of Earth, so it feels a stronger-than-average gravitational attraction toward the

> The Moon's gravity pulls harder on the side of Earth facing the Moon.

Moon. In contrast, the side of Earth facing away from the Moon is *farther* than average from the Moon, so it feels a weaker-than-average attraction toward the Moon. Evaluating these effects numerically, we find that the pull on the near side of Earth is about 7 percent larger than the pull on the far side of Earth.

You can perhaps better visualize how this difference in gravitational attraction works if you carry out a "thought experiment." Imagine holding three rocks at different altitudes high above the surface of the Moon, as shown in **Figure 10.4a**. If you let them go at the same time, they would all fall toward the Moon. But rock 1, located closest to the Moon and thus subject to greater acceleration, would fall faster than rocks 2 and 3. Rock 3, located farthest from the Moon, would fall more slowly than either of the other two rocks. As the rocks fell toward the Moon, the separation between them would grow. A person falling along with rock 2 would see both rocks 1 and 3 moving away in opposite directions. So far so good. Now suppose the three rocks were connected by springs, as in **Figure 10.4b**. As the rocks fell toward the Moon, the differences in the gravitational forces they felt would stretch the springs. To someone falling along with rock 2, it would seem as if there were forces pulling rocks 1 and 3 in opposite directions. Exactly the same thing happens with Earth. If we replace the three rocks with different parts of Earth, as shown in **Figure 10.4c**, we see that differences in the Moon's gravitational attraction on different parts of Earth try to stretch Earth out along a line pointing toward the Moon. ▶‖ **AstroTour: Tides and the Moon**

FIGURE 10.4 (a) As three objects fall toward the Moon, they experience different gravitational accelerations. (b) If the three objects were connected by springs, the springs would be stretched as if there were forces pulling outward on each end of the chain. (c) The difference in the Moon's gravitational attraction between the near and far sides of Earth is the cause of Earth's tides.

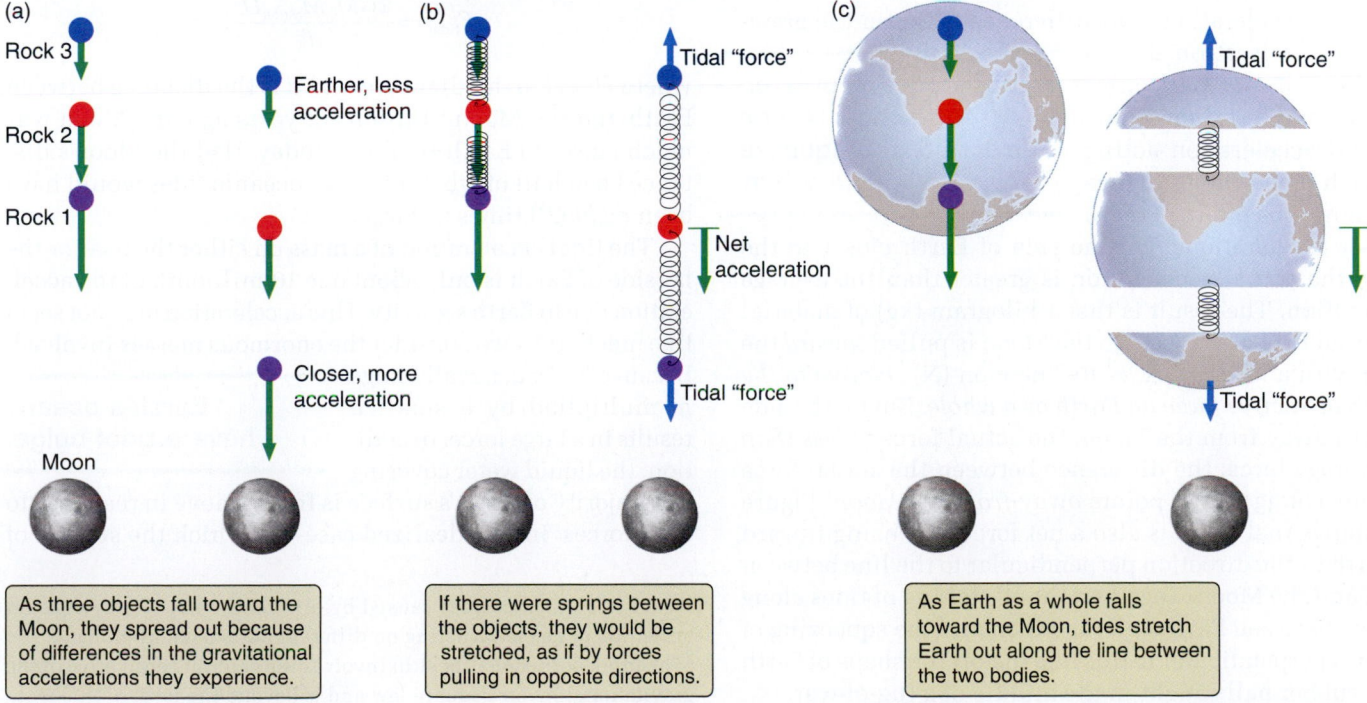

(a)

Rock 3
Rock 2
Rock 1
Moon

Farther, less acceleration

Closer, more acceleration

As three objects fall toward the Moon, they spread out because of differences in the gravitational accelerations they experience.

(b)

Tidal "force"

Net acceleration

Tidal "force"

If there were springs between the objects, they would be stretched, as if by forces pulling in opposite directions.

(c)

Tidal "force"

Tidal "force"

As Earth as a whole falls toward the Moon, tides stretch Earth out along the line between the two bodies.

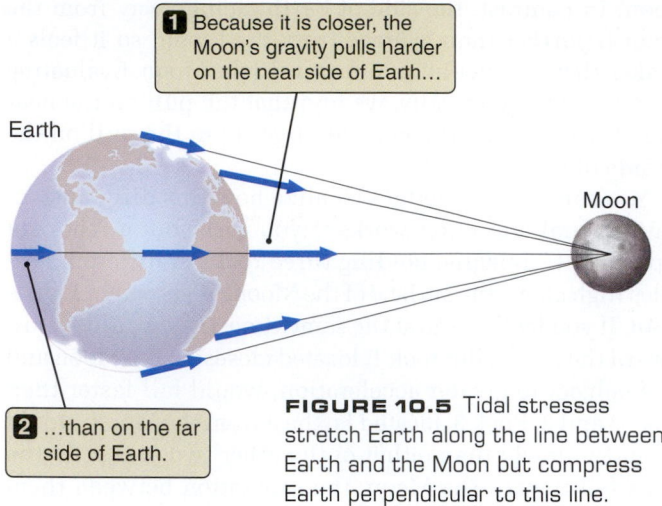

① Because it is closer, the Moon's gravity pulls harder on the near side of Earth...

Earth

Moon

② ...than on the far side of Earth.

FIGURE 10.5 Tidal stresses stretch Earth along the line between Earth and the Moon but compress Earth perpendicular to this line.

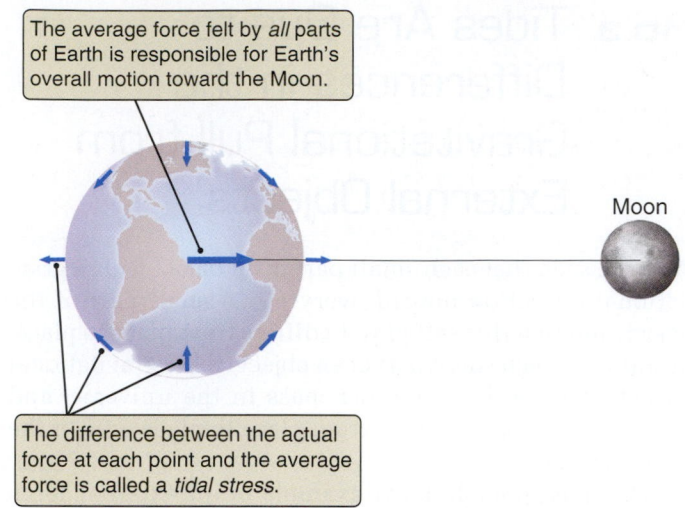

The average force felt by *all* parts of Earth is responsible for Earth's overall motion toward the Moon.

Moon

The difference between the actual force at each point and the average force is called a *tidal stress*.

We can also think about this situation by applying the idea of relative motion from our discussions of Newton's laws (see Chapter 3). Earth as a whole is constantly falling toward the Moon. The acceleration that Earth experiences is not very great—only about 3.3×10^{-5} meter per second per second (m/s²). The gravitational acceleration that we feel toward the center of Earth is 300,000 times greater. Yet the gravitational pull of the Moon on Earth is a substantial influence on our planet, causing it to shift more than 9,300 kilometers (km) back and forth about their common center of gravity over the course of a month (see Figure 6.14 for a similar effect involving the Sun and Jupiter). However, we do not personally perceive any sensation of movement from the gravitational attraction of the Moon on Earth. Just as everything in a traveling car shares the motion of the car, everything on Earth falls *together* toward the Moon.

What is *not* exactly the same everywhere on Earth is the "leftover" acceleration—the difference between the gravitational acceleration due to the Moon at any particular location and the average gravitational acceleration acting on Earth as a whole. **Figure 10.5** shows the effect of these leftover accelerations. On the side of Earth closer to the Moon, the actual acceleration is greater than the average acceleration. The result is that 1 kilogram (kg) of material on the side of Earth closer to the Moon is pulled *toward* the Moon with a force of 1.1×10^{-6} newton (N) *relative to the Moon's attractive force on Earth as a whole*. But on the side of Earth away from the Moon, the actual force is *less than* the average force; the difference between the actual force and the average force points *away from the Moon*! Figure 10.5 shows that there is also a net force squeezing inward on Earth in the direction perpendicular to the line between Earth and the Moon. Together, the stretching of tides along the line between Earth and the Moon and the squeezing of the tides perpendicular to this line distort the shape of Earth like a rubber ball caught in the middle of a tug-of-war.

> Tidal stresses stretch out Earth in one direction and squeeze it in the other.

Out of convenience we speak of **tidal stresses**[1] as if there were actually a force pulling the far side of Earth away from Earth's center. However, it is important to remember that the Moon is not pushing the far side of Earth away. Rather, it simply is not pulling on the far side of Earth as hard as it is on the planet as a whole. The far side of Earth is "left behind" as the rest of the planet falls more rapidly toward the Moon.

In Chapter 3 we learned that the strength of the gravitational force between two bodies is proportional to their masses and inversely proportional to the square of the distance between. But the strength of *tidal* forces caused by one body acting on another is far more complicated. Tidal influence is also proportional to the mass of the tide-raising body, but it is inversely proportional to the *cube* of its distance. Consider, for example, the tidal force acting on Earth by the Moon:

$$F_{\text{tidal}} = \frac{2GM_{\oplus}M_{\text{Moon}}D_{\oplus}}{d^3},$$

where $D_{\oplus}$ is Earth's diameter and d is the distance between Earth and the Moon.[2] Billions of years ago, the Moon was much closer to Earth than it is today. Had the Moon's distance been half of what it is now, oceanic tides would have been *eight* (2^3) times as large.

The tidal acceleration of a mass on either the near or the far side of Earth is only about one 10-millionth of the acceleration due to Earth's gravity. This acceleration may not seem like much until we consider the enormous masses involved. Because $F = ma$, a really huge m multiplied by a small a results in a large force. In addition, the liquid water covering the majority of Earth's surface is free to move in response to tidal forces. In the idealized case—in which the surface of

> Earth's oceans have a tidal bulge.

---

[1]Tidal stresses are forces caused by differences in the gravitational attraction of one mass acting on different parts of another mass.
[2]The proof of this relationship involves the calculus derivative of the gravitational inverse square law and is beyond the level of this text.

Earth is perfectly smooth and covered with a uniform ocean, and Earth does not rotate—the **lunar tides** (tides on Earth due to the gravitational pull of the Moon) would pull our oceans into an elongated **tidal bulge** like that in **Figure 10.6a**. The water would be at its deepest on the side toward the Moon and on the side away from the Moon, and at its shallowest at points midway between. Of course, our Earth is *not* a perfectly smooth, nonrotating body covered with perfectly uniform oceans, and many effects complicate this simple picture. One complicating effect is Earth's rotation. As a point on Earth rotates through the ocean's tidal bulges, that point experiences the ebb and flow of the tides. In addition, friction between the spinning Earth and its tidal bulge drags the oceanic tidal bulge around in the direction of Earth's rotation, as shown in **Figure 10.6b**.

Follow along in **Figure 10.6c** as we ride our planet through the course of a day. We begin as the rotating Earth carries us through the tidal bulge on the Moonward side of the planet. Because the tidal bulge is not exactly aligned with the

Tides rise and fall twice each day.

Moon, the Moon is not exactly overhead but is instead high in the western sky. When we are at the high point in the tidal bulge, the ocean around us is deeper than average—a condition referred to as "high tide." About 6¼ hours later, probably somewhat after the Moon has settled beneath the western horizon, the rotation of Earth carries us through a point where the ocean is shallowest. It is "low tide." If we wait another 6¼ hours, it is again high tide. We are now passing through the region where ocean water is "pulled" (relative to Earth as a whole) into the tidal bulge on the side of Earth that is away from the Moon. The Moon, which is responsible for the tides we see, is itself hidden from view on the far side of Earth. About 6¼ hours later, probably sometime after the Moon has risen above the eastern horizon, it is low tide. About 25 hours after we started this journey—the amount of time the Moon takes to return to the same point in the sky from which it started—we again pass through the tidal bulge on the near side of the planet. This is the age-old pattern by which mariners have lived their lives for millennia: the twice-daily coming and going of high tide, shifting through the day in lockstep with the passing of the Moon. ▶Ⅱ **AstroTour:** Tides and the Moon

Those who live near the ocean may have noticed that our description of daily tides is often simpler than what you actually observe. The shapes of Earth's shorelines and ocean basins complicate our simple picture of tides seen at any particular point, even in midocean.[3] As they respond to the tidal stresses from the Moon, Earth's oceans flow around the various landmasses that break up the water covering our planet. Some places, like the Mediterranean Sea and the Baltic Sea, are protected from tides by their relatively

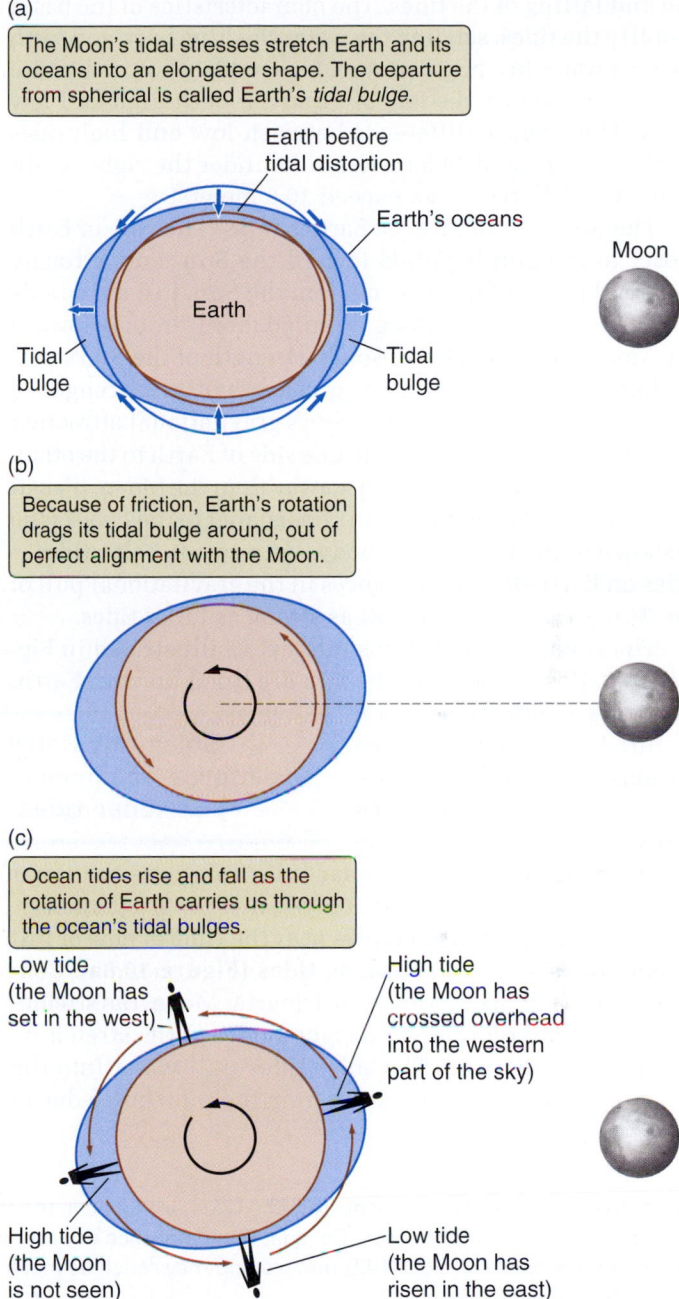

(a)

The Moon's tidal stresses stretch Earth and its oceans into an elongated shape. The departure from spherical is called Earth's *tidal bulge.*

Earth before tidal distortion

Earth's oceans

Moon

Earth

Tidal bulge

Tidal bulge

(b)

Because of friction, Earth's rotation drags its tidal bulge around, out of perfect alignment with the Moon.

(c)

Ocean tides rise and fall as the rotation of Earth carries us through the ocean's tidal bulges.

Low tide (the Moon has set in the west)

High tide (the Moon has crossed overhead into the western part of the sky)

High tide (the Moon is not seen)

Low tide (the Moon has risen in the east)

FIGURE 10.6 (a) Tidal stresses pull Earth and its oceans into a tidal bulge. (b) Earth's rotation pulls its tidal bulge slightly out of alignment with the Moon. (c) As Earth's rotation carries us through these bulges, we experience the well-known ocean tides.

[3]High tide does not necessarily occur when the Moon is near its highest point above the horizon. The reason is that, in addition to local shoreline effects, there are oceanwide oscillations similar to water sloshing around in a basin.

small sizes and the narrow passages connecting these bodies of water with the larger ocean. In other places, the shape of the land funnels the tidal surge from a large region of ocean into a relatively small area, concentrating its effect. The Bay of Fundy lies between the Canadian provinces of Nova Scotia and New Brunswick. This bay, along with the Gulf of Maine, forms a great basin in which water naturally rocks back and forth with a period of about 13 hours. This amount of time is close to the 12½-hour period of the rising and falling of the tides. The characteristics of the basin amplify the tides, sending the water sloshing back and forth like the water in a huge bathtub. **Figure 10.7** shows a picture of one location on the Bay of Fundy at both high and low tides. The average difference between low and high tides on the bay is about 14.5 meters, and under the right conditions this difference can exceed 16.6 meters!

The Sun also influences Earth's tides. The side of Earth closer to the Sun is pulled toward the Sun more strongly than is the side of Earth away from the Sun, just as the side of Earth closest to the Moon is pulled more strongly toward the Moon. Although the absolute strength of the Sun's pull on Earth is nearly 200 times greater than the strength of the Moon's pull on Earth, the Sun's gravitational attraction does not change by much from one side of Earth to the other, because the Sun is much farther away than the Moon. (Recall that tidal effects are inversely proportional to the *cube* of the distance to the tide-raising body.) As a result, **solar tides**—tides on Earth due to differences in the gravitational pull of the Sun—are only about half as strong as lunar tides.

The solar and lunar tides interact as illustrated in **Figure 10.8**. If the Moon and the Sun are lined up with Earth, as occurs at either new Moon or full Moon, then the tides on Earth due to the Sun are in the same direction as the tides on Earth due to the Moon. At these times the solar tides reinforce the lunar tides, so the tides are about 50 percent stronger than average. The especially strong tides near the time of new or full Moon are referred to as **spring tides (Figure 10.8a)**. Conversely, around the first and third quarter Moon, the stretching due to the solar tide is at right angles to the stretching due to the lunar tide. The solar tides pull water into the dip in the lunar tides and away from the tidal bulge due to

> Solar tides may reinforce or diminish lunar tides.

the Moon. This stretching in perpendicular directions has the effect of making low tides higher and high tides lower. Such tides, when lunar tides are diminished by solar tides, are called **neap tides (Figure 10.8b)**. Neap tides are only about half as strong as average tides and only a third as strong as spring tides.

The daily ebb and flow of the tides allows beachgoers to experience the wonders of tide pools, the natural aquariums teeming with marine life. As the tides roll in and out, they recharge the tide pools with fresh seawater while sometimes exchanging the pool's inhabitants. As you gaze into these microenvironments, think back a billion years or so: tide pools hold an ancient connection to all plants and creatures now living on solid ground, as described in **Connections 10.1**.

10.4 Tidal Effects on Solid Bodies

So far in our discussion we have focused on the sometimes dramatic movements of the liquid of Earth's oceans in response to the tidal stresses from the Moon and Sun. But these tidal stresses also affect the *solid* body of Earth. As we pointed out earlier, Earth is somewhat elastic (like a rubber ball), and tidal stresses cause a vertical displacement of about 30 centimeters (cm) between high tide and low tide, or roughly a third of the displacement of the oceans. As Earth rotates through its tidal bulge, the solid body of our planet is constantly being deformed. It takes energy to deform the shape of a solid object. (If you want a practical demonstration of this fact, hold a rubber ball in your hand, and squeeze and release it a few dozen times. While you are shaking your

FIGURE 10.7 The world's most extreme tides are found in the Bay of Fundy, located between Nova Scotia and New Brunswick, Canada. The difference in water depth between low tide (a) and high tide (b) is typically about 14.5 meters and may reach as much as 16.6 meters.

(a)

(b)

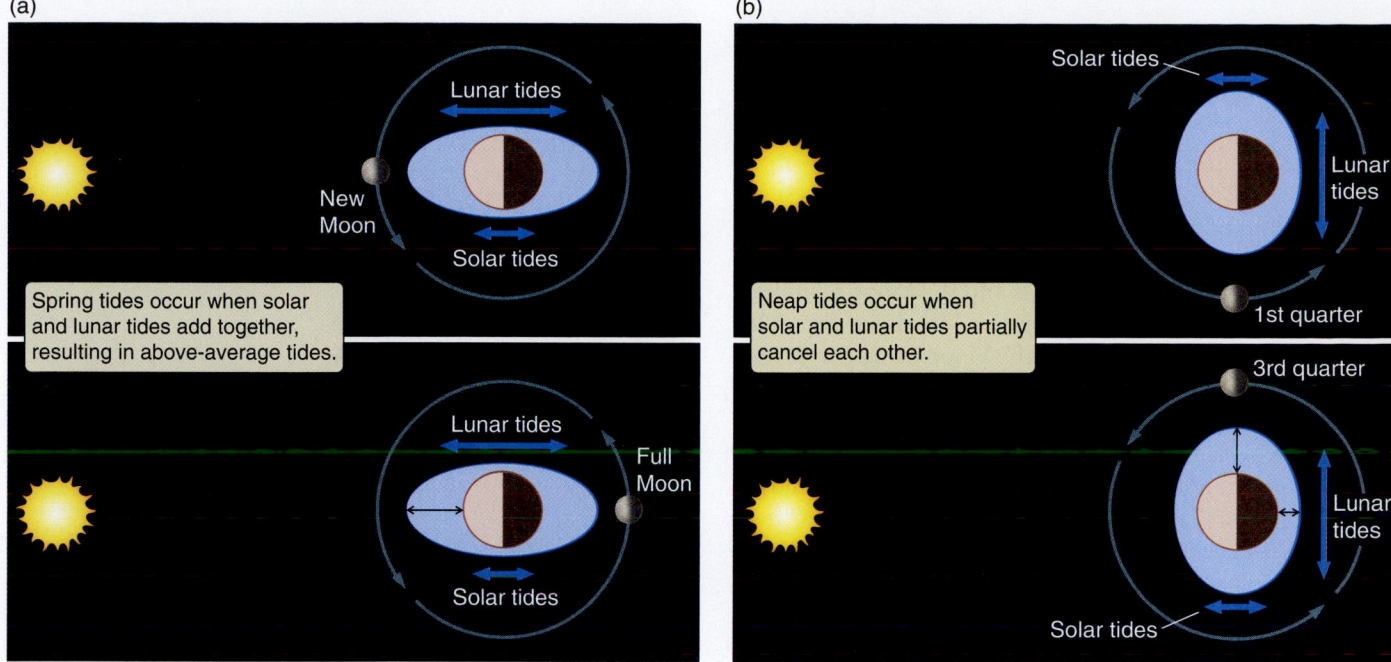

FIGURE 10.8 Earth experiences solar tides that are about half as strong as tides due to the Moon. The interactions of solar and lunar tides result in either (a) *spring tides* when they are added together or (b) *neap tides* when they partially cancel one another.

CONNECTIONS 10.1

Lunar Tides and Life

To ancient Romans and Greeks she was known as Diana or Artemis, goddess of the Moon. From the earliest times to the present, the Moon has occupied a special place in the minds of romantics and scientists alike. Whether we look upon the Moon as an object of nighttime beauty or as our nearest planetary neighbor, few of us are aware that lunar tides played a crucial early role in the drama of life. Directly and indirectly, the Moon imposed its influence on the emergence and evolution of advanced life-forms. Life's first home was the oceans, and instant death was in store for any sea creature having the misfortune to be tossed onto barren land by an errant wave. But there were places of relative safety where life could be exposed briefly to air and land before retreating to the refuge of its natural aquatic environment. These were the tide pools.

The twice-daily rise and fall of sea level swept traces of sea life into tide pools, where some of these creatures spent a few adventurous hours on solid ground. Many probably perished, but a few organisms were successful in making the difficult transition to dry land. The first to take this enormous step were the plants, establishing their dominance on land about 500 million years ago. Animal life followed soon after.

Today, all creatures of land and air, including us, may well owe their existence to the tides created by the Moon's gravity. Does this mean, as some have claimed, that advanced life can arise only on planets that have a large moon? Perhaps. No one really knows.

The Moon's gravitational grip on Earth has had another probable effect on terrestrial life, albeit one that is a bit more difficult to quantify. The Moon's gravity has held Earth's axial tilt, or *obliquity*, relatively steady between 21.5° and 24.5° over at least the past half billion years. This stability has maintained a smooth and steady transition in climate between our frigid polar regions and the warmth of the tropics, allowing for a wide diversity of life. Not so for Mars, though, which has no stabilizing companion. Gravitational tugs from the Sun and planets, among other forces, cause its axial tilt to change chaotically from as little as 0° to as much as 60° over a few tens of millions of years. Without restraint from our Moon, Earth might also have gyrated widely, possibly resulting in an unstable environment for life.

So the next time you gaze at the Moon, look upon it with respect. It may be one of the reasons you are here!

sore hand, imagine the energy that it must take to compress Earth by a third of a meter twice a day!) This energy is converted into thermal energy by *friction* in Earth's interior. This friction opposes and takes energy from the rotation of Earth, causing it to gradually slow. Earth's internal friction adds to the even greater slowing caused by friction between Earth and its oceans as the planet rotates through the oceans' tidal bulge. As a result, Earth's days are currently lengthening by about 0.0015 second every century. We'll turn now to tidal stresses on other solid bodies beyond Earth.

The Moon Is Tidally Locked to Earth

The Moon has no bodies of liquid to make tidal stresses obvious, but it is distorted in the same manner as Earth. In fact, because of Earth's much greater mass and the Moon's smaller radius, the tidal effects of Earth on the Moon are about 20 times as great as the tidal effects of the Moon on Earth. Given that the average tidal deformation of Earth is about 30 cm, the average tidal deformation of the Moon should be about 6 meters. However, what we actually observe on the Moon is a tidal bulge of about 20 meters! This unexpectedly large displacement exists because the Moon's tidal bulge was "frozen" into its relatively rigid crust at an earlier time when the Moon was closer to Earth and tidal stresses were much stronger than they are today. Planetary scientists sometimes refer to this deformation as the Moon's "fossil tidal bulge."

This extreme deformation is the reason that the Moon always keeps the same face toward Earth. Early in its history, the period of the Moon's rotation was almost certainly different from its orbital period. As the Moon rotated through its extreme tidal

> Tidal forces lock the Moon's rotation to its orbit around Earth.

bulge, however, friction within the Moon's crust was tremendous, rapidly slowing the Moon's rotation. After a fairly short time, the period of the Moon's rotation equaled the period of its orbit. When its orbital and rotation periods became equalized, the Moon no longer rotated *with respect to its tidal bulge*. Instead, the Moon and its tidal bulge rotated *together* in lockstep with the Moon's orbit about Earth. With the frictional forces within the Moon gone,[4] the Moon's rotation no longer slowed but instead remained equal to the period of the Moon's orbit about Earth. This scenario is how things remain today as the tidally distorted Moon orbits Earth, always keeping the same face and the long axis of its tidal bulge toward Earth (**Figure 10.9**). The **synchronous rotation** of the Moon discussed in Chapter 2 is a result of the **tidal locking** of the Moon to Earth.

Tidal forces affect not only the rotations of the Moon and Earth, but also their orbits. Because of its tidal bulge, Earth is not a perfectly spherical body. Therefore, the material in Earth's tidal bulge on the side nearer the Moon pulls on the Moon more strongly than does material in the tidal bulge on the back side of Earth. Because the tidal bulge on the Moonward side of Earth "leads" the Moon somewhat, as shown in **Figure 10.10**, the gravitational attraction of the bulge causes the Moon to accelerate slightly along the direction of its orbit about Earth. It is as if the rotation of Earth were dragging the Moon along with it, and in a sense this

---

[4]A small amount of tidal friction remains in the Moon. It is caused by the Moon's noncircular orbit and *libration*, as discussed in the next subsection.

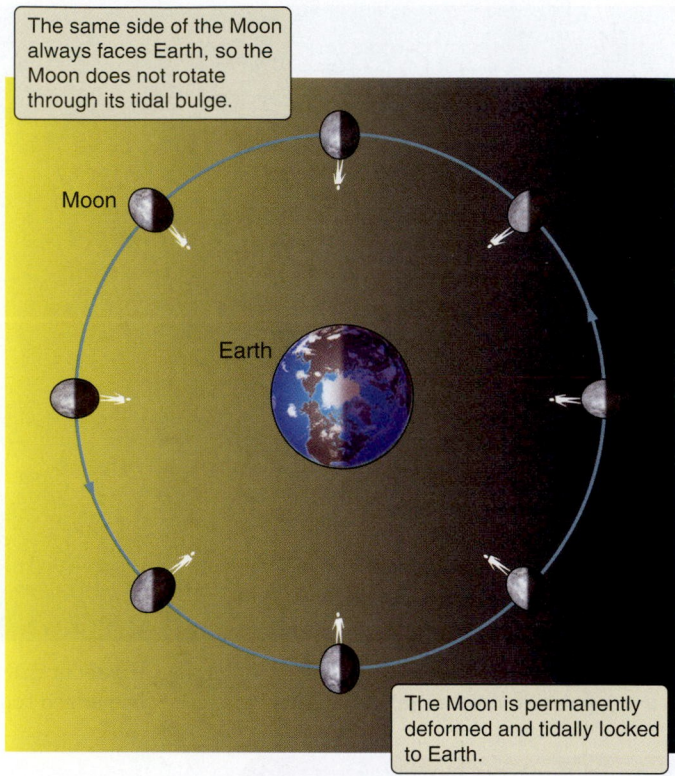

The same side of the Moon always faces Earth, so the Moon does not rotate through its tidal bulge.

Moon

Earth

The Moon is permanently deformed and tidally locked to Earth.

FIGURE 10.9 Tidal forces due to Earth's gravity lock the Moon's rotation to its orbital period.

is exactly what is happening. As we can see in **Connections 10.2**, the angular momentum lost by Earth as its rotation slows is exactly equal to the angular momentum gained by the Moon as it accelerates in its orbit.

The acceleration of the Moon in the direction of its orbital motion causes the orbit of the Moon to grow larger. At present, the Moon is drifting away from Earth at a rate of 3.83 cm per year. That's a very small number when compared with the Moon's distance, so how are we able to make such a precise measurement? Going to the Moon made it possible. In 1969, the *Apollo* astronauts placed an array of special reflectors on the lunar surface. Laser beams pointed at the Moon from Earth are reflected by these arrays and directed back to Earth, and by precise timing of a laser pulse's round-trip, astronomers can measure the distance between the centers of Earth and the Moon with an accuracy of better than one part in 10 billion!

As the Moon grows more distant, the length of the lunar month increases by about 0.014 second each century. At this rate, within slightly over a billion years the Moon will be far enough away from Earth that a total solar eclipse will no longer be possible. If this were to continue long enough (about 50

> The Moon's orbit is growing larger and Earth's rotation is slowing.

billion years), Earth would become tidally locked to the Moon, just as the Moon is now tidally locked to Earth. At that point the period of rotation of Earth, the period of rotation of the

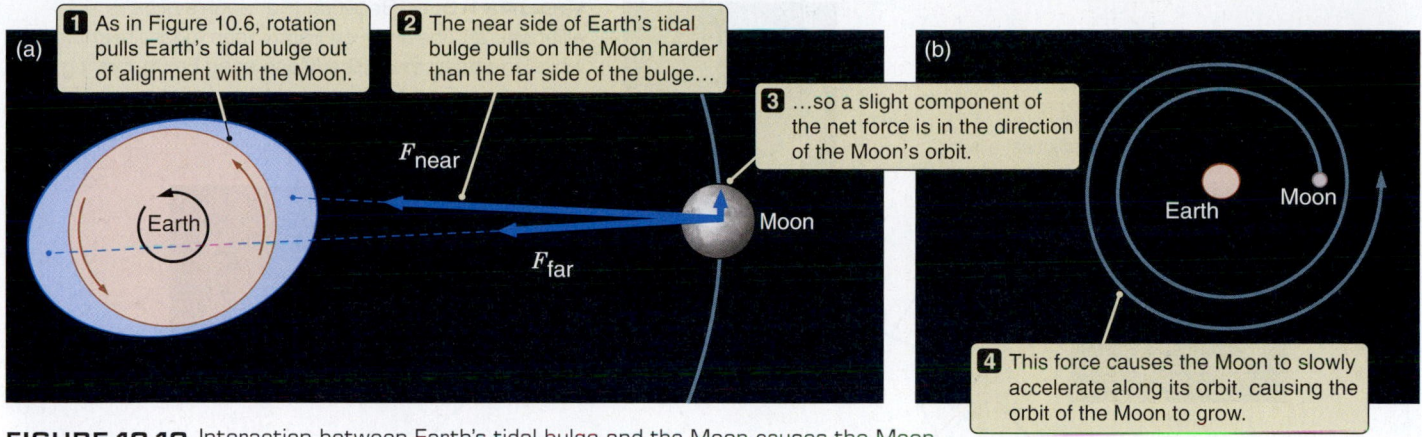

FIGURE 10.10 Interaction between Earth's tidal bulge and the Moon causes the Moon to accelerate in its orbit and the Moon's orbit to grow.

CONNECTIONS 10.2

Earth, the Moon, and Conservation of Angular Momentum

Earlier in this chapter we learned that the tidal interaction between Earth and the Moon is causing Earth's rotation to slow while it causes the Moon's orbit to grow larger. Given enough time, Earth and the Moon will become tidally locked to one another, with each keeping the same face toward the other. Looking at Figure 10.10, you might imagine that it would be rather difficult to calculate the exact size of the net force causing the Moon's orbit to grow, or the net frictional forces within Earth that slow its rotation. You would be correct. However, we can understand the relationship between these two effects, as well as their end result, without having to worry about any of those difficult details.

As Earth's rotation slows, Earth loses angular momentum. How can this be? When we discussed the formation of stars and planetary systems in Chapter 6, we found that angular momen-

tum is conserved. Where does the angular momentum that Earth loses as tidal forces slow it down go? The answer is that angular momentum is transferred to the Moon. The angular momentum that the Moon picks up as its orbit grows balances exactly the angular momentum that Earth loses as its rotation about its axis slows down. This may seem like a grand conspiracy, but it is not. The frictional forces in Earth slow its rotation, but they also cause its tidal bulge to be pulled around out of alignment with the Moon. This misalignment in turn causes the Moon's orbit to grow. In the end it *must be* the case that these effects work together to balance Earth's angular-momentum loss and the Moon's angular-momentum gain. Conservation of angular momentum is a grand pattern that *always* holds true, whether we are talking about the formation of stars, rotations and orbits of planets, spinning of a top, or pinwheel motion of the galaxy itself.

Moon, and the orbital period of the Moon would all be exactly the same—about 47 of our present days—and the Moon would be about 43 percent farther from Earth than it is today. However, this situation will never actually occur—at least not before the Sun itself has burned out.

Tides and Noncircular Orbits: An Imperfect Match

The effects of tidal forces can be seen throughout the Solar System. Most of the moons in the Solar System are tidally locked to their parent planets, and in the case of Pluto and its largest moon, Charon, each is tidally locked to the other. The tidal locking between a moon and a planet (or between any two objects) can be perfect only if both objects

are in circular orbits. Even though *on average* an object in an elliptical orbit may rotate at exactly the same rate at which it moves in its orbit, as the object follows its orbit, it is constantly speeding up and slowing down. This is the case with the Earth-Moon system, in which the Moon is in an elliptical, not circular, orbit about Earth.

Like a spinning top, the Moon rotates on its axis at a steady rate that equals the *average* rate of the Moon's progress around Earth. When the Moon is closest to Earth and moving fastest in its orbit, however, the Moon's orbital motion "gains" a little on its rotation. Conversely, when the Moon is farthest from Earth and moving most slowly on its orbit, the Moon's orbital motion lags behind its steady rotation. From our perspec-

> **Tidal stresses make orbits more circular in shape.**

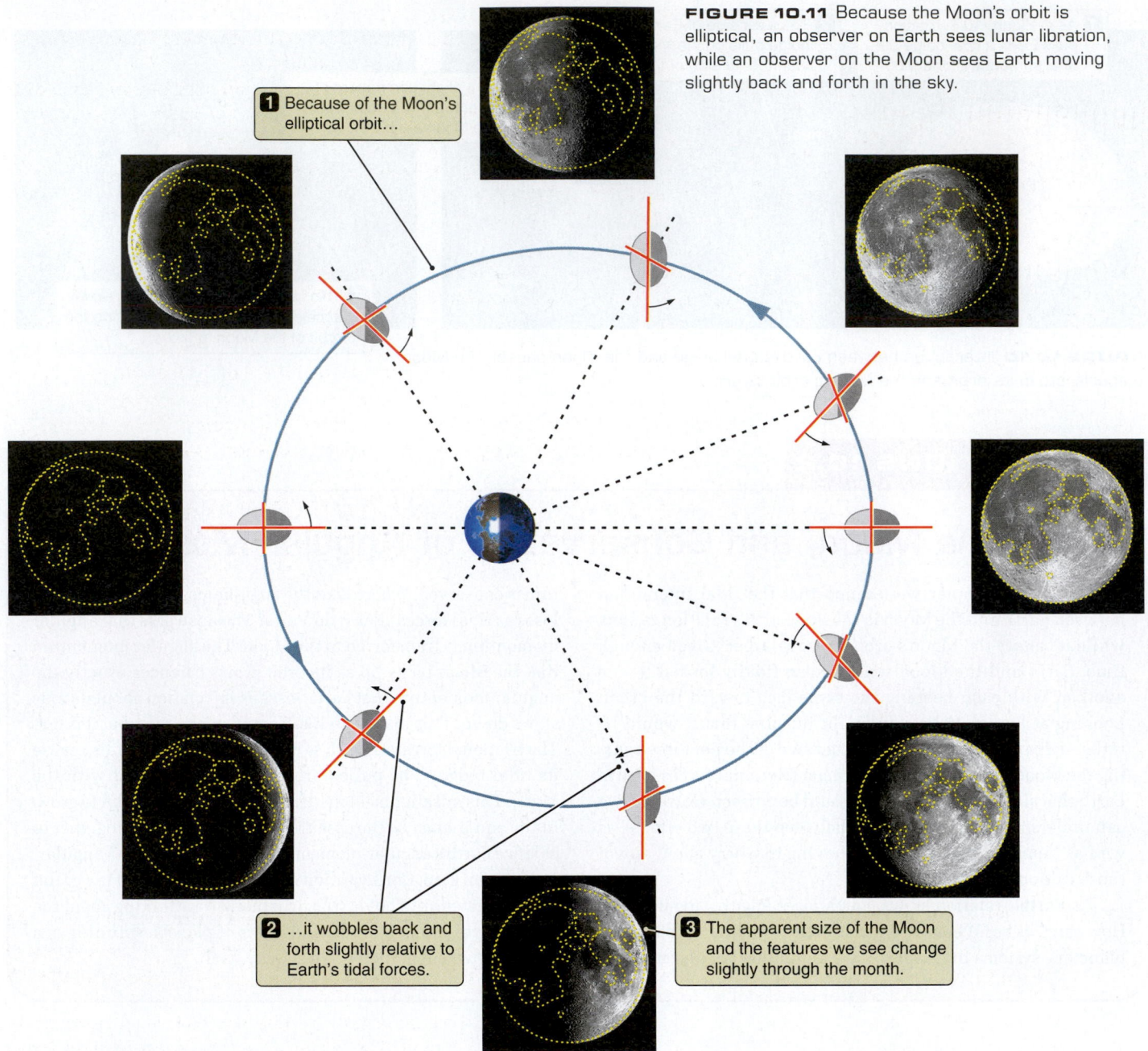

FIGURE 10.11 Because the Moon's orbit is elliptical, an observer on Earth sees lunar libration, while an observer on the Moon sees Earth moving slightly back and forth in the sky.

1 Because of the Moon's elliptical orbit…

2 …it wobbles back and forth slightly relative to Earth's tidal forces.

3 The apparent size of the Moon and the features we see change slightly through the month.

tive on Earth, the face of the Moon appears to be rocking back and forth. This effect is referred to as lunar **libration**. The motions responsible for lunar libration are illustrated in **Figure 10.11**. One consequence of these motions is that the Moon's frozen gravitational bulge is constantly swinging back and forth with respect to the direction of Earth's gravitational pull. Friction within the Moon, working together with the gravitational interaction between the Moon's tidal bulge and Earth's gravity, is slowly causing the Moon's orbit to become more circular.

The masses of the giant planets are far greater than the mass of Earth, so the tidal stresses they exert on their moons are correspondingly stronger. The tidal effect of Jupiter on its innermost moon, Io, is 250 times greater than the effect of Earth's tidal stresses on the Moon. As with most of the

moons of the giant planets, Io is tidally locked to Jupiter, rotating once on its axis in exactly the same amount of time that it takes Io to complete one orbit about the planet. Normally, we would have expected Io long ago to have settled into a perfectly circular orbit. But the fact that its orbit is constantly being perturbed by the gravity of Jupiter's other moons has prevented it from achieving a steady circular orbit. As a result, Io experiences libration, just as the Moon does, but with much more dramatic consequences. As Io's libration forces it to rock back and forth through the moon's tremendous tidal bulge, there is enough frictional heating to keep the interior of Io molten, powering the perpetually active volcanism seen on this small world (see the opening photograph for this chapter and **Figure 10.12**). In Chapter 11 we will learn that Io is the most volcanically active

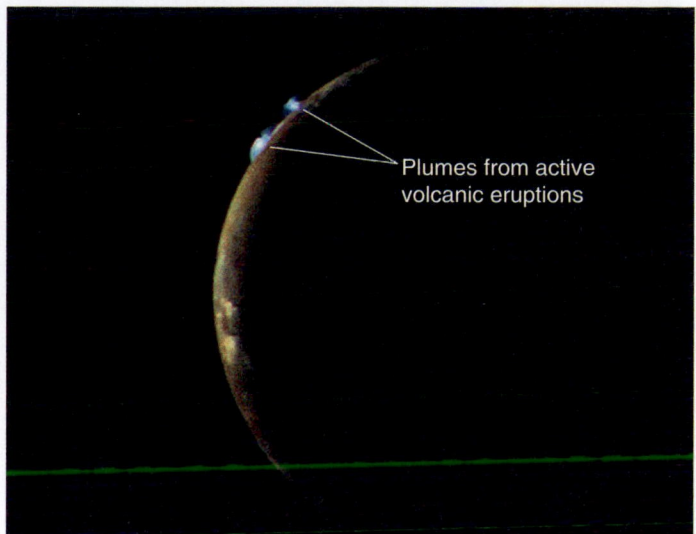

FIGURE 10.12 Volcanoes are constantly erupting on Jupiter's moon Io. The energy to power these volcanoes comes from tides due to Jupiter's gravity.

Plumes from active volcanic eruptions

object in the Solar System. The powerhouse behind that activity is tides.

Tidal locking, in which an object's rotation period exactly equals its orbital period, is only one example of **spin-orbit resonance**.[5] Other types of spin-orbit resonance are also possible. For example, Mercury is in a very elliptical orbit about the Sun. As with the Moon, tidal stresses have coupled Mercury's rotation to its orbit. Yet unlike the Moon with its synchronous rotation, Mercury rotates in a 3:2 spin-orbit resonance, spinning on its axis three times for every two trips around the Sun. The period of Mercury's orbit—87.97 Earth days—is exactly 1½ times the 58.64 days that it takes Mercury to spin once on its axis. Thus, each time Mercury comes to **perihelion**—the point in its orbit that is closest to the Sun—first one hemisphere and then the other faces the Sun.

Tides Can Be Destructive

We normally think of the effects of tidal forces as small compared with the force of gravity holding an object together, yet this is not always the case. Under certain conditions, tidal effects can be extremely destructive. Consider for a moment the fate of a small moon, asteroid, or comet that wanders too close to a massive planet such as Jupiter or Saturn. All objects in the Solar System larger than about a kilometer are held together by their self-gravity. However, the self-gravity of a small object such as an asteroid, a comet, or a small moon is feeble. In contrast, the tidal stresses close to a massive object such as Jupiter can be fierce. If the tidal

stresses trying to tear an object apart become greater than the self-gravity trying to hold the object together, the object will begin to break into pieces.

Astronomers have a special name for the distance from the center of a planet at which tidal stresses on a second, smaller body just equal that body's self-gravity. It is called the planet's **Roche limit**. For a body having the same density as the planet and having no internal mechanical strength, the Roche limit is about 2.45 times the planet's radius. Such an object bound together solely by its own gravity can remain intact when it is outside a planet's Roche limit, but not when it is inside the limit.[6] Tidal disruption of small bodies is thought to be the source of the particles that make up the rings of the giant planets. Comparison of the major ring systems around the giant planets shows that most rings lie within their respective planets' Roche limits.

> Tidal stresses exceed self-gravity inside the Roche limit.

In the mid-1990s, astronomers were treated to a rare look at tidal disruption in action. Comet Shoemaker-Levy 9 was discovered in March 1993. As astronomers studied the comet, they realized that it had been captured by Jupiter and that in July 1992 the comet had passed within 1.4 Jupiter radii of the planet's center, well within the planet's Roche limit. The comet was broken into pieces by this encounter, so pictures taken by the Hubble Space Telescope showed not a single comet, but instead a chain of orbiting fragments (**Figure 10.13a**). In the summer of 1994, Comet Shoemaker-Levy 9 treated the world to an even more unusual spectacle: one after another, the pieces of the comet slammed into Jupiter's atmosphere, creating enormous fireballs and scars larger than Earth. **Figure 10.13b** shows the shattered remains of another comet, Comet West, which was broken apart by the Sun's tidal stresses as it passed close to the Sun in 1976.

> Comet Shoemaker-Levy 9 was shattered by Jupiter's tidal force.

We have seen that tidal forces are at work throughout the Solar System. In **Connections 10.3** we find they are at work throughout the universe as well.

10.5 More Than Two Objects Can Join the Dance

We have seen that tidal interactions can lead to simple relationships or resonances between the orbital period of an object and the period of the object's rotation. Resonances can

---

[5]In a broader sense, spin-orbit resonance is a relationship between the spin and orbital periods of an object such that the ratio of the two periods can be expressed as simple integers.

[6]Small objects generally have mechanical strength that is much stronger than their own self-gravity. This is why the International Space Station and other Earth satellites are not torn apart even though they orbit well within Earth's Roche limit.

Tides on Many Scales

This section of the book concerns itself with the objects that make up our Solar System and the processes that influence them. Therefore, it is fitting that we have concentrated on the role that tidal forces play in the Solar System. On the other hand, we could have discussed tides and their effects at almost any point in the book. Tidal stresses result from differences in gravitational forces from place to place, and such differences are a general consequence of the inverse square law of gravity. *Any time* two objects of significant size or two collections of objects interact gravitationally, the gravitational forces will differ from one place to another within the objects, giving rise to tidal effects.

We might easily have included a discussion of tidal forces in Chapter 13, where we will find that stars are often members of binary pairs in which two stars orbit each other. Strong tidal interactions between such stars can tidally lock the rotation of each star to its orbital period about its companion, and can even pull material from one star onto the other. When we study this process in Chapter 16, we will learn of the *Roche lobes* of a binary star. These are named after Édouard A. Roche (1820–1883), the French scientist who did pioneering work on the disruption of satellites and for whom the Roche limit is named. Under some circumstances in the universe, tidal stresses become unimaginably powerful. In the vicinity of very massive, very dense objects the pull of gravity can change by huge amounts over minute distances. In Chapter 17 we will find that the tidal effects near the surface of a black hole can be billions of times stronger than the force of gravity holding us to the surface of Earth. Such tidal stresses are enough to rip normal matter apart atom by atom.

Tides continue to play a role at even larger scales. Tidal effects can strip stars from clusters consisting of thousands of stars. Beginning in Chapter 18, we will also learn about *galaxies*—vast collections of hundreds of billions of stars that all orbit each other under the influence of gravity. Often two galaxies pass close enough together to strongly interact gravitationally. When this happens, as in **Figure 10.14**, both galaxies taking part in the interaction can be grossly distorted by tidal effects. Tidal forces even play a role in shaping huge collections of galaxies—the largest known structures in the universe.

The next time you are at the beach "watching the tide roll in," you might take a moment to think about tides and the role they play throughout the universe. Vast collections of billions of stars strewn about millions of light-years of space, matter shredded on its way into a black hole, gas stripped from a burgeoning star, comets and moons pulled to pieces, planets "flipped" like pancakes—all are the result of the same gravitational effect that is responsible for the gentle twice-daily rise and fall of the waterline on an oceanside dock.

1 Tidal interactions between galaxies draw stars into long tails…

2 …and distort the inner parts of the galaxies as well.

FIGURE 10.14 Tidal interactions distort the appearance of two galaxies. The tidal "tails" seen here are characteristic of tidal interactions between galaxies.

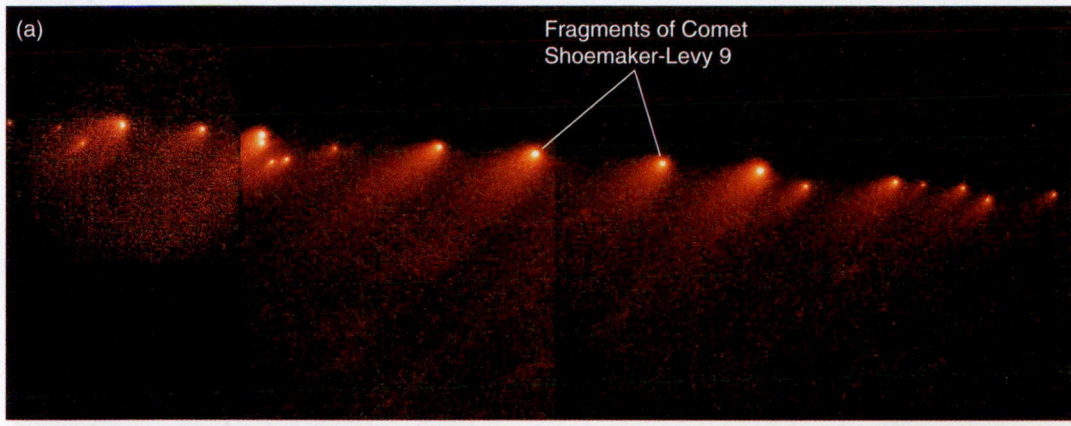

(a)

Fragments of Comet
Shoemaker-Levy 9

FIGURE 10.13 (a) Comet Shoemaker-Levy 9 was shattered into pieces as it passed within Jupiter's Roche limit in July 1992. (b) Comet West was shattered by solar tidal stresses during its 1976 pass through the inner Solar System.

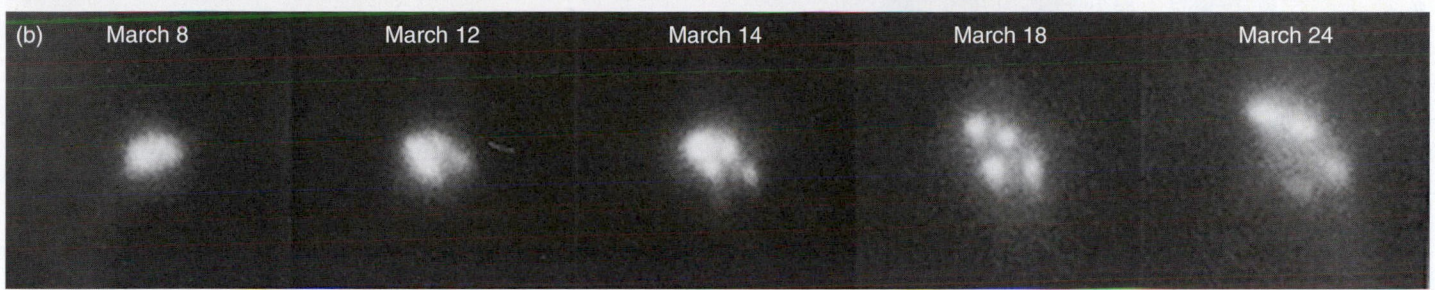

(b)   March 8 | March 12 | March 14 | March 18 | March 24

also exist between the orbital periods of two or more bodies orbiting about the Sun or a planet. We'll now see how tidal interactions affect the orbits of asteroids and the multitude of particles that make up planetary rings.

Gaps in the Asteroid Belt

In Chapter 12 we will study the smaller objects that orbit about the Sun, including the small rocky or metallic worlds called *asteroids*. Most of the asteroids in the Solar System are located in a "belt" between the orbits of Mars and Jupiter. You might imagine that asteroids would be distributed randomly throughout this region. However, if you look at the distribution of the sizes of the orbits` of the asteroids in the main belt, shown in **Figure 10.15**, you will see an amazing thing. Rather than asteroids having orbits of all sizes in this region, there are certain-sized orbits that are seldom occupied. These gaps in the asteroid belt are referred to as the **Kirkwood gaps** after Daniel Kirkwood (1814–1895), the American astronomer who first recognized them. Astronomers studying the Kirkwood gaps found a simple relationship between the "missing" orbits and the orbit of Jupiter. For example, none of the asteroids orbiting the Sun have an orbital period equal to half of Jupiter's orbital period. In fact, all of the Kirkwood gaps in the asteroid belt correspond to orbits that are related to the orbital period of Jupiter by the ratio of two small integers. Such a simple relationship between the periods of the orbits of two or more objects is referred to as an **orbital reso-**

> Orbital resonances with Jupiter cause the Kirkwood gaps.

nance. (Note that this is different from spin-orbit resonance, which we discussed in Section 10.4.)

The easiest way to understand the origin of the Kirkwood gaps is to imagine what would happen to an asteroid that started out with an orbital period exactly half that of Jupiter. Follow along in **Figure 10.16** as we watch the orbit of such an asteroid, beginning when the asteroid and Jupiter are at their closest to one another. Because the asteroid is closer to the Sun than Jupiter is, its orbital velocity is greater than that of Jupiter, so it leaves Jupiter behind as they continue on their respective orbits. By the time the asteroid has completed half of its orbit, Jupiter has completed only a fourth of its orbit. By the time the asteroid has gone around its orbit once, Jupiter is halfway. When the asteroid has completed one and one-half orbits, Jupiter has been through three-quarters of its orbit. Only after the asteroid has made two complete orbits and Jupiter has been around the Sun once do the two objects again line up. Thanks to the relationship between the periods of their two orbits, when Jupiter and the asteroid line up they are in the same place where they started. As Jupiter and the asteroid continue in their orbits, they line up at this same location again and again, every 11.86 years (the orbital period of Jupiter).

The gravitational force that Jupiter exerts on an asteroid at its closest approach is tiny compared with the gravitational force of the Sun on the asteroid, which is over 360 times stronger. A single close pass between Jupiter and the asteroid does very little to the asteroid's orbit. For an asteroid that is *not* in orbital resonance with Jupiter, the tiny nudges from Jupiter come at a different place in its orbit each time. The effects of these random nudges average out, and as a result even multiple passes close to Jupiter have little over-

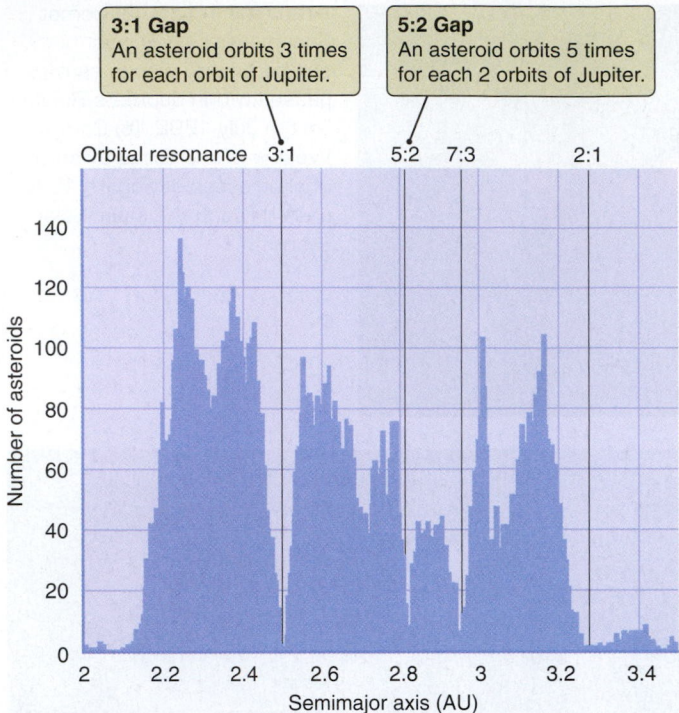

FIGURE 10.15 This plot shows the number of asteroids in the main belt with a given orbital period. The gaps in the distribution of asteroids, called *Kirkwood gaps*, are caused by orbital resonances with Jupiter.

all effect on the asteroid's orbit. As we just saw, however, for an asteroid that has a 2:1 orbital resonance with Jupiter (in other words, it has an orbital period equal to half that of Jupiter), the nudge from Jupiter comes at the *same place* in its orbit *every time*.

Although no single tug has much influence on the asteroid, the *repeated* tugs from Jupiter at the same place add up, changing the asteroid's orbit. This is why there are no asteroids with orbital periods equal to half the orbital period

of Jupiter. If we were to put an asteroid in such an orbit, it would not stay there long. The same phenomenon works for other combinations of orbital periods as well. With a piece of paper and a pencil, you ought to be able to convince yourself that other orbital resonances, such as a 3:1 resonance (in which an asteroid completes three orbits about the Sun in the time that it takes Jupiter to complete one orbit), will have a similar effect. The reason asteroids are not found in the Kirkwood gaps is that their gravitational interaction with Jupiter prevents them from staying there.

Gravity and Ring Systems

Asteroids are not the only objects in the Solar System that are affected by orbital resonances. Exactly the same phenomenon occurs in the ring systems around the giant planets. These ring systems consist of countless numbers of small particles—varying in size from tiny grains to house-sized boulders—that orbit individually around the giant planets like so many tiny moons. Kepler's laws dictate that the speed and orbital periods of all ring particles must vary with their distance from the planet, with the closest moving the fastest and having the shortest orbital periods. The orbital periods of particles in Saturn's bright rings, for example, range from 5 hours 45 minutes (5^h45^m) at their inner edge to 14^h20^m at the outer one. In the densely packed rings of Saturn, the orbits of all of these particles are perfectly circular and in precisely the same plane.

> Rings are swarms of tiny moons orbiting according to Kepler's laws.

To understand why the orbits of particles in dense rings must be so orderly, imagine yourself riding along on a large particle in the middle of Saturn's bright rings. Around you is a swarm of particles of all shapes and sizes, many of them close enough for you to reach out and touch. They

FIGURE 10.16 An asteroid with an orbital period exactly half that of Jupiter would line up with Jupiter at exactly the same place in every second orbit. The repeated gravitational pull from Jupiter at the same point in the asteroid's orbit would nudge the asteroid enough to change its orbital period. No asteroids have a 2:1 orbital resonance with Jupiter.

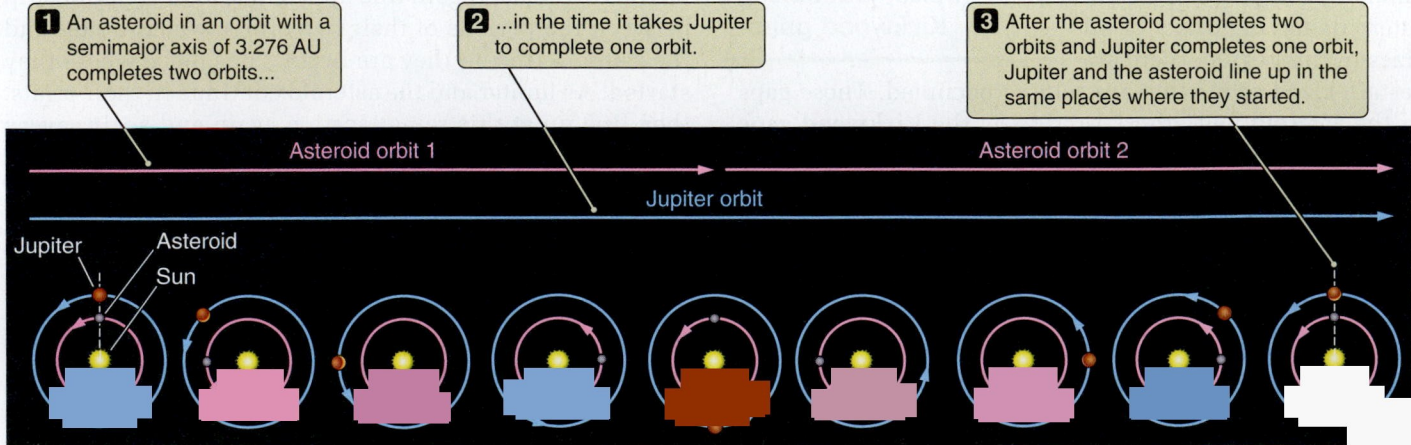

are all moving along in almost perfect step with your own, like members of a well-disciplined marching band. We say "almost" because the particles slightly closer to the planet are moving just a bit faster than you are, and those a little farther out are moving just a bit slower, as Kepler's laws demand. The differences in speed are very small. The particle orbiting 1 meter inward from you is moving only a tenth of a millimeter per second faster than you are—a relative speed difference slower than a snail's pace.

Now consider what would happen if that particle were moving in a slightly noncircular orbit, perhaps because it was bumped by one of its other neighbors. In response to its orbital eccentricity, it would drift alternately a bit outward and then inward as it completed each orbit. But as it moved outward, the particle would eventually nudge up against you and be unable to go any farther. With no room for maneuvering, the orbits of errant particles would (and do) quickly become circular. The same would be true for any particle in an orbit with even the smallest inclination: During each orbit, the deviant ring fragment would be carried alternately above and below the ring plane. As it approached your particle from, say, above, the mild encounter would quickly remove its out-of-plane motion and bring it into a coplanar (in the same plane) orbit. Saturn's densely packed rings have no place for nonconformists. Like the members of a well-disciplined marching band, all particles must march along together in unison.

> **Particles in dense rings must move in circular orbits and travel in unison with nearby particles.**

In addition to these interactions with each other, the orbits of these small particles deep within a ring can be influenced by the planet's larger moons. All that is necessary is for the moon to be massive enough to create a significant gravitational tug on the particles and be in orbital resonance with them. Such is the case with Saturn's moon Mimas, which causes the famous gap in the rings around Saturn called the **Cassini Division** (**Figure 10.17**). This gap is so prominent that it can be seen easily from Earth, even with small telescopes. The Cassini Division corresponds to a 2:1 orbital resonance with Mimas. That is to say, a ring particle located in the Cassini Division would have an orbital period about the planet equal to half the orbital period of Mimas. Just as Jupiter's gravity pulls asteroids out of resonant orbits, leading to the Kirkwood gaps, Mimas pulls ring particles out of the resonant orbits found within the Cassini Division. Such resonances with other moons are known to produce some of the gaps that appear in Saturn's bright rings.

Many planetary rings have sharp edges with no visible material appearing in the space beyond. For these rings, something is holding most of the particles in place, preventing the rings from quickly dissipating. The key to the sharp edges and stability of the rings lies with the same sort of orbital resonance that causes the Cassini Division. As moons orbit a planet, their gravity can nudge ring particles that are in resonant orbits, keeping them in line. Consider, for example, the abrupt outer edge of Saturn's outermost bright ring. The ring particles at the edge of this ring are in a 7:6 orbital resonance with Saturn's co-orbital moons Janus and Epimetheus, meaning that the ring particles make *precisely* seven orbits for every six orbits of its two co-orbital moons. ("Co-orbital" moons are moons that occupy the same orbit.)

> **Moons maintain sharp ring edges and create gaps in the rings.**

Now picture a particle suffering a collision and being forced a little outside the edge of the ring. Each time Janus passes the particle, once out of every six Janus orbits and every seven particle orbits, it does so at the same position in the particle's orbit. And each time the moon passes, the particle receives a small gravitational nudge from the moon in the same direction, causing the particle to lose a bit of its orbital energy. The effects build up, and eventually the particle loses enough energy to fall back into the ring. In short, the sharp edge of the ring itself is just like the sharp edge of a resonance gap such as the Cassini Division.

Each time a moon nudges a ring particle inward, the minute amount of energy lost by the particle is absorbed by

FIGURE 10.17 Saturn and its rings. The Cassini Division is easily seen here as the widest gap in Saturn's bright ring system.

the more massive moon, causing it to move imperceptibly farther from the ring and the planet. Over time, a moon may move so far from the ring edge that it can no longer provide stability to the ring. The ring is then free to dissipate. In the case of Saturn's bright rings, the situation is even more complicated because Janus itself is also in resonance with other moons of Saturn, meaning that *all* of these moons must be pushed away before the rings can dissipate. Thus, the bright rings of Saturn may be much more stable than most rings.

> Orbital resonances shape Saturn's rings.

Other kinds of orbital resonances are possible in ring systems. For example, most narrow rings are caught up in a periodic gravitational tug-of-war with nearby moons. These are called **shepherd moons** in recognition of the way they shepherd a flock of ring particles. Shepherd moons are usually small, are located close to a narrow ring, and often come in pairs, one orbiting just inside and the other just outside the narrow ring. The shepherding mechanism is much like the resonances discussed earlier. A shepherd moon just outside a ring will rob orbital energy from any particles that drift outward beyond the edge of the ring, causing them to move back inward. A shepherd moon just inside a ring will give up orbital energy to a ring particle that has drifted too far in, nudging it back in line with the rest of the ring. In some cases, narrow rings are trapped between two shepherd moons in slightly different orbits.

> Moons that keep narrow rings from spreading are called shepherd moons.

Planetary rings are a virtual laboratory for studying such orbital resonances, as well as tidal stresses and similar gravitational interactions. Tidal stresses, for example, are thought to be largely responsible for much of the material found in planetary rings. If a moon or other planetesimal held together by gravity alone happens to be within the Roche limit of a giant planet, it will be pulled apart by tidal stresses. The location of Saturn's bright rings inside the planet's Roche limit, for example, suggests that the rings formed from the breakup of one or more moons or captured planetesimals that were perturbed by larger moons into orbits that took them too close to the planet. If rings are made of material that originated on moons, we might expect to find the composition of a planet's rings mimicking the makeup of the moons that surround the planet. Such is indeed the case, as we will see in the next chapter.

Jupiter's ring system turns out to be largely the product of the strong gravity of the planet itself, combined with the presence of a handful of small, rocky moons close to the planet. As interplanetary meteoroids are pulled toward Jupiter by its strong gravity, a few of them strike the surface of one of Jupiter's four innermost moons. These moons are so tiny that some of the dust from these impacts is kicked off with speeds exceeding the escape velocities of the moons. This dust provides a steady supply of material for the rings.

Gravity Affects the Orbits of Three Bodies

One of the most interesting examples of orbital resonance occurs when two massive bodies (such as a planet and the Sun, or a moon and a planet) move about their common center of mass in circular or nearly circular orbits. In this situation, there are five locations where the combined gravity of the two bodies adds up in such a way that a third, lower-mass object will orbit in lockstep with the other two. These five locations, called **Lagrangian equilibrium points** or simply "Lagrangian points," are named for the Italian-born French mathematician Joseph-Louis Lagrange (1736–1813), who first called attention to them.

> Lagrangian points are orbital resonances formed by the gravity of two objects.

The exact locations of the Lagrangian points depend on the ratio of the masses of the two primary bodies. The Lagrangian points for the Earth-Moon system are shown in **Figure 10.18a**. (Remember that the pattern shown in the figure rotates like a turntable as the two massive objects orbit each other.) Three of the Lagrangian points lie along the line between the two principal objects. These points are designated L_1, L_2, and L_3. Although they are equilibrium points, these three points are unstable in the same way that the top of a hill is unstable. A ball placed on the top of a hill will sit there if it is perfectly perched, but give it the slightest bump and it goes rolling down one side. Similarly, an object displaced slightly from L_1, L_2, or L_3 will move away from that point. Thus, although L_1, L_2, and L_3 are equilibrium points, they do not capture and hold objects in their vicinity. Even so, they can be useful. For example, as of this writing, the *SOHO* (Solar and Heliospheric Observatory) and *ACE* (Advanced Composition Explorer) spacecraft sit near the L_1 point of the Sun-Earth system, where tiny nudges from their onboard jets are enough to keep them orbiting directly between Earth and the Sun, giving a constant, unobscured view of the side of the Sun facing Earth.

The situation is different for the other two Lagrangian points: L_4 and L_5. These two points are located 60° in front of and 60° behind the less massive of the two main bodies in its orbit about the center of mass of the system. These are *stable* equilibrium points, like the stable equilibrium that exists at the bottom of a bowl. If you bump a marble sitting at the bottom of a bowl, it will roll around a bit but it will stay in the bowl. In like fashion, objects near L_4 or L_5 follow elongated, tadpole-shaped paths relative to these gravitational "bowls."

Under the correct circumstances, L_4 and L_5 are able to capture and hold on to passing objects. Objects are often found orbiting about the stable L_4 and L_5 Lagrangian points of two-body systems. For example, the stable Lagrangian points of the Sun-Jupiter system are home to a collection of

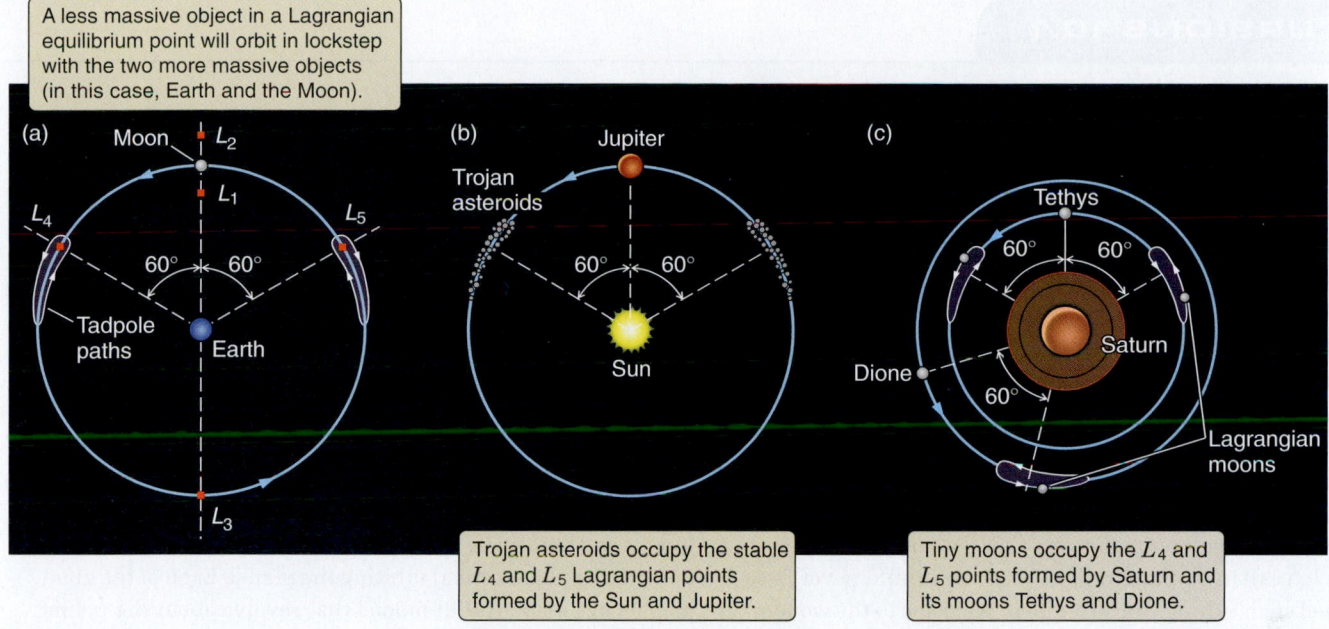

A less massive object in a Lagrangian equilibrium point will orbit in lockstep with the two more massive objects (in this case, Earth and the Moon).

(a) Moon L_2 L_1 L_5 L_4 60° 60° Tadpole paths Earth L_3

(b) Jupiter Trojan asteroids 60° 60° Sun

(c) Tethys 60° 60° Dione Saturn 60° Lagrangian moons

Trojan asteroids occupy the stable L_4 and L_5 Lagrangian points formed by the Sun and Jupiter.

Tiny moons occupy the L_4 and L_5 points formed by Saturn and its moons Tethys and Dione.

FIGURE 10.18 (a) The pattern of Lagrangian equilibrium points of a two-body system, in this case composed of Earth and the Moon. This entire pattern of Lagrangian points rotates like a turntable. (b) Trojan asteroids share Jupiter's orbit. (c) Small moons occupy Lagrangian points formed by Saturn and Tethys and by Saturn and Dione.

planetesimals called the **Trojan asteroids** (**Figure 10.18b**). There are also small moons in the L_4 and L_5 Lagrangian points of Tethys and Dione, two moons of Saturn (**Figure 10.18c**). (A group that formed in the 1970s and advocated human colonization of space called itself the "L5 Society" in recognition of the fact that the L_4 and L_5 Lagrangian points of the Earth-Moon system would make excellent locations for large orbital colonies.)

10.6 Such Wondrous Complexity Comes from Such a Simple Force

The tidal forces and resonances that we have discussed in this chapter are but a small sample of the tremendous wealth and variety of phenomena that result from the gravitational interactions between extended objects or collections of more than two objects.

> Gravity causes the precession of Earth's axis.

We might have mentioned many other such phenomena. The 26,000-year precession of Earth's axis, for example, is a result of the Sun's gravity causing Earth's axis to wobble like a top's. How a gravitational tug in one direction can cause the axis of a planet (or a top) to move in another direction is a truly fascinating story.

It is with regret that we leave behind the gravitational ballet, but move on we must. Before departing, however, it is worth pointing out just how broad and deep the general problem of gravitational interactions is. Although the motion of *two* masses orbiting about each other was solved centuries ago by Isaac Newton, the general problem of the motion of *three* bodies remains unsolved today! More than a few PhD theses, scholarly journal articles, and monographs in mathematics, astronomy, and physics have been filled with years of work on various aspects of this seemingly simple problem. Even in this day of powerful computers, which can crank out brute-force solutions to problems that seemed unapproachable a few decades ago, we remain humble in the face of gravity. As discussed in **Excursions 10.1**, even the gravitational interactions between a few objects such as the planets in our Solar System are *chaotic*. Unbelievably tiny differences in the initial positions and velocities of the planets—differences that are far too tiny to keep track of with the best computers—can, over time, lead to dramatically different end results.

As we leave this interlude about gravity behind, we can only marvel at the extraordinary complexity and diversity that arise in nature from this deceptively simple force. What an amazing array of implications, insights, puzzles, and surprises there are within Newton's elegant statement that the force of gravity is proportional to the product of the masses of two objects and inversely proportional to the square of the distance between them!

A World of Chaos

Imagine that you head for class but you leave 1 second later than you intended. That single second causes you to get caught at a traffic light, delaying you enough that you wind up first in line waiting for a passing train. While you await the train, an accident occurs up ahead, and as you sit in the resulting traffic jam, your car's engine overheats. Eventually, while your classmates are listening to an inspiring lecture about orbital motion, you find yourself sitting in a mechanic's waiting room learning the hard way about how much expensive damage you can do to an engine by letting it get too hot. You do not realize it, but the only thing that separated you from a normal day in class was the extra sip of coffee that put you out the door a second later than you might have been. You could never have predicted such a thing would happen. Welcome to the wonderful world of **chaos**.

Chaos is the technical term to describe interrelated systems in which tiny differences at one point in time lead to large differences at later times. Any system capable of exhibiting chaotic behavior is referred to as a **complex system**. Complex systems need not be extremely complicated, like the example in the previous paragraph. They can be seemingly very simple. Newton's law of gravity is simplicity itself, and as long as only two objects are involved, the resulting motions are simple as well. When we think of a planet orbiting the Sun, we think of the regular, repeating elliptical orbits described by Kepler's laws. When more than two objects are involved, however, the resulting motions can be anything but simple and regular.

Figure 10.19a shows a calculation of the orbit of Jupiter's moon Pasiphae over an interval of approximately 38 years. Pasiphae orbits far enough from Jupiter that its orbit is strongly perturbed by the Sun's gravity. Rather than a simple ellipse, Pasiphae's orbit looks more like what a 5-year-old might produce if handed a crayon and asked to quickly trace out a circle again and again. Pasiphae's orbit is an example of chaos. Chaotic orbits are complex, irregular motions in which extremely small differences in an object's position or speed grow into large differences in the object's subsequent motion.

Capture of moons by planets is another example of chaos at work. **Figure 10.19b** shows the path of a small planetesimal, such as a comet, as it approaches a planet like Jupiter. Very tiny differences in the position of the comet lead to very different outcomes. In the illustration there is a gnat's eyelash of difference between the starting point of a path in which the comet impacts the planet, a path in which the comet swings around Jupiter and continues to orbit the Sun, and a path that winds up with the planetesimal orbiting the planet. Each of the giant planets has a handful of moons that revolve about the planet in the wrong direction, indicating that they must have been captured in such chaotic events.

Chaos also affects the orbits of planets themselves. Each planet in our Solar System moves under the combined gravitational influence not only of the Sun, but of all the other planets as well. Although these extra influences are small, they are not negligible. Over millions of years they lead to significant differences in the locations of planets in their orbits. Although analysis of our Solar System indicates that the orbits of all eight classical planets are fairly stable, at least over times of a few billion years, that is not always the case. In many possible planetary systems, chaotic interactions among planets might cause planets to dramatically change their orbits or even be ejected from the system entirely.

In Chapter 6 we discussed the fact that planetary systems have been found around many other stars. Some of these systems contain giant planets that seem much too close to their parent stars to have formed there. Models suggest that these planets might have formed farther out in these systems and then

Seeing the Forest for the Trees

In Chapter 3 we learned about the birth of modern science. We saw how a few straightforward ideas about forces and the inverse square law of gravity opened a new window on the universe. When Newton used his laws of motion and gravitation to explain the motions of the planets about the Sun, he changed the face of science and the course of history.

On this leg of our journey we have pursued the implications of Newton's insight even further. We have seen how that same inverse square law of gravitation explains not only the simple elliptical orbits of planets about the Sun, but also an amazing wealth of different phenomena. The coming and going of the tides, the synchronous rotation of the Moon, the exquisite structure of Saturn's rings—all of these and more are the logical consequences of the fact that gravity is one-fourth as strong when two objects are twice as far apart. The insights that we have arrived at in this chapter will have profound implications again and again as we continue our journey. For example, when we follow the birth, evolution, and death of stars like our Sun, much of what we will see will be determined by the sizes and masses of the cores of those stars. Because of what we have learned in this chapter, we

moved inward toward the stars as a result of chaotic interactions. Indeed, events such as the collision between Earth and a Mars-sized body that formed the Moon tell of a more chaotic period early in the history of our own Solar System.

A key characteristic of chaotic systems is that although they are *deterministic*, they are not *predictable*. The law of gravity is well known, and there is in principle no problem with calculating the path that Earth will follow as a result of its ongoing interactions with other planets. It is a perfectly well-determined situation. However, even Earth's orbit is subject to chaos. An uncertainty of only 1 *centimeter* in our knowledge of the position of Earth along its orbit today is enough to make it completely impossible to predict where Earth will be in its orbit 200 million years from now.

Scientists studying the orbits of celestial bodies were among the first to recognize the existence of chaotic behavior, and in the last decades of the 20th century they began to develop the mathematical tools needed to describe and even exploit chaos. These tools are finding more and more common applications in our lives. The beating human heart, weather patterns, and traffic control are all examples of complex systems to which these tools have been applied. Pacemakers use these techniques to recognize when the regular beating of a heart is about to give way to the deadly, uncoordinated fluttering called "fibrillation." A tiny nudge on the part of the pacemaker, slightly changing the timing of a handful of heartbeats before fibrillation begins, is enough to prevent disaster. Chaos—the physics of complexity—is but one of many examples in which studies of the universe have spilled over into better understanding of the immediate world in which we live.

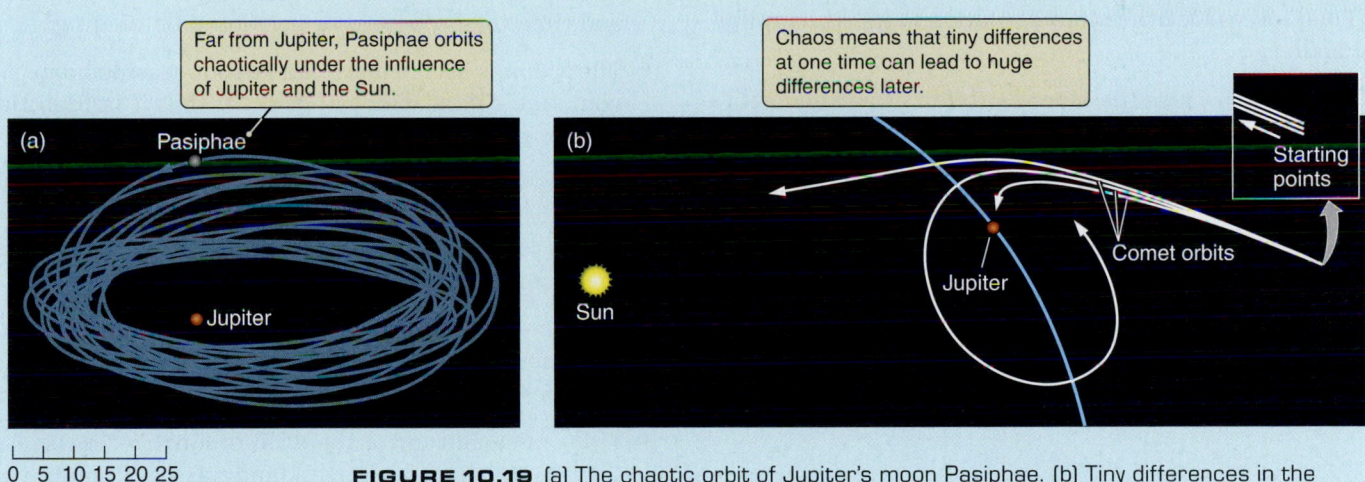

Far from Jupiter, Pasiphae orbits chaotically under the influence of Jupiter and the Sun.

Chaos means that tiny differences at one time can lead to huge differences later.

(a) Pasiphae · Jupiter

0 5 10 15 20 25
Millions of km

(b) Sun · Jupiter · Comet orbits · Starting points

FIGURE 10.19 (a) The chaotic orbit of Jupiter's moon Pasiphae. (b) Tiny differences in the motion of a comet lead to dramatic differences in the comet's fate.

will be able to say that the force of gravity at the surface of a stellar core is determined by the core alone, regardless of what the rest of the star is doing.

Even more important, this leg of our journey has shown us again the aesthetics of science. If you ask a physicist today about the ultimate goal of physics, you will hear about the desire to explain the way the universe behaves in terms of the fewest and simplest possible physical laws. In Chapter 3 we discovered Newton's laws of motion. In a sense, once we know those laws, everything—from the flight of a bumblebee, to the swirl of a hurricane, to the motions of an Olympic gymnast—is just arithmetic. We began this chapter with the single statement about the inverse square law of gravitation, and went on to tell stories ranging from the ebb and flow of Earth's tides to the shredding of distant galaxies.

At the turn of the 21st century, physicists have come to believe that even the basic physical laws that we are using on our journey—Newton's laws, relativity, gravitation, and quantum mechanics—are themselves embodiments of patterns in nature that are more fundamental still. Toward the end of our journey we will learn that today's frontiers of science are located at the two extremes. Theoretical physicists and cosmologists push ever backward in time toward the beginning of the universe in search of the single theory—the "Theory of Everything"—from which all else follows. Some scientists believe we may be just a few decades (or perhaps only a few years) from knowing the shape of such an all-encompassing theory. We may even come to know whether it was possible for the universe to become other than it is.

At the same time, physicists, astronomers, planetary scientists, biologists, and others are pushing forward with their understanding of the complex world around us. They are developing new tools and approaches, such as the those describing chaos, to enable them to find the hand of simple physical law in all that we see. So the next time you find yourself looking up at the Moon, gazing on the same face of our sister world that our ancestors saw from time immemorial, think about gravity and tides and the interplay between sister worlds. But at the same time, think about what they represent—the glorious complexity of this fascinating universe in which we live, and the beautiful simplicity of physical law from which it all derives.

Summary

- Planets and moons are held together by self-gravity.

- Both the Moon and the Sun create tides on Earth.

- As Earth rotates, tides rise and fall twice each day.

- Tidal forces lock the Moon's rotation to its orbit around Earth.

- Tidal forces cause the Moon's orbit to grow and Earth's rotation rate to slow.

- Tidal forces make orbits circular.

- Tidal stresses are stronger than self-gravity inside the Roche limit.

- Tidal forces create gaps in the distribution of asteroid orbits.

- Tidal forces control the orbits of particles in planetary rings.

SmartWork, Norton's online homework system, includes algorithmically generated versions of these questions, plus additional conceptual exercises. If your instructor assigns questions in SmartWork, log in at **smartwork.wwnorton.com.**

Student Questions

THINKING ABOUT THE CONCEPTS

1. Define what we mean by *self-gravity*.

2. In our daily lives we encounter many types of symmetry, including rotational, reflection, and bilateral (the case in which left and right sides are identical). List some ordinary examples of each of these types of symmetry.

3. If you could live in a house at the center of Earth, you could float from room to room as though you were living in the International Space Station. Explain why this statement is true.

4. What determines the strength of gravity at various radii between Earth's center and its surface?

5. The best time to dig for clams along the seashore is when the ocean tide is at its lowest. What phases of the Moon and times of day would be best for clam digging?

6. The Moon is on the meridian at your seaside home, but your tide calendar does not show that it is high tide. What might explain this apparent discrepancy?

**7. We may have an intuitive feeling for why lunar tides raise sea level on the side of Earth facing the Moon, but why is sea level also raised on the side facing away from the Moon?

8. Both the Sun and the Moon raise tides on Earth. Explain how the phase of the Moon affects the total strength of the tides.

9. Tides raise and lower the level of Earth's oceans. Can they do the same for Earth's landmasses? Explain your answer.

10. If lunar tides cannot raise the ocean surface more than 1 meter above mean sea level, how can tides as large as 5–10 meters occur?

11. If the Moon spins on its axis once every 29½ days relative to Earth, why are there not Earth tides on the Moon rising and falling over this same interval?

12. Tidal friction slows Earth's rotation and causes the Moon to orbit ever farther from Earth. Is this an example of conservation of angular momentum or conservation of orbital energy? Explain your answer.

13. What is meant by *tidal locking*?

*14. In 1959, a spacecraft launched by the Soviet Union photographed the far side of the Moon, showing us a large part of the Moon's surface that had never before been seen. Yet more than half (59 percent) of the Moon's surface had already been mapped prior to the space age. If the Moon always keeps the same face toward Earth, how was this possible?

*15. Could there be tidal heating effects on a moon located in a perfectly circular orbit around a planet? Explain your answer.

16. How is tidal stress responsible for powering volcanic activity on Jupiter's moon Io?

17. Small bodies orbiting large bodies within the Roche limit can be torn apart. Explain why this happens.

18. Most commercial satellites orbit Earth well inside the Roche limit. Why are they not torn apart?

19. There is a region in the main asteroid belt (about 3.28 astronomical units [AU] from the Sun) where the orbital period is exactly half that of Jupiter's. Explain why no asteroids are found there.

20. Explain why all of the particles in Saturn's bright rings remain in circular orbits and in the same plane.

21. Explain why Saturn's bright ring system ends abruptly at its outer edge.

22. What are shepherd moons, and what is their relationship to planetary rings?

*23. Explain the significance of the five Lagrangian points.

24. Which of Earth's five Lagrangian points can capture and hold passing objects?

**25. Even our most powerful computers have limited ability to predict the orbits and locations of the various bodies in our Solar System in the far distant future, and no matter how powerful we make our computers there will always be uncertainties. Why?

APPLYING THE CONCEPTS

26. Assume for the moment that Earth is homogeneous (having constant density throughout) and spherical. (We know, of course, from Chapter 7 that neither of these assumptions is true.) If you were in a deep well halfway to Earth's center, how would your weight there compare with your weight at Earth's surface?

27. If the Moon had twice the mass that it does, how would the strength of lunar tides change?

*28. At some time in the past, the Moon was only two-thirds its present distance from Earth. How would lunar tides at that time compare with those of today?

29. Saturn's small moon Hyperion is in a 4:3 orbital resonance with its largest moon, Titan, which orbits closer to

Saturn than does Hyperion. The orbital period of Titan is approximately 16 days. What is the orbital period of Hyperion?

*30. Tidal influence is proportional to the mass of a disturbing body and is inversely proportional to the *cube* of the distance to that body. Some astrologers claim that your destiny is determined by the "influence" of the planets that are rising above the horizon at the moment of your birth. Compare the tidal influence of Jupiter (mass 1.9×10^{27} kg; distance 7.8×10^{11} meters) with that of the doctor in attendance at your birth (mass 80 kg, distance 1 meter).

31. What is Earth's Roche limit, expressed in kilometers, for a body having the same density as Earth and no internal strength? (Earth's mean radius is 6,371 km.) Compare this distance with the orbital radius of the International Space Station, which circles Earth at an average altitude of 350 km.

32. Saturn's radius is approximately 60,000 km, and the outer radius of its bright ring system is approximately 137,000 km. Show that Saturn's bright ring system falls within Saturn's Roche limit.

**33. The mass of the Sun is 30 million times greater than the mass of the Moon, but the Sun is about 400 times farther away. Show why solar tides on Earth are only about half as strong as lunar tides.

34. Calculate the orbital radius of the Kirkwood gap that is in a 3:1 orbital resonance with Jupiter. (Hint: Refer to the discussion of Kepler's third law in Chapter 3. If you cannot do the calculation, show how you would set it up.)

35. The radius of the outer edge of Saturn's bright ring system is 136,755 km, and the radius of the orbit of Janus is 151,460 km. Using tools learned in Chapter 3, show that particles at the edge of the bright ring system are in a 7:6 orbital resonance with Janus.

36. The L_4 and L_5 Lagrangian points form equilateral triangles with the Sun and Earth (see Figure 10.18). If we were ever to colonize L_5, how far would the space pioneers making a home at L_5 be living from their previous homes on Earth?

StudySpace is a free and open Web site that provides a Study Plan for each chapter of *21st Century Astronomy*. Study Plans include animations, reading outlines, vocabulary flashcards, and multiple-choice quizzes, plus links to premium content in SmartWork and the ebook. Visit **www.wwnorton.com/studyspace**.

Although we are mere sojourners on the surface of the planet, chained to a mere point in space, enduring but for a moment in time, the human mind is not only enabled to number worlds beyond the unassisted ken of mortal eye, but to trace the events of indefinite ages before the creation of our race.

SIR CHARLES LYELL (1797–1875)

Complex structure in Saturn's three bright rings, as seen by the *Cassini* spacecraft.

CHAPTER 11

Planetary Adornments— Moons and Rings

11.1 Galileo's Moons

In 1610 the Italian astronomer Galileo Galilei observed that Jupiter was accompanied by four "stars" that changed their positions nightly. He quickly realized that, like Earth, Jupiter has moons of its own. We honor his discovery by calling them the "Galilean moons" of Jupiter. Galileo showed that Jupiter and its moons resemble a miniature Copernican planetary system: just as planets revolve around the Sun, so do the many moons of Jupiter revolve around it. By the end of the 17th century, astronomers had found five moons orbiting around Saturn as well, and this discovery further strengthened their belief in the Copernican system (see Chapter 3). Today we realize that the Solar System abounds with moons. As of August 2009, we count more than 170 observed moons in orbit around the classical and dwarf planets, and there are probably many others—especially in the outer Solar System—that we have not yet found. Some of these moons are unique worlds of their own, exhibiting geological processes similar to those on the terrestrial planets, and some may even harbor forms of life.

The moons of our Solar System are not distributed equally among the planets. In the inner part of the Solar System there are only three moons; Earth has one, Mars has two, and Mercury and Venus have none. Among the dwarf planets, Pluto possesses three known moons, Haumea has two, and Eris has one. All of the remaining moons belong to the giant planets. We might ask why Mercury and Venus failed to form or capture any moons of their own. But is this really surprising? Mercury and Venus, after all, have only one less moon than Earth has. Mars apparently captured two small asteroids, but Mars is situated adjacent to the main asteroid belt. (We will return to the two martian moons, Phobos and Deimos, when we discuss asteroids in

KEY CONCEPTS

For centuries, celestial wonders such as Saturn's rings and the Galilean moons of Jupiter delighted those who looked through telescopes. But with the dawn of the space age, robotic explorers traveling through the Solar System have shown us much more than points of light, revealing wondrous, diverse families of worlds orbiting other planets. Among them we will find

- Scores of small worlds composed of rock and solid ice, some of which formed with their planets and others of which were captured later.

- Geologically active moons freckled with volcanoes and geysers, and geologically dead moons covered with impact craters.

- Moons that may harbor deep liquid oceans beneath their ice-covered surfaces and may conceivably provide a home for extraterrestrial life.

- Exquisite, delicate structure in ring systems resulting from subtle gravitational interactions among planets, moons, and ring particles.

- The fate of most planetary rings, short-lived compared to the planets they surround.

FIGURE 11.1 Images of the major moons in the Solar System obtained by various spacecraft. The images are shown to scale. The planet Mercury and dwarf planet Pluto are shown for comparison. The martian moons, Phobos and Deimos, are too small to be shown. (The pictures of Pluto and Charon are artists' conceptions rendered from groundbased and space-based images.)

Chapter 12.) Finally, as we learned in Chapter 7, Earth would be without a moon if not for a cataclysmic collision when the planet was young. What seems important is that while the larger planets were forming, they had greater attracting mass and greater amounts of debris around them, which is what gave rise to their greater number of moons.

Figure 11.1 displays images of some of the major moons in the Solar System, all shown to scale. In many ways moons resemble smaller versions of the terrestrial planets. Some, such as our own, are made of rock. Others, especially in the outer Solar System, are mixtures of rock and water ice, with densities intermediate between the two. A few seem to be made almost entirely of ice. Several moons are comparable in

> Moons are made of rock, ice, or mixtures of both.

size to or even larger than Mercury, and the smallest known moons would fit within the expanse of a large metropolitan airport. Although most moons are airless, one has an atmosphere denser than Earth's, and several have very low-density atmospheres. Some of the larger moons appear to have differentiated chemically, and three are known to have active volcanoes or geysers. (Our own Moon is not covered in this chapter; we discussed it in Chapter 7 because of its close similarity to the four terrestrial planets of the inner Solar System.)

As discussed in Chapter 6, we believe that many moons formed at the same time as the planets they orbit, and that they accreted in much the same way that the planets themselves grew—from the accumulation of planetesimals orbiting the Sun. For example, Ganymede, Jupiter's largest moon, grew from the accretion of ice and dust grains

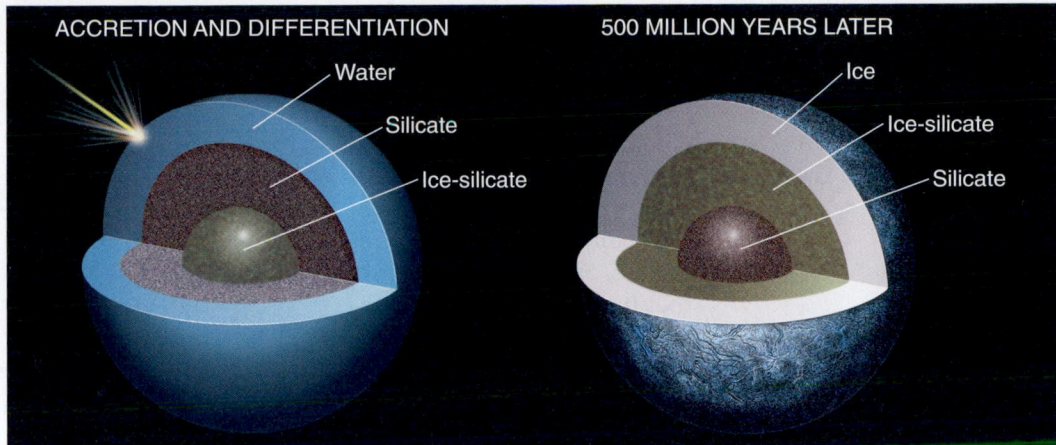

FIGURE 11.2 The evolution of Ganymede's interior. In the final stages of accretion, Ganymede was covered with a global ocean of liquid water, underlain by a zone of silicates and a mixed core of ice and silicates. Five hundred million years later the dense silicates had sunk gravitationally to form a core, and much of the ocean had frozen.

that were orbiting in a disk surrounding the young, hot proto-Jupiter. At the distance of Ganymede's orbit from the glowing proto-Jupiter core, temperatures were low enough for grains of water ice to survive, and along with silicate materials, they coalesced to form planetesimals. It took less than a half million years for these planetesimals to accrete and form the moon Ganymede. Heating generated by accretion melted parts of the moon to form an outer water layer, an inner silicate zone, and an ice-silicate core. These layers, however, were not stable. As cooling took place, much of the outer water layer froze to form a dirty ice crust. Most of the denser silicate materials sank to the center to form a core, leaving an intermediate ice-silicate zone (**Figure 11.2**).

We call Ganymede and other moons that formed, together with their parent planets, **regular moons**. Regular moons revolve around their planets in the same direction as the planets rotate and in orbits that lie nearly in the planets' equatorial planes. These characteristic orbits arose because the debris from which the regular moons formed was orbiting in the planets' equatorial planes and in the same direction as the evolving planets were rotating. With few exceptions, regular moons are tidally locked to their parent planets. Recall from Chapter 10 that tidal locking causes a body to rotate synchronously with respect to its orbit, as does Earth's Moon. A moon in synchronous rotation around its planet has fixed leading and trailing hemispheres: one hemisphere permanently faces the direction the moon is traveling in its orbit around the planet, and the other permanently looks backward. These two surfaces can appear very different from one another, because the leading hemisphere is always flying directly into any local debris surrounding the planet.

An especially strange regular moon is Saturn's Hyperion, seen in **Figure 11.3**. Hyperion has an unusual appearance, but that is not the only thing peculiar about it. It rotates in a

> Moons that formed together with their planets are called regular moons.

way that is spectacularly different from the rotation of any other known moon in the Solar System. Hyperion's rotation is chaotic (see Chapter 10), meaning that it tumbles in its orbit with a rotation period and a spin-axis orientation that are constantly and unpredictably changing!

Some moons revolve in a direction that is opposite the rotation of their planets, and some are situated in distant, unstable orbits. These are almost certainly bodies that

> Some captured, or "irregular," moons have retrograde orbits.

FIGURE 11.3 Saturn's Hyperion, certainly among the strangest-looking moons, rotates chaotically, with its rotation period and spin axis constantly changing. The 250-kilometer (km) moon's low density and spongelike texture, seen in this *Cassini* image, suggest that its interior houses a vast system of caverns.

formed elsewhere and were later captured by the planets. We call them **irregular moons**. The largest irregular moon is Neptune's Triton. It orbits Neptune in a **retrograde**, or "backward," direction. Other moons, such as Saturn's Phoebe, also have retrograde orbits. Most of the recently discovered moons of the outer planets are irregular, and many are only a few kilometers across.

11.2 Moons as Small Worlds

There are several ways we could group the moons of our Solar System. Some schemes are based on the sequence of the moons in their orbits around the parent planets; others are based on the sizes or compositions of the moons. For example, we have already noted that a few moons are predominantly rocky objects, some are mostly ices, and most appear to be mixtures of ice and rock.

The scheme we will use here is based on the amount and timing of the moons' geological activity as expressed by the features we see on their surfaces. As we learned when looking at the terrestrial planets and Earth's Moon, surface features observed on planets and moons provide critical clues to their geological history. For example, water ice is a common surface material among the moons of the outer Solar System, and the freshness of that ice tells us something about the ages of those surfaces. Meteorite dust darkens the icy surfaces of moons just as dirt darkens snow late in the season in our own urban areas. In other words, a bright surface often (but not always) means a fresh surface. (As we will see later in this section, the *oldest* surfaces on Jupiter's Europa are bright, whereas the *youngest* are dark.)

> Brightness, structure, and crater density of moon terrains give clues to geological activity.

As we discussed in Chapter 7, the size and number of impact craters gives us the relative timing of events such as volcanism, and this timing allows us to gauge if and when a moon may have been active in the past. Terrains having a large number of craters are older than those having only a few. Observations of erupting volcanoes, as on Io and Enceladus, are direct evidence that some moons are geologically active today.

What happens when we apply these age-dating techniques to the moons in the Solar System? We find an immense diversity: some moons have been frozen in time since their formation during the early development of the Solar System, while others are even more geologically active than Earth. In our classification scheme of moons, we include four categories of geological activity: (1) definitely active today, (2) probably or possibly active today, (3) active in the past but not today, and (4) apparently not active at any time since their formation.

Geologically Active Moons: Io, Enceladus, and Triton

One of the more spectacular surprises in Solar System exploration was the discovery of active volcanoes on Io, the innermost of the four large Galilean moons of Jupiter. Yet in one of those rare events that happen in science, planetologists predicted Io's volcanism just 2 weeks before its discovery, basing their conclusions on the very same tidal stresses discussed in Chapter 10. Did you ever take a piece of metal and bend it back and forth, eventually breaking it in half? Touch the crease line and you can burn your fingers! Just as bending metal in your hands creates heat, the continual flexing of Io's crust caused by the changing strength and direction of the tides generates enough energy to melt parts of the crust. In this way, Jupiter's gravitational energy is converted into thermal energy, powering the most active volcanism in the Solar System.

> Io is the Solar System's most volcanically active body.

As *Voyager* approached Jupiter, images showed that Io's surface is literally covered by volcanic features, including vast lava flows, volcanoes, and volcanic craters (see the opening photograph for Chapter 10). Amazingly, however, pictures from *Voyager* (and later from *Galileo*) failed to show a single impact crater, making Io unique among all the solid planetary bodies seen so far in planetary exploration. With a surface so young, Io must be volcanically active indeed, with lava flows and volcanic ash burying impact craters as quickly as they form. Scientists working on the *Voyager* mission discovered just how volcanically active Io is in spectacular and undeniable fashion when postencounter images looking back toward this moon showed explosive volcanic eruptions sending debris hundreds of kilometers above Io's surface!

As of this writing, Io has more than 300 known volcanic vents, with more than 150 active volcanoes observed by the *Voyager*, *Galileo*, and *New Horizons* spacecraft. The most vigorous eruptions, with vent velocities of up to 1 kilometer per second (km/s), spray sulfurous gases and solids as high as 300 km above the surface. Ash and other particles rain onto the surface as far as 600 km from the vents, as can be seen in the opening photograph in Chapter 10. The moon is so active that several huge eruptions are often occurring at the same time. One look at an image such as **Figure 11.4** leaves little doubt about the source of the material supplying the plasma torus and Io flux tube that we discussed in Chapter 9.

Io's surface, shown in **Figure 11.5** and the opening photograph for Chapter 10, displays a wide variety of colors—pale shades of red, yellow, orange, and brown. Mixtures of sulfur, sulfur dioxide frost, and sulfurous salts of sodium and potassium on the moon's surface are the likely cause of the colors. Bright patches may be fields of sulfur diox-

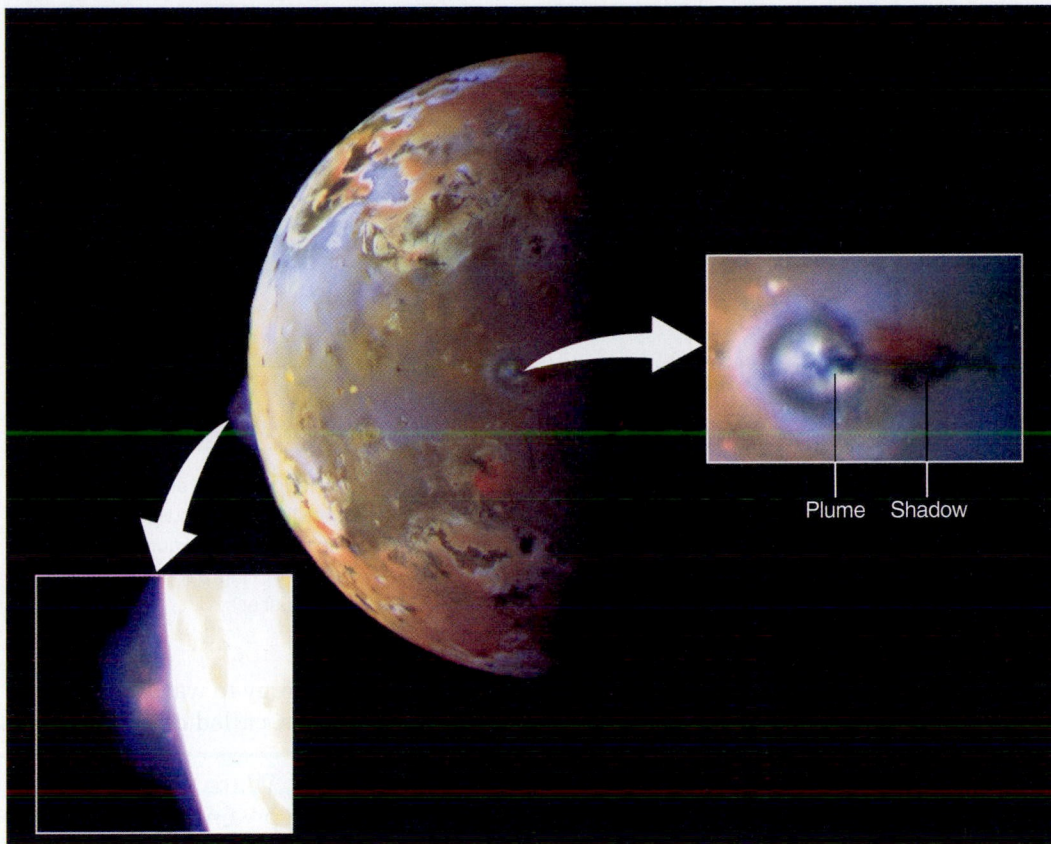

ide snow. You may well wonder how snow can fall from Io's nearly nonexistent atmosphere. According to current understanding, liquid sulfur dioxide must flow beneath Io's surface, held at high pressure by the weight of overlying material. Like water from an artesian spring, this pressurized sulfur dioxide can be pushed out though fractures in the crust, producing sprays of sulfur dioxide snow crystals that travel for up to hundreds of kilometers before settling back to the moon's surface. (To see a similar process in action, just operate a carbon dioxide fire extinguisher. These fire extinguishers contain liquid carbon dioxide at high pressure that immediately turns to "dry ice" snow as it leaves the nozzle.)

Voyager and *Galileo* images show parts of the surface of Io at high resolution. They reveal a variety of plains, irregular craters, and flows, all related to eruption of mostly silicate magmas onto the surface of the moon. They also show high-standing mountains, some nearly twice the height of Mount Everest, Earth's tallest mountain. Huge structures, some 65 km across, show multiple calderas and other complex structures telling of a long history of repeated eruptions followed by collapse of the partially emptied magma chambers. Many of the floors are very hot (**Figure 11.6**) and might still contain molten material similar to magnesium-rich lavas that erupted on

Silicate magmas dominate Io's volcanism.

FIGURE 11.5 An image of Jupiter's volcanically active moon Io, constructed of images obtained by *Galileo*.

FIGURE 11.6 A *Galileo* image of Io showing regions where lava has erupted. A molten lava flow is shown in false color to make it more visible.

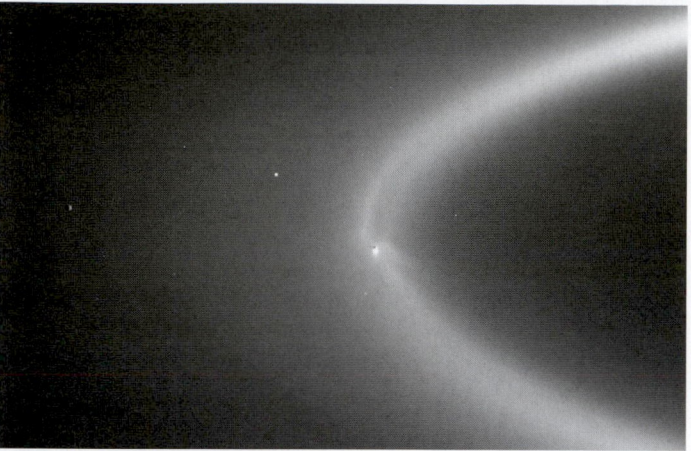

FIGURE 11.7 Saturn's moon Enceladus is the source of material in Saturn's E Ring. Note the distortion in the distribution of ring material in the immediate vicinity of the moon caused by its gravitational influence on the orbits of ring particles. Other bright objects in this *Cassini* image are stars.

Earth more than 1.5 billion years ago. Note that volcanoes on Io are spread around the moon in a much more random fashion than on Earth, where tectonic patterns influence the location of volcanism.

Io probably formed at about the same time as the other giant planets' moons. Io's current volcanic activity suggests that its entire mass has recycled, or turned inside out, more than once in the past, leading to chemical differentiation. Volatiles such as water and carbon dioxide probably escaped into space long ago, with most heavier materials sinking to the interior to form a core. Sulfur and various sulfur compounds, aided by silicate magmas, are constantly being recycled, forming the complex surface we see today.

> Volcanism may have turned Io inside out several times.

As we move farther out in the Solar System, we find volcanism of a different type. Enceladus, one of Saturn's icy moons, is only 500 km in diameter. Even so, it shows a wide variety of ridges, faults, and smooth plains—evidence of tectonic processes unexpected for such a small body. Some impact craters, rather than showing crisp features, appear "softened," as though they were "relaxed" by the viscous flow of warm ice. Parts of the moon show no craters at all, indicating recent resurfacing. As we will see in the next section, the orbit of Enceladus is located in the densest part of Saturn's E Ring, and astronomers had long suspected that the moon was providing the *ring* with a continuous supply of tiny particles, most probably ice crystals (**Figure 11.7**).

The joint NASA–European Space Agency *Cassini* mission confirmed the existence of tectonic activity on Enceladus. Close-up images of terrain near the moon's south pole reveal a cracked and twisted "taffylike" appearance (**Figure 11.8a**). Temperature measurements of these geologically young "cracks" show them to be warmer than their surroundings. A combination of tidal heating and the decay of radioactive elements within the moon's rocky component

could generate enough thermal energy to warm the ice and drive it to the surface. This process, called **cryovolcanism**, is similar to normal volcanism but is driven by subsurface low-temperature liquids rather than molten rock. *Cassini* showed that active cryovolcanic plumes, like those seen in **Figure 11.8b**, are energetic enough to overcome the moon's low gravity, sending tiny ice crystals into space to repopulate particles that are being continually lost from Saturn's E Ring.[1] We might logically ask why Enceladus is so active while Mimas, a neighboring moon of about the same size and also subject to tidal heating, appears to be quite dead. The answer, unfortunately, remains a mystery.

> Water volcanoes on Enceladus are an example of low-temperature cryovolcanism.

Cryovolcanism also occurs in the Neptune system. Triton, the largest moon in the Neptune system, is an irregular moon that must have been captured by its planet. To achieve its present circular, synchronous orbit, it must have experienced extreme tidal stresses from Neptune following its capture. Such stresses would be similar to those on Io and would have generated large amounts of thermal energy. The interior might even have melted, allowing Triton to become chemically differentiated.

Like all other moons in the outer reaches of the Solar System, Triton is a cold place. Its surface temperature is only about 38 kelvins (K). Triton has a thin atmosphere, with a surface composed mostly of ices and frosts of methane and

---

[1] Saturn's diffuse G Ring also contains a tiny "moonlet" embedded in a *ring arc*. This moonlet appears to be the source of the G Ring particles, although it is much too small to be geologically active. The small icy particles that make up the ring are easily dislodged from the moonlet's surface by impacts of micrometeorites.

(a)
(b)

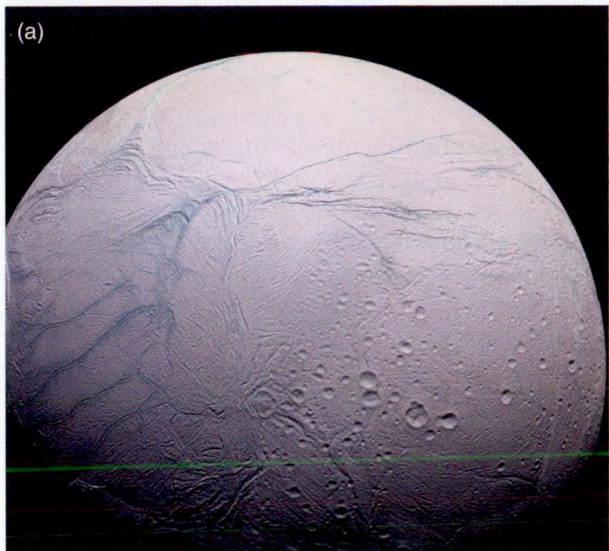
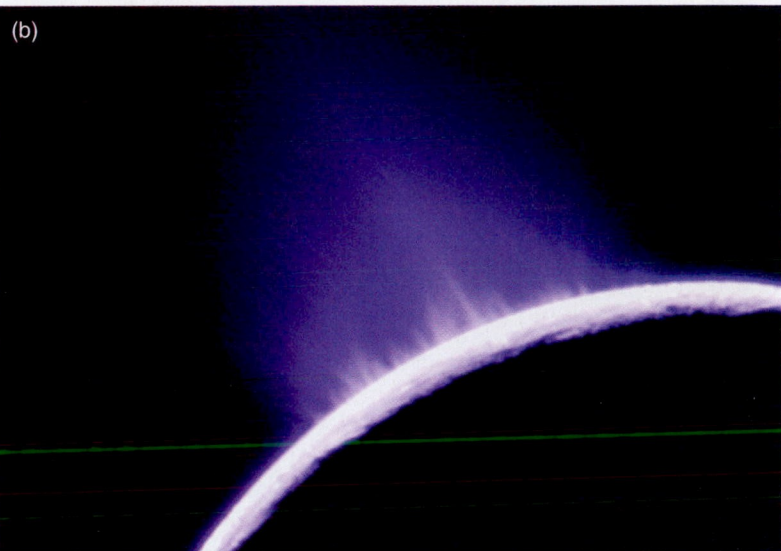

FIGURE 11.8 *Cassini* images of Enceladus. (a) The twisted and folded surface of deformed ice cracks near the moon's south pole (shown blue in false color) are warmer than surrounding terrain and were found to be the sources of cryovolcanism. (b) Cryovolcanic plumes in the south polar region are seen spewing ice particles into space.

nitrogen. From the relative lack of craters, we know that the surface is geologically young. Part of Triton is covered with "cantaloupe terrain," so named because it looks like the skin of a cantaloupe (**Figure 11.9**). Irregular pits and hills may represent surface deformation and extrusion of slushy ice onto the surface from the interior. Veinlike features include grooves and ridges that could result from the extrusion of ice along fractures. The rest of Triton is covered with smooth plains of volcanic origin. Irregularly shaped depressions as wide as 200 km (**Figure 11.10**) formed when

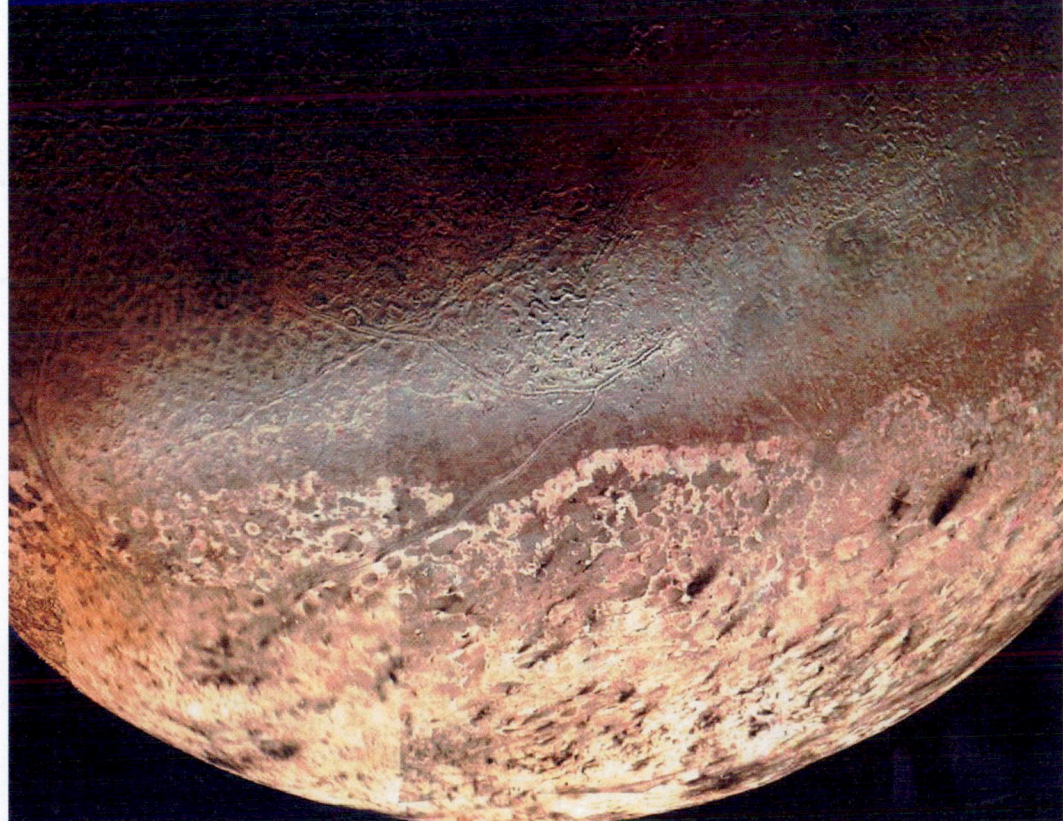

FIGURE 11.9 This *Voyager* mosaic shows various terrains on the Neptune-facing hemisphere of Triton. "Cantaloupe terrain" is visible at the top; its lack of impact craters indicates a geologically younger age than the bright, cratered terrain at the bottom.

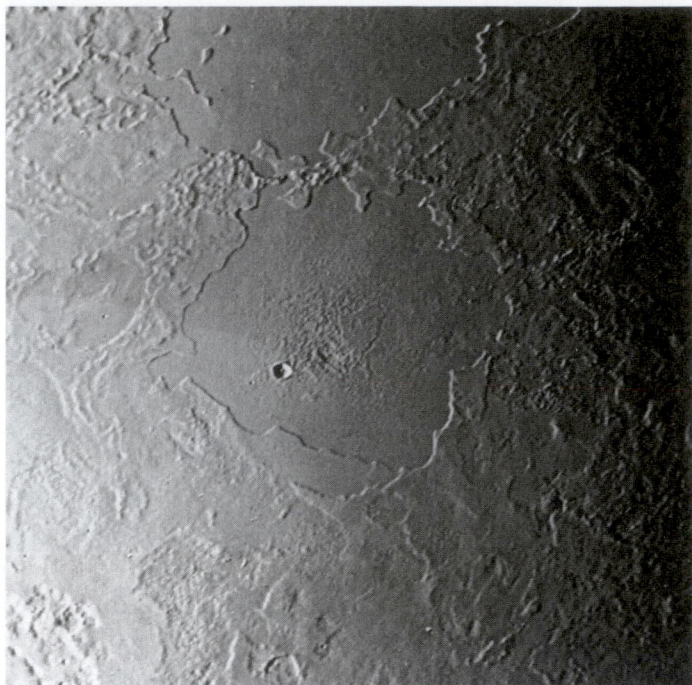

FIGURE 11.10 This irregular basin on Triton has been partly filled with frozen water, forming a relatively smooth ice surface. The state of New Jersey could just fit within the basin's boundary.

mixtures of water, methane, and nitrogen ice melted in the interior of Triton and erupted onto the surface, much as rocky magmas erupted onto the lunar surface and filled impact basins on Earth's Moon.

Voyager 2 found four active geyserlike cryovolcanoes on Triton. Each consisted of a plume of gas and dust as much as a kilometer wide rising 8 km above the surface, where the plume was caught by upper atmospheric winds and carried for hundreds of kilometers downwind. How do the eruptions on Triton work? Their association with Triton's southern hemispheric ice cap may provide some clues to the process. We know that nitrogen ice in its pure form is transparent and allows easy passage of sunlight. Thus, clear nitrogen ice could create a localized greenhouse effect (see Chapter 8), in which solar energy trapped beneath the ice raises the temperature at the base of the ice layer. A temperature increase of only 4°C would be enough to vaporize the nitrogen ice. As gas is formed, the expanding vapor exerts very high pressures beneath the ice cap. Eventually the ice ruptures and vents the gas explosively into the low-density atmosphere. Dark material, perhaps silicate dust or radiation-darkened methane ice grains, is carried along with the expanding vapor into the atmosphere, from which it subsequently settles to the surface, forming dark patches streaked out by local winds, as seen near the bottom of Figure 11.9.

> **Nitrogen propels Triton's geyserlike volcanism.**

Possibly Active Moons: Europa and Titan

One of the more fascinating objects in the outer Solar System is Jupiter's moon Europa, a rock world slightly smaller than our Moon but with an outer shell of water. We know that the surface of the water is frozen, but what lies beneath the ice? Like Io, Europa experiences continually changing tidal stresses from Jupiter that generate internal energy and possibly volcanism. Some calculations show that the tidal heating may be too small to produce volcanism, but large enough to keep much of the subsurface water in a liquid state. Beneath its cap of solid ice—estimated to be anywhere from a few hundred meters to tens of kilometers thick—Europa may possess a global ocean.

Some planetologists believe this subsurface sea could be 100 km deep and contain more water than all of Earth's oceans! Such an ocean would have some interesting characteristics. We might expect it to be salty because of dissolved minerals. Supporting evidence for Europa's salty sea came when physicists took a close look at *Galileo*'s magnetometer data and found that Europa's magnetic field is variable, indicating an internal electrically conducting fluid. It may also contain an abundance of organic material brought in by impacting comet nuclei. In fact, what we may have on Europa is an environment that is not so different from some of our terrestrial ecological niches that support "extremophiles" (see **Connections 11.1**). The universal conditions necessary to create and support life—liquid water, heat, and organic material—could all be present in Europa's oceans. As such, Europa is a high-priority target in the search for extraterrestrial life.

Europa's surface is young, with few impact craters, but it is deformed by tectonic activity driven by heating from its interior. Regions of chaotic terrain, as shown in **Figure 11.11**, are places where the icy crust has been broken into slabs that have been rafted into new positions. These slabs could be fitted back together like pieces of a jigsaw puzzle. In other areas the crust split apart and has been filled in with new dark material forced to the surface from the interior, in a process somewhat resembling seafloor spreading on Earth (see Chapter 7). With time, frosts cover the dark material, causing a general brightening of the surface. Of the handful of large impact structures preserved on Europa, all are shallow features resembling the patterns formed when a rock is dropped into stiff mud. The chaotic terrain, spreadinglike features, impact structures, and other surface features all suggest that the icy crust of Europa consisted of a thin brittle shell overlying either liquid water or warm, slushy ice at the time when the features formed. Although we do not know whether these are the conditions on Europa today, the geologically young surface holds open that possibility.

> **Europa is covered with broken slabs of ice.**

FIGURE 11.11 A high-resolution *Galileo* image of Jupiter's moon Europa, showing where the icy crust has been broken into slabs that, in turn, have been rafted into new positions. These areas of chaotic terrain are characteristic of a thin, brittle crust of ice floating atop a liquid or slushy ocean.

The case for current geological activity on Saturn's large moon Titan appears to be stronger than for Jupiter's Europa. Titan is larger than Mercury and has a density that suggests a composition of about 45 percent water and 55 percent rocky material. What makes Titan especially remarkable is a thick atmosphere, which historically has obscured our view of its surface. As we learned in Chapter 8, Mercury's secondary atmosphere has been lost to space, while Titan's greater mass and distance from the Sun has allowed it to retain an atmosphere that is 30 percent denser than Earth's. Like Earth, Titan has an atmosphere that is mostly nitrogen, but views of Titan's limb show atmospheric layers that are most likely photochemical hazes, much like smog. Titan's atmosphere is reminiscent of the smog over Los Angeles on a bad day.

> A dense, hazy atmosphere obscures Titan's surface.

CONNECTIONS 11.1

Extreme Environments and an Organic Deep Freeze

During a visit in 1835 to the Galápagos Islands in the eastern Pacific Ocean, British naturalist Charles Darwin (1809–1882) observed variations in animal and plant life that eventually led to his now legendary theory of evolution. Off the coast of the Galápagos, 2,500 meters beneath the ocean's surface, is a form of life that Darwin could never have imagined. Here the Nazca and Pacific plates (see Figure 7.19) grind furiously against one another, creating friction, high temperatures, and seafloor volcanism. Mineral-rich superheated water pours out of "hydrothermal vents." The surrounding water contains very little dissolved oxygen. No sunlight reaches these depths. Yet in the total darkness of the ocean bottom, life abounds. From tiny bacteria to shrimp to giant clams and tube worms, sea life not only survives but thrives in this severe environment. In the complete absence of sunlight, the small single-celled organisms at the bottom of the local food chain get their energy from "chemosynthesis," a process by which inorganic materials are converted into food through the use of chemical energy.

Biologists call these life-forms "extremophiles." Robust types of bacteria are found flourishing in the scalding waters of Yellowstone's hot springs; in the bone-dry oxidizing environment of Chile's Atacama Desert; and in the Dead Sea, where salt concentrations run as high as 33 percent. Bacteria have even been found in core samples of ancient ice 3,600 meters below the surface of the East Antarctic ice sheet. When it comes to harsh habitats, life is amazingly adaptable. If life can exist under such extreme conditions on Earth, then it might also exist on those moons of the giant planets that have the necessary ingredients for life: liquid water, an energy source, and the presence of organic compounds.

As we have learned in this chapter, Jupiter's moon Europa may contain a shell of liquid water beneath a surface layer of water ice. Europa is not the only moon suspected of having a subsurface ocean. Like Europa, Callisto also shows magnetic variability, the signature of a salty ocean. In our search for life elsewhere in the Solar System, Jupiter's large moons should be among our prime targets. So, too, should Enceladus, the moon of Saturn that spews water-ice crystals.

The presence of comet-borne organic material in the oceans of Europa and Callisto cannot yet be confirmed, but there is one place where we *can* find such materials—in Titan's massive atmosphere. Here, organic gases reveal themselves through the analytical eyes of the spectrograph. We know that Titan's nitrogen atmosphere contains methane and traces of ethane, propane, ethylene, hydrogen cyanide, carbon monoxide, acetylene, and other compounds of biological interest. For example, five molecules of hydrogen cyanide (HCN) will spontaneously combine to form adenine ($C_5H_3N_4NH_2$), one of the four primary components of DNA and RNA. HCN is also a building block of amino acids, which in turn combine to form proteins.

Photodissociation and recombination of these various gases produce complex organic molecules that then rain out onto Titan's surface as a frozen tarry sludge. Biochemists believe that many of these substances are biological precursors, similar to the organic molecules that preceded the development of life on Earth. For now, these possible clues to the origins of terrestrial life remain locked up in Titan's deep freeze, quietly awaiting analysis by future space missions.

As Titan heated up and differentiated chemically, various ices, including methane (CH_4) and ammonia (NH_3), emerged from the interior to form an early atmosphere. Ultraviolet photons from the Sun have enough energy to break apart ammonia and methane molecules—a process called **photodissociation**. Photodissociation of ammonia is the likely source of Titan's abundant supply of atmospheric nitrogen. Photodissociation of methane breaks the molecules into fragments, which can recombine to form complex hydrocarbons such as ethane and other organic compounds. Organic compounds tend to form in tiny particles, creating an organic smog and giving Titan's atmosphere its characteristic orange hue (**Figure 11.12a**). The *Cassini* spacecraft, which began orbiting Saturn in 2004, gave astronomers their first close-up views of the moon's surface (**Figure 11.12b**). Haze-penetrating infrared imaging shows broad regions of dark and bright terrain (**Figures 11.12c and d**).

Titan's northern hemisphere (currently its winter hemi-

> Titan's atmosphere and surface are rich in organic compounds.

sphere) is now in darkness and not accessible to infrared imaging. Radar imaging, however, easily pierces the moon's thick cloud cover and reveals its surface features without the aid of solar illumination. Radar images have revealed irregularly shaped features in Titan's northern hemisphere that appear to be widespread lakes of methane, ethane, and other hydrocarbons (**Figure 11.13**). However, the abundance of methane—both on the surface and in the atmosphere of Titan—presents a problem. The photodissociative process by sunlight should have removed all atmospheric methane within a geologically brief period of 50 million years or so. Therefore, there must be some process for renewing the methane that is being destroyed by solar radiation. A likely candidate is cryovolcanism. As on Earth, radioactive decay is an important source of internal heating, and Titan's internal thermal energy could cause the release of "new" methane from underground through cryovolcanism in much the same way that terrestrial volcanism releases water vapor and carbon dioxide into Earth's atmosphere (see Chapter 8). So far, there has been no direct evidence of active cryovolcanism on Titan, but the presence of abundant

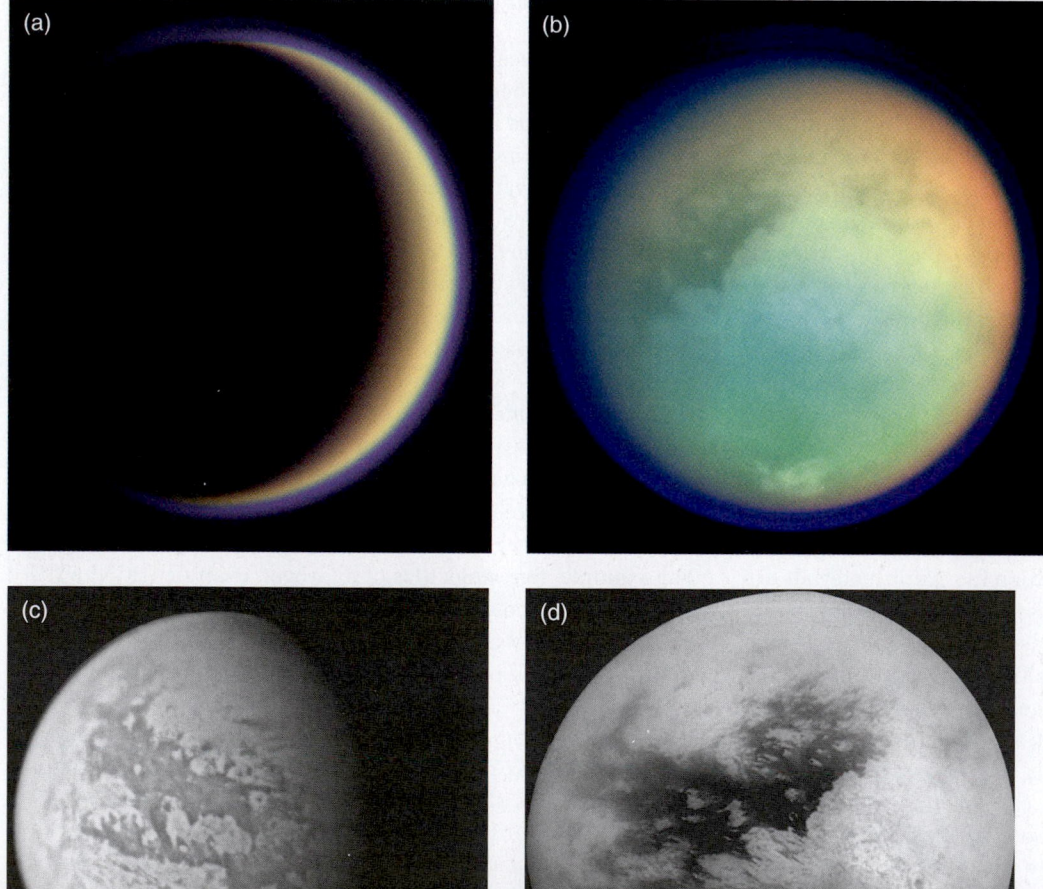

(a) (b) (c) (d)

FIGURE 11.12 *Cassini* images of Saturn's largest moon, Titan. (a) Visible-light imaging shows Titan's orange atmosphere, which is caused by the presence of organic smoglike particles. (b) This infrared- and ultraviolet-light combined image shows surface features and a bluish atmospheric haze caused by scattering of ultraviolet sunlight by small atmospheric particles. Bright clouds in Titan's lower atmosphere are seen near the moon's south pole. (c) Infrared-light imaging penetrates Titan's smoggy atmosphere and reveals surface features. (d) This infrared image covers the same general area of Titan as seen in (b). The large dark area, called Xanadu, is about the size of Australia.

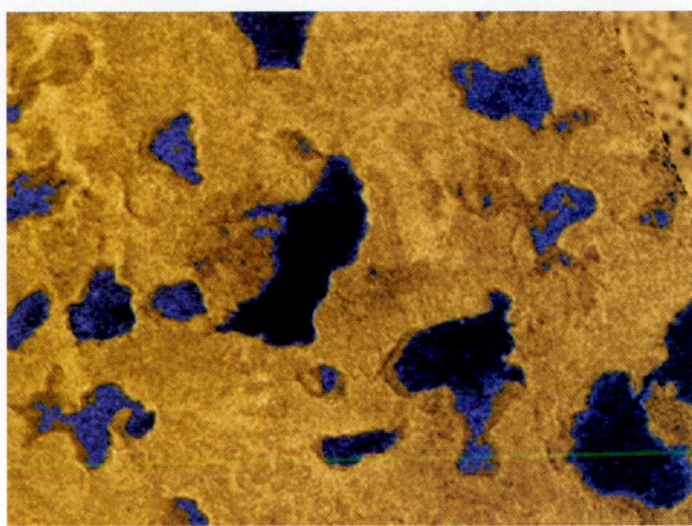

FIGURE 11.13 Radar imaging (false color) near Titan's north pole shows what appear to be lakes of liquid hydrocarbons covering 100,000 square kilometers (km²) of the moon's surface. Features commonly associated with terrestrial lakes, such as islands, bays, and inlets, are clearly visible in many of these radar images.

atmospheric methane and methane lakes strongly suggests that Titan is indeed geologically active.

Shortly after arriving at Saturn, the *Cassini* spacecraft released a 320-kilogram (kg) probe, *Huygens*, which plunged through Titan's atmosphere measuring composition, temperature, pressure, and winds and taking pictures as it descended. *Huygens* confirmed that cloud particles contain nitrogen-bearing organic compounds—key components in

the production of terrestrial proteins. During its descent, *Huygens* encountered 120-meter-per-second (m/s) winds and temperatures as low as 88 K. As it reached the surface, though, winds had died down to less than 1 m/s and the temperature had warmed to 112 K. Images taken during the descent (**Figure 11.14a**) show terrains reminiscent of those on Earth, with networks of channels, ridges, hills, and flat areas that may be dry lake basins. These terrains suggest a sort of "hydrological cycle" in which methane rain falls to the surface, washes the ridges free of the dark hydrocarbons, and then collects into drainage systems that empty into low-lying, liquid methane pools.[2] Stubby, dark channels appear to be springs where liquid methane emerges from the subsurface; bright, curving streaks could be water ice that has oozed to the surface to feed glaciers. Although no liquid methane rain or surface pools were seen in the *Huygens* images, the near absence of impact craters on the surface would point to recent—if not current—hydrological activity. Radar views also indicate an active surface, showing features that resemble terrestrial sand dunes and channels, but only an occasional impact crater (**Figure 11.14b**).

Once on Titan's surface, *Huygens* continued to take pictures and make physical and compositional measurements. *Huygens* scientists described the surface as having characteristics similar to wet or dry sand or lightly packed snow. The surface was wet with liquid methane, which evaporated as the probe—heated to 2,000 K during its pas-

[2]We are a bit loose with our terminology here. Technically, *hydrological* refers to *water*, as in the terrestrial hydrological (water) cycle discussed in Chapter 8. On Titan, methane assumes the role that water plays on Earth. Some planetologists prefer to use the term *methanological cycle* when applied to Titan.

FIGURE 11.14 (a) The surface of Titan viewed from the *Huygens* probe during its descent to the surface. The dark drainage patterns resemble river systems on Earth. (b) This radar image of Titan taken from the *Cassini* orbiter shows a rare impact crater and its ejecta blanket.

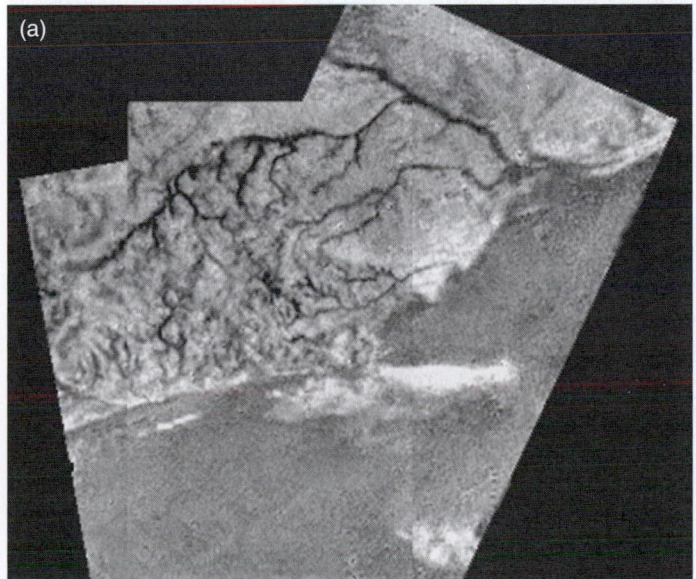
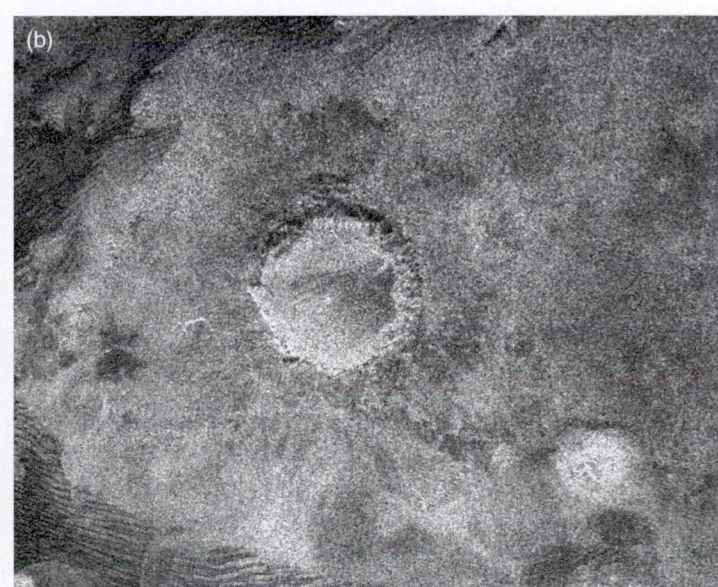

FIGURE 11.15 View of the surface of Titan obtained from the *Huygens* probe, showing a relatively flat surface littered with water-ice "rocks." The two rocks just below the center of the image are about 85 centimeters (cm) from the camera and roughly 15 and 4 cm across, respectively. The dark "soil" is probably made up of hydrocarbon ices.

sage through the atmosphere—landed in the frigid soil. The surface was also rich with other organic compounds, such as cyanogen $(CN)_2$ and ethane (C_2H_6). As shown in **Figure 11.15**, the surface around the landing site is relatively flat and littered with rounded "rocks" of water ice. The dark "soil" is probably a mixture of water ice and of hydrocarbon ices that have precipitated from the atmospheric haze seen in the background.

In many ways, Titan resembles a primordial Earth, albeit at much lower temperatures. The presence of organic compounds that could be biological precursors in the right environment makes Titan another high-priority target for continued exploration (see Connections 11.1).

Formerly Active Moons: Ganymede and Some Moons of Saturn and Uranus

Some moons show clear evidence of past ice volcanism and tectonic deformation, but no current geological activity. Ganymede is the largest moon in the Solar System, even larger than the planet Mercury. The surface is composed of two prominent terrains: a

> Ganymede is the Solar System's largest moon.

dark, heavily cratered (and therefore ancient) terrain, and a bright terrain characterized by ridges and grooves. Ganymede's low density (1.9 times that of water) indicates that its bulk composition is about half water and half rocky materials. Overall, the moon's surface is relatively bright. Even the so-called dark terrains are brighter than the bright

areas on Earth's Moon. The high number of impact craters on the dark terrain reflects the period of intense bombardment during the early history of the Solar System. The most extensive region of ancient dark terrain includes a semicircular area more than 3,200 km across (about the size of Australia) on the leading hemisphere. Furrowlike depressions occurring in many dark areas are among Ganymede's oldest surface features. They may represent surface deformation from internal processes, or they may be relics of impact-cratering processes.

Impact craters range up to hundreds of kilometers in diameter, with the larger craters being proportionately more shallow. With time, the icy crater rims deform by viscous (very slow) flow, as might a lump of soft clay, and they can ultimately lose nearly all of their varying topography. Such features are seen as flat circular patches of bright terrain called "palimpsests," which are characteristic of Ganymede's icy lithosphere (**Figure 11.16a**). Palimpsests are found principally in the dark terrain of Ganymede and are believed to be scars left by early impacts onto a thin, icy crust overlying water or slush (**Figure 11.16b**).

When first viewed in *Voyager* images, the bright terrain was thought to represent regions that had been flooded by water or slush erupted from the interior of Ganymede. *Galileo* images, however, failed to show any indications of such flooding, other than in a few local places. How, then, did the bright terrain form? The answer seems to be related to a style of surface formation by tectonic processes not previously considered. In Chapter 7 we discussed how planetary surfaces can be fractured by faults or folded by compression resulting from movements such as those initiated in the mantle. On Ganymede, the tectonic processes have been so intense that the fracturing and faulting have completely deformed the icy crust, destroying all signs of older features, such as impact craters.

We might ask why Ganymede is no longer geologically active. The energy that powered Ganymede's early activity was liberated during a period of differentiation when the moon was very young. Once the differentiation process was complete, that source of internal energy ran out and geological activity ceased.

Many other moons show evidence of an early period of geological activity that has resulted in a dazzling array of terrains. A 400-km impact crater scars Saturn's moon Tethys, covering 40 percent of the diameter of the moon itself, and an enormous canyonland wraps at least three-fourths of the way around the moon's equator. Saturn's moon Dione shows bright ice cliffs up to several hundred meters high, created by tectonic fracturing. The trailing hemisphere of Saturn's Iapetus is bright, reflecting half the light that falls on it, while much of the leading hemisphere is as black as tar (**Figure 11.17**). These dark deposits appear *only* in the leading hemisphere of Iapetus, suggesting that they might be debris that was blasted off small retrograde moons of Saturn by micrometeoritic impacts and swept up by Iapetus as it moved along in its prograde orbit around Saturn.

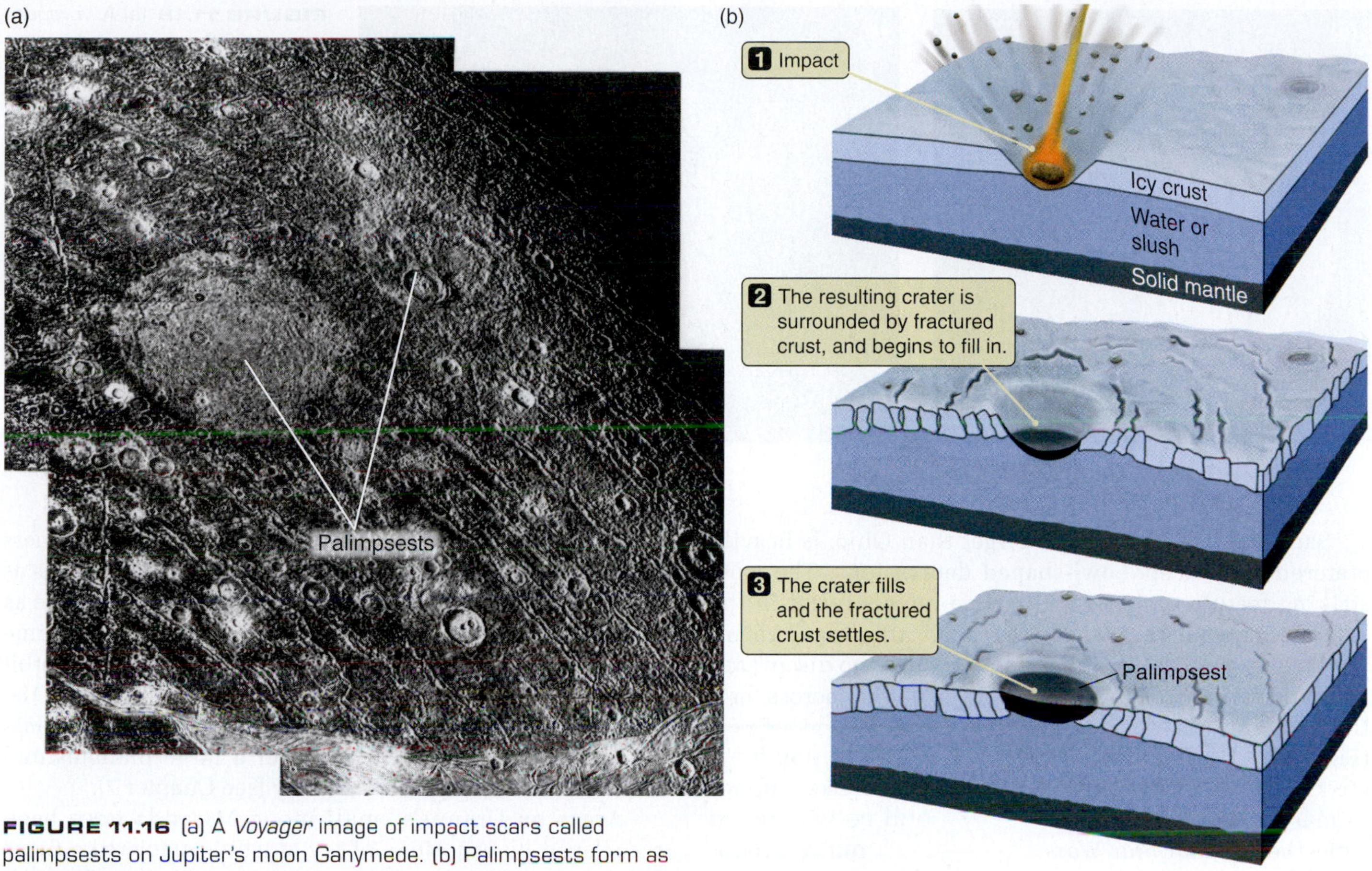

(a)

(b)

1 Impact

Icy crust

Water or slush

Solid mantle

2 The resulting crater is surrounded by fractured crust, and begins to fill in.

3 The crater fills and the fractured crust settles.

Palimpsest

Palimpsests

FIGURE 11.16 (a) A *Voyager* image of impact scars called palimpsests on Jupiter's moon Ganymede. (b) Palimpsests form as viscous flow smooths out structures left by impacts on icy surfaces.

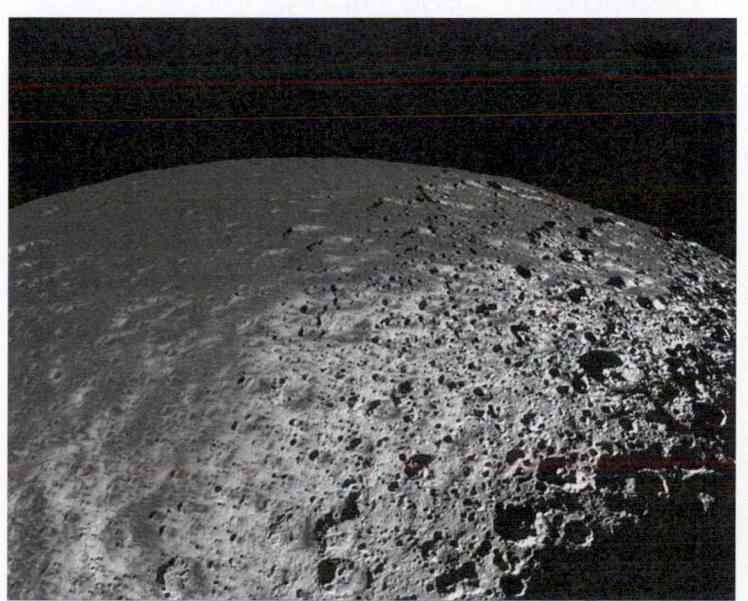

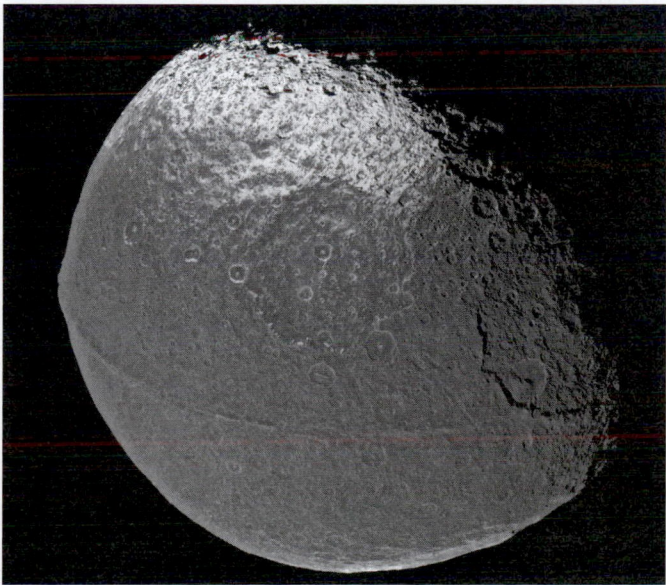

FIGURE 11.17 *Cassini* images of Iapetus show the boundary between its bright and dark hemispheres.

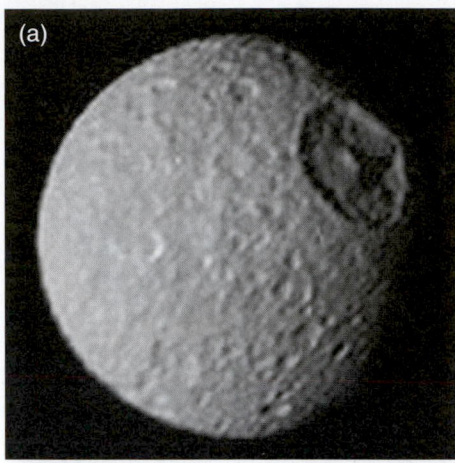

FIGURE 11.18 (a) A *Voyager* image of Saturn's moon Mimas and the crater Herschel. (b) The Death Star from *Star Wars*. (The movie was released 3 years before the image of Mimas was taken.)

Saturn's moon Mimas, no larger than Ohio, is heavily cratered with deep, bowl-shaped depressions. The most striking feature on Mimas is a huge impact crater in the leading hemisphere. Named "Herschel" after the German-English astronomer Sir William Herschel, who discovered many of Saturn's moons, the crater is 130 km across, or a third the size of Mimas itself. (**Figure 11.18** illustrates the striking similarity between Mimas and the "Death Star" from the original *Star Wars*, which was released in 1977, 3 years before the image of Mimas was taken.) It is doubtful that Mimas could have survived the impact of a body much

> **Mimas may have been broken apart and reassembled many times.**

larger than the one that created Herschel. Some astronomers believe that Mimas (and perhaps other small, icy moons as well) was hit many times in the past by objects so large as to fragment the moon into many small pieces. Each time this happened, the individual pieces still in Mimas's orbit would coalesce to re-form the moon, perhaps in much the same way that our own Moon coalesced from fragments that remained in Earth orbit after a large planetesimal impacted Earth early in its history (see Chapter 7).

Areas on Uranus's small moon Miranda have been resurfaced by eruptions of icy slush or glacierlike flows (**Figure 11.19**). Other Uranus moons—Oberon, Titania, and Ariel—are covered with faults and other signs of early tectonism. On Ariel, in particular, very old, large craters appear to be missing, perhaps obliterated by earlier volcanism.

FIGURE 11.19 A *Voyager* image of fault zones, high cliffs, and cratered terrain on Miranda, a 472-km-diameter moon of Uranus.

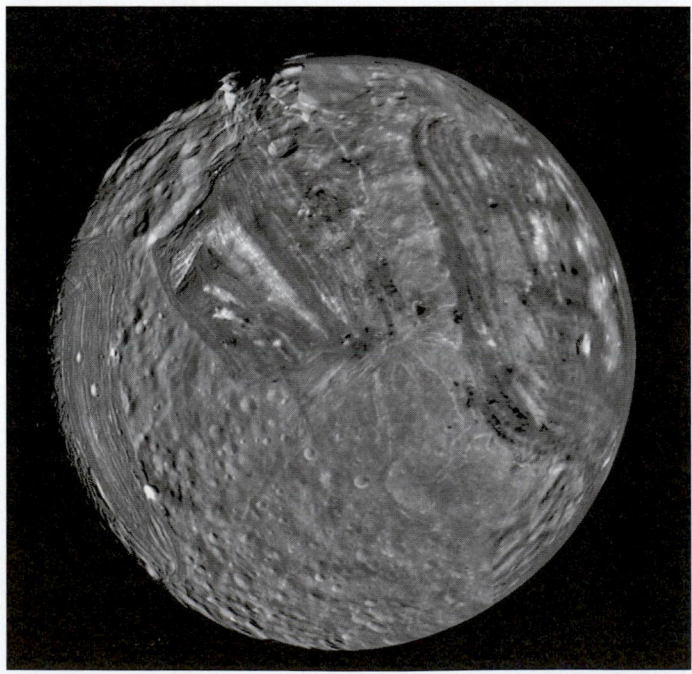

Geologically Dead Moons

Geologically dead moons, including Jupiter's Callisto, Uranus's Umbriel, and a large assortment of irregular moons, are those for which there is little or no evidence of internal activity having occurred at any time since their formation. Their surfaces are heavily cratered and show no modification other than the cumulative degradation caused by a long history of impacts.

Callisto (**Figure 11.20**) is about the size of Mercury. It is also the darkest of the Galilean moons, yet it is still twice as reflective as Earth's Moon. This brightness indicates that Callisto is rich in water ice, but with a mixture of dark, rocky materials. Except for terrains that experienced large impact events, the surface is essentially uniform, consisting of relatively dark, heavily cratered terrain. High-resolution images reveal that Callisto's surface has been modified by local landslides and places where the small craters have been erased by an unknown process. Its most prominent feature is a 2,000-km, multiringed structure of impact origin named Valhalla (the largest bright feature visible on Callisto's face in Figure 11.1). The impact

FIGURE 11.20 A *Galileo* image of Jupiter's second-largest moon, Callisto. Its ancient surface is dominated by impact craters and shows no sign of early internal activity.

may have occurred in a relatively thin, rigid crust overlying a fluidlike interior. The fluid mass then rapidly filled the initial crater bowl, leaving only a trace of the impact scar. Geophysical measurements obtained from the *Galileo* spacecraft suggest that Callisto is not differentiated, implying that it never went through a molten phase. On the other hand, the magnetometer aboard *Galileo* returned results suggesting that a liquid ocean could exist beneath the heavily cratered surface, implying that some sort of differentiation has occurred. These observations simply point out that many times in science we are faced with conflicting ideas that usually can be resolved only with additional measurements or observations.

Umbriel, the darkest and third-largest of the Uranus moons, appears uniform in color, reflectivity, and general surface features, indicative of an ancient surface. The real puzzle posed by Umbriel is why it has been geologically dead, while the surrounding large moons of Uranus have been active, at least at some time in their past.

11.3 Rings Surround the Giant Planets

If Galileo's discovery of Jupiter's moons was personally satisfying, his other important discovery was decidedly less so. When he first pointed his telescope at Saturn in 1610, the tiny disk of the planet seemed to be accompanied by smaller companions on both sides. Unlike the moons of Jupiter that he had found a few months earlier, these features did not move. Galileo was troubled by the lack of movement because it was like nothing else he had observed. Two years later the "companions" had vanished. Their unexpected disappearance upset the Italian astronomer greatly; he feared his earlier observations had been in error. A few years later, the mysterious features reappeared.

For more than four decades astronomers puzzled over Galileo's discovery. In 1655 a 26-year-old Dutch instrument maker, Christiaan Huygens (1629–1695), pointed a superior telescope of his own design at Saturn and saw what the astronomers of his day had failed to see. Saturn is surrounded by an apparently continuous, flat **ring**, and as Huygens correctly deduced, the variations in its visibility are caused by changes in the apparent tilt of the ring as Saturn orbits the Sun. What was the nature of this strange ring encircling Saturn? Astronomers assumed it was a solid disk spinning around the planet, an interpretation that persisted for more than a century after Huygens's discovery. That notion began to weaken in 1675 when the great Italian-French astronomer Jean-Dominique Cassini (1625–1712) found a gap in the planet's seemingly solid ring. Saturn now appeared to have two rings rather than one, and the gap that separated them became known as the Cassini Division that we learned of in the previous chapter.

In 1850 a fainter ring, located just inside the two bright rings, was found independently by English and American observers, giving Saturn a total of three known rings. For illustrators and cartoonists, Saturn had become the iconic image of a planet. Yet over the centuries following the discovery of Saturn's rings, exhaustive searches failed to detect rings around any other planet. Papers published by more than one distinguished astronomer explained why, theoretically, only Saturn could have rings. Then, in the latter part of the 20th century, a new search technique became available: observation of stellar occultations (see Chapter 9). In 1977 a team of American astronomers using the occultation technique to study the atmosphere of Uranus saw brief, minute changes in the brightness of a star as it first approached and then receded from the planet. The interpretation was immediately obvious: Uranus has rings! Over the next several years stellar occultations revealed a total of nine rings surrounding the planet. In 1986 *Voyager 2* imaged two additional Uranus rings, and in 2005 the Hubble Space Telescope recorded two more, bringing the total to 13.

Stellar occultations not only show the existence of rings; they also tell us something about the rings themselves. The duration of an occultation event is a measure of the width of the ring. The observed decrease in the brightness of a star is an indication of the ring's transparency and therefore of the amount of material it contains. The very brief interruption of starlight as the Uranus rings passed in front of the star showed that they are far too narrow to have been seen in the earlier, unsuccessful searches by more conventional methods.

More ring discoveries were to follow from still another technology: close-up studies by planetary probes. In 1979, cameras on *Voyager 1* recorded a faint ring around Jupiter; and in the same year, *Pioneer 11* found a narrow ring just outside the bright rings of Saturn.

For a while Neptune seemed to be the only giant planet devoid of rings. Then, in the early to mid-1980s, occultation searches by teams of American and French astronomers began yielding positive but confusing results. Several occultation events that appeared to be due to rings were seen on only one side of the planet. The astronomers concluded that Neptune was surrounded not by complete rings but rather by several arclike ring segments. Only when *Voyager 2* reached Neptune in 1989 did we learn that its rings are indeed complete, and that the **ring arcs** are merely high-density segments within one of its narrow rings. All of Neptune's rings are faint and, with the exception of the ring arcs, they contain too little material to be detected by the stellar occultation technique.

> One of Neptune's rings contains higher-density segments called ring arcs.

Saturn's Magnificent Rings—A Closer Look

All of the giant planets are now known to have ring systems, although each is unique (**Figure 11.21**). Moreover, it turns out that the giant planets are the only planets in our Solar System that have rings; none of the terrestrial planets do. The most complex ring system belongs to what was once thought to be the only ringed planet—spectacular Saturn. Huygens, with his mid–17th century understanding of physics, believed Saturn's ring to be a solid disk surrounding the planet. This view was challenged in later years, but it was not until the middle of the 19th century that the brilliant Scottish mathematician James Clerk Maxwell showed that solid rings would be unstable and would quickly break apart. As we saw in Chapter 10, orbital resonances and other gravitational interactions maintain the stability of these rings.

Figure 11.21b shows the individual components of Saturn's bright ring system and its major divisions and gaps. The most conspicuous features are its expansive bright rings, which dominate all photographs of Saturn (**Figure 11.22**). Among the four giant planets, only Saturn has rings so wide and so bright.

Photographs often show only the two outer and brighter rings, separated by the Cassini Division. The outermost ring, the A Ring, is the narrowest of the three bright rings. It has a sharp outer edge and contains several narrow gaps. The conspicuous Cassini Division is so wide (4,700 km) that the planet Mercury would almost fit within it. Astronomers once thought that it was completely empty. In fact, space scientists planning the 1979 encounter of *Pioneer 11*

with Saturn gave serious consideration to flying the spacecraft through the Cassini Division to get as close to the planet as possible. Had they carried out this plan, the *Pioneer 11* mission would certainly have ended right there. Images taken the following year by *Voyager 1* show the Cassini Division to be filled with material. Why, then, did astronomers think it was empty? The Cassini Division and many other smaller "gaps" in the bright rings are simply regions with less ring material. In contrast with their denser surroundings, they appear darker and thus empty.

> "Divisions" and "gaps" in Saturn's rings are not truly empty.

The B Ring is the brightest of Saturn's rings. With a width of 25,500 km, two Earths could fit side by side between its inner and outer edges. Strangely, the B Ring seems to have no gaps at all, at least on the scale of those seen in the other bright rings. The C Ring often fails to show up in normally exposed photographs because of the limited ability to display a wide range of brightnesses. At the eyepiece of the telescope, though, this beautifully translucent ring appears like delicate gossamer and hence is often called the *Crepe Ring*. There is no known gap between the C Ring and either of the adjacent rings. Only an abrupt change in brightness marks the boundary between them. What could cause such a sharp change in the amount of ring material remains an unanswered question. Too dim to be seen next to Saturn's bright disk, the D Ring is a fourth wide ring that was unknown until imaged by *Voyager 1*. It shows less structure than any of the bright rings, and it does not appear to have a definable inner edge. The D Ring may extend all the way down to the top of Saturn's atmosphere, where its ring particles would enter and burn up as meteors.

Saturn's bright rings are far from homogeneous. The A and C rings contain hundreds, and the B Ring thousands, of individual **ringlets**, some only a few kilometers wide (**Figure 11.23**). Each of these ringlets is a narrowly confined concentration of ring particles bounded on both sides by regions of relatively little material. Spacecraft images of Saturn's rings turned up many other surprises, as we will see later.

> Saturn's rings contain thousands of individual narrow rings called ringlets.

Each time the plane of Saturn's rings lines up with Earth, as it does about every 15 years, the rings all but vanish for a day or so in even the largest telescopes. With the glare of the rings temporarily gone, astronomers can search for undiscovered moons or other faint objects close to Saturn. In 1966, American astronomer Walter A. Feibelman (1930–2004) was looking for moons when he found weak but compelling evidence for a faint ring near the orbit of Saturn's moon Enceladus. In 1980, *Voyager 1* confirmed the existence of this faint ring, now called the E Ring, and found another closer one known as the G Ring. The E Ring and the G Ring are both examples of

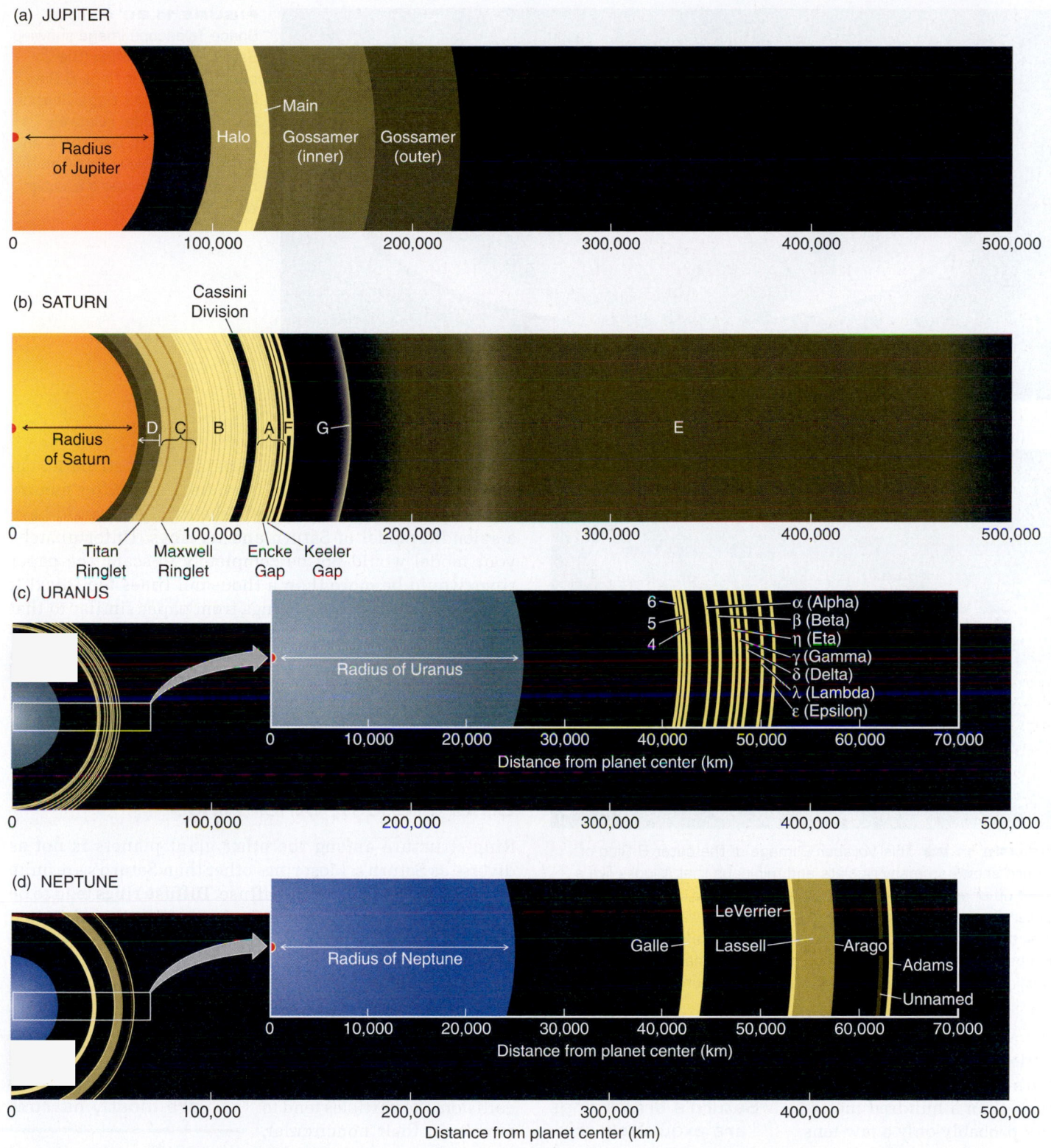

FIGURE 11.21 A comparison of the ring systems of the four giant planets.

what astronomers call a *diffuse ring*, a ring with no distinct boundaries. We'll say more about diffuse rings in the next subsection.

Although Saturn's bright rings are very wide—more than 62,000 km from the inner edge of the C Ring to the outer edge of the A Ring—they are extremely thin. From our previous discussion, you might guess that they could be no thicker than the diameter of the larger ring particles. But in these densely packed rings, there is simply not enough room to jam all of the particles into the same plane, so they

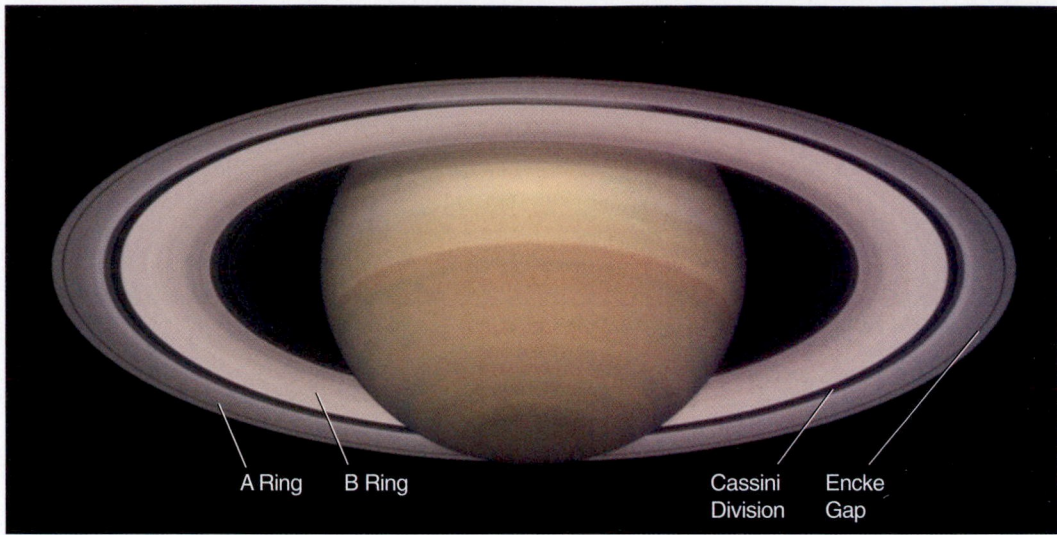

FIGURE 11.22 A Hubble Space Telescope image showing Saturn and its A Ring (the outermost bright ring), B Ring (the middle bright ring), Cassini Division (the wide, relatively dark division between the A and B rings), and Encke Gap (the narrow division near the outer edge of the A Ring)

A Ring   B Ring   Cassini   Encke
                                          Division   Gap

FIGURE 11.23 This *Voyager 2* image of the outer B Ring of Saturn shows so many ringlets and minigaps that it looks like a close-up of an old-fashioned phonograph record. The narrowest features are only 10 km across, at the limit of resolution. Even finer structure was noted during stellar occultations by the rings, as observed by the *Voyager* photometer and imaged later by *Cassini*. The cause of most if this structure is unknown.

the Cassini Division, you could paint a dark stripe 1.5 cm wide and about 12 cm from the outer edge. After mounting the paper ring around the basketball, you would have a splendid model of Saturn and its rings. Unfortunately, your model would not be completely to scale; the paper rings would be more than a thousand times too thick! If you wanted to make your rings from paper similar to that used in this book, the planet would have to be a ball 250 meters in diameter, and the paper rings would have to extend over the length of six football fields! The diameter of Saturn's bright ring system is 10 million times the thickness of the rings.

Other Planets, Other Rings

Ring structure among the other giant planets is not as diverse as Saturn's. Most rings other than Saturn's are quite narrow, although a few are diffuse. **Diffuse rings** tend to be sparsely populated and have characteristics quite different from Saturn's tightly packed bright rings. The separations between particles in diffuse rings are generally very large, and the occasional collisions between particles can cause their individual orbits to become eccentric, inclined, or both. Because it is unlikely that these disturbed orbits will experience a restoring collision, the particles tend to remain in their noncircular, non-coplanar orbits. For this reason, diffuse rings spread out horizontally and thicken vertically, sometimes without any obvious boundaries. These characteristics make diffuse rings difficult to detect at most viewing angles. When seen under *backlit* conditions however, diffuse rings appear to light up because of the strong forward scattering of very small ring particles by sunlight (see Excursions 8.2 and **Excursions 11.1**).

> **Rings around the other giant planets are mostly narrow.**

settle down as close as they can get to the ring plane. Saturn's bright rings are thus no more than a hundred meters and probably only a few tens of meters from their lower to upper surfaces. The extremes between their width and their thickness can be difficult to picture, but let's try.

> **Saturn's bright rings are exquisitely thin.**

Suppose you want to make a scale model of Saturn and its rings using a basketball to represent Saturn. The basketball is about 20 cm in diameter. You could make the three bright rings out of paper by cutting a circle 45 cm in diameter, with a 25-cm hole cut from the center. To represent

EXCURSIONS 11.1

The Backlighting Phenomenon

Diffuse rings contain tiny particles that show up best when we are looking backward toward the Sun. You can witness this same effect by observing a familiar phenomenon here on Earth. Pebbles and boulders are easiest to see when the light illuminating them is coming from behind. Dust and other small particles, however, stand out most strongly when you look *into* the light. For example, notice how dust particles on your windshield appear brightest when you are driving toward the Sun.

Photographers call this effect **backlighting**. A photographer will often place his subject in front of a bright light to highlight her hair (**Figure 11.24**). Under backlit conditions, individual strands of hair shine brightly, creating a halo effect around the person's face. This effect happens when light falls on very small objects—those with dimensions a few times to several dozen times the wavelength of light. Human hair is near the upper end of this range. Light falling on strands of hair is not scattered uniformly in all directions; rather, it tends to continue in the direction away from the source of illumination. Very little of the light is scattered off to the side, and almost none is scattered back toward the source.

Some of the dustier planetary rings are filled with particles that are just a few times larger than the wavelength of visible light. To a spacecraft approaching from the direction of the Sun, such rings may be difficult or even impossible to see. These tiny

FIGURE 11.24
Backlighting of a person's hair creates a halo effect. The light coming from behind the subject is scattered in a direction toward the photographer. This effect is caused by forward scattering of small particles. Typical human hair has a thickness only a few times the wavelength of visible light.

ring particles scatter very little sunlight back toward the Sun and the approaching spacecraft. When the spacecraft passes by the planet and looks backward in the general direction of the Sun, these same dusty rings suddenly appear as a circular blaze of light, much like a halo surrounding the nighttime hemisphere of the planet. Many planetary rings are best seen with backlighting, and some have been observed only under these conditions. Backlighting has also been remarkably successful in revealing tiny ice crystals erupting from cryovolcanoes on Enceladus, as shown dramatically in Figure 11.8b.

When *Voyager* scientists looked at Jupiter's ring system with the Sun behind the camera, all they saw was a narrow, faint strand. But when they looked back toward the Sun while in the shadow of the planet, Jupiter's rings suddenly blazed into prominence. **Figure 11.25a** shows a nearly edge-on, backlit view of Jupiter's rings taken by the *Galileo* spacecraft. Most of the material in Jupiter's rings is made up of fine dust dislodged by meteoritic impacts on the surfaces of Jupiter's small inner moons, shown orbiting among the rings in **Figure 11.25b**.

Nine of the 13 rings of Uranus are very narrow and widely spaced relative to their widths. Most are only a few kilometers wide, but they are many hundreds of kilometers apart (see Figure 11.21c and **Figure 11.26a**). Two rings, discovered by the Hubble Space Telescope in 2005, are much wider and more distant than the narrow rings. With the outermost nearly twice as far from the planet as the outermost of the previously known rings, they constitute what is being called an "outer ring system." The most prominent Uranus ring, the Epsilon Ring, is eccentric and the widest of the planet's inner narrow rings, varying in width between 20 and 100 km. The innermost ring is wide and diffuse, with an undefined inner edge. As with Saturn's D

Ring, material in this ring may be spiraling into the top of the Uranus atmosphere. When viewed under backlit conditions by *Voyager 2* (**Figure 11.26b**), the space *between* the Uranus rings turned out to be filled with dust, much as in Jupiter's ring system.

Four of Neptune's six rings are very narrow, similar to the 11 narrow rings surrounding Uranus. The other two have widths of a few thousand kilometers (see Figure 11.21d). Neptune's rings are named for 19th century astronomers who made major contributions to Neptune's discovery. Among Neptune's six rings, it is Adams Ring that attracts our greatest attention (**Figure 11.27**). Much of the material in the Adams Ring is clumped together into several ring arcs.[3] These ring segments extend over lengths of 4,000–10,000 km yet are only about 15 km wide. When first discovered, the ring arcs were a puzzle, because mutual collisions among their particles should cause them to be spread more or less uniformly around their orbit. Most astronomers now attribute this clumping to orbital resonances (see Chapter 10)

---

[3]The names of the three most prominent ring arcs—Liberté, Égalité, and Fraternité—are taken from the famous motto of the French Revolution.

(a)

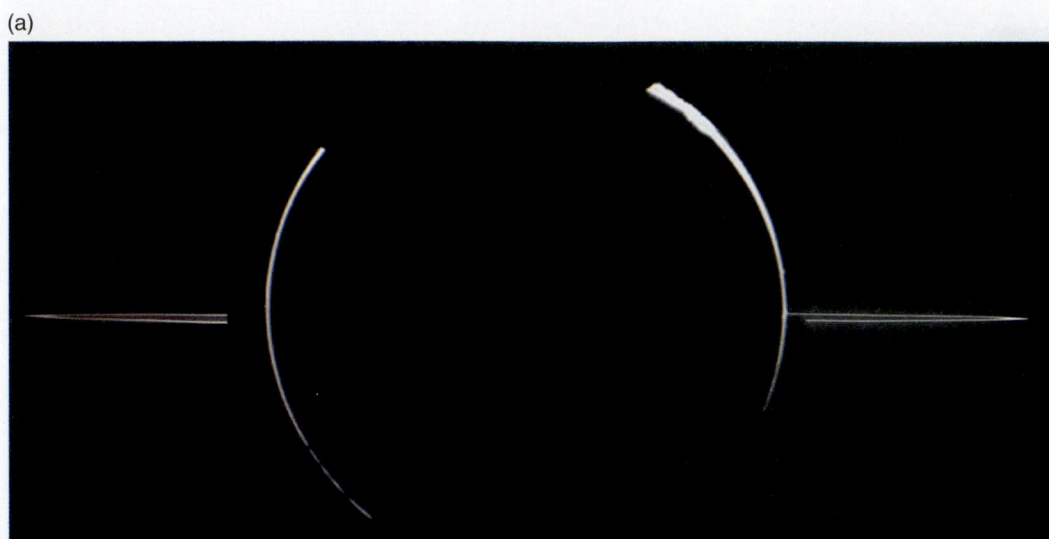

FIGURE 11.25 (a) A backlit *Galileo* view of Jupiter's rings. Note also the forward scattering of sunlight by tiny particles in the upper layers of Jupiter's atmosphere. (b) A diagram of the Jupiter ring system and the small moons that form them.

(b)

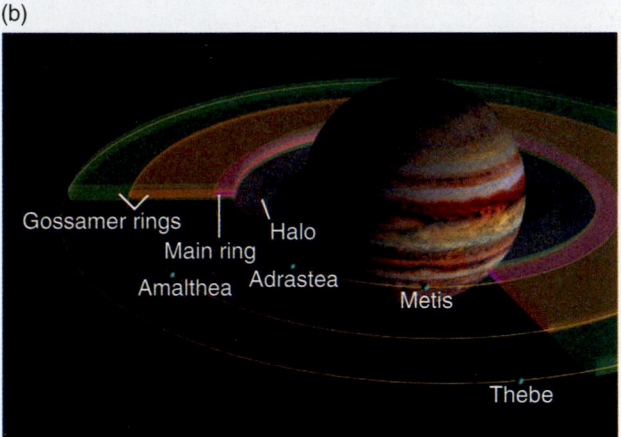

Gossamer rings
Halo
Main ring
Amalthea Adrastea
Metis
Thebe

(a)   (b)

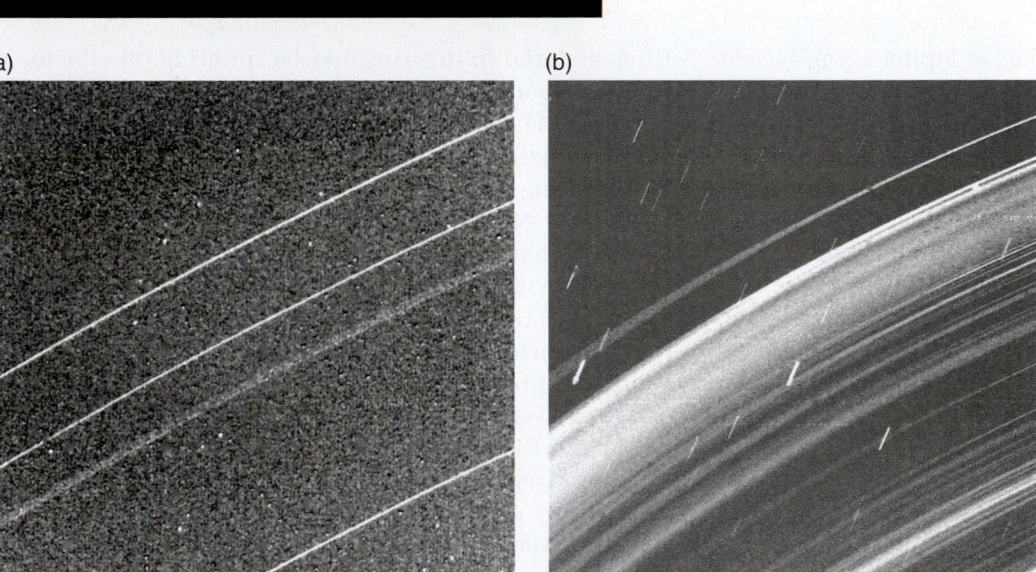

FIGURE 11.26 The appearance of rings depends dramatically on lighting conditions. Here the rings of Uranus appear as narrow, faint bands when viewed with the Sun at our back (a) but burst into dazzling brilliance when illuminated from behind (b). (Bright lines in the images are stars "streaked out" by the spacecraft motion during exposures.)

with the moon, Galatea, that orbits just inside the Adams Ring. Recent groundbased imaging suggests that some parts of the Neptune rings may be unstable. Images obtained early in 2002–2003 show decay in the ring arcs when compared to images taken by *Voyager* in 1989. One of the ring arcs, Liberté, may disappear entirely before the end of the century. The Adams Ring is not unique. Uranus's Lambda Ring and Saturn's G Ring also show ring arcs.

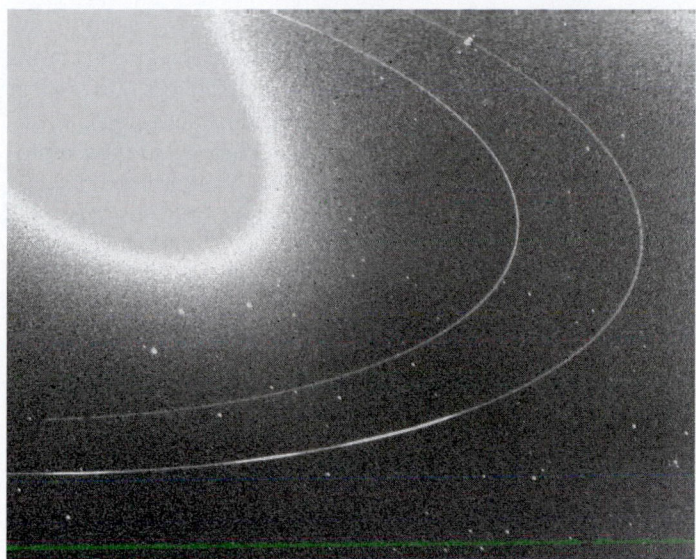

FIGURE 11.27 A *Voyager 2* image of the three brightest arcs in Neptune's Adams Ring. Neptune itself is very much overexposed in this image.

The Composition of Ring Material

As we learned in Chapter 10, some planetary rings are created when a moon or planetesimal comes within a planet's Roche limit and is torn apart by tidal stresses. The fragmented pieces of the disrupted body are then distributed around the planet in the form of a ring. The composition of the ring will, of course, be the same as that of the fated body. Such is the case for Saturn's bright ring system. Saturn's bright rings appear bright because they reflect about 60 percent of the sunlight falling on them. We might suspect from their brightness alone that they are made of water ice, and spectral observations confirm this suspicion, clearly showing the distinct signature of water. A slight reddish tint to the rings tells us that they are not made of pure ice but must be contaminated with other materials, such as silicates. The icy moons around Saturn or the frozen comets that prowl the outer Solar System could easily provide this material.

Saturn's bright rings are the brightest in the Solar System and are the only ones that we know are composed of water ice. In stark contrast, the rings of Uranus and Neptune are among the darkest objects known in our Solar System. Only 2 percent of the sunlight falling on them is reflected back into space, making the ring particles blacker than coal or soot. No silicates or similar rocky materials are this dark.[4] The rings of Uranus and Neptune are likely composed of

Saturn's rings are composed primarily of water ice.

---

[4]A number of carbon-rich meteorites are nearly this dark, as is the organic-rich material that makes up the nucleus of Comet Halley.

organic materials and ices that have been radiation-darkened by high-energy, charged particles in the planets' magnetospheres. (Radiation processing of organic ices, such as methane, blackens the ice by releasing carbon from the ice's molecular structure.) Jupiter's rings are neither as bright as Saturn's nor as dark as those of Uranus and Neptune, suggesting that they may be rich in dark silicate materials, like the innermost of Jupiter's small moons.

The rings of Uranus and Neptune are darker than coal.

The jumble of fragments that make up Saturn's rings is easily understood as a product of tidal disruption of a moon or planetesimal (see Chapter 10), but moons can contribute material to rings in other ways as well, as we see in the case of Jupiter's ring system. The last view that *Voyager* had at Jupiter's rings carried hints of things unseen, and so matters would remain for 20 years. It is ironic that for so many years we knew less about Jupiter's rings than about any other ring system in the Solar System; yet as a result of *Galileo*'s arrival at the planet, Jupiter's ring system is now among the best observed (see Figures 11.21a and 11.25).

The brightest of Jupiter's rings is a narrow strand only 6,500 km across, consisting of material from Metis and Adrastea (see Figure 11.25). These two moons orbit in Jupiter's equatorial plane, and the ring they form is thin. Beyond the main ring, however, are the very different gossamer rings, so called because they are extremely tenuous. The gossamer rings are supplied by dust from the moons Amalthea and Thebe. Unlike the main ring, the gossamer rings are rather thick; the inner gossamer ring, associated with Amalthea, is actually located within the outer gossamer ring formed of material from Thebe. These rings are so thick because the orbits of the moons that supply the ring material are slightly tilted with respect to Jupiter's equatorial plane. The orbital planes of these moons wobble, as does the orbital plane of Earth's Moon; but instead of taking almost 19 years to complete one wobble (as our Moon does), these moons complete a wobble in only a few months. The gossamer rings, made up of material from these wandering moons, are spread as far below and above Jupiter's equatorial plane as the orbits of the satellites that form them.

The innermost ring in Jupiter's system, called the Halo Ring, consists mostly of material from the main ring. As the dust particles in the main ring drift slowly inward toward the planet, they pick up electric charges and are pulled into this rather thick torus, or doughnut-shaped ring, by **electromagnetic forces** associated with Jupiter's powerful magnetic field.

Finally, moons may contribute ring material through volcanism. Volcanoes on Jupiter's moon Io continually eject sulfur particles into space, many of which drift inward under the influence of pressure from sunlight and find their way into the Jupiter ring. The particles in Saturn's E Ring are ice crystals ejected by cryovolcanism from the moon

Enceladus, which is located in the very densest part of the E Ring (see Figure 11.7).

Moons Maintain Order and Create Gaps

In Chapter 10 we discussed the 2:1 orbital resonance between Saturn's moon Mimas and particles in Saturn's rings that creates the Cassini Division. Jupiter creates the Kirkwood gaps in the asteroid belt by this same mechanism. Similarly, we know that one of the gaps in Saturn's C Ring is caused by a 4:1 resonance between the ring particles and Mimas. Unfortunately, the cause of many—perhaps we should say most—other gaps in Saturn's rings remains unexplained. If they also are produced by resonances, we have yet to identify the source. One possibility is that these gaps are the result of collisions between ring particles. Any collision between two ring particles will cause one of the particles to move to an orbit farther out, and the other to an orbit farther in. Over time, this process could sweep some areas clean of ring particles, forming gaps while piling those ring particles up in narrow ringlets between the gaps.

There are also important gravitational interactions between ring particles themselves. These interactions determine the ring shapes at the edges of gaps. In fact, analysis of the shapes of ring edges enables us to estimate the masses of the rings. Even though planetary rings can be large and prominent, they account for only the tiniest fraction of the mass of the material around a giant planet. Saturn's bright rings are by far the most massive rings in the Solar System. In fact, they contain more material than all other planetary rings combined. Even so, their total mass is estimated to be less than

> The mass of all Saturn's rings combined is about the same as the mass of a small icy moon.

that of Mimas, Saturn's small icy moon that is less than 400 km in diameter. The amount of material in the narrow rings is, of course, much less. All of the particles in the largest ring of Uranus, the Epsilon Ring, could be compressed into a single body no more than 20 km across. All of the material in both the Neptune rings and the Jupiter ring could fit into single objects only a few kilometers in diameter.

Strange Things among the Rings

Among the many strange rings imaged by *Voyager*, the archetype is clearly Saturn's F Ring, shown in **Figure 11.28**. Images of the ring taken by *Pioneer 11* a year earlier had shown nothing out of the ordinary, but the first high-resolution images of the F Ring made by *Voyager* had spacecraft scientists staring in disbelief. The ring was separated into several strands that appeared to be intertwined. Some media reporters quickly claimed that the F Ring was "disobeying the laws of physics." This, of course, was not the case; but ready explanations for this seemingly non-Keplerian behavior of the ring particles were not immediately forthcoming. And as if the multiple strands were not perplexing enough, the ring also displayed what appeared to be a number of knots and kinks.

Saturn's F Ring is now understood to be a dramatic example of the action of a pair of shepherd moons (see Chapter 10). The F Ring is flanked by Prometheus (a moon that orbits 860 km inside the ring) and Pandora, a second moon, which orbits 1,490 km on the outside (**Figure 11.29a**). Both

> Rings can be distorted by the gravitational influence of nearby moons.

moons are irregular in shape, with average diameters of 85 and 80 km, respectively. Because of their relatively large size and proximity, they exert significant gravitational forces on nearby ring particles. The resulting tug-of-war between Prometheus pulling ring particles in its vicinity

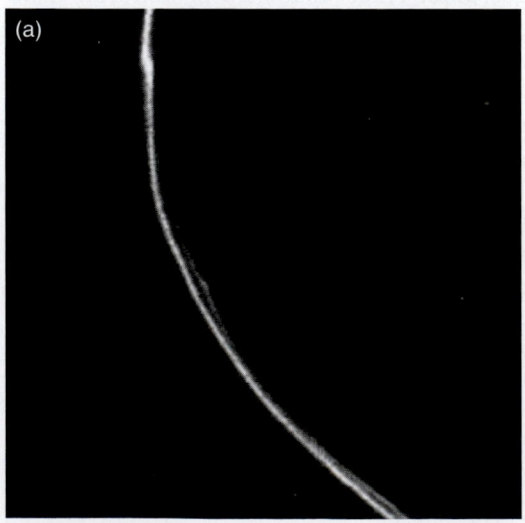
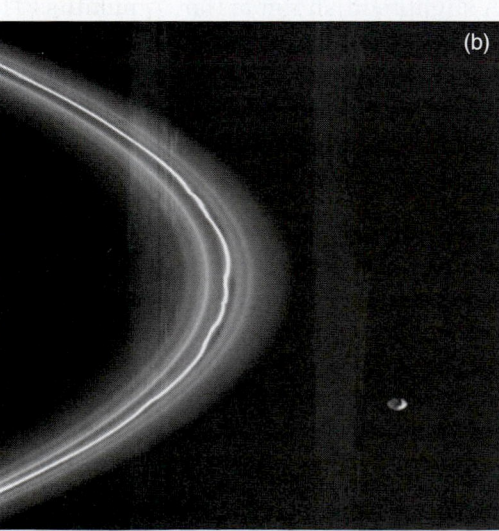

FIGURE 11.28 (a) A *Voyager 1* image of Saturn's "twisted" F Ring. (b) A *Cassini* view of the F Ring and the 80-km-diameter moon Pandora.

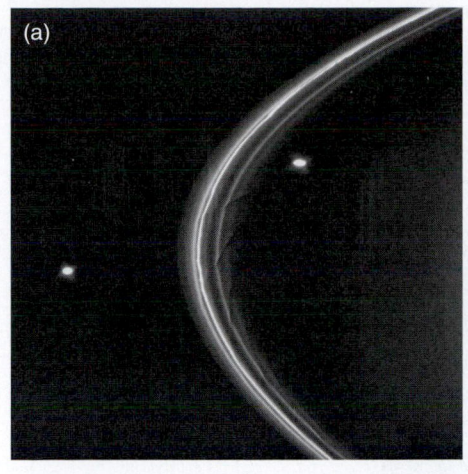

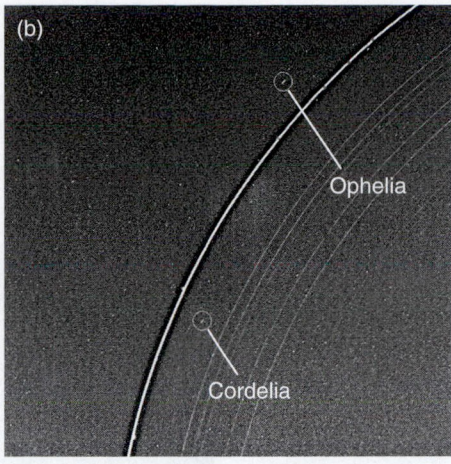

FIGURE 11.29 (a) *Cassini* image of Saturn's F Ring and its shepherd moons: 80-km-diameter Pandora (left) and 85-km Prometheus (right). (b) *Voyager 2* image of Uranus's Epsilon Ring with its 40-km shepherds, Cordelia and Ophelia.

to larger orbits and Pandora drawing its neighbors into smaller orbits is the cause of the bizarre structure that originally baffled scientists and reporters alike.

The F Ring is not an isolated case. The 360-km-wide Encke Gap in the outer part of Saturn's A Ring contains two narrow rings that show knots and gaps (**Figure 11.30a**)—a structure that must be related to a 20-km-diameter moon, named Pan, that orbits within the gap. The rings in the Encke Gap are unusual but not unique, for they bear some resemblance to the arcs in Neptune's Adams Ring.

If shepherd moons, such as those shown in **Figure 11.29b**, are in eccentric or inclined orbits, they cause the

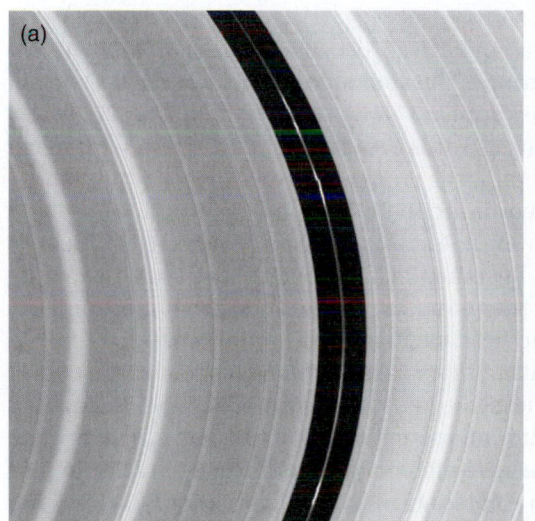

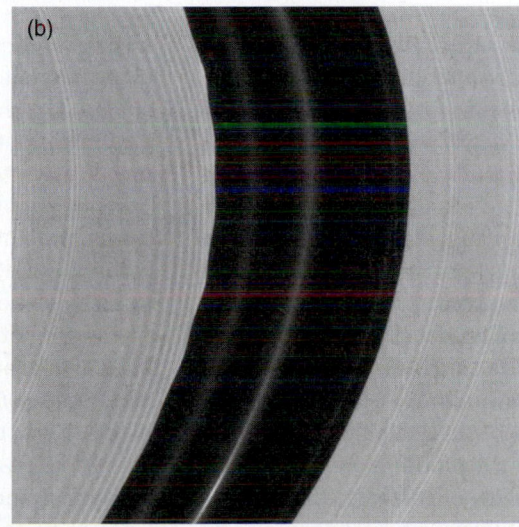

FIGURE 11.30 (a) One of two discontinuous and knotted rings within Saturn's Encke Gap. (b) A higher-resolution view of the Encke Gap reveals a scalloped pattern along the division's inner edge caused by the small moon Pan. (c) The tiny moon Daphnis disrupts both edges of Saturn's Keeler Gap.

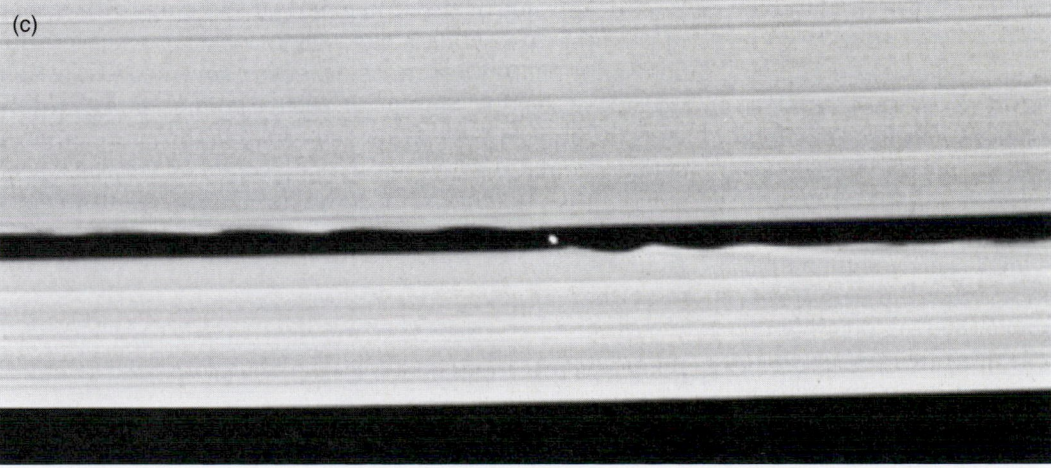

(a)

(b)

FIGURE 11.31 Spokes in Saturn's B Ring appear dark in normal viewing (a) but bright with backlighting (b), indicating that the spoke particles are very small.

confined ring to also be eccentric or inclined, as is the case for Uranus's Epsilon Ring. Because shepherd moons can be so small, they often escape detection. According to current theories of ring dynamics, a number of still unknown shepherd moons must be interspersed among the ring systems of the outer Solar System.

Small moons orbiting within ring gaps can also disturb ring particles along the edges of the gaps. **Figure 11.30b** shows the scalloped pattern, caused by Pan, that is found along the inner edge of the Encke Gap. **Figure 11.30c** catches the 7-km-diameter moon Daphnis in the act of disrupting the inner and outer edges of Saturn's 35-km-wide Keeler Gap, located near the outer edge of the A Ring.

Gravitational interactions do not hold a monopoly on strange behavior among ring systems. One of the more puzzling discoveries made by *Voyager 1* at Saturn was the appearance of dozens of dark spokelike features in the outer part of the B Ring (**Figure 11.31**). As they formed, the **spokes** tended to grow in a more or less radial direction, but they were distorted by "Keplerian shear," which causes the outer tips to rotate more slowly than the inner tips. No single spoke was seen to last for more than half an orbit. This half-orbit survival tells us that the particles in the spokes must be suspended above the ring plane, probably by electrostatic forces. Why is this obvious? Any particle that is not in the ring plane must be in an inclined orbit, and it thus has to pass through the plane twice during each orbit of the planet. As the spoke particles try to pass through the densely packed B Ring, they run into the

> Saturn's B Ring has transient radial features called spokes.

B Ring particles and are absorbed. Such a model can nicely explain what causes the spokes to disappear, but it does not tell us *why* they appear.

One hypothesis links the origin of the spokes with meteoroid impacts on large ring particles. As they strike these particles, meteoroids can collide with so much energy that they create an ionized cloud of tiny charged particles—a plasma—that becomes briefly suspended above the ring plane. Before the cloud can return to the ring plane, Saturn's magnetic field causes the charged particles to drift outward, creating the spokelike features. Still, important questions remain. For example, the spokes seem to form primarily in the "morning" sector of the rings, which tends to argue against the meteorite hypothesis. In addition, our model has no explanation for the fact that *Voyager* and *Cassini* have imaged spokes only in the outer part of the B Ring and not in the inner part of the ring or in either of the other two wide rings. When Voyager observed the spokes in 1980–81, Saturn was near equinox. But when Cassini arrived at Saturn in 2004, 5 years prior to equinox, the spokes were at first not visible. So there is even some speculation that Saturn's ring spokes may be a seasonal phenomenon. As of now, three decades after their discovery, the spokes in Saturn's rings remain as mysterious as ever.

Planetary Rings Are Ephemeral

Planetary rings do not have the long-term stability of most Solar System objects. Ring particles are constantly colliding with one another in their tightly packed environment, either gaining or losing orbital energy as they do so. This redistribution of orbital energy can cause particles at the ring edges to leave the rings and drift away, aided by non-gravitational influences such as the pressure of sunlight. Although various orbital resonances with moons may help guide the orbits of ring particles and delay the dissipation of the rings themselves, at best this can be only a temporary holding action. Most planetary rings eventually face their inevitable fate—total dissipation. At least one ring, however, seems immune from this eventual demise: as we learned earlier, volcanic emissions from Saturn's moon Enceladus are constantly supplying icy particles to Saturn's E Ring, replacing those that drift away. The E Ring will survive for as long as Enceladus remains geologically active.

Rings don't last forever.

It is quite unlikely that any of the planetary rings we see today have existed in their current form since the Solar System's beginning. Indeed, it is far more likely that many ring systems have come and gone over the history of the Solar System. Even our own planet has probably had several short-lived rings at various times during its long history.

Earth's lack of shepherd moons explains why it has no rings.

Any number of comets or asteroids must have passed within Earth's Roche limit[5] and disintegrated catastrophically into a swarm of small fragments, thereby creating a temporary ring. Yet Earth lacks shepherd moons to provide orbital stability to rings. Interactions between ring particles would have caused such a ring to spread out and dissipate, and the inner parts of a ring around Earth would feel the drag of Earth's extended atmosphere and spiral inward, creating spectacular meteor displays as they fell. A similar absence of small inner moons also prevents Venus and Mercury from keeping rings over geological timescales. We leave Mars off the list for now. Although we know of no ring around Mars, its two tiny moons—Phobos and Deimos—could, in principle, serve to shepherd a collection of orbiting debris.

Planetary rings in the outer Solar System will continue to come and go over the eons ahead for as long as the giant planets maintain the small moons that provide temporary ring stability. Whether any will rival the splendor of Saturn's bright ring system that we see today is, of course, unknown. We may be living in a fortuitous time when Saturn is putting on its best face.

---

[5] Earth's Roche limit is about 25,000 km for rocky bodies and more than twice that for icy bodies.

Seeing the Forest for the Trees

We normally think of the Solar System as consisting of the planets we learned about in grade school. Yet there are far more moons in our Solar System than there are planets, and each moon is a unique world in its own right. There was no way in 1610 for Galileo to know that the four points of light he had discovered circling giant Jupiter were the flagships of a vast armada of strange and amazing worlds. The diversity of the solid worlds of the inner Solar System, while remarkable, is nothing compared with the diversity of the worlds that surround Jupiter, Saturn, Uranus, and Neptune. Some of these worlds are frozen remnants of the time long ago when the giant planets formed at the centers of their own swirling disks. Others are testament to chaotic events in which passing objects were captured in gravitational interactions between planets and other moons. At the other extreme are moons such as Io, a world so geologically active that its surface is remaking itself through volcanism as we watch.

These moons are not the uniform worlds, with iron-and-nickel cores and silicate mantles, found close to the Sun. Instead they are assembled from the diverse mix of metals, rocks, and ices that existed in solid form in the outer reaches of the disk that surrounded the newly formed Sun nearly 4.6 billion years ago. As such, these bodies offer important clues about the history of our Solar System, allowing us to test our ideas of what that young Solar System must have been like. These worlds may even give us clues about the history of life in the universe. The fractured ice floes that make up the surface of Europa may well cover a vast, deep ocean, warmed and enriched by geothermal vents not unlike those that dot the floors of Earth's oceans. When scientists finally realized that Mars and Venus were not as Earth-like as once imagined, prospects for finding life elsewhere in the Solar System seemed dim. Now, in the opening decades of the 21st century, we are turning our attention with fresh excitement and hope to these small worlds that circle far from the Sun.

Systems of moons may offer a fascinating future for geologists; but they also represent gravitational playgrounds in which the simple two-body interactions described by Kepler's and Newton's laws give way to complex, chaotic interactions among many different objects. The tidal stresses we learned about in Chapter 10 also rip and pull at these worlds, locking them into synchronous orbits and in some cases churning their interiors into seething cauldrons of geological activity. Sometimes these tidal stresses are so great that a moon or passing comet pays the ultimate price and is ripped into tiny fragments. Born from these violent events are the majestic rings that circle Saturn and the less spectacular but equally fascinating systems of rings surrounding the other giant planets. The extraordinary complexity of these rings is also a testament to Newton's remarkable inverse square law of gravitation as planets, moons, and rings do their gravitational dance. These interactions can even distort rings into structures so remarkable that they seemed all but impossible until revealed by the eyes of the *Voyager* and *Cassini* spacecraft.

It is with sadness that we leave this realm of moons and rings so quickly; these worlds deserve much more attention than we can give them on our journey. As we might do on a package tour promising to show us Europe in 3 days, we look at the marvels through the window, glance at our guidebooks, and promise to return someday when we have more time. Planetary scientists also share our wistful regret. Our glances from passing spacecraft at the moons and rings surrounding the giant planets are really only enough to whet our appetites. No doubt, lifetimes of wonder and surprise remain as we continue to explore the smaller worlds of our Solar System.

There may be more moons than there are planets in the Solar System, but planets, moons, and rings do not complete the accounting of objects adrift within the gravitational realm of the Sun. Still remaining are the dwarf planets and the uncounted swarms of asteroids and comets that orbit within the domain of the planets and also stretch halfway to the nearest star. In the next stage of our journey we will discover that, far from being insignificant, this flotsam and jetsam carries with it the most direct record of the history of our Solar System, providing essential pieces to the puzzle of the origin and history of life on Earth.

Summary

- The moons of the outer Solar System are composed of rock and ice.

- Some moons were formed along with their parent planets; others were captured later.

- A few moons are geologically active; most are dead.

- Jupiter's moon Io is the most volcanically active body in the Solar System.

- Saturn's moon Enceladus is erupting water-ice crystals into space.

- Jupiter's moon Europa may contain an enormous subsurface ocean.

- All four giant planets are surrounded by rings.

- Saturn's ring system is the most complex.

- Some rings may be transient features held in place by moons.

- Planetary rings are ephemeral.

SmartWork, Norton's online homework system, includes algorithmically generated versions of these questions, plus additional conceptual exercises. If your instructor assigns questions in SmartWork, log in at **smartwork.wwnorton.com.**

Student Questions

THINKING ABOUT THE CONCEPTS

1. Which among the planets in our Solar System do not have moons?

2. Explain the difference between regular and irregular moons.

3. Identify the three moons known to be geologically active.

*4. Explain the process that drives volcanism on Jupiter's moon Io.

5. Describe cryovolcanism and explain its similarities and differences with respect to terrestrial volcanism.

6. Which moons show evidence of cryovolcanism?

7. Europa and Titan may both be geologically active. What leads astronomers to suspect such activity?

8. Discuss evidence supporting the idea that Europa might have a subsurface ocean of liquid water.

9. Titan contains abundant amounts of methane. What process destroys methane in this moon's atmosphere?

10. In certain ways, Titan resembles a frigid version of Earth. Explain the similarities.

11. Some moons display signs of geological activity in the past. Identify some of the evidence for past activity.

12. Name three geologically dead moons.

13. Name the planets of the Solar System known to have rings.

14. Describe a groundbased technique that has led to the discovery of rings around the outer planets.

15. What are ring arcs and where are they found?

*16. Identify and explain two possible mechanisms that can produce planetary ring material.

17. Will the particles in Saturn's bright rings eventually stick together to form one solid moon orbiting at the mean distance of all the ring particles? Explain your answer.

18. How does the mass of a planet's rings compare with the mass of its individual moons?

*19. Explain two mechanisms that create gaps in Saturn's bright ring system.

20. Under what lighting conditions are the tiny dust particles found in some planetary rings best observed?

21. Explain ways in which diffuse rings differ from other planetary rings.

22. Describe and explain a mechanism that keeps planetary rings from dissipating.

23. Astronomers believe that most planetary rings eventually dissipate. Explain why they do not last forever.

24. Name one ring that might continue to exist indefinitely, and explain why it could survive when others might not.

25. Why does Earth not have a ring?

APPLYING THE CONCEPTS

26. Io has a mass of 8.9×10^{22} kg and a radius of 1,820 km.
    a. Using the formula provided in Chapter 3, calculate Io's escape velocity.
    b. How does Io's escape velocity compare with the 1-km/s vent velocities from the moon's volcanoes?

27. Enceladus has a mass of 8.4×10^{19} kg and a radius of 250 km. Calculate the minimum speed at which ice crystals from the moon's cryovolcanoes must be traveling in order to escape to Saturn's E Ring.

*28. Planetary scientists have estimated that Io's extensive volcanism could be covering the moon's surface with lava and ash to an average depth of up to 3 millimeters (mm) per year.
    a. Io's radius is 1,815 km. If we model Io as a sphere, what are its surface area and volume?
    b. What is the volume of volcanic material deposited on Io's surface each year?
    c. How many years would it take for volcanism to perform the equivalent of depositing Io's entire volume on its surface?
    d. How many times might Io have "turned inside out" over the age of the Solar System?

29. Particles at the very outer edge of Saturn's A Ring are in a 7:6 orbital resonance with the moon Janus. If the orbital period of Janus is 16 hours 41 minutes (16^h41^m), what is the orbital period of the outer edge of Ring A?

*30. The inner and outer diameters of Saturn's B Ring are 184,000 and 235,000 km, respectively. If the average thickness of the ring is 10 meters and the average density is 150 kilograms per cubic meter (kg/m^3), what is the mass of Saturn's B Ring?

31. The mass of Saturn's small icy moon Mimas is 3.8×10^{19} kg. How does this mass compare with the mass of Saturn's B Ring, as calculated in the previous question? Why is this comparison meaningful?

StudySpace is a free and open Web site that provides a Study Plan for each chapter of *21st Century Astronomy*. Study Plans include animations, reading outlines, vocabulary flashcards, and multiple-choice quizzes, plus links to premium content in SmartWork and the ebook. Visit **www.wwnorton.com/studyspace**.

Almost in the center of it, above the Prechistenka Boulevard, surrounded and sprinkled on all sides by stars ... shone the enormous and brilliant comet of 1812—the comet which was said to portend all kinds of woes and the end of the world.

LEO TOLSTOY (1828–1910), *WAR AND PEACE*

Comet Hyakutake approached close to Earth in 1996.

Dwarf Planets and Small Solar System Bodies

12.1 Leftover Material: From the Small to the Tiniest

We began our discussion of the Solar System in Chapter 6 with the story of its history. We learned that very early on—at the same time our Sun was becoming a star—tiny grains of primitive material were sticking together to produce swarms of small bodies called **planetesimals**. Those that formed in the hotter, inner part of the Solar System were composed mostly of rock and metal, while those in the colder, outer parts were made up of ice, organic compounds, and rock. We also learned what happened to many of these planetesimals. Some were consumed during the era of large-body building to become planets and moons, and others were ejected from the Solar System by gravitational encounters with larger bodies. However, not all of these primitive bodies suffered the same fate. Many of them are still around, and they remain a small, but scientifically important, component of our present-day Solar System. As we now conclude our discussion of the Solar System, we find that we have come full circle. These remaining planetesimals, and the fragments they continually create, provide planetologists with the opportunity to look back to the physical and chemical conditions of the earliest moments in the history of the Solar System.

Among the larger surviving planetesimals are five bodies known as **dwarf planets** (Table 12.1). Four of them—the former planet Pluto and the more recently discovered Eris, Haumea, and Makemake—are an icy-organic-rocky type of planetesimal. All four reside in the frigid region beyond the orbit of Neptune. The fifth, the former *asteroid* Ceres, orbits within the *main asteroid belt* between Mars

KEY CONCEPTS

Dwarf planets are among the lesser worlds in the Solar System. Asteroids and comets are even smaller; yet these objects, and their fragments that fall to Earth as meteorites, have told us much of what we know about the early history of the Solar System. As we explore these pieces of interplanetary flotsam and jetsam, we will discover

- The littlest of planetary bodies, which we call dwarfs.
- Small, irregular worlds called asteroids that are made of rock and metal.
- Different types of meteorites that are fragments of these varieties of asteroids.
- That some asteroids are primitive while others are differentiated or are pieces of larger, differentiated bodies.
- That most asteroids orbit between Mars and Jupiter, but some have orbits that cross Earth's.
- Comet nuclei—pristine, icy planetesimals—adrift in the frozen outer reaches of the Solar System.
- Spectacular active comets that are warmed by the Sun as they dive through the inner Solar System.
- World-jarring meteorite impacts that still occur today and have played a vital role in shaping the history of life on Earth.
- Meteor showers that occur when Earth passes through a comet's trail of debris.

TABLE 12.1

Physical Properties of the Dwarf Planets

	Ceres	Pluto	Haumea	Makemake	Eris
Orbital radius (AU)	2.8	39.5	43.1	45.7	67.7
Orbital period (years)	4.6	248.1	283.3	309.9	557
Orbital velocity (km/s)	17.9	4.7	4.5	4.4	3.4
Mass ($M_\oplus = 1$)	0.00016	0.0021	0.00067	~0.0007	0.0028
Equatorial diameter (km)	975	2,306	~1,400	~1,500	~2,400
Equatorial diameter ($D_\oplus = 1$)	0.07	0.18	0.11	~0.12	0.19
Density (water = 1)	2.1	2.0	~3	~2	~2
Rotation period (hours)	9.1	153.3	3.9	7.8	>8?
Obliquity (degrees)	~3	119.6	?	?	?
Surface gravity (m/s^2)	0.27	0.81	0.44	~0.5	~0.6
Escape speed (km/s)	0.51	1.27	0.84	~0.8	~1.3

and Jupiter. As such, it is a member of the rock-metal group of planetesimals.

Smaller than dwarf planets, yet respectable in size, are the rocky **asteroids** and the icy **comet nuclei** (the nuclei of *comets* that orbit beyond Neptune and have yet to come into the inner Solar System). These remote icy planetesimals are known as **Kuiper Belt objects** (**KBOs**) or sometimes as "trans-Neptunian objects" (TNOs.)

At the very smallest scale are *meteoroids* and interplanetary dust—the widespread debris created by collisions among asteroids and the disintegration of comet nuclei by solar heating.

12.2 Dwarf Planets: Pluto and Others

Pluto has been an enigma since its discovery. The story begins with what appeared to be discrepancies between the observed and predicted orbital positions of Uranus and Neptune throughout the 19th century. In the early 20th century, astronomers began a search for the unseen body that they believed was responsible for perturbing the orbits of Uranus and Neptune. They called it Planet X and estimated that it had six times Earth's mass and was located somewhere beyond Neptune's orbit. Planet X was finally found by the American astronomer Clyde W. Tombaugh (1906–1997) in 1930, not far from its predicted position. It became the Solar System's ninth planet and was named Pluto for the Roman god of the underworld. In the years that followed, however, observational evidence began to indicate that the mass of Pluto was far too small to have produced the presumed perturbations in the orbits of Uranus and Neptune. When astronomers reanalyzed the 19th century observations, they found that the orbital "discrepancies" were in fact erroneous. Pluto's discovery thus turned out to be a strange and improbable coincidence based on faulty data.

Pluto's largest moon, Charon, was first observed in 1978. By applying Kepler's and Newton's laws to observations of the moon's motion around Pluto, astronomers were able to accurately "weigh" the Pluto-Charon system. Its total mass turned out to be only about 1/400 that of Earth (**Figure 12.1**). Pluto is only two-thirds as large as the Moon and only twice as big as its largest moon, Charon. Because of the relative similarity in size between Pluto and Charon, we might think of the Pluto-Charon system as a "double planet." Moreover, Pluto and Charon are a dynamically locked pair—the only known example in the Solar System. The two are in synchronous rotation with one another, so each has one hemisphere that always faces the other body and another that never sees the other body (**Figure 12.2**). If immobile inhabitants were living on opposite sides of Pluto and Charon, they would never know of each other's existence. In 2005, two smaller moons were found orbiting Pluto. All three of its moons appear

> Pluto has a moon half as big as it is.

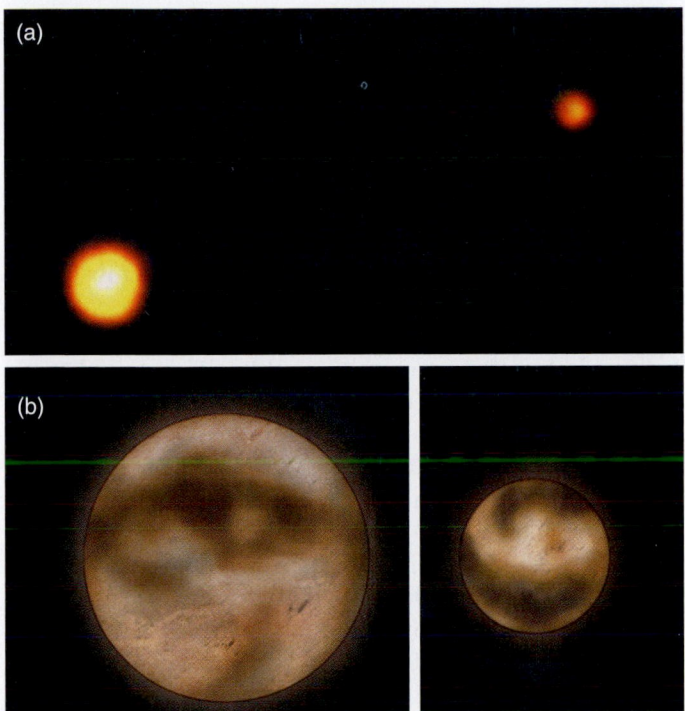

FIGURE 12.1 The dwarf planet Pluto and its moon Charon in a Hubble Space Telescope image (a) and an artist's rendition based on analysis of HST images (b). Pluto and Charon are so similar in size they could be considered a "double planet."

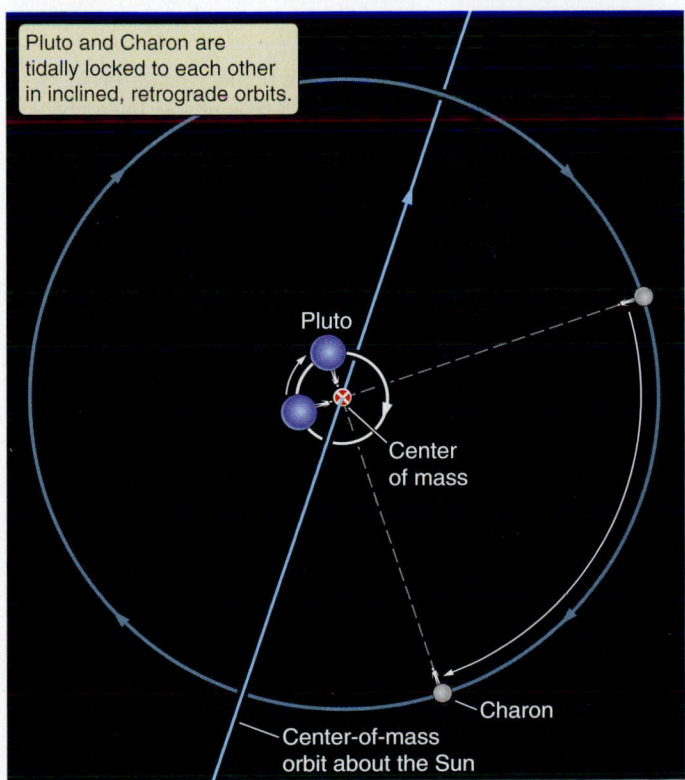

Pluto and Charon are tidally locked to each other in inclined, retrograde orbits.

Pluto

Center of mass

Charon

Center-of-mass orbit about the Sun

FIGURE 12.2 A diagram showing the doubly synchronous rotation and revolution in the Pluto-Charon system. The two bodies permanently face one another.

to be in orbital resonance with one another. As with Uranus, the plane of Pluto's equator is tipped at almost right angles to its orbital plane.

Pluto's eccentric orbit periodically brings it inside Neptune's orbit, which is nearly circular, and from 1989 to 1999 Pluto was closer to the Sun than Neptune was. More than 248 Earth years are required for Pluto to complete one orbit. Since its discovery, Pluto has traveled less than a third of its

> **Pluto is a mixture of rock and ice and has a thin methane atmosphere.**

way around the Sun. With densities nearly twice that of water, both Pluto and Charon probably consist of a rocky core that makes up about 70 percent of their mass, surrounded by a water-ice mantle. Because of their great distance from Earth, Pluto and Charon were not included in the great program of spacecraft exploration that took place late in the 20th century. Therefore, we have limited information about their surface properties, and we know nothing about their geological history. Spectra obtained with ground-based telescopes and the Hubble Space Telescope tell us that Pluto's surface contains an icy mixture of frozen water, carbon dioxide, nitrogen, methane, and carbon monoxide. Unlike Pluto, Charon's surface seems to be made up primarily of dirty water ice. Pluto has a temporary, low-density atmosphere composed mostly of nitrogen, methane, and carbon monoxide. As Pluto moves farther from the Sun and temperatures drop, its atmosphere will gradually freeze and fall to the ground as snow. We should soon know much more about Pluto and Charon. In 2006, NASA launched a planetary spacecraft called *New Horizons* that will reach the "double planet" in 2015.

As more was learned about Pluto and other bodies beyond Neptune's orbit, some astronomers called into question Pluto's status as a planet, and a heated debate ensued. The discovery of a distant body, later named Eris, added fuel to the already fiery argument. Eris (**Figure 12.3**) was found to be even larger and more massive than Pluto, but it was similar to Pluto in other respects. Each is attended by a relatively large moon, although Pluto has two other smaller moons. Both Pluto and Eris have frozen methane on their surfaces. At this point, the inevitable question emerged: should astronomers consider Eris to be the Solar System's tenth planet—or should neither Pluto nor Eris be called planets?

The International Astronomical Union (IAU) made its decision in August 2006. "To be a planet, a body must be massive enough to (1) pull itself into a round shape and (2) be able to clear the neighborhood around it." (Details of the IAU resolution can be found in Appendix 8.) Pluto *is* round like the classical planets, but it is *not* able to clear its neighborhood. With that simple pronouncement, Pluto lost its status as the "ninth planet." Although the IAU is considered by professional astronomers to be the final authority on such matters, not all were happy with its decision, and the question of Pluto's classification remains controversial. We should make it very clear, however, that regardless of the label that the IAU has put on it, Pluto is in no way a less

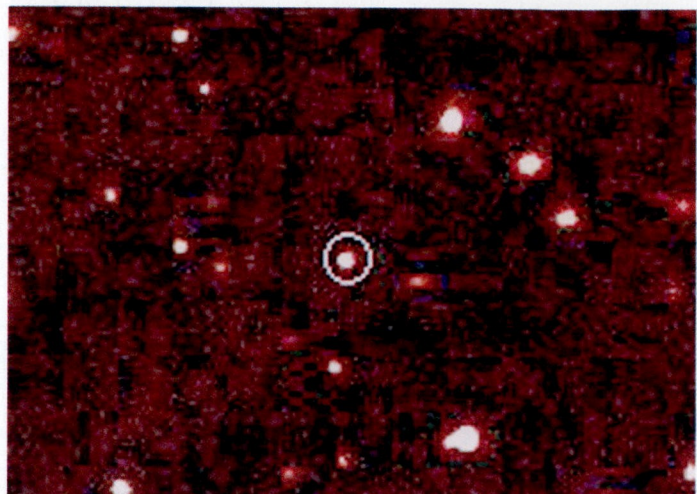

FIGURE 12.3 The distant dwarf planet Eris (shown within the white circle) is slightly larger than Pluto and has similar physical characteristics.

interesting body to the scientific community, as is evidenced by NASA's funding of the *New Horizons* mission.

What, then, can we say about Eris, Haumea, and Makemake? Observations by the Hubble Space Telescope confirmed earlier estimates that the diameter of Eris is approximately 2,600 kilometers (km), making it about 13 percent larger than Pluto. In 2005, a year before the IAU decision, astronomers had discovered Dysnomia, Eris's only known moon. As observations of Charon did for Pluto, observations of Dysnomia's orbital motion around Eris yielded an accurate mass for Eris, which turned out to be about 28 percent greater than Pluto's mass. Eris thus has sufficient mass to pull itself into a round shape, but not enough mass to clear its surroundings. Along with Pluto, Eris was designated as a dwarf planet.

Its highly eccentric orbit carries Eris from 37.8 astronomical units (AU)—its closest point to the Sun—out to 97.6 AU, with an orbital period of 557 years. By chance, Eris was found near the most distant point in its orbit, making it currently the most remote known object in the Solar System.[1] Because of its great distance at this time, Eris now appears about 100 times fainter than Pluto. When we combine the observed brightness of Eris with its diameter, we arrive at an albedo of 0.86. This is an astonishing result. With the single exception of Saturn's moon Enceladus, the surface of Eris is more highly reflecting than that of any other major Solar System body. Like Enceladus, Eris must have a coating of pristine ice. The surface of Enceladus is water ice, while Eris is covered with methane ice. At its

> **Dwarf planet Eris is currently the most distant known Solar System object.**

present location, the average surface temperature on Eris is about 43 kelvins (K), easily cold enough to freeze out any atmospheric methane. As is now the case for Pluto, Eris will probably develop a methane atmosphere when it comes closest to the Sun, in the year 2257.

The most recent KBOs to be classified as dwarf planets are Haumea and Makemake. Both are smaller than Pluto and both orbit slightly farther from the Sun than does Pluto (see Table 12.1). Haumea has two moons of its own—Hi'iaka and Namaka—enabling astronomers to calculate the system's mass as they did for Pluto and Eris. Although Haumea has sufficient mass to pull itself into a spherical shape (one of the IAU requirements for classification as a dwarf planet), it spins so rapidly on its axis that its shape is distorted into a flattened oval shape, with an equatorial radius that is approximately twice its polar radius. As we learned in Chapter 9, this difference between the equatorial and polar radii would give Haumea an **oblateness** of 0.5, by far the largest of any classical planet or dwarf planet. No moons have been discovered orbiting around Makemake, so we know less about this dwarf planet than about Pluto, Haumea, or Eris.

On New Year's Day 1801, Sicilian astronomer Giuseppe Piazzi found a bright object between the orbits of Mars and Jupiter. Piazzi named the new object Ceres. When Piazzi discovered Ceres, he thought he might have found a hypothetical "missing planet." But as

> **Dwarf planet Ceres was previously the largest asteroid.**

FIGURE 12.4 Ceres, once known as the largest asteroid, is now called a dwarf planet. Note its spherical shape, one of the criteria that gives Ceres its dwarf planet status. The nature of the bright spot is unknown, but it reveals the rotation of Ceres as seen in this set of four images taken by HST over an interval of 2 hours and 20 minutes. Time increases from left to right and top to bottom.

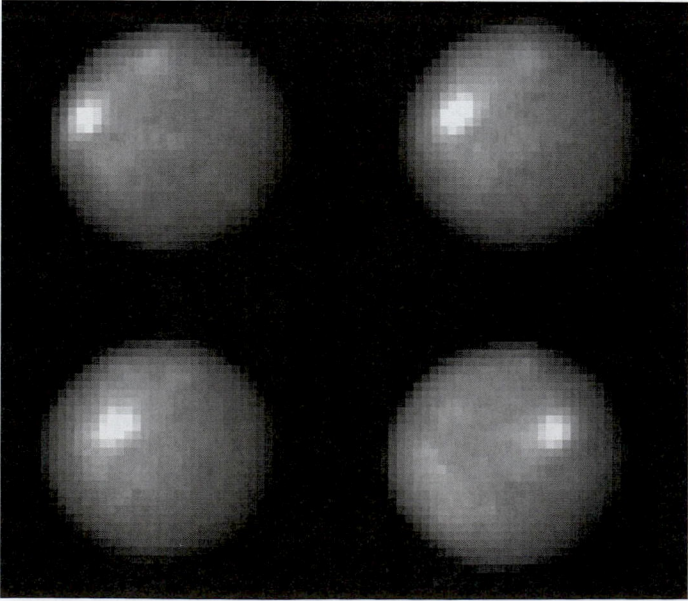

---

[1] Eris's remote status is temporary. Other Solar System bodies, such as the Kuiper Belt object Sedna, have eccentric orbits that will eventually carry them as far as a thousand astronomical units from the Sun.

more objects were discovered orbiting the region between Mars and Jupiter, astronomers classified Ceres as a new kind of Solar System object, which they called asteroids.

With a diameter of about 975 km, Ceres (**Figure 12.4**) is larger than most moons but smaller than any planet. It contains about a third of the total mass in the asteroid belt, but only about 4 percent of the mass of our Moon. Ceres rotates on its axis with a period of about 9 hours, typical of many asteroids. As a large planetesimal, Ceres seems to have survived largely intact, although there are indications that it underwent differentiation at some point in its early history. Spectra show that the surface of Ceres contains hydrated minerals such as clays and carbonates, indicating the presence of significant amounts of water in its interior. Perhaps as much as a quarter of its mass exists in the form of a water-ice mantle, which surrounds a rocky inner core.

For more than two centuries, Ceres would be known as the largest asteroid—but no longer. Curiously, the same IAU resolution that demoted Pluto from planet to dwarf planet has raised the status of Ceres from asteroid to dwarf planet—Ceres is round, but it is unable to clear its neighborhood. NASA's *Dawn* mission, launched in 2007, will explore a large asteroid named Vesta in 2011–12 and then go on to Ceres in 2015.

12.3 Ghostly Apparitions and Rocks Falling from the Sky

Like an ocean filled with flotsam and jetsam, our Solar System swarms with smaller members that might escape our attention until one of them washes ashore. It would be a mistake to assume that this planetary debris is unimportant, however. Comets, asteroids, and *meteorites* provide astronomers with some of their most important clues about how our Solar System formed and evolved.

Throughout human history, the sudden and unexpected appearance of a bright **comet** has elicited fear, wonder, and superstition. Early cultures viewed these spectacular visitors as omens. Comets were often seen as dire warnings of disease, destruction, and death, but sometimes as portents of victory in battle or as heavenly messengers announcing the impending birth of a great leader. The earliest records of comets date from as long ago as the 23rd century B.C. Mediterranean and Far Eastern literatures are especially full of references and popular superstitions about comets.

Meteorites have also engendered fascination and awe for as long as humankind has watched the heavens. Nearly all ancient cultures venerated these rocks from the sky. Early Egyptians preserved meteorites along with the remains of their pharaohs, Japanese placed them in Shinto shrines, and ancient Greeks worshipped them. In 1492—the same year that Columbus voyaged to the New World—the townsfolk of Ensisheim (now in France) watched as a 172-kilogram (kg) stone fell to Earth. A woodcut artist recorded the famous event (**Figure 12.5**), and for a while the meteorite was enshrined in a local church. Despite numerous eyewitness accounts of meteorite falls, however, many people were slow to accept that these peculiar rocks actually came from far beyond Earth. In 1807, two Yale professors investigating the meteorite that landed in Weston, Connecticut, concluded that the object truly had dropped from the sky. Thomas Jefferson, third president of the United States and an enlightened scientist, is rumored to have responded, "I would rather believe that two Yankee professors would lie, than that stones fall from heaven." But as the evidence continued to mount, it became impossible to ignore. By the early 1800s, scientists had documented so many meteorite falls that their true origin was indisputable. Today, hardly a year passes without a recorded meteorite fall, including some that have smashed into cars, houses, pets, and (occasionally) people (see Chapter 7).

A third type of interplanetary debris was not discovered until the age of telescopes, when Piazzi found Ceres in 1801. Subsequently, a number of similar objects were discovered in the region between the orbits of Mars and Jupiter. Because

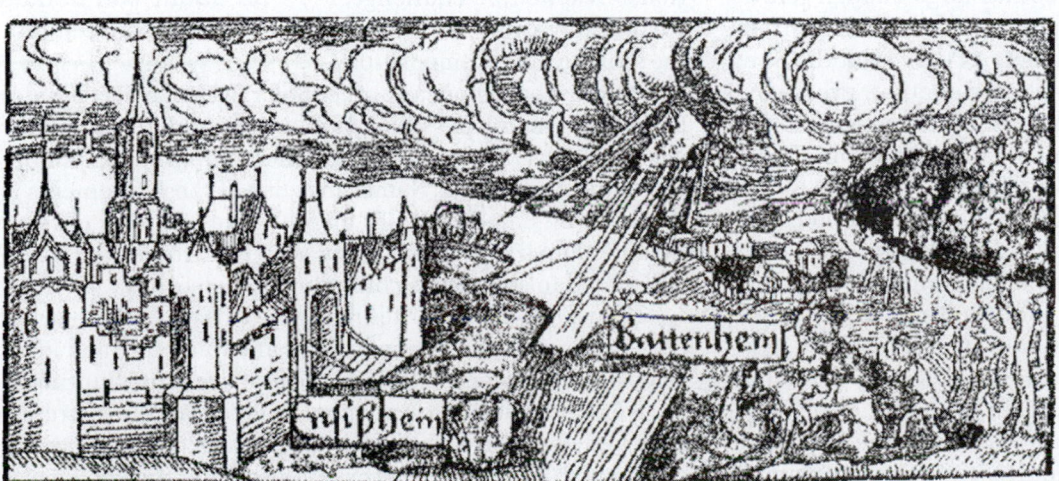

FIGURE 12.5 A woodcut of the meteorite fall at Ensisheim in 1492. More than three centuries would pass before scientists were convinced that such objects come from beyond Earth.

these new objects appeared in astronomers' eyepieces as nothing more than faint points of light, William and Caroline Herschel (the brother-sister pair of English astronomers who discovered Uranus) named them *asteroids*, a Greek word meaning "starlike." As the years went by, more asteroids were discovered, and by the first decade of the 21st century we have found nearly 450,000 of them, of which about 200,000 have well-determined orbits. Discoveries continue at the rate of nearly 4,000 per month.

Apart from their role as fascinating curiosities and reminders that there is more to heaven and Earth than meets the eye, these types of inter-planetary debris play a very significant role in our quest to discover humankind's cosmic origins. We now know that meteorites, asteroids, and comets are fragments of our Solar System's ancient past. Like archaeologists who assemble the history of civilizations from fossil bones and shards of pottery, planetary scientists work to piece together the story of our Solar System from these relics of a time when the Sun, Earth, and planets were young. We begin by describing what the completed puzzle looks like and then talk about the pieces and how they fit together.

> Comets, asteroids, and meteorites provide scientific clues to our past.

up into tiny clumps of dusty ice. When Earth passes through swarms of dusty ice from a disintegrating comet nucleus or pebbles from the breakup of an asteroid, the particles, called **meteoroids**, can burn up in our atmosphere, creating an atmospheric phenomenon we refer to as a **meteor**. Excursions 12.1 describes and differentiates among meteoroids, meteors, meteorites, and comets.

> Comet nuclei are remnants of icy planetesimals that formed far from the Sun.

Planetesimals that became part of the planets or their moons were so severely modified by planetary processes that nearly all information about their original physical condition and chemical composition has been hopelessly lost. By contrast, asteroids and comet nuclei constitute an ancient and far more pristine record of what the early Solar System was like. If you visit your local planetarium or science museum, you may find a meteorite on display that you can walk right up to and touch. Do so with respect. The meteorite under your hand may be older than Earth itself. Some meteorites contain tiny grains of material that pre-date the formation of our Solar System. These grains include diamonds and carbon compounds that originated as material ejected from dying stars.

12.4 Asteroids and Comets: Pieces of the Past

As we learned in Chapter 6, when the planet-building process finally ended, some of the original planetesimals remained in orbit around the Sun. Most asteroids are relics of rocky or metallic planetesimals that originated in the region between the orbits of Mars and Jupiter. Although early collisions between these planetesimals created several bodies large enough to differentiate, Jupiter's tidal disruption prevented them from forming a single Moon-sized planet. As they orbit the Sun, asteroids continue to collide with one another, producing small fragments of rock and metal. Most meteorites are pieces of these asteroidal fragments that have found their way to Earth and crashed to its surface.

> Asteroids are relics of rocky planetesimals that formed between the orbits of Mars and Jupiter.

Icy planetesimals surviving today form the solid cores, or "nuclei," of comets. Although asteroids can survive in the inner Solar System, comet nuclei cannot. Today most comet nuclei are preserved in the frigid regions of space beyond the planets. Sometimes, however, something happens to cause one of these icy bodies to change direction and dive inward toward the domain of the inner planets. When a comet nucleus approaches too close to the Sun, solar heating unglues its surface structure, gradually breaking it

12.5 Meteorites: A Chip Off the Old Asteroid Block

Most asteroids are so small compared with planets that they appear as nothing but unresolved points of light through telescopes on Earth. (They reveal themselves by their motion among the fixed stars.) Historically, their small size has made it difficult to learn much about them directly. Even determining gross properties such as size, shape, and rate of spin has required sophisticated analysis of telescopic, spacecraft, and radar observations. Despite this challenge, we probably know more about the structure and composition

> Meteorites are pieces of asteroids. They tell us about the bodies they came from.

of asteroids than about any other Solar System object besides Earth! How is this possible? As asteroids orbit the Sun, they occasionally collide with each other, chipping off smaller rocks and bits of dust. Sometimes one of these fragments is captured by Earth's gravity and survives its fiery descent through Earth's atmosphere as a meteor, to be picked up and added to someone's collection of meteorites.

Thousands of meteorites reach the surface of Earth every day, but only a tiny fraction of these are ever found and identified as such. Antarctica offers the best meteorite hunting in the world. Meteorites are no more likely to fall in Antarctica than anywhere else, but in Antarctica they are far easier to distinguish from their surroundings. Antarctic snowfields

EXCURSIONS 12.1

Meteors, Meteoroids, Meteorites, and Comets

Meteors are familiar to all of us. Just stand outside for a few minutes on a starry night, away from bright city lights, and you will almost certainly be rewarded by a glimpse of a "falling star," a meteor. Most of us realize that we are not witnessing the demise of a real star but are instead seeing a piece of Solar System debris entering our atmosphere. Few are aware that what we see is seldom the actual debris, but rather a trail of heated atmospheric gas glowing from the friction caused by a small unseen object entering at speeds of up to about 70 kilometers per second (km/s). A meteor, then, is an "atmospheric phenomenon."

The small[2] solid body that creates the meteor is called a **meteoroid**. These small objects are of either cometary or asteroidal origin. Cometary fragments are typically less than a centimeter across and have about the same density as cigarette ash. What they lack in size and mass, they make up for in speed.

[2]By definition, meteoroids are objects whose size lies within a range of 100 micrometers (μm) to 100 meters. Particles smaller than this are considered "interplanetary dust"; larger objects are called planetesimals. Some astronomers place a meteoroid's upper size limit at 10 meters.

A 1-gram meteoroid (about half the mass of a dime) entering Earth's atmosphere at 50 km/s has a kinetic energy comparable to that of an automobile cruising along at the fastest highway speeds. Before plunging into our atmosphere, the meteoroid may have been orbiting the Sun for millions of years after being chipped from an asteroid or left behind by a disintegrating comet nucleus. Most meteoroids are so small and fragile that they burn up completely before reaching Earth's surface.

A meteoroid large enough to survive all the way to the ground is called a **meteorite**. All meteoroids are pieces of Solar System debris. The larger pieces that survive the plunge through Earth's atmosphere are usually fragments of asteroids. Meteorites likely litter the ground of all solid planets and moons. Most of the smaller pieces that burn up in the atmosphere before reaching the ground are cometary fragments.

Comets appear as fuzzy patches of light in the nighttime sky. They are bodies of ice, dust, and gas that shine by reflected sunlight. Comets frequently remain visible in our skies for weeks at a time, with tails that can be 100 million km long and stretch halfway across our sky. In contrast, a meteor may streak across 100 km of our atmosphere and last at most a few seconds.

---

serve as a huge net for collecting meteorites. Glaciers then carry the meteorites along, concentrating them in regions where the glacial ice is eroded by wind. The meteorites are left lying on the surface for collectors to pick up. The big advantage of hunting on Antarctic ice is that in many places the *only* stones to be found on the ice are meteorites. Antarctic meteorites have also spent their time on Earth literally in the deep freeze; many are thus relatively well preserved, showing little weathering or contamination from terrestrial dust or organic compounds.

Meteorites are extremely valuable because they are samples of the same relatively pristine material from which asteroids are made. In addition, unlike planets or asteroids themselves, we can take meteorites into the laboratory and study them using sophisticated equipment and techniques. Scientists compare them to rocks found on Earth and the Moon and contrast their structure and chemical makeup with those of rocks studied by spacecraft that have landed on Mars and Venus. Meteorites are also compared with asteroids and other objects on the basis of what colors of sunlight they reflect and absorb.

Meteorites are normally grouped into three categories based on the kinds of materials they are made of and the degree of differentiation they experienced within their parent bodies. Over 90 percent of meteorites are **stony meteorites (Figures 12.6a and b)**, which are similar to terrestrial

silicate rocks. A stony meteorite can be recognized by the thin coating of melted rock that forms as it passes through the atmosphere. Many stony meteorites contain small round spherules called **chondrules** (see Figure 12.6a), which range in size from sand grains to marble-sized objects. Laboratory experiments simulating formation show that chondrules formed at higher temperatures than the finer-grained minerals that surround them. Chondrules must once have been molten droplets that rapidly cooled to form the crystallized spheres we see today.

> Meteorites tell us the age of the Solar System.

Chondrules are all held together by a matrix of finer-grained material, much as the gravel in concrete is held together by a matrix of cement. Stony meteorites containing chondrules are called **chondrites** (see Figure 12.6a). One type of chondrite stands out with particular interest: **carbonaceous chondrites**, chondrites that are rich in carbon, are thought to be the very building blocks of the Solar System. Although these meteorites cannot be dated directly, because they are composed of aggregates of many individual grains, indirect measurements suggest that they are about 4.56 billion years old—our best measurement of the time that has passed since the formation of the Solar System. By contrast, the stony meteorites that do not contain chondrules are known as **achondrites** (see Figure 12.6b). Unlike chondrites, achon-

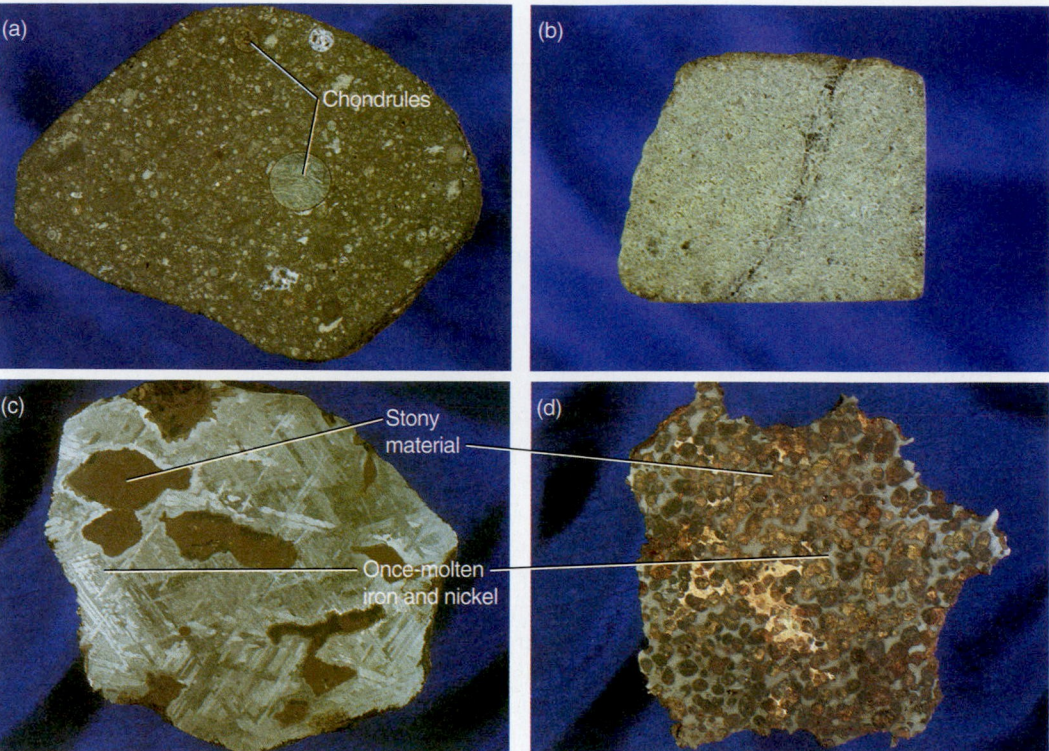

FIGURE 12.6 Cross sections of several kinds of meteorites: (a) a chondrite (a stony meteorite with chondrules), (b) an achondrite (a stony meteorite without chondrules), (c) an iron meteorite, and (d) a stony-iron meteorite.

drites have crystals that appear to have formed in the same place and at the same time.

The second major category of meteorites, **iron meteorites** (**Figure 12.6c**), is the easiest to recognize. The surface of an iron meteorite has a melted and pitted appearance generated by frictional heating as it streaked through the atmosphere. (The meteorite in your local museum is probably an iron meteorite.) Even so, many are never found, either because they land in water or simply because no one happens to notice and recognize them for what they are. As an interesting side note, one of the more pleasant surprises from the Mars rover *Opportunity* was the discovery of an iron meteorite lying in plain sight on the martian surface (**Figure 12.7**). Both its appearance—typical of iron meteorites found on Earth—and its position on the smooth, featureless plains made it instantly recognizable. Chemical analysis from the rover instruments showed it to be composed mostly of iron and nickel. The final category is the **stony-iron meteorites** (**Figure 12.6d**), which consist of a mixture of rocky material and iron-nickel alloys. Stony-iron meteorites are relatively rare.

> There are three classes of meteorites: stony (including chondrites and achondrites), iron, and stony-iron.

Meteorites come from asteroids, which in turn derive from rocky-iron planetesimals. Refer to **Figure 12.8** as we follow the possible fates of a planetesimal. Recall from Chapter 6 that, during the growth of the terrestrial planets, large amounts of thermal energy were released as larger planetesimals accreted smaller objects. Additional thermal energy was released as radioactive elements inside the planetesimals decayed. Some planetesimals never reached the temperatures needed to melt their interiors. These planetesimals—aggregations of the primordial material from which they formed—simply cooled off, and they have since remained pretty much as they were when they formed. These planetesimals are known as **C-type asteroids**, which are composed of primitive material that we believe has largely been unmodified since the origin of the Solar System, almost 4.6 billion years ago.

> C-type asteroids never differentiated.

In contrast to the C-type asteroids, some planetesimals were heated enough by impacts and radioactive decay to cause them to melt and differentiate, with denser matter such as iron sinking to the center of the planetesimal. Lower-density material—such as compounds of calcium, silicon, and oxygen—floated toward the surface and combined to form a mantle and crust of silicate rocks. **S-type asteroids** (and the stony achondritic meteorites that come from them) may be pieces of the mantles and crusts of such differentiated planetesimals. They are chemically more similar to igneous rocks found on Earth than to primitive chondrites and C-type asteroids. That is, they were hot enough at some point to lose their carbon compounds and other volatile materials to space. The large interlocking

> S- and M-type asteroids may be pieces of larger differentiated bodies.

FIGURE 12.7 A basketball-sized iron meteorite lying on the surface of Mars, imaged by the Mars exploration rover *Opportunity*.

crystals that make up achondrites are also like those seen in rocks that slowly cooled from molten material. Similarly, **M-type asteroids** (from which iron meteorites come) are fragments of the iron- and nickel-rich cores of one or more differentiated planetesimals that shattered into small pieces during collisions with other planetesimals. Indeed, slabs cut from iron meteorites show large, interlocking crystals characteristic of iron that cooled very slowly from molten metal. The rare stony-iron meteorites may come from the transition zone between the stony mantle and the metallic core of such a planetesimal.

A few planetesimals in the region between the orbits of Mars and Jupiter seem to have evolved toward becoming tiny planets before being shattered by collisions. They even became volcanically active, complete with eruption of lava flows onto their surfaces. Vesta (now the largest known asteroid, with a diameter of 525 km) seems to be a piece of such a highly evolved planetesimal. Vesta does not fit well into any of the main asteroid groups, but its spectrum is strikingly similar to the reflection spectrum of a peculiar group of meteorites. These meteorites, which are probably pieces of Vesta, look like rocks taken from iron-rich lava flows on Earth and the Moon. The main difference is that they are *only* 4.4 billion to 4.5 billion years old, a bit younger than most meteorites. It is hard to escape the conclusion that early in the history of the Solar System, some planetesimals were large enough to become differentiated and to develop volcanism. But rather than going on to form planets, these planetesimals broke into pieces in collisions with other planetesimals.

The story of how planetesimals, asteroids, and meteorites are related is one of the great successes of planetary science. A vast wealth of information about this diverse collection of objects has come together into a single con-

FIGURE 12.8 The fate of a rocky planetesimal in the young Solar System depends on whether it gets large and hot enough to melt and differentiate, as well as on the impacts it experiences. Different histories lead to the different types of asteroids and meteorites found today.

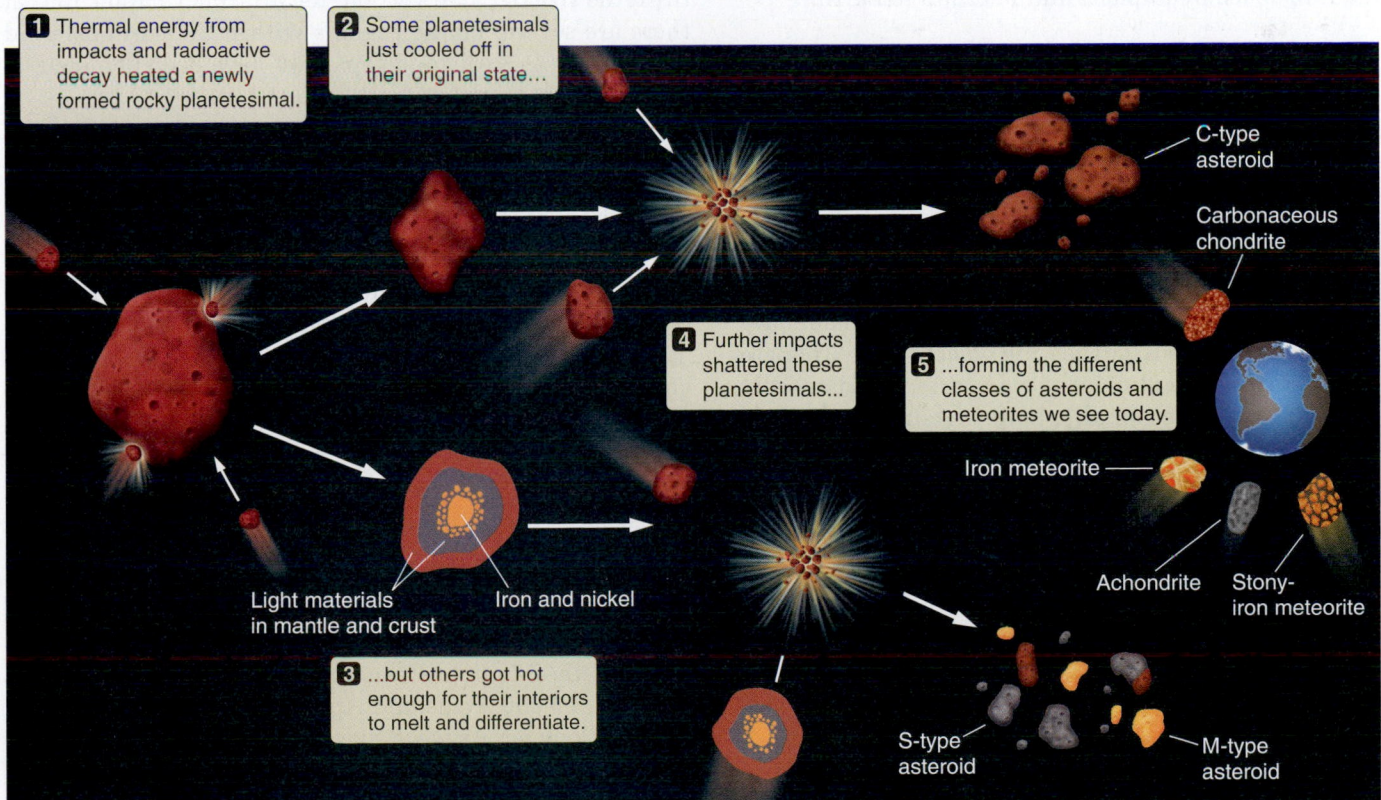

sistent story of planetesimals growing and possibly becoming differentiated, and then being shattered by subsequent collisions. The story is even more satisfying because it, in turn, fits so well with the even grander story of how most planetesimals were accreted into the planets and their moons.

As we stress throughout this book, patterns are extremely important in science—not only when they are followed, but also when they are broken. Some types of meteorites fail to follow the patterns just discussed. Whereas most achondrites have ages in the range

> To hold certain meteorites is to hold a piece of Mars.

of 4.5 billion to 4.6 billion years, some members of one group are less than 1.3 billion years old. Some are chemically and physically similar to the soil and the atmospheric gases that our lander instruments have measured on Mars. The similarities are so strong that most planetary scientists believe these meteorites are pieces of Mars that were knocked into space by asteroidal impacts.[3] This means we have pieces of another planet we can study in laboratories here on Earth. The martian meteorites support the general belief that much of the surface of Mars is covered with iron-rich volcanic materials.

If pieces of Mars have reached Earth from its orbit almost 80 million km beyond our own, we might expect that pieces of our companion Moon would find their way to Earth as well. Indeed, meteorites of another group bear striking similarities to samples returned from the Moon. Like the meteorites from Mars, these are chunks of the Moon that were blasted into space by impacts and later fell to Earth.

12.6 Asteroids Are Fractured Rock

With a few exceptions, asteroids are too small for their self-gravity to have pulled them into a spherical shape. Some asteroids imaged by spacecraft or by Earth-based radar have highly elongated shapes, suggesting objects that either are fragments of larger bodies or were created haphazardly from collisions between smaller bodies. Astronomers have measured the masses of a number of asteroids by noting the effect of their gravity on spacecraft passing nearby. Knowing the mass and the size of an asteroid enables us to determine its density. The densities of these asteroids range between 1.3 and 3.5 times the density of water. Those at the lower end of this range are considerably less dense than the meteorite fragments they create. How can

asteroids be less dense than the rock that likely came from them? Planetologists think that some of them are shattered heaps of rubble, with large voids between the fragments. Once again, this is what we would expect of objects that were assembled from smaller objects and then suffered a history of violent collisions.

Asteroids rotate just as planets and moons do, although the rotation of irregularly shaped asteroids is more of a tumble than a spin. A day on a typical asteroid is about 9 hours long, although the rotation periods of some asteroids are as short as 2 hours, and

> A day on a typical asteroid is only 9 hours long.

others are longer than 40 Earth days. How can we measure the rotation period of an object that appears starlike in a telescope? Unless the asteroid is perfectly round, and we know that very few are, we measure rotation periods by watching changes in their brightness as they alternately present their broad and narrow faces to Earth.

Asteroids are found throughout the Solar System. Most orbit the Sun in several distinct zones, with the majority residing between the orbits of Mars and Jupiter in the **main asteroid belt**. The main belt contains at least 1,000 objects larger than 30 km in diameter (about the size of Washington DC), of which about 200 are larger than 100 km. We estimate that there may be as many as 10 million asteroids larger than 1 km in the main asteroid belt. However, although there are a great number of asteroids, they account for only a tiny fraction of the matter in the Solar System. If all of the asteroids were combined into a single body, it would be about a third the size of Earth's Moon. Recall from Chapter 10 that there are several empty regions within the main asteroid belt, known as Kirkwood gaps, that have been depleted by perturbations from Jupiter.

Figure 12.9 shows orbits of several asteroids that are representative of their classes. In Chapter 10 we discussed the Trojan asteroids that occupy the L_4 and L_5 Lagrangian points of the Sun-Jupiter system. Asteroids whose orbits bring them close to Earth's orbit are called **near-Earth asteroids**. Members of one group, called the **Amors**, cross the orbit of Mars but do not cross Earth's orbit. Members of two other groups, the **Atens** and the **Apollos**, have orbits that cross Earth's orbit (see Figure 12.9). The difference between these two populations is that the Apollos also cross the orbit of Mars, while the Atens remain inside Mars's orbit. Atens and Apollos, along with a few degenerate comet nuclei, are known collectively as **near-Earth objects** (**NEOs**) and occasionally collide with

> Apollo and Aten asteroids cross Earth's orbit and could collide with us one day.

Earth or the Moon. Astronomers estimate that there are more than 3,500 Atens and Apollos with diameters larger than a kilometer. As we discovered in Chapter 7, collisions between Earth or the Moon and such objects are geologically important and some have dramatically altered the history of life on Earth.

[3]In 1996 a NASA research team announced that a particular martian meteorite (ALH84001) showed physical and chemical evidence of early life on Mars. These conclusions have since been challenged by many scientists and are now generally discredited.

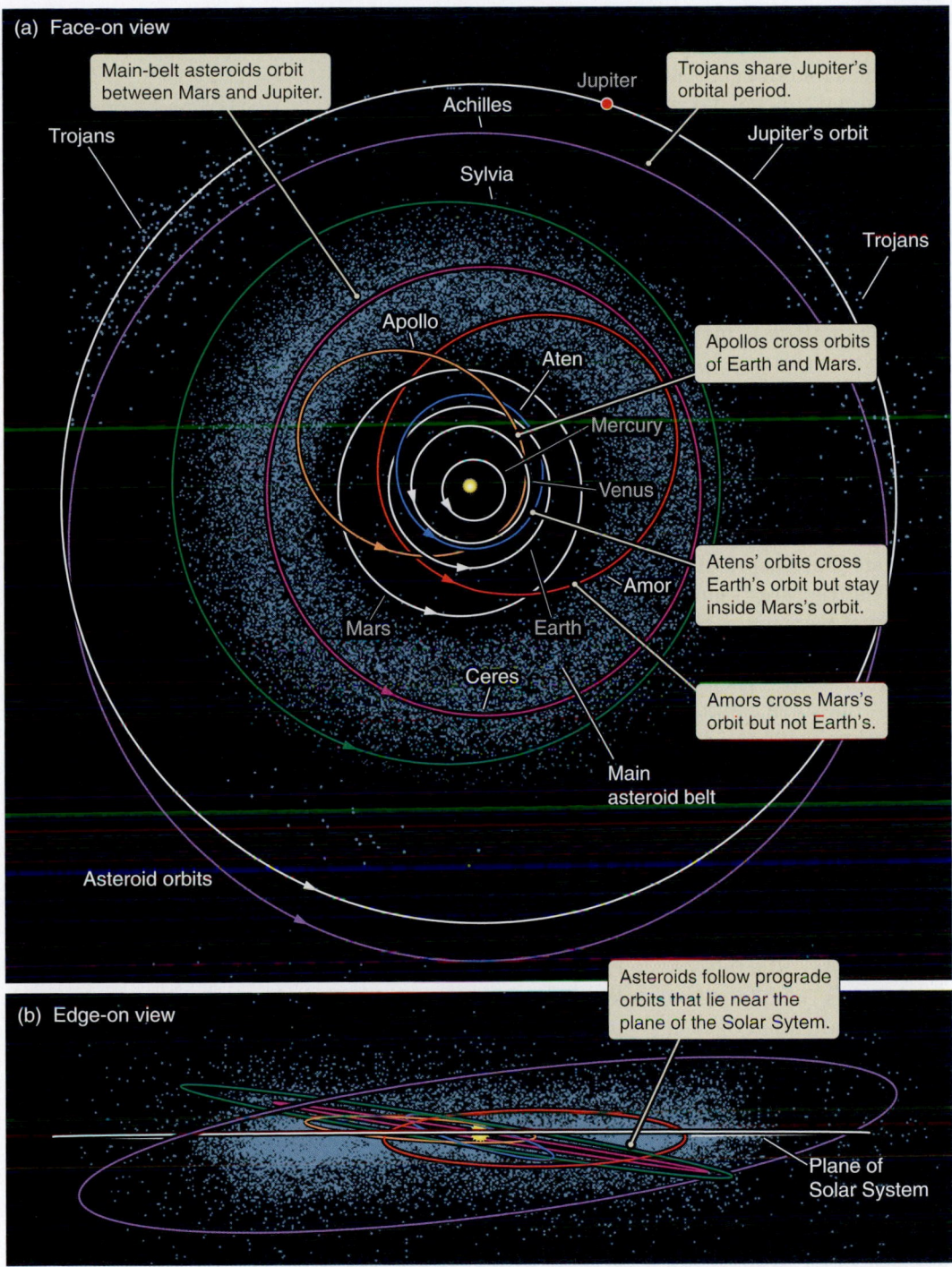

(a) Face-on view

Main-belt asteroids orbit between Mars and Jupiter.

Trojans

Jupiter

Achilles

Trojans share Jupiter's orbital period.

Sylvia

Jupiter's orbit

Trojans

Apollo

Aten

Apollos cross orbits of Earth and Mars.

Mercury

Venus

Mars

Earth

Amor

Atens' orbits cross Earth's orbit but stay inside Mars's orbit.

Ceres

Amors cross Mars's orbit but not Earth's.

Main asteroid belt

Asteroid orbits

Asteroids follow prograde orbits that lie near the plane of the Solar Sytem.

(b) Edge-on view

Plane of Solar System

FIGURE 12.9 Blue dots show the locations of known asteroids at a single point in time. Most families of asteroids take their names from prototype members of the family. For example, Apollo-family asteroids cross the orbits of Earth and Mars, like the asteroid Apollo. Most asteroids, such as Sylvia, are main-belt asteroids. Achilles was the first Trojan asteroid to be discovered.

12.7 Asteroids Viewed Up Close

Until the space age, scientists had no good idea of what asteroids looked like. As the 1980s drew to a close, planetary scientists were about to have their theories put to the test. Would the first close-up views of asteroids pull the rug out from under our carefully constructed story of asteroids, meteorites, and the early Solar System? (If so, it would not be the first time a closer examination of a particular phe-

nomenon revealed that people's ideas about it were wrong!) Fortunately, as spacecraft began to return detailed information about the asteroids they visited, the data they collected placed our ideas about the history of the Solar System on even more solid footing.

In 1991 the *Galileo* spacecraft was following a circuitous path through the inner Solar System, gaining momentum from encounters with Earth and Venus to help send it on its journey to Jupiter. While flying through the main asteroid belt, it got close enough to the asteroid Gaspra to obtain dozens of good images. Gaspra is an S-type asteroid, about

$9 \times 10 \times 20$ km. The *Galileo* images show it to be cratered and irregular in shape. Faint, groovelike patterns on its surface may be fractures resulting from the same impact that chipped Gaspra from a larger planetesimal. Images taken in several different wavelengths of light show distinctive colors on Gaspra's surface, implying that Gaspra is covered with a variety of different types of rock.

Later in its mission, *Galileo* made another pass through the main belt, and this time it sent back images of the asteroid Ida (**Figure 12.10**). Ida orbits in the outer part of the main asteroid belt, and it, too, is an S-type asteroid. *Galileo* flew so close to Ida that it could see details as small as 10 meters across—about the size of a small house or cottage. Ida is shaped like a croissant, 54 km long and ranging from 15 to 24 km in diameter. Like Gaspra, Ida shows the scars of a long history of impacts with smaller bodies. Just as planetary scientists use the number and sizes of craters to estimate the ages of the surfaces of planets and moons (as we saw in Chapter 7), the cratering on Ida indicates that its surface is a billion years old, twice the age estimated for Gaspra. Also like Gaspra, Ida contains fractures. The fractures seen in the two asteroids indicate that they must be made of relatively solid rock. (You can't "crack" a loose pile of rubble.) This is an important confirmation of the idea—crucial to our story of asteroids and meteorites—that some asteroids must be pieces chipped from larger, solid objects.

The *Galileo* images also revealed a tiny moon orbiting the asteroid. Ida's moon, called Dactyl, is only 1.4 km across, and like nearly all Solar System objects with solid surfaces, it is cratered from impacts. Although planetary scientists had long speculated that asteroids might themselves have moons, Dactyl was the first such object to be discovered. But is Dactyl really a moon? To planetary scientists, any significant Solar System body orbiting a larger body other than the Sun is called a moon. Dactyl may not be the stuff of poetry, but by the planetary scientist's definition of a moon, it counts. As it turns out, Dactyl is hardly unique. Moons have now been found around nearly 100 asteroids, and at least three asteroids are known to have two moons.

> **Ida's feeble gravity holds a tiny moon called Dactyl.**

Since *Galileo*'s visits to Gaspra and Ida, spacecraft have visited four other asteroids: Mathilde, Eros, Braille, and Itokawa. Braille was seen only briefly, in 1999 during the flyby of *Deep Space 1*. It was found to have a highly reflective surface, similar to Vesta's, leading to speculation that Braille could be a chunk knocked from Vesta by a collision sometime in the past.

The *NEAR Shoemaker*[4] spacecraft was designed to do more than make a brief encounter with an asteroid—in

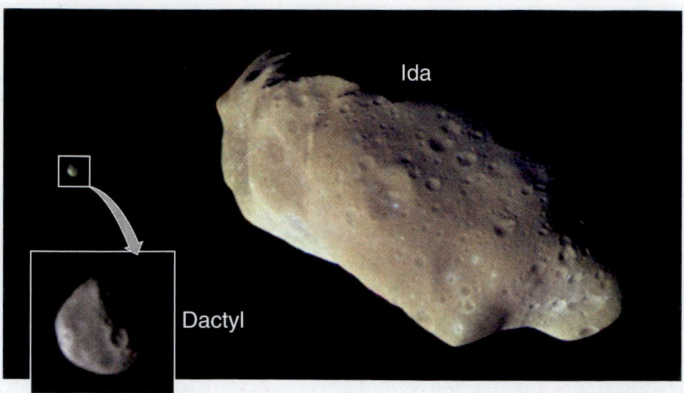

FIGURE 12.10 A *Galileo* spacecraft image of the asteroid Ida with its tiny moon, Dactyl (also shown enlarged in the inset).

early 2000 it was placed into orbit around the asteroid Eros to begin long-term observations of this S-type object (**Figure 12.11**). As a near-Earth asteroid, Eros is one of more than 5,000 known objects whose orbits bring them within 1.3 AU of the Sun. Eros measures about $34 \times 11 \times 11$ km and, like its counterparts Gaspra and Ida, shows a surface with grooves, rubble, and impact craters, including a crater 8.5 km across. However, the scarcity of smaller craters suggests that its surface could be younger than Ida's. A year after being placed in orbit, the spacecraft was gently crash-landed onto the asteroid's surface. Chemical analyses confirmed the similarity in composition between Eros and primitive meteorites.

On its way to Eros, *NEAR Shoemaker* flew past the asteroid Mathilde and provided our first information about a C-type body. Mathilde was found to measure $66 \times 48 \times 46$ km and to have a surface only half as reflective as charcoal. Its color properties suggest a composition of materials such as carbonaceous chondrules. However, estimates of the overall density of Mathilde are about 1,300 kilograms per cubic meter, or kg/m³ (1.3 times the density of water), only half that of carbonaceous chondrite meteorites measured in the laboratory. This density estimate implies that Mathilde is porous inside, as we would expect if it is composed of chunks of rocky material stuck together by the accretion process with open spaces between the chunks. The surface of Mathilde is dominated by craters, the largest of which is more than 33 km across. The great number of craters suggests that Mathilde likely dates back to the very early history of the Solar System.

The spectrum of sunlight reflected by Mars's tiny moons, Phobos and Deimos (**Figure 12.12**), is similar to that of C-type asteroids; and many scientists believe these moons must be asteroid-like objects that were somehow captured by Mars. But are Phobos and Deimos really asteroids? Controversy about this question has never been put to rest. Some scientists argue that it is unlikely

> **Phobos and Deimos may be asteroids captured by Mars.**

---

[4]*NEAR*, the original name, stands for *Near Earth Asteroid Rendezvous*. The spacecraft was renamed *NEAR Shoemaker* in March 2000 in honor of Gene Shoemaker, a pioneer of the study of comets and asteroids who died in an automobile accident in 1997 while studying impact craters in the Australian outback.

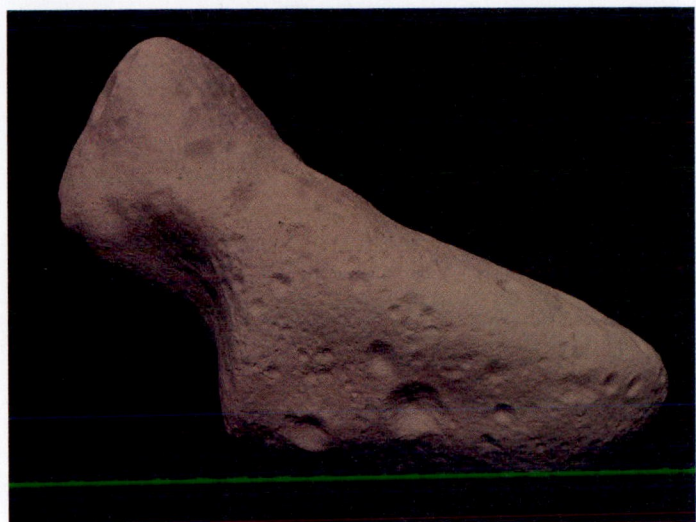

FIGURE 12.11 Asteroid Eros viewed by the *NEAR Shoemaker* spacecraft. This image was produced by high-resolution scanning of the asteroid's surface by the spacecraft's laser range finder. *NEAR Shoemaker* became the first spacecraft to land on an asteroid, when in 2001 the orbiter was gently crashed onto the surface of Eros.

12.8 The Comets: Clumps of Ice

Close observations and more refined analyses have redefined our view of asteroids. Likewise, our concept of comets changed markedly during the 16th and 17th centuries, when we looked upon nature with an increasingly refined and rational eye. In much earlier times comets were regarded simply as ghostly apparitions, pale luminous patches or streaks of light in the nighttime sky that would mysteriously appear, remain for a few days or weeks, and then vanish. Until the end of the Middle Ages, the Aristotelian view of comets prevailed, regarding them as atmospheric phenomena rather than astronomical objects. The popular beliefs that comets were very nearby and that their tails contained poisonous vapors contributed to the fear and panic that comets frequently generated.

A major change in our view came in the 16th century, thanks to Tycho Brahe. Tycho reasoned that if comets were atmospheric phenomena like clouds, as Aristotle had supposed, then their appearance and location in the sky should be very different to observers located many miles away

> Comets are astronomical objects, *not* atmospheric phenomena.

from each other. But when Tycho compared sightings of comets made by observers at several different sites, he found no evidence of such differences, and he concluded that comets must be at least as far away as the Moon. Lest we become arrogant in our faith in humankind's rationality, it is worth noting that the proof that comets are far outside Earth's atmosphere did not completely dispel the fear they evoke. As recently as 1910, the news that Earth would pass through the tail of Comet Halley (popularly known as Halley's Comet) was accompanied by widespread apprehension throughout Europe and the United States. And during the approach of Comets Kohoutek in 1973 and Hale-Bopp in

Mars could have "captured" two asteroids and that Phobos and Deimos must somehow have evolved together with Mars. A third possibility is that the bodies were parts of a much larger body that was fragmented by a collision early in the history of Mars.

Once called the "vermin of the skies," asteroids have now achieved considerable respectability in scientific circles. Not only do they hold fascinating and unique clues about how the Solar System formed, but they also play a part in the continuing (albeit slow) evolution of the Solar System. And if an asteroid were to get *too* close to Earth, as happens from time to time, it would quickly become a *very* respectable topic for conversation!

(a)

(b)

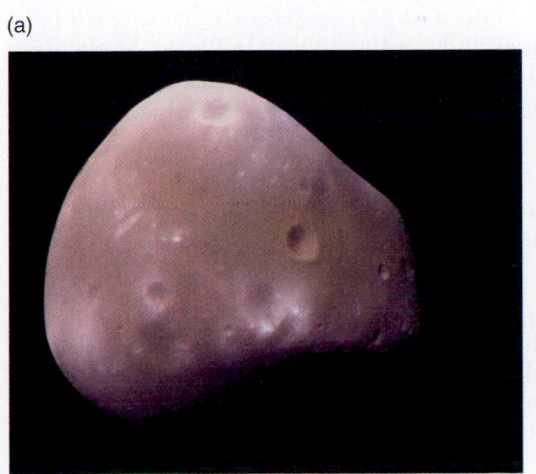

FIGURE 12.12 *Mars Reconnaissance* orbiter and *Mars Express* images of Mars's two tiny moons: (a) Deimos and (b) Phobos.

1997,[5] many otherwise rational people experienced similar anxieties, regrettably fueled by certain opportunistic writers who exploit human fear and superstition for power or profit.

The Abode of the Comets

When we think of a comet, we think of a spectacular display of the sort produced by Comet Hyakutake in 1996, as seen in the opening photograph of this chapter. But for most of their lives, comets are merely icy planetesimals—comet nuclei formed from primordial material that managed

> Swarms of comets spend most of their lives far beyond the planets.

to escape the planet-building process—adrift in the frigid outer reaches of the Solar System. Drifting in these distant regions, most are much too small and far away to be seen and counted by telescopes on Earth, so no one really knows their total number. Estimates range as high as a trillion (10^{12}) comet nuclei—more than all the stars in the Milky Way Galaxy. Nevertheless, we have seen hardly more than a thousand, perhaps only a billionth of the presumed total. Comet nuclei put on a show only when they come deep enough into the inner Solar System to suffer destructive heating from the Sun. When they are close enough to show the effects of solar heating, we call them **active comets**.

We know where comets reside by observing their orbits as they pass through the inner Solar System. These studies suggest that comets fall into two distinct groups: the "*Kuiper Belt* comets" and the "*Oort Cloud* comets." These two populations of comet

> The Kuiper Belt and the Oort Cloud are reservoirs of comets.

nuclei are named for scientists Gerard Kuiper (1905–1973) and Jan Oort (1900–1992), who first proposed their existence in the mid–20th century.[6]

The extent of the realm of comets is illustrated in **Figure 12.13**. Comet nuclei from the **Kuiper Belt** orbit the Sun within a disk-shaped region that begins just beyond the orbit of Neptune and extends outward to perhaps a thousand astronomical units from the Sun. The innermost part of the Kuiper Belt, which seems to end somewhat abruptly at about 50 AU, appears to contain tens of thousands of icy planetesimals, which we call Kuiper Belt objects. The larger KBOs are well over 1,000 km across, making them

Kuiper Belt objects are shown in red.

Oort Cloud comets are shown in green.

500 AU
Eris's orbit
Plane of Solar System
Sun
Neptune's orbit (diameter 60 AU)

Zoom out by 20×

10,000 AU

The Kuiper Belt is an extension of the Solar System disk.

Zoom out by 20×

200,000 AU

The Oort Cloud is a larger spherical cloud of comet nuclei.

Distance to nearest star Proxima Centauri approximately 270,000 AU

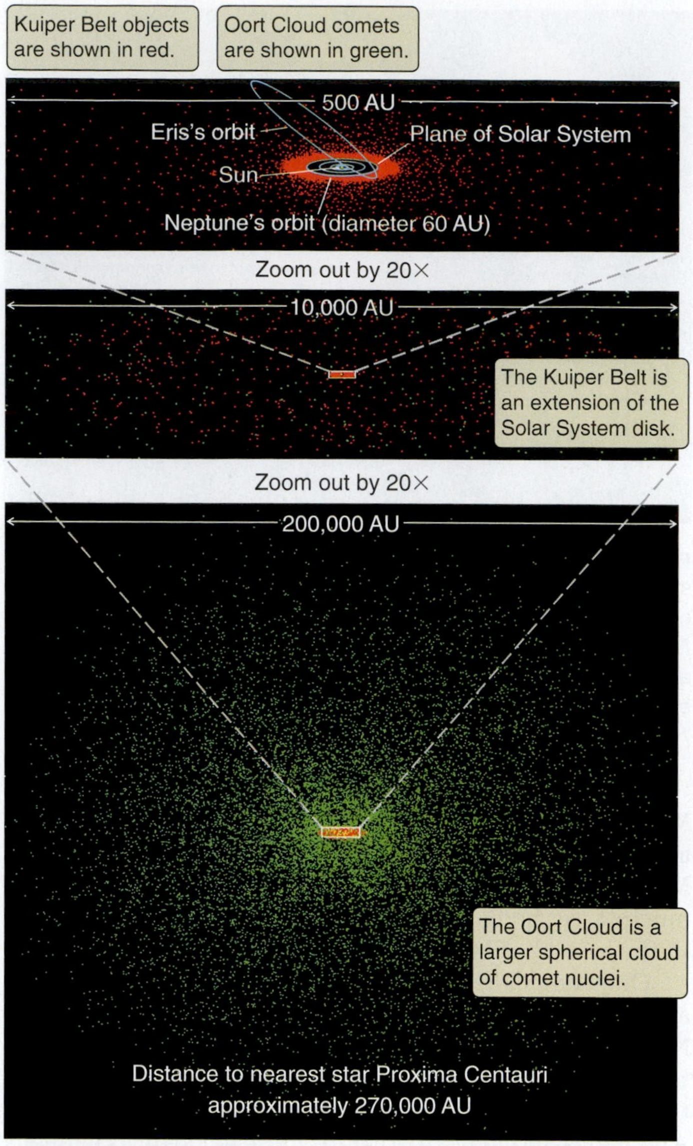

FIGURE 12.13 Most comet nuclei near the inner Solar System form an extension to the disk of the Solar System called the Kuiper Belt. The spherical Oort Cloud is far larger and contains many more comets.

larger than Ceres and similar in size to Pluto.[7] Some planetologists would classify Pluto not as a dwarf planet but as a large member of the family of Kuiper Belt objects. Yet one characteristic sets Pluto aside from all the known KBOs and makes it seem more planetlike: *Pluto has an atmosphere.* Like the asteroids, some KBOs are attended by moons, and at least one has as many as three moons. Astronomers have

---

[5]Believing that Comet Hale-Bopp was accompanied by a UFO that would carry their souls away to a "better place," 39 members of the cult Heaven's Gate committed mass suicide in March 1997.

[6]Some planetary scientists refer to the Kuiper Belt as the Edgeworth-Kuiper Belt in honor of Kenneth Edgeworth (1880–1972), an Irish engineer who predicted a reservoir of comets beyond the planets in a paper published in 1943, several years before the predictions of Kuiper and Oort.

[7]With a few exceptions, the sizes of KBOs cannot be measured directly and must be estimated. Although brightness and approximate distance are known, their albedos are uncertain, making estimates of their size correspondingly uncertain. Nevertheless, reasonable limits for the albedos of KBOs can set maximum and minimum values for their size.

already found more than 1,000 KBOs, but many smaller ones must also be there, just beyond the reach of our largest telescopes. Although they are too far from the Sun to be "active," it seems likely that these planetesimals are comet nuclei.

One of the larger known KBOs—Quaoar (pronounced "kwa-whar")—is also one of the few whose size we have been able to measure. Hubble Space Telescope observations show a tiny disk with a diameter of about 1,250 km, slightly larger than Pluto's Charon. By knowing its apparent brightness, distance, and size, we can easily calculate Quaoar's albedo. It turns out to be 0.12, which is more reflective than the nuclei of those comets that have entered the inner Solar System but far less reflective than Pluto. Quaoar's remote location and pristine condition have apparently allowed some bright ices to survive on its surface. Quaoar orbits between 42 and 45 AU from the Sun, and in this frigid region of the Solar System its surface temperature is only 50 K.

Sedna is an interesting object that remains outside the inner Kuiper Belt. Its highly elliptical orbit takes it from 75 AU out to 1,000 AU from the Sun. In such an extended orbit, Sedna requires more than *10,000 years* to make a single trip around the Sun. When discovered in 2003, Sedna was about 90 AU from the Sun and getting closer. It will reach its closest point in 2076. Because of its great distance, Sedna's size cannot be measured directly, even by the Hubble Space Telescope. If its albedo were the same as Quaoar's, Sedna would be the largest KBO.

Right now our knowledge of the chemical and physical properties of KBOs remains meager at best. Because of their great distance, we know very little about their composition and surface properties and absolutely nothing about their geology. This lack of information may change in the not-too-distant future, however. Following its encounter with Pluto in 2015, the *New Horizons* spacecraft is scheduled to continue outward into the Kuiper Belt, where it will be maneuvered to fly close to one or more KBOs.

Unlike the flat disk of the Kuiper Belt, the **Oort Cloud** is a spherical distribution of comet nuclei that are much too remote to be seen by even the most powerful telescopes. We know the size and shape of the Oort Cloud because comet nuclei from the Oort Cloud approach the inner Solar System from seemingly random directions in the sky, following orbits that bring them in from as far as 100,000 AU from the Sun, or nearly halfway to the nearest stars!

We can think of the Kuiper Belt and Oort Cloud as enormous reservoirs of icy planetesimals that now and then fall into the inner Solar System. But why do they sometimes leave their frigid haven and take what will probably be a fatal plunge inward toward the Sun? It seems likely that gravitational perturbations by objects beyond the Solar System—such as stars and interstellar clouds—disturb the orbits of comet nuclei, kicking the nuclei in toward the Sun. Astronomers estimate that every 5 million to 10 million years, one star or another, traveling in nearly the same orbit around the Milky Way as does the Sun, passes within about 100,000 AU of the Sun, perturbing the orbits

of comet nuclei. The gravitational attraction from huge clouds of dense interstellar gas concentrated in the plane of the Milky Way Galaxy[8] might also stir up the Oort Cloud as the Sun passes back and forth through the galactic plane during its 220-million-year orbital journey around the center of our galaxy.

These disturbances from beyond the Solar System have little effect on the orbits of the planets and other bodies in the inner Solar System. Inner Solar System objects are close enough to the Sun that external disturbances never exert more than a tiny fraction of the gravitational force of the Sun. In the distant Oort Cloud, however, things are quite different. Oort Cloud comet nuclei are so far from the Sun, and the Sun's gravitational force on them is so feeble, that they are barely bound to the Sun at all. The tug of a slowly passing star or interstellar cloud can compete with the Sun's gravity, significantly changing the orbits of Oort Cloud objects. If the interaction adds to the orbital energy of a comet nucleus, it may move outward to an even more distant orbit, or perhaps escape from the Sun completely to begin an eons-long odyssey through interstellar space. On the other hand, a comet nucleus that loses orbital energy falls inward. Some of these comet nuclei come all the way into the inner Solar System, where they may appear briefly in our skies as active comets before returning once again to the Oort Cloud.

Unlike comet nuclei in the Oort Cloud, those in the Kuiper Belt are packed together close enough to interact gravitationally from time to time. In such events, one nucleus gains energy while the other loses it. The "winner" may gain enough energy to be sent into an orbit that reaches far beyond the boundary of the Kuiper Belt. It seems likely that the Oort Cloud was populated in just this way (by comet nuclei ejected from the inner edge of the Kuiper Belt in what is now the Uranus-Neptune region). The experience could prove fatal for the "loser" if enough orbital energy is lost for it to fall inward toward the Sun.

Anatomy of an Active Comet

Unlike asteroids, which have been through a host of chemical and physical changes as a result of collisions, heating, and differentiation, most comet nuclei have been preserved over the past 4.6 billion years by the "deep freeze" of the outer Solar System. Comet nuclei are made of the most nearly pristine material remaining from the formation of the Solar System.

The small object at the center of the comet—the icy planetesimal itself—is the comet nucleus. The **nucleus** is by far the smallest component of a comet, but it is the source of all the mass that we see stretched across the skies as the comet nears the Sun (**Figure 12.14**). Comet nuclei are anywhere from a few dozen meters to several hundred kilometers across. We

---

[8]Interstellar clouds and the structure of the Milky Way Galaxy will be discussed in Chapters 15 and 20, respectively.

might formally describe comet nuclei as "planetesimals of modest size composed of a mixture of various ices of volatile compounds, organics, and dust grains loosely packed together to form a porous conglomerate"; but, mercifully, a shorter description is possible. In the middle of the 20th century, astronomer Fred Whipple (1906–2004) offered an elegant way to sum all of this up in just two words: comets are "dirty snowballs."

> An active comet has a tiny nucleus at its heart.

As a comet nucleus nears the Sun, sunlight heats its surface, turning volatile ices into gases, which then stream away from the nucleus, carrying embedded dust particles along with them. This process of conversion from solid to gas is called **sublimation**. An example much closer at hand illustrates this process. Dry ice (frozen carbon dioxide) does not melt like water ice but instead turns directly into carbon dioxide gas. Water ice melts.[9] Dry ice sublimates— that is why we call it "dry." Set a piece of dry ice out in the sun on a summer day, and you will get a pretty good idea of what happens to a comet.

The gases and dust driven from the nucleus of an active

---

[9]Water ice also sublimates in certain environments. For example, an ice cube left in a frost-free freezer will eventually disappear as water molecules slowly break free from its surface.

comet form a nearly spherical atmospheric cloud around the nucleus called the **coma** (plural *comae*—Figure 12.14a). The nucleus and the inner part of the coma are sometimes referred to collectively as the comet's **head**. Pointing from the head of the

> The coma is an extended cometary atmosphere surrounding the nucleus of a comet.

comet in a direction more or less away from the Sun are long streamers of dust, gas, and ions called the **tail**.

Most naked-eye comets seem to exhibit similar behavior— developing first a coma and then an extended tail as they approach the inner Solar System. Comet McNaught in 2007 was such a comet (**Figure 12.15a**). But there are exceptions. Comet Holmes was a very faint telescopic object when it reached its closest point to the Sun just beyond the orbit of Mars on May 4, 2007. Then, several months later, on October 21, as it was well on its way outward toward Jupiter's orbit, the comet suddenly became a half-million times brighter in just 42 hours! This abrupt outburst made Comet Holmes a bright naked-eye comet that graced Northern Hemisphere skies for several months (**Figure 12.15b**). Astronomers remain puzzled over the cause of this dramatic eruption. Explanations range from a meteoroid impact to a sudden (but unexplained) buildup of subsurface gas. ▶❙❙ **Astro-Tour: Cometary Orbits**

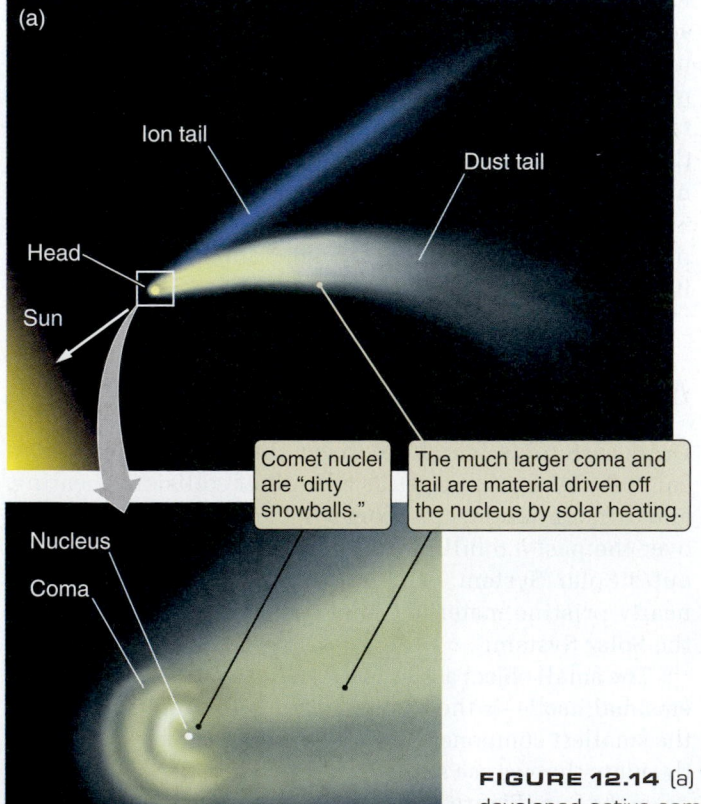

FIGURE 12.14 (a) The principal components of a fully developed active comet are the nucleus, the *coma*, and two types of tails, called the *dust tail* and the *ion tail*. Together, the nucleus and the coma are called the *head*. (b) Comet Hale-Bopp, the great comet of 1997.

(a)

(b)

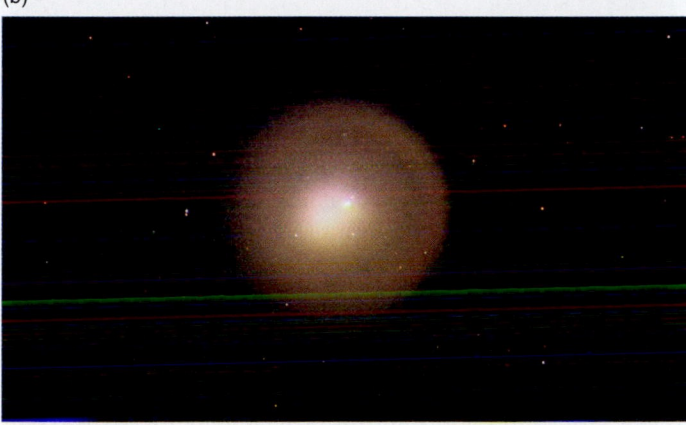

FIGURE 12.15 (a) Comet McNaught, in 2007, was the brightest comet to appear in decades, but its true splendor was visible only to observers in the Southern Hemisphere. (b) Months after its closest approach to the Sun in 2007, Comet Holmes suddenly became a half-million times brighter within hours, turning into a naked-eye comet with an angular diameter larger than that of the full Moon. Comet Holmes favored Northern Hemisphere observers.

A Visit to Comet Halley

Typical comet nuclei at typical distances appear so small that even the most powerful telescopes see them only as points of light. Even when a rare comet happens to approach close to Earth, we are still out of luck because the nucleus remains hidden within the dusty coma that surrounds it. Although astronomers have studied the gases that make up the comae and tails of comets with great interest for centuries, direct observations of comet nuclei remained an elusive prize.

All this changed in 1986, when Comet Halley made its "once in a lifetime" trip into the inner Solar System, providing a rare opportunity to send space probes to observe a nucleus at close range. This opportunity was so rare because,

despite the fact that even relatively pristine comet nuclei are frequent visitors to the inner Solar System, we seldom have enough advance warning of a comet's visit or sufficiently detailed knowledge of its orbit to mount a successful mission to intercept it. But although planetary scientists knew the orbit of Halley extremely well, sending probes to the comet still turned out to be a difficult task as space missions go. The problem was that the orbit of Comet Halley is *retrograde*—it goes around the Sun in a direction opposite to that of Earth and the other planets. As a result, the closing speed between an Earth-launched spacecraft and the comet would be 68 km/s, or about 75 times the speed of a bullet from a high-powered rifle. Observations close to the nucleus would have to be made very quickly, and there would be the always-present danger of high-speed collisions with debris from the nucleus.

Despite the difficulties, working in cooperation with each other, three different space agencies rose to the challenge of sending a five-spacecraft armada to meet Comet Halley as it passed through the inner Solar System. The whole operation was rather like driving down the highway at 70 miles per hour (mph) and trying to thread a needle held out the window by someone in the oncoming lane, but in the end the mission was a glorious success. Much of what we know about comet nuclei and the innermost parts of the coma was learned from data sent back by this armada of Soviet, European, and Japanese spacecraft.

Images taken by these spacecraft showed the nucleus of Halley to be an irregular peanut shell–shaped object measuring $8 \times 8 \times 14$ km (**Figure 12.16a**). As the Soviet *VEGA* and the European *Giotto* spacecraft passed close by the nucleus, they observed jets of gas and dust moving at speeds of up to 1,000 meters per second (m/s) from its surface, far exceeding the ability of its feeble gravity to contain the material. Here was dramatic evidence that the gas and dust that make up a comet's coma and tail come directly from the nucleus. The spacecraft images showed, however, that instead of streaming out from the whole surface of the nucleus, material was coming from several hot spots, or jets, that covered only about a tenth of the surface. Ninety percent of the surface was "quiet" at the time of the observations. The escaping gas emerged from beneath the crust through a number of small cracks or fissures.

By observing the jets of material streaming away from the nucleus of Halley, planetary scientists estimated that 20,000 kg of gas and 10,000 kg of dust were being lost by the nucleus each second. During its 1986 passage around the Sun, Comet Halley must have lost at least 100 billion (10^{11}) kg of material. But these jets also indirectly enabled planetary scientists to measure another fundamental property of this comet. Just like the engines on a jet airplane, the jets of material streaming away from the nucleus push on it, subtly altering its orbit. By observing the jets flowing away from the nucleus, scientists could tell how hard the jets

> Comet Halley loses 100 billion kg of mass each time it swings around the Sun.

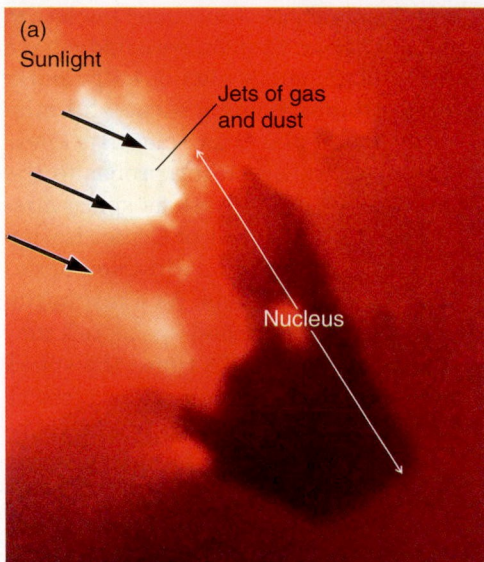

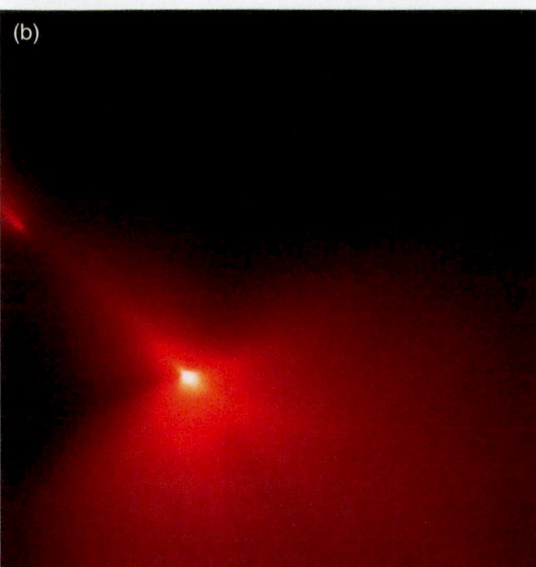

FIGURE 12.16 (a) The nucleus of Comet Halley as imaged in false color by the *Giotto* spacecraft in 1986. (b) An HST image of the coma of Comet Hyakutake taken in red light. A jet of material is seen at the upper left pointing in the antisunward direction.

were pushing on the nucleus. And by observing the comet's orbit, they learned how the nucleus was responding to these shoves. Using these observations, scientists were then able to apply Newton's laws of motion, which require the change in the momentum of the nucleus to balance the change in momentum of the material streaming away. They thus estimated the mass of the nucleus to be about 2.2×10^{14} kg.

This mass, divided by the volume as measured from the images of the nucleus, tells us that the nucleus is only about a fourth as dense as water. This is more like a loosely packed powdery snow than normal ice. As suggested by Whipple years earlier, the nucleus of Comet Halley must be a very fragile object that is porous throughout. Even with a mass loss of 300 billion kg of material during each encounter, the comet is losing only about 0.1 percent of its total mass, so Comet Halley should still be entertaining terrestrial onlookers hundreds of generations from now. It will return to the inner Solar System again in the summer of 2061.

The surface of the nucleus is also much darker than expected. It reflects only 3 percent of the sunlight falling on it, making it even blacker than coal or black velvet. The "dirty snowball" at the heart of Comet Halley is not only dirty; it is among the *darkest known objects* in the Solar System. This low reflectivity means that comet nuclei, or at least the nucleus of Halley, must be rich in organic matter that is far more complex than simple hydrocarbons such as methane, propane, butane, or octane. Simple organic compounds such as these form clear liquids and ices at low temperatures. In contrast, more complex hydrocarbons such as tar are usually very dark. The implication is that very complex organic matter was present as dust in the disk around the young Sun, perhaps even in the interstellar cloud from which the Solar System formed.

The coma of a comet is a tenuous and temporary atmo-

> The nucleus of Comet Halley is blacker than coal.

sphere surrounding the nucleus, consisting of gas and dust that are driven from the nucleus by solar heating. **Figure 12.16b** shows an image of the coma of Comet Hyakutake obtained with the Hubble Space Telescope. As a comet nucleus falls from the outer Solar System inward toward the Sun, the coma begins to form at about the time the nucleus passes within the orbit of Jupiter. (Although from our vantage point so close to the heart of the Solar System, Jupiter may seem far removed from the Sun's warmth, from the comet's perspective the solar warming at 5 AU is hundreds to many millions of times greater than it is in the Kuiper Belt or the Oort Cloud.) As the nucleus approaches still closer to the Sun, rapid sublimation of ices near its surface can cause the coma to swell to a diameter of a million kilometers, three times the distance between Earth and the Moon. But although the coma is huge compared with the nucleus, it is still more tenuous than the very best vacuums we can produce on Earth.

VEGA and *Giotto* entered the coma of Comet Halley when they were still nearly 300,000 km from the nucleus. The spacecraft measured the size and chemical composition of the dust particles in Halley's coma and found that the dust grains escaping from the nucleus were generally between 0.01 and 10 millimeters (mm) in size. This dust is very fine—finer even than the dust responsible for that annoying thin film that collects on your car and ruins its shine. Experiments found that the dust from Comet Halley consists of a mixture of light organic substances (combinations of hydrogen, carbon, nitrogen, and oxygen) and heavier rocky material (combinations of magnesium, silicon, iron, and oxygen). Instruments on the spacecraft designed to study the gases surrounding the nucleus found that it consisted of about 80 percent water (H_2O) and 10 percent carbon monoxide (CO), with smaller amounts of carbon dioxide (CO_2), methane (CH_4), ammonia (NH_3), and hydrogen cyanide (HCN).

Visits to Comets Borrelly, Wild 2, and Tempel 1

The early years of the 21st century have witnessed dramatic progress in our understanding of comet nuclei. NASA sent three spacecraft to visit three Jupiter-family comets: Borrelly, Wild 2, and Tempel 1. Such comets are escapees from the Kuiper Belt that have been captured by Jupiter into short-period orbits. It may come as no surprise to learn that all three comet nuclei appear vastly different when studied up close. In 2001, NASA's *Deep Space 1* spacecraft flew within 2,200 km of Comet Borrelly's nucleus. Comet Borrelly's tar-black surface is among the darkest seen on any Solar System object and, much to the surprise of scientists, showed no evidence of water ice. In 2004, NASA's *Stardust* spacecraft flew within 235 km of the nucleus of Comet Wild 2 (pronounced "vilt too"). Astronomers were especially interested in this comet because it is a newcomer to the inner Solar System. Comet Wild 2 had previously resided in the frigid region between the orbits of Jupiter and Uranus, but a close encounter with Jupiter in 1974 perturbed its orbit, bringing this relatively pristine body closer to the Sun as it travels back and forth in its new orbit between Jupiter and Earth. At the time of *Stardust*'s encounter with Wild 2, the comet had made only five trips around the Sun in its new orbit, compared with more than 100 visits by Comet Halley.

Wild 2's nucleus is about 5 km across, and unlike other comets visited earlier by spacecraft (Comets Halley and Borrelly), it has a more spherical shape. Images taken by *Stardust* show at least 10 active gas jets. Some were as large as 100 meters in diameter and carried within them surprisingly large chunks of surface material. (A few particles as large as a bullet penetrated the outer layer of the spacecraft's protective shield.) The surface of Wild 2 appears covered with craterlike features (**Figure 12.17**). Although these depressions do not have the normal impact characteristics, they may in fact be impact craters that have been modified by ice sublimation, small landslides, and erosion by jetting gas. The larger depressions are more than 1.5 km across. Some show flat floors, suggesting a relatively solid interior beneath a porous surface layer.

How do we know what really lies beneath a comet's surface? To find the answer, astronomers came up with a novel approach. They proposed firing a massive bullet at very high speed into the nucleus of a comet and observing what happens. In 2005 a 370-kg impacting projectile, launched by NASA's *Deep Impact* spacecraft, hit the nucleus of Comet Tempel 1 (**Figure 12.18**) at a speed of more than 10 km/s. The kinetic energy of the impact was equivalent to 5 tons of TNT, sending 10,000 tons of water and dust flying off into space at speeds of 50 m/s (**Figure 12.19**). A camera mounted on the projectile snapped photos of its target until both were vaporized by the impact. Observations of the event were made locally by the *Deep Impact* spacecraft and back at Earth by a multitude of orbiting and groundbased telescopes.

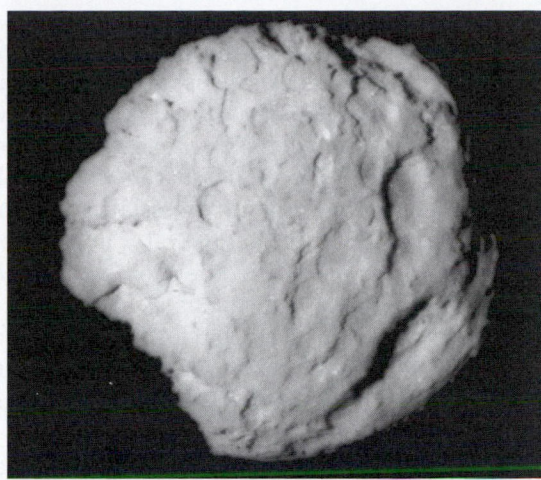

FIGURE 12.17 The nucleus of Comet Wild 2, imaged by the *Stardust* spacecraft.

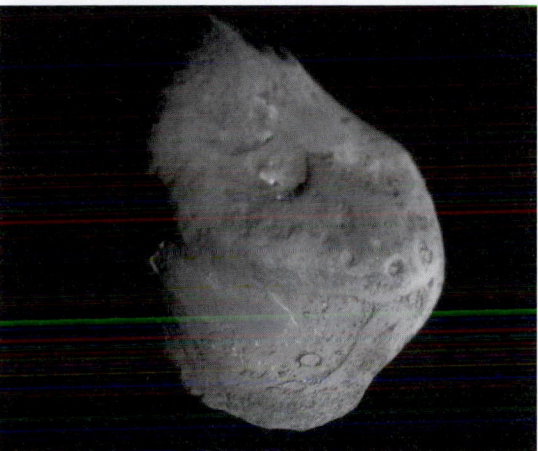

FIGURE 12.18 The surface of Comet Tempel 1 seen just before impact by the *Deep Impact* projectile. The impact occurred between the two 370-meter-diameter craters located near the bottom of the image. The smallest features appearing in this image are about 5 meters across.

FIGURE 12.19 A 370-kg projectile impacts Comet Tempel 1 at a speed of 10.2 km/s. This image, taken 67 seconds after impact, shows subsurface cometary material being ejected into space by the force of the impact.

Spectra indicated the presence of water, carbon dioxide, hydrogen cyanide, iron-bearing minerals, and a host of complex organic molecules. Analysis of the debris showed that the comet's outer layer is composed of fine dust with a consistency of talcum powder. No large chunks were seen flying from the impact site. Beneath the dust are layers made up of water ice and organic materials. The nucleus has a density only 0.6 times that of water and is perhaps best described as a porous rubble pile weakly held together by self-gravity rather than the mechanical strength of its constituent materials. One surprise for scientists was the presence of well-formed impact craters (see Figure 12.18), which had been absent in close-up images of Comets Borrelly and Wild 2. Why should some comet nuclei have fresh impact craters and others none? This remains a question that planetary scientists have yet to answer.

Comets Have Two Types of Tails

The tail, which is the largest and most spectacular part of a comet, is also the "hair" for which comets are named. (*Comet* comes from the Greek word *kometes*, which means "hairy one.") Active comets have two different types of tails, as shown in Figure 12.14a. One type of tail is called the **ion tail**. Many of the atoms and molecules that make up a comet's coma are ions. Because they are electrically charged, ions in the coma feel the effect of the solar wind, the stream of charged particles that blows continually away from the Sun. The solar wind pushes on these ions, rapidly accelerating them to speeds of more than 100 km/s—far greater than the orbital velocity of the comet itself—and sweeps them out into a long wispy structure. Because the particles that make up the ion tail are so quickly picked up by the solar wind, ion tails are usually very straight and point from the head of the comet directly away from the Sun.

Dust particles in the coma can also have a net electric charge and feel the force of the solar wind. In addition, sunlight itself exerts a force on cometary dust. But dust particles are much more massive than individual ions, so they are accelerated more gently and do not reach such high relative speeds as the ions do. As a result, the dust particles are unable to keep up with the comet, and the **dust tail** often appears to curve gently away from the head of the comet as the dust particles are gradually pushed from the comet's orbit in the direction away from the Sun.

Figure 12.20 shows the tails of a comet at various points in its orbit. Remember that both types of tails always point *away from the Sun*, regardless of the direction in which the comet is moving. As it approaches the Sun, the comet's two tails trail behind its nucleus. But like the flag on a sailboat sailing with the wind, the comet's tails extend *ahead* of the nucleus as it moves outward from the Sun.

Comet tails are fascinating structures that vary greatly

> Comet tails *always* point away from the Sun.

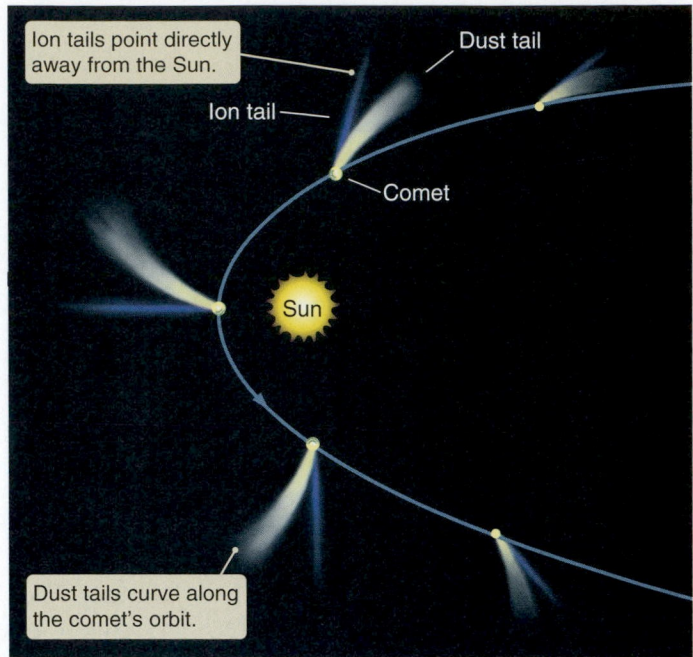

FIGURE 12.20 The orientation of the dust and ion tails at several points in a comet's orbit. The ion tail points directly away from the Sun while the dust tail curves along the comet's orbit.

from one comet to another. Some comets, such as Comets West and Halley, display both types of tails simultaneously; while, for reasons that we do not understand, some comets produce no tails at all! A tail often forms as the comet crosses the orbit of Mars, where the increase in solar heating drives gas and dust away from the nucleus at a prodigious rate. In October 1965, early-morning risers were held spellbound by the sight of Comet Ikeya-Seki's glowing 150 million km-long tail (**Figure 12.21**) stretching across the sky—a length as great as the distance between Earth and the Sun. For a brief few weeks, Comet Ikeya-Seki achieved temporary splendor as the longest object in the entire Solar System.

The gas in a comet's tail is even more tenuous than the gas in its coma, with densities of no more than a few hundred molecules per cubic centimeter. Compare this density with that of Earth's atmosphere, which at sea level contains more than 10^{19} molecules per cubic centimeter. Dust particles in the tail are typically about a micrometer in diameter, about the size of smoke particles. One astronomer's often-repeated quip that "all of the material in a long comet tail could be packed into a suitcase" is less an exaggeration than you might imagine. When Earth passed through the tail of Comet Halley in 1910, there were no observable effects. In contrast to the widespread predictions of disaster, it was all "much ado about nothing."

The Orbits of Comets

How long a dying nucleus will survive depends on its orbital period and **perihelion** (closest point to the Sun) distance—in

FIGURE 12.21 Comet Ikeya-Seki as it appeared in the predawn hours in October 1965. Soon afterward, just before it passed behind the Sun, the head of the comet became so bright that it was visible in broad daylight. When it reappeared a few days later, it had split into two separate components.

other words, how frequently it passes by the Sun and how close it comes. Each passage takes its toll of ice and dust. By convention, comet orbits are generally referred to as *short-period* or *long-period* orbits. The division between the two is somewhat arbitrarily set at a period of 200 Earth years. Comets with periods of less than 200 years are called **short-period comets**. The total number of short-period comets known today is about 400. Comets with orbital periods longer than 200 years are termed **long-period comets**. The total number of long-period comets observed to date is about 3,000, with an average of six new ones being discovered each year.

The orbits of nearly all comets are highly elliptical, with one end of the orbit close to the Sun and the other in the distant parts of the Solar System. As a result, we might expect all comets passing through the inner Solar System to have extremely long orbital periods that carry them back to the Oort Cloud or

> The orbits of most comets are very elliptical.

the Kuiper Belt. The surprise is that more than 100 short-period comets remain relatively close to the Sun. How do these denizens of the frigid hinterlands become resident aliens in the realm of the planets?

The key to solving this mystery lies in an interesting difference between the orbits of long- and short-period comets. **Figure 12.22** shows the orbits of a number of comets. Whereas long-period comets come diving into the inner Solar System from all directions, short-period comets have orbits that are strongly concentrated in the ecliptic plane. Because their orbits cross the orbits of many of the planets, short-period comets frequently get close enough to a planet for the planet's gravity to change the comet's orbit about the Sun in the sorts of chaotic encounters discussed in Chapter 10. Presumably, short-period comets originated in the Kuiper Belt; but as they fell in toward the Sun, they were forced into their present short-period orbits by gravitational encounters with planets.

Short-Period Comets **Figure 12.22a** shows the orbits of a number of short-period comets in the inner Solar System. Nearly all short-period comets have **prograde** orbits. That is, they orbit the Sun in the same direction as the planets. (Comet Halley, with its retrograde orbit, is an exception.) This tendency toward prograde orbits, along with the concentration of short-period comets in the plane of the Solar System, is important evidence supporting our theory of the origin of short-period comets and how they were captured. The disklike Kuiper Belt appears to share the prograde rotation of the rest of the Solar System (except the Oort Cloud). Because Kuiper Belt comet nuclei start out on prograde orbits, they are more likely to be captured into short-period prograde orbits as well. In addition, the encounter between a planet and a comet nucleus on a prograde orbit will last longer than the encounter between a planet and a comet nucleus on a retrograde orbit. This prolonged contact means that a prograde comet's orbit will be more strongly influenced by such an encounter than will the orbit of a retrograde comet. Jupiter, the most massive of the planets, is the one most likely to affect the orbit of a Kuiper Belt comet. There are about 270 known short-period comets, referred to as the Jupiter family, that all have their **aphelion** (farthest point from the Sun) near Jupiter's orbit and travel in prograde orbits around the Sun.

Comet Halley is the brightest and most famous of the short-period comets. It also has a very special place in the history of science because it was the first comet whose return was predicted. In 1705, Edmund Halley was studying the orbits of comets and employing the gravitational laws that had recently been discovered by his friend and colleague Isaac Newton. (Halley was instrumental in encouraging Newton to publish his work on orbits and gravitation—work that became the foundation for the science of physics.) Halley noted that a bright comet that had appeared in 1682 had an orbit remarkably similar to those of comets seen in 1531 and 1607. He concluded that all three were actually one and the same, and he made the daring prediction that this

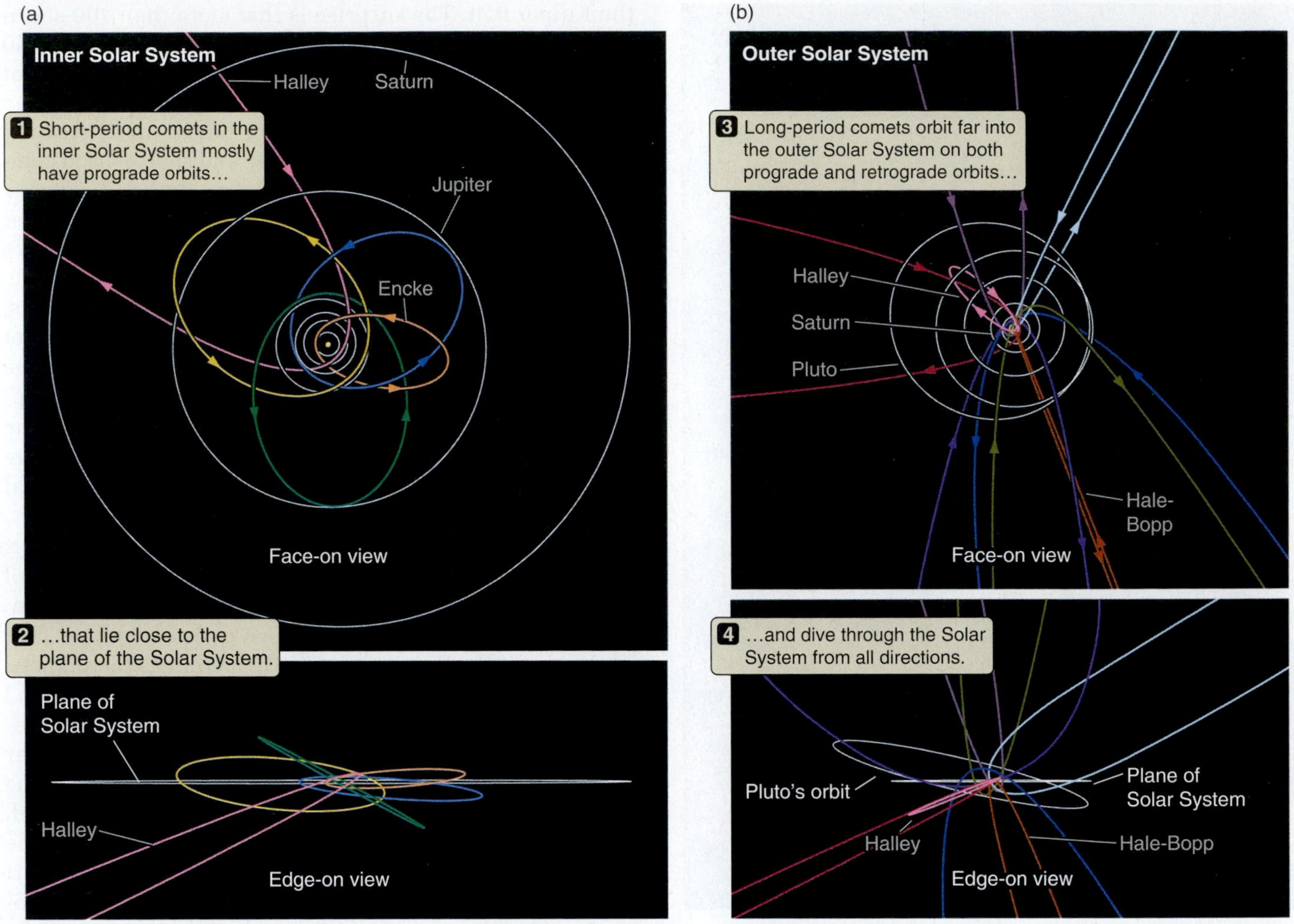

(a)

Inner Solar System

— Halley   Saturn

1 Short-period comets in the inner Solar System mostly have prograde orbits…

Jupiter

Encke

Face-on view

2 …that lie close to the plane of the Solar System.

Plane of
Solar System

Halley

Edge-on view

(b)

Outer Solar System

3 Long-period comets orbit far into the outer Solar System on both prograde and retrograde orbits…

Halley
Saturn
Pluto

Hale-
Bopp

Face-on view

4 …and dive through the Solar System from all directions.

Pluto's orbit

Plane of
Solar System

Halley   Hale-Bopp

Edge-on view

FIGURE 12.22 Orbits of a number of comets in face-on and edge-on views of the Solar System. Populations of short-period comets (a) and long-period comets (b) have very different orbital properties. Comet Halley, which appears in both diagrams for comparison, is a short-period comet.

comet would return in 1758. Unfortunately, Halley did not live to see it. His comet reappeared on Christmas Eve in the very year he had predicted. Astronomers quickly named it Halley's Comet and heralded it as a triumph for the genius of both Newton and Halley.

Comet Halley's highly elongated orbit takes it from perihelion, about halfway between the orbits of Mercury and Venus, out to aphelion beyond the orbit of Neptune. Astronomers and historians have now identified possible sightings of the comet that go back to 467 B.C. Comet Halley has an average period of 76 years, and many of us mark the "once in our lifetime" when we are fortunate enough to see it. Mark Twain is famous for saying that he came in with the comet in 1834 and he would go out with the comet in 1910—a promise he kept. For the authors of this book, our opportunity to see Halley's Comet came in 1986. Although that appearance was not especially spectacular compared to the one in 1910—in 1986 Comet Halley

Halley's Comet comes close to Earth every 76 years.

and Earth were on opposite sides of the Sun when the comet put on its display—seeing the ghostly sight of Halley is an experience we will long remember. Comet Halley will reach aphelion in 2024 and then begin its long journey back to the inner Solar System, becoming visible to the naked eye once again in the summer of 2061.

Long-Period Comets   Of the known long-period comets, more than 600 have well-determined orbits. Some have orbital periods of hundreds of thousands or even millions of years, and their nuclei spend almost all their time in the frigid, outermost regions of the Solar System. Orbits of a few long-period comets are shown in **Figure 12.22b**. Unlike the mostly prograde, ecliptic-plane orbits of the short-period comets, long-period comets split about evenly into prograde and retrograde and fall into the inner Solar System from all directions. These are the comets that tell us of the existence

Some long-period comets come by only once in a million years.

of the Oort Cloud. Because of their very long orbital periods, these comets cannot make more than a single appearance throughout the course of recorded history and, in most cases, in all of human history.

Long-period comets differ from short-period comets in another way as well. With a few exceptions, Comet Halley among them, the nuclei of short-period comets have been badly "worn out" by their repeated exposure to heating by the Sun. As the volatile ices are driven from a nucleus, some of the dust and organics are left behind on the surface. The buildup of this covering slows down cometary activity. (Envision how, as a pile of urban—and therefore dirty—snow melts, the dirt left behind is concentrated on the surface of the snow.) That is why most short-period comets create little excitement. In contrast, long-period comets are usually relatively pristine. More of their supply of volatile ices still remains close to the surface of the nucleus, and they can produce a truly magnificent show. A half dozen or so long-period comets arrive each year. Most pass through the inner Solar System at relatively large distances from Earth or the Sun and never become bright enough to attract much public attention.

Comet Ikeya-Seki (see Figure 12.21) is a member of a family of comets called **sungrazers**, comets whose perihelia are located very close to the surface of the Sun. Many sungrazers fail to survive even a single pass by our local star. Ikeya-Seki became so bright as it neared perihelion in 1965 that it was visible in broad daylight, only two solar diameters away from the noontime Sun; and when it reappeared from behind the Sun, it had been split into two pieces. Sungrazers generally come in groups, with successive comets following in nearly identical orbits. Each member of such a group started as part of a single larger nucleus that broke into pieces during an earlier perihelion passage.

The closing years of the 20th century witnessed two spectacular long-period comets sporting long, beautiful tails: Hyakutake in 1996 (see the opening photograph of this chapter) and Hale-Bopp in 1997 (see Figure 12.14b). Both were widely seen by the viewing public, Comet Hale-Bopp being perhaps the most observed comet ever. It is a huge comet, with a nucleus perhaps as large as 60 km in diameter. It was bright enough to spot even in urban nighttime skies, and it dazzled those who were fortunate enough to see it far from city lights. Comet Hale-Bopp is an especially important scientific object for professional astronomers because it was discovered far from the Sun near Jupiter's orbit 2 years before its perihelion passage. This early discovery extended the total time available to study its development as it approached the Sun and provided ample opportunity to plan the important observations as it neared perihelion. Warmed by the Sun, the nucleus produced large quantities of gas and dust, and as much as 300 tons of water per second, with lesser amounts of carbon monoxide, sulfur dioxide, cyanogen, and other gases. Comet Hale-Bopp is still being observed with large telescopes. It remains active and continues to show a tail, even though it is now far from the Sun

and approaching the orbit of Neptune. Comet Hale-Bopp will continue its outward journey for well over 1,000 years. It will not return to the inner Solar System until sometime around the year 4530.

The most spectacular comet yet seen in the 21st century was Comet McNaught. Known as the "Great Comet of 2007," it was the brightest to appear in over 40 years and was perhaps the second brightest in nearly a century. The comet's nucleus and coma were visible in broad daylight as its orbit carried it within 25 million km of the Sun. When Comet McNaught next appeared in the evening skies to observers in Australia and elsewhere in the Southern Hemisphere, its tail had grown to a length of more than 160 million km and stretched 35° across the sky (see Figure 12.15a). Comet McNaught came into the inner Solar System from the Oort Cloud, but it left on a slightly hyperbolic orbit (see Chapter 3), meaning that this is one comet that will never return.

When will the next bright comet like Hyakutake, Hale-Bopp, or McNaught come along? On average, a spectacular comet appears about once per decade, but it is all a matter of chance. It might be many years from now—or it could happen tomorrow.

12.9 Collisions Still Happen Today

As we studied the inner planets and the moons of the outer planets, we learned that almost all hard-surfaced objects in the Solar System still bear the scars of a time when the Sun and planets were young and tremendous impact events were common occurrences. The collision of Comet Shoemaker-Levy 9 with Jupiter in 1994 (see **Excursions 12.2**) focused attention on the fact that although such impacts are far less frequent today than they once were, they still happen.[10] Having observed the effects of Shoemaker-Levy 9's collision with Jupiter, we can easily imagine the global firestorms on Earth that accompanied the impact that occurred at the Cretaceous-Tertiary boundary 65 million years ago (see Excursions 7.1). **Connections 12.1** discusses several ways that such impacts affected the course of life on Earth. The "Jupiter comet crash" was also a landmark event in the history of the public's access to fast-breaking scientific events. Occurring when the Internet was growing in popularity, the impact's latest images were downloaded daily by millions of people around the world, scientists and laypeople alike.

> In 1994, much of the world watched as a comet crashed into Jupiter.

[10] In 2009, observers saw evidence of yet another comet hitting Jupiter, the sudden appearance of a dark atmospheric scar nearly as large as Earth.

What we learned from the crash of Shoemaker Levy-9 into Jupiter also helps explain an event that has puzzled scientists for the past century. The Tunguska River flows through a remote region of western Siberia, inhabited mostly by reindeer and a few reindeer herders. In the summer of 1908 the

> In 1908 a small asteroid or comet smashed into Siberia.

region was blasted with the energy equivalent of 2,000 times the atomic bomb dropped over Hiroshima. **Figure 12.25** (on p. 353) shows a map of the region, along with a photo of the devastation caused by the blast. Eyewitness accounts detailed the destruction of dwellings, the incineration of reindeer (including one herd of 700), and the deaths of at least five people. Although trees were burned or flattened for over 2,150 square kilometers (km^2)—an

EXCURSIONS 12.2

Comet Shoemaker-Levy 9

From a wealth of ground- and space-based observations, planetary scientists have pieced together the events leading to the collision of Comet Shoemaker-Levy 9 with Jupiter. Early in the 20th century this comet nucleus from the Kuiper Belt was perturbed from its path and sent on an orbital journey that carried it close to Jupiter. In 1992 it passed so close to the planet that tidal stresses broke it into two dozen major fragments, which

subsequently spread out along more than 7 million km of its orbit. The trajectory carried the fragments around for one more 2-year orbit about the planet.

Then, in July 1994, the entire string of fragments crashed into Jupiter. Over a week's time, one after another of the fragments, each traveling at 60 km/s, plunged through Jupiter's stratosphere. Even though the impacts occurred just behind

FIGURE 12.23 Events following the impact of a fragment of Comet Shoemaker-Levy 9 on Jupiter.

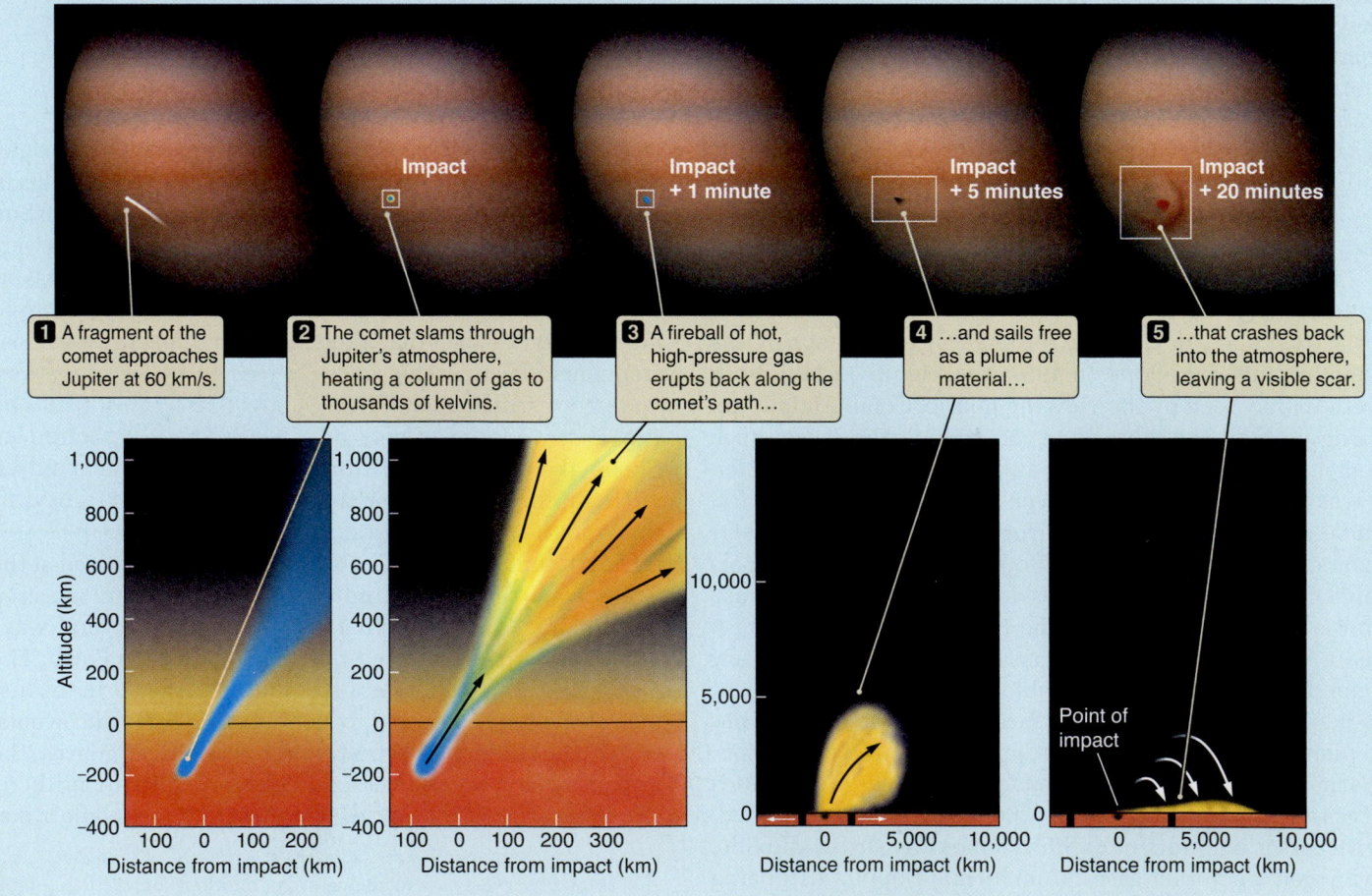

Impact

Impact + 1 minute

Impact + 5 minutes

Impact + 20 minutes

1 A fragment of the comet approaches Jupiter at 60 km/s.

2 The comet slams through Jupiter's atmosphere, heating a column of gas to thousands of kelvins.

3 A fireball of hot, high-pressure gas erupts back along the comet's path...

4 ...and sails free as a plume of material...

5 ...that crashes back into the atmosphere, leaving a visible scar.

area greater than metropolitan New York City—no crater was left behind!

For many years, scientists and nonscientists alike speculated about the cause of the Tunguska event. Ideas included such fanciful notions as a collision with a mini–black hole that passed through Earth, a collision with an object made of antimatter, or an act of aggression by aliens (a favorite of the supermarket tabloids). It now seems fairly clear that the Tunguska event was the result of a tremendous high-altitude explosion that occurred when a small asteroid or comet nucleus hit Earth's atmosphere, ripped apart, and formed a fireball before reaching Earth's surface. This is just the sort of event seen on Jupiter (when Shoemaker-Levy 9 crashed into it) and hypothesized to account for the "splotches" seen on Venus. Recent expeditions to the Tunguska area have recovered resin from the trees blasted by the event. Chemi-

the limb of the planet, where they could not be observed from Earth, the *Galileo* spacecraft was on its way to Jupiter and was able to image some of the impacts. In addition, astronomers using groundbased telescopes and the Hubble Space Telescope could see immense plumes rising from the impacts to heights of more than 3,000 km above the cloud tops at the limb. The debris in these plumes then rained back onto Jupiter's strato-sphere, causing ripple effects like pebbles thrown into a pond. **Figure 12.23** shows the sequence of events that took place as each fragment of the comet slammed into the giant planet. Sulfur and carbon compounds released by the impacts formed giant, Earth-sized scars in the atmosphere that persisted for months (**Figure 12.24**) and were visible even through small amateur telescopes.

FIGURE 12.24 (a) Groundbased infrared images of Jupiter, showing the hot, glowing scars left by fragments of Comet Shoemaker-Levy 9. (b) Although the fragments struck Jupiter on the back side, these HST images show the fireball from one fragment rising above the limb of the planet. (c) HST images of the evolution of the scar left by one fragment of the comet.

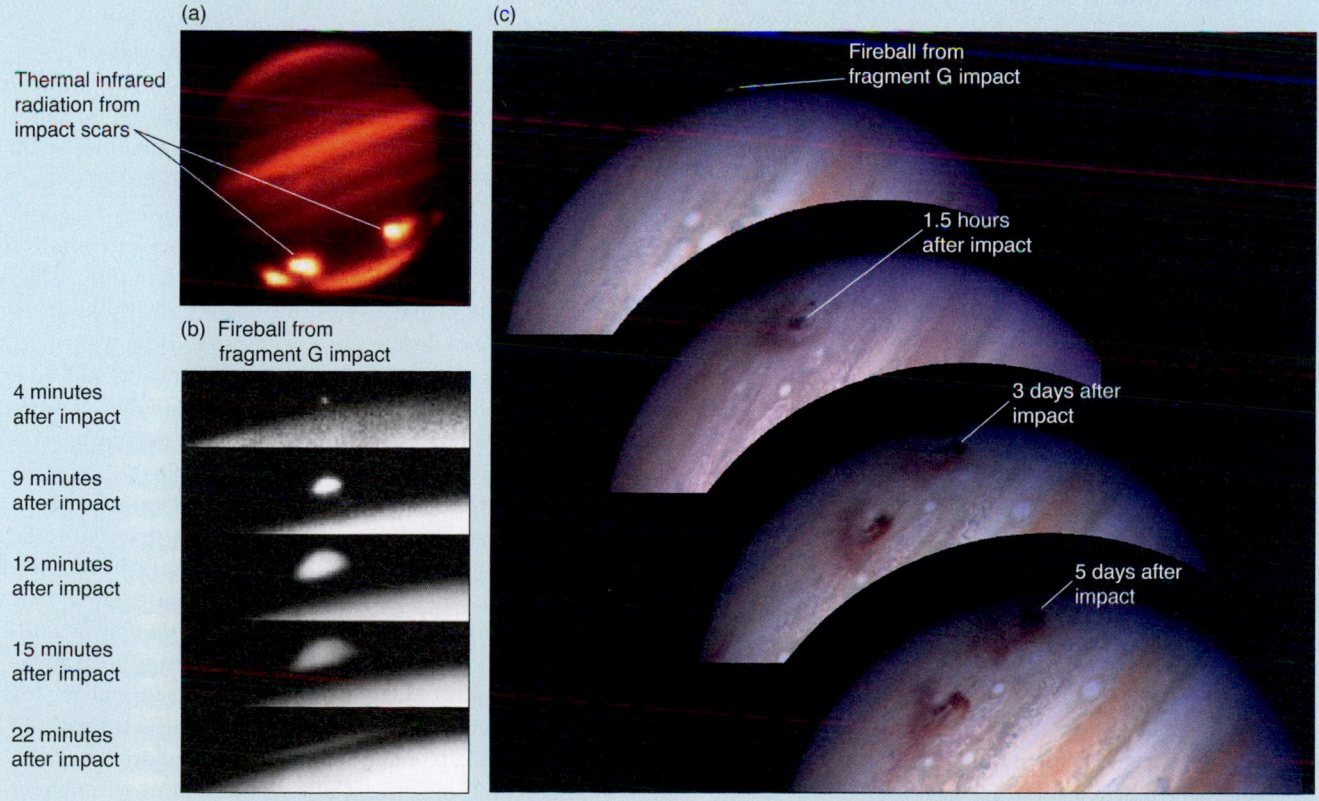

(a)

Thermal infrared radiation from impact scars

(b) Fireball from fragment G impact

4 minutes after impact

9 minutes after impact

12 minutes after impact

15 minutes after impact

22 minutes after impact

(c)

Fireball from fragment G impact

1.5 hours after impact

3 days after impact

5 days after impact

cal traces in the resin suggest that the impacting object may have been a stony asteroid.

The Tunguska impact and the collision of Shoemaker-Levy 9 with Jupiter are truly sobering events. We know of several impacts occurring in the Solar System within a human lifetime, and more than one of them involved our planet![11] We know enough about the distribution of relatively large asteroids in the inner Solar System to say that it is highly improbable a populated area on Earth will experience a major collision with an asteroid within our lifetimes (see Chapter 7). Comets and smaller asteroids, however, are less predictable. There may be as many as 10 million asteroids larger than a kilometer across, but only about 130,000 have well-known orbits, and most of the unknowns are too small

---

[11]At 10:38 A.M. on February 12, 1947, yet another planetesimal struck Earth, this time in the Sikhote-Alin region of eastern Siberia. Composed mostly of iron, it had an estimated diameter of about 100 meters and broke into a number of fragments before hitting the ground, leaving a cluster of craters and widespread devastation. Witnesses reported a fireball brighter than the Sun and sound that was heard 300 km away.

to see until they come very close to Earth. Recently, the US government—along with the governments of several other nations—became aware of the risk posed by NEOs. Although the probability of a collision between a small asteroid and Earth is quite small, the consequence can be catastrophic. In response to this recognition, NASA has been given a congressional mandate to catalog all NEOs and to scan the skies for those that remain undiscovered.

Comets, however, present a more serious problem. Several previously unknown long-period comets enter the inner Solar System each year. If one happens to be on a collision course with Earth, we might not notice it until just a few weeks or months before

> Comet or asteroid impacts are infrequent but devastating.

impact. For example, Comet Hyakutake was discovered only 2 months before passing near Earth, and a potentially destructive asteroid that just missed Earth in 2002 was not discovered until 3 days *after* its closest approach! Although this has become a favorite theme of science fiction disaster stories (*Lucifer's Hammer*, by Larry Niven and Jerry Pournelle, being a favorite of some of the authors

---

CONNECTIONS 12.1

Comets, Asteroids, and Life

One of the great questions of modern astronomy involves understanding how life on Earth is connected with processes at work in the greater cosmos. In a certain sense, this quest is not so different from that pursued by many of the world's great religions and philosophies. At most of the stops along our journey through the universe, we find threads of the cosmic tapestry of which terrestrial life is a part, and this stop is no exception.

For example, were it not for impacts by comet nuclei, water on Earth might have been less plentiful than it is today. We think that some of Earth's water was contributed during impacts of icy planetesimals (in other words, comet nuclei) early in the history of the Solar System. How could this have come about? The icy planetesimals likely condensed from the protoplanetary disk surrounding the young Sun and grew to their present size near the orbits of the giant planets. These planetesimals subsequently suffered strong orbital disturbances from those same planets. In such interactions, about half of the planetesimals would be flung outward to form the Kuiper Belt and Oort Cloud and half inward toward the Sun. Some of the objects flung toward the Sun would likely have hit Earth, the largest planet in the inner Solar System. Because most of the mass in comet nuclei appears to be in the form of water ice, it is likely that some of Earth's current water supply came from this early bombardment. Water, as we know, is essential to life.

Yet the existence of comets can also threaten life on Earth. It seems certain that occasional collisions of comet nuclei and

asteroids with Earth have resulted in widespread devastation of Earth's ecosystem and the extinction of many species. Passing stars or the passage of the Sun through the "galactic plane" may send showers of comet nuclei into the inner Solar System every 10 million to 100 million years—intervals similar to those between mass extinctions found in fossil records of life on our planet. Although such events certainly qualify as global disasters for the plants and animals alive at the time, they also represent global opportunities for new life-forms to evolve and fill the niches left by species that did not survive. Such a collision with an asteroid or comet probably played a central role in ending the 180-million-year reign of dinosaurs and provided our ancient mammalian ancestors an opportunity for world dominance (see Excursions 7.1).

In our study of comets, we may also have found a key to the chemical origins of life on Earth. Comets turn out to be rich in complex organic material—the chemical basis for terrestrial life—and cometary impacts on the young Earth may have played a role in chemically seeding the planet. In addition, the fact that comets are believed to be pristine samples of the material from which the Sun and planets formed means that organic material must be widely distributed throughout interstellar space. This conclusion, which is supported by radio telescope observations of vast interstellar clouds throughout our galaxy, might have significant implications as we consider the possibility of life elsewhere in the universe.

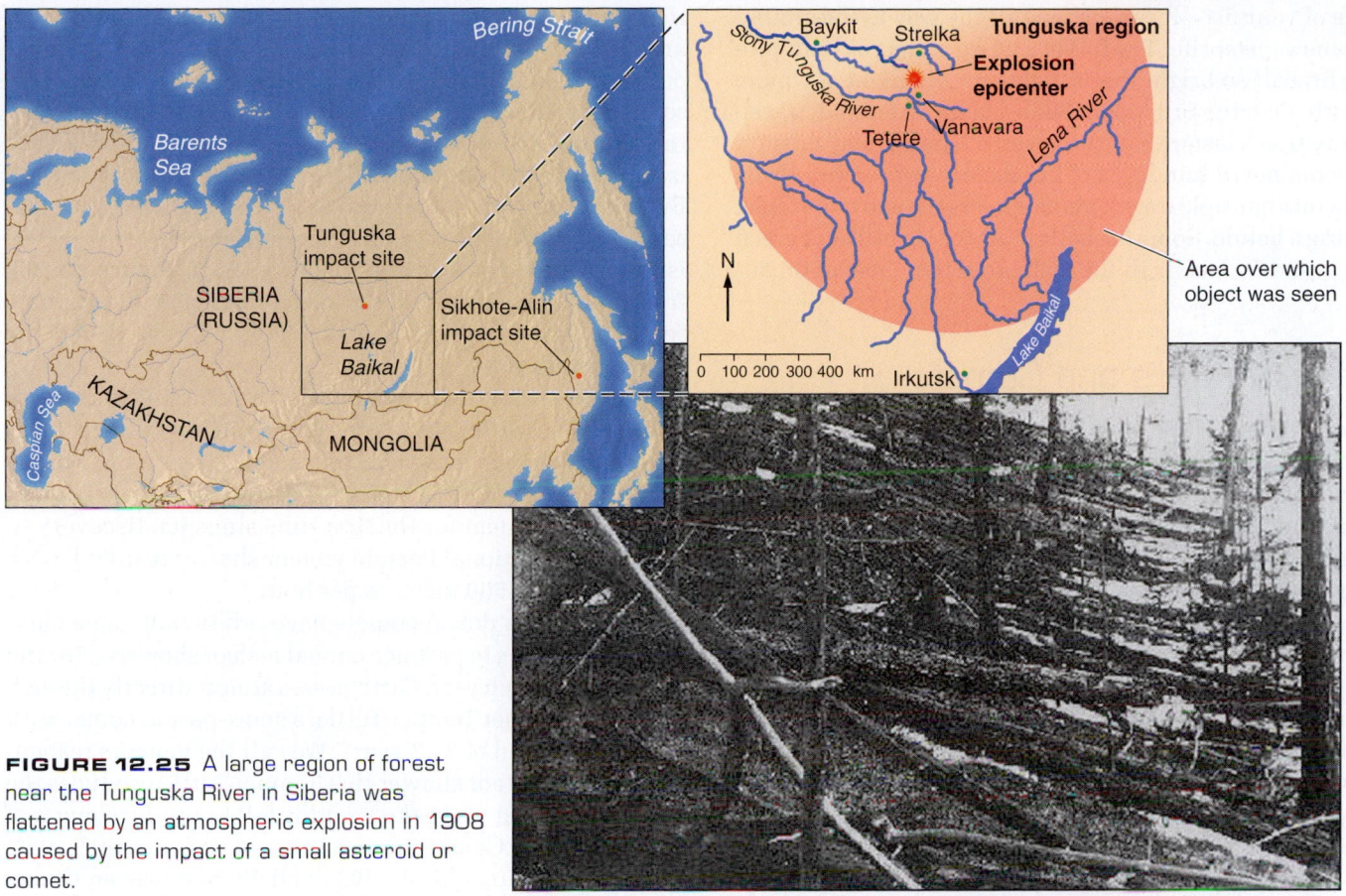

FIGURE 12.25 A large region of forest near the Tunguska River in Siberia was flattened by an atmospheric explosion in 1908 caused by the impact of a small asteroid or comet.

of this text), Earth's geological and historical record suggests that actual impacts by large bodies are infrequent events. Does this mean we need not lose any sleep over a possible collision with a large comet or asteroid? Probably, but remember that even though the probability may be small, the consequences of such an event are enormous. Just ask the dinosaurs.

12.10 Solar System Debris

Although some comets and asteroids meet the spectacular fate of Comet Shoemaker-Levy 9 or the Tunguska planetesimal, most go out with more of a fizzle than a bang. Comet nuclei that enter the inner Solar System generally disintegrate within a few hundred thousand years as a result of their repeated passages close to the Sun. Asteroids have much longer lives but still are slowly broken into pieces from occasional collisions with each other. Disintegration of comet nuclei and asteroid collisions are the sources of most of the debris that fills the inner part of the Solar Sys-

> Comet nuclei and asteroids are the sources of Solar System debris.

tem. As Earth and other planets move along in their orbits, they continually sweep up this fine debris. This debris is the source of most of the meteoroids that Earth encounters. When they burn up in our atmosphere, these meteoroids become the meteors that you can see streaking across the sky on any clear, dark night. As discussed in Excursions 12.1, only when they are large enough to reach Earth's surface do we call them meteorites.

Meteoroids and Meteors

Some 100,000 kg of meteoritic debris is swept up by Earth every day, and what does not burn up (mostly particles smaller than 100 μm) eventually settles to the ground as fine dust. We can measure meteor heights with radar because radio waves bounce off the ionized gas in meteor trails just as they bounce off the metal in an airplane or an automobile. Using these techniques, we find that the altitudes of meteors are between 50 and 150 km. For comparison, commercial jet aircraft typically fly at heights of about 10 km.

> Meteors are atmospheric phenomena.

Fragments of asteroids are much denser than cometary meteoroids. If an asteroid fragment is large enough—about the size of your fist—it can survive all the way to the ground to become a meteorite. The fall of a 10-kg meteoroid can produce a fireball so bright that it lights up the night sky more brilliantly than the full Moon. Such a large meteoroid, traveling many times faster than the speed of sound, may create a sonic boom heard hundreds of kilometers away. It may even explode into multiple fragments as it nears the end of its flight, becoming a **bolide**. Some fireballs glow with a brilliant green color, caused by metals in the meteoroid that created them.

Meteor Showers and Comets

Standing under a dark nighttime sky with a clear horizon, you can expect to see about a dozen meteors per hour on any night of the year. These are called **sporadic meteors**, and they occur as Earth sweeps up random bits of cometary and asteroidal debris in its annual path around the Sun. However, if you happen to be meteor watching on the nights of August 11–13, for instance, you may see four to five times this number. If you pay close attention, you might also notice that nearly all of the meteors seem to be coming from the same region of the sky. This phenomenon is called a **meteor shower** (**Table 12.2**). We call the particular meteor shower that peaks in mid-August the **Perseids** because the trails left behind all point back to the constellation Perseus (**Figure 12.26a**).

Meteor showers happen when Earth's orbit crosses the orbit of a comet. Bits of dust and other debris released by a comet nucleus as it rounds the Sun remain in orbits of their own that are similar to the orbit of the nucleus itself. When Earth passes through this concentration of cometary debris, the result is a meteor shower. Because the meteoroids that are being swept up are all in similar orbits, they all enter our atmosphere moving in the same direction. As a result, the paths of all shower meteors are parallel to one another. But just as the parallel rails of a railroad track appear to vanish to a single point in the distance, as in **Figure 12.26b**, from our perspective all the meteors appear to originate from the same point in the sky. This point is called the shower's **radiant**.

> Meteor showers occur when Earth passes through cometary debris.

The Perseids are the result of Earth's crossing the orbit of Comet Swift-Tuttle. Although spread out along the comet's orbit, the debris is more concentrated in the vicinity of the comet itself. In 1992, Comet Swift-Tuttle returned to the inner Solar System for the first time since its discovery in 1862. An exceptional Perseid meteor shower resulted, with counts of up to 500 meteors per hour.

More than a dozen comets have orbits that come close enough to Earth's to produce annual meteor showers. Around November 16 each year, Earth passes almost directly through the orbit of Comet Tempel-Tuttle, a short-period comet with an orbital period of 33.2 years. We call the meteors responsible for the meteor shower that Tempel-Tuttle produces the **Leonids**. In most years the Leonids fail to produce much of a show because Comet Tempel-Tuttle distributes little of its debris around its orbit. In 1833 and 1866, however, Comet Tempel-Tuttle was not far away when Earth passed through its orbit. The Leonid showers in those 2 years were so intense that meteors filled the sky with as many as 100,000 per hour (**Figure 12.27**). In 1900, one comet orbit later, nothing out of

TABLE 12.2

Selected Meteor Showers

Shower	Approximate Date	Parent Object
Quadrantids	January 3–4	?
Lyrids	April 21–22	Comet Thatcher
Eta Aquariids	May 5–6	Comet Halley
Perseids	August 12–13	Comet Swift-Tuttle
Draconids	October 8–9	Comet Giacobini-Zinner
Orionids	October 21–22	Comet Halley
Taurids	November 5–6	Comet Encke
Leonids	November 17–18	Comet Tempel-Tuttle
Geminids	December 13–14	Asteroid Phaethon
Ursids	December 22–23	Comet Tuttle

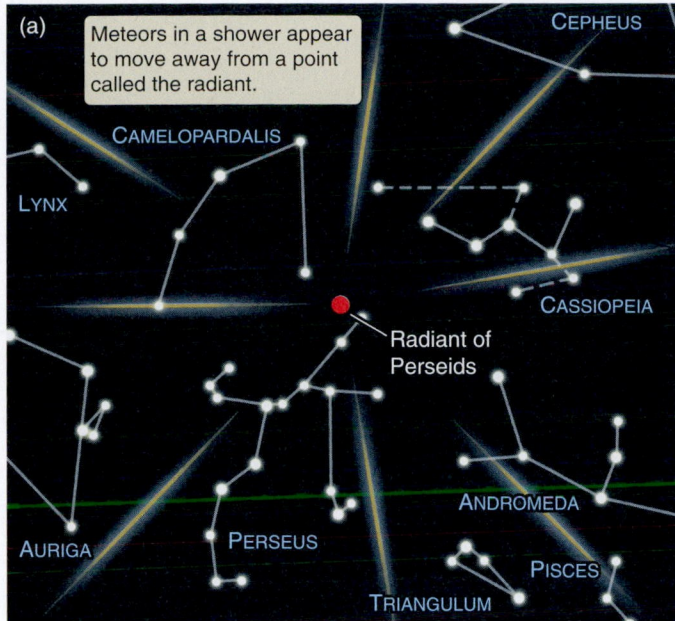

Meteors in a shower appear to move away from a point called the radiant.

CEPHEUS

CAMELOPARDALIS

LYNX

CASSIOPEIA

Radiant of Perseids

ANDROMEDA

AURIGA

PERSEUS

PISCES

TRIANGULUM

Actually the meteor paths are parallel. The radiant is a vanishing point.

VISUAL ANALOGY FIGURE 12.26 Meteors appear to stream away from the radiant of the Perseid meteor shower (a). These streaks are actually parallel paths that appear to emerge from a vanishing point, like the railroad tracks shown in (b).

FIGURE 12.27 An engraving of the 1833 Leonid shower seen in France.

the ordinary happened. Again in 1933, the Leonids were disappointing. Perturbations of Comet Tempel-Tuttle by Jupiter had moved the orbit of the comet's nucleus slightly away from Earth's, causing a sharp decrease in shower strength. What Jupiter took away, though, it gave back. Further perturbations of the comet's orbit caused a spectacular Leonid shower in 1966—which one of the authors of this text witnessed—that may have produced as many as a half-million meteors per hour! The Leonid shower put on less spectacular but still impressive shows between 1999 and 2003, when several thousand meteors per hour were seen (**Figure 12.28**).

Zodiacal Dust

In the same way that we can "see" sunlight streaming through an open window by observing its reflection from motes of dust drifting in the air, we can see the sunlight reflected off tiny **zodiacal dust** particles that fill the inner parts of the Solar System close to the plane of the ecliptic. On a clear, moonless night, not long after the western sky has grown dark, this dust is visible as a faint column of light slanting upward from the western horizon along the path of the ecliptic. This band, called the **zodiacal light**, can also be seen in the eastern sky before dawn (**Figure 12.29**). With good eyes and an especially dark night, you may be able to follow the zodiacal dust band all the way across the sky. In its brightest parts, the zodiacal light can be several times brighter than the Milky Way, for which it is sometimes mistaken.

Like meteoroids, zodiacal dust is a mixture of cometary debris and ground-up asteroidal material. The dust grains are roughly a millionth of a meter in diameter—the size of smoke particles. In the vicinity of Earth there are only a few particles of zodiacal

FIGURE 12.28 Leonid meteors seen in 2001. This image is a summation of eight individual exposures with a total exposure time of about an hour.

FIGURE 12.29 Zodiacal light shines in the eastern sky before dawn.

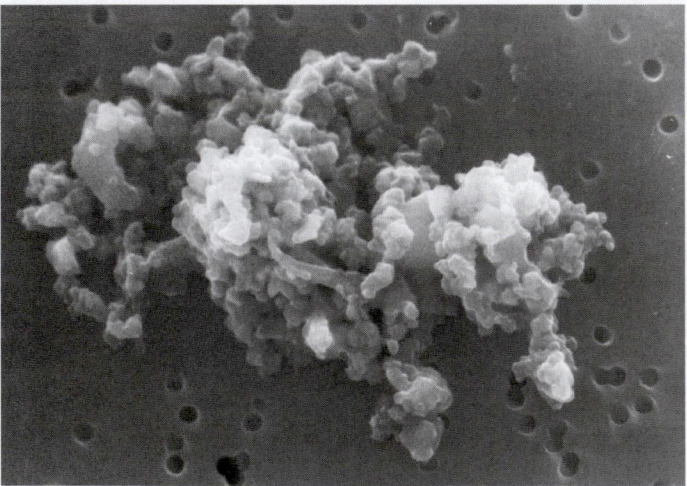

FIGURE 12.30 A 10-µm-diameter cluster of interplanetary dust grains collected in Earth's stratosphere by a NASA U-2 airplane.

dust in each cubic kilometer of space. The total amount of zodiacal dust in the entire Solar System is estimated to be 10^{16} kg, equivalent to a solid body about 25 km across, or roughly the size of a large comet nucleus. Grains of zodiacal dust are constantly being lost as they are swept up by planets or pushed out of the Solar System by the pressure of sunlight. Such interplanetary dust grains have been recovered from Earth's upper atmosphere by aircraft flying very high (**Figure 12.30**). If not replaced by new dust from comets, all zodiacal dust would be gone within a brief span of 50,000 years.

> **Zodiacal dust is cometary debris and ground-up asteroidal material.**

In the infrared region of the spectrum, thermal emission from the band of warm zodiacal dust makes it one of the brightest features in the sky. It is so bright that astronomers wanting to observe faint infrared sources are frequently hindered by its foreground glow.

Seeing the Forest for the Trees

Long before their true nature was understood, comets were granted great significance by humankind. These spectacular celestial displays were taken as omens of great events and, more often than not, as harbingers of the end of the world. How ironic that they have turned out instead to be messengers from the time when our world was born.

Anyone who has read a mystery novel knows that sometimes things that seem least significant at first glance turn out to hold the crucial clues to the biggest questions. In our study of the Solar System, two questions rise above all others: how did the Solar System form, and what is its history? Planets may dominate the environment around the Sun, but planets keep imperfect records of the earliest days of the Solar System. The violence of the planet formation process and eons of geological activity together effectively erase most clues to their origin. The best samples of the early Solar System that have survived to the present day are instead the smallest bodies: asteroids and especially comets, frozen denizens of the outer reaches of the Sun's influence. Buried within these dirty snowballs are grains that pre-date even the Solar System itself.

These tiny bits of solid material formed in the atmospheres of stars and in material blasted into space in tremendous stellar explosions, survived for a time in the vast reaches between the stars, participated in the collapse of the interstellar cloud that would become the Solar System, and wound up embedded within the small bodies that are the most numerous citizens of our planetary system. The discovery of such grains provides a direct link between our existence and our origins in the stars. How remarkable and fortuitous it is that these very pieces of our Solar System's past are delivered to our doorstep, falling to Earth as meteorites to be picked up from a cornfield in the Midwest or a glacier in the Antarctic, and then deciphered using the most advanced tools modern science has to offer.

Grains that pre-date the Sun make up only the tiniest fraction of the material found in meteorites. Most of this material was instead formed along with the Solar System itself. This material was vaporized in the violence that accompanied the formation of the disk around the young Sun and then condensed again into solid form. Some asteroids and all comets are formed directly from this pristine material. Meteorites broken off from such bodies provide a window into the conditions that existed at that time. Other asteroids instead carry clues about how planets formed. Iron and stony-iron meteorites, for example, are pieces broken off from what were once larger, differentiated bodies. Pick up a stony-iron meteorite and you hold in your hand a snapshot, frozen in time, of the processes that shaped the world.

Comets and asteroids are far more than scientific curiosities; they have played a major role in shaping the history of life on our planet. Your body is largely water, and it is quite likely that much of that water (along with the water that covers the surface of our world) arrived on Earth billions of years ago as volatile-rich comets slammed into the surface of our young world. Across the ages, occasional impacts of comets and asteroids on Earth have dramatically altered the planet's climate for a time and redirected the course of life's flow. Intelligent life on Earth descended from mammals only because of a cosmic fluke—the impact of a comet, 65 million years ago, in the region that is now the Yucatán Peninsula. The awe-inspiring fireballs that accompanied the impact of Comet Shoemaker-Levy 9 on Jupiter and the devastation of a remote corner of Siberia by a small piece of a comet or asteroid that hit Earth's atmosphere in 1908 are reminders that there is a grain of truth in humankind's deep-seated superstitions about these objects. Sometimes the appearance of a comet *does* mean the end of the world as we know it. It has happened in the past, and unless we develop technology to prevent such events (not an easy task), it *will* happen again in the future.

However, as we look to the future of human exploration and utilization of the Solar System, asteroids and comets may play a vital and positive role as ready-made way stations and caches of raw materials that we would need if we were to expand beyond our home planet. The history and destiny of our kind, as well as the course of our intellectual journey through the universe, are inextricably tied to this flotsam and jetsam adrift in interplanetary space.

So far on our journey, we have spent our time digging in our own backyard. The Solar System may dwarf the scales of our everyday lives, but it is vanishingly small compared with the universe. Just as we found that we could understand our own planet and its history only by putting it within the context of the Solar System, we are unable to understand the Sun and the worlds of our Solar System without placing them within the context of the broader universe. The next leg of our journey begins as we gaze at the myriad stars that fill the night sky and wonder about what we see there.

Summary

- Pluto, Eris, Haumea, Makemake, and Ceres are classified as dwarf planets.

- Asteroids are small Solar System bodies made of rock and metal.

- The orbits of most asteroids are located between the orbits of Mars and Jupiter.

- Some asteroids cross Earth's orbit and are potentially dangerous.

- Comets are small, icy planetesimals that reside in the frigid regions beyond the planets.

- Comets that venture into the inner Solar System are warmed by the Sun, producing an atmospheric coma and a long tail.

- Very large asteroids or comets striking Earth create enormous explosions that can wipe out most of terrestrial life.

- Meteoroids are small fragments of asteroids and comets.

- When a meteoroid enters Earth's atmosphere, frictional heat causes the air to glow, producing a phenomenon called a meteor.

- Meteor showers occur when Earth passes through a trail of cometary debris.

- A meteoroid that survives to a planet's surface is called a meteorite.

SmartWork, Norton's online homework system, includes algorithmically generated versions of these questions, plus additional conceptual exercises. If your instructor assigns questions in SmartWork, log in at **smartwork.wwnorton.com.**

Student Questions

THINKING ABOUT THE CONCEPTS

1. This chapter deals with leftover planetesimals. What became of the others?

2. Pluto's designation as a dwarf planet continues to be a source of controversy among astronomers. Discuss arguments for and against its being listed as a planet. What is your opinion and why?

3. Describe ways in which Pluto differs significantly from the classical Solar System planets.

4. If the dwarf planet Eris is larger than Pluto, why is it nearly 100 times fainter?

5. How does the composition of an asteroid differ from that of a comet nucleus?

6. Define *meteoroid*, *meteor*, and *meteorite*.

7. What are the differences between a comet and a meteor in terms of their size, distance, and how long they remain visible?

*8. Most meteorites (pieces of S-type and M-type asteroids) are 4.54 billion years old. Carbonaceous chondrites (pieces of C-type asteroids), however, are 20 million years older. What determines the time of "birth" of these pieces of rock? What does this information tell you about the history of their parent bodies?

9. Most meteorites are pieces of asteroids, but a few came from elsewhere. What is the origin of these nonasteroidal meteorites?

10. Most asteroids are found between the orbits of Mars and Jupiter, but astronomers are especially interested in the relative few whose orbits cross that of Earth. Why?

*11. How could you and a friend, armed only with your cell phones and a knowledge of the night sky, prove conclusively that meteors are an atmospheric phenomenon?

*12. Suppose you find a rock that has all of the characteristics of a meteorite. You take it to a physicist friend who confirms that it is a meteorite but says that radioisotope dating indicates an age of only a billion years. What might be the origin of this meteorite?

13. Describe differences between the Kuiper Belt and the Oort Cloud as sources of comets.

14. Name the three parts of a comet. Which part is the smallest? Which is the most massive?

15. Kuiper Belt objects (KBOs) are actually comet nuclei. Why do they not display comae and tails?

16. Comets contain substances closely associated with the development of life, such as water (H_2O), ammonia (NH_3), methane (CH_4), carbon monoxide (CO), and hydrogen cyanide (HCN). Which of these are organic compounds?

17. In 1910, Earth passed directly through the tail of Halley's Comet. Among the various gases in the tail was hydrogen cyanide, deadly to humans. Yet nobody became ill from this event. Why?

18. Picture a comet leaving the vicinity of the Sun and racing toward the outer Solar System. Does its tail point backward toward the Sun or forward in the direction the comet is moving? Explain your answer.

19. Comets have two types of tails. Describe them and explain why they sometimes point in different directions.

20. Explain the orbital terms *perihelion* and *aphelion*.

21. What is the ultimate fate of every short-period comet?

22. If collisions of comet nuclei and asteroids with Earth are rare events, why should we be concerned about the possibility of such a collision?

23. What is the source of meteors we see during a meteor shower?

*24. During a meteor shower, all meteors trace back to a single region in the sky known as the radiant. Explain why.

25. What is zodiacal light, and what is its source?

APPLYING THE CONCEPTS

26. Our Moon has a diameter of 3,474 km and orbits at an average distance of 384,400 km. At this distance it subtends an angle just slightly larger than half a degree in our sky. Pluto's moon Charon has a diameter of 1,186 km and orbits at a distance of 19,600 km from the dwarf planet.
    a. Compare the appearance of Charon in Pluto's skies with the Moon in our own skies.
    b. Describe where in the sky Charon would appear as seen from various locations on Pluto.

*27. One recent estimate concludes that nearly 800 meteorites with mass greater than 100 grams (massive enough to cause personal injury) strike the surface of Earth each day. Assuming that you present a target of 0.25 square meter (m^2) to a falling meteorite, what is the probability that you will be struck by a meteorite during your 100-year lifetime? (Note that the surface area of Earth is approximately 5×10^{14} m^2.)

28. Electra is a 182-km-diameter asteroid accompanied by a small moon orbiting at a distance of 1,350 km in a circular orbit with a period of 3.92 days. Refer back to Chapter 3 to answer the following questions.
    a. What is the mass of Electra?
    b. What is Electra's density?

29. The orbital periods of Comets Encke, Halley, and Hale-Bopp are 3.3 years, 76 years, and 2,530 years, respectively.
    a. What are the semimajor axes (in astronomical units) of the orbits of these comets?
    b. Assuming negligible perihelion distances, what are the maximum distances from the Sun (in astronomical units) reached by Comets Halley and Hale-Bopp in their respective orbits?
    c. Which would you guess is the most pristine comet among the three? Which is the least? Explain your reasoning.

*30. Comet Halley has a mass of approximately 2.2×10^{14} kg. It loses about 3×10^{11} kg each time it passes the Sun.
    a. The first confirmed observation of the comet was made in 230 B.C. Assuming a constant period of 76.4 years, how many times has it reappeared since that early sighting?
    b. How much mass has the comet lost since 230 B.C.?
    c. What percentage of its total mass does this amount represent?

31. A cubic centimeter of the air you breathe contains about 10^{19} molecules. A cubic centimeter of a comet's tail may typically contain 10 molecules. Calculate the size of a cubic volume of comet tail material that would hold 10^{19} molecules.

*32. The total number of asteroids larger than 1,000 meters in diameter that cross Earth's orbit (Aten and Apollo asteroids) is currently estimated to be about 3,500, with five times as many having diameters larger than 500 meters. Assuming this progression of number versus size remains constant for still smaller asteroids, how many would there be with diameters larger than 125 meters? (Note that the impact on Earth of any asteroid larger than 100 meters in diameter could cause major, widespread damage.)

*33. A 1-megaton hydrogen bomb releases 4.2×10^{15} joules (J) of energy. Compare this amount of energy with that released by a 10-km-diameter comet nucleus ($m = 5 \times 10^{14}$ kg) hitting Earth at a speed (v) of 20 km/s. Recall from Chapter 4 that $E_K = \frac{1}{2}mv^2$ (where E_K is the kinetic energy in joules, m is the mass in kilograms, and v is the speed in meters per second).

34. The estimated amount of zodiacal dust in the Solar System remains constant at approximately 10^{16} kg. Yet zodiacal dust is constantly being swept up by planets or removed by the pressure of sunlight.
    a. If all the dust would disappear (at a constant rate) over a span of 30,000 years, what would the average production rate, in kilograms per second, have to be to maintain the current content?
    b. Is this an example of static or dynamic equilibrium? Explain your answer.

StudySpace is a free and open Web site that provides a Study Plan for each chapter of *21st Century Astronomy*. Study Plans include animations, reading outlines, vocabulary flashcards, and multiple-choice quizzes, plus links to premium content in SmartWork and the ebook. Visit **www.wwnorton.com/studyspace**.

To man, that was in th' evening made,
Stars gave the first delight;
Admiring, in the gloomy shade,
Those little drops of light.

EDMUND WALLER (1606–1687)

Stars on the Hertzsprung-Russell (H-R) diagram.

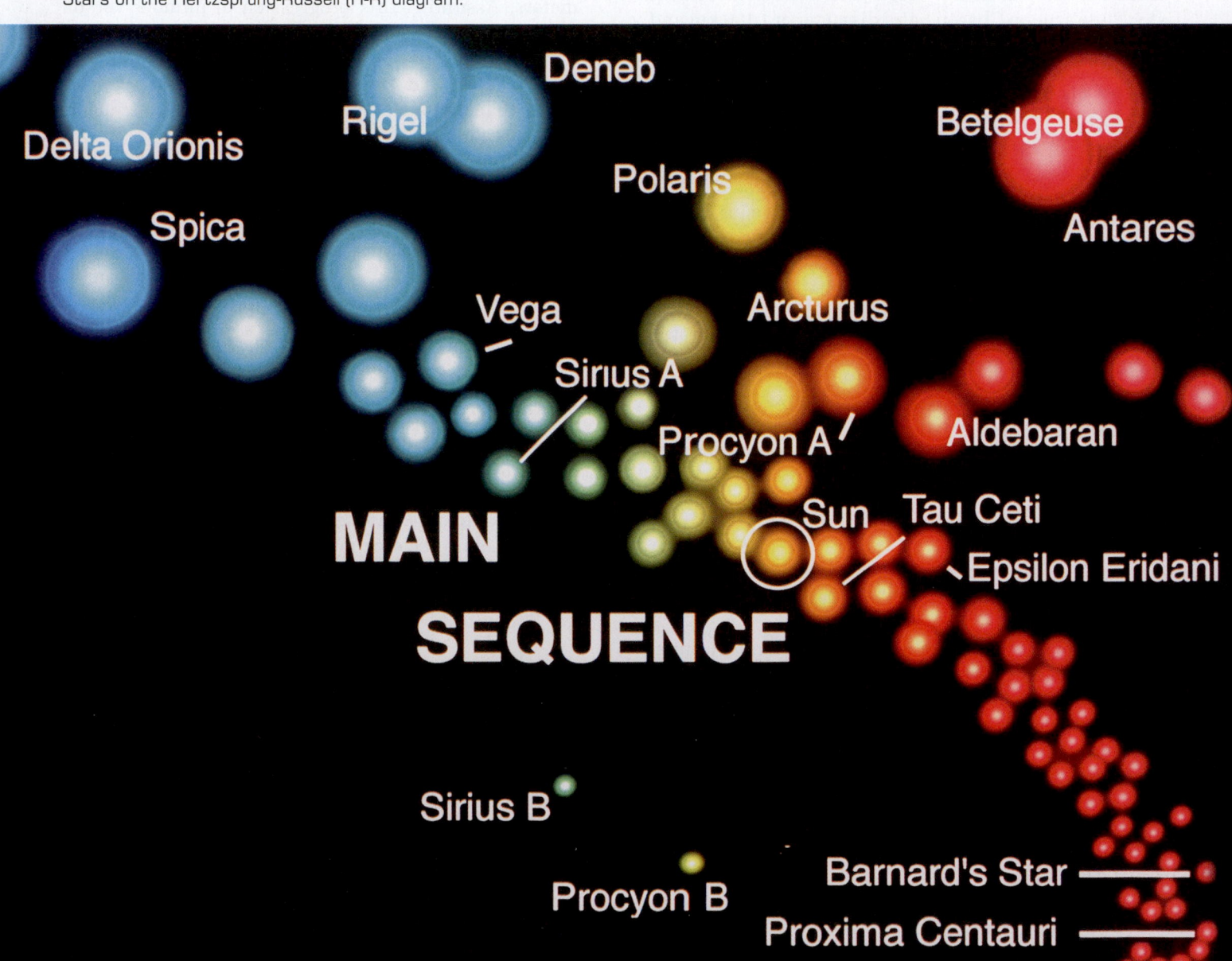

CHAPTER 13

Taking the Measure of Stars

13.1 Twinkle, Twinkle, Little Star, How I Wonder What You Are

As children, we look at the sky, see the *stars*, and may wonder, What are those points of light? How far away are they? How bright are they? How hot is a star? How big is a star? How much does a star "weigh"? What are stars made of? How long does a star last? What makes a star shine? These questions have always been a part of the human experience.

For most of human history, **stars**, like so much of nature, have seemed mysterious and unknowable. Even so, their passage follows the daily and annual rhythms that shape our lives. It is no wonder that in the absence of any real understanding of the stars, ancient humans turned their imaginations loose and viewed the stars as the province of gods and magic, with power over our lives. Today we understand that the patterns of the constellations have no more mystical influence over the course of events than does the random toss of a coin (even though some people persist in believing the ramblings of astrologers).

The story of how we know what we know about the stars is a wonderful tale. It starts with simple but ingenious observations of the sky, borrows from lessons learned about the behavior of matter on Earth and in the Solar System, and ends with clear, straightforward answers to the very questions that a child might ask while gazing at the sky.

Unlike our exploration of the Solar System, we cannot send space probes to a star to take close-up pictures or land on its surface! Instead, we study the stars by observing their light, by using some of the laws of physics discussed in earlier chapters, and by finding patterns in subgroups of stars

KEY CONCEPTS

To all but the most powerful of telescopes, a star is just a point of light in the night sky. But by applying our understanding of light, matter, and motion to what we see, we are able to build a remarkably detailed picture of the physical properties of stars. As we take this first step beyond our local Solar System, we will

- Apply a form of stereoscopic vision, extended by Earth's orbit, to measure distances to nearby stars.

- Use the brightness of stars and their distances from Earth to discover how luminous they are.

- Use our knowledge of thermal radiation to infer the temperatures and sizes of stars from their colors.

- Measure the composition of stars from spectra.

- Study the orbits of binary stars and use Kepler's laws to calculate their masses.

- Classify stars, and organize this information on a plot of luminosity versus temperature called the Hertzsprung-Russell, or H-R, diagram.

- Discover that 90 percent of the stars we see lie along a well-defined "main sequence" in the H-R diagram.

- Discover that the mass and composition of a main-sequence star determine its luminosity, temperature, and size.

- Explore the range of stellar properties, learning how our Sun compares to other stars.

that enable us to extrapolate to most other stars. A bit of geometry, a bit about radiation, a bit about orbits—all things that we studied in earlier chapters—and we begin to find ourselves with solid answers to age-old questions.

13.2 The First Step Is Measuring the Distance, Brightness, and Luminosity of Stars

In a sense, one of the most amazing feats that a human can perform is to catch a fly ball. Here comes the ball, traveling at speeds of 150 kilometers per hour (km/h) or more. To put a glove on the ball, the fielder must judge not only the direction to the ball but also its distance. How does the fielder do it? Like most predators, we have two forward-looking eyes, one on each side of the face. Our two eyes have somewhat different views of the world. Exactly how different these perspectives are depends on the distance to the object you are looking at. If you are looking at a house down the street and you blink back and forth between your two eyes, your view changes very little. But if you are looking at a nearby object—perhaps a finger held up at arm's length—the perspectives of your two eyes differ quite a lot. This difference in our eyes' perspectives on objects at different distances is the basis of our **stereoscopic vision**. Put another way, stereoscopic vision is the major key to the way we perceive

distances. (**Figure 13.1** shows that if each eye is shown a different view, the brain can even be fooled into perceiving distance where none exists.)

Our stereoscopic vision allows us to judge the distances of objects as far away as a few hundred meters, but beyond that it is of little use. Your right eye's view of a mountain several kilometers away is indistinguishable from the view seen by your left eye. Comparing the two views tells your brain only that the mountain is too far away for you to judge its distance. (Evolution provided us with only enough depth perception to judge distances to things that we might be trying to eat or that might be trying to eat us!) The distance over which our stereoscopic vision works is limited by the separation between our two eyes, which is only about 6 centimeters (cm). If you wanted to increase the differences between the views your two eyes see, the obvious thing to do would be to somehow move them farther apart. If you could separate your eyes by several meters, their perspectives would be different enough to allow you to judge the distances to objects that were kilometers away.

Of course, we cannot literally take our eyes out of our heads and hold them apart at arm's length, but we can compare pictures taken from two widely separated locations. The greatest separation we can get without leaving Earth is to let Earth's orbital motion carry us from one side of the Sun to the other. If we take a picture of the sky tonight and then wait 6 months and take another picture, our point of view between the two pictures will have changed by the diameter of Earth's orbit, or 2

> We measure distances to nearby stars by comparing the view from opposite sides of Earth's orbit.

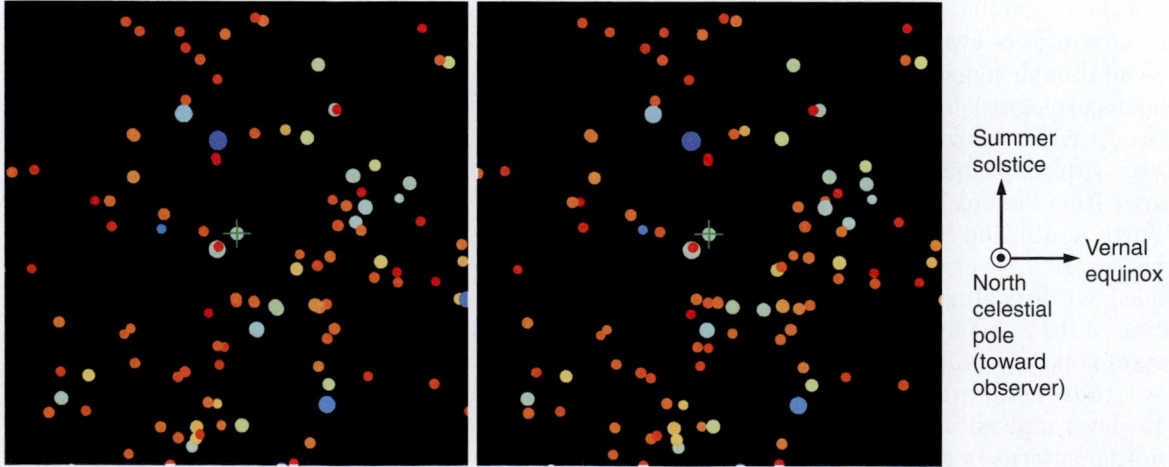

FIGURE 13.1 Your brain uses the slightly different views offered by your two eyes to "see" the distances and three-dimensional character of the world around you. This stereoscopic pair shows the stars in the neighborhood of the Sun as viewed from the direction of the north celestial pole. The field shown is 40 *light-years* on a side. The Sun is at the center, marked with a green cross. The observer is 400 light-years away and has "eyes" separated by about 30 light-years. To get the stereoscopic view, hold a card between the two images and look at them from about a foot and half away. Relax and look straight at the page until the images merge into one.

Summer solstice

Vernal equinox

North celestial pole (toward observer)

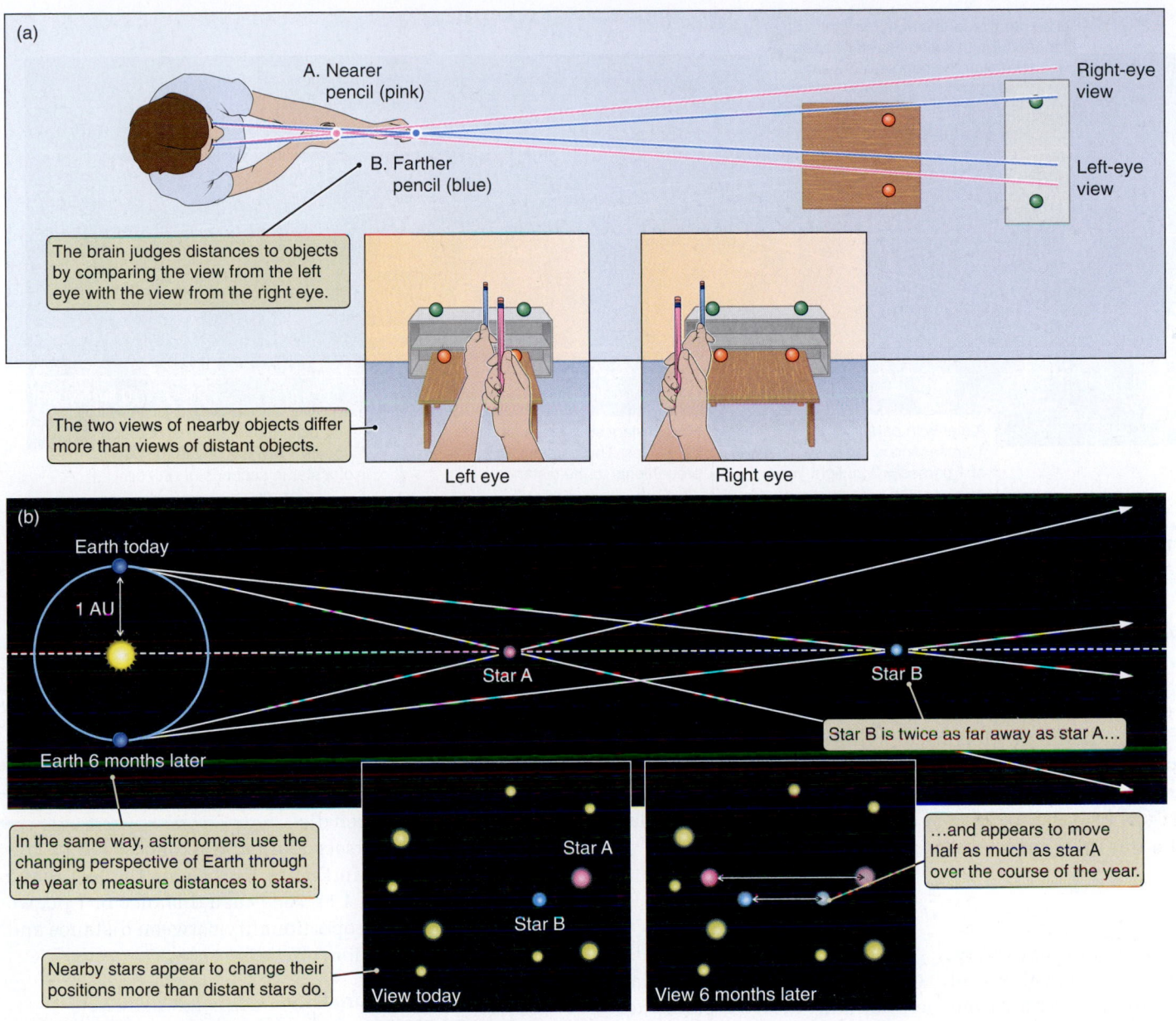

FIGURE 13.2 As we move around the Sun, the apparent positions of nearby stars change more than the apparent positions of more distant stars. (The diagram is not to scale.) This is the starting point for measuring the distances to stars.

astronomical units (AU). With 2 AU separating our two "eyes," we should have very powerful stereoscopic vision indeed. **Figure 13.2** shows how our view of a field of stars changes as our perspective changes during the year. This change in perspective is what enables us to measure the distances to nearby stars.

We Use Parallax to Measure Distances to Nearby Stars

The eye cannot detect the changes in position of a nearby star throughout the year, but telescopes can reveal these small shifts relative to the background stars. **Figure 13.3**

shows Earth, the Sun, and three stars. Look first at star 1. When Earth, the Sun, and the star are in this position, they form a long, skinny right triangle. The short leg of the triangle is the distance from Earth to the Sun, or 1 AU. The long leg of the triangle is the distance from the Sun to the star. The small angle at the end of the triangle is called the "parallactic angle," or simply **parallax**, of the star. Over the course of a year, the star's position in the sky appears to shift back and forth, returning to its original position 1 year later. The amount of this shift—the angle between one extreme in the star's apparent motion and the other—is equal to twice the parallax.

The more distant the star, the longer and skinnier the triangle that it forms, and the smaller the star's parallax.

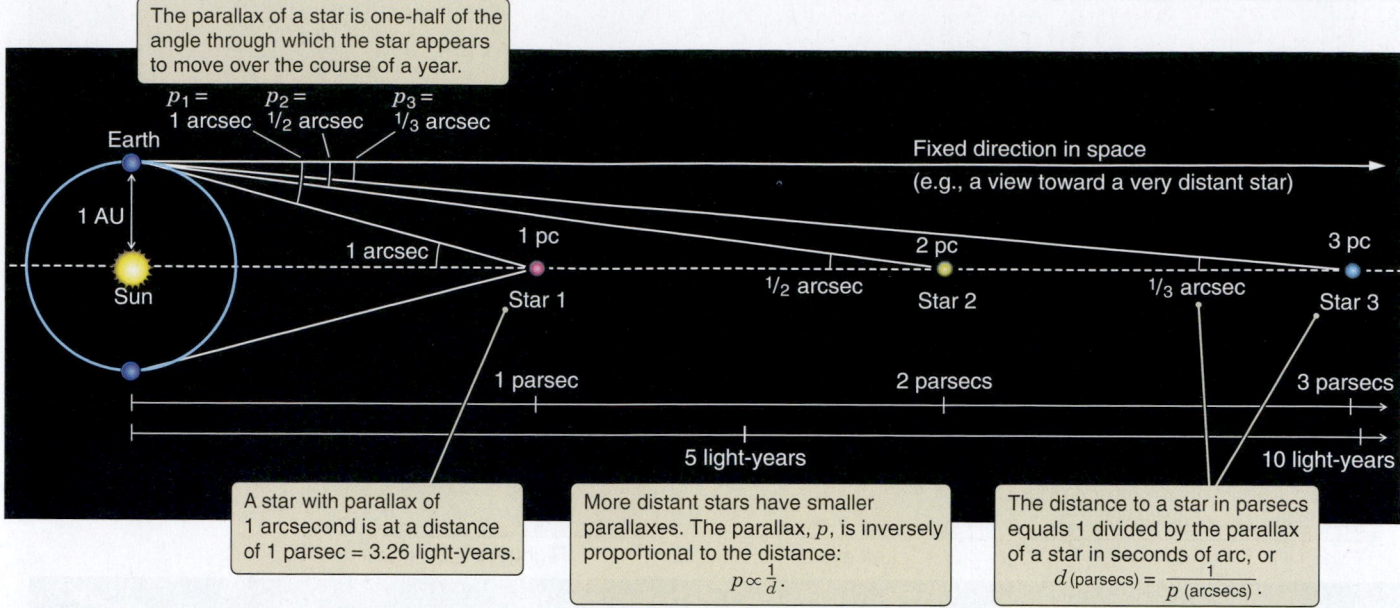

The parallax of a star is one-half of the angle through which the star appears to move over the course of a year.

$p_1 =$ 1 arcsec $p_2 =$ 1/2 arcsec $p_3 =$ 1/3 arcsec

Earth

Fixed direction in space (e.g., a view toward a very distant star)

1 AU

Sun

1 arcsec

1 pc
Star 1

1/2 arcsec

2 pc
Star 2

1/3 arcsec

3 pc
Star 3

1 parsec   2 parsecs   3 parsecs

5 light-years   10 light-years

A star with parallax of 1 arcsecond is at a distance of 1 parsec = 3.26 light-years.

More distant stars have smaller parallaxes. The parallax, p, is inversely proportional to the distance: $p \propto \frac{1}{d}$.

The distance to a star in parsecs equals 1 divided by the parallax of a star in seconds of arc, or $d \text{(parsecs)} = \frac{1}{p \text{(arcsecs)}}$.

FIGURE 13.3 The parallax of three stars at different distances. (The diagram is not to scale.) Parallax is inversely proportional to distance.

Look again at Figure 13.3. Star 2 is twice as far away as star 1, and its parallax is only half as great. If you were to draw a number of such triangles for different stars, you would find that increasing the distance to the star always reduces the star's parallax. Move a star 3 times farther away, as with star 3 in Figure 13.3, and you reduce its parallax to ⅓ of its original value. Move a star 10 times farther away, and you reduce its parallax to ¹⁄₁₀ of its original value. The parallax of a star (p) is inversely proportional to its distance (d):[1]

$$p \propto \frac{1}{d} \text{ or } d \propto \frac{1}{p}.$$

The parallaxes of real stars are tiny. Rather than talking about parallaxes of 0.0000028° or 4.8×10^{-8} radian (1 **radian** = 57.3°), astronomers normally measure parallaxes in units of seconds of arc. Just as an hour on the clock is divided into minutes and seconds of time, a degree can be divided into minutes and seconds of arc. A minute of arc (or **arcminute**, abbreviated **arcmin**) is ¹⁄₆₀ of a degree, and a second of arc (or **arcsecond**, abbreviated **arcsec**) is ¹⁄₆₀ of a minute of arc.[2] That makes a second of arc 1/3,600 of a degree, or 1/1,296,000 of a complete circle. (The apparent diameter of the full Moon in the sky varies from 29 to 34 arcmin, averaging just over half a degree.) An arcsecond is about equal to the angle formed by the diameter of a golf ball at the distance of 5 miles.

[1]Note that parallax is inversely proportional to distance only if the parallax is tiny, as it is for stars. For nearby objects, it is the *tangent* of the parallax that is inversely proportional to the distance.
[2]Arcseconds are often denoted by the symbol ″ and arcminutes by ′; one second of arc is written as 1″, and one minute of arc is written as 1′. In this text, however, we will spell out these units.

If the angle at the apex of a triangle is 1 arcsec, and the base of the triangle is 1 AU, then the length of the triangle is 206,265 AU (see Appendix 1). This distance, which corresponds to 3.09×10^{16} meters, or 3.26 light-years, is referred to as a **parsec** (abbreviated **pc**). The relationship between distance measured in parsecs and parallax measured in arcseconds is illustrated in Figure 13.3. Using the fact that a star with a parallax of 1 arcsec is at a distance of 1 pc, we can turn the inverse proportionality between distance and parallax into an equation:

> A parsec is defined as the distance at which the parallax equals 1 arcsecond.

$$\left(\begin{array}{c}\text{Distance measured} \\ \text{in parsecs}\end{array}\right) = \frac{1}{\left(\begin{array}{c}\text{Parallax measured} \\ \text{in arcseconds}\end{array}\right)}.$$

If you measure the parallax of a star to be 0.5 arcsec, then you know the star is located at a distance of 1/0.5 = 2 pc. A star with a parallax of 0.01 arcsec is located at a distance of 1/0.01 = 100 pc.

In this book we will usually use units of **light-years** to indicate distances to stars and galaxies. One light-year is the distance that light travels in 1 year—about 9 trillion kilometers (km). We use this unit because it is the unit you are most likely to see in a newspaper article or a popular book about astronomy. When astronomers discuss distances to stars and galaxies, however, the unit they often use is the parsec. (You can always convert between the two units. One parsec equals 3.26 light-years.)

The star closest to us (other than the Sun) is Proxima Centauri. Located at a distance of 4.22 light-years, or 1.3 pc, Proxima Centauri is a faint member of a system of three stars called Alpha Centauri. This star has a parallax of only about ¾ arcsec. It is no wonder that ancient astronomers were unable to detect the apparent motions of the stars over the course of a year!

When astronomers began to apply this technique, they discovered that stars are very distant objects indeed. The first successful measurement of the parallax of a star was made by F. W. Bessel, who in 1838 reported a parallax of 0.314 arcsec for the star 61 Cygni. This finding implied that 61 Cygni was 3.2 pc away, or 660,000 times as far away as the Sun. With this one measurement, Bessel increased the known size of the universe 10,000-fold! Today we know of 65 stars in 50 single-, double-, or triple-star systems within 5 pc (16.3 light-years) of the Sun. A sphere with a radius of 16.3 light-years has a vol-

Stars are few and far between in our neighborhood.

ume of about 18,000 cubic light-years, corresponding to a local density of 50 systems per 18,000 cubic light-years. That is about 0.003 star system per cubic light-year. Stated another way, in the neighborhood of the Sun each system of stars has, on average, about 360 cubic light-years of space (a volume about 4.4 light-years in radius) all to itself.

Knowledge of our stellar neighborhood took a tremendous step forward during the 1990s with the completion of the Hipparcos mission. The Hipparcos satellite measured the positions and parallaxes of 120,000 stars. These measurements, taken from a satellite well above Earth's obscuring atmosphere, are better than the measurements that can typically be made from telescopes located on the surface of Earth. But even this catalog has its limits. The accuracy of any given Hipparcos parallax measurement is about ±0.002 arcsec. Because of this **observational uncertainty**, our measurements of the distances to stars are not perfect. For exam-

---

MATH TOOLS 13.1

The Magnitude System

The **magnitude** system of brightness for celestial objects can be traced back 2,000 years to the Greek astronomer Hipparchus, who classified the brightest stars he could see as being "of the first magnitude" and the faintest as being "of the sixth magnitude." Later, astronomers defined Hipparchus's 1st magnitude stars as being exactly 100 times brighter than his 6th magnitude stars. With five steps between 1st and 6th magnitudes, each step is thus equal to the fifth root of 100, or $100^{1/5}$, which is a factor of approximately 2.512. To put it another way, $(2.512)^5 = 100$.

Let's see how this works. The magnitude system says that 5th magnitude stars are 2.512 times brighter than 6th magnitude stars, and that 4th magnitude stars are $2.512 \times 2.512 = 6.310$ times brighter than 6th magnitude stars. Continuing with this progression, we find that 3rd and 2nd magnitude stars are, respectively, $(2.512)^3$ and $(2.512)^4 = 15.85$ and 39.81 times brighter than 6th magnitude stars. We can now put this progression into a more useful expression. The brightness ratio between any two stars is equal to $(2.512)^N$, where N is the magnitude difference between them. Note that this is really a backward system, in which a larger magnitude refers to a fainter object.

Hipparchus must have had typical eyesight, because an average person under dark skies can see stars only as faint as 6th magnitude. How does this visual limit compare with that of our most powerful telescopes? Hubble Space Telescope can detect stars as faint as 30th magnitude. Since $N = 30 - 6 = 24$, HST can detect stars that are $(2.512)^{24} = 4 \times 10^9$, or 4 billion, times fainter than the naked eye can see!

So far so good, but what do we do with stars that are brighter than 1st magnitude? The answer in this backward system is that we go first to zero magnitude and then to negative num-

bers. For example, Sirius, the visually brightest star in the sky, has a magnitude of −1.46. Venus can be as bright as magnitude −4.4, or about 15 times brighter than Sirius and bright enough to cast a shadow. The magnitude of the full Moon is −12.6, and that of the Sun is −26.7. Thus, the Sun is 14.1 magnitudes—or more than 400,000 times—brighter than the full Moon.

The magnitude of a star, as we have discussed it, is called the star's **apparent magnitude** because it reflects the brightness of the star as it *appears* to us in our sky. As we know, however, stars are found at different distances from us, so a star's apparent brightness does not provide a clue to its luminosity (see Chapter 4). But suppose all stars *were* located at the same distance from us. The brightness of each star would then be representative of its luminosity. In principle, astronomers make exactly that assumption. If the distance to a star is known, they assign a brightness to each star as if it were located at a standard distance—in this case, 10.0 pc (32.6 light-years). Astronomers refer to this fundamental property of a star as its **absolute magnitude**.

The brightness of astronomical objects generally varies with spectral region (color), and astronomers use special symbols to represent magnitudes in various colors. For example, they use V and B, respectively, to represent magnitudes in the visual (green) and blue regions of the spectrum. Recall 51 Pegasi from Chapter 6—the first solar-type star found to have an orbiting planet. This star has a visual magnitude, V, of 5.49. Its blue magnitude, B, is 6.16 (a larger number), indicating that it is less bright in blue light than in visual (green) light. With these numbers, we can assign the star's *color index*, B-V, as $6.16 − 5.49 = 0.67$ magnitude (see Appendix 6 and the second footnote accompanying Table 13.3).

ple, a star with a parallax measured by Hipparcos of 0.004 arcsec might really have a parallax of anywhere between 0.002 and 0.006 arcsec. So instead of knowing that the distance to the star is exactly 250 pc (1 divided by 0.004 arcsec), we know only that the star is probably between about 170 pc (1/0.006 arcsec) and 500 pc (1/0.002 arcsec). With current technology, parallax becomes useless as a way of measuring stellar distances for stars more than a few hundred parsecs away. If you are measuring stellar distances using parallax, and you need to know the distance to an accuracy of 10 percent or better, you are restricted to stars that are less than about 50 pc (160 light-years) away.

Once We Know Distance and Brightness, We Can Calculate Luminosity

When we talk about the brightness of an object, we are making a statement about how that object appears to us. In Chapter 4 we saw that brightness corresponds to the amount of energy falling on a square meter of area each second in the form of electromagnetic radiation. (When astronomers talk about the brightness of stars, they usually use a system called *magnitude*, discussed in **Math Tools 13.1** and Appen-

dix 6.) But although the brightness of a star is directly measurable, it does not immediately tell us much about the star itself. As illustrated in **Figure 13.4**, a bright star in the night sky may in fact be a dim one, appearing bright only because it is nearby. Conversely, a faint star may be a powerful beacon, still visible despite its tremendous distance.

> Brightness depends on the observer's perspective; luminosity does not.

To learn about the stars themselves, we need to know the total energy radiated by a star each second—the star's **luminosity**. In Chapter 4 we studied the relationships between the brightness, luminosity, and distance of objects. Borrowing from that earlier work, recall that the brightness of an object that has a known luminosity and is located at a distance d is given by the following equation:

$$\text{Brightness} = \frac{\text{Total light emitted per second}}{\text{Area of a sphere of radius } d}$$

$$= \frac{\text{Luminosity}}{4\pi d^2}.$$

We can rearrange this equation, moving the quantities we know how to measure (distance and brightness) to the right-hand side and the quantity we would like to know (luminosity) to the left, giving us

(a)

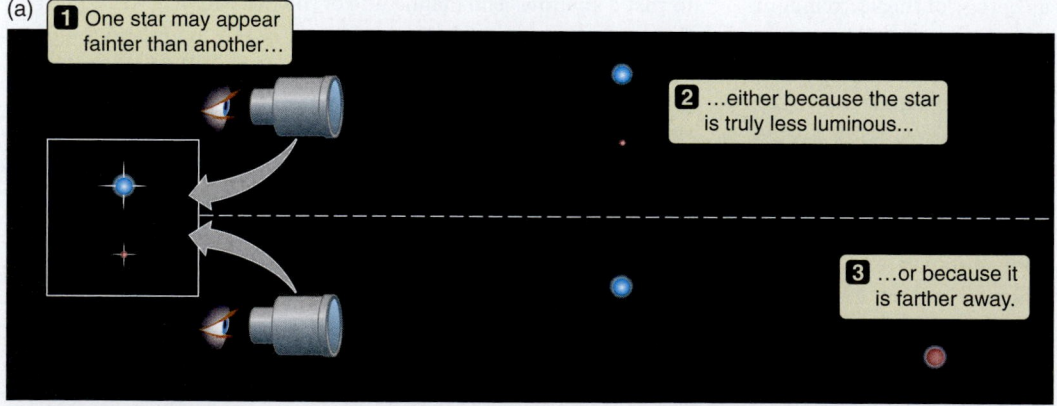

1 One star may appear fainter than another...

2 ...either because the star is truly less luminous...

3 ...or because it is farther away.

(b)

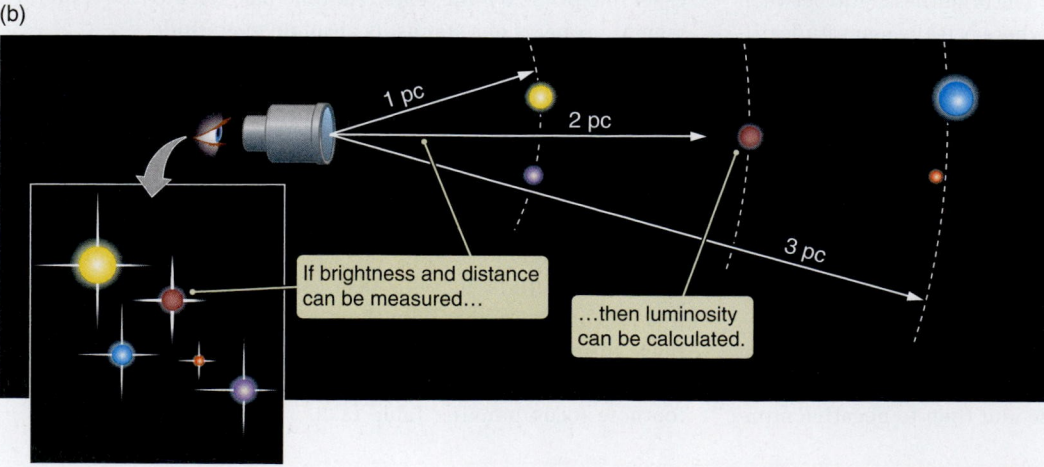

If brightness and distance can be measured...

...then luminosity can be calculated.

1 pc

2 pc

3 pc

FIGURE 13.4 The brightness of a visible star depends on both its luminosity and its distance. If brightness and distance are measured, luminosity can be calculated.

$$\text{Luminosity} = 4\pi d^2 \times \text{Brightness.}$$

This equation answers the question, How much total light must a star be giving off to appear as bright as it does at its distance from us? In the process, we take two measurable quantities that depend on our particular perspective (distance and brightness) and from them obtain a quantity that is a property of the star itself—namely, luminosity.

When measuring the luminosity of stars as just described, we find that stars vary tremendously in the amount of light they give off. The Sun provides a convenient yardstick for measuring the properties of stars, including their luminosity. The luminosity of the Sun is written as $L_\odot$. The most luminous stars can exceed 1,000,000 $L_\odot$, or 10^6 $L_\odot$ (a million times the luminosity of the Sun). The least luminous stars have luminosities less than 0.0001 $L_\odot$, or 10^{-4} $L_\odot$ (1/10,000 that of the Sun). The most luminous stars are therefore over 10 billion (10^{10}) times more luminous than the least luminous stars. Only a very small fraction of stars are near the upper end of this range of luminosities. The vast majority of stars are at the faint end of this distribution, less luminous even than our Sun. **Figure 13.5** shows the relative number of stars compared to their luminosity in solar units. (Distances for the nearest stars are obtained from their parallax; other methods—to be discussed later—are used for the more distant stars.)

> Some stars are 10 billion times more luminous than the least luminous stars.

> There are many more low-luminosity stars than high-luminosity stars.

FIGURE 13.5 The distribution of the luminosities of stars. Both axes are plotted logarithmically, as explained in Section 13.5.

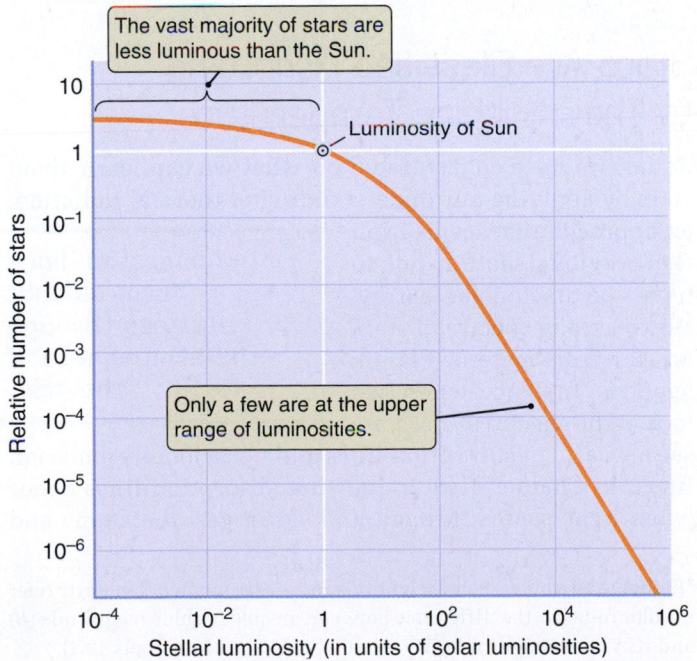

The vast majority of stars are less luminous than the Sun.

Luminosity of Sun

Only a few are at the upper range of luminosities.

Relative number of stars

Stellar luminosity (in units of solar luminosities)

13.3 Radiation Tells Us the Temperature, Size, and Composition of Stars

Two everyday concepts—stereoscopic vision, and the fact that the closer an object is the brighter it appears—have given us the tools we need to measure the distance and luminosity of relatively nearby stars. In these two steps, stars have gone from being merely faint points of light in the night sky to being extraordinarily powerful beacons located at distances almost impossible for the mind to comprehend. To continue on this journey of discovery, we again turn to the laws of radiation that we studied in Chapter 4.

Stars are gaseous, but they are fairly dense—dense enough that the radiation from a star comes close to obeying the same laws as the radiation from objects like the heating element on an electric stove or the filament in a lightbulb. That means we can use our understanding of Planck radiation—results such as the Stefan-Boltzmann law (hotter at same size means more luminous) and Wien's law (hotter means bluer)—to understand the radiation from stars. In particular, these two laws enable us to measure the temperatures and sizes of our stellar neighbors.

Wien's Law Revisited: The Color and Surface Temperature of Stars

The hotter the surface of an object, the bluer the light that it emits. Stars with especially hot surfaces are blue, stars with especially cool surfaces are red, and our Sun is a middle-of-the-road yellow. More formally, Wien's law states that

> Measuring the color of a star tells us its surface temperature.

$$T = \frac{2{,}900\ \mu\text{m K}}{\lambda_{\text{peak}}},$$

where λ_{peak} is the wavelength, in micrometers (μm, 10^{-6} meter), at which the electromagnetic radiation from a star is most intense. Obtain a spectrum of a star, measure the wavelength at which the spectrum peaks, and Wien's law will tell you the temperature, T, of the star's surface.[3]

In practice, it is usually not necessary to obtain a complete spectrum of a star to determine its temperature. Instead, stellar temperatures are often measured in a way that is similar to the way that your eyes see color. Your eyes and brain distinguish color by comparing how bright an

[3] It is important to remember that the color of a star tells us only about the temperature at the star's *surface*, because the surface is the part of the star giving off the radiation that we see. Star surfaces may be hot, but later in our journey we will discover that star interiors are far hotter still.

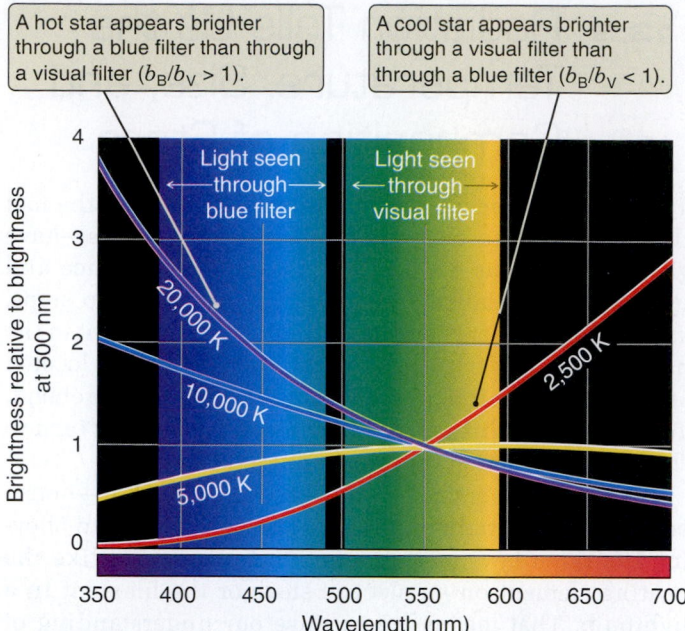

A hot star appears brighter through a blue filter than through a visual filter ($b_B/b_V > 1$).

A cool star appears brighter through a visual filter than through a blue filter ($b_B/b_V < 1$).

FIGURE 13.6 A star's color depends on its temperature. The Planck spectra (blackbody curves) shown here are adjusted so that they have the same brightness at 550 nm. A wider range of blackbody curves is shown in Figure 4.21.

object is at one wavelength with how bright it is at another wavelength. The human eye contains nearly 100 million light-sensitive cells, and among them are certain color-sensitive cells called cones. Cones come in three color-discriminating types, with their peak responses in red, green, and blue regions of the spectrum, respectively.[4] If you look at a very red lightbulb, the red-sensitive cones in your eye report a bright light, and the blue- and green-sensitive cones see less light. If you look at a yellow light, the green- and red-sensitive cones are strongly stimulated, but those with blue sensitivity are not. White light stimulates all three kinds of color-sensitive cells. By combining the signals from cells that are sensitive to different wavelengths of light, your brain can distinguish among fine shades of color.

Astronomers often measure the colors of stars in much the same way. The brightness of a star is usually measured through a **filter**—sometimes just a piece of colored glass—that lets through only a certain range of wavelengths. Two of the most common filters used by astronomers are a blue filter that transmits light with wavelengths of about 440 nanometers (nm), and a yellow-green filter that passes light with wavelengths of about 550 nm. The first of these filters is (sensibly enough) called a "blue" filter. By convention, however, the second of these filters is referred to as a "visual" filter rather than a "yellow-green" filter because it

transmits roughly the range of wavelengths to which our eyes are most sensitive.

Figure 13.6 shows four Planck spectra with temperatures of 2,500–20,000 kelvins (K), adjusted so that they are all the same brightness at 550 nm (the wavelength of the center of the range transmitted by the visual filter). A hot star, with a spectrum like the 20,000-K Planck spectrum shown, gives off more light in the blue part of the spectrum than in the visual part of the spectrum. If we were to divide the brightness of the star as seen through the blue filter by the brightness of the star as seen through the visual filter, the blue/visual brightness ratio (b_B/b_V) would be *greater* than 1. In contrast, a cool star with a spectrum more like the 2,500-K Planck spectrum shown is much fainter in the blue part of the spectrum than in the visual part of the spectrum. The ratio of blue light to visual light is *less* than 1 for the cool star. This ratio of brightness between the blue and visual filters is referred to as the **color index** of the star.[5] By convention, astronomers usually adjust the way they measure the blue and visual brightnesses of stars so that the star Vega (which has a surface temperature of about 10,000 K) has a *defined* color index of 1.00. On this scale, the Sun has a color index of 0.56.

The fact that the color of a star depends on the star's temperature is an extremely handy tool for astronomers. It means that a single pair of snapshots of a group of stars, each taken through a different filter, provides a measurement of the surface temperature of every star in our camera's field of view! This technique makes it possible to measure the temperatures of hundreds or even thousands of stars at once. When we do, we find that just as there are many more low-luminosity stars than high-luminosity stars, there are also many more cool stars than hot stars. We also discover that most stars have surface temperatures lower than that of the Sun. ▶‖ **AstroTour:** Stellar Spectrum

Stars Are Classified According to Their Surface Temperature

So far, we have concentrated on what we can learn about stars by applying our understanding of thermal radiation, an approach that seems from the previous section not to have led us too far astray. However, the spectra of stars are far from perfect Planck spectra. Instead, when we look at the spectra of stars, we

Absorption lines form as light escapes through the atmosphere of the star.

see a wealth of absorption lines and occasionally emission lines. In Chapter 4 we found that absorption lines occur when light passes through a cloud of gas: the atoms and

---

[4]Cones sensitive to red and green light have a rather broad spectral response and, in fact, partially overlap one another. This overlap is the cause of red-green color blindness in some people.

[5]Rather than a blue/visual brightness ratio, astronomers generally refer to color index as the difference between an object's blue magnitude (*B*) and its visual magnitude (*V*)—that is, *B-V* (see Math Tools 13.1).

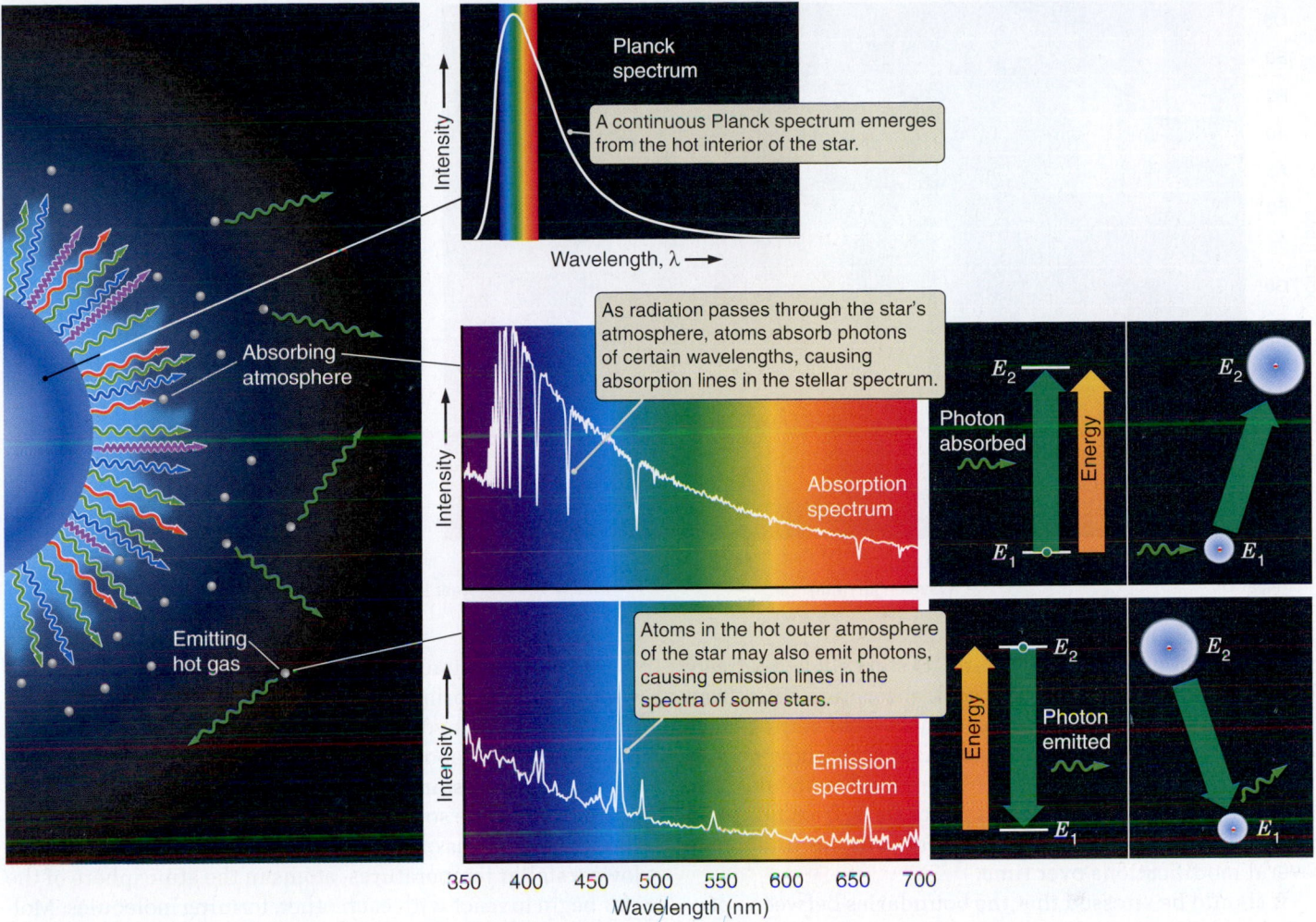

Planck spectrum

A continuous Planck spectrum emerges from the hot interior of the star.

Intensity

Wavelength, λ →

As radiation passes through the star's atmosphere, atoms absorb photons of certain wavelengths, causing absorption lines in the stellar spectrum.

Absorbing atmosphere

Intensity

Absorption spectrum

Photon absorbed

E_2

E_1

Energy

E_2

E_1

Emitting hot gas

Atoms in the hot outer atmosphere of the star may also emit photons, causing emission lines in the spectra of some stars.

Intensity

Emission spectrum

Energy

E_2

E_1

Photon emitted

E_2

E_1

350   400   450   500   550   600   650   700
Wavelength (nm)

FIGURE 13.7 The formation of absorption and emission lines in the spectra of stars.

molecules of the gas absorb light of certain specific wavelengths, characteristic of the kind of atom or molecule that is doing the absorbing. Similarly, the atoms and molecules in diffuse hot gas will emit light of specific wavelengths as well. Both of these processes are at work in stars.

Although the hot "surface" of a star emits radiation with a spectrum very close to a smooth Planck curve, this light must then escape through the outer layers of the star's atmosphere. The atoms and molecules in the cooler layers of the star's atmosphere leave their absorption-line fingerprints in this light, as shown in **Figure 13.7**. The atoms and molecules in the star's atmosphere and any gas that might be found in the vicinity of the star can also produce emission lines in stellar spectra. Although absorption and emission lines complicate how we use the laws of Planck radiation to interpret light from stars, spectral lines more than make up for this trouble by providing a wealth of information about the state of the gas in a star's atmosphere.

The spectra of stars were first classified during the late 1800s, long before stars, atoms, or radiation were well understood. Stars were classified not on the basis of their physical properties, but on the appearance of the dark bands (now known as absorption lines) seen in their spectra. The original ordering of this classification was arbitrarily based on the prominence of particular absorption lines known to be associated with the element hydrogen. Stars with the strongest hydrogen lines were labeled "A stars," stars with somewhat weaker hydrogen lines were labeled "B stars," and so on.

Stars are classified by the appearance of their spectra.

This classification scheme was refined and turned into a real sequence of stellar properties in the early 20th century. The new system, originally proposed in 1901, was the work of Annie Jump Cannon (1863–1941), who led an effort at the Harvard College Observatory to systematically examine and classify the spectra of hundreds of thousands of stars. She dropped many of the earlier spectral types, keeping only seven, which she reordered into a sequence that was no longer arbitrary, but instead was based on surface temperatures. Spectra of stars of different types are shown in **Figure 13.8**. The hottest stars, with surface temperatures

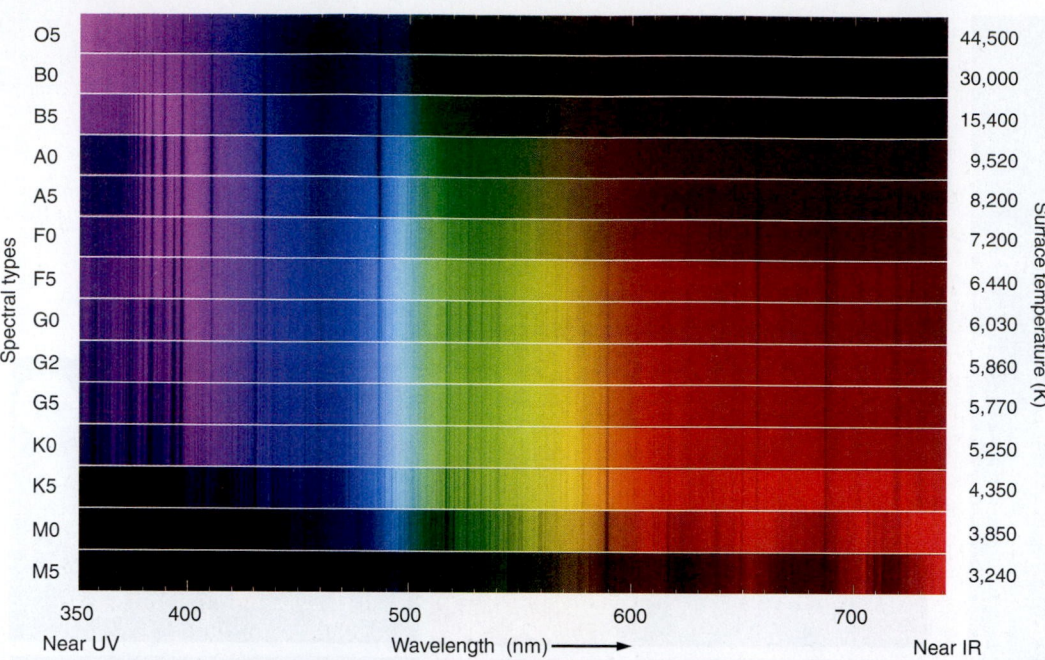

FIGURE 13.8 Spectra of stars with different spectral types, ranging from hot blue O stars to cool red M stars. Hotter stars are brighter at shorter wavelengths. The dark lines are absorption lines.

over 30,000 K, are labeled "O stars." O stars show relatively featureless spectra, with only weak absorption lines from hydrogen and helium. The coolest stars—"M stars"—have temperatures as low as about 2,800 K. M stars show myriad lines from many different types of atoms and molecules. The complete sequence of **spectral types** of stars, from hottest to coolest, is O, B, A, F, G, K, M. This sequence has undergone several modifications over time.

It should be stressed that the boundaries between spectral types are not precise. A hotter-than-average G star is very similar to a cooler-than-average F star. Astronomers break down the main spectral types into a finer sequence of subclasses by adding numbers to the letter designations. For example, the hottest B stars are called B0 stars, slightly cooler B stars are called B1 stars, and so on. The coolest B stars are B9 stars, which are only slightly hotter than A0 stars. The Sun is a G2 star.

Returning to Figure 13.8, we can see that not only are hot stars bluer than cool stars, but the absorption lines in their spectra are quite different as well. The reason is that differences in the temperature of the gas in the atmosphere of a star affect the state of the atoms in that gas, which in turn affects the energy level transitions available to absorb radiation. At the hot end of the spectral classification scheme, the temperature in the atmospheres of O stars is so high that most atoms have had one or more electrons stripped from them by energetic collisions within the gas. There are few transitions available in these ionized atoms that cause absorption lines in the visible part of the electromagnetic spectrum, so the visible spectrum of an O star is relatively featureless. At lower temperatures there are more atoms in energy states that can absorb light in the visible part of the spectrum, so the visible spectra of cooler stars are far more complex than are the spectra of O stars.

Most absorption lines have an optimal temperature at which they are formed most strongly. For example, absorption lines from hydrogen are most prominent at surface temperatures of about 10,000 K, which is the surface temperature of an A star. (This should be no surprise. Spectral-type A stars were so named because they are the stars with the strongest lines of hydrogen in their spectra.) At the very lowest stellar temperatures, atoms in the atmosphere of the star begin to react with each other, forming molecules. Molecules such as titanium oxide (TiO) are responsible for much of the absorption in the atmospheres of cool M stars.

Because different spectral lines are formed at different temperatures, we can use these absorption lines to measure a star's temperature directly. The surface temperatures of stars measured in this way agree extremely well with the surface temperatures of stars measured using Wien's law, again confirming a prediction of the cosmological principle: the physical laws that apply on Earth apply to stars as well. ▶‖ **AstroTour: Stellar Spectrum**

Stars Consist Mostly of Hydrogen and Helium

The most obvious differences in the lines seen in stellar spectra are due to temperature, but the details of the absorption and emission lines found in starlight carry a wealth of other information as well. By applying our knowledge of the physics of atoms and molecules to the study of stellar absorption lines, we can accurately determine not only surface tem-

> Spectral lines are used to measure many properties of stars, including chemical composition.

peratures of stars, but also pressures, chemical composi-tions, magnetic-field strengths, and other physical properties of stars. In addition, by making use of the Dop-pler shift of emission and absorption lines, we can measure rotation rates, motions of the atmosphere, expansion and contraction, "winds" driven away from stars, and other dynamic properties of stars.

As we continue to learn about stars, we will come to appreciate that one of the most interesting and important things about a star is its chemical composition. We already saw in Chapter 4 that if we look at a source of thermal radia-tion through a cloud of gas, the strength of various absorp-tion lines tells us what kinds of atoms are present in the gas and in what abundance. Although we must take great care in interpreting spectra to properly account for the temperature and density of the gas in the atmosphere of a star, in most respects stars are ready-made laboratories for carrying out just such an experiment.

When analyzed in this way, most stars are found to have atmospheres that consist predominantly of the least massive elements. Hydrogen typically makes up over 90 percent of the atoms in the atmosphere of a star, with helium account-ing for most of what remains. All of the other chemical

elements, which are collectively referred to as **heavy ele-ments** or (more properly) as "massive elements," are present in only trace amounts.[6] (This would be a good time to go back and review the "astronomer's periodic table" in Fig-ure 9.5.) **Table 13.1** shows the chemical composition of the atmosphere of the Sun, which is fairly typical for stars in our vicinity. On the other hand, chemical composition can vary tremendously from star to star. In particular, some stars show even smaller amounts of elements other than hydrogen and helium in their spectra. The existence of such stars, all but devoid of more massive elements, provides important clues about the origin of chemical elements and the chemi-cal evolution of the universe.

The Stefan-Boltzmann Law Revisited: Finding the Sizes of Stars

Once we know the temperature of a star, we also know how much radiation each square meter of the star is giving off each second. As shown in **Figure 13.9b**, each square meter of the surface of a hot blue star gives off more radiation per second than a square meter of the surface of a cool red star. A hot star will be more luminous than a cool star of the same size. A small hot star might even be more luminous than a larger cool star. As noted in **Figure 13.9c**, we can use the relationship between temperature and luminosity of each square meter of a surface to infer the sizes of stars.

According to the Stefan-Boltzmann law (see Chapter 4), the amount of energy radiated each second by each square meter of the surface of a star is equal to the constant σ multiplied by the surface temperature of the star raised to the fourth power. Written as an equation, this relationship says

$$\text{Energy radiated each second by 1 m}^2 \text{ of surface} = \sigma T^4.$$

To find the total amount of light radiated each second by the star, we need to multiply the radiation per second from each square meter by the number of square meters of the star's surface:

$$\begin{array}{c}\text{Total energy radiated} \\ \text{by the star} \\ \text{each second}\end{array} = \begin{array}{c}\text{Energy radiated} \\ \text{by a square meter} \\ \text{each second}\end{array} \times \begin{array}{c}\text{Number of square} \\ \text{meters of surface} \\ \text{of the star}\end{array}.$$

The left-hand item in this equation—the total energy emitted by the star per second (in units of joules per sec-ond, abbreviated J/s)—is the star's luminosity, L. The middle item—the energy radiated by each square meter of the star in a second (in units of joules per square meter per second, or $J/m^2/s$)—can be replaced with the σT^4 factor from the Stefan-Boltzmann law. The remaining item—the number of square

TABLE 13.1

The Chemical Composition of the Sun's Atmosphere*

Element	Percentage by Number[†]	Percentage by Mass[††]
Hydrogen	92.5	74.5
Helium	7.4	23.7
Oxygen	0.064	0.82
Carbon	0.039	0.37
Neon	0.012	0.19
Nitrogen	0.008	0.09
Silicon	0.004	0.09
Magnesium	0.003	0.06
Iron	0.003	0.16
Sulfur	0.001	0.04
Total of others	0.001	0.03

*The relative amounts of different chemical elements in the atmo-sphere of the Sun.
[†]The percentage of atoms accounted for by the listed element.
[††]The percentage of mass consisting of the listed element.

(a)

Wien's Law:
Blue stars are hot, red stars are cool.

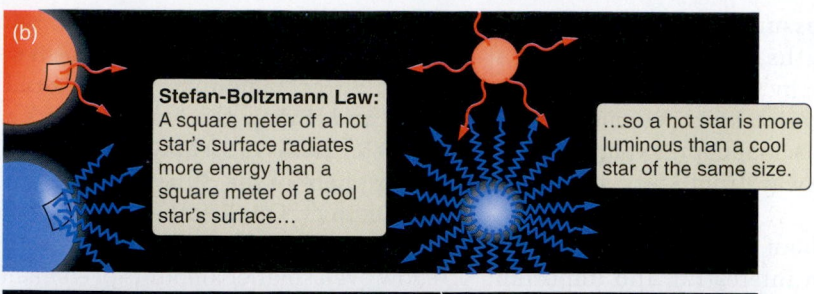

(b)

Stefan-Boltzmann Law:
A square meter of a hot star's surface radiates more energy than a square meter of a cool star's surface…

…so a hot star is more luminous than a cool star of the same size.

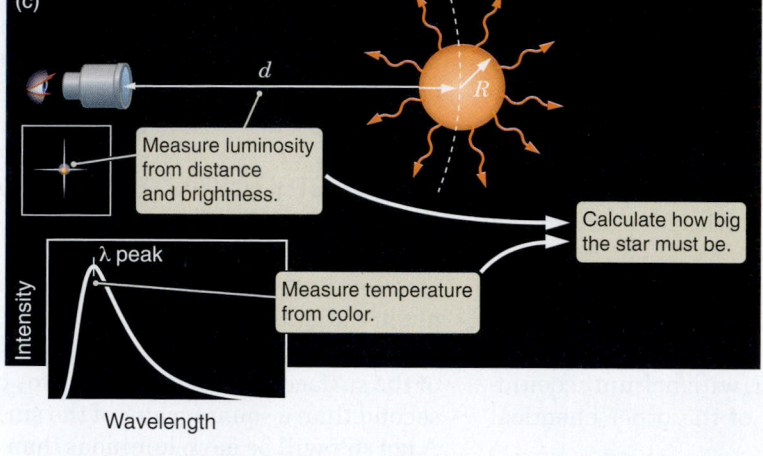

(c)

d

R

Measure luminosity from distance and brightness.

Calculate how big the star must be.

λ peak

Measure temperature from color.

Intensity

Wavelength

FIGURE 13.9 We measure the temperature and size of stars by applying our understanding of Planck radiation.

meters covering the surface of the star—is the surface area of a sphere, $A_{sphere} = 4\pi R^2$ (in units of square meters, or m²), where R is the radius of the star. If we replace the words in the equation with the appropriate mathematical expressions for the Stefan-Boltzmann law and the area of a sphere, our equation for the luminosity of a star looks like this:

$$\text{Luminosity}\left(\frac{J}{s}\right) = \left(\frac{J}{m^2 s}\right) \times m^2$$

$$= (\sigma T^4) \times (4\pi R^2).$$

Combining the right sides of these two equations, we get

$$L = 4\pi R^2 \sigma T^4, \text{ in units of J/s (watts).}$$

It is worth thinking about this equation for a minute to see the sense behind it. Because the constants (4, π, and σ) do not change, the luminosity of a star is proportional only to $R^2 T^4$. The R^2 makes sense: The surface area of a star is proportional to the square of the radius of the star. Make a star three times as large, and its surface area becomes $3^2 = 9$ times as large. There is nine times as much area to radiate, so there is nine times as much radiation. The effect of T^4 is understandable as well: make a star twice as hot, and each square meter of the star's surface radiates $2^4 = 16$ times as much energy. (As we have seen before, an apparently complex mathe-

A larger and hotter star is also more luminous.

matical expression is really a shorthand way of writing our understanding of what is going on physically—in this case the commonsense result that larger, hotter stars are more luminous than smaller, cooler stars.)

The previous equation answers the question, How luminous will a star of a given size and a given temperature be? Now turn this question around and ask, How large does a star of a given temperature need to be to have a total luminosity of L? The star's luminosity (L) and temperature (T) are quantities that we can measure, and the star's radius (R) is what we want to know. Rearrange the previous equation, moving the things that we know how to measure (temperature and luminosity) to the right-hand side of the equation, and the thing that we would like to know (the radius of the star) to the left side. After a couple of steps of algebra, we find

If the temperature and luminosity of a star are known, its size can be calculated.

$$R = \sqrt{\frac{L}{4\pi\sigma}} \times \frac{1}{T^2}.$$

Again, the right-hand side of the equation contains only things that we know or can measure. The constants 4, π, and σ are always the same. We can find L, the luminosity of the star, by combining measurements of the star's brightness and parallax (although only for those nearer stars with known parallax). T is the surface temperature of the star, which can be measured from its color. From the relationship of these measurements we now know something new: the size of the star. We will refer to this last equation as the **luminosity-temperature-radius relationship** for stars.

The luminosity-temperature-radius relationship has been used to measure the sizes of many thousands of stars. When we talk about the sizes of stars, we again use our Sun as a yardstick. The radius of the Sun, written $R_\odot$, is 696,000 km, or about 700,000 km. When we look at stars around us, we find that the smallest stars we see, called white dwarfs, have radii that are only about 1 percent of the Sun's radius ($R = 0.01\,R_\odot$). The largest stars that we see, called red supergiants, can have radii more than 1,000 times that of the Sun. There are many more stars toward the small end of this range, smaller than our Sun, than there are giant stars.

> There are many more small stars than large stars.

13.4 We Measure Stellar Masses by Analyzing Binary Star Orbits

The most important property of stars that we have yet to investigate is their mass. Determining the mass of an object can be a tricky business. We certainly cannot rely on the amount of light from an object or the object's size as a measure of its mass. Massive objects can be large or small, faint or luminous. The only thing that *always* goes with mass is gravity. Mass is responsible for gravity, and gravity in turn affects the way masses move.

> To measure mass, astronomers look for the effects of gravity.

When astronomers are trying to determine the masses of astronomical objects, they almost always wind up looking for the effects of gravity.

When discussing the orbits of the planets in Chapter 3, we found that Kepler's laws of planetary motion are the result of gravity. We even went so far as to show that the properties of the orbit of a planet about the Sun can be used to measure the mass of the Sun. If we could find a planet orbiting a distant star, we could apply the same technique to determine the mass of that star. Unfortunately, today's telescopes are barely powerful enough to directly see planets orbiting other stars, but we can watch as two *stars* orbit about each other. About half of the higher-mass stars in the sky are actually multiple systems consisting of several stars moving about under the influence of their mutual gravity. Most of these are **binary stars** in which two stars orbit each other in elliptical orbits as predicted by Kepler's laws. However, most low-mass stars are single, and low-mass stars far outnumber their higher-mass brethren. This means that *most stars are single*.

Binary Stars Orbit a Common Center of Mass

When discussing the motions of the planets, we usually say that the planets orbit around the Sun and we don't worry about the effect of their gravity on the Sun's motion. On the other hand, if two objects—say, two stars—are closer to the same mass, we have to abandon this simple mental picture. Each object's motion will be noticeably affected by the gravitational force from the other, and we must worry about what happens when the two objects *orbit each other*. (In fact, as we learned in Chapter 6, our knowledge of planets beyond the Solar System comes primarily from the wobbles they cause in the motions of stars.)

Think about what happens if we set two unequal masses—m_1 and m_2—adrift in space with no motion of one mass relative to the other. As soon as we release the two masses, gravity begins to pull them together. The force of mass 1 on mass 2 equals the force of mass 2 on mass 1, and each mass begins to fall toward the other. But even though the forces on each mass are equal, the accelerations experienced by the two masses are not. Acceleration equals force divided by mass, so the *less massive object experiences the greater acceleration*.

Suppose, for the moment, that m_1 is three times as massive as m_2. That means that the acceleration of m_2 will be three times as great as the acceleration of m_1. At any given point in time, m_2 will be moving toward m_1 three times as fast as m_1 is moving toward m_2. When the two objects collide, m_2 will have fallen three times as far as m_1. The point where the two objects meet, called the **center of mass** of the two objects, will be three times as far from the original position of the less massive m_2 as from the original position of the more massive m_1. (If the two objects were sitting on a seesaw in a gravitational field, the support of the seesaw would have to be directly under the center of mass for the objects to balance, as shown in **Figure 13.10**.)

If we give our two masses some motion perpendicular to the line between them, then instead of simply falling into each other they will orbit around each other. When Newton applied his laws of motion to the problem of orbits, he found that two objects move in elliptical orbits with their common center of mass at one focus of both of the ellipses, as shown in **Figure 13.11**. The center of mass, which lies along the line between the two objects, remains stationary. The two objects will always be found on exactly opposite sides of the center of mass, and the elliptical orbit of the more massive object is just a smaller version of the elliptical orbit of the less massive object.

> Each star in a binary system follows an elliptical orbit around the system's center of mass.

Because the perimeter of the orbit of the less massive star is larger than that of the more massive star's orbit, the less

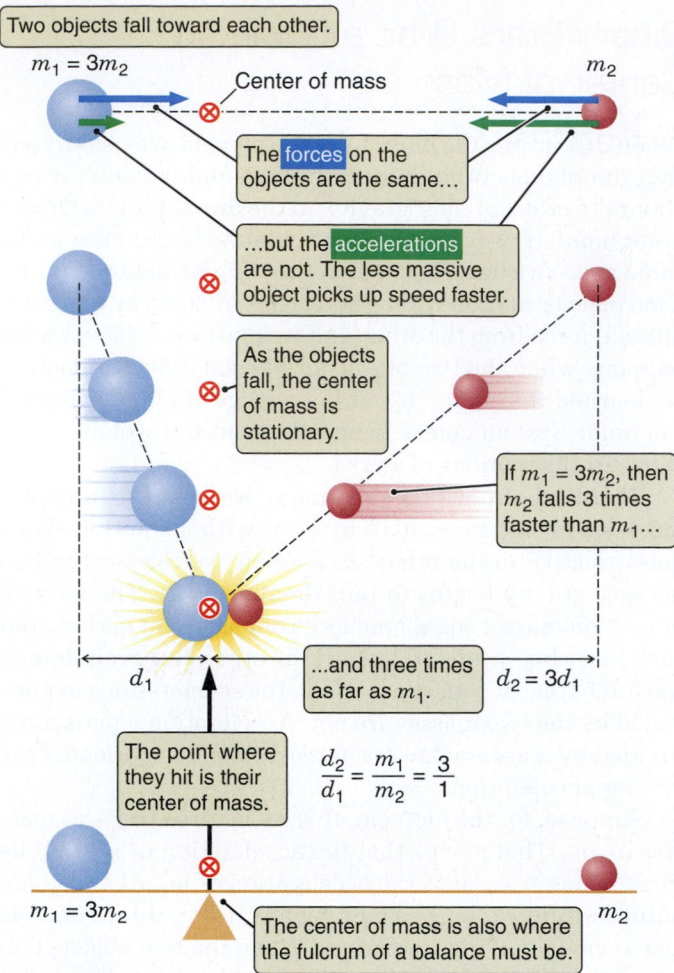

FIGURE 13.10 The center of mass of two objects is the "balance" point on a line joining the centers of two masses.

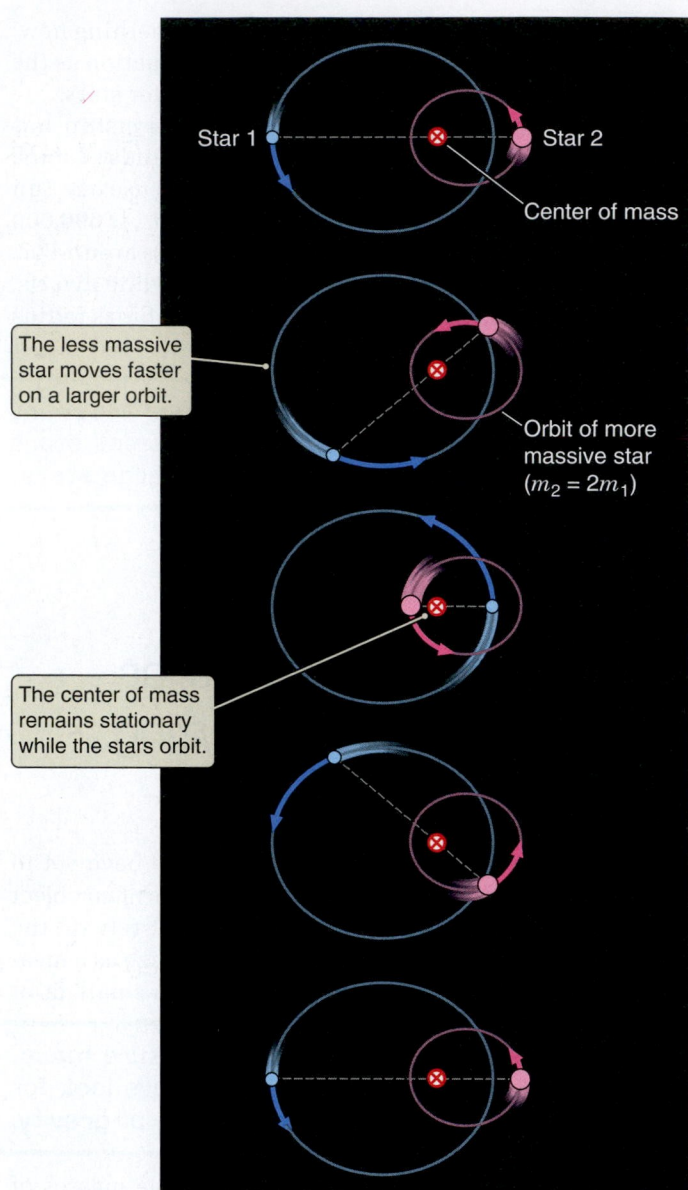

FIGURE 13.11 In a binary star system, the two stars orbit on elliptical paths about their common center of mass. In this case, star 2 has twice the mass of star 1. The eccentricity of the orbits is 0.5. There are equal time steps between the frames.

massive star must be moving *faster* than the more massive star. The less massive star has farther to go than the more massive star, but it must cover

The less massive star has a larger orbit and so must move faster.

that distance in the same amount of time. As a result, the velocity (*v*) of a star in a binary system is inversely proportional to its mass:

$$\frac{v_1}{v_2} = \frac{m_2}{m_1}.$$

This relationship between the velocities and the masses of the stars in a binary system is a crucial part of how binary stars are used to measure stellar masses. Imagine that you are watching a binary star edge-on, as shown in **Figure 13.12**. As one star is moving toward you, the other star will be moving away from you; and vice versa. If you were to look at the spectra of the two stars, you might see

that the absorption lines from star 1 are redshifted relative to the overall center-of-mass velocity, while the absorption lines from star 2 are blueshifted. If you were to look again half an orbital period later, the situation would be reversed: lines from star 2 would be redshifted and lines from star 1 would be Doppler-shifted to the blue. Because the two stars are always exactly opposite one another about their common center of mass, they are always moving in opposite directions. We can measure the ratio of the masses of the two stars by comparing the size of the Doppler shift in the spectrum of star 1 with the size of the Doppler shift in the spectrum of star 2.

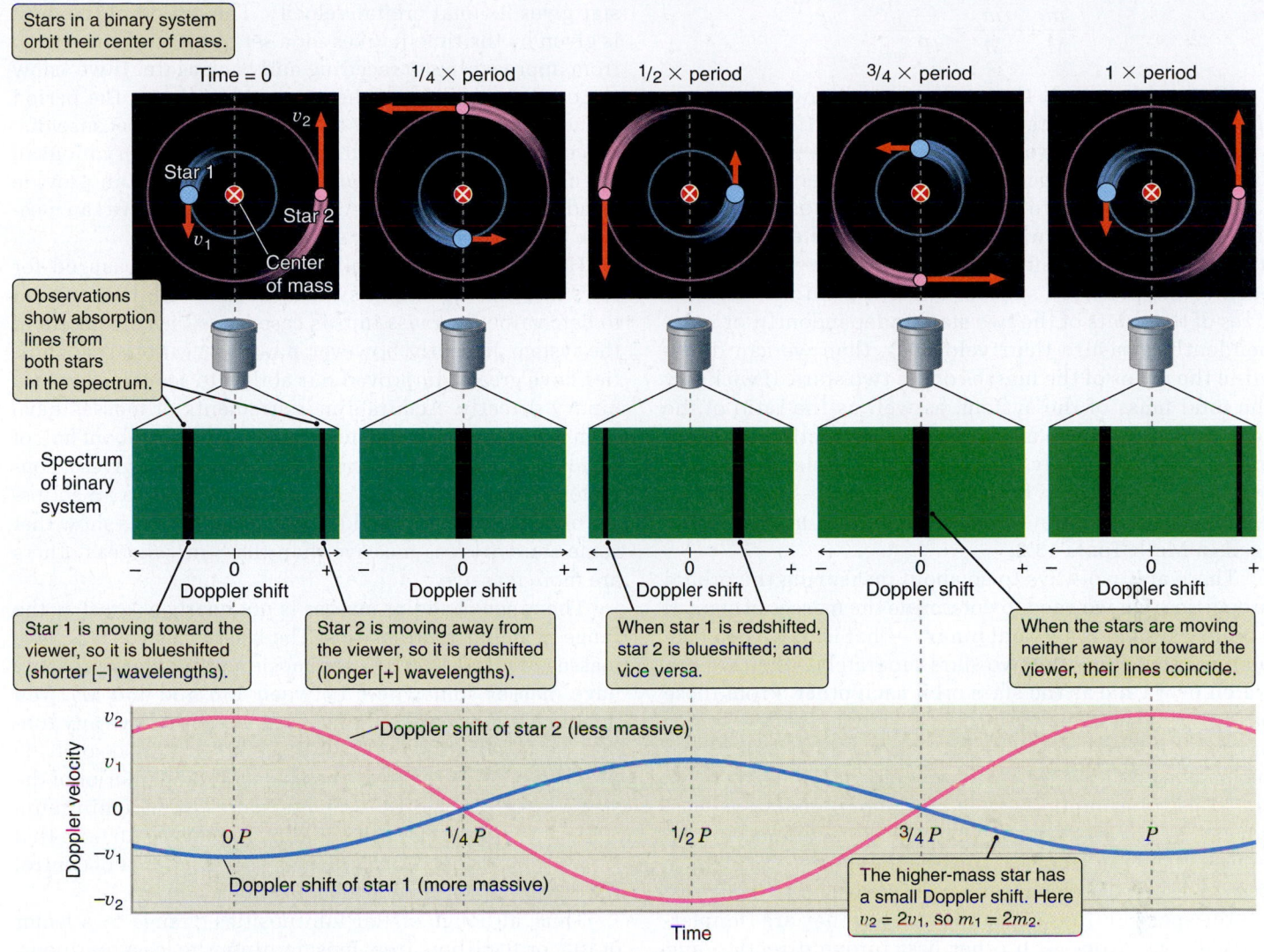

FIGURE 13.12 The orbits of two stars in a binary system are shown here, along with the Doppler shifts of the absorption lines in the spectrum of each star. Star 1 is twice as massive as star 2 and so has half the Doppler shift.

Kepler's Third Law Gives the Total Mass of a Binary System

In Chapter 3 we conveniently ignored all of this complexity having to do with the motion of two objects about their common center of mass. Now, however, it is this very complexity that enables us to measure the masses of the two stars in a binary system. In his derivation of Kepler's laws, Newton showed that if two objects with masses m_1 and m_2 are in orbit about each other, then the period of the orbit, P, is related to the average distance between the two masses, A,[7] by the equation

$$P^2 = \frac{4\pi^2 A^3}{G(m_1 + m_2)}.$$

Rearranging this a bit turns it into an expression for the sum of the masses of the two objects:

$$m_1 + m_2 = \frac{4\pi^2}{G} \times \frac{A^3}{P^2}.$$

We can cleverly ignore the value of $4\pi^2/G$ by using what we know about Earth's orbit around the Sun, and then expressing m, A, and P as ratios. If $A = 1$ AU and $P = 1$ year, then we know that $m_1 + m_2 = M_\odot + M_\oplus \approx M_\odot$ (because $M_\odot$ is much larger than $M_\oplus$). So if we express the masses m_1 and m_2, A, and P in that equation in terms of Solar System units—such as $m_1 = m_1/(1\ M_\odot)$, $A_{AU} = A/(1\ \text{AU})$, and $P_{years} = P/(1\ \text{year})$—then this equation simplifies to

[7] Here, A is equivalent to the semimajor axis, which we discussed in Chapter 3.

$$\frac{m_1}{M_\odot} + \frac{m_2}{M_\odot} = \frac{(A_{\mathrm{AU}})^3}{(P_{\mathrm{years}})^2}.$$

We now know all that we need to know to measure the masses of two stars in a binary system. If we can measure the period of the binary system and the average separation between the two stars, then Kepler's third law gives us the total mass. Referring back to the previous subsection, if we can measure the sizes of the orbits of the two stars independently, or independently measure their velocities, then we can determine the ratios of the masses of the two stars. If we know the total mass of the system, as well as the ratio of the two masses, we have all we need to determine the mass of each star separately. To express these relationships in symbols, if we know both m_1/m_2 and $m_1 + m_2$, a bit of algebra yields separate values for m_1 and m_2, as can be seen in Math Tools 13.2.

> **Kepler's third law gives the total mass of a binary system.**

There are two ways to go about measuring the orbital properties that we need to determine the masses of stars. If a binary system is a **visual binary**—that is, if we can take pictures that show the two stars separately—then we can watch over time as the stars orbit each other. From these observations we can measure the shapes and period of the orbits of the two stars. When combined with Doppler measurements of the line-of-sight velocities of the stars, this is all the information we need. In many binary systems, however, the two stars are so close together and so far away from us that we cannot actually see the stars separately. We know these stars belong to binary systems only because we see the spectral lines of the two stars as they are Doppler-shifted away from each other first in one direction and then in the other; these are called **spectroscopic binary** stars. Yet even those Doppler shifts alone can sometimes be used to measure the masses of such stars.

If a binary system is an **eclipsing binary**, in which we see a dip in brightness as one star passes in front of the other (**Figure 13.13**), then we know we are looking at the system nearly edge-on. In this case, the peak Doppler shift of each star gives its total orbital velocity. The period of the orbit is given by the time it takes for a set of spectral lines to go from approaching to receding and back again. If we know the orbital velocities of the stars and we know the period of the orbit, we also know the size of the orbit because distance equals velocity multiplied by time. (Observations of the changing brightness of eclipsing binaries can provide an additional bonus. Under certain conditions, we can measure the size [radii] of the stars as well.)

Historically, most stellar masses were measured for stars in eclipsing binary systems because all that we need to determine the mass in this case is a series of spectra of the system. Recently, however, new observational capabilities have greatly improved our ability to see the stars in a binary directly. Accurate measurements of masses have been obtained for several hundred binary stars, about half of which are eclipsing binaries. (**Math Tools 13.2** gives a concrete example of how the masses of two stars in an eclipsing binary are determined.) These measurements show that some stars are less massive than the Sun, whereas others are more massive.

The range of stellar masses is not nearly as great as the range of stellar luminosities. The least massive stars have masses of about 0.08 $M_\odot$; the most massive stars probably have masses somewhere between 130 and 150 $M_\odot$. You might wonder why the mass of a star should have any limits. These lower and upper limits are determined solely by the physical processes that go on deep in the interior of the star. We will learn in the chapters ahead that a minimum stellar mass is necessary to ignite the nuclear furnace that keeps a star shining, but the furnace can run out of control if the stellar mass is too great.

Thus, although stellar luminosities change by a factor of 10^{10}, or 10 billion, from most luminous to least luminous, the most massive stars are only about 10^3, or a thousand, times more massive than the least massive stars.

13.5 The H-R Diagram Is the Key to Understanding Stars

We have come a long way in our effort to measure the physical properties of stars. **Table 13.2** summarizes the techniques we have used for certain types of stars, such as nearby stars and binary stars. However, just knowing some of the basic properties of stars does not mean that we understand stars. The next step in our journey involves looking for patterns in the properties we have determined. The first astronomers to take this step were a Dane by the name of Ejnar Hertzsprung (1873–1967) and the American astronomer Henry Norris Russell (1877–1957). In the early part of the 20th century (from 1906 to 1913, to be precise), Hertzsprung and Russell were independently studying the

FIGURE 13.13 The shape of the light curve of an eclipsing binary can reveal information about the relative size and surface brightness of the two stars.

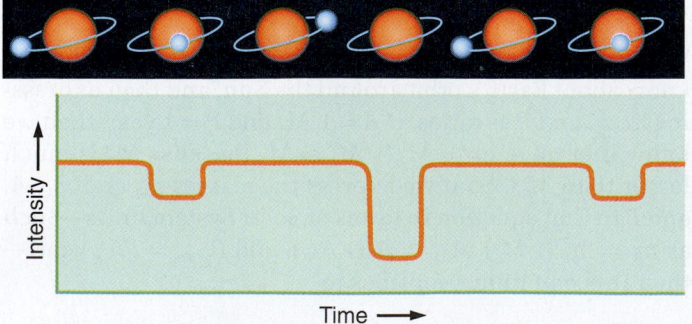

MATH TOOLS 13.2

Using a Celestial Bathroom Scale to Measure the Masses of an Eclipsing Binary Pair

It is worth looking at an example to see how the relationships discussed in the text are actually put to work. Suppose you are an astronomer studying a binary star system. After observing the star for several years, you accumulate the following information about the binary system:

1. The star is an eclipsing binary.
2. The period of the orbit is 2.63 years. You learned this by observation.
3. Star 1 has a Doppler velocity that varies between +20.4 and −20.4 kilometers per second (km/s).
4. Star 2 has a Doppler velocity that varies between +6.8 and −6.8 km/s.
5. The stars are in circular orbits. You know this because the Doppler velocities about the star are symmetric.

These data are summarized in **Figure 13.14**. You begin your analysis by noting that the star is an eclipsing binary, which tells you that the orbit of the star is edge-on to your line of sight. The Doppler velocities tell you the total orbital velocity of each star, and you determine the size of the orbits using the relationship Distance = Speed × Time. In one orbital period, star 1 travels a distance of

$$\left(20.4\ \frac{\text{km}}{\text{s}}\right) \times (2.63\ \text{years}) \times \left(3.156 \times 10^7\ \frac{\text{s}}{\text{year}}\right)$$
$$= 1.69 \times 10^9\ \text{km}.$$

This distance is 2π times the radius of the star's orbit, so star 1 is following an orbit with a radius of 2.7×10^8 km, or 1.8 AU. A similar analysis of star 2 shows that its orbit has a radius of 0.6 AU.

Next you apply Kepler's third law, which says that

$$\frac{m_1}{M_\odot} + \frac{m_2}{M_\odot} = \frac{(A_\text{AU})^3}{(P_\text{years})^2}.$$

where $m_1/M_\odot$ refers to the star's mass in terms of solar masses and A is the average distance between the two stars. Here you have $A = 1.8\ \text{AU} + 0.6\ \text{AU} = 2.4\ \text{AU}$. Because you know A and the period P, you can calculate the total mass of the two stars:

$$\frac{m_1}{M_\odot} + \frac{m_2}{M_\odot} = \frac{(A_\text{AU})^3}{(P_\text{years})^2} = \frac{(2.4)^3}{(2.63)^2} = 2.0,$$

so you have learned that the combined mass of the two stars is twice the mass of the Sun.

To sort out the individual masses of the stars, you use the fact that the more massive the star is, the less it moves in response to the gravitational attraction of its companion:

$$\frac{m_2}{m_1} = \frac{v_1}{v_2} = \frac{20.4\ \text{km/s}}{6.8\ \text{km/s}} = 3.0,$$

so star 2 is three times as massive as star 1. Algebra gets you the rest of the way. Substituting $m_2 = 3 \times m_1$ into the equation $m_1 + m_2 = 2.0\ M_\odot$ gives you

$$m_1 + (3 \times m_1) = 4 \times m_1 = 2.0\ M_\odot,$$

or $m_1 = 0.5\ M_\odot$. Substituting $m_1 = 0.5\ M_\odot$ into the equation $m_2 = 3 \times m_1$ gives you $m_2 = 1.5\ M_\odot$.

Star 1 has a mass of 0.5 $M_\odot$, and star 2 has a mass of 1.5 $M_\odot$. You may tell your stars to step off the scale now.

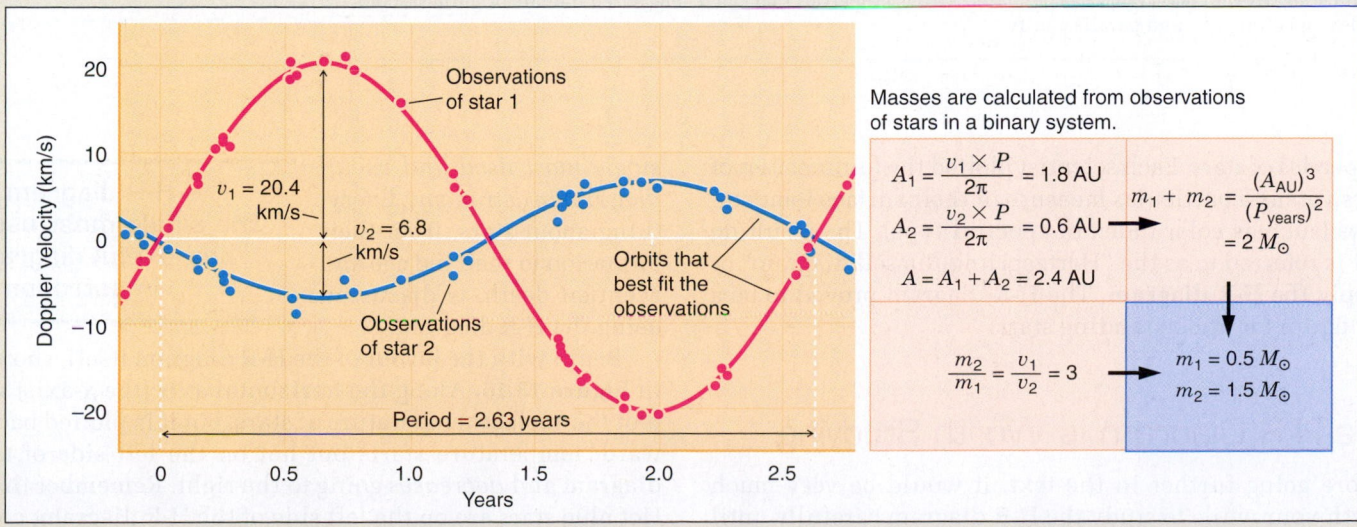

FIGURE 13.14 Doppler velocities of the stars in an eclipsing binary are used to measure the masses of the stars.

TABLE 13.2

Taking the Measure of Stars*

Property	Methods
Luminosity	• For a star with a known distance, measure the brightness, then apply the inverse square law of radiation: $$\text{Luminosity} = 4\pi \times \text{Distance}^2 \times \text{Brightness}$$ • For a star without a known distance, take a spectrum of the star to determine its spectral and luminosity classes, plot them onto the H-R diagram, and read the luminosity from the diagram.
Temperature	• Measure the color index of the star using blue and visual filters. Use Wien's law to relate the color to a temperature. • Take a spectrum of the star and estimate the temperature from its spectral class by noting which spectral lines are present.
Distance	• For relatively nearby stars (within a few hundred parsecs), measure the parallax shift of the star over the course of the year. • For more distant stars, find the luminosity using the H-R diagram as noted above, and then use the spectroscopic parallax method to relate luminosity, distance, and brightness.
Size	• For a few of the largest stars, measure the size directly or by the length of eclipse in eclipsing binary stars. • From the width of the star's spectral lines, estimate the luminosity class (supergiant, giant, or main-sequence). • For a star with known luminosity and temperature, use the Stefan-Boltzmann law to compute how large the star must be (the luminosity-temperature-radius relationship).
Mass	• Measure the motions of the stars in a binary system, and use these to determine the orbits of the stars; then apply Newton's form of Kepler's laws. • For nonbinary stars, use the mass-luminosity relation to estimate the mass.
Composition	• Analyze the lines in the star's spectrum to measure chemical composition.

*A brief summary of the methods used to measure basic properties of stars. Of the properties listed here, only temperature, distance, and composition can be *measured*. Luminosity must be *inferred* from the H-R diagram, and size and mass must be *calculated*. Other properties that can be measured include brightness, color index, spectral type, and parallax shift.

---

properties of stars. Each scientist plotted the luminosities of stars versus a particular measure of their surface temperatures (such as color index or spectral type). The resulting plot is referred to as the "Hertzsprung-Russell diagram" or simply the **H-R diagram**. The H-R diagram proved to be a key figure for understanding stars.

The H-R Diagram Is Worth Studying

Before going further in the text, it would be very much worth your while to study the H-R diagram carefully until you understand what it can tell you. *The H-R diagram is the*

single most used and useful diagram in astronomy. Everything about stars, from their formation to their old age and eventual death, is discussed using the H-R diagram.

> The H-R diagram is the single most used and useful diagram in astronomy.

Begin with the layout of the H-R diagram itself, shown in **Figure 13.15**. Along the horizontal axis (the *x*-axis) we plot the surface temperature of stars, but it is plotted backward: temperature starts out hot on the left side of the diagram and *decreases* going to the right. Remember that. Hot blue stars are on the left side of the H-R diagram; cool red stars are on the right. This sequence is plotted *loga-*

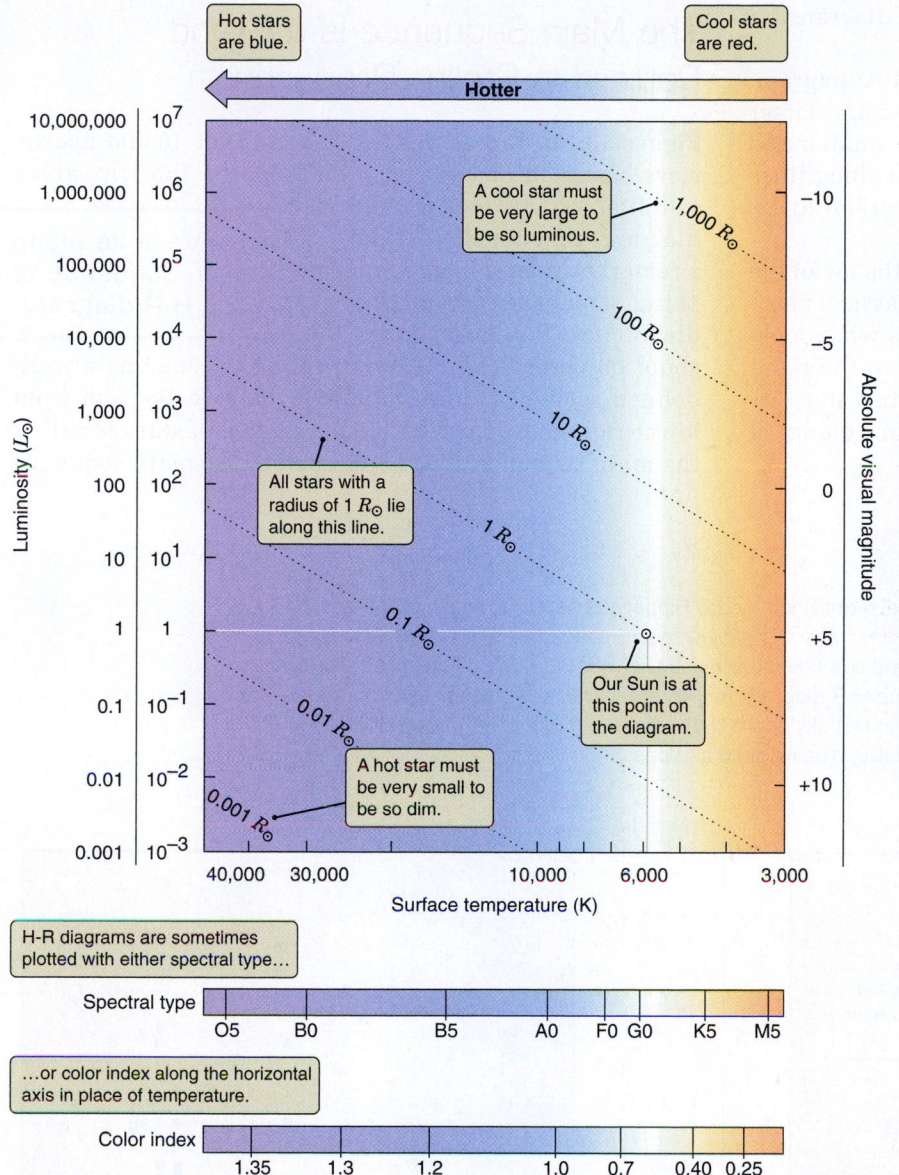

FIGURE 13.15 The layout of the Hertzsprung-Russell, or H-R, diagram used to plot the properties of stars. More luminous stars are at the top of the diagram. Hotter stars are on the left. Stars of the same radius (*R*) lie along the dotted lines moving from upper left to lower right. Absolute magnitudes are discussed in Math Tools 13.1 and Appendix 6.

this case with each step along the axis corresponding to a multiplicative factor of 10 in the luminosity. To understand why the plotting is done this way, recall that the most luminous stars are 10 billion times more luminous than the least luminous stars, yet all of these stars must fit on the same plot.

It turns out that the H-R diagram tells us about more than just the surface temperatures and luminosities of stars. Earlier in the chapter we discovered that if the temperature and the luminosity of a star are known, we can calculate its radius. Because each point in the H-R diagram is specified by a surface temperature and a luminosity, we can use the luminosity-temperature-radius relationship to find the size of a star at that point as well. We can think through how this works using what we know about radiation. A star in the upper right corner of the H-R diagram is very cool, which, according to the Stefan-Boltzmann law, means that each square meter of its surface is radiating only a small amount of energy. But at the same time, this star is extremely luminous. Such a star must be huge to account for its high overall luminosity despite the feeble radiation coming from each square meter of its surface. Conversely, a star in the lower left corner of the H-R diagram is hot, which means that a large amount of energy is coming from each square meter of its surface. However, this star has a very low overall luminosity. The conclusion is that its surface area cannot be very

rithmically, which is to say that a step along the axis from a point representing a star with a surface temperature of 30,000 K to one with a surface temperature of 10,000 K (that is, a temperature change by a factor of 3) is the same as a step between points representing a star with a temperature of 9,000 K and a star with a temperature of 3,000 K (also a factor-of-3 change). The temperature axis can also be labeled with the spectral types or color indices of stars. Hot blue O stars with large color indices are on the left side of the diagram; cool red M stars with small color indices are on the right.

Along the vertical axis (the *y*-axis) we plot the luminosity of stars—the total amount of energy that a star radiates each second. This time the plot is the "right way" around: more luminous stars are toward the top of the diagram, and less luminous stars are toward the bottom. As with the temperature axis, luminosities are plotted logarithmically, in

large. Stars in the lower left corner of the H-R diagram are small. ▶‖ **AstroTour:** H-R Diagram

This result persists everywhere in the H-R diagram. Moving up and to the right takes you to larger and larger stars. Moving down and to the left takes you to smaller and smaller stars. All stars of the same radius lie along lines running across the H-R diagram from the upper left to the lower right, as shown in Figure 13.15.

Like an experienced surveyor who "sees" the lay of the land when looking at a topographic map, or a classical musician who "hears" the rhythms and harmonies when looking at a piece of sheet music, you should get to the point where you automatically "see" the properties of a star—its temperature, color, size, and luminosity—from a glance at its position on the H-R diagram.

The Main Sequence Is a Grand Pattern in Stellar Properties

Figure 13.16a shows the H-R diagram for 16,600 nearby stars based on observations obtained by the Hipparcos satellite. A quick look at this diagram immediately reveals a remarkable fact. Instead of being strewn all about the diagram, we find instead that

> Most stars lie along the main sequence of the H-R diagram.

about 90 percent of the stars in the sky lie along a well-defined sequence running across the H-R diagram from lower right to upper left. This sequence of stars is called the **main sequence**. On the left end of the main sequence

FIGURE 13.16 (a) An H-R diagram for 16,600 stars obtained by the Hipparcos satellite. Most of the stars lie in a band running from the upper left of the diagram toward the lower right, called the main sequence. (Dot color represents number of stars, with blue indicating the fewest and red the most.) (b) An H-R diagram for two different samples of stars. The red symbols show the H-R diagram for 46 stars that are especially close to the Sun. The blue symbols show the 97 brightest stars in the sky. Note that because these are observational H-R diagrams, they are plotted against an observed quantity: their spectral type.

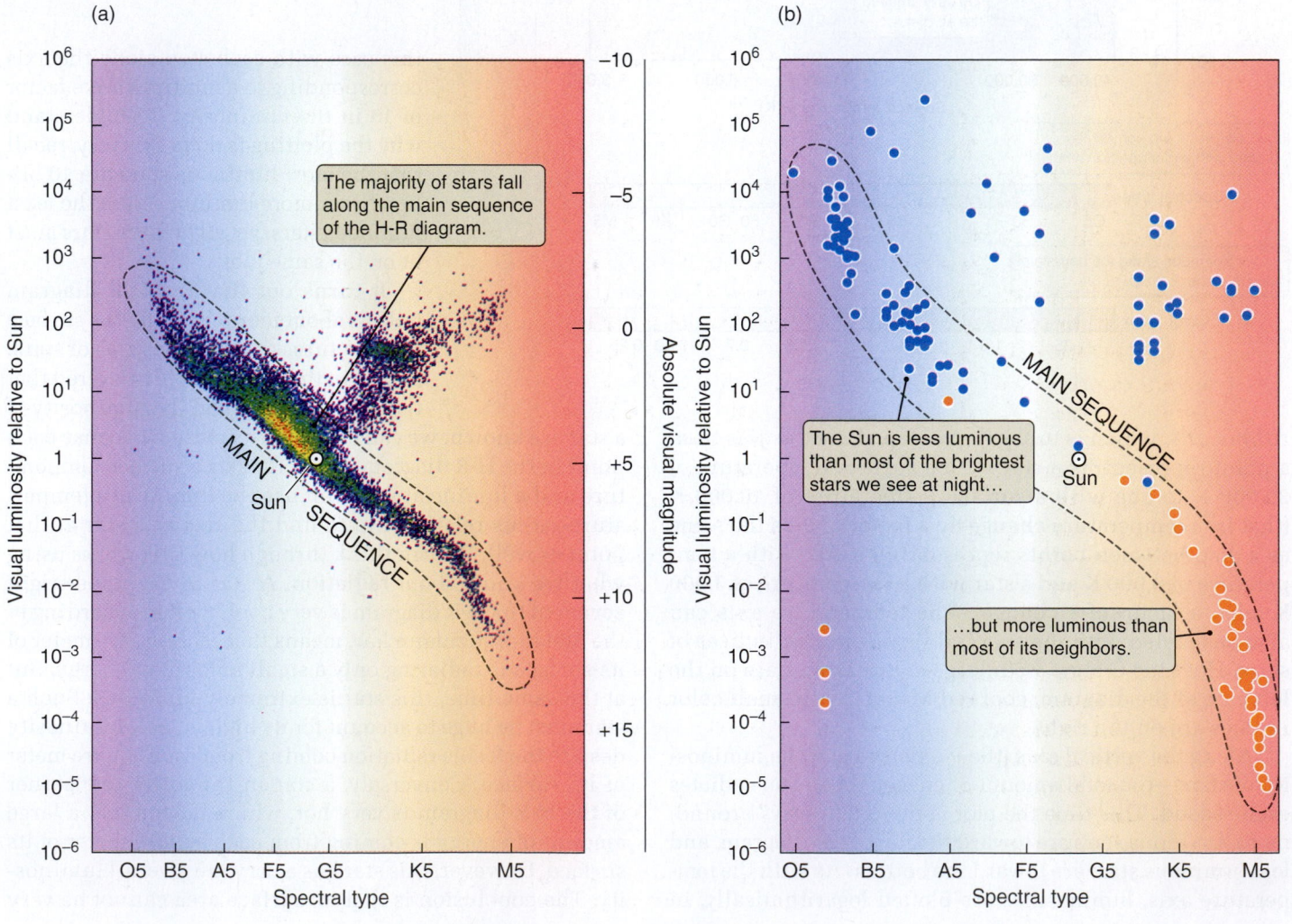

are the O stars: hotter, larger, and more luminous than the Sun. On the right end of the main sequence are the M stars: cooler, smaller, and fainter than the Sun. If you know where a star lies on the main sequence, then you know its approximate luminosity, surface temperature, and size.

From a practical standpoint, the main sequence is extremely handy. For example, you may have wondered in our discussion of stellar parallax how we measure the distances to stars that are farther away than a few hundred light-years. The main sequence comes to the rescue. We can determine whether a star is a main-sequence star by looking at the absorption lines in its spectrum. If a star is on the main sequence, then by virtue of its location on the H-R diagram we know its luminosity. If we know its luminosity and we measure its brightness, we can use the inverse square law of radiation to find its distance. (How far away must a star of that luminosity be to have the brightness we measure?) This method of determining distances to stars is called **spectroscopic parallax**, and is discussed further in Appendix 6. Parallax and spectroscopic parallax are the first two steps along a chain of reasoning that will let us build our knowledge of distances all the way to the edge of the observable universe.

Figure 13.16b makes another important point for astronomy, as well as for all other sciences. The red symbols represent 46 of the nearest stars to Earth; the blue symbols represent the 97 brightest stars as seen in our sky. There are many more cool, low-luminosity stars in the sky than there are hot, high-luminosity stars. If we restrict our attention to stars that are near the Sun, these are all that we see. Yet if we look at the brightest stars in the sky, we get a very different picture. Even though these stars are farther away, they are so luminous that they dominate what we see. Think about how different your impression of the properties of stars would be if all you knew about were the nearest stars, or if all you knew about were the brightest stars. Neither of these groups alone accurately represents what stars as a whole are like, but the nearest stars provide a much fairer sample because luminous stars, which dominate the brightest-stars group, are rare. Whether you are an astronomer studying the properties of stars or a political pollster measuring the sense of the people, the validity of your results depends on the care with which you choose the sample (of stars or people) that you study. If astronomers were conducting a poll, they would be much better off interviewing the nearest stars rather than the brightest.

When we add mass to the picture, the main sequence of the H-R diagram gets even more interesting. Stellar mass increases smoothly as we go from the lower right to the upper left along the main sequence. The faint, cool stars on the right-hand side of the main sequence have low masses; the luminous, hot stars on the left side of the

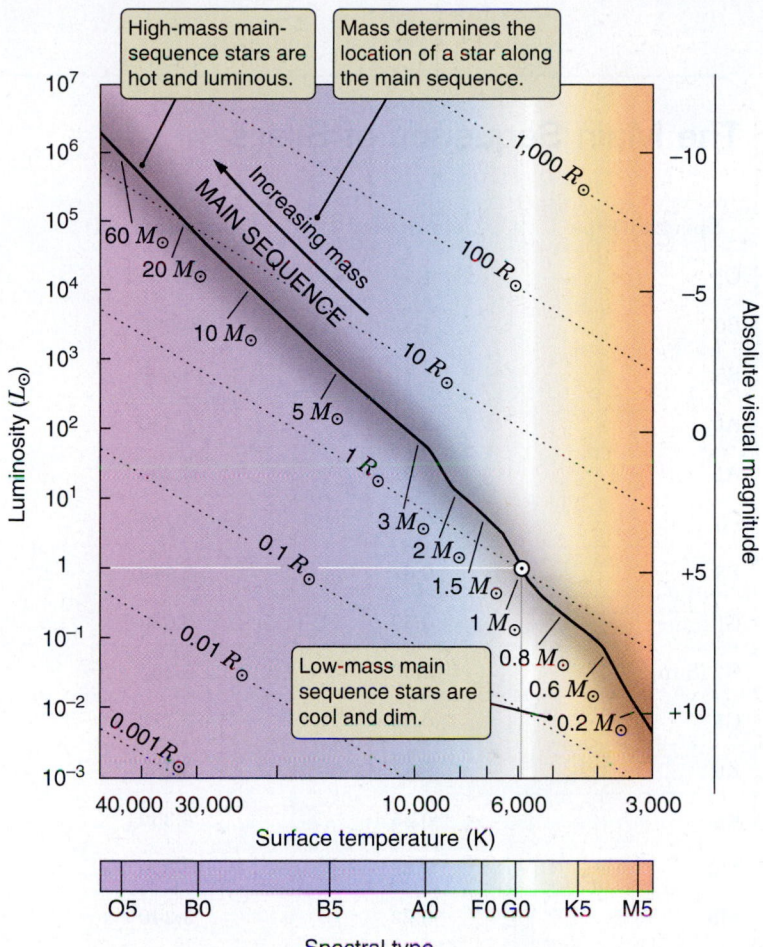

FIGURE 13.17 The main sequence of the H-R diagram is a sequence of masses.

main sequence are **high-mass stars**. Stated another way, if we know the mass of a main-sequence star, then we know where on the main sequence the star is, which means we know the star's approximate temperature, size, and luminosity. If a main-sequence star is less massive than the Sun, it will be smaller, cooler, redder, and less luminous than the Sun; it will be located to the lower right of the Sun on the main sequence. On the other hand, if a star is more massive than the Sun, it will be larger, hotter, bluer, and more luminous than the Sun; it will be located to the upper left of the Sun on the main sequence. *The mass of a star determines where on the main sequence the star will lie.*

To see the relationship between a star's mass and its position on the main sequence directly, look at **Table 13.3** and **Figure 13.17**, which show the properties of stars along the main sequence. If, for example, a main-sequence star has a mass of 18 $M_\odot$, it will be a B0 star. It will have a sur-

> Low-mass main-sequence stars are faint and cool. High-mass main-sequence stars are hot and luminous.

TABLE 13.3

The Main Sequence of Stars*

Spectral Type	Color Index†	Temperature (K)	Mass ($M_\odot$)	Radius ($R_\odot$)	Luminosity ($L_\odot$)
O5	1.36	44,500	60	17.8	794,000
B0	1.32	30,000	18	9.3	52,500
B5	1.17	15,400	5.9	3.8	832
A0	1.02	9,520	2.9	2.5	54
A5	0.87	8,200	2.0	1.74	14
F0	0.76	7,200	1.6	1.35	6.5
F5	0.67	6,440	1.3	1.2	3.2
G0	0.59	6,030	1.05	1.05	1.5
G2 (Sun)	0.56	5,860	1.00	1.00	1.0
G5	0.53	5,770	0.92	0.93	0.8
K0	0.47	5,250	0.79	0.85	0.4
K5	0.35	4,350	0.67	0.74	0.15
M0	0.27	3,850	0.51	0.63	0.08
M5	0.22	3,240	0.21	0.32	0.011
M8	0.19	2,640	0.06	0.13	0.0012

*Approximately 90 percent of the stars in the sky fall on the main sequence of the H-R diagram. This table gives the properties of stars along the main sequence.
†Normally, astronomers refer to the color index (or blue/visual brightness ratio) of a star as the difference between the B and V magnitudes of the star. (See Math Tools 13.1 and Appendix 6 for more information on calculating the color index.)

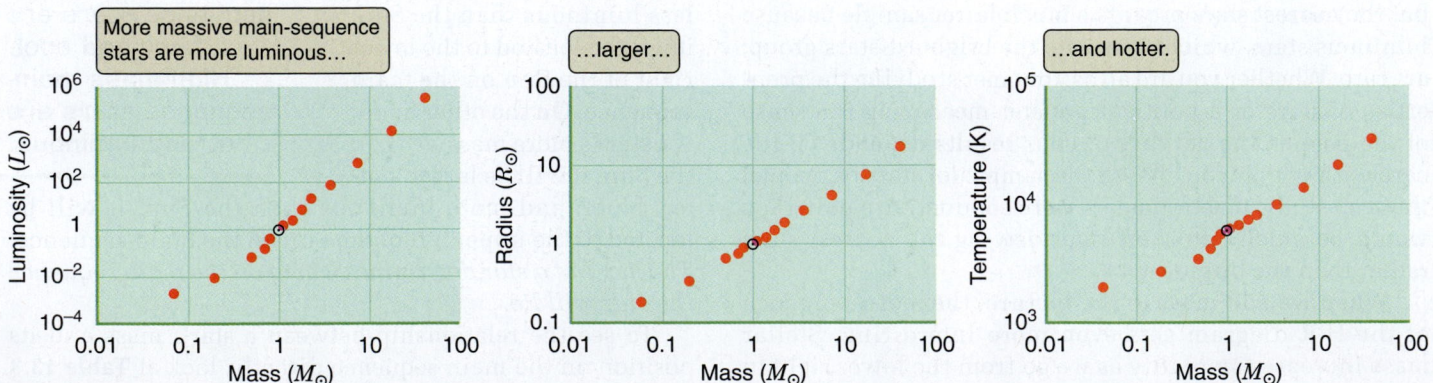

FIGURE 13.18 Plots of luminosity, radius, and temperature versus mass for stars along the main sequence. The mass (and chemical composition) of a main-sequence star determines all of its other properties.

face temperature of 30,000 K, a radius of over 9 $R_\odot$, and a luminosity about 50,000 times that of the Sun. If a main-sequence star instead has a mass of 0.21 $M_\odot$, it will be an M5 star. It will have a surface temperature of 3,240 K, a radius of about 0.32 $R_\odot$, and a luminosity of about 0.01 $L_\odot$. A main-sequence star with a mass of 1 $M_\odot$ will be a G2 star like the Sun and will have the same surface temperature, size, and luminosity as the Sun.

It is difficult to overemphasize the importance of this result, so we state it again for clarity. For stars of similar chemical composition, *the mass of a main-sequence star alone determines all of its other characteristics*. **Figure 13.18** illustrates this point. Knowing the mass (and chemical composition) of a main-sequence star tells us how large it is, what its surface temperature is, how bright it is, what its internal structure is, how long it will live, how it will evolve, and what its final fate will be!

> The mass of a main-sequence star determines what its fate will be.

Upon reflection, this relationship is both sensible and possibly the most important and fundamental relationship in all of astrophysics. If you have a certain amount and type of material to make a star, there is only one kind of star you can make. As we go on to discuss the physical processes that give a star its structure, this statement will make even more sense. We will find that a star is a "battle" between gravity (which is trying to pull the star together) and the energy released by nuclear reactions in the interior of the star (which is trying to blow it apart). The mass of the star determines the strength of its gravity, which in turn determines how much energy must be generated in its interior to prevent it from collapsing under its own weight. The mass of a star determines where the balance is struck.

Not All Stars Are Main-Sequence Stars

Before leaving our discussion of the observed properties of stars, we need to point out that although most stars in the sky are main-sequence stars, some are not. Some stars are found in the upper right portion of the H-R diagram, well above the main sequence. From their position we know that they must be bloated, luminous, cool giants, with radii hundreds or thousands of times the radius of the Sun. If the Sun were such a star, its atmosphere would swallow the orbits of the inner planets, including Earth. At the other extreme are stars found in the far lower left corner of the H-R diagram. These stars must be tiny, with sizes comparable to the size of Earth. Their small surface areas explain why they have such low luminosities despite having temperatures that can rival or even exceed the surface temperature of the hottest main-sequence O stars.

How can astronomers tell which stars are not members of the main sequence? Stars that lie off the main sequence on the H-R diagram can be identified by their luminosities (determined by their distance), or by slight differences within their spectral type. The width of a star's spectral lines is an indicator of the density and "surface pressure" of gas in a star's atmosphere. Those puffed-up stars above the main sequence have atmospheres with lower density and lower surface pressure compared to main-sequence stars.

It is important for astronomers to know where stars fall on the H-R diagram. For example, when using the H-R diagram to estimate the distance to a star by the spectroscopic parallax method, they must know whether the star is on, is above, or falls below the main sequence in order to get the star's luminosity. Stars both on and off the main sequence have a property called **luminosity class**, which tells us the *size* of the star. Luminosity classes are defined as follows: supergiant stars are luminosity class I, bright giants are class II, giants are class III, subgiants are class IV, main-sequence stars are class V, and white dwarfs are class WD. Luminosity classes I though IV lie above the main sequence; class WD falls below and to the left of the main sequence (**Figure 13.19**). Thus, the complete spectral classification of a star includes both its spectral type, which tells us temperature and the color; and its luminosity class, which indicates size.

Another point to consider is that some non-main-sequence stars vary in luminosity, and through the lumi-

FIGURE 13.19 Stellar luminosity classes.

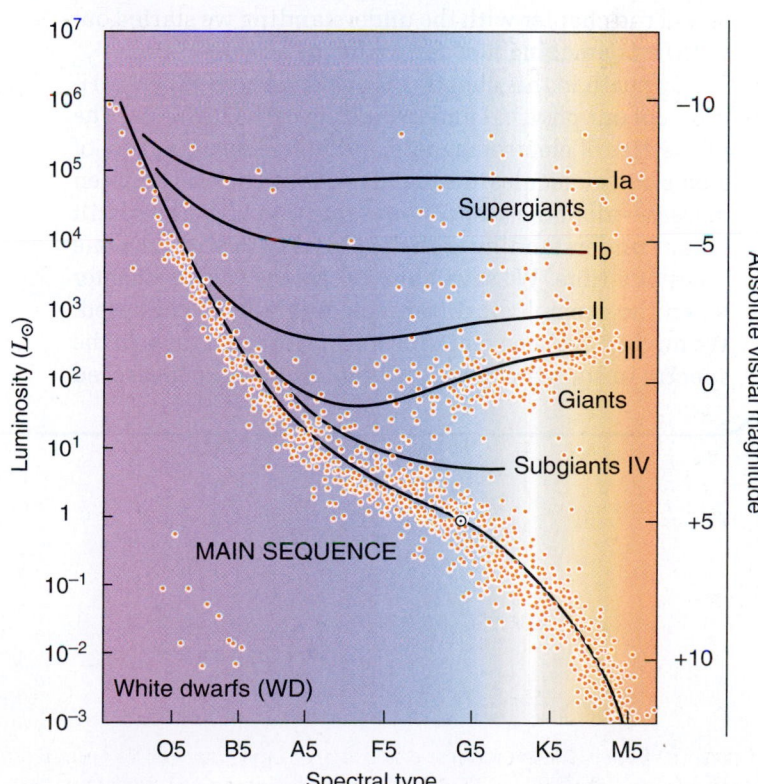

nosity-temperature-radius relationship we see that their temperature and size (radius) must be changing as well. We will return to these stars in later chapters as we look at stellar evolution.

The existence of the main sequence, together with the fact that the mass of a main-sequence star determines where on the main sequence it will lie, is a grand pattern that points to the possibility of a deep understanding of what stars are and what makes them tick. By the same token, the existence of stars that do *not* follow this grand pattern raises yet more questions. What is it about a star that determines whether or not it is part of the main sequence?

In the decades that followed the discovery of the main sequence, few problems in astronomy attracted more attention than these questions. Their answers turned out to be as fundamental as stellar pioneers Annie Jump Cannon, Ejnar Hertzsprung, and Henry Norris Russell could ever have hoped. The existence of the main sequence holds the essential clue to what stars are and how they work. The properties of stars that are not on the main sequence point to an understanding of how stars form, how they evolve, and how they die. Much of the rest of the text discussing stars will be spent on the lessons learned from patterns in the H-R diagram.

Seeing the Forest for the Trees

When you walk down a path, each step you take covers only a short distance. But by persistently putting one foot in front of the other, you find after a time that you have gotten somewhere. The same is true of the road we traveled in this chapter. Each step along the way was relatively small and understandable—involving the application of a physical principle that we see at work in the world around us, or the use of a tool from algebra or geometry. But when we compare our understanding of stars reached by the end of the chapter with the understanding we started out with, it is amazing how far we have come.

Our path in this chapter followed the course of the triumph of our physical understanding of the universe. The properties of electromagnetic radiation, the structure of atoms and molecules, the gravitational attraction between masses—all of these and more came into play as we built up our understanding of the physical nature of stars one piece at a time. We saw many occasions in this chapter when the cosmological principle might have collapsed. We might have seen emission and absorption lines in the spectra in stars that were different from those measured

in terrestrial laboratories. We might have found that the temperatures of stars inferred from their absorption spectra disagreed with the temperatures of stars inferred from their peaks in blackbody emission. We might have found that measurements of the sizes of stars based on the Stefan-Boltzmann law disagreed with measurements of the sizes of stars in eclipsing binaries. We might have found that the motions of stars in binary systems failed to follow the predictions of Newton's physics. We might have found any or all of these things, *but we did not!* The successes of this chapter give us strong reason to believe that the same physical laws at work right here on Earth and in our Solar System also describe the fundamental character and behavior of matter and energy throughout the rest of the universe.

With an understanding of the basic physical properties of stars in place, we are now ready to ask much more fundamental questions about stars. We are ready, figuratively speaking, to "lift the hood" and see what lies within. How do stars work? How do they form? How do they evolve? How do they die? We will begin to address these questions by investigating the star that serves as the standard by which we measure other stars: the star we know best, our Sun.

Summary

- The distances to nearby stars are measured stereoscopically by their parallax.

- The nearest star (other than the Sun) is about 4 light-years away.

- When stars are too remote for trigonometric parallax measurements, we must use their spectra and the H-R diagram to estimate their distances.

- Radiation tells us the temperature, size, and composition of stars.

- The color index of a star reveals its surface temperature.

- A star's luminosity class and temperature tells us its size.

- Small, cool stars greatly outnumber large, hot stars.

- Binary stellar systems enable us to measure the masses of stars.

- The H-R diagram is the single most useful diagram in astronomy, plotting stellar evolution and showing the relationship among the various physical properties of stars.

- The mass and composition of a main-sequence star determine its luminosity, temperature, and size.

SmartWork, Norton's online homework system, includes algorithmically generated versions of these questions, plus additional conceptual exercises. If your instructor assigns questions in SmartWork, log in at **smartwork.wwnorton.com.**

Student Questions

THINKING ABOUT THE CONCEPTS

1. Distances to stars can be measured in inches, miles, kilometers (km), astronomical units (AU), light-years, and parsecs (pc). Why do many stellar astronomers prefer to use parsecs?

2. Even our best measurements always have experimental uncertainties. Using the measurement uncertainties given in the book, how accurately can we know the distance to a star using trigonometric parallax?

3. The distances of nearby stars are determined by their parallaxes. Why is the uncertainty in our knowledge of a star's distance greater for stars that are farther from Earth?

4. What would happen to our ability to measure stellar parallax if we were on the planet Mars? What about Venus or Jupiter?

5. To know certain properties of a star, you must first determine the star's distance. For other properties, knowledge of distance is not necessary. Into which of these two categories would you place each of the following properties: size, mass, temperature, color, spectral type, and chemical composition? In each case, state your reason(s).

*6. The light from stars passes through dust in our galaxy before it reaches us, making stars appear dimmer than they actually are. How does this phenomenon affect stellar trigonometric parallax? How does it affect spectroscopic parallax?

7. Compare the temperature, luminosity, and radius of stars at the lower left and upper right of the H-R diagram.

8. Albireo, a star in the constellation of Cygnus, is a visual binary system whose two components can be seen easily with even a small amateur telescope. Viewers describe the brighter star as "golden" and the fainter one as "sapphire blue."
   a. What does this description tell you about the relative temperatures of the two stars?
   b. What does it tell you about their respective sizes?

9. Very cool stars have temperatures around 2,500 K and emit Planck spectra with peak wavelengths in the red part of the spectrum. Explain whether or not these stars emit any blue light.

*10. It is possible for two stars to have different temperatures and different sizes but the same luminosity. Explain how.

11. The stars Betelgeuse and Rigel are both in the constellation Orion. Betelgeuse appears red in color and Rigel is bluish white. To the eye, the two stars appear equally bright. If you can compare the temperature, luminosity, or size from just this information, do so. If not, explain why.

12. Capella (in the constellation Auriga) is the sixth brightest star in the sky. With a high-power telescope, Capella is actually two pairs of binary stars: the first pair are G-type giants; the second pair are M-type main-sequence stars. From this information, what color do you expect Capella to appear as, and why?

13. Two stars have the same luminosity but one is larger. Compare their temperatures. Now suppose that the two stars have the same size but one is more luminous. Again, compare their temperatures.

*14. In this chapter, we learned that the spectra of stars contain many absorption features, yet we treat them as blackbodies when discussing their properties. Why is this approach justified?

15. Explain why the stellar spectral types (O, B, A, F, G, K, M) are not in alphabetical order. Also explain the sequence of temperatures defined by these spectral types.

*16. Other than the Sun, the only stars whose mass we can measure *directly* are those in eclipsing or visual binary systems. Explain why.

17. In Math Tools 13.2, we show how one can calculate the masses of both stars in an eclipsing binary system using their orbital details. Implicit in these calculations was the assumption that the binary system is seen edge-on. What would happen to our final masses if the system were tilted (or "inclined") instead?

18. Once we determine the mass of a certain spectral type of star that is located in a binary system, we assume that all other stars of the same spectral type and luminosity class also have the same mass. Why is this a safe assumption?

*19. How do we estimate the mass of stars that are not in eclipsing or visual binary systems?

20. Although we tend to think of our Sun as an "average" main-sequence star, it is actually hotter and more luminous than average. Explain.

21. Many times throughout this chapter, we have commented that there are many more low-mass stars than high-mass stars. Why do you think this is so? (We will address the reasons in chapters to come).

22. Star masses range from 0.08 to about 100 $M_\odot$.
    a. Why are there no stars with masses less than 0.08 $M_\odot$?
    b. Why are there no stars with masses much greater than 100 $M_\odot$?

23. Very old stars often have very few heavy elements, while very young stars have much more. What does this difference imply about the chemical evolution of the Universe?

**24. Explain whether we would still be able to identify the spectral types of stars if there were no elements other than hydrogen and helium in their atmospheres.

25. Logarithmic (log) plots show major steps along an axis scaled to represent equal factors, most often factors of 10. Why, in astronomy, do we sometimes use a log plot instead of the more conventional linear plot?

APPLYING THE CONCEPTS

26. Sketch a circle and mark an arc along the circumference equal to the radius of the circle. The size of the angle subtended by the arc, measured in *radians*, is given by the length of the arc divided by the radius (*r*) of the circle.

Because the circumference of a circle is $2\pi r$, there must be $2\pi r/r$ or 2π radians in a circle.
    a. How many degrees are there in 1 radian?
    b. How many arcseconds are there in 1 radian? How does this value compare to the number of astronomical units in 1 pc (206,265) that we cite in the text?
    c. What angle would a round object that has a diameter of 1 AU make in our sky if we see it at a distance of 1 pc? A distance of 10 pc?
    d. What do your answers to parts (b) and (c) of this question tell you about how the actual size of an object is related to its angular size measured in units of radians?

27. Our eyes are typically 6 cm apart. Suppose your eye separation is average and you see an object jump from side to side by half a degree as you blink back and forth between your eyes. How far away is that object?

28. Sirius, the brightest star in the sky, has a parallax of 0.379 arcsec. What is its distance in parsecs? In light-years?

**29. Sirius is actually a binary pair of two A-type stars. The brighter of the two stars is called the "Dog Star," and the fainter is called the "Pup Star" because Sirius is in the constellation Canis Major (meaning "Big Dog"). The Pup Star appears about 6,800 times fainter than the Dog Star. Compare the temperatures, luminosities, and sizes of both stars.

30. Sirius and its companion orbit around a common center of mass with a period of 50 years. The mass of Sirius is 2.35 times the mass of the Sun.
    a. If the orbital velocity of the companion is 2.35 times greater than that of Sirius, what is the mass of the companion?
    b. What is the semimajor axis of the orbit?

31. Sirius is 22 times more luminous than the Sun, and Polaris (the "North Pole Star") is 2,350 times more luminous than the Sun. Sirius appears 23 times brighter than Polaris. How much farther away from us is Polaris than Sirius? What is the distance of Polaris in light-years?

32. Proxima Centauri, the star nearest to Earth other than the Sun, has a parallax of 0.772 arcsec. How long does it take light to reach us from Proxima Centauri?

*33. Betelgeuse (in Orion) has a parallax of 0.00763 ± 0.00164 arcsec, as measured by the Hipparcos satellite. What is the distance to Betelgeuse, and the uncertainty in that measurement?

34. Rigel (also in Orion) has a Hipparcos parallax of 0.00412 arcsec. Given that the two stars appear equally bright in the sky, which star is actually more luminous? Knowing that Betelgeuse appears reddish while Rigel appears bluish white, as stated in question 11, which star would you say is larger and why?

35. The Sun is about 16 trillion (1.6×10^{13}) times brighter than the faintest stars visible to the naked eye.
    a. How far away (in astronomical units) would an identical solar-type star be if it were just barely visible to the naked eye?
    b. What would be its distance in light-years?

*36. Our galaxy (the Milky Way) contains over 100 billion stars. If you assume that the average density of stars is the same as in the solar neighborhood, how much volume does the Milky Way take up? If the galaxy were a sphere, what would its radius be?

37. Find the peak wavelength of blackbody emission for a star with a temperature of about 10,000 K. In what region of the spectrum does this wavelength fall? What color will this star appear to us?

38. About 1,470 watts (W) of solar energy hits each square meter of the Earth's surface. Use this and our distance to the Sun to calculate its luminosity.

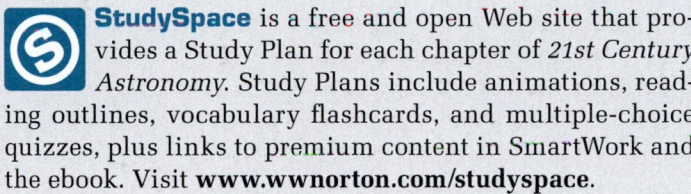 **StudySpace** is a free and open Web site that provides a Study Plan for each chapter of *21st Century Astronomy*. Study Plans include animations, reading outlines, vocabulary flashcards, and multiple-choice quizzes, plus links to premium content in SmartWork and the ebook. Visit **www.wwnorton.com/studyspace**.

It is stern work, it is perilous work to thrust your hand in the sun
And pull out a spark of immortal flame to warm the hearts of men.

JOYCE KILMER (1886–1918)

Sunset in New York City.

CHAPTER 14

A Run-of-the-Mill G-Type Star—Our Sun

14.1 The Sun Is More Than Just a Light in the Sky

How wonderful, after a long cold night, to see the light and feel the warmth of the rays of the Sun. Energy from the Sun is responsible for daylight, for our weather and seasons, and for terrestrial life itself. No object in nature has been more revered or more worshipped than the Sun. In fact, the modern symbol for the Sun, ☉, is nothing other than the ancient Egyptian hieroglyph for the Sun god Ra. Yet although the Sun may have dominated human consciousness since the dawn of our species, the discussion in Chapter 13 offers a very different perspective. To an astronomer of the 21st century, the Sun is the prototype for main-sequence stars. With a middle-of-the-road spectral type of G2, the Sun is all but indistinguishable from billions of other stars in our galaxy. It is also the star against which all other stars are measured. The mass of the Sun, the size of the Sun, the luminosity of the Sun—these basic properties provide the yardsticks of modern astronomy.

The Sun may be run-of-the-mill as far as stars go, but that makes it no less awesome an object on a human scale. The mass of the Sun, 1.99×10^{30} kilograms (kg), is over 300,000 times that of Earth. The Sun's radius, 696,000 kilometers (km), is over 100 times that of Earth. At a luminosity of 3.85×10^{26} watts (W), the Sun produces more energy in a second than all of the electric power plants on Earth could generate in a half-million years. The Sun is also the only star we can study at close range. Much of the detailed information that we know about stars has come only from studying our local star.

KEY CONCEPTS

To most humans the Sun is the most important object in the heavens. It lights our days, warms our planet, and provides the energy for life. But to astronomers the Sun is a typical main-sequence star, located conveniently nearby for detailed study. In this chapter, as we take a closer look at our local star, we will learn about

- The balances between pressure and gravity, and between energy generation and loss, that determine the structure of the Sun.
- Fusion of hydrogen to helium, and how mass is efficiently converted into energy in the Sun's core.
- The different ways that energy moves outward from the Sun's core toward its surface.
- Physical models of the Sun's interior, and how observations of solar neutrinos and seismic vibrations on the surface of the Sun test these models.
- The structure of the Sun's atmosphere, from its 5,780-kelvin photosphere to its 1-million-kelvin corona.
- The solar wind streaming away from the Sun.
- Sunspots, flares, coronal mass ejections, and other consequences of magnetic activity on the Sun.
- Solar-activity cycles of 11 and 22 years.
- How solar activity affects Earth.

In the previous leg of our journey, we looked at the gross physical properties of distant stars, including their mass, luminosity, size, temperature, and chemical composition. Now, as we turn our attention toward our own local star, we face more fundamental questions: How does the Sun work? Where does it get its energy? Why does it have the size, temperature, and luminosity that it does? How has it been able to remain so constant over the billions of years since the Solar System formed? In short, we now confront the question, What is a star?

14.2 The Structure of the Sun Is a Matter of Balance

In Chapter 7 we tackled the question of how we know about the interior of Earth even though no machine has ever done more than scratch the surface of our planet. The answer was a combination of physical understanding, detailed computer models, and clever experiments that test the predictions of those models. The task of exploring the interior of the Sun is much the same. As with Earth, the structure of the Sun is governed by a number of physical processes and relationships. Using our understanding of physics, chemistry, and the properties of matter and radiation, we express these processes and relationships as mathematical equations. High-speed computers are then used to solve these equations and arrive at a model of the Sun. One of the great successes of 20th century astronomy was the successful construction of a physical model of the Sun that agrees with our observations of the mass, composition, size, temperature, and luminosity of the real thing.

Our current model of the interior of the Sun is the culmination of decades of work by thousands of physicists and astronomers. Understanding the details that lie within this model is the work of lifetimes. Even so, the essential ideas underlying our understanding of the structure of the Sun are found in a few key insights. In turn, these insights can be summed up in a single statement: *The structure of the Sun is a matter of balance.*

The first key balance within the Sun is the balance between pressure and gravity illustrated in **Figure 14.1**. The Sun is a huge ball of hot gas. If gravity were stronger than pressure within the Sun, the Sun would collapse. Conversely, if pressure were stronger than gravity, the Sun would blow itself apart. We saw this balance in Chapter 7 and gave it a name: *hydrostatic equilibrium.* Hydrostatic equilibrium sets the pressure at each point within a planet and determines the atmospheric pressure at Earth's surface. Hydrostatic equilibrium says

> The Sun is in hydrostatic equilibrium.

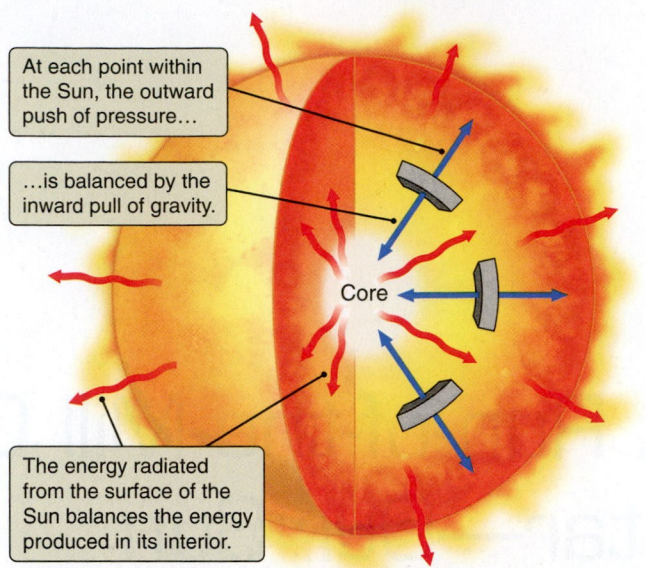

FIGURE 14.1 The structure of the Sun is determined by the balance between the forces of pressure and gravity, and the balance between the energy generated in its core and energy radiated from its surface.

that the pressure at any point within the Sun's interior must be just enough to hold up the weight of all the layers above that point. If the Sun were not in hydrostatic equilibrium, then forces within the Sun would not be in balance, so the surface of the Sun would *move.* The Sun today is the same as it was yesterday and the day before. That is all the observation we require to infer that the interior of the Sun is in hydrostatic equilibrium.

Hydrostatic equilibrium becomes an even more powerful concept when combined with what we know about the way gases behave. As we move deeper into the interior of the Sun, the weight of the material above us becomes greater, and hence the pressure must increase. As we learned in Foundations 8.1, in a gas higher pressure means higher density and/or higher temperature. **Figure 14.2a** shows how conditions vary as distance from the center of the Sun changes. As we go deeper into the Sun, the pressure climbs; and as it does, the density and temperature of the gas climb as well.

A second fundamental balance within the Sun is a balance of energy (see Figure 14.1). Stars like the Sun are remarkably stable objects. Geological records show that the luminosity of the Sun has remained nearly constant for billions of years.[1] In fact, the very existence of the main sequence says that stars do not change much over the main

> Solar energy production must balance what is radiated away.

---

[1] Astrophysicists believe that the Sun increases its luminosity about 5 percent every billion years.

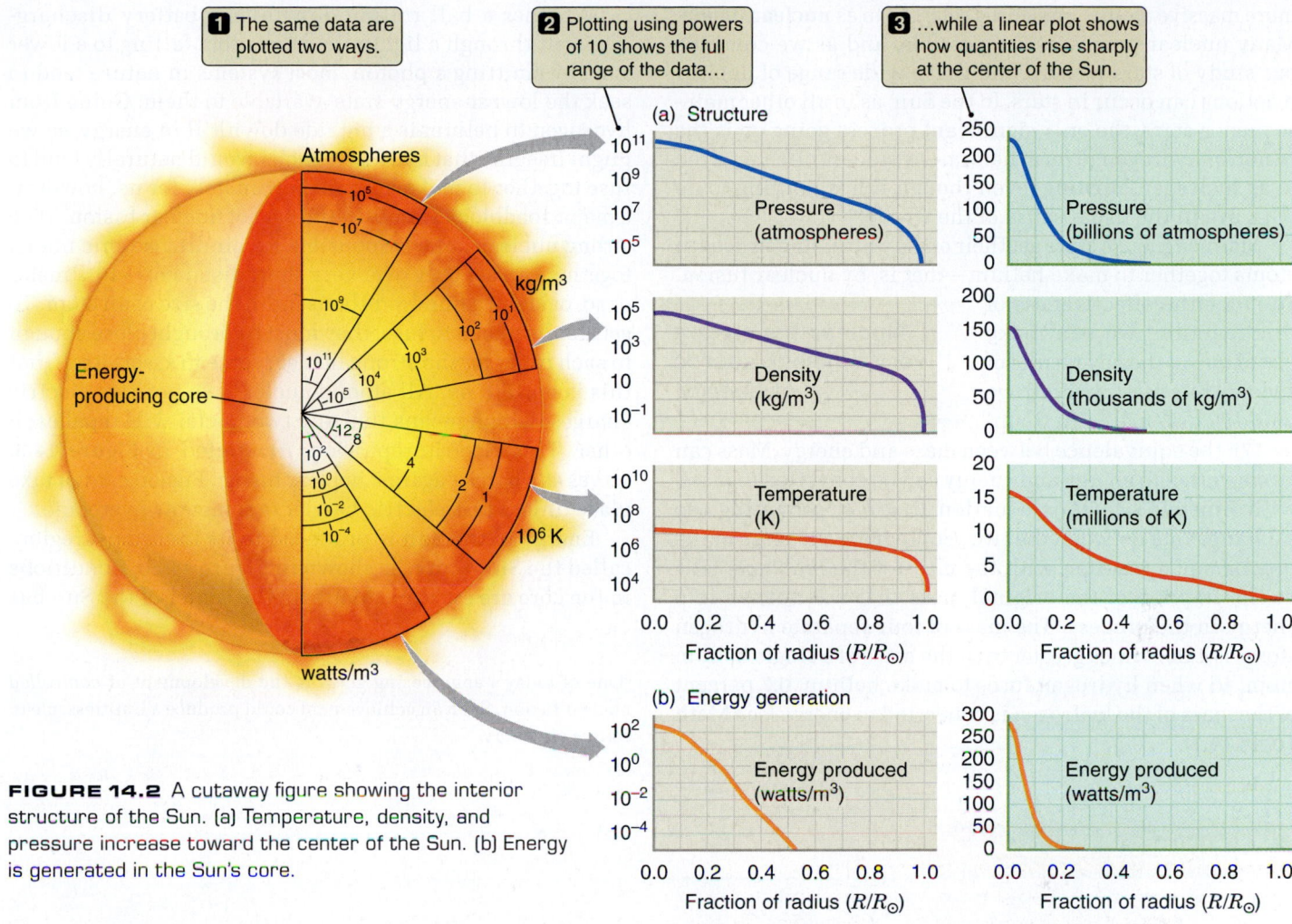

1 The same data are plotted two ways.

2 Plotting using powers of 10 shows the full range of the data...

3 ...while a linear plot shows how quantities rise sharply at the center of the Sun.

(a) Structure

Atmospheres

Energy-producing core

kg/m³

10⁶ K

watts/m³

FIGURE 14.2 A cutaway figure showing the interior structure of the Sun. (a) Temperature, density, and pressure increase toward the center of the Sun. (b) Energy is generated in the Sun's core.

(b) Energy generation

part of their lives. To remain in balance, the Sun must produce just enough energy in its interior each second to replace the energy that is radiated away by its surface each second. This is a new type of balance, one we have not dealt with before. Understanding the balance of energy within the Sun requires thinking about how energy is generated in the interior of the Sun (**Figure 14.2b**), and how that energy finds its way from the interior to the Sun's surface, where it is radiated away.

The Sun Is Powered by Nuclear Fusion

One of the most basic questions facing the pioneers of stellar astrophysics was where the Sun and stars get their energy. The answer to this question came not from astronomers' telescopes, but from theoretical work and the laboratories of nuclear physicists. At the heart of the Sun lies a nuclear furnace capable of powering the star for billions of years.

The nucleus of most hydrogen atoms consists of a single proton. Nuclei of all other atoms are built from a mixture of protons and neutrons. Most helium nuclei, for example,

consist of two protons and two neutrons. Most carbon nuclei consist of six protons and six neutrons. Protons have a positive electric charge, and neutrons have no net electric charge. Like charges repel, so all of the protons in an atomic nucleus must be pushing away from each other with a tremendous force because they are so close to each other. If electric forces were all there was to it, the nuclei of atoms would rapidly fly apart—yet atomic nuclei *do* hold together. We conclude that there must be another force in nature, even stronger than the electric force, that "glues" the protons and neutrons in a nucleus together. That force, which acts only over extremely short distances, is called the **strong nuclear force**.

The strong nuclear force is indeed a very powerful force. It would take an enormous amount of energy to pull apart the nucleus of an atom such as helium into its constituent parts. The reverse of this process says that when you assemble an atomic nucleus from its component parts, this same enormous amount of energy is released. The process of combining two less massive atomic nuclei into a single

> Atomic nuclei are held together by the strong nuclear force.

more massive atomic nucleus is referred to as **nuclear fusion**. Many nuclear processes are possible, and as we continue our study of stars we will find that a wide range of nuclear reactions can occur in stars. In the Sun, as in all other main-sequence stars, the only significant process going on is the fusion of hydrogen to form helium—a process often referred to as **hydrogen burning** (even though it has nothing to do with fire in the usual sense of the word).

Main-sequence stars get their energy by fusing hydrogen atoms together to make helium—that is, by nuclear fusion. To judge the effectiveness of this reaction, we can make use of one of the key results of special relativity (which we will discuss further in Chapter 17): the equivalence between mass and energy. Mass can be converted to energy and energy can be converted to mass, with Einstein's famous equation $E = mc^2$ providing the exchange rate between the two. Comparing the mass of the *products* of a reaction with the mass of the *reactants* tells us the fraction of the original mass that was turned into energy in the process. The mass of four separate hydrogen atoms is 1.007 times greater than the mass of a single helium atom; so when hydrogen fuses to make helium, 0.7 percent of the mass of the hydrogen is converted to energy (see **Math Tools 14.1**).

> Nuclear fusion is a very efficient source of energy.

Whether a ball rolling downhill, a battery discharging itself through a lightbulb, or an atom falling to a lower state by emitting a photon, most systems in nature tend to seek the lowest-energy state available to them. Going from hydrogen to helium is a big ride downhill in energy, so we might imagine that hydrogen nuclei would naturally tend to fuse together to make helium. Fortunately for us, however, a major roadblock stands in the way of nuclear fusion.[2] The strong nuclear force responsible for binding atomic nuclei together can act only over very short distances—10^{-15} meter or so, or about a hundred-thousandth the size of an atom. To get atomic nuclei to fuse, they must be brought close enough to each other for the strong nuclear force to assert itself, but this is hard to do. All atomic nuclei have positive electric charges, which means that any two nuclei will repel each other. This electric repulsion, illustrated in **Figure 14.3**, serves as a barrier against nuclear fusion. Fusion cannot take place unless this electric barrier is somehow overcome.

Energy in the Sun is produced in its innermost region, called the Sun's **core**, as shown in Figure 14.2b. Conditions in the core are extreme. Matter at the center of the Sun has

---

[2]One of today's engineering goals is the development of *controlled* nuclear fusion. Such an achievement could produce a limitless, clean source of energy.

MATH TOOLS 14.1

The Source of the Sun's Energy

We have seen that the conversion of four hydrogen nuclei (protons) into a single helium nucleus results in a loss of mass. The mass of a single proton is 1.6726×10^{-27} kg. So, four protons add up to a mass of 6.6904×10^{-27} kg. But the mass of a helium nucleus is 6.6465×10^{-27} kg. Doing the subtraction, we see that this is 0.0439×10^{-27} kg (or 4.39×10^{-29} kg) less than the mass of the four protons—a loss of about 0.7 percent. Conversion of 0.7 percent of the mass of the hydrogen into energy might not seem very efficient—until we compare it with other sources of energy and discover that it is millions of times more efficient than even the most efficient chemical reactions.

Using Einstein's equation $E = mc^2$, we can see that the energy released by this mass-to-energy conversion is

$$E = (4.39 \times 10^{-29} \text{ kg})(3.00 \times 10^8 \text{ m/s})^2 = 3.95 \times 10^{-12} \text{ J}.$$

Fusing a *single gram* of hydrogen into helium releases about 6×10^{11} joules (J) of energy, which is enough to boil all of the water in about 10 average backyard swimming pools. Scale this up to converting roughly 600 million metric tons (1 metric ton = 10^3 kg) of hydrogen into helium every second (with 4 mil-

lion metric tons of matter converted to energy in the process), and you have our Sun. The sunlight falling on Earth may be responsible for powering almost everything that happens on our planet, but it amounts to only about a hundred-billionth of the energy radiated by our local star.

The Sun has been burning hydrogen at this prodigious rate for 4.6 billion years, so how much of its available fuel has already been spent? We estimate that the amount of hydrogen available for fusion in the young Sun's core was about 10 percent of the total solar mass, or 2×10^{29} kg. At a rate of 600 million metric tons per second, the Sun consumes 6×10^{11} kg per second, multiplied by 3.16×10^7 seconds per year, or 1.90×10^{19} kg per year. Over the Sun's age of 4.6 billion years, it has consumed 1.9×10^{19} kg per year multiplied by 4.6×10^9 years, or a total of 8.7×10^{28} kg of hydrogen. This is about 44 percent of the Sun's original supply of available hydrogen fuel, confirming that Old Sol is now approaching middle age. A favorite theme of science fiction is the fate that awaits Earth when the Sun dies, but we need not worry anytime soon. The Sun is only about halfway through its 10-billion-year lifetime as a main-sequence star.

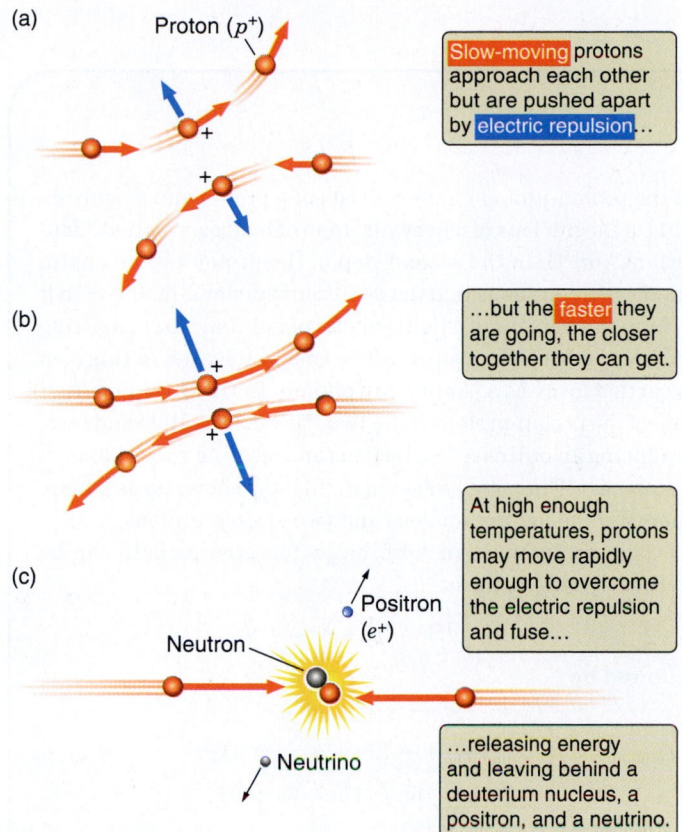

(a) Proton (p⁺)

Slow-moving protons approach each other but are pushed apart by electric repulsion…

(b)

…but the faster they are going, the closer together they can get.

(c)

At high enough temperatures, protons may move rapidly enough to overcome the electric repulsion and fuse…

Positron (e⁺)

Neutron

Neutrino

…releasing energy and leaving behind a deuterium nucleus, a positron, and a neutrino.

FIGURE 14.3 Atomic nuclei are positively charged and so repel each other. (a, b) If two nuclei are moving toward each other, the faster they are going, the closer they will get before veering away. (c) At the temperatures and densities found in the centers of stars, thermal motions of nuclei are energetic enough to overcome this electric repulsion, so fusion takes place.

a density about 150 times the density of water (which is 1,000 kilograms per cubic meter [kg/m³]), and the temperature at the center of the Sun is about 15 million kelvins (K). The thermal motions of atomic nuclei in the Sun's core contain tens of thousands times more kinetic energy than the thermal motions of atoms at room temperature. As illustrated in **Figure 14.3c**, under these conditions atomic nuclei slam into each other hard enough to overcome the electric repulsion between them and allow short-range nuclear forces to act. The hotter and denser a gas is, the more of these energetic collisions will take place each second. For this reason, the rate at which nuclear fusion reactions occur is extremely sensitive to the temperature and the density of the gas. Half of the energy produced by the Sun is generated within the inner 9 percent of the Sun's radius, or less than 0.1 percent of the volume of the Sun (see Figure 14.2b).

There are several reasons why hydrogen burning is the most important source of energy in main-sequence stars.

Hydrogen fuses to helium in the core of the Sun.

Hydrogen is the most abundant element in the universe, so it offers the most abundant source of nuclear fuel at the beginning of a star's lifetime. Hydrogen burning is also the most efficient form of nuclear fusion, converting a larger fraction of mass into energy than does any other type of reaction. But the most important reason why hydrogen burning is the dominant process in main-sequence stars is that hydrogen is also the easiest type of atom to fuse. Hydrogen nuclei—protons—have an electric charge of +1. The electric barrier that must be overcome to fuse protons is the repulsion of a single proton against another. Compare this to the force required, for example, to get two carbon nuclei close enough to fuse. To fuse carbon we must overcome the repulsion of six protons in one carbon nucleus pushing against the six protons in another carbon nucleus. The resulting force is proportional to the product of the charges of the two atomic nuclei, making the repulsion between two carbon nuclei 36 times stronger than that between two protons. For this reason, hydrogen fusion occurs at a much lower temperature than does any other type of nuclear fusion. In the cores of low-mass stars such as the Sun, hydrogen burns primarily through a process called the *proton-proton chain*. The dominant branch of the proton-proton chain is discussed in **Foundations 14.1**. ▶❙❙ **AstroTour:** The Solar Core

Hydrogen burns mostly via the proton-proton chain.

Energy Produced in the Sun's Core Must Find Its Way to the Surface

Some of the energy released by hydrogen burning in the core of the Sun escapes directly into space in the form of neutrinos (see Foundations 14.1), but most of the energy goes instead into heating the solar interior. To understand the structure of the Sun we must understand how thermal energy is able to move outward through the star. The nature of **energy transport** within a star is one of the key factors determining the star's structure.

Thermal energy can be transported from one place to another by a number of methods. Pick up a bucket of hot coals and carry it from one side of the room to the other, and you have transported thermal energy. A common way in which energy is transported in our everyday lives is **thermal conduction**. For example, hold one end of a metal rod while you put the other end into a fire. Soon the end of the rod that is in your hand becomes too hot to hold. Thermal conduction occurs as the energetic thermal vibrations of atoms and molecules in the hot end of the rod cause their cooler neighbors to vibrate more rapidly as well. However, although thermal conduction is the most important way energy is transported in *solid* matter, it is typically ineffective in a *gas* because the atoms and molecules are too far apart to transmit vibrations to one another efficiently. Thermal conduction is unimportant in the transport of energy

The Proton-Proton Chain

In the Sun and in other low-mass stars, hydrogen burning takes place through a series of nuclear reactions called the **proton-proton chain**. The proton-proton chain has three different branches. The most important of these, responsible for about 85 percent of the energy generated in the Sun, consists of three steps, illustrated in **Figure 14.4**. In the first step, two hydrogen nuclei fuse. In the process, one of the protons is transformed into a neutron by emitting a positively charged particle called a **positron** and another type of elementary particle called a **neutrino**. The conversion of a proton into a neutron by the emission of a positron and a neutrino is one variety of a process referred to as **beta decay**.

The positron is expelled at a great velocity, carrying away some of the energy released in the reaction. Electrons and positrons have opposite electric charges, so they attract each other. As a result, our expelled positron will soon collide with one of the many electrons moving freely about in the center of the Sun. But the positron is the **antiparticle** of the electron and, as we'll learn in Chapter 21, when particle and antiparticle collide they annihilate each other, with their total mass being converted into energy. Thus, the annihilation of electrons and positrons in the Sun's core produces energy in the form of gamma-ray photons. These photons carry part of the energy released when the two protons fused, thereby helping to heat the surrounding gas. The neutrino, on the other hand, is a very elusive particle. Its interactions with matter are so feeble that its most likely fate is to escape from the Sun without further interactions with any other particles.

Follow along in Figure 14.4 as we step through the proton-proton chain. The new atomic nucleus formed by the first step in the proton-proton chain consists of a proton and a neutron. This is the nucleus of a heavy isotope of hydrogen called "deuterium," or ^{2}H. In the second step of the proton-proton chain, another proton slams into the deuterium nucleus, fusing with it to form the nucleus of a light isotope of helium, ^{3}He, consisting of two protons and a neutron. The energy released in this step is carried away as a gamma-ray photon. In the third and final step of the proton-proton chain, two ^{3}He nuclei collide and fuse, producing an ordinary ^{4}He nucleus and ejecting two protons in the process. The energy released in this step shows up as kinetic energy of the helium nucleus and two ejected protons.

This dominant branch of the proton-proton chain can be written symbolically as

$$^1H + {}^1H \rightarrow {}^2H + e^+ + \nu,$$

followed by

$$e^+ + e^- \rightarrow \gamma + \gamma$$
$$^2H + {}^1H \rightarrow {}^3He + \gamma$$
$$^3He + {}^3He \rightarrow {}^4He + {}^1H + {}^1H$$

Here the symbols are e^+ for a positron, e^- for an electron, ν (the Greek letter nu) for a neutrino, and γ (gamma) for a gamma-ray photon.

The rate at which the proton-proton chain reaction takes place depends on both temperature and pressure. At the temperature and pressure that exist within the Sun's core, the reaction rate is relatively slow—in fact, extremely slow compared to a nuclear bomb explosion. The Sun's slow **nuclear burning** is fortunate for us. Otherwise the Sun would have exhausted its supply of hydrogen long ago.

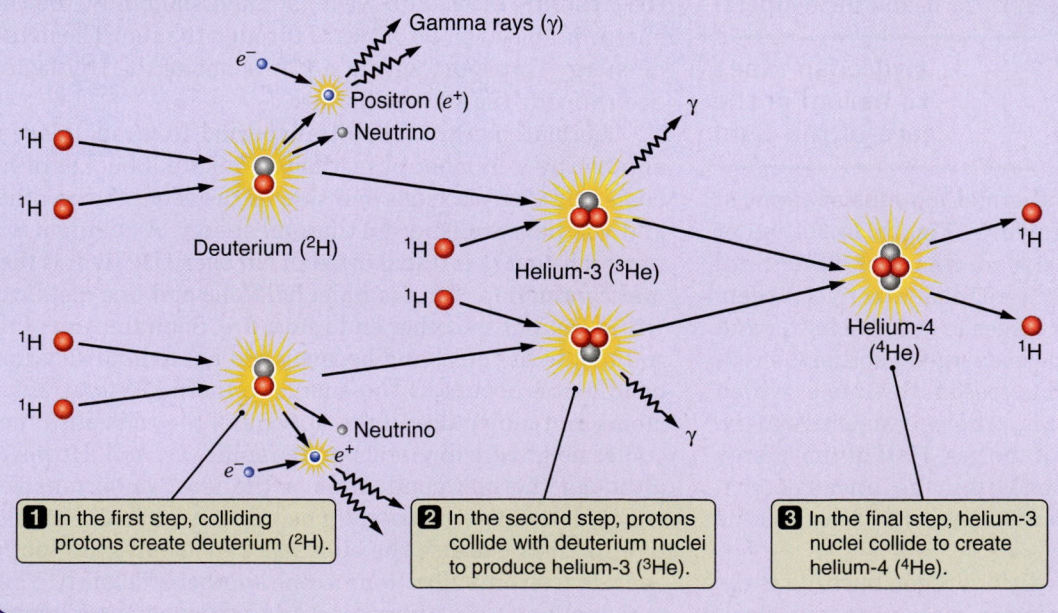

Gamma rays (γ)

e^-

Positron (e^+)

^{1}H   Neutrino

Deuterium (^{2}H) ^{1}H

Helium-3 (^{3}He)

^{1}H γ

^{1}H ^{1}H

^{1}H   Helium-4 (^{4}He) ^{1}H

^{1}H

Neutrino e^+

e^- γ

1 In the first step, colliding protons create deuterium (^{2}H).

2 In the second step, protons collide with deuterium nuclei to produce helium-3 (^{3}He).

3 In the final step, helium-3 nuclei collide to create helium-4 (^{4}He).

FIGURE 14.4 The Sun and all other main-sequence stars get their energy by fusing the nuclei of four hydrogen atoms together to make a single helium atom. In the Sun, about 85 percent of the energy produced comes from the branch of the proton-proton chain shown here.

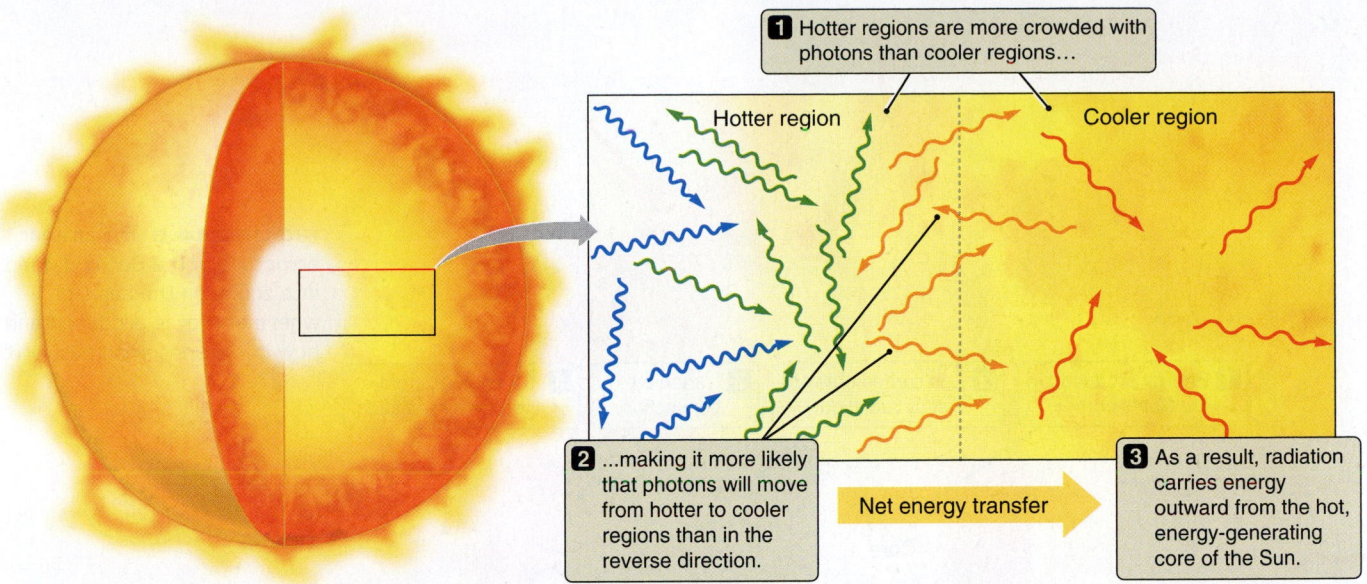

1 Hotter regions are more crowded with photons than cooler regions...

Hotter region   Cooler region

2 ...making it more likely that photons will move from hotter to cooler regions than in the reverse direction.

Net energy transfer

3 As a result, radiation carries energy outward from the hot, energy-generating core of the Sun.

FIGURE 14.5 Higher-temperature regions deep within the Sun produce more radiation than do lower-temperature regions farther out. Although radiation flows in both directions, more radiation flows from the hot regions to the cooler regions than from the cooler regions to the hot regions. Therefore, radiation carries energy outward from the inner parts of the Sun.

from the core of the Sun to its surface. Thermal energy is instead carried outward from the center of the Sun by two other mechanisms: radiation and convection.

In **radiative transfer**, energy is transported from hotter to cooler regions by photons, which carry the energy with them. Imagine a hotter region of the Sun sitting next to a cooler region, as shown in **Figure 14.5**. Recall from our study of radiation in Chapter 4 that the hotter region will contain more (and more energetic) photons than the cooler region. (There will be a Planck spectrum of photons in both regions, so the total energy carried by photons will be proportional to the fourth power of the temperature in each region.) More photons will move by chance from the hotter ("more crowded") region to the cooler ("less crowded") region than in the reverse direction. Thus, there is a net transfer of photons and photon energy from the hotter region to the cooler region, and in this way radiative transfer carries energy outward from the Sun's core.

Radiation carries energy from hotter regions to cooler regions.

If the temperature differs by a large amount over a short distance, then the concentration of photons will differ sharply as well, favoring rapid radiative energy transfer. The transfer of energy from one point to another by radiation also depends on how freely radiation can move from one point to another within a star. The degree to which matter impedes the flow of photons through it is referred to as **opacity**. The opacity of a mate-

Opacity impedes the outward flow of radiation.

rial depends on many things, including the density of the material, its composition, its temperature, and the wavelength of the photons moving through it.

Radiative transfer is most efficient in regions where opacity is low. In the inner part of the Sun, where temperatures are high and atoms are ionized, opacity comes mostly from the interaction between photons and free electrons (electrons not attached to any atom). Here opacity is relatively low, and radiative transfer is capable of carrying the energy produced in the core outward through the star. The region in which radiative transfer is responsible for energy transport extends 71 percent of the way out toward the surface of the Sun. This region is referred to as the Sun's **radiative zone** (**Figure 14.6**). Even though the opacity of the radiative zone is relatively low, photons are still able to travel only a short distance before interacting with matter. The path that a photon follows is so convoluted that, on average, it takes the energy of a gamma-ray photon produced in the interior of the Sun about 100,000 years to find its way to the outer layers of the Sun. Opacity serves as a blanket, holding energy in the interior of the Sun and letting it seep away only slowly.

From a peak of 15 million K at the center of the Sun, the temperature falls to about 100,000 K at the outer margin of the radiative zone. At this temperature, atoms are no longer completely ionized, and the opacity is therefore higher. As the opacity increases, radiation becomes less efficient in carrying energy from one place to another. The energy that is flowing outward through the Sun "piles up." The physical sign that energy is piling up is that the *temperature gradient*—that is, how rapidly temperature drops with

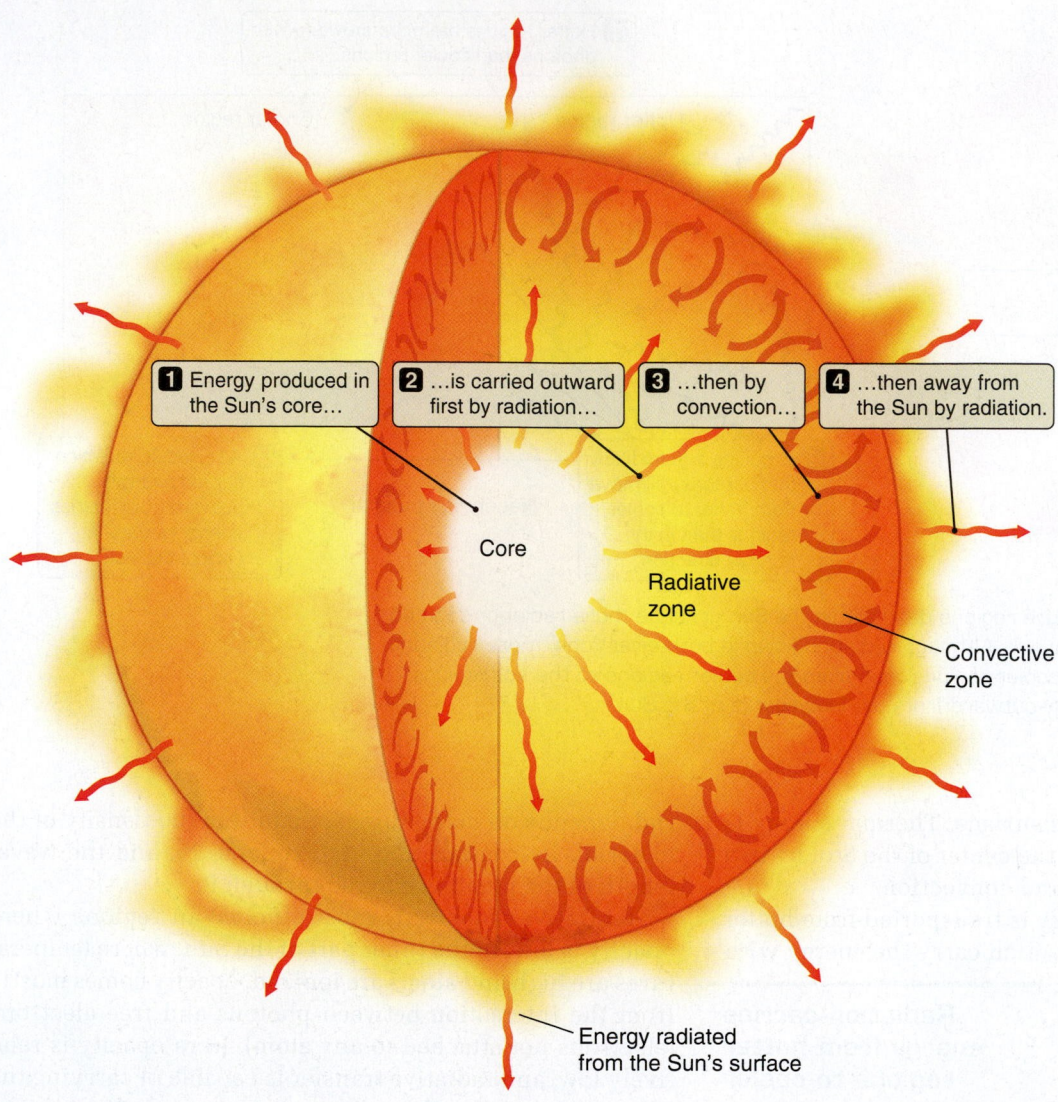

1 Energy produced in the Sun's core…

2 …is carried outward first by radiation…

3 …then by convection…

4 …then away from the Sun by radiation.

Core

Radiative zone

Convective zone

Energy radiated from the Sun's surface

FIGURE 14.6 The interior structure of the Sun is divided into zones on the basis of where energy is produced and how it is transported outward.

increasing distance from the center of the Sun—becomes very steep. (Radiative transfer carries energy from hotter regions to cooler regions, smoothing out temperature differences between them. As the opacity increases, radiation becomes less effective in smoothing out temperature differences, so temperature differences between one region and another become greater.)

As we move farther toward the surface of the Sun, radiative transfer becomes so inefficient (and the temperature gradient so steep) that a different way of transporting energy takes over. Like a hot-air balloon, cells (or packets) of hot gas become buoyant and rise up through the lower-temperature gas above them, carrying energy with them. Thus, **convection** begins. Just as convection carries energy from the interior of planets to their surfaces, or from the Sun-heated surface of Earth upward through Earth's *atmosphere*, convection also plays an important role in the transport of energy outward from the interiors of many stars, including the Sun. The solar

> In the outer part of the Sun, energy is carried by convection.

convective zone (see Figure 14.6) extends from the outer boundary of the radiative zone out to just below the visible surface of the Sun.

In the outermost layers of stars, radiation again takes over as the primary mode of energy transport. (This must be the case. After all, it is radiation that transports energy from the outermost layers of a star off into space.) Even so, the effects of convection can be seen as a perpetual roiling of the visible surface of the Sun.

What If the Sun Were Different?

At this point we have seen the pieces that go into calculating a model of the interior of the Sun. To put these pieces together, we again stress that the key is *balance*. The temperature and density in the core of the model Sun must be just right to produce the same amount of energy as would be radiated away by a star of the same size and surface temperature as the Sun. The temperature and density at each point within the model Sun must be just right so that transport of energy away from the core by radiation and convection just balances the amount of energy produced by fusion in the core. The density, temperature, and pressure of the model Sun must vary from point to point in such a way that the outward push of pressure is everywhere balanced by the inward pull of gravity. Finally, the whole model must depend on only two things: the total mass of gas from which the star is made, and the chemical composition of that gas.

Let us restate that last idea to drive the point home. *In order to be successful, our model of the Sun must start with no more information than the known mass and chemical composition of the real Sun, and from these it must correctly predict all the other observed properties of the real Sun.* The amazing thing is that our model does exactly that. Computer modeling shows that a ball of gas containing one solar mass with the same composition as our Sun can have only one structure and still satisfy all the different balances *at the same time.* Our model predicts what the size, temperature, and luminosity of the resulting star should be—predictions that agree remarkably well with the observed properties of the real Sun.

To better understand *why* there is only one possible structure that a 1-$M_\odot$ star (a star with a mass equal to the Sun's mass) can have, we can "what if" the Sun. For example, we might ask, "What if a star had the same mass, surface temperature, and composition as the Sun, but was somehow larger than the Sun? What properties would such a hypothetical star have? Specifically, what would happen to the balance between the amount of energy generated within such a hypothetical star and the amount of energy that it radiates away into space?" Follow along in **Figure 14.7** as we consider what would happen if the Sun were "too large."[3]

We begin with the second part of this balance. Because our hypothetical star would have more surface area than the Sun, it would be able to more effectively radiate its energy into space. To keep a one-solar-mass star inflated to a size larger than the Sun would require that the star be more luminous than the Sun.

Now let's consider what is going on in the interior of our hypothetical star. Because the star is larger than the Sun but contains the same amount of mass as the Sun, the force of gravity at any point within our hypothetical star would be less than the force of gravity at the corresponding location within the Sun. (This is a result of the inverse square law of gravitation: if the radius R is larger in our hypothetical star, then $1/R^2$ must be smaller.) Weaker gravity means that

---

[3]Our approach here is admittedly simplified. We have not taken into consideration certain subtle factors, such as the Sun's *differential rotation.*

FIGURE 14.7 A star like the Sun can have only the structure that it has. Here we imagine the fate of a Sun with too large a radius.

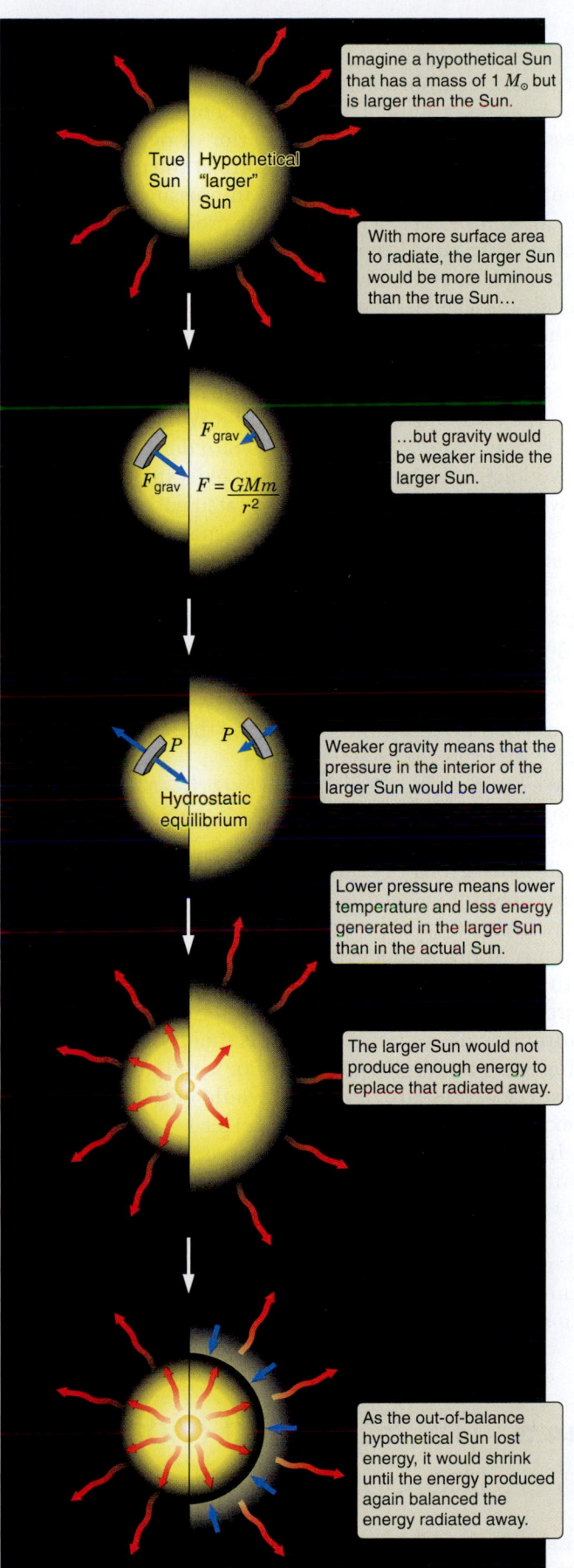

Imagine a hypothetical Sun that has a mass of 1 $M_\odot$ but is larger than the Sun.

With more surface area to radiate, the larger Sun would be more luminous than the true Sun…

…but gravity would be weaker inside the larger Sun.

$$F = \frac{GMm}{r^2}$$

Weaker gravity means that the pressure in the interior of the larger Sun would be lower.

Lower pressure means lower temperature and less energy generated in the larger Sun than in the actual Sun.

The larger Sun would not produce enough energy to replace that radiated away.

As the out-of-balance hypothetical Sun lost energy, it would shrink until the energy produced again balanced the energy radiated away.

the weight of matter pushing down on the interior of our hypothetical star would be less than in the Sun. Because hydrostatic equilibrium means that the pressure at any point within a star is equal to the weight of overlying matter, the pressure at any point in the interior of our hypothetical star would be less than the pressure at the corresponding point in the Sun. This reduction in pressure would affect the amount of energy that our star produced. The proton-proton chain runs faster at higher temperatures and densities, so the lower pressure in the interior of our hypothetical star means that less energy would be generated there than in the core of the Sun.

Wait a minute—there is a contradiction here. We found that our hypothetical star would have to be more luminous than the Sun, but at the same time it would be producing less energy in its interior than the Sun does. This discrepancy violates the balance that must exist in any stable star between the amount of energy generated within the star and the amount of energy radiated into space. Our hypothetical star cannot exist! Stated another way, even if we could somehow magically pump up the Sun to a size larger than it actually is, it would not remain that way. Less energy would be generated in its core, while more energy would be radiated away at its surface. The Sun would be out of balance. As a result, the Sun would lose energy, the pressure in the interior of the Sun would decline, and the Sun would shrink back toward its original (true) size.

We could do the same thought experiment the other way around, asking what would happen if the Sun were smaller than it actually is. With less surface area, it would radiate less energy. At the same time the Sun's mass would be compacted into a smaller volume, driving up the strength of gravity and therefore the pressure in its interior. Higher pressure implies higher density and temperature, which in turn would cause the proton-proton chain to run faster, increasing the rate of energy generation. Again there is a contradiction—an imbalance. This time, with more energy being generated in the interior than is radiated away from the surface, pressure in the Sun would build up, causing it to expand toward its original (true) size.

In the end, there is only one possible Sun. If a star contains one solar mass of material of solar composition, only one structure allows that star to maintain all of the balances that must be maintained. Our stable, reliable Sun is the result.

> If the Sun were any different, it would not be in balance.

14.3 The Standard Model of the Sun Is Well Tested

The standard model of the Sun correctly predicts such global properties of the Sun as its size, temperature, and luminosity. This is a remarkable feat, but the model predicts much more than these properties. In particular, the standard model of the Sun predicts exactly what nuclear reactions should be occurring in the core of the Sun, and at what rate. The nuclear reactions that make up the proton-proton chain produce copious quantities of neutrinos. As we have noted, neutrinos are extremely elusive beasts—so elusive that almost all of the neutrinos produced in the heart of the Sun travel freely through the outer parts of the Sun and on into space as if the Sun were not there. The core of the Sun lies buried beneath 700,000 km of dense, hot matter, seemingly buried forever away from our view. Yet the Sun is *transparent* to neutrinos.

> Neutrinos escape freely from the core of the Sun.

It may take thermal energy produced in the heart of the Sun 100,000 years to find its way to the Sun's surface, but the solar neutrinos streaming through you as you read these words were produced by nuclear reactions in the heart of the Sun only $8\frac{1}{3}$ minutes ago. If we could find a way to capture and analyze these neutrinos, think of what we might learn! In principle, neutrinos offer a direct window into the very heart of the Sun's nuclear furnace.

We Use Neutrinos to Observe the Heart of the Sun

Transforming the promise of neutrino astronomy into reality turns out to be a formidable technical challenge. The same property of neutrinos that makes them so exciting to astronomers—the fact that their interaction with matter is so feeble that they can escape unscathed from the interior of the Sun—also makes them notoriously difficult to observe. Suppose we wanted to build a neutrino detector capable of stopping half of the neutrinos falling on it. Our hypothetical detector would need the stopping power of a piece of lead a light-year thick! Yet despite the difficulties, neutrinos offer a unique window into the Sun so powerful that they are worth going to great lengths to try to detect.

Fortunately, the Sun produces a truly enormous number of neutrinos. As you lie in bed at night, about 400 trillion solar neutrinos pass through your body each second, having already passed through Earth. With this many neutrinos about, a neutrino detector does not have to be very efficient to be useful. Several methods have been devised to measure neutrinos from the Sun and from other astronomical sources, such as supernova explosions, and a number of such experiments are under way. These experiments have successfully detected neutrinos from the Sun, and in so doing they have provided crucial confirmation that nuclear fusion reactions indeed are responsible for powering the Sun.

As with many good experiments, however, measurements of solar neutrinos raised new questions while answering others. After their initial joy at confirming that the Sun really is a nuclear furnace, astronomers became troubled that there seemed to be only about a third to a half as many

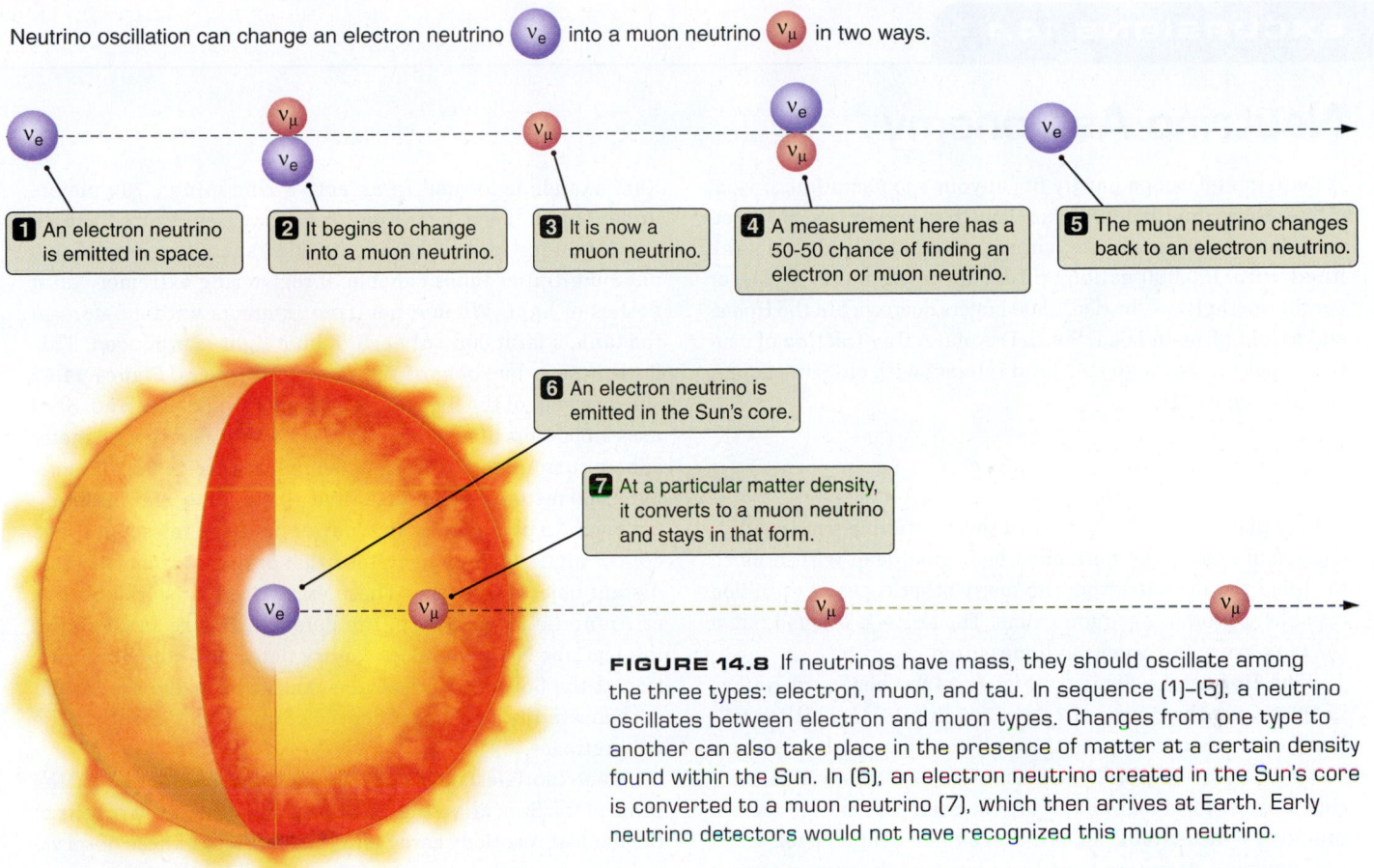

Neutrino oscillation can change an electron neutrino v_e into a muon neutrino v_μ in two ways.

1 An electron neutrino is emitted in space.

2 It begins to change into a muon neutrino.

3 It is now a muon neutrino.

4 A measurement here has a 50-50 chance of finding an electron or muon neutrino.

5 The muon neutrino changes back to an electron neutrino.

6 An electron neutrino is emitted in the Sun's core.

7 At a particular matter density, it converts to a muon neutrino and stays in that form.

FIGURE 14.8 If neutrinos have mass, they should oscillate among the three types: electron, muon, and tau. In sequence (1)–(5), a neutrino oscillates between electron and muon types. Changes from one type to another can also take place in the presence of matter at a certain density found within the Sun. In (6), an electron neutrino created in the Sun's core is converted to a muon neutrino (7), which then arrives at Earth. Early neutrino detectors would not have recognized this muon neutrino.

solar neutrinos as predicted by solar models. The difference between the predicted and measured flux of solar neutrinos was referred to as the **solar neutrino problem**.

Understanding the solar neutrino problem was an area of very active research. One possible explanation was that our understanding of the structure of the Sun was somehow wrong. This possibility seemed unlikely, however, because of the many other successes of our solar model. A second possibility was that our understanding of the neutrino itself was incomplete. The neutrino was long thought to have zero mass (like photons) and to travel at the speed of light. But if neutrinos actually do have a tiny amount of mass, then theories from particle physics predict that solar neutrinos should *oscillate* (alternate back and forth) among three different kinds, or "flavors"—the *electron*, *muon*, and *tau* neutrinos, as shown in **Figure 14.8**. Because only one of these, the electron neutrino, could interact with the atoms in the earlier neutrino detectors (described in **Excursions 14.1**), neutrino oscillations provided a convenient explanation for why we saw only about a third of the expected number of neutrinos. And, as seen in Figure 14.8, electron neutrinos should also change flavor as they interact with solar material during their escape from the Sun.

After several decades of work on the solar neutrino problem, this last idea has won out. Work currently under way at high-energy physics labs, nuclear reactors, and neutrino telescopes around the world is showing that neutrinos *do* have a nonzero mass, and this work has uncovered evidence of neutrino oscillations.

Solving the solar neutrino problem is a good example of how science works—how a better model of the neutrino showed that the solar neutrino problem was real and not merely an experimental mistake, and how a single set of anomalous observations was later confirmed by other more sophisticated experiments. All of this has led to a better understanding of basic physics.

Helioseismology Enables Us to Observe Echoes from the Sun's Interior

In Chapter 7 we found that models of Earth's interior predict how density and temperature change from place to place within our planet. These differences affect the way pressure waves travel through Earth, bending the paths of these waves. We test models of Earth's interior by comparing measurements of seismic waves from earthquakes with model predictions of how seismic waves should travel through the planet.

The same basic idea has now been applied to the Sun. Detailed observations of motions of material from place

Neutrino Astronomy

A neutrino telescope hardly fits anyone's expectation of what a telescope should look like. The first apparatus designed to detect solar neutrinos consisted of a cylindrical tank filled with 100,000 gallons of dry-cleaning fluid—C_2Cl_4, or perchloroethylene—buried 1,500 meters deep within the Homestake Gold Mine in Lead, South Dakota. A tiny fraction of neutrinos passing through this fluid interact with chlorine atoms, causing the reaction

$$^{37}Cl + \nu \rightarrow {}^{37}Ar + e^-$$

to take place. The ^{37}Ar formed in the reaction is a radioactive isotope of argon. The tank must be buried deep within Earth to shield the detector from the many other types of radiation capable of producing argon atoms. The argon is flushed out of the tank every few weeks and measured.

The Homestake detector (**Figure 14.9a**) began operations in 1965. Over the course of 2 days, roughly 10^{22} (10 billion trillion) solar neutrinos pass through the Homestake detector. Of these, on average only *one* neutrino interacts with a chlorine atom to form an atom of argon. Even so, this interaction produces a measurable signal. Since then, over a dozen and a half neutrino detectors have been built, each using different reactions to detect neutrinos of different energies. For example, the Soviet-American Gallium Experiment (SAGE) and the European Gallium Experiment (GALLEX) use reactions involving the conversion of gallium atoms into germanium atoms ($^{71}Ga + \nu \rightarrow {}^{71}Ge + e^-$) to detect solar neutrinos. SAGE uses 60 tons of metallic gallium. GALLEX uses 30 tons of gallium in the form of gallium chloride. These two experiments account for most of the world's current supply of gallium.

Among the more ambitious detectors capable of detecting all three flavors of neutrinos are the Super-Kamiokande detector and the Sudbury Neutrino Observatory (SNO). Super-Kamiokande is located in an active zinc mine 2,700 meters under Mount Ikena, near Kamioka, Japan (see Figure 21.14.) It is a 50,000-ton tank of ultrapure water, surrounded by 13,000 photomultiplier tubes capable of registering extremely faint flashes of light. When a neutrino interacts with an atom in the tank, a faint conical flash of blue light is produced. This flash is seen by some of the photomultipliers. **Figure 14.9b** shows a map of the flash of light from a single neutrino. SNO uses 1,000 tons of heavy water (D_2O) contained in a 12-meter sphere surrounded by light detectors (**Figure 14.9c**) and buried 2,000 meters deep in an active nickel mine near Sudbury, Ontario. An objective of still newer neutrino telescopes is to collect higher-energy neutrinos that originate from the most distant objects in space. These experiments use the oceans or Antarctic ice as part of the detector. **Figure 14.9d** shows the IceCube Neutrino Observatory detector under construction at the South Pole. IceCube's optical sensors are buried far beneath the surface, at depths of up to 2.5 km within the Antarctic ice.

Neutrino telescopes observe neutrinos produced in the heart of the Sun, allowing us to directly observe the results of the nuclear reactions going on there. Although these observations have provided crucial confirmation that stars are powered by nuclear reactions, they have also challenged our models of the solar interior and led us to change our ideas about the nature of the neutrino itself (as is evident from the discussion of the solar neutrino problem in the text). In addition to solar neutrinos, a number of experiments detected neutrinos from Supernova 1987A. As we will see in Chapter 17, this was an explosion marking the end of the life of a massive star located 160,000 light-years away in a small galaxy called the Large Magellanic Cloud. Neutrino astronomy was one of the great innovations of 20th century astronomy, and it is certain to have many applications in the 21st century.

---

to place across the surface of the Sun show that the Sun vibrates or "rings," something like a struck bell. Compared to a well-tuned bell, however, the vibrations of the Sun are very complex, with many different frequencies of vibrations occurring simultaneously. These motions are echoes of what lies below. Just as geologists use seismic waves from earthquakes to probe the interior of Earth, solar physicists use the surface oscillations of the Sun to test our understanding of the solar interior. This science, new to the later years of the 20th century, is called **helioseismology**. Like neutrino astronomy, helioseismology has created quite a stir among astronomers by letting us "see" into the invisible heart of the Sun (**Figure 14.10**).

Observing the "ringing" motion of the surface of the Sun is a difficult business. To detect the disturbances of helioseismic waves on the surface of the Sun, astronomers must use instruments capable of measuring Doppler shifts of less than 0.1 meter per second (m/s) while detecting changes in brightness of only a few parts per million at any given location on the Sun. In addition, tens of millions of different wave motions are possible within the Sun. Some waves travel around the circumference of the Sun, providing information about the density of the upper convection zone. Other waves travel through the interior of the Sun, revealing the density structure of the Sun close to its core. Still others travel inward toward the center of

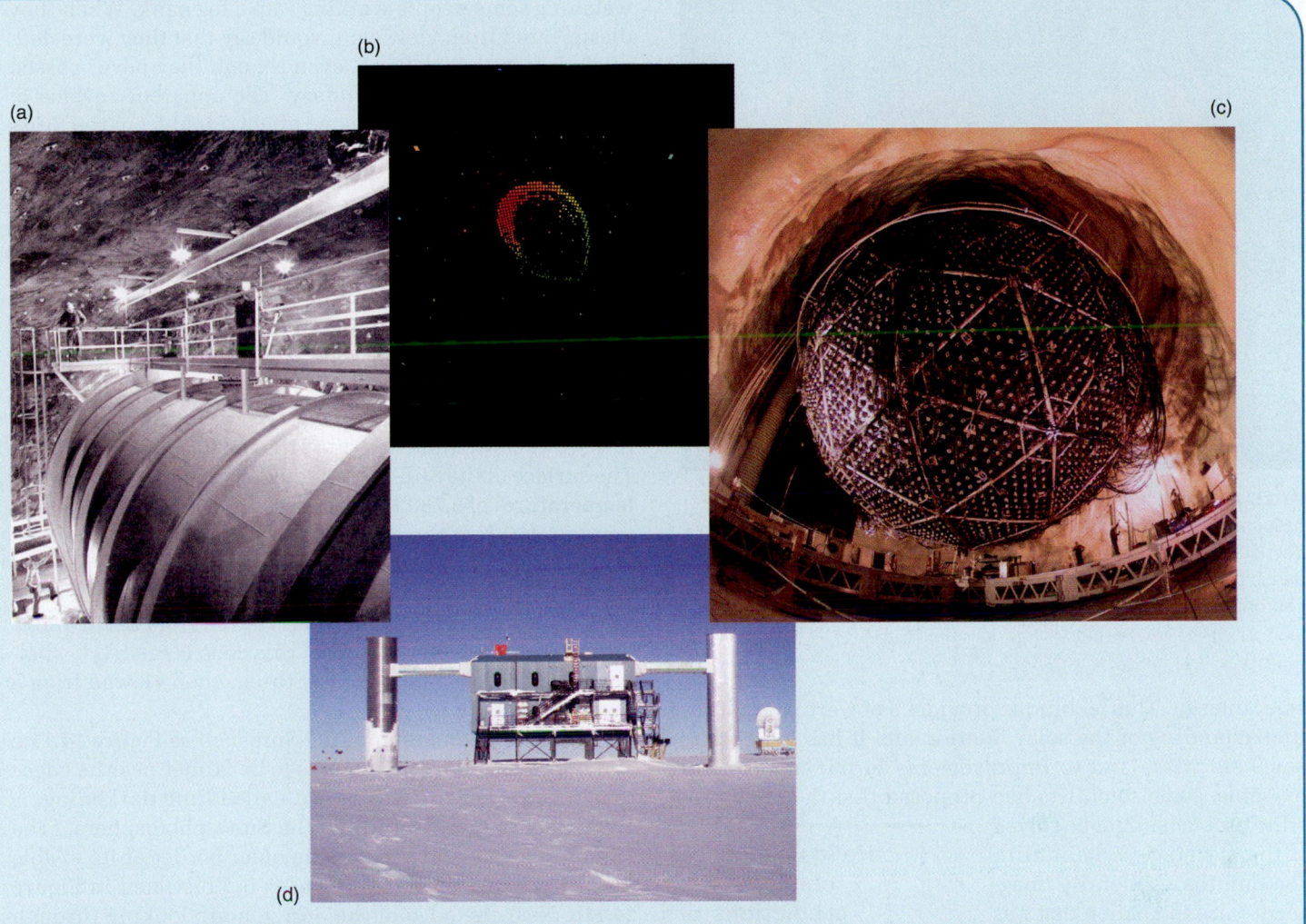

FIGURE 14.9 Neutrino "telescopes" do not look much like visible-light telescopes. (a) The Homestake neutrino detector is a 100,000-gallon tank of dry-cleaning fluid located deep in a mine in South Dakota. (b) A map of the flash of light from a single neutrino caught by a neutrino detector. (c) The Sudbury Neutrino Observatory, buried 2 km deep in a Canadian nickel mine. (d) The IceCube Neutrino Observatory at the South Pole.

the Sun, until they are bent by the changing solar density and return to the surface. All of these wave motions are going on at the same time. Sorting out this jumble requires computer analysis of long, unbroken strings of solar observations.

Helioseismology studies of the Sun got a huge boost in the closing years of the 20th century with two very successful projects. The Global Oscillation Network Group, or GONG, is a network of six solar observation stations spread around the world. With this network, solar astronomers are able to observe the surface of the Sun approximately 90 percent of the time. The other project is the Solar and Heliospheric Observatory, or *SOHO* spacecraft, which is

a joint mission between NASA and the European Space Agency. By orbiting at the L_1 Lagrangian point of the Sun-Earth system (see Chapter 10), *SOHO* moves in lockstep with Earth at a location approximately 1,500,000 km (0.01 astronomical unit, or AU) from Earth almost directly in line between Earth and the Sun. *SOHO* carries a complement of 12 scientific instruments designed to monitor the Sun and measure the *solar wind* upstream of Earth. *SOHO* observations have dramatically improved our detailed knowledge of the Sun.

To interpret helioseismology data, scientists compare the strength, frequency, and wavelengths of observed vibrations with predicted vibrations calculated from models of the

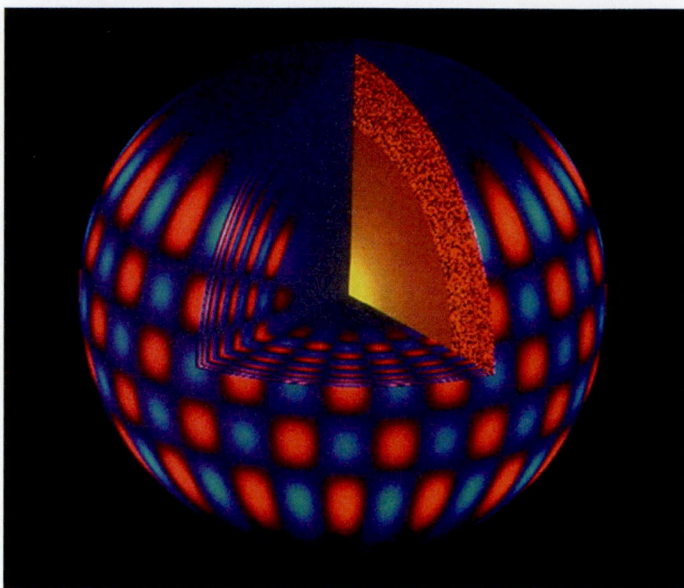

FIGURE 14.10 The interior of the Sun rings like a bell as helioseismic waves move through it. This figure shows one particular "mode" of the Sun's vibration. Red shows regions where gas is traveling inward; blue, where gas is traveling outward. We can observe these motions by using Doppler shifts.

solar interior. This technique provides a powerful test of our understanding of the solar interior, and it has led both to some surprises and to improvements in our models. For example, some scientists had proposed that the solar neutrino problem might be solved if there were less helium in the Sun than generally imagined—an explanation that was ruled out by analysis of the waves that penetrate to the core of the Sun. Helioseismology also showed that the value for opacity used in early solar models was too low. This realization led astronomers to recalculate the location of the bottom of the convective zone. Both theory and observation now put the base of the convective zone at 71.3 percent of the way out from the center of the Sun, with an uncertainty in this number of less than half a percent! The amazing agreement between the predictions of computer models of the Sun and the temperature and density structure within the Sun measured by helioseismology is truly remarkable.

> Helioseismology confirms the predictions of solar models.

14.4 The Sun Can Be Studied Up Close and Personal

The Sun is a large ball of gas, so it has no solid surface in the sense that Earth does. Instead, it has the kind of surface that a cloud on Earth does, or a "surface" like that of one of the Jovian planets in our Solar System. To help you under-

stand such a surface, imagine a fog bank. The surface of the fog bank is a gradual thing—an illusion really. Imagine watching some people walking into a fog bank. When they disappeared from view, you would say that they were definitely inside the fog bank, even though they never passed through a noticeable boundary. The apparent surface of the Sun is defined by the same effect. Light from the Sun's surface can escape into space, so we can see it. Light from below the Sun's surface cannot escape directly into space, so we cannot see it.

The Sun's surface is referred to as the solar **photosphere**. (*Photo* means "light"; the photosphere is the place light comes from.) There is no instant when you can say that you have suddenly crossed the surface of a fog bank, and by the same token there is no instant when we suddenly cross the photosphere of the Sun. The surface of the Sun—the photosphere—has an **effective temperature**[4] of 5,780 K and ranges from 6,600 K at the bottom to 4,400 K at the top. It is a zone about 500 km thick, across which the density and opacity of the Sun increase sharply. The reason the Sun appears to have a well-defined surface and a sharp outline when viewed from Earth (*never* look at the Sun directly!) is that this zone is relatively shallow; 500 km does not look very thick when viewed from a distance of 150 million km.

> The apparent surface of the Sun is called the photosphere.

Look at a photograph of the Sun such as **Figure 14.11a**, and notice that the Sun appears to be fainter near its edges than near its center. This effect, called **limb darkening**, is an artifact of the structure of the Sun's photosphere. (The "limb" of a celestial body is the outer border of its visible disk.) The cause of limb darkening is illustrated in **Figure 14.11b**. Near the edge of the Sun you are looking through the photosphere at a steep angle. As a result, you do not see as deeply into the interior of the Sun as when you are looking directly down through the photosphere near the center of the Sun's disk. The light you see coming from near the limb of the Sun is from a layer in the Sun that is shallower and hence cooler and fainter.

The Solar Spectrum Is Complex

In one sense the surface of the Sun may be an illusion, but in another sense it is not. The transition between "inside" the Sun and "outside" the Sun is quite abrupt. In the outermost part of the Sun, the density of the gas drops very rapidly with increasing altitude. This is the region we refer to as the Sun's **atmosphere**. **Figure 14.12** shows how the pressure and temperature change across the atmosphere of the Sun. The Sun's atmosphere is where all visible solar phenomena take place.

---

[4]The effective temperature of an object is the temperature of a blackbody (Chapter 4) that would emit the same amount of electromagnetic radiation as the object.

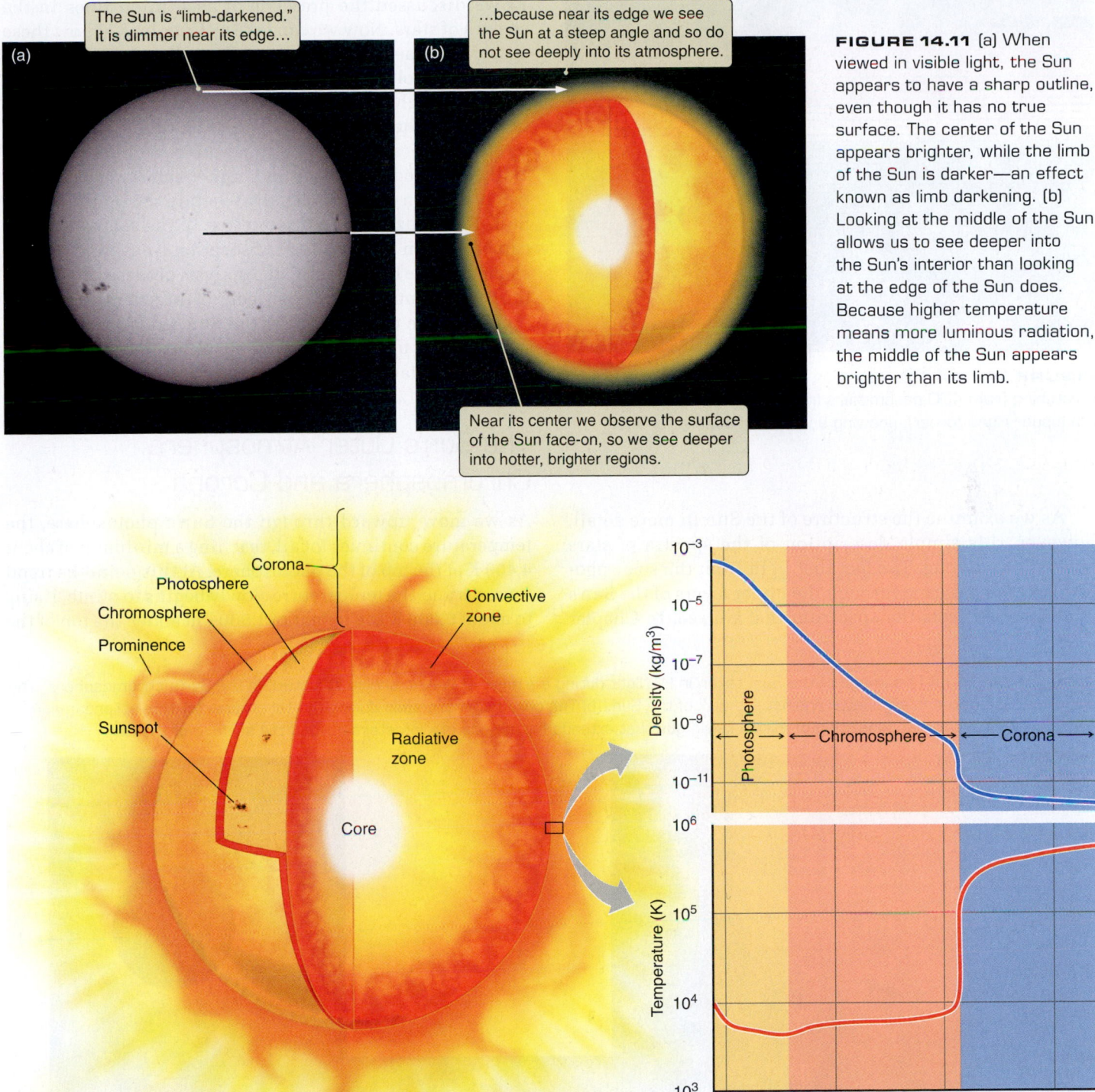

The Sun is "limb-darkened." It is dimmer near its edge…

…because near its edge we see the Sun at a steep angle and so do not see deeply into its atmosphere.

(a)

(b)

Near its center we observe the surface of the Sun face-on, so we see deeper into hotter, brighter regions.

FIGURE 14.11 (a) When viewed in visible light, the Sun appears to have a sharp outline, even though it has no true surface. The center of the Sun appears brighter, while the limb of the Sun is darker—an effect known as limb darkening. (b) Looking at the middle of the Sun allows us to see deeper into the Sun's interior than looking at the edge of the Sun does. Because higher temperature means more luminous radiation, the middle of the Sun appears brighter than its limb.

Corona

Photosphere

Chromosphere

Prominence

Convective zone

Sunspot

Radiative zone

Core

Density (kg/m^3)

Temperature (K)

Photosphere

Chromosphere

Corona

Height (km)

FIGURE 14.12 The components of the Sun's atmosphere, along with plots of how the temperature and density change with height at the base of the Sun's atmosphere.

Very nearly all the radiation from below the Sun's photosphere is absorbed by matter and cannot escape—it is trapped. This is exactly our definition from Chapter 4 for the conditions under which blackbody radiation is formed. We should not then be surprised that the radiation able to leak out of the Sun's interior has a spectrum very close to being a Planck (blackbody) spectrum. This is why, in Chapter 13, we were able to understand much about the physical properties of stars by applying our understanding of blackbody radiation.

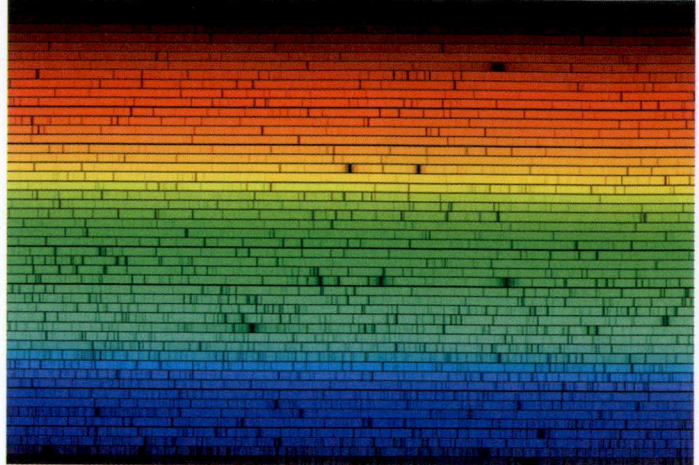

FIGURE 14.13 A high-resolution spectrum of the Sun, stretching from 400 nanometers (nm) (lower left corner) to 700 nm (upper right corner), showing a wealth of absorption lines.

13 we discussed the presence of absorption lines in the spectra of stars. Now we can take a closer look at how these absorption lines form. As photospheric light travels upward through the solar atmosphere, atoms in the solar atmosphere absorb the light at discrete wavelengths. Because from our perspective the Sun is so much brighter than any other star, its spectrum can be studied in far more detail. Today, specially designed telescopes and high-resolution spectrometers have been built specifically to study the light from the Sun. The amazing structure present in the solar spectrum can be seen in **Figure 14.13**. Absorption lines from over 70 elements have been identified. Analysis of these lines forms the basis for much of our knowledge of the solar atmosphere, including the composition of the Sun, and is the starting point for our understanding of the atmospheres and spectra of other stars.

The Sun's Outer Atmosphere: Chromosphere and Corona

As we examine the structure of the Sun in more detail, however, this simple description of the spectra of stars begins to go wrong. The fact that light from the solar photosphere must escape through the upper layers of the Sun's atmosphere affects the spectrum that we see. In Chapter

As we move upward through the Sun's photosphere, the temperature continues to fall, reaching a minimum of about 4,400 K at the top of the photosphere. At this point the trend reverses and the temperature slowly begins to climb, rising to about 6,000 K at a height of 1,500 km above the top of the

FIGURE 14.14 (a) This image of the Sun, taken in Hα light during a transit of Venus, shows structure in the Sun's chromosphere. The planet Venus is seen in silhouette against the disk of the Sun. (b) The chromosphere seen during a total eclipse. (c) This eclipse image shows the Sun's corona, consisting of million-kelvin gas that extends for millions of kilometers beyond the surface of the Sun.

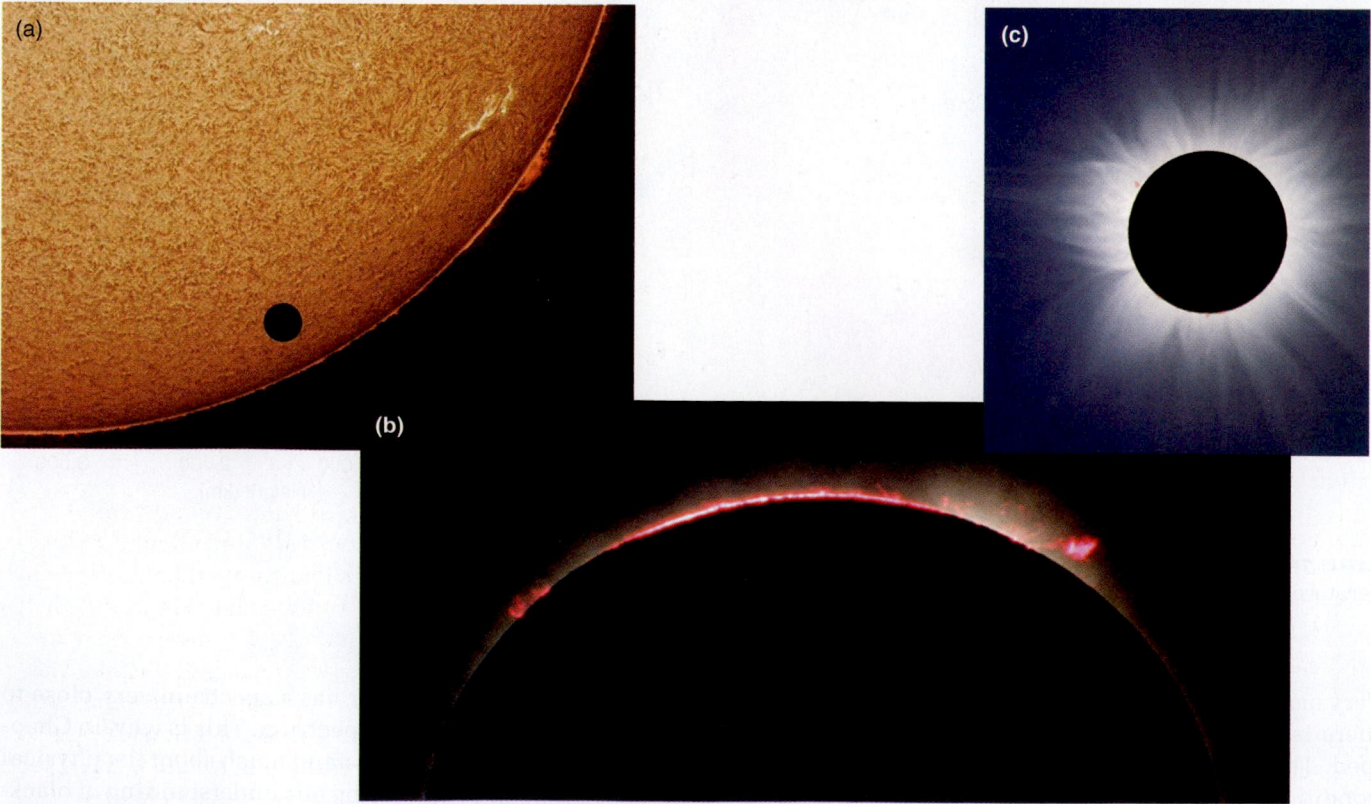

photosphere. This region above the photosphere is called the **chromosphere** (**Figure 14.14a**). The chromosphere was discovered in the 19th century during observations of total solar eclipses (**Figure 14.14b**). The chromosphere is seen most strongly as a source of emission lines, especially the Hα line from hydrogen (the "hydrogen alpha line"). In fact, the deep red color of the Hα line is what gives the *chromosphere* ("the place where color comes from") its name. It was also from a spectrum of the chromosphere that the element helium was discovered in 1868. Helium is named after *helios*, the Greek word for "Sun."[5] The reason for the chromosphere's temperature reversal with increasing height is not well understood, but it may be caused by magnetic waves propagating through the region.

> The Sun's chromosphere lies above the photosphere.

At the top of the chromosphere, across a transition region that is only about 100 km thick, the temperature suddenly soars (see Figure 14.12). Above this transition lies the outermost region of the Sun's atmosphere, called the **corona**, in which temperatures reach 1 million–2 million K. The corona is probably heated by magnetic waves and magnetic fields in much the same way the chromosphere is, but why the temperature changes so abruptly at the transition between the chromosphere and the corona is not at all clear. Since ancient times the Sun's corona has been known—visible during total solar eclipses as an eerie outer glow stretching a distance of several solar radii beyond the Sun's surface (**Figure 14.14c**). Because it is so hot, the solar corona is a strong source of X-rays. Atoms in the corona are also highly ionized. The spectrum of the corona shows emission lines from ionic species such as Fe^{13+} (iron atoms from which 13 electrons have been stripped) and Ca^{14+} (calcium atoms from which 14 electrons have been stripped).

> The corona has a temperature of millions of kelvins.

Solar Activity Is Caused by Magnetic Effects

Virtually all of the structure seen in the atmosphere of the Sun is imposed on the gas by the Sun's magnetic field. High-resolution images of the Sun, such as **Figure 14.15**, show "coronal loops" that make the Sun look as though it were covered with matted, tangled hair. This fibrous or rope-like texture in the chromosphere is the result of magnetic structures called flux tubes, much like the flux tube that connects Io with Jupiter (see Chapter 9). Magnetic fields are responsible for much of the structure of the corona as well.

FIGURE 14.15 A close-up image of the Sun, showing the tangled structure of coronal loops.

Recall from Chapter 8 that "hot" atoms are able to escape from the tops of planetary atmospheres. Shouldn't this happen on the Sun as well? In this case, the answer is yes and no. The corona is far too hot to be held in by the Sun's gravity, but over most of the surface of the Sun coronal gas is confined instead by magnetic loops with both ends firmly anchored deep within the Sun. The magnetic field in the corona acts almost like a network of rubber bands that coronal gas is free to slide along but cannot cross. In contrast, about 20 percent of the surface of the Sun is covered by an ever-shifting pattern of **coronal holes**. These are apparent in X-ray images of the Sun as dark regions (**Figure 14.16**), indicating that they are cooler and lower in density than their surroundings. Coronal holes are large regions where the magnetic field points outward, away from the Sun, and where coronal material is free to stream away into interplanetary space.

We have encountered this flow of coronal material away from the Sun before. This is the same **solar wind** responsible for shaping the magnetospheres of planets (see Chapters 8 and 9) and for blowing the tails of comets away from the Sun (see Chapter 12). The relatively steady part of the solar wind consists of lower-speed flows, with velocities of about 350 kilometers per second (km/s), and higher-speed flows, with velocities up to about 700 km/s. The higher-speed flows originate in coronal holes. Depending on their speed, particles in the solar wind take about 2–5 days to reach Earth. Frequently, 2–5 days after a coronal hole passes across the center of the face of the Sun, there is an increase in the speed and density of the solar wind reaching Earth. The solar wind drags the Sun's magnetic field along with it. The magnetic field in the solar wind gets "wound up" by the Sun's rotation, as shown in **Figure 14.17**.

> A "wind" blows away from the Sun.

[5]It is rather remarkable that helium, the second most common element in the universe, was discovered in the Sun before it was identified on Earth!

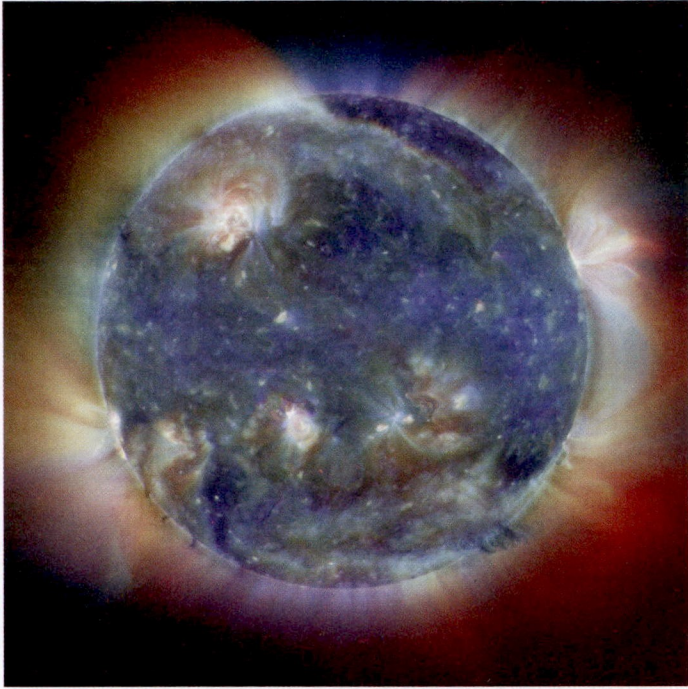

FIGURE 14.16 X-ray images of the Sun show a very different picture of our star than do images taken in visible light. The brightest X-ray emission comes from the base of the Sun's corona, where gas is heated to temperatures of millions of kelvins. This heating is most powerful above magnetically active regions of the Sun. The dark areas are coronal holes, regions where the Sun's corona is cooler and has lower *plasma* density than in their surroundings.

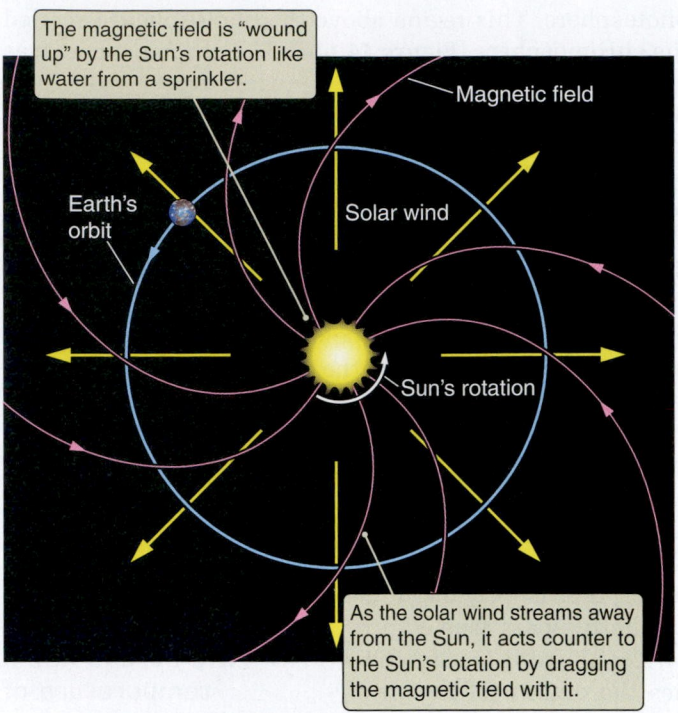

The magnetic field is "wound up" by the Sun's rotation like water from a sprinkler.

As the solar wind streams away from the Sun, it acts counter to the Sun's rotation by dragging the magnetic field with it.

FIGURE 14.17 The solar wind streams away from active areas and coronal holes on the Sun. As the Sun rotates, the solar wind picks up a spiral structure, much like the spiral of water that streams away from a rotating lawn sprinkler.

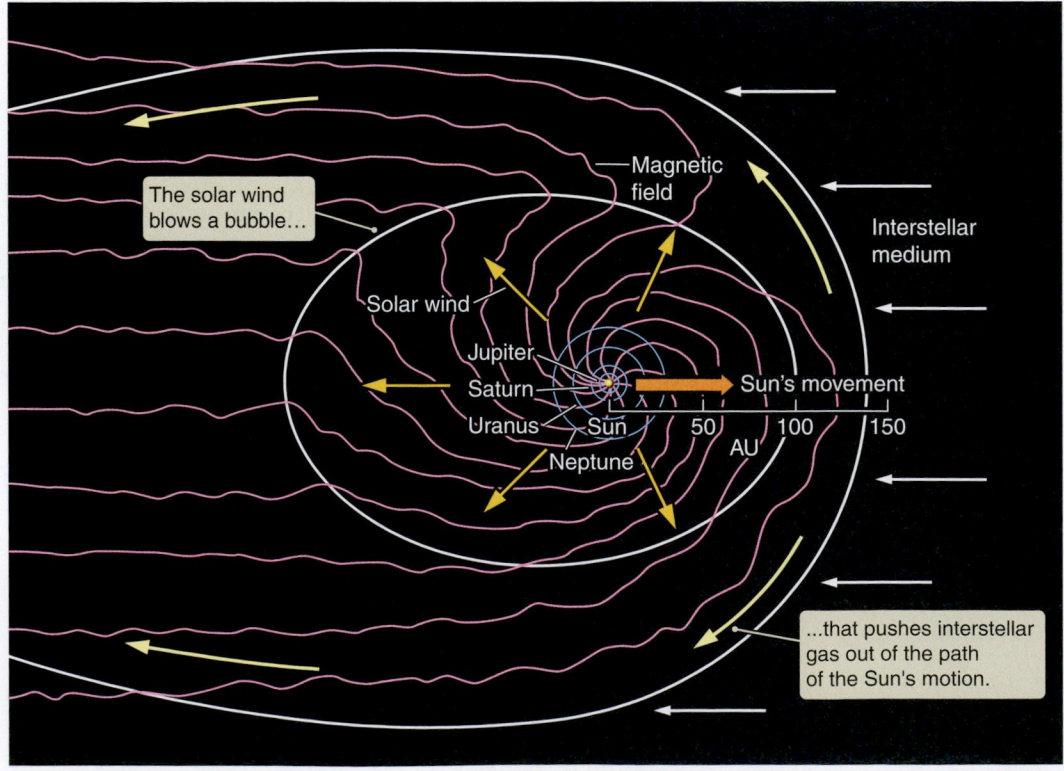

The solar wind blows a bubble…

…that pushes interstellar gas out of the path of the Sun's motion.

FIGURE 14.18 The solar wind streams away from the Sun for about 100 AU, until it finally piles up against the pressure of the interstellar medium through which the Sun is traveling. The *Voyager 1* spacecraft is now passing through this boundary and is expected to cross over into the true interstellar medium within the coming decade.

Consequently, the solar wind has a spiral structure, something like a stream of water from a rotating lawn sprinkler.

The effects of the solar wind are felt throughout the Solar System. As we have seen, the solar wind causes the tails of comets, shapes the magnetospheres of the planets, and provides the energetic particles that power Earth's spectacular auroral displays. Using space probes, we have been able to observe the solar wind extending out to 100 AU from the Sun. But the solar wind does not go on forever. The farther it gets from the Sun, the more it has to spread out. Just like radiation, the density of the solar wind follows an inverse square law. At a distance of about 100 AU from the Sun, the solar wind is assumed no longer to be powerful enough to push the **interstellar medium** (the gas and dust that lie between stars in a galaxy and that surround the Sun) out of the way. There the solar wind will stop abruptly, "piling up" against the pressure of the interstellar medium. **Figure 14.18** shows this region of space over which the wind from the Sun holds sway. Sometime within the next decade the *Voyager 1* spacecraft is expected to cross the very outer edge of this boundary and begin sending back our first "on the scene" measurements of true interstellar space.[6]

The best-known features on the surface of the Sun are relatively dark blemishes in the solar photosphere, called **sunspots**, that come and go over time. Sunspots appear dark, but only in contrast to the brighter surface of the Sun (see **Math Tools 14.2**). Early telescopic observations of sunspots made during the 17th century led to the discovery of the Sun's rotation, which has an average period of about 27

---

[6]The truth is, we really don't know just how far from the Sun this boundary lies, but we should know it when we see it.

MATH TOOLS 14.2

Sunspots and Temperature

Sunspots are about 1,500 K cooler than their surroundings. What does this tell us? Think back to the Stefan-Boltzmann law in Chapter 4. The flux from a blackbody is proportional to the fourth power of the temperature. We write this relationship as

$$\mathcal{F} \propto T^4.$$

Let's take round numbers for the temperature of a typical sunspot and the surrounding photosphere: 4,500 and 6,000 K, respectively. The Stefan-Boltzmann law informs us that the surface brightness of a sunspot is about

$$\left(\frac{4,500}{6,000}\right)^4 = 0.32$$

or about $\frac{1}{3}$ the brightness of the surrounding photosphere.

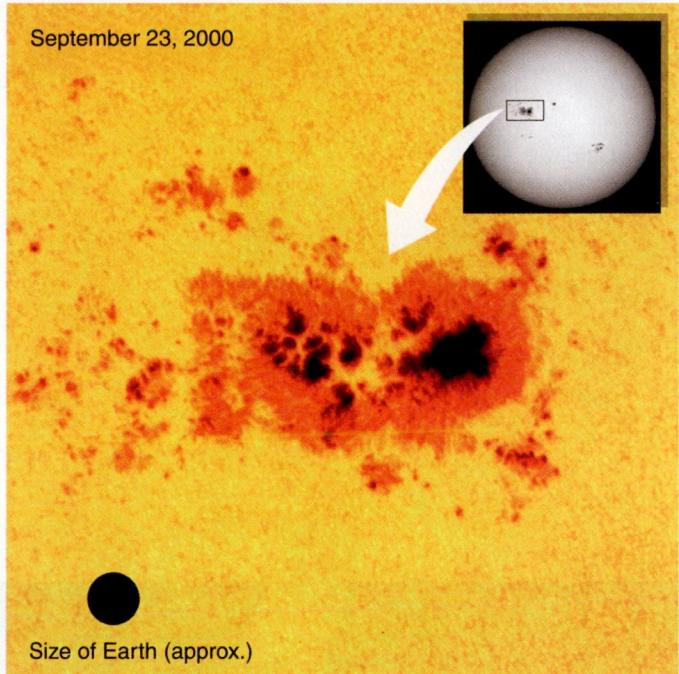

September 23, 2000

Size of Earth (approx.)

FIGURE 14.19 These *SOHO* images show a large sunspot group. Sunspots are magnetically active regions that are cooler than the surrounding surface of the Sun.

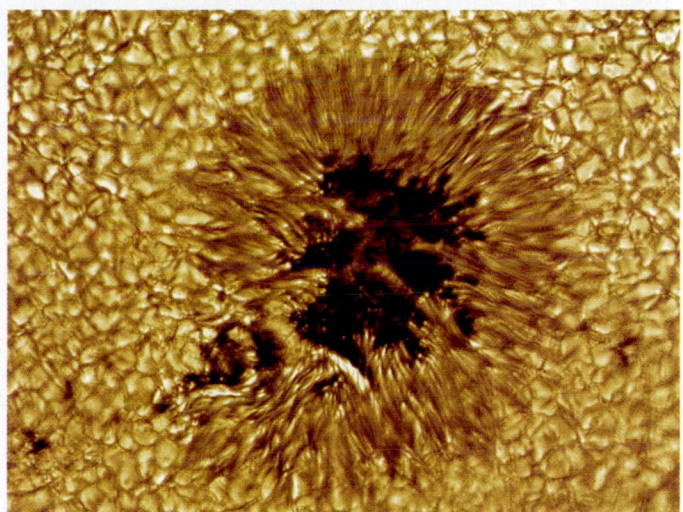

FIGURE 14.20 A very high-resolution view of a sunspot, showing the dark *umbra* surrounded by the lighter *penumbra*. The solar surface around the sunspot bubbles with separate cells of hot gas called "granules." The smallest features are about 100 km across.

days as seen from Earth and 25 days relative to the stars. Observations of sunspots also show that the Sun, like Saturn, rotates more rapidly at its equator than it does at higher latitudes. This effect, referred to as **differential rotation**, is possible only because the Sun is a large ball of gas rather than a solid object.

Figure 14.19 is a photograph of a large sunspot group, and **Figure 14.20** shows the remarkable structure of one of

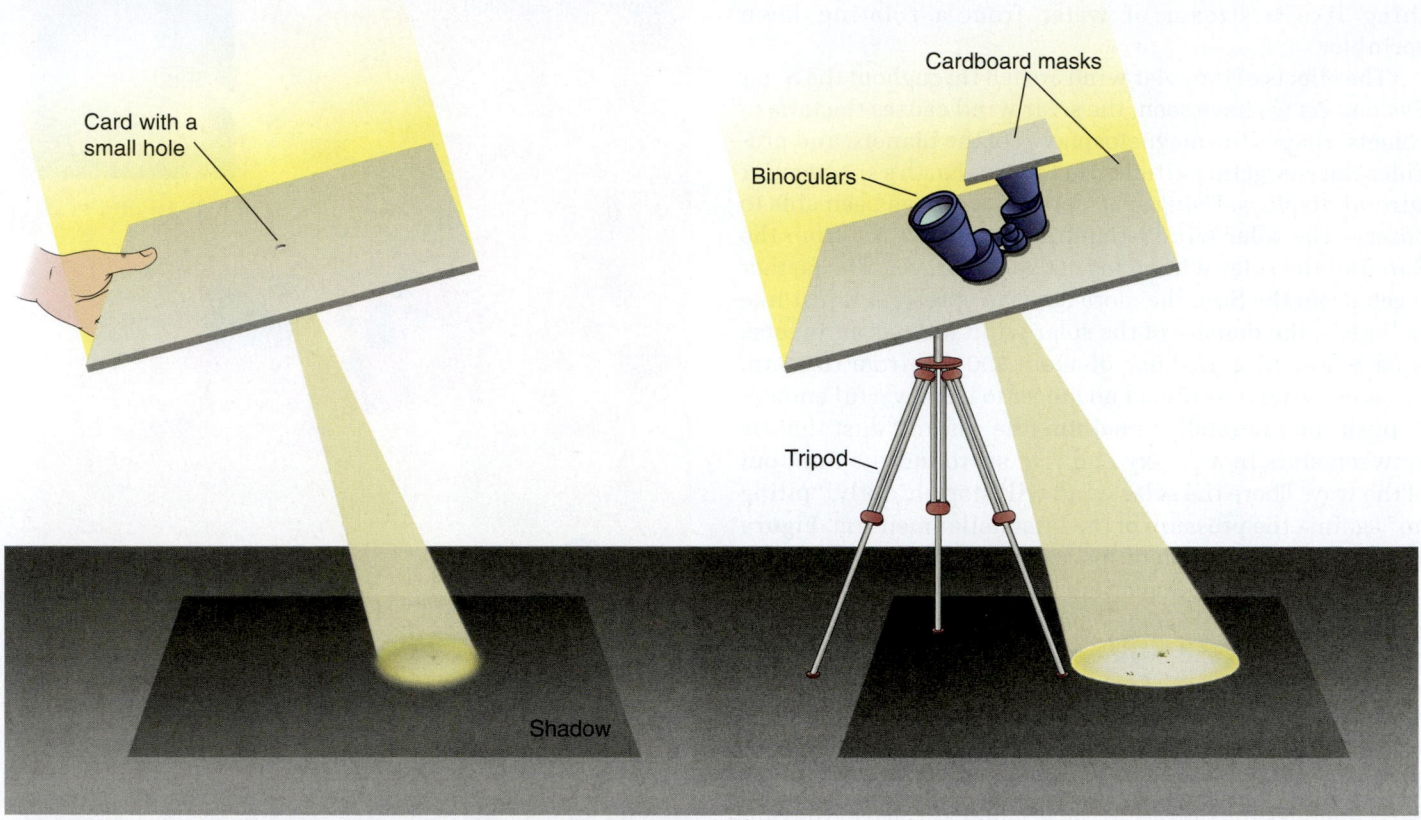

Card with a small hole

Cardboard masks

Binoculars

Tripod

Shadow

FIGURE 14.21 There are a number of ways of safely viewing sunspots on the Sun or observing a partial solar eclipse. Two methods are illustrated here. Never look at the Sun directly (except during the totality phase of a total solar eclipse)!

these blemishes on the surface of the Sun. Each sunspot consists of an inner dark core called the **umbra**, which is surrounded by a less dark region called the **penumbra**. The penumbra shows an intricate radial pattern, reminiscent of the petals of a flower. Observations of sunspots show that they are magnetic in origin, with magnetic fields thousands of times greater than the magnetic field at Earth's surface. Sunspots occur in pairs that are connected by loops in the magnetic field, much like the shape of a horseshoe. Sunspots range in size from about 1,500 km across up to complex groups that may contain several dozen individual spots and span 50,000 km or more. The largest sunspot groups are so large that they can be seen with the naked eye and have been noted since antiquity. *But do not look directly at the Sun!* More than one casual observer of the Sun has paid the price of blindness for a glimpse of a naked-eye sunspot. **Figure 14.21** shows two ways to look at the Sun in safety. Direct viewing through welder's glass is also safe.

Individual sunspots are ephemeral. Although sunspots occasionally last 100 days or longer, half of all sunspots

> **Magnetic fields cause sunspots.**

come and go in about 2 days, and 90 percent are gone within 11 days. The amount of sunspot activity and the locations on the Sun where sunspots are seen to change over time in a pronounced 11-year pattern called the **sunspot cycle** (**Figure 14.22a**). At the beginning of a cycle, sunspots begin to appear at solar latitudes of about 30° to the north and south of the solar equator. Over the following years the regions where most sunspots are seen move toward the equator as the number of sunspots increases and then declines. As the last few sunspots near the equator are seen, sunspots again begin appearing at middle latitudes and the next cycle begins. A diagram showing the number of sunspots at a given latitude plotted against time has the appearance of a series of opposing diagonal bands and is often referred to as the sunspot "butterfly diagram" (**Figure 14.22b**).

> **Sunspots come and go in an 11-year cycle.**

Telescopic observations of sunspots date back almost 400 years, and there were naked-eye reports of sunspots even before that. **Figure 14.23** shows the historical record of sunspot activity. Although astronomers often speak of the 11-year sunspot cycle, the cycle is neither perfectly periodic nor especially reliable. The time between peaks in the num-

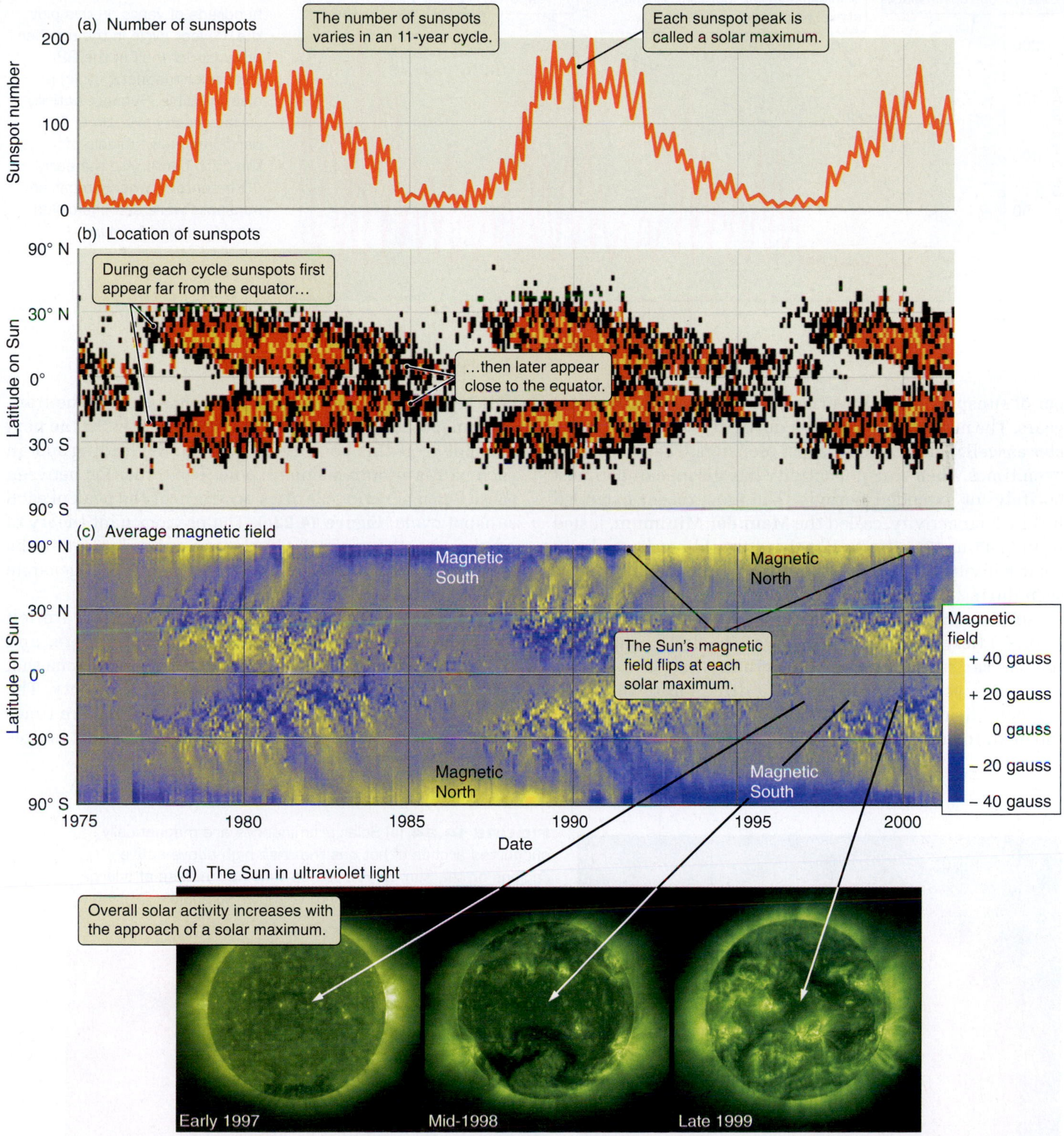

(a) Number of sunspots

The number of sunspots varies in an 11-year cycle.

Each sunspot peak is called a solar maximum.

(b) Location of sunspots

During each cycle sunspots first appear far from the equator…

…then later appear close to the equator.

(c) Average magnetic field

Magnetic South

Magnetic North

Magnetic North

Magnetic South

The Sun's magnetic field flips at each solar maximum.

Magnetic field

+ 40 gauss

+ 20 gauss

0 gauss

− 20 gauss

− 40 gauss

Date

(d) The Sun in ultraviolet light

Overall solar activity increases with the approach of a solar maximum.

Early 1997

Mid-1998

Late 1999

FIGURE 14.22 (a) The number of sunspots versus time for the last few solar cycles. (b) The solar butterfly diagram, showing the fraction of the Sun covered at each latitude. (c) The Sun's magnetic field flips every 11 years. (d) The approach of a solar maximum is apparent in these *SOHO* images taken in ultraviolet light.

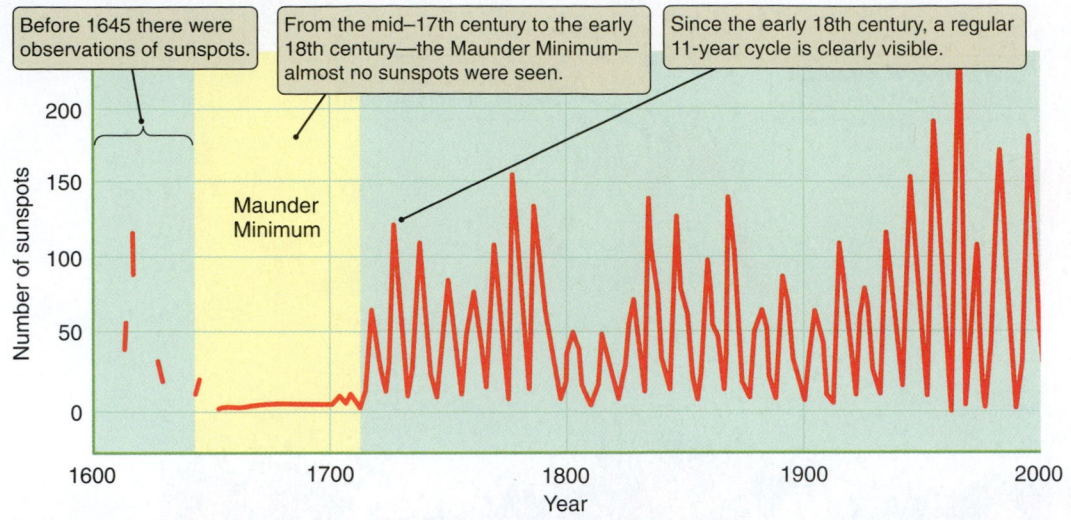

Before 1645 there were observations of sunspots.

From the mid–17th century to the early 18th century—the Maunder Minimum—almost no sunspots were seen.

Since the early 18th century, a regular 11-year cycle is clearly visible.

Maunder Minimum

FIGURE 14.23 Sunspots have been observed for hundreds of years. In this plot the 11-year cycle in the number of sunspots (half of the 22-year solar magnetic cycle) is clearly visible. Sunspot activity varies greatly over time. The period from the middle of the 17th century to the early 18th century, when almost no sunspots were seen, is called the *Maunder Minimum*.

ber of sunspots actually varies between about 9.7 and 11.8 years. The number of spots seen during a given cycle fluctuates as well. Some cycles are real monsters. There have also been times when sunspot activity has disappeared almost entirely for extended periods. The most recent extended lull in solar activity, called the **Maunder Minimum**, lasted from 1645 to 1715. Normally there would be six peaks in solar activity in 70 years, but virtually no sunspots were seen during the Maunder Minimum.

In the early 20th century, American solar astronomer George Ellery Hale (1868–1938) was the first to show that the 11-year sunspot cycle was actually half of a 22-year magnetic cycle during which the direction of the Sun's magnetic field reverses from one 11-year sunspot cycle to the next. In one sunspot cycle the leading sunspot in each

pair tends to be a north magnetic pole, whereas the trailing sunspot tends to be a south magnetic pole. In the next sunspot cycle this polarity is reversed: the leading spot in each pair is a south magnetic pole. The transition between these two magnetic polarities occurs near the peak of each sunspot cycle (**Figure 14.22c**). The predominant theory of what causes this magnetic cycle involves a *dynamo* in the interior of the Sun, much like the dynamos that generate the magnetic fields of the planets.

The effects of magnetic activity on the Sun are felt throughout the Sun's photosphere, chromosphere, and corona. Sunspots are only one of a host of phenomena that follow the Sun's 22-year cycle of magnetic activity. The peaks of the cycle, referred to as **solar maxima**, are times of intense activity, as can be seen in the ultraviolet images

(a)

FIGURE 14.24 (a) Solar prominences are magnetically supported arches of hot gas that rise high above active regions on the Sun. (b) A close-up view at the base of a large prominence. An image of Earth is shown to scale.

(b)

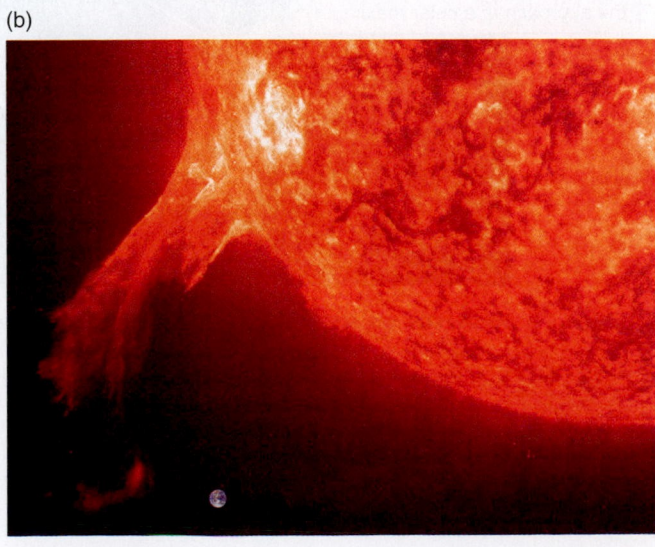

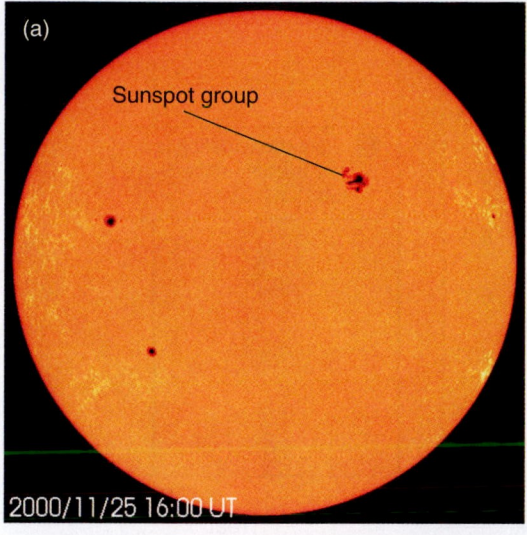

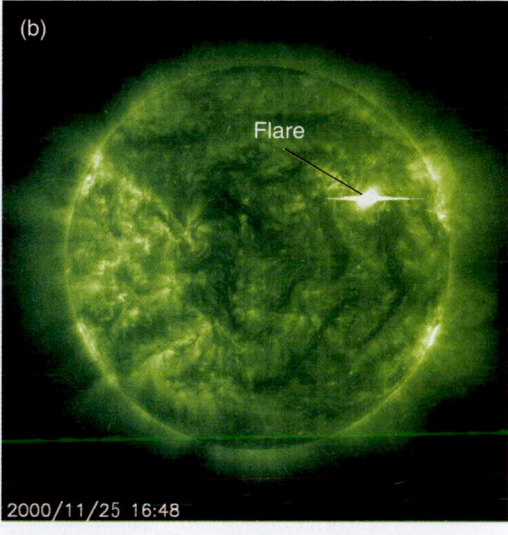

FIGURE 14.25 A large sunspot group (a) emits a powerful flare, shown (b) in ultraviolet light.

of the Sun in **Figure 14.22d**. Although sunspots are darker-than-average features, they are often accompanied by a brightening of the solar chromosphere that is seen most clearly in the light of emission lines such as Hα. The magnificent loops seen arching through the solar corona are also anchored in solar active regions. These include solar **prominences** such as those shown in **Figure 14.24**. Prominences are magnetic flux tubes of relatively cool (5,000–10,000 K) gas extending through the million-kelvin gas of the corona. Although most prominences are relatively quiescent, others can erupt out through the corona, towering a million kilometers or more over the surface of the Sun and ejecting material into the corona at velocities of 1,000 km/s.

FIGURE 14.26 A *SOHO* image of a coronal mass ejection (upper right), with a simultaneously recorded ultraviolet image of the solar disk superimposed.

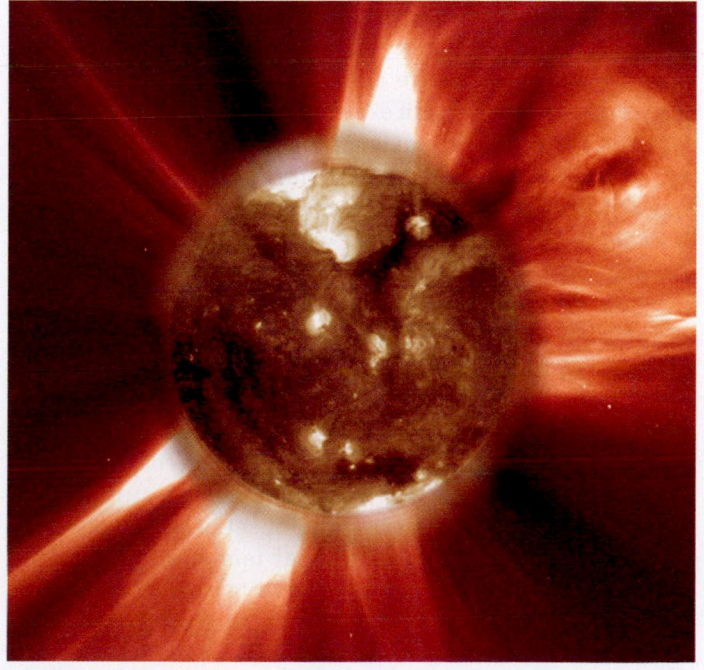

Figure 14.25 shows images of a **solar flare** erupting from a sunspot group. Solar flares are the most energetic form of solar activity. Flares are violent eruptions in which enormous amounts of magnetic energy are released over the course of a few minutes to a few hours. Solar flares can heat gas to temperatures of 20 million K, and they are the source of intense X-ray and gamma-ray radiation. Hot **plasma** (consisting of atoms stripped of some of their electrons) moves outward from flares at speeds that can reach 1,500 km/s. Magnetic effects can then accelerate subatomic particles to almost the speed of light. Such events, called **coronal mass ejections (Figure 14.26)**, send powerful bursts of energetic particles outward through the Solar System, affecting us in many ways when they reach Earth (see the next subsection). Coronal mass ejections occur about once per week during the minimum of the solar activity cycle, but they can reach bursts of two or more per day near the maximum of the cycle.

Solar Activity Affects Earth

The amount of solar radiation at Earth's distance from the Sun is, on average, 1.35 kilowatts per square meter (kW/m²).[7] Satellite measurements of the amount of radiation coming from the Sun (**Figure 14.27**) show that this value can vary by as much as 0.2 percent over periods of a few weeks as dark sunspots and bright spots in the chromosphere move across the disk. Overall, however, the increased radiation from active regions on the Sun more than makes up for the reduction in radiation from sunspots. On average, the Sun seems to be about 0.1 percent brighter during the peak of a solar cycle than it is at its minimum.

Solar activity has many effects on Earth. Solar active regions are the source of most of the Sun's extreme ultraviolet and X-ray emission. This energetic radiation heats

---

[7]Photons from the Sun provide an important alternative source of energy.

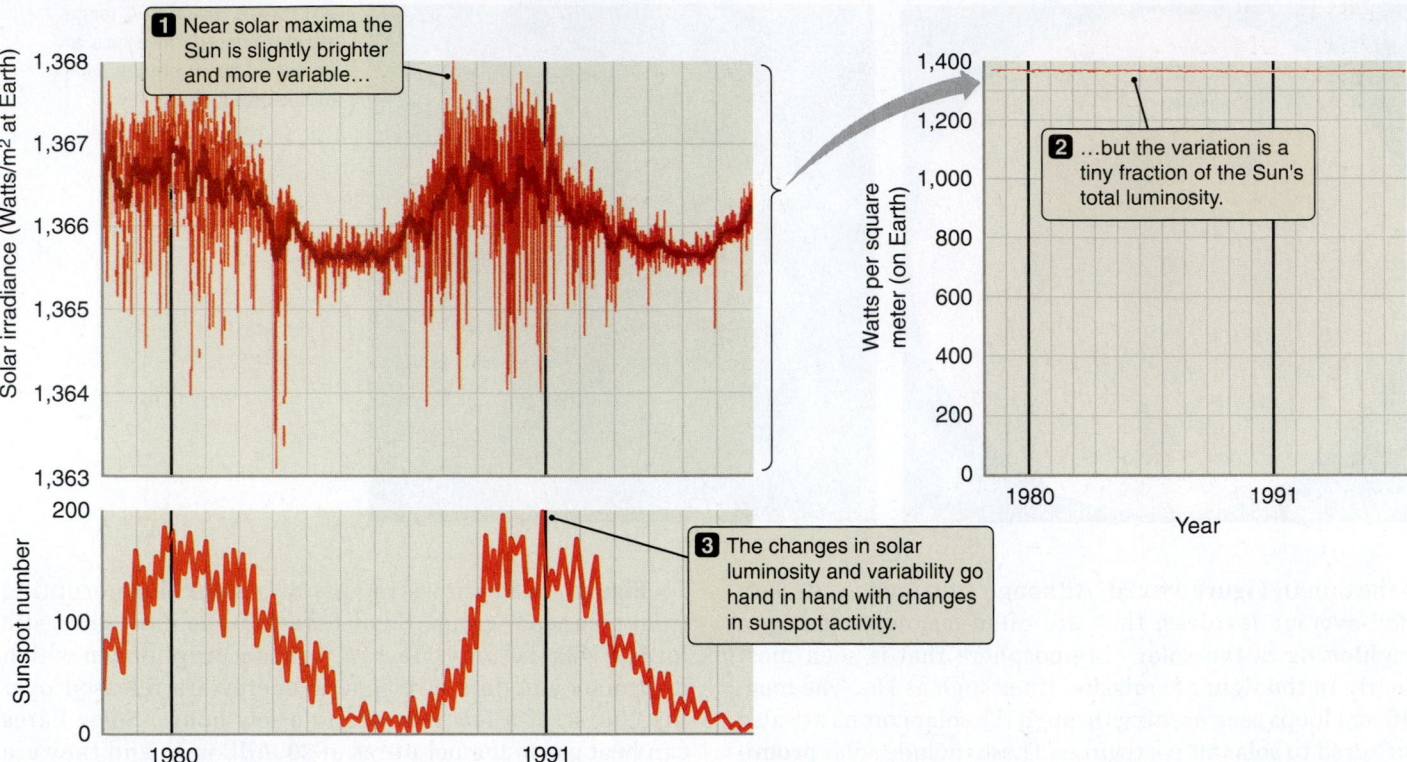

FIGURE 14.27 Measurements taken by satellites above Earth's atmosphere show that the amount of light from the Sun changes slightly over time.

Earth's upper atmosphere and, during periods of solar activity, causes Earth's upper atmosphere to expand. When this happens, the swollen upper atmosphere can significantly increase the atmospheric drag on spacecraft orbiting near Earth, causing their orbits to decay. (Prior to the launch of the Hubble Space Telescope, many people working on the project were concerned that, because it was to be launched into the teeth of a solar maximum, the telescope might not survive very long.) One reason for repeated shuttle visits to the HST was to "reboost" it to its original orbit to make up for the slow decay in its orbit caused by drag in the rarefied outer parts of Earth's atmosphere. NASA gave a final shuttle "reboost" to the HST in 2009, keeping the space telescope in operation until at least 2014, when it will be replaced by the larger James Webb Space Telescope. Earth's atmospheric drag will eventually bring the HST back to Earth—although not in one piece.

Earth's magnetosphere is the result of the interaction between Earth's magnetic field and the solar wind. Increases in the solar wind accompanying solar activity, especially coronal mass ejections directed at Earth, can disrupt Earth's magnetosphere in ways that are obvious even to nonscientists. Spectacular auroras can accompany such events, as can magnetic storms that have been known to

> **Solar storms cause auroras and disrupt electric power grids on Earth.**

disrupt electric power grids, causing blackouts across large regions. Coronal mass ejections also have adverse effects on radio communication and navigation, and they can damage sensitive satellite electronics, including communication satellites. Energetic particles accelerated in solar flares pose one of the greatest dangers to human exploration of space.

Over the years, people have attempted to relate sunspot activity to phenomena ranging from the performance of the stock market to the mating habits of Indian elephants. Few of these supposed relationships have withstood serious scrutiny, however. An especially interesting idea is that variations in Earth's climate might be related to solar activity. We have certainly seen that solar activity affects Earth's upper atmosphere, and it might not be too much of a reach to imagine that it could affect weather patterns as well. It has also been suggested that variations in the amount of radiation from the Sun might be responsible for variations in Earth's climate in the past. Although the causes of the current global climatic change remain controversial, models suggest that observed variations in the Sun's luminosity could account for only about 0.1-K differences in Earth's average temperature—much less than the effects due to the ongoing buildup of carbon dioxide in Earth's atmosphere. On the other hand, triggering the onset of an ice age may require a sustained drop in global temperatures of only about 0.2–0.5 K, so the quest for a definite link between solar variability and changes in Earth's climate persists.

Seeing the Forest for the Trees

The universe contains more stars than there are grains of sand on all of Earth's beaches. Even "nearby" stars are so distant that at best we see them as faint points of light in the night sky. Only one star is close enough for its complex structure to be studied in great detail. That star is a maelstrom of fierce activity, fueled by nuclear fires burning deep within its interior. Storms rage across its surface. Giant flares burst forth, arching millions of kilometers into space before collapsing back again. That star heats a huge and rarefied corona to millions of kelvins and blows a bubble in interstellar space large enough to swallow our entire Solar System. That star—a main-sequence G2 star, different in no way from main-sequence G2 stars throughout our galaxy and the universe—is the star we call our Sun.

No machine of human manufacture has ever visited the interior of the Sun, just as no machine has more than scratched the surface of Earth. Even so, our understanding of the interior of the Sun is remarkably complete. Like everything else in the universe, the Sun is governed by the laws of physics. Computer models based on those physical laws show that there is only one possible structure that a 1-$M_\odot$ ball of gas with the chemical composition of the Sun can have while remaining in balance. According to those models, the Sun draws its energy from the fusion of hydrogen to make helium in its core. That energy, carried from the Sun's core to its surface by radiation and convection, arrives at just the rate needed to replace the energy radiated away into space. All of this takes place while a delicate standoff is maintained in the battle between the inward pull of gravity and the outward push of pressure fueled by the nuclear fires in the Sun's interior. Models of the Sun make a host of detailed predictions, ranging from the density and temperature at every point within the Sun's interior to the nuclear reactions that take place in its core. Over the past decades the predictions of models of the Sun have been tested in ever more sophisticated and detailed ways, and those predictions have been confirmed. One decade into the 21st century, it is fair to say that we *know* what the interior of the Sun is like.

Our knowledge of the Sun also forms the cornerstone of our knowledge of all other stars everywhere. Starting with a successful model of the Sun, we can now go on to ask a host of other questions: What happens if we increase the mass of the model star? What happens if we change its chemical abundances? What happens if we run the model forward in time until all of the hydrogen at the star's center is gone? How do we assemble something that looks like the Sun in the first place? Our study of the Sun has set the stage for the next leg of our journey as we turn our attention to how stars form, how they differ from one another, how they change with time, and how they die.

Summary

- Nuclear reactions converting hydrogen to helium are the source of the Sun's energy.

- Energy created in the Sun's core moves outward to the surface, first by radiation and then by convection.

- Neutrinos are elusive, almost massless particles that interact only very weakly with other matter.

- The apparent surface of the Sun is called the photosphere.

- The temperature of the Sun's atmosphere ranges from about 6,000 K near the bottom to about 1 million K at the top.

- Material streaming away from the Sun's corona creates the solar wind.

- Sunspots are photospheric regions that are cooler than their surroundings.

- The Sun's outermost region is called the corona.

- Sunspots reveal the 11- and 22-year cycles in solar activity.

- Solar storms can produce terrestrial auroras and disrupt power grids.

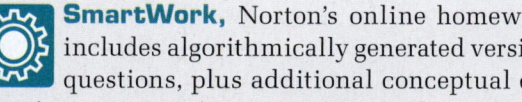 **SmartWork,** Norton's online homework system, includes algorithmically generated versions of these questions, plus additional conceptual exercises. If your instructor assigns questions in SmartWork, log in at **smartwork.wwnorton.com.**

Student Questions

THINKING ABOUT THE CONCEPTS

1. The Sun's stability depends on hydrostatic equilibrium and energy balance. Describe how both of these work.

*2. Using the "what if" discussion in Section 14.2 as a guide, explain how hydrostatic equilibrium acts as a safety valve to keep the Sun at its constant size, temperature, and luminosity.

3. What is the strong nuclear force, and how does it provide stability for atomic nuclei?

4. Explain nuclear fusion and how it relates to the Sun's source of energy.

5. Two of the three atoms in a molecule of water (H_2O) are hydrogen. Why are Earth's oceans not fusing hydrogen into helium and setting Earth ablaze?

6. Engineers and physicists dream of solving the world's energy supply problem by constructing power plants that would convert globally plentiful hydrogen to helium. Our Sun seems to have solved this problem. On Earth, what is a major obstacle to this potentially environmentally clean solution?

7. Name the part of the Sun where most of its energy is produced, and explain why the Sun's primary source of energy is located there.

*8. Explain the proton-proton chain process through which the Sun generates energy by converting hydrogen to helium.

9. On Earth, nuclear power plants use *fission* to generate electricity. Here, a heavy element like uranium is broken into many atoms, where the total mass of the fragments is less than the original atom. Explain why fission could not be powering the Sun today.

10. In the proton-proton chain process, the mass of four protons is slightly greater than the mass of a helium nucleus. Explain what happens to this difference in mass.

11. What experiences have you had with energy transmitted to you by radiation alone? By convection alone? By conduction alone? (Hint: You are exposed to all three every day.)

12. Is gas an efficient conductor of thermal energy? Explain why or why not.

*13. Suppose an abnormally large amount of hydrogen suddenly burned in the core of the Sun. What would happen to the rest of the Sun? Would we see anything as a result?

14. The Sun has a radius equal to about 2.3 light-seconds. Explain why a gamma ray produced in the Sun's core does not emerge from the Sun's surface 2.3 seconds later.

15. A high-energy photon and a neutrino are created in the Sun's core at the same instant. Which one will reach Earth first? Explain your answer.

16. Why are neutrinos so difficult to detect?

*17. Discuss the "solar neutrino problem" and how this problem was solved.

18. What technique do you find in common between how we probe the internal structure of the Sun and the internal structure of Earth?

19. The Sun's visible "surface" is not a true surface, but a feature we call the photosphere. Explain why the photosphere is not a true surface.

20. The second most abundant element in the universe was not discovered here on Earth. What is it and how was it discovered?

21. Describe the solar corona. Under what circumstances can we see it without using special instruments?

22. The solar corona has a temperature of millions of degrees; the photosphere has a temperature of only about 6,000 K. Why isn't the corona much, much brighter than the photosphere?

23. What is the solar wind?

24. Sunspots are dark splotches on the Sun. Why are they dark?

25. What have sunspots told us about the Sun's rotation?

26. Explain the relationship between the 11-year sunspot cycle and the 22-year magnetic cycle.

27. What is a solar flare?

28. How is the fate of the Hubble Space Telescope tied to solar activity?

29. Solar flares and magnetic storms can have an adverse effect on certain things important to our daily lives. Name some of the things that are affected.

APPLYING THE CONCEPTS

30. Assume that the Sun's mass is about 300,000 Earth masses and that its radius is about 100 times that of Earth. The density of Earth is about 5,500 kg/m^3.
   a. What is the density of the Sun?
   b. How does this compare with the density of water?

31. The Sun shines by converting mass into energy according to Einstein's well-known relationship $E = mc^2$. Show that if the Sun produces 3.85×10^{26} J of energy per second, it must convert 4.3 million metric tons (4.3×10^9 kg) of mass per second into energy.

32. Assume that the Sun has been producing energy at a constant rate over its lifetime of 4.5 billion years (1.4×10^{17} seconds).

a. How much mass has it lost creating energy over its lifetime?

b. The present mass of the Sun is 2×10^{30} kg. What fraction of its present mass has been converted into energy over the lifetime of the Sun?

33. Imagine that the source of energy in the interior of the Sun changed abruptly.

a. How long would it take before a neutrino telescope detected the event?

b. When would a visible-light telescope see evidence of the change?

34. On average, how long does it take particles in the solar wind to reach Earth from the Sun if they are traveling at an average speed of 400 km/s?

35. A sunspot appears only 70 percent as bright as the surrounding photosphere. The photosphere has a temperature of approximately 5,780 K. What is the temperature of the sunspot?

36. The hydrogen bomb represents humankind's efforts to duplicate processes going on in the core of the Sun. The energy released by a 5-megaton hydrogen bomb is 2×10^{16} J.

a. How much mass did Earth lose each time a 5-megaton hydrogen bomb was exploded?

b. This textbook, *21st Century Astronomy*, has a mass of about 1.6 kg. If we converted all of its mass into energy, how many 5-megaton bombs would it take to equal that energy?

37. Verify the claim made at the start of this chapter that the Sun produces more energy per second than all the electric power plants on Earth could generate in a half-million years. Estimate or look up how many power plants there are on the planet, and how much energy an average power plant produces. Be sure to account for different kinds of power—for example, coal, nuclear, wind.

*38. Let's examine how we know that the Sun cannot power itself by chemical reactions. Using Math Tools 14.1 and the fact that an average chemical reaction between two atoms releases 1.6×10^{-19} J (1 electron-volt, or eV), estimate how long the Sun could emit energy at its current luminosity. Compare that estimate to the known age of Earth.

*39. How much distance must a photon actually cover in its convoluted path from the center of the Sun to the outer layers?

StudySpace is a free and open Web site that provides a Study Plan for each chapter of *21st Century Astronomy*. Study Plans include animations, reading outlines, vocabulary flashcards, and multiple-choice quizzes, plus links to premium content in SmartWork and the ebook. Visit **www.wwnorton.com/studyspace**.

There are no whole truths; all truths are half truths. It is
trying to treat them as whole truths that plays the devil.

ALFRED NORTH WHITEHEAD (1861–1947)

A star-forming region 3,000 light-years away in the constellation Cepheus.

CHAPTER 15

Star Formation and the Interstellar Medium

15.1 Whence Stars?

To mortal humans, the stars seem fixed and unchanging. Indeed, our local star, the Sun, has remained much the same for long enough to allow life on Earth to arise, evolve, and begin to contemplate its own origin and fate. But our perspective is distorted by our brief existence. All of recorded human history amounts to less than a millionth of the age of the Sun and Earth. In Chapter 6 we told the story of how our Solar System formed some 4.6 billion years ago out of a swirling disk of gas and dust that surrounded the protostellar Sun. We are now ready to return to that story, but with a different focus. First we will step back and look at the interstellar environment that gave birth to the Sun and the Solar System, and all the other stars, both large and small. Then we will focus our attention on the **protostar** at the center of the **circumstellar disk** and watch as it becomes a star.

The birth of a star—from a cloud of gas and dust to nuclear burning—is a process that can happen within a few tens of thousand of years for the most massive stars, or require hundreds of millions of years for the least massive. Such evolutionary times are far too long for us to witness changes in an individual protostar, but astronomers have pulled the larger picture together by observing many different stars at various stages of stellar development.

The root of the word *astronomy* may mean "star," but if we ask what fraction of space is filled with stars, we get a very different picture of their importance. The volume of our Sun is about 1.3 million times the volume of Earth. Though this volume might seem enormous to us, it is almost incomprehensibly tiny compared with the volume of space. Our Sun is the only star in a region of space with a volume of over 300 cubic light-years. If our Sun were the size of a golf

KEY CONCEPTS

When you look at the night sky it is stars that you notice, yet stars fill only the tiniest fraction of the volume of space. On this leg of our journey we turn our attention to the tenuous medium between the stars, and we discover that

- There are various phases of the interstellar medium, ranging from cold, relatively dense molecular clouds to hot, tenuous intercloud gas.
- Different phases of the interstellar medium emit various types of radiation and can be observed at different wavelengths, ranging from radio waves to X-rays.
- Dust in the interstellar medium blocks visible light but glows with infrared radiation.
- Gas in the interstellar medium is heated and ionized by energy from stars and stellar explosions.
- Stars form in clusters from dense cores buried within giant molecular clouds.
- Forming stars are seen by their infrared emission and by the effects they have on their surroundings.
- We can follow the progress of an evolving protostar by its track across the H-R diagram.
- Protostars collapse, radiating away their gravitational energy until fusion starts in their cores and they settle onto the main sequence.

ball sitting on an author's office desk in Tempe, Arizona, its nearest stellar neighbor would be a similar golf ball sitting on another author's desk in Santa Cruz, California, 1,000 kilometers (km) away. The rest of the volume of our galaxy is filled with the *interstellar medium*. It is here that the story of stars begins and ends. Stars are formed from the gas in the interstellar medium, they live their lives there, and when they die they bequeath some of their chemical and energetic legacy back to the interstellar medium.

Before taking a look at the interstellar medium, however, we need to put it into another context as well. Remember that galaxies are *crowded* in comparison with the vast expanses of intergalactic space. Returning to the golf ball analogy, if we ask the whereabouts of a star in the spiral galaxy nearest to our own, that star would be equivalent to a golf ball somewhere on Jupiter. But that is a story for later in our journey. For now, we turn our attention to our "local neighborhood" and the interstellar medium that fills the volume of our own galaxy.

15.2 The Interstellar Medium

The Sun formed from the **interstellar medium**, so it is not too surprising that the chemical composition of the interstellar medium in our region of the Milky Way is similar to the chemical composition of the Sun (see the "astronomer's periodic table" in Figure 9.5). In the interstellar medium about 90 percent of the atomic nuclei are hydrogen, and the remaining 10 percent are almost all helium. More massive elements together account for only 0.1 percent of the atomic nuclei, or about 2 percent of the mass in the interstellar medium. Roughly 99 percent of that interstellar matter is gaseous, consisting of individual atoms or **molecules** moving freely about, as do the molecules in the air around us.

However, interstellar gas is far more tenuous than the thick soup that we breathe. Each cubic centimeter of the air around you contains about 2.5×10^{19} molecules. A good terrestrial vacuum pump can reduce this density down to about 10^{10} molecules per cubic centimeter (cm^3)—about a billionth as dense. By comparison, the interstellar medium has an average density of less than 1 atom/cm^3—one ten-billionth as dense still! To understand this concept more clearly, imagine a square tube 1 centimeter (cm) on a side and a meter long, filled with air and sealed at both ends. Also imagine that you can stretch the tube without changing its cross section. As you do so, the air in the tube has to spread out to fill the larger volume and so is less dense. When the tube is stretched to a length of 10 meters, the air inside will be a tenth as dense as the air around you. When the tube is stretched to 100 meters, the air inside will be a hundredth as dense. For the air in the tube to match the

Interstellar gas is extremely tenuous.

average density of interstellar gas—that is, for it to have the same amount of mass per unit volume as the interstellar medium has—you would have to stretch it into a tube 25,000 light-years long! Stated another way, there is about as much material in a column of air between your eye and the floor beside you as there is interstellar gas in a column with the same cross section stretching from the Solar System to the center of the galaxy.

The Interstellar Medium Is Dusty

About 1 percent of the material in the interstellar medium is in the form of solid grains, referred to as **interstellar dust**. "Interstellar soot" might be a better name. Ranging in size from little more than large molecules up to particles about 300 nanometers (nm) across, these solid grains more closely resemble the particles of soot from a candle flame than they do the dust that collects on a windowsill. (It would take several hundred "large" interstellar grains to span the thickness of a single human hair.) Interstellar dust begins to form when **refractory materials** (see Chapter 6) such as iron, silicon, and carbon stick together to form grains in dense, relatively cool environments like the outer atmospheres and "stellar winds" of cool, red giant stars; or in dense material thrown into space by stellar explosions. (More about these topics later.) Once these grains are in the interstellar medium, other atoms and molecules may stick to them. This process is remarkably efficient: about half of all interstellar matter more massive than helium (1 percent of the total mass of the interstellar medium) is found in interstellar grains.

About 1 percent of the mass of the interstellar medium is in grains.

If there is only as much material between us and the center of the galaxy as there is between your eye and the floor, then you might expect our view of distant objects to be as clear as your view of your own big toe. But nothing could be further from the truth. It turns out that interstellar dust is extremely effective at blocking light. If the air around you contained as much dust as a comparable mass of interstellar material, it would be so dirty that you would be hard-pressed to see your hand held up 10 cm in front of your face. Go out on a dark summer night in the Northern Hemisphere (or a dark winter night in the Southern Hemisphere) and look closely at the Milky Way, visible as a faint band of diffuse light running through the constellation Sagittarius. You will see a dark "lane" running roughly down the middle of this bright band, splitting it in two. This dark band is the result of vast expanses of interstellar dust blocking our view of distant stars (see, for example, Figure 20.1).

Interstellar dust obscures our view.

When interstellar dust gets in the way of radiation from distant objects, the effect is called **interstellar extinction**.

(a)

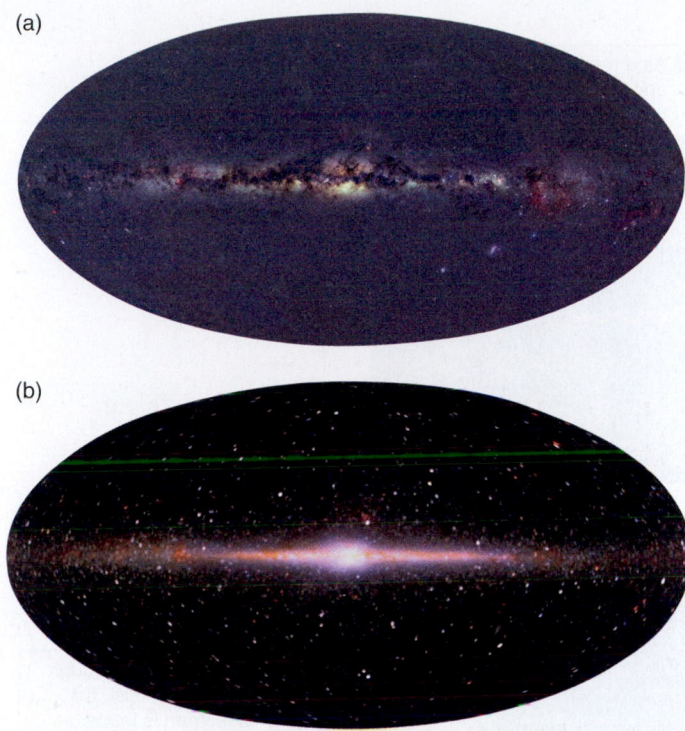

(b)

FIGURE 15.1 (a) An all-sky picture of the Milky Way taken in visible light. The dark splotches blocking our view are dusty interstellar clouds. (b) The same field of view as seen in the near infrared. Infrared radiation penetrates interstellar dust, giving us a clearer view of the stars in our galaxy.

Not all electromagnetic radiation suffers equally from interstellar extinction, however. **Figure 15.1** shows two images of our galaxy: one taken in visible light, the other taken in the infrared (IR). The dark clouds that block the shorter-wavelength visible light seem to have vanished in the longer-wavelength IR image (**Figure 15.1b**), enabling us to see through the clouds to the center of our galaxy and beyond.

As a starting point for understanding why short-wavelength radiation is blocked by dust while long-wavelength radiation is not, think about a different kind of wave—waves on the surface of the ocean. Imagine you are on the ocean in a strong swell. If you are floating in a small boat—one much smaller than the wavelength of ocean waves—the swell causes you to bob gently up and down. But that is about all; there is no other interaction between the waves and the small boat. The story is quite different if you are on a larger boat that is about half as long as the wavelength of the ocean waves. Now the bow of the boat may be on a wave crest while the stern of the boat is in a trough, or vice versa. The boat tips wildly back and forth as the waves go by. If the size of the boat and the wavelength of the waves are a good match, even fairly modest waves will rock the boat. (You might have noticed this if you were ever in a canoe or rowboat when the wake from

a speedboat came by.) Now imagine viewing these two situations from the perspective of the wave. The wave is hardly affected by the small boat, but it is strongly affected by the large boat. The energy to drive the wild motions of the boat comes from the wave, so the motion of the wave is affected by the interaction.

The interaction of electromagnetic waves with matter is more involved than that of a boat rocking on the ocean, but the same basic idea often applies. Tiny interstellar dust grains interact most strongly with ultraviolet and blue light, which have wavelengths comparable to the typical size of dust grains. For this reason, ultraviolet and blue light are effectively blocked by interstellar dust. Short wavelengths suffer heavily from interstellar extinction. Infrared and radio radiation, on the other hand, have wavelengths that are too long to interact strongly with the tiny interstellar dust grains, so they can travel largely unimpeded across great interstellar distances. To sum up, at visible and ultraviolet wavelengths, most of the galaxy is hidden from our view by dust. In the infrared and radio portions of the spectrum, however, we get a far more complete view.

> **Long-wavelength radiation penetrates interstellar dust.**

In the visible part of the spectrum, short-wavelength blue light suffers more from extinction than does longer-wavelength red light. As a result, as shown in **Figure 15.2**, an object viewed through dust looks redder (actually less blue) than it really is. This effect is called **reddening**. Correcting for the fact that stars and other objects appear both fainter and redder than they would in the absence of dust can be one of the most difficult parts of interpreting astronomical observations, often adding to our uncertainty in the measurement of an object's properties.

Interstellar extinction may not be much of a concern at infrared wavelengths, but dust still plays an important role in what we see when we look at the sky in the infrared. Like any other solid object, grains of dust glow at wavelengths determined by their temperature. In Chapter 4 we learned that the equilibrium between absorbed sunlight and emitted thermal radiation determines the temperatures of the terrestrial planets (see Figure 4.23). A similar equilibrium is at work in interstellar space, where dust is often heated by starlight and by the gas in which it is immersed to temperatures of tens to hundreds of kelvins (**Figure 15.3**). At a temperature of 100 K, Wien's law says that dust will glow most strongly at a wavelength of 29 micrometers (μm):

> **Dust glows in the infrared.**

$$\lambda_{max} = \frac{2{,}900 \ \mu m \ K}{T} = \frac{2{,}900 \ \mu m \ K}{100 \ K} = 29 \ \mu m.$$

Cooler dust—say, at a temperature of 10 kelvins (K)—glows most strongly at a wavelength of 290 μm. Both of these wavelengths are in the far-infrared part of the electromagnetic spectrum. Much of what we see when looking at the sky

(a)

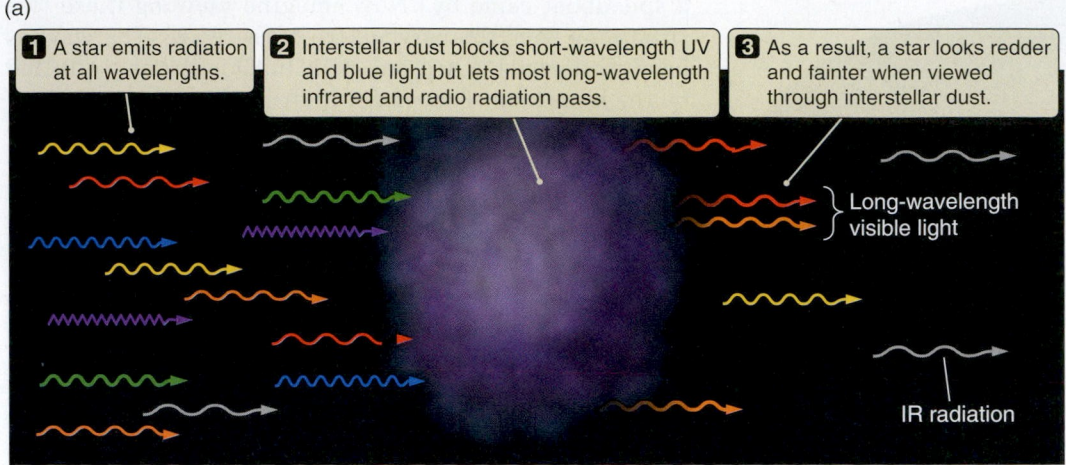

1 A star emits radiation at all wavelengths.

2 Interstellar dust blocks short-wavelength UV and blue light but lets most long-wavelength infrared and radio radiation pass.

3 As a result, a star looks redder and fainter when viewed through interstellar dust.

Long-wavelength visible light

IR radiation

(b)   (c)

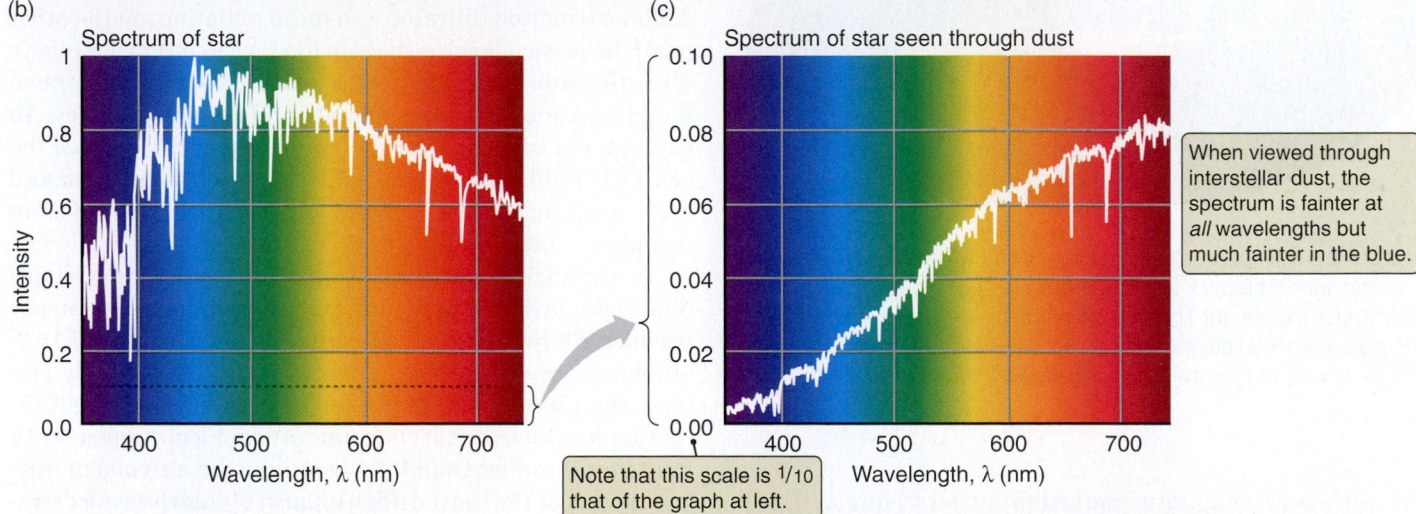

Spectrum of star

Spectrum of star seen through dust

Intensity

Wavelength, λ (nm)

When viewed through interstellar dust, the spectrum is fainter at *all* wavelengths but much fainter in the blue.

Note that this scale is $^{1}/_{10}$ that of the graph at left.

FIGURE 15.2 (a) The wavelengths of ultraviolet and blue light are close to the size of interstellar grains, so the grains effectively block this light. Grains are less effective at blocking longer-wavelength light. As a result, the spectrum of a star (b) when seen through an interstellar cloud (c) appears fainter and redder.

with infrared "eyes" is not really starlight at all, but thermal radiation from dust. When NASA launched the Infrared Astronomical Satellite (IRAS) in 1983, astronomers got their first look at the sky at wavelengths out to 100 μm, and the views (like **Figure 15.4a**) were unlike any seen before. IRAS showed the Milky Way's dark clouds glowing brilliantly in infrared radiation from dust. More recently, the Japanese infrared satellite AKARI provided a similar view (**Figure 15.4b**) at much shorter wavelengths (9 μm).

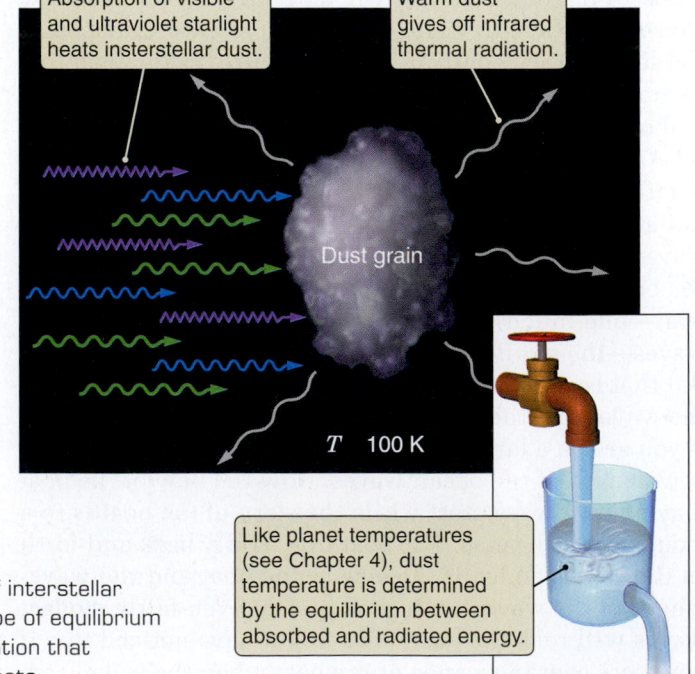

Absorption of visible and ultraviolet starlight heats insterstellar dust.

Warm dust gives off infrared thermal radiation.

Dust grain

T 100 K

Like planet temperatures (see Chapter 4), dust temperature is determined by the equilibrium between absorbed and radiated energy.

FIGURE 15.3 The temperature of interstellar grains is determined by the same type of equilibrium between absorbed and emitted radiation that determines the temperatures of planets.

(a)

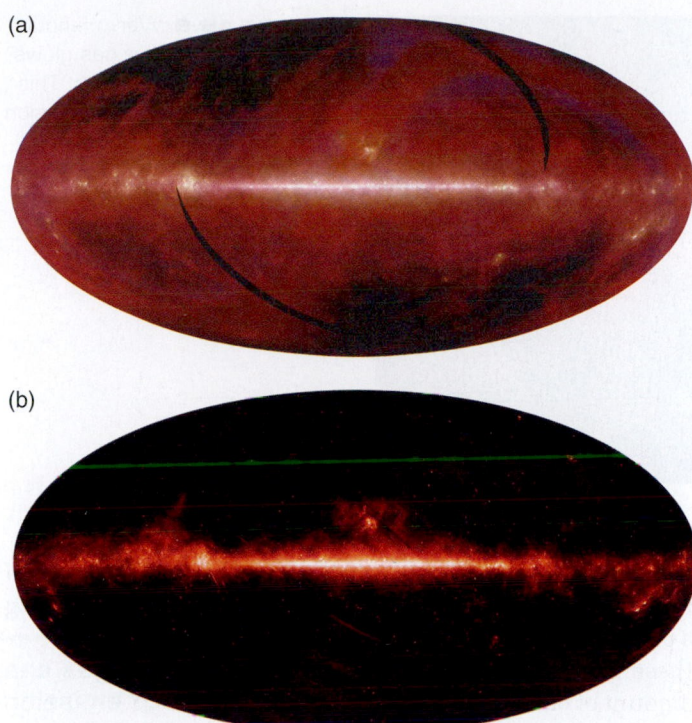

(b)

FIGURE 15.4 (a) Far-infrared (100 μm) IRAS images of the sky show clouds of warm, glowing interstellar dust. (The black bands here and in Figure 15.5 are due to missing data.) (b) This AKARI image shows the galactic plane at shorter infrared wavelengths (9 μm).

Interstellar Gas Has Different Temperatures and Densities

The gas and dust that fill interstellar space within our galaxy are not spread out evenly. About half of all interstellar gas is concentrated into only about 2 percent of the volume of interstellar space. These relatively dense regions are referred to as **interstellar clouds**. Interstellar clouds are *not* like terrestrial clouds, which are locations where water vapor has condensed to form liquid droplets. Rather, interstellar clouds are places where the interstellar gas is more concentrated than in surrounding regions. The other half of the interstellar gas is spread out through the remaining 98 percent of the volume of interstellar space. This is called **intercloud gas**, meaning gas that is found between the clouds.

> Most interstellar material is concentrated in relatively dense clouds.

The properties of intercloud gas vary from place to place. Some intercloud gas is extremely hot, with temperatures in the millions of kelvins, close to those found in the centers of stars. Even so, were you to find yourself adrift in an expanse of hot intercloud gas, your first concern would be freezing to death! The gas is hot, which means that the atoms that make it up are moving about very rapidly; so when one of these atoms runs into you, it hits you very hard. But the gas is also extremely tenuous. Typically, you would have to search a liter (1,000 cm³) or more of hot intercloud gas to find a single atom. To gather up 1 gram of hot intercloud gas—about the mass found in a liter of air—you would have to collect all of the gas in a cube over 8,000 km on an edge! There are so few atoms in a given volume of hot intercloud gas that atoms would very rarely run into you, so this million-kelvin gas would do little to keep you warm. You would radiate energy away and cool off much faster than the gas around you could replace the lost energy.

Extremely hot intercloud gas is thought to occupy about half the volume of interstellar space. This gas is heated mostly by the energy of tremendous stellar explosions called **supernovae**. (We will return to the story of supernovae in Chapters 16 and 17. For now, it is enough to know that these are *very* energetic events!) Our Solar System is located inside a rarefied bubble of hot intercloud gas that has a density of about 0.005 hydrogen atom per cubic centimeter and is at least 650 light-years across. Some astronomers have suggested that this is the remnant of the hot bubble produced by a supernova explosion 300,000 years ago. Like the million-kelvin gas in the corona of the Sun, hot intercloud gas glows faintly in the energetic X-ray portion of the electromagnetic spectrum. X-rays cannot penetrate Earth's atmosphere, so it is necessary to get above our atmosphere to observe the radiation coming from hot intercloud gas. For rocket- and satellite-borne X-ray telescopes, however, the entire sky is aglow with faint X-rays coming from the bubble of million-kelvin hot gas in which our Solar System is immersed (**Figure 15.5**).

> We live in a bubble of million-kelvin intercloud gas.

Not all intercloud gas is as hot as our local bubble. Most of the remaining intercloud gas has a temperature of about 8,000 K and a density ranging from about 0.01 to 1 atom per

FIGURE 15.5 The faint X-ray glow that fills the sky is due largely to emission from the bubble of million-kelvin gas that surrounds the Sun. Bright spots are more distant sources, including objects such as bubbles of very hot, high-pressure gas surrounding the sites of recent supernova explosions.

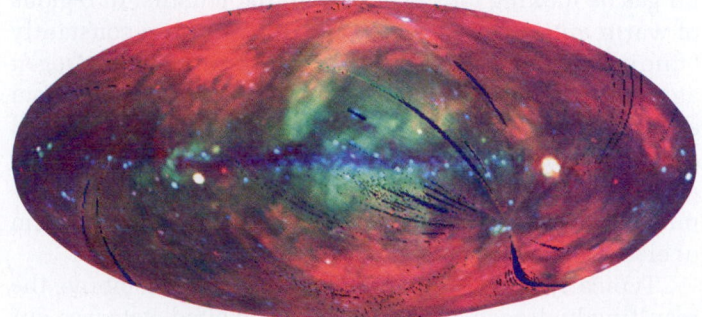

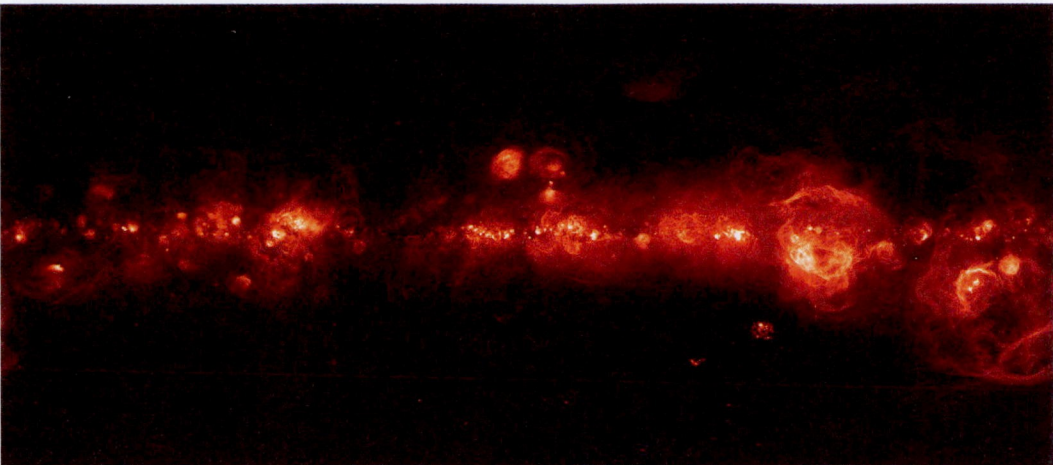

FIGURE 15.6 Warm (about 8,000 K) interstellar gas glows in the Hα line of hydrogen. This image, showing the Hα emission from much of the northern sky, reveals how complex the structure of the interstellar medium is.

cubic centimeter. To gather up a gram of this warm intercloud gas, we would need to collect all of the gas in a cube of interstellar space "only" 800 km on an edge.

Interstellar space is awash with the light from stars. Some of this is ultraviolet light with wavelengths shorter than 91.2 nm. Each photon of this ultraviolet light has enough energy that, if absorbed by a hydrogen atom, it will **ionize** the atom, kicking its electron free of the nucleus. If enough of these energetic photons are around, then atoms in the interstellar gas cannot remain neutral (because any neutral hydrogen atom will soon be ionized by starlight). In this way about half of the volume of warm intercloud gas is kept ionized. But if there is a large enough expanse of warm intercloud gas, then the ionizing photons in the starlight get "used up." About half of the volume of warm intercloud gas is mostly neutral. This gas is usually surrounded by regions of ionized intercloud gas that shield the neutral gas from starlight, much as Earth's ozone layer shields the surface of Earth from harmful ultraviolet radiation from the Sun.

One way to look for interstellar gas, including both warm and hot interstellar gas, is to study the spectra of distant stars. As starlight passes through interstellar gas, absorption lines form. Just as absorption lines that formed in the atmosphere of a star tell us much about the star's temperature, density, and chemical composition, interstellar absorption lines provide similar information about the gas the light has passed through. We can also study interstellar gas by looking for the radiation that it emits. In regions of warm *ionized* gas, protons and electrons are constantly "finding each other" and recombining to make hydrogen atoms. Because it takes energy to tear a hydrogen atom apart, conservation of energy says that when a proton and an electron combine to form a hydrogen atom, this same amount of energy must be given up. This freed-up energy must go somewhere, and that somewhere is into the form of electromagnetic radiation.

Typically, when a proton and an electron combine, the resulting hydrogen atom is left in an excited state (see our discussion of energy states of atoms in Chapter 4). The atom then drops down to lower and lower energy states, emitting a photon at each step. Together these photons carry away an amount of energy equal to the ionization energy of hydrogen plus whatever energy of motion the electron and proton had before recombining. The photons emitted in this process have wavelengths corresponding to the energy differences between the allowed energy states of hydrogen. In other words, warm ionized interstellar gas glows in emission lines characteristic of hydrogen (as well as other elements). Usually, the strongest emission line given off by warm interstellar gas in the visible part of the spectrum is the Hα (hydrogen alpha) line, which is seen in the red part of the spectrum at a wavelength of 656.3 nm.

> Interstellar gas can produce both emission and absorption lines.

Figure 15.6 shows an image of a portion of the sky taken in the light of Hα emission. The faint diffuse emission in the figure comes mostly from warm, ionized intercloud gas. The especially bright spots, on the other hand, are quite different. These are regions where intense ultraviolet radiation from massive, hot, luminous O and B stars is able to ionize even relatively dense interstellar clouds. These are called **H II** ("H two") **regions**, signifying that they are made up of the second, or ionized, form of hydrogen. In the chapters to come we will address the question of how long stars live, and we will discover that O stars live only a few million years. As a result, they usually do not move very far from where they formed. The glowing clouds seen as H II regions are the very clouds from which the stars were born. Thus, H II regions are signposts to regions where *active star formation* is taking place.

> H II regions are ionized by hot, luminous O and B stars.

One of the closest H II regions to the Sun is the Orion Nebula, a **nebula** in the constellation Orion that was formerly known as the Great Nebula (**Figure 15.7**). In fact, the

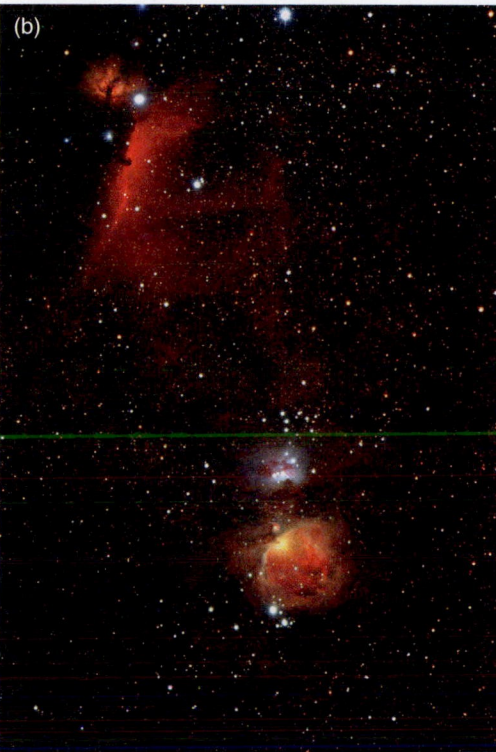

FIGURE 15.7 (a) The Orion Nebula is seen here as a glowing region of interstellar gas surrounding a cluster of young, hot stars. New stars are still forming in the dense clouds surrounding the nebula. (b) The Orion Nebula (at bottom) is only a small part of the larger Orion star-forming region. The dark Horsehead Nebula is seen at the upper left of part (b) and in Figure 15.10a.

Orion Nebula is tiny as H II regions go. It is "great" only because it happens to be very close to us. Almost all of the ultraviolet light that powers the nebula comes from a single hot star, and all told only a few hundred stars are forming in its immediate vicinity. In contrast, **Figure 15.8** shows a giant H II region called 30 Doradus (also known as the Tarantula Nebula). The 30 Doradus H II region is located in the Large Magellanic Cloud, a small companion galaxy to the Milky Way, 160,000 light-years distant. This enormous cloud of ionized gas is powered by a dense star cluster containing thousands of hot, luminous stars. If 30 Doradus were as close as the Orion Nebula, it would be bright enough in the nighttime sky to cast shadows!

Warm, *neutral* hydrogen gas gives off radiation in a different way than warm, ionized hydrogen does. Many subatomic particles, including protons and electrons, are magnetized. It is as if each of these particles had a bar magnet, with a north and a south pole, built into it. According to the rules of quantum mechanics, a hydrogen atom can exist in only two different configurations. Either the magnetic "poles" of the proton and electron point in the *same* direction, or they point in *opposite* directions. The electron's "magnet" and the proton's "magnet" push on each other, so the way they line up affects the energy that the atom has. When the two "magnets" point in opposite directions, the atom has slightly more energy than when they point in the same direction.

It takes energy to change a hydrogen atom from the lower-energy, magnetically aligned state to the higher-energy, magnetically unaligned state. This is not a problem,

FIGURE 15.8 This European Southern Observatory image shows the 30 Doradus Nebula, located in a small companion galaxy to the Milky Way, 160,000 light-years from Earth. The ultraviolet radiation coming from the dense cluster of thousands of young, massive stars causes interstellar gas in this region of intense star formation to glow.

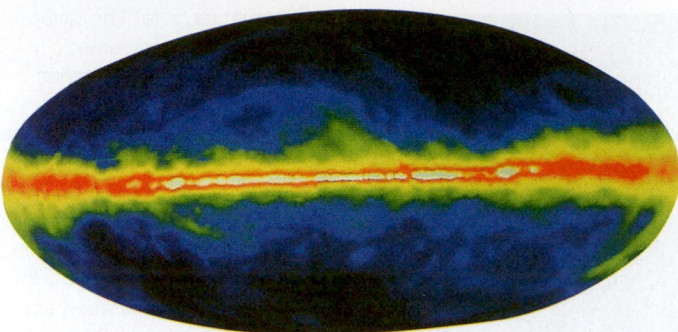

FIGURE 15.9 This radio image of the sky taken at a wavelength of 21 cm shows clouds of neutral hydrogen gas throughout our galaxy. Because radio waves penetrate interstellar dust, 21-cm observations are a crucial probe of the structure of our galaxy.

because in warm, neutral interstellar gas, interactions between atoms easily supply enough energy to do the trick. But once a hydrogen atom is in its higher-energy (magnetically unaligned) state, getting back into its lower-energy (magnetically aligned) state requires nothing but time. If left undisturbed long enough, a hydrogen atom in the higher-energy state will spontaneously jump to the lower-energy state, emitting a photon in the process. The energy difference between the two magnetic states of a hydrogen atom is extremely small: less than half a millionth of the energy needed to tear the electron away from the proton completely. The radiation given off by these low-energy transitions has a wavelength of 21 cm, which lies in the radio portion of the spectrum.

> Neutral hydrogen emits 21-cm radiation.

The tendency for hydrogen atoms to emit 21-cm radiation is extremely weak. On average, you would have to wait about 11 million years for a hydrogen atom in the higher-energy state to spontaneously jump to the lower-energy state and give off a photon. Even so, there is a *lot* of hydrogen in the universe. In **Figure 15.9**, the sky is aglow with 21-cm radiation from neutral hydrogen. Because of its long wavelength, 21-cm radiation freely penetrates dust in the interstellar medium, enabling us to see neutral hydrogen throughout our galaxy, while measurements of the Doppler shift of the line indicate how fast the source of the radiation is moving. These two attributes make the 21-cm line of neutral hydrogen perhaps the most important kind of radiation there is for understanding the structure of our galaxy.

Regions of Cool, Dense Gas Are Called Clouds

As mentioned earlier, intercloud gas fills 98 percent of the volume of interstellar space but accounts for only half of the mass of interstellar gas. The remaining 50 percent of all interstellar gas is concentrated into much denser interstellar clouds that occupy only 2 percent of the volume of the galaxy. Most interstellar clouds are composed primarily of neutral atomic hydrogen (that is, isolated neutral hydrogen atoms). These clouds are much cooler and denser than the warm intercloud gas. They have temperatures of about 100 K and densities in the range of about 1–100 atoms per cubic centimeter. But even at a density of 100 atoms/cm^3, you would still have to gather up all of the gas in a box 180 km on an edge to collect a single gram of this material.

On Earth it is uncommon to find atoms in isolation—most atoms are tied up in molecules. (For example, in Earth's atmosphere only nonreactive gases such as argon are typically found in their atomic form.) Yet the interstellar gas that we have encountered so far consists of isolated atoms rather than molecules. The reason is that throughout most of interstellar space, including most interstellar clouds, molecules do not survive for long. If interstellar gas is too hot, then any molecules that do exist will soon collide with other molecules or atoms that have enough energy to break the molecules back down into their constituent atoms. The temperature in a neutral hydrogen cloud may be low enough for some molecules to survive, but even here a different process will soon destroy any molecules that might form. Photons of starlight with enough energy to break molecules apart can penetrate neutral hydrogen clouds. Only in the hearts of the densest of interstellar clouds, where dust effectively blocks even the relatively low-energy photons required to shatter molecules, does interstellar chemistry get its chance. For obvious reasons, these dark clouds are referred to as **molecular clouds**.

In images such as those in **Figure 15.10**, molecular clouds are evident from their silhouettes against a background of stars. Inside such clouds it is dark and usually very cold, with a typical temperature of only about 10 K—that is, 10 kelvins above absolute zero. Most of these clouds have densities of about 100–1,000 molecules/cm^3, but densities as high as 10^{10} molecules/cm^3 have been observed. Even at 10^{10} molecules/cm^3 this gas is still less than a billionth as dense as the air around you, making it an extremely good vacuum by terrestrial standards. In this cold, relatively dense environment, atoms combine to form a wide variety of molecules.

> Molecular clouds are dense and cold.

By far the most common component of molecular clouds is molecular hydrogen. Molecular hydrogen (H_2) consists of two hydrogen atoms and is the smallest possible molecule. Molecules radiate mostly in the radio and infrared portions of the electromagnetic spectrum. Molecular emission lines are useful to astronomers in much the same way that the spectral lines from atoms are useful: Each type of molecule is unique in its properties, and the energies of its allowed states are unique as well. The wavelengths of emission lines from molecules are an unmistakable fingerprint of the kind of molecule responsible for them.

(a)

(b)

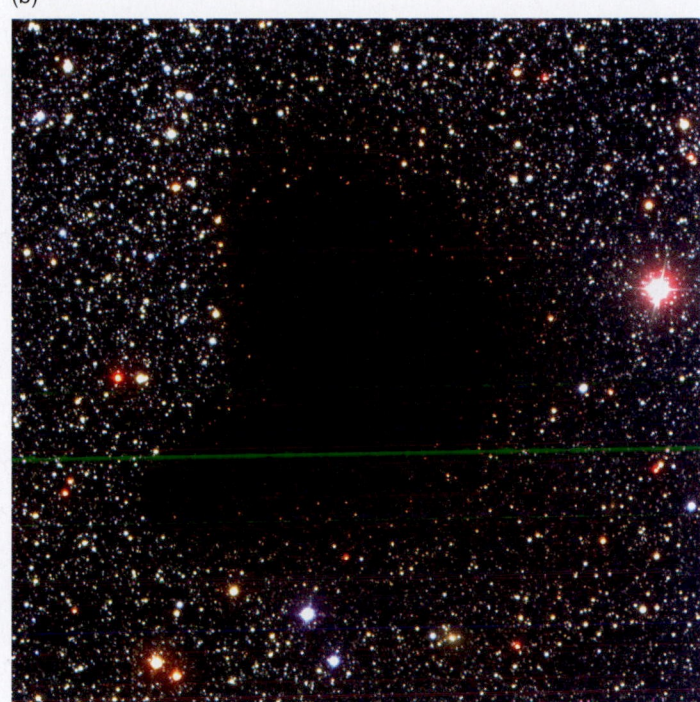

(c)

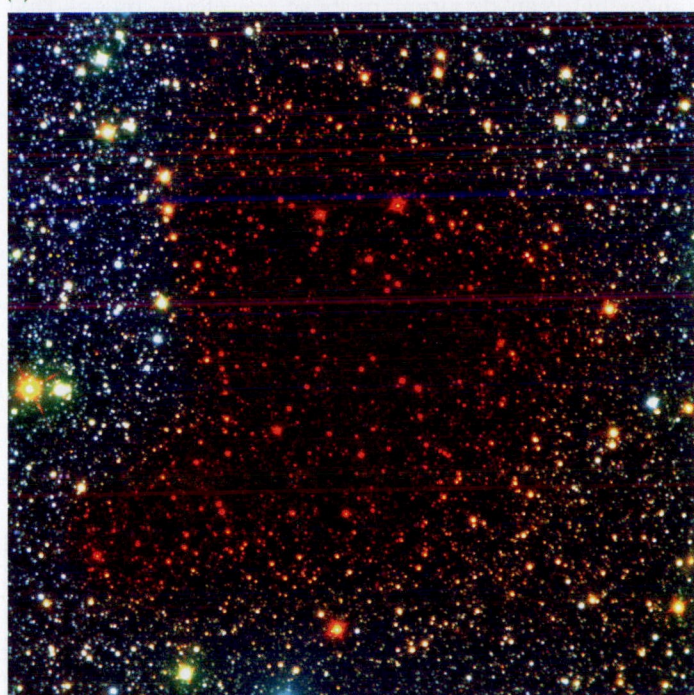

FIGURE 15.10 In visible light, interstellar molecular clouds are seen in silhouette against a background of stars and glowing gas. (a) The Horsehead Nebula in Orion is one of the most identifiable of dark nebulae because of its recognizable shape. The circular halos around the bright stars in this image are a photographic artifact. (b) Light from background stars is blocked by dust and gas in nearby Barnard 68, a dense, dark molecular cloud. (c) Infrared wavelengths can penetrate much of this gas and dust, as seen in this false-color image of Barnard 68.

In addition to molecular hydrogen, approximately 150 other molecules have been discovered in interstellar space. These molecules range from very simple structures such as carbon monoxide (CO), one of the more common molecules, to complex organic compounds such as methanol (CH_3OH) and acetone [$(CH_3)_2CO$], to carbon chains as large as $HC_{11}N$ (cyanopolyyne). Very large carbon molecules, made of hundreds of individual atoms, bridge the gap between large interstellar molecules and small interstellar grains. As mentioned earlier, visible light cannot escape from the molecular clouds where interstellar molecules are concentrated. Fortunately, however, the radio waves emitted by molecules are unaffected by interstellar dust, so they escape easily from dark molecular clouds. Observations of radio waves from molecules give us a remarkable look at the innermost workings of the densest and most opaque interstellar clouds. Among the more important species is CO. Many astronomers believe that the ratio of CO to H_2 is relatively constant, and because molecular hydrogen is so difficult to detect even by infrared and radio observations, carbon monoxide is often used to reveal the amounts and distribution of molecular hydrogen.

Molecular clouds have masses ranging from a few times the mass of our Sun to 10 million solar masses. The smallest molecular clouds may be less than half a light-year across, whereas the largest molecular clouds may be over a thousand light-years in size. The largest molecular clouds qualify for the title **giant molecular cloud**. These behemoths typically have masses of a few hundred thousand times that of our Sun and on average are about 120 light-years across. Our galaxy contains about 4,000 giant molecular clouds and a much larger number of smaller ones. Molecular clouds fill only a tiny fraction of the volume of our galaxy—probably only about 0.1 percent of interstellar space. These clouds may be rare, but they are extremely important to our story because they are the cradles of star formation.

Having looked at the many different phases of the interstellar medium, ranging from vast expanses of tenuous million-kelvin intercloud gas to the cold dense interiors of molecular clouds, an obvious question to ask is, How does it all fit together? Discussing the structure of the interstellar medium is a lot like discussing the nature of weather on Earth: it is extremely complex, often difficult to understand, and all but impossible to predict precisely.

In the interstellar medium, the energy to power the "weather" comes from stars. Warm gas heated by ultraviolet radiation from massive, hot stars pushes outward into its surroundings. Blast waves from supernova explosions sweep out vast, hot "cavities," piling up everything in their path like snow in front of a snowplow. Interstellar clouds are destroyed under the onslaught of these violent events, but they are also formed by these same forces. Swept-up intercloud gas becomes the next generation of clouds. Hot bubbles of high-pressure interstellar gas crush molecular clouds, driving up their densities and perhaps triggering the formation of new generations of stars. The interstellar medium is so well stirred that *on average*, interstellar clouds are moving around with random velocities of approximately 20 kilometers per second (km/s). The fastest motions of interstellar material are those of very hot gas and are measured in thousands of kilometers per second.

The story of the wide range of phenomena that shape the interstellar medium is fascinating. We could easily stay here and look around for quite some time, but instead we must press on in our journey. The time has come to watch as a molecular cloud begins the slow process of collapsing under its own weight, beginning a chain of events that will culminate with the ignition of nuclear fires within a new generation of stars.

15.3 Molecular Clouds Are the Cradles of Star Formation

As we have seen, objects such as planets and stars are held together by gravity. The mutual gravitational attraction between each part of a planet or star and every other part of the object results in a net inward force that pulls all parts of the object toward its center. If an object is stable, then this inward force of the object's **self-gravity** must be balanced by something—often the outward push of pressure within the object. In our discussion of planets and stars we referred to the balance between gravity and pressure in a stable object as **hydrostatic equilibrium**.

All of these same concepts can be applied to clouds in the interstellar medium. As shown in **Figure 15.11**, interstellar clouds also have self-gravity: each part of the cloud feels a gravitational attraction for every other part of the cloud. However, interstellar clouds are rarely in hydrostatic equilibrium. In most interstellar clouds, internal pressure is much stronger than self-gravity. (Because gravity follows an inverse square law, the more spread out an object's mass is, the weaker is its self-gravity.) The outward push of pressure in these clouds would cause them to expand if not for the opposing pressure of the surrounding intercloud medium. The intercloud medium is much less dense than interstellar clouds, but it is also much hotter, so its pressure is high enough to confine the clouds. (Pressure is proportional to both density and temperature.)

Some molecular clouds, on the other hand, violate hydrostatic equilibrium in the other direction. These clouds are massive enough and dense enough—that is, there is enough mass packed together in a small enough volume of space—that their self-gravity becomes important. (More mass, along with smaller distances between different parts of the cloud, means that gravity is stronger.) Furthermore, these clouds are cool enough that their internal pressure is relatively low despite their high density. In such clouds, self-gravity is much greater than pressure, so the clouds collapse under their own weight.

> Some molecular clouds collapse under their own weight.

If self-gravity in a molecular cloud is much more significant than pressure, we might expect that gravity should win outright, and that the cloud should rapidly collapse toward its center. In practice, the process goes very slowly because several other effects stand in the way of the collapse. One

FIGURE 15.11 Self-gravity causes a molecular cloud to collapse.

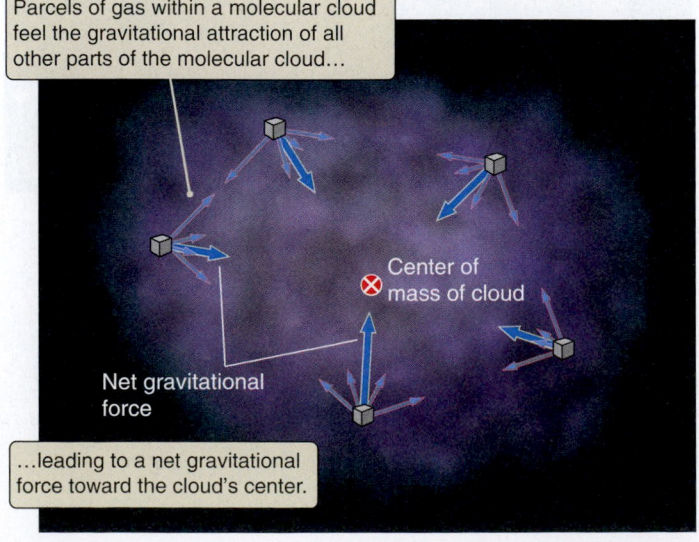

Parcels of gas within a molecular cloud feel the gravitational attraction of all other parts of the molecular cloud…

Center of mass of cloud

Net gravitational force

…leading to a net gravitational force toward the cloud's center.

such effect that slows the collapse of a cloud is **conservation of angular momentum**. (We discussed the effects of angular momentum and the flattening of a collapsing cloud in Chapter 6.) Other properties of molecular clouds that prevent them from collapsing rapidly are **turbulence** and the effects of magnetic fields. Even though these effects may slow the collapse of a molecular cloud, in the end gravity will win. Magnetic fields in the cloud can slowly die away. One part of the cloud can lose angular momentum to another part of the cloud, allowing the part of the cloud with less angular momentum to collapse further. Turbulence ultimately fades away. The details of these processes are complex and the subject of much current research, but the crucial point is this: the effects preventing the collapse of a molecular cloud are temporary, and gravity is persistent. As the forces that oppose the cloud's self-gravity gradually fade away, the cloud slowly becomes smaller. ▶❚❚ **AstroTour:** Star Formation

> Angular momentum, turbulence, and magnetic fields slow the collapse of a molecular cloud, but gravity wins in the end.

Molecular Clouds Fragment As They Collapse

Crucial to the process of star formation is the fact that as a cloud becomes smaller, the gravitational forces trying to pull it together grow stronger. This effect is a result of the inverse square law of gravitation discussed in Chapter 3. Suppose a cloud starts out being 4 light-years across. By the time the cloud has collapsed to 2 light-years across, the different parts of the cloud are, on average, only half as far away from each other as they were when they started. As a result, the gravitational attraction they feel toward each other will be four times stronger than it was when the cloud was 4 light-years across. When the cloud is $\frac{1}{4}$ as large as when it started out (or 1 light-year across), the force of gravity will be 16 times as strong. As a collapsing cloud becomes smaller, gravity becomes stronger. As gravity becomes stronger, the collapse picks up speed. And as the collapse picks up speed, gravity increases even more rapidly. In other words, the collapse "snowballs."

Molecular clouds are never uniform. Some regions within the cloud are invariably denser than others and so collapse within themselves more rapidly than do surrounding regions. As these regions collapse, their self-gravity becomes stronger because they are more compact, so they collapse even faster. **Figure 15.12** shows the result of this process. What started out as slight variations in the density of the cloud grow to become very dense concentrations of gas. Instead of collapsing into a single object, the molecular cloud fragments into a number of very dense **molecular-cloud cores**. A single molecular cloud may form hundreds or thousands of molecular-cloud cores. It is from these dense cloud cores, typically a few light-months in size, that stars may form.

> Stars form in molecular-cloud cores.

As a molecular-cloud core becomes smaller, the gravitational forces that are trying to crush the cloud grow stronger. Eventually, gravity is able to overwhelm the forces due to pressure, magnetic fields, and turbulence that have been resisting it. This happens first near the center of the cloud because that's where the cloud material is most strongly

FIGURE 15.12 As a molecular cloud collapses, denser regions within the cloud collapse more rapidly than less dense regions. As this process continues, the cloud fragments into a number of very dense molecular-cloud cores that are embedded within the large cloud. These cloud cores may go on to form stars.

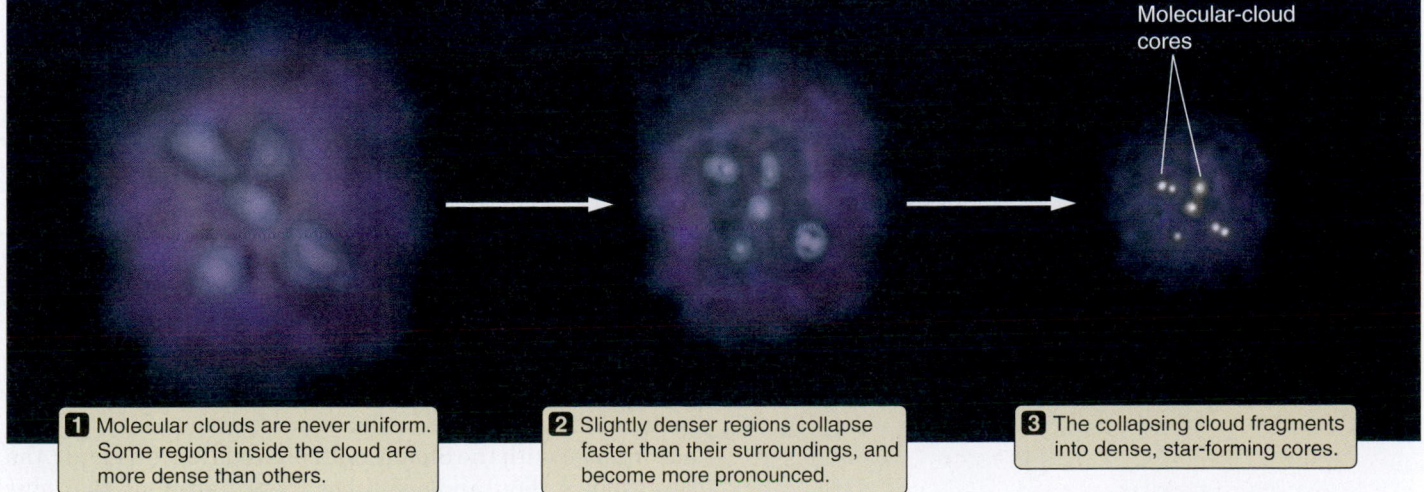

Molecular-cloud cores

1 Molecular clouds are never uniform. Some regions inside the cloud are more dense than others.

2 Slightly denser regions collapse faster than their surroundings, and become more pronounced.

3 The collapsing cloud fragments into dense, star-forming cores.

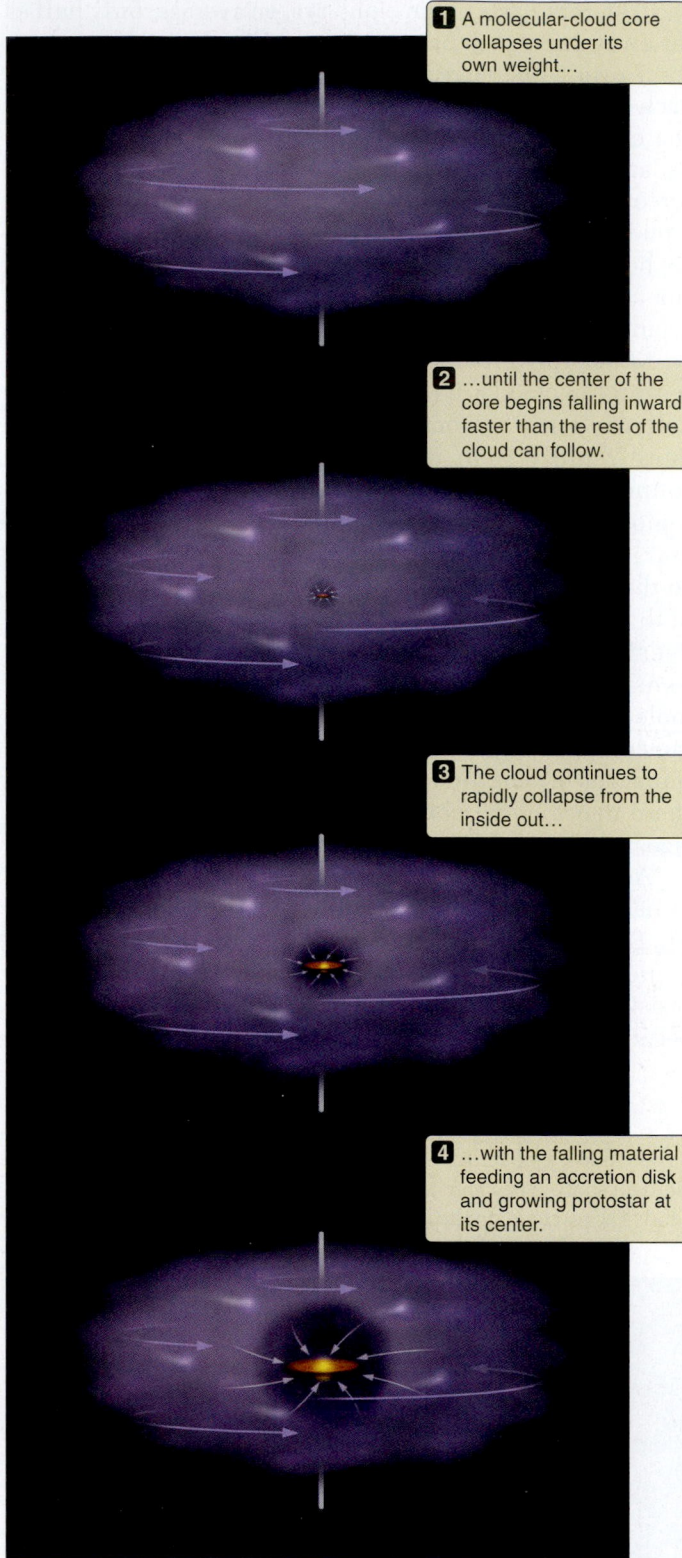

1 A molecular-cloud core collapses under its own weight...

2 ...until the center of the core begins falling inward faster than the rest of the cloud can follow.

3 The cloud continues to rapidly collapse from the inside out...

4 ...with the falling material feeding an accretion disk and growing protostar at its center.

FIGURE 15.13 When a molecular-cloud core gets very dense, it collapses from the inside out. Conservation of angular momentum causes the infalling material to form an accretion disk that feeds the growing protostar.

concentrated. The inner parts of the cloud core start to fall rapidly inward, starting a sequence of events much like a chain of dominoes in which one domino topples the next, which topples the next. The pressure from the central part of the cloud core had been supporting the weight of the layers above it. When the center of the cloud collapses, this support is suddenly removed. Without this "pressure support," the next-outer layer begins to fall freely inward toward the center as well. But what about the layers farther out? Now there is nothing to hold them up either, so the process continues: each layer of the cloud core falls inward in turn, thereby removing support from the layers still farther out. As shown in **Figure 15.13**, the cloud core collapses like a house of cards when the bottom layer is knocked out. The whole structure comes crashing down.

We have been here before. It was at this point in the story of star formation that our discussion of the formation of the Solar System in Chapter 6 began. **Figure 15.14** shows an overview of the process. Material from the collapsing molecular-cloud core falls inward. Because of its angular momentum, this material lands on a flat, rotating accretion disk. Figure 6.2 shows several such disks. The dark bands in these images are the disks seen edge-on; the bright regions are starlight reflected from the surfaces of the disks. Most of this material eventually finds its way inward onto the growing protostar at the center of the disk, but a small fraction of it remains to become the stuff planets are made of. The last time we passed this way, we followed the evolution of the gas and dust left behind in the disk and saw how it leads naturally to a planetary system with the properties of our Solar System. This time through, we will instead follow the story of the protostar as it becomes a star.

> Our Solar System began in a molecular cloud.

15.4 The Protostar Becomes a Star

We pick up the story of our protostar at a time when the cloud is still collapsing, and more and more material is falling onto the disk. At this point the surface of the protostar has been heated to a temperature of thousands of kelvins as the gravitational energy of the original molecular cloud has been converted into thermal energy. The protostar is also huge—hundreds of times larger than our Sun—which means that the surface of the protostar is tens of thousands of times larger than the surface of the Sun. Each square meter of that enormous surface is radiating energy away in accordance with the Stefan-Boltzmann law. As a result, the protostar is thousands of times more luminous than our

> Protostars are large and luminous.

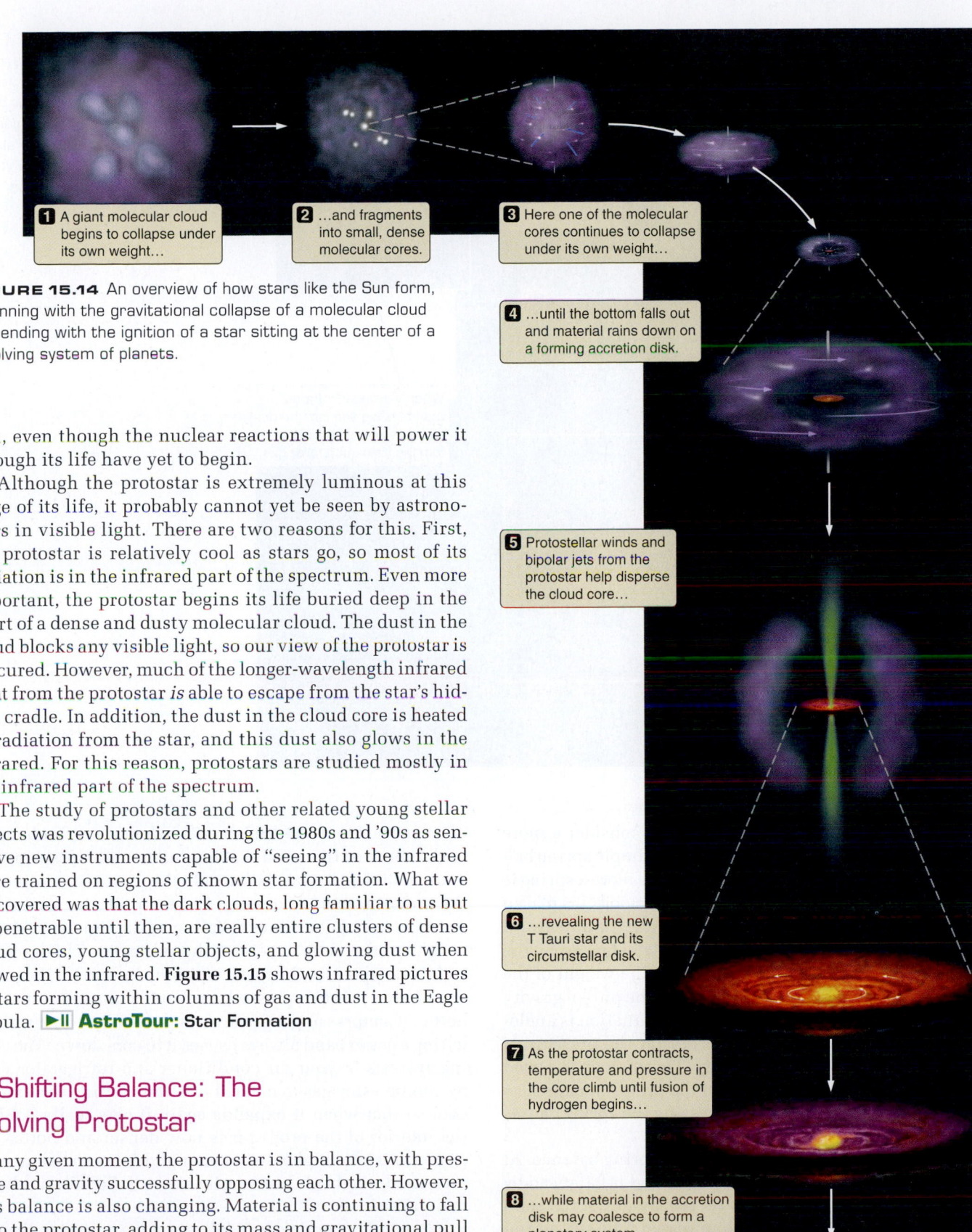

1 A giant molecular cloud begins to collapse under its own weight…

2 …and fragments into small, dense molecular cores.

3 Here one of the molecular cores continues to collapse under its own weight…

FIGURE 15.14 An overview of how stars like the Sun form, beginning with the gravitational collapse of a molecular cloud and ending with the ignition of a star sitting at the center of a revolving system of planets.

4 …until the bottom falls out and material rains down on a forming accretion disk.

5 Protostellar winds and bipolar jets from the protostar help disperse the cloud core…

6 …revealing the new T Tauri star and its circumstellar disk.

7 As the protostar contracts, temperature and pressure in the core climb until fusion of hydrogen begins…

8 …while material in the accretion disk may coalesce to form a planetary system.

Sun, even though the nuclear reactions that will power it through its life have yet to begin.

Although the protostar is extremely luminous at this stage of its life, it probably cannot yet be seen by astronomers in visible light. There are two reasons for this. First, the protostar is relatively cool as stars go, so most of its radiation is in the infrared part of the spectrum. Even more important, the protostar begins its life buried deep in the heart of a dense and dusty molecular cloud. The dust in the cloud blocks any visible light, so our view of the protostar is obscured. However, much of the longer-wavelength infrared light from the protostar *is* able to escape from the star's hidden cradle. In addition, the dust in the cloud core is heated by radiation from the star, and this dust also glows in the infrared. For this reason, protostars are studied mostly in the infrared part of the spectrum.

The study of protostars and other related young stellar objects was revolutionized during the 1980s and '90s as sensitive new instruments capable of "seeing" in the infrared were trained on regions of known star formation. What we discovered was that the dark clouds, long familiar to us but impenetrable until then, are really entire clusters of dense cloud cores, young stellar objects, and glowing dust when viewed in the infrared. **Figure 15.15** shows infrared pictures of stars forming within columns of gas and dust in the Eagle Nebula. ▶❚❚ **AstroTour:** Star Formation

A Shifting Balance: The Evolving Protostar

At any given moment, the protostar is in balance, with pressure and gravity successfully opposing each other. However, this balance is also changing. Material is continuing to fall onto the protostar, adding to its mass and gravitational pull and therefore increasing the weight that underlying layers of the protostar must support. The protostar is also slowly losing its internal thermal energy by radiating it away. How can an object be in perfect balance and yet be changing at

HST image in visible light shows columns of molecular gas illuminated by nearby stars.

When viewing in infrared radiation, we see into the dusty columns where new stars are forming from interstellar gas.

FIGURE 15.15 This HST image of the Eagle Nebula shows dense columns of molecular gas and dust at the edge of an H II region. Infrared images of the same field, also taken with the HST, show young stars forming within these columns.

the same time? To answer this question, consider a more everyday example. **Figure 15.16a** shows a simple spring balance, which works on the principle that the more a spring is compressed, the harder it pushes back. If an object is placed on the balance, it compresses the spring until the downward force of the weight of the object is exactly balanced by the upward force of the spring. We measure the weight of the object by determining the point at which the pull of gravity and the push of the spring are equal. The situation is analogous to our protostar, but in the protostar the pressure of the gas takes the place of the spring. (The analogy is a good one because compressing the gas in the protostar causes its pressure to increase, just as compressing the spring causes the spring to push back harder.)

Let's now slowly pour sand onto our spring balance. At any point the downward weight of the sand is balanced by the upward force of the spring. As the weight of the sand increases, the spring is slowly compressed. The spring and the weight of the sand are always in balance, but this balance is *changing with time* as more and more sand is added. In just the same way, the protostar is always in balance, with the inward force of gravity matched by the outward force

of pressure (**Figure 15.16b**). But its balance, too, is changing with time. Just like sand that has fallen onto the spring balance, material that has fallen onto the protostar compresses the protostar more and more, and the protostar evolves.

As material continues to fall onto the protostar, it adds weight to the outer layers of the protostar. This growing weight compresses the material in the interior of the protostar, and as it is compressed, the protostar's interior grows hotter. (Compressing a gas always causes it to heat up, and letting a gas expand always causes it to cool down. The cooling systems in your air conditioner and refrigerator work by compressing gas to make it hot, then letting this hot gas cool so that when it expands again it gets really cold.) If the interior of the protostar is now denser and hotter, the pressure is also higher—just enough higher to balance the increased weight of the material above it. Balance is always maintained.

Even after material stops falling on the protostar, the protostar keeps evolving in much the same way. Earlier we saw that the protostar is radiating away many times as much energy as the Sun. The source of the energy being radiated away is the thermal energy trapped in the interior of the protostar—the same thermal energy that is responsible for supporting the protostar against the forces of gravity. So, as thermal energy leaks out of the protostar, gravity gains the upper hand and the protostar slowly contracts. As the protostar shrinks, the forces of gravity become greater (because the parts of the protostar are closer together—remember the inverse square law of gravity). Meanwhile, as the protostar shrinks, its interior is compressed, so it grows hotter. As density and temperature climb, so does the pressure—*just enough to continue to counteract the growing force of gravity*. Balance between gravity and pressure is always maintained. This process continues, with the star becoming smaller and smaller and its interior grow-

> The protostar radiates away thermal energy and shrinks.

ing hotter and hotter, until the center of the star is finally hot enough for hydrogen to begin fusing into helium.

Something strange is going on here. When an object radiates energy, it normally gets cooler. When you turn off the electric coil on your stove, the coil cools as it loses the thermal energy within it. Yet we just concluded that as a protostar radiates thermal energy away, it actually grows hotter. How can that be? Once again the answer lies with the concept of conservation of energy. As the protostar contracts, every part of the protostar is slowly falling toward its center, which means that the protostar is *losing gravitational energy*. This gravitational energy has to show up in another form, which in this case is thermal energy. Some of this thermal energy is radiated away, but not all of it—so the interior of the star grows hotter.

VISUAL ANALOGY **FIGURE 15.16** (a) A spring balance comes to rest at the point where the downward force of gravity is matched by the upward force of the compressed spring. As sand is added, the location of this balance point shifts. (b) Similarly, the structure of a protostar is determined by a balance between pressure and gravity. Like the spring balance, the structure of the protostar constantly shifts as the protostar radiates energy away and additional material falls onto its surface.

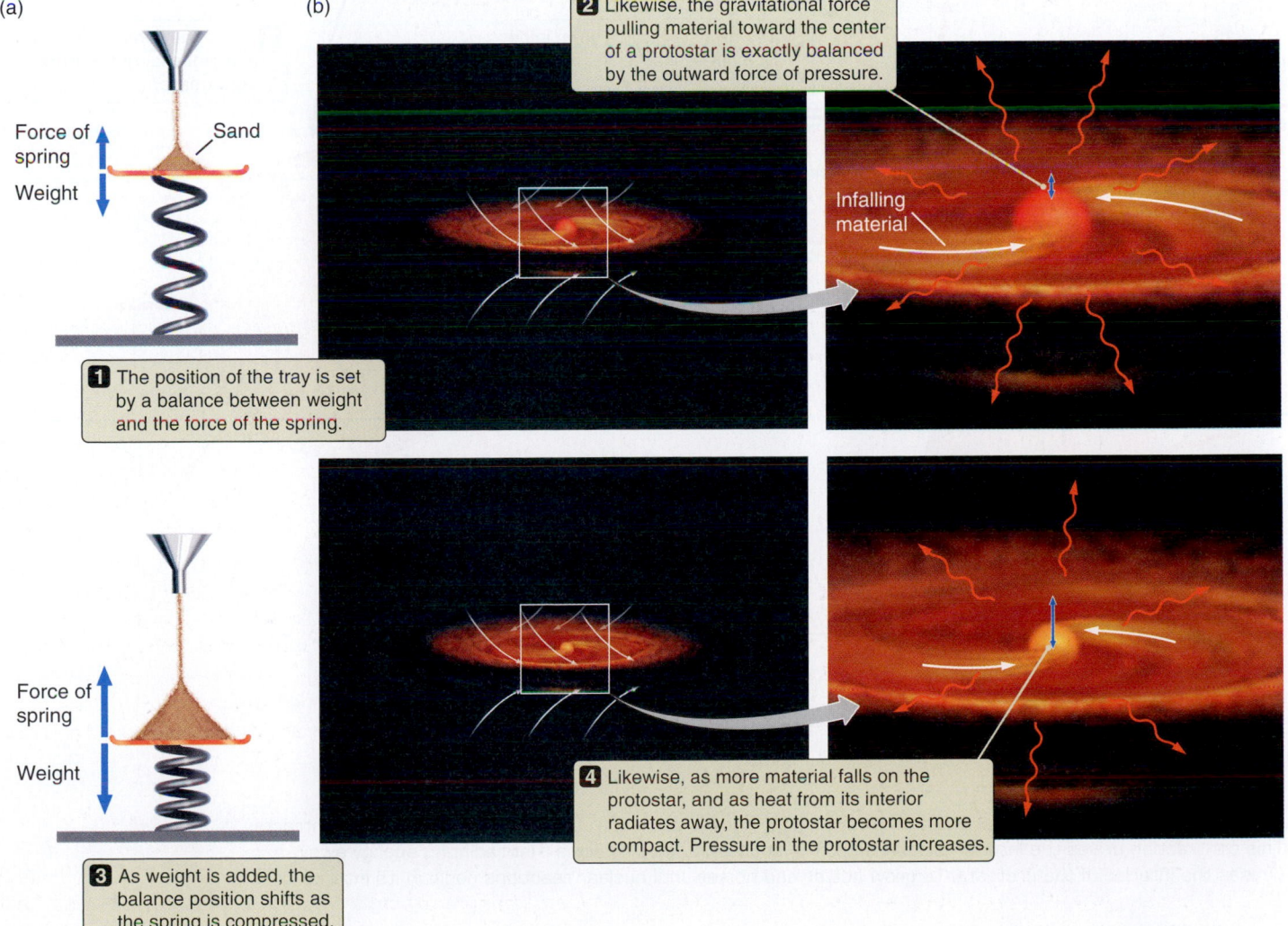

(a)

(b)

Force of spring

Sand

Weight

1 The position of the tray is set by a balance between weight and the force of the spring.

2 Likewise, the gravitational force pulling material toward the center of a protostar is exactly balanced by the outward force of pressure.

Infalling material

Force of spring

Weight

3 As weight is added, the balance position shifts as the spring is compressed.

4 Likewise, as more material falls on the protostar, and as heat from its interior radiates away, the protostar becomes more compact. Pressure in the protostar increases.

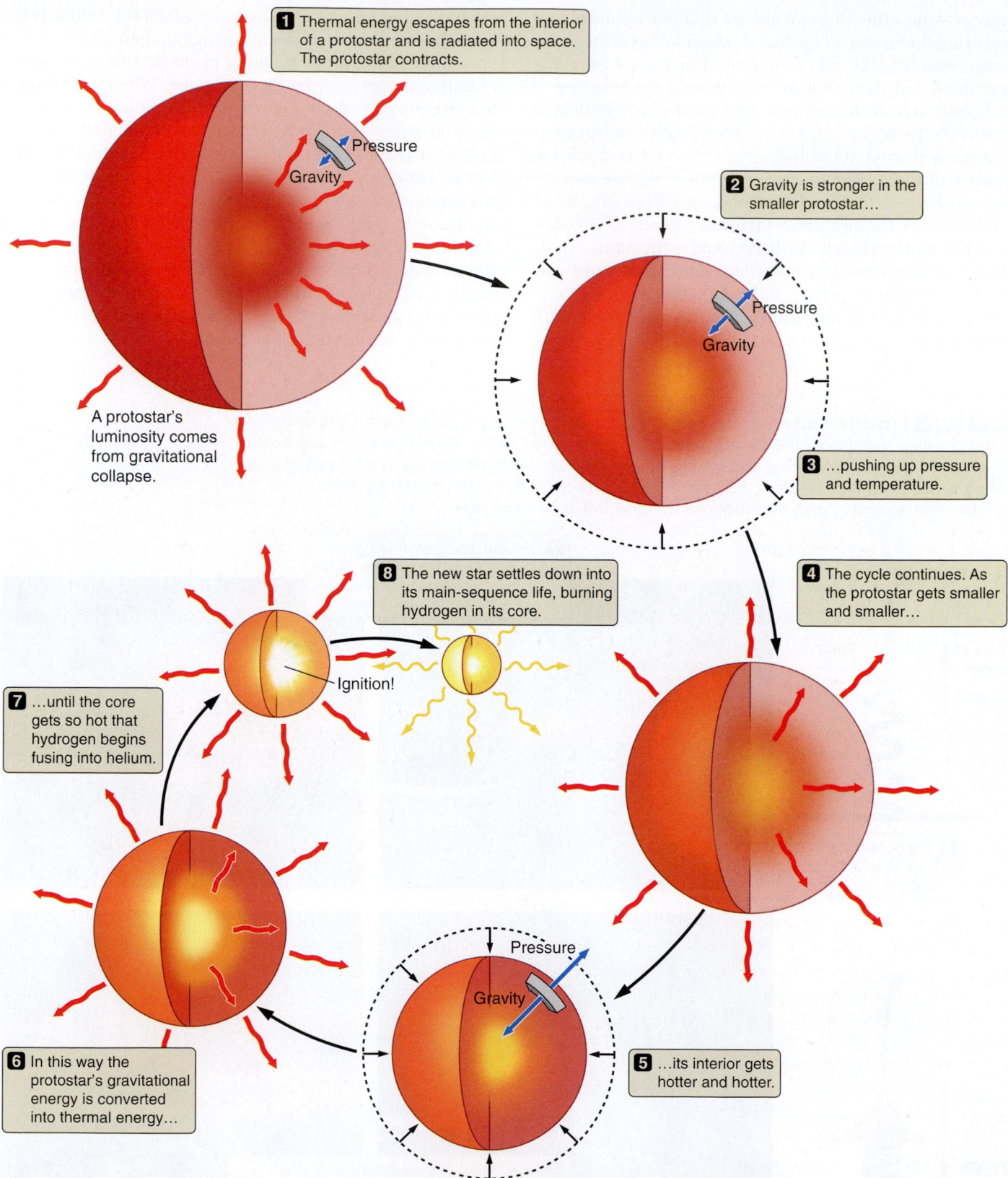

1 Thermal energy escapes from the interior of a protostar and is radiated into space. The protostar contracts.

Pressure
Gravity

A protostar's luminosity comes from gravitational collapse.

2 Gravity is stronger in the smaller protostar…

Pressure
Gravity

3 …pushing up pressure and temperature.

4 The cycle continues. As the protostar gets smaller and smaller…

8 The new star settles down into its main-sequence life, burning hydrogen in its core.

Ignition!

7 …until the core gets so hot that hydrogen begins fusing into helium.

6 In this way the protostar's gravitational energy is converted into thermal energy…

Pressure
Gravity

5 …its interior gets hotter and hotter.

FIGURE 15.17 As a protostar radiates away energy, it loses pressure support in its interior and contracts. This contraction drives the interior pressure up. The counterintuitive result is that radiating energy away causes the interior of the protostar to grow hotter and hotter until nuclear reactions begin in its interior.

Figure 15.17 shows this chain of events in a nutshell. The protostar radiates away thermal energy and, in the process, loses pressure support. Deprived of pressure support, the protostar contracts. As the protostar contracts, gravitational energy is converted to thermal energy, driving up the temperature in the protostar's interior. If the protostar is massive enough, the temperature in its interior will eventually become hot enough for nuclear fusion to begin. This is the point at which the transition from protostar to star takes place. The distinction between the two is that a *protostar* draws its energy from gravitational collapse, whereas a *star* draws its energy from thermonuclear reactions in its interior.

> Fusion begins, and the protostar becomes a star.

The protostar's mass determines whether it will ever become a star. As the protostar slowly collapses, the temperature at its center gets higher and higher. If the protostar is more massive than about 0.08 times the mass of the Sun ($0.08\ M_\odot$), the temperature in its core will eventually reach the 10-million-K mark, and fusion of hydrogen into helium will begin. At this onset of fusion, the newly born star will once again adjust its structure until it is radiating energy away from its surface at just the rate that energy is being liberated in its interior. As it does so, it "settles" onto the main sequence of the H-R diagram, where it will spend the majority of its life.

> At least $0.08\ M_\odot$ is needed for a protostar to become a star.

If the mass of the protostar is less than $0.08\ M_\odot$, it will never reach the point where nuclear burning takes place. Such a failed star is called a **brown dwarf**. A brown dwarf is neither star nor planet, but something in between—what we might call a "substellar object." The International Astronomical Union (IAU) sets the boundary between a brown dwarf and a supermassive giant planet at 13 Jupiter masses (M_J), although there is still some debate as to whether this is really the appropriate criterion.

> Brown dwarfs are failed stars.

A brown dwarf forms the same way a star forms, yet in many respects it is like a giant Jupiter. Brown-dwarf candidates are seen as especially cool M stars, and stars that are even cooler than spectral type M.[1] A brown dwarf never burns hydrogen in its core but instead glows by continually cannibalizing its own gravitational energy.[2] As the years pass, a brown dwarf gets progressively smaller and fainter.

---

[1] Relatively new spectral types, L and T, have been added to represent stars that are even cooler than M-type stars.

[2] The cores of brown dwarfs can get hot enough to burn deuterium, a heavy isotope of hydrogen; and those with a mass greater than 65 Jupiters can burn lithium. But both of these energy sources are in very short supply, and after a brief period of deuterium or lithium fusion, brown dwarfs shine only by the energy of their own gravitational contraction.

The smallest known brown dwarf has a mass of about 8 M_J and a temperature of about 1,350 K.

Since the first brown dwarfs were identified in the mid-1990s, many hundreds have been found. The cooler among them have methane and ammonia in their atmospheres, similar to what we find in the atmospheres of the giant planets of the Solar System. Winds can be very high, though, producing weather far more violent than storms observed in the atmospheres of our own giant planets.

Evolving Stars and Protostars Follow "Evolutionary Tracks" on the H-R Diagram

Within the evolving protostar it is convection rather than radiation that carries energy outward, keeping the protostar's interior well stirred. Interestingly, although the interior of the protostar grows hotter and hotter as it contracts, its *surface* stays at about the same temperature through most of this phase of its evolution. This distinction is important. The surface temperature of a star or protostar is *not* the same as the temperature deep in its interior. For example, in Chapter 14 we found that the temperature of the surface of the Sun is about 5,780 K while the temperature of its interior is millions of kelvins. As the protostar contracts, the temperature deep within the star grows hotter and hotter, but the temperature of the star's *surface* remains nearly constant.

In the 1960s the Japanese theoretician Chushiro Hayashi explained why this is so. Hayashi pointed out that the atmospheres of stars and protostars contain a natural thermostat: the H⁻ ion. (An H⁻, or "H minus," ion is a hydrogen atom that has acquired an extra electron and therefore has a negative charge.) The amount of H⁻ in the atmosphere of a protostar is highly sensitive to the temperature at the surface of the protostar. The cooler the atmosphere of a star, the more slowly atoms and electrons are moving, and the easier it is for a hydrogen atom to hold on to an extra electron. As a result, the cooler the atmosphere of the star, the more H⁻ there is.

The H⁻ ion, in turn, helps control how much energy is radiated away by a star or protostar. The more H⁻ there is in the atmosphere of a star or protostar, the more opaque the atmosphere is, and the more effectively the thermal energy of the protostar is trapped in its interior. Imagine that the surface of the protostar is too cool. "Too cool" means that extra H⁻ forms in the atmosphere, making the atmosphere of the protostar more opaque. The atmosphere thus traps more of the radiation that is trying to escape, and the trapped energy heats up the star. As the temperature climbs, H⁻ ions are destroyed (that is, changed to neutral H atoms). Now imagine the other possibility—that the protostar is too hot.

> H⁻ is a stellar thermostat.

Then, H⁻ in its atmosphere will be destroyed, so the atmosphere will become more transparent, allowing radiation to more freely escape from the interior. Because the protostar cannot hold on to enough of its energy to stay warm, the surface cools. In either case—too cold or too hot—H⁻ is formed or destroyed until the star's atmosphere once again traps just the right amount of escaping radiation.

The H⁻ ion is basically doing the same thing that you do with your bedcovers at night. If you get too cold, you pile on extra covers to trap your body's thermal energy and keep you warm (more H⁻ ions). If you get too hot, you kick off some covers to cool off (fewer H⁻ ions). It is a shame that we have no such "automatic cover" to maintain our body temperature at night as effectively as the H⁻ ion controls the surface temperature of a protostar.

The amount of H⁻ in the atmosphere of a protostar keeps the surface temperature of the protostar somewhere between about 3,000 and 5,000 K, depending on the mass and age of the protostar. Because the temperature of the star is not changing much, the amount of energy per unit time (power) radiated away by each square meter of the surface of the star does not change much either. Recall the Stefan-Boltzmann law from Chapter 4, which says that the amount of power radiated by each square meter of the star's surface is determined by its temperature. But as the star shrinks, the area of its surface shrinks as well. There are fewer square meters of surface to radiate, so the luminosity of the protostar drops. As viewed from the outside, the protostar stays at nearly the same temperature and color but gradually gets fainter as it evolves toward its eventual life as a main-sequence star.

In Chapter 13 we introduced the H-R diagram and used it to begin to understand how the properties of stars differ. For the next several chapters we will also use the H-R diagram to keep track of how stars change as they evolve through their lifetimes. The path across the H-R diagram

FIGURE 15.18 (a) The evolution of pre–main sequence stars can be followed on the H-R diagram. Protostars in the upper right portion of the diagram are large and cool. (b) The roughly vertical, constant-temperature parts of the evolutionary tracks of low-mass protostars are referred to as Hayashi tracks.

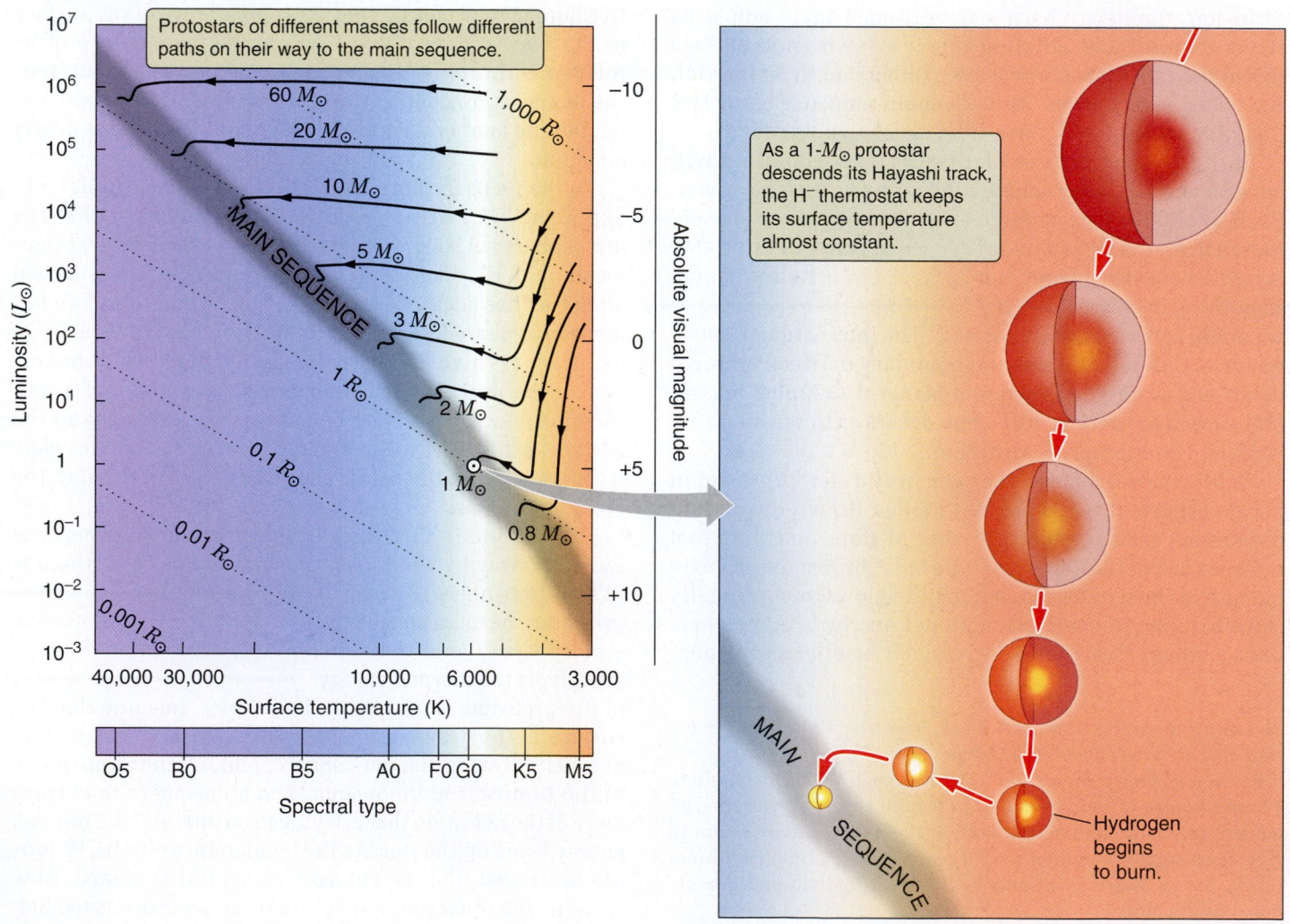

that a star follows as it goes through the different stages of its life is called the star's **evolutionary track (Figure 15.18a)**. The particular path that an evolving protostar follows as it approaches the main sequence is called its **Hayashi track (Figure 15.18b)**. The protostar is brighter than it will be as a true star on the main sequence, so a protostar's Hayashi track is located above the main sequence on the H-R diagram. Because the surface temperature of the protostar stays nearly constant as the protostar contracts, the protostar's Hayashi track prior to the start of hydrogen burning is an almost vertical line on the H-R diagram. Figure 15.18 shows the pre–main sequence evolutionary tracks of stars of several different masses. Astronomers say that an evolving protostar "descends the Hayashi track."

Not All Stars Are Created Equal

In Chapter 13 we found that stars can have a wide range of masses, varying from less than $1/10$ the mass of the Sun up to perhaps 100 times the mass of the Sun. What determines how massive a star will be? The answer to this question is unclear and is a topic of a great deal of ongoing research. One obvious possibility is that a forming star grows until it uses up all of the gas around it; it becomes no larger simply because it has run out of material. Although this explanation would be easy to understand, it does not match our observations of how stars actually form. At this point in our story the contracting protostar is at the center of a collapsing molecular-cloud core, which in turn is a denser-than-average region inside a molecular cloud whose total mass may be hundreds of thousands of times greater than the mass of the Sun. Observations indicate that under most circumstances star formation is a very inefficient process. Only a small fraction of the material in a molecular cloud—perhaps a few percent—ends up as part of the stars forming within it. Something must prevent most of the material in a molecular cloud from ever actually falling onto protostars. There are a number of ideas about what this something might be. One intriguing possibility is that forming stars control their own masses.

FIGURE 15.19 (a) Material falls onto an accretion disk around a protostar and then moves inward, eventually falling onto the star. In the process, some of this material is driven away in powerful jets that stream perpendicular to the disk. (b) This infrared Spitzer Space Telescope image shows jets streaming outward from a young, developing star. Note the nearly edge-on, dark accretion disk surrounding the young star.

(a)

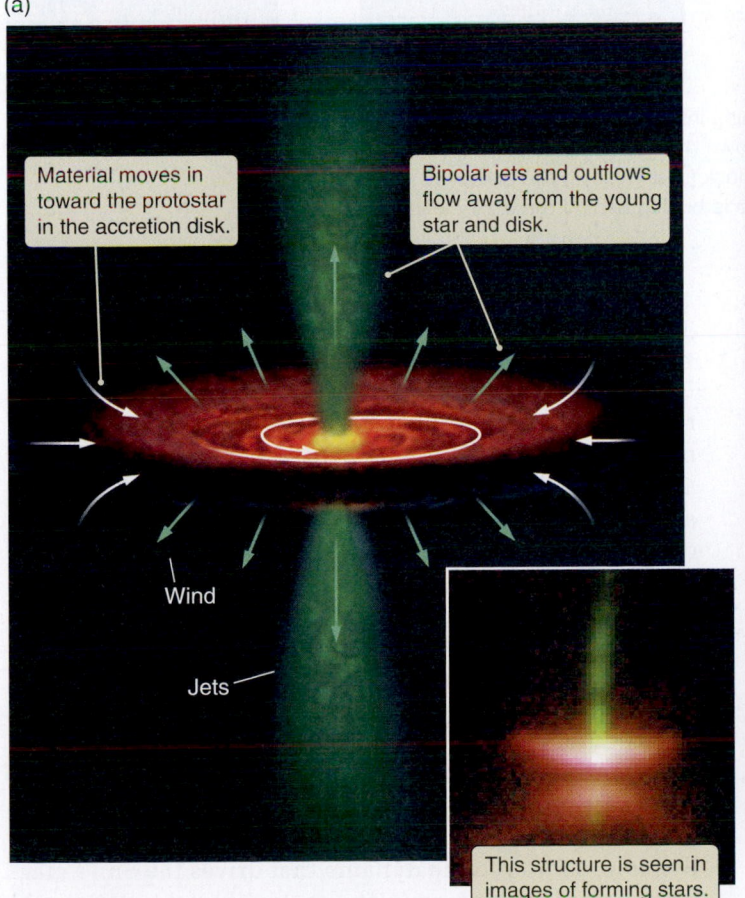

Material moves in toward the protostar in the accretion disk.

Bipolar jets and outflows flow away from the young star and disk.

Wind

Jets

This structure is seen in images of forming stars.

(b)

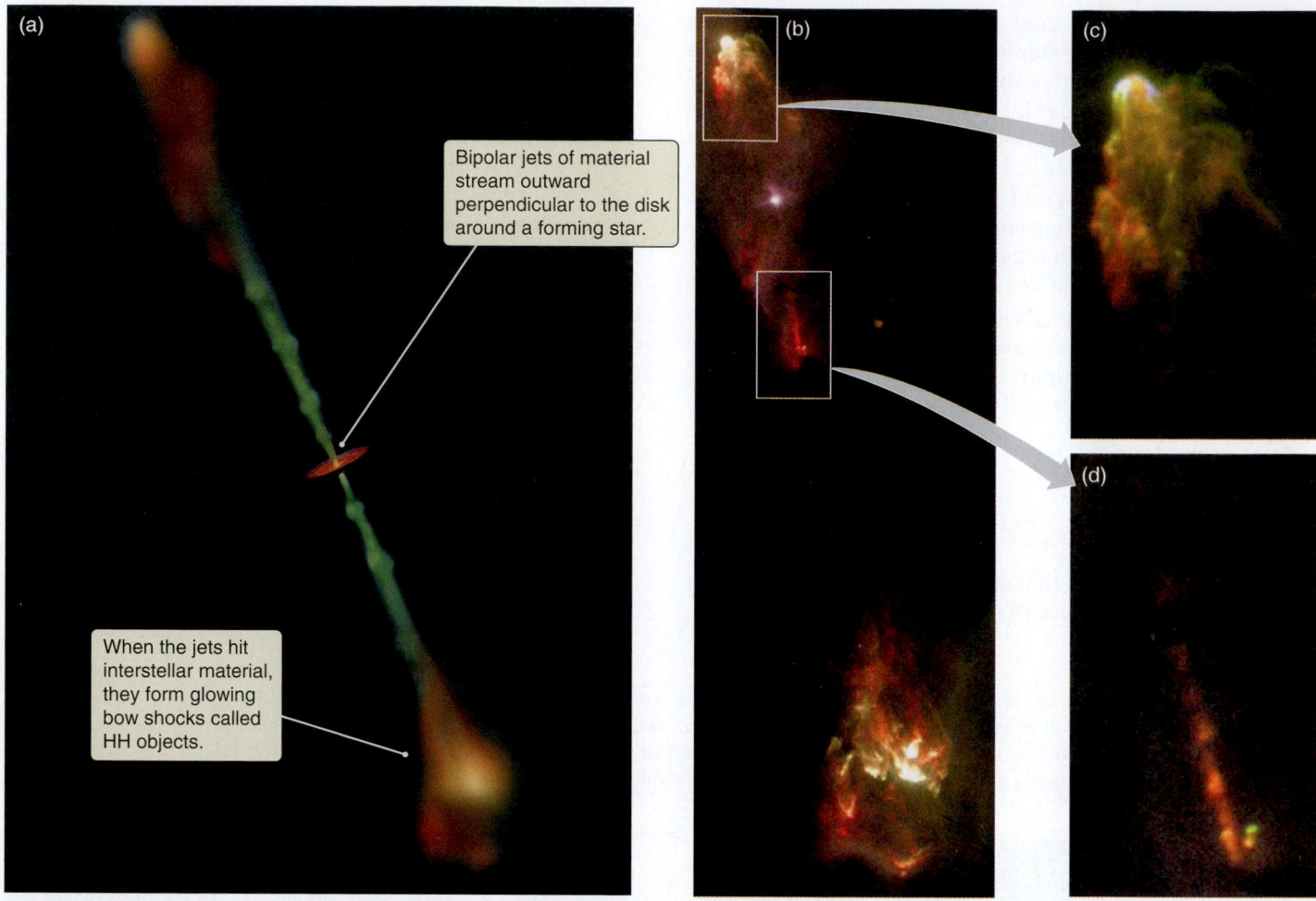

FIGURE 15.20 (a) Artist's view of jets from a protostar slamming into surrounding interstellar gas, heating the gas and causing it to glow. (b) This HST image shows the bow shocks formed at the ends of a bipolar jet from a protostar. Enlargements of the bow shock (c) and jet (d) are shown at right. Only one side of the jet itself is visible because the other side is hidden behind the dark cloud in which the star is embedded.

When we observe young stellar objects, we often find clear signs that material is falling onto the central protostar and accretion disk, as discussed earlier. However, it is also common to observe powerful flows of material moving *away* from forming stars at the same time that material is being accreted *onto* the stars. How can this be? As **Figure 15.19a** shows, material falls onto the accretion disk and moves inward toward the equator of the star, while at the same time other material is blown away from the protostar and disk in the two opposite directions away from the plane of the disk. The resulting stream of material away from the protostar is called a **bipolar outflow**.

Some bipolar outflows from young stellar objects are slow and fairly disordered, but others produce remarkable **jets** of material that move away from the central protostar and disk at velocities of hundreds of kilometers per second (**Figure 15.19b**). The

Protostars drive powerful bipolar outflows.

material in these jets flows out into the interstellar medium, where it heats, compresses, and pushes away surrounding interstellar gas. Knots of glowing gas accelerated by jets are referred to as **Herbig-Haro objects** (or **HH objects** for short), named after the two astronomers who first identified them and associated them with star formation. **Figure 15.20** shows a Hubble Space Telescope image of the first HH objects discovered: HH 1 and HH 2. These two HH objects are the two sides of the bipolar outflow from a single source.

The origin of outflows from protostars is not as well understood as we would like, but current models suggest that they are the result of magnetic interactions between the protostar and the disk. The interior of a protostar on its Hayashi track is convective. Great cells of hot gas are rising from the interior of the star, while other cells of cooler gas are falling toward the center. This convection, coupled with the protostar's rapid rotation, can lead to the formation of a dynamo, similar to the dynamo that drives the Sun's magnetic field. The dynamo in the center of a protostar would

(a)

(b)

FIGURE 15.21 If you have good eyes (or a pair of binoculars), on a clear winter night you can see the brightest of the stars in this cluster—a tight bunch in the constellation Taurus called the Pleiades or Seven Sisters. The diffuse blue light around the stars is starlight scattered by interstellar dust. (b) A Spitzer infrared image of the Pleiades shows more detail in the structure of the interstellar dust through which the cluster is traveling.

view of the protostar: the protostar is "revealed." Once the contracting protostar makes its appearance, it is referred to as a **T Tauri star**. This name comes from the first recognized member of this class of objects, the star labeled T in the constellation Taurus.

Astronomers have long known that stars are often found together in closely knit collections called **star clusters**. Figure 15.21 shows one such star cluster: a group called the Pleiades or Seven Sisters. Star clusters gave astronomers their first evidence that many stars of all different masses can form together at the same place and at about the same time. When we look around the galaxy at large, we see a hodgepodge of stars—some very old and others very young. If these were the only stars we had to study, it would be extremely difficult to learn much about how stars evolve. Star clusters, on the other hand, are large collections of stars that all formed *at the same time, in the same place, and from the same material*. They provide us with tailor-made samples to study star formation.

Even though the few brightest and most massive stars in a cluster dominate what we see, the vast majority of stars in a cluster are much less massive, like our Sun. In fact, some star-forming regions do not seem to form any especially massive stars at all. Astronomers are very interested in how and why molecular clouds subdivide themselves into low- and high-mass stars. The details of this division—specifically, what fraction of newly formed stars will be of what masses—are crucial if we are to use observations of the stars around us today to untangle the history of star formation in our galaxy. Unfortunately, we are still far from a detailed understanding of why some cloud cores become 1-$M_\odot$ stars while others become 10-$M_\odot$ stars.

After a cloud core collapse, the evolution of a protostar is determined largely by its mass. Calculations suggest that

be much more powerful than the Sun's dynamo, however. The protostar's resulting strong magnetic field might cause the protostar to begin blowing a powerful wind. It might also act something like the blade in a blender, tearing at the inner edge of the accretion disk and flinging material off into interstellar space.

Powerful protostellar winds, jets, and other outflows could disrupt the cloud core and accretion disk from which the protostar formed, shutting down the flow of material onto the protostar. Until the protostellar wind begins, the protostar is enshrouded in the dusty molecular-cloud core from which it was born. As the wind from the protostar disperses this obscuring envelope, we get our first direct, visible-light

a star with the mass of our Sun probably takes about 10 million years or so to descend its Hayashi track and become a star on the main sequence. If we look at the entire history, including the collapse and fragmentation of the molecular cloud itself, the total time is probably more like 30 million years. More massive stars go through this process much faster. A 10-$M_\odot$ star might go from the stage of being a molecular-cloud core to burning hydrogen in its interior in only 100,000 years. A 100-$M_\odot$ star might make the journey in less than 10,000 years. By comparison, a 0.1-$M_\odot$ star might take 100 million years to finally reach the main sequence.

> Star formation may take millions of years.

The 30 million years or so that it took for our Sun to form is a long time, but it is a tiny fraction of the 10 billion years during which the Sun will steadily fuse hydrogen into helium as a main-sequence star. It is no wonder that so few among the many stars we see in the sky are such young objects. But every star that we see was young *at one time*, including our own Sun.

Not surprisingly, many questions about star formation remain. For example, how must we modify our story to account for the formation of binary stars or other multiple-star systems? When we observe the sky, we find that about half of the stars we see are part of multiple-star systems.

At what point during star formation is it determined that a collapsing cloud core will form several stars instead of just one? Some models suggest that this split may happen early in the process, during the fragmentation and collapse of the molecular cloud. The advantage of these ideas is that they provide a natural way of dealing with much of the angular momentum of the cloud core: it goes into the orbital angular momentum of the stars about each other. Other models suggest that additional stars may form from the accretion disk around an initially single protostar.

> Collapsing cloud cores often form multiple stars.

The picture of star formation presented here is remarkably complete, considering that we have never visited a protostar or watched a star form. Instead, astronomers have observed many different stars at different stages in their formation and evolution, and then used their knowledge of physical laws to tie these observations together into a coherent, consistent description of how, why, and where stars form. This two-pronged attack—using observations to see what things exist in the universe, and using physics to understand how they work and the relationships between them—is how all astronomy (and in a certain sense all physical science) works.

Seeing the Forest for the Trees

We began our discussion of the Solar System in Chapter 6 by describing its formation. Nearly 5 billion years ago a vast interstellar cloud collapsed under its own weight to form a swirling disk of gas and dust. Within that disk, small objects stuck together to become parts of larger objects, culminating in the formation of the planets and other solid bodies of our Solar System. We now can see that earlier story—and in some sense our entire discussion of the Sun, Earth, and Solar System—as a "sidebar" to a larger story, the thread of which we picked up again on this leg of our journey. The tenuous expanse of gas and dust that fills the vast reaches of interstellar space sets the stage for the ongoing formation of generations upon generations of stars and planetary systems, of which ours is but one among thousands of billions of billions in the universe.

The interstellar medium is a difficult topic, even for experts in the field. The time that we have for our journey is limited. To go beyond the basic description of the interstellar medium here would be a long and arduous excursion. It would also quickly move on to more speculative ground. For now we suggest you think of the interstellar medium as you might a complex and subtle ecosystem. It is one thing to catalog the inhabitants of an ecosystem, but it is quite another to understand the interrelationships and dependencies among those organisms well enough to say that you know how the ecosystem works. In like fashion, astronomers at the turn of the 21st century have a fairly complete picture of what makes up the interstellar medium, but we do not fully understand the complex interplay between the phases of the interstellar medium and the stars that are born and die there. Even so, we have learned some important lessons and have developed some important tools. For example, 21-cm radiation from ubiquitous interstellar hydrogen—radiation that easily penetrates the dust that obscures our view of the universe at visible wavelengths—gives us a tool to see out to the far reaches of our galaxy and map out our home in the universe.

Fortunately, our understanding of stars is far more complete. In fact, the workings of stars represent such a nicely posed physics problem that we would probably include them on our journey even if this were a pure physics text rather than an astronomy text. Many of the physical principles employed in our discussion of star formation and the evolution of protostars in this chapter should have sounded familiar. The balance between gravity and pressure that gives a protostar its structure is the same balance at work within the Sun. The chain of physical reasoning that we used to understand the collapse of a protostar (gravitational energy that is converted to thermal energy and then radiated away) applies almost unmodified to understanding the ongoing (albeit slow) collapse of Jupiter.

As we go forward on our journey, the physical insight that we have developed will continue to serve us well. Stars are temporal objects. They are born from interstellar gas, they shine by the light of nuclear fires deep within their hearts, and when they exhaust their fuel they die. The changing balance of the protostar is only the first chapter in a process of evolution that continues throughout the star's life. We found it convenient to follow the changes taking place within an evolving protostar by tracking its progress across the face of the H-R diagram. In doing so, we discovered the best way that astronomers have found to draw a road map of a star's life. That road map will be our guide for the next stretch of our journey.

Summary

- The interstellar medium is a complex region, ranging from cold, relatively dense molecular clouds to hot, tenuous intercloud gas.

- Dust and gas in the interstellar medium blocks visible light but becomes more transparent at longer, infrared wavelengths.

- Neutral hydrogen cannot be detected at visible and infrared wavelengths, but it is revealed by its 21-cm microwave emission.

- Stars form in clusters within fragmented collapsing cores inside giant molecular clouds.

- Our Solar System began within a molecular-cloud core.

- Protostars glow with thermal energy released from converted gravitational energy as they collapse.

- A protostar lands on the main sequence when nuclear reactions begin in its core.

- A protostar must have at least 0.08 $M_\odot$ to become a true star.

- Brown dwarfs are neither stars nor planets, but something in between.

- What we know about the evolution of the birth of stars comes from observations of many protostars at various stages of their development.

SmartWork, Norton's online homework system, includes algorithmically generated versions of these questions, plus additional conceptual exercises. If your instructor assigns questions in SmartWork, log in at **smartwork.wwnorton.com.**

Student Questions

THINKING ABOUT THE CONCEPTS

1. The interstellar medium is approximately 99 percent gas and 1 percent dust. Why is it the dust and not the gas that blocks our visible-light view of the galactic center?

*2. In Chapter 13 we learned that we can measure the temperature of a star by comparing its brightness in blue and yellow light. Does reddening by interstellar dust affect the temperature we measure of a star? If so, how?

3. Using your own experience of walking or riding in a dense fog, explain how the existence of large amounts of interstellar dust may have influenced the views held by 19th century astronomers regarding the structure of our galaxy.

4. Explain how the development of infrared astronomy has made it possible for astronomers to study the detailed processes of star formation.

5. How does the material in interstellar clouds and intercloud gas differ in density and distribution?

6. When a star forms inside a molecular cloud, what happens to the cloud? Is it possible for the molecular cloud to remain cold and dark with one or more stars inside it? Explain your answer.

7. If you placed your hand in boiling water (100°C) for even one second, you would get a very serious burn. If you placed your hand in a hot oven (200°C) for a second or two, you would hardly feel the heat. Explain this difference and how it relates to million-kelvin regions of the interstellar medium.

8. What is the "local bubble" and how was it formed?

9. What is the primary source of heating of hot intercloud gas?

10. Interstellar gas typically cools by colliding with other gas atoms or grains of dust; during the collision, the atom loses energy and hence its temperature is lowered. How does this effect explain why tenuous gases are generally so hot, while dense gases tend to be so cold?

11. Stellar radiation can convert atomic hydrogen (H I) to ionized hydrogen (H II).
    a. Why does a B8 main-sequence star ionize far more interstellar hydrogen in its vicinity than does a K0 giant of the same luminosity?
    b. What properties of a star are important in determining whether it can ionize large amounts of nearby interstellar hydrogen?

12. When a hydrogen atom is ionized, a single particle becomes two particles.
    a. Identify the two particles.
    b. If both particles have the same kinetic energy, which moves faster?

*13. What causes a hydrogen atom to radiate a photon of 21-cm radio emission?

*14. Explain how the important discovery of 21-cm radio emission has enabled us to detect interstellar clouds of neutral hydrogen (H I), even when large amounts of interstellar dust are in the way.

15. Molecular hydrogen is very difficult to detect from the ground, but we can easily detect carbon monoxide (CO) by observing its 2.6-cm microwave emission. Describe how observations of CO might help astronomers infer the amounts and distribution of molecular hydrogen within giant molecular clouds.

16. Our galaxy contains several thousand giant molecular clouds. Describe a giant molecular cloud and the role it plays in star formation.

17. What two forces establish hydrostatic equilibrium in an evolving protostar?

18. As a cloud collapses to form a protostar, the forces of gravity felt by all parts of the cloud (which follow an inverse square law) become stronger and stronger. One might argue that under these conditions, the cloud should keep collapsing until it has zero size (a black hole). Why doesn't this happen?

*19. The internal structure of a protostar maintains hydrostatic equilibrium even as more material is falling onto it. Explain how this can be.

20. What determines whether a protostar will become a true star or a brown dwarf?

21. You can think of a brown dwarf as a failed star—that is, one lacking sufficient mass for nuclear reactions to begin. What similarities and differences do you see between a brown dwarf and a giant planet such as Jupi-

ter? Would you classify a brown dwarf as a supergiant planet? Explain your answer.

*22. The H^- ion acts as a thermostat in controlling the surface temperature of a protostar. Explain how.

*23. The Hayashi track is a nearly vertical (constant-temperature) evolutionary track on the H-R diagram. Why?

24. What do astronomers mean by *bipolar outflow*?

25. Describe a Herbig-Haro object.

26. What distinguishes a T Tauri star from a protostar?

27. Why are star clusters so important to our understanding of how stars of different mass evolve?

28. Of the many stars that astronomers see in the sky, why are so few of them protostars?

29. What is the single most important property of a star that will determine its evolution?

APPLYING THE CONCEPTS

30. Estimate the typical density of dust grains (grains per cubic centimeter) in the interstellar medium. A typical grain has a mass of about 10^{-17} kg. (Hint: You know the typical density of gas, and what fraction of the interstellar medium, by mass, is made of dust.)

31. Using Figure 15.2, estimate the blackbody temperature of the star as shown in part (b) (without dust) and part (c) (with dust). How significant are the effects of interstellar dust when we are using observed data to determine the properties of a star?

32. A typical temperature of intercloud gas is 8,000 K. Using Wien's law (see Chapter 4), calculate the wavelength at which this gas would radiate.

33. The mass of a proton is 1,850 times the mass of an electron. If a proton and an electron have the same kinetic energy, $E_K = \frac{1}{2} mv^2$, how many times greater is the velocity of the electron than that of the proton?

34. If a typical hydrogen atom in a collapsing molecular-cloud core starts at a distance of 10,000 astronomical units, or AU (1.5×10^{12} km) from the core's center and falls inward at an average velocity of 1.5 km/s, how many years does it take to reach the newly forming protostar? Assume that a year is 3×10^7 seconds.

35. From Table 13.1 we can see that the ratio of hydrogen atoms (H) relative to carbon atoms (C) in the Sun's atmosphere is approximately 2,400 to 1. It would be reasonable to assume that this ratio also applies to molecular clouds. If 2.6-cm radio observations indicate 100 $M_\odot$ of carbon monoxide (CO) in a giant molecular cloud, what is the implied mass of molecular hydrogen (H_2)

in the cloud? (Carbon represents $\frac{3}{7}$ of the mass of a CO molecule.)

36. Neutral hydrogen emits radiation at a radio wavelength of 21 cm when an atom drops from a higher-energy magnetic state to a lower-energy magnetic state. On average, each atom remains in the higher-energy state for 11 million years (3.5×10^{14} seconds).
    a. What is the probability that any given atom will make the transition in 1 second?
    b. If there are 6×10^{59} atoms of neutral hydrogen in a 500-$M_\odot$ cloud, how many photons of 21-cm radiation will the cloud emit each second?
    c. How does this number compare with the 1.8×10^{45} photons emitted each second by a solar-type star?

37. The Sun took 30 million years to evolve from a collapsing cloud core to a star, with 10 million of those years spent on its Hayashi track. It will spend a total of 10 billion years on the main sequence. Suppose we were to compress the Sun's main-sequence lifetime into just a single day.
    a. How long would the total collapse phase last?
    b. How long would the Sun spend on its Hayashi track?

*38. Assume that the Sun, starting out as a protostar, had a diameter 100 times that of the present-day Sun and a surface temperature of 3,300 K.

a. Compare the surface area of the protostellar Sun with today's Sun.
b. Using the Stefan-Boltzmann law, compare the luminosity of the protostellar Sun with today's Sun. (Note: The surface temperature of the Sun is 5,780 K.)

*39. A protostar with the mass of the Sun starts out with a temperature of about 3,500 K and a luminosity about 200 times larger than the Sun's current value. Estimate this protostar's size and compare it to the size of the Sun today.

40. The star-forming region 30 Doradus is 160,000 light-years away in the Large Magellanic Cloud and appears about $\frac{1}{6}$ the brightness of the faintest stars visible to the naked eye. If it were located at the distance of the Orion Nebula (1,300 light-years) how much brighter than the faintest visible stars would it appear?

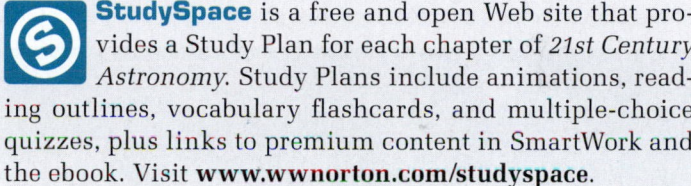

StudySpace is a free and open Web site that provides a Study Plan for each chapter of *21st Century Astronomy*. Study Plans include animations, reading outlines, vocabulary flashcards, and multiple-choice quizzes, plus links to premium content in SmartWork and the ebook. Visit **www.wwnorton.com/studyspace**.

It is said an Eastern monarch once charged his wise men to invent him a sentence to be ever in view, and which should be true and appropriate in all times and situations. They presented him the words "And this, too, shall pass away."

ABRAHAM LINCOLN (1809–1865), SEPTEMBER 30, 1859

Shells of dust reflect successive outbursts of the supergiant star V838 Monocerotis.

CHAPTER 16

Stars in the Slow Lane— Low-Mass Stellar Evolution

16.1 This, Too, Shall Pass Away

If you go outside tonight and look up, you will not see the same sky that you would have seen a year ago. The stars themselves have not changed, but *the way you see them* has changed. A year ago the sky was probably full of points of light, called stars, that were "somewhere off in space." Now you see an expanse of distant suns, each farther away than the mind can easily comprehend, and each shining with a devastating brilliance powered by a nuclear inferno deep in its heart. When you look at the dark splotches that mar the Milky Way's eerie glow, you see vast clouds of interstellar gas and dust. In your mind's eye you look inside these clouds at the new suns and new solar systems being born there, and imagine the birth of our own planet and the star we orbit, 4.6 billion years ago. The sky is like a page from a book. To those who cannot read, written words are so many chicken scratches; but to those who can, thoughts and words and ideas spring off the page to touch our hearts and stimulate our minds and imaginations. The sky is a book worth reading, as is the rest of nature. At this stage in our journey we are well on our way to becoming literate.

So far on our journey, we have seen interstellar clouds collapse under the force of gravity to form immense protostars surrounded by swirling disks of gas and dust. We have watched as dust left behind in the disk accumulates into planets and moons, asteroids and comets. We have followed along as protostars continue their collapse until the nuclear fuel in their cores ignites, and we have recognized a grand pattern—the main sequence—in the stars that those protostars go on to become. We have even looked

KEY CONCEPTS

Within its core, the Sun fuses over 4 billion kilograms of hydrogen to helium each second; and although the Sun may seem immortal by human standards, eventually it will run out of fuel. When it does, some 5 billion years from now, the Sun's time on the main sequence will come to an end. As we examine what happens when a low-mass star like the Sun nears the end of its life, we will find that

- The more massive a star is, the shorter is its lifetime.
- When the Sun runs out of fuel at its center, it will grow into a larger, more luminous red giant star.
- We can follow the post–main sequence evolution of stars by tracing their paths on the H-R diagram.
- Red giants, and some other evolved stars, are shaped by dense "degenerate" cores in which atoms have been crushed by gravity.
- Evolving low-mass stars go through a series of stages, eventually burning helium to carbon and building up a core of carbon ash.
- In the end, a low-mass star will eject its outer layers, possibly forming a planetary nebula, and leaving behind a tiny degenerate white dwarf.
- Low-mass stars in binary systems may experience more exciting fates as novae or supernovae.

inside one of these stars, our Sun, and come to appreciate the battle between gravity and pressure that gives our Sun its structure. Our understanding of that battle provides our understanding of all stars along the main sequence. A star's mass determines the strength of its gravity, and the need to balance that gravity in turn determines the pressure in the star's interior. The more massive the star, the higher the pressure that is needed to hold it up, and the more rapidly the star must burn its nuclear fuel to support its own weight. From luminous O stars on the hot end of the main sequence to faint M stars on the cool end, mass is the fundamental and overriding property that makes a main-sequence star what it is.

We have seen stars born and seen how they live. Now the time has come to watch them die. Like all main-sequence stars, the Sun gets its energy by converting hydrogen to helium in its core. This is what *defines* the main sequence: being on the main sequence *means* that a star is burning hydrogen in its core. But a star cannot remain a main-sequence star forever. Hydrogen at the core of a star is a consumable resource. Any star eventually exhausts this resource—it "runs out of gas"—and when it does, its structure begins to change dramatically. Just as the balance between pressure and gravity within a protostar constantly changes as it descends the Hayashi track toward the main sequence, new balances must also constantly be found as a star evolves beyond the main sequence, until at last no balance is possible at all.

> Stars eventually exhaust their nuclear fuel.

The mass and composition of a star control the star's life on the main sequence, and they remain at center stage in the closing acts of the star's story as well. The evolutionary course followed by each of the hundred billion or so stars that make up the Milky Way Galaxy (and each of the stars in every other galaxy throughout the universe) is locked in place when the star forms, determined foremost by the seemingly incidental fact of the amount of mass incorporated into the star at the time of its birth, and secondarily by the chemical composition of the material from which it formed.

> Mass and composition determine a star's fate.

Each star is unique. Relatively minor differences in the masses and chemical compositions of two stars can sometimes result in significant, and possibly even dramatic, differences in their fates. The course followed by a star with a mass of 1.1 times the mass of the Sun (1.1 $M_\odot$) is not identical to the fate of a star with a mass of 0.9 $M_\odot$. Nevertheless, stars can be divided roughly into two broad categories whose members evolve in qualitatively different ways. Massive, luminous O and B stars follow a course fundamentally different from that of the cooler, fainter, less massive stars found toward the lower right end of the main sequence (see Figure 13.17). These

> Low-mass and high-mass stars evolve differently.

stars, which have masses less than about 3 $M_\odot$, are referred to as **low-mass stars** and are typified by our Sun. In this chapter we begin our discussion of stellar evolution by examining the stages through which low-mass stars progress. What better place to start than by asking what fate awaits our own Sun?

16.2 The Life and Times of a Main-Sequence Star

In Chapter 14 we learned that the structure of the Sun is determined by a balance between the inward force of gravity and the outward force of pressure. The pressure within the Sun is, in turn, maintained by energy released by nuclear fusion in the heart of the star. If you were to add mass to the Sun, the weight of material pushing down on the inner regions of the star would increase. Gravity would gain an advantage. The inner parts of the Sun would be compressed by the added weight, driving up the temperature and density there. This increase in temperature and density would in turn accelerate the pace of the nuclear reactions occurring there.

We have followed this chain of reasoning before, but it is so crucial to what is to come that it is worth reviewing here. Increasing the temperature and pressure at the center of a star means several things. For one, it means that more atomic nuclei are packed together into a smaller volume. You know that you are far more likely to bump into another person while strolling through a crowded shopping mall than through an empty park. Similarly, packing atomic nuclei more tightly together increases the likelihood that they will run into each other and fuse. Higher density means more frequent collisions between atomic nuclei, and a higher number of collisions means faster burning. Higher temperature also drives up the rate of nuclear reactions: higher temperature means that atomic nuclei are moving faster, so they are more likely to bump into each other. More important, higher temperature means that atomic nuclei collide with each other more *violently*, making it more likely they will overcome the electric repulsion that pushes the positively charged nuclei apart. As a result of the combined effects of temperature and density, modest increases in pressure within a star can sometimes lead to dramatic increases in the amount of energy released by nuclear burning.

> Increases in temperature and pressure speed up nuclear burning.

Here is the key to understanding why the main sequence is primarily a sequence of masses, with low-mass stars on the faint end and high-mass stars on the luminous end. More mass means stronger gravity, stronger gravity means higher temperature and pressure in the star's interior, higher temperature and pressure mean faster nuclear reactions, and faster nuclear reactions mean a more luminous star. If the Sun were more massive, it would necessarily have a different balance between gravity and pressure—a balance in which the Sun would burn its nuclear fuel more rapidly and thus would be more luminous. In other words, if the Sun were more massive, it would be located at a different position on the H-R diagram: farther up and to the left on the main sequence. Mass determines the structure of a star and its place on the main sequence. This, in a nutshell, is the heart of our understanding of stellar structure.

Higher Mass Means a Shorter Lifetime

A main-sequence star can live only so long. But how long is long? The question "How long can a main-sequence star continue to burn hydrogen in its core?" is much like the question "How long can you drive your car before it runs out of gas?" In the case of the car, the answer depends in part on how much gas your tank holds. The larger the gas tank, the more fuel you have and the longer your motor might run. But the answer also depends on the size and efficiency of your motor. A gas-guzzling eight-cylinder SUV drinks fuel a lot faster than a motor scooter. The amount of time your motor will run is determined by a competition between these two effects. The larger motor might run out of gas first, even if it is attached to a much larger tank.

The competition between these two effects—tank size and motor size—is most readily expressed as a ratio. How long your motor runs is given by the amount of gas in the tank, divided by how quickly the motor uses it:

$$\text{Lifetime of tank of gas (hours)} = \frac{\text{Amount of fuel (gallons)}}{\text{Rate at which fuel is used (gallons/hour)}}.$$

If you have a 15-gallon tank and your motor is burning fuel at a rate of 3 gallons each hour, then your motor will use up all of the gas in just 5 hours.

The same principle works for main-sequence stars. The amount of fuel is determined by the mass of the star. The more massive the star is, the more hydrogen is available to power nuclear burning. The rate at which fuel is used is measured by the luminosity of the star. Main-sequence stars are "in balance," so energy is radiated into space from the surface of the star at the same rate at which energy is being generated in its core. (This balance between energy generation and luminosity remains true at almost every stage of a star's evolution.) If one main-sequence star has twice the luminosity of another, then it must be burning hydrogen at twice the rate of the other star.

> How quickly a star runs out of fuel depends on its mass and luminosity.

An expression for the **main-sequence lifetime** of the star looks very similar to our previous expression for the time it takes your car to run out of fuel:

$$\text{Lifetime of star} = \frac{\text{Amount of fuel } (\propto \text{ mass of star})}{\text{Rate fuel is used } (\propto \text{ luminosity of star})}.$$

As **Math Tools 16.1** shows us, this relationship among mass, luminosity, and lifetime indicates that lower-mass stars will have much longer lives than their higher-mass cousins.

The lifetimes of stars should be of more than passing interest to us. We don't yet know if life exists on planets orbiting other stars, but we might assume that life would be unlikely to evolve on a planet orbiting a massive star with a stable life of only a few million years (see Chapter 23). Likewise, it would be unlikely for life to develop on a planet surrounding a very low-mass star that had yet to initiate the stable nuclear burning phase.

The Structure of a Star Changes As It Uses Its Fuel

To say that a main-sequence star is stable is not to say that it does not continually change. When the Sun formed, about 90 percent of its atoms were hydrogen atoms. Since then, the Sun has produced its energy by converting hydrogen into helium via the proton-proton chain. As the composition of a star changes, so must its structure. When we discussed the collapse of a protostar toward the main sequence in Chapter 15, we considered the idea of a changing balance between gravity and pressure. The protostar was always in balance; but as the star radiated away thermal energy, this balance was constantly changing, shifting toward that of a smaller and denser object.

The same concept applies here. As a main-sequence star uses the fuel in its core, its structure must continually shift in response to the changing core composition. At any given point in its lifetime, a main-sequence star like the Sun is in balance, but the balance in the Sun today is slightly different from the balance the Sun had 4 billion years ago, and slightly different from the balance it will have 4 billion years from now. Between the time the Sun was born and the time it will leave the main sequence, its luminosity will roughly double, with most of this change occurring during the last billion years of its life on the main sequence. Stars evolve even as they "sit" on the main sequence, although this evolution is slow and modest in comparison with the events that follow.

> Helium ash builds up in the core of a main-sequence star.

What happens to the helium that is produced in the core of a low-mass main-sequence star like our Sun? We might imagine that it begins to burn, with helium atoms fusing to form heavier elements; but such is not the case. For fusion to occur, atomic nuclei must be slammed together with enough energy to overcome the electric repulsion between them. For two helium nuclei to get close enough to fuse, the two protons in one nucleus must overcome the repulsive force of the two protons in the other nucleus—four times the repulsive force of one hydrogen nucleus acting on another hydrogen nucleus. At the temperature found at the center of a

MATH TOOLS 16.1

Estimating Main-Sequence Lifetimes

Astronomers can determine the lifetime of main-sequence stars either observationally or by modeling the evolution of stars of a given composition. One simple method is to employ what we might call a "rule-of-thumb approach." If we use what we know about how much hydrogen must be converted into helium each second to produce a given amount of energy, as well as the fraction of its hydrogen that a star burns, we can come up with a relationship that says the main-sequence lifetime of a star, τ_{MS}, can be expressed as

$$\tau_{MS} \propto \frac{M_{MS}}{L_{MS}},$$

where M is mass (amount of fuel) and L is luminosity (the rate fuel is used). We can put this relationship in quantitative terms by introducing the constant of proportionality, 1.0×10^{10}, which is the computed lifetime (in years) of a 1-$M_\odot$ star:

$$\tau_{MS} = (1.0 \times 10^{10}) \times \frac{M_{MS}/M_\odot}{L_{MS}/L_\odot} \text{ years.}$$

Now, let's see how the lifetime of a more massive star compares with that of the Sun. The relationship between the mass and the luminosity of stars is very sensitive. Relatively small differences in the masses of stars result in large differences in their main-sequence luminosities. One method for estimating luminosities of main-sequence stars is known as the **mass-luminosity relationship**, $L \propto M^{3.5}$, which is based on observed luminosities of stars of known mass,[1] as illustrated in **Figure 16.1**. As in the preceding example, we can express this relationship relative to the Sun's mass and luminosity:

$$\frac{L_{MS}}{L_\odot} = \left(\frac{M_{MS}}{M_\odot}\right)^{3.5}$$

Substituting the mass-luminosity relationship into the lifetime equation gives us

$$\tau_{MS} = (1.0 \times 10^{10}) \times \frac{M_{MS}/M_\odot}{(M_{MS}/M_\odot)^{3.5}} = (1.0 \times 10^{10}) \times \left(\frac{M_{MS}}{M_\odot}\right)^{-2.5} \text{ years.}$$

[1]The exponent 3.5 is not a hard-and-fast number. Observational evidence shows that it can vary from 2.5 to 5.0, depending on the mass of the star (see Figure 16.1).

As an example, let's look at a main-sequence O5 star. According to **Table 16.1**, which is based on computer models, an O5 star has a mass that is about equal to 60 times that of the Sun:

$$\tau_{O5} = (1.0 \times 10^{10}) \times (60)^{-2.5} = 3.6 \times 10^5 \text{ years.}$$

Instead of the 10-billion-year life span of the Sun, an O5 star has a main-sequence lifetime of *less than a half-million years*! Several generations of O stars have lived and died in the time that hominids have walked the surface of Earth. Even though the 60-$M_\odot$ star starts out with 60 times as much fuel as the Sun, it burns that fuel so much faster that it uses it up in less than a ten-thousandth of the Sun's lifetime. While low-mass stars live their lives in the slow lane, high-mass stars live fast and die young.

FIGURE 16.1 The mass-luminosity relationship for main-sequence stars: $L \propto M^{3.5}$. The exponent (3.5) is an average value over the wide range of main-sequence star masses. Observational data show that the deviation of stars from the average relationship depends on their mass.

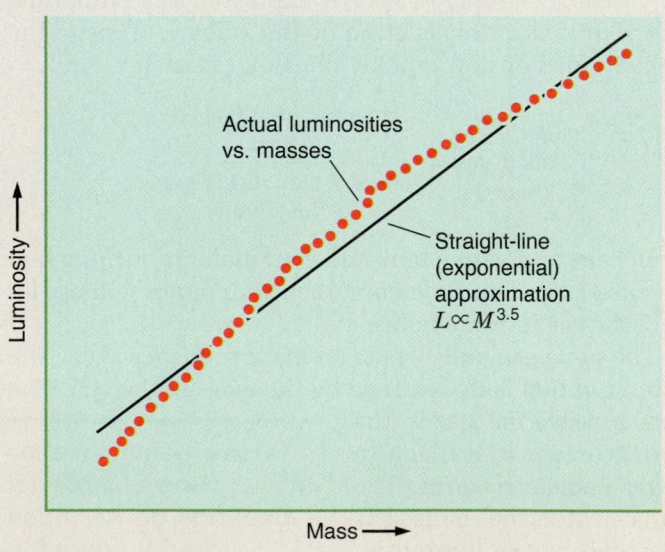

low-mass main-sequence star, collisions are not energetic enough to overcome the electric repulsion between helium nuclei.[2] As hydrogen burns in the core of a low-mass star,

the resulting helium collects there, building up like the nonburning ash in the bottom of a fireplace.

[2]You may be puzzled, remembering from Chapter 14 that one step of the proton-proton chain is the fusion of two ^{3}He nuclei, which have as strong an electric repulsion as two ^{4}He nuclei do. The answer is

that the strong-nuclear-force interaction between the two ^{3}He nuclei is much more powerful than the strong-nuclear-force interaction between two ^{4}He nuclei, so two ^{3}He nuclei do not have to get as close together as do two ^{4}He nuclei in order to fuse.

TABLE 16.1

Main-Sequence Lifetimes*

Spectral Type	Mass ($M_\odot$)	Luminosity ($L_\odot$)	Main-Sequence Lifetime (Years)
O5	60	794,000	3.6×10^5
B0	17.5	52,500	1.0×10^7
B5	5.9	832	7.2×10^7
A0	2.9	54	3.9×10^8
A5	2.0	14	1.1×10^9
F0	1.6	6.5	2.1×10^9
F5	1.3	3.2	3.5×10^9
G0	1.05	1.5	8.3×10^9
G2 (our Sun)	1.0	1.0	1.0×10^{10}
G5	0.92	0.8	1.5×10^{10}
K0	0.79	0.4	3.7×10^{10}
K5	0.67	0.15	5.3×10^{10}
M0	0.51	0.077	$6.6 \times 10^{10}*$
M5	0.21	0.011	$1.9 \times 10^{11}*$
M8	0.06	0.0012	$5.0 \times 10^{12}*$

*Lifetimes marked with an asterisk (*) are based on the "rule of thumb" discussed in Math Tools 16.1. The other lifetimes are based on computer models.

Helium Ash Builds Up in the Center of the Star

Helium does not build up at the same rate throughout the interior of a star. In Chapter 7 we found that the temperature and pressure within Earth *must* be highest at the center of the planet. No other configuration makes sense. Exactly the same arguments apply equally well to stars. The fact that the temperature and pressure are highest at the center of a main-sequence star means that hydrogen burns most rapidly there as well. As a result, nonburning helium "ash" accumulates most rapidly at the center of the star.

If we could cut a star open and watch as it evolves, we would see the chemical composition of the star changing most rapidly at its center and less rapidly as we move outward through the star. **Figure 16.2** shows how the chemical composition inside a star like the Sun changes throughout its main-sequence lifetime. When the Sun formed, it had a uniform composition throughout, with hydrogen accounting for about 70 percent of the mass in the Sun and helium accounting for most of the remaining 30 percent. With time, the helium fraction in the center of the Sun climbed. Today, roughly 5 billion years later, only about 35 percent of the mass at the center of the Sun is hydrogen.

16.3 A Star Runs Out of Hydrogen and Leaves the Main Sequence

Hydrogen burning in the core of a star cannot continue forever. Eventually—about 5 billion years from now in the case of the Sun—a star exhausts all of the hydrogen fuel available at its center. At this point, the innermost core of the

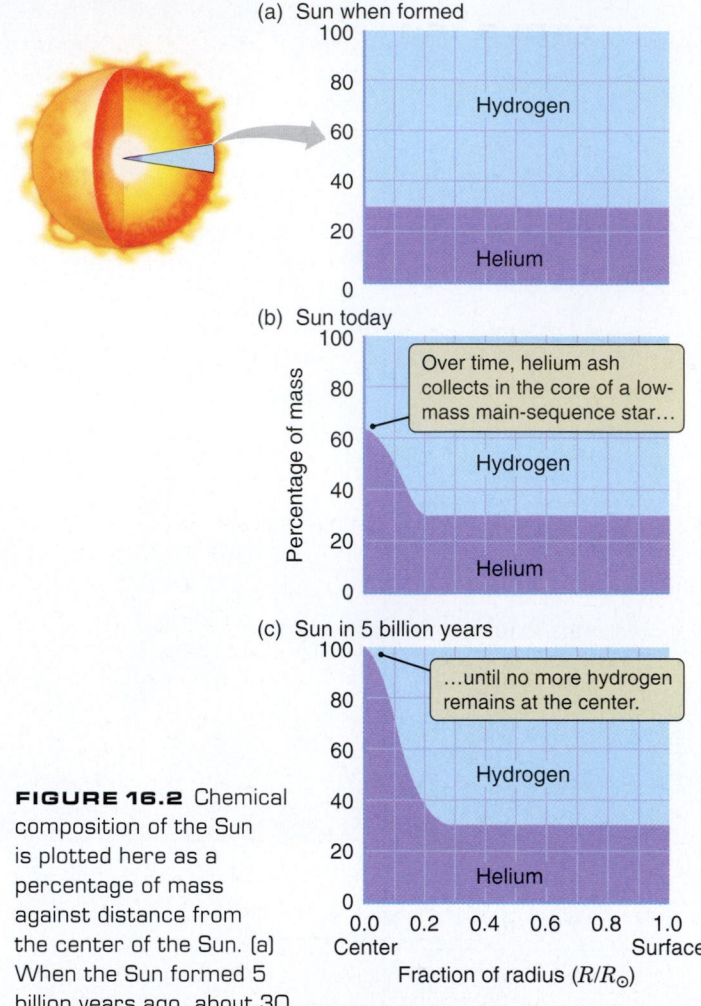

(a) Sun when formed

(b) Sun today

Over time, helium ash collects in the core of a low-mass main-sequence star...

(c) Sun in 5 billion years

...until no more hydrogen remains at the center.

FIGURE 16.2 Chemical composition of the Sun is plotted here as a percentage of mass against distance from the center of the Sun. (a) When the Sun formed 5 billion years ago, about 30 percent of its mass was helium and 70 percent was hydrogen throughout. (b) Today the material at the center of the Sun is about 65 percent helium and 35 percent hydrogen. (c) The Sun's main-sequence life will end in about 5 billion years, when all of the hydrogen at the center of the Sun is gone.

star is composed entirely of helium ash. As thermal energy leaks out of the helium core into the surrounding layers of the star, no more energy is generated within the core to replace it. The balance that has maintained the structure of the star throughout its life is now broken. The star's life on the main sequence has come to an end.

The Helium Core Is Degenerate

Throughout our lives, all of the matter we directly experience is mostly *empty space*. Just as the Solar System is mostly empty except for the tiny bit of space occupied by the Sun and the planets, an atom is mostly empty except for the tiny bit of space occupied by the nucleus and the electrons. The same is true for the matter within the Sun. At the enormous temperatures within the Sun the electrons have almost all been stripped away from their atoms by

energetic collisions. (In other words, the gas is completely ionized.) So the gas inside the Sun is a mixture of electrons and atomic nuclei all flying about freely. Even so, the gas that makes up the Sun is still mostly empty space, with the electrons and atomic nuclei filling only a tiny fraction of the volume.

When a low-mass star like the Sun exhausts the hydrogen at its center, the situation changes. As gravity begins to win its shoving match against pressure, the helium core is crushed to an ever-smaller size and an ever-greater density, but there is a limit to how dense the core can get. The rules of quantum mechanics (the same rules that say that atoms can have only certain discrete amounts of energy and that light comes in packets called photons) limit the number of electrons that can be packed into a given volume of space at a given pressure.[3] As the matter in the core of the star is compressed further and further, it finally bumps up against this limit. The space occupied by the core of the star is no longer mostly empty, but is now effectively "filled" with electrons that are smashed tightly together. Matter at the center of the star is now so dense that a single cubic centimeter of this material can have a mass of a metric ton (1,000 kilograms [kg]) or more. Matter that has been compressed to this point is said to be **electron-degenerate**.

> The crushed helium core is electron-degenerate.

Hydrogen Burns in a Shell Surrounding a Core of Helium Ash

Once a low-mass star exhausts the hydrogen at its center, nuclear burning may end there, but the story is very different outside the core. The layers surrounding the degenerate core still contain hydrogen, and this hydrogen continues to burn. Astronomers speak of **hydrogen shell burning** because the hydrogen is burning in a shell surrounding a core of helium, like the flesh of a fruit around its pit.

The electron-degenerate core of a star has a number of fascinating properties. For example, as more and more helium ash piles up on the degenerate core, the core *shrinks* in size. It does so because the added mass increases the strength of gravity and therefore the weight bearing down on the core, which means that the electrons can be smashed together into a smaller volume. The presence of the degenerate core triggers a chain of events that will dominate the evolution of our 1-$M_\odot$ star for the next 50 million years. Follow along as we step through the chain of cause and effect that takes center stage in the post–main sequence evolution of a low-mass star:

---

[3]If you have taken high school or college chemistry, you may remember the "Pauli exclusion principle," which limits the number of electrons that can go into a single orbital in an atom. This principle also limits the number of electrons that can be packed into the energy states available in the center of a star.

1. From our discussion in Chapter 10 (see Figure 10.3), we know that just outside the star's degenerate core, the gravitational acceleration g_{core} is given by

$$g_{core} = \frac{GM_{core}}{r^2_{core}},$$

where M_{core} is the mass of the helium core and r_{core} is its radius. As the helium core grows, its larger mass (bigger M_{core}) and its shrinking size (smaller r_{core}) both cause the strength of gravity at the surface of the core to increase. As more helium is added to the core, the strength of gravity at its surface increases dramatically.

2. Increasing the strength of gravity around the core increases the weight of the overlying material pushing down on the hydrogen-burning shell surrounding the core.

3. This increase in weight must be balanced by an increase in pressure in the inner parts of the star. In particular, the pressure in the hydrogen-burning shell must increase.

4. Increasing the pressure in the hydrogen-burning shell drives up the rate of the nuclear reactions occurring in the shell.

5. Faster nuclear reactions release more energy, so the luminosity of the star increases.

This is a very counterintuitive result! We might have imagined that when a star like the Sun used up the nuclear fuel at its center, it would grow fainter. Yet just the opposite happens. A degenerate core means stronger gravity, stronger gravity means higher pressure, higher pressure means faster nuclear burning, and faster nuclear burning means a more luminous star. When the low-mass star "runs out of gas" at its center, it responds by getting more luminous!

> Running out of fuel makes the star grow more luminous.

Tracking the Evolution of the Star on the H-R Diagram

The changes that occur in the heart of a star with a degenerate helium core are reflected in changes in the overall structure of the star. With time, the mass of the degenerate helium core grows as more and more hydrogen is converted into helium ash in the surrounding shell. And as the mass of the degenerate helium core grows, so does the rate of energy generation in the surrounding hydrogen-burning shell. This increase in energy generation heats the overlying layers of the star, causing them to expand. The star becomes a bloated, luminous giant. As illustrated

> A bloated luminous giant surrounds a tiny degenerate core.

in **Figure 16.3**, the internal structure of the star is now fundamentally different from the structure when the star was on the main sequence. The giant can grow to have a luminosity hundreds of times the luminosity of the Sun and a radius of over 50 solar radii (50 $R_{\odot}$). Yet at the same time the core of the giant star is far more compact than that of the Sun, with much of the star's mass becoming concentrated into a volume that is only a few times the size of Earth.

From our vantage point outside the star, we are not privy to the changes taking place deep within its interior. All we can see directly is that the star becomes larger and more luminous, and perhaps surprisingly, *cooler and redder* as well. The enormous expanse of the star's surface allows it to cool very efficiently. Even though its *interior* grows hotter and its luminosity increases, the *surface* temperature of the star actually begins to drop.

Just as the H-R diagram helped us follow the changes in a protostar on its way to the main sequence, it is a handy device for keeping track of the changing luminosity and surface temperature of the star as it evolves away from the main sequence (**Figure 16.4**). As soon as the star exhausts the hydrogen in its core, it leaves the main sequence and begins to move upward and to the right on the H-R diagram, growing more luminous but cooler. We refer to such a star, which is somewhat brighter and larger than it was on the main sequence, as a **subgiant** star. As the subgiant continues to evolve, it grows larger and cooler, but after a time its progress to the right on the H-R diagram hits a roadblock: the H⁻ thermostat. When the surface temperature of the subgiant star has dropped by about 1,000 kelvins (K) relative to its temperature on the main sequence, H⁻ ions start to form in great abundance in its atmosphere. We have encountered the formation of H⁻ ions before. In our discussion of protostars in Chapter 15, it was the H⁻ ion in the protostar's atmosphere that acted as a thermostat, regulating the star's temperature. The H⁻ ion serves exactly the same role here, regulating how much radiation can escape from the star and preventing it from becoming any cooler.

> An evolving low-mass star moves up and to the right on the H-R diagram.

Because the star can cool no further, it begins to move almost vertically upward on the H-R diagram, growing larger and more luminous but remaining about the same temperature. The star has become a **red giant**—an obvious name for a star that is now both redder and larger than it was on the main sequence. We can think of the path that a star follows on the H-R diagram as it leaves the main sequence as being a tree "branch" growing out of the "trunk" of the main sequence. Astronomers refer to these tracks as the **subgiant branch** and the **red giant branch** of the H-R diagram. Interestingly, the path that a red giant follows on the H-R diagram closely parallels the path that it followed earlier as a collapsing protostar on its way toward the main sequence—except, of course, in reverse: this time the star is moving up that path rather than coming down it. This similarity is not a coincidence. The same physical processes

FIGURE 16.3 The structure of a star near the top of the red giant branch is compared with the structure of the Sun. Left panels compare the size of the Sun with the size of the red giant. Right panels compare the size and structure of the Sun with the core of the red giant. The panels at right are magnified about 50 times compared to the panels on the left.

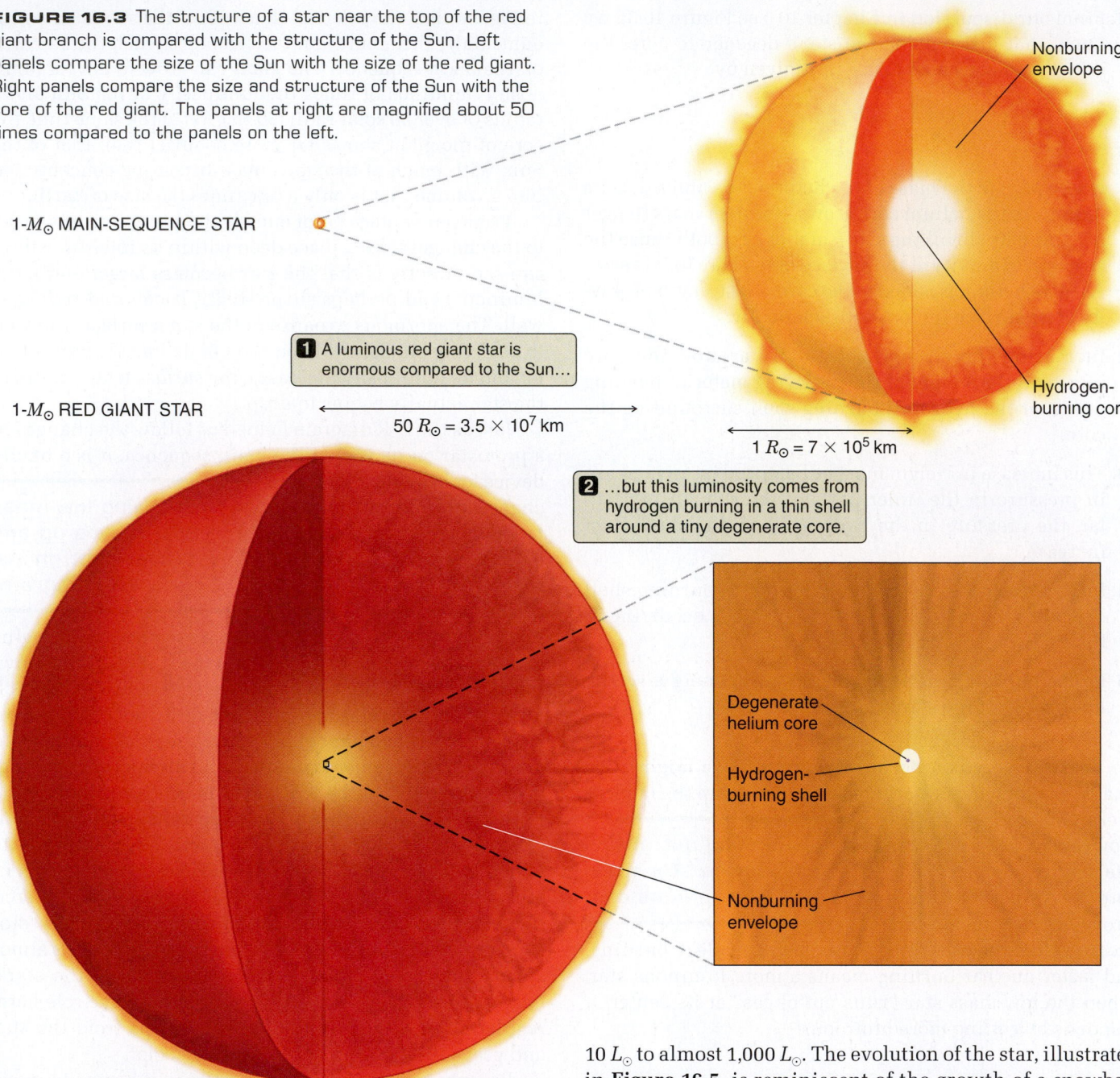

1-$M_\odot$ MAIN-SEQUENCE STAR

Nonburning envelope

1-$M_\odot$ RED GIANT STAR

1 A luminous red giant star is enormous compared to the Sun…

50 $R_\odot$ = 3.5 × 10^7 km

Hydrogen-burning core

1 $R_\odot$ = 7 × 10^5 km

2 …but this luminosity comes from hydrogen burning in a thin shell around a tiny degenerate core.

Degenerate helium core

Hydrogen-burning shell

Nonburning envelope

(such as the H⁻ thermostat) that give rise to the Hayashi track followed by a collapsing protostar also control the relationship of luminosity, size, and surface temperature in an expanding red giant.

As the star leaves the main sequence, the changes in its structure occur sluggishly at first, but then they pick up steam as the star moves up the red giant branch faster and faster. It takes 200 million years or so for a star like the Sun to go from the main sequence to the top of the red giant branch. Roughly the first half of this period is spent on the subgiant branch as the star's luminosity increases to about 10 times the luminosity of the Sun ($L_\odot$). During the second half of this time the star's luminosity skyrockets from

10 $L_\odot$ to almost 1,000 $L_\odot$. The evolution of the star, illustrated in **Figure 16.5**, is reminiscent of the growth of a snowball rolling downhill. The larger the snowball becomes, the faster it grows; and the faster it grows, the larger it becomes. "Growth" and "size" feed off each other, and what began as a bit of snow at the top of the mountain soon becomes a huge ball that could trigger an avalanche.

The analogy between the evolution of a red giant star and the growth of a snowball is actually not a bad one. The helium core of the star grows in mass (but not in radius!) as hydrogen is converted to helium in the hydrogen-burning shell. The increasing mass of the ever–more compact helium core drives up the force of gravity in the heart of the star. Stronger gravity means higher pressure, and higher pressure accelerates nuclear burning in the shell. But faster nuclear reactions in the shell convert hydrogen into helium more quickly, so the part of the star's mass that is in its core grows

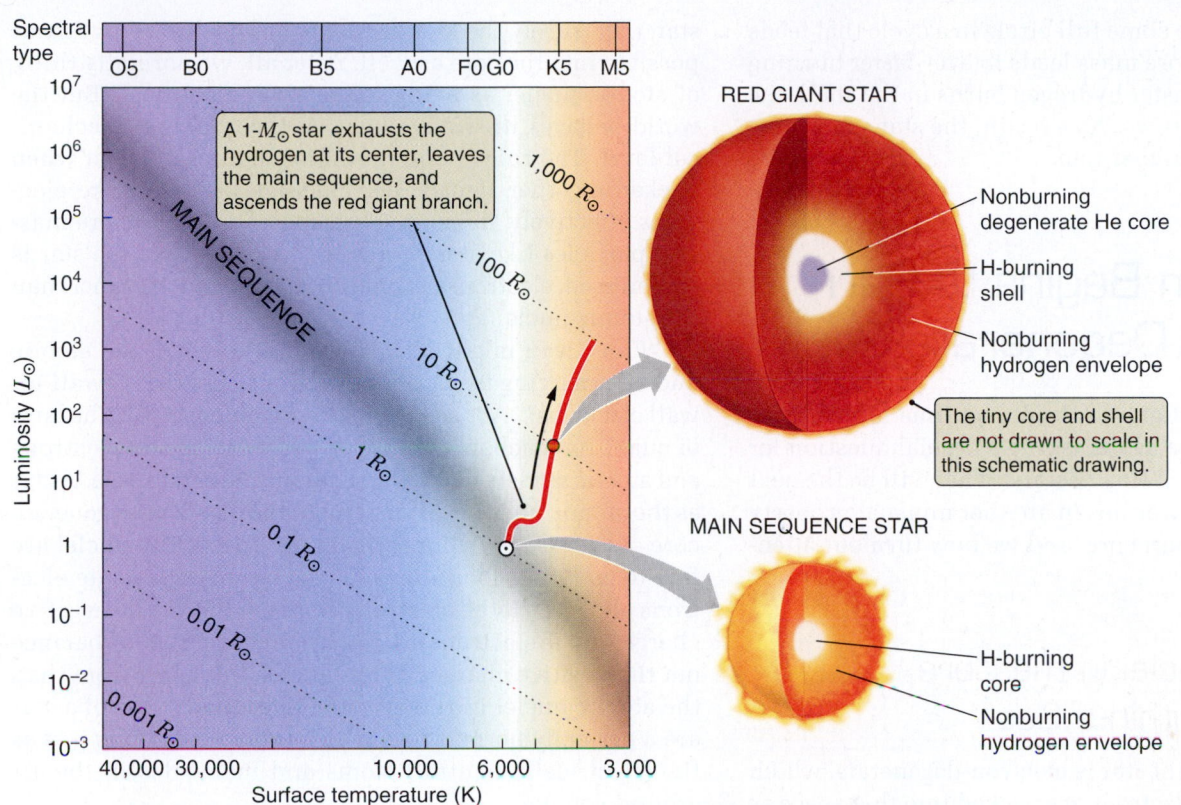

Spectral type
O5
B0
B5
A0
F0
G0
K5
M5
A 1-$M_\odot$ star exhausts the hydrogen at its center, leaves the main sequence, and ascends the red giant branch.
MAIN SEQUENCE
1,000 $R_\odot$
100 $R_\odot$
10 $R_\odot$
1 $R_\odot$
0.1 $R_\odot$
0.01 $R_\odot$
0.001 $R_\odot$
Luminosity ($L_\odot$)
10^7
10^6
10^5
10^4
10^3
10^2
10^1
1
10^{-1}
10^{-2}
10^{-3}
40,000
30,000
10,000
6,000
3,000
Surface temperature (K)
RED GIANT STAR
Nonburning degenerate He core
H-burning shell
Nonburning hydrogen envelope
The tiny core and shell are not drawn to scale in this schematic drawing.
MAIN SEQUENCE STAR
H-burning core
Nonburning hydrogen envelope

FIGURE 16.4 The evolution of a red giant star on the H-R diagram. The structure of a red giant star consists of a degenerate core of helium ash surrounded by a hydrogen-burning shell. As the star moves up the red giant branch, it comes close to retracing the Hayashi track that it followed when it was a protostar collapsing toward the main sequence.

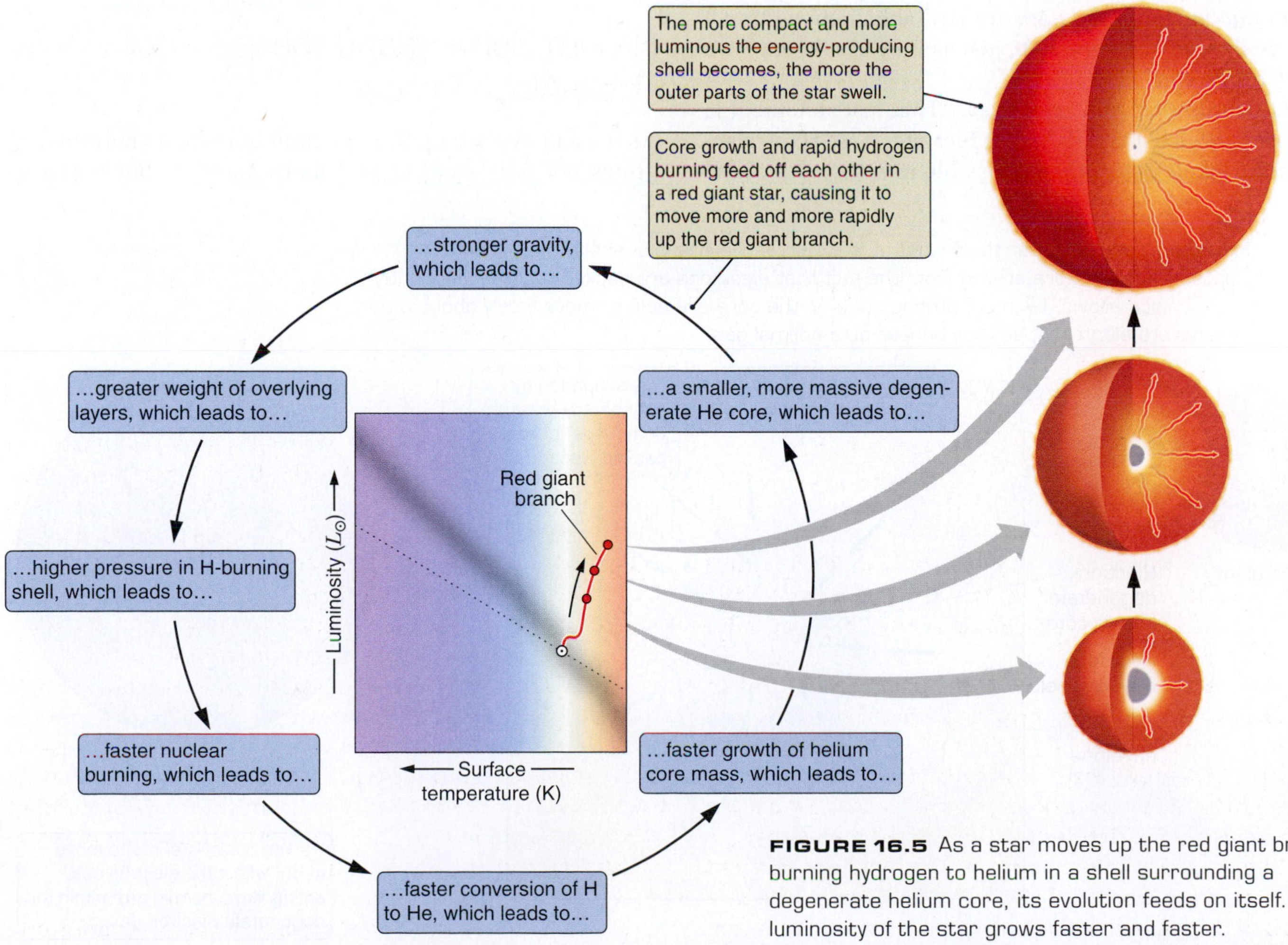

The more compact and more luminous the energy-producing shell becomes, the more the outer parts of the star swell.
Core growth and rapid hydrogen burning feed off each other in a red giant star, causing it to move more and more rapidly up the red giant branch.
…stronger gravity, which leads to…
…greater weight of overlying layers, which leads to…
…higher pressure in H-burning shell, which leads to…
…faster nuclear burning, which leads to…
…faster conversion of H to He, which leads to…
…faster growth of helium core mass, which leads to…
…a smaller, more massive degenerate He core, which leads to…
Red giant branch
Luminosity ($L_\odot$)
Surface temperature (K)

FIGURE 16.5 As a star moves up the red giant branch, burning hydrogen to helium in a shell surrounding a degenerate helium core, its evolution feeds on itself. The luminosity of the star grows faster and faster.

more rapidly. We have come full circle in a cycle that feeds on itself. Increasing core mass leads to ever-faster burning in the shell; and the faster hydrogen burns in the shell, the faster the core mass grows. As a result, the star's luminosity climbs at an ever-higher rate.

16.4 Helium Begins to Burn in the Degenerate Core

The growth of the red giant cannot continue forever, and we find ourselves once again asking a crucial question for understanding the evolution of stars: What will be the next thing to give? The answer lies in another unusual property of the degenerate helium core, and we now turn our attention there.

The Atomic Nuclei in the Core Form a "Gas within a Gas"

The core of the red giant star is electron-degenerate, which means that as many *electrons* are packed into that space as the rules of quantum mechanics allow at that pressure. However, *atomic nuclei* in the core are still able to move freely about, as shown in **Figure 16.6**, just as they are throughout the rest of the star.

"Wait a minute!" you might say. "That last statement is nonsense for at least two different reasons. First, if the electrons are packed as tightly as possible into the core of the

star, then surely the atomic nuclei are packed as tightly as possible into the core as well. After all, we normally think of atomic nuclei as being larger than electrons." But the world behaves in strange ways at the quantum mechanical level. The rules of quantum mechanics say that when packed together tightly, less massive particles like electrons effectively "take up more space" than do more massive particles like atomic nuclei. As the core of the star is compressed, electrons become degenerate much sooner than the atomic nuclei do.

"Fine," you might continue, "but how can the atomic nuclei go moving freely about within a core that is wall-to-wall electrons?" Actually, this is no problem at all: The laws of quantum mechanics place few restrictions on electrons and atomic nuclei occupying the same physical space. As far as the atomic nuclei are concerned, the electron-degenerate core of the star is still mostly empty space. The nuclei are free to go flying about through the sea of degenerate electrons almost as if the electrons were not there. The negative charges of the electrons are important because they balance out the positive charges of the nuclei; but apart from that, the atomic nuclei in the electron-degenerate core of a star are a perfectly normal "gas within a gas," behaving just as the (electrically neutral) atoms and molecules in the air around you do.

Helium Burning and the Triple-Alpha Process

As the star evolves up the red giant branch, its helium core grows not only smaller and more massive, but hotter as

FIGURE 16.6 In a red giant star the weight of the overlying layers is supported by electron degeneracy pressure in the core arising from the fact that electrons are packed together as tightly as quantum mechanics allows. Even so, atomic nuclei in the core are able to move freely about within the sea of degenerate electrons, so they behave as a normal gas.

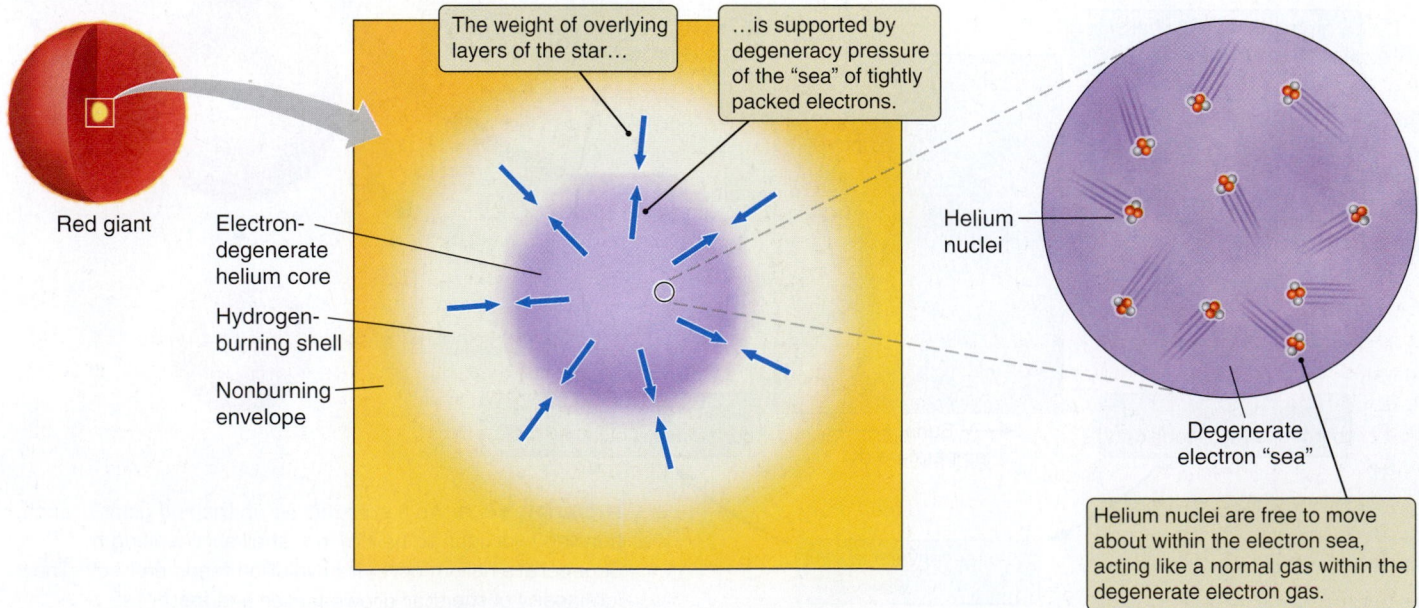

Red giant

Electron-degenerate helium core

Hydrogen-burning shell

Nonburning envelope

The weight of overlying layers of the star…

…is supported by degeneracy pressure of the "sea" of tightly packed electrons.

Helium nuclei

Degenerate electron "sea"

Helium nuclei are free to move about within the electron sea, acting like a normal gas within the degenerate electron gas.

well. This increase in temperature is partly because of the gravitational energy released as the core shrinks (just as the protostar's core grew hotter as it collapsed) and partly due to the energy released by the ever-faster pace of hydrogen burning in the surrounding shell. The climbing temperature of the core means that the thermal motions of the atomic nuclei in the core become more and more energetic. Eventually, at a temperature of about 10^8 K (a hundred million kelvins), the collisions between helium nuclei in the core become energetic enough to overcome the electric repulsion between them. Helium nuclei are slammed together hard enough for the strong nuclear force to act, and helium burning begins.

Helium burns in a two-stage process, referred to as the **triple-alpha process**, illustrated in **Figure 16.7**. First, two helium-4 nuclei (^{4}He) fuse to form a beryllium-8 nucleus (^{8}Be) consisting of four protons and four neutrons. The ^{8}Be nucleus is extremely unstable. Left on its own, it would break apart again after only about a trillionth (10^{-12}) of a second. But if, in that short time, it collides with another ^{4}He nucleus, the two nuclei will fuse into a stable nucleus of carbon-12 (^{12}C) consisting of six protons and six neutrons. The triple-alpha process takes its name from the fact that it involves the fusion of three ^{4}He nuclei, which are traditionally referred to as **alpha particles**.

> Helium burns via the triple-alpha process.

The Helium Core Ignites in a Helium Flash

In the next phase of the star's evolution, the helium in the core begins burning, as **Figure 16.8** illustrates. Degenerate material is a very good conductor of thermal energy, so any differences in temperature within the core are quickly evened out. As a result, the degenerate core of a red giant star is at almost exactly the same temperature throughout. When helium burning begins at the center of the core, the energy released quickly heats the entire core. Within a few minutes the entire core is burning helium into carbon by the triple-alpha process.

In a normal gas like the air around you, the pressure of the gas comes from the random thermal motions of the atoms. Increasing the temperature of such a gas means that the motions of the atoms become more energetic, so the pressure of the gas increases. If the helium core of a red giant star were a normal gas, the increase in temperature that accompanies the onset of helium burning would lead to an increase in pressure. The core of the star would expand; the temperature, density, and pressure would decrease; nuclear reactions would slow; and the star would settle down into a new balance between gravity and pressure. These are exactly the sorts of changes that are steadily occurring within the core of a main-sequence star like the Sun, as the structure of the star steadily and smoothly shifts in response to the changing composition in the star's core.

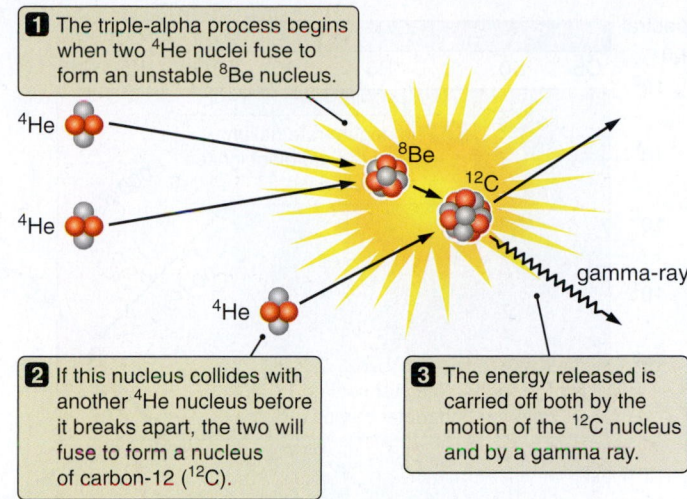

1 The triple-alpha process begins when two ^{4}He nuclei fuse to form an unstable ^{8}Be nucleus.

2 If this nucleus collides with another ^{4}He nucleus before it breaks apart, the two will fuse to form a nucleus of carbon-12 (^{12}C).

3 The energy released is carried off both by the motion of the ^{12}C nucleus and by a gamma ray.

FIGURE 16.7 The triple-alpha process. Two helium-4 (^{4}He) nuclei fuse to form an unstable beryllium-8 (^{8}Be) nucleus. If this nucleus collides with another ^{4}He nucleus before it breaks apart, the two will fuse to form a stable nucleus of carbon-12 (^{12}C). The energy produced is carried off both by the motion of the ^{12}C nucleus and by a high-energy gamma ray emitted in the second step of the process.

However, the degenerate core of a red giant is not a normal gas. In a certain sense the degenerate core behaves more like a rock than like a normal gas. A rock is hard to crush because of how closely the atoms within it are packed together. Heating a rock does not cause it to swell up like a hot-air balloon, and cooling a rock does not cause it to deflate. Similarly, the pressure in a red giant's degenerate core comes from how tightly the electrons in the core are packed together. Heating the core does not change the number of electrons that can be packed into its volume, so the core's pressure does not respond to changes in temperature. And if the pressure does not increase, then there is nothing to cause the core to expand.

Take another run at that to be sure it clicks. When helium begins to burn in the degenerate core, the temperature of the core goes up but the pressure does not. So the onset of helium burning in the degenerate core of a red giant does not cause the core to expand! Yet even though the higher temperature does not change the pressure, it does cause the helium nuclei to be slammed together with more frequency and greater force, so the nuclear reactions become more vigorous. The process begins to snowball again. More vigorous reactions mean higher temperature, and higher temperature means even more vigorous reactions. Thermonuclear burning in the degenerate core runs away with itself, wildly out of control as increasing temperature and increasing reaction rates feed each other. This is the **helium flash**.

> The helium flash is a thermonuclear runaway—an explosion within the star.

It is difficult to imagine the drama of the thermonuclear runaway that takes place during the helium flash. Helium

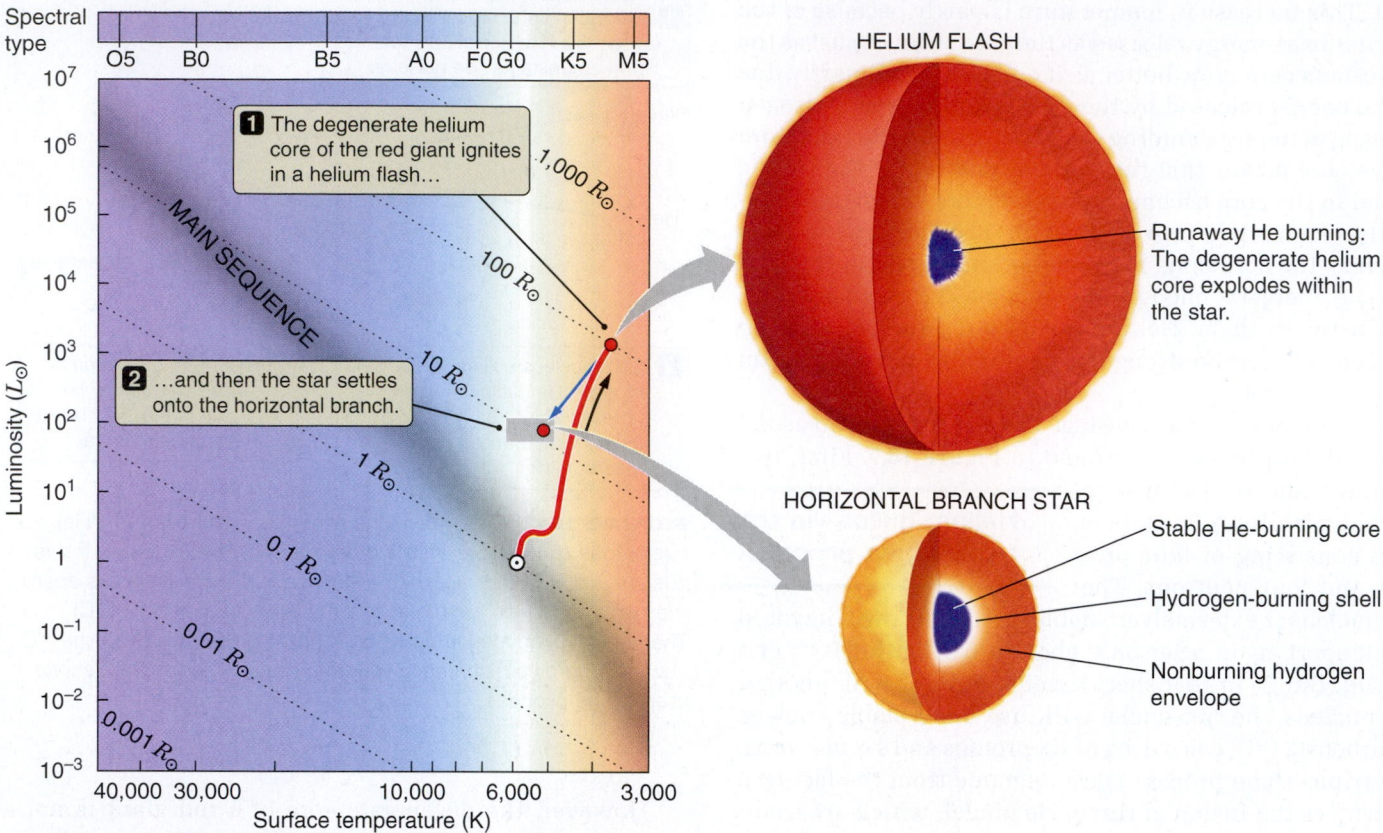

FIGURE 16.8 When the core temperature of a red giant reaches about 10^8 K, helium begins to burn explosively in the degenerate core, leading to a helium flash. After a few hours the core of the star begins to inflate, ending the helium flash. Over about the next 100,000 years (a relatively short time) the star settles onto the *horizontal branch*, where it burns helium in its core and hydrogen in a surrounding shell.

burning begins at a temperature of about 100 million K. By the time the temperature has climbed by a mere 10 percent, to 110 million K, the rate of helium burning has increased to 40 times what it was at 100 million K. By the time the core's temperature reaches 200 million K, the core is burning helium 460 million times faster than it was at 100 million K! As the temperature in the core grows higher and higher, the thermal motions of the electrons and nuclei become more energetic, and the pressure due to these thermal motions becomes greater and greater. Within seconds of ignition, the thermal pressure increases until it is no longer smaller than the degeneracy pressure, at which point the core literally explodes. Because the explosion is contained within the star, however, we do not see it. The energy released in this runaway thermonuclear explosion lifts the overlying layers of the star, and as the core expands, the electrons are able to spread out. The drama is over within a few hours. The expanded helium-burning core is no longer degenerate, and the star is on its way toward a new equilibrium.

Following the helium flash, the star once again does something counterintuitive. You might imagine that helium burning in the core would cause the star to grow more luminous, but it does not. The tremendous energy released during the helium flash goes into fighting gravity and puffing up the core. After the helium flash, the core (which is no

longer degenerate) is much larger, so the acceleration due to gravity within it and the surrounding shell is much smaller. (Again, $g = GM/r^2$, so a larger core radius means smaller values of g.) Weaker gravity means less weight pushing down on the core and the shell, which means lower pressure. Lower pressure, in turn, slows the nuclear reactions. The net result is that following the helium flash, core helium burning keeps the core of the star puffed up, and the star becomes less luminous than it was as a red giant.

The star spends the next 100,000 years or so settling into a stable structure in which helium burns to carbon in a normal, nondegenerate core while hydrogen burns to helium in a surrounding shell. The star is now about a hundredth as luminous as it was at the time of the helium flash.

Horizontal branch stars burn helium in the core and hydrogen in a shell.

The lower luminosity means that the outer layers of the star are not as puffed up as they were as a red giant. The star shrinks, and as it does so its surface temperature climbs. (This is just the reverse of the sequence of events that caused the red giant to become larger and redder as it grew more luminous.) At this point in their evolution, low-mass stars with chemical compositions similar to that of the Sun will bunch up on the H-R diagram just to the left of the red giant branch. Stars that contain much less iron than the Sun tend to distribute themselves away from the red giant branch along a nearly horizontal line on the H-R diagram. This stage of stellar evolution takes its name from this horizontal band. The star is now referred to as a **horizontal branch** star (see Figure 16.8).

16.5 The Low-Mass Star Enters the Last Stages of Its Evolution

The evolution of a solar-type star from the main sequence to the helium flash and horizontal branch is fairly well understood. Just as our understanding of the interior of the Sun comes from computer models of the physical conditions within our local star, our understanding of the evolution of a red giant comes from computer models that look at the changes in structure as the star's degenerate helium core grows. These models show that any star with a mass of about 1 $M_\odot$ will follow the march from main sequence to helium flash, and then drop down onto the horizontal branch. But when we try to use computer models to push our understanding to what happens next, the road that we follow gets a bit trickier. We already noted that differences in chemical composition between stars significantly affect where they fall on the horizontal branch. From this point on, small changes in the properties of a star—mass, chemical composition, strength of the star's magnetic field, or even the rate at which the star is rotating—can lead to qualitative differences in how the star evolves. The farther we go in time beyond the helium flash, the more possible it is that divergent evolutionary paths are open to a low-mass star.

With this caution in mind, we continue our story of the evolution of a 1-$M_\odot$ star with solar composition, following what we currently believe to be the most likely sequence of events awaiting our Sun.

The Star Moves Up the Asymptotic Giant Branch

The structure of a horizontal branch star is much like the structure of a main-sequence star in many respects. The biggest difference is that now instead of burning hydrogen into helium in a stable, nondegenerate core, the horizontal branch star is burning helium into carbon in a stable, nondegenerate core. (The other difference, of course, is that hydrogen is continuing to burn in a shell surrounding the core.) The behavior of a star on the horizontal branch is remarkably similar to that of a star on the main sequence. This similarity offers the key to understanding what comes next. In our account of what happened as the star evolved off the main sequence, replacing the word *hydrogen* with *helium* and then

helium with *carbon* yields a pretty good description of how the star evolves as it leaves the horizontal branch.

The star's life on the horizontal branch, however, is much shorter than its life on the main sequence. For one thing, there is now less fuel to burn in its core. In addition, the star is more luminous, so it must be consuming fuel more rapidly. Finally, helium is a much less efficient nuclear fuel than hydrogen. Even so, for 100 million years the horizontal branch star remains stable, burning helium to carbon in its core and hydrogen to helium in a shell.

The temperature at the center of a horizontal branch star is not high enough for carbon to burn,[4] so carbon ash builds up in the heart of the star, just as helium ash accumulates at the center of a main-sequence star. When the horizontal branch star has burned all of the helium at its core, gravity once again begins to win. The nonburning carbon ash core is crushed by the weight of the layers of the star above it until once again the electrons in the core are packed together as tightly as the laws of quantum mechanics allow at its pressure. The carbon core is now electron-degenerate, with physical properties much like those of the degenerate helium core at the center of a red giant.

> The star forms a degenerate carbon core as it leaves the horizontal branch.

The small, dense electron-degenerate carbon core drives up the strength of gravity in the inner parts of the star, which in turn drives up the pressure, which speeds up the nuclear reactions, which causes the degenerate core to grow more rapidly—we have heard this story before. The internal changes occurring within the star are similar to the changes that took place at the end of the star's main-sequence lifetime, and the path the star follows as it leaves the horizontal branch echoes that earlier phase of evolution as well. Just as the star accelerated up the red giant branch as its degenerate helium core grew, the star now leaves the horizontal branch and once again begins to grow larger, redder, and more luminous as its degenerate carbon core grows. The path that the star follows on the H-R diagram (**Figure 16.9**) closely parallels the path it followed as a red giant, getting closer to the red giant branch as the star grows more luminous. That is why this phase of evolution is referred to as the **asymptotic giant branch** (**AGB**) of the H-R diagram. An AGB star burns helium and hydrogen in nested concentric shells surrounding a degenerate carbon core.

Giant Stars Lose Mass

Building on our analogy between AGB stars and red giants, you might imagine that the next step in the evolution of an AGB star should be a "carbon flash," when carbon burning begins in the star's degenerate core. Yet a carbon flash never happens. Before the temperature in the carbon core

---

[4] You may recall from Chapter 14 that *burn* as used here refers not to fire but to nuclear fusion.

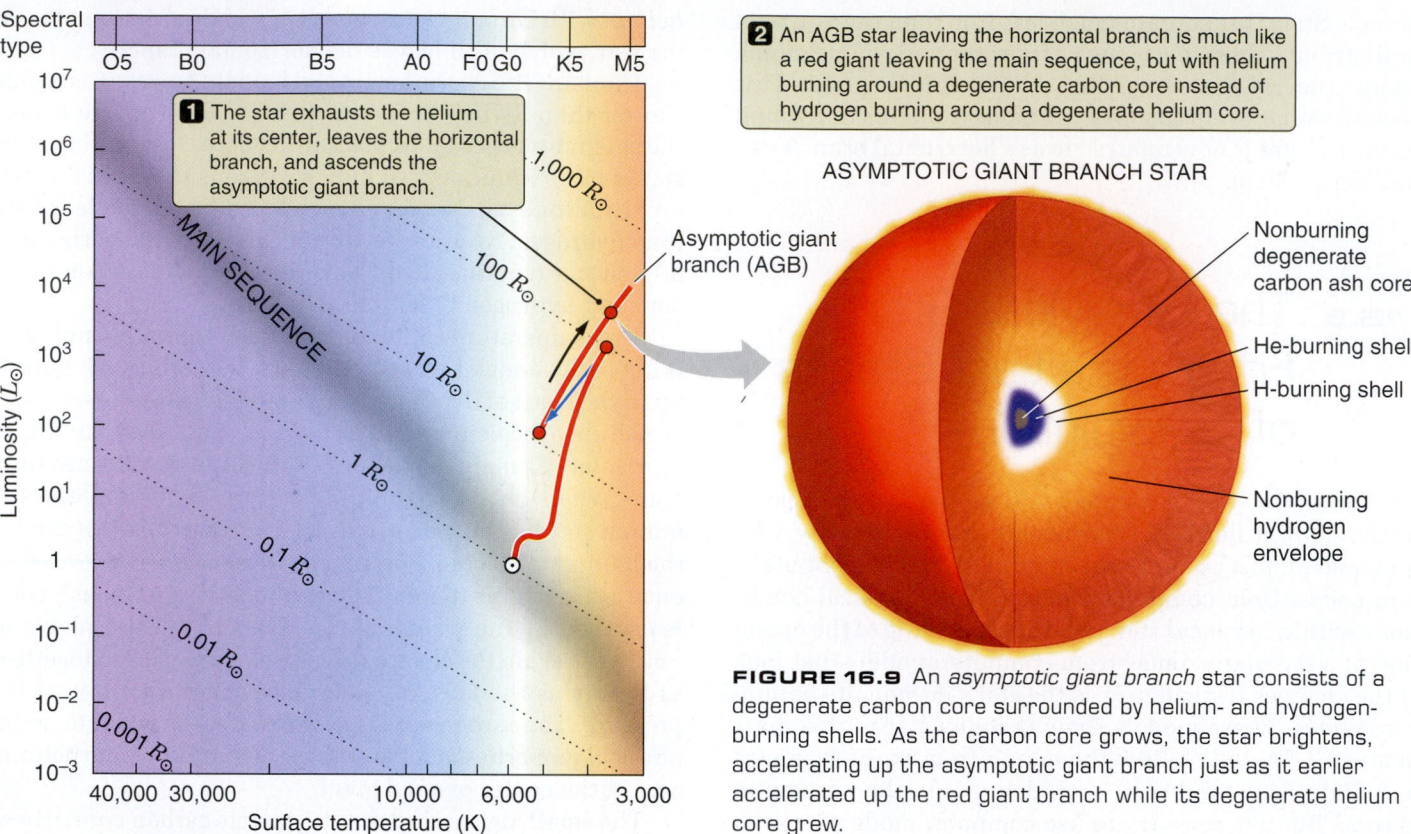

1 The star exhausts the helium at its center, leaves the horizontal branch, and ascends the asymptotic giant branch.

2 An AGB star leaving the horizontal branch is much like a red giant leaving the main sequence, but with helium burning around a degenerate carbon core instead of hydrogen burning around a degenerate helium core.

ASYMPTOTIC GIANT BRANCH STAR

Asymptotic giant branch (AGB)

Nonburning degenerate carbon ash core

He-burning shell

H-burning shell

Nonburning hydrogen envelope

FIGURE 16.9 An *asymptotic giant branch* star consists of a degenerate carbon core surrounded by helium- and hydrogen-burning shells. As the carbon core grows, the star brightens, accelerating up the asymptotic giant branch just as it earlier accelerated up the red giant branch while its degenerate helium core grew.

becomes high enough for carbon to burn, the star loses its gravitational grip on itself and expels its outer layers into interstellar space.

Red giant and AGB stars are huge objects. The AGB star into which a 1-$M_\odot$ main-sequence star evolves can grow to a radius hundreds of times the radius of our Sun. When our Sun becomes an AGB star, its outer layers will swell to the point that they engulf the orbits of the inner planets, possibly including Earth. (Do not lose much sleep over this eventuality, however. It will be 5 billion years before Earth is engulfed by the burgeoning Sun, and well before then Earth will be well toasted by the thousandfold increase in the Sun's luminosity.) When a star expands to such a size, its hold on its outer layers becomes tenuous indeed.

Once again, an understanding of what happens to the star rests with Newton's universal law of gravitation. Recalling that $g = GM/r^2$, we know that the acceleration due to gravity (g) at the surface of a star with a mass of 1 $M_\odot$ and a radius of 100 $R_\odot$ is only 1/10,000 as strong as the gravity at the surface of the Sun. It takes little in the way of a kick from below to push material near the surface of such a giant star over the edge, driving it completely away from the star. The process of **stellar mass loss** begins when the star is still on the red giant branch; by the time a 1-$M_\odot$ main-sequence star reaches the horizontal branch, it may have lost 10–20 percent of its total mass. As

> Red giants and AGB stars lose mass from their outer layers.

the star ascends the asymptotic giant branch, it loses another 20 percent or even more of its total mass. By the time it is well up on this branch, a star that began as a 1-$M_\odot$ star probably has a mass less than about 0.7 $M_\odot$, and it may even have lost more than half of its original mass. Mass loss on the asymptotic giant branch can be spurred on by a lack of stability in the star's interior. The extreme sensitivity of the triple-alpha process to temperature can lead to episodes of rapid energy release, which can provide the extra kick needed to expel material from the star's outer layers. It also means that stars that are initially quite similar can behave very differently when they reach this stage in their evolution.

The Post-AGB Star May Cause a Planetary Nebula to Glow

Toward the end of an AGB star's life, mass loss itself becomes a runaway process. When a star loses a bit of its outer layers, the weight pushing down on the underlying layers of the star is reduced. Without this weight holding them down, the outer layers of the star puff up even larger than they were before. The star, which is now both less massive and larger, is even less tightly bound by gravity, so even less energy is needed to push outer layers away from it. The situation is a bit like taking the lid off a pressure cooker. Mass loss leads to weaker gravity, which leads to faster mass loss, which leads to weaker gravity, and so on. When the end comes, much of

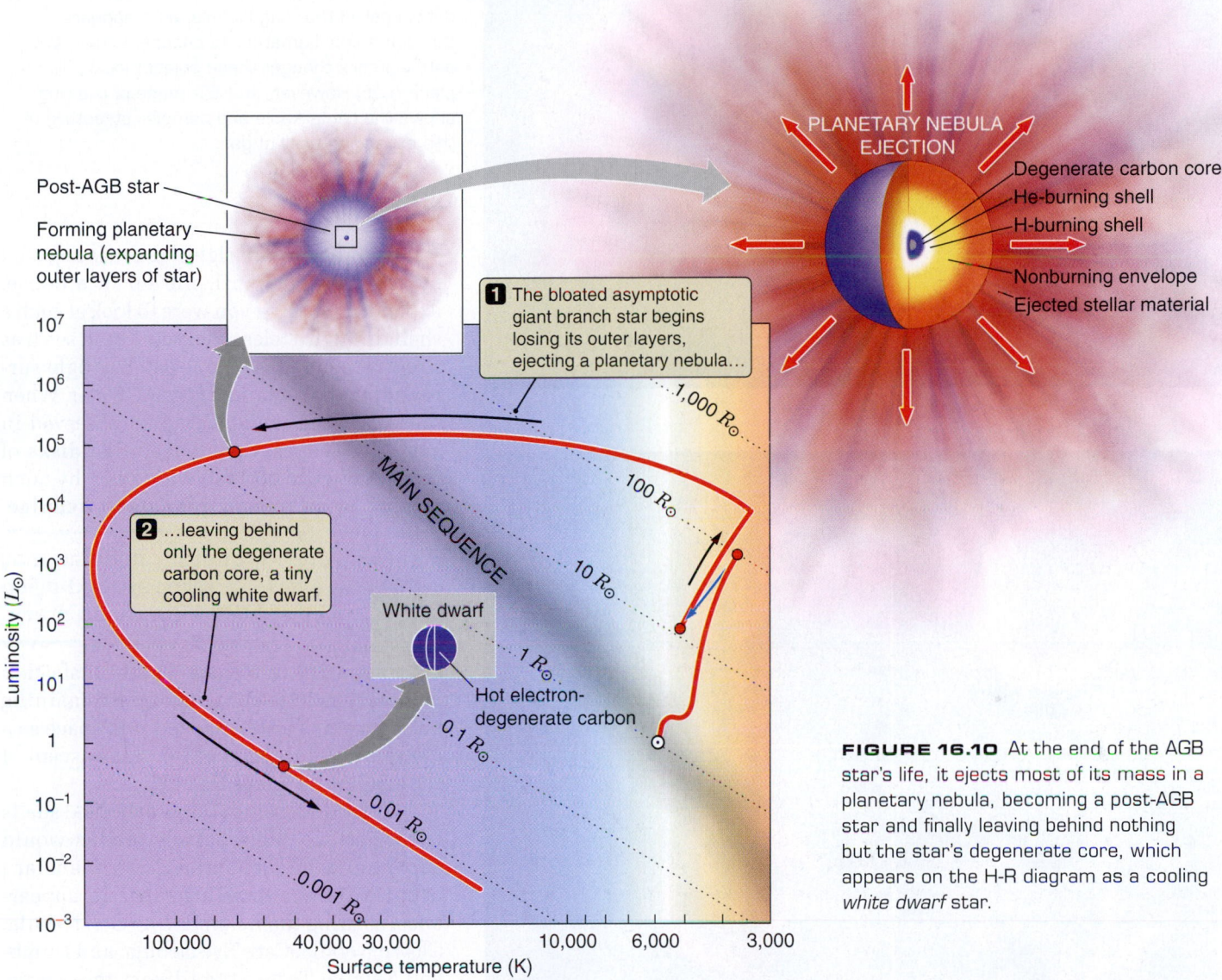

Post-AGB star

Forming planetary nebula (expanding outer layers of star)

1 The bloated asymptotic giant branch star begins losing its outer layers, ejecting a planetary nebula...

2 ...leaving behind only the degenerate carbon core, a tiny cooling white dwarf.

White dwarf

Hot electron-degenerate carbon

MAIN SEQUENCE

Luminosity ($L_\odot$)

Surface temperature (K)

1,000 $R_\odot$
100 $R_\odot$
10 $R_\odot$
1 $R_\odot$
0.1 $R_\odot$
0.01 $R_\odot$
0.001 $R_\odot$

PLANETARY NEBULA EJECTION

Degenerate carbon core
He-burning shell
H-burning shell
Nonburning envelope
Ejected stellar material

FIGURE 16.10 At the end of the AGB star's life, it ejects most of its mass in a planetary nebula, becoming a post-AGB star and finally leaving behind nothing but the star's degenerate core, which appears on the H-R diagram as a cooling *white dwarf* star.

the remaining mass of the star is ejected into space, typically at speeds of 20–30 kilometers per second (km/s).

After ejection of its outer layers, all that is left of the low-mass star itself is a tiny, very hot electron-degenerate carbon core, surrounded by a thin envelope in which hydrogen and helium are still burning. This star is now somewhat less luminous than when it was at the top of the asymptotic giant branch, but it is still much more luminous than a horizontal branch star. The remaining hydrogen and helium in the star rapidly burn to carbon, and as more and more of the mass of the star ends up in the carbon core, the star itself shrinks and becomes hotter and hotter. Over the course of only 30,000 years or so following the beginning of runaway mass loss, the star moves very rapidly from

right to left across the top of the H-R diagram, as shown in **Figure 16.10**.

The surface temperature of the star may eventually reach 100,000 K or hotter. Wien's law says that at such temperatures, most of the light from the star is in the hard (high-energy) UV part of the spectrum:

$$\lambda_{\text{peak}} = \frac{2{,}900 \ \mu\text{m K}}{100{,}000 \ \text{K}} = 0.029 \ \mu\text{m}.$$

The intense UV light from what remains of the star heats and ionizes the expanding shell of gas that was recently ejected by the star. The ultraviolet light causes the shell of gas to glow in the same way that UV light from an O star causes an H II region to glow.

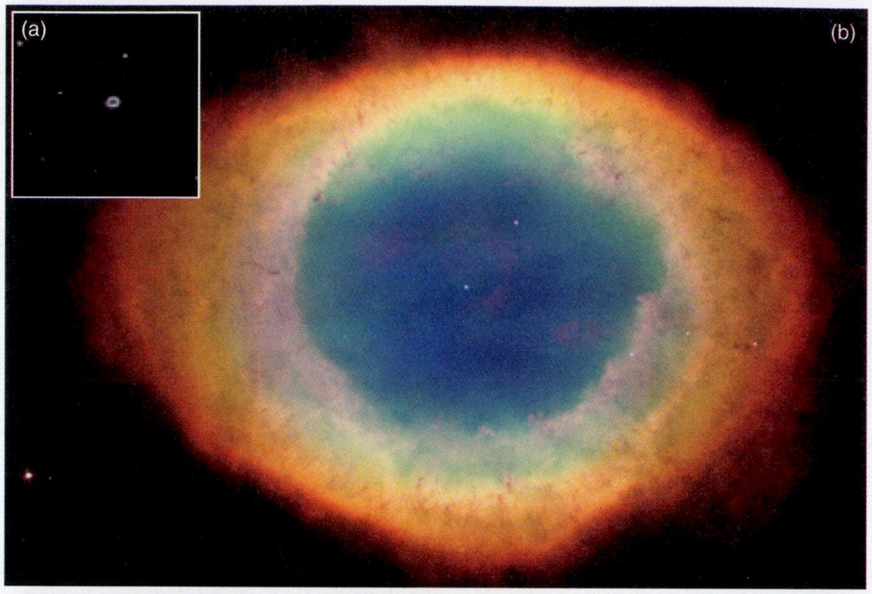

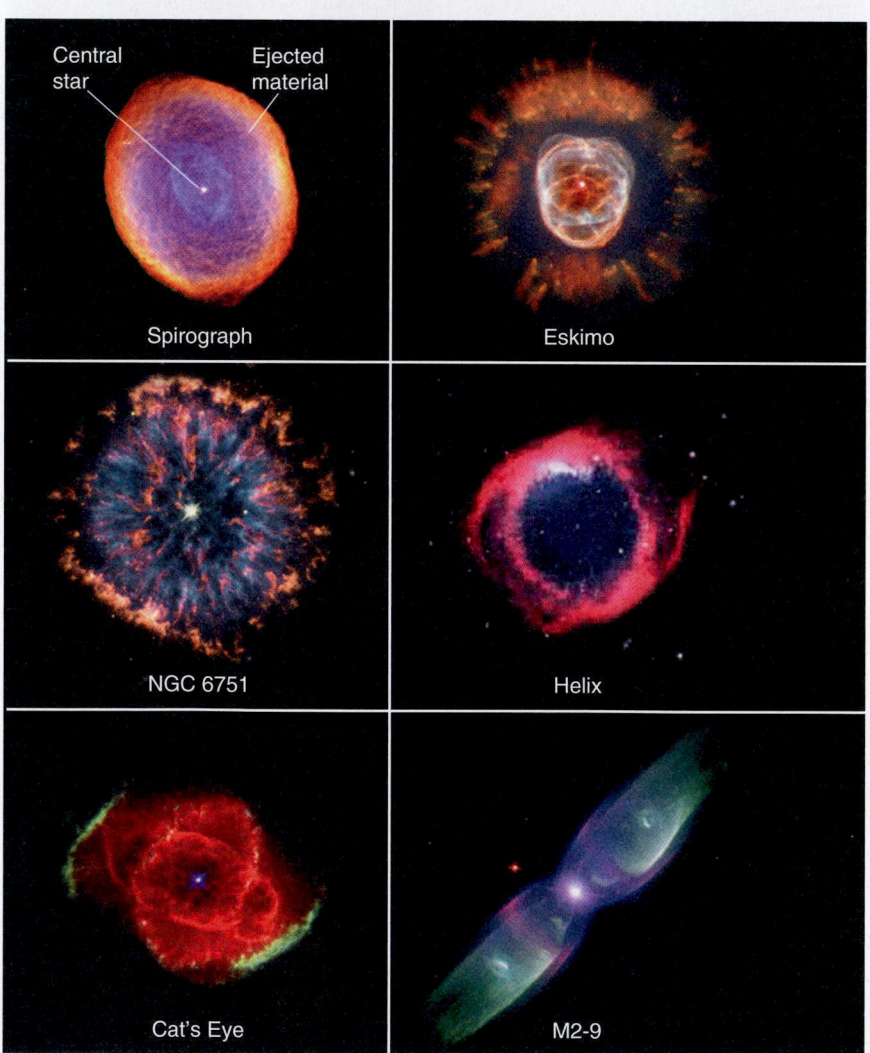

(a) (b)

FIGURE 16.11 At the end of its life, a low-mass star ejects its outer layers and may form a "planetary nebula" consisting of an expanding shell of gas surrounding the white-hot remnant of the star. (a) This picture of a famous planetary nebula called the Ring Nebula, as it appears through a small amateur telescope, shows why astronomers thought these objects looked like planets. (b) However, an HST image of the ring shows the remarkable and complex structure of this expanding shell of gas.

If conditions are right,[5] the mass ejected by the AGB star will pile up in a dense, expanding shell. If you were to look at such a shell through a telescope, you would see it as a round or perhaps oblong patch of light surrounding the remains of the AGB star. When these glowing shells were first observed in small telescopes, they looked like disks of planets (**Figure 16.11a**), which is why such objects were named **planetary nebulae**. But there is nothing truly "planetary" about a planetary nebula,

> Planetary nebulae may form around dying low-mass stars.

as is apparent in **Figure 16.11b**. Instead, a planetary nebula consists of the remaining outer layers of a star, ejected into space as a dying gasp at the end of the star's ascent of the asymptotic giant branch.

Rather than being simple spherical shells of the sort we might naïvely predict would surround a nice spherical star, planetary nebulae show a dazzling range in appearance, earning them names like Owl Nebula, Clown Nebula, Cat's Eye Nebula, and Dumbbell Nebula. This extraordinary menagerie, illustrated in **Figure 16.12**, serves to rub astronomers' noses in the complexity of the late stages of a star's evolution, telling of eras when mass loss from the star was slower or faster, and of times when mass was ejected primarily from the star's equator or its poles. The gas in the planetary nebula also contains chemical elements that were produced by nuclear burning in the star, offering us

Central star   Ejected material

Spirograph   Eskimo

NGC 6751   Helix

Cat's Eye   M2-9

FIGURE 16.12 Planetary nebulae are not all simple spherical shells around their parent stars. These HST images of planetary nebulae show a wealth of structure resulting from the complex processes by which low-mass stars eject their outer layers.

[5] Not all stars form planetary nebulae. Massive stars go through the post-AGB stage too quickly. Stars with insufficient mass take too long, so their envelope evaporates before they can illuminate it. Some astronomers believe that our own Sun will not retain enough mass during its post-AGB phase to form a planetary nebula.

CONNECTIONS 16.1

Seeding the Universe with New Chemical Elements

One of the grand themes of 20th and 21st century astronomy is the origin of the chemical elements. When we discuss the properties of the early universe, we will see that the only chemical elements that were formed in abundance when the universe was young were hydrogen and helium, along with trace amounts of lithium, beryllium, and boron.

Astronomers studying red giant and AGB stars see direct evidence that more massive chemical elements formed deep within the interiors of low-mass stars find their way to the "outside" world. In Chapter 14 we learned that if the change in temperature within a star becomes too steep, convection will begin. Like the boiling motion in a pot of water, convection is a process of mixing. Material from the inner parts of the star is carried upward and mixed with material in the outer parts of the star. As a star ascends first the red giant and later the asymptotic giant branches of the H-R diagram, the core of the star grows hotter and hotter, and the temperature gradient within the star grows steeper. Convection spreads through more and more of the star. Under certain circumstances, convection can spread so deep into a star that it is able to dredge up chemical elements formed by nuclear burning within the star and carry them to the star's surface. Many AGB stars, known as **carbon stars**, show an overabundance of carbon and other by-products

of nuclear burning in their spectra. This extra carbon originated in the star's helium-burning shell and then was carried to its surface by convection.

Mass loss from giant stars carries the chemical elements enriching the stars' outer layers off into interstellar space. Planetary nebulae—the ejected atmospheres of low-mass stars—often show an overabundance of elements such as carbon, nitrogen, and oxygen, which are by-products of nuclear burning. Once this chemically enriched material leaves the star, it mixes with the interstellar gas there, enriching the chemical diversity of the universe. *Novae* and *Type I supernovae*, discussed at the end of this chapter, also do their part. The nuclear burning that occurs in these explosive events goes far beyond the formation of elements such as carbon. Novae and Type I supernovae help seed the universe with much more massive atoms, such as iron and nickel.

As we continue to discuss the evolution of stars and the universe in the chapters to come, we will learn more about the origin of the chemical elements. Today, when scientists study the patterns of the abundance of different chemical elements on Earth and in our Solar System, they recognize the patterns they see. Those patterns are clear signatures of nuclear reactions in stars.

---

our first look at the processes responsible for the chemical evolution of the universe (see **Connections 16.1**). The planetary nebula is visible for 50,000 years or so before the gas ejected by the star disperses so far that the nebula is too faint to be seen.

The Star Becomes a White Dwarf

Within 50,000 years or so, a post-AGB star burns all of the fuel remaining on its surface, leaving nothing behind but a cinder—a nonburning ball of carbon with a remaining mass that is probably less than 70 percent of the mass of the original star. In the process, the star plummets down the left side of the H-R diagram, becoming smaller and fainter. Within a few thousand years the burned-out core shrinks to about the size of Earth, at which point it has become fully degenerate and so can shrink no further. The degenerate stellar cinder is now referred to as a **white dwarf**. The white dwarf continues to radiate energy away into space, and as it does so it cools, just as the filament of a lightbulb cools when the switch is turned off. Because the white dwarf is electron-degenerate, its size does not change much as it

> A newly formed white dwarf is hot but tiny.

cools, so it moves down and to the right on the H-R diagram, following a line of constant radius. Even though the white dwarf may remain very hot for 10 million years or so, its luminosity may now be only 1/1,000 that of a main-sequence star like our Sun. Many white dwarfs are known, but all are much too faint to be seen without the aid of a telescope. Yet when you look at Sirius, you are also looking at a white dwarf. The brightest star in Earth's sky has a faint white dwarf as a binary companion.

Here we have the fate of our Sun, some 6 billion or so years from now. It will become a white dwarf that, bereft of all sources of nuclear fuel, will shine ever more feebly as it continues to cool, radiating its thermal energy away into space. It is amazing to recall that this superdense ball—with a density of a ton per teaspoonful—actually began its life billions of years earlier as a cloud of interstellar gas billions of times more tenuous than the vacuum in the best vacuum chamber on Earth.

Figure 16.13 recaps the evolution of a solar-type 1-$M_\odot$ main-sequence star through to its final existence as a 0.6-$M_\odot$ white dwarf. We point out once again that what we have described is representative of the fate of low-mass stars. Although all low-mass stars form white dwarfs at the end points of their evolution, the exact path a low-mass star follows from core hydrogen burning on the main sequence to

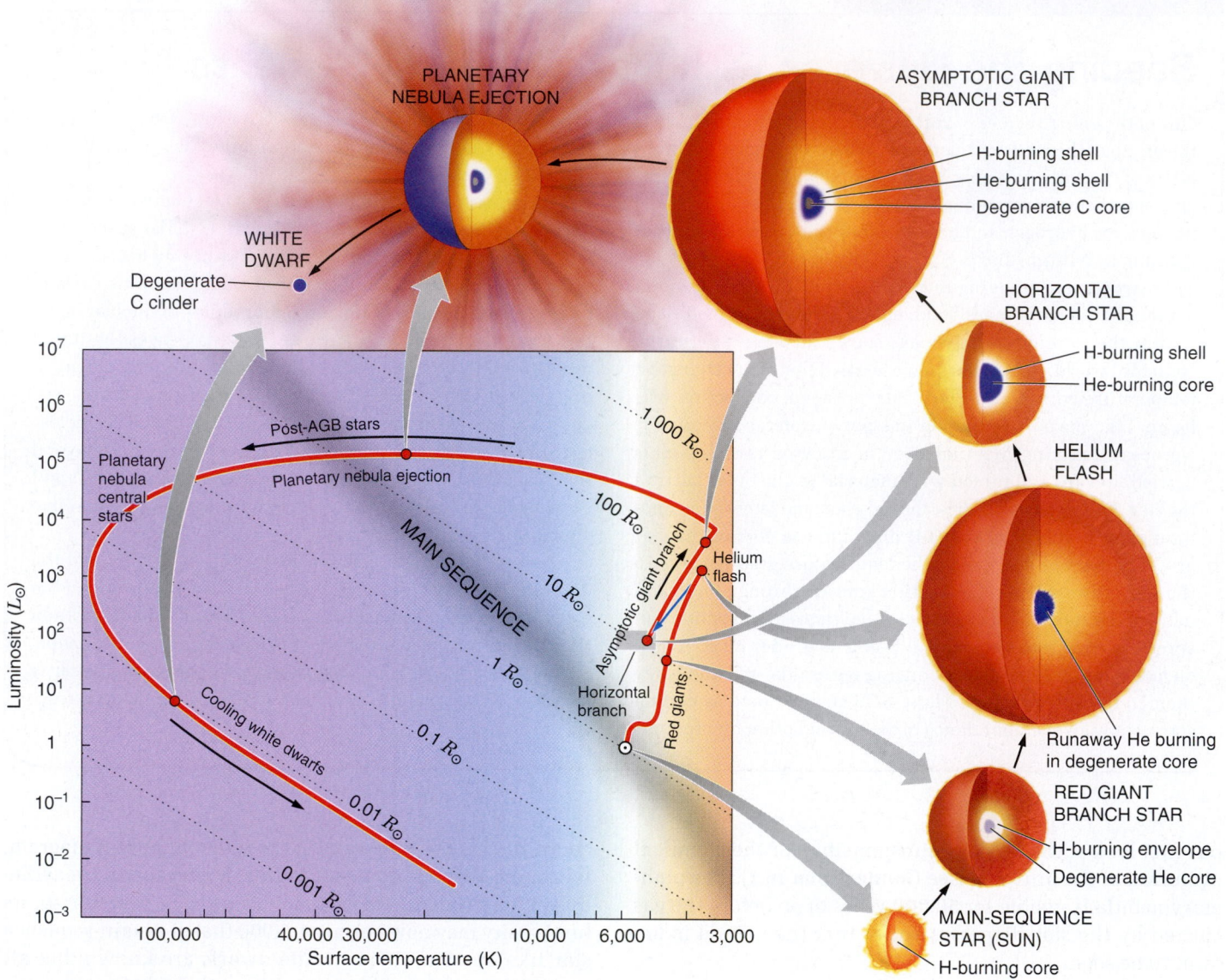

FIGURE 16.13 This H-R diagram summarizes the stages in the post–main sequence evolution of a 1-$M_\odot$ star.

white dwarf depends on many details particular to the star. Some stars less massive than the Sun may become white dwarfs composed largely of helium rather than carbon. Temperatures in the cores of evolved 2-$M_\odot$ to 3-$M_\odot$ stars are high enough to allow additional nuclear reactions to occur, leading to the formation of somewhat more massive white dwarfs composed of materials such as oxygen, neon, and magnesium. Differences in chemical composition of a star can also lead to dramatic differences in its post–main sequence evolution. Finally, we note that studies of the evolution of stars whose main-sequence masses are less than about 0.8 $M_\odot$ or so are pure theory. These stars are able to burn hydrogen in their cores for longer than the current age of the universe. No star with a mass lower than that has ever evolved beyond the main sequence except in the virtual world of a computer's calculations.

In our discussion of stellar evolution we have focused on what happens after a star leaves the main sequence. It is important, however, to remind ourselves that the spectacle of a red giant or AGB star is ephemeral. The Sun will travel the path from subgiant to white dwarf in less than one-tenth of the time that it spends on the main sequence steadily burning hydrogen to helium in its core. Stars spend most of their luminous lifetimes on the main sequence, which is why most of the stars that we see in the sky are main-sequence stars. (Were it otherwise, the "main sequence" would not be "main" at all!) In the end, though, white dwarfs will carry the day. Too faint to be noticed as we look at the sky on a clear summer evening, white dwarfs still constitute the final resting place for the vast majority of stars that have been or ever will be formed.

16.6 Many Stars Evolve as Pairs

Possibly the most significant complication in our picture of the evolution of low-mass stars arises from the fact that many stars are members of binary systems. Binary stars were quite useful to us when we were uncovering the properties of stars. We used them as a sort of celestial bathroom scale, applying Kepler's laws to the orbits of two stars about each other to determine their masses. Although both members of a binary pair are on the main sequence, they usually have little effect on each other. When one of the stars evolves off the main sequence, however, it can swell to hundreds of times its original size, and its outer layers can literally cross the line between what belongs to the evolved star and what belongs to its companion. Shortly we will see the consequences of this behavior.

Mass Flows from an Evolving Star onto Its Companion

Think for a moment about what would happen if you were to travel in a spacecraft from Earth toward the Moon. When you are still near Earth, the force of Earth's gravity is far stronger than that of the Moon. There is no question that you "belong" to Earth, just as there is no question that you "belong" to Earth as you sit in your chair reading this book. Yet, as you move away from Earth and closer to the Moon, this terrestrial dominance becomes less secure: the gravitational attraction of Earth weakens, and the gravitational attraction of the Moon becomes stronger. At a certain point you reach an intermediate zone where neither body has the upper hand. If you continue beyond this point, the lunar gravity begins to successfully assert itself until you find yourself firmly in the grip of the Moon.

Exactly the same situation exists between two stars. Gas near each star clearly "belongs" to that star. But what happens if, as a star swells up, its outer layers cross that gravitational dividing line separating the star from its companion? Any material that crosses this line no longer belongs to the first star, but instead can be pulled toward the companion. A star reaches this point when it fills up its portion of an imaginary figure eight–shaped volume of space, as shown in **Figure 16.14**. These regions surrounding the two stars—their gravitational domains—are referred to as the **Roche lobes** of the system. Once the first star has expanded to fill its Roche lobe, material begins to pour through the "neck" of the figure eight and fall toward the other star. Astronomers refer to this exchange of material between the two stars as **mass transfer**.

> An evolving star loses mass to its companion.

You may recall the name Édouard A. Roche from Chapters 10 and 11, when we discussed how the tidal disruption of moons and comets can lead to the formation of planetary rings. The Roche limit describes how close a small object might come to a planet before being torn apart by the planet's tides. Similarly, mass transfer between stars in a binary can be thought of as the "tidal stripping" of one star by the other. The physical principles at work in the two situations are much the same, and Roche is responsible for early calculations of both.

Evolution of a Binary System

The best way to understand how mass transfer affects the evolution of stars in a binary system is to apply what we have learned from the evolution of single low-mass stars. **Figure 16.14a** shows a binary system consisting of two low-mass stars. We will rather unimaginatively call the more massive of the two stars "star 1" and the less massive of the two "star 2." This is an ordinary binary system, and each of these stars is an ordinary main-sequence star for most of the system's lifetime. When one of the stars begins to evolve, however, things start to get interesting.

The more massive a main-sequence star is, the more rapidly it evolves. Therefore, star 1, which is more massive, will be the first to use up the hydrogen at its center and begin to evolve off the main sequence (**Figure 16.14b**). If the two stars are close enough together, star 1 will eventually grow to fill its Roche lobe, and the overflow from star 1 will start falling onto star 2 (**Figure 16.14c**). A number of interesting things can happen at this point. For example, the transfer of mass between the two stars can result in a sort of "drag" that causes the orbits of the two stars to shrink, bringing the stars closer together and further enhancing mass loss. The two stars can even reach the point where they are effectively two cores sharing the same extended envelope of material.

> The more massive star evolves first.

Despite these complexities, star 2 probably remains a basically normal main-sequence star throughout this process, burning hydrogen in its core. However, it does not remain the same main-sequence star that it started out to be. The mass of star 2 increases, thanks to the largesse of its companion. As it does so, the structure of star 2 must change to accommodate its new status as a higher-mass star. If we plotted star 2's position on the H-R diagram during this period, we would see it move up and to the left along the main sequence, becoming larger, hotter, and more luminous.

While star 2 gains from the interaction, star 1 suffers. Star 1 never gets to experience the full glory of being an isolated red giant or AGB star, because star 1 can never grow larger than its Roche lobe. The presence of star 2 prevents star 1 from swelling to the size of the behemoths that populate the top of the H-R diagram. Yet star 1 continues to evolve, spend-

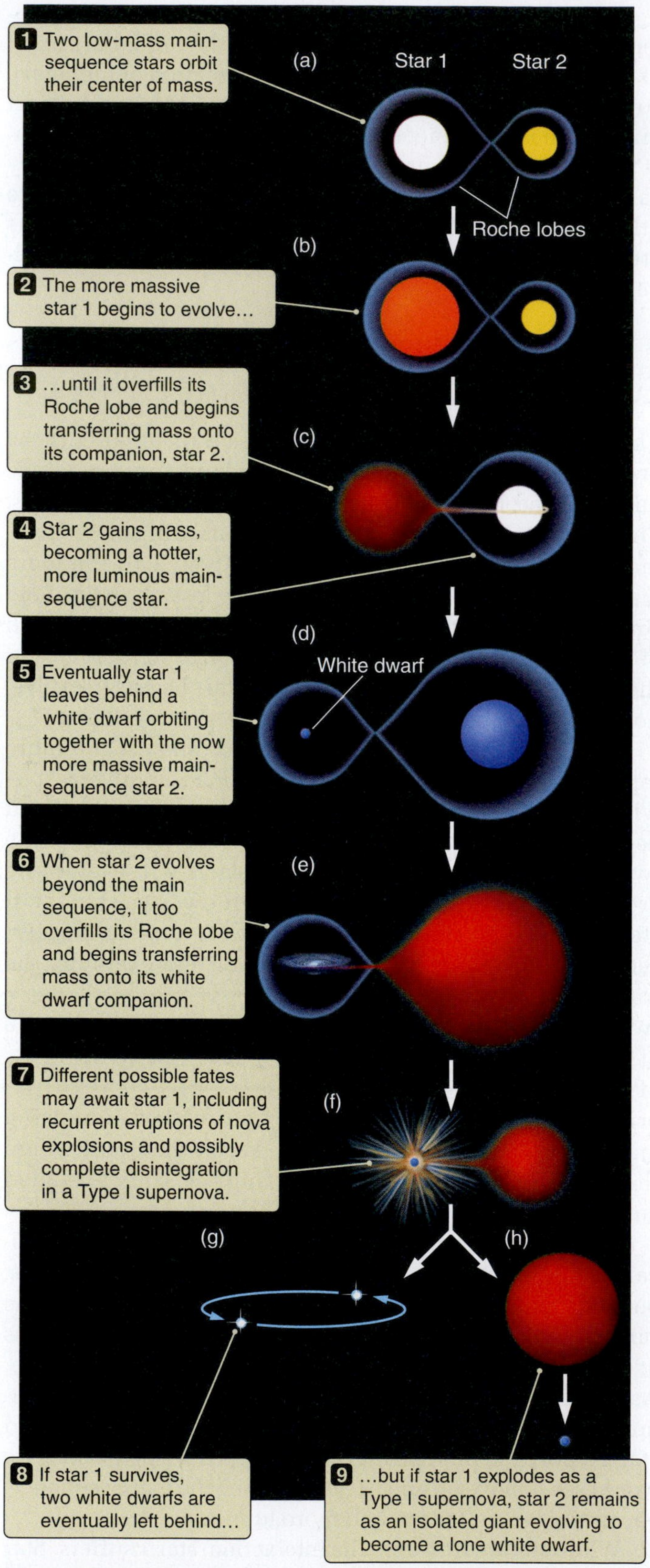

1 Two low-mass main-sequence stars orbit their center of mass.

(a)　Star 1　Star 2

Roche lobes

(b)

2 The more massive star 1 begins to evolve...

3 ...until it overfills its Roche lobe and begins transferring mass onto its companion, star 2.

(c)

4 Star 2 gains mass, becoming a hotter, more luminous main-sequence star.

(d)

White dwarf

5 Eventually star 1 leaves behind a white dwarf orbiting together with the now more massive main-sequence star 2.

6 When star 2 evolves beyond the main sequence, it too overfills its Roche lobe and begins transferring mass onto its white dwarf companion.

(e)

7 Different possible fates may await star 1, including recurrent eruptions of nova explosions and possibly complete disintegration in a Type I supernova.

(f)

(g)　(h)

8 If star 1 survives, two white dwarfs are eventually left behind...

9 ...but if star 1 explodes as a Type I supernova, star 2 remains as an isolated giant evolving to become a lone white dwarf.

FIGURE 16.14 A compact binary system consisting of two low-mass stars passes through a sequence of stages as the stars evolve and mass is transferred back and forth. The evolution of the binary system can lead to novae or even to a Type I supernova that destroys the initially more massive star 1.

ing its allotted time burning helium in its core on the horizontal branch, proceeding through a stage of helium shell burning, and finally losing its outer layers and leaving behind a white dwarf. **Figure 16.14d** shows the binary system after star 1 has completed its evolution. All that remains of star 1 is a white dwarf, orbiting about its bloated main-sequence companion, star 2.

> The first star becomes a white dwarf orbiting a main-sequence star.

Fireworks Occur When the Second Star Evolves

The stage is now set for a sequence of events that can lead to one of nature's most spectacular displays. **Figure 16.14e** picks up the evolution of the binary system as star 2 begins to evolve off the main sequence. Like star 1 before it, star 2 grows to fill its Roche lobe; as it does, material from star 2 begins to pour through the "neck" connecting the Roche lobes of the two stars. However, this time the mass is not being added to a normal star but is drawn toward the tiny white dwarf left behind by star 1. Because the white dwarf is so small, the infalling material generally misses the star. Instead of landing directly on the white dwarf, the infalling mass forms an **accretion disk** around the white dwarf, similar in some ways to the accretion disk that forms around a protostar. As in the process of star formation, the accretion disk serves as a way station for material that is destined to find its way onto the white dwarf but that starts out with too much angular momentum to hit the white dwarf directly.

We have already seen that a white dwarf has a mass comparable to that of the Sun, but a size comparable to that of Earth. A large mass and a small radius mean strong gravity. In Chapter 6 we saw how the gravitational energy of material falling toward a forming protostar is converted

> Material from the second star gets hot as it falls onto the white dwarf.

into thermal energy when the material hits the accretion disk around the protostar. The same principle applies here, but with a vengeance. A kilogram of material falling from space onto the surface of a white dwarf releases a hundred times more energy than a kilogram of material falling from the outer Solar System onto the surface of the Sun. The material streaming toward the white dwarf in the binary system falls down into an incredibly deep gravitational "well." All of this energy has to go somewhere, and that somewhere is thermal energy. The spot where the stream of material from star 2 hits the accretion disk can be heated to millions of kelvins, where it glows in the far ultraviolet and X-ray parts of the electromagnetic spectrum.

Eventually the infalling material accumulates on the surface of the white dwarf (**Figure 16.15a**), where it is compressed by the enormous gravitational pull of the white dwarf to a density close to that of the white dwarf itself. As more and more material builds up on the surface of the white dwarf, the white dwarf shrinks (just as the core of the red giant shrinks as it grows more massive). The density increases more and more, and at the same time the release of gravitational energy drives the temperature of the white dwarf higher and higher. Now recall that the infalling material is from the outer, unburned layers of star 2, so it is composed mostly of hydrogen. We have a dangerous situation here. Hydrogen, the best nuclear fuel around, is being compressed to higher and higher densities and heated to higher and higher temperatures on the surface of the white dwarf.

The mental picture you should have at this point is of gasoline pooling on the floor of a match factory. Once the temperature at the base of the layer of hydrogen reaches about 10 million K, hydrogen begins to burn. But this is not the contained hydrogen burning that takes place in the center of the Sun. Rather, this is explosive hydrogen burning in a degenerate gas. Energy

Runaway burning of hydrogen on a white dwarf causes a nova.

released by hydrogen burning drives up the temperature, and the higher temperature drives up the rate of hydrogen burning. This runaway thermonuclear reaction is much like the runaway helium burning that takes place during the helium flash, except now there are no overlying layers of a star to keep things contained. The result is a tremendous explosion—a **nova** (plural: *novae*)—that blows part of the layer covering the white dwarf out into space at speeds of thousands of kilometers per second (see **Figures 16.14f and 16.15b**).

About 50 novae are thought to occur in the Milky Way Galaxy each year, but because our view is restricted by interstellar extinction, we can see only a few of these, typically two or three each year. Novae get bright in a hurry—typically reaching their peak brightness in only a few hours—and for a brief time they can be almost half a million times more luminous than the Sun. Although the brightness of a nova sharply declines in the weeks following the outburst, it can sometimes still be seen for years. During this time the glow from the expanding cloud of material ejected by the explosion is powered by the decay of radioactive isotopes created in the explosion.

The explosion of a nova does not destroy the underlying white dwarf star. In fact, much of the material that had built up on the white dwarf can remain behind after the explosion. The nova leaves the binary system in much the same configuration it was in previously—the configuration shown in Figure 16.15a—with material from star 2 still pouring onto the white dwarf. This cycle can repeat itself many times, with material building up and igniting over and over again on the surface of the white dwarf. In most cases, outbursts are separated by thousands of years, so most novae have been seen only once in historical times. Some novae, however, are known to be recurrent, erupting anew every decade or so.

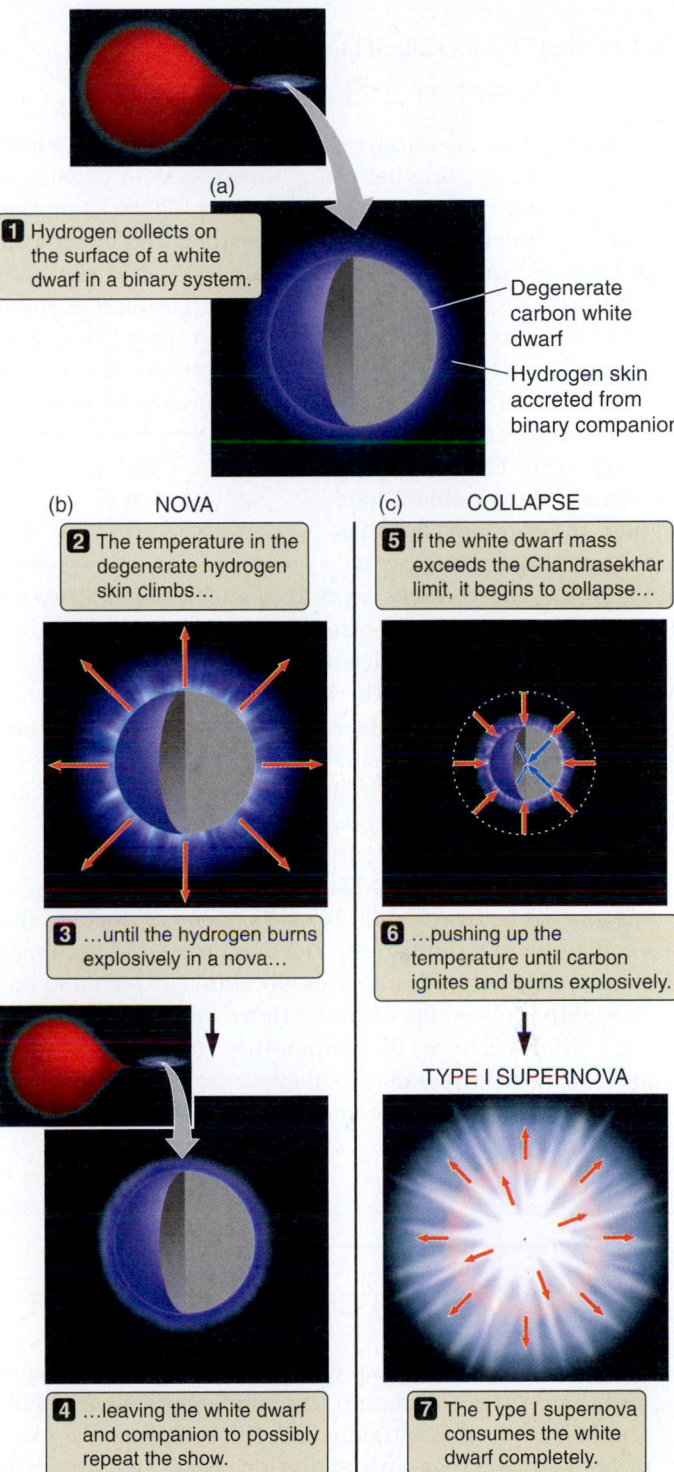

(a)

1 Hydrogen collects on the surface of a white dwarf in a binary system.

Degenerate carbon white dwarf

Hydrogen skin accreted from binary companion

(b) NOVA

2 The temperature in the degenerate hydrogen skin climbs…

3 …until the hydrogen burns explosively in a nova…

4 …leaving the white dwarf and companion to possibly repeat the show.

(c) COLLAPSE

5 If the white dwarf mass exceeds the Chandrasekhar limit, it begins to collapse…

6 …pushing up the temperature until carbon ignites and burns explosively.

TYPE I SUPERNOVA

7 The Type I supernova consumes the white dwarf completely.

FIGURE 16.15 (a) In a binary system in which mass is transferred onto a white dwarf, a skin of hydrogen builds up on the surface of the degenerate white dwarf. (b) If hydrogen burning ignites on the surface of the white dwarf, the result is a nova. (c) If enough hydrogen accumulates to force the white dwarf to begin to collapse, carbon ignites and the result is a Type I supernova.

A Stellar Cataclysm May Await the White Dwarf in a Binary System

It is possible that eventually star 2 will simply go on to form a white dwarf, leaving behind a binary system consisting of two white dwarfs, as in **Figure 16.14g**. There is another possibility, however. Novae may be spectacular events, but they pale in comparison with an alternate fate that may await the white dwarf in a binary system. Through millions of years of mass transfer from star 2 onto the white dwarf, and possibly through countless nova outbursts, the mass of the white dwarf slowly increases. As it does, the white dwarf shrinks, and the grip of self-gravity on this already extremely dense object gets tighter and tighter. The mass of a white dwarf can be pushed

> The mass of a white dwarf cannot exceed 1.4 $M_\odot$.

only so far. There comes a point when even the pressure supplied by degenerate electrons is no longer enough to balance gravity. This precipice occurs at a mass of about 1.4 $M_\odot$, a value referred to as the **Chandrasekhar limit**, named for Subrahmanyan Chandrasekhar (1910–1995), the Indian-born American astrophysicist who derived it. It is impossible for a star to increase in mass above 1.4 $M_\odot$ and remain a white dwarf. Raise its mass above the Chandrasekhar limit, and it will begin to collapse into even more extreme forms of matter (**Figure 16.15c**).

As the white dwarf in our binary system approaches this limit, it grows smaller and smaller and hotter and hotter. As the star crosses the Chandrasekhar limit, it begins to collapse and the conversion of gravitational energy into thermal energy quickly drives the temperature in the white dwarf past the 6×10^8 K needed to slam carbon nuclei together with enough force to overcome their electric repulsion and fuse them. Once again, the onset of nuclear burning in a degenerate gas leads to a thermonuclear runaway, with ever-higher temperatures and ever-faster nuclear burning feeding off each other. But whereas the thermonuclear runaway in a nova involves only a thin shell of material on the surface of the white dwarf, runaway carbon burning involves the entire white dwarf. Within about a second, the whole white dwarf is consumed in the resulting conflagration. In this single instant, 100 times more energy is liberated than will be given off by the Sun over its entire 10-billion-year lifetime on the main sequence. Runaway fusion reactions convert a large fraction of the mass of the star into elements such as iron and nickel, and the explosion blasts the shards of the white dwarf into space at top speeds in excess of 20,000 km/s. The explosion completely destroys star 1, leaving star 2 behind as a lone giant to continue its evolution toward becoming a white dwarf (**Figure 16.14h**).

Explosive carbon burning in a white dwarf is a leading theory used to explain colossal events called **Type I supernovae**. Type I supernovae[6] occur in a galaxy the size of the Milky Way about once a century. For a brief time they can shine with a luminosity

> The carbon in a white dwarf can ignite in a Type I supernova.

10 billion times that of our Sun, possibly outshining the galaxy itself.

---

[6] Strictly speaking, we should refer here to Type Ia supernovae. Technically, a "Type I" supernova is a supernova that shows no emission lines from hydrogen. This category includes some variants of the supernovae from massive stars that will be discussed in the next chapter. We hope the purists will forgive our simplification of referring to all supernovae from white dwarfs as Type I and all supernovae from massive stars as Type II.

Seeing the Forest for the Trees

Stars are born of the tenuous matter of interstellar space, and then settle down onto the main sequence as stable and steady cosmic citizens. But the universe is a place of constant change and evolution. Stars consume the fuel they were born with, and when that fuel is gone they must die.

In the world of humans, it is wealth and power and birthright that determine our lifestyle. In the lives of stars, birthright is spelled "mass." Mass determines the weight bearing down on the interior of a star, which in turn controls the rate of the nuclear reactions taking place there. High-mass stars are spectacular but ephemeral, burning out the fuel in their cores in only a few million years. Our Sun, however, is typical of the much longer-lived low-mass stars. Yet eventually, even these steady performers must exhaust the hydrogen at their centers.

In this chapter we have followed the evolution of a low-mass star like our Sun as it evolves beyond the main sequence, passing through a fascinating series of evolutionary stages. Some of these stages are dramatic and violent, such as the runaway thermonuclear reactions of the helium flash or the ejection of a planetary nebula that carries the chemical legacy of nuclear burning deep within the dying AGB star. Other phases are far more quiescent, such as the star's time on the horizontal branch burning helium to carbon in its core and hydrogen to helium in a surrounding shell. Eventually, even an undistinguished star like our Sun may reach a luminosity rivaling all but the most luminous main-sequence O stars, and swell to a size that would swallow much of our inner Solar System. Curiously, such a behemoth is supported by frenzied nuclear reactions in a region surrounding the tiniest of cores—a superdense inert cinder crushed by the weight of the overlying star to a size not much larger than Earth. At a density of a thousand kilograms per cubic centimeter, the electron-degenerate core defies our normal notions of matter.

All of these stages are signposts along the star's inexorable march toward its eventual fate, in which the outer layers of the star will have been ejected into space and nothing will remain but a tiny cooling cinder—a degenerate white dwarf that will eventually fade from the notice of the rest of the universe. Yet if it happens to orbit about a binary companion, the white dwarf may live again. Unburned nuclear fuel pouring onto the surface of the white dwarf may trigger the episodic nuclear outbursts of novae or even a cataclysmic Type I supernova.

We have never witnessed the life cycle of a low-mass star from birth to death, nor will we. Not even a hundred million human lifetimes would be enough to see the saga through from beginning to end. Yet we are confident of most of what we have reported here, just as we are confi-dent of our knowledge of the internal structure of the Sun and Earth. Physics works, and the closing decades of the 20th century saw steady improvements in the quality of our calculations of the structure and evolution of stars. And although we cannot witness the life cycle of a single star, when we look at the sky we find many examples of red giants, AGB stars, white dwarfs, and the other members of this evolutionary menagerie. Later on our journey we will also see how clusters of stars that form together provide snapshots that enable us to test our theories of stellar evolution in exquisite detail.

What we have seen is representative of the fate awaiting the vast majority of stars in the universe. In the next chapter we turn our attention to the evolution of the small fraction of stars that fall into the category of high-mass stars. There we will find that the extreme and counterintuitive behavior of white dwarfs presages a whole zoo of bizarre and exotic objects. Where low-mass stars have bent our notions about matter, high-mass stars will shatter those notions, leading us to consider forms of matter so extreme that they cease to be matter at all.

Summary

- All stars eventually exhaust their nuclear fuel.

- The more massive a star is, the shorter will be its lifetime.

- Helium accumulates like ash in the cores of main-sequence stars.

- After exhausting its hydrogen, a low-mass star leaves the main sequence and swells to become a red giant.

- A red giant burns helium via the triple-alpha process.

- In their dying stages, some stars form planetary nebulae.

- All low-mass stars eventually become white dwarfs.

- A white dwarf is very hot but very small, and cannot exceed 1.4 $M_\odot$.

- White dwarfs cool to become cold, dark cinders—the stellar graveyard.

- Transfer of mass within a binary system can lead to novae or supernovae.

 SmartWork, Norton's online homework system, includes algorithmically generated versions of these questions, plus additional conceptual exercises. If your instructor assigns questions in SmartWork, log in at **smartwork.wwnorton.com.**

Student Questions

THINKING ABOUT THE CONCEPTS

1. Why do most nearby stars turn out to be low-mass, low-luminosity stars?

2. The Sun is slowly becoming more luminous as it ages on the main sequence. How do you think this change will affect terrestrial life?

3. Is it possible for a star to skip the main sequence and immediately begin burning helium in its core? Explain your answer.

4. What is the primary reason that the most massive stars have the shortest lifetimes? (Note: The answer can be expressed in just a few words.)

5. Consider a hypothetical star having the same mass as our Sun being one of the first stars that formed when the universe was still very young. Would you expect it to be surrounded by planets? Explain your answer.

6. Describe some possible ways in which a star might increase the temperature within its core while at the same time lowering its density.

7. Suppose a main-sequence star were to suddenly start burning hydrogen at a faster rate in its core. How would the star react? Discuss changes in size, temperature, and luminosity.

8. We typically say that the mass of a newly formed star determines its destiny from birth to death. However, there is a frequent environmental circumstance for which this statement is not true. Identify this circumstance and explain why the birth mass of a star might not fully account for its destiny.

9. What physical effect or explanation do you think could be responsible for the "break" in slope in the mass-luminosity relationship between low- and high-mass stars, as shown in Figure 16.1? Justify your answer.

**10. Is it fair to assume that stars do not change their structure while on the main sequence? Why or why not?

11. Suppose Jupiter were not a planet, but instead were a G5 main-sequence star with a mass of 0.8 $M_\odot$.
  a. How do you think terrestrial life would be affected, if at all?
  b. How would the Sun be affected as it came to the end of its life?

12. When a star runs out of nuclear fuel at the core, why does it become more luminous? Why does the surface temperature of the star cool down?

13. When a star leaves the main sequence, its luminosity increases tremendously. What does this increase in luminosity imply about the amount of time the star has left to live, or the amount of time the star spends on any subsequent part of its evolutionary path?

14. Why are the paths along the H-R diagram that a star follows as it forms (from a protostar) and as it leaves the main sequence (climbing the red giant branch) so similar?

**15. When compressed, ordinary gas heats up but degenerate gas does not. Why, then, does a degenerate core heat up as the star continues shell burning around it?

16. Suppose you were an astronomer making a survey of the observable stars in our galaxy. What would be your chances of seeing a star undergoing the helium flash? Explain your answer.

*17. Why is a horizontal branch star (which burns helium at a high temperature) less luminous than a red giant branch star (which burns hydrogen at a lower temperature)?

*18. Suppose a star is able to heat its core temperature high enough to begin fusing oxygen. Predict how the star will continue to evolve. Include how you think the star will evolve on the H-R diagram.

19. As an AGB star evolves into a white dwarf, it runs out of nuclear fuel, and one might argue that the star should cool off and move to the right of the H-R diagram. Why does the star move instead to the left?

20. Why *must* a white dwarf move down and to the right along the H-R diagram?

*21. Why does fusion in degenerate material *always* lead to a runaway reaction?

22. The intersection of the Roche lobes in a binary system is the equilibrium point between the two stars where the gravitational attraction from both stars is equally strong and opposite in direction. Is this an example of stable or unstable equilibrium? Explain.

23. Suppose the more massive red giant star in a binary system engulfs its less massive main-sequence companion, and their nuclear cores combine. What structure do you think the new star will have? Where will the star lie along the H-R diagram?

24. In Latin, *nova* means "new." Novae, as we now know, are not new stars. Explain how novae might have gotten their name and why they are really not new stars.

25. T Corona Borealis is a well-known recurrent nova.
  a. Is it a single star or a binary system? Explain.
  b. What mechanism causes a nova to flare up?
  c. How can a nova flare-up happen more than once?

APPLYING THE CONCEPTS

26. For most stars on the main sequence, luminosity scales with mass as $M^{3.5}$ (see Math Tools 16.1). What luminosity does this relationship predict for (a) 0.5-$M_\odot$ stars, (b) 6-$M_\odot$ stars, and (c) 60-$M_\odot$ stars? Compare these numbers to values given in Table 16.1.

27. Compute the main-sequence lifetimes for (a) 0.5-$M_\odot$ stars, (b) 6-$M_\odot$ stars, and (c) 60-$M_\odot$ stars. Compare them to the values given in Table 16.1.

28. Notice in Figure 16.1 that low-mass and high-mass stars have different mass-luminosity relationships with the average value of $M^{3.5}$. What does this fact imply about the actual main-sequence lifetimes of a 0.5- or 6-$M_\odot$ star, as compared to the predictions you made in question 27?

29. The escape velocity ($v_{esc} = \sqrt{2\,GM_\odot/r_\odot}$) from the Sun's surface today is 618 km/s.
  a. What will the escape velocity be when the Sun becomes a red giant with a radius 50 times greater and a mass only 0.9 times that of today?

b. What will it be when the Sun becomes an AGB star with a radius 200 times greater and a mass only 0.7 times that of today?

c. How would these changes in escape velocity affect mass loss from the surface of the Sun as a red giant, and later as an AGB star?

30. Each form of energy generation in stars depends on temperature.

a. The rate of hydrogen fusion (proton-proton chain) near 10^7 K increases with temperature as T^5. If you raise the temperature of the hydrogen-burning core by 10 percent, how much does the hydrogen fusion energy increase?

b. Helium fusion (the triple-alpha process) at 10^8 K increases with an increase in temperature at a rate of T^{38}. If you raise the temperature of the helium-burning core by 10 percent, how much does the helium fusion energy increase?

31. After leaving the main sequence, low-mass stars "seed" the interstellar medium with heavy elements such as carbon, nitrogen, and oxygen. Assume that the universe is approximately 1.4×10^{10} (14 billion) years old and that the Sun is 4.6 billion years old. Refer to Table 16.1 for the masses and main-sequence lifetimes of stars. How many generations of each type of star preceded the birth of our Sun and Solar System?

32. Assume that there are twice as many F0 stars as A0 stars, and that each type contributes heavy elements to the interstellar medium in approximate proportion to its main-sequence mass. What was the relative contribution from the two types of stars before the Sun was born?

33. Roughly how large does a planetary nebula grow before it disperses? Use an expansion rate of 20 km/s and a lifetime of 50,000 years.

34. In typical binary systems, a red giant can transfer mass onto a white dwarf at rates of about 10^{-9} $M_\odot$ per year. Roughly how long after mass transfer begins will the white dwarf undergo a Type I supernova? How does this length of time compare to the typical lifetime of a low-mass star? (Hint: Assume that a typical white dwarf starts with a mass of 0.6 $M_\odot$.)

**35. How fast does material in an accretion disk orbit around a white dwarf? (Hint: Use Kepler's third law.)

36. A white dwarf has a density of approximately 10^9 kilograms per cubic meter (kg/m^3). Earth has an average density of 5,500 kg/m^3 and a diameter of 12,700 kilometers (km). If Earth were compressed to the same density as a white dwarf, how large would it be?

37. What is the density of degenerate material? Calculate how large the Sun would be if all of its mass were degenerate.

38. Recall from Chapter 4 that the luminosity of a spherical object at temperature T is given by $L = 4\pi R^2 \sigma T^4$, where R is the object's radius. If the Sun became a white dwarf with a radius of 10^7 meters, what would be its luminosity at temperatures of (a) 10^8 K, (b) 10^6 K, (c) 10^4 K, and (d) 10^2 K?

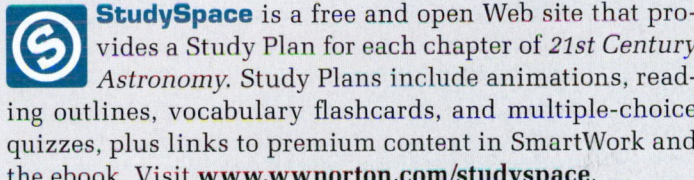 **StudySpace** is a free and open Web site that provides a Study Plan for each chapter of *21st Century Astronomy*. Study Plans include animations, reading outlines, vocabulary flashcards, and multiple-choice quizzes, plus links to premium content in SmartWork and the ebook. Visit **www.wwnorton.com/studyspace**.

It does not do to leave a live dragon out of your calculations, if you live near him.

J. R. R. TOLKIEN (1892–1973), *THE HOBBIT*

The inner Crab Nebula seen in X-ray (blue) and visible (red) light.

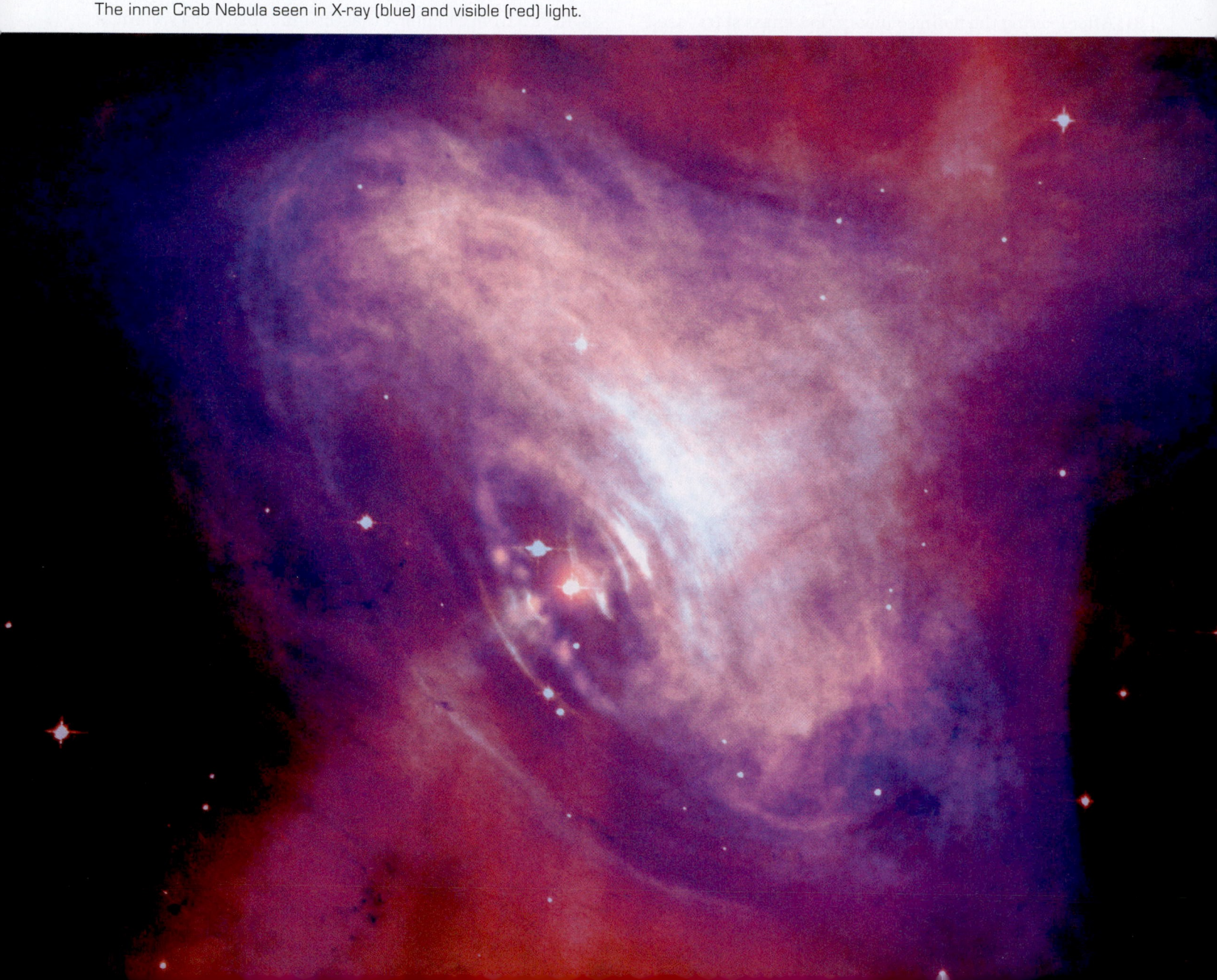

CHAPTER 17

Live Fast, Die Young— High-Mass Stellar Evolution

17.1 Here Be Dragons

So far in our discussion of the lives of stars, we have concentrated on what happens to low-mass stars like our Sun. These are stars that steadily burn hydrogen to helium in their cores and produce a fairly constant output of energy for billions of years—long enough for our Earth to cool from a molten ball of rock and for life to evolve to the point of contemplating its place in the universe. High-mass stars— stars with masses greater than about 8 solar masses ($M_\odot$)— are very different beasts. These are objects that burn with luminosities thousands or even millions of times as great as the luminosity of our Sun and squander their generous allotments of nuclear fuel in less time than mammals have walked on Earth.

The differences between the ways in which high-mass stars and low-mass stars evolve follow from the same relationship among gravity, pressure, and the rate of nuclear burning that determines the balance in a star like our Sun. The litany should be familiar by now: More mass means stronger gravity and therefore more weight bearing down on the inner parts of the star. The increase in weight means higher pressure; higher pressure means faster reaction rates; and faster reaction rates mean greater luminosity. There are many differences in the evolution of low- and high-mass stars, but in the end they all trace back to the greater gravitational force bearing down on the interior of a high-mass star.

KEY CONCEPTS

Our Sun is but a pale ember, compared with the brilliance of more massive stars. These stars consume their fuel in a cosmic blink of the eye, and then in their death they dazzle us with brilliant fireworks, seed the universe with the elements of life, and force us to once again alter our concept of space and time. As our exploration of stars draws to a close, we will discover

- How the structure of high-mass stars differs from that of low-mass stars.

- The stages of nuclear burning that evolving high-mass stars experience as they form progressively more massive chemical elements.

- That massive stars die as Type II supernovae.

- The origin of the most massive chemical elements.

- A menagerie of bizarre objects, including neutron stars, pulsars, and X-ray binary stars.

- How H-R diagrams enable us to measure the ages of stars and test our theories of stellar evolution.

- That the speed of light is a universal constant, which forces us to completely rethink classical physics.

- That gravity is a consequence of the way mass distorts the very shape of spacetime.

- The bottomless pits in spacetime that we call black holes.

(a) CNO cycle

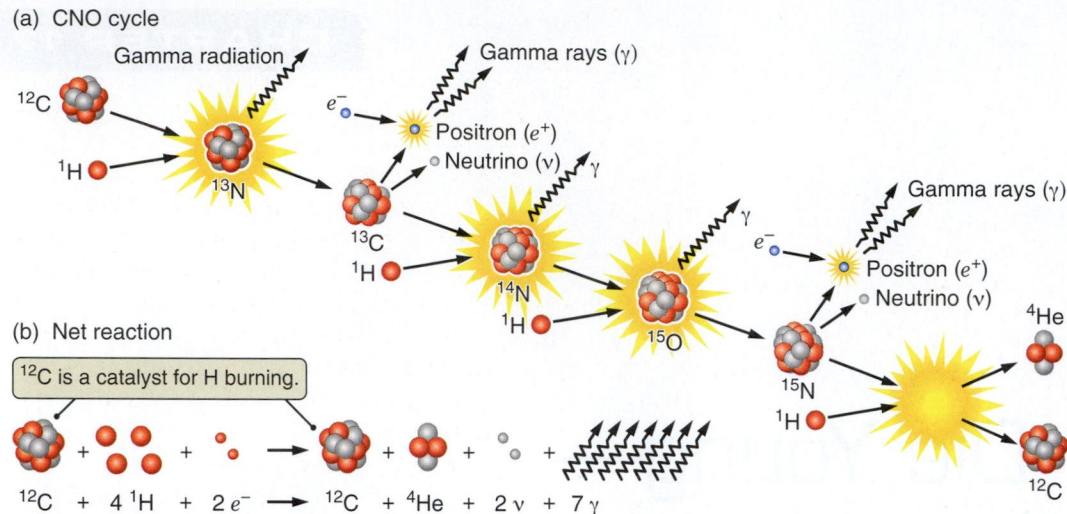

(b) Net reaction

^{12}C is a catalyst for H burning.

$$^{12}C \;+\; 4\,^{1}H \;+\; 2\,e^{-} \;\longrightarrow\; ^{12}C \;+\; ^{4}He \;+\; 2\,\nu \;+\; 7\,\gamma$$

FIGURE 17.1 In high-mass stars, carbon serves as a catalyst for fusion of hydrogen to helium. This process is called the carbon-nitrogen-oxygen (or CNO) cycle.

17.2 High-Mass Stars Follow Their Own Path

Recall from Chapter 14 that the first step in the hydrogen-burning proton-proton chain is the collision and fusion of two protons. Protons have only a single positive charge, so they are the easiest atomic nuclei to force close enough together to fuse. This is a great advantage for low-mass stars, in which hydrogen burns at temperatures as low as a few million kelvins. However, the proton-proton chain suffers from a large disadvantage as well. The "affinity" between two protons is not very great. Even when two protons are slammed together hard enough for the strong nuclear force to act on them, the likelihood that they will fuse is low. The lack of vigor of this first step in the proton-proton chain limits how rapidly the entire process can run.

At the much higher temperatures at the center of a high-mass star, additional nuclear reactions become possible. In particular, hydrogen nuclei are able to interact with the nuclei of more massive elements such as carbon. It takes a lot of energy to get past the electric barrier that repels a carbon nucleus, with its six protons, from a hydrogen nucleus. But if this barrier can be overcome, the strong-nuclear-force interaction between the carbon and the hydrogen is much more favorable than the interaction between two hydrogen nuclei. This reaction, in which a carbon-12 (^{12}C) nucleus and a proton combine to form a nitrogen-13 (^{13}N) nucleus consisting of seven protons and six neutrons, is the first reaction in a chain of events that is referred to as the **carbon-nitrogen-oxygen (CNO) cycle**. The remaining steps in the CNO cycle are shown in **Figure 17.1**. Note that carbon is not consumed by the CNO cycle but instead is a catalyst. The ^{12}C nucleus that started the cycle is there again at the end of the cycle, but now instead of the four hydrogen nuclei that took part in the chain of reactions, we have a helium nucleus. **Figure 17.2** shows that the CNO cycle becomes far more efficient than the proton-proton chain in stars more massive than 1.5 $M_\odot$.

> The **CNO** cycle burns hydrogen in massive stars.

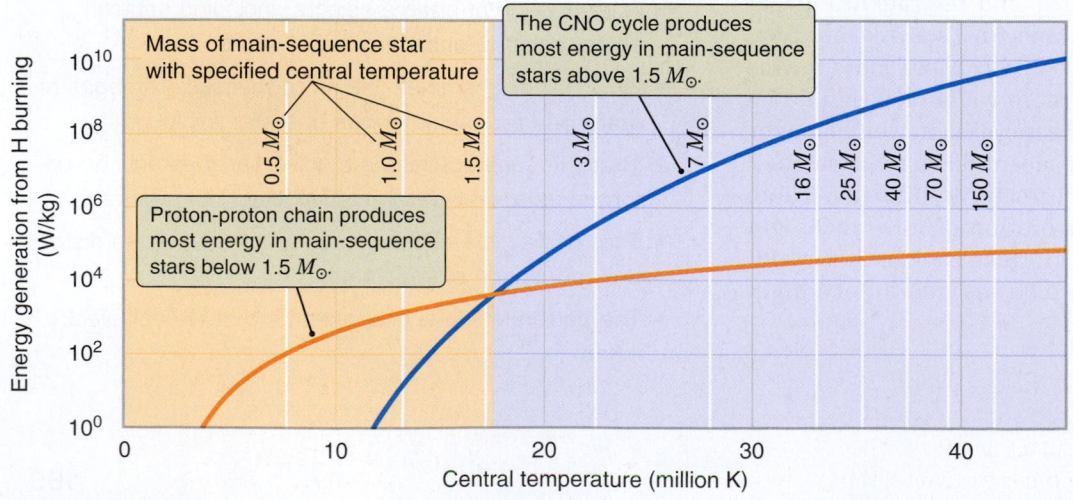

FIGURE 17.2 Plots of the rate of energy generation as a function of temperature for the proton-proton chain and the CNO cycle. At the higher central temperatures of stars more massive than 1.5 $M_\odot$, the CNO cycle more efficiently fuses hydrogen into helium.

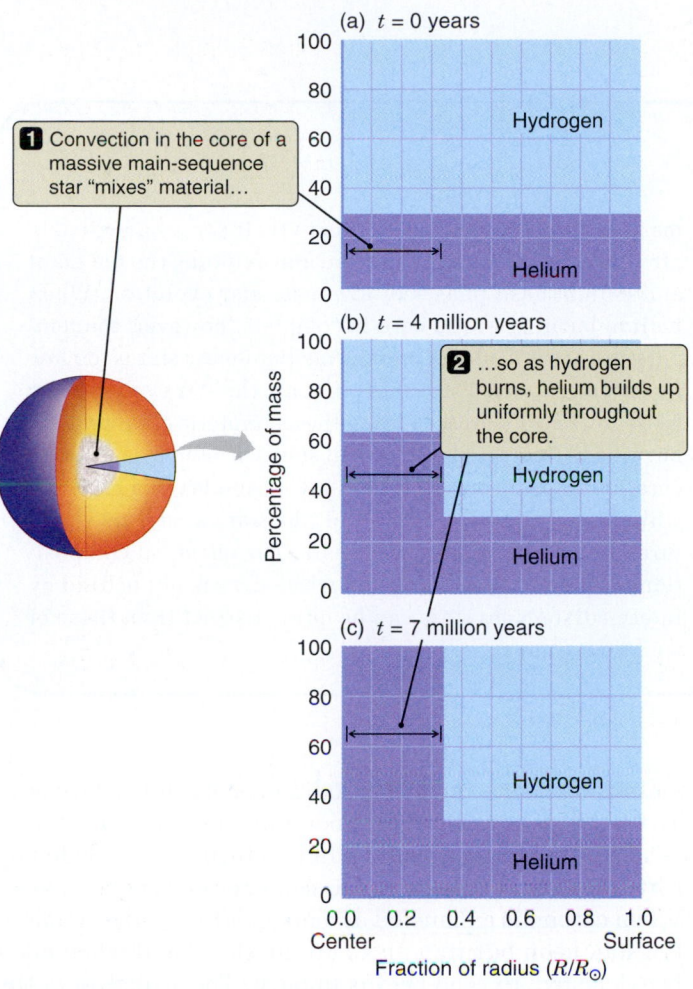

(a) $t = 0$ years

1 Convection in the core of a massive main-sequence star "mixes" material...

Hydrogen

Helium

(b) $t = 4$ million years

2 ...so as hydrogen burns, helium builds up uniformly throughout the core.

Hydrogen

Helium

(c) $t = 7$ million years

Hydrogen

Helium

Percentage of mass

Center   Surface
Fraction of radius ($R/R_\odot$)

FIGURE 17.3 Convection keeps the core of a high-mass main-sequence star well mixed, so the composition remains uniform throughout the evolving core. (Evolution times are for a 25-$M_\odot$ star.)

The different ways that hydrogen burning takes place in high- and low-mass stars are reflected in the different structures of their cores. The temperature gradient in the core of a high-mass star is so steep that convection sets in within the core itself, "stirring" the core like the water in a boiling pot. Compare **Figure 17.3** with Figure 16.2. Rather than building up from the center outward, helium ash is spread uniformly throughout the core of a high-mass star as the star consumes its hydrogen.

> **Convection stirs the core of a high-mass star.**

The High-Mass Star Leaves the Main Sequence

These distinctions between high- and low-mass stars might seem a bit esoteric, but as the high-mass star's life on the main sequence comes to an end, the

visible differences in its structure and evolution become far more pronounced. As the high-mass star runs out of hydrogen in its core, the weight of the overlying star compresses the core, just as it does in a low-mass star. Yet long before the core of the high-mass star becomes electron-degenerate, the pressure and temperature in the core reach the 10^8-kelvin (K) point needed for helium burning to begin. There will be no growing degenerate core in the high-mass star. There will be no accelerating ascent up the red giant and asymptotic giant branches on the H-R diagram. Instead, the star makes a fairly smooth transition from hydrogen burning to helium burning. The overall structure of the star responds to the changes taking place in its interior, but its luminosity will change relatively little.

When a low-mass star leaves the main sequence, the path that it follows on the H-R diagram is largely vertical (see Figure 16.4), going to higher and higher luminosities. As the high-mass star leaves the main sequence, on the other hand, it grows in size while its surface temperature falls, so it moves mostly off to the right on the H-R diagram (**Figure 17.4**). The massive star now has the same structure as a low-mass horizontal branch star, burning helium in its

> **High-mass stars skip the red giant phase.**

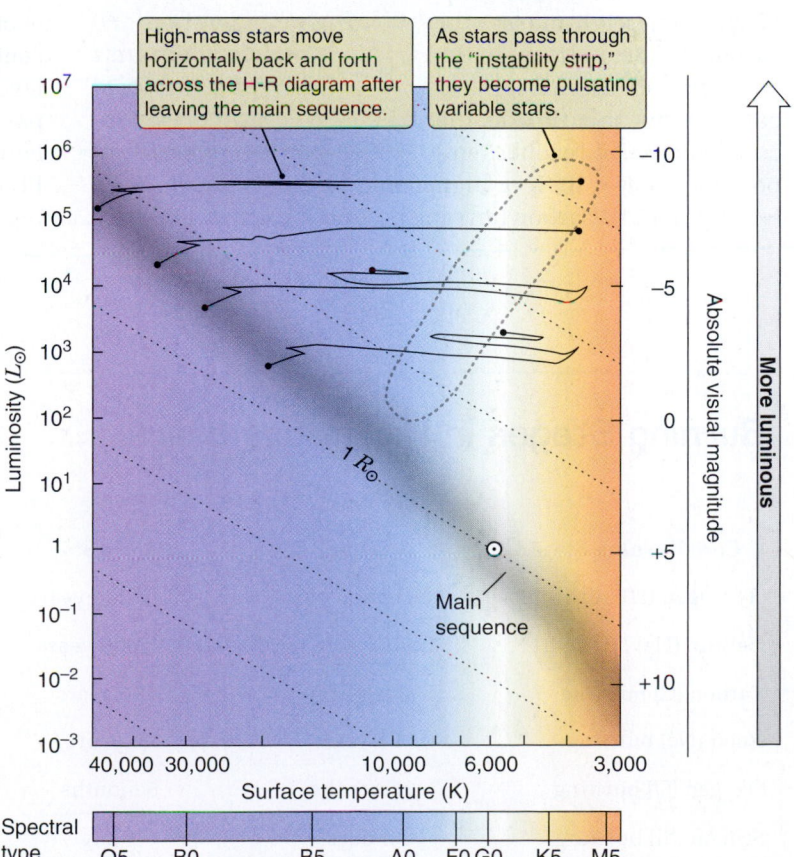

FIGURE 17.4 When massive stars leave the main sequence, they move horizontally across the H-R diagram.

High-mass stars move horizontally back and forth across the H-R diagram after leaving the main sequence.

As stars pass through the "instability strip," they become pulsating variable stars.

$1 R_\odot$

Main sequence

Luminosity ($L_\odot$)

Surface temperature (K)

Absolute visual magnitude

More luminous

Spectral type

O5   B0   B5   A0   F0 G0   K5   M5

EXCURSIONS 17.1

Each Star Is Unique

The distinction made in this book between low-mass and high-mass stars is a useful, convenient way to think about stellar evolution. After all, there is a fundamental and qualitative difference between stars that end their lives quietly as dying cinders and those that die in spectacular explosions. Even so, this distinction is an oversimplification. In reality, each star is unique, with its own individual mass and composition and its own particular circumstances.

Stars with masses between about 3 and 8 $M_\odot$ exist in something of a gray area between the low-mass stars discussed in Chapter 16 and the high-mass stars discussed in this chapter. Like massive stars, these **intermediate-mass stars**, despite being on the main sequence, burn hydrogen via the CNO cycle. They also leave the main sequence as massive stars do, burning helium in their cores immediately after their hydrogen is exhausted and skipping the red giant and helium flash phases of low-mass star evolution. When helium burning in the core is complete, however, the temperature at the center of an intermediate-mass star is too low for carbon to burn. From this point on, the star evolves more like a low-mass star, ascending the asymptotic giant branch, burning helium and hydrogen in shells around a degenerate core, and then ejecting its outer layers and leaving behind a white dwarf. Yet even at the end, the star carries the signature of its early "high-mass" years. The chemical compositions of planetary nebulae and white dwarfs left behind by intermediate-mass stars can be quite distinct from those of truly low-mass stars.

---

core and hydrogen in a surrounding shell, but it has skipped the red giant phase of low-mass star evolution. (Some less massive stars skip the red giant phase as well. See **Excursions 17.1** for a more nuanced view of the distinction we have drawn between low-mass and high-mass stars.)

The next stage in the evolution of a high-mass star has no analog in low-mass stars. When the high-mass star exhausts the helium in its core, the core again begins to collapse, but this time as the core collapses it reaches temperatures of 8×10^8 K or higher, and carbon begins to burn (see **Table 17.1**). Carbon burning produces a number of more

> Progressively more massive elements burn as massive stars evolve.

massive elements, including sodium, neon, and magnesium. The star now consists of a carbon-burning core surrounded by a helium-burning shell, which in turn is surrounded by a hydrogen-burning shell. The sequence does not end here. When carbon is exhausted as a nuclear fuel at the center of the star, neon burning picks up the slack; and when neon is exhausted, oxygen begins to burn. The structure of the evolving high-mass star, shown in **Figure 17.5**, is reminiscent of an onion, with layer surrounding layer surrounding layer. As we move inward toward the center of the star, we pass through a layer of hydrogen burning, then a layer of helium burning, then a layer of carbon burning, and so on. Finally, we reach the most advanced stages of nuclear burning in the core of the star.

TABLE 17.1

Burning Stages in High-Mass Stars

Core Burning Stage	9-$M_\odot$ Star	25-$M_\odot$ Star	Typical Core Temperatures
Hydrogen (H) burning	20 million years	7 million years	$(3–10) \times 10^7$ K
Helium (He) burning	2 million years	700,000 years	$(1–7.5) \times 10^8$ K
Carbon (C) burning	380 years	160 years	$(0.8–1.4) \times 10^9$ K
Neon (Ne) burning	1.1 years	1 year	$(1.4–1.7) \times 10^9$ K
Oxygen (O) burning	8 months	6 months	$(1.8–2.8) \times 10^9$ K
Silicon (Si) burning	4 days	1 day	$(2.8–4) \times 10^9$ K

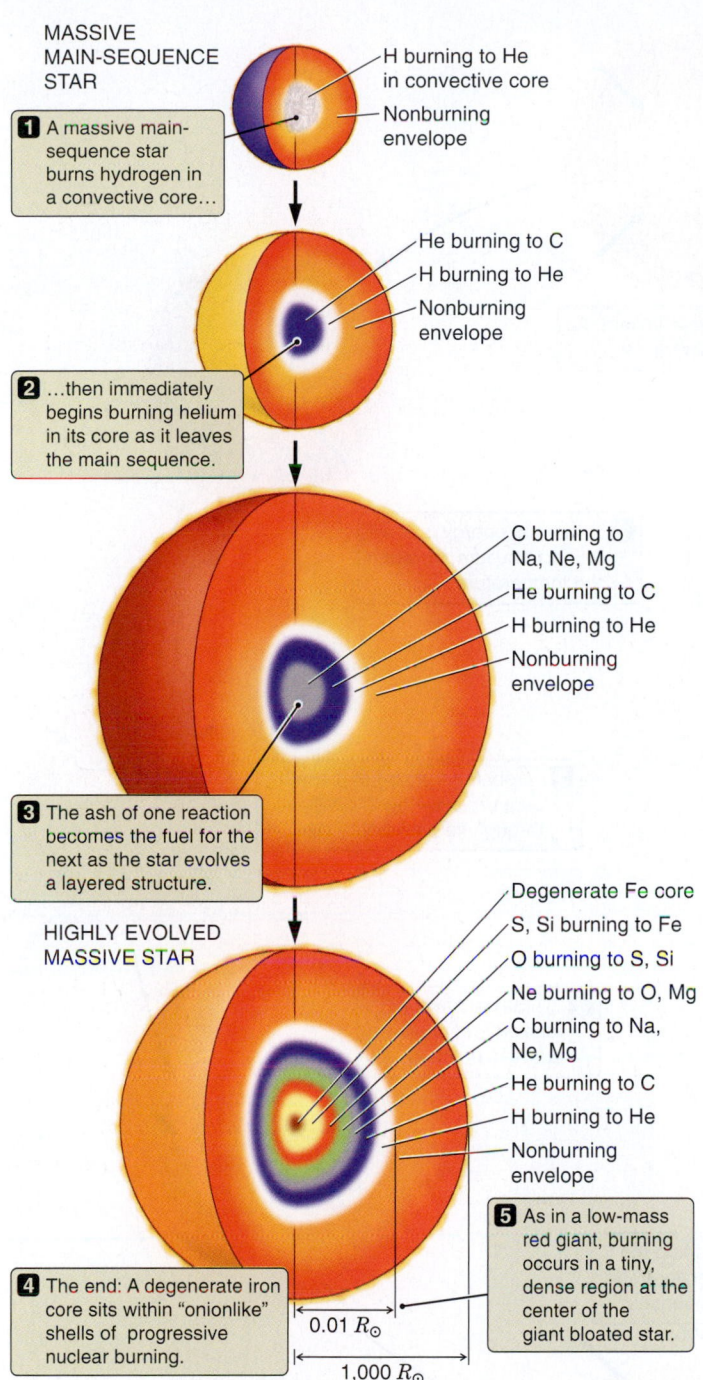

MASSIVE MAIN-SEQUENCE STAR

H burning to He in convective core

Nonburning envelope

1 A massive main-sequence star burns hydrogen in a convective core…

He burning to C

H burning to He

Nonburning envelope

2 …then immediately begins burning helium in its core as it leaves the main sequence.

C burning to Na, Ne, Mg

He burning to C

H burning to He

Nonburning envelope

3 The ash of one reaction becomes the fuel for the next as the star evolves a layered structure.

HIGHLY EVOLVED MASSIVE STAR

Degenerate Fe core

S, Si burning to Fe

O burning to S, Si

Ne burning to O, Mg

C burning to Na, Ne, Mg

He burning to C

H burning to He

Nonburning envelope

4 The end: A degenerate iron core sits within "onionlike" shells of progressive nuclear burning.

$0.01\ R_\odot$

$1{,}000\ R_\odot$

5 As in a low-mass red giant, burning occurs in a tiny, dense region at the center of the giant bloated star.

FIGURE 17.5 As a high-mass star evolves, it builds up a layered structure like that of an onion, with progressively more advanced stages of nuclear burning found deeper and deeper within the star. Note the change in scale for the bottom image.

Not All Stars Are Stable

When we discussed the structure of the Sun in Chapter 14 and a low-mass star in Chapter 16, we asserted that a single stable solution always exists. Given one solar mass with a mix of 70 percent hydrogen and 30 percent helium, that mass *will* form a stable star with the structure of our Sun. As we

"what if" the Sun, considering the consequences of forcing the Sun to be larger or smaller than it is, we find that changes in the star would cause it to settle back into this unique balance between the inward pull of gravity and the outward push of pressure. This relationship remains true for high-mass stars on the main

> Evolved stars pulsate while passing through the instability strip.

sequence as well. However, it is *untrue* for many evolved stars. As a star undergoes post–main sequence evolution, it may make one or more passes through a region of the H-R diagram known as the **instability strip** (see Figure 17.4). Rather than finding a steady balance, stars in the instability strip pulsate, alternately growing larger and smaller.

Stars lying within the instability strip of the H-R diagram are heat engines, powering their pulsation by tapping into the thermal energy flowing outward through them. Changes in the ionization state of the gas within the star alternately trap thermal energy, forcing the pressure in the star up and causing it to expand, and then they release this energy, allowing the star to contract again. These changes serve the role of the valves in an engine: they rob the star of pressure support at one part of its cycle, allowing it to shrink too far, and then pump the pressure in the star back up at a different point in its cycle, puffing up the star to a size larger than it can support. The process is illustrated in **Figure 17.6**. Rather than settling at a constant radius, like the Sun, the star alternately expands and shrinks, moving out and in like the piston of a steam engine. The pulsations in the outer parts of the star have very little effect on the nuclear burning in the star's interior. However, the pulsations do affect the light escaping from the star. Recall from Chapter 13 the **luminosity-temperature-radius relationship** for stars: both the luminosity and the color of the star change as the star expands and shrinks. The star is at its brightest and bluest while it expands through its equilibrium size, and at its faintest and reddest while it falls back inward. Such stars are referred to as **pulsating variable stars**.

The most massive and luminous pulsating variable stars are named **Cepheid variables**, after Delta Cephei, the first recognized member of this class. A Cepheid variable completes one cycle of its pulsation in anywhere from about 1 to 100 days, depending on its luminosity. The more luminous the star, the longer it takes it to complete its cycle. This **period-luminosity relationship** for Cepheid variables, first discovered experimentally by Henrietta Leavitt (1868–1921) in 1912, is the basis for the use of Cepheid variables as indicators of the distances to galaxies beyond our own.

Cepheid variables are not the only type of **variable star**. The horizontal branch of the evolutionary tracks of low-mass stars (see Figure 16.8) may pass through the instability strip as well. These unstable horizontal branch stars are known as **RR Lyrae variables** after their prototype. They pulsate by the same mechanism as Cepheid variables but are typically hundreds of times less luminous. The instability strip also intersects the main sequence around spectral type A, and many A stars do show significant variability (although

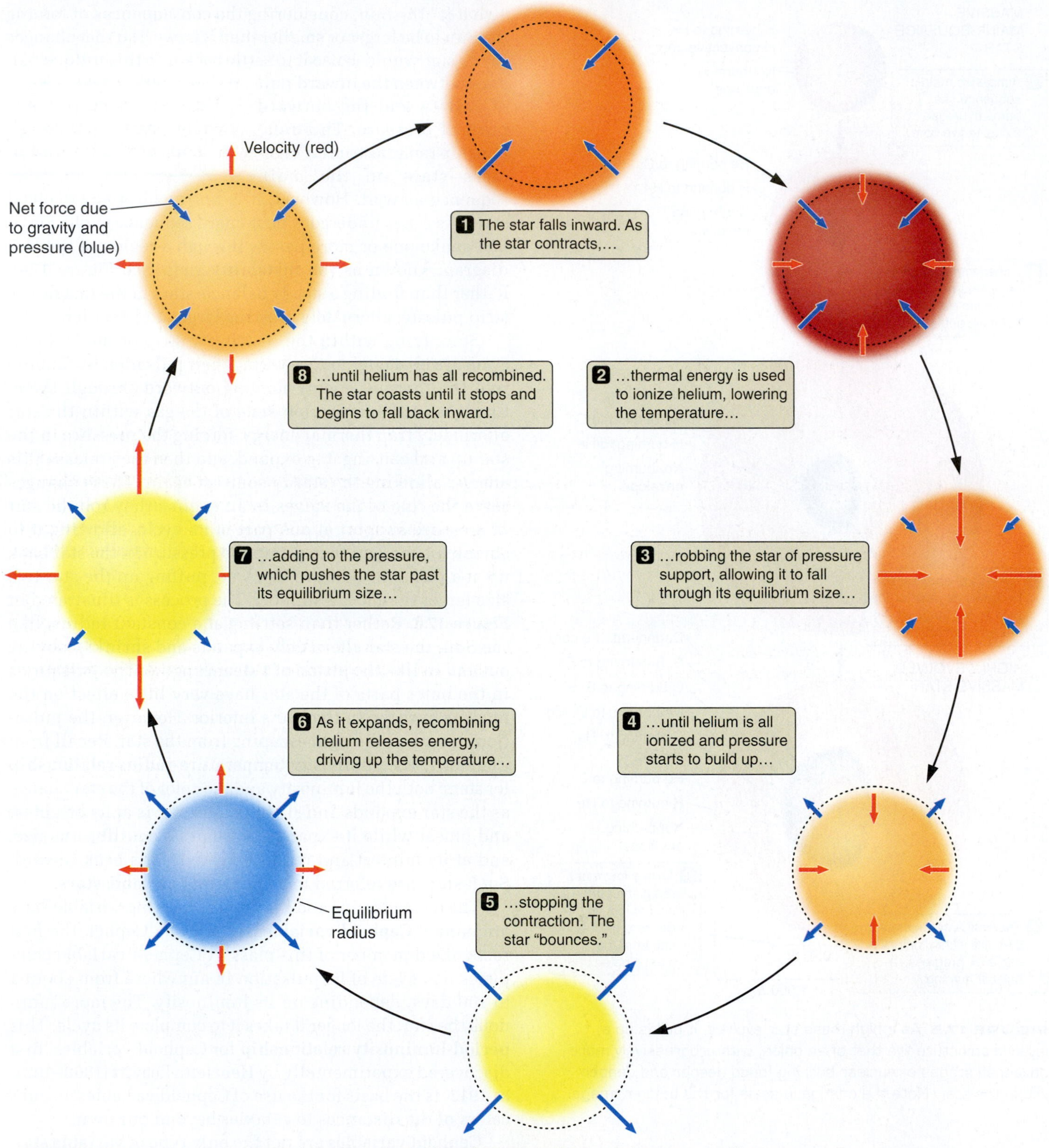

Velocity (red)

Net force due to gravity and pressure (blue)

1 The star falls inward. As the star contracts,…

2 …thermal energy is used to ionize helium, lowering the temperature…

8 …until helium has all recombined. The star coasts until it stops and begins to fall back inward.

7 …adding to the pressure, which pushes the star past its equilibrium size…

3 …robbing the star of pressure support, allowing it to fall through its equilibrium size…

6 As it expands, recombining helium releases energy, driving up the temperature…

4 …until helium is all ionized and pressure starts to build up…

Equilibrium radius

5 …stopping the contraction. The star "bounces."

FIGURE 17.6 In a pulsating Cepheid variable, ionization of atoms in the star's interior acts like the valves in a steam engine, alternately allowing the surface of the star to fall inward, and then pushing it back out again. (Color changes shown here are greatly exaggerated.)

it is much less pronounced than the variability in Cepheid or RR Lyrae variables). A number of other kinds of variable stars are seen elsewhere in the H-R diagram, each driven by its own variety of "engine."

High-mass stars also change their composition by violently expelling a significant percentage of their mass back into space throughout their existence. Even while on the main sequence, massive O and B stars drive away tenuous winds at velocities that can reach 3,000 kilometers per second (km/s). These winds are pushed outward by the pressure of the radiation from the star. We do not normally think of light as "pushing" on something, but the pressure of the intense radiation at the surface of a massive star is enough to overcome the star's gravity and drive away material in its outermost layers. Main-sequence O and B stars lose mass at rates ranging from about 10^{-7} to 10^{-5} $M_\odot$ of material per year, with the greatest mass loss occurring in the most massive stars.

> Massive stars lose mass as they generate high-velocity winds.

These numbers may sound tiny, but added up over millions of years they mean that mass loss plays a prominent role in the evolution of high-mass stars. O stars with masses of 20 $M_\odot$ or more may lose about 20 percent of their mass while on the main sequence, and possibly more than 50 percent of their mass over their entire lifetimes. Even an 8-$M_\odot$ star may lose 5–10 percent of its mass. An extreme example of mass loss in a massive star is seen in Eta Carinae (**Figure 17.7**), a 100-$M_\odot$ star with a luminosity (summed over all wavelengths) of 5 million Suns. Currently, Eta Carinae is losing mass at a rate of about 10^{-3} $M_\odot$ per year (or 1 $M_\odot$ every 1,000 years). However, during a 19th century eruption, when Eta Carinae became the second-brightest star in the sky, its mass loss must have reached the amazing rate of 0.1 $M_\odot$ per year, shedding 2 $M_\odot$ of material over a mere 20 years. Eta Carinae is expected to explode as a supernova in the astronomically near future.

17.3 High-Mass Stars Go Out with a Bang

We have already seen how a low-mass star approaches the end of its life—relatively slowly and gently, ejecting its outer parts into nearby space (sometimes forming a planetary nebula) and leaving behind a degenerate core. In stark contrast, for a high-mass star the end comes suddenly and amid considerable fury. An evolving high-mass star builds up its onionlike structure (see Figure 17.5) as nuclear burning in its interior proceeds to more and more advanced stages. Hydrogen burns to helium, helium burns to carbon and oxygen, carbon burns to magnesium, oxygen burns to sulfur and silicon, and then silicon and sulfur burn to iron. Many different types of nuclear

> Iron is the most massive element formed by fusion.

FIGURE 17.7 This Hubble Space Telescope image shows an expanding cloud of dusty material ejected by the luminous blue variable star Eta Carinae. The star itself, which is largely hidden by the surrounding dust, has a luminosity 5 million times that of the Sun and a mass probably in excess of 100 $M_\odot$. Dust is created when volatile material ejected from the star condenses.

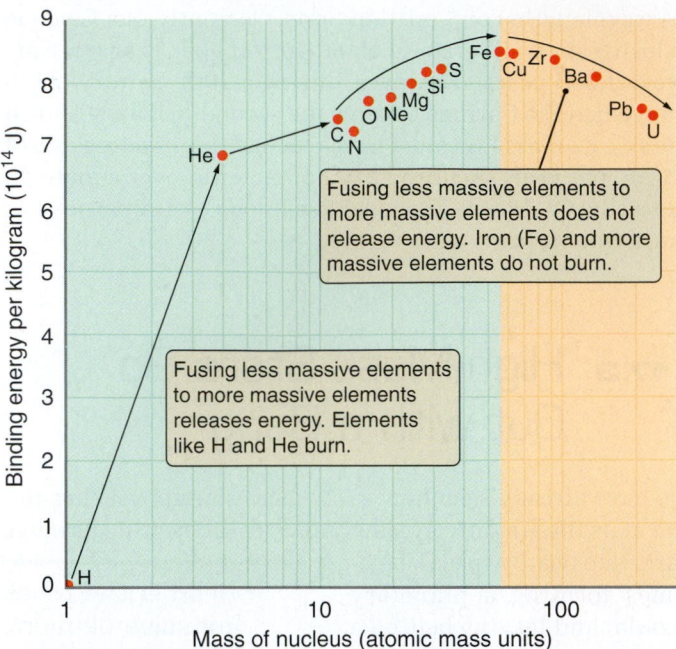

Fusing less massive elements to more massive elements does not release energy. Iron (Fe) and more massive elements do not burn.

Fusing less massive elements to more massive elements releases energy. Elements like H and He burn.

FIGURE 17.8 The nuclear binding energy of a kilogram of material is plotted against the mass of the atomic nuclei. This is the energy that it would take to break a kilogram of the material apart into protons and neutrons. Energy is released by nuclear fusion only if the binding energy of the products is greater than the binding energy of the reactants.

reactions occur up to this point, forming almost all of the different stable isotopes of **elements** less massive than iron. But the essential point is this: *With iron, the chain of nuclear fusion stops.*

Why does hydrogen burn while iron does not? More generally, what must be true of a material for it to serve as fuel? Gasoline burns because when it chemically combines with oxygen, energy is released. This extra energy pushes up the temperature, which then causes the chemical reaction between gasoline and oxygen to go faster. The reaction is "self-sustaining," meaning that the reaction itself is the source of thermal energy needed to cause the reaction to go. The same is true of nuclear fusion reactions in the interiors of stars. When four hydrogen atoms combine to form a helium atom, the resulting helium atom has less energy than the four hydrogen atoms had separately. This difference in energy is converted to thermal energy, which maintains the temperature of the gas at the high levels needed to sustain the reaction.

How much energy is available from a nuclear reaction of combining, say, three helium nuclei to form a ^{12}C nucleus (the triple-alpha process)? We can answer the question in steps. First, how much energy would it take to break down each of the three helium nuclei into its constituent six neutrons and six protons? Next, how much energy would be released if these six protons and neutrons combined to form a ^{12}C nucleus? The net energy produced by the reaction is just the difference between these two amounts.

The energy it would take to break up an atomic nucleus into its constituent parts is called the **binding energy** of the nucleus. The net energy released by a nuclear reaction is the difference between the binding energy of the products and the binding energy of the reactants:

> **Different nuclei have different binding energies.**

$$\text{Net energy} = \begin{pmatrix} \text{Binding energy} \\ \text{of products} \end{pmatrix} - \begin{pmatrix} \text{Binding energy} \\ \text{of reactants} \end{pmatrix}$$

Figure 17.8 shows a plot of the nuclear binding energy per kilogram for a number of different atomic nuclei.

It is worth making a brief digression to see how this works in practice. Using these values to work through the previous example of the triple-alpha process, we find that the binding energy of a helium nucleus is 6.822×10^{14} joules (J) per kilogram of helium. The binding energy of a ^{12}C nucleus is 7.410×10^{14} J per kilogram of carbon. To find the amount of energy available from fusing 1 kilogram (kg) of helium nuclei into carbon, take the difference between the two numbers:

$$\begin{pmatrix} \text{Net energy from} \\ \text{burning 1 kg He} \end{pmatrix} = \begin{pmatrix} \text{Binding energy} \\ \text{of C formed} \end{pmatrix} - \begin{pmatrix} \text{Binding energy} \\ \text{of He burned} \end{pmatrix}$$

$$= 7.410 \times 10^{14} \text{ J} - 6.822 \times 10^{14} \text{ J}$$

$$= 5.88 \times 10^{13} \text{ J}$$

This release of net energy indicates that helium is a good nuclear fuel, as Figure 17.8 shows directly. Moving from helium to carbon on the plot increases the binding energy of each nucleon,[1] so fusing helium to carbon releases energy. But what if we instead try to fuse iron to form more massive elements? Because iron is at the peak of the binding-energy curve, the products of iron burning will have *less* binding energy than the reactants (going from iron to more massive elements takes us down on the binding-energy curve in Figure 17.8), so the net energy in the reaction will be *negative*. Rather than producing energy, fusion of iron *uses* energy. Iron does not burn.[2]

The Final Days in the Life of a Massive Star

The nuclear reactions following hydrogen burning are energetically much less favorable than conversion of hydrogen to helium. A look at Figure 17.8 shows that conversion of 1 kg of helium to carbon produces less than $\frac{1}{10}$ as much energy

---

[1] A "nucleon" is a constituent of an atomic nucleus—that is, a proton or a neutron.

[2] Iron does not burn in a nuclear sense, but iron does burn *chemically*. For example, combining iron and oxygen increases the *chemical* binding energy of the atoms involved in the reaction. However, chemical burning plays no role in the interior of a star.

as conversion of 1 kg of hydrogen to helium. In order to support the star against gravity, this less efficient nuclear fuel must be consumed more rapidly. Although conversion of hydrogen into helium can provide the energy needed to support the high-mass star against the force of gravity for millions of years, helium burning can support the star for only a few hundred thousand years.

Following helium burning, the nature of the balance within the star becomes qualitatively different. There is almost as much energy available from burning a kilogram of carbon, neon, oxygen, or silicon as there is from burning a kilogram of helium. We might expect, then, that each of these stages would last about as long as helium burning. Yet when we look at Table 17.1, we see that the star proceeds from helium burning to the end of its life in a cosmic blink of the eye. So where is the catch?

You can think of the balance in a star as something like trying to keep a leaky balloon inflated. The larger the leak, the more rapidly you have to pump air into the balloon. A star that is burning hydrogen or helium is like a balloon with a slow leak. At the temperatures generated by hydrogen or helium burning, energy leaks out of the interior of a star primarily by radiation and convection. Neither of these processes is very efficient, because the outer layers of the star act like a thick, warm blanket. Nuclear fuels need to burn at only a relatively modest rate to support the weight of the outer layers of the star while keeping up with the energy escaping outward.

Beginning with carbon burning, this balance shifts in a dramatic and fundamental way. Rather than being carried by radiation and convection, energy now begins to escape from the core primarily in the form of neutrinos produced by the many nuclear reactions occurring there. When this happens, the floodgates are opened. Like air pouring out through a huge rent in the side of a balloon, neutrinos produced in the interior of the star stream through the overlying layers of the star as if they were not even there, carrying the energy from the stellar interior out into space. Keeping the balloon inflated becomes difficult indeed. As thermal energy pours out of the interior of the star, the outer layers of the star push inward, driving up the density and temperature, and forcing nuclear reactions to run at the furious rate necessary to replace the energy escaping in the form of neutrinos.

Once this process of **neutrino cooling** becomes significant, the star begins evolving much more rapidly. Carbon burning is capable of supporting the star for only something less than about a thousand years. Oxygen burning holds the star up for only about a year. Once a massive star reaches the point that it begins to burn silicon, it is within a few days of the end. The luminosity of a silicon-burning star in electromagnetic radiation may be not much more than it was when the star was burning helium in its core. But if our eyes could sense neutrinos, we would see that a

> Stages of burning after helium burning are progressively shorter-lived.

star burning silicon is actually giving off about 200 million times more energy per second than its former self did!

Following silicon burning, we come to the end of the line. Once a star forms its iron core, no source of nuclear energy remains to replenish the energy that is being taken away by escaping neutrinos. The high-mass star's life as a balancing act between gravity and a controlled thermonuclear furnace is over. Gravity will have its way.

The Core Collapses and the Star Explodes

Throughout the story of stellar evolution, we have seen that although gravity can be held off for a time—perhaps a very long time—it is infinitely patient, and in the end it generally wins the battle. Bereft of support from thermonuclear fusion, the iron core of the massive star begins to collapse. Refer to **Figure 17.9** as we follow the dramatic sequence of events that comes next.

The early stages of the collapse of the iron ash core of an evolved massive star are much the same as in the collapse of a nonburning core in a low-mass star. As the core collapses, its density and temperature skyrocket, and the force of gravity becomes even stronger. The gas in the core is once again compressed beyond the realm of normal matter, becoming electron-degenerate when it reaches the approximate size of Earth. Unlike the electron-degenerate core of a low-mass red giant, however, the weight bearing down on the interior of the iron ash core is too great to be held up by electron degeneracy. Here, gravity is too strong even for the quantum mechanical rules that limit how tightly electrons can be packed together. As the collapse continues, the core reaches temperatures of 10 billion K (10^{10} K) and higher, while the density exceeds 10 metric tons per cubic centimeter, or 10 times the density of an electron-degenerate white dwarf.

These phenomenal temperatures and pressures trigger fundamental changes in the makeup of the core. The laws describing thermal radiation say that at these temperatures the nucleus of the star will be awash in extremely energetic thermal radiation. This radiation is so energetic that thermal gamma-ray photons are produced with enough energy to break iron nuclei apart into helium nuclei. This process, known as "photodisintegration," literally begins undoing the results of nuclear fusion—a task that uses up a tremendous amount of the thermal energy of the core. At the same time, the density of the core is so great that electrons are squeezed into atomic nuclei, where they combine with protons to produce neutron-rich isotopes in the core of the star. This process uses up thermal energy as well, robbing the core of even more of its pressure support. All the while, neutrinos continue to pour out of the core of the dying star. These events take place within a second! The collapse of the core accelerates, reaching velocities of 70,000 km/s, or almost one-fourth the speed of light, on its inward fall.

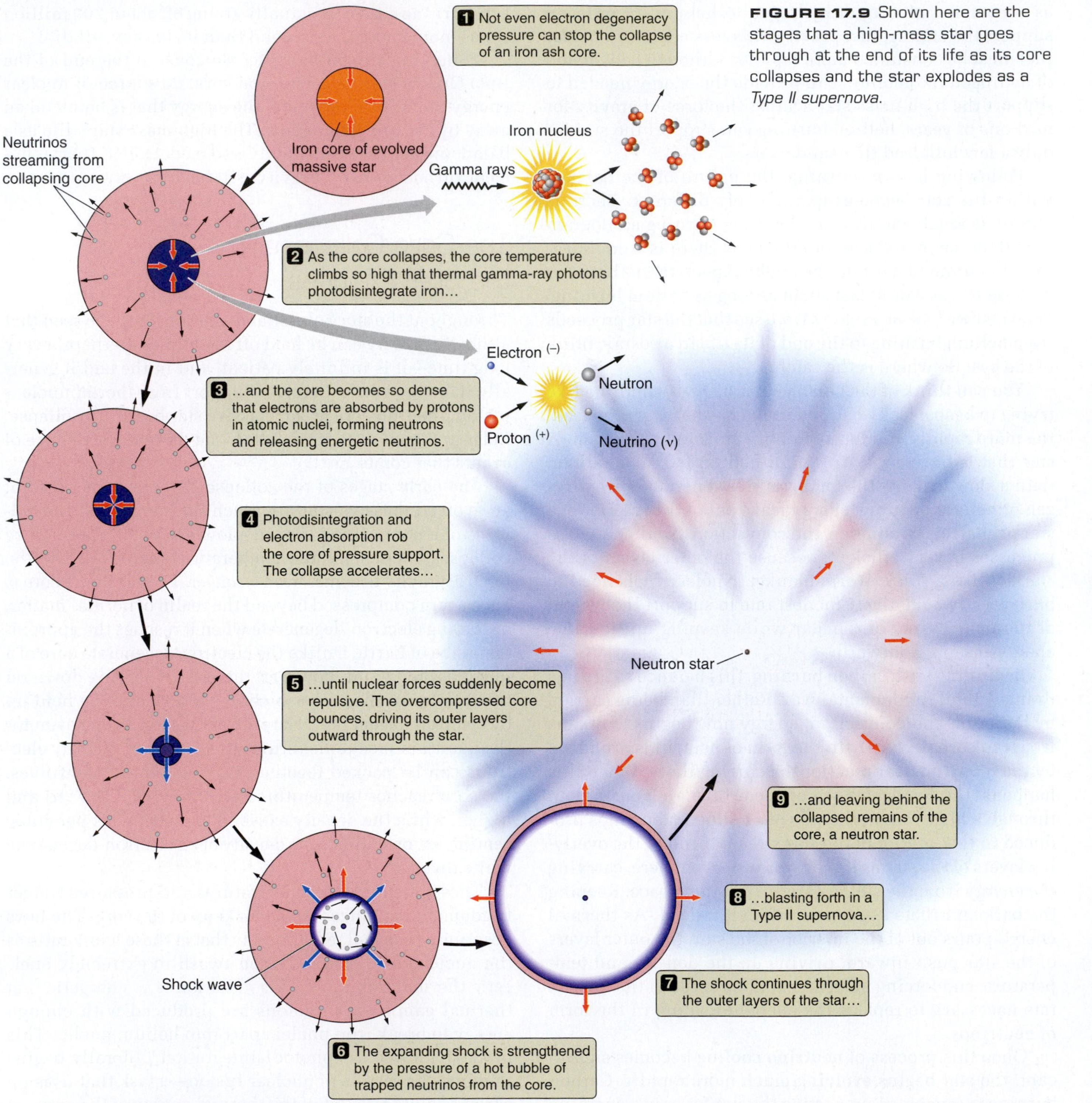

1 Not even electron degeneracy pressure can stop the collapse of an iron ash core.

Iron core of evolved massive star

Iron nucleus

Gamma rays

Neutrinos streaming from collapsing core

2 As the core collapses, the core temperature climbs so high that thermal gamma-ray photons photodisintegrate iron…

Electron (−)

Neutron

Proton (+)

Neutrino (ν)

3 …and the core becomes so dense that electrons are absorbed by protons in atomic nuclei, forming neutrons and releasing energetic neutrinos.

4 Photodisintegration and electron absorption rob the core of pressure support. The collapse accelerates…

5 …until nuclear forces suddenly become repulsive. The overcompressed core bounces, driving its outer layers outward through the star.

Neutron star

9 …and leaving behind the collapsed remains of the core, a neutron star.

8 …blasting forth in a Type II supernova…

7 The shock continues through the outer layers of the star…

Shock wave

6 The expanding shock is strengthened by the pressure of a hot bubble of trapped neutrinos from the core.

FIGURE 17.9 Shown here are the stages that a high-mass star goes through at the end of its life as its core collapses and the star explodes as a *Type II supernova.*

The next hurdle standing in the way of the collapsing core is the force that holds atomic nuclei together. As material in the collapsing core reaches and exceeds the density of an atomic nucleus, the strong nuclear force actually becomes repulsive. Computer models say that about half of the collapsing core suddenly slows its inward fall. The remaining half slams into the innermost part of the star at a significant fraction of the speed of light and "bounces," sending a tremendous shock wave back out through the star.

Under the extreme conditions in the center of the star, neutrinos are being produced at an enormous rate. Over the next second or so, almost a fifth of the mass of the material in the core is converted into neutrinos and their energy. Most of these neutrinos pour outward through the star; but at the phenomenal densities found in the collapsing core of the massive star, not even neutrinos pass with complete freedom. A few tenths of a percent of the energy of the neutrinos streaming out of the core of the dying star is trapped by the dense material behind the expanding shock wave. The energy of these trapped neutrinos drives the pressure and temperature in this region ever higher, inflating a bubble of extremely hot gas and intense radiation around the core of the star. The pressure of this bubble adds to the strength of the shock wave moving outward through the star. Within about a minute the shock wave has pushed its way out through the helium shell within the star. Within a few hours it reaches the surface of the star itself, heating the stellar surface to 500,000 K and blasting material outward at velocities of up to about 30,000 km/s. Our evolved massive star has exploded in an event referred to as a **Type II supernova**.

> The high-mass star explodes as a tremendous Type II supernova.

The death of a star as a Type I or Type II supernova should mean more to us than just another example of nature's spectacular fireworks. As we will see in the following section, supernovae leave behind a chemical legacy that was critical in the formation of next-generation stars like our Sun and its Solar System. Were it not for supernovae, it is highly likely that either we or our planet would not exist.

17.4 The Spectacle and Legacy of Supernovae

A Type II supernova is comparable in its spectacle to a Type I supernova. For a brief time a Type II supernova can shine with the light of 100 billion suns, yet the energy that comes out of the supernova in the form of light is but a tiny fraction of the energy of the explosion itself. About a hundred times more energy comes out as the kinetic energy of the outer parts of the star that are blasted into space by the explosion. This ejected material contains about 10^{47} J of kinetic energy—enough energy to accelerate the mass of our Sun to a speed of 10,000 km/s. The kinetic energy of the ejecta from both Type I and Type II supernovae is responsible for heating the hottest

phases of the interstellar medium and pushing around the clouds in the interstellar medium. Yet even this amount of energy is a pittance in comparison with the energy carried away from the supernova explosion by neutrinos—an amount of energy at least a hundred times larger still!

One of the most important astronomical events in the last part of the 20th century was the explosion of a massive star in the small companion galaxy to the Milky Way known as the Large Magellanic Cloud. Even at a distance of 160,000 light-years, Supernova 1987A was so bright that it dazzled sky gazers in the Southern Hemisphere (**Figure 17.10**). While astronomers working in all parts of the electromagnetic spectrum scrambled to point their telescopes at the new supernova, solar astronomers had already quietly and unknowingly captured one of the true scientific prizes of SN 1987A. Solar neutrino telescopes recorded a burst of neutrinos passing through Earth—neutrinos that originated not in the Sun but in the tremendous stellar explosion that occurred beyond the bounds of our galaxy itself. The detection of neutrinos from SN 1987A provided us with a rare and crucial glimpse of the very heart of a massive star at the moment of its death, confirming a fundamental prediction of our theories about the collapse of the core and its effects.

The Energetic and Chemical Legacy of Supernovae

Type II supernova explosions leave a rich and varied legacy to the universe. We have already mentioned their energetic legacy: the energy from these explosions is responsible for driving much of the dynamics of the interstellar medium. When we look at the sky over a wide range of wavelengths, we see huge expanding bubbles of million-kelvin gas, like that in **Figure 17.11a**, aglow in X-rays and driving visible shock waves (**Figure 17.11b**) into the surrounding interstellar medium. These bubbles are the still-powerful blast waves of supernova explosions that took place thousands of years ago. In fact, in many cases supernova explosions are thought to be responsible for compressing nearby clouds (like the

(a)

(b)

SN 1987A

FIGURE 17.10 SN 1987A was a supernova that exploded in a small companion galaxy of the Milky Way called the Large Magellanic Cloud (LMC). These images show the LMC before the explosion (a) and while the supernova was near its peak (b).

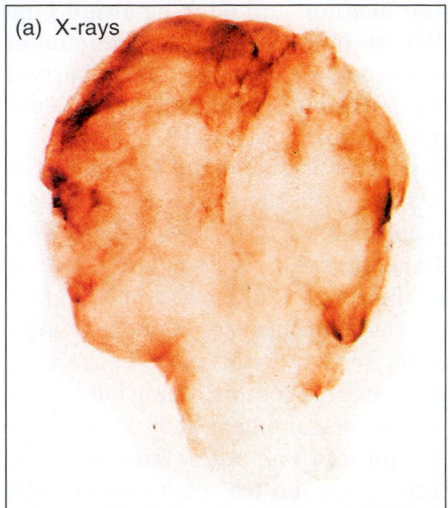

(a) X-rays

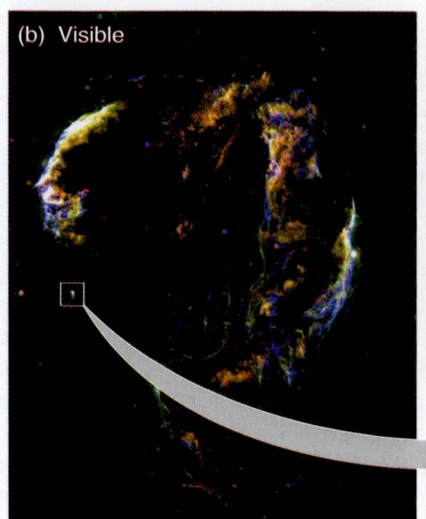

(b) Visible

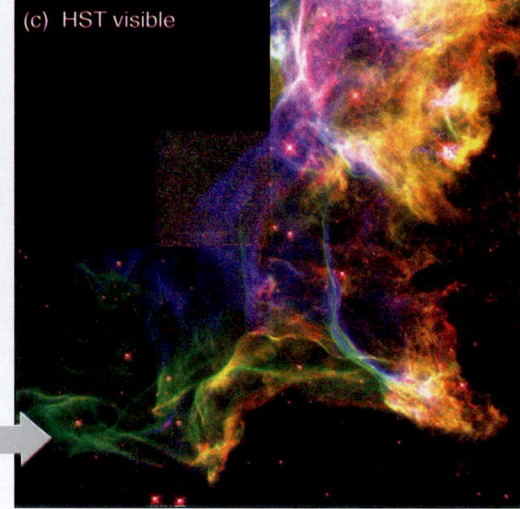

(c) HST visible

FIGURE 17.11 The Cygnus Loop is a supernova remnant—an expanding interstellar blast wave caused by the explosion of a massive star. Gas in the interior with a temperature of millions of kelvins glows in X-rays (a), while visible light comes from locations where the expanding blast wave pushes through denser gas in the interstellar medium (b). (c) A Hubble Space Telescope image of a location where the blast wave is hitting an interstellar cloud.

one in **Figure 17.11c**) enough to trigger their collapse toward the formation of new generations of stars.

Perhaps even more important to us is the chemical legacy left behind by supernova explosions. Later in our journey we will turn our attention to the earliest moments after the birth of the universe; we will find that the only chemical elements that formed at that time were the least massive ones: hydrogen, helium, and trace amounts of lithium, beryllium, and boron. All of the rest of the chemical elements, including a large fraction of the atoms of which we are made, were formed in the hearts of stars and then returned to the interstellar medium. This process, which is responsible for the progressive chemical enrichment of the universe, is called **nucleosynthesis**. But although mass loss from both low- and high-mass stars enriches the interstellar medium with massive elements formed in their interiors, Types I and II supernovae are the true champions of nucleosynthesis. A look at Figure 17.5 should offer a clue as to why this is so. Whereas low-mass stars are able to form elements only as massive as carbon and oxygen, nuclear burning in high-mass stars produces elements as massive as iron. Yet this is not the whole story. A look at a table of elements that occur in nature (see Appendix 3) shows that many elements, up through uranium (the most massive among the naturally occurring elements), are far more massive than iron. If iron is the most massive element that can be formed by nuclear burning, then where do these even more massive elements come from?

Answering this question takes us back to the reason that high temperatures are necessary for fusion to occur. Under

> Supernovae eject newly formed massive elements into interstellar space.

normal circumstances, electric repulsion keeps positively charged atomic nuclei far apart. Extreme temperatures are needed to slam nuclei together hard enough to overcome this electric repulsion. Free neutrons, on the other hand, are not subject to these ground rules: they have no net electric charge, so there is no electric repulsion to prevent them from simply running into an atomic nucleus, regardless of how many protons that nucleus contains. Under normal circumstances in nature, free neutrons are very rare. In the interiors of evolved stars, however, a number of nuclear reactions produce free neutrons, and under some circumstances—including those shortly before and during a Type II supernova—free neutrons can be produced in very large numbers. Free neutrons are easily captured by atomic nuclei and later decay to become protons. In this way, elements with higher and higher mass are formed.

Calculations of nucleosynthesis in stars make clear predictions about which nuclei should be formed in abundance and which should not. These same patterns are found in measurements of the abundances of nuclei on Earth, in meteorite material, and in the atmospheres of stars. Although some of these patterns are subtle, we can appreciate others by comparing what we have learned about nucleosynthesis with a plot of the relative abundance of elements found in nature (**Figure 17.12**). Notice, first of all, that less massive elements are far more abundant than more massive elements—a consequence of the way more massive elements are progressively built up from less massive elements. An exception to this pattern is the dip in the abundances of the light elements lithium (Li), beryllium (Be), and boron (B). These light elements are easily destroyed by nuclear burning, but their production is mostly bypassed by the main reactions involved

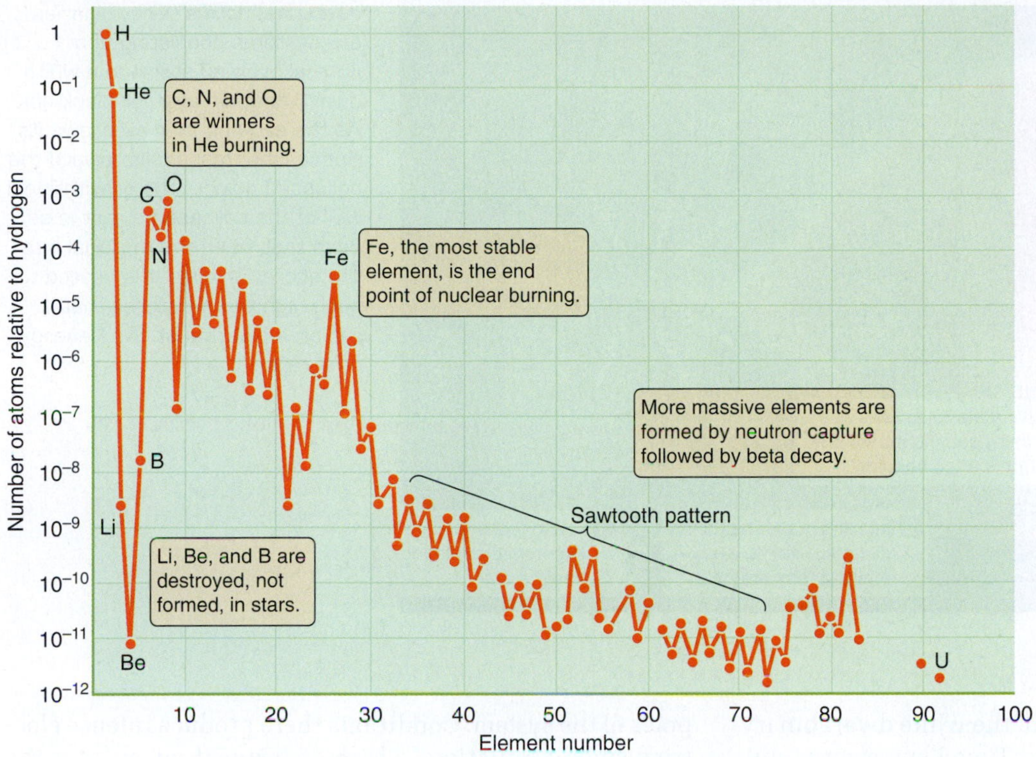
Number of atoms relative to hydrogen
Element number
C, N, and O are winners in He burning.
Fe, the most stable element, is the end point of nuclear burning.
More massive elements are formed by neutron capture followed by beta decay.
Sawtooth pattern
Li, Be, and B are destroyed, not formed, in stars.
H
He
C
O
N
Fe
B
Li
Be
U

FIGURE 17.12 The relative abundances of different elements on Earth are plotted against the element number of the nucleus. This pattern can be understood as a result of the process of nucleosynthesis in stars. The periodic table of elements in Appendix 3 identifies individual elements by their element number (the number of protons).

in burning hydrogen (H) and helium (He). Conversely, carbon (C), nitrogen (N), and oxygen (O) are big winners in the CNO cycle of hydrogen burning and the triple-alpha process of helium burning, and their high abundances reflect this fact. The spike in the abundances of the "iron peak" elements is evidence of the nucleosynthetic processes, including those in Type I and Type II supernovae, that favor these most tightly bound of all nuclei. Even the sawtooth pattern in the abundances of even- and odd-numbered elements can be understood as a consequence of stellar nucleosynthesis. (To complete our discussion of nucleosynthesis, we will need to consider the formation of elements that took place not in stars, but during the formation of the universe itself. We will take up this discussion in Chapter 18.)

Once again we find connections and insights in places that, prior to our study of the universe, we might never have imagined looking. Our understanding of the processes at work within the interiors of dying stars is being confirmed by analysis of the chemical composition of the very Earth under our feet. At the same time, we can make a pretty good stab at identifying the kinds of stars responsible for forming the atoms that make up our own bodies. Our growing understanding of the chemical evolution of the universe and our connection to it represents one of the triumphs of modern astronomy.

Neutron Stars and Pulsars

We have now seen the fate of the outer parts of the star, which are blasted back into interstellar space by the explosion of a Type II supernova. But what remains of the core that was left behind? Picking up our story where we left off, the matter at the center of the massive star has collapsed to the point where it has about the same density as the nucleus of an atom. As long as the mass of the core left behind by the explosion is no more than about 3 $M_{\odot}$, this collapse will be halted by quantum mechanical rules similar to those responsible for holding up a white dwarf. But in this case, instead of electrons it is neutrons that are forced together as tightly as the rules of quantum mechanics allow. The neutron-degenerate core left behind by the explosion of a Type II supernova is referred to as a **neutron star**. It has a radius of perhaps 10 kilometers (km), making it roughly the size of a small city; but into that volume is packed a mass from 1.4 to about 3 times that of our Sun. At a density of about a billion metric tons per cubic centimeter, the neutron star is a billion times more dense than a white dwarf and a thousand trillion (10^{15}) times more dense than water! That density is roughly what we would get by crushing the entire Earth down to an object the size of a football stadium.

The Type II supernova leaves behind a neutron-degenerate core.

As if neutron stars were not extraordinary enough in their own right, they also form the hearts of a number of other exotic objects. If the massive star responsible for the formation of a neutron star is part of a binary system, then the neutron star will be left with a binary companion.

X-ray binaries arise from the accretion of mass onto neutron stars.

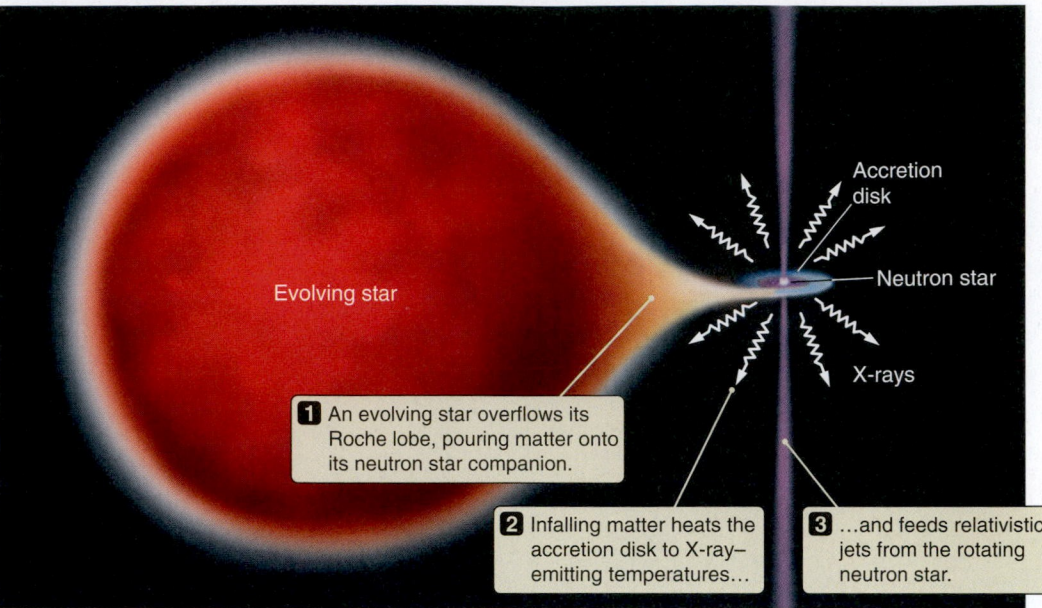

FIGURE 17.13 X-ray binaries are systems consisting of a normal evolving star and a white dwarf, neutron star, or black hole. As the evolving star overflows its Roche lobe, mass falls toward the collapsed object. The gravitational well of the collapsed object is so deep that when the material hits the accretion disk, it is heated to such high temperatures that it radiates away most of its energy as X-rays.

Evolving star

Accretion disk

Neutron star

X-rays

1 An evolving star overflows its Roche lobe, pouring matter onto its neutron star companion.

2 Infalling matter heats the accretion disk to X-ray–emitting temperatures…

3 …and feeds relativistic jets from the rotating neutron star.

Such a system might remind you of the white dwarf binary systems responsible for novae and Type I supernovae discussed in Chapter 16. As the second star in such a binary system evolves and overfills its Roche lobe, matter plummets down the deep gravitational well of the tiny but massive neutron star. This matter slams into the accretion disk around the neutron star with enough energy to heat the disk to temperatures of millions of kelvins, and the accretion disk glows brightly in X-rays. Such an object, illustrated in **Figure 17.13**, is known as an **X-ray binary**. Many fascinating phenomena occur in X-ray binaries, including the formation of powerful jets that blast away from the neutron star in directions perpendicular to its accretion disk and at speeds approaching the speed of light.

Besides phenomenal density, a neutron star has a number of other extraordinary properties. The same principle of conservation of angular momentum that requires a collapsing molecular cloud to spin faster as it grows smaller also says that as the core of a massive star collapses, it must spin faster as well. As a main-sequence O star, a massive star rotates perhaps once every few days. As a neutron star, it might instead rotate tens or even hundreds of times each second! We also saw in our discussion of a collapsing interstellar cloud that the magnetic field in the cloud is carried along and concentrated by the collapse. This phenomenon also occurs in the collapse of a star, amplifying its magnetic field to values that are trillions of times greater than the magnetic field at Earth's surface. A neutron star has a magnetosphere, just as Earth and several other planets do, except that the neutron star's magnetosphere is unimaginably stronger and is whipped around many times a second by the spinning star.

Energetic subatomic particles such as electrons and positrons move along the magnetic-field lines of the neutron star and are "funneled" by the field toward the magnetic

poles of the system. Conditions there produce intense electromagnetic radiation, which is beamed away from the magnetic poles of the neutron star as shown in **Figure 17.14**. As the neutron star rotates, these beams of radiation sweep through space much like the rotating beams of a lighthouse. When we are located in the paths of these beams, we detect a pattern resembling what the beams from a lighthouse look like to sailors entering a harbor at night. The neutron star appears to flash on and off with a regular period equal to the period of rotation of the star (or half the rotation period, if we see both beams).

Rapidly pulsing objects were first discovered by observers working with radio wavelengths in 1967. These

> **Pulsars are rapidly spinning, magnetized neutron stars.**

objects, which blinked like regularly ticking clocks, puzzled astronomers. One of the early tongue-in-cheek names given to these objects was "LGMs," which stood for "little green men." Today these objects are referred to by the less flamboyant but more accurately descriptive term **pulsar**. As of this writing, well over a thousand pulsars are known, and more are being discovered all the time.

The Crab Nebula—Remains of a Stellar Cataclysm

In A.D. 1054, Chinese astronomers recorded the presence of a "guest star" in the part of the sky that we call the constellation Taurus. The new star was so bright that it could be seen during the daytime for 3 weeks, and it did not fade from visibility altogether for many months. On the basis of the Chinese description of the changing brightness and color of the object, we can say today that the guest star of

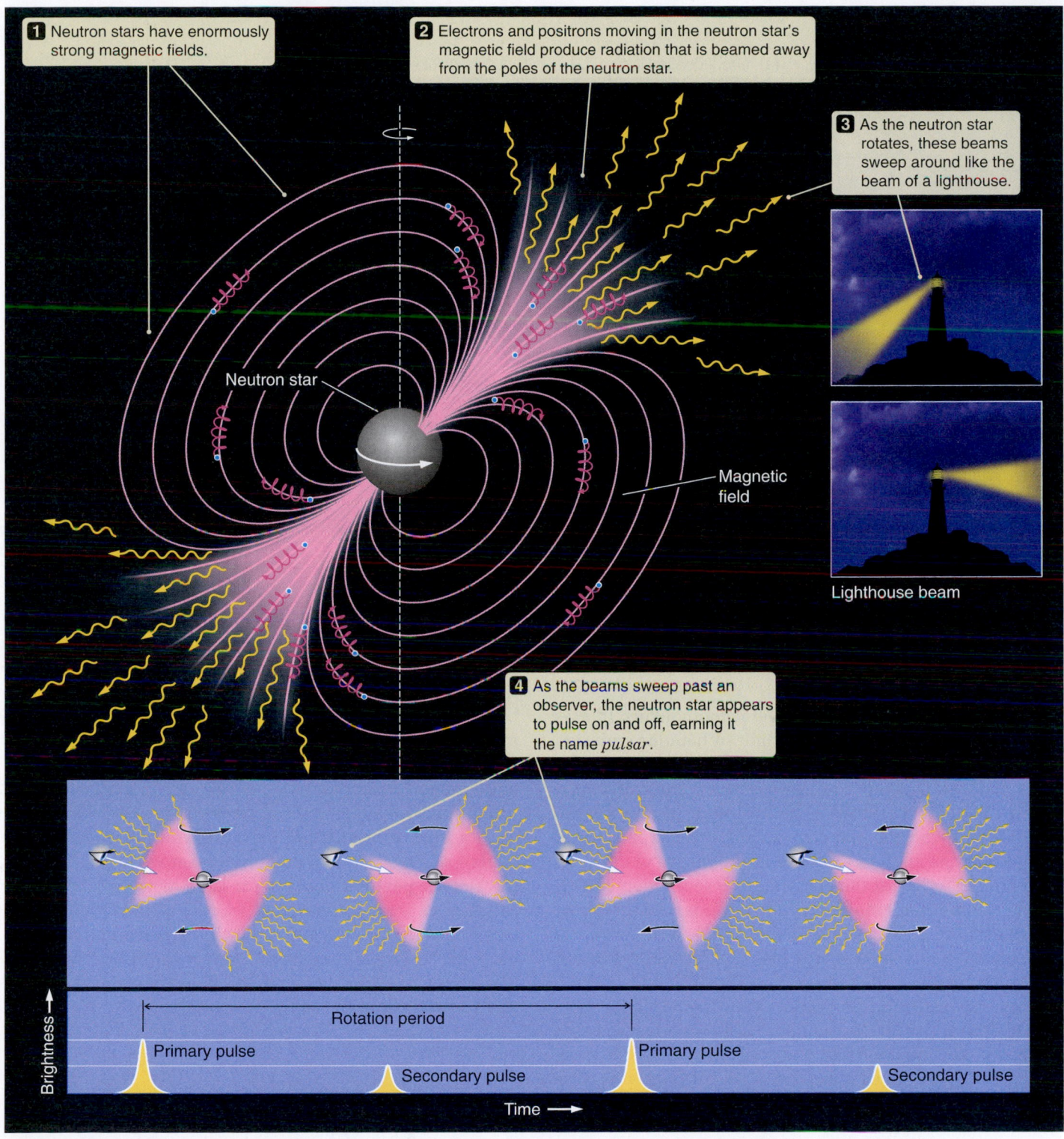

1 Neutron stars have enormously strong magnetic fields.

2 Electrons and positrons moving in the neutron star's magnetic field produce radiation that is beamed away from the poles of the neutron star.

3 As the neutron star rotates, these beams sweep around like the beam of a lighthouse.

Neutron star

Magnetic field

Lighthouse beam

4 As the beams sweep past an observer, the neutron star appears to pulse on and off, earning it the name *pulsar*.

Rotation period

Brightness

Primary pulse

Secondary pulse

Primary pulse

Secondary pulse

Time →

VISUAL ANALOGY **FIGURE 17.14** As a highly magnetized neutron star rotates rapidly, light is given off, much like the beams from a rotating lighthouse lamp. From our perspective, as these beams sweep past us the star will appear to pulse on and off, earning it the name *pulsar*.

HST image

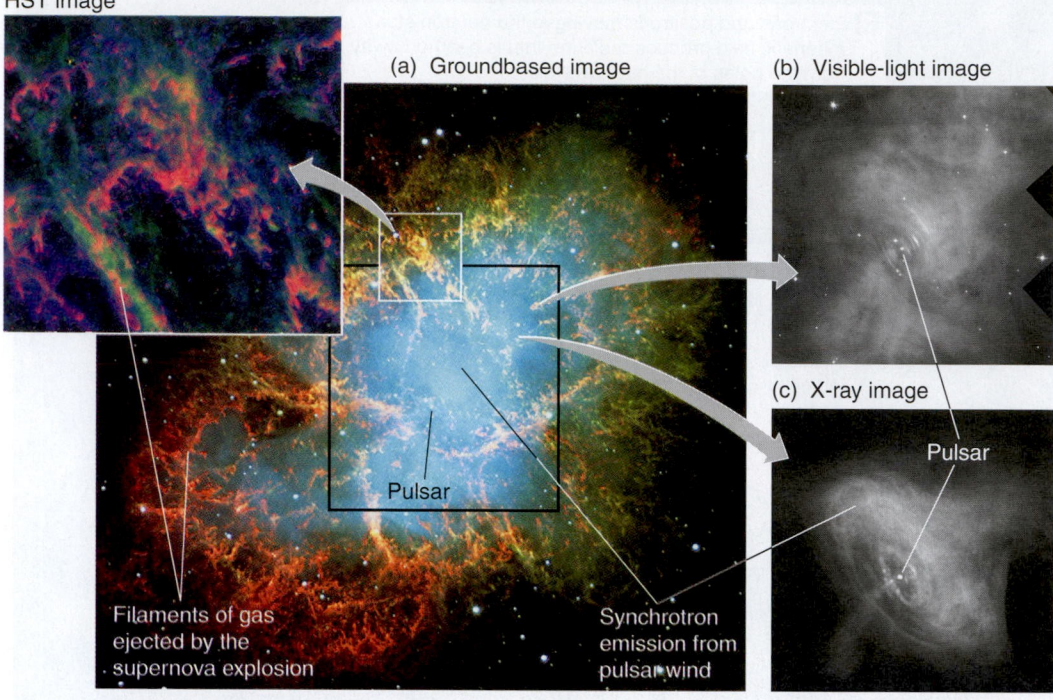

(a) Groundbased image

(b) Visible-light image

(c) X-ray image

Pulsar

Pulsar

Filaments of gas ejected by the supernova explosion

Synchrotron emission from pulsar wind

FIGURE 17.15 The Crab Nebula is the remnant of a supernova explosion witnessed by Chinese astronomers in A.D. 1054. (a) The object we see today is an expanding cloud of "shrapnel" from that earlier cataclysm. The spinning pulsar at the heart of the Crab Nebula sends off a "wind" of electrons and positrons moving at close to the speed of light. The synchrotron radiation from these particles is shown in visible light (b) and in X-rays (c).

1054 was a fairly typical Type II supernova. When we look at this spot in the sky today, we see an expanding cloud of debris from this explosion—an extraordinary object called the **Crab Nebula** (**Figure 17.15**).

The Crab Nebula consists of several components. Images of the Crab Nebula taken in the light of nebular emission lines show filaments of glowing gas. Doppler shift measurements of these filaments reveal a pattern much like that seen in planetary nebulae—the hallmark of an expanding shell. But whereas planetary nebulae are expanding at 20–30 km/s, the shell of the Crab is expanding at closer to 1,500 km/s. Studies of the spectra of these filaments show that they contain anomalously high abundances of helium and other more massive chemical elements—the products of the nucleosynthesis that took place in the supernova and its progenitor star.

There is a pulsar at the center of the Crab Nebula. This was the first pulsar to be seen at visible wavelengths as it flashed on and off 30 times a second. Actually, the Crab pulsar flashes 60 times a second: first with a main pulse associated with one of the "lighthouse" beams, then with a fainter secondary pulse associated with the other beam. Today, the Crab Nebula has been observed in all parts of the electromagnetic spectrum, from low-energy radio waves to high-energy X-rays and even higher-energy gamma rays (see the chapter opening photograph).

As the Crab pulsar spins 30 times a second, it whips its powerful magnetosphere around with it. At a distance from the pulsar about equal to the radius of the Moon, material in the magnetosphere must move at almost the speed of light to keep up with this rotation. Like a tremendous sling-shot, the rotating pulsar magnetosphere flings elementary particles—probably mostly electrons and positrons—away from the neutron star in a powerful wind moving at nearly the speed of light. Material from this wind fills the space between the pulsar and the expanding shell. The Crab Nebula is almost like a big balloon; but instead of being filled with hot air, it is filled

> A pulsar powers the Crab's eerie synchrotron glow.

with a mix of *relativistic* particles and strong magnetic fields—an environment more like what we might find in a physicist's particle accelerator than what we normally think of as interstellar space. Images of the Crab Nebula (see Figures 17.15b and c) show this bizarre bubble as an eerie glow. This glow is synchrotron radiation from the relativistic electrons and positrons as they spiral around the magnetic field in the Crab.

17.5 Star Clusters Are Snapshots of Stellar Evolution

Over the course of this and the previous chapters we have told a remarkable tale about the evolution of stars of different masses, presenting it as a well-corroborated theory. In other words, we have presented the story of stellar evolution as fact. Yet even the most massive stars take hundreds

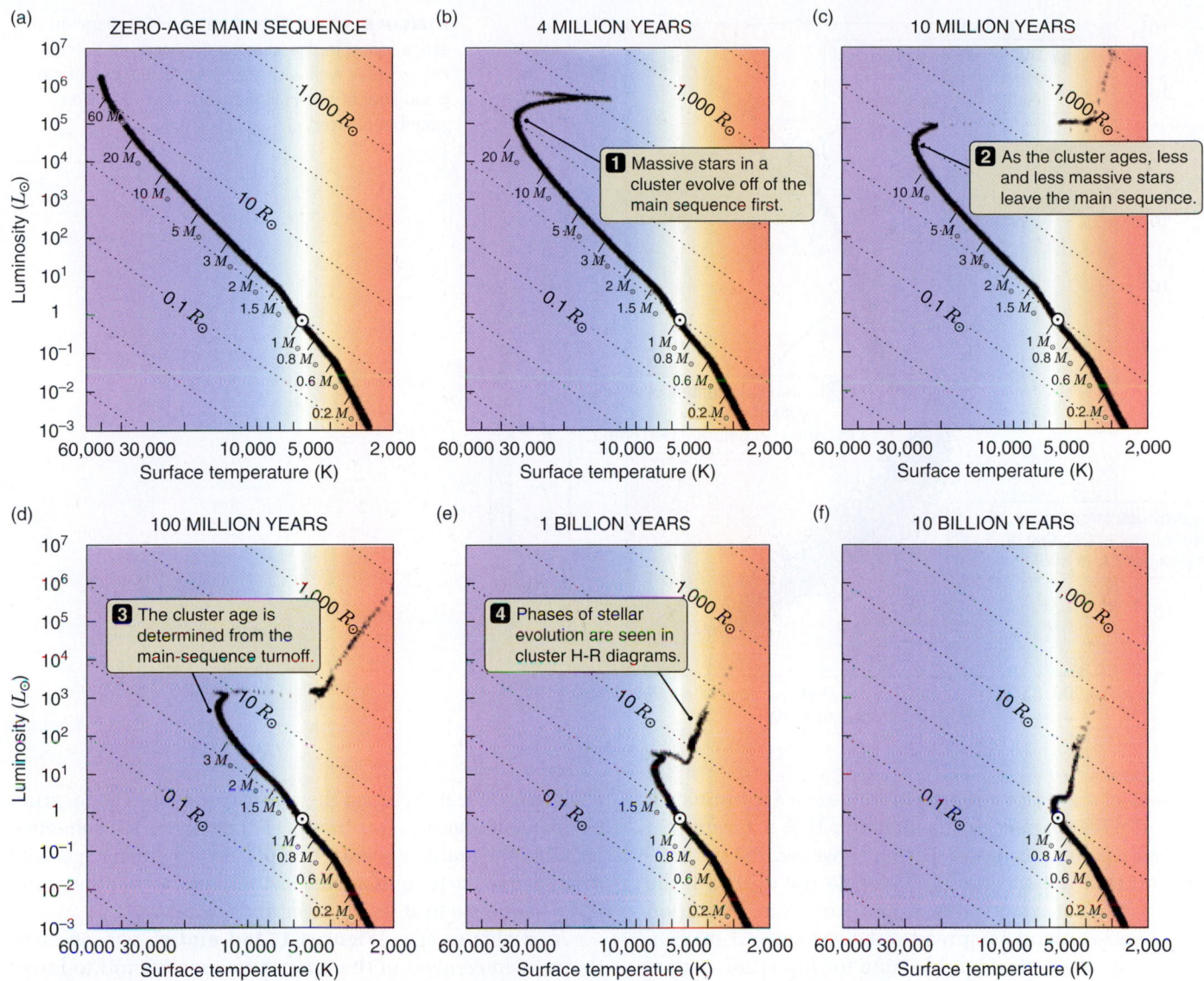

FIGURE 17.16 H-R diagrams of star clusters are snapshots of stellar evolution. These are H-R diagrams of a simulated cluster of 40,000 stars of solar composition seen at different times following the birth of the cluster. Note the progression of the *main-sequence turnoff* to lower and lower masses.

of thousands of years to evolve, which is far longer than the handful of decades we have spent studying their ways. Upon what testable predictions of our theories of stellar evolution do we base such bold claims of knowledge?

As we learned in Chapter 15, when an interstellar cloud collapses, it fragments into pieces, forming not one star but many stars of different masses. We see many such **star clusters** around us today, containing anywhere from a few dozen to millions of stars. The fact that all of the stars in a cluster formed together at nearly the same time means that clusters are snapshots of stellar evolu-

> Stars in clusters formed together at about the same time.

tion. A look at a cluster that is 10 million years old shows us what stars of all different masses evolve into during the first 10 million years after they are formed. A look at a cluster 10 *billion* years after it formed shows us what becomes of stars of different masses after 10 billion years pass.

This basic result—the fact that high-mass stars in a cluster evolve more rapidly than low-mass stars that formed at the same time—provides the key to our knowledge of stellar evolution. **Figure 17.16** shows the H-R diagram of a simulated cluster of 40,000 stars as it would appear at several different ages. In **Figure 17.16a** stars of all masses are located on the zero-age main sequence, showing where they begin their lives as main-sequence stars. The increas-

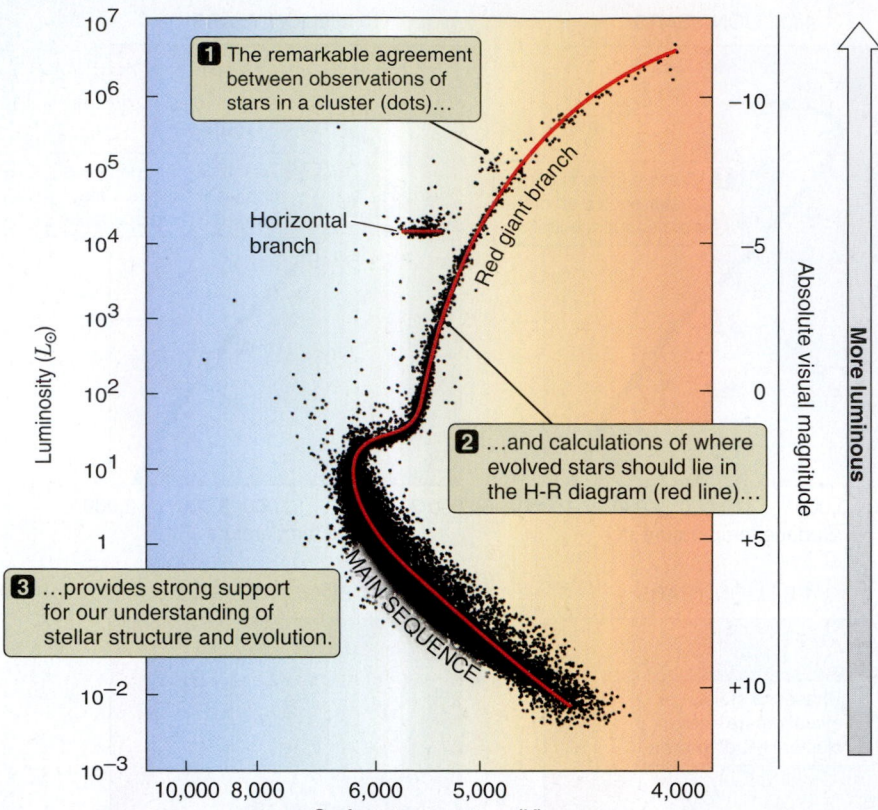

On the H-R diagram (luminosity vs. surface temperature):

1. The remarkable agreement between observations of stars in a cluster (dots)…

Horizontal branch

Red giant branch

2. …and calculations of where evolved stars should lie in the H-R diagram (red line)…

MAIN SEQUENCE

3. …provides strong support for our understanding of stellar structure and evolution.

Luminosity ($L_\odot$)

Surface temperature (K)

Absolute visual magnitude

More luminous

FIGURE 17.17 The observed H-R diagram of stars (dots) in the cluster 47 Tucanae agrees remarkably well with the theoretical calculation (solid line) of the H-R diagram of a 12-billion-year-old cluster.

ing masses of stars along the main sequence are indicated. We would never expect to see a cluster H-R diagram that looks like the one in Figure 17.16a, however, for the simple reason that the stars in a cluster do not all reach the main sequence at exactly the same time. Star formation in a molecular cloud is spread out over several million years, and it takes considerable time for lower-mass stars to contract to reach the main sequence. The H-R diagram of a very young cluster normally shows many lower-mass stars located well above the main sequence, still descending their Hayashi tracks.

The more massive a star is, the shorter its life on the main sequence will be. After only 4 million years (**Figure 17.16b**), all stars with masses greater than about 20 $M_\odot$ have evolved off the main sequence and are now spread out across the top of the H-R diagram. The most massive stars have already disappeared from the H-R diagram entirely, having vanished in supernovae.

As time goes on, stars of lower and lower mass evolve off the main sequence, and the turnoff point moves toward the bottom right in the H-R diagram. By the time the cluster is 10 million years old (**Figure 17.16c**), only stars with masses less than about 15 $M_\odot$ remain on the main sequence. The cluster H-R diagram looks as if we grabbed the top of the band of cluster stars and gradually peeled it away from

The main-sequence turnoff shifts to lower-mass stars as a cluster grows older.

its original location along the main sequence. The location of the most massive star that is still on the main sequence is called the **main-sequence turnoff**. As the cluster ages, the main-sequence turnoff moves farther and farther down the main sequence to stars of lower and lower mass.

As a cluster ages (**Figures 17.16d and e**), we see more than the movement of the main-sequence turnoff to lower and lower masses; we see the details of all stages of stellar evolution. By the time the star cluster is 10 *billion* years old (**Figure 17.16f**), stars with masses of only 1 $M_\odot$ are beginning to pull away from the main sequence. Stars slightly more massive than this are seen as giant stars of various types. In this cluster the horizontal branch appears as a knot of stars sometimes referred to as the "red clump." Note how few giant stars are present in any of the cluster H-R diagrams. The giant, horizontal, and asymptotic giant branch phases in the evolution of low- and intermediate-mass stars pass so quickly in comparison with a star's main-sequence lifetime that even though the cluster started with 40,000 stars, only a handful of stars are seen in these phases of evolution at any given time. Similarly, it takes a newly formed white dwarf only a few tens of millions of years to cool to the point that it disappears off the bottom of these figures. Even though the majority of evolved stars in an old cluster are white dwarfs, all but a few of these stars will have cooled and faded into obscurity at any given time.

The cluster H-R diagrams in Figure 17.16 are theoretical calculations of what clusters of different ages *should* look

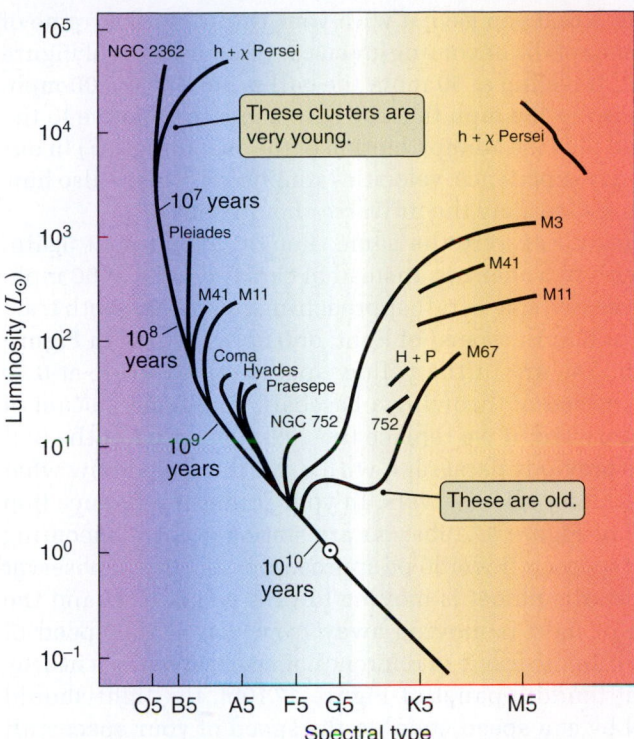

FIGURE 17.18 H-R diagrams for clusters having a range of different ages. The ages associated with the different main-sequence turnoffs are indicated.

like. The crucial point is that this is what H-R diagrams of real clusters *must* look like if our theories of stellar evolution are correct. Fortunately, this is also what H-R diagrams of real clusters *do* look like. **Figure 17.17** shows the observed H-R diagram for the cluster 47 Tucanae, along with a theoretical calculation of the H-R diagram for a 12-billion-year-old cluster. The quality of the agreement speaks for itself. The fact that observed star clusters have H-R diagrams that agree so well with the predictions of models is strong support for our theories of stellar evolution.

Armed with confidence in the reliability of our ideas about stellar evolution, we can turn cluster evolution models into a powerful tool for studying the history of star formation. When we observe a star cluster, the location of the main-sequence turnoff immediately tells us the age of the cluster. If shown any of the cluster H-R diagrams in Figure 17.16, we should have no difficulty estimating the age of the cluster. **Figure 17.18** traces the observed H-R diagrams for several real star clusters. Once we know what to look for, the difference between young and old clusters is obvious. NGC 2362 is clearly a young cluster. Its complement of massive, young stars shows it to be only a few million years old. In contrast, 47 Tucanae (see Figure 17.17) has a main-sequence turnoff of about 0.85 $M_\odot$, signifying a cluster age of about 12 billion years. (Note that the evolution of stars depends on their chemical composition as well as their masses. Detailed comparisons between models and observed cluster H-R diagrams must account for the abun-

dances of massive elements in the atmospheres of cluster stars, as well as the cluster age.)

We can apply our understanding of stellar evolution even when groups of stars are so far away that we cannot see each star individually. Although many fewer high-mass stars form in a cluster than low-mass stars, higher-mass stars are *far* more luminous than their lower-mass siblings. Likewise, giant, evolved low-mass stars are far more luminous than less massive stars that remain on the main sequence. As a result, the most massive, most luminous stars that are present will dominate the light from a star cluster. If the cluster is young, then most of the light we see will usually come from luminous hot, blue stars. If the cluster is old, then the light from the cluster will have the color of red giants and relatively cool, low-mass stars.

As always, there are caveats to such general statements. Young clusters may contain very luminous red supergiants, and stars with lower abundances of massive elements in their atmospheres often look significantly bluer than their more chemically enriched counterparts. Even so, we can usually get some idea about the properties of a group of stars by looking at its overall color. We will put this knowledge to good use in the chapters that follow, as we turn our attention to the large collections of stars called galaxies. A group of stars with similar ages and other characteristics is referred to as a **stellar population**. An especially bluish color to a galaxy or a part of a galaxy often signifies that the galaxy contains a young stellar population that still includes hot, luminous, blue stars that must have formed recently. In contrast, if a galaxy or part of a galaxy has a reddish color, we expect that it is composed primarily of an old stellar population.

> **Colors reveal the ages of stellar populations.**

17.6 Beyond Newtonian Physics

Having discussed the evolution of clusters of stars of all masses, we now return to the stellar remnants left behind by the evolution of the most massive stars. You might imagine that, surely, the neutron star represents the final extreme of stellar evolution; but there is one more step a star can take. The physics of a neutron star is much like the physics of a white dwarf, except that it is neutrons rather than electrons that cause it to be degenerate. A white dwarf can have a mass of no more than about 1.4 $M_\odot$. This is the Chandrasekhar limit, discussed in Chapter 16. If the mass of the object exceeds this limit, then gravity will be able to overcome electron degeneracy, and the white dwarf will begin collapsing again.

Just as there is a Chandrasekhar limit for white dwarfs, there is a Chandrasekhar limit for the mass of neutron stars as well. If the mass of a neutron star exceeds about 3 $M_\odot$, then gravity will begin to win out over matter once again.

The neutron star grows smaller, and gravity becomes stronger and stronger at an ever-accelerating pace. However, this time there is no force in nature powerful enough to prevent gravity's final victory. The collapsing object quickly crosses a threshold where the escape velocity from its surface exceeds the speed of light, and not even light can escape its gravity. From this point on, nothing can escape from the collapsing object and find its way back into the universe of which it was once a part. The object is now a **black hole**. A black hole will form if the stellar core left behind by a Type II supernova exceeds about 3 $M_\odot$. Alternatively, a neutron star will collapse to become a black hole if it accretes enough matter from a binary companion to push it over the 3-$M_\odot$ limit. Regardless of how it formed, any collapsed object with a mass greater than 3 $M_\odot$ must be a black hole.

> If a neutron star's mass exceeds 3 $M_\odot$, it will collapse to a black hole.

At this point we must pause. What sort of object is this? It turns out that black holes are so strange, so far from our common understanding of reality, that the laws of Newtonian physics (which we explored in Chapter 3) are inadequate to describe them. To do so, we must first take a step back and—as Einstein did—question our intuitive assumptions about the very nature of space and time.

The Speed of Light Is a Very Special Value

Even though we have seen that light behaves like a wave, it turns out that there is a serious flaw in our Newtonian description of that wave. To understand this flaw, we need to think back to Chapter 2, where we used a moving car as an example of a moving frame of reference. Imagine that you are sitting in a moving car and there is a ball on the seat beside you. In your frame of reference the ball is at rest. But if the car is moving at 50 miles per hour (mph) down the highway, someone standing by the road will say that the ball is moving at 50 mph. To someone in oncoming traffic moving at 50 mph, the relative speed of both your car and the ball would be 100 mph. There really is no difference among these three perspectives. The laws of physics are the same in *any* inertial frame of reference.

As a variant of the "ball in the car" experiment, imagine two cars approaching one another, each traveling at 50 mph, as in **Figure 17.19a**. As your car (the red car) moves down the highway at 50 mph, you pitch a fastball forward at 100 mph. In *your* frame of reference (top panel of Figure 17.19a), you are stationary, the oncoming (green) car is approaching you at 100 mph, an observer standing still by the side of the road is moving toward you at 50 mph, and the ball is moving at 100 mph. Now consider frame of reference of the observer (middle panel of Figure 17.19a). Both cars are approaching at 50 mph, and the ball is moving at 150 mph. (The ball has the original 50-mph speed of your car plus the additional 100 mph that you gave it with your throw.) In the frame of reference of the oncoming green car (bottom panel of Figure 17.19a) traveling at 50 mph, the ball is moving at 200 mph. (This is the 150 mph that the ball is moving relative to the ground plus the 50-mph motion of the oncoming car.) In our everyday experience, velocities simply add. This is also how Newton's laws say the universe should behave.

Now do exactly the same thought experiment again, but with two changes. Instead of cars traveling at 50 mph, imagine two spacecraft approaching each other, both traveling at half the speed of light, or 0.5c, as shown in **Figure 17.19b**. You are in the yellow spaceship traveling at 0.5c and, instead of throwing a baseball, you shine a beam of light forward. If we replace the 100-mph speed of the ball in the previous paragraph with c, we think we know what to expect for all observers. In your frame of reference (top panel of Figure 17.19b) you are stationary, the oncoming (blue) spacecraft would be approaching at 1.0c, an observer on a nearby planet is moving toward you at 0.5c, and the beam of light is moving away from you at the speed of light, c. In the frame of reference of an observer on a nearby planet (middle panel of Figure 17.19b), the light should travel by at a speed equal to the speed of your spacecraft plus the speed of light, or 1.5c. Similarly, in the reference frame of the oncoming blue spacecraft (bottom panel of Figure 17.19b) traveling at 0.5c, the light should appear to travel at a speed of 2.0c. (This is the 1.5c that the light is moving relative to the nearby planet plus the 0.5c motion of the oncoming spacecraft.) But something is wrong here. The beam of light appears to be traveling at 1.5c and 2.0c to, respectively, observers on the nearby planet and the approaching blue spacecraft, and the two spacecraft appear to be approaching one another at the speed of light. We have already learned in Chapter 4 that the speed of light has a constant value, c; and, as we will see later in this chapter, objects cannot move at a speed equal to that of light. In the relativistic world, it turns out, speeds do *not* simply add. The relative speed between the two spacecraft in the top and bottom panels of Figure 17.19b (0.5c + 0.5c) adds to 0.8c, not 1.0c! At relativistic speeds, our everyday experience no longer works for us.

Although we would like to compare the speed of light measured by observers in the spacecraft and by a stationary observer, we do not have the luxury of performing this experiment while traveling through space at half the speed of light. However, physicists are ingenious folk. During the closing years of the 19th century and the early years of the 20th century, physicists were conducting laboratory experiments that were the functional equivalent of our thought experiment. What they found puzzled them greatly. Rather than the speed of the beam of light differing from one observer to the next, as expected on the basis of Newton's physics and "common sense," they found

> Surprisingly, experiments showed that the speed of light is the same for all observers.

instead that *all observers measure exactly the same value for the speed of the beam of light, regardless of their motion!*

As you ride in your spaceship, you measure the speed of the beam of light to be c, or 3×10^8 meters per second (m/s). That is as expected because you are holding the source of the light. But the observer on the planet *also* measures the speed of the passing beam of light to be 3×10^8 m/s. Even the passenger in the oncoming spacecraft finds that the

beam from your light is traveling at exactly c in her own frame of reference. In fact, it turns out that *every observer always finds that light in a vacuum travels at exactly the same speed, c, regardless of his or her own motion or the motion of the light source.*

If at this point you are feeling uneasy and saying to yourself, "This is very bizarre," then you probably have followed what the last few paragraphs state. And if this discus-

FIGURE 17.19 The rules of motion that apply in our daily lives break down when speeds approach the speed of light. The fact that light itself always travels at the same speed for any observer is the basis of special relativity. (Note that relativity also affects the relative speeds of the two spacecraft.)

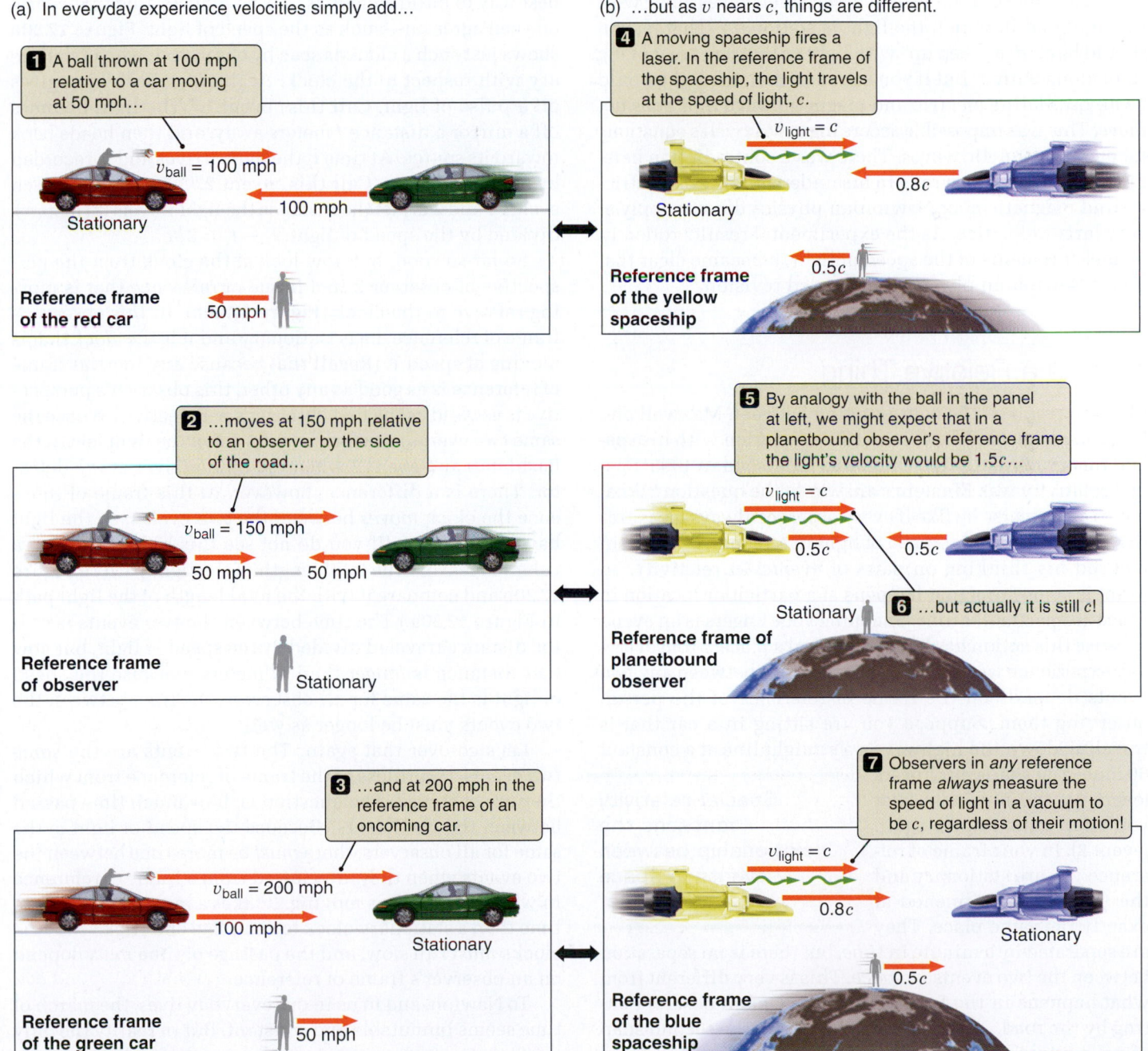

sion bothers you, imagine the reaction of those physicists! Newton's laws of motion had been the bedrock of science for 200 years, standing up to every experimental challenge that came their way. Now, suddenly that bedrock seemed to turn to sand. How could light have the same speed for *all* observers, regardless of their own velocity? Preposterous! And yet that was the inescapable experimental result. Despite all its spectacular successes, Newtonian physics seemed to be in serious trouble.

Enter a young German-born Swiss patent clerk named Albert Einstein. As a 16-year-old schoolboy Einstein had already realized that there was trouble afoot. Light travels in a straight line at a constant speed. Einstein reasoned that according to Newton's laws of motion, there should be a perfectly good inertial frame of reference that moves along with the light and in which the light is stationary. That is, you should be able to "keep up" with light so that you are moving right along with it. But if you could do that, the light would be an oscillating electric and magnetic wave *that does not move*. This was impossible according to Maxwell's equations for electromagnetic waves. There was a contradiction here. Either Maxwell was wrong in his understanding of electricity and magnetism, or Newtonian physics did not apply at very large velocities. As the experimental results rolled in on measurements of the speed of light, it became clear that it was Newtonian physics that needed revision.

Time Is a Relative Thing

Einstein resolved the contradiction between Maxwell and Newton and ushered in a scientific revolution with his **special theory of relativity**, which was published in 1905. Special relativity was Einstein's answer to the question "What must the universe be like if every observer always measures the same value for the speed of light in a vacuum?" Einstein focused his thinking on pairs of *events*. In relativity, an **event** is something that happens at a particular location in space at a particular time. Snapping your fingers is an event, because this action has both a time and a place. From everyday experience we know that the distance between any two events depends on the frame of reference of the person observing them. Suppose you are sitting in a car that is traveling down the highway in a straight line at a constant 60 mph. You snap your fingers (event 1), and a minute later you snap your fingers again (event 2). In your frame of reference *you* are stationary and the two events happened at exactly the same place. They are separated by a minute in *time*, but there is no separation between the two events in *space*. This is very different from what happens in the frame of reference of an observer sitting by the road. This observer agrees that the second snap of your fingers (event 2) occurred a minute after the first snap of your fingers (event 1), but to this observer the two

> Special relativity concerns the relationship between events in space and time.

events were separated from each other in space by a mile. In this everyday, "Newtonian" view, the *distance* between two events depends on the motion of the observer, but the *time* between the two events does not.

Einstein questioned why there was such a distinction between the way Newton treated space and the way Newton treated time. Einstein realized that the *only* way the speed of light can be the same for all observers is if *the passage of time is different from one observer to the next!* This is a *very* counterintuitive idea, but it is so central to our modern understanding of the universe that it is worth wrestling with a bit. Hang on to your hat while we reconstruct some of the reasoning that led Einstein to this remarkable conclusion.

To measure time, the first thing we need is a clock. The best way to build a clock is to base it on a value that everyone can agree on—such as the speed of light. **Figure 17.20a** shows just such a clock as seen by observer 1, who is stationary with respect to the clock. At time t_1 a flashlamp gives off a pulse of light. Call this "event 1." The light bounces off a mirror a distance l meters away, and then heads back toward its source. At time t_2 the light arrives and is recorded by a photodetector. Call this "event 2." The time between events 1 and 2 is just the distance the light travels ($2l$ meters), divided by the speed of light: $t_2 - t_1 = 2l/c$.

So far so good, but now look at the clock from the perspective of observer 2 in a frame of reference that is moving relative to the clock (**Figure 17.20b**). In *this* observer's frame of reference, he is stationary and it is the *clock* that is moving at speed *v*. (Recall that because any inertial frame of reference is as good as any other, this observer's perspective is as valid as the first observer's perspective.) We see the same two events as before: event 1 when the light leaves the flashlamp and event 2 when the light arrives at the detector. There is a difference, however. In this frame of reference the clock *moves* between the two events, so the light has *farther to go*. (If you do not see this right away, use a ruler to measure the total length of the light path in Figure 17.20b and compare it with the total length of the light path in Figure 17.20a.) The time between the two events is still the distance traveled divided by the speed of light, but now that distance is *longer* than $2l$ meters. Because the speed of light is the same for all observers, the time between the two events must be longer as well!

Let's go over that again. The two events are the *same two events*, regardless of the frame of reference from which they are observed. The question is, how much time passed between the two events? Because the speed of light is the same for all observers, there *must* be more time between the two events when they are viewed from a frame of reference in which the clock is moving. It takes a moving clock more time than a stationary clock to complete one "tick." Moving clocks *must* run slow, and the passage of time *must* depend on an observer's frame of reference.

To Newton, and to us in our everyday lives, the march of time seems immutable and constant. But in reality the only thing that is truly constant is the speed of light, and even time itself flows differently for different observers.

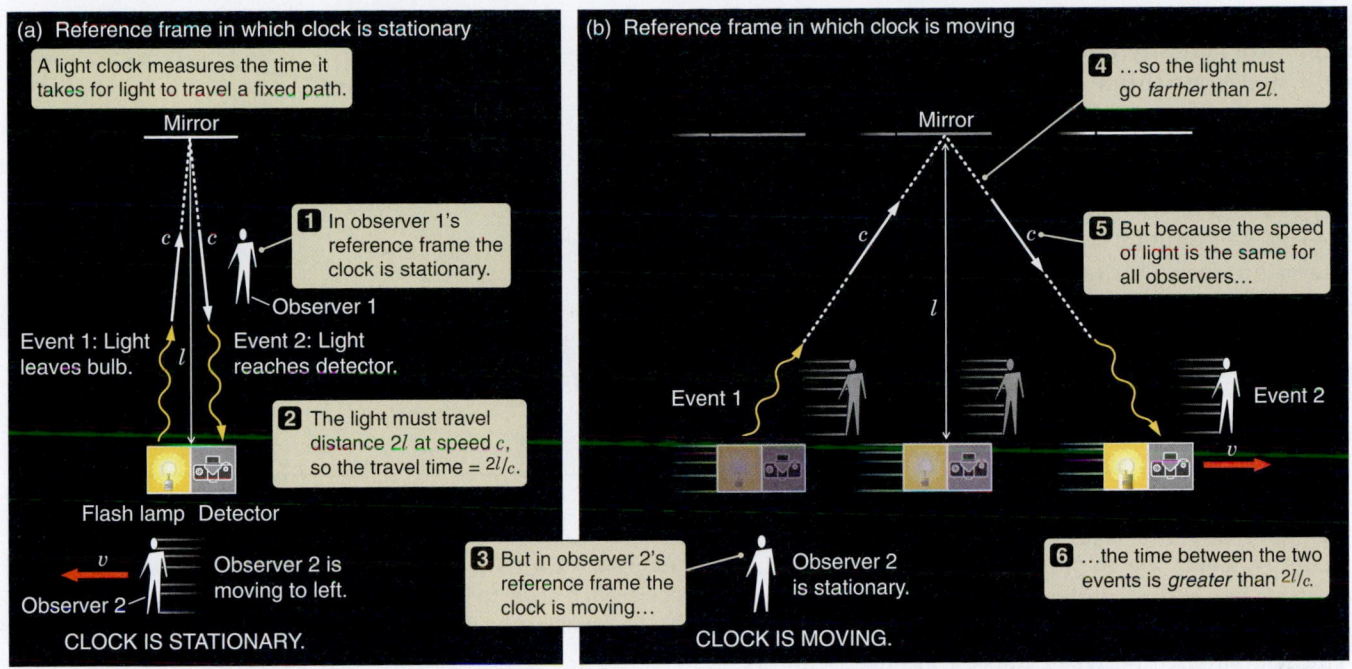

(a) Reference frame in which clock is stationary

A light clock measures the time it takes for light to travel a fixed path.

Mirror

c ↕ *c*

1 In observer 1's reference frame the clock is stationary.

Observer 1

Event 1: Light leaves bulb.

l

Event 2: Light reaches detector.

2 The light must travel distance $2l$ at speed c, so the travel time = $2l/c$.

Flash lamp   Detector

v

Observer 2 is moving to left.

Observer 2

3 But in observer 2's reference frame the clock is moving…

CLOCK IS STATIONARY.

(b) Reference frame in which clock is moving

4 …so the light must go *farther* than $2l$.

Mirror

c   *c*

5 But because the speed of light is the same for all observers…

Event 1   *l*   Event 2

v

Observer 2 is stationary.

6 …the time between the two events is *greater* than $2l/c$.

CLOCK IS MOVING.

FIGURE 17.20 The "tick" of a light clock as seen in two different reference frames: stationary (a) and moving (b). As Einstein's thought experiment demonstrates, if the speed of light is the same for every observer, then moving clocks *must* run slow.

Here, in a nutshell, is the heart of Einstein's special theory of relativity. In our everyday Newtonian view of the world, we live in a three-dimensional space through which time marches steadily onward. Events occur in space at a certain time. By the time Einstein finished working out the implications of his insight, he had reshaped this three-dimensional universe into a four-dimensional **spacetime**. Events occur at specific locations within this four-dimensional spacetime, but how this spacetime translates into what we perceive as "space" and what we perceive as "time" depends on our frame of reference.

> Space and time together form a four-dimensional spacetime.

It is very important to state that Einstein did not throw out Newtonian physics. We were not wasting our time in Chapter 3 when we studied Newton's laws of motion. Instead, Einstein found that Newtonian physics is *contained within* special relativity. In our everyday experience we never encounter speeds that approach that of light. Even the breakneck speed of the space shuttle is only about 0.000025c. When Einstein's special relativity is restricted to cases in which velocities are much less than the speed of light, then Einstein's equations become the very equations that describe Newtonian physics! In our everyday lives we experience a Newtonian world. Only when relative velocities approach that of light do things begin to depart from the predictions of Newtonian physics. When great velocities cause something to turn out

> Newtonian physics is contained within special relativity.

differently from what Newtonian physics would lead us to expect, we call this a **relativistic** effect.

The Implications of Relativity Are Far-Ranging

The story of special relativity is another case study of how science works. Newton's laws had proven for a long time to be an extraordinarily powerful way of viewing the world. But as science turned its attention to a different phenomenon—light—difficulties arose. Newton's theory of motion, Maxwell's theory of electromagnetic radiation, and empirical measurements of the speed of light met head-on. Such conflicts are what scientists live for; they point the way to new knowledge and new understanding. Einstein was able to step in and reconcile this conflict, and in the process he changed the way we think about the universe. Einstein's ideas remained controversial well into the 20th century. But as one experiment after another confirmed the strange and counterintuitive predictions of relativity, scientists came to accept its validity. Today, special relativity is an integral and indispensable part of all of physics, shaping our thinking about the motions of both the tiniest subatomic particles and the most distant galaxies. Puzzling out relativity is time well spent, but here we will explore only a few of the essential insights that come from Einstein's work:

> Conflicts between theory and observation point the way to new knowledge.

1. **What we think of as "mass" and what we think of as "energy" are actually two manifestations of the same thing.** Usually we think of the energy of an object as depending on its speed. The faster it moves, the more energy it has. But Einstein's famous equation $E = mc^2$ says that even a *stationary* object has an intrinsic "rest" energy that equals the mass (m) of the object multiplied by the speed of light (c) squared. The speed of light is a very large number. This relationship between mass and energy says that a single tablespoon of water has a rest energy equal to the energy released in the explosion of over 300,000 tons of TNT! All reactions that produce energy do so by converting some of the mass of the reactants into other forms of energy. But even the most efficient chemical or nuclear reactions release only tiny fractions of the total energy available. Exploding TNT, for example, converts less than a trillionth of its mass into energy. Even the explosion of a hydrogen bomb releases far less than 1 percent of the energy contained in the mass of the bomb.

    The equivalence between mass and energy points both ways. In Chapter 3 we defined *mass* as the property of matter that resists changes in motion. Does the energy of an object really increase its resistance to changes in motion? Yes. Even adding to the energy of motion of an object increases its inertia. For example, a proton in a high-energy particle accelerator may approach the speed of light so closely that its total energy is 1,000 times greater than its rest energy. Such an energetic proton is, indeed, harder to "push around" (in other words, it has more inertia) than a proton at rest.

2. **The speed of light is the ultimate speed limit.** There are several ways to think about this. We already discussed the insight that led Einstein to relativity in the first place. Were it possible to travel at the speed of light, then in that frame of reference light would cease to be a traveling wave, and all of the laws of physics would come tumbling down around us. We can also think about this limit in terms of the equivalence of mass and energy just discussed. As the speed of an object gets closer and closer to the speed of light, its energy, and therefore its mass, become greater and greater, so it becomes increasingly resistant to further changes in its motion. We can continue to push on it all we like, making it go faster, but we face diminishing returns.

    The situation is like trying to get from 0 to 1 by halving the remainder again and again. The resulting sequence—0, ½, ¾, ⅞, ¹⁵⁄₁₆, ³¹⁄₃₂, ⁶³⁄₆₄, . . .—gets arbitrarily close to 1 but never actually reaches it. In the same way, a continuous force applied to an object will cause its velocity to get closer and closer to the speed of light, but it will never actually reach the speed of light. You just cannot get there. It would take an *infinite* amount of energy to accelerate an object with a nonzero rest mass to the speed of light. In short, all the energy in the entire universe is inadequate to accelerate a single electron to

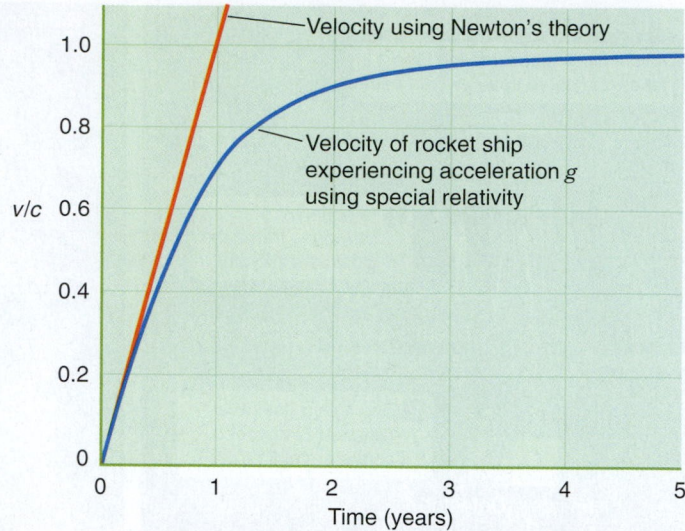

FIGURE 17.21 The speed of a rocket ship experiencing an acceleration equal to Earth's acceleration of gravity. The rocket ship approaches the speed of light but never gets there.

the speed of light. We can get the electron arbitrarily close to that number—0.99999999999999999999 . . . $\times$ c is no problem, at least in principle—but there is no getting over the hump. In **Figure 17.21** we show how a rocket ship, which experiences a constant acceleration equal to that of gravity on Earth (so that its occupants will feel at home), moves faster and faster but never reaches the speed of light. Faster-than-light travel may be a mainstay of science fiction, but this is one of those cases where wishing that something is physically possible does not necessarily make it so.

3. **Time passes more slowly in a moving reference frame.** This phenomenon is referred to as **time dilation** because time is "stretched out" in the moving reference frame. Were you to compare clocks with an observer moving at ⁹⁄₁₀ the speed of light (0.9c), you would find that the other observer's clock was running less than half as fast as your clock (about 0.44 times as fast).[3] You might guess that to the other observer your clock would be fast, but actually the other observer would find instead that it is *your* clock that was running slow! A bit of thought shows why it must be this way. To you, the other observer may be moving at 0.9c, but to the other observer, *you* are moving. Either frame of reference is equally valid, so it stands to reason that if a clock in a moving reference frame runs slow, you would each find the other's clock to be slow.

---

[3] The factor by which time is dilated and space is contracted is given by

$$\frac{1}{\sqrt{1 - \dfrac{v^2}{c^2}}}.$$

This is called the Lorentz factor, named for Dutch physicist Hendrik Lorentz (1853–1928), and is often symbolized as γ.

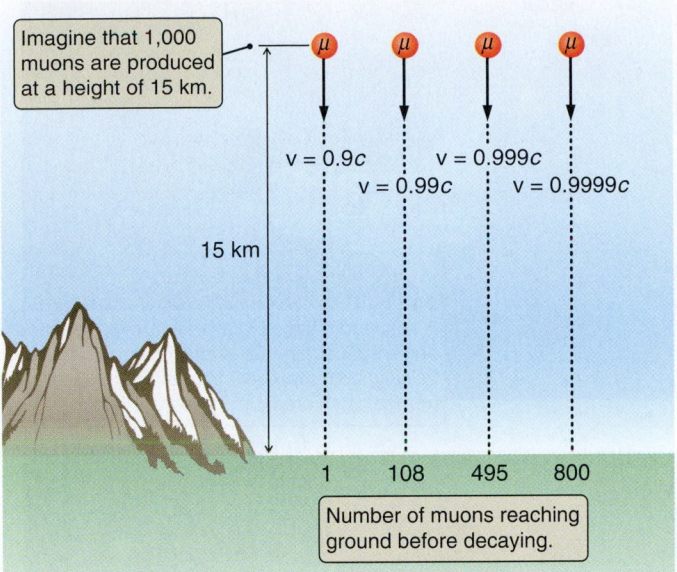

Imagine that 1,000 muons are produced at a height of 15 km.

$v = 0.9c$ $v = 0.999c$
$v = 0.99c$ $v = 0.9999c$

15 km

1   108   495   800

Number of muons reaching ground before decaying.

FIGURE 17.22 Muons created by cosmic rays high in Earth's atmosphere would decay long before reaching the ground if they were not traveling at nearly the speed of light. An observer on the ground sees a longer lifetime because of time dilation. Here we show what happens to 1,000 muons produced at an altitude of 15 km for a variety of speeds.

A straightforward scientific observation demonstrates this effect. As illustrated in **Figure 17.22**, fast particles called "cosmic-ray muons" are produced at about the 15-km level in Earth's atmosphere when high-energy primary cosmic rays strike atmospheric atoms or molecules. Muons at rest decay very rapidly into other particles. Within 2.2 microseconds (μs), half of all muons will have changed their identity. This happens so quickly that, even if they *could* move at the speed of light, virtually all muons would have decayed long before traveling the 15 km to reach Earth's surface. However, the time dilation effect causes the muons' clocks to run slower, so the particles live longer and can travel farther. That is why we are able to detect cosmic-ray muons on the ground.

4. **"At the same time" is a relative concept.** Two events that occur at the *same* time for one observer may occur at *different* times for a different observer. Hold out your arms and snap the fingers on both hands at the same time. For you, the two snaps were simultaneous. But to an observer moving by you from right to left at nearly the speed of light, you snapped the fingers of your left hand first and the fingers of your right hand later.

5. **An object is shorter in motion than it is at rest.** More specifically, moving objects are compressed in the direction of their motion. A meter stick moving at 0.9c is only 43.6 centimeters (cm) long.[4]

[4]See footnote 3.

These consequences of relativity can be combined in what is often called the "twin paradox." Suppose you head off on a trip to the center of the Milky Way Galaxy, roughly 25,000 light-years distant. Your spectacularly powerful star drive accelerates your ship up to 0.9999999992c. To you, the surrounding galaxy is moving by at this speed, so the distance of 25,000 light-years to the center of the galaxy is compressed by a factor of 25,000, to a distance of a single light-year (see number 5 in the previous list). At your speed, you cross this distance in a single year. You snap a picture of the gas swirling around the black hole at the center of the galaxy and then turn around to head home and show it to your twin. Again, the return trip takes only a year. So, in 2 years you have traveled to the center of the galaxy and back again. (Who says interstellar travel is such a big deal?) But when you return, you find that your twin died 50,000 years ago. In the reference frame of Earth, your spacecraft crossed the 25,000–light-year distance to the center of the galaxy moving at just under the speed of light. The only reason you survived the journey, according to an Earth-bound observer, is that in your moving frames of reference, clocks (including biological clocks) ran extremely slow (number 3 in our list). Each leg of the two-way journey took 25,000 years to observers on Earth, and your twin just could not wait that long.

> The twin paradox illustrates many aspects of relativity.

You might puzzle over the twin paradox a bit. Both on the way out and on the way back, in your reference frame it is the clocks on Earth that are running slowly, so you are aging *faster* than your twin. Yet when you return, more time has passed for your twin than for you. How can this be? The answer is that, unlike your twin, you *changed reference frames* during your trip. Event 1 is when you left Earth, and event 2 is when you returned to Earth. Your twin went from one event to the other while riding along in Earth's frame of reference. You, on the other hand, changed reference frames when you left Earth, changed again when you stopped at the center of the galaxy, changed a third time when you left the galactic center to return home, and changed reference frames one final time when you arrived back at Earth. It happens that the path through spacetime that you followed between the two events involved the passage of only 2 years of what you experienced as time, while your twin's path involved 50,000 years of what your twin experienced as time.

Another way to view this is that the key difference between you and your twin is that you experienced acceleration during your trip, while your twin did not. When two observers are in *uniform motion* relative to one another, *neither* of them can lay claim to being in a unique frame of reference. However, *acceleration is a real phenomenon.* You *feel* acceleration when you are riding in a car, and you would surely *feel* the acceleration of the spaceship in this example. It is the fact that you experienced an acceleration that enabled you to "outlive" your twin.

17.7 Gravity Is a Distortion of Spacetime

Our exploration of special relativity began with the observation that the speed of light is always the same (regardless of the motion of an observer or the source), and ended by shattering our everyday notions of space and time. Now, as we confront the properties of black holes—indeed, of all massive objects in the universe—our concepts of space and time will be pulled even further from the comfortable absolutes of Newtonian physics. First we learned that what we traditionally called (three-dimensional) space and time are actually just a result of our particular, limited perspective on a four-dimensional spacetime that is different for each observer. Now we shall discover that this four-dimensional spacetime is warped and distorted by the masses it contains. One of the consequences of this deformation is the gravity that holds you to Earth. This realization, called the **general theory of relativity**, is another of Einstein's great contributions to science.

> Mass warps the fabric of spacetime.

A crucial clue to the fundamental connection between gravity and spacetime has been with us since Chapter 3, where we found that the *inertial mass* of an object—the mass appearing in Newton's $F = ma$—is *exactly* the same as the object's gravitational mass. Another clue is that, left on their own, any two objects at the same location and moving with the same velocity will follow the same path through spacetime, regardless of their masses. The space shuttle astronaut falls around Earth, moving in lockstep with the space shuttle itself. A feather dropped by an *Apollo* astronaut standing on the Moon falls toward the surface of the Moon at exactly the same rate as a dropped hammer does. In some sense, rather than thinking of gravity as a "force" that "acts on" objects, it is more accurate to think of it as a consequence of the path through spacetime that objects will follow in the absence of other forces. *Gravitation is the result of the shape of the spacetime terrain through which objects move.*

> Inertial mass and gravitational mass are the same.

Free Fall Is the Same as Free Float

The essence of special relativity is that any inertial reference frame is as good as any other. There is no experiment you can do to distinguish between sitting in an enclosed spaceship floating stationary in deep space (**Figure 17.23a**) and sitting in an enclosed spaceship traveling through our galaxy at 0.99999 times the speed of light (**Figure 17.23b**). You cannot tell any difference between these two cases—they do not *feel* any different—because there *is* no difference between them.

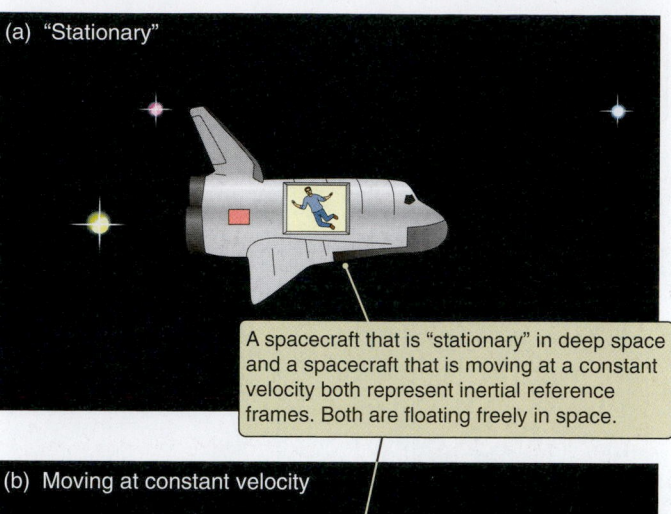

(a) "Stationary"

A spacecraft that is "stationary" in deep space and a spacecraft that is moving at a constant velocity both represent inertial reference frames. Both are floating freely in space.

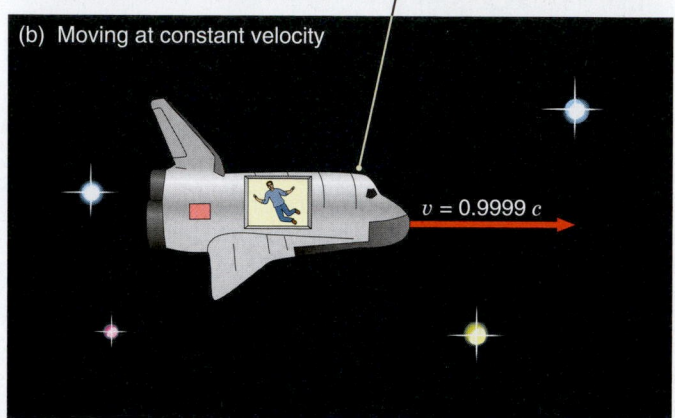

(b) Moving at constant velocity

$v = 0.9999\ c$

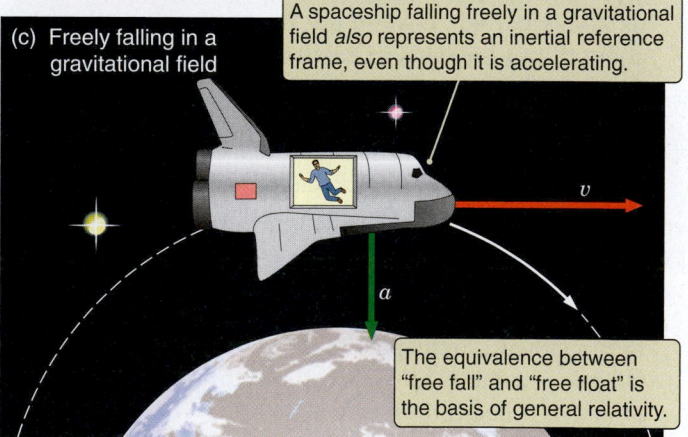

(c) Freely falling in a gravitational field

A spaceship falling freely in a gravitational field *also* represents an inertial reference frame, even though it is accelerating.

v

a

The equivalence between "free fall" and "free float" is the basis of general relativity.

FIGURE 17.23 Special relativity says that there is no difference between (a) a reference frame that is floating "stationary" in space and (b) one that is moving through the galaxy at constant velocity. General relativity adds that there is no difference between these inertial reference frames and (c) an inertial reference frame that is falling freely in a gravitational field. Free fall is the same as free float, as far as the laws of physics are concerned.

Each of these reference frames is an equally valid inertial reference frame. As long as nothing acts to *change* the motion—that is, nothing pushes on either spaceship—the laws of physics are exactly the same inside both spacecraft.

General relativity begins by the application of this same idea to an astronaut inside the space shuttle orbiting Earth, as shown in **Figure 17.23c**. As long as we restrict our attention to a small enough volume of space and a short enough period of time that we can ignore changes in the strength and direction of gravity from place to place, our astronaut again has no way to tell the difference between being inside the space shuttle as it falls around Earth and being inside a spaceship coasting through interstellar space. Close your eyes and jump off a diving board. For the brief time that you are falling freely through Earth's gravitational field, the sensation you feel is exactly the same as the sensation that you would feel adrift in the abyss of interstellar space! The implications of this result are somewhat startling. Even though its velocity is constantly changing as it falls, *the inside of a space shuttle orbiting Earth is as good an inertial frame of reference as that of an object drifting along a straight line through interstellar space.* This principle—which can be simply stated as "free fall is the same as free float"—is called the **equivalence principle**.

> A freely falling object defines an inertial reference frame.

The equivalence principle says that a falling object is simply following its "natural" path through spacetime—it is going where its inertia carries it—every bit as much as an object that drifts along a straight line at a constant speed through deep space. The natural path that an object will follow through spacetime in the absence of other forces is referred to as the object's "world line" or **geodesic**. In the absence of a gravitational field, the geodesic of an object is a straight line—hence Newton's statement of inertia that an object, unless acted on by an unbalanced external force, will move at a constant speed in a constant direction. However, in the presence of mass the shape of spacetime becomes distorted, so an object's geodesic becomes curved.

> Falling objects follow curved paths through curved spacetime.

The equivalence principle gives us a way of understanding why gravitational mass and inertial mass are one and the same thing. When we discussed Newton's law, $F = ma$, we found it useful to state it instead as $a = F/m$. When you apply a force F to an object, you move it away from its natural path and its inertial mass m tells you how strongly it resists the change. General relativity says that when you are standing still on the surface of Earth, your natural path through spacetime—your geodesic—is actually a path falling inward toward the center of Earth. In the absence of any external forces, this is what you would do. Of course, the surface of Earth gets in your way. Put another way, the surface of Earth exerts an external force on your feet, and that force causes you to accelerate continuously away from your natural path through spacetime.

This idea leads to a different, equally valid, way of stating the equivalence principle. Another thought experiment, shown in **Figure 17.24**, demonstrates the point. Imagine you are in a box inside a rocket ship that is accelerating through deep space at a rate of 9.8 meters per second per second (m/s²) in the direction of the arrow shown in **Figure 17.24b**. The floor of the box exerts enough of a force on you to overcome your inertia and cause you to accelerate at 9.8 m/s², so you feel as though you are being pushed into the floor of the box. Now imagine instead that you are sitting in a closed box on the surface of Earth. Again the floor of the box exerts enough upward force on you to overcome your inertia, causing you to accelerate at 9.8 m/s². You feel as though you are being pushed into the floor of the box.

According to the equivalence principle, *the two cases are identical.* There is no difference between sitting in an armchair in a rocket ship traveling through deep space with an acceleration of 9.8 m/s² and sitting in an armchair on the surface of Earth reading this book. In the first case, the force of the rocket ship is pushing you away from your "floating" straight-line geodesic through spacetime. In the second case, Earth's surface is pushing you away from your curved "falling" geodesic through a spacetime that has been distorted by the mass of Earth. An acceleration is an acceleration, regardless of whether you are being accelerated off a straight-line geodesic through deep space or being accelerated off a "falling" geodesic in the gravitational field of Earth. And in all cases, it is the same mass—the mass that gives an object inertia—that resists the change. Gravitational mass and inertial mass are the same thing!

> Being stationary in a gravitational field is the same as being in an accelerated reference frame.

There is an important caveat to the equivalence principle. In an accelerated reference frame such as an accelerating rocket ship, the *same* acceleration is experienced *everywhere*. In contrast, the curvature of space by a massive object changes from place to place. Tides are one result of changes in the curvature of space from one place to another. A more careful statement of the equivalence principle is that the effects of gravity and acceleration are indistinguishable *locally*—that is, as long as we restrict our attention to small enough volumes of space that changes in gravity can be ignored.

Spacetime as a Rubber Sheet

General relativity is a *geometric* theory. It describes how mass distorts the *geometry* of spacetime. You can get a sense for how mass distorts spacetime by imagining the surface of a tightly stretched rubber sheet. The rubber sheet is flat. If you roll a marble across the sheet, it will roll in a straight line. All of the Euclidean geometry that you learned in high school applies on the

> Mass distorts the geometry of spacetime.

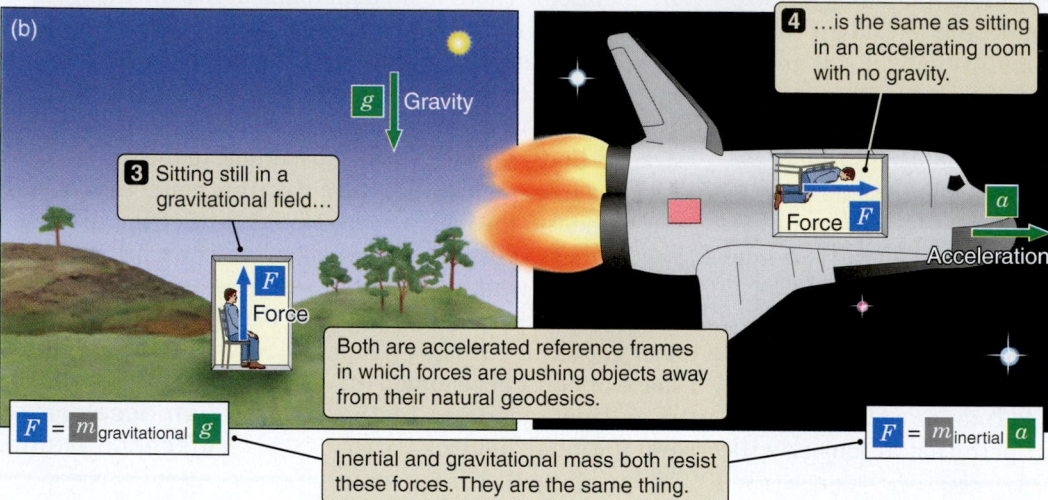

FIGURE 17.24 According to the equivalence principle, an object falling freely in a gravitational field is in an inertial reference frame (a), and an object at rest in a gravitational field is in an accelerated frame of reference (b).

surface of the sheet as well: The angles in a triangle on the sheet add up to 180°. Right triangles obey the Pythagorean theorem. If you draw a circle on the sheet, you will find that the circumference of the circle is equal to 2π times its radius. If you draw a line on this sheet and a point off to one side, there will be exactly one line that passes through that point but never intersects the first line.

Now, though, think about what happens if you place a bowling ball in the middle of the rubber sheet, as in **Figure 17.25**. The surface of the sheet will be stretched and distorted. If you roll a marble across the sheet, its path will dip and curve (**Figure 17.25a**). You might even find that you can roll the marble so that it moves around and around the bowling ball, like a planet orbiting about the Sun. Next you might revisit the relationship between the radius of a circle and its circumference. If you draw a circle around the bowling ball, measure the distance around that circle, and then compare that distance with the distance from the circle to its center along the surface of the sheet (**Figure 17.25b**), you will find that the circumference of the circle is less than $2\pi r$. Finally, you might try to draw a triangle on the surface of the sheet, connecting three points around the bowling ball with the straightest and shortest lines you can draw on the surface of the sheet, as in **Figure 17.25c**. If

you do this and then look at the sheet from above, you will be amazed to see that rather than adding to 180°, the angles in this new triangle sum to more than 180°. The surface of the sheet is no longer flat, and Euclid's geometry ("plane geometry") no longer applies.

Mass has an effect on the fabric of spacetime that is analogous to the effect of the bowling ball on the fabric of the rubber sheet. The bowling ball stretches the sheet, changing the distances between any two points on the surface of the sheet. (Think of the deep depression in the rubber sheet as a "well." We will frequently use this language.) Similarly, mass distorts the shape of spacetime, changing the "distance" between any two locations or events in that spacetime.

It is easy to understand how the two-dimensional surface of the sheet is distorted by the bowling ball, because we can visualize how the sheet is stretched through a third spatial dimension. It is virtually impossible for us to "see" in our mind's eye what a curved four-dimensional spacetime would "look like." Once again we have run into a limitation in how our brains are wired. Yet certain experiments, much like those done on the surface of the rubber sheet, demonstrate that the geometry of our four-dimensional spacetime is distorted much like the rubber sheet.

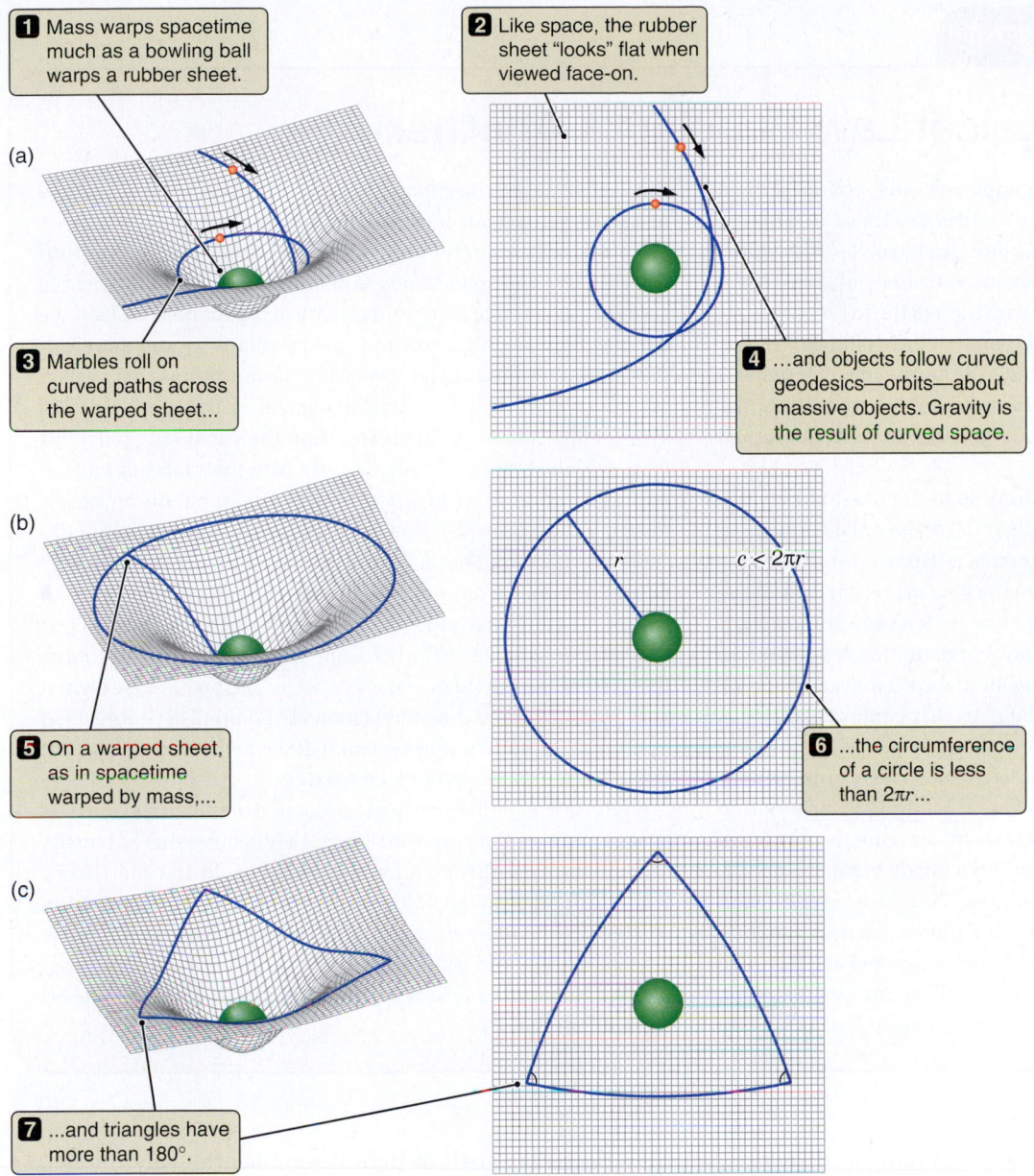

1 Mass warps spacetime much as a bowling ball warps a rubber sheet.

2 Like space, the rubber sheet "looks" flat when viewed face-on.

3 Marbles roll on curved paths across the warped sheet...

4 ...and objects follow curved geodesics—orbits—about massive objects. Gravity is the result of curved space.

5 On a warped sheet, as in spacetime warped by mass,...

6 ...the circumference of a circle is less than $2\pi r$...

7 ...and triangles have more than 180°.

r $c < 2\pi r$

VISUAL ANALOGY FIGURE

17.25 Mass warps the geometry of spacetime in much the same way that a bowling ball warps the surface of a stretched rubber sheet. This distortion of spacetime has many consequences; for example, (a) objects follow curved paths or geodesics through curved spacetime, (b) the circumference of a circle around a massive object is less than 2π times the radius of the circle, and (c) angles in triangles need not add to exactly 180°.

Before going any further in our discussion of general relativity, it is important to point out that general relativity does not mean that Newton's law of gravitation is "wrong." See **Connections 17.1** for a discussion of what happens when one physical law supplants another.

The Observable Consequences of General Relativity

Curved spacetime has many observable consequences. You can imagine, at least in principle, stretching a rope all the way around the circumference of Earth's orbit about the Sun, and then comparing the length of that rope with the length of a rope taken from the orbit of Earth to the center of the Sun. Having studied geometry in high school, you might expect to find that the circumference of Earth's orbit is equal to 2π times the radius of Earth's orbit, just like a circle drawn on a flat piece of paper.[5] If you could carry out this experiment, however, you would find instead that *the rope around the circumference of Earth's orbit was shorter by 10 km than 2π times the length of the rope stretched from Earth to the center of the Sun*—just as the circumference of the circle on the stretched rubber sheet is less than 2π times the radius of the circle.

It is not practical to stretch a rope from Earth to the Sun. But we *can* do an experiment that is conceptually very similar. The long axis of Mercury's elliptical orbit about the Sun is slowly "precessing"; that is,

> The consequences of curved spacetime include precession of Mercury's orbit.

[5]As we learned in Chapter 3, Earth's orbit is actually an ellipse, but it is close enough to a circle for purposes of this thought experiment.

CONNECTIONS 17.1

When One Physical Law Supplants Another

If you have been reading this chapter closely, you might feel some justifiable annoyance with this business of gravity. Throughout the book until this point, we have described gravity as a force that obeys Newton's universal law of gravitation: $F = Gm_1m_2/r^2$. Now we suddenly introduce the ideas of general relativity and in the process ask you to totally change the way you think about gravity. So which is the real deal? If general relativity is right, does that not imply that Newton's formulation of gravity is wrong? And if so, then why have we continued to use Newton's law?

The answers to these questions go to the heart of how science progresses and how our conception of the universe evolves. Under most circumstances there is virtually no difference between the predictions made using general relativity and the predictions made using Newton's law. As long as a gravitational field is not too strong, Newton's law of gravitation is a very close *approximation* to the results of a calculation using general relativity. The meaning of "too strong" in this context is relative. For example, in most ways the enormous gravitational field near the core of a massive main-sequence star would be considered "weak." Had we used a general relativistic formulation of gravity rather than Newton's laws to compute the structure of a main-sequence star, it would have made virtually no difference in the results of the calculation. Similarly, even though spacetime is curved by gravity, this curvature near Earth is very slight, so over small regions it can be ignored entirely. The flat Euclidean geometry is a good "local" approximation even to curved spacetime—and is a lot easier to use. This is exactly the kind of approximation we use when, despite the curvature of Earth, we navigate a city using a flat road map.

Similarly, we now know that Newton's laws of motion are actually *approximations* to the more generally applicable rules of special relativity and quantum mechanics. In fact, we can *derive* Newton's laws from special relativity and quantum mechanics by making the "everyday" assumptions that speeds of objects are much less than the speed of light and that the sizes of objects are much greater than the subatomic particles from which atoms are made. We use Newton's laws of motion and gravitation most of the time because they are far easier to apply than the relativistically and quantum mechanically "correct" laws, and because any inaccuracies we introduce by using Newtonian approximations are usually far too tiny to matter. It is only in conditions very different from those of our everyday lives (such as the behavior of an electron in an atom or the gravitational field of a black hole), or in special cases when very high accuracy is needed (such as the precise timing used by the global positioning system [GPS] satellite network), that the more general laws must be used.

This is a general feature of new scientific theories. If a new theory is to replace an earlier, highly successful scientific theory, the new theory must normally "hold the old theory within it"—it must be able to reproduce the successes of the earlier theory—just as special relativity contains Newton's laws of motion and general relativity holds within it the successful Newtonian description of gravity that we have used throughout the book.

---

the axis of Mercury's orbit is slowly changing its direction. Even after allowing for the perturbations caused by the gravity of the other planets, this precession is not predicted by Newton's inverse square law of gravity, working in a flat Euclidean space. However, it is *just* what is predicted for the path of a planet that is moving alternately deeper and outward within the stretched-out non-Euclidean fabric of spacetime that has been warped by the Sun.

The real-life equivalent of the triangle with more than 180° is probably easier to understand. A straight line in space is *defined* by the path followed by a beam of light. This is the shortest distance between any two points. A beam of light moving through the distorted spacetime around a massive object is bent by gravity, just as the lines in Figure 17.25c are bent by the curvature of the sheet. This phenomenon is called **gravitational lensing** because the way the curvature of space-

> Gravitational lensing can displace and distort an object's image.

time bends the path of light resembles the way the lenses bend light in a pair of eyeglasses.

The first measurement of gravitational lensing came during the solar eclipse of 1919. Prior to the eclipse, English astrophysicist Sir Arthur Stanley Eddington (1882–1944) measured the positions of a number of stars in the part of the sky where the eclipse would occur. Eddington then repeated his measurement during the solar eclipse and found that the apparent positions of the stars had been deflected outward by the presence of the Sun. The light from the stars followed a bent path through the curved spacetime around the Sun, causing the stars to appear farther apart in Eddington's measurement, as illustrated in **Figure 17.26**. During the eclipse, the triangle formed by Earth and the two stars contained more than 180°—just like the triangle on the surface of our rubber sheet. The results of Eddington's measurements were just as predicted by Einstein's theory. Eddington's result was the first experimental test of a prediction of general relativity and is considered to be one of

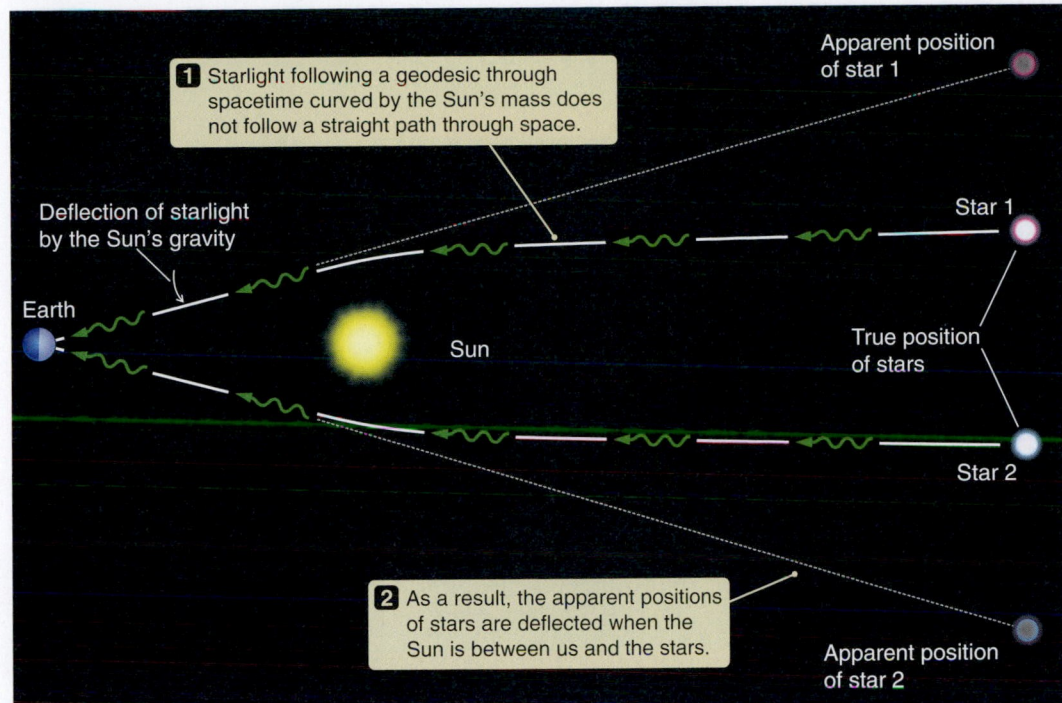

① Starlight following a geodesic through spacetime curved by the Sun's mass does not follow a straight path through space.

Apparent position of star 1

Deflection of starlight by the Sun's gravity

Earth

Sun

Star 1

True position of stars

Star 2

② As a result, the apparent positions of stars are deflected when the Sun is between us and the stars.

Apparent position of star 2

FIGURE 17.26
Measurements obtained by Sir Arthur Eddington during the total solar eclipse of 1919 found that the gravity of the Sun bends the light from distant stars by the amount predicted by Einstein's general theory of relativity. This is an example of gravitational lensing. Note that the "triangle" formed by Earth and the two stars contains more than 180°, just like the triangle in Figure 17.25c.

the landmark experiments of 20th century physics. More recently, gravitational lensing has been used to search for unseen massive objects adrift in space. These objects do not give off light, but their gravity can distort the light from background stars that they happen to pass in front of. (See the discussion of MACHOs in Chapter 20.)

General relativity also affects spacetime in ways that have no direct comparison with a rubber sheet because mass distorts not only the geometry of space, but the geometry of time as well. The deeper we descend into the gravitational field of a massive object, the more slowly our clocks appear to run from the perspective of a distant observer. This effect is called **general relativistic time dilation**. To understand one consequence of general relativistic time dilation, suppose a light is attached to a clock sitting on the surface of a neutron star. The light is timed so that it flashes once a second. Because time near the surface of the star is "stretched out," however, an observer far from the neutron star perceives the light to be pulsing less frequently than once a second. The frequency of the flashing is lowered. Now suppose we have an emission line source on the surface of the neutron star. Because time is running slowly on the surface of the neutron star, at least from our distant perspective, the light that reaches us will have a lower frequency as well. Remember that lower frequency means longer wavelength. So the light from the source will be seen at a longer, redder wavelength than the wavelength at which it was emitted.

This phenomenon, shown in **Figure 17.27**, is called the **gravitational redshift** because the wavelengths of light from

> Time runs more slowly near massive objects.

objects deep within a gravitational well are shifted to longer wavelengths. Gravitational redshift is similar in its effect to the **Doppler redshift** we saw earlier. In fact, there is no way to tell the difference between light that is redshifted by gravity and the Doppler-shifted light from an object moving away from us. Astronomers often describe the gravitational redshift of an object as an "equivalent velocity." The gravitational redshift of lines formed on the surface of the Sun is equivalent to a Doppler shift of 0.6 km/s. The gravitational redshift of light from the surface of a white dwarf is equivalent to a Doppler shift of about 50 km/s. The gravitational redshift from the surface of a neutron star is equivalent to a Doppler shift of about a tenth the speed of light. Sometimes astronomers get sloppy and talk about the gravitational redshift as if it truly were a Doppler shift. We might say, for example, that the "gravitational redshift of the surface of a particular white dwarf is 57.1 km/s." However, this analogy does not mean that the surface of the white dwarf is moving away from us at 57.1 km/s. It means that time is running so slowly on the surface of the white dwarf that the light reaching us from the white dwarf *looks as though* it were coming from an object moving away from us at 57.1 km/s.

Bringing this phenomenon a bit closer to home, a clock on the top of Mount Everest gains about 80 nanoseconds (ns—80 billionths of a second) a day compared with a clock at sea level. The difference between an object on the surface of Earth and an object in orbit is even greater. A global positioning system (GPS) receiver uses the results of sophisticated calculations of the effects of general relativistic time dilation to help you accurately find your position on the surface of Earth. Even after allowing for slowing due

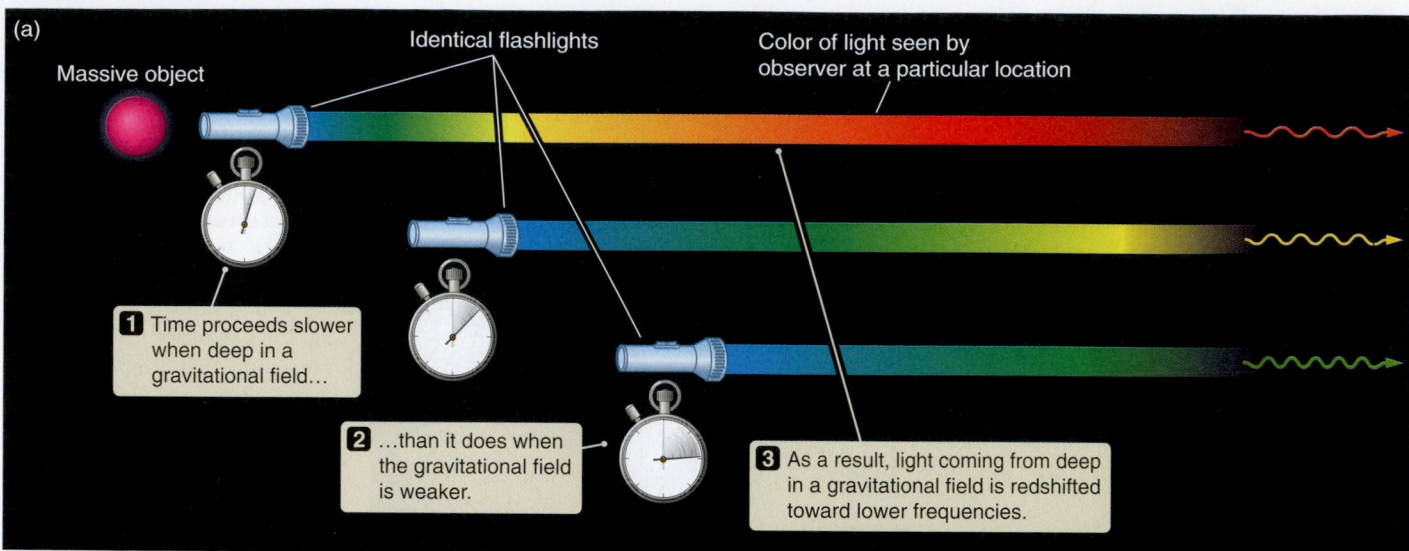

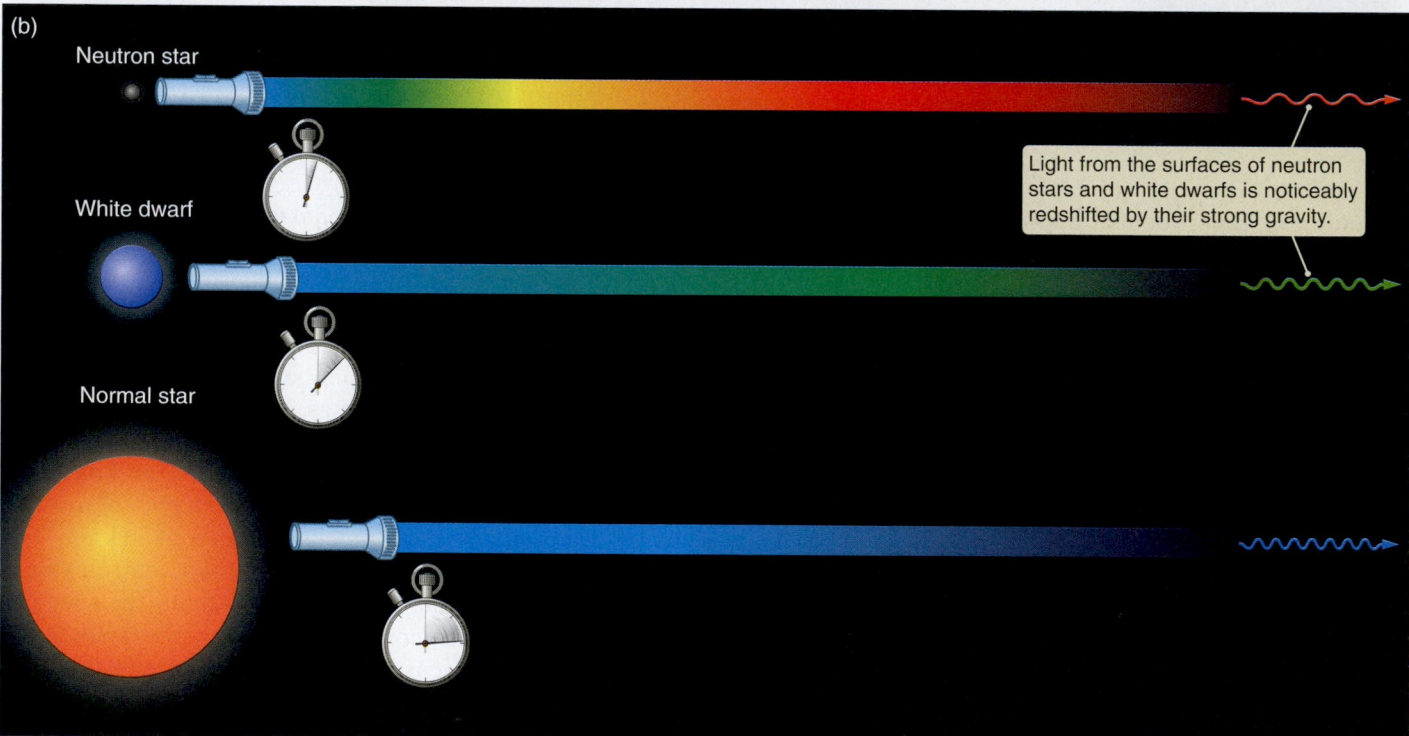

FIGURE 17.27 Time passes more slowly near massive objects because of the curvature of spacetime. As a result, to a distant observer light from near a massive object will have a lower frequency and longer wavelength. The closer the source of radiation is to the object (a) or the more massive and compact the object is (b), the greater the gravitational redshift will be.

to special relativity, the clocks on the satellites that make up the GPS run faster than clocks on the surface of Earth. If the satellite clocks *and* your GPS receiver did not correct for this and other effects of general relativity, then the position your GPS receiver reported would be in error by up to half a kilometer. The fact that the GPS works is actually a strong experimental confirmation of a number of predictions of general relativity, including general relativistic time dilation.

We could easily fill the rest of this book with fascinating tales about general relativity. It is pretty heady stuff if you think about it—discussing the fabric of the universe itself as though it were a substance in a test tube to be poked and prodded (see **Connections 17.2**). One final phenomenon that we should mention before moving on is the phenomenon of **gravitational**

Gravitational waves travel through the fabric of spacetime.

CONNECTIONS 17.2

General Relativity and the Structure of the Universe

Our discussion of gravity as the warping of spacetime by the mass it contains was motivated by the events that accompany the collapse of a massive star. If general relativity applied only to black holes, it might not have been appropriate to devote so much effort in this chapter to a discussion of it. However, this is not the last time we will run into general relativity. Many of the same physical processes we have seen at work here also shape events in the larger universe. In our discussion of galaxies, for example, we will find that some galaxies contain supermassive black holes at their centers—objects with masses millions of times that of our Sun, which grow by consuming entire stars and interstellar clouds. Material falling into such monsters is a leading candidate to explain the powerful radiation from *quasars*—beacons at the edge of the observable universe—which we will explore in greater detail in Chapter 19. We will also find that just as the Sun bends the path of light passing near it, as observed by Eddington during the total eclipse of 1919, the curved spacetime around distant galaxies can act as a lens, magnifying and distorting the appearance of even more distant objects behind them.

The grandest application of general relativity will come as we consider the history and fate of the universe itself. We live in an expanding universe—one that is the result of a singular event that took place about 14 billion years ago. This event, called the **Big Bang**, was not the result of mass coming into existence *within* the spacetime of the universe. Rather, *it was the coming into existence of the spacetime of the universe itself.* When we say that the universe is expanding, we do not mean that the stars and the galaxies in the universe are flying away from each other through space. Instead, the spacetime of the universe itself is expanding with time. Just as the spacetime around a black hole has a shape, so does the spacetime of the universe, and that shape is determined at least in part by the mass that our universe contains. What was the universe like when it was very young? Will the universe expand forever, or is there enough mass in the universe to warp spacetime into a closed shape that will eventually collapse back in on itself? General relativity gives astronomers the tools they need to address such basic questions about the nature of existence itself.

---

waves. If you thump the surface of our rubber sheet, waves will move away from where you thump it, something like ripples spreading out over the surface of a pond. Similarly, the equations of general relativity predict that if you "thump" the fabric of spacetime (for example, with the catastrophic collapse of a high-mass star or the formation of a black hole), then ripples in spacetime will move outward at the speed of light. These gravitational waves are like electromagnetic waves in some respects. Accelerating an electrically charged particle gives rise to an electromagnetic wave. Accelerating a massive object gives rise to gravitational waves.

Gravitational waves have never been observed in a laboratory or anywhere else, but there is strong circumstantial evidence for their existence. In 1974, astronomers discovered a binary system consisting of two neutron stars, one of which is an observable pulsar. By using the pulsar as a precise clock, astronomers have been able to very accurately measure the orbits of both stars. The stars themselves are 2.8 solar radii apart, but their orbits are gradually decaying, which means that they are losing energy *somewhere*. Calculations show that the energy being lost by the system is just what general relativity predicts the system should be losing in the form of gravitational waves. More recently, astronomers discovered a similar binary pair separated by only 1.0 solar radius ($R_\odot$),

and they found once again that the orbital energy loss was consistent with the radiation of gravitational waves. Still, both systems provide only indirect evidence. These measurements very strongly suggest that gravitational waves exist, but they do not prove it.

Let's stop for a minute and think about this. Here we are talking about a phenomenon that has *never been observed*, at least not directly. At the moment, gravitational waves exist only as a scientific theory. Is this so different from the pseudoscientific theory of "intelligent design," which claims that life on Earth was deliberately designed by an intelligent agent? Yes, very much so. The prediction that gravitational waves exist is *falsifiable*. By this we mean that the gravitational wave theory *can be shown to be false*. We have already noted that the theory predicts that certain events, such as the catastrophic collapse of a high-mass star or the formation of a black hole, will generate gravitational waves. As we learned in Chapter 5, astrophysicists have a new kind of "telescope," called the Laser Interferometer Gravitational-Wave Observatory (LIGO), that will be able to detect the predicted gravitational waves emanating from such events. When this happens, either we will see them or we will not. The theory that predicts gravitational waves can be tested because it is falsifiable! Pseudoscientific theories, on the other hand, are *not* falsifiable; they cannot be put to a straightforward "yes or no" test.

17.8   Back to Black Holes

We began our digression into relativity when we encountered black holes, and we return to the nature of black holes now. When we placed an object on the surface of our rubber sheet, it caused a funnel-shaped distortion that is analogous to the distortion of spacetime by a mass. Now imagine such a funnel-shaped distortion in the rubber sheet that is *infinitely* deep—a funnel that keeps getting narrower and narrower as we go deeper and deeper, but that has no bottom. This is the rubber sheet analog to a black hole. With black holes, the mathematics describing the shape of spacetime fail in the same way that

> Black holes are bottomless pits in spacetime.

the mathematical expression $1/x$ fails when $x = 0$. Such a mathematical anomaly is called a **singularity**. Black holes are singularities in spacetime (**Figure 17.28**).

A black hole has only three properties: mass, electric charge, and angular momentum. The amount of mass that falls into a black hole determines the extent of its distortion of spacetime. The electric charge of a black hole is the net electric charge of the matter that fell into it. The angular momentum of a black hole causes the spacetime around the black hole to be twisted around much like the water around an eddy in a river. Apart from these three properties, all information about the material that fell into the black hole is lost. Nothing of its former composition, structure, or history survives.

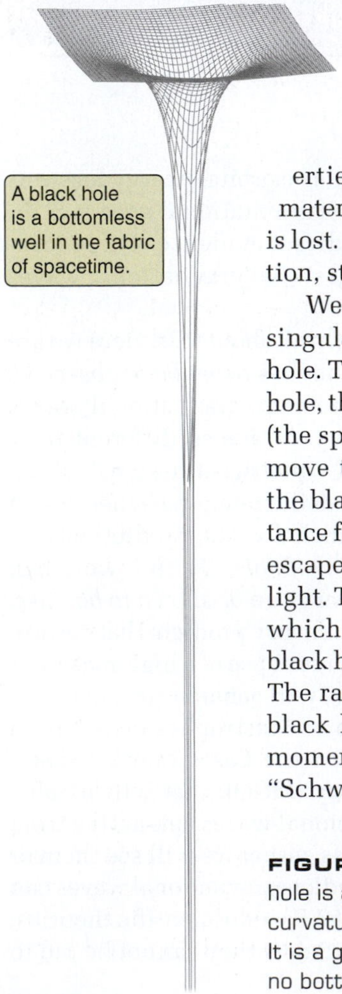

We can never actually "see" the singularity at the center of a black hole. The closer an object is to a black hole, the greater is its escape velocity (the speed at which it would have to move to escape from the gravity of the black hole). There is a radial distance from the black hole at which the escape velocity reaches the speed of light. This point of no return, beyond which even light is trapped by the black hole, is called its **event horizon**. The radius of the event horizon of a black hole that has neither angular momentum nor charge is called the "Schwarzschild radius," named for

A black hole is a bottomless well in the fabric of spacetime.

FIGURE 17.28 A black hole is a singularity in the curvature of spacetime. It is a gravitational well with no bottom.

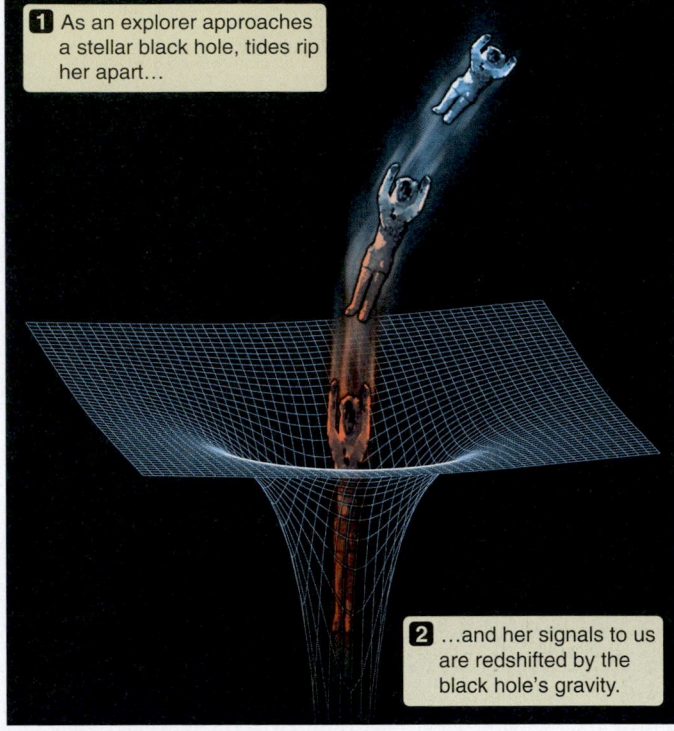

1 As an explorer approaches a stellar black hole, tides rip her apart...

2 ...and her signals to us are redshifted by the black hole's gravity.

FIGURE 17.29 Journey into a black hole.

German physicist Karl Schwarzschild (1873–1916), and it is proportional to the mass of the black hole:

$$R_s = \frac{2GM_{BH}}{c^2},$$

where R_S is the Schwarzschild radius, G is the universal gravitational constant, M_{BH} is the mass of the black hole, and c is the speed of light. A black hole with a mass of $1\ M_\odot$ has an event horizon of about 3 km. A black hole with a mass equivalent to that of Earth would have an event horizon of only about a centimeter—the mass of our planet compressed into a volume equal to that of a pecan!

Let's consider what would happen if an adventurer were willing to journey into a black hole (**Figure 17.29**). From our perspective outside the black hole, we would see our adventurer fall toward the event horizon; but as she did, her watch would run more and more slowly and her progress toward the event

> The event horizon is the boundary of no return.

horizon would get slower and slower as well. Like Achilles in Zeno's famous paradox, even though our adventurer got closer and closer to the event horizon, she would never quite make it, from our perspective. The event horizon is where the gravitational redshift becomes infinite and where clocks stop altogether. Yet our adventurer's own experience would be very different. From her standpoint, there would be nothing special about the event horizon at

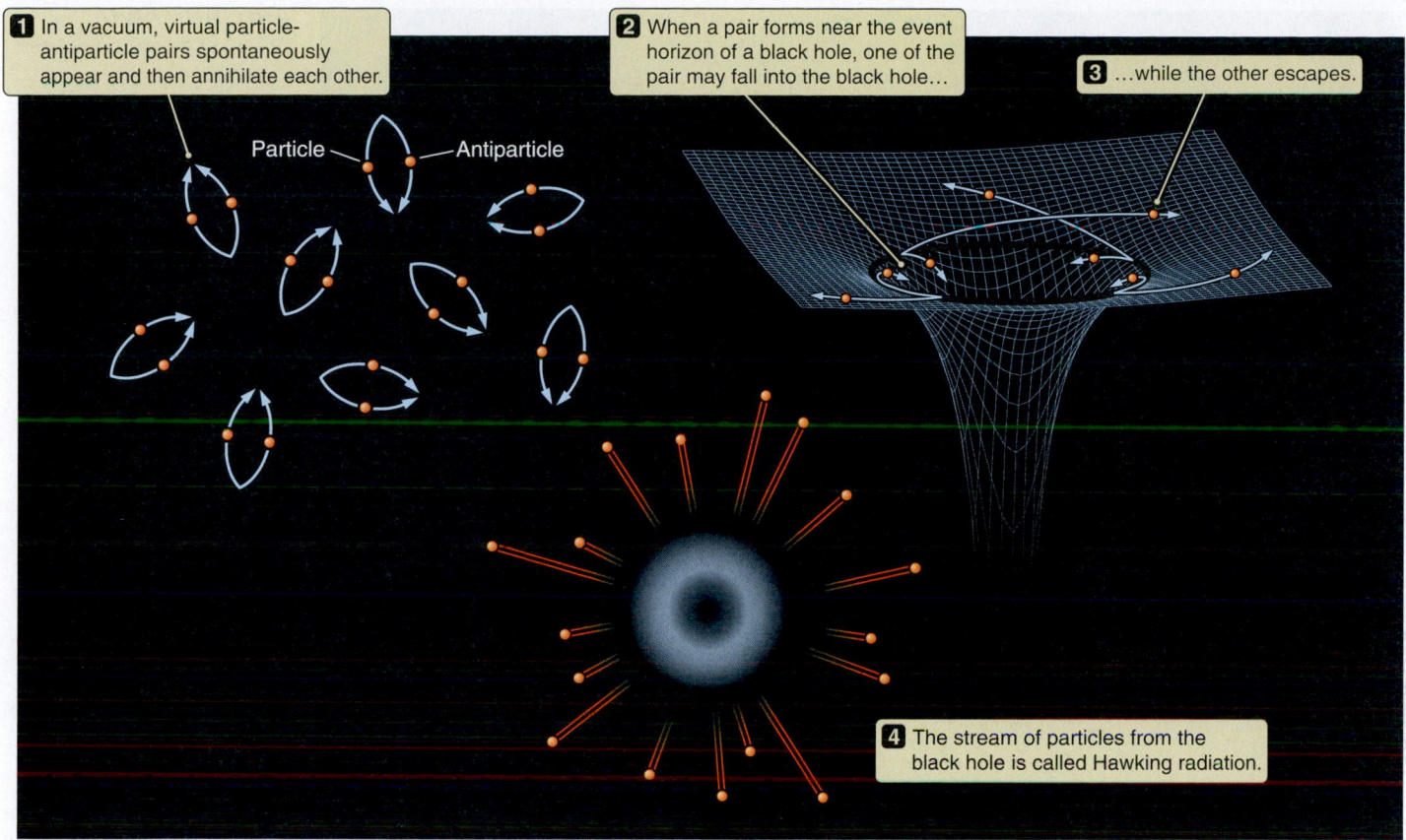

1 In a vacuum, virtual particle-antiparticle pairs spontaneously appear and then annihilate each other.

Particle   Antiparticle

2 When a pair forms near the event horizon of a black hole, one of the pair may fall into the black hole...

3 ...while the other escapes.

4 The stream of particles from the black hole is called Hawking radiation.

FIGURE 17.30 In the vacuum of empty space, particles and antiparticles are constantly being created and then annihilating each other. Near the event horizon of a black hole, however, one particle may cross the horizon before it recombines with its partner. The remaining particle leaves the black hole as *Hawking radiation*.

all. She would fall past the event horizon and on, deeper into the black hole's gravitational well. However, she would now have entered a region of spacetime cut off from the rest of the universe. The event horizon is like a one-way door: once our adventurer has passed through, she can never again pass back into the larger universe to which she once belonged.

Actually, we have overlooked a rather crucial fact. Our intrepid explorer would have been torn to shreds long before she reached the black hole. Near the event horizon of a 3-$M_\odot$ black hole, the difference in gravitational acceleration between our explorer's feet and her head—the tidal "force" pulling her apart—would be about a billion times her weight on the surface of Earth. Obviously, this is not an experiment we would ever want to perform! Although scientific theories *must* produce testable predictions, it is not required that *all* individual predictions be directly testable.

"Seeing" Black Holes

In 1974 the British physicist Stephen Hawking (1942–) realized that black holes should actually be *sources* of radiation.

In the ordinary vacuum of empty space, quantum theory says that particles and their antiparticle "mates" spontaneously spring into existence and then quickly annihilate each other and disappear. These particle pairs typically live for less than about 10^{-21} second, but their effects are seen in sensitive measurements of atomic transitions. If such a pair of **virtual particles** comes into existence near the event horizon of a very small black hole, as shown in **Figure 17.30**, then one of the particles might wind up falling into the black hole while the other particle is able to escape. Some of the gravitational energy of the black hole will have been used up in making one of the pair of virtual particles real. Taking into account all of the esoteric physics, Hawking was able to show that a black hole should actually emit a Planck spectrum, and that the effective temperature of this spectrum would increase as the black hole became smaller. Although this phenomenon, called **Hawking radiation**, is of considerable interest to physicists and astronomers, in a practical sense it is usually negligible. A 3-$M_\odot$ black hole should emit radiation at a whopping temperature of only 2×10^{-8} K, which means that the black hole should radiate with a power of 1.6×10^{-29} watts (W)—very feeble indeed.

FIGURE 17.31
An artist's depiction of the Cygnus X-1 binary system, showing material from the B0 supergiant being pulled off and falling onto an accretion disk surrounding the black hole, thereby producing X-ray emission.

Hawking radiation is hardly a useful way to "see" a black hole. For all intents and purposes, black holes remain true to their name. Nonetheless, by the end of the 20th century astronomers had found strong circumstantial evidence for black holes in two very different kinds of systems. In Chapter 19 we will find that there is strong evidence for supermassive black holes at the very centers of galaxies, but the first and perhaps strongest evidence for black holes comes from X-ray binary stars in our own galaxy. In 1972, astronomers did not yet have a good understanding of the X-ray emission from stars. In that year, the Uhuru X-ray satellite made a puzzling discovery. The brightest X-ray source in the constellation Cygnus was found to be rapidly flickering. We now know that the brightness of the X-ray emission from this object, called **Cygnus X-1**, can change in as little as 0.01 second. For reasons we will cover in more detail in our discussion of quasars in Chapter 19, this means that the source of the X-rays must be smaller than the distance that light travels in 0.01 second, or 3,000 km. Thus, the source of X-rays in Cygnus X-1 must be smaller than Earth!

When astronomers began to study this object in other parts of the electromagnetic spectrum, Cygnus X-1 was identified with both a radio star and with an already cataloged optical star called HD 226868. The spectrum of HD 226868 shows that it is a normal B0 supergiant star with a mass of about 30 $M_\odot$. Such a star is far too cool to explain the X-ray emission from Cygnus X-1. But HD 226868 was

> Black holes are found through the effects of their gravity.

also discovered to be part of a binary system. The wavelengths of absorption lines in the spectrum of HD 226868 are Doppler-shifted back and forth with a period of 5.6 days. Using the same techniques we used to measure the masses of stars in Chapter 13 (namely, analyzing the orbits of binaries), astronomers found that the mass of the unseen compact companion of HD 226868 must be at least 6 $M_\odot$. (Because the tilt of the orbit of the binary is not known, only a lower limit can be determined.) The companion to HD 226868 is too compact to be a normal star, yet it is much more massive than the Chandrasekhar limit for a white dwarf or a neutron star. According to our understanding of the laws of physics, such an object can only be a black hole. Astronomers believe that the X-ray emission from Cygnus X-1 arises when material from the B0 supergiant falls onto an accretion disk surrounding the black hole, as illustrated in **Figure 17.31**.

Since 1972, a number of other good candidates for stellar-mass black holes have been discovered. One such object is a rapidly varying X-ray source in our companion galaxy, the Large Magellanic Cloud. Called LMC X-3, this X-ray source orbits a B3 main-sequence star every 1.7 days, and the data show that the compact source must have a mass of at least 9 $M_\odot$. Although the evidence that these systems contain black holes is circumstantial, the arguments that lead to this conclusion seem airtight. With dozens of compelling examples of such objects on the books, the evidence is in. Black holes, once regarded as nothing more than a bizarre quirk of the mathematics describing gravitation and space-time, exist in nature!

Seeing the Forest for the Trees

The story of low-mass stars like our Sun is by and large a story of longevity and stability. No one has ever seen a star less massive than about 0.8 $M_\odot$ evolve off the main sequence, because all the time in the universe has literally not been enough for this to happen even once. What a contrast on this leg of our journey to look instead at the high-mass end of the family of stars! Rather than stability, these stars offer spectacle. Living for only a short while, they blaze with the light of thousands or even millions of Suns. Such stars are few and far between, but the role they play in the life (and in our understanding) of the universe is significant far beyond their numbers. Indeed, when we move on to study galaxies and the universe, massive stars such as Cepheid variables will provide the signposts that we use to gauge the scale of our universe.

Contemplating the lives of massive stars has also taken us even further afield into the extremes to which matter, space, and time can be molded. White dwarfs may have seemed incomprehensibly bizarre when we encountered them in our discussion of low-mass stars. Objects with densities of a ton per teaspoonful are so far beyond our everyday experience as to be all but unimaginable. Now we have encountered neutron stars—objects a billion times more dense yet. The matter in a neutron star is to the electron-degenerate matter in a white dwarf as the matter in a white dwarf is to the air in a summer breeze. But even the neutron star is but a way station on our journey to the limits of matter and mind. As we contemplate massive stars, we also come face to face with the most extreme object of all—if *object* is even an appropriate term for the bottomless pit in the fabric of the universe that we call a black hole.

This part of our journey has not only taken us into the realm of tortured spacetime but has also shown us the answers to immediate questions about our own existence. We have watched as high-mass stars begin their lives by tapping the same reservoir of nuclear fuel as their low-mass counterparts do, fusing hydrogen into helium and helium into elements such as carbon. But the alchemy of high-mass stars does not end with a cinder of carbon and oxygen. Instead, it goes on, fusing massive elements into more massive elements, and those elements into more massive elements yet.

Finally, we have seen the spectacle of a Type II supernova. As gravity has its final victory, pulling the inner parts of a massive star down into a neutron star or black hole, the outer layers of the star blaze forth, casting into the universe the seeds of future generations of planets and, in all likelihood, life.

We now come to the end of this leg of our journey, having filled out the family album of stars, following them from their origins in the vast reaches of interstellar space to their final places of rest. But we are not done with stars—far from it. All that we have seen has happened on the stage of galaxies, and the universe is the grand hall in which that stage resides. Knowledge of stars and the interstellar medium is the astronomers' starting point as we endeavor to tease from galaxies and the universe their secrets. At the same time, knowledge of the theater will give us further insight into the players on the stage.

Thus, we again head outward, on the last major segment of our journey: an exploration of our expanding universe and the galaxies and other structures it contains.

Summary

- The CNO cycle burns hydrogen in massive stars.

- Iron is the most massive element formed by fusion.

- More massive elements are created in successive burning stages.

- Massive stars eventually explode as supernovae, leaving behind neutron stars.

- Supernovae eject newly formed massive elements into interstellar space.

- Neutron stars contain from 1.4 to about 3 $M_\odot$ packed into a 10-km-diameter sphere.

- Pulsars are rapidly spinning magnetized neutron stars.

- Special relativity concerns the relationship between events in space and time.

- Inertial mass and gravitational mass are the same.

- Mass warps the fabric of spacetime so that objects move on the shortest path in this warped geometry.

- Time runs more slowly near massive objects.

- Objects deep in a gravitational well appear redshifted.

- Black holes are bottomless pits in spacetime.

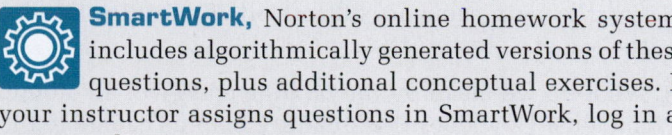
SmartWork, Norton's online homework system, includes algorithmically generated versions of these questions, plus additional conceptual exercises. If your instructor assigns questions in SmartWork, log in at **smartwork.wwnorton.com.**

Student Questions

THINKING ABOUT THE CONCEPTS

1. Explain the differences between the ways that hydrogen is converted to helium in a low-mass star (proton-proton chain) and in a high-mass star (CNO cycle). What is the catalyst in the CNO cycle, and how does it take part in the reaction?

2. How does a low-mass star begin burning helium in its core? What about a high-mass star? How are these processes different or similar?

*3. Why does the core of a high-mass star not become degenerate, as in the cases of low-mass stars?

4. For what two reasons does each post-helium-burning cycle for high-mass stars (carbon, neon, oxygen, silicon, and sulfur) become shorter than the preceding cycle?

5. Cepheids are highly luminous, variable stars in which the period of variability is directly related to luminosity. Explain why Cepheids are good indicators for determining stellar distances that lie beyond the limits of accurate parallax measurements.

6. Identify and explain two important ways in which supernovae influence the formation and evolution of new stars.

7. Is a Type II supernova really an explosion, or is it an implosion? Explain your answer.

8. Describe what we will observe on Earth when Eta Carinae explodes.

9. Recordings show that SN 1987A was detected by neutrinos on February 23, 1987. About 3 hours later it was detected in optical light. Why was there this time delay?

10. Why can the accretion disk around a neutron star release so much more energy than the accretion disk around a white dwarf, even though both stars have approximately the same mass?

11. In Section 17.3, we told you that Type II supernovae blast material outward at 30,000 km/s, while the material in the Crab Nebula (see Section 17.4) is expanding at only 1,500 km/s. What explains the difference?

12. An experienced astronomer can take one look at the H-R diagram of a star cluster and immediately estimate its age. How is this possible?

*13. Can the main-sequence turnoff tell us anything besides the age of a cluster? If so, what?

14. What do we mean by the *binding energy* of an atomic nucleus? How does this quantity help us calculate the energy given off in nuclear fusion reactions?

15. An astronomer sees a redshift in the spectrum of an object. With no other information available, can she determine whether this is an extremely dense object (gravitational redshift) or one that is receding from us (Doppler redshift)? Explain your answer.

*16. Einstein's special theory of relativity tells us that no object can travel faster than, or even at, the speed of light. We know that light is an electromagnetic wave, but we know that it is also a particle called a photon. If it acts as a particle, how can a photon travel at the speed of light?

17. Imagine a future cosmonaut traveling in a spaceship at 0.866 times the speed of light. Special relativity says that the length of the spaceship along the direction of flight is only half of what it was when it was at rest on Earth. The cosmonaut checks this prediction with a meter stick that he brought with him. Will his measurement confirm the contracted length of his spaceship? Explain your answer.

18. Of the four forces in nature (strong nuclear, electromagnetic, weak nuclear, and gravity), gravity is by far the weakest. Why, then, is gravity such a dominant force in stellar evolution? (Note: Although not explicitly discussed so far in this text, the weak nuclear force is involved in certain decay processes within the nucleus.)

19. Suppose astronomers discover a 3-$M_\odot$ black hole located a few light-years from Earth. Should we be concerned that its tremendous gravitational pull will lead to our untimely demise?

20. If you could watch a star falling into a black hole, how would the color of the star change as it approached the event horizon?

21. Why don't we detect the effects of special and general relativity in our everyday lives here on Earth?

22. Many movies and TV programs (like *Star Wars*, *Star Trek*, or *Battlestar Galactica*) rely on faster-than-light travel. How likely is it that we will ever develop such technology in the future?

APPLYING THE CONCEPTS

23. Our galaxy has about 50,000 stars of average mass (0.5 $M_\odot$) for every main-sequence star of 20 $M_\odot$. But 20-$M_\odot$ stars are about 10^4 times more luminous than the Sun, and 0.5-$M_\odot$ stars are only 0.08 times as luminous as the Sun.
    a. How much more luminous is a single massive star than the total luminosity of the 50,000 less massive stars?
    b. How much mass is in the lower-mass stars compared to the single high-mass star?

c. Which stars—lower-mass or higher-mass stars—contain more mass in the galaxy and which produce more light?
Explain your answers.

24. In 1841 the 100-$M_\odot$ star Eta Carinae was losing mass at the rate of 0.1 $M_\odot$ per year. Let's put that into perspective.
   a. The mass of the Sun is 2×10^{30} kg. How much mass (in kilograms) did Eta Carinae lose each minute?
   b. The mass of the Moon is 7.35×10^{22} kg. How does Eta Carinae's mass loss per minute compare with the mass of the Moon?

*25. Using values given in Section 17.2, verify that an O star can lose 20 percent of its mass during its main-sequence lifetime.

**26. The approximate relationship between the luminosity and the period of Cepheid variables is L_{star} ($L_\odot$ units) = 335 P (days). Delta Cephei has a cycle period of 5.4 days and a parallax of 0.0033 arcseconds (arcsec). A more distant Cepheid variable appears 1/1,000 as bright as Delta Cephei and has a period of 54 days.
   a. How far away (in parsecs) is the more distant Cepheid variable?
   b. Could the distance of the more distant Cepheid variable be measured by parallax? Explain.

27. If the Crab Nebula has been expanding at an average velocity of 3,000 km/s since A.D. 1054, what was its average radius in the year 2006? (Note: There are approximately 3×10^7 seconds in a year.)

28. Use Einstein's famous mass-energy equivalence formula ($E = mc^2$) to verify that 5.88×10^{13} J of energy is released from fusing 1 kg of helium.

29. According to Einstein, mass and energy are equivalent. So which weighs more on Earth: a cup of hot coffee or a cup of iced coffee? Why? Do you think the difference is measurable?

30. We know that pulsars are rotating neutron stars. For a pulsar that rotates 30 times per second, at what radius in the pulsar's equatorial plane would a co-rotating satellite (rotating about the pulsar 30 times per second) have to be moving at the speed of light? Compare this to the pulsar radius of 1 km.

31. Verify the claim in Section 17.4 that Earth would be roughly the size of a football stadium if it were as dense as a neutron star.

32. The Moon has a mass equal to 3.74×10^{-8} $M_\odot$. Suppose the Moon suddenly collapsed into a black hole.
   a. What would be the radius of the event horizon (the "point of no return") around the black-hole Moon?
   b. What effect would this collapse have on tides raised by the Moon on Earth? Explain.
   c. Do you think this event would generate gravitational waves? Explain.

33. If a spaceship approaching us at 0.9 times the speed of light shines a laser beam at Earth, how fast will the photons in the beam be moving when they arrive at Earth?

34. Suppose we discover signals from an alien civilization coming from a star that is 2,500 light-years away, and we send you to visit it using the same spaceship as discussed in the twin paradox of Section 17.6.
   a. How long will it take you to reach that planet, according to your clock? According to our clock on Earth? According to the aliens on the other planet?
   b. How likely is it that someone you know will be there to greet you when you return to Earth?

StudySpace is a free and open Web site that provides a Study Plan for each chapter of *21st Century Astronomy*. Study Plans include animations, reading outlines, vocabulary flashcards, and multiple-choice quizzes, plus links to premium content in SmartWork and the ebook. Visit **www.wwnorton.com/studyspace**.

A man said to the universe: "Sir, I exist!" "However,"
replied the universe, "The fact has not created in me a
sense of obligation."

STEPHEN CRANE (1871–1900)

A hemispheric sky map showing tiny fluctuations in the cosmic background radiation (CBR).

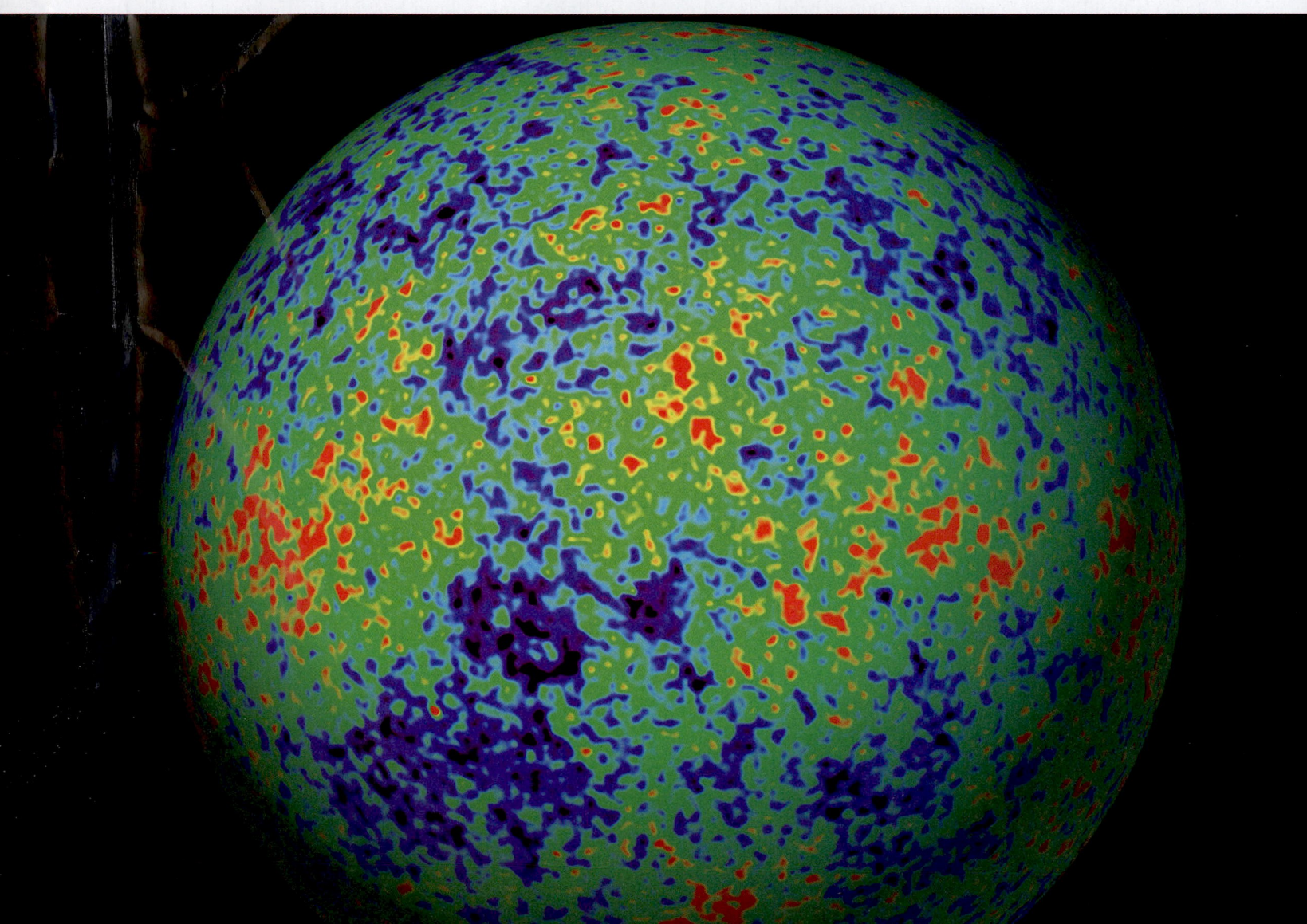

CHAPTER 18

Our Expanding Universe

18.1 Twentieth Century Astronomers Discovered the Universe of Galaxies

We have come a long way on our journey of discovery, building an understanding of stars and the planetary systems that surround them. Yet cosmically speaking, everything we have come across so far is in our own back-yard. We will now greatly expand our vision to regions far beyond our Solar System and Milky Way. Let's look at a deep-space image of a piece of "dark" sky, such as the Hubble Space Telescope image shown in **Figure 18.1**. It reveals myriad faint smudges of light filling the gaps among a sparse smattering of nearby stars. It has long been known that the sky contains faint, misty patches of light. These objects were originally referred to as **nebulae** (singular: *nebula*) because of their nebulous (fuzzy) appearance. Prior to the 1780s, observers using telescopes and naked-eye observations had found only about a hundred of these smudges of light. In 1784 the French comet hunter Charles Messier (1730–1817) published a catalog of 103 nebulous objects, mostly as a warning to other comet hunters not to waste their time on these objects. Yet 20 years later, courtesy of the remarkable observations of the astronomers William Herschel and his sister Caroline Herschel, that number jumped to 2,500. From this time on, astronomers were aware of systematic differences in the appearance of nebulae. Although some of the Herschels' nebulae looked diffuse and amorphous, most were round or elliptical or resembled spiraling whirlpools. These dis-

KEY CONCEPTS

Just as stars are the building blocks of the structure of our own galaxy, so too are galaxies the building blocks of the structure of the universe. Observations of the motions of galaxies led to one of the most remarkable discoveries in the history of our species: the universe is expanding! As we investigate this discovery we will learn

- That the universe contains hundreds of billions of galaxies.

- That the cosmological principle predicts isotropy and homogeneity of the observed distribution and velocity of galaxies.

- About the discovery of Hubble's law, which relates the redshift of a galaxy to its distance.

- How Hubble's law is used to map the universe in space and to look back in time.

- That galaxies are not flying apart through space but rather "ride along" as space itself expands in accordance with the general theory of relativity.

- That spacetime itself was born approximately 13.7 billion years ago in an event called the Big Bang.

- That each of the major predictions of Big Bang theory, from the glow of the cosmic background radiation to the types of atoms that fill the universe, is confirmed by observation.

- What observations of the cosmic background tell us about the properties of the young universe.

FIGURE 18.1 A deep image of a section of "dark" sky made with the Hubble Space Telescope (HST). When we look hard enough, the sky seems literally to be covered with faint galaxies. Nearly every object in this image is a galaxy. The faintest smudges are images of galaxies being formed more than 10 billion years ago.

instead galaxies located far beyond the bounds of our local galaxy. Each tiny smudge in Figure 18.1 is such a galaxy. The universe contains billions upon billions of galaxies—perhaps even an infinite number of galaxies—but certainly more galaxies than there are stars in our own galaxy! Each galaxy is a collection ranging from millions to hundreds of billions of stars, thereby rivaling our own cosmic home. Most of these galaxies are located at such astonishing distances from us that they appear too small and faint to see with any but the most powerful telescopes. We will study distant galaxies and the Milky Way in Chapters 19 and 20.

As has often been the case in astronomy, the nature of galaxies was, at its heart, a question about size and distance. Early attempts to understand the size of the Milky Way were confounded by interstellar dust, which blocks the passage of visible light and thus limits our view. Early astronomers, not knowing of the existence or consequences of this dust, assumed that what they could see in visible light was all there is. Unable to see past this obscuring shroud, they concluded that we live in a system of stars some 6,000 light-years across. It was not until the beginning of the 20th century that American astronomer Harlow Shapley (1885–1972), of the Harvard College Observatory, estimated that our galaxy is over 300,000 light-years in size.[1] Shapley based his estimate on observations of globular clusters, and these observations greatly expanded the previously held views about the extent of space.

In an interesting historical twist, the same insight that brought Shapley to the correct conclusion about the approximate size of the Milky Way also led him to an erroneous conclusion about the nature of the so-called spiral and elliptical nebulae. In 1920, Shapley and fellow astronomer Heber D. Curtis (1872–1942) from California's Lick Observatory met in Washington DC to publicly debate these issues. Historians call this meeting astronomy's "Great Debate." When the Great

tinctions formed the original three categories—diffuse, elliptical, and spiral—used to classify nebulae.

Discovering the existence of nebulae was one thing, but uncovering their true nature was quite another. Speculations about the nature of these objects abounded for the next 140 years. It was suggested that spiral nebulae might be planetary systems in various stages of formation. The great 18th century German philosopher Immanuel Kant (1724–1804) had a very different idea. He speculated that nebulae were instead "island universes"—distant realms of existence separate from our own Milky Way. Herschel himself shared this belief but realized that no telescope of his day would ever be able to resolve the issue. At the time, many astronomers thought that the sum of existence—the *universe*—consisted solely of the swarm of stars to which our Sun belongs. It was not until the first third of the 20th century that the technological tools became available to turn Kant's philosophical musing into scientific knowledge.

Today we know that Kant was correct. Our Milky Way is only one of Kant's island universes, which were renamed **galaxies** (from the Greek *gala*, meaning "milky") to reflect this change in understanding. (The term **universe** is now used to refer to the full expanse of space and all that it contains.) Whereas most diffuse nebulae are nearby clouds of gas and dust in our own Milky Way Galaxy, what Messier and others thought of as elliptical and spiral nebulae are

---

[1]This is actually the size of the galactic *halo* (see Chapter 20). The galactic disk is smaller, about 100,000 light-years across.

FIGURE 18.2 The Andromeda Galaxy, our nearest large galactic neighbor, is about 2.5 million light-years away. Observations of variable stars as standard candles enabled Hubble to estimate its distance. His measurement provided the first observational evidence of the vastness of the universe.

Debate was held, Curtis defended the earlier, smaller model of our galaxy, but the tide against that picture was already turning. The question about the nature of the observed spiral nebulae, however, was still wide open. In Shapley's opinion, his far larger Milky Way was ample enough to encompass everything in the universe. Having worked to show that the galaxy is 50 times larger than previously thought, Shapley balked at the idea that the whole universe was hugely larger still. But Curtis favored the idea that the spiral nebulae were really galaxies separate from our own and that the universe was, indeed, far larger than our own galaxy.

> The Great Debate focused attention on the size and distance of nebulae.

Unlike questions of politics and law, scientific questions are not resolved by the rhetorical skills of partisans. Instead, they are settled by the results of well-crafted and carefully conducted experiments and observations. However, scientific debates do help bring issues into sharper focus, leading scientists to concentrate their attention and efforts on key questions. The reason we call that 1920 meeting the Great Debate is that it clearly marked the final steps that would lead to a correct understanding of nebulae. The 1920 debate set the stage and pointed the direction for the subsequent work of Edwin P. Hubble (1889–1953), whose name was to become forever entwined with our modern understanding of the universe.

Using the newly finished 100-inch telescope on Mount Wilson, high above the then-small city of Los Angeles, Hubble was able to find some variable stars in the large neighboring galaxy of Andromeda (**Figure 18.2**). He recognized that these stars were very similar to the Cepheid variable stars studied by Henrietta Leavitt in the Milky Way

and the nearby Magellanic Cloud galaxies, though his stars appeared much fainter. Using the period-luminosity relation for Cepheid variable stars discussed in Chapter 17, Hubble turned his observations of these stars into measurements of the distances to these objects. The results showed that the distances to these nebulae are far greater than even the galaxy size that Shapley measured. No doubt remained: spiral and elliptical nebulae are really galaxies in their own right, similar in size to our own galaxy but located at truly immense distances. Hubble may not have shared the stage during the Great Debate, but when he spoke through his results, those earlier questions were answered once and for all. Shapley's Milky Way, itself vast beyond comprehension, is but a speck adrift in a universe full of galaxies.

> The nondebater Hubble settled the Great Debate by measuring the distance to galaxies.

18.2 The Cosmological Principle Shapes Our View of the Universe

Hubble's discovery greatly expanded our concept of the universe, and as we look back on our journey of the mind—and on the history of our species—it is remarkable how far we have come. Beginning with Copernicus's realization some 500 years ago that Earth is not the center of all things, we

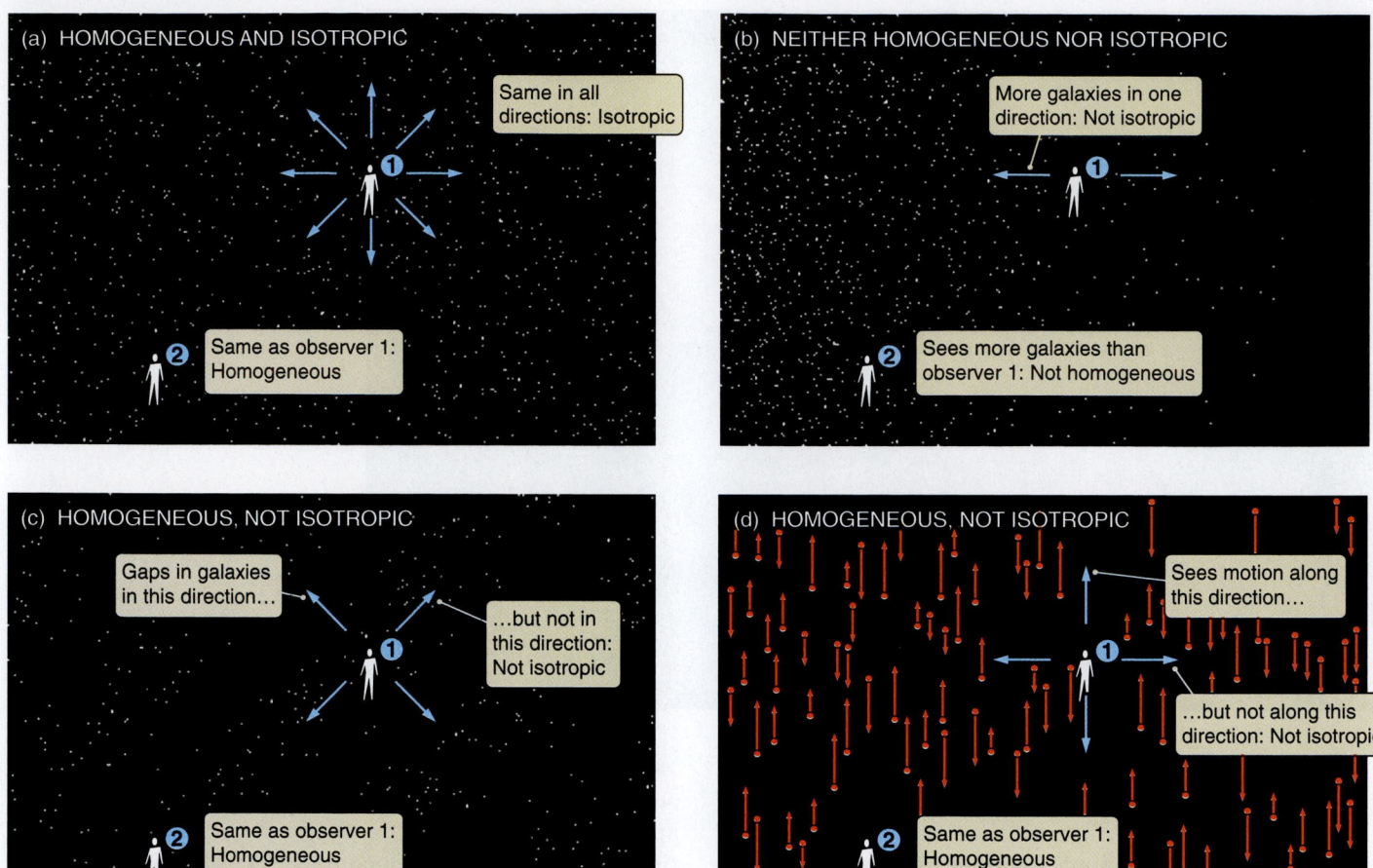

FIGURE 18.3 Homogeneity and isotropy in four different theoretical models of a universe. Blue arrows indicate the direction of view. (a) The distribution of galaxies is uniform, so this universe is both homogeneous and isotropic. (b) The density of galaxies is decreasing in one direction, so this universe is neither homogeneous nor isotropic. (c) The bands of galaxies lie along a unique axis, making this universe anisotropic. (d) The distribution of galaxies is uniform, but galaxies move along only one direction, so this universe also is not isotropic.

have shared the insights of Galileo and Newton and their intellectual heirs as they have torn down the conceptual barriers separating terrestrial existence from that of the heavens. Casting aside the aura of mysticism and magic, we can now go out and look at the night sky and begin to see it for what it is. We now know that Earth sits within an enormous spiral galaxy consisting of hundreds of billions of stars. This galaxy, in turn, is but one of at least hundreds of billions of galaxies that fill a universe vastly larger than our ancestors might have imagined. Through it all, the **cosmological principle** has been at the center of our conceptual understanding. No progress has been possible without the enabling realization that *the rules that apply to one part of our universe apply everywhere.*

The time has come to put the cosmological principle to work in its namesake field: **cosmology**. The science of cosmology is the study of the universe itself, including its structure, history, origins, and fate. As we stressed in Chapter 1, the cosmological principle is not an article of faith. Rather, it is a testable scientific theory. An important prediction of the

cosmological principle is that the conclusions we reach about our universe should be more or less the same, regardless of whether we live in the Milky Way or in a galaxy billions of light-years away at the limits of the observable universe. In other words, we predict that if the cosmological principle is correct, then our universe will be **homogeneous**.

According to one popular dictionary, the word *homogeneous* means "having at all points the same composition and properties." Clearly the universe is not truly homogeneous in an absolute sense of the word. The conditions we encounter at the surface of

> **Observers everywhere should see the same universe.**

Earth are very different from those we would encounter in deep space or in the heart of the Sun. (Even homogenized milk varies at the molecular level from place to place.) When we speak as cosmologists of homogeneity of the universe, we mean instead that stars and galaxies in our part of the universe are much the same, and behave in the same manner, as stars and galaxies in remote corners of the universe.

We also mean that stars and galaxies everywhere are distributed in space in much the same way as they are in our cosmic neighborhood, and that observers in those galaxies see the same properties for our universe that we do. In other words, when we speak of a homogeneous universe, we are applying the term *homogeneous* with a very broad brush.

It is not easy to verify the prediction of homogeneity directly. We do not have the luxury of traveling from our galaxy to a galaxy in the remote universe to see whether conditions are the same. However, we can observe light arriving from the distant universe and see the ways in which features look the same or different. For example, we can look at the way galaxies are distributed in space (see Figure 18.1) and ask whether that distribution is homogeneous.

In addition to predicting that the universe is homogeneous, the cosmological principle requires that all observers (including us) have the same impression of the universe, regardless of the *direction* in which they are looking. If something is the same in all directions, then it is **isotropic**. This prediction of the cosmological principle is much easier to test directly than is homogeneity. For example, if galaxies were lined up in rows—a violation of the cosmological principle—we would get very different impressions, depending on the direction in which we looked. In most instances, isotropy goes hand in hand with homogeneity, and the cosmological principle requires them both. **Figure 18.3** shows examples of how the universe could have violated the cosmological principle by being either inhomogeneous or anisotropic, as well as examples of how the universe might satisfy the cosmological principle.

The isotropy and homogeneity of the distribution of galaxies in the universe are predictions of the cosmological principle that we can go to the telescope and test directly. As you read this, you can benefit from the experience of decades of astronomers who have used the world's most powerful telescopes to test these predictions. These observations show that the properties of the universe are basically the same, regardless of the direction in which we look. If we look on very large scales, the universe appears homogeneous as well. Had these observations turned out otherwise, they would have proven the cosmological principle false, but the cosmological principle withstood the test and is now a bedrock of modern cosmology.

> The distribution of galaxies is isotropic and homogeneous, as predicted.

18.3 We Live in an Expanding Universe

As we have seen, one of the great pioneers of cosmology was Edwin Hubble, pictured in **Figure 18.4**. In the 1920s, Hubble and his coworkers were studying the properties of a large collection of galaxies. Vesto Slipher (1875–1969), one of

FIGURE 18.4 Edwin Hubble, shown on the cover of the February 9, 1948, issue of *Time* magazine.

Hubble's colleagues, was taking images of these galaxies at Lowell Observatory in Flagstaff, Arizona, to obtain their spectra. Unsurprisingly, Slipher's galaxy spectra looked like the spectra of ensembles of stars with a bit of glowing interstellar gas mixed in. The surprise, however, was that the emission and absorption lines in the spectra of these galaxies were seldom seen at the same wavelengths in laboratory-generated spectra. The lines were almost always shifted to longer wavelengths, as shown in **Figure 18.5**.

> Light from distant galaxies is redshifted to longer wavelengths.

Slipher characterized the observed shifts in galaxy spectra as **redshifts**. That is, the spectral lines in these galaxies were shifted to longer, or redder, wavelengths. Recall from Chapter 4 that the Doppler shift causes the observed wavelengths of objects moving away from us to shift toward the red end of the spectrum. The wavelength at which a line is observed in an object that is stationary relative to the observer is called the **rest wavelength** of the line, written λ_{rest}. The redshift of a galaxy, written z, is defined as the difference between the observed wavelength, $\lambda_{observed}$, and the rest wavelength, divided by the rest wavelength:

$$z = \frac{\lambda_{observed} - \lambda_{rest}}{\lambda_{rest}}.$$

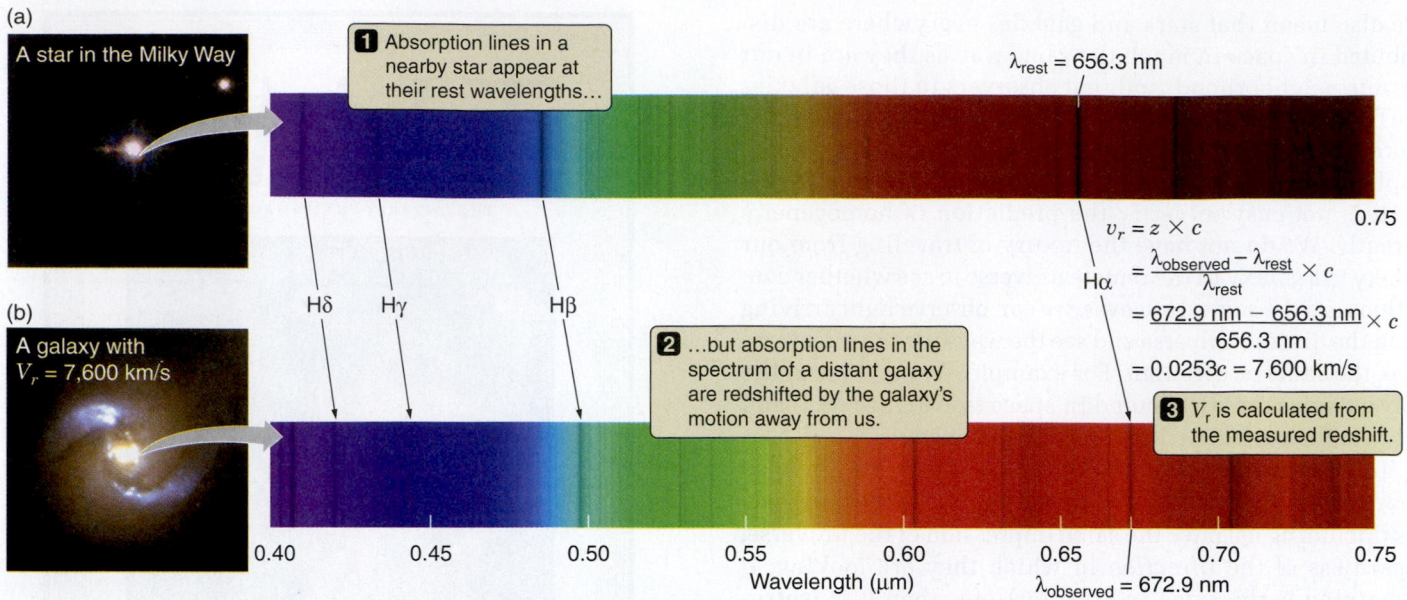

(a)

A star in the Milky Way

1 Absorption lines in a nearby star appear at their rest wavelengths…

$\lambda_{rest} = 656.3$ nm

0.75

$$v_r = z \times c$$
$$= \frac{\lambda_{observed} - \lambda_{rest}}{\lambda_{rest}} \times c$$

Hα

$$= \frac{672.9 \text{ nm} - 656.3 \text{ nm}}{656.3 \text{ nm}} \times c$$
$$= 0.0253c = 7,600 \text{ km/s}$$

(b)

A galaxy with $V_r = 7,600$ km/s

Hδ Hγ Hβ

2 …but absorption lines in the spectrum of a distant galaxy are redshifted by the galaxy's motion away from us.

3 V_r is calculated from the measured redshift.

0.40   0.45   0.50   0.55   0.60   0.65   0.70   0.75

Wavelength (μm)

$\lambda_{observed} = 672.9$ nm

FIGURE 18.5 (a) A star in our galaxy, shown with its spectrum. (b) A distant galaxy, shown with its spectrum at the same scale as that of the star. Note that lines in the galaxy spectrum are redshifted to longer wavelengths.

Hubble's insight was to interpret Slipher's redshifts as Doppler shifts, and he concluded that almost all of the galaxies in the universe are moving away from the Milky Way (see **Math Tools 18.1**). When he combined these measurements of galaxy recession velocities with his own estimates of the distances to these galaxies, he made one of the greatest discoveries in the history of astronomy. Hubble found that distant galaxies are moving away from us more rapidly than are nearby galaxies. Specifically, *the velocity at which a galaxy is moving away from us is proportional to the distance of that galaxy.* This simple relationship between distance and recession velocity has become known as **Hubble's law**. ▶❚❚ **AstroTour: Hubble's Law**

> Hubble's law says that a galaxy's recession velocity is proportional to its distance.

When talking about the distances of remote galaxies in the universe, we need a suitable yardstick. The yardstick we choose is the **mega-light-year** (**Mly**), equal to a million light-years. According to Hubble's law, a galaxy located 100 Mly away is, on average, moving away from us at twice the speed of a galaxy at a distance of 50 Mly.

We usually write Hubble's law as

$$v_r = H_0 \times d_G,$$

where d_G is the distance to the galaxy and v_r is the recession velocity of the galaxy. H_0 (which astronomers pronounce as "H naught") is a constant of proportionality and is called the **Hubble constant**. When thinking of Hubble's law, just remember that if you double the distance d_G, you also double the recession velocity v_r.

> Hubble's constant relates a galaxy's recession velocity to its distance.

All Observers See the Same Hubble Expansion

Hubble's law is a remarkable observation about the universe that has far-reaching implications. For one thing, Hubble's law helps us test the prediction that the universe is homogeneous and isotropic. When we look at galaxies in one direction in the sky, we find that they obey the same Hubble law that galaxies observed in other directions in the sky obey. But although Hubble's law corroborates the prediction that our view of the universe is isotropic, the law appears at first glance to contradict the prediction of the cosmological principle that the universe is homogeneous. Hubble's law might seem to imply that we are sitting in a very special place—at the *center* of a tremendous expansion of space, with everything else in the universe streaming away from us. This initial impression, however, is incorrect. *Hubble's law actually says that we are sitting in a uniformly expanding universe and that the expansion looks the same, regardless of the location of the galaxy from which we view it!* To understand why this is the case, we now turn to a useful model that you can build for yourself with materials you can probably find in your desk.

> Hubble's law says that the universe is expanding uniformly.

Figure 18.6 shows a long rubber band with paper clips attached along its length. If you stretch the rubber band, the paper clips, which represent galaxies in an expanding universe, get farther and farther apart. Imagine what this expansion would look like if you were an ant riding on paper clip A. As the rubber band is stretched, you notice that all of the paper clips are moving away from you. When you look at clip B, which is the next clip over, you see it moving away

MATH TOOLS 18.1

Redshift: Calculating the Recession Velocity and Distance of Galaxies

In our discussion of an expanding universe we defined the redshift, z, of a galaxy as

$$z = \frac{\lambda_{observed} - \lambda_{rest}}{\lambda_{rest}},$$

where $\lambda_{observed}$ and λ_{rest} are, respectively, the observed and rest wavelengths of a spectral line.

Because lines from distant galaxies have wavelengths shifted to the red, the galaxies must be moving away from us. Note that the redshift of a galaxy is independent of the wavelength of the line used to measure it. Note also that this equation is identical to the Doppler formula (see Chapter 4) if we replace z with v_r/c, where v_r is the speed of an object moving away from us and c is the speed of light. So, with a little rearranging, we can rewrite the preceding equation as

$$v_r = \frac{\lambda_{observed} - \lambda_{rest}}{\lambda_{rest}} \times c$$

or

$$v_r = z \times c$$

(Be aware, however, that this correspondence works only for velocities much slower than the speed of light. See Foundations 18.1.)

Suppose we observe a hydrogen line, with a rest wavelength of 122 nanometers (nm), in the spectrum of a distant galaxy. If the observed wavelength of the hydrogen line is 124 nm, then its redshift is

$$z = \frac{\lambda_{observed} - \lambda_{rest}}{\lambda_{rest}} = \frac{124 - 122}{122} = 0.016.$$

We can now calculate the recession velocity from this redshift:

$$v_r = z \times c = 0.016 \times 300{,}000 \text{ km/s} = 4{,}800 \text{ km/s}.$$

How far away, though, is our distant galaxy? This is where *Hubble's law* and the *Hubble constant* ($H_0 = 22$ kilometers per second per mega-light-year, or km/s/Mly) come in. Hubble's law relates a galaxy's recession velocity to its distance ($v_r = H_0 \times d_G$), where d_G is the distance to a galaxy measured in millions of light-years, or *mega-light-years* (*Mly*). Again, a little shifting of variables gives us

$$d_G = \frac{v_r}{H_0} = \frac{4{,}800 \text{ km/s}}{22 \text{ km/s/Mly}} = 220 \text{ Mly}.$$

From a simple measurement of the wavelength of a hydrogen line, we have learned that the distant galaxy is approximately 220 million light-years away.

---

slowly. When you look at clip C, located twice as far down the line from you as B, you see it moving away twice as fast as B. By the time you work your way down the line to clip E (located four times as far away as B), you see it moving away four times as fast as B. In other words, from the perspective of an ant riding on clip A, all of the other paper clips on the rubber band are moving away with a speed that is proportional to their distance. The paper clips located along the rubber band obey a Hubble-like law.

This demonstration of Hubble's law is a handy result, but the key insight to our analogy comes from realizing that there is nothing special about the perspective of paper clip A. If, instead, the ant were riding on clip E, clip D would be the one moving away slowly and clip A would be moving away four times as fast. Repeat this experiment for any paper clip along the rubber band, and you will arrive at the same result: the speed at which other clips are moving away from the ant is proportional to their distance. Bringing our cosmological terminology to bear, the stretching rubber band is "homogeneous." The same Hubble-like law applies, regardless of the paper clip selected as a vantage point.

The observation that nearby paper clips move away slowly and distant paper clips move away more rapidly does not say that we are at the center of anything. Instead, it says that the rubber band is being stretched uniformly along its length. In like fashion, Hubble's law for galaxies does not mean that our galaxy is at the center of an expanding universe. *Hubble's law means that the universe is expanding uniformly.* Any observer viewing our universe from any galaxy will see nearby galaxies moving away slowly and more distant galaxies moving away more rapidly. Observers anywhere else will find that the same Hubble law applies from their vantage point as applies from our vantage point on Earth. The expansion of the universe is homogeneous.

We Must Build a Distance Ladder to Measure the Hubble Constant

The discovery of Hubble's law for galaxies marked a fundamental change in our perception of the universe. It also marked the beginning of a quest that has taken center stage in astronomy for nearly a century. The form of Hubble's

VISUAL ANALOGY **FIGURE 18.6** In this analogy of Hubble's law, a rubber band with paper clips evenly spaced along its length is stretched. As the rubber band stretches, an ant riding on clip A will observe clip C moving away twice as fast as clip B. Similarly, an ant riding on clip E will see clip C moving away twice as fast as clip D. Any ant will see itself as stationary, regardless of which paper clip it is riding, and it will see the other clips moving away with speed proportional to distance.

law—as a proportionality between recession velocity and distance—tells us that our universe is expanding. But to know the present *rate* of the expansion, we need a good value for the Hubble constant, H_0.

How shall we go about obtaining the value for the Hubble constant? In principle, the answer is straightforward. If we can measure the redshifts and distances of a number of galaxies and plot velocity versus distance, then H_0 will be the slope of the resulting line. With a large enough telescope, measuring the redshifts of galaxies is easy; however, measuring the actual distances to galaxies is much, much more difficult.

The difficulty in determining the Hubble constant comes from the fact that we must measure the distances not only to nearby galaxies, but to galaxies that are very far away. To see why, think about the motion of the water in a river. All of the water in a flowing river moves downstream, but even very uniform and steady rivers contain eddies and crosscurrents that disturb the uniform flow. If you want to get a good overall picture of river flow, you need to look at a large portion of the river, not just the motion of a single leaf or two drifting downstream.

Similarly, there are eddies and crosscurrents in the motions of galaxies that make up our universe. The over-

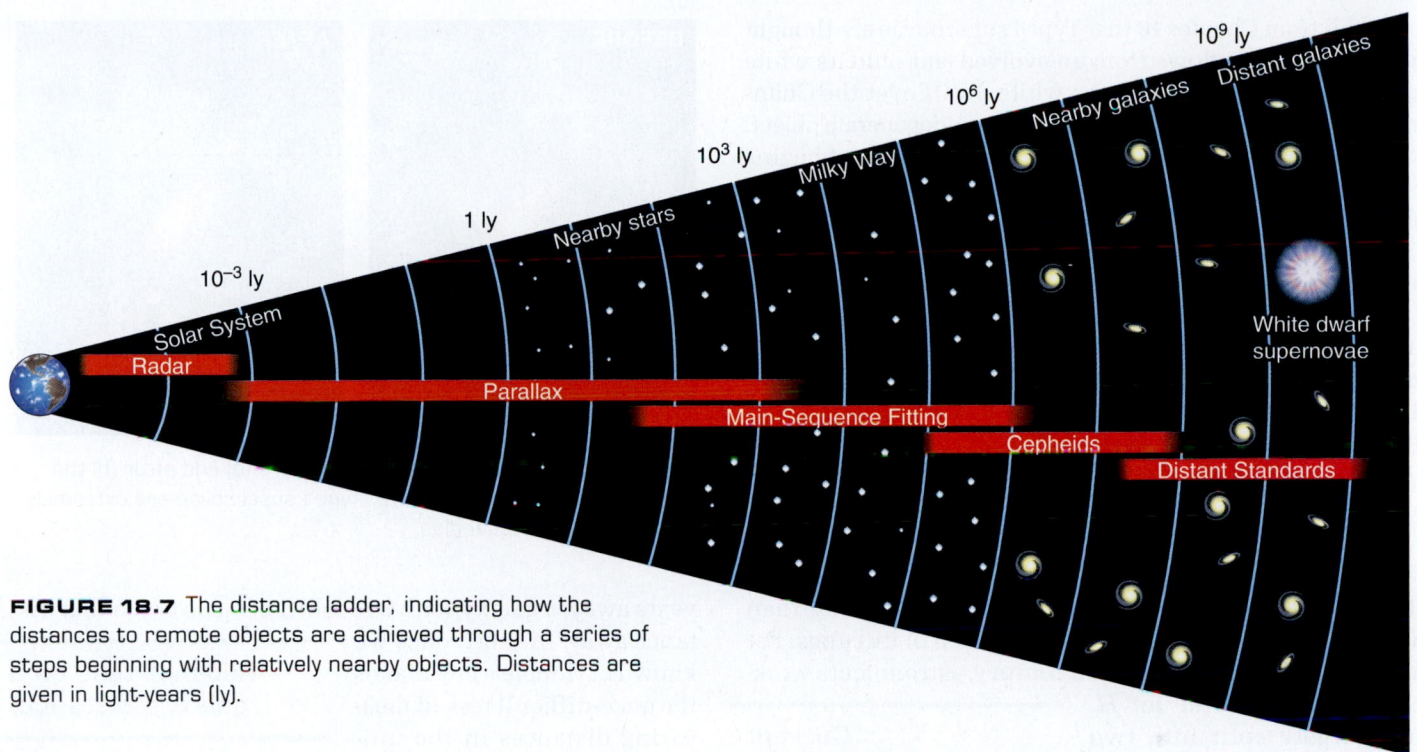

FIGURE 18.7 The distance ladder, indicating how the distances to remote objects are achieved through a series of steps beginning with relatively nearby objects. Distances are given in light-years (ly).

all motion of galaxies in accord with Hubble's law is often referred to as "Hubble flow." Departures from a smooth Hubble flow, referred to as **peculiar velocities**, are the result of gravitational attractions that cause galaxies to fall toward their neighbors or toward large concentrations of mass scattered throughout the universe. If we look only at nearby galaxies, their motions due to Hubble's law will be small, so most of the motion that we see will instead be due to their peculiar velocities. If we want to measure the Hubble flow itself to obtain a reliable value for H_0, we need to study galaxies that are far enough away that most of their velocity comes from the expansion or Hubble flow and relatively little of their observed motion is due to peculiar velocity or gravity.

Peculiar velocities of galaxies are typically a few hundred kilometers per second, so to determine H_0 we need to measure accurately the distances to galaxies with Hubble velocities of several thousand kilometers per second or more (that is, galaxies with redshift z greater than about 0.01). The distances to such galaxies—150 Mly—are far too great to measure using the same techniques that enable us to measure the distances of objects in our own galaxy. Instead, we must find recognizable objects of known luminosity within our own galaxy that are also luminous enough to be seen at greater distances. We refer to these objects as **standard candles**.[2] We will come back to the use of standard candles when we turn to our own Milky Way Galaxy in Chapter 20.

Distances of remote objects are measured in a series of steps, referred to as the **distance ladder**, which relates distances on a variety of scales, as illustrated in **Figure 18.7**.

The distance ladder begins with the size of the astronomical unit, measured by radar and telemetry from space probes. It then steps outward with stellar parallax (see Chapter 13), which enables us to measure distances to nearby stars and thereby to build up the H-R diagram. For more distant stars, astronomers use their spectral and luminosity classifications to determine their position on the H-R diagram. Its position on the H-R diagram provides a star's luminosity, which in turn enables us to estimate its distance by comparing its *apparent* brightness with its known luminosity, as described in Appendix 6. Moving farther out, we can measure the distance to relatively nearby galaxies by identifying recognizable luminous objects, such as O stars, globular clusters, planetary nebulae, novae, and variable stars. (In so doing, we must assume that the luminosity of each object is the same as in our galaxy.) For example, the Magellanic Clouds (see Figure 20.20) are located 160,000 light-years away and contain large numbers of Cepheid variable stars (we learned about Cepheid variables in Chapter 17). Cepheid variables take us another step up the distance ladder and enable us to accurately measure distances to galaxies as far away as 100 Mly. Even this is not far enough to determine a reliable value for the Hubble constant, but within that volume of space are many galaxies that we can scour for yet more powerful distance indicators. Among the best of these are Type I supernovae.[3]

> Measuring H_0 requires measuring distances to remote galaxies.

---

[2]This term is borrowed from an old unit of light intensity that was based on actual candles.

[3]Formally, a specialist in the field would refer to these as "Type Ia supernovae," as noted in Chapter 16.

Recall from Chapter 16 that Type I supernovae are thought to occur when gas flows from an evolved star onto its white dwarf companion, pushing the white dwarf over the Chandrasekhar limit for the mass of an electron-degenerate object. When this happens, the overburdened white dwarf begins to collapse and then explodes. Because all Type I supernovae occur in white dwarfs of the same mass, we might expect all such explosions to have

> Type I supernovae are very luminous standard candles.

about the same luminosity. This prediction is borne out by observations of Type I supernovae in galaxies with known distances. With a peak luminosity that outshines a hundred billion Suns (**Figure 18.8**), Type I supernovae can be seen and measured with modern telescopes that can reach almost to the edge of the observable universe.

It is often said that a chain is no stronger than its weakest link. In constructing the cosmic distance ladder, the situation is even worse: the final ladder is no better than the combination of the weaknesses in each of its rungs. For the last few decades of the 20th century, astronomers working to obtain a value for H_0 were largely split into two camps. One group favored a Hubble constant of about 18 km/s/Mly; a second camp

> Current measurements put H_0 at 22 km/s/Mly.

viewed the data as supporting a value of about 35 km/s/Mly. A virtual war raged between these two groups for years, until HST observations in the mid-1990s began to converge on an intermediate value of about 22 km/s/Mly.[4] The most recent (2008) analysis of satellite data using other measures of distance has yielded results consistent with that value.

Figure 18.9 plots the measured recession velocities of galaxies against their measured distances. Notice how well the universe follows Hubble's law. Today, most astronomers believe that we have measured the Hubble constant to an accuracy of 10 percent—and perhaps even better than that. Therefore, we can be confident that it lies between 20 and 25 km/s/Mly and that it is likely to be further refined in the years to come.

Hubble's Law Maps the Universe in Space and Time

Why all the fuss about Hubble's law and the Hubble constant? What do these tell us about the universe? First, Hubble's law gives us a practical tool for measuring distances to remote objects. Once we know the value of H_0, Hubble's law enables us to turn a straightforward measurement of the redshift of a galaxy into knowledge of its distance. For example, a galaxy with a redshift of 0.1 is 1.4 billion light-

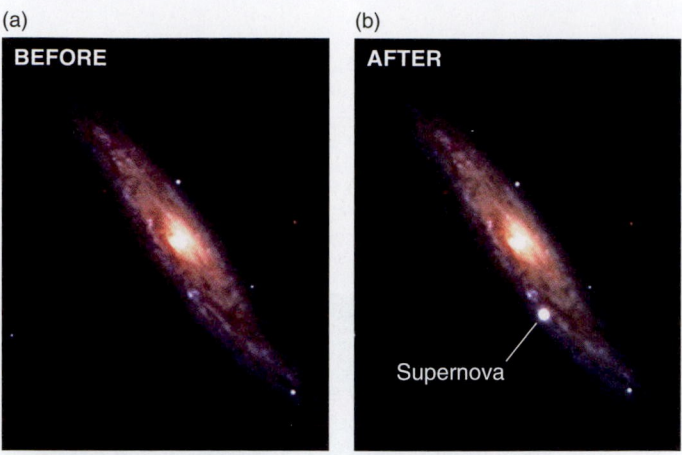

(a) BEFORE   (b) AFTER

Supernova

FIGURE 18.8 Galaxy NGC 3877 before (a) and after (b) the explosion of a Type I supernova. Type I supernovae are extremely luminous standard candles.

years away; a galaxy with a redshift of 0.2 is twice that distance away. In short, once we know H_0, Hubble's law makes the once-difficult task of measuring distances in the universe relatively easy, providing us with a tool to literally map the structure of the universe (see Math Tools 18.1). We will put this tool to good use in Chapter 22 when we turn our attention to the large-scale structure of the universe.

> Redshift tells us a galaxy's distance.

Hubble's law does more than place galaxies in space. It also places galaxies in time. Light travels at a huge but finite speed. Remember that when we look at the Sun, we see it as it existed 8 minutes ago. When we look at Alpha Centauri, the nearest stellar system beyond the Sun, we see it as it existed 4.3 years ago. When we look at the center of our galaxy, the picture we see is

> When observing the universe, elsewhere is "elsewhen."

27,000 years old. When looking at a distant object, we speak of its **look-back time**—the time it has taken for the light from that object to reach our telescope. As we look into the distant universe, look-back times become very great indeed. The distance to a galaxy whose redshift $z = 0.1$ is 1.4 billion light-years (assuming $H_0 = 22$ km/s/Mly), so the look-back time to that galaxy is 1.4 billion years. The look-back time to a galaxy where $z = 0.2$ is 2.7 billion years. As we look at objects with greater and greater redshifts, we are seeing increasingly younger versions of our universe.

18.4 The Universe Began in the Big Bang

Hubble's law provides a very powerful and practical tool for mapping the distribution of galaxies throughout the universe. Yet the most significant aspect of Hubble's law is what it tells us about the structure and origin of the uni-

---

[4]Astronomers normally express H_0 in units of kilometers per second per million parsecs (km/s/Mpc) rather than kilometers per second per mega-light-years. In these units, 22 km/s/Mly corresponds to 72 km/s/Mpc.

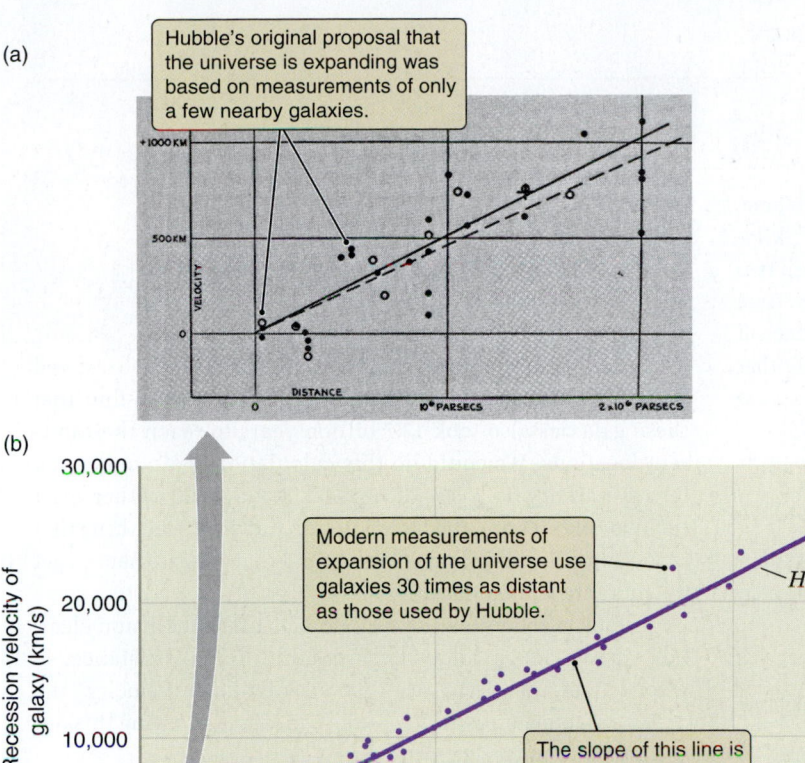

(a)

Hubble's original proposal that the universe is expanding was based on measurements of only a few nearby galaxies.

(b)

Modern measurements of expansion of the universe use galaxies 30 times as distant as those used by Hubble.

$H_0 = 22$ km/s/Mly

The slope of this line is Hubble's constant H_0.

Hubble's original plot

FIGURE 18.9 (a) Hubble's original figure illustrating that more distant galaxies are receding faster than less distant galaxies. (b) Modern data on galaxies up to 30 times farther away than those studied by Hubble show that recession velocity is proportional to distance.

verse itself. We know that all galaxies in the universe are moving away from each other, but if we could run the "movie" of this separation backward in time, we would find the galaxies getting closer and closer together. Using Hubble's law, we find that about 6.8 billion years ago, when the universe was half its present age, all of the galaxies in the universe were separated from each other by half their present distances. Twelve billion years ago, all of the galaxies in the universe must have been separated from each other by about a tenth of their present distances. Assuming that galaxies have been moving apart at the same speed that we see today, then 13.7 billion years ago (a time equal to $1/H_0$), *all the stars and galaxies that make up today's universe must have been concentrated together at the same location!* (See **Math Tools 18.2**.) The value of 1 divided by the Hubble constant is referred to as the **Hubble time**. If our line of reasoning is correct, then today's universe is hurtling outward from a tremendous expansion of space that took place approximately 13 billion to 14 billion years ago. This colossal event, which marked the beginning of our universe, is referred to as the **Big Bang (Figure 18.11)**. ▶❙❙ **AstroTour: Hubble's Law**

The idea of the Big Bang greatly troubled many astronomers in the early and middle years of the 20th century. Several different suggestions were put forward to explain the

> **Expansion started with a Big Bang.**

observed fact of Hubble expansion without resorting to the idea that the universe came into existence in an extraordinarily dense fireball billions of years ago. However, as more and more observations have come in and more discoveries about the structure of the universe have been made, the Big Bang theory has only grown stronger. Today only a very few serious workers in this field question whether the Big Bang took place. Virtually all the major predictions of the Big Bang theory (expansion of the universe being but one of them) have proven to be correct. As we will see, the Big Bang theory for the origin of our universe is now such a well-corroborated theory that most astronomers would probably say it has crossed into the realm of scientific fact.

The implications of Hubble's law are striking. This single discovery forever changed our concept of the origin, history, and possible future of the universe in which we live. At the same time, Hubble's law has pointed to many new questions about the universe. To address them, we next need to consider exactly what we mean by the term *expanding universe.*

Galaxies Are *Not* Flying Apart through Space

The mental picture that you have of the expanding universe at this point in our discussion is probably one of a

MATH TOOLS 18.2

Expansion and the Age of the Universe

We can use Hubble's law to estimate the age of the universe. Consider two galaxies located 100 Mly ($d_G = 9.5 \times 10^{20}$ kilometers [km]) away from each other (**Figure 18.10**). If these two galaxies are flying apart from each other, then at some time in the past they *must* have been together in the same place at the same time. According to Hubble's law, and assuming that $H_0 = 22$ km/s/Mly, the distance between these two galaxies is increasing at the rate

$$v_r = H_0 \times d_G = 22 \text{ km/s/Mly} \times 100 \text{ Mly} = 2{,}200 \text{ km/s}.$$

Knowing the speed at which they are traveling, we can calculate the time it took for the two galaxies to become separated by 100 Mly:

$$\text{Time} = \frac{\text{Distance}}{\text{Speed}} = \frac{9.5 \times 10^{20} \text{ km}}{2{,}200 \text{ km/s}} = 4.32 \times 10^{17} \text{ s} = 1.37 \times 10^{10} \text{ yr.}$$

In other words, *if* expansion of the universe has been constant, two galaxies that today are 100 Mly apart started out at the same place 13.7 billion years ago.

Now let's do the same calculation with two galaxies that are twice as far apart: 200 Mly, or 19×10^{20} km (see Figure 18.10). These two galaxies are twice as far apart, but the distance between them is increasing twice as rapidly:

$$v_r = H_0 \times d_G = 4{,}400 \text{ km/s}.$$

Therefore,

$$\text{Time} = \frac{19 \times 10^{20} \text{ km}}{4{,}400 \text{ km/s}} = 4.32 \times 10^{17} \text{ s} = 1.37 \times 10^{10} \text{ yr.}$$

Again, we calculate time as distance divided by speed (twice the distance divided by twice the speed) to find that these galaxies also took 13.7 billion years to reach their current locations. We could do this calculation again and again for any pair of galaxies in the universe today. The farther apart the two galaxies are, the faster they are moving. The thing that is the same for all galaxies is the time that it took them to get to where they are today.

A look at the following math makes this conclusion clear. Velocity equals Hubble's constant multiplied by distance, so when we calculate a time as "distance divided by velocity," the distance factors on top and bottom cancel out. Writing this out as an equation, we get

$$\text{Time} = \frac{\text{Distance}}{\text{Velocity}} = \frac{\text{Distance}}{H_0 \times \text{Distance}} = \frac{1}{H_0}.$$

We define $1/H_0$ as look-back time, or *Hubble time*. It is one way of estimating the age of the universe.

FIGURE 18.10 Assuming that the velocities of galaxies remain constant, the time needed for any two galaxies to reach their current locations is the same, regardless of their separation.

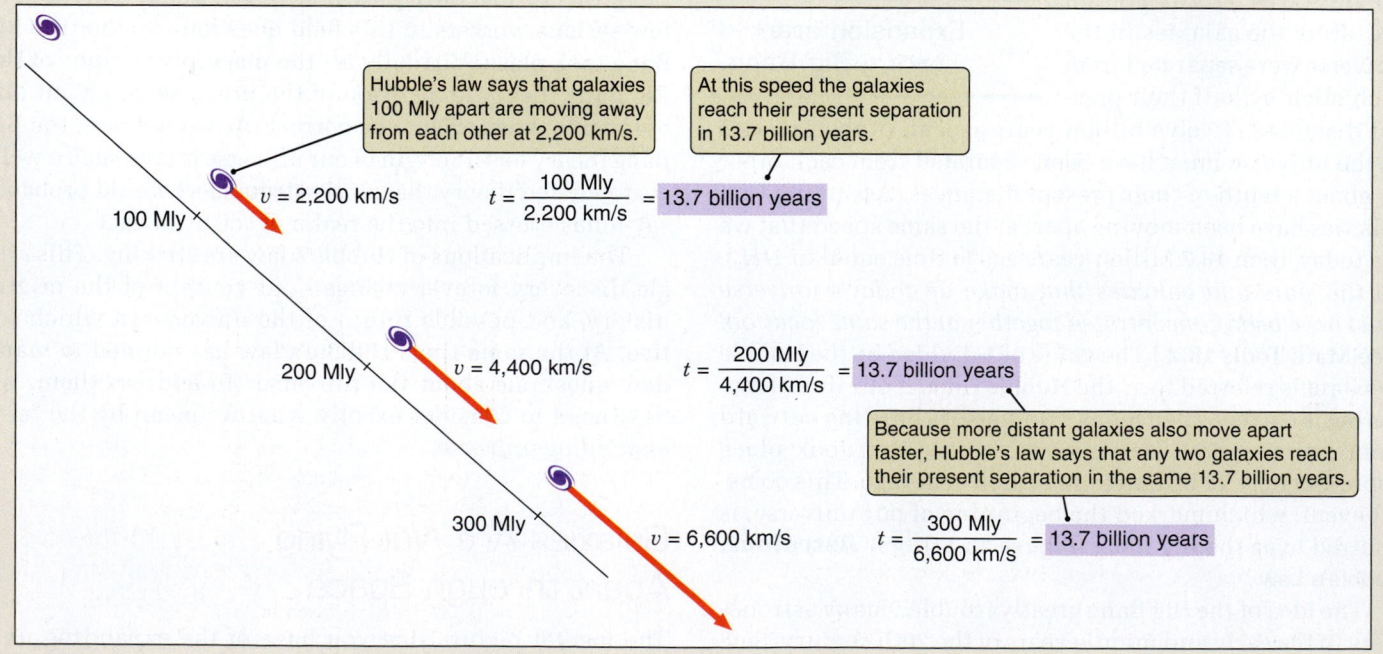

Hubble's law says that galaxies 100 Mly apart are moving away from each other at 2,200 km/s.

At this speed the galaxies reach their present separation in 13.7 billion years.

$$v = 2{,}200 \text{ km/s} \qquad t = \frac{100 \text{ Mly}}{2{,}200 \text{ km/s}} = 13.7 \text{ billion years}$$

100 Mly

$$v = 4{,}400 \text{ km/s} \qquad t = \frac{200 \text{ Mly}}{4{,}400 \text{ km/s}} = 13.7 \text{ billion years}$$

200 Mly

Because more distant galaxies also move apart faster, Hubble's law says that any two galaxies reach their present separation in the same 13.7 billion years.

$$v = 6{,}600 \text{ km/s} \qquad t = \frac{300 \text{ Mly}}{6{,}600 \text{ km/s}} = 13.7 \text{ billion years}$$

300 Mly

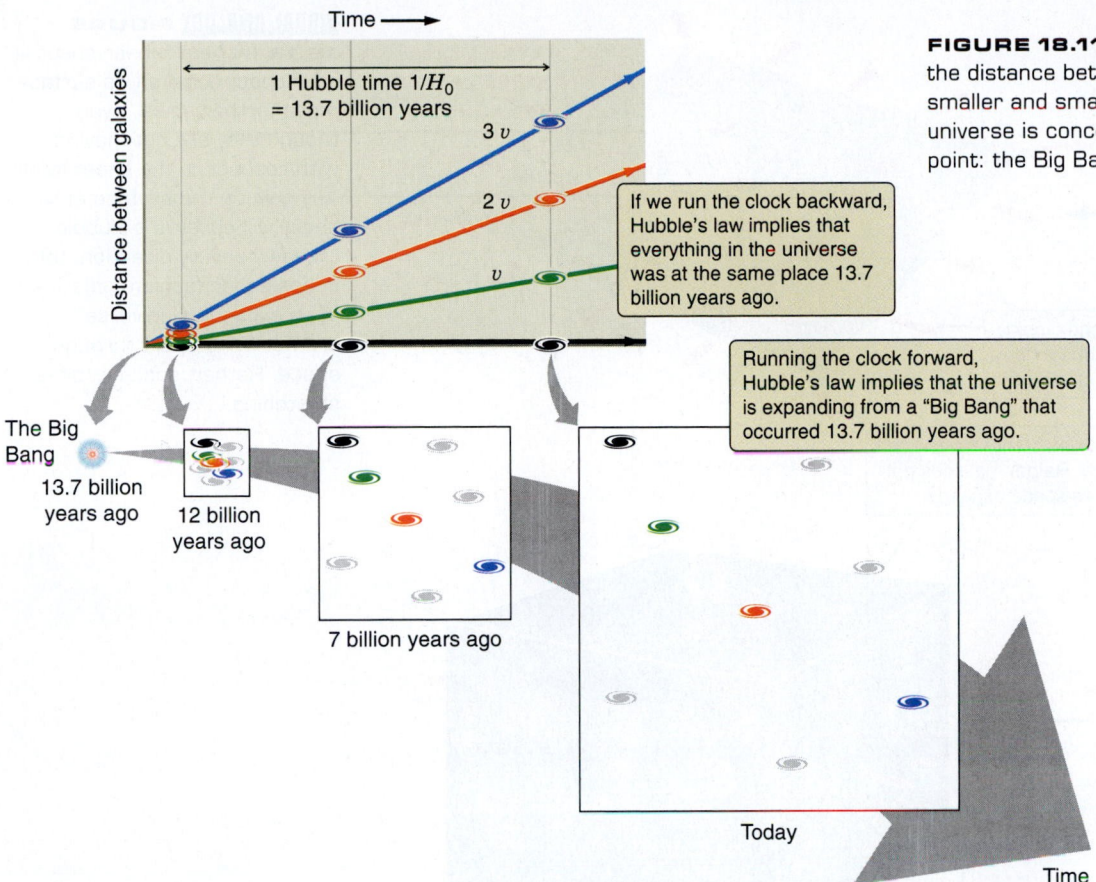

FIGURE 18.11 Looking backward in time, the distance between any two galaxies is smaller and smaller, until all matter in the universe is concentrated together at the same point: the Big Bang.

Time ⟶

1 Hubble time $1/H_0$ = 13.7 billion years

Distance between galaxies

$3\,v$

$2\,v$

v

If we run the clock backward, Hubble's law implies that everything in the universe was at the same place 13.7 billion years ago.

Running the clock forward, Hubble's law implies that the universe is expanding from a "Big Bang" that occurred 13.7 billion years ago.

The Big Bang

13.7 billion years ago

12 billion years ago

7 billion years ago

Today

Time

cloud of debris from an explosion flying outward through surrounding space. As we will see, however, this is not an explosion in the usual sense of the word, and in fact there is no surrounding space.

One of the first questions students usually ask about the Big Bang is, "Where did it take place?" The answer to this question, amazing as it seems, is that the Big Bang took place *everywhere*. Wherever you are in the universe today, you are sitting at the site of the Big Bang. The reason is that galaxies are not flying apart through space at all. Rather, *space itself* is expanding, carrying the stars and galaxies that populate the universe along with it.

The Big Bang happened everywhere.

This may seem an incredible notion, but we have already dealt with the basic ideas that enable us to understand the expansion of space. In our discussion of neutron stars and black holes in Chapter 17, we encountered Einstein's general theory of relativity. General relativity says that space is distorted by the presence of mass, and that the consequence of this distortion is gravity. For example, the mass of the Sun, like any other object, distorts the geometry of spacetime around it; so Earth, coasting along in its inertial frame of reference, follows a curved path around the Sun. We illustrated this phenomenon in Figure 17.25 with

The expansion of space itself is similar to stretching a two-dimensional rubber sheet.

the analogy of a ball placed on a stretched rubber sheet, showing how the ball distorted the surface of the sheet.

The surface of a rubber sheet can be distorted in other ways as well. Imagine a number of coins placed on a rubber sheet, as shown in **Figure 18.12**. Suppose we grab the edges of the sheet and begin pulling them outward. As the rubber sheet stretches, each coin remains at the same location on the surface of the sheet, but the distances between the coins increase. Two coins sitting close to each other move apart only slowly, while coins farther apart move away from each other more rapidly. In other words, the distances and relative motions of the coins on the surface of a rubber sheet will obey a Hubble-like relationship as the sheet is stretched.

Analogously, this movement is what is happening in the universe, with galaxies taking the place of the coins and space itself taking the place of the rubber sheet. Obviously, in the case of coins on a rubber sheet there is a limit to how far we can stretch the sheet before it breaks. With space and the real universe, there is no such limit. The fabric of space can, in principle, go on expanding forever. Hubble's law is the observational consequence of the fact that the space making up the universe is expanding.

How will this expansion of the universe behave in the future? In the next chapter we will learn that most of the mass in the universe consists of invisible dark matter, and in Chapter 21 we will see that gravity due to this dark matter has the effect of slowing down the expansion of the

(a) Coins on a rubber sheet

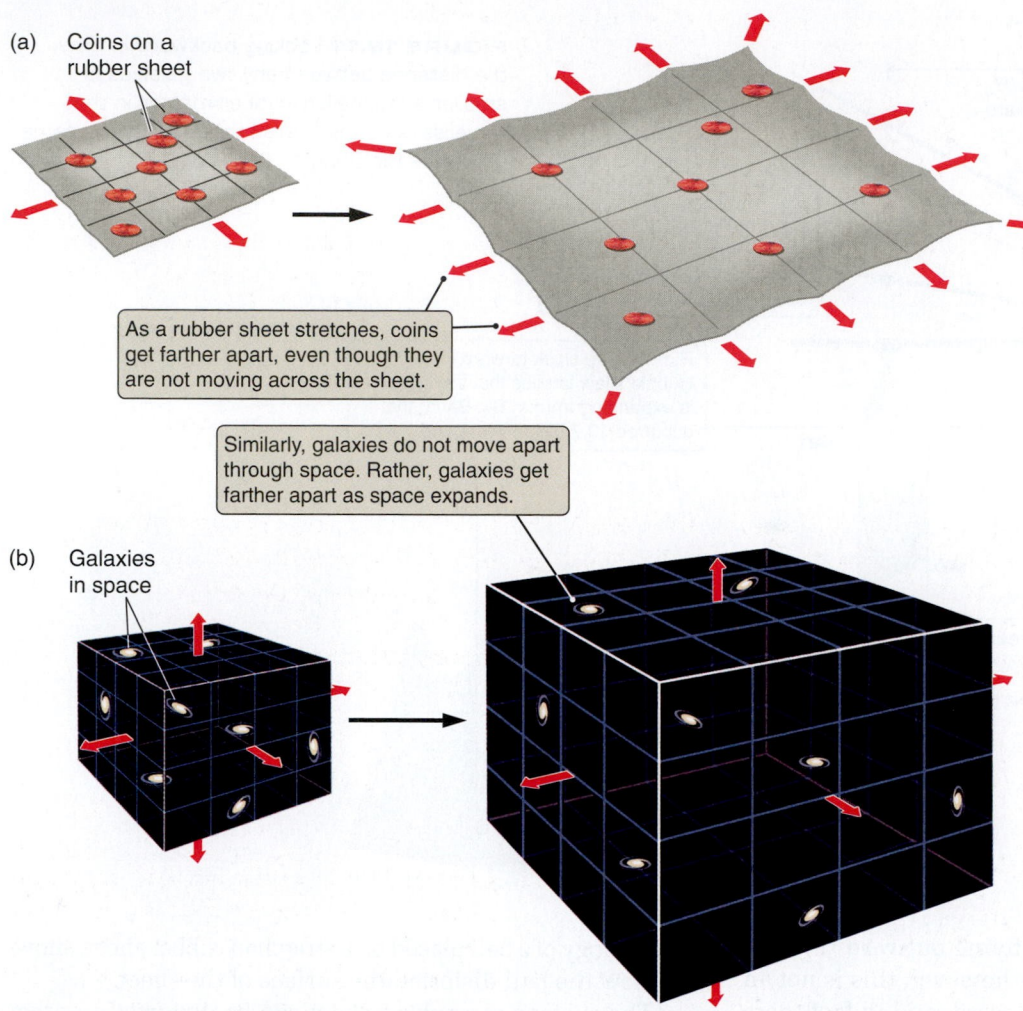

As a rubber sheet stretches, coins get farther apart, even though they are not moving across the sheet.

Similarly, galaxies do not move apart through space. Rather, galaxies get farther apart as space expands.

(b) Galaxies in space

universe. But now it becomes more complicated. We will also learn that there is another unseen constituent of the universe, called dark energy, and this constituent causes the expansion of the universe instead to *accelerate*. At the current stage in the expansion of the universe, the accelerating effect of dark energy is beginning to dominate over the slowing effect of dark matter.

Expansion Is Described with a Scale Factor

When astronomers discuss the expansion of the universe, they talk in terms of the **scale factor** of the universe. To understand this concept, we return to our analogy of the rubber sheet. Suppose we place a ruler on the surface of the sheet and draw a tick mark every centimeter, as in **Figure 18.13a**. If we want to know the distance between two points on the sheet, all we need to do is count the marks between the two points and multiply by 1 centimeter (cm) per tick mark.

As the sheet is stretched, however, the distance between the tick marks does not remain 1 cm. When the sheet is stretched to 150 percent of the size that it had when we

drew our ruler, each tick mark is separated from its neighbors by 1½ times the original distance, or 1.5 cm. If we wanted to know the distance between two points, we could still count the marks, but we would have to scale up the distance between tick marks by 1.5 to find the distance in centimeters. The scale factor of the sheet is now 1.5. If the sheet were twice the size that it was when we drew the ruler (**Figure 18.13b**), each mark would correspond to 2 cm of actual distance; the scale factor of the sheet would now be 2. The scale factor tells us the size of the sheet relative to its size at the time when we drew our ruler. The scale factor also tells us how much the distance between points on the sheet has changed.

We can apply this same idea to the universe. Suppose we choose today to lay out a "cosmic ruler" on the fabric of space, placing an imaginary tick mark every 10 Mly. We define the scale factor of the universe at this time to be 1. In the past, when the universe

> The scale factor, R_U, increases as the universe expands.

was smaller, distances between the points in space marked by our cosmic ruler would have been less than 10 Mly. The scale factor of that younger, smaller universe would have

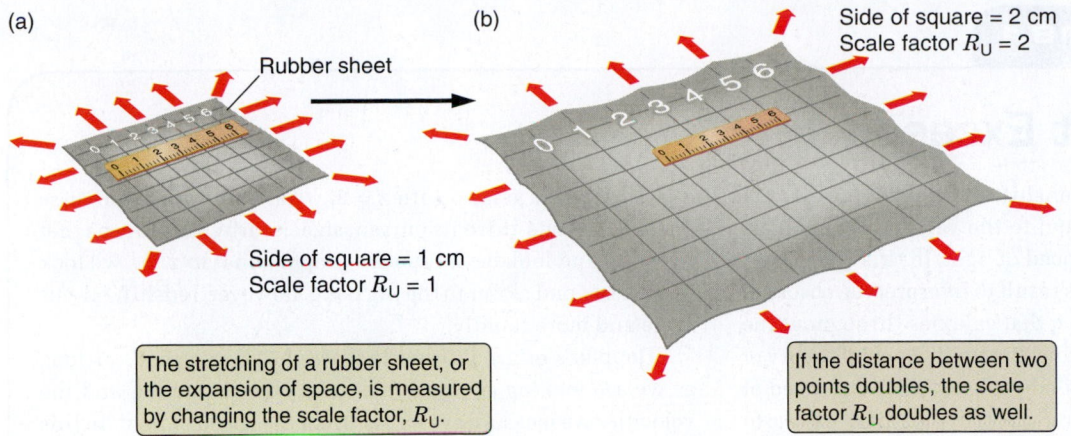

(a)

Rubber sheet

Side of square = 1 cm
Scale factor $R_U = 1$

(b)

The stretching of a rubber sheet, or the expansion of space, is measured by changing the scale factor, R_U.

Side of square = 2 cm
Scale factor $R_U = 2$

If the distance between two points doubles, the scale factor R_U doubles as well.

VISUAL ANALOGY FIGURE

18.13 (a) On a rubber sheet, tick marks are drawn 1 cm apart. As the sheet is stretched, the tick marks move farther apart. (b) When the spacing between the tick marks is 2 cm, or twice the original value, we say that the scale factor of the sheet, R_U, has doubled. A similar scale factor, R_U, is used to describe the expansion of the universe.

been less than 1 compared to the scale factor today. In the future, as the universe continues to expand, the distances between the tick marks on our cosmic ruler will grow to more than 10 Mly, and the scale factor of the universe will be greater than 1. In this way we can use the scale factor, usually written as R_U, to keep track of the changing scale of the universe.

It is important to remember, when thinking about the expanding universe, that the laws of physics are themselves unchanged by the changing scale factor. For example, when we stretched out the rubber sheet, we did not change the properties of the coins on its surface. In like fashion, as the universe expands, the sizes and other physical properties of atoms, stars, and galaxies also remain unchanged.

Let's return now to the question of locating the center of expansion. Looking back in time, we see that the scale factor of the universe gets smaller and smaller, approaching zero as we get closer and closer to the Big Bang. The fabric of space that today spans billions of light-years spanned much smaller distances when the universe was young. When the universe was only a day old, all of the space that we see today amounted to a region only a few times the size of our Solar System. When the universe was a 50th of a second old, the vast expanse of space that makes up today's observable universe (and all the matter in it) occupied a volume only the size of today's Earth. As we continue to approach the Big Bang itself, going backward in time, the space that makes up today's observable universe becomes smaller and smaller— the size of a grapefruit, a marble, an atom, a proton. Every point in the fabric of space that makes up today's universe was right there at the beginning, a part of that unimaginably tiny, dense universe that emerged from the Big Bang.

These points bear repetition: *Where is the center of the Big Bang?* There is no center. The Big Bang did not occur at a specific point in space, because space itself came into existence with the Big Bang. *Where did the Big Bang happen?* It happened everywhere, including right where you

Expansion does not affect local physics— stars, atoms, or anything else.

are sitting. This is an important concept. If a particular point in today's universe marked the site of the Big Bang, that would be a very special point indeed. But there is no such point. The Big Bang happened everywhere. A Big Bang universe is homogeneous and isotropic, consistent with the cosmological principle.

Redshift Is Due to the Changing Scale Factor of the Universe

General relativity gives us a powerful tool for interpreting Hubble's great discovery of the expanding universe. It also forces us to rethink just what we mean when we talk about the redshift of distant galaxies. Although it is true that the distance between galaxies is increasing as a result of the expansion of the universe, and that we can use the equation for Doppler shifts to measure the redshifts of galaxies, these redshifts are not due to Doppler shifts at all! As light comes toward us from distant galaxies, the scale factor of the space through which the light travels is constantly increasing; and as it does so, the distance between adjacent wave crests increases as well. The light is "stretched out" as the space it travels through expands. (See **Foundations 18.1**.)

Let's return to our rubber sheet analogy. If we draw a series of bands on the rubber sheet to represent the crests of an electromagnetic wave, as in **Figure 18.15**, we can watch what happens to the wave as the sheet is stretched out. By the time the sheet is stretched to twice its original size— that is, by the time the scale factor of the sheet is 2—the distance between wave crests has doubled. When the sheet has been stretched to three times its original size (a scale factor of 3), the wavelength of the wave will be three times what it was originally.

We apply this idea to light coming from a distant galaxy. When the light left the galaxy of its origin, the scale factor of the universe was smaller than it is today. The universe expanded while the light was in transit, and as it did so, the wavelength of the light grew longer in proportion to the increasing scale factor of the universe. The redshift of

When Redshift Exceeds 1

In our discussion of the Doppler shift in Chapter 4, we found that $(\lambda_{observed} - \lambda_{rest})/\lambda_{rest}$ is equal to the velocity of an object away from us, divided by the speed of light. In this chapter we see how Edwin Hubble used this result to interpret the observed redshifts of galaxies as evidence that galaxies throughout the universe are moving away from us. Einstein's special theory of relativity says that nothing can move faster than the speed of light. Hubble's initial assumption was that redshifts are due to the Doppler effect. The resulting relation, $z = v_r/c$, would then seem to imply that no object can have a redshift (z) greater than 1. Yet that is not the case. Astronomers routinely observe redshifts significantly in excess of 1. As of this writing, the most distant objects known have redshifts greater than 8! What is wrong here? It is worth taking a brief digression to consider how redshifts can exceed 1.

The first thing to note is that, to arrive at the expression for the Doppler effect—$v_r/c = (\lambda_{observed} - \lambda_{rest})/\lambda_{rest}$—we have to *assume* that v_r is much less than c. If that were not the case—that is, if v_r were close to c—we would have to take into account more than just the fact that the waves from an object are stretched out by the object's motion away from us. We would also have to consider relativistic effects, including the fact that moving clocks run slowly. When combining these effects, we would find that as the speed of an object approaches the speed of light, its redshift becomes arbitrarily large (**Figure 18.14**).

A second source of redshift is the gravitational redshift discussed in Chapter 17. As light escapes from deep within a gravitational well, it loses energy, so photons are shifted to longer and longer wavelengths. If the gravitational well is deep enough, then the observed redshift of this radiation can be boundlessly large. In fact, the event horizon of a black hole—that is, the surface around the black hole from which not even light can escape—is where the gravitational redshift becomes infinite.

Cosmological redshift, which is most relevant to this chapter, results from the amount of "stretching" that space has undergone during the time the light from its original source has been *en route* to us. The amount of stretching that has occurred is given by the factor $1 + z$. When we look at a distant galaxy whose redshift $z = 1$, we are seeing the universe at a time when it was half the size that it is today. When we

see light from a galaxy with $z = 2$, we are seeing the universe when it was one-third its current size. Nearby, this means that distance and look-back time are proportional to z. As we look back closer and closer to the Big Bang, however, redshift climbs more and more rapidly.

Doppler's original formula is essentially correct—as long as we are looking at shifts due to an object's motion and the velocities we measure are far less than the speed of light. In this case, $v_r/c = (\lambda_{observed} - \lambda_{rest})/\lambda_{rest}$. When we look at the motions of orbiting binary stars or the peculiar velocities of galaxies relative to the Hubble flow, this equation works just fine. But any time you have a redshift of 1 or greater, you should immediately recognize that you are now in the realm of relativity.

FIGURE 18.14 Plot of the redshift (z) of an object versus its recession velocity (v_r) as a fraction of the speed of light. According to special relativity, as v_r approaches c, the redshift becomes large without limit.

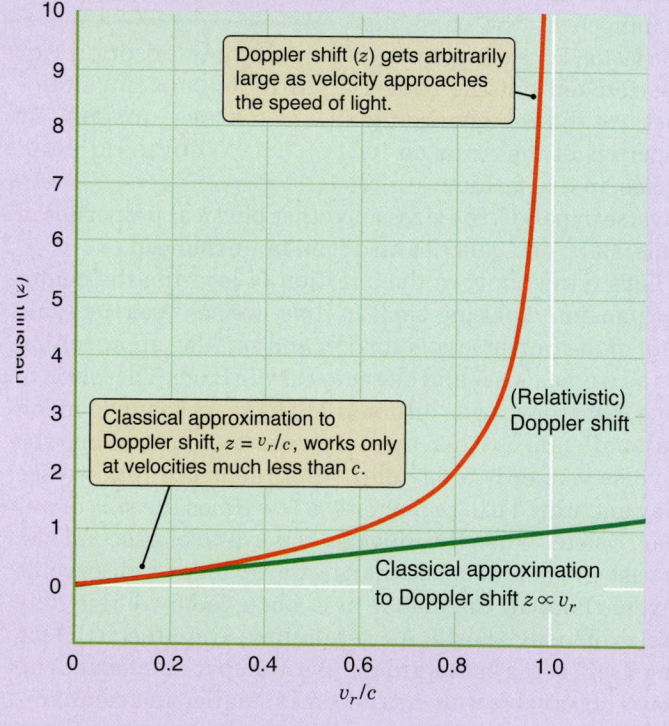

Doppler shift (z) gets arbitrarily large as velocity approaches the speed of light.

(Relativistic) Doppler shift

Classical approximation to Doppler shift, $z = v_r/c$, works only at velocities much less than c.

Classical approximation to Doppler shift $z \propto v_r$

light from distant galaxies is therefore a direct measure of how much the universe has expanded since the time when the radiation left its source. Redshift measures how much the scale factor of the universe, R_U, has changed since the light was emitted.

If we see radiation with a redshift of 1, then its wavelength is twice as long as when that radiation left its source.

When the light left its source, R_U was equal to half of today's value. If we see radiation with a redshift of 2, then the wavelength of the radiation is three times its original wavelength. The radiation was emitted when the scale factor of the universe was a third of what it is today. This wonderfully direct relationship lets us use the observed redshift of a galaxy to calculate the size of the universe at the look-back time to

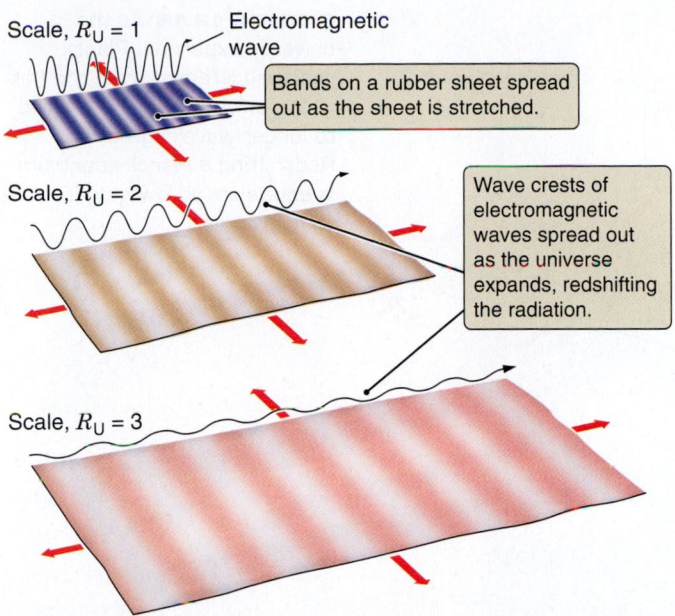

Scale, $R_U = 1$ — Electromagnetic wave

Bands on a rubber sheet spread out as the sheet is stretched.

Scale, $R_U = 2$

Wave crests of electromagnetic waves spread out as the universe expands, redshifting the radiation.

Scale, $R_U = 3$

VISUAL ANALOGY **FIGURE 18.15** Bands drawn on a rubber sheet represent the positions of the crests of an electromagnetic wave in space. As the rubber sheet is stretched—that is, as the universe expands—the wave crests get farther apart. The light is redshifted.

that galaxy. Written as an equation, the scale factor of the universe we see when looking at a distant galaxy is equal to 1, divided by 1 plus the redshift of the galaxy:

$$R_U(z) = \frac{1}{1+z}.$$

18.5 The Major Predictions of the Big Bang Theory Are Resoundingly Confirmed

The questions we are grappling with when discussing ideas such as the origin of the universe are some of the most fundamental questions humankind can ask. Throughout human history, answers to these same questions have been among the most prized goals of philosophers and theologians. It is quite remarkable that we live in a time when we are finding real, testable answers to these questions by appealing not to divine inspiration or to the blind logic of philosophy, but rather to the empirical methods of science. It is essential, then, that we place extraordinary demands on the quality of the evidence we rely on to support the theory of the Big Bang. What evidence, apart from the observed expansion of the universe itself, requires us to accept that the Big Bang actually took place?

We See Radiation Left Over from the Big Bang

The story of the single most important confirmation of the Big Bang theory begins in the mid-1940s, when Russian-American cosmologist George Gamow (1904–1968) and his student Ralph Alpher (1921–2007) were thinking about the implications of Hubble expansion. Their logical reasoning about the expansion of the universe connected it to changes in temperature: When a gas is compressed, it grows hotter. By contrast, as a gas expands it becomes cooler. So, reasoned Gamow and Alpher, since the universe is expanding, it must also be cooling. Consequently, when the universe was very young and small, it must have consisted of an extraordinarily hot, dense gas. As with any hot, dense gas, this early universe would have been awash in the same kind of radiation that we have encountered so many times before on our journey of discovery: radiation from a blackbody, which exhibits a **Planck spectrum**.

Gamow and Alpher took this idea a step further. As the universe expanded, they reasoned, this radiation would have been redshifted to longer and longer wavelengths. Recall Wien's law in Chapter 4, which states that the temperature associated with Planck radiation is inversely proportional to the peak wavelength: $T = (2{,}900 \ \mu m)/\lambda_{peak}$. Shifting the wavelength of Planck radiation to longer and longer wavelengths is therefore the equivalent of shifting the temperature of the radiation to lower and lower values. As illustrated in **Figure 18.16**, doubling the wavelength of the photons in a Planck spectrum by doubling the scale factor of the universe is equivalent to cutting the temperature of the Planck spectrum in half.

On April Fools' Day, 1948, Alpher and Gamow published a paper asserting that this radiation should still be visible today and should have a Planck spectrum with a temperature in the neighborhood of 5–10 kelvins (K). The paper actually listed as authors not only Alpher and Gamow, but also Hans Bethe (1906–2005), who—though an esteemed physicist in his own right— did not actively participate in this research. Gamow, being a renowned jokester, added

> Alpher, Bethe, and Gamow predicted the glow from a young hot universe in 1948.

Bethe's name between his and Alpher's. "Alpher, Bethe, and Gamow" is a play on the first three letters of the Greek and Hebrew alphabets—an appropriate authorship for a paper making predictions about the very early universe!

This prediction languished until the early 1960s, when two physicists at Bell Laboratories, Arno Penzias (1933–) and Robert Wilson (1936–), were trying to bounce radio signals off the newly launched Echo satellites. This hardly seems much of a feat today, when we routinely use cell phones and handheld GPS systems that communicate directly with satellites. At the time, however, the effort pushed radio technology to its limits. Penzias and Wilson needed a very sensitive microwave telescope for their work. Any spurious signals coming from the telescope itself might wash out the

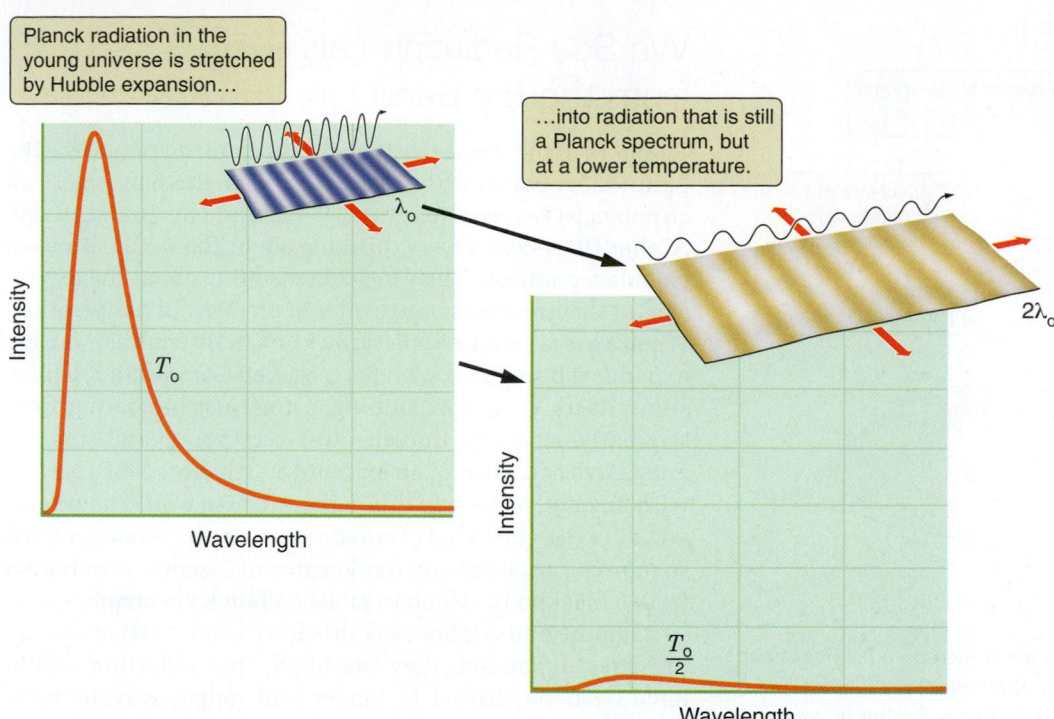

Planck radiation in the young universe is stretched by Hubble expansion…

…into radiation that is still a Planck spectrum, but at a lower temperature.

λ_0

T_0

Intensity

Wavelength

Intensity

$\dfrac{T_0}{2}$

Wavelength

$2\lambda_0$

FIGURE 18.16 As the universe expanded, Planck radiation left over from the hot young universe was redshifted to longer wavelengths. Redshifting a Planck spectrum is equivalent to lowering its temperature.

faint signals bounced off a satellite. Pictured in **Figure 18.17** along with their radio telescope, Penzias and Wilson worked tirelessly to eliminate all possible sources of interference originating from within their instrument. This work included such endless and menial tasks as keeping the telescope free of bird droppings and other extraneous material. Even so, Penzias and Wilson found that no matter how hard they tried to eliminate sources of extraneous noise, they could still detect a faint microwave signal when they pointed the telescope at the sky. Eventually, they came to accept that the signal they were detecting was real. The sky faintly glows in microwaves.

In the meantime, physicist Robert Dicke (1916–1997) and his colleagues at Princeton University also predicted a hot early universe, arriving independently at the same basic conclusions that Alpher and Gamow had reached two decades earlier. When Dicke and colleagues heard of the signal that Penzias and Wilson had found, they interpreted it as the radiation left behind by the hot early universe. The strength of the detected signal was consistent with the glow from a blackbody with a temperature of about 3 K, very close to the predicted value. Their results, published in 1965, reported the discovery of the glow left behind by the Big Bang. Penzias and Wilson shared the 1979 Nobel Prize in physics for their remarkable discovery.

> Penzias and Wilson discovered the cosmic background radiation.

It is worth noting that timing can be everything in science. Alpher, who first predicted the existence of a faint glow from the Big Bang, had searched unsuccessfully for the signal 10 years before Penzias and Wilson made their discovery. Unfortunately, however, the technology of the late 1940s and early 1950s was simply not up to the task.

This radiation left over from the early universe is called the **cosmic background radiation (CBR)**. Today, the cosmic background

FIGURE 18.17 Penzias and Wilson next to the Bell Labs radio telescope antenna with which they discovered the cosmic background radiation. This antenna is now a US National Historic Landmark.

radiation and the conditions in the early universe are much better understood than they were in the early 1960s. The origin of the CBR is illustrated in **Figure 18.18**. When the universe was young, it was hot enough that all of the atoms in the universe were ions. In our discussion of the structure of the Sun and stars, we found that radiation does not travel well through an ionized plasma. Free electrons in a plasma interact strongly with the radiation, blocking its progress. At this time in the early universe, the conditions within the universe were much like the conditions within a star: the universe was an opaque blackbody.

> The CBR is thermal radiation that arose when the universe was hot and ionized.

As the universe expanded, the gas filling it cooled. By the time the universe was about a thousandth of its current size, the temperature had dropped to a few thousand kelvins, so protons and electrons were able to combine to form hydrogen atoms. This event, called the **recombination** of the universe, occurred when the universe was several hundred thousand years old.

Hydrogen atoms are much less effective at blocking radiation than free electrons are; so when recombination occurred, the universe suddenly became transparent to radiation. Since that time, the radiation left behind from the Big Bang has been able to travel largely unimpeded throughout the universe. At the time of recombination, when the temperature of the universe was a few thousand kelvins, the wavelength of this radiation peaked at about 1 μm, according to Wien's law (see Chapter 4). As the universe expanded, this radiation was redshifted to longer and longer wavelengths. Today, the scale of the universe has increased a thousandfold since recombination, and the peak wavelength of the cosmic background radiation has increased by a thousandfold as well, to a value close to 1 mm. The spectrum of the CBR still has the shape of a Planck spectrum, but with a characteristic temperature of 2.73 K—only a thousandth what it was at the time of recombination.

> Since recombination, the CBR has traveled freely and cooled by a factor of 1,000.

1 In the ionized early universe, light was trapped by free electrons. Radiation had a Planck spectrum.

(a)

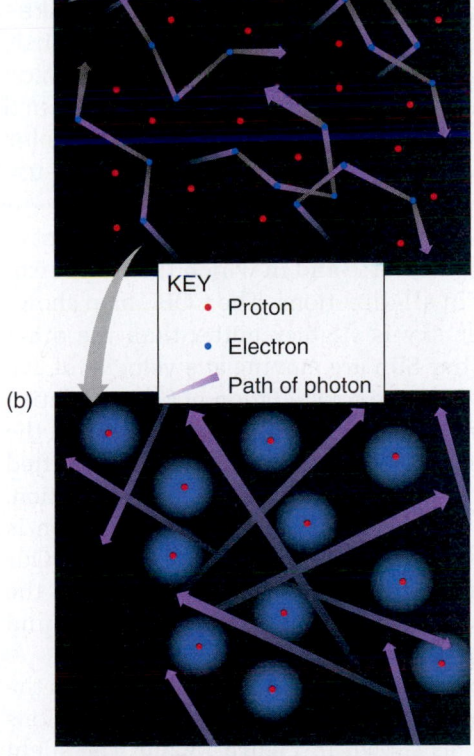

KEY
• Proton
• Electron
➤ Path of photon

(b)

VISUAL ANALOGY FIGURE 18.18 The origin of the cosmic background radiation. (a) Prior to *recombination*, the universe was like a foggy day, except that the "fog" was a sea of hydrogen atoms. Radiation interacted strongly with free electrons and so could not travel far. The trapped radiation had a Planck spectrum. (b) When the universe recombined, the fog cleared and this radiation was free to travel unimpeded.

2 When the universe recombined, it became transparent, and the Planck radiation went its own way.

3 Recombination was like the fog suddenly clearing.

Satellite Data Confirmed That the Cosmic Background Radiation Is Real

The presence of cosmic background radiation with a Planck spectrum is a very strong prediction of the Big Bang theory. Penzias and Wilson had confirmed that a signal with the correct strength was there, but they could not say for certain whether the signal they saw had the spectral shape of a Planck spectrum. From the late 1960s to the 1980s, most experiments at different wavelengths supported these same conclusions. Yet it was not until the end of the 1980s that the predictions of Big Bang cosmology for the CBR were put to the ultimate test. The year 1989 saw the launch of a satellite called the Cosmic Background Explorer, or COBE. Instruments carried on COBE were capable of making extremely precise measurements of the CBR at many wavelengths, from a few micrometers out to 1 cm. In January 1990, hundreds of astronomers gathered in a large conference room in Washington DC, at the winter meeting of the American Astronomical Society, to hear the COBE team present its first results. Security surrounding the new findings had been tight, so the atmosphere in the room was electric. The tension did not last long; presentation of a single viewgraph brought the room's occupants to their feet in a spontaneous ovation.

The data shown on that viewgraph are reproduced in **Figure 18.19**. The small dots in the figure are the COBE measurements of the CBR at different frequencies. The uncertainty in each measurement is far less than the size of each dot. The line in the figure, which runs perfectly through the data points, is a Planck spectrum with a temperature of 2.73 K. The agreement between theoretical prediction and observation is truly remarkable. The observed spectrum so perfectly matches the one predicted by Big Bang cosmology that there can be no real doubt we are seeing the residual radiation left behind from the primordial fireball of the early universe.

> COBE unambiguously showed the CBR to have a Planck spectrum at 2.73 K.

The CBR Measures Earth's Motion Relative to the Universe Itself

COBE provided us with much more than a measurement of the spectrum of the cosmic background radiation. **Figure 18.20a** shows a map obtained by COBE of the CBR from the entire sky. The different colors in the map correspond to variations in the temperature of the CBR. The differences are not as extreme as the colors might suggest, however. The range in temperature in the map corresponds to a variation of only about 0.1 percent in the temperature of the CBR. Most of this range of temperature is present because one side of the sky looks slightly warmer than the opposite side of the sky. This difference has nothing to do with the large-scale

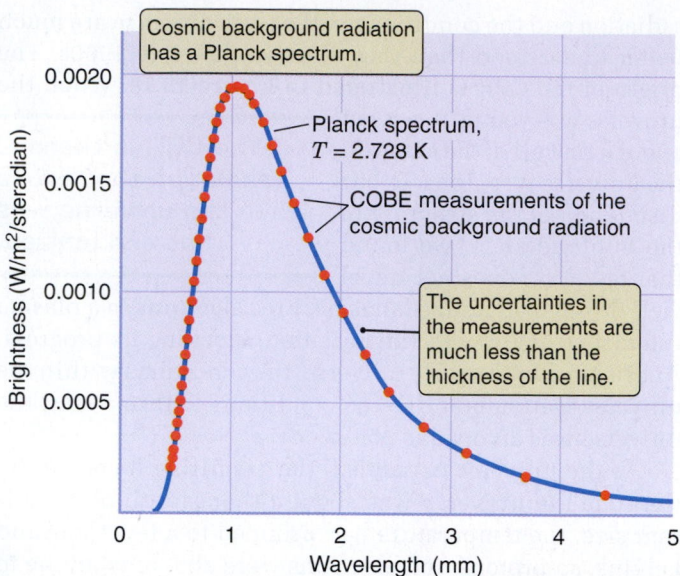

FIGURE 18.19 The spectrum of the cosmic background radiation as measured by the Cosmic Background Explorer (COBE) satellite (red dots). The uncertainty in the measurement at each wavelength is much less than the size of a dot. The line running through the data is a Planck spectrum with a temperature of 2.728 K.

structure of the universe itself, but rather is the result of the motion of Earth with respect to the CBR.

Time and again on our journey we have stressed that there is no preferred frame of reference. The laws of physics are the same in *any* inertial reference frame, so none is better than any other. Yet in a certain sense there is a preferred frame of reference at every point in the universe. This is the frame of reference that is at rest with respect to the expansion of the universe and in which the CBR is isotropic, or the same in all directions. The COBE map shows that one side of the sky is slightly hotter than the other because Earth and our Sun are moving at a velocity of 368 km/s in the direction of the constellation of Crater, relative to this cosmic reference frame. Radiation coming from the direction in which we are moving is slightly blueshifted (shifted to a higher characteristic temperature) by our motion, whereas radiation coming from the opposite direction is Doppler-shifted toward the red (or cooler temperatures). Our motion is due to a combination of factors, including the motion of our Sun around the center of the Milky Way and the motion of our galaxy relative to the CBR.

> Our motion makes the CBR slightly hotter in the direction we are moving toward and slightly cooler behind us.

If we subtract from the COBE map this asymmetry in the CBR caused by the motion of Earth, only slight variations in the CBR remain, as shown in **Figure 18.20b**. The slight variations seen in this map have an amplitude that is only about 1/100,000 the brightness of the CBR. This means that the brighter parts of this image are only about 1.00001 times

(a)

(b)

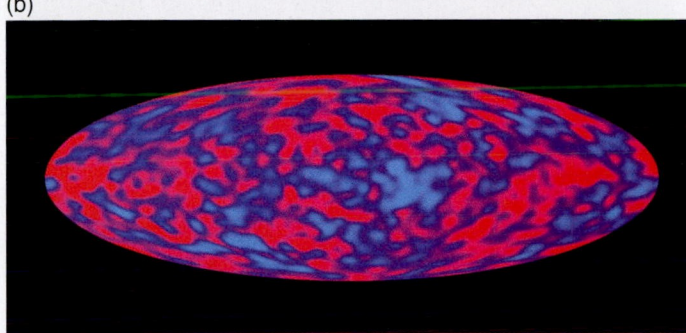

(c)

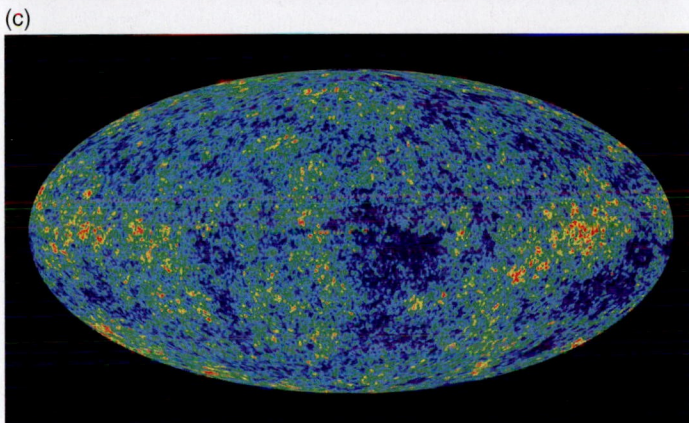

FIGURE 18.20 (a) The COBE map of the cosmic background radiation. The CBR is slightly hotter (by about 0.003 K) in one direction in the sky than in the other direction. This difference is due to Earth's motion relative to the CBR. (b) The COBE map with Earth's motion removed, showing tiny ripples remaining in the CBR. (c) WMAP has provided the highest resolution yet of the CBR. The radiation seen in this image was emitted less than 400,000 years after the Big Bang.

brighter than the fainter parts. These slight variations might not seem like much, but they are actually of crucial importance in the history of the universe. Recall from Chapter 17 that gravity itself can create a redshift. These tiny fluctuations in the cosmic background radiation are the result of gravitational redshifts caused by concentrations of mass that existed in the early

> COBE found, and WMAP further revealed, tiny variations in the CBR resulting from the formation of structure during early times.

universe. These concentrations later gave rise to galaxies and the rest of the structure that we see in the universe today.

Subsequent observations from Antarctica and from instruments carried aloft by balloons support the COBE findings. Beginning in 2001, more precise measurements of the variations of the CBR have been carried out by a satellite called the Wilkinson Microwave Anisotropy Probe, or WMAP. **Figure 18.20c** shows the ripples measured by WMAP with much higher resolution than could be detected by COBE. The much higher-resolution maps obtained by WMAP have profound implications for our understanding of the origin of structure in the universe and enable us to determine several cosmological parameters. For example, the value of the Hubble constant we use in this book is identical to the value inferred from the WMAP experiment.

The Big Bang Theory Correctly Predicts the Abundance of the Least Massive Elements

The next confrontation between the predictions of the Big Bang theory and observations of the universe came from a very different direction. When the universe was only a few minutes old, its temperature and density were high enough for nuclear reactions to take place. Just as we can use our knowledge of nuclear physics to calculate the nuclear reactions occurring in the interiors of stars, we can use this knowledge to calculate the nuclear reactions that took place in the early universe. Collisions between protons in the early universe built up low-mass nuclei, including deuterium (heavy hydrogen) and isotopes of helium, lithium, beryllium, and boron. The formation of new elements in this early nuclear brew, called **Big Bang nucleosynthesis**, determined the final chemical composition of the matter that emerged from the hot phase of the Big Bang. ▶❚❚ **AstroTour:** Big Bang Nucleosynthesis

We have already discussed how differences in the abundances of the products of *stellar* nucleosynthesis help us track the chemical evolution of the universe and the history of star formation. These ideas will play an important role in our discussion of the Milky Way in Chapter 20, and they will come to the fore again in our later discussion of galaxy formation. Here we focus on the products of Big Bang nucleosynthesis.

The amounts of various elements that formed from Big Bang nucleosynthesis depended in detail on the temperature and density of normal matter in the early universe. In turn, the density in the early universe at appropriate nucleosynthesis temperatures accounts for the density of normal matter found in the

> Big Bang theory predicts a universe that is 24 percent helium by mass, which observations confirm.

universe today. **Figure 18.21** shows the calculated predictions of element abundances from Big Bang nucleosynthesis,

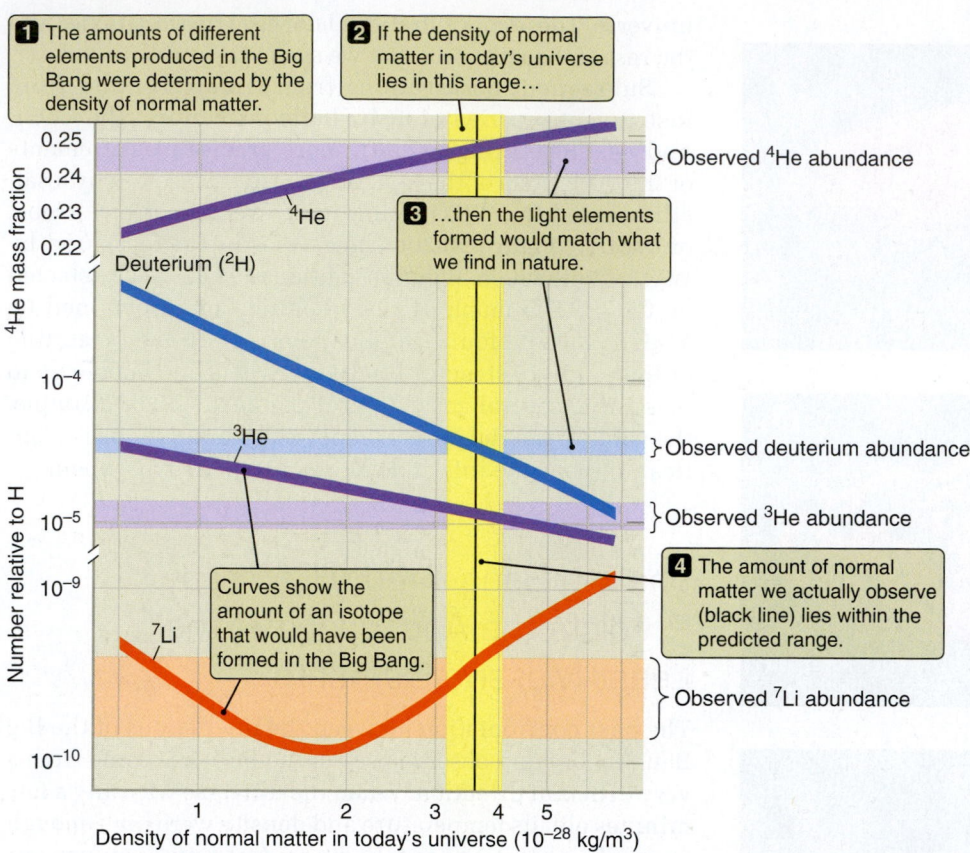

1. The amounts of different elements produced in the Big Bang were determined by the density of normal matter.

2. If the density of normal matter in today's universe lies in this range…

3. …then the light elements formed would match what we find in nature.

4. The amount of normal matter we actually observe (black line) lies within the predicted range.

Curves show the amount of an isotope that would have been formed in the Big Bang.

Observed ⁴He abundance

Observed deuterium abundance

Observed ³He abundance

Observed ⁷Li abundance

⁴He mass fraction
Number relative to H

⁴He
Deuterium (²H)
³He
⁷Li

Density of normal matter in today's universe (10^{-28} kg/m³)

FIGURE 18.21 Observed and calculated abundances of the products of Big Bang nucleosynthesis, plotted against the density of normal matter in today's universe. Big Bang nucleosynthesis correctly predicts the amounts of these isotopes found in the universe today.

plotted as a function of the present-day density of normal (luminous) matter in the universe. The first thing we see from this figure is that about 24 percent of the mass of the normal matter formed in the early universe should have ended up in the form of the very stable isotope ⁴He, regardless of exactly what the density of matter in the universe was. Indeed, when we look about us in the universe today, we find that about 24 percent of the mass of normal matter in the universe is in the form of ⁴He, in complete agreement with the prediction of Big Bang nucleosynthesis.

Unlike helium, the abundances of most of the isotopes formed in the Big Bang depended sensitively on the density of normal matter in the universe. Beginning with the amounts of isotopes such as deuterium (²H) and ³He found in the universe (shown as roughly horizontal bands in Figure 18.21), we can ask what the density of normal matter in today's universe must be for these isotopes to have been formed in the Big Bang. This prediction is shown as the vertical band in the figure. We have talked a great deal about the ways astronomers measure the amount of normal matter in the universe. The best current measurements give a value of about 3.9×10^{-28} kilograms per cubic meter (kg/m³) for the average density of normal matter in the universe today. This value lies well within the range predicted by the observations shown in

Abundances of other low-mass elements are also consistent with the Big Bang model.

Figure 18.21. Once again, the agreement is remarkable. Turning this around, we can begin with an observation of the amount of normal matter in and around galaxies and then use our understanding of the Big Bang to calculate what the chemical composition emerging from the Big Bang should have been. When we do this, the answer we get agrees remarkably well with the amounts of these elements we actually find in nature.

Two other points are worth noting here. The first is that no elements more massive than boron could have been formed in the Big Bang. Reactions such as the triple-alpha process, which forms carbon in the interiors of stars (see Chapter 16) simply would not work under the conditions existing in the early universe. That is how we know that all the more massive elements in the universe, including the atoms making up the bulk of our planet and ourselves, must have formed in subsequent generations of stars. The second point is that the agreement between the observed density of normal matter in the universe and the abundances of light elements also provides a powerful constraint on the nature of the dark matter dominating the mass in the universe. As we will see in the next chapter, dark matter cannot consist of normal matter made up of neutrons and protons; if it did, the density of neutrons and protons in the early universe would have been much higher, and the resulting abundances of light elements in the universe would have been much different from what we actually observe.

Seeing the Forest for the Trees

You never know when you get up in the morning if this will be one of those rare days when you happen upon an extraordinary experience that leaves you looking at the world in a different way. So it was for Edwin Hubble. Hubble, who also was the first to prove the true nature of galaxies, was certainly at the right place at the right time in his career. He was doing as scientists do, using the cutting-edge tools of his day and following his instincts about what questions might turn out to have interesting answers. His investigation into the basic properties of the newly recognized class of objects called galaxies took an amazing and unexpected turn. With one graph, plotting the redshift of galaxies against his estimates of their distances, he forever changed our view of the universe. What must it have felt like to be Edwin Hubble, looking at his data and realizing for the first time what the implications of those data were? What would it mean to be the first human to glimpse the true history and potential future of the universe?

From the early discoveries of Hubble and of those who followed in his footsteps, we have come to marvel at how vast the universe truly is. Yet, from our vantage point here in our home galaxy, much of what we see is not the actual universe as it exists today; instead, what we see is how it appeared billions of years ago. How, then, do we know what the entirety of today's universe is really like? When we travel from the city to the countryside, we see completely different surroundings as we look about. This, of course, is what we would expect. But suppose we were transported suddenly to another galaxy billions of light-years away. Would the universe appear to be quite different from the way we see it at home? The cosmological principle tells us *it would not*. We have learned that the universe is both homogeneous and isotropic, meaning that at the largest scales it appears the very same in every direction and from every location.

We could possibly take some comfort in knowing that no matter where we might find ourselves in the universe, everything would look and work pretty much the same as it does right here. But that doesn't mean change isn't happening. It is happening, and on a grand scale. No matter what our location in space might be, we would find that galaxies appear to be flying apart as space itself expands.

At one time, everything we see in this vast universe—including space itself—was contained in a volume that was as small compared to the nucleus of an atom as you are compared to today's enormous universe. From a single moment 13.7 billion years ago, space has been growing into a colossus so spread out that even light takes nearly 30 billion years to cross it.

There is still much more to learn about the origin, structure, and destiny of the universe, but we will put the universe itself aside briefly while we take a close look at the largest of its individual components. The next step in our journey brings us to the galaxies themselves, these immense collections of stars, gas, and dust that were once called "island universes."

Summary

- Our observable universe contains hundreds of billions of galaxies, each with millions to hundreds of billions of stars.

- Observations suggest that our universe is homogeneous and isotropic, in agreement with the cosmological principle.

- The expansion of the universe is governed by Hubble's law: a galaxy's recession velocity is proportional to its distance.

- Expansion of the universe produces the observed redshifts, but it does not affect the local physics or structure of objects.

- The universe is expanding uniformly from a Big Bang, which occurred nearly 14 billion years ago.

- Observed redshifts of distant galaxies are a result of the increasing scale factor of the universe.

- We observe thermal radiation at 2.73 K, which is the cooled remnant of the radiation present in the universe when it was much smaller and hotter.

- The CBR enables us to measure our velocity with respect to the background radiation, and it also shows evidence of the ripples that grew to become large-scale structure in the universe.

- Nuclear reactions of normal matter in the hot early universe produced all the helium we see today, as well as trace amounts of other light elements.

SmartWork, Norton's online homework system, includes algorithmically generated versions of these questions, plus additional conceptual exercises. If your instructor assigns questions in SmartWork, log in at **smartwork.wwnorton.com.**

Student Questions

THINKING ABOUT THE CONCEPTS

1. Explain what astronomers mean by the term *universe*.

2. What was the subject of the Great Debate, and why was it important to our understanding of the scale of the universe?

3. How did Cepheid variable stars finally settle the Great Debate?

4. Describe the cosmological principle in your own words.

5. We see individual stars and galaxies. What, then, do astronomers mean when they say the universe is "homogeneous"?

*6. The universe could be homogeneous without being isotropic. Explain what is meant by *isotropic*.

7. Imagine that you are standing in the middle of a dense fog.
   a. Would you describe your environment as isotropic?
   b. Would you describe it as homogeneous?

   Explain your answers.

8. Early in the 20th century, astronomers discovered that most galaxies are moving away from our own (that is, they are redshifted.) What was the significance of this discovery?

9. Edwin Hubble later made an even more important discovery: that the speed with which galaxies are receding is proportional to their distance. Why was this among the more important scientific discoveries of the 20th century?

10. If you lived in one of the remote galaxies pictured in Figure 18.1, would you observe distant galaxies receding from your own galaxy, or would they appear to be approaching? Explain your answer.

11. Why is the Milky Way Galaxy not expanding together with the rest of the universe?

12. Why can we not use the measured radial velocities of nearby galaxies, such as the Andromeda Galaxy, to evaluate the Hubble constant (H_0)?

13. What are peculiar velocities, and how do they affect our ability to measure H_0?

14. Explain what astronomers mean by a *distance ladder*.

15. As astronomers extend their distance ladder beyond 100 Mly, they change their measuring standard from Cepheid variable stars to Type I supernovae. Why is this necessary?

16. What is meant by *Hubble time*?

*17. As the universe expands from the Big Bang, we know that galaxies are not actually flying apart from one another. What is really happening?

18. The scale factor can be used as a "cosmic ruler" in describing the expansion of the universe. Explain the scale factor.

19. Knowing that you are studying astronomy, a curious friend asks where the center of the universe is located. You smile and answer, "Right here and everywhere." Explain in detail why you would give this answer.

*20. The general relationship between radial velocity (v_r) and redshift (z) is $v_r = cz$. This simple relationship fails, however, for very distant galaxies with large redshifts. Explain why.

21. What is the origin of the cosmic background radiation (CBR)?

*22. Why is it significant that the CBR displays a Planck spectrum?

23. As the sensitivity of our instrumentation increases, we are able to look ever farther into space and, therefore, ever further back in time. When we reach the era of recombination, however, we run into a wall and can see no further back in time. Explain why.

24. What is the significance of the tiny brightness variations that are observed in the CBR?

25. What important characteristics of the early universe are revealed by today's observed abundances of various isotopes, such as ^{2}H and ^{3}He?

APPLYING THE CONCEPTS

26. Hubble time ($1/H_0$) represents the age of a universe that has been expanding at a constant rate since the Big Bang. Assuming an H_0 value of 22 km/s/Mly and a constant rate of expansion, calculate the age of the universe in years. (Note: 1 year = 3.16×10^7 seconds, and 1 light-year = 9.46×10^{12} km.)

27. Throughout the latter half of the 20th century, estimates of H_0 ranged from 18 to 35 km/s/Mly. Calculate the age of the universe in years for each of these estimated values of H_0.

28. Assume that the most distant observable galaxies have a redshift $z = 10$. (None have yet been found at this distance.)
   a. What was the scale factor (R_U) of the universe at the time light was leaving those hypothetical distant galaxies?
   b. How did the volume of the universe at that time compare with today's volume?

29. One of the more distant galaxies has a redshift $z = 5.82$ and a recession velocity of 287,000 km/s (about 96 percent of the speed of light).
    a. If $H_0 = 22$ km/s/Mly and if Hubble's law remains valid out to such a large distance, then how far away is this galaxy?
    b. What was the scale factor of the universe when that galaxy emitted the light we see today?

30. The spectrum of a distant galaxy shows the Hα line of hydrogen ($\lambda_{rest} = 656.28$ nm) at a wavelength of 750 nm. Assume that $H_0 = 22$ km/s/Mly.
    a. What is the redshift (z) of this galaxy?
    b. What is its recession velocity in kilometers per second?
    c. What is the distance of the galaxy in mega-light-years?

31. The rest wavelength of the Ly α line of hydrogen is in the extreme ultraviolet region of the spectrum at 121.6 nm. ("Ly" stands for Theodore Lyman [1874–1954], the American physicist who discovered this far-ultraviolet series of lines in the hydrogen spectrum.)
    a. What would be the wavelength of this line in the spectrum of a galaxy with a redshift $z = 6.4$?
    b. In what region of the spectrum would this redshifted line be located?

32. One of the most distant known galaxies has a redshift $z = 7.6$.
    a. What would be the observed wavelength of the Hα line ($\lambda_{rest} = 656.28$ nm)?
    b. In what region of the spectrum would this line be located?

33. The emission of the Ly α line of hydrogen (121.6 nm) from a distant galaxy is observed at the same wavelength as the rest wavelength of the Hα line of hydrogen (656.28 nm). Calculate the redshift of this galaxy.

34. Suppose we observe two galaxies: one at a distance 35 Mly with a radial velocity of 580 km/s, and another at a distance of 1,100 Mly with a radial velocity of 25,400 km/s.
    a. Calculate the Hubble constant (H_0) for each of these two observations.
    b. Which of the two calculations would you consider to be more trustworthy? Why?

c. Estimate the peculiar velocity of the closer galaxy.
    d. If the more distant galaxy had this same peculiar velocity, how would your calculated value of the Hubble constant change?

35. The temperature of the CBR is 2.73 K. What is the peak wavelength of its Planck spectrum expressed both in micrometers and in millimeters?

36. COBE observations show that our Solar System is moving in the direction of the constellation Crater at a speed of 368 km/s relative to the cosmic reference frame. What is the blueshift (negative value of z) associated with this motion?

*37. If new observations suggested that the mass fraction of helium remains at 24 percent but the mass fraction of deuterium (D) is really 10^{-6}, how would our estimates of the density of normal matter in the universe be affected? Refer to Figure 18.21.

38. The average density of normal matter in the universe is 4×10^{-28} kg/m^3. The mass of a hydrogen atom is 1.66×10^{-27} kilograms (kg). On average, how many hydrogen atoms are there in each cubic meter in the universe?

39. To get a feeling for the emptiness of the universe, compare its density (4×10^{-28} kg/m^3) with that of Earth's atmosphere at sea level (1.2 kg/m^3). How much denser is our atmosphere? Write this ratio using standard notation.

*40. Assume that the most distant galaxies have a redshift $z = 10$. The average density of normal matter in the universe today is 4×10^{-28} kg/m^3. What was its density when light was leaving those distant galaxies? (Hint: Keep in mind that volume is proportional to the cube of the scale factor.)

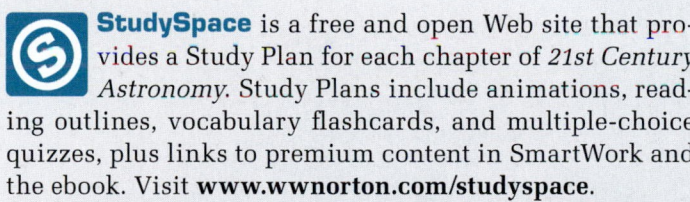 **StudySpace** is a free and open Web site that provides a Study Plan for each chapter of *21st Century Astronomy*. Study Plans include animations, reading outlines, vocabulary flashcards, and multiple-choice quizzes, plus links to premium content in SmartWork and the ebook. Visit **www.wwnorton.com/studyspace**.

It may not be amiss to point out some other very remarkable Nebulae which cannot well be less, but are probably much larger than our own system; and being also extended, the inhabitants of the planets that attend the stars which compose them must likewise perceive the same phenomena. For which reason they may also be called milky ways.

SIR WILLIAM HERSCHEL (1738–1822)

The large, barred spiral galaxy NGC 1300 is 70 million light-years away in the constellation Eridanus.

CHAPTER 19

Galaxies

19.1 Galaxies Come in Many Flavors

It is still less than a century since we came to realize that the universe is filled with "spiral nebulae," objects that closely resemble our own galaxy, the Milky Way. But just as stars come in many flavors based on their mass or on their stage of evolution, galaxies also come in many forms. After realizing that galaxies are huge collections of stars and dust, astronomers found ways to classify these galaxies into various types and make sense of the differences among them. One of their goals is to understand the origin of each flavor of galaxy, just as we interpret the origins of the different types of stars.

Galaxies Are Classified According to Their Appearance

Imagine what you would see if you took a handful of coins and threw them into the air as shown in **Figure 19.1**. You know that all of these objects are very much the same: dimes, pennies, nickels, quarters—all flat and circular. When you look at the objects falling through the air, however, they do not appear to be all the same. Some coins you see face on, and they appear circular. Some coins are instead seen edge on and appear as nothing but thin lines. Most coins are seen from an angle between these two extremes and appear with various degrees of "ellipticity," or flattening.

In principle, we can learn a lot about the properties of coins by looking at a picture like Figure 19.1. If we began

KEY CONCEPTS

Stars are not spread uniformly through space but are instead grouped into what Immanuel Kant referred to as "island universes," which today we call "galaxies." As we look beyond stars to the galaxies they belong to, we will find that

- Galaxies are classified into different types according to their shapes, which are determined by the orbits of the stars they contain.

- The arms of spiral galaxies, which form whenever the disk of a spiral galaxy is disturbed, are sites of star formation.

- Stars and gas account for only a small fraction of the mass of a galaxy.

- Galaxies are composed mostly of dark matter, which interacts through gravity but does not emit or absorb noticeable amounts of electromagnetic radiation.

- Most—perhaps all—large galaxies have supermassive black holes at their centers.

- When these supermassive black holes are "fed," possibly during encounters between galaxies, they may blaze forth with the light of thousands of normal galaxies.

- The light emerging from active galactic nuclei comes from a region no larger than the orbit of Neptune in our own Solar System.

535

by assuming that coins have a particular three-dimensional shape, we could predict what we should see in Figure 19.1. We could then compare our prediction with what is actually observed. In this example we would probably have little trouble convincing ourselves that coins must be disks. As astronomers, we play exactly this game in our efforts to discover the true three-dimensional shapes of galaxies. **Figure 19.2** shows a set of galaxies seen from various viewing angles, from face on to edge on. We can infer from images of the sky that, just like the coins in Figure 19.1, galaxies often have a disklike shape and are randomly oriented on the sky.

A quick look at a group of galaxies imaged by the Hubble Space Telescope (**Figure 19.3**), shows that galaxies come in a wide range of apparent sizes and shapes. The first advance in understanding galaxies came from sorting these different shapes into categories. The classifications we use today date back to the 1930s, when Edwin Hubble devised a scheme much like that shown in **Figure 19.4**. Hubble grouped all

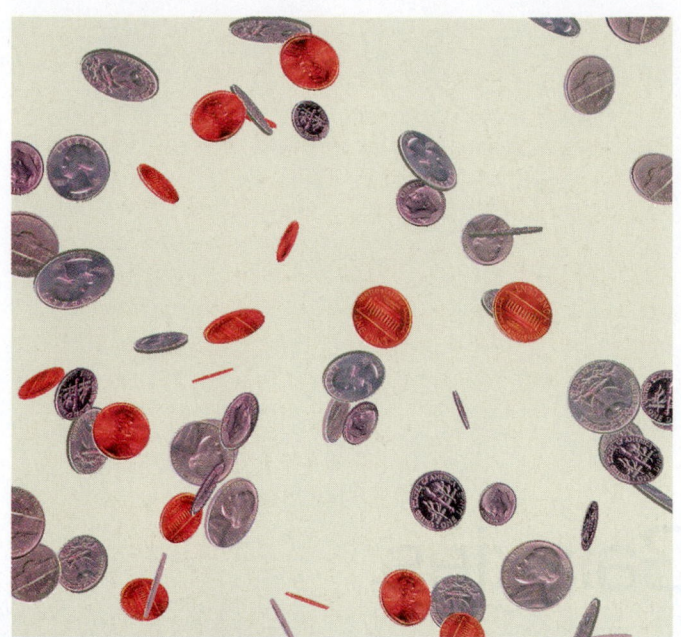

VISUAL ANALOGY FIGURE

19.1 A handful of coins thrown in the air provides a helpful analogy for the difficulties in identifying the shapes of certain types of galaxies. We see some face on, some edge on, and most somewhere in between.

VISUAL ANALOGY FIGURE

19.2 Disk-shaped galaxies seen from various perspectives or angles. The variety of angles we see for galaxies corresponds to the range of perspectives for the coins in Figure 19.1.

FIGURE 19.3 This Hubble Space Telescope image of a small group of galaxies called Hickson Compact Group 87 illustrates something of the range of shapes and sizes found among galaxies.

galaxies according to appearance and positioned them on a diagram that resembles the tuning fork used in the tuning of a musical instrument.

On the bottom (or "handle") of this **tuning fork diagram** sit oval-shaped objects called **elliptical galaxies** (labeled "E" on the diagram); these galaxies can have either round or elliptical shapes, and they show little evidence of the disk-like structure that we see in other types of galaxies.

On the two "tines" of the fork are the **spiral galaxies**, designated with an initial "S." The defining feature of a spiral galaxy is a flattened, rotating disk. Spiral galaxies also contain the spiral arms that give these galaxies their name. In addition to disks and arms, spiral galaxies have central **bulges**, which look like elliptical galaxies.

Hubble recognized that the distinction between spiral and elliptical galaxies is not always clear-cut. Some galaxies seem to be a cross between the two types, having stellar disks but no spiral arms. Hubble called these intermediate types of galaxies **S0 galaxies** and placed them near the junction of his tuning fork. Today, the distinction between elliptical and S0 galaxies is even more blurred. Modern telescopic observations have revealed that many, if not most, elliptical galaxies contain small rotating disks at their centers. Elliptical and S0 galaxies share another similarity: as of now, both have stopped producing any new stars.

Hubble noticed that the bulges of roughly half of the spiral galaxies are bar-shaped (see the opening photograph of this chapter). He called these galaxies **barred spirals** ("SB")

and placed them along the right-hand tine of the tuning fork, as shown in Figure 19.4. Spirals that lack a barlike bulge lie along the left-hand tine. Notice that Hubble positioned the spiral galaxies vertically along the tines of the fork according to the prominence of the central bulge and how tightly the spiral arms are wound. For example, Sa and SBa galaxies have the largest bulges and display tightly wound and smooth spiral arms. Sc and SBc galaxies have small central bulges and more loosely woven spiral arms, often very knotty in appearance.

Galaxies that fall into none of these classes are called **irregular galaxies** ("Irr"). As their name implies, irregular galaxies are often without symmetry in shape or structure, and they do not fit neatly on Hubble's tuning fork.

Originally, Hubble thought that his tuning fork diagram might do for galaxies what the H-R diagram had done for stars. This hope turned out to be incorrect, but his classification scheme did succeed in bringing order to the study of these objects. **Table 19.1** summarizes the criteria that Hubble used to classify galaxies.

Stellar Motions Give Galaxies Their Shapes

A galaxy is not a solid object like a coin, but a collection of stars, gas, and dust orbiting under the influence of its overall gravitational field. The shapes of elliptical galaxies are determined from the almost random orbits of their stars. For example, in any region in an elliptical galaxy, some stars are falling in while others are climbing out—in fact, there are stars moving in all possible directions. Unlike planets, which move on simple elliptical orbits about the Sun, stars in an elliptical galaxy follow orbits with a wide range of different shapes, as shown in **Figure 19.5**. These orbits are more complex than the orbits of planets because the gravitational field within an elliptical galaxy does not come from a single central object.

> The collective orbits of all its stars give an elliptical galaxy its shape.

Taken together, all of these stellar orbits are what give an elliptical galaxy its shape. Orbital speeds are also a factor. The faster the stars are moving, the more spread out the galaxy is. (After all, if the stars were not moving at all, they would all clump together at the center of the galaxy.) If the stars in an elliptical galaxy are moving in *truly* random directions, the galaxy will have a spherical shape. However, if stars tend to move faster in one direction than in others, the galaxy will be more spread out in that direction, giving it an elongated shape. These differences in stellar orbits are responsible for the variety of shapes that Hubble noted—namely, that some elliptical galaxies (those on the bottom of the tuning fork handle—see Figure 19.4) are round, while others (those on the top of the handle) are elongated. One difficult problem faced by astronomers studying elliptical

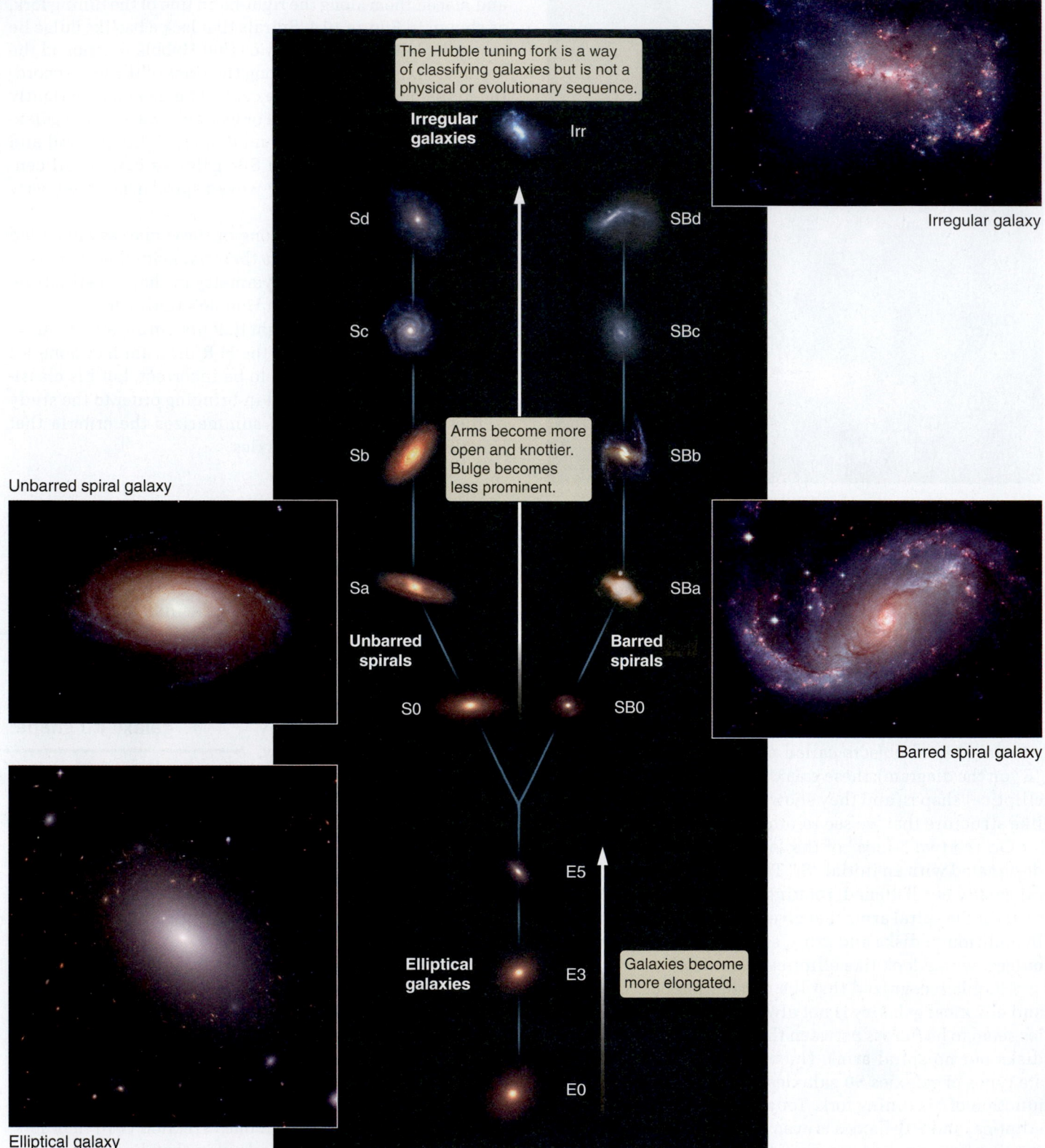

Irregular galaxy

Unbarred spiral galaxy

Barred spiral galaxy

Elliptical galaxy

The Hubble tuning fork is a way of classifying galaxies but is not a physical or evolutionary sequence.

Irregular galaxies   Irr

Sd   SBd

Sc   SBc

Arms become more open and knottier. Bulge becomes less prominent.

Sb   SBb

Sa   SBa

Unbarred spirals   **Barred spirals**

S0   SB0

E5

Elliptical galaxies   Galaxies become more elongated.

E3

E0

FIGURE 19.4 *Tuning fork diagram* showing Edwin Hubble's scheme for classifying galaxies according to their appearance. *Elliptical galaxies* form the "handle" of this tuning fork. Unbarred and barred *spiral galaxies* lie along the left and right tines of the fork, respectively. *S0 galaxies* lie along the bottom left and right tines. *Irregular galaxies* are not placed on the tuning fork.

TABLE 19.1

The Hubble Sequence of Galaxies

A Morphological Classification Scheme Based on the Gross Properties of Galaxies

Category/Criteria	Abbreviation	Range of Features			
Ellipticals	E0	Rounder			
Mostly bulge	E1	↑			
Old, red stellar population	E2				
Smooth-appearing	E3				
	E4				
	E5				
	E6	↓			
	E7	More elongated			
S0 (unbarred/barred)	S0/SB0	Smooth disk and bulge			
Bulge and disk with no arms					
Bulge and disk contain mostly old, red stars					
Spirals (unbarred/barred)	Sa/SBa	More bulge	Tightly wound arms	Smooth arms	
Bulge and disk with arms	Sb/SBb	↕	↕	↕	
Bulge has old, red stars	Sc/SBc				
Disk has both old, red stars and young, blue stars	Sd/SBd	Little bulge	Open arms	Knotty arms	
Spirals (S) have roundish bulges					
Barred spirals (SB) have elongated or barred bulges					
Irregulars	Irr				
No arms					
No bulge					
Some old stars, but mostly young stars, giving a knotty appearance					

galaxies, however, is that their appearance in the sky does not necessarily tell us their true shape. For example, a galaxy might actually be shaped like an American football, but if we happen to see it end on, it will look instead round like a baseball.

In contrast, the orbits of stars in the disks of spiral galaxies are quite different from those of stars in elliptical galaxies. The components of a spiral galaxy are shown in **Figure 19.6**. The defining feature of a spiral galaxy is that it has a flattened, rotating disk.

> **Spiral galaxies have a rotating disk and a central bulge.**

Like the planets of our Solar System, most of the stars in the disk of a spiral galaxy follow nearly circular orbits and travel in the same direction about a concentration of mass at the center of the galaxy. But the stellar orbits in a spiral galaxy's central bulge are quite different from those in the galaxy's disk. As with elliptical galaxies, the gravitational field within the bulge does not come from a single object, and the stars therefore follow random orbits. The bulges of spiral galaxies are thus roughly spherical in shape.

Other Differences among Galaxies

In addition to the differences in their stellar orbits, there is another important distinction between spiral and elliptical galaxies. Most spiral galaxies contain large amounts of dust

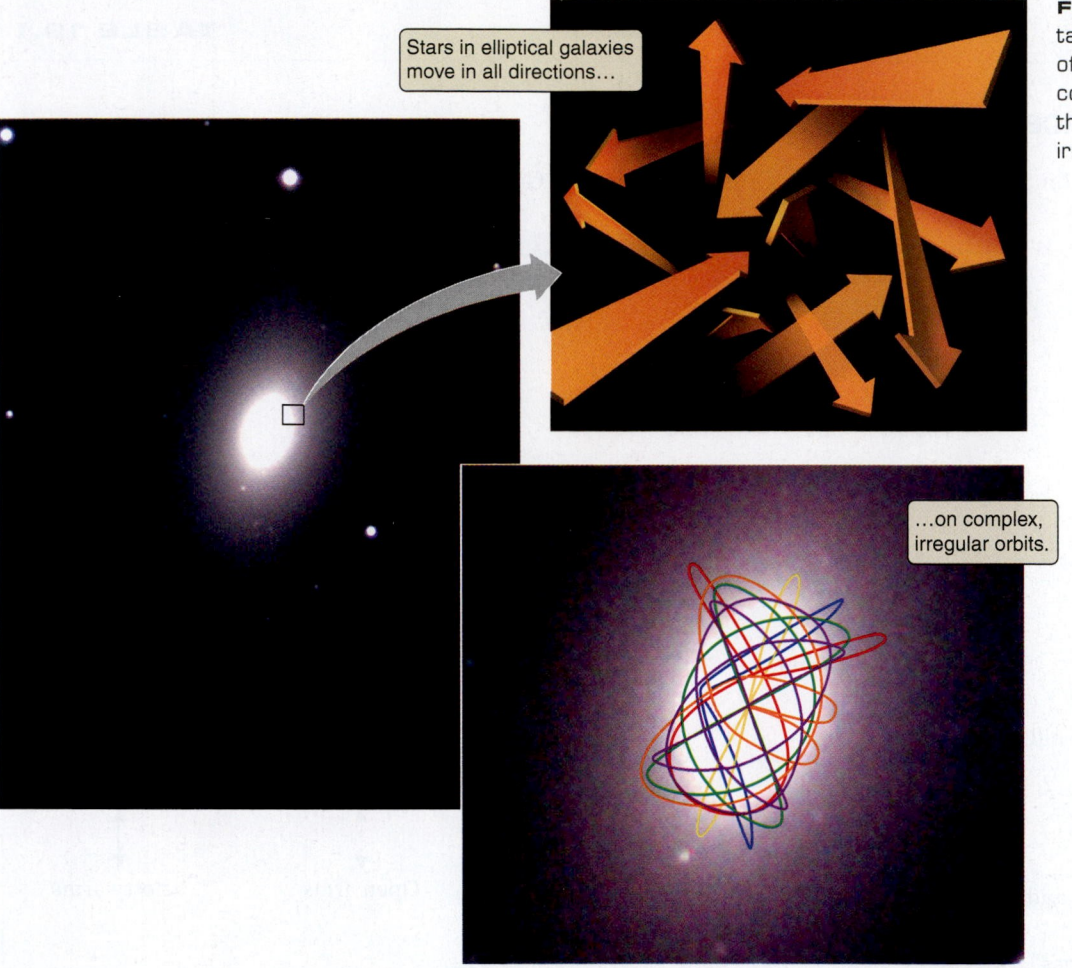

Stars in elliptical galaxies move in all directions…

…on complex, irregular orbits.

FIGURE 19.5 Elliptical galaxies take their shape from the orbits of the stars they contain. The colored lines superimposed on the galaxy represent the complex, irregular orbits of its stars.

and cold, dense molecular gas concentrated in the midplanes of their disks. Just as the dust in the disk of our own galaxy can be seen on a clear summer night as a dark band slicing the Milky Way in two (see Figure 15.1), the dust in an edge-on spiral galaxy appears as a dark, obscuring band running down the midplane of the disk (**Figure 19.7**). The cold molecular gas that accompanies the dust can also be seen in radio observations of spiral galaxies. In contrast, elliptical galaxies contain large amounts of very hot gas that we see primarily by observing the X-rays it emits.

> **Ellipticals contain mostly hot gas, and spirals contain mostly cold gas.**

The difference in shape between elliptical and spiral galaxies offers some insight into why the gas in ellipticals is hot, while spirals contain a large amount of cold, dense gas. Just as gas settles into a disk around a forming star, so, too, does cold gas settle into the disk of a spiral galaxy as a result of conservation of angular momentum. In contrast, the only place in an elliptical galaxy where cold gas could collect is at the center. However, the density of stars in elliptical galaxies is so high that Type I supernovae continually reheat this gas, preventing most of it from cooling off and forming cold clouds.

The colors of spiral and elliptical galaxies tell us a great deal about their star formation histories. As we learned in Chapter 14, stars form from dense clouds of cold molecular gas. Because the gas we see in elliptical galaxies is very hot, we would not expect star formation to be taking place today. The red colors of elliptical and S0 galaxies tell us that we are indeed looking at old stellar populations and that there has been little or no star formation for quite some time. The stars in these galaxies are an older population of lower-mass stars. The blue colors of the disks of spiral galaxies, on the other hand, tell us that we are looking at regions of ongoing star formation where massive, young hot stars are being born in the cold molecular clouds contained within the disk. Even though *most* of the stars in a spiral disk are old, the massive, young stars are so luminous that their blue light dominates what we see. When it comes to star formation, most irregular galaxies are like spiral galaxies. Some irregular galaxies are currently forming stars at prodigious rates, given their relatively small sizes.

> **Stars are still forming in spiral galaxies but not in elliptical galaxies.**

There is not a straightforward relationship between luminosity and size among the different types of galaxies.

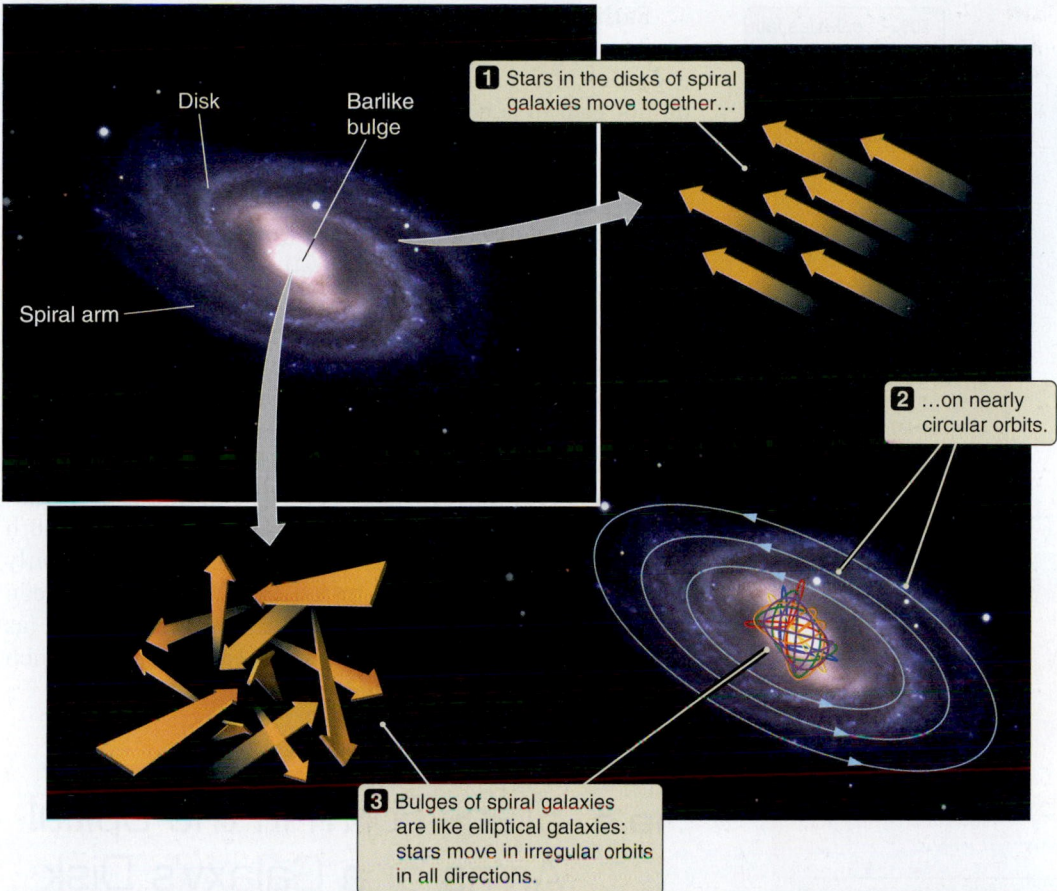

Disk

Barlike bulge

1 Stars in the disks of spiral galaxies move together...

Spiral arm

2 ...on nearly circular orbits.

3 Bulges of spiral galaxies are like elliptical galaxies: stars move in irregular orbits in all directions.

FIGURE 19.6 The components of a barred spiral galaxy. The orbits of stars in the rotating disk and the ellipse-like bulge are indicated.

FIGURE 19.7 The dust in the plane of the nearly edge-on spiral galaxy M104 is seen as a dark, obscuring band in the midplane of the galaxy. Note the bright halo made up of stars and globular clusters. Compare this image with Figure 15.1, which shows the dust in the plane of the Milky Way.

Galaxies range in luminosity from about a million up to a million million solar luminosities (10^6–10^{12} $L_\odot$) and in size from about a thousand up to hundreds of thousands of light-years. There is no strict

All types of galaxies come in a wide range of sizes.

size difference between elliptical and spiral galaxies. About half of both types of galaxies fall within a similar range of sizes. Although it is true that the most luminous elliptical galaxies are more luminous than the most luminous spiral galaxies, there is considerable overlap in the range of luminosities among all Hubble types (**Figure 19.8**).

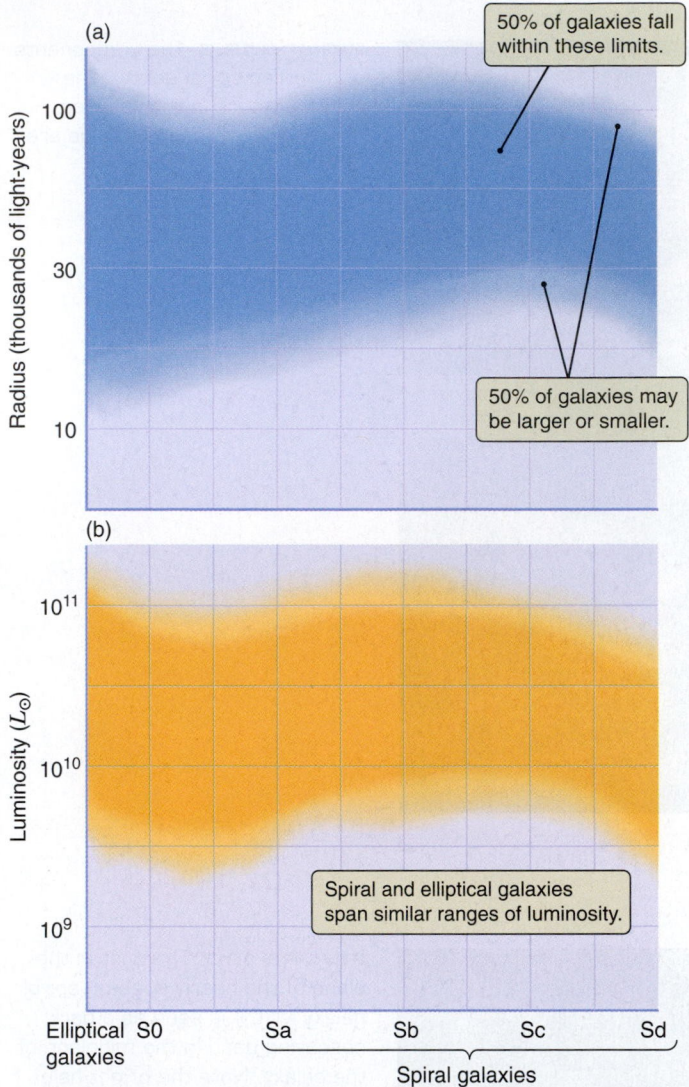

(a)

50% of galaxies fall within these limits.

50% of galaxies may be larger or smaller.

Radius (thousands of light-years)

100

30

10

(b)

10^{11}

10^{10}

10^{9}

Luminosity ($L_\odot$)

Spiral and elliptical galaxies span similar ranges of luminosity.

Elliptical S0
galaxies

Sa   Sb   Sc   Sd

Spiral galaxies

FIGURE 19.8 Galaxies span an enormous range in both size (a) and luminosity (b). Note the great overlap in luminosity and size among galaxies of different Hubble types.

Earlier in our journey we found that mass is the single most important parameter in determining the properties and evolution of a star. In contrast, differences in mass and size do not lead to such obvious differences in galaxies. Only subtle differences in color and concentration exist between large and small galaxies, making it difficult to distinguish which are large and which are small. Even when a smaller, nearby spiral galaxy is seen next to a larger, distant spiral (**Figure 19.9**), it can be hard to tell which is which. If you were to look at a photo of a nearby child appearing to stand next to a more distant adult, you could almost certainly tell which one was the child. There would be obvious differences in appearance. But that is not the case with spiral galaxies. Still, astronomers prefer to call galaxies that have relatively low luminosity (less than 1 billion $L_\odot$) **dwarf galaxies** and those that are more luminous than this **giant galaxies**. Only elliptical and irregular galaxies come in both types. In fact, among spiral and S0 galaxies, we find only giants. It is relatively easy to tell the difference between a dwarf elliptical galaxy and a giant elliptical galaxy (as shown in **Figure 19.10**). Giant elliptical galaxies have a much higher density of stars than dwarf ellipticals have.

19.2 Stars Form in the Spiral Arms of a Galaxy's Disk

Spiral galaxies take their name from the spiral arms they contain, but what is a spiral arm? From pictures of spiral galaxies outside of our own spiral Milky Way Galaxy, we might have guessed that stars in the disk of a spiral galaxy are concentrated in the spiral arms. This turns out not to be the case. **Figure 19.11** shows images of the Andromeda Galaxy taken in ultraviolet and visible light. Notice that whereas the spiral arms are relatively

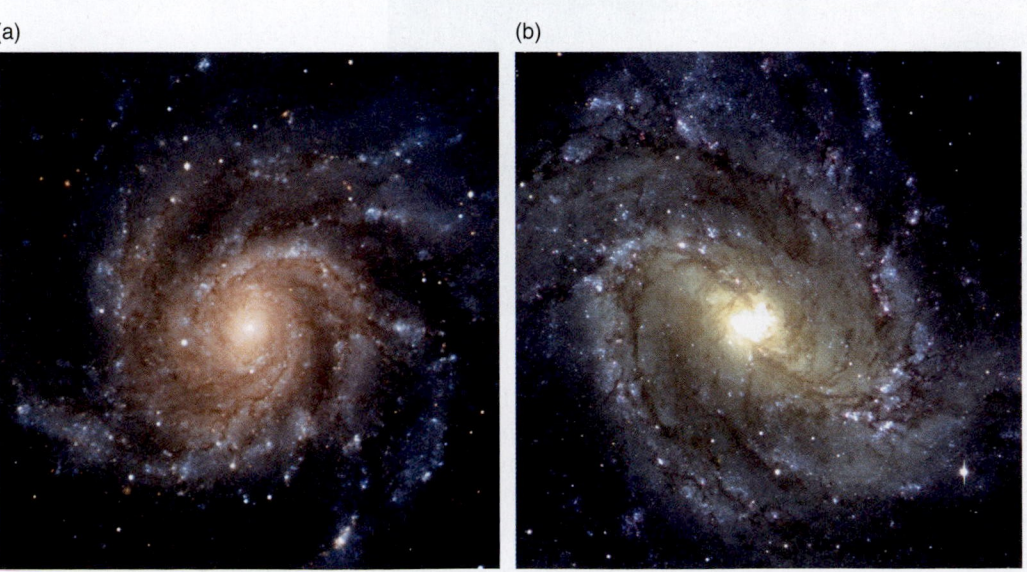

(a)   (b)

FIGURE 19.9 The mass or size of a spiral galaxy does not determine its appearance. Even though these galaxies appear to be similar in size and luminosity, the larger galaxy (a) is 4 times more distant and 10 times more luminous than the smaller galaxy (b).

(a)   (b)

FIGURE 19.10 Dwarf elliptical galaxies (a) differ in appearance from giant elliptical galaxies (b).

(a)   (b)

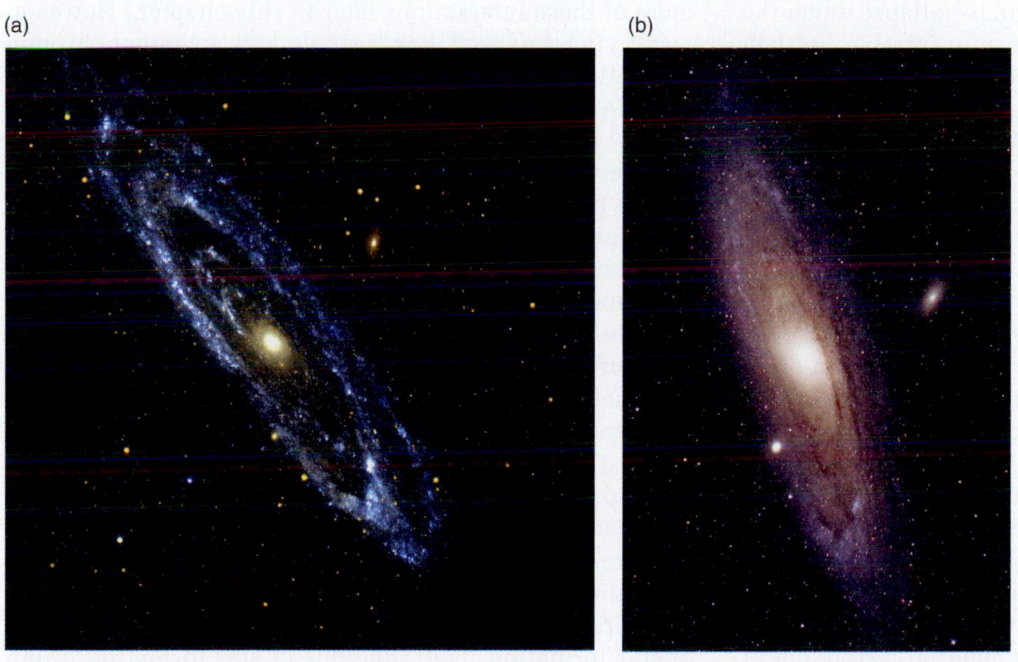

FIGURE 19.11 The Andromeda Galaxy in ultraviolet (a) and visible (b) light. Note that the spiral arms are most prominent in ultraviolet light, which is dominated by young hot stars, and in emission from interstellar clouds that are ionized by the radiation from young hot stars. The spiral arms are less prominent in visible light.

prominent in the UV image (**Figure 19.11a**), they are less prominent when viewed in visible light (**Figure 19.11b**). If we carefully trace the actual numbers of stars rather than just their brightness, we find that although stars are slightly concentrated in spiral arms, this concentration is not strong—certainly not strong enough to account for the prominence of the spiral arms we see. In fact, the concentration of stars in the disks of spiral galaxies varies quite smoothly as it decreases outward from the center of the disk to the edge of the galaxy.

Spiral arms look so prominent when viewed in blue or UV light because that is where we find significant concentrations of young, massive, luminous stars. In other words, what is strongly concentrated in the arms of spiral galaxies is ongoing star formation. H II regions, molecular clouds, associations of O and B stars, and other structures that we

(a)

Dust lane

Spiral arms are traced by concentrations of gas and dust, seen here as dark absorbing lanes…

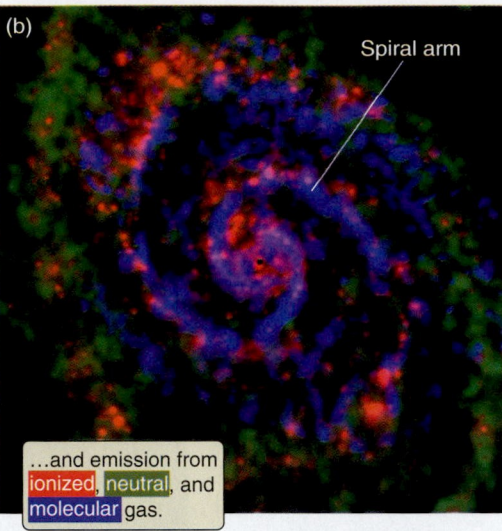

(b)

Spiral arm

…and emission from ionized, neutral, and molecular gas.

FIGURE 19.12 Two images of a face-on spiral galaxy showing the spiral arms. (a) This visible-light image also shows dust absorption. (b) This image of 21-cm emission shows the distribution of neutral interstellar hydrogen, CO emission from cold molecular clouds, and Hα emission from ionized gas.

have learned to associate with star formation are all found predominantly in the spiral arms of galaxies.

We can say a lot about what spiral arms must be like just by applying what we already know about star formation. Stars form when dense interstellar clouds become so massive and concentrated that they begin to collapse under the force of their own gravity. If stars form in spiral arms, then spiral arms must be places where clouds of interstellar gas pile up and are compressed. Such is indeed the case. There are many ways to trace the presence of gas in the spiral arms of galaxies. Pictures of face-on spiral galaxies, such as the one featured in **Figure 19.12a**, show dark lanes where clouds of dust block starlight. These lanes provide one of the best tracers of spiral arms. Spiral arms also show up in other tracers of concentrations of gas, such as 21-centimeter (cm) radiation from neutral hydrogen or radio emission from carbon monoxide (**Figure 19.12b**).

> Gas, dust, and young stars are concentrated in spiral arms.

Spiral arms are concentrations of gas where stars form, but why do spiral arms exist at all? Part of the answer is that any disturbance in the disk of a spiral galaxy will naturally be made into a spiral pattern by the disk's rotation. Material that is closer to the center takes less time to complete a revolution around the galaxy than does material farther out in the galaxy. **Figure 19.13** illustrates the point. We begin with a single linear arm through the center of a model galaxy and then watch what happens as the model galaxy rotates. In the time it takes for objects in the inner part of the galaxy to complete several rotations, objects in the outer parts of the galaxy may not have completed even a single revolution. In the process, the originally straight arms are slowly made into the spiral structure shown.

> Rotation in a disk galaxy naturally produces spiral structure.

As Figure 19.13 shows, the way disk galaxies rotate implies that any disturbances, or "kicks," to a galaxy's disk will naturally lead to spiral structure in that disk. A spiral galaxy can be kicked in several ways. One way is by gravitational interactions with other galaxies. (We will see much more of these interactions later in this chapter.) However, just giving a galaxy disk a single kick (whether through gravitational interactions or star formation) will not produce a stable spiral-arm pattern. Spiral arms produced from a "one-shot" kick will wind themselves up completely in two or three rotations of the disk and then disappear. In addition, some types of kicks are repetitive, so they are capable of sustaining spiral structure indefinitely. Some kicks come from within a galaxy itself. When the bulge in the center of a spiral galaxy is not spherically symmetric (as indeed seems to be the case for most spiral galaxies), then the bulge produces a gravitational disturbance in the disk. As the disk rotates through this disturbance, it is subjected to repeated kicks that, in turn, trigger star formation and the formation of spiral structure.

> Gravitational interactions and star formation are processes that kick disks.

Spiral structure can also be created by the process of star formation itself. Regions of star formation dump considerable energy into their surroundings in the form of UV radiation, stellar winds, and supernova explosions. All of this energy drives up the pressure in the region, compressing clouds of gas and triggering more star formation. Many massive stars typically form in the same region at about the same time; their combined mass outflows and supernova explosions occur one after another in the same region of space over the course of only a few million years. The result can be large, expanding bubbles of hot gas that sweep out cavities in the interstellar medium and concentrate the swept-up gas into dense,

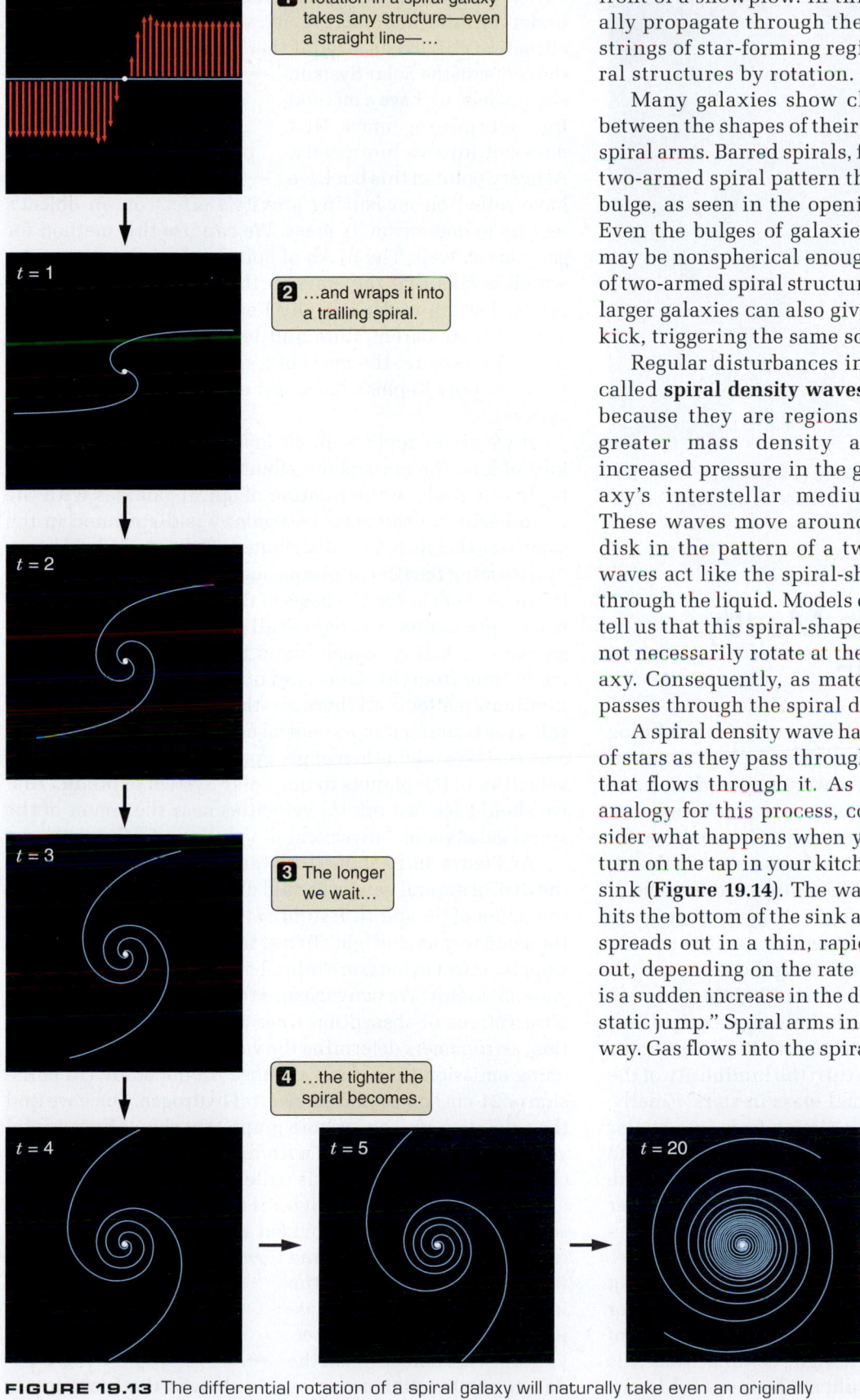

1 Rotation in a spiral galaxy takes any structure—even a straight line—...

2 ...and wraps it into a trailing spiral.

3 The longer we wait...

4 ...the tighter the spiral becomes.

FIGURE 19.13 The differential rotation of a spiral galaxy will naturally take even an originally linear structure and wrap it into a progressively tighter spiral as time (t) goes by.

star-forming clouds, much like the snow that piles up in front of a snowplow. In this way, star formation can actually propagate through the disk of a galaxy. The resulting strings of star-forming regions can then be swept into spiral structures by rotation.

Many galaxies show clear evidence of a relationship between the shapes of their bulges and the structure of their spiral arms. Barred spirals, for example, have a characteristic two-armed spiral pattern that is connected to the elongated bulge, as seen in the opening photograph for this chapter. Even the bulges of galaxies that are not obviously barred may be nonspherical enough to contribute to the formation of two-armed spiral structure. Smaller galaxies in orbit about larger galaxies can also give rise to a periodic gravitational kick, triggering the same sort of two-armed structure.

Regular disturbances in the disks of spiral galaxies are called **spiral density waves**. We call them "density" waves because they are regions of greater mass density and increased pressure in the galaxy's interstellar medium. These waves move around a disk in the pattern of a two-armed spiral. Spiral density waves act like the spiral-shaped blade in a blender slicing through the liquid. Models of how spiral density waves form tell us that this spiral-shaped, two-armed wave pattern does not necessarily rotate at the same rate as the rest of the galaxy. Consequently, as material in the disk orbits about, it passes through the spiral density waves.

> Regular disturbances lead to two-armed spirals.

A spiral density wave has very little effect on the motions of stars as they pass through it, but it does compress the gas that flows through it. As an analogy for this process, consider what happens when you turn on the tap in your kitchen sink (**Figure 19.14**). The water hits the bottom of the sink and spreads out in a thin, rapidly moving layer. A few inches out, depending on the rate at which water is flowing, there is a sudden increase in the depth of the water called a "hydrostatic jump." Spiral arms in galaxies work in much the same way. Gas flows into the spiral density wave and piles up like water in a hydrostatic jump. Stars form in the resulting compressed gas. Massive stars have such short lives (typically 10 million years or so) that they never get the chance to drift far from the spiral arms where they were born, so that is where we see them. Less massive stars, on the other hand, have plenty of time to move away from their places of birth, forming a smooth underlying disk.

> Spiral density waves compress gas, triggering star formation.

FIGURE 19.14 Water from a tap flows in a thin layer along the bottom of a kitchen sink. The sudden increase in the depth of the water is called a hydrostatic jump.

19.3 Galaxies Are Mostly Dark Matter

We have seen the importance of stellar motions in defining the shape of a galaxy, and we now turn to another important property of a galaxy: its mass. As we found earlier, mass plays the dominant role in determining the properties of stars. However, mass does not play such a clear-cut role for galaxies. Even so, efforts to measure the masses of galaxies during the last decades of the 20th century led to some of the most remarkable and surprising findings in the history of astronomy. To understand these results, we first need to ask how astronomers go about measuring the mass of a galaxy. One way is to look at the amount of light it gives off. We can use the spectrum of starlight from a galaxy to determine the types of stars the galaxy contains. We then use our knowledge of stellar evolution to turn the luminosity of the galaxy into an estimate of the total mass in stars. Finally, our knowledge of the physics of radiation from interstellar gas at X-ray, infrared, and radio wavelengths enables us to estimate the mass of these other components. Together, the stars, gas, and dust in a galaxy are called **luminous matter** (or simply **normal matter**) because this matter emits electromagnetic radiation.

However, looking at the light from a galaxy does not allow us to determine a galaxy's total mass. Imagine, for example, if we were to replace the Sun with a black hole of the same mass. We would see no light coming from this black hole, and thus its mass would not be included in our estimate based on the luminosity of starlight from our galaxy. Yet the planets would continue on their orbits, moving under the influence of its gravity, showing that gravitational attraction still exists, even if there is no light coming from the center of the Solar System. Fortunately, we have a method for determining mass that does not involve luminosity. At every point in this book we have relied on measuring gravity's effect on an object's motion to determine its mass. We can use this method for galaxies as well. The disks of spiral galaxies are rotating, which means that the stars in those disks are following orbits that are much like the Keplerian orbits of planets around their parent stars and binary stars around each other. To measure the mass of a spiral galaxy, all we need to do is apply Kepler's laws, just as we did for those other systems.

> **Astronomers use Kepler's laws to measure galaxy mass.**

If we are to apply Kepler's laws, we should have some idea of how the mass is distributed in a galaxy. We might begin our study of the rotation of spiral galaxies with the hypothesis that the mass in a galaxy is distributed in the same way that its light is distributed. That is, we could begin by assuming that the luminous mass in these galaxies is all the mass there is. On the basis of this hypothesis, we would make a prediction. The light of all galaxies, including spiral galaxies, is highly concentrated toward their centers. We could infer from this fact—and our initial assumption that luminous matter is all there is—that all the mass of a spiral galaxy is contained in its central concentration of stars, gas, and dust. We might then apply our knowledge of the orbital velocities of the planets in our Solar System to predict that we should see fast orbital velocities near the center of the spiral galaxy and slower orbital velocities farther out.

As **Figure 19.15** shows, orbital velocities of material in the disk of a spiral galaxy should change with distance from the center of the spiral, assuming that mass is distributed in the same way as starlight. To test this prediction, we use the Doppler effect to measure orbital motions. There are several ways to do this. We can measure the velocities of stars from observations of absorption lines in their spectra. In addition, astronomers determine the velocities of interstellar gas using emission lines such as those produced by Hα emission or 21-cm emission from neutral hydrogen. Once we find the velocities, we can create a graph that shows how orbital velocity in a galaxy varies with distance from the galaxy's center. This kind of graph is called a **rotation curve**.

Astronomers had to wait until the mid-1970s before telescope instrumentation provided reliable measurements of orbital velocities of stars and interstellar gas outside the inner, bright regions of galaxies. After making these observations, astronomers used the data to create rotation curves that showed that, contrary to

> **Rotation curves of spiral galaxies are remarkably flat.**

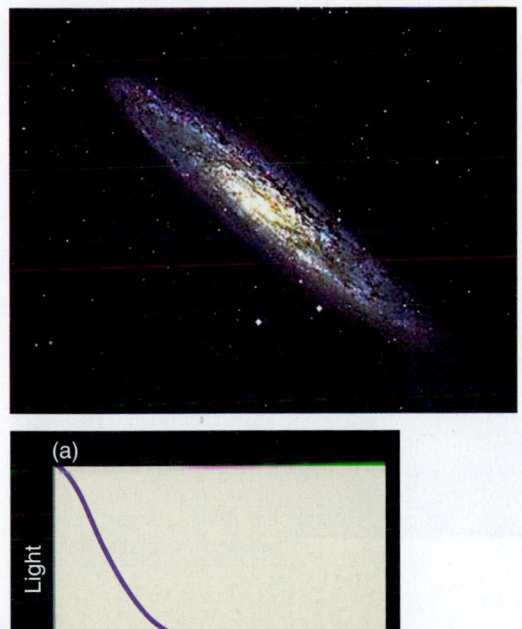

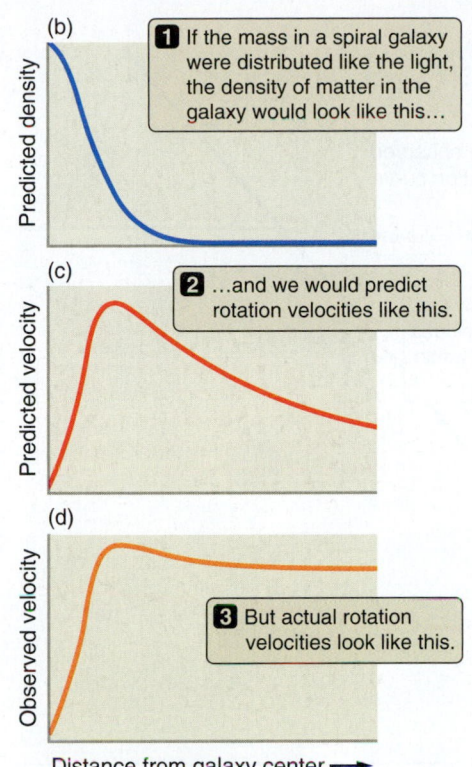

(b)

1 If the mass in a spiral galaxy were distributed like the light, the density of matter in the galaxy would look like this…

(c)

2 …and we would predict rotation velocities like this.

(d)

3 But actual rotation velocities look like this.

Distance from galaxy center ⟶

FIGURE 19.15 (a) The profile of visible light in a typical spiral galaxy. (b) The mass density of stars and gas located at a given distance from the galaxy's center. If stars and gas accounted for all of the mass of the galaxy, then the galaxy's rotation curve would be as shown in (c). However, galaxies have observed rotation curves more like the curve shown in (d).

(a)

Light

Distance from galaxy center

prediction, the rotation velocities of spiral galaxies remain about the same out to the most distant measured parts of the galaxies. As shown in Figure 19.15, the rotation curves of spiral galaxies appear level, or "flat," in their outer parts, rather than sloping downward as predicted. (These flat curves are quite logically referred to as **flat rotation curves**.) Observations of 21-cm radiation from neutral hydrogen show that the rotation curves remain flat even well outside the extent of the visible disks. These discoveries came as a shock. They meant that the long-held idea that mass in galaxies is distributed in the same way as their visible light was wrong.

The rotation curve of a spiral galaxy enables us to directly determine how the mass in that galaxy is distributed. We can apply Kepler's laws to these rotation curves and ask, "How much mass must be present inside a given radius to account for the orbital velocity we measure at that radius?" Recall from Chapter 10 that only the mass inside a given radius contributes to the net gravitational force felt by an object. Strictly speaking, this is true only for spherically symmetric objects, but a spiral galaxy is symmetric enough for this to be a good approximation. **Figure 19.16a** shows the result of such a calculation. In addition to the centrally concentrated luminous matter, these galaxies must have a second component consisting of matter that does not show up in our census of stars, gas, and dust. This material, which reveals itself only by the influence of its gravity, is called **dark matter**. ▶❚❚ **AstroTour:** Dark Matter

We "see" the presence of dark matter in spiral galaxies only because it is distributed differently from the starlight.

Had the dark matter been distributed just like the stars, we might not have realized that galaxies have these two major components. In their centers, galaxies have a higher concentration of luminous matter than dark matter. The rotation curves of the inner parts of spiral galaxies match fairly well what would be predicted by their luminous matter, indicating that normal luminous matter accounts for most of the matter in the inner parts of spiral galaxies. Within the entire *visual* image of a galaxy, the mix of dark and luminous matter is about half and half. However, rotation curves measured with 21-cm radiation from neutral hydrogen show that the outer parts of spiral galaxies are mostly dark matter. Astronomers currently estimate that as much as 95 percent of the total mass in some spiral galaxies consists of a greatly extended **dark matter halo (Figure 19.16b)**, far larger than the visible spiral portion of the galaxy located at its center. This is a startling statement. A spiral galaxy illuminates only the inner part of a much larger distribution of mass that is dominated by some type of matter we cannot see!

Like spiral galaxies, elliptical galaxies contain a large fraction of dark matter. To calculate an elliptical galaxy's total mass, we begin by measuring its luminous matter. Since elliptical galaxies do not rotate, however, we cannot use the same approach to measuring their masses that we used for spiral galaxies. Our discussion of planetary atmo-

> **Most of the mass comprising spiral galaxies is dark matter.**

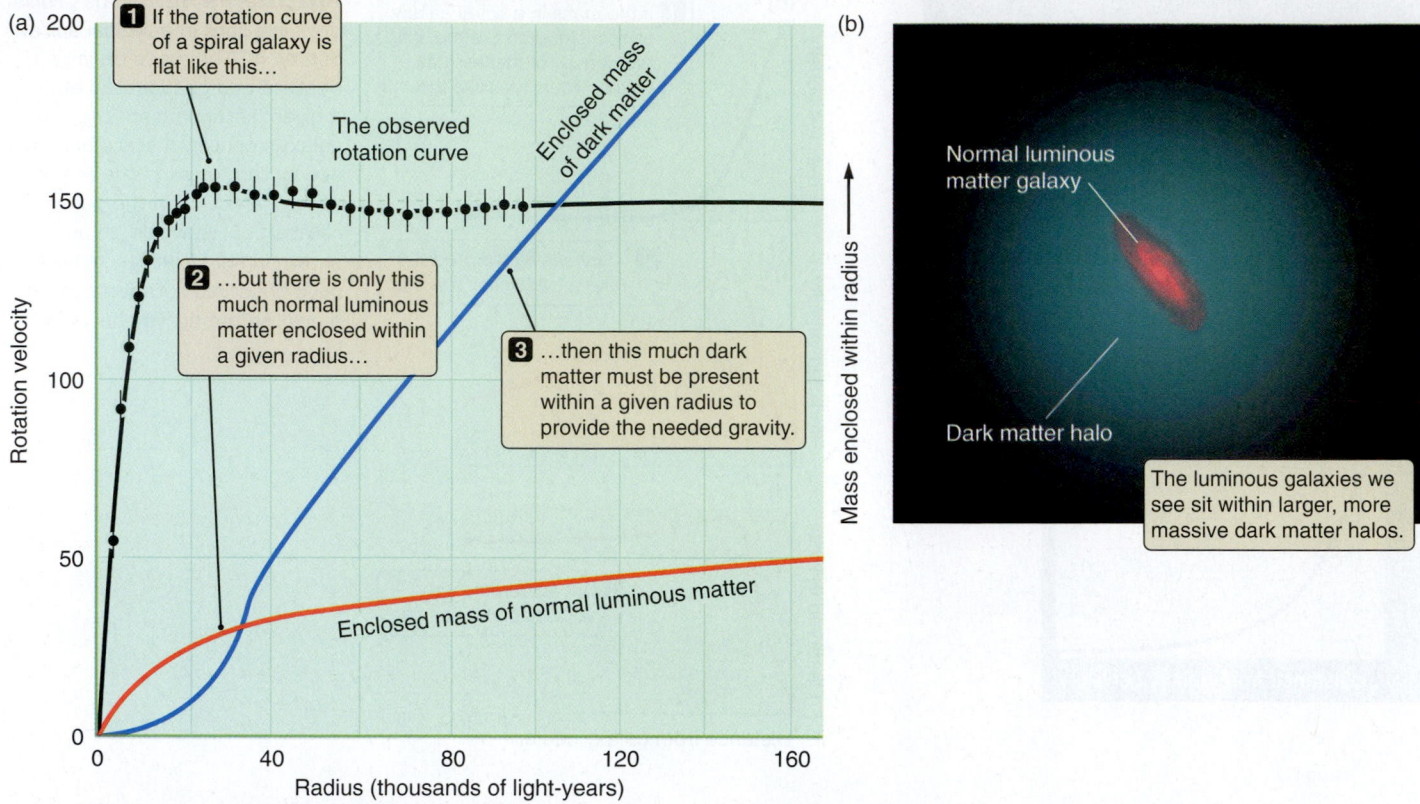

(a)

1 If the rotation curve of a spiral galaxy is flat like this…

The observed rotation curve

Enclosed mass of dark matter

2 …but there is only this much normal luminous matter enclosed within a given radius…

3 …then this much dark matter must be present within a given radius to provide the needed gravity.

Enclosed mass of normal luminous matter

Rotation velocity

Radius (thousands of light-years)

(b)

Mass enclosed within radius

Normal luminous matter galaxy

Dark matter halo

The luminous galaxies we see sit within larger, more massive dark matter halos.

FIGURE 19.16 (a) We can use the flat rotation curve of the spiral galaxy NGC 3198 to determine the total mass within a given radius. Notice that the normal mass that can be accounted for by stars and gas accounts for only part of the needed gravity. Extra dark matter is needed to explain the rotation curve. (b) In addition to the matter we can see, galaxies must be surrounded by halos containing a large amount of dark matter.

spheres in Chapter 8 provides just the tool we need. A planet's ability to hold on to its atmosphere depends on its mass. In like fashion, an elliptical galaxy's ability to hold on to its hot, X-ray-emitting gas depends on its mass. First we infer the total amount of gas from X-ray images, such as the blue/purple halo seen in **Figure 19.17**. Next we calculate the mass that is needed to hold on to the gas. We then compare that mass with the amount of luminous matter in the center of the galaxy (the whitish region of the galaxy pictured in Figure 19.17). The amount of dark matter is the difference between what is needed to hold on to the gas and the observed luminous matter.

> The dark matter of elliptical galaxies enables them to hold on to their hot gases.

When we do this, we discover the same thing that we found from the rotation curves of spirals. Some elliptical galaxies contain up to 20 times as much mass as can be accounted for by their stars and gas alone, so they also must be dominated by dark matter. As with spirals, the luminous matter in ellipticals is more centrally concentrated than is the dark matter. The transition from the inner parts of galaxies (where luminous matter dominates) to the outer

FIGURE 19.17 Combined visible-light and X-ray images of elliptical galaxy NGC 1132. The false-color blue/purple halo is X-ray emission from hot gas surrounding the galaxy. The hot gas extends well beyond the visible light from stars.

parts (which are dominated by dark matter) is remarkably smooth. As we will see in Chapter 22, this seamless combination of dark and luminous matter is an important piece of evidence that any successful theory of galaxy formation must explain. Of course, some galaxies may contain less dark matter than others, but it is probably safe to conclude that about 90 percent of the total mass in a typical galaxy is in the form of dark matter.

What, then, is this dark matter that makes up most of a galaxy? We do not yet know. A number of suggestions have been made over the years: Jupiter-like objects, swarms of black holes, copious numbers of white dwarf stars, and exotic unknown elementary particles. (The last of these is currently the favored explanation.) Given that galaxies are formed mostly of dark matter, have we been wasting our time paying so much attention to the small fraction of mass tied up in stars and the interstellar medium? No. For one thing, stars and gas are the only parts of galaxies that we can see directly. Stars are also of undeniable importance from our human perspective. Stars formed the atoms in our bodies and are the kernels around which planetary systems form. The star we call our Sun supplies the energy that makes life on Earth possible. The attention we have paid to stars is well placed indeed!

19.4 There Is a Beast at the Heart of Most Galaxies

Galaxies are remarkable objects, each shining with the light of hundreds of billions of stars. Galaxies themselves change over time as the universe expands and as their stars evolve. In the past, the rates of star formation in galaxies were higher and the galaxies were closer together, increasing the chance of galaxy collisions. We will return to these points and to the subject of dark matter in galaxies when we discuss galaxy formation in Chapter 22. But galaxies pale in comparison with the most brilliant beacons of all: **quasars**. *Quasar* is short for "quasi-stellar radio source," so named because astronomers first observed these mysterious objects as unresolved points at radio wavelengths. The story of the discovery of quasars provides an interesting insight into the discovery of new phenomena in astronomy and how conventional thinking can sometimes hinder progress.

In the late 1950s, radio surveys had detected a number of bright, compact objects that at first seemed to have no optical counterparts. Eventually, astronomers found the optical counterparts when improved radio positions revealed that the radio sources coincided with faint, very blue, stellarlike objects. Unaware of the true nature of these objects,

> Quasars are phenomenally luminous.

astronomers called them "radio stars." Obtaining spectra of the first two radio stars was a laborious task, requiring 10-hour exposures with the only recording technique available in those days: slow photographic plates (see Chapter 5). Hardly prepared for what they would see in these painstakingly obtained spectra, astronomers were greatly puzzled by the results. Rather than displaying the expected absorption lines that are characteristic of blue stellar objects, the spectra showed only a single pair of emission lines that were broad—indicating very rapid motions within these objects—and that did not seem to correspond to the lines of any known substances. Puzzling indeed!

For several years astronomers believed they had discovered a new type of star. After all, they reasoned, "If it looks like a star, it must be a star." Finally, one astronomer, Maarten Schmidt (1929–), realized what no one else had considered: that these broad spectral lines were, in fact, the highly redshifted lines of ordinary hydrogen. The implications were astounding! These "stars" were not stars. They were extraordinarily luminous objects at enormous distances.

Other "quasars," as they came to be known, were soon found by the same techniques. Many were relatively easy to identify because of their unusual blue color. As still more were found, astronomers did what astronomers always do: they began cataloging them. (The young daughter of two of the catalogers misunderstood what her parents termed "quasi-stellar objects," hearing it instead as "crazy-stellar objects"—a not altogether unfitting description of this strange new phenomenon, and a source of much amusement in that household!)

Quasars are phenomenally powerful, pouring forth the luminosity of a trillion to a thousand trillion (10^{12}–10^{15}) Suns! Quasars are objects of the distant universe. The nearest quasar to us is approximately 1 billion light-years away. Literally billions of galaxies are closer to us than the nearest quasar. Recall that the distance to an object also tells us the amount of time that has passed since the light from that object left its source. The fact that the nearest quasar is seen as it existed about a billion years ago tells us that quasars are quite rare in the universe today. Quasars were once much more common. The discovery that quasars existed in the distant and therefore earlier universe provided one of the first pieces of evidence that the universe has evolved over time.

Quasars are not the isolated beacons they were once thought to be. Instead, they are centers of violent activity in the hearts of large galaxies, as seen in **Figure 19.18**. Astronomers now recognize that quasars are only the most extreme form of activity that can occur

> There are several types of active galactic nuclei.

in the nuclei of galaxies. In the universe we see today, approximately 3 percent of galaxies contain brilliant points of light in their centers that may outshine all of the stars in the galaxies that host them. Together, quasars and their less luminous but still active cousins are called **active galactic nuclei**,

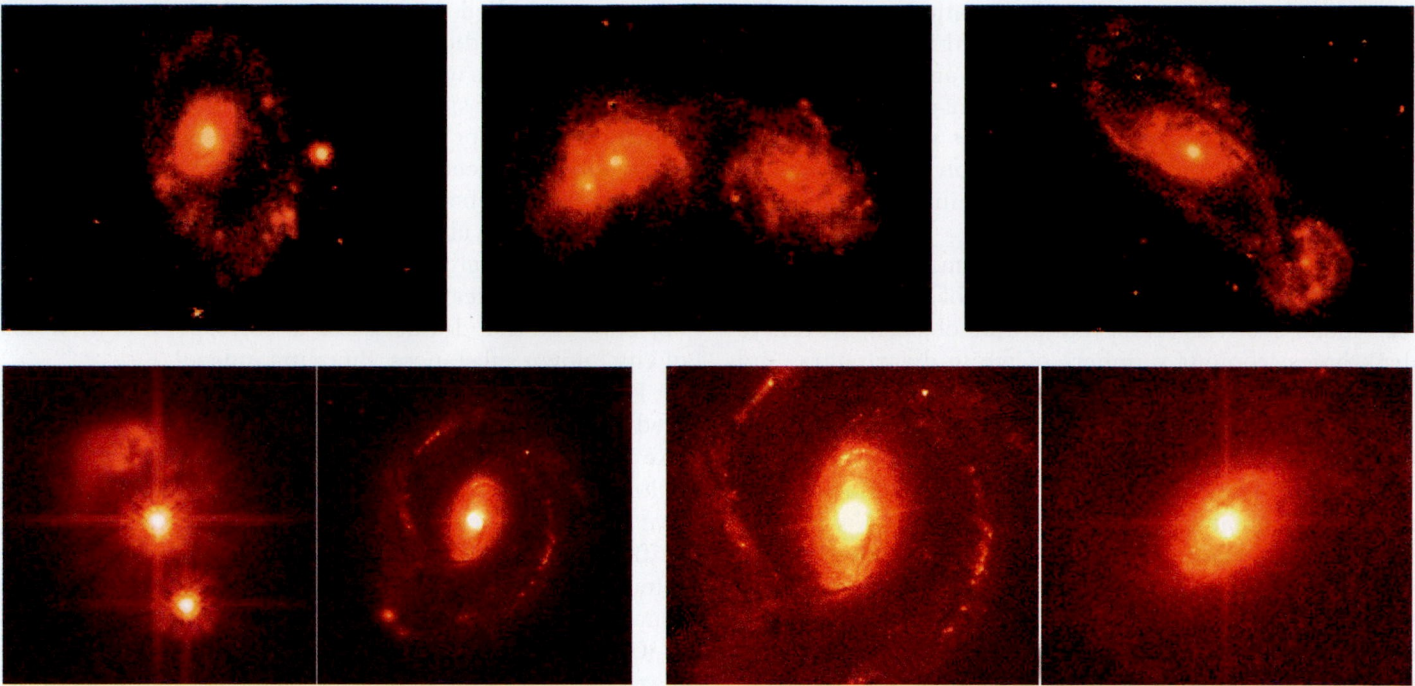

FIGURE 19.18 The environments around quasars, which are found in the centers of galaxies. Those galaxies often show evidence of interactions with other galaxies.

or simply **AGNs**. Today, we know that there are several distinct types of active nuclei within various types of galaxies.

Seyfert galaxies, named after American astronomer Carl Seyfert (1911–1960), who discovered them in 1943, are spiral galaxies whose centers contain AGNs discernible in visible light as a distinct bright spot. The luminosity of a typical Seyfert nucleus can be 10 billion to 100 billion $L_\odot$, comparable to the luminosity of the rest of the galaxy as a whole. The luminosities of AGNs found in elliptical galaxies are similar to those of Seyfert nuclei (10 billion to 100 billion $L_\odot$). Unlike Seyfert nuclei, however, AGNs in elliptical galaxies are usually most prominent in the radio portion

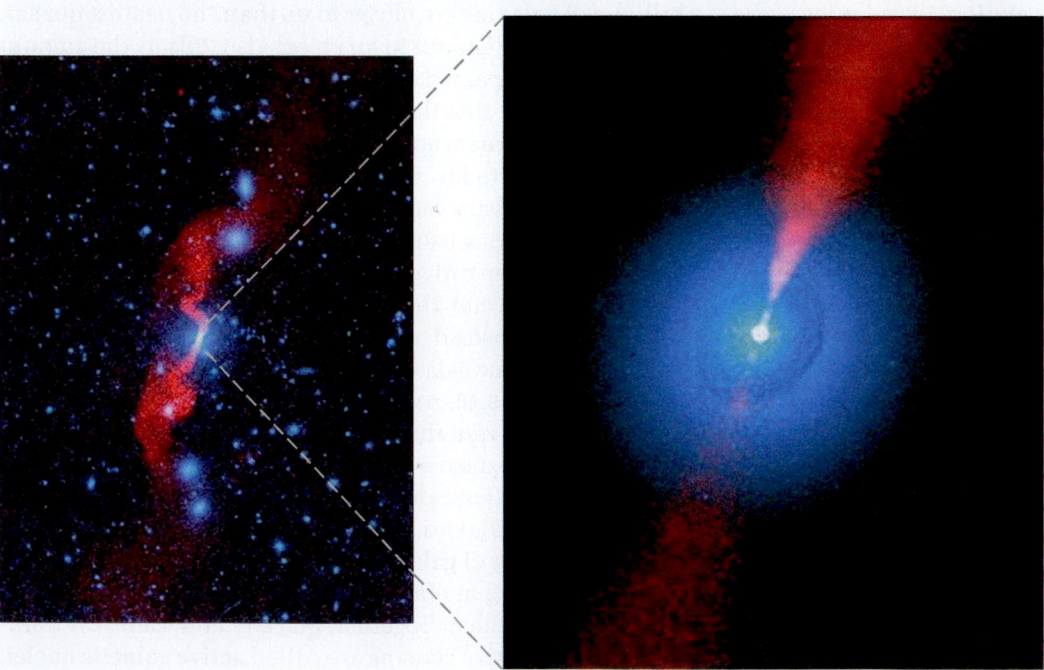

FIGURE 19.19 Radio emission from a double-lobed radio galaxy (shown in red) is superimposed over an image of visible starlight from the galaxy (shown in blue). The lobes, powered by a relativistic jet streaming outward from the nucleus of the galaxy, extend to more than 3 million light-years from the galaxy's center.

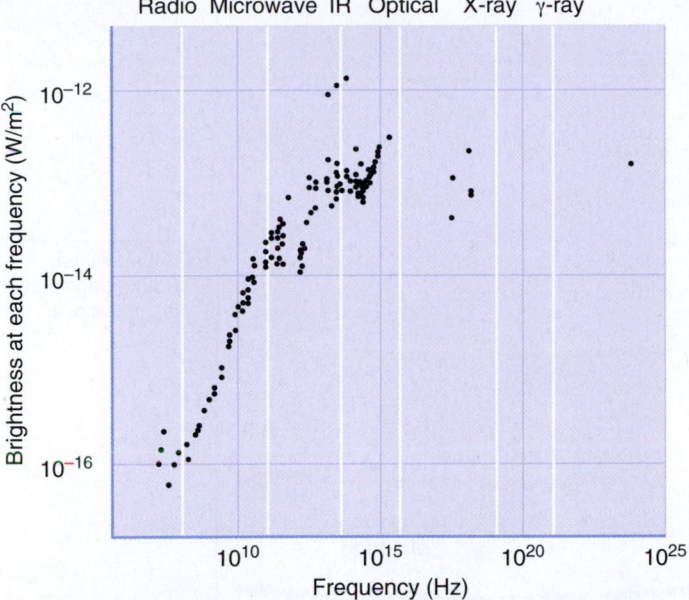

Radio Microwave IR Optical X-ray γ-ray

FIGURE 19.20 Electromagnetic energy spectrum of the well-studied and bright quasar 3C 273 from low-frequency radio to high-energy gamma rays. Most of the energy emitted by this quasar lies between infrared and gamma-ray frequencies.

of the electromagnetic spectrum, earning them the name **radio galaxies**. Radio galaxies, and their distant, extremely luminous cousins the quasars, are often the sources of slender jets that extend outward millions of light-years from the galaxy, powering twin lobes of radio emission, such as those seen in **Figure 19.19**.

What could power such phenomenal cosmic beacons as quasars, Seyfert galaxies, and radio galaxies? Clues lie in the light they emit. **Figure 19.20** shows the electromagnetic spectrum, ranging from radio frequencies to high-energy gamma rays, of the well-studied quasar 3C 273. This quasar is about 100 times as luminous as our entire Milky Way Galaxy, and it exhibits much stronger radio and gamma-ray emission than is typical for AGNs. Much of the light from AGNs is synchrotron radiation. This is the same type of radiation that we first encountered coming from Jupiter's magnetosphere and later saw again in such extreme environments as the Crab Nebula. Recall from Chapter 9 that synchrotron radiation comes from relativistic charged particles spiraling around in the direction of a magnetic field (see Figure 9.20). The fact that AGNs accelerate large amounts of material to nearly the speed of light indicates that they are very violent objects indeed. In addition to the continuous spectrum of synchrotron emission, the spectra of many quasars and Seyfert nuclei also show emission lines that are smeared out by the Doppler effect across a wide range of wavelengths. This observation implies that gas in AGNs is swirling around the centers of these galaxies at speeds of thousands or even tens of thousands of kilometers per second.

> **AGNs emit synchrotron radiation.**

AGNs Are the Size of the Solar System

The enormous radiated power and mechanical energy of active galactic nuclei are mind-boggling on their own, but they are made even more spectacular by the fact that all of this power emerges from a region that can be no larger than a light-day or so across. That is comparable in size to our own Solar System! How can we even make such a claim? Quasars and other AGNs appear only as unresolved points of light, even in our most powerful telescopes.[1] What information in the light that we receive from AGNs tells us they are such compact objects? For an answer, we turn not to the sky but to the halftime show at a local football game.

Figure 19.21 illustrates a problem that faces every marching-band director. When a band is all together in a tight formation at the center of the field, the notes you hear in the stand are clear and crisp; the band plays together beautifully. But as the band spreads out across the field, its sound begins to get mushy. This is not because the marchers are poor musicians. Rather, it is a consequence of the fact that sound travels at a finite speed. On a cold, dry, December day sound travels at a speed of about 330 meters per second (m/s). At this speed, it takes sound approximately $\frac{1}{3}$ of a second to travel from one end of the football field to the other. Even if every musician on the field played a note at exactly the same instant in response to the director's cue, in the stands you would hear the instruments close to you first but would have to wait longer for the sound from the far end of the field to arrive.

If the band is spread from one end of the field to the other, then the beginning of a note will be smeared out over about $\frac{1}{3}$ of a second, or the difference in sound travel time from the near side of the field to the far side. If the band were spread out over two football fields, it would take about $\frac{2}{3}$ of a second for the sound from the most distant musicians to arrive at your ear. If our marching band were spread out over a kilometer, then it would take roughly 3 seconds—the time it takes sound to travel a kilometer—for us to hear a crisply played note start and stop. Even with our eyes closed, it would be easy to tell whether the band was in a tight group or spread out across the field.

Exactly the same principle works for AGNs; but here we are working with the speed of light, not the speed of sound. We observe that quasars and other AGNs can change their brightness dramatically over the course of only a day or two—and in some cases as briefly as a few hours! This rapid

---

[1] HST and large groundbased telescopes show faint "fuzz" around the images of some quasars and AGNs. This fuzz is light from the surrounding galaxy. The quasars and AGNs themselves, however, remain as unresolved points of light.

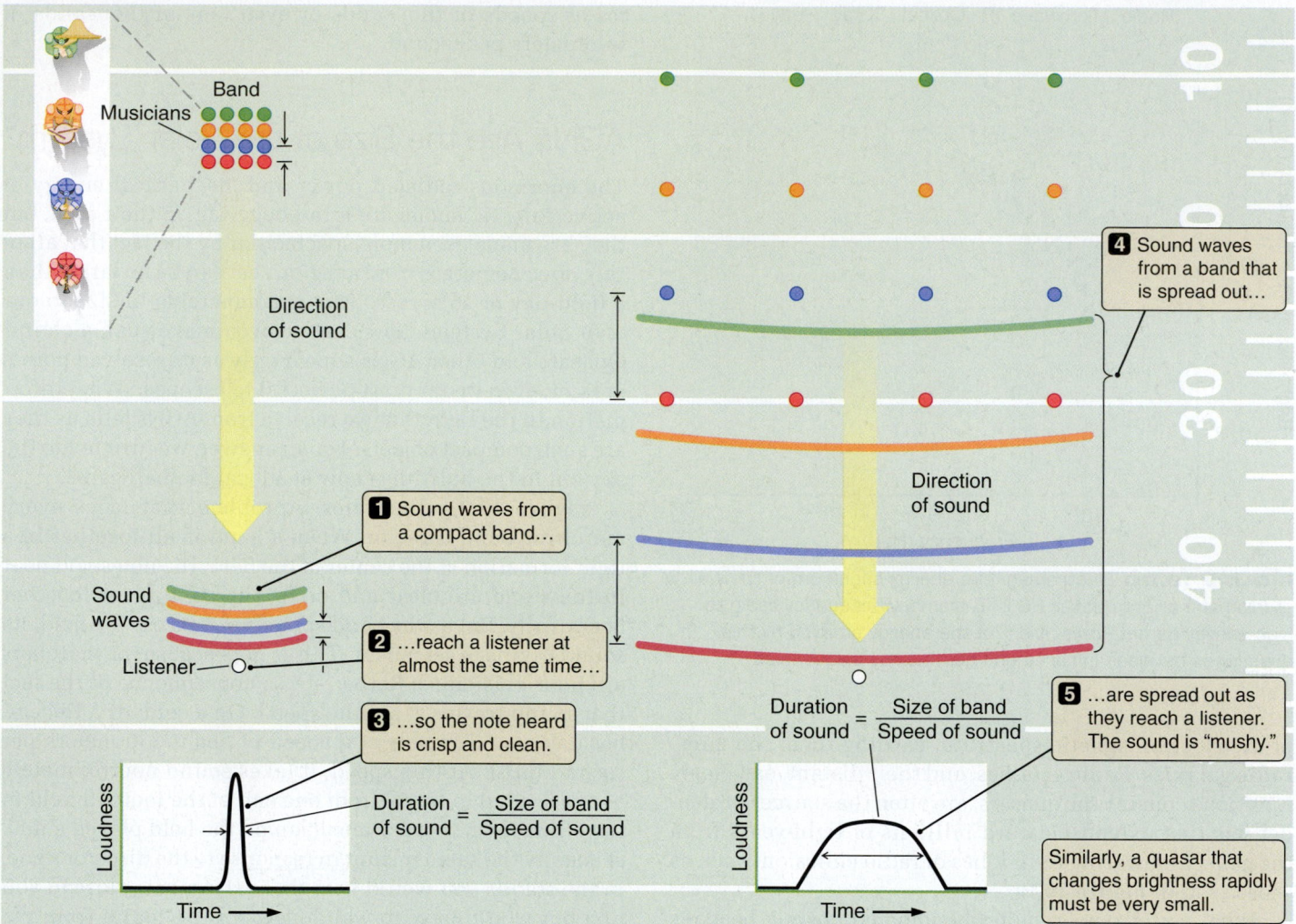

Band

Musicians

Direction of sound

1 Sound waves from a compact band...

Sound waves

Listener

2 ...reach a listener at almost the same time...

3 ...so the note heard is crisp and clean.

Loudness

$$\text{Duration of sound} = \frac{\text{Size of band}}{\text{Speed of sound}}$$

Time

4 Sound waves from a band that is spread out...

Direction of sound

5 ...are spread out as they reach a listener. The sound is "mushy."

$$\text{Duration of sound} = \frac{\text{Size of band}}{\text{Speed of sound}}$$

Loudness

Time

Similarly, a quasar that changes brightness rapidly must be very small.

FIGURE 19.21 A marching band spread out across a field cannot play a clean note. Similarly, AGNs must be very small to explain their rapid variability (see the text).

variability sets an upper limit on the size of the AGN, just as hearing clear music from the band tells us that the band musicians are close together. The AGN powerhouse must therefore be no more than a light-day or so across because if the powerhouse were larger, what we see could not possibly change in a day or two.

AGNs vary rapidly and so must be relatively small.

Here is the image that should come to mind when you think about active galactic nuclei: *the light of 10,000 galaxies pouring out of a region of space that would come close to fitting within the orbit of Neptune!* ▶‖ **AstroTour: Active Galactic Nuclei**

Supermassive Black Holes and Accretion Disks Run Amok

When astronomers first discovered AGNs, they put for-ward a variety of ideas to explain them. But as observa-tions revealed the tiny sizes and incredible energy densities of AGNs, only one answer seemed to make sense: violent accretion disks surrounding **supermassive black holes**—black holes with masses from thousands to tens of billions of solar masses—power AGNs. We have run across accretion disks on a number of occasions during our journey. Accretion disks surround young stars, pro-viding the raw material for solar systems. Accretion disks around white dwarfs, fueled by material torn from their bloated evolving com-panions, lead to novae and

AGNs are powered by accretion onto supermassive black holes.

Type I supernovae. Accretion disks around neutron stars and star-sized black holes a few kilometers across are seen as X-ray binary stars. Now take these examples and scale them up to a black hole with a mass of a billion solar masses and a radius comparable in size to the orbit of Neptune. Furthermore, imagine an accretion disk fed by

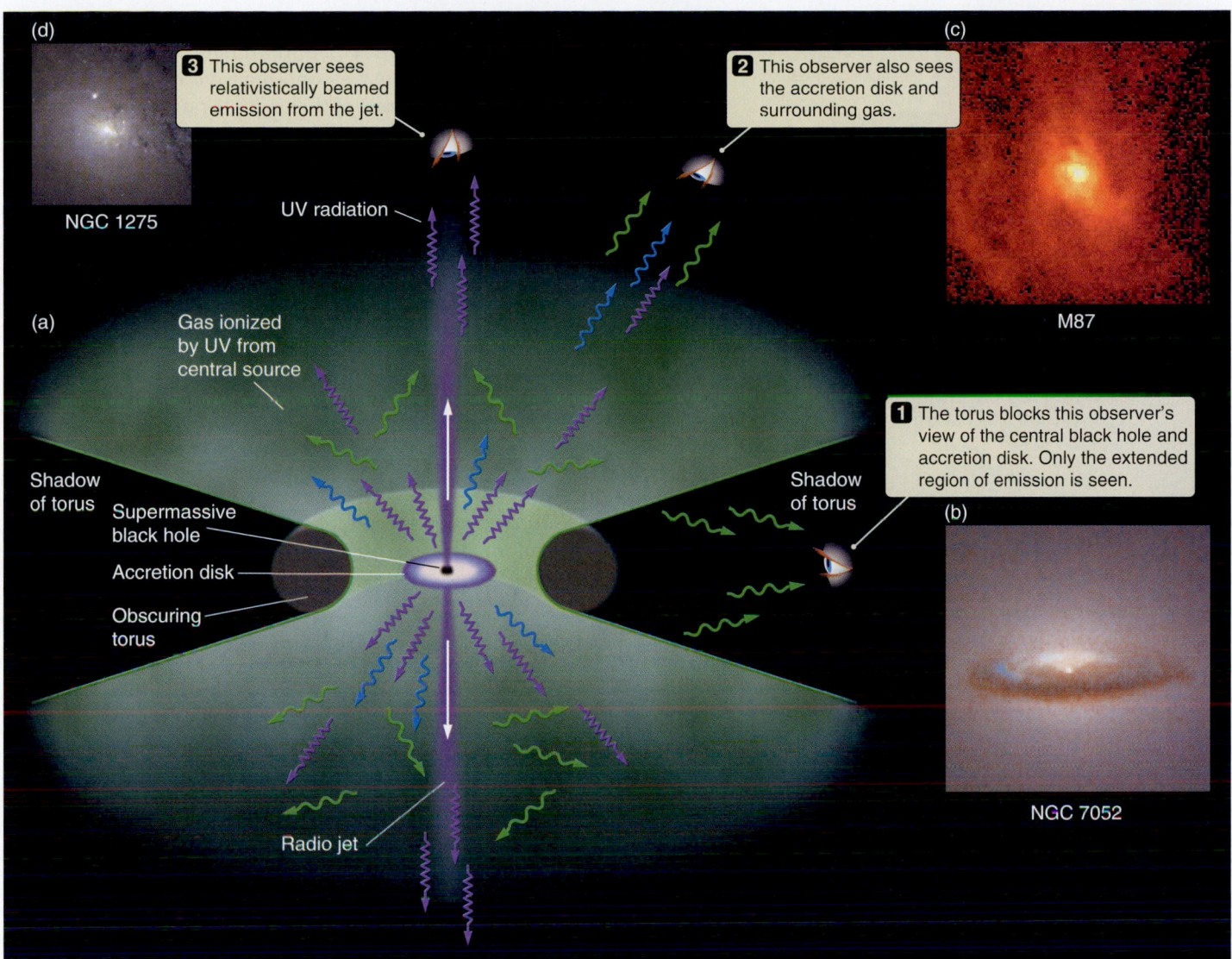

(d)

3 This observer sees relativistically beamed emission from the jet.

NGC 1275

(c)

2 This observer also sees the accretion disk and surrounding gas.

M87

UV radiation

(a)

Gas ionized by UV from central source

Shadow of torus

Shadow of torus

1 The torus blocks this observer's view of the central black hole and accretion disk. Only the extended region of emission is seen.

(b)

Supermassive black hole

Accretion disk

Obscuring torus

Radio jet

NGC 7052

FIGURE 19.22 (a) The unified model of active galactic nuclei. The appearance of the object changes when the disk and torus are viewed from the edge (b), at higher inclination (c), and close to face on (d). Astronomers believe that the viewing angle, the mass of the central black hole, and the rate at which it is being fed determine the properties of any AGN that we see.

substantial *fractions of entire galaxies* rather than the small amounts of material being siphoned off a star. *That* is an active galactic nucleus.

Astronomers have developed this basic picture of a supermassive black hole surrounded by an accretion disk into a more complete physical description called the **unified model of AGNs**. The unified model takes its name from the fact that it attempts to unify our understanding of all AGNs—quasars, Seyfert galaxies, and radio galaxies— within the same framework of understanding.

Figure 19.22a shows the various components of the unified model of AGNs. In this model, an accretion disk surrounds a supermassive black hole. Much farther out lies a large torus (doughnut) of gas and dust consisting of material that is feeding the central engine. Each of the differ-

ent components of the unified model accounts for different observed properties of AGNs.

In our discussion of star formation, we learned that gravitational energy is converted to thermal energy as material moves inward toward the growing protostar. As material moves inward toward a supermassive black hole, conversion of gravitational energy heats the accretion disk to

> AGN accretion disks are phenomenally efficient at converting mass to energy.

hundreds of thousands of kelvins, causing it to glow brightly in visible and ultraviolet light. Conversion of gravitational energy to thermal energy as material falls onto the accretion disk is also a source of X-rays, UV radiation, and other energetic emission. When we discussed the Sun, we mar-

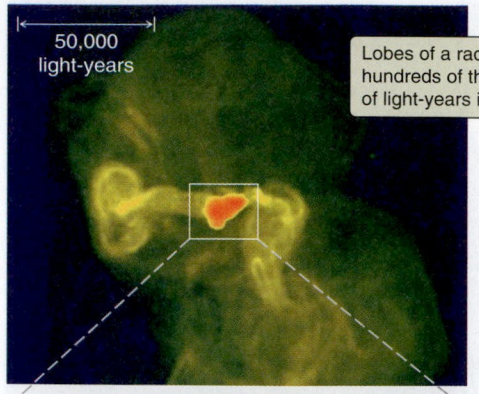

50,000 light-years

Lobes of a radio galaxy hundreds of thousands of light-years in size...

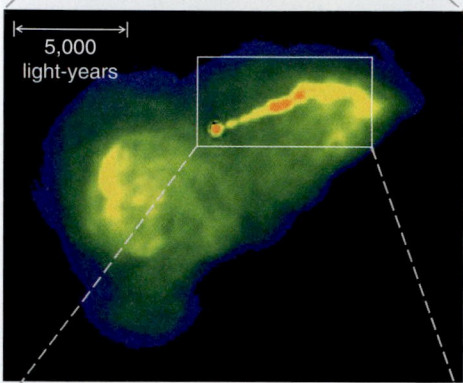

5,000 light-years

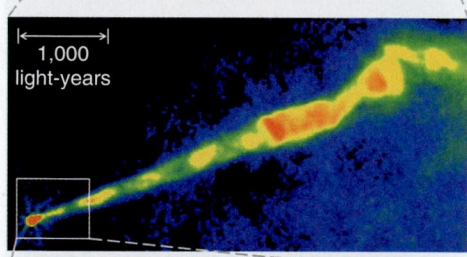

1,000 light-years

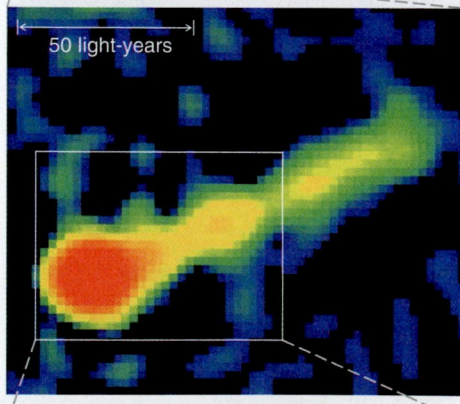

50 light-years

veled at the efficiency of fusion, which converts 0.7 percent of the mass of hydrogen into energy. In contrast, approximately 50 percent of the mass of infalling material around a supermassive black hole is converted to luminous energy. The rest of that mass is pulled into the black hole itself, causing it to grow even more massive.

The interaction of the accretion disk with the black hole gives rise to powerful radio jets—superpowerful analogs to the Herbig-Haro jets formed by accretion disks around young stellar objects (see Chapter 15). Throughout, twisted magnetic fields accelerate charged particles such as electrons and protons to relativistic speeds, accounting for the observed synchrotron emission. Gas in the accretion disk or in nearby clouds orbiting the central black hole at high speeds gives off emission lines that are smeared out by the Doppler effect into the broad lines seen in AGN spectra. This accretion disk surrounding a supermassive black hole is the "central engine" that leads to AGNs.

The outer torus plays a somewhat different role in the unified model. Located far from the inner turmoil of the accretion disk, and far larger than the central engine, some of the outer torus is ionized by UV light from the AGN much as an H II region is ionized by the UV light from O stars. This ionization provides one possible source for the H II region–like emission lines seen in many AGNs. The most important conceptual role of the outer torus is that it allows a single model to

The outer torus of an AGN determines what we can see.

explain many of the differences observed among various AGNs. The outer torus obscures our view of the central engine in different ways, depending on the angle from which we happen to see it. By invoking variations in the viewing angle, the mass of the black hole, and the rate at which it is being fed, the unified model of AGNs can account for a wide range of AGN properties, including the average spectra of AGNs, an example of which is shown in Figure 19.20.

What AGN We See Depends on Our Perspective

When we view the unified model edge on, we see emission lines from the surrounding torus and other surrounding gas. We can also some-

FIGURE 19.23 The visible jet from the galaxy M87 extends over a hundred thousand light-years but originates in a tiny volume at the heart of the galaxy (far lower right).

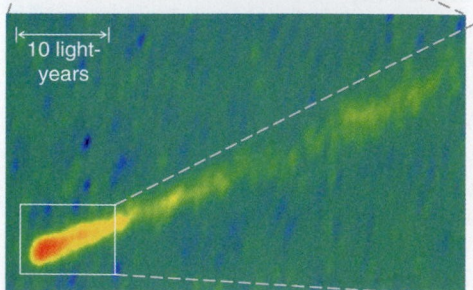

10 light-years

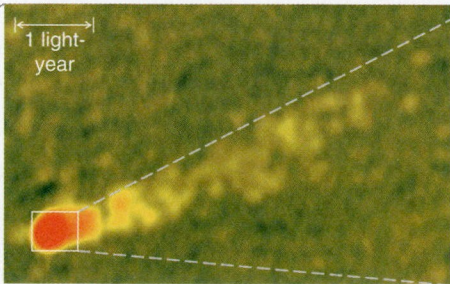

1 light-year

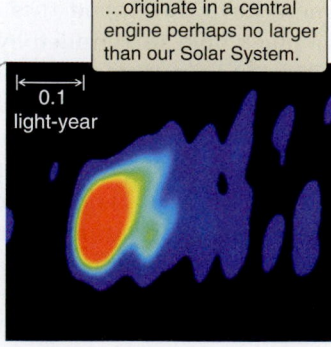

...originate in a central engine perhaps no larger than our Solar System.

0.1 light-year

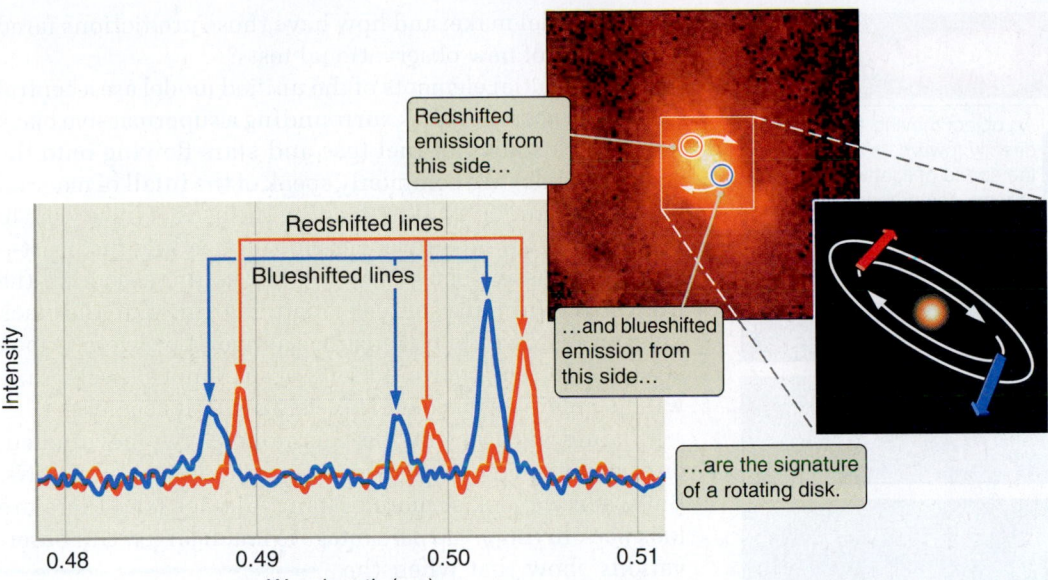

FIGURE 19.24 An HST image of the nearby radio galaxy M87. The high velocities observed provide direct evidence of rotation about a supermassive black hole at this galaxy's center.

times see the torus in absorption against the background of the galaxy. **Figure 19.22b** is a Hubble Space Telescope image of the inner part of the galaxy NGC 7052, showing what appears to be just such a shadow. From this nearly edge-on orientation, we cannot see the accretion disk itself, so we do not expect to see the Doppler-smeared lines that originate closer to the supermassive black hole. If jets are present in the AGN, however, we should be able to see these emerging from the center of the galaxy.

> Viewed edge on, an AGN's accretion disk is not visible.

If we look at the inner accretion disk somewhat more face on (**Figure 19.22d**), we can see over the edge of the torus, and thus we get a more direct look at the accretion disk and the location of the black hole. In this case, we should see more of the synchrotron emission from the region around the black hole and the Doppler-broadened lines produced in and around the accretion disk. **Figure 19.22c** shows a Hubble Space Telescope image of one such object, called M87,[2] at an intermediate inclination. M87 is a source of powerful jets that continue outward for a hundred thousand light-years, but originate in the tiny engine at the heart of the galaxy (**Figure 19.23**). Spectra of the disk at the center of this galaxy (**Figure 19.24**) show the rapid rotation of material around a central black hole with a mass of 3 billion solar masses ($3 \times 10^9 M_\odot$).

> When we view it more face on, we see the AGN's accretion disk.

The material in an AGN jet travels very close to the speed of light. As a result, what we see is strongly influenced by relativistic effects. One of these is an extreme form of the

Doppler effect called **relativistic beaming**. Matter traveling at close to the speed of light concentrates any radiation it emits into a tight beam pointed in the direction in which it is moving. As a result of relativistic beaming, an AGN jet coming toward us will look much brighter than its twin moving away from us. The unified model clearly predicts that AGN jets should be two-sided, but most of the time what we actually see appear to be one-sided jets. The reason is that we see only the portion of the jet that is being beamed toward us. The emission from the other portion of the jet always exists. We come to this conclusion because the radio lobes of radio galaxies are always two-sided. We don't see the jet that is moving away from us because it is beaming its radiation in the other direction.

> Relativistic effects influence what part of an AGN we see.

In rare instances we happen to see the accretion disk in a quasar or radio galaxy almost directly face on. In such cases, relativistic beaming dominates what we see. Rather than seeing emission lines and other light coming from hot gas in the accretion disk, we are blinded by the bright glare of jet emission beamed directly at us (see Figure 19.22d).

Relativistic beaming is not the only thing that complicates what we see when we look down the barrel of an AGN jet. The material in an AGN jet is moving so close to the speed of light that the radiation it emits is barely able to outrun its source. As observers, we see all of the light emitted by the jet over thousands of years arrive at our telescopes over the course of only a few years. Time appears to be compressed. From our perspective, the jet seems to travel great distances in brief periods of time. In extreme cases, such as the jet in M87 (**Figure 19.25**), features in the jet

---

[2]Object number 87 in Messier's catalog (see Chapter 18).

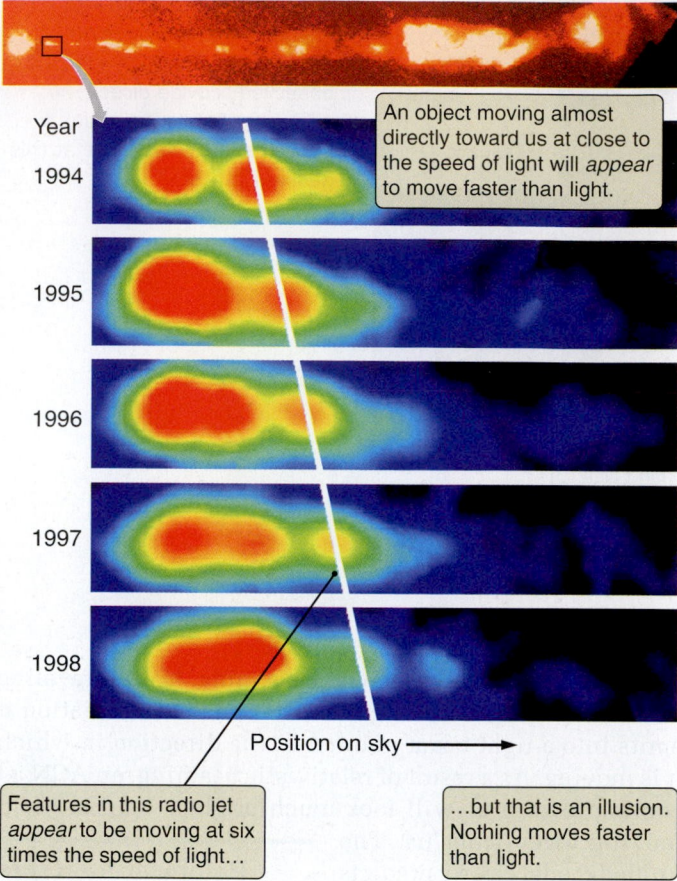

An object moving almost directly toward us at close to the speed of light will *appear* to move faster than light.

Year

1994

1995

1996

1997

1998

Position on sky ⟶

Features in this radio jet *appear* to be moving at six times the speed of light…

…but that is an illusion. Nothing moves faster than light.

FIGURE 19.25 Because the jet material in M87 is moving toward us at relativistic speeds, we see an optical illusion: the radio blobs appear to be moving apart at more than six times the speed of light.

appear to be moving across the sky faster than the speed of light. We stress the word *appear* because this phenomenon, referred to as **superluminal motion**, is an optical illusion. Despite the name, nothing in these jets is actually traveling through space faster than the speed of light. Einstein's special theory of relativity remains safe.

It is worth noting that we have used the same galaxy, M87, as an example of several of the phenomena discussed in this chapter. The presence of many different phenomena predicted by the unified model in the same object is important support for the view that the model truly is *unified*.

Normal Galaxies and AGNs— A Question of Feeding the Beast

Astronomers began developing the unified model of AGNs in the 1980s, and it has been modified frequently to account for new and better data. What testable predictions does the

unified model make, and how have those predictions fared in the light of new observational tests?

The essential elements of the unified model are a central engine (an accretion disk surrounding a supermassive black hole) and a source of fuel (gas and stars flowing onto the accretion disk). We commonly speak of the infall of material onto the accretion disk as "feeding the beast." If we were to shut off this infall—if we were to stop feeding the beast— the supermassive black hole would remain, absent all the fireworks. Without a source of matter falling onto the black hole, an AGN would no longer be active. If we were to look at such an object, we would see a normal (not active) galaxy with a supermassive black hole sitting in its center.

Should such objects be common? As we have noted, only about 3 percent of present-day galaxies contain AGNs. But when we look at more distant galaxies (and therefore look back in time), the percentage is much larger. Our observations show that when the universe was younger, there were many more AGNs than there are today. If the unified model of AGNs is correct, then all the supermassive black holes that powered those dead AGNs should still be around. If we combine what we know of the number of AGNs in the past with ideas about how long a given galaxy remains in an active phase, we are led by the unified model to predict that many— perhaps even *most*—normal galaxies today contain supermassive black holes!

> The unified model predicts that normal galaxies contain supermassive black holes.

This is a somewhat startling prediction—that quiescent collections of stars, gas, and dust like our own Milky Way should have slumbering beasts at their centers. It is a bit like suggesting that all of our dignified, soft-spoken grandfathers were once members of biker gangs. Yet here is a prediction of the unified model that can be tested.

If supermassive black holes are present in the centers of normal galaxies, they should reveal themselves in a number of ways. For one thing, such a concentration of mass at the center of a galaxy should draw surrounding stars close to it. The central region of such a galaxy would be much brighter than could be explained if stars alone were responsible for the gravitational field in the inner part of the galaxy. Stars feeling the gravitational pull of a supermassive black hole in the center of a galaxy should also orbit at very high velocities. We should therefore see large Doppler shifts in the light from stars near the centers of normal galaxies. Astronomers have, in fact, found evidence of this sort in every normal galaxy with a substantial bulge in which a careful search has been conducted. The masses inferred for these black holes range from 10,000 $M_\odot$ (for a "small" black hole) to 5 billion $M_\odot$ (a "gargantuan"

> Supermassive black holes have been discovered in the nuclei of many nearby galaxies.

FIGURE 19.26 These tidally interacting galaxies show severe distortions, including stars and gas drawn into long tidal tails.

one)! The mass of the supermassive black hole seems to be related to the mass of the elliptical-galaxy or spiral-galaxy bulge in which it is found. At the beginning of the 21st century we have come to accept that all large galaxies probably contain supermassive black holes. These observations confirm the most fundamental prediction of the unified model. They also tell us something remarkable about the structure and history of normal galaxies.

Apparently the only difference between a normal galaxy and an active galaxy is whether the supermassive black hole at its center is being fed at the time we see that galaxy. The fact that only 3 percent of present-day galaxies have AGNs does not indicate which galaxies have the potential for AGN activity. Rather, it indicates which galaxy centers are being lit up at the moment. If we were to drop a large amount of gas and dust directly into the center of any large galaxy, this material would fall inward toward the central black hole, forming an accretion disk and a surrounding torus. The predicted result of this process is that the nucleus of this galaxy would change into an AGN.

Mergers and Interactions Make the Difference

Galaxies do not exist in isolation. Even our own Milky Way has several neighbors. In Chapter 10 we discussed tidal interactions between galaxies. **Figure 19.26** shows the havoc that such interactions can cause, pulling interacting galaxies into distorted shapes in which stars and gas are drawn out into sweeping arcs and tidal tails (see also Figures 5.31 and 10.14). Sometimes when galaxies interact, they pass by each other and go their separate ways. Such interactions can trigger the formation of spiral structure, and they can also slam clouds of interstellar gas together at high speeds, triggering additional star formation. Tidally distorted or interacting galaxies often contain regions of vigorous ongoing star formation. Sometimes when two galaxies interact, they merge to form a single, larger galaxy. This turns out to be an important part of the process of galaxy formation, as we will learn in Chapter 22. ▶❚❚ **AstroTour:** Galaxy Interactions and Mergers

To account for the many large galaxies that we see today, interactions and mergers must have been much more prevalent in the past. The prevalence of interacting galaxies when the universe was younger explains the large number of AGNs that existed

> Galaxy-galaxy interactions fuel AGN activity.

in the past. Computer models show that galaxy-galaxy interactions can cause gas located tens of thousands of light-years from the center of a galaxy to fall inward toward the center, where it can provide fuel for an AGN. During mergers, a significant fraction of a cannibalized galaxy might wind up being fed to the beast. HST images of quasars, such as those in Figure 19.18, often show that quasar host galaxies are tidally distorted or are surrounded by other visible matter that is probably still falling into the galaxies. The most violent forms of AGN activity were likely most common in the early universe because that was the time when galaxies

were forming, and large amounts of matter were constantly being drawn in by the gravity of newly formed galaxies. This process is still at work today. Galaxies that show evidence of recent interactions with other galaxies are far more likely to house AGNs in their centers.

There are still many puzzles. For example, the unified model has not been developed to the point that it predicts how long an outburst of AGN activity will last, or how often galaxies will undergo episodes of AGN activity. To answer these questions, we will have to combine the unified model of AGNs with much better models of galaxy formation and evolution than we currently have. In addition, some observations are not yet explained by the unified model of AGNs. For instance, the unified model does not account for why one quasar can be a very powerful radio source while another, identical in all other respects, is radio-quiet, even when observed with the most sensitive radio telescopes.

Our understanding of AGNs is far from complete. Even so, the unified model has had many successes—enough for us to say with confidence that any large galaxy, including our own, might be only a chance encounter away from becoming an AGN. How different our own sky might be if, a few tens of millions or hundreds of millions of years from now, our descendants look toward the center of our galaxy and see the brilliant light of a powerful Seyfert nucleus blazing forth.

Seeing the Forest for the Trees

A century is but a blink of an eye compared with the age of Earth or even the age of our species. Such is the story of 20th century astronomy. A hundred years ago, humankind knew virtually nothing of the true answers to the most basic questions we might ask about the universe. Virtually all that we know of these matters we have learned within the past century. The discovery and study of galaxies is one of the major threads running throughout the story of that century.

From the time that Copernicus first dislodged Earth from the center of creation, each new discovery has pushed us further and further from the human-centered conceptions that shaped our view of the world for millennia (and persist in many ways to this day). The discovery of the almost unthinkable size of our own Milky Way and the realization that the universe contains countless more galaxies comparable to our own were huge steps in humanity's expanding awareness of the universe. The work of Hubble and others did far more than merely answer a few arcane and esoteric scientific questions. These scientists forever changed the way we must view everything, including ourselves.

The scale and distance of galaxies may have shattered preconceptions about the extent of the universe, but galaxies themselves seemed at first to be composed of familiar objects. In essence, the zoo grew far larger but the animals themselves seemed the same. Stars in distant galaxies shine in accordance with the same laws that govern the Sun. Galaxies are held together by the same force of gravity that set the course of the cannonball in Newton's famous thought experiment. But as our understanding grew and new observations were made, observations of galaxies forced astronomers to push our knowledge of physics in extreme and unexpected ways.

When we first learned of black holes, we were faced with the thought of several solar masses compressed into a region only a few kilometers across, and we had to confront ideas of general relativity that turn our everyday notions of space and time inside out. Now we discover that such star-sized black holes are the merest of specks compared with the monsters residing in the centers of large galaxies. We have witnessed the consequences when these behemoths, with masses as great as billions of times that of the Sun, are fed with gas supplied by collisions between galaxies. Radio observations of the sky show us relativistic jets stretching across millions of light-years of intergalactic space. Yet the source of such a jet is tiny. The light of a quasar outshines a thousand galaxies—a beacon that can be seen from the very edge of the universe—yet originates from a region no larger than our Solar System. As we continue our journey, we will find that such a monster lurks at the heart of our own mundane Milky Way, waiting (perhaps forever) for its next meal.

Active galaxies stir the imagination, but one of the most startling and fundamental results of our study of galaxies comes from the simple application of Newton's derivation of Kepler's laws to the motions of stars in galaxies. We have used this technique over and over on our journey to measure the mass of planets and stars. When we apply this comfortable tool to galaxies, however, the results are shocking. The matter that we see in the stars, gas, and dust is but the tip of a much larger iceberg, the rest of which is composed of a substance known only as dark matter. Most of the matter in the universe consists of we know not what.

As we have seen, in the 20th century astronomers made dramatic progress toward building a physical understanding of the formation, evolution, and death of stars. We now can use the laws of physics to peel back the layers of a star, peer into the heart of a supernova, or run the clock forward and see our Sun's ultimate fate. Astronomers cannot claim a similar understanding of galaxies, because the study of galaxies is still in its infancy. Remember how little time has passed since galaxies were even recognized for what they are. It is unlikely that there will ever be a simple scheme that does for galaxies what the H-R diagram did for stars. On the other hand, astronomers are beginning to better understand what the "ecology" within galaxies is like and to piece together something of how they form and evolve. In the next chapter we take a step down this road by looking in more detail at the galaxy we know best: our own Milky Way.

Summary

- The shapes of galaxies and the types of orbits of their stars determine their Hubble classification.

- Currently, stars are forming in the disks of spiral galaxies but not in ellipticals or S0 galaxies.

- Spiral arms are regions of intense star formation, and the arms are visible because of the concentration of bright young stars.

- Most of the mass in galaxies does not reside in gas, dust, or stars; rather, about 90 percent of a galaxy's mass is in the form of dark matter, which does not emit or absorb light to any significant degree.

- Most—perhaps all—large galaxies have supermassive black holes at their centers.

- When gas accretes onto one of these supermassive black holes, the center of the galaxy becomes an active galactic nucleus, which can emit as much as a thousand times the light of the whole galaxy, all coming from a region the size of our Solar System.

SmartWork, Norton's online homework system, includes algorithmically generated versions of these questions, plus additional conceptual exercises. If your instructor assigns questions in SmartWork, log in at **smartwork.wwnorton.com.**

Student Questions

THINKING ABOUT THE CONCEPTS

1. Name and describe a common, everyday object that appears so dissimilar when viewed from different angles that you might think from these different views that it is actually a different object.

2. Name and describe three types of galaxies.

3. What are the principal morphological (structural) differences between spiral and elliptical galaxies?

4. How do E0 and E7 elliptical galaxies differ in shape?

5. What would be the shape of an elliptical galaxy whose stars were all traveling in random directions in their orbits? Explain your answer.

6. How does molecular-gas temperature differ between elliptical and spiral galaxies?

7. Explain why star formation in spiral galaxies takes place mostly in the spiral arms.

8. Some galaxies have regions that are relatively blue in color, while other regions appear redder. Aside from color, what can you say about the differences between these regions?

9. Describe the characteristics of irregular galaxies.

10. Galaxies come in a large range of both luminosity and size. Which of these two properties varies more among galaxies?

*11. Describe the spiral arms in a galaxy and explain at least one of the mechanisms that create them.

12. In describing galaxies, what do astronomers mean by *luminous* (or *normal*) matter?

13. With regard to its interaction with electromagnetic radiation, how does dark matter differ from normal matter?

14. What evidence do we have that most galaxies are composed largely of dark matter?

15. Contrast the rotation curve for a galaxy containing only normal matter with one containing dark matter.

16. How does a spiral galaxy's dark matter halo differ from its visible spiral component?

*17. Name some of the candidates for the composition of dark matter.

18. How would you explain a quasar to a relative or friend?

19. Which is more luminous: a quasar or a galaxy with a hundred billion solar-type stars? Explain you answer.

*20. The nearest quasar is about a billion light-years away. Why do we not see any that are closer?

21. What distinguishes a "normal" galaxy from one we call "active"—that is, one that contains an AGN?

22. Contrast the size of a typical AGN with the size of our own Solar System. How do we know how big an AGN is?

23. Describe what must be happening at the centers of galaxies that contain AGNs.

**24. The material in some AGN jets appears to be moving faster than the speed of light (superluminal motion). Why is this not possible, and what contributes to this illusion?

25. It is likely that most galaxies contain supermassive black holes, yet in many galaxies there is no obvious evidence for their existence. Why do some black holes reveal their presence while others do not?

APPLYING THE CONCEPTS

26. Assume that there are 1 trillion (10^{12}) galaxies in the universe, that the average galaxy has a mass equivalent to 100 billion (10^{11}) average stars, and that an average star has a mass of 10^{30} kilograms (kg).
    a. Ignoring dark matter, how much mass (in kilograms) does the universe contain?
    b. If the mass of an average particle of normal matter is 10^{-27} kg, how many particles are there in the entire universe?

27. Suppose the number density of galaxies in the universe is, on average, 3×10^{-68} galaxies per cubic meter (m^3). If astronomers could observe all galaxies out to a distance of 10^{10} light-years, how many galaxies would they find?

28. The Keplerian speed of a distant star orbiting a galaxy is twice the speed that the galaxy's visible mass would suggest. What is the ratio of dark matter to normal matter within the star's orbit?

29. The nearest known quasar is 3C 273. It is located in the constellation of Virgo and is bright enough to be seen in a medium-sized amateur telescope. With a redshift of 0.158, what is the distance of 3C 273 in light-years?

30. The quasar 3C 273 has a luminosity of 10^{12} $L_\odot$. Assuming that the total luminosity of a large galaxy, such as the Andromeda Galaxy, is 10 billion times that of the Sun, compare the luminosity of 3C 273 with that of the entire Andromeda Galaxy.

31. Consider a hypothetical star orbiting 3C 273 at a distance of 100,000 astronomical units (AU) with a period of 1,080 years. What is the mass of 3C 273 in solar masses?

32. Stars at a distance of 50 light-years from the center of a galaxy are orbiting the galactic center at a speed of 200 kilometers per second (km/s). Use Newton's laws to calculate the mass of the supermassive black hole at the galaxy's center.

33. A quasar has the same brightness as a foreground galaxy that happens to be 6 million light-years distant. If the quasar is 1 million times more luminous than the galaxy, what is the distance of the quasar?

34. You read in the newspaper that astronomers have discovered a "new" cosmological object that appears to be flickering with a period of 83 minutes. Having read *21st Century Astronomy*, you are able to quickly estimate the maximum size of this object. How large can it be?

35. A solar-type star ($M_\odot = 2 \times 10^{30}$ kg), accompanied by its retinue of planets, approaches a supermassive black hole. As it crosses the event horizon, half of its mass falls into the black hole, while the other half is completely converted to luminous energy. As it signals its demise in a burst of electromagnetic radiation, how much energy (in joules) does the dying solar system send out to the rest of the universe?

*36. A quasar has a luminosity of 10^{41} watts (W), or joules per second (J/s), and 10^8 $M_\odot$ to feed it. Assuming constant luminosity and 50 percent conversion efficiency, what would be your estimate of the quasar's lifetime?

37. Assume that Earth, whose mass ($M_\oplus$) is 5.97×10^{24} kg, fell into a supermassive black hole with a 50 percent energy conversion.

a. How much energy (in joules) would be radiated by the black hole?

b. Compare that with the energy radiated by the Sun each second: 3.85×10^{26} joules (J).

38. If a luminous quasar has a luminosity of 2×10^{41} W (or J/s), how many solar masses ($M_\odot = 2 \times 10^{30}$ kg) per year does this quasar consume to maintain its average energy output?

39. Material ejected from the supermassive black hole at the center of the galaxy M87 extends outward from the galaxy to a distance of approximately 100,000 light-years. M87 is approximately 50 million light-years away.

a. If this material were visible to the naked eye, how large would it appear in our nighttime sky? Give your answer in degrees, recalling from Chapter 13 that 1 radian = 57.3°.

b. Compare this with the angular size of the Moon.

40. A lobe in a visible jet from the galaxy M87 is observed at a distance of 5,000 light-years from the galaxy's center and moving outward at a speed of 0.99 times the speed of light (0.99c). Assuming constant speed, how long ago was the lobe expelled from the supermassive black hole at the galaxy's center?

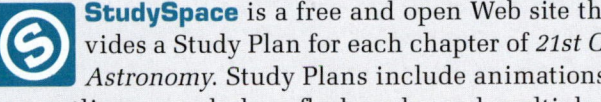

StudySpace is a free and open Web site that provides a Study Plan for each chapter of *21st Century Astronomy*. Study Plans include animations, reading outlines, vocabulary flashcards, and multiple-choice quizzes, plus links to premium content in SmartWork and the ebook. Visit **www.wwnorton.com/studyspace**.

O Milky Way, sister in whiteness
To Canaan's rivers and the bright
Bodies of lovers drowned,
Can we follow toilsomely
Your path to other nebulae?

GUILLAUME APOLLINAIRE (1880–1918)

The star fields and dark dust clouds of the Milky Way Galaxy, with a Lyrid meteor and ancient Bristlecone pines in the foreground.

CHAPTER 20

The Milky Way— A Normal Spiral Galaxy

20.1 We Look Up and See Our Galaxy

We live in a universe full of galaxies of many sizes and types, visible in our most powerful telescopes all the way to the edge of the observable universe. Yet when we go outside at night away from city lights and look up, it is not this universe of galaxies that we see. Rather, the night sky is filled with a single galaxy—our home, the galaxy we call the Milky Way. With what we know of galaxies from the previous chapter, we can learn a great deal about our local galaxy by looking at the night sky. **Figure 20.1a** shows what our Milky Way Galaxy looks like as seen in our own skies; **Figure 20.1b** is an image of an edge-on spiral galaxy. Comparison of the two can leave little doubt that we live in the disk of such a spiral. The flattened disk of the Milky Way is obvious when we look at the sky, once we appreciate what we are looking at. We can even see dark bands where clouds of interstellar gas and dust obscure much of the central plane of our galaxy. There are many things that we know about galaxies only from our experience with the Milky Way. We see it from a much closer perspective than we do any other galaxy. We see it from the inside!

At the same time, there are real disadvantages to our perspective on the Milky Way Galaxy. Buried within the disk of the Milky Way, we lack a bird's-eye view of our galactic home. The situation is further complicated by the dust in the surrounding interstellar medium, which also limits our view. **Figure 20.2a** shows a model of the structure

> We live in a barred spiral galaxy called the Milky Way.

KEY CONCEPTS

Of the hundreds of billions of galaxies in the universe, the one that means the most to us is our cosmic home, the Milky Way. The Milky Way may be just another galaxy, but it is the only galaxy we can study at close range. As we focus our attention on our galaxy, we will learn

- How variable stars in globular clusters are used as standard candles, enabling us to measure the size of the Milky Way.

- How Doppler-shifted radio emission from gas throughout the rotating disk of the Milky Way enables us to map the galaxy's structure.

- That the Milky Way is a typical giant barred spiral galaxy, and like all such galaxies it is composed mostly of dark matter.

- How the chemical composition of the Milky Way has evolved with time.

- What differences in the age and chemical composition of groups of stars tell us about the history of star formation in our galaxy.

- About the environment within the disk of the Milky Way, and the halo of stars, globular clusters, and dark matter that surrounds and permeates our galaxy.

- About the black hole at the center of our galaxy.

FIGURE 20.1 (a) We see the Milky Way as a luminous band stretching across our night sky. Note the prominent dark lanes caused by interstellar dust that obscures the light from more distant stars. (b) The edge-on spiral galaxy NGC 891, whose disk greatly resembles the Milky Way.

of the Milky Way proposed in the late 1700s by William Herschel. Bearing little resemblance to our modern understanding of the Milky Way, Herschel's model was based simply on counting the number and brightness of stars seen in different directions. Herschel did not know about the interstellar extinction of starlight, the properties of stars, or the relationship between our galaxy and the "spiral nebulae" that he studied throughout his life.

> Dust obscures our view of much of our own galaxy.

Early models of our galaxy are of interest today mostly as historical curiosities. Even so, today's astronomers must cope with the same suite of difficulties that stood in the way of 19th and early 20th century astronomers. What type of spiral galaxy do we live in? Are the spiral arms prominent? Is the bulge barlike, and is it large or small relative to the disk? The questions that are easier to ask and answer for distant galaxies are far more difficult to address for our own galaxy. Yet at this stage in our journey, we have already learned far more about our galaxy than about any other, as seen in **Figure 20.2b**. Stars, planets, and the interstellar medium—almost

(a)

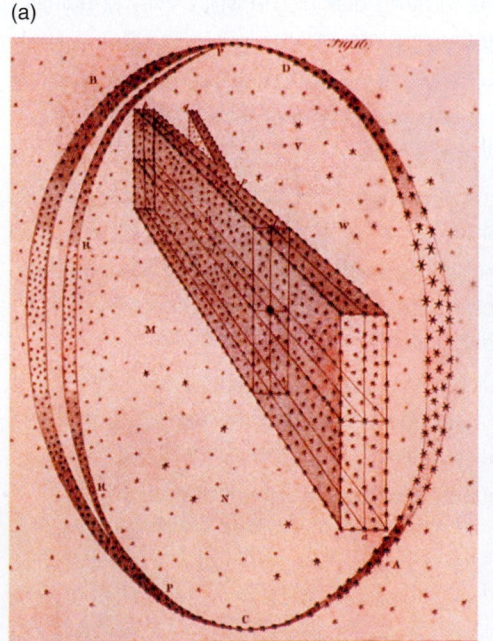

(b)

FIGURE 20.2 (a) An early model proposed by William Herschel in the late 1700s in which he depicts the Milky Way as a dense slab of stars. (b) Infrared and radio observations contribute to a modern model of the Milky Way Galaxy. The galaxy's two major arms (Scutum-Centaurus and Perseus) are seen attached to the ends of a thick central bar.

everything we know about them we have learned within the context of our own galactic home. It is time to merge the perspective of our look outward at the universe of other galaxies with our knowledge of our own locality to better understand our Milky Way as a spiral galaxy.

20.2 Measuring the Milky Way

As noted in earlier chapters, one of the more difficult practical issues faced by astronomers is determining the distances to objects in the sky. Such distance measurements are crucial to determining the sizes of objects. In Chapter 13 we learned how trigonometric parallax is used to measure the distances to nearby stars, but this method got us out to distances of only a few hundred light-years. In Chapter 18 we mentioned that Harlow Shapley determined the size of the Milky Way and we discussed the effect that his measurement had on astronomy. We also discussed Hubble's discovery of Cepheid variables in other galaxies, and then we learned how distances to galaxies are estimated through the use of standard candles. We turn now to how some of these standard candles were developed and how they have been used in the study of the Milky Way.

Globular Clusters and the Size of the Milky Way Galaxy

The key to determining the size of the Milky Way Galaxy turned out to be our finding a type of object, a standard candle, whose distance we can measure throughout our galaxy. Astronomers found that **globular clusters** were just what they needed. Recall from Chapter 17 that we observe many stars in large groups called clusters. A globular cluster, such as the one in **Figure 20.3**, is a large spheroidal group of stars held together by gravity. Many clusters can be seen through small telescopes. At first glance they look something like small elliptical galaxies, and the analogy is not a bad one. The motions of stars within a globular cluster are much like the motions of stars within an elliptical galaxy. However, globular clusters are quite different from elliptical galaxies in size and in concentration of stars.

There are more than 150 cataloged globular clusters in our galaxy (and there are likely many more—dust in the disk of our galaxy may hide them from view). The known globular clusters have luminosities ranging from a low of 400 solar luminosities (400 $L_\odot$) to a high of about 1 million $L_\odot$. A typical globular cluster consists of 500,000 stars packed into a volume of space with a radius of only 15 light-years. Let's put this density into

> Globular clusters are very luminous and easy to identify at great distances within the galaxy.

FIGURE 20.3 A Hubble Space Telescope image of the globular cluster M80.

perspective: we find only about 50 stars within that same distance from our own Sun. (You might imagine how our sky would appear if our Solar System were located at the center of a globular cluster, as portrayed so dramatically in **Excursions 20.1**.) Globular clusters are therefore much denser concentrations of stars than occur on average throughout our galaxy. Still, they contain only a small percentage of the Milky Way's stars; our galaxy has over 100,000 times as many stars as a typical globular cluster has.

About one-fourth of the globular clusters in the Milky Way reside in or near the disk of our galaxy. The rest of them occupy a large volume of space surrounding the disk and bulge, referred to as the **halo** of the Milky Way. Globular clusters offer two distinct advantages that make them relatively easy to study. First, they are very luminous and so can be easily seen at great distances. Second, because many globular clusters lie outside the disk of the Milky Way, we can see them at great distances without encountering much absorption due to the obscuring dust within the disk. However, simply being able to see globular clusters does not necessarily put us any closer to measuring their distances. To do that, we need to look at the properties of the stars they contain.

This is not the first time that we have been confronted with the problem of measuring the distance to remote objects. Recall that the inverse square law (discussed at length in Chapter 4) enables us to determine distance when (1) we know the luminosity of a star and (2) we can measure its brightness. Rearranging that law for our present purpose, we can calculate the star's distance:

$$d = \sqrt{\frac{L_{\text{star}}}{4\pi b_{\text{star}}}}.$$

Translated into words, this equation tells us that the distance of a star is proportional to the square root of the ratio of its luminosity (L_{star}) to its brightness (b_{star}) as seen from Earth.

Some types of stars are especially useful for determining distances. These are stars that are very luminous (so that they can be seen at great distances) and have *known* luminosities. Globular clusters, located in the Milky Way and other galaxies, contain good standard candles (Cepheid variables) and are also good standard candles themselves. **Figure 20.4** shows the H-R diagram for the stars in globular cluster M92. Recall from Chapter 17 that star clusters offer snapshots of stellar evolution. The main-sequence turnoff in this cluster's H-R diagram occurs for stars with

> We use stars of known luminosity as standard candles to measure distances.

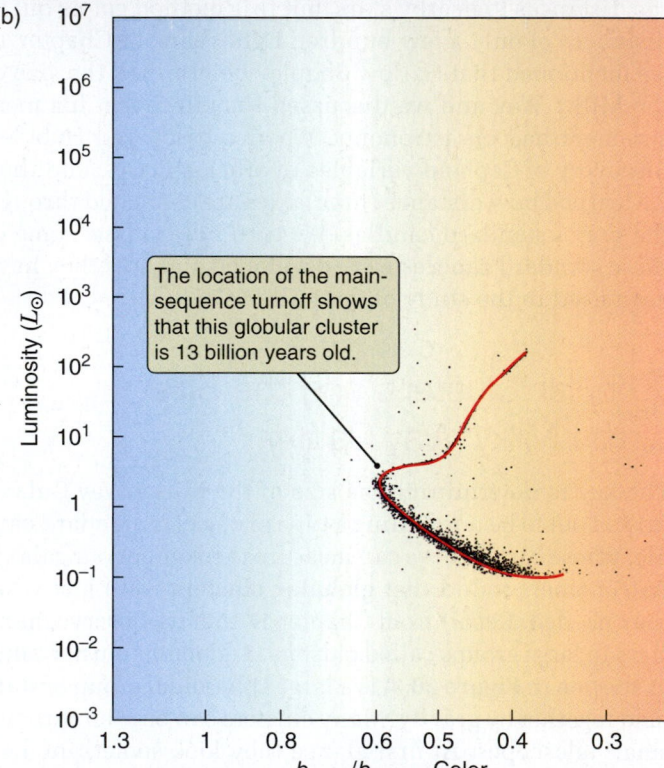

The location of the main-sequence turnoff shows that this globular cluster is 13 billion years old.

FIGURE 20.4 (a) The globular cluster M92 and (b) an H-R diagram of the stars it contains. (The color plotted on the diagram is based on the ratio of visible to infrared light.) The main-sequence turnoff at 0.8 solar mass (0.8 $M_\odot$) indicates that the cluster is about 13 billion years old.

EXCURSIONS 20.1

Nightfall

Isaac Asimov (1920–1992), the famous writer of science and science fiction, imagined what might happen to a civilization on a planet orbiting within a system of six stars located in the heart of a giant globular cluster. His short story "Nightfall" has become one of the more famous works of science fiction. On Asimov's fictional planet Lagash, at least one of its six stars is almost always above the horizon. "Nightfall" occurs on Lagash only once every 2,049 years. The story tells of the great madness that afflicts the inhabitants on this one night as they recoil in fear from a sky filled with hundreds of thousands of bright stars.

Asimov's story leads us to consider how our perspectives as human beings are formed by the circumstances in which we live. Yet we need not go to science fiction to ask these kinds of questions. Our perspective on the universe changes with time. At the moment, we are traversing an open, rather dust-free part of the Milky Way. On a moonless night we see a dark sky and gaze with our telescopes into a universe of galaxies, but that will not always be the case. Just "down the road," astronomically speaking, are dark interstellar clouds and star-forming regions filled with glowing gas and obscuring dust through which the Sun and its entourage of planets will occasionally pass. How different would our view of the universe be if, instead of a dark sky, we looked up each night and saw a sky filled with a soft green glow, punctuated by a few points of intense light? How much different would our history be if, at some point during the rise of our civilization, we had suddenly emerged over the course of just a few years from within a molecular cloud and gotten our first look at the larger universe?

masses of about 0.8 $M_\odot$, which corresponds to a main-sequence lifetime of close to 13 billion years. The age of this globular cluster is similar to those of many others, making them the oldest objects known in our galaxy or in any nearby galaxy. Globular clusters must have formed when the universe and our galaxy were very young. Compared to globular-cluster stars, our Sun, at about 5 billion years old, is a relatively young member of our galaxy.

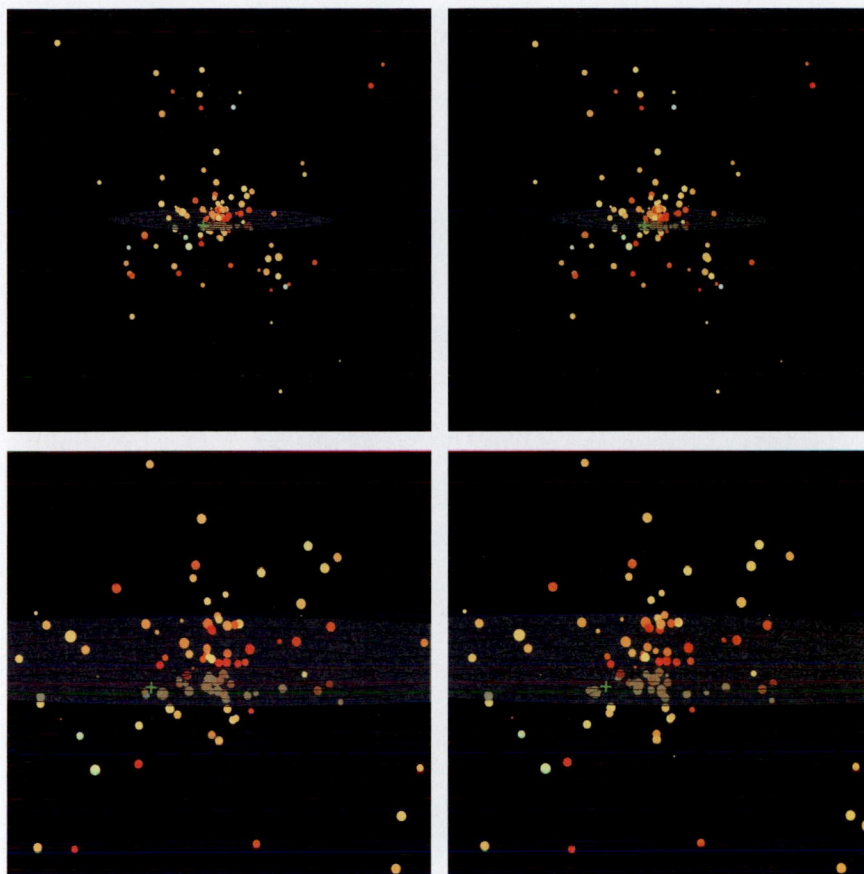

FIGURE 20.5 Two stereoscopic views of the distribution of globular clusters in the Milky Way Galaxy. Colors represent the average colors of stars in each cluster. The green cross shows the location of the Sun. The light gray area represents the plane of the galaxy. (See Figure 13.1 for viewing instructions.)

a roughly spherical region of space with a diameter of about 300,000 light-years! These globular clusters trace out the halo of the Milky Way Galaxy, as shown in **Figure 20.5**, which reflects the modern view of the globular-cluster distribution.

> Harlow Shapley used globular clusters to determine the size of our galaxy.

The globular clusters around the Milky Way are moving about under the gravitational influence of the galaxy just as stars in an elliptical galaxy move about under the gravitational influence of that kind of galaxy. Symmetry therefore requires that the center of the distribution of globular clusters coincide with the gravitational center of the galaxy itself. Shapley realized that, because he could determine the distance to the center of this distribution, he had actually determined the Sun's distance from the center of the Milky Way, as well as the size of the galaxy itself.

Figure 20.6 identifies the disk, bulge, and halo of the Milky Way. Modern determinations indicate that the Sun is located about 27,000 light-years from the center of the galaxy, or roughly halfway out toward the edge of the disk. Armed with strong evidence, in 1915 Shapley confidently presented the then-new view of our greatly expanded galaxy. It took all of human history up to 1610 to go from an Earth-centered universe to a Sun-centered Solar System. It took 305 years to go from a small galaxy to a large galaxy. Yet it was barely a decade later that Edwin Hubble, using Cepheid and RR Lyrae stars and the period-luminosity relationship, proved that Shapley's enormous Milky Way is but one of billions of galaxies in the universe.

In Chapter 17 we also found that there is a region in the H-R diagram, called the *instability strip*, in which stars pulsate. As they do so, their luminosity changes. In an old cluster like a globular cluster, the horizontal branch of the H-R diagram crosses the instability strip. Recall that horizontal branch stars that lie in the instability strip are variable stars called **RR Lyrae variables**. RR Lyrae stars in globular clusters are easy to spot because they are very luminous (horizontal branch stars are giant stars) and because their periodic changes in brightness signal their identity. As with Cepheid variables, the time it takes for an RR Lyrae star to undergo one pulsation is related to the star's luminosity. Harlow Shapley used Henrietta Leavitt's determination of this period-luminosity relationship to calculate the luminosities of RR Lyrae stars in globular clusters. He then used the inverse square law of radiation to combine these luminosities with measured brightnesses to determine the distances to globular clusters. Finally, Shapley cross-checked his results by noting that more distant clusters (as measured with his standard candle) also tended to appear smaller in the sky, as expected.

> Globular clusters contain RR Lyrae standard candles.

Knowing both the distances to globular clusters and where they appear in the sky, Shapley was able to make a three-dimensional map of the distribution of globular clusters in space. This map showed that globular clusters occupy

We Use 21-cm Radiation to Measure Rotation of the Galaxy

There is much we would like to know about the Milky Way apart from its size and the fact that it is a spiral galaxy. But as we've noted many times already, there is a problem: because we live inside the dusty disk of the galaxy itself, our visible-light view of the Milky Way is badly obscured. If you go out on a dark night, away from any streetlights, and look in the

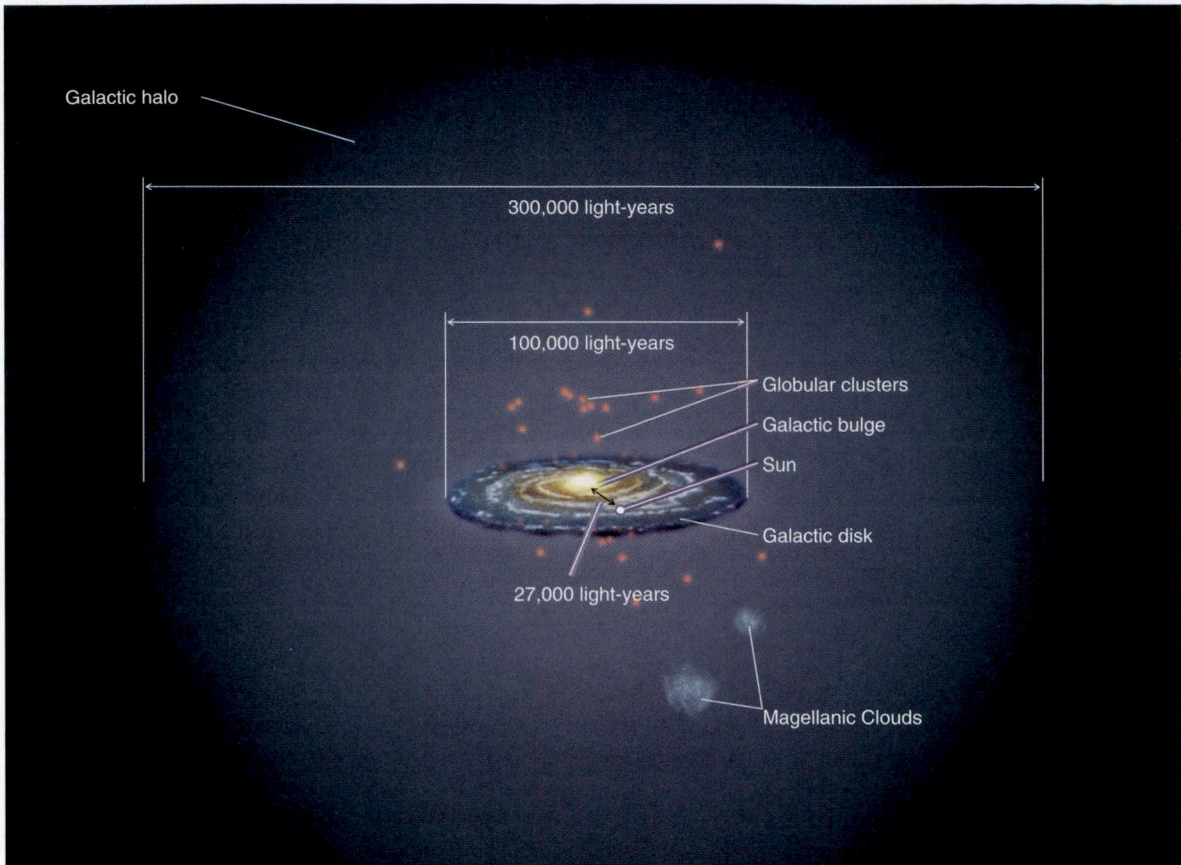

FIGURE 20.6
This diagram of the disk, bulge, and halo of the Milky Way Galaxy also shows the Magellanic Clouds and the location of the Sun within the Milky Way disk.

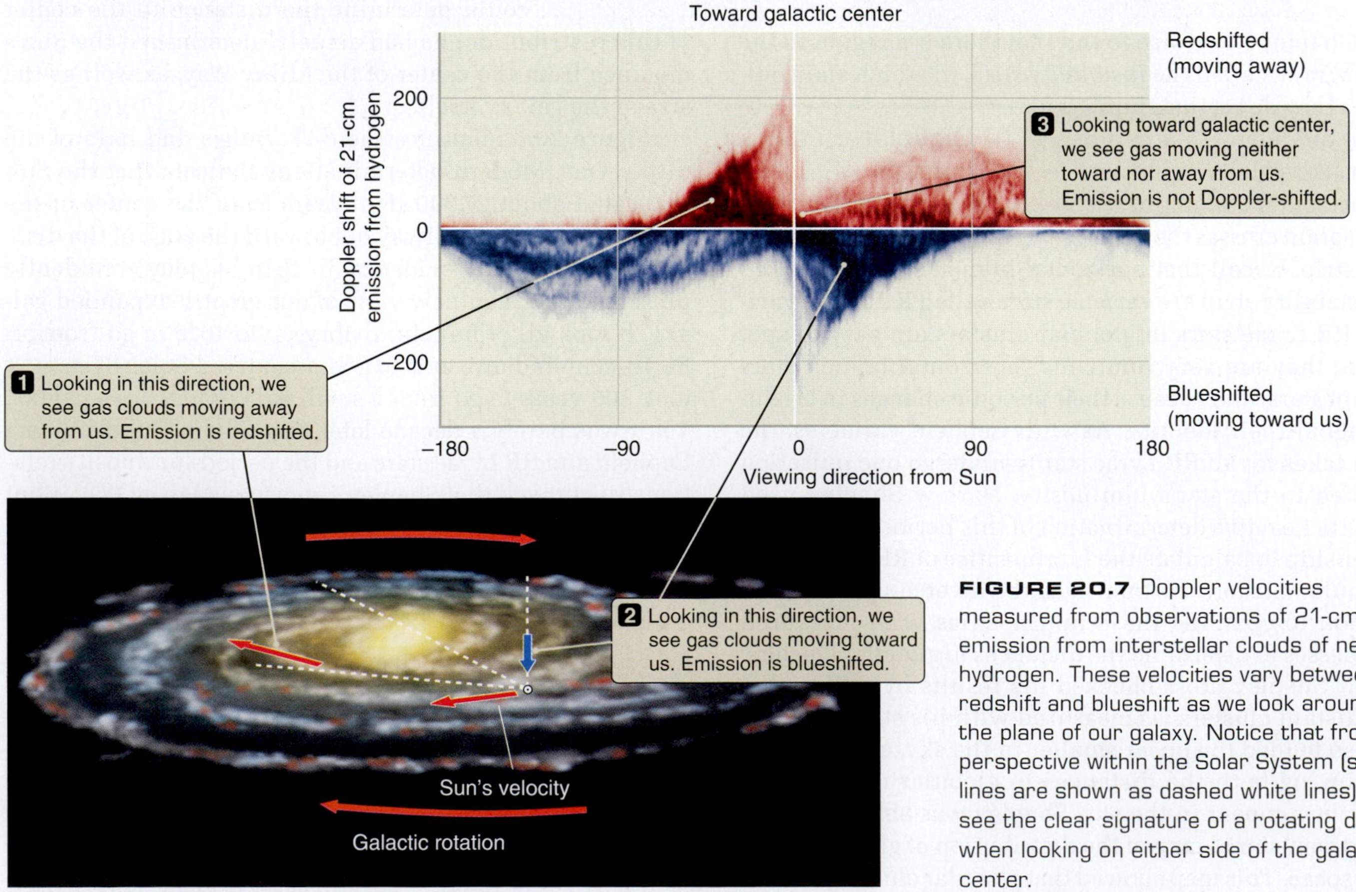

Galactic halo

300,000 light-years

100,000 light-years

Globular clusters

Galactic bulge

Sun

Galactic disk

27,000 light-years

Magellanic Clouds

Toward galactic center

Redshifted (moving away)

Doppler shift of 21-cm emission from hydrogen

200

0

−200

3 Looking toward galactic center, we see gas moving neither toward nor away from us. Emission is not Doppler-shifted.

1 Looking in this direction, we see gas clouds moving away from us. Emission is redshifted.

Blueshifted (moving toward us)

−180   −90   0   90   180

Viewing direction from Sun

2 Looking in this direction, we see gas clouds moving toward us. Emission is blueshifted.

FIGURE 20.7 Doppler velocities measured from observations of 21-cm emission from interstellar clouds of neutral hydrogen. These velocities vary between redshift and blueshift as we look around in the plane of our galaxy. Notice that from our perspective within the Solar System (sight lines are shown as dashed white lines), we see the clear signature of a rotating disk when looking on either side of the galactic center.

Sun's velocity

Galactic rotation

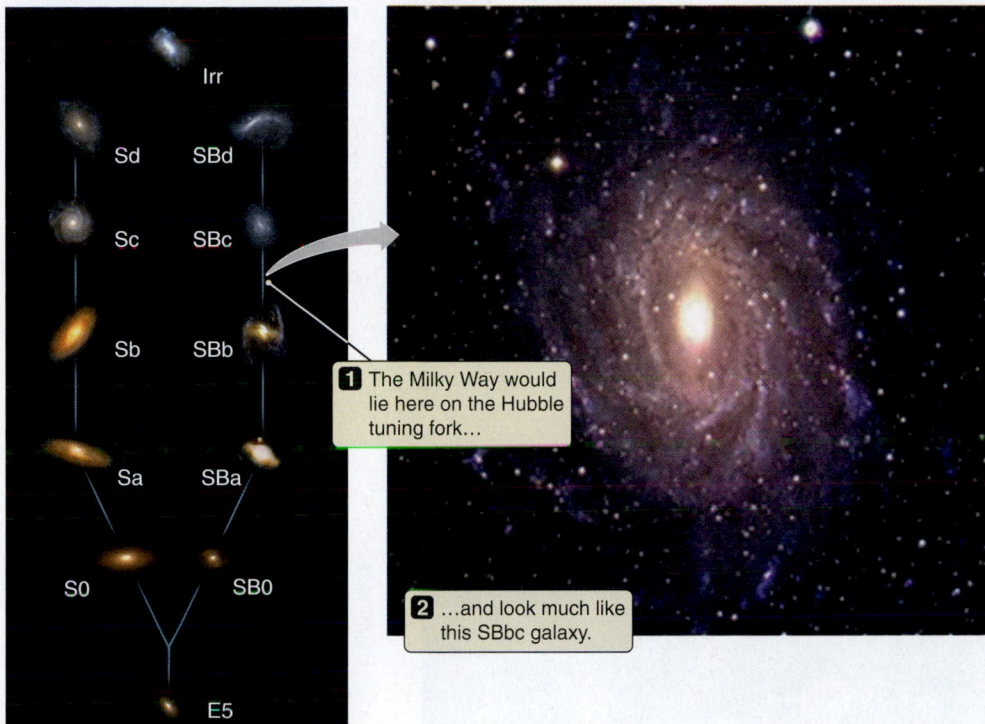

1 The Milky Way would lie here on the Hubble tuning fork…

2 …and look much like this SBbc galaxy.

direction of the center of our galaxy (located in the constellation Sagittarius), instead of a bright spot you will see a dark lane of dusty clouds, as shown in Figure 20.1a. To probe the structure of our galaxy, we must therefore use long-wavelength infrared and radio radiation that can penetrate the disk without being affected much by dust. The most powerful tool for this work is the same 21-centimeter (cm) line from neutral interstellar hydrogen that we used in the previous chapter to measure the rotation of other galaxies.

The velocities of interstellar hydrogen measured from 21-cm radiation are plotted in **Figure 20.7** as a function of the direction in which we are looking. Looking in the region around the center of the galaxy, we see that on one side hydrogen clouds are moving toward us while on the other side clouds are moving away from us. This is a pattern that we have seen before—it is the pattern of the rotation velocity of gas in a disk (see Figures 19.15 and 19.16). The only difference is that instead of looking at it from outside, we see our own galaxy's rotation curve from a vantage point located within—and rotating with—the galaxy. In other directions, the velocities we see are complicated by our moving vantage point within the disk and so

> Doppler velocities measured from 21-cm radiation show that we live inside a rotating galaxy.

are more difficult to interpret at a glance. Even so, observed velocities of neutral hydrogen enable us to measure our galaxy's rotation curve and even determine the structure present throughout the disk of our galaxy.

Figure 20.2b shows an artist's reconstruction of the Milky Way based on Spitzer Space Telescope infrared observations and 21-cm radio data. Our galaxy has a substantial bar with a modest bulge at its center, as indicated by infrared studies of the distribution and motions of stars toward the center of our galaxy. Two major spiral arms—Scutum-Centaurus and Perseus—connect to the ends of the central bar and sweep through the galaxy's disk, just like the arms we saw in external spiral galaxies. Putting all the available information together, we conclude that the Milky Way is a middle-of-the-road giant barred spiral. From outside, our galaxy probably looks much like the galaxy shown in **Figure 20.8**, and we would place the Milky Way about halfway along the right tine of the Hubble tuning fork diagram (see Figure 19.4). In fact, we classify the Milky Way as an SBbc galaxy.

The Milky Way Is Mostly Dark Matter

As we saw in the previous chapter, observations of rotation curves have led us to conclude that the masses of spiral galaxies consist mostly of dark matter. We can use the rotation of our own galaxy to determine whether the Milky Way is also dominated by dark matter. **Figure 20.9** shows the rotation curve of the Milky Way as inferred primarily from 21-cm observations. The orbital motion of the nearby dwarf galaxy called the Large Magellanic Cloud provides data for the outermost point in the rotation curve, at a dis-

> Our galaxy has a flat rotation curve.

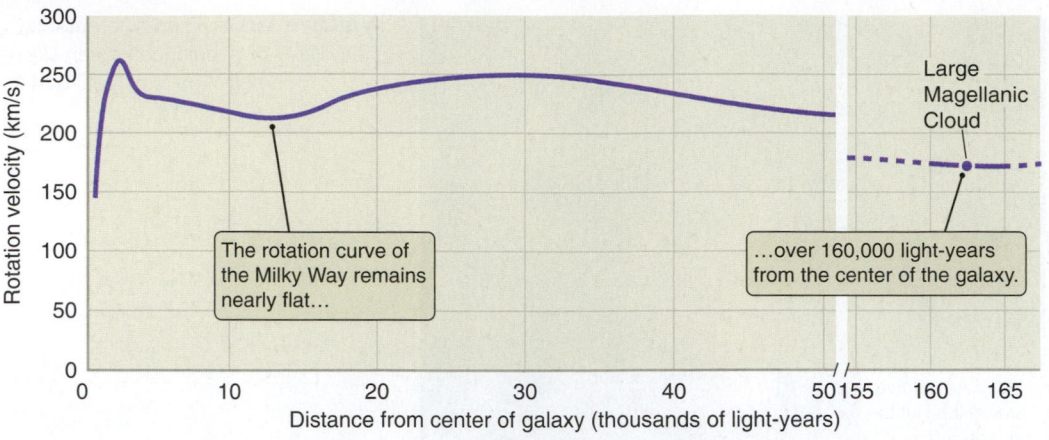

FIGURE 20.9 A plot showing rotation velocity versus distance from the center of the Milky Way. The most distant point comes from measurements of the orbit of the Large Magellanic Cloud. The nearly flat rotation curve indicates that dark matter dominates the outer parts of our galaxy.

The rotation curve of the Milky Way remains nearly flat…

…over 160,000 light-years from the center of the galaxy.

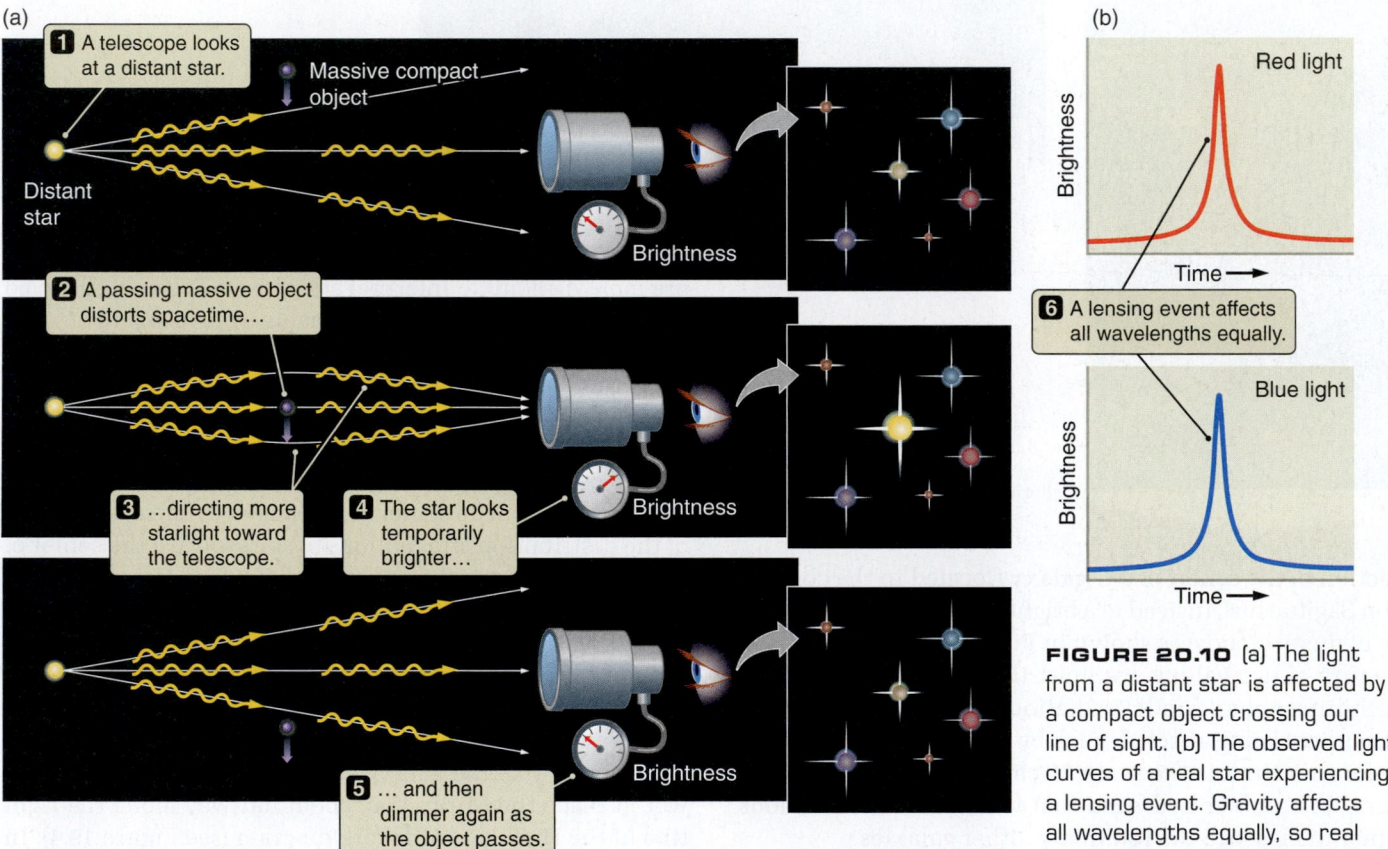

(a)
1 A telescope looks at a distant star.
Massive compact object
Distant star
Brightness

2 A passing massive object distorts spacetime…

3 …directing more starlight toward the telescope.
4 The star looks temporarily brighter…
Brightness

5 … and then dimmer again as the object passes.
Brightness

(b)
Red light
Brightness
Time →

6 A lensing event affects all wavelengths equally.

Blue light
Brightness
Time →

FIGURE 20.10 (a) The light from a distant star is affected by a compact object crossing our line of sight. (b) The observed light curves of a real star experiencing a lensing event. Gravity affects all wavelengths equally, so real "lensing events" should look the same in all colors.

tance of roughly 160,000 light-years from the center of the galaxy. As we have come to expect for spiral galaxies, the Milky Way has a fairly flat rotation curve.

By applying the same techniques laid out in the previous chapter to the rotation curve of the Milky Way, we infer that the galaxy's mass must be about 6×10^{11} to 10×10^{11} $M_\odot$. However, if we instead estimate the mass of the Milky Way by measuring the light from its stars and knowing how much stellar mass is needed to produce that much light (using infrared to see through the dust), we find a much lower value. As in other spiral galaxies, we can therefore infer that the Milky Way is mostly dark matter. The spatial distribution of dark and normal matter within the Milky Way is also much like what we have seen in other galaxies. Visible matter dominates the inner part of our galaxy. Dark matter dominates its outer parts, at least to a distance of 150,000 light-years from the galaxy's center.

What is dark matter? Is it exotic, or could it perhaps consist of very dim, compact objects of the kind we have discussed so far? There is a very clever way to search for dark matter within the halo of our own galaxy, when the dark matter consists of compact objects such as low-mass stars (for example, small main-sequence M stars), planets, white dwarfs, neutron stars, or black holes. We refer to such dark matter candidates as **MACHOs**, which stands for "massive compact halo objects." If the dark matter in our halo consists of MACHOs, there must be a lot of these objects, and they must each exert gravitational force but not emit much light.

How might we detect MACHOs? Because of their gravity, MACHOs can gravitationally deflect light according to Einstein's general theory of relativity. If we were observing a distant star and a MACHO passed between us and the star, the star's light would be deflected and perhaps focused by the intervening MACHO as it passed across our line of sight, as illustrated in **Figure 20.10a**. Because gravity affects all wavelengths equally, such "lensing events" should look the same in all colors—a fact that rules out other causes of variability.

We would be remarkably lucky if such an event occurred just as we were observing a single distant star. However, astronomers monitored the stars in the Large and Small Magellanic Clouds (two of the small companion galaxies of the Milky Way Galaxy), observing tens of millions of stars for several years. They found a number of examples of events of the sort shown in **Figure 20.10b**, some believed to be small main-sequence M stars, but not nearly enough to account for the amount of dark matter in the halo of our galaxy. Thus, it was concluded that the dark matter in our galaxy is probably *not* composed primarily of MACHOs. (MACHOs are not to be confused with another dark matter candidate called "weakly interacting massive particles," or **WIMPs**. Who says astronomers have no sense of humor?)

20.3 Studying the Milky Way Galaxy Up Close and Personal

Earlier in the chapter we concentrated on the disadvantages of our perspective that make it hard to see the Milky Way as a galaxy. However, our perspective also has advantages. For example, we can study the stellar content of the Milky Way at very close range, star by star, looking at subtle aspects of the populations of stars that give us direct clues about how spiral galaxies like ours form. For example, stellar orbits determine the shapes of the different parts of our galaxy, and it is much easier to measure stellar orbits in our own galaxy than it would be to measure them in other galaxies. The stars in the Milky Way's disk rotate about the center of the galaxy, just like the gas and dust in the disk. The stars in the halo move in orbits similar to those of stars in elliptical galaxies. The bar in the bulge of our galaxy is shaped primarily by stars and gas moving both in highly elongated orbits up and down the long axis of the bar and in short orbits aligned perpendicular to the bar.

Our Sun is a middle-aged disk star located among other middle-aged stars that orbit around the galaxy within the disk. Yet near the Sun are other stars, usually much older, that are a part of the galactic halo and whose orbits are carrying them

through the disk. Using the ages, chemical abundances, and motions of nearby stars, we can differentiate between disk and halo stars to learn more about the galaxy's structure.

We See Stars of Different Age and Different Chemical Composition

Stellar ages and their chemical abundances provide the most fundamental categories into which populations of stars can be grouped. Conveniently, some stars come prepackaged into distinct groups that split up along just these two lines. Recall from Chapter 17 that there are two dif-

FIGURE 20.11 (a) The open star cluster NGC 6530, located 5,200 light-years away, in the disk of our galaxy. (b) The H-R diagram of stars in this cluster shows that it has an age of a few million years or less.

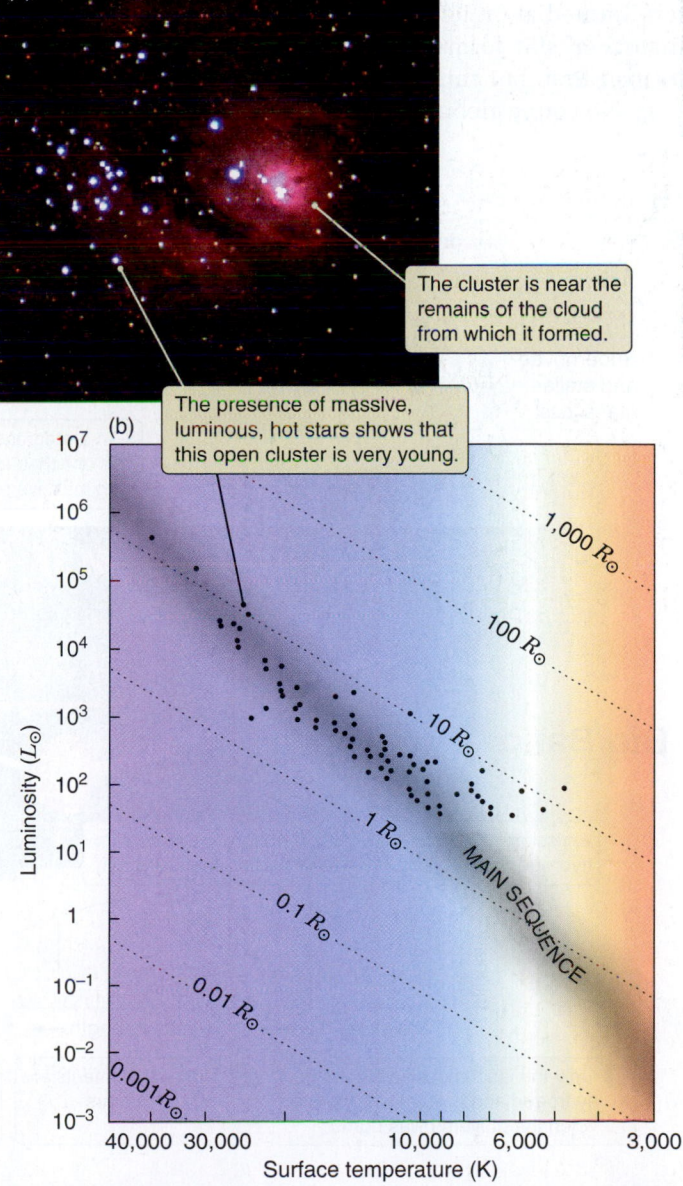

(a)

The cluster is near the remains of the cloud from which it formed.

(b)

The presence of massive, luminous, hot stars shows that this open cluster is very young.

ferent varieties of star clusters in the Milky Way. Globular clusters—the densely packed collections of millions of stars studied by Shapley—are mostly halo objects. With ages of up to 13 billion years, they are among the oldest objects known. In contrast, **open clusters**, like the one shown in **Figure 20.11a**, are much less tightly bound collections of a few dozen to a few thousand stars that are found orbiting in the disk of our galaxy. As with globular clusters, the stars in an open cluster all formed in the same region at about the same time. When we study the H-R diagrams of open clusters (**Figure 20.11b**), we find a wide range of ages. Some open clusters contain the very youngest stars known. Other open clusters contain stars that are somewhat older than the Sun. There is no overlap in age between open clusters and globular clusters, however. Even the youngest globular clusters are several billion years older than the oldest open clusters. Open clusters do not survive long in the disk of our galaxy, because they are loosely bound together and easily disrupted by the gravitational tug from nearby objects.

The differences in ages between globular and open clusters immediately tell us something interesting about the history of star formation in our galaxy. Stars in the halo formed first, but this epoch of star formation did not last long. No young globular clusters are seen. In the disk of the galaxy, star formation seems not to have gotten started until later, but it has been continuing ever since. The process that formed the stars in the massive, compact globular clusters must have also been much different from the more sedate process responsible for creating stars in the less massive, more scattered open clusters.

It is obvious why we would be so interested in grouping stars according to their ages, but why would we focus on the chemical composition of stars? We have discussed the chemical evolution of the universe on a number of occasions. When the universe was very young, only the least massive of elements existed. All elements more massive than boron must have formed by nucleosynthesis in stars. For this reason, the abundance of massive elements in the interstellar medium provides a record of the cumulative amount of star formation that has taken place up to the present time. Gas that shows large abundances of massive elements must have gone through a great deal of stellar processing (see Chapters 13 and 17), whereas gas with low abundances of massive elements is more pristine.

In turn, the abundance of massive elements in the atmosphere of a star provides a snapshot of the chemical composition of the interstellar medium *at the time the star formed*. (In main-sequence stars, material from the core

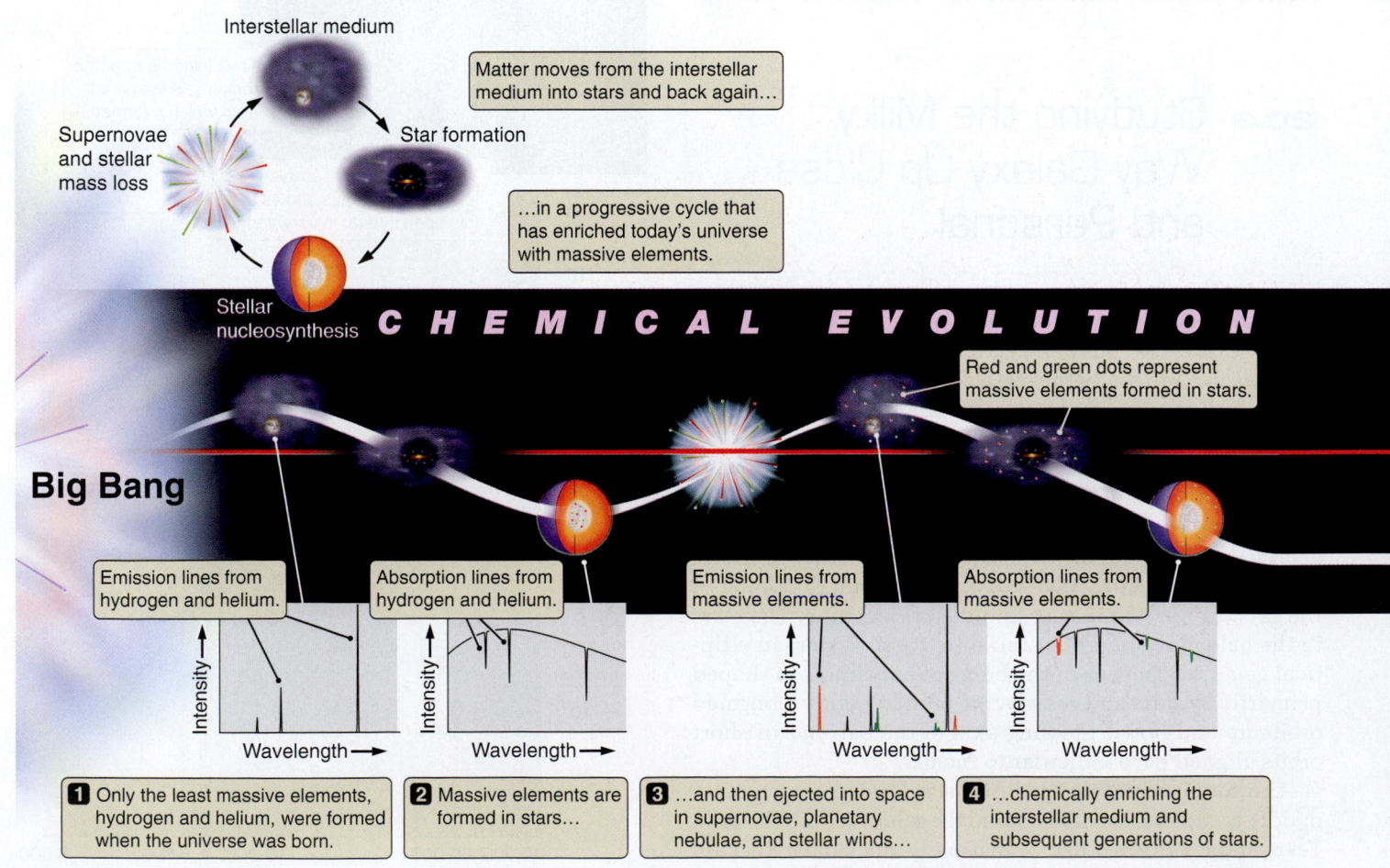

Interstellar medium

Matter moves from the interstellar medium into stars and back again…

Supernovae and stellar mass loss

Star formation

…in a progressive cycle that has enriched today's universe with massive elements.

Stellar nucleosynthesis

C H E M I C A L   E V O L U T I O N

Red and green dots represent massive elements formed in stars.

Big Bang

Emission lines from hydrogen and helium.

Absorption lines from hydrogen and helium.

Emission lines from massive elements.

Absorption lines from massive elements.

Intensity → / Wavelength →

1 Only the least massive elements, hydrogen and helium, were formed when the universe was born.

2 Massive elements are formed in stars…

3 …and then ejected into space in supernovae, planetary nebulae, and stellar winds…

4 …chemically enriching the interstellar medium and subsequent generations of stars.

does not mix with material in the atmosphere, so the abundances of chemical elements inferred from the spectra of a star are the same as the abundances in the interstellar gas from which the star formed.) As illustrated in **Figure 20.12**, the chemical composition of a star's atmosphere reflects the cumulative amount of star formation that has occurred up to that moment.

> Abundances of massive elements record the cumulative history of star formation.

If our ideas about the chemical evolution of the universe are correct, we expect to see large differences in massive-element abundances between globular and open clusters. Stars in globular clusters, being among the earliest stars to form, should contain only very small amounts of massive elements. And that is exactly what we see. Some globular-cluster stars contain only 0.5 percent as much of these massive elements as our Sun has. This relationship between age and abundances of massive elements is evident throughout much of the galaxy. The chemical evolution of the Milky Way has continued within the disk as generation

> Younger stars typically have higher massive-element abundances than older stars have.

after generation of disk stars have further enriched the interstellar medium with the products of their nucleosynthesis. Within the disk, younger stars typically have higher abundances of massive elements than do older stars. Similarly, older stars in the outer parts of our galaxy's bulge have lower massive-element abundances than do young stars in the disk. Such lower abundances of heavy elements characterize not only globular-cluster stars, but all of the stars in our galaxy's halo, where globular-cluster stars comprise only a minority among the total number of stars in the galactic halo.

Within the galaxy's disk, we can even see differences in abundances of massive elements from place to place that are related to the rate of star formation in different regions. Star formation is generally more active in the inner part of the Milky Way than in the outer parts. This higher level of activity is a result of the denser concentrations of gas that are found in the inner galaxy. If such activity has continued throughout the history of our galaxy, we might predict massive elements to be more abundant in the inner parts of our galaxy than in the outer parts. Observations of chemical abundances in the interstellar medium, based both on interstellar absorption lines in the spectra of stars and on emission lines in glowing H II regions, confirm this predic-

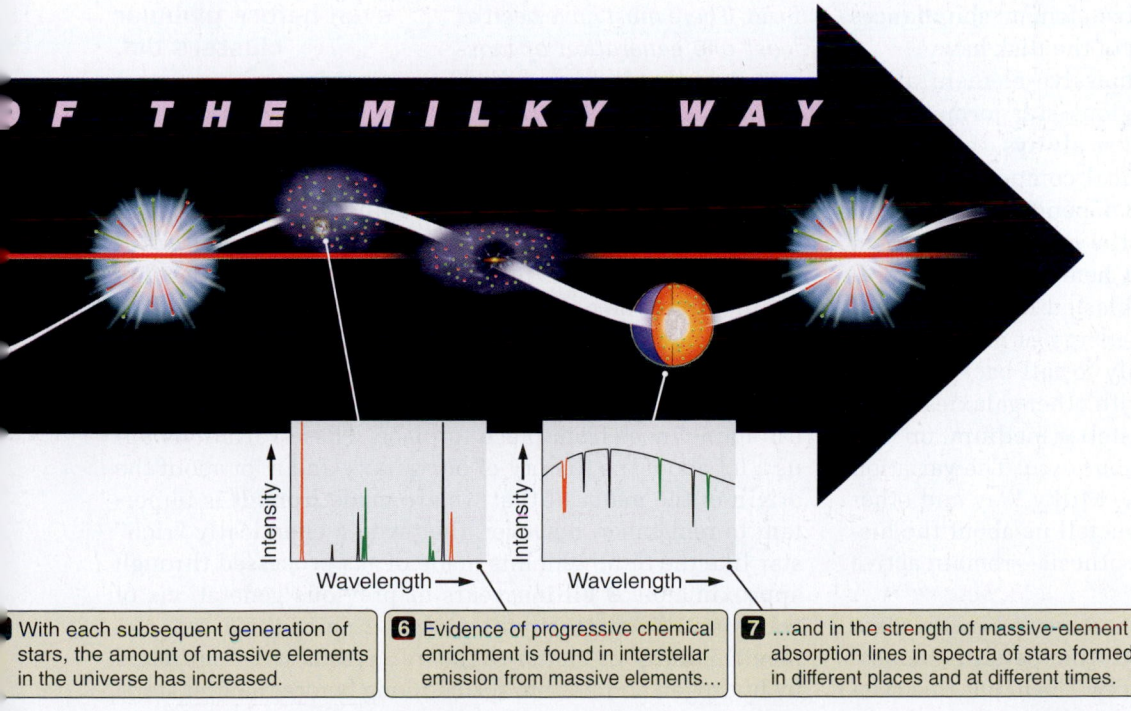

FIGURE 20.12 As subsequent generations of stars form, live, and die, they enrich the interstellar medium with massive elements—the products of stellar nucleosynthesis. The chemical evolution of our galaxy and other galaxies can be traced in many ways, including by the strength of interstellar emission lines and stellar absorption lines.

With each subsequent generation of stars, the amount of massive elements in the universe has increased.

6 Evidence of progressive chemical enrichment is found in interstellar emission from massive elements…

7 …and in the strength of massive-element absorption lines in spectra of stars formed in different places and at different times.

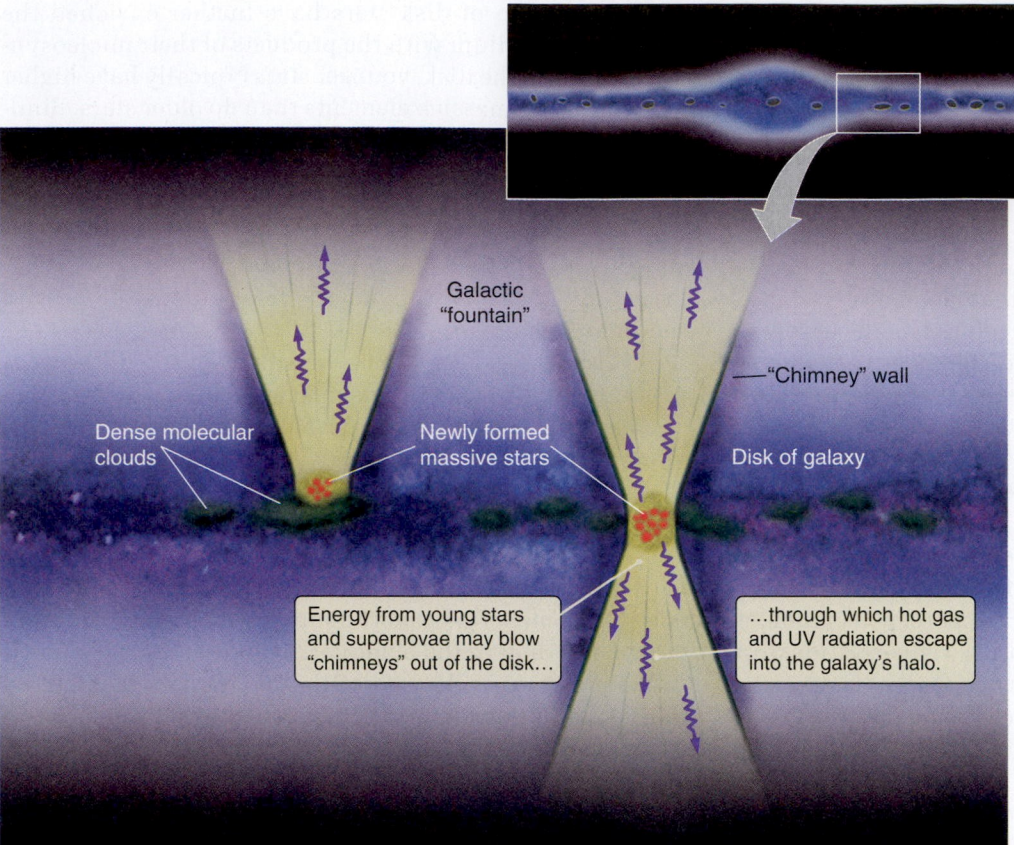

FIGURE 20.13 In the "galactic fountain" model of the disk of a spiral galaxy, gas is pushed away from the plane of the galaxy by energy released by young stars and supernovae and then falls back onto the disk.

Galactic "fountain"

"Chimney" wall

Dense molecular clouds

Newly formed massive stars

Disk of galaxy

Energy from young stars and supernovae may blow "chimneys" out of the disk…

…through which hot gas and UV radiation escape into the galaxy's halo.

tion. As expected, there is a smooth decline in abundances of massive elements from the inner to the outer parts of the disk. Astronomers have documented similar trends in other galaxies. These trends can be seen with stars as well. Within a galactic disk, relatively old stars near the center of a galaxy often have greater massive-element abundances than young stars in the outer parts of the disk have.

Our basic idea about higher massive-element abundances following the more prodigious star formation in the inner galaxy seems correct; but as always, the full picture is not this simple. The chemical composition of the interstellar medium at any location depends on a wealth of factors. New material falling into the galaxy might affect interstellar chemical abundances. Chemical elements produced in the inner disk might be blasted into the halo in great "fountains" powered by the energy of massive stars (as illustrated in **Figure 20.13**), only to fall back onto the disk elsewhere. Past interactions with other galaxies might have stirred the Milky Way's interstellar medium, mixing gas from those other galaxies with our own. The variation of chemical abundances within the Milky Way and other galaxies—and what these variations tell us about the history of star formation and nucleosynthesis—remain active topics of research.

Although the details are complex, there are several clear and important lessons to be learned from patterns in massive-element abundances in the galaxy. The first is that even the very oldest globular-cluster stars contain *some* amount of massive chemical elements. The implication is clear: globular-cluster stars and other halo stars were not the first stars in our galaxy to form. *There must have been at least one generation of massive stars that lived and died, ejecting newly synthesized massive elements into space, before even the oldest globular clusters formed.* Further, every star less massive than about 0.8 $M_\odot$ that ever formed is still around today. Even so, we find *no* disk stars with exceptionally low massive-element abundances. We would have found these stars by now if they existed. The gas that wound up in the plane of the Milky Way must have seen a significant amount of star formation *before* it settled into the disk of the galaxy and made stars.

> Generations of stars must have formed even before globular clusters did.

We have placed great emphasis on variations in chemical abundances from place to place. These variations tell us a lot about the history of our galaxy and a lot about the origin of the material that we are made from. It is important to remember, however, that even a chemically "rich" star like the Sun, which is made of gas processed through approximately 9 billion years of previous generations of stars, is still composed of less than 2 percent massive elements. Luminous matter in the universe is still dominated by hydrogen and helium formed long before the first stars.

Looking at a Cross Section through the Disk

The youngest stars in our galaxy are most strongly concentrated in the galactic plane, defining a disk about 1,000 light-years thick (but over 100,000 light-years across—thin indeed). The older population of disk stars, distinguishable by lower abundances of massive elements, has a much "thicker" distribution, about 12,000 light-years thick. **Figure 20.14** illustrates how the population of stars changes with distance from the galactic plane. Not too surprisingly, the youngest stars are concentrated closest to the plane of the galaxy for the simple reason that this is where the molecular clouds are. Older stars make up the thicker parts of the disk. There are two hypotheses for the origin of this thicker disk. One suggests that these stars formed in the midplane of the disk long ago but have since been kicked up out of the plane of the galaxy, primarily by gravitational interactions with massive molecular clouds (see Figure 20.14). The other hypothesis suggests that these stars were acquired from the merging process that formed our Milky Way Galaxy.

When we discussed the formation of accretion disks in Chapters 6 and 15, we found that a rotating cloud of gas naturally collapses into a thin disk as a consequence of gas falling from one direction (above the disk) running into gas falling from the other direction (below the disk.) Clouds of gas cannot pass through each other, so the colliding gas clouds instead settle into a disk. The same process applies to clouds of gas that are pulled by gravity toward the midplane of the disk of a spiral galaxy. Although stars are free to pass back and forth from one side of the disk to the other, cold, dense clouds of interstellar gas settle down into the central plane of the disk. These clouds are seen as the concentrated dust lanes that slice the disks of spiral galaxies (as shown in Figure 20.1), like a layer of bologna in a very thin, flat sandwich. This thin lane slicing through the midplane of the disk is the place both where new stars form and where new stars are found.

> **Gas tends to concentrate into thin sheets.**

The interstellar medium is a dynamic place—energy from star-forming regions can shape it into impressively large structures. We mentioned earlier that energy from regions of star formation can impose interesting structure on the interstellar medium, clearing out large regions of gas in the disk of a galaxy. Many massive stars forming in the same region can blow "chimneys" out through the disk of the galaxy via a combination of supernova explosions and strong stellar winds. If enough massive stars are formed together, sufficient energy may be deposited to blast holes all the way through the plane of the galaxy. In the process, dense interstellar gas can be thrown high above the plane of the galaxy (see Figure 20.13). Maps of the 21-cm emission from neutral hydrogen in our galaxy and visible-light images of hydrogen emission from some edge-on external galaxies show a wealth of vertical structures in the interstellar medium of disk galaxies. These vertical structures are often interpreted as the "walls" of chimneys.

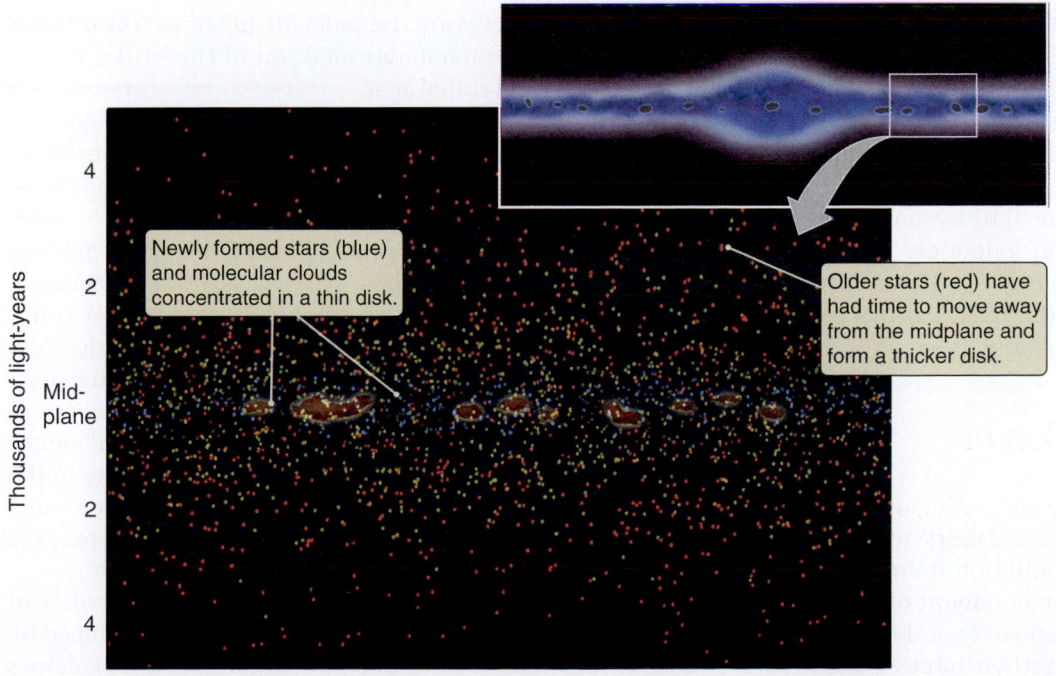

FIGURE 20.14 A vertical profile of the disk of the Milky Way Galaxy. Gas and young stars are concentrated in a thin layer in the center of the disk. Older populations of stars make up the thicker portions of the disk.

Thousands of light-years

4

2

Mid-plane

2

4

Newly formed stars (blue) and molecular clouds concentrated in a thin disk.

Older stars (red) have had time to move away from the midplane and form a thicker disk.

The Halo Is More Than Globular Clusters

Earlier, when discussing the halo of our galaxy, we concentrated our attention on its globular clusters. These are very important to astronomy, in part because they tell us a great deal about the history of star formation in the halo. Yet globular clusters account for only about 1 percent of the total mass of stars in the halo. As halo stars fall through the disk of the Milky Way, some pass close to the Sun, giving us a sample of the halo that we can study at close range.

Because most of the stars near our Sun are, like the Sun, disk stars, you might be wondering how we can tell them apart from nearby halo stars. They are distinguishable in two ways. First, most halo stars have much lower abundances of massive elements than disk stars have. Second, halo stars appear to be whizzing by us at high velocities. Actually, it is often not the halo stars that are moving fast, but we who are moving fast relative to them. The reason is that halo stars do not rotate about the center of the galaxy in the same way that disk stars do, but rather are in the same kind of odd-shaped orbits as stars in elliptical galaxies. In contrast, the disk stars near the Sun are moving, mostly together, in a 220-kilometer-per-second (km/s) rotation about the center of our galaxy. Just as it is easy to tell the difference between a person sitting beside you on a bus and the people who (in your frame of reference) are whizzing by outside the window, it is easy to tell the difference between disk stars that share our motion and halo stars that, to us, are just passing through.

Reflecting once again the bias of the observer, astronomers call the halo stars **high-velocity stars**, even though it is usually the disk stars that are moving faster relative to the galaxy as a whole! Enough halo stars are near us that we can measure their distances and map out their orbits, presenting a very detailed look at the kinds of orbits that stars in a halo (or in an elliptical galaxy) have. The picture we get is complex. About half the halo stars are orbiting in the same direction as the disk rotation, while the other half are moving in the *opposite* direction. We can deduce the complete orbits of the halo stars near us by measuring their motions. Such orbits suggest that halo stars fill a volume of space similar to that occupied by the globular clusters in the halo. Halo stars and globular clusters both paint the same picture of the halo of our galaxy.

Magnetic Fields and Cosmic Rays Fill the Galaxy

Almost all that we know about the galaxy we learn from analyzing the electromagnetic radiation it emits—almost, but not quite, all. An important component of the galaxy's disk, whose energy we can measure directly on Earth, is made up of **cosmic rays**. Despite their name, cosmic rays are not a form of electromagnetic radiation. (They were named before their true nature was known.) Rather, they are charged particles moving at nearly the speed of light. Cosmic rays are continually hitting Earth.

Most cosmic-ray particles are protons, but some are nuclei of helium, carbon, and other elements produced by nucleosynthesis. A few are high-energy electrons and other subatomic particles. What makes cosmic rays especially fascinating is that they span an enormous range in particle energy. We can observe the lowest-energy cosmic rays using interplanetary spacecraft. These cosmic rays have energies as low as about 10^{-11} joules (J), which corresponds to the energy of a proton moving at a velocity of a few tenths the speed of light. In contrast, the most energetic cosmic rays are 10 trillion (10^{13}) times as energetic as the lowest-energy cosmic rays. These high-energy cosmic rays are known from the showers of elementary particles that they cause when crashing through Earth's atmosphere.

Astronomers hypothesize that cosmic rays are accelerated to these incredible energies by the shock waves produced in supernova explosions. The highest-energy cosmic rays are more difficult to explain. These are as much as a billion times more energetic than any particle ever produced in a particle accelerator on Earth. These cosmic rays have energies up to 100 J, corresponding to the energy of a proton moving at a velocity that is 99.9999999999999999999 percent of the speed of light. To get a better sense of just how much energy we're talking about, consider this: if you were to drop your copy of *21st Century Astronomy* (this irreplaceable textbook) from the ceiling, the energy it would have when it hit the floor is about the same as the energy concentrated into a *single* high-energy cosmic-ray proton!

> **Some cosmic rays have enormously high energies.**

Most cosmic rays are trapped within our galaxy by magnetic fields. The interstellar medium of the Milky Way is laced with measurable magnetic fields that are wound up and strengthened by the rotation of the galaxy's disk. The interstellar magnetic field, however, is weak compared to Earth's magnetic field—a million times weaker. When we discussed Earth's magnetosphere in Chapter 8, we saw that charged particles spiral around magnetic fields, with a net motion along the field rather than across it. The other side of this coin is that magnetic fields cannot freely escape from a cloud of gas containing charged particles. Earlier we mentioned that dense clouds of interstellar gas are concentrated by gravity in the midplane of the Milky Way. The mass of these clouds anchors the galaxy's magnetic field and its charged-particle cosmic rays to the disk, as shown in **Figure 20.15**.

> **Cosmic rays are trapped by the Milky Way's magnetic field.**

Are all cosmic rays confined to our galaxy? The disk of our galaxy glows from **synchrotron radiation** produced by cosmic rays (mostly electrons) spiraling around the galaxy's

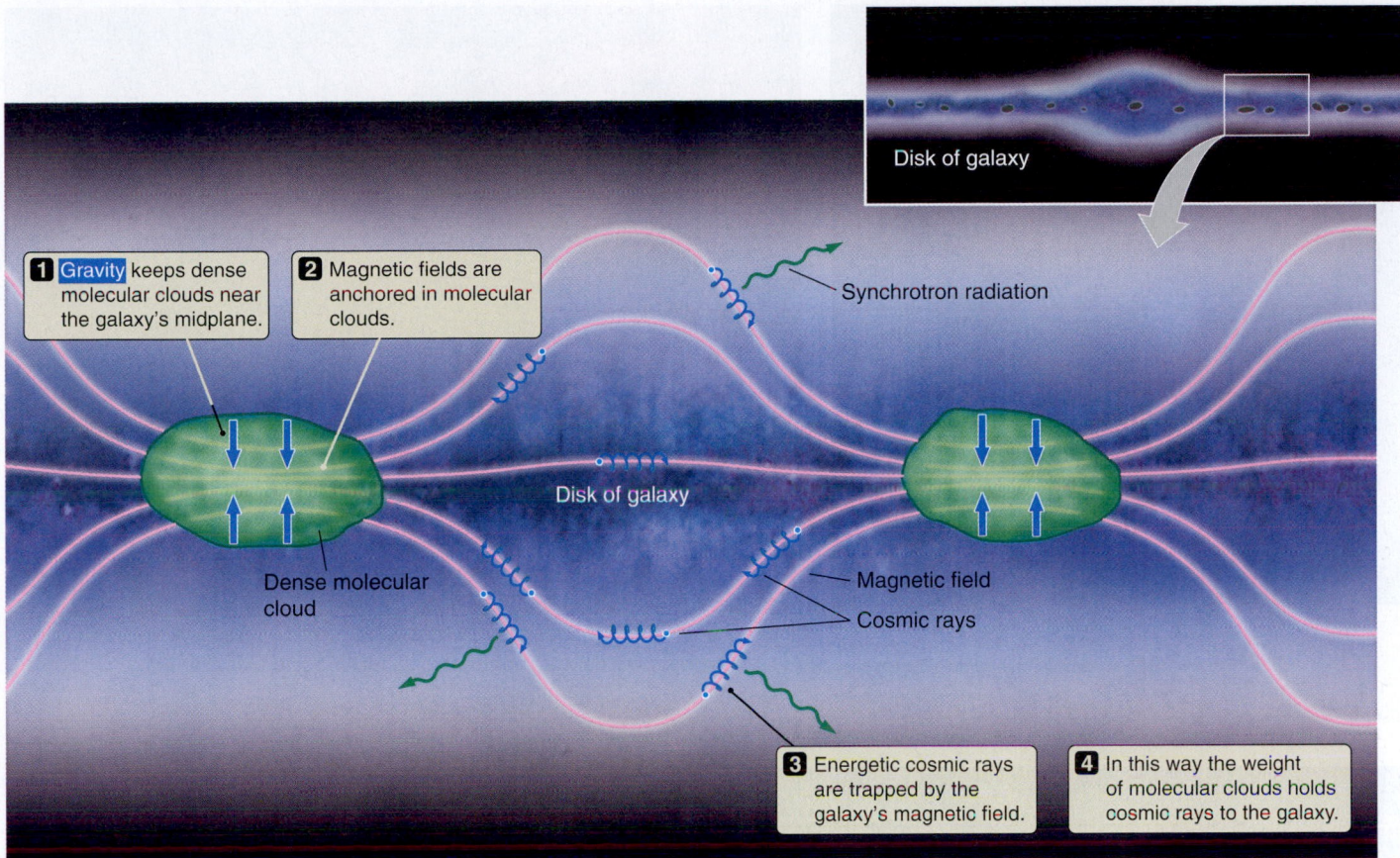

1 Gravity keeps dense molecular clouds near the galaxy's midplane.

2 Magnetic fields are anchored in molecular clouds.

Synchrotron radiation

Disk of galaxy

Dense molecular cloud

Magnetic field

Cosmic rays

3 Energetic cosmic rays are trapped by the galaxy's magnetic field.

4 In this way the weight of molecular clouds holds cosmic rays to the galaxy.

Disk of galaxy

FIGURE 20.15 The weight of interstellar clouds anchors the magnetic field of the Milky Way to the disk of the galaxy. The magnetic field in turn traps the galaxy's cosmic rays, much as a planetary magnetosphere traps charged particles.

magnetic field. Such synchrotron emission is seen in the disks of other spiral galaxies as well, telling us that they, too, have magnetic fields and populations of energetic cosmic rays. Even so, the very highest-energy cosmic rays are moving much too fast to be confined to our galaxy. Any such cosmic rays formed in the Milky Way soon stream away from the galaxy into intergalactic space. Thus, although we do not currently understand their origin, it is likely that a fraction of the energetic cosmic rays reaching Earth originated in energetic events outside our galaxy.

The total energy of all of the cosmic rays in the galactic disk can be estimated from the energy of the cosmic rays reaching Earth. The strength of the interstellar magnetic field can be measured in a variety of ways, including the effect that it has on the properties of radio waves passing through the interstellar medium. These measurements indicate that in our galaxy, the magnetic-field energy and the cosmic-ray energy are about equal to each other. Both are comparable to the energy present in other energetic components of the galaxy, including the motions of interstellar gas and the total energy of electromagnetic radiation within the galaxy. Magnetic fields and cosmic rays are *not* bit players in the production we call the Milky Way.

20.4 The Milky Way Hosts a Supermassive Black Hole

We began this chapter by pointing out that dense clouds of dust and gas hide our visible-light view of the Milky Way's center. Yet it is here that we might expect to find a supermassive black hole, as is often found in other galaxies with central bulges. Fortunately, we can overcome the obscuration problem by utilizing the penetrating ability of X-ray, infrared, and radio radiation. **Figure 20.16** shows images of the Milky Way's center taken with the Chandra X-ray Observatory and the Spitzer Space Telescope. The X-ray view (**Figure 20.16a**) shows the location of a strong radio source called Sagittarius A* (abbreviated Sgr A*), which astronomers believe lies at the exact center of the Milky Way. The infrared image (**Figure 20.16b**) cuts through the dust to reveal the crowded, dense core of the galaxy containing hundreds of thousands of stars.

Studies of the motions of stars closest to the Sgr A* source suggest a central mass very much greater than

FIGURE 20.16 (a) An X-ray view of the Milky Way's central region showing the active source, Sgr A*, as the brightest spot at the middle of the image. Lobes of superheated gas (shown in red) are evidence of recent, violent explosions happening near Sgr A*. (b) This wide infrared view (measuring 640 × 890 light-years) of the central core of the Milky Way shows hundreds of thousands of stars. The bright white spot at the far right marks the galaxy's center, home of a supermassive black hole.

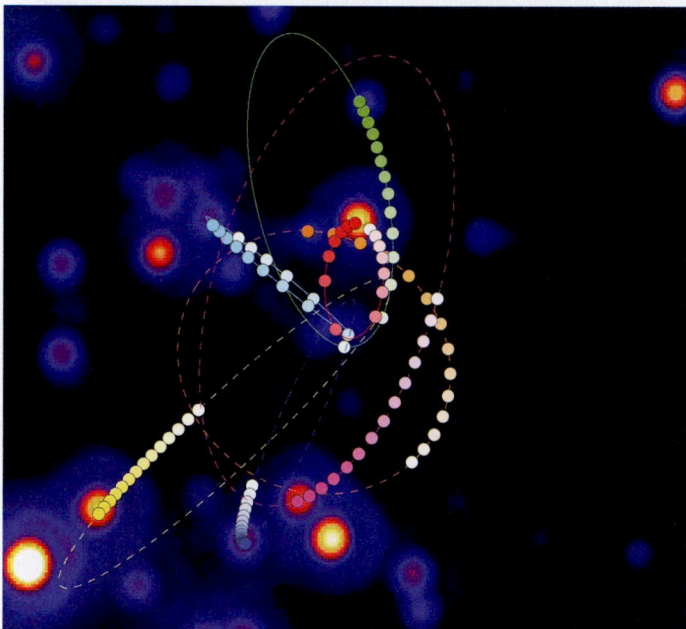

FIGURE 20.17 Orbits of several stars within 0.1 light-year (about 6,000 astronomical units, or AU) of the Milky Way's center. The Keplerian motions of these stars reveal the presence of a 3.7-million-$M_\odot$ supermassive black hole at the galaxy's center. Colored dots show the measured positions of the stars over a 12-year interval.

that of the few hundred stars orbiting there. Furthermore, observations of our galaxy's rotation curve also show rapid rotation velocities very close to its center. This is the same sort of evidence that led us to surmise the presence of supermassive black holes in other galaxies. More

recently, astronomers have studied the orbits of stars less than 0.1 light-year from the galaxy's center. These stars follow Kepler's laws as they orbit about the center, and they are so close that it takes them only a matter of a dozen or so years to complete an orbit! This means we can observe the positions of these stars changing noticeably over time, and we can see them speed up as they whip around what can only be a supermassive black hole at the focus of their elliptical orbits (**Figure 20.17**). Using Kepler's third law, we can then estimate that the black hole at the center of our own galaxy is a relative lightweight, having a mass of "only" $3.7 \times 10^6\ M_\odot$.

Although invisible to our eyes, the central region of our Milky Way Galaxy is awash in radio, infrared, X-ray, and gamma radiation. Radio observations (**Figure 20.18**) reveal synchrotron emission from fascinating wisps and loops of material distributed throughout the region. This is reminiscent of the synchrotron emission seen from active galactic nuclei (AGNs), except that this emission is seen at far lower levels.

> The center of our galaxy shows rapid rotation and AGN-like emission.

Clouds of interstellar gas at the galaxy's center are heated to millions of degrees by shock waves from supernova explosions and colliding stellar winds blown outward by young massive stars. Superheated gas produces X-rays, and the Chandra X-ray Observatory has detected more than 2,000 X-ray sources within the central region of the galaxy. These include frequent, short-lived X-ray flares near Sgr A* (**Figure 20.19**), which provide direct evidence that matter falling toward the supermassive black hole fuels the energetic activity at the galaxy's center. Huge lobes of 20-million-K

FIGURE 20.18 Radio observations of the center of the Milky Way reveal wispy molecular clouds (purple) glowing from strong synchrotron emission. Cold dust (20–30 kelvins, or K) associated with molecular clouds is shown in orange. Diffuse infrared emission appears in blue-green. The galactic center (Sgr A*) lies within the bright area to the right of center.

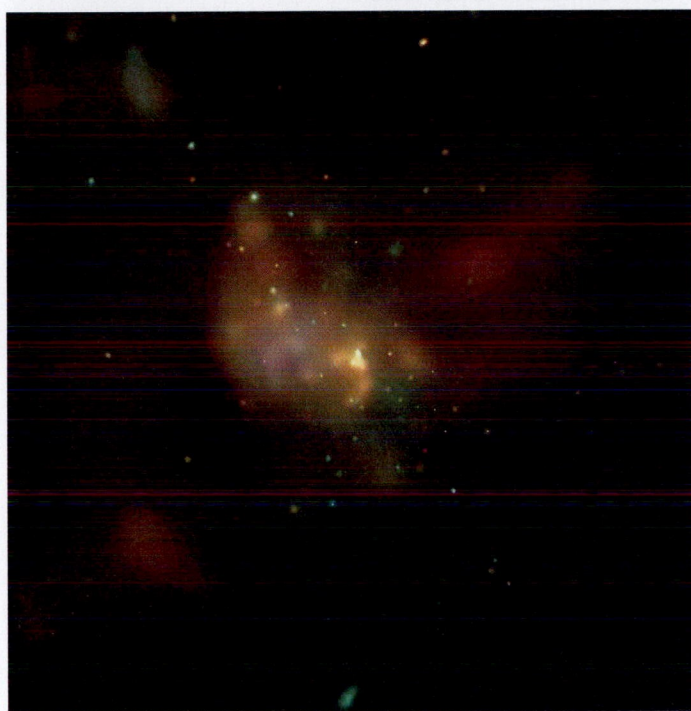

FIGURE 20.19 The central region of the Milky Way is also a source of strong X-rays, with evidence of the "beast" at its very center. The bright, pointlike feature in the middle is a huge X-ray flare that occurred within 1 AU of the supermassive black hole. The flare developed within a matter of minutes and declined to previous levels after 3 hours.

gas surrounding Sgr A* (see Figure 20.16a) bear witness to several such explosions that have occurred within the past 10,000 years. Nevertheless, this activity is not as intense as that seen in other galaxies with central black holes. What does this mean? It tells us that today our supermassive black hole is starving. Without a source of large amounts of material to feed the beast, the Milky Way's center is relatively quiet. Like steam rising from the cauldron of a dormant vol-

cano, however, what we are seeing in the inner Milky Way is a reminder that our galaxy was almost certainly "active" in the past and, like an episodic volcano, the beast could become active once again.

20.5 The Milky Way Offers Clues about How Galaxies Form

One of the fundamental goals of stellar astronomy is to understand the life cycle of stars, including how stars form from clouds of interstellar gas. In Chapter 15 we were able to tell a fairly complete story of this process, at least as it occurs today, and tie this story strongly to observations of our galactic neighborhood. Galactic astronomy has the same basic goal. That is, astronomers would like very much to have a complete and well-tested theory of how our galaxy formed. Unfortunately, such a complete theory is not yet in hand. Even so, what we have seen so far offers us many clues.

Important among these clues are the properties of globular clusters and high-velocity stars in the halo of the galaxy. For reasons discussed earlier, these objects must have been among the first stars formed that still exist today. The fact that they are not concentrated in the disk or bulge of the galaxy says that they formed from clouds of gas well before those clouds had settled into the galaxy's disk. The observations that globular clusters are very old and that the youngest cluster is older than the oldest disk stars agree with this hypothesis. The presence of small amounts of massive elements in the atmospheres

> At least one generation of stars had to exist before today's halo stars were formed.

of halo stars also tells us that at least one generation of stars must have lived and died *before* the formation of the halo stars we see today. We have yet to find any stars from that first generation in our galaxy today.

From these and other clues, we have come to understand that our galaxy must have formed when the gas within a huge "clump" of dark matter collapsed into a large number of small protogalaxies. Some of these smaller clumps are still around today in the form of small, satellite dwarf galaxies near our own. The larger among them are the Large and Small Magellanic Clouds (**Figure 20.20**), which are easily seen by the naked eye and appear much like detached pieces of the Milky Way. Among the closest is the Sagittarius Dwarf, which at this moment is plowing through the disk of the Milky Way on the other side of the bulge. It seems likely that at some point the Sagittarius Dwarf will become incorporated into the Milky Way—an indication that our galaxy is still growing. There are more than 20 of these satellite dwarf galaxies, although we can't be sure that all are gravitationally bound to the Milky Way. Many were discovered only very recently, because of their low luminosity and the possibility that they are dominated by an even greater fraction of invisible dark matter than we find in other known galaxies. Observations of the motions and speeds of the dwarf galaxies about the Milky Way can lead to new estimates of the dark matter mass within the Milky Way itself.

The remainder of these protogalaxies merged to form the barred spiral galaxy we call the Milky Way. In this process, stars were formed in the halo, in the bulge, and in the disk. The first stars to form ended up in the halo—some in globular clusters, but many not. Gas that settled into the disk of the Milky Way quickly formed several generations of stars. This process resulted in the formation of a dense, concentrated mass that quickly grew into the supermassive black hole at the center of the galaxy. The details of this process are sketchy, but some calculations indicate that so much mass was concentrated in this small region that almost any sequence of events would have led to the formation of a massive black hole.

> Our galaxy formed from the mergers of many smaller protogalaxies.

We need to be careful not to get ahead of ourselves. The Milky Way offers many clues about the way galaxies form, but much of what we know of the process comes from looking beyond our local system. Images of distant galaxies (which we see as they existed billions of years ago), as well as observations of the glow left behind by the formation of the universe itself, provide equally important pieces of the puzzle. We will have to leave the story of galaxy formation as we currently understand it unfinished, for now, as we turn our attention to the immensely larger structure of the universe as a whole.

FIGURE 20.20 Our Milky Way is surrounded by more than 20 dwarf companion galaxies, the largest among them being the Large and Small Magellanic Clouds. The Magellanic Clouds were named for Ferdinand Magellan (c. 1480–1521), who headed the first European expedition that ventured far enough into the Southern Hemisphere to see them.

Seeing the Forest for the Trees

Many books of puzzles contain optical illusions in which familiar patterns lie hidden in jumbles of lines and shapes. When looking at such a picture, at first we see nothing but confusion. Then in a flash of insight we suddenly see the pattern! Once the pattern is recognized, our perception shifts: a moment earlier all we saw was a mess; but now when we look at the picture, the "totally obvious" pattern jumps off the page at us. Such is the case with observing our galaxy. Go outside on a dark, moonless summer night, away from the city lights, and stare at the sky. In addition to the dazzling jewels of individual stars that fill your view, you will see a faint, patchy river of light running from horizon to horizon. This river—this "Milky Way"—has been known since earliest prehistory. Even so, its nature was unknown until a remarkably short time ago. As recently as the early 20th century, it was unclear how this river of light fit into models of our stellar system based on counts of individual stars. Yet once the answer is known, there it is for all to see—how could we ever have been so blind as to miss the obvious? Stretched across our sky lies an unmistakable spiral galaxy, as viewed edge on from within the disk.

Our conclusion from this flash of insight is confirmed by the results of countless observational studies. Whether we look at the distribution of globular clusters, 21-cm observations of the motions of hydrogen gas, or any number of other tracers of structure, the answer is always the same. The Milky Way is a giant "plain vanilla," dark matter–dominated, barred spiral galaxy with a supermassive black hole at its center, just like countless others. The Sun and Earth are located in a relatively open and clear region of interstellar space partway out in the galaxy's disk.

Our vantage point within the Milky Way is hardly the best from which to answer questions about global properties of our galaxy. If ever there was a problem in astronomy of not being able to see the forest for the trees, this is it. For example, we can determine the morphological classification of the Andromeda Galaxy by glancing at a single picture. In contrast, obtaining a reliable and complete classification of our own Milky Way is a task that has taken decades of detailed study. The classification of the Milky Way as an SBbc galaxy was difficult to arrive at and remains at least somewhat uncertain today.

On the other hand, the Milky Way gives us the ultimate insider's view. For example, stars in the Milky Way are not just part of a large blur, but are instead individuals that can be studied as such. Here is where we learn about the stars in globular and open clusters. These collections of stars not only tell us about the Milky Way as a galaxy, but also provide the data we use to test and refine our ideas about the way stars everywhere evolve. In the Milky Way we see stars in the bulge, stars in the thin and thicker portions of the disk, and stars in the halo that are falling through the disk near us. We measure their properties—luminosity, size, temperature, and abundance of the chemical elements of which they are made—and we learn what each variety of stars in a spiral galaxy is like. Spiral arms in the Milky Way are not just bright swirls against the background of a spiral disk. Rather, they are collections of young stars, old stars, and clouds of gas and dust that can be studied at close range. We know that stars form in the arms of other spiral galaxies because we see the brightest of those stars, along with the generally blue color of the arms. We know that stars form in the spiral arms of the Milky Way because that is where we see dense molecular-cloud cores containing individual young stars surrounded by their accretion disks.

It is difficult to study the Milky Way as a galaxy, but it is worth the investment. In our study of the grand patterns within the Milky Way, we are guided by what we know of other galaxies; conversely, in the Milky Way we can learn the details of processes that are at work shaping all of those other islands of activity spread out across the expanse of the universe. These details are beginning to help us clarify our understanding of some of the "big" questions about the way galaxies form and evolve.

Our location in the Milky Way may not give us the best perspective from which to study our galaxy, but it is far from the worst. Fate has been kind. Since the Sun and Earth formed, they have made numerous trips around the galaxy, passing back and forth through the disk and encountering a wide range of environments. What if our species had become aware at a moment when Earth was in the depths of a molecular cloud and the sky held nothing of interest other than the Sun and planets? That would have brought a whole new and very literal meaning to the notion of the "dark ages."

In some sense this portion of our journey has been about context. We have learned to see stars within the context of the history of the Milky Way. We have seen star-forming molecular clouds within the context of the sweep of spiral arms. We have learned to see the Milky Way within the context of its similarities with other galaxies. On the next leg of our journey we will pose the ultimate question of context as we learn to see galaxies, including our Milky Way, within the context of the universe itself.

Summary

- We live in the disk of a barred spiral SBbc galaxy called the Milky Way, which is 100,000 light-years across.

- We use variable stars of known luminosity to find the distances to globular clusters, which enable us to measure the size of the Milky Way's extended halo.

- The Sun is about 27,000 light-years from the Milky Way's center.

- The Doppler velocities of radio lines show that the rotation curve of the Milky Way is flat, like those of other galaxies, and that the mass of our galaxy is mostly in the form of dark matter.

- The chemical composition of the Milky Way has evolved with time, and there must have been a generation of stars before the oldest halo and globular-cluster stars we see today formed.

- Star formation is actively occurring in the disk of our galaxy, leading to complex structures within the disk.

- At the center of the galaxy is a massive black hole, which produces rapid orbital velocities nearby.

SmartWork, Norton's online homework system, includes algorithmically generated versions of these questions, plus additional conceptual exercises. If your instructor assigns questions in SmartWork, log in at **smartwork.wwnorton.com.**

Student Questions

THINKING ABOUT THE CONCEPTS

1. What can you see for yourself in the night sky (unaided by a telescope) that suggests we live in a spiral galaxy?

2. Why is it so difficult for astronomers to get an overall picture of the structure of our galaxy?

3. What do astronomers mean by *standard candle*?

4. Why are globular clusters so useful as standard candles in determining the size of the galaxy?

5. Describe the distribution of globular clusters within the galaxy and what that implies about the size of the galaxy and our distance from its center.

6. How do we know that the stars in globular clusters are among the oldest stars in our galaxy?

7. Compare globular and open clusters.
   a. What are the main differences in the gas out of which globular and open clusters formed?
   b. Why do globulars have such high masses while open clusters have low masses?
   c. Are these issues related? Explain.

*8. Old stars in the inner disk of our galaxy have higher abundances of massive elements than do young stars in the outer disk. Shouldn't old stars *always* have a lower abundance of massive elements than young stars have? Explain your reasoning.

9. Do you think it is likely that astronomers will find a young star that has no heavy elements in it? Why or why not? How about an old star?

10. How do 21-cm radio observations reveal the rotation of our galaxy?

11. What does the rotation curve of our galaxy say about the presence of dark matter in the galaxy?

12. Explain the observational evidence that shows we live in a spiral galaxy, not an elliptical galaxy.

13. In the Hubble scheme for classifying galaxies, how is the Milky Way classified?

14. What does the abundance of a star's massive elements tell us about the age of the star?

15. Where do we find the youngest stars in our galaxy?

16. Halo stars are found in the vicinity of the Sun. What observational evidence distinguishes them from disk stars?

17. Are cosmic rays a form of electromagnetic radiation? Explain your answer.

18. Can a cosmic ray travel at the speed of light? Why or why not?

*19. What is one source of synchrotron radiation in our galaxy, and where is it found?

20. Why must we use X-ray, infrared, and 21-cm radio observations to probe the center of our galaxy?

21. What is Sgr A* and how was it detected?

22. Explain the evidence for a supermassive black hole at the center of the Milky Way.

23. How does the mass of the supermassive black hole at the center of our galaxy compare with that found in most other spiral galaxies?

24. To observers in Earth's Southern Hemisphere, the Large and Small Magellanic Clouds look like detached pieces of the Milky Way. What are these "clouds," and why is it not surprising that they look so much like pieces of the Milky Way?

*25. What is the origin of the Milky Way's satellite galaxies?

26. What has been the fate of most of the Milky Way's satellite galaxies?

27. Why are most of the Milky Way's satellite galaxies so difficult to detect?

28. Use your imagination to describe how our skies might appear if our Sun and Solar System were located
    a. Near the center of the galaxy.
    b. Near the center of a large globular cluster.
    c. Near the center of a large, dense molecular cloud.

APPLYING THE CONCEPTS

29. The Sun completes one trip around the center of the galaxy in approximately 230 million years. How many times has our Solar System made the circuit since its formation 4.6 billion years ago?

30. The Sun is located about 27,000 light-years from the center of the galaxy, and the galaxy's disk probably extends another 30,000 light-years farther out from the center. Assume that the Sun's orbit takes 230 million years to complete.
    a. With a truly flat rotation curve, how long would it take a globular cluster located near the edge of the disk to complete one trip around the center of the galaxy?
    b. How many times has that globular cluster made the circuit since its formation about 13 billion years ago?

31. Parallax measurements of the variable star RR Lyrae indicate that it is located 750 light-years from the Sun. A similar star observed in a globular cluster located far above the galactic plane appears 160,000 times fainter than RR Lyrae.
    a. How far from the Sun is this globular cluster?
    b. What does your answer to part (a) tell you about the size of the galaxy's halo compared to the size of its disk?

32. Although the flat rotation curve indicates that the total mass of our galaxy is approximately $8 \times 10^{11} \, M_\odot$, electromagnetic radiation associated with normal matter suggests a total mass of only $3 \times 10^{10} \, M_\odot$. Given this information, calculate the fraction of our galaxy's mass that is made up of dark matter.

*33. Compare the H-R diagram for the young cluster NGC 6530 (see Figure 20.11) with H-R diagrams in previous chapters to determine how soon the most massive stars will explode, and how massive the protostars are that are still contracting onto the main sequence.

34. Given what you have learned about the distribution of massive elements in the Milky Way, and what you know about the terrestrial planets, where do you think such planets are most likely and least likely to form?

35. A cosmic-ray proton is traveling at nearly the speed of light (3×10^8 meters per second [m/s]).
    a. Using Einstein's familiar relationship between mass and energy ($E = mc^2$), show how much energy (in joules) the cosmic-ray proton would have if m were based only on the proton's rest mass (1.7×10^{-27} kilograms [kg]).
    b. The actual measured energy of the cosmic-ray proton is 100 J. What, then, is the relativistic mass of the cosmic-ray proton?
    c. How much greater is the relativistic mass of this cosmic-ray proton than the mass of a proton at rest?

**36. One of the fastest cosmic rays ever observed had a speed of $(1.0 \text{ minus } 10^{-24}) \times c$. Assume that the cosmic ray and a photon left a source at the same instant. To a stationary observer, how far behind the photon would the cosmic ray be after traveling for 100 million years?

37. Consider a black hole with a mass of 5 million solar masses ($M_\odot = 2 \times 10^{30}$ kg.) A star orbiting the black hole has a semimajor axis of 0.02 light-year (1.9×10^{14} meters). Using the universal gravitational constant, $G = 6.67 \times 10^{-11}$ newton times square meters per square kilogram (N m²/kg²), calculate the star's orbital period. (Hint: For this and the questions that follow, you may want to refer back to Chapter 3.)

38. A star in a circular orbit about the black hole at the center of our galaxy ($M_{BH} = 7.36 \times 10^{36}$ kg) has an orbital radius of 0.0131 light-year (1.24×10^{14} meters). What is the average speed of this star in its orbit?

39. How large is the black hole at the center of our galaxy (that is, where is its event horizon)?

40. A star is observed in a circular orbit about a black hole with an orbital radius of 1.5×10^{11} kilometers (km) and an average speed of 2,000 km/s. What is the mass of this black hole in solar masses?

There is a theory which states that if ever anyone discovers exactly what the universe is for and why it is here, it will instantly disappear and be replaced by something even more bizarre and inexplicable. There is another which states that this has already happened.

DOUGLAS ADAMS (1952–2001)

To understand the large-scale universe, we must understand everything from the smallest subatomic particles to the largest structures we see.

CHAPTER 21

Modern Cosmology

21.1 The Universe Has a Destiny and a Shape

Cosmology is the study of the universe on the very grandest of scales. During our journey we have already learned many things about the universe and its evolution. We learned that the universe is expanding, having originated in a Big Bang nearly 14 billion years ago; that it was once very hot, filled with thermal radiation that has now cooled to a temperature of 2.7 kelvins (K); that light elements in the universe were produced within the first few minutes after the Big Bang. But there is still more to learn. In this chapter we take a closer look at the nature of the universe and how it has evolved over time, and we contemplate its ultimate fate. We also develop an understanding of particle physics, which is necessary for describing the very smallest pieces of the universe.

At first glance, particle physics and cosmology might seem to have almost nothing in common. Whereas particle physics is the study of the quantum mechanical world that exists on the tiniest scales imaginable, cosmology is the study of the changing structure of a universe that extends for billions of light-years and probably much farther. Yet the last quarter of the 20th century saw the boundary between these two fields fade and eventually disappear as cosmologists and particle physicists came to realize that the structure of the universe and the fundamental nature of matter are two sides of the same scientific coin.

We know we live in an expanding universe, but will that expansion continue forever? This is clearly one of the great questions of modern cosmology. What is the fate of

KEY CONCEPTS

Cosmology is the study of our large-scale universe, including its nature, origin, evolution, and ultimate destiny. As we study and try to understand our universe on the largest possible scales, we will learn

- That gravity acts to slow the expansion of the universe.

- How the universe can have a shape or geometry different from our standard Euclidean geometry.

- How the history, shape, and fate of the universe are determined by how much mass it contains and by a poorly understood cosmological constant.

- That the expansion of our universe appears to be accelerating.

- What observations of the cosmic background radiation tell us about the properties of the young universe.

- The universe may have undergone a period of huge expansion that we call inflation.

- That matter and the fundamental forces of nature "froze out" of the uniformity of the expanding and cooling universe moments after the Big Bang.

the universe? The answer depends in part on the amount of distributed mass the universe contains on very large scales. The gravitational effect of this distributed matter is only one of the factors—and the first one we discuss—that determines how the universe evolves.

To see how gravity affects the expansion of the universe, it will help to recall gravity's effects on the motion of projectiles. Think back to our discussion of escape velocity in Chapter 3 and take the case of the Moon as an example. The fate of a projectile fired straight up from the surface of the Moon depends on its speed. As long as the speed is less than the Moon's escape velocity (2.4 kilometers per second [km/s]), gravity will eventually stop the rise of the projectile and pull it back to the Moon's surface. But if the speed of the projectile is greater than the Moon's escape velocity, then gravity will lose. Although the projectile will slow down, it will never stop. It will escape from the Moon entirely.

Just as the gravity of the Moon pulls on a projectile, slowing its climb, the gravity arising from the mass contained in the universe slows its expansion. If there is enough mass in the universe, then gravity will be strong enough to stop the expansion. And in that case, the universe will slow, stop, and eventually collapse in on itself in a catastrophic "Big Crunch." But if there is not enough mass, then the expansion of the universe may slow, but it will never stop. The universe will expand forever.

> **Gravity slows the expansion of the universe.**

A planet's mass and radius determine the escape velocity from its surface. The "escape velocity" of the universe is also determined by its mass and size—specifically, its average *density*. If the universe is denser on average than a particular value, called the **critical density**, then we expect gravity will be strong enough to eventually stop and reverse the expansion. If the universe is less dense than this value, we expect gravity will be too weak and the universe will expand forever.

The faster the universe is expanding, the more mass is needed to turn that expansion around. For that reason, the critical density depends on the value of the Hubble constant, H_0. Assuming that $H_0 = 22$ kilometers per second per mega-light-year (km/s/Mly) and that gravity is the only thing we have to worry about, the universe's critical density has a value of 8×10^{-27} kilograms per cubic meter (kg/m^3). Rather than trying to keep track of such awkward numbers, we will instead talk about the *ratio* of the actual density of the universe to its critical density. We call this ratio Ω_{mass} (pronounced "omega sub mass"). Note that Ω_{mass} is a dimensionless pure number (a number that has no units).

We can follow the expansion of different possible universes in **Figure 21.1**, which shows a plot of the scale factor R_U (see Chapter 18) versus time for different values of Ω_{mass}. (In these plots we have assumed that gravity alone controls the fate of the universe—an assumption that we will revisit in the following subsection.) If Ω_{mass} in a universe controlled

by gravity alone is greater than 1, then gravity is strong enough to turn the expansion around. Like a projectile fired from the Moon at less than the escape velocity, the expansion of such a universe will slow and eventually stop, and the universe will then fall back in on itself, culminating in the "Big Crunch." Conversely, if Ω_{mass} in a possible universe is less than 1, that universe will expand forever, but slowed down to some extent by gravity. This is like a projectile fired at more than the escape velocity from the Moon. The dividing line, where Ω_{mass} equals 1, corresponds to a universe that expands more and more slowly—continuing forever, but never quite stopping. Such a universe is expanding at exactly the escape velocity.

> **Ω_{mass} determines the fate of a universe governed exclusively by gravity.**

Look again at the three plots in Figure 21.1 and you will see that they are not straight lines. Curvature in these plots represents the effect that gravity is having on expansion. As the value of Ω_{mass} increases (corresponding to greater density in the universe), the amount of curvature also increases. In the case where Ω_{mass} is greater than 1, the scale factor begins to decrease and gravity wins.

Until the closing years of the 20th century, most astronomers thought that this straightforward application of gravity to the universe was all there was to the question of expansion and collapse. Researchers focused great efforts on carefully measuring the mass of galaxies and assemblages of galaxies in the belief that this study would reveal the density and therefore the fate of the universe. As we have seen time and again, measuring masses of objects in the universe is tricky. When we calculate Ω_{mass} using just the luminous matter that we see in galaxies and groups of galaxies, we get a value of about 0.02. In Chapter 19 we saw that galaxies contain about 10 times as much dark matter as normal matter, so adding in the dark matter in galaxies pushes the value of Ω_{mass} up to about 0.2. Finally, when we include the mass of dark matter *between* galaxies (a subject we will return to in the next chapter), Ω_{mass} could increase to 0.3 or higher. In other words, by our current accounting there is only about a third as much mass in the universe as is needed to stop the universe's expansion.

Reviving Einstein's "Biggest Blunder"

As they were closing in on a good value for Ω_{mass}, it seemed to astronomers that our understanding of the expansion of the universe was almost complete. Then the other shoe dropped.

If the expansion of the universe has been slowing with time, as these simple models using gravity predict, then when the universe was young it must have been expanding more rapidly than it is today. That is a prediction that we can go to our telescopes and check. Objects that are very far away (so that we see them as they were long ago) should have larger velocities than our local Hubble's law would lead us

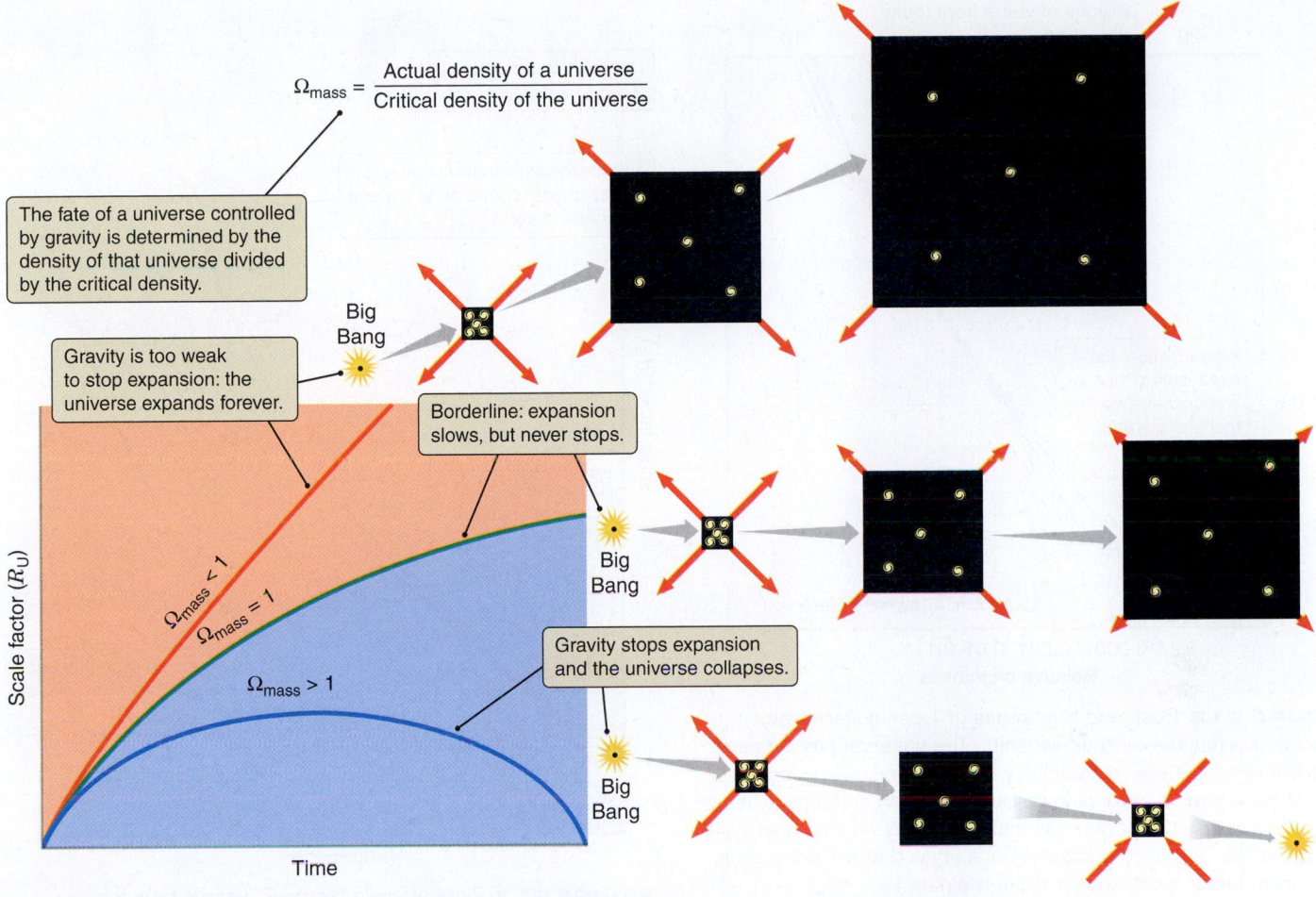

$$\Omega_{mass} = \frac{\text{Actual density of a universe}}{\text{Critical density of the universe}}$$

The fate of a universe controlled by gravity is determined by the density of that universe divided by the critical density.

Gravity is too weak to stop expansion: the universe expands forever.

Borderline: expansion slows, but never stops.

Gravity stops expansion and the universe collapses.

Big Bang

Big Bang

Big Bang

Scale factor (R_U)

$\Omega_{mass} < 1$

$\Omega_{mass} = 1$

$\Omega_{mass} > 1$

Time

FIGURE 21.1 Three possible scenarios for the fate of the universe based on the critical density of the universe, but ignoring any cosmological constant.

to expect. This is the same as saying that the plots in Figure 21.1 are all curving away from a straight line. (If gravity were not present, the Hubble expansion would not be changing, and the plots would instead be straight lines.)

During the 1990s, some groups of astronomers began using tools such as the Hubble Space Telescope and the giant Keck Observatory telescopes in Hawaii to test this prediction. They measured the brightnesses of standard candles (in particular, Type I supernovae) in very distant galaxies and compared those brightnesses with expected brightness based on the redshifts of those galaxies. The findings of these studies sent a wave of excitement through the astronomical community. Rather than showing that the expansion of the universe has slowed down over time, the data indicated that it is *speeding up!*

The observational data plotted in **Figure 21.2** indicate that the expansion of the universe is speeding up. The observations on which this conclusion is based are extremely difficult to carry out, and we must take into account many different factors when interpreting the data. Even so, the

Evidence suggests that the expansion of the universe is accelerating.

fact that similar findings have been obtained independently by different groups of astronomers lends credence to the results. In addition, results obtained early in the 21st century by the WMAP experiment (see Chapter 18) provide independent confirmation of this increasing rate of expansion of the universe.

How could it be that the rate of expansion of the universe has *increased* over time? For this to be the case, there would have to be some force at work to push the universe outward in opposition to normal gravity. Physicists have some ideas about how such a hypothetical repulsive force might originate. These ideas are related to theories from particle physics concerning the nature of what we call the **vacuum**. Normally we think of a vacuum as "empty space," but it turns out that even empty space has some very interesting physical properties. Discussion of these ideas (which are related to Hawking radiation from black holes, discussed in Chapter 17) will have to wait until later in the chapter, when we take up fundamental questions related to the nature of the interactions of matter. For now we turn our attention to the effects that such a force would have.

The idea of a repulsive force opposing the attractive force of gravity is not new. When Einstein used his newly

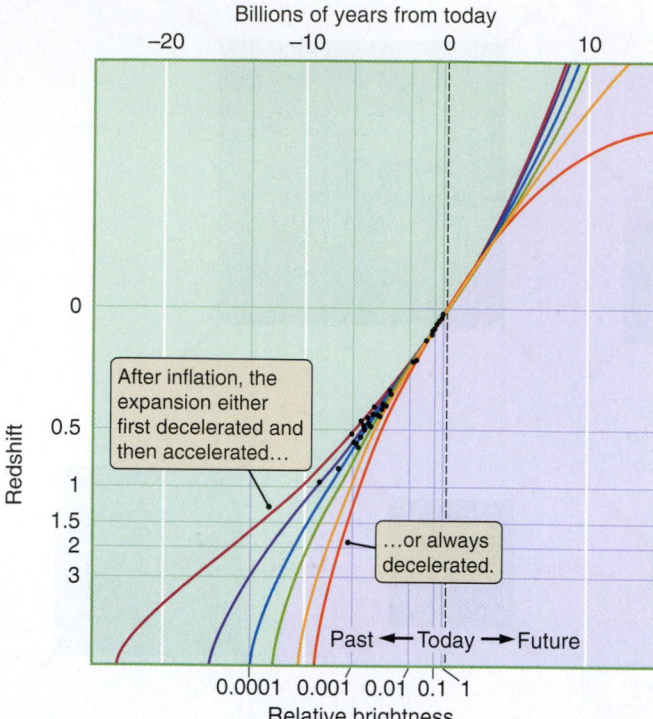

FIGURE 21.2 Observed brightness of Type I supernovae, plotted as a function of their redshift. The observations indicate that the redshifts are too small for their distances—evidence that the universe is expanding faster today than in the past. (Each point represents the brightness and redshift of an individual Type 1 supernova. Overall, the observations show that the universe is expanding faster today than it did in the past.)

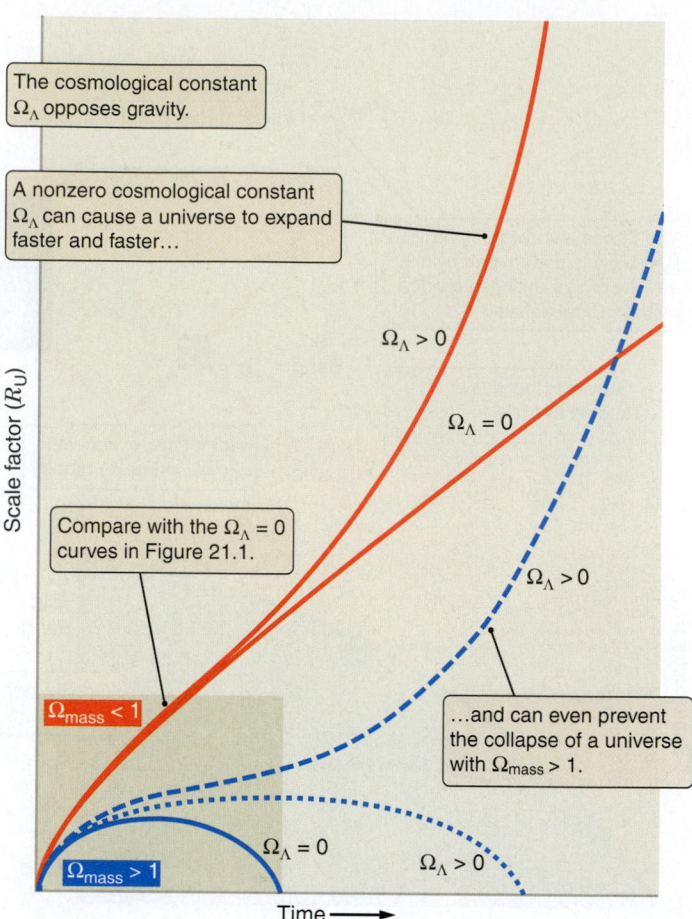

FIGURE 21.3 Plots of scale factor R_U versus time for cosmologies with and without a cosmological constant, Ω_Λ. If there is enough mass in a universe, gravity could still overcome the cosmological constant and cause that universe to collapse. Any universe without enough mass to eventually collapse will instead end up expanding at an ever-increasing rate.

formulated equations of general relativity to calculate the structure of spacetime in the universe, he was greatly troubled. The equations clearly indicated that any universe containing mass could not be static, any more than a ball can hang motionless in the air. He found the same result that Figure 21.1 shows—namely, that gravity always makes the universe move toward slower expansion or even collapse. However, Einstein's formulation of spacetime came more than a decade before Hubble did his epic work on the expansion of the universe, and the conventional wisdom at the time was that the universe was indeed static—that it neither expands nor collapses.

In order to force his new general theory of relativity to allow for a static universe, Einstein inserted a "fudge factor" into his equations. He called this fudge factor the **cosmological constant**, which we write as Ω_Λ (pronounced "omega sub lambda").[1] (Just as we did with Ω_{mass}, we define

> Einstein invented a cosmological constant to oppose gravity in a static universe.

Ω_Λ relative to the critical density.) The cosmological constant acts as a repulsive force in the equations, opposing gravity and allowing galaxies to remain stationary despite their mutual gravitational attraction. The cosmological constant was Einstein's version of the magician's trick that allows a subject to hang apparently unsupported in midair.

When Hubble announced his discovery that the universe is expanding, Einstein realized his mistake. The unmodified equations of general relativity demand that the structure of the universe be dynamic. Instead of doctoring them with the inclusion of Ω_Λ, Einstein realized he should have *predicted* that the universe must either be expanding or contracting with time. What a coup it would have been for his new theory to successfully predict such an amazing and previously unsuspected result. He called the introduction of his fudge factor, the cosmological constant, the "biggest blunder" of his career as a scientist. It is ironic that with the new results on the brightness of Type I supernovae, Einstein's biggest blunder has returned to center stage. The

---

[1] Einstein probably never wrote Ω_Λ, referring instead to the related quantity as Λ.

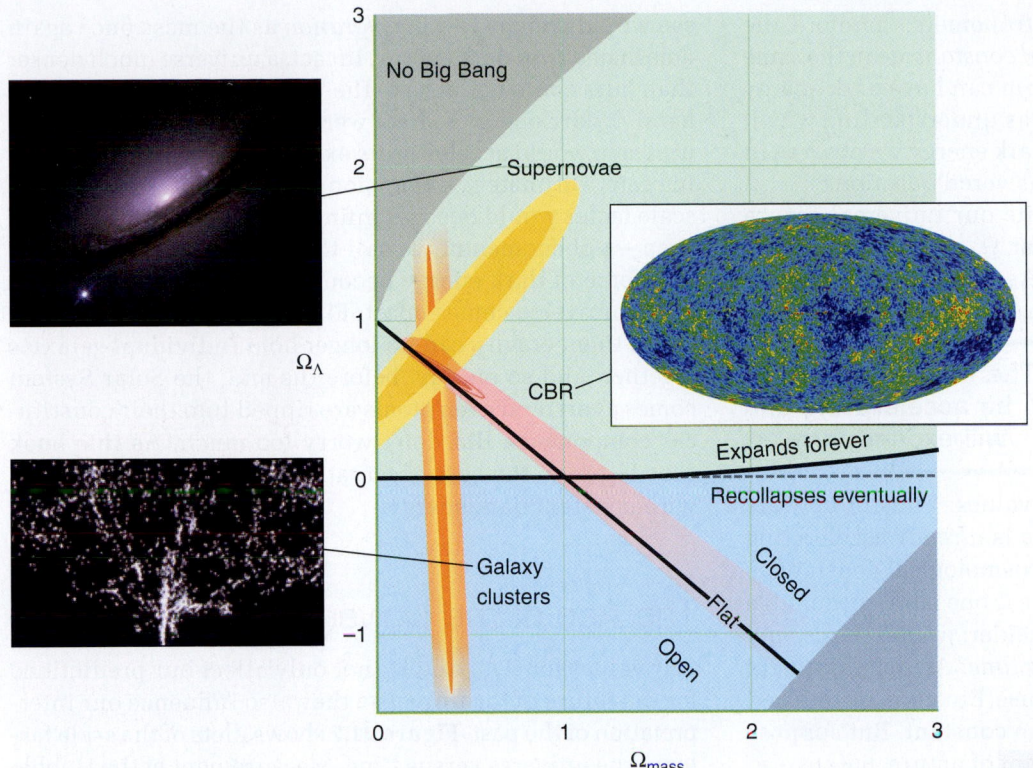

FIGURE 21.4 Current observations from different sources—Type I supernovae, measurements of mass in galaxies and clusters, and detailed observations of the structure of the cosmic background radiation—suggest that the best current estimate for Ω_Λ is about 0.7 and about 0.3 for Ω_{mass}, which means that the expansion of the universe is accelerating.

repulsive force represented by the infamous Ω_Λ in Einstein's equations is just what is needed to describe a universe that is expanding at an ever-accelerating rate.

How does the possibility of a nonzero value for Ω_Λ affect the possible fates that await our universe? If Ω_Λ is not zero, the fate of the universe is no longer controlled exclusively by Ω_{mass}. If something is effectively pushing outward from within the universe, adding to its expansion, then gravity will have a harder time turning the expansion around. In that case, the mass needed to halt the expansion of the universe will be greater than the critical mass we already discussed.

Figure 21.3 shows plots of the scale factor R_U versus time that are similar to those shown in Figure 21.1; but now we have included the effects of a nonzero cosmological constant. The evolution of a universe that collapses back on itself (below the horizontal line in **Figure 21.4**) is similar, regardless of whether Ω_Λ is zero. In contrast, if a universe with a nonzero cosmological constant expands forever, its evolution will look drastically different from that of any universe in which Ω_Λ is zero. As a universe expands, gravity gets weaker and weaker because the mass is more and more spread out. Unlike gravity, however, the effect of the cosmological constant becomes increasingly *greater*. While a universe is young and compact, gravity is strong enough to dominate the effect of the cosmological constant. Unless gravity is able to turn the expansion around, however, the cosmological constant wins out in the end, causing the expansion to continue accelerating forever. Indeed, even for a universe that otherwise would have collapsed back

Λ as the source of cosmic expansion: DeSitter in 1930

FIGURE 21.5 Cartoon from physicist Willem de Sitter showing that the cosmological constant is the force blowing up the balloon that is the universe. Note that the cartoon figure looks like lambda.

on itself, a large enough cosmological constant could make that same universe expand forever. One of the first scientists to work with the cosmological constant was Willem de Sitter (1872–1934). He provided an early cartoon (**Figure 21.5**) showing that it is the cosmological constant that drives the expansion of the universe.

When Einstein added the cosmological constant to his equations of general relativity, he introduced what he considered a new fundamental constant, in many ways similar to Newton's universal gravitational constant G. Today we realize that the vacuum can have distinct physical properties of its own. For example, the vacuum can have a nonzero energy even in the total absence of matter. We call this energy **dark energy**. It turns out that dark energy produces exactly the same kind of repulsive force that Einstein's cosmological constant does. To astronomers, therefore, the terms *dark energy* and *cosmological constant* mean the same thing. But knowing that the vacuum can have dark energy associated with it is not the same as understanding why it must have exactly the amount of dark energy we observe in the universe. That is still an unanswered question.

What fate, then, actually awaits our universe? Figure 21.4 shows the range of values for Ω_{mass} and Ω_Λ that are allowed by current observations. Assuming these observations are correct, the values for Ω_{mass} and Ω_Λ are about 0.3 and 0.7, respectively. Indeed, the data from Type I supernovae, from WMAP, and from clusters of galaxies (which will be discussed in Chapter 22) are all consistent with these values. Thus, it appears that the expansion of our universe is *already* accelerating under the dominant effect of the cosmological constant.

> The universe appears to be accelerating and will expand forever.

Before leaving this subject, there is one more speculative but fascinating idea we should consider: *what if the cosmological constant is not constant with time?* At first, this might seem a strange supposition, because Einstein introduced the cosmological constant as a true constant. But suppose dark energy was not really a constant of nature, but instead was an effect of the vacuum that arises from fundamental physics. If this were the case, dark energy's effect could be either increasing or decreasing with time. In truth, we do not yet understand the origin of dark energy, so it is entirely possible that its effects *could* vary with time.

Such a possibility would significantly change the future of the universe, as illustrated in **Figure 21.6**. For example, if dark energy were to decrease rapidly enough with time, the accelerating expansion of the universe that we now

see would change to a *deceleration* as the mass once again dominated over dark energy. In fact, a universe much denser than ours could collapse to the "Big Crunch." On the other hand, if dark energy's effect were to increase with time, the universe would accelerate its expansion at an ever-increasing rate. Ultimately, expansion could be so rapid that the scale factor would become infinite within a finite period of time—a phenomenon we call the **Big Rip**. Here the repulsive force of dark energy becomes so dominating that the entire universe comes apart. First, clusters of galaxies rip apart; then, gravity can no longer hold individual galaxies together; and so on. Just before the end, the Solar System comes apart, and even atoms are ripped into their constituent components. But don't worry too much. As this book goes to press, the best observational data seem consistent with constant dark energy.

The Age of the Universe

Our values for Ω_{mass} and Ω_Λ not only affect our predictions for the future of the universe; they also influence our interpretation of the past. **Figure 21.7** shows plots of the scale factor of the universe versus time. Measurement of the Hubble constant (H_0) tells us how fast the universe is expanding *today*. That is, it tells us the *slope* of the curves in Figure 21.7 *at the current time*. As we saw earlier, if the expansion of the universe has not changed in time, then the plot of R_U versus time is the straight red line in Figure 21.7. The age of the universe in this case is equal to the Hubble time: $1/H_0$. If the expansion of the universe has been slowing down with time (the green line in Figure 21.7), then the universe is actually *younger* than the Hubble time. (The

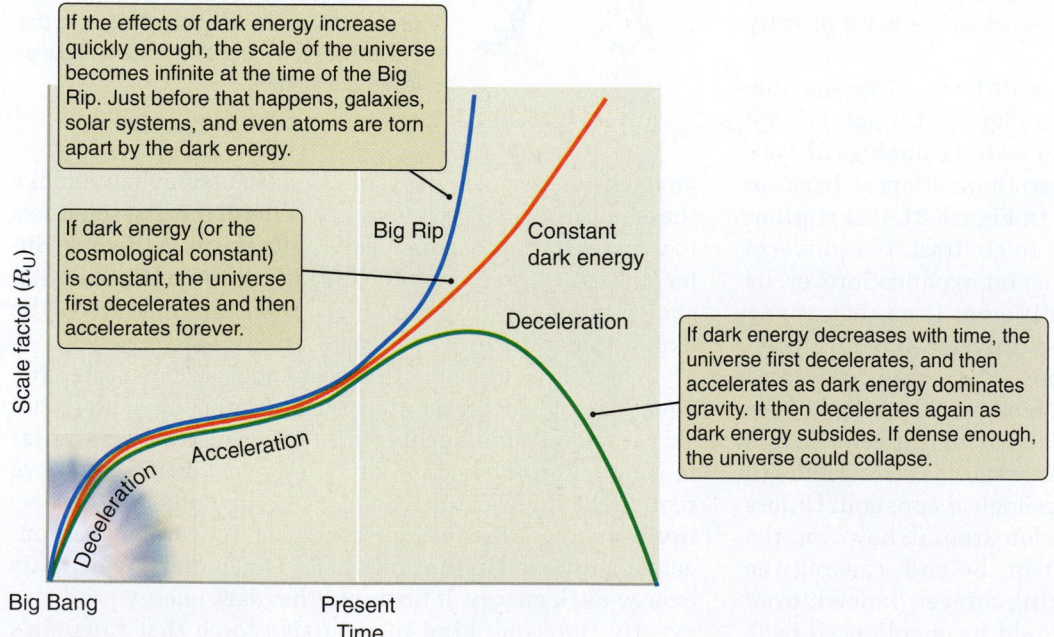

If the effects of dark energy increase quickly enough, the scale of the universe becomes infinite at the time of the Big Rip. Just before that happens, galaxies, solar systems, and even atoms are torn apart by the dark energy.

If dark energy (or the cosmological constant) is constant, the universe first decelerates and then accelerates forever.

Big Rip

Constant dark energy

Deceleration

If dark energy decreases with time, the universe first decelerates, and then accelerates as dark energy dominates gravity. It then decelerates again as dark energy subsides. If dense enough, the universe could collapse.

Acceleration

Deceleration

Scale factor (R_U)

Big Bang

Present

Time

FIGURE 21.6 How the scale factor R_U of the universe varies depending on how dark energy changes with time. If dark energy is constant, then the universe first decelerates and then accelerates its expansion as dark energy dominates gravity. If dark energy decreases with time, then dark energy can cause acceleration for some period of time, but mass density later dominates gravity and the universe decelerates again, and could even collapse for high enough density. When the effect of dark energy increases with time, the acceleration of the universe gets faster and faster, and the scale factor can become infinite at a time called the Big Rip.

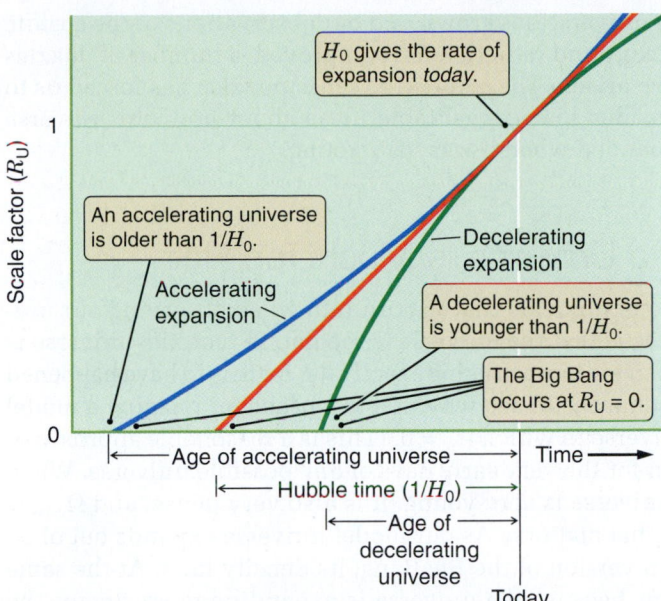

FIGURE 21.7 A plot of the scale factor R_U versus time for three possible universes. If the universe has expanded at a constant rate, then its age is equal to the Hubble time, $1/H_0$. If the expansion of the universe has slowed with time, it is younger than the Hubble time. If the expansion has sped up with time, then the universe is older than $1/H_0$.

curve crosses $R_U = 0$ at a point more recent than $1/H_0$.) If, on the other hand, the expansion of the universe has been speeding up with time (blue line), then the true age of the universe is greater than the Hubble time.

The current measured value for H_0 (22 km/s/Mly) corresponds to a Hubble time ($1/H_0$) of about 13.6 billion years. If expansion of the universe has slowed over time, the universe is actually younger than 13.6 billion years. Having a younger universe is a problem if the measured ages of globular clusters—13 billion years—is correct. (Globular clusters

clearly cannot be older than the universe that contains them!) On the other hand, if the expansion of the universe has sped up with time, as suggested by the observations of Type I supernovae and by WMAP, then the universe is about 13.7 billion years old—comfortably older than globular clusters.

The Universe Has a Shape

We have already discussed such properties of the universe as density, dark energy, and age. But the universe also has another key property: its shape in spacetime. By this time the concept of spacetime as described by general relativity is getting to be a familiar friend. Space is a "rubber sheet" that has stretched outward from the Big Bang. In Chapter 17 we saw that the rubber sheet of space is also curved by the presence of mass. We saw how the shape of space around a massive object can be detected through changes in simple geometric relationships, such as the ratio of the circumference of a circle to its radius, or the sum of the angles in a triangle. If the mass of a star, planet, or black hole causes a distortion in the shape of space, then should not the mass of everything in the universe—including galaxies, dark matter, and dark energy—also distort the shape of the universe *as a whole*? The answer is yes.

Three basic shapes are possible for our universe. Which shape actually describes the universe is determined by the total amount of mass and energy—in other words, the sum of Ω_{mass} and Ω_Λ. Continuing with the rubber sheet analogy, the first possibility, corresponding to $\Omega_{mass} + \Omega_\Lambda = 1$, is that we live in a **flat universe**. A flat universe is described overall by the rules of Euclidean geometry. As shown in **Figure 21.8a**, circles in a flat universe have a circumference of 2π times their radius, and triangles contain angles whose sum is 180°. A flat universe stretches on forever.

The second possibility is that the universe is shaped something like the surface of a saddle (**Figure 21.8b**). This

FIGURE 21.8 Two-dimensional representations of the possible geometries that space can have in a universe. In a flat universe (a), Euclidean geometry holds: triangles have angles that sum to 180°, and the circumference of a circle equals 2π times the radius. In an open universe (b) or a closed universe (c), these relationships are no longer correct over very large distances.

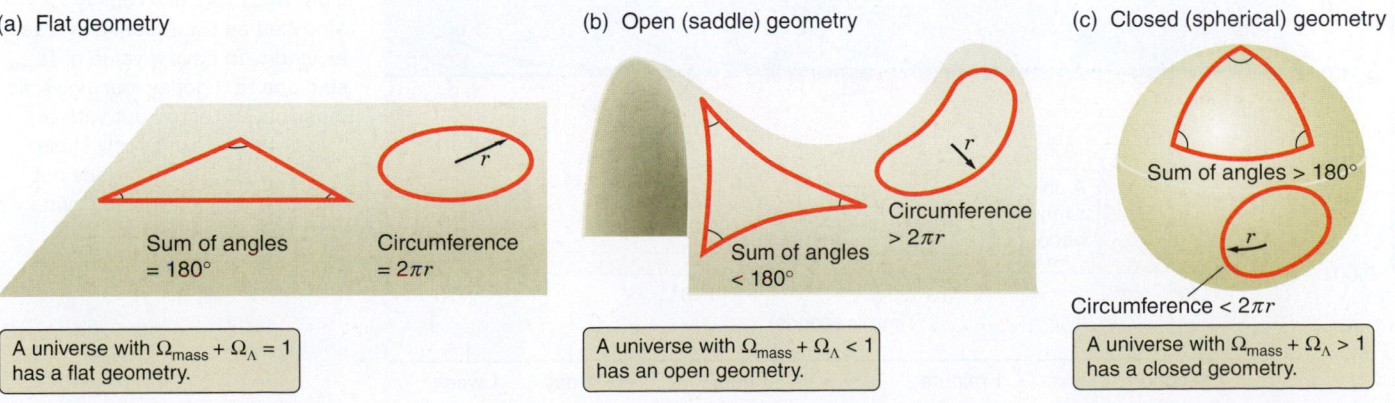

(a) Flat geometry

Sum of angles = 180°
Circumference = $2\pi r$

A universe with $\Omega_{mass} + \Omega_\Lambda = 1$ has a flat geometry.

(b) Open (saddle) geometry

Sum of angles < 180°
Circumference > $2\pi r$

A universe with $\Omega_{mass} + \Omega_\Lambda < 1$ has an open geometry.

(c) Closed (spherical) geometry

Sum of angles > 180°
Circumference < $2\pi r$

A universe with $\Omega_{mass} + \Omega_\Lambda > 1$ has a closed geometry.

type of universe, in which $\Omega_{mass} + \Omega_\Lambda < 1$, is also infinite and is referred to as an **open universe**. In this type of universe the circumference of a circle is greater than 2π times its radius, and triangles contain less than 180°.

The final possibility, in which $\Omega_{mass} + \Omega_\Lambda > 1$, is a universe shaped like the surface of a sphere (**Figure 21.8c**). The geometric relationships on a sphere are similar to those in the vicinity of a massive object, as discussed in Chapter 17. The circumference of a circle on a sphere is less than 2π times its radius, and triangles contain more than 180°. This possibility is called a **closed universe** because space is finite and closes back on itself.

Again we face the question "Which of these shapes describes the universe in which we live?" The measurements are difficult. Even so, as seen in Figure 21.4, $\Omega_{mass} + \Omega_\Lambda$ is close to 1 (0.3 + 0.7), meaning that our universe is very nearly flat.

> Recent evidence suggests that our universe is remarkably flat.

21.2 Inflation

Even with our understanding of the expansion, acceleration, and shape of the universe, it is remarkable that Big Bang cosmology makes so many correct predictions about the properties of the universe in which we live. A century ago, astronomers were struggling just to get a handle on the size of the universe. Today we have a comprehensive theory that ties together many diverse facts about nature: the constancy of the speed of light, the properties of gravity, the motions of galaxies, and even the origins of the very atoms of which we are made. The case for the Big Bang is compelling. Even so, as our knowledge of the expansion of

the universe has grown and our observations of the cosmic background radiation have improved, a number of puzzles have arisen. The solution to these puzzles has forced us to consider some remarkable ideas about how our universe expanded when it was very young.

The Universe Is Much Too Flat

The first puzzle that we run into when observing our universe is that the universe is too flat. In fact, the universe is much too close to being exactly flat for this to have happened by chance. To see why this is a problem, imagine a model universe in which $\Omega_\Lambda = 0$. (This is a reasonable approximation for the very early days of *any* possible universe. When a universe is very young, it is also very dense, and Ω_{mass} is all that matters.) As our model universe expands out of its own version of the Big Bang, its density falls. At the same time, because the universe is expanding more slowly, the critical density needed to eventually stop the expansion falls as well. If Ω_{mass} is *exactly* 1, the decline in the actual density and the decline in the critical density go hand in hand: the ratio between the two, Ω_{mass}, remains 1 for all time, as shown by the middle curve in **Figure 21.9**. A universe that starts out perfectly flat *stays* perfectly flat.

On the other hand, a universe that does *not* start out perfectly flat has a very different fate. If a universe started out with Ω_{mass} even slightly greater than 1, its expansion would slow more rapidly than that of the flat universe, meaning that less and less density would be required to stop the expansion. At the same time, the actual density would be falling less rapidly than in the flat universe. This disparity between the actual density of the universe and the critical density would increase, causing the ratio between the two, Ω_{mass}, to skyrocket. This condition is shown by the blue curves that climb toward the top of

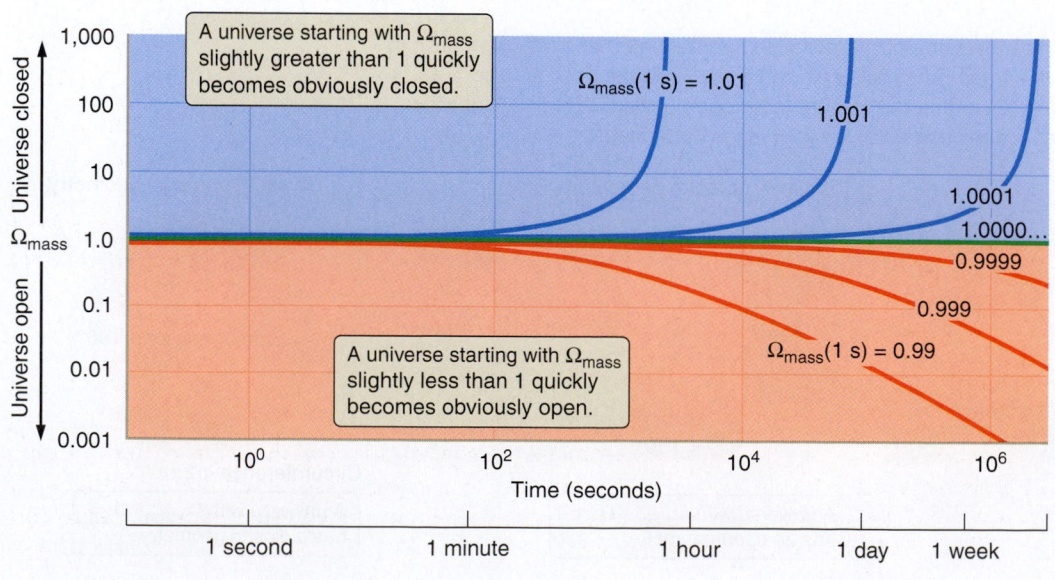

FIGURE 21.9 Universes in which Ω_{mass} has slightly different values at an age of 1 second. Notice that even tiny differences from $\Omega_{mass} = 1$ are rapidly amplified as the model universe expands. To have a value of Ω_{mass} so close to 1 today, our universe must have started out with a value of Ω_{mass} exquisitely close to 1. (The value of Ω_Λ does not affect the calculations shown.)

Figure 21.9. A universe that starts out even *slightly* closed rapidly becomes *obviously* closed, and would collapse long before stars could form.

Conversely, if a universe started with Ω_{mass} even a tiny bit less than 1, the expansion would slow *less* rapidly than in a flat universe. As time passed, more and more mass would be required for gravity to stop the too-rapidly-expanding universe. At the same time, the actual density of the universe would be dropping faster than in a flat universe. In this case, Ω_{mass} (the ratio between the actual density and the critical density) would plummet, leading to the red curves that dive toward the bottom of Figure 21.9.

Adding Ω_Λ to the picture makes the math a bit more complex, but it does not change the basic results. Try balancing a razor blade on its edge. If the blade is tipped just a tiny bit in one direction, it quickly falls that way. If the blade is tipped just a tiny bit in the other direction, it quickly falls in the other direction instead. By all rights, we would expect our universe to be either *obviously* open or *obviously* closed—analogous to the tipped razor blade. We find ourselves instead in a universe in which $\Omega_{mass} + \Omega_\Lambda$ is so close to 1 that we have difficulty telling which way the razor blade is tipped at all! Discovering that $\Omega_{mass} + \Omega_\Lambda$ is extremely close to 1 after more than 13 billion years is like balancing a razor blade on its edge and coming back 10 years later to find that it still has not tipped over!

For the present-day value of $\Omega_{mass} + \Omega_\Lambda$ to be as close to 1 as it is, $\Omega_{mass} + \Omega_\Lambda$ could not have differed from 1 by more than one part in 100,000 when the universe was 2,000 years old. When the universe was 1 second old, it had to be flat by at least one part in 10 billion. At even earlier times, it had to be much flatter still. This is simply too special a situation to be the result of chance—a fact referred to in cosmology as the **flatness problem**. *Something* about the early universe must have *forced* $\Omega_{mass} + \Omega_\Lambda$ to have a value incredibly close to 1.

The Cosmic Background Radiation Is Much Too Smooth

The second problem faced by our cosmological models is that the cosmic background radiation is surprisingly smooth. Following the discovery of the CBR in the 1960s, many observers turned their attention to mapping this background glow. At first they were reassured as result after result showed that the temperature of the CBR is remarkably constant, regardless of where one looks in the sky. Yet over time this strong confirmation of Big Bang cosmology turned instead into a puzzle that challenged our view of the early universe. Once we remove our motion relative to the CBR from the picture, the CBR is not just smooth—it is *too* smooth.

Why should we expect the CBR to be less uniform than it is? To understand the answer, we need to shift our attention from the very large to the very small. In Chapter 4 we discovered the bizarre world of quantum mechanics that shapes the world of atoms, light, and elementary particles. When the universe was extremely young, it was so small that quantum mechanical effects played a role in shaping the structure of the universe as a whole. In particular, the early universe was subject to the quantum mechanical **uncertainty principle**. The

> We expect nonuniformities in the early universe.

uncertainty principle says that as we look at a system at smaller and smaller scales, the properties of that system become less and less well determined. This principle applies whether we are talking about the properties of an electron in an atomic orbital, or about our entire universe at the time when it would have fit within the size of an atom.

Let's look at a simple analogy of how the uncertainty principle applies to the universe. Imagine sitting on the beach looking out across the ocean. Off in the distance, the surface of the ocean appears smooth and flat. The horizon looks almost like a geometric straight line. Yet the apparent smoothness of the ocean as a whole hides the tumultuous structure present at smaller scales, where waves and ripples upon waves fluctuate dramatically from place to place. In similar fashion, quantum mechanics says that as we look at smaller and smaller scales in the universe, conditions *must* fluctuate in unpredictable ways. In particular, quantum mechanics says that the smaller the universe we consider (that is, the earlier in the history of our universe we go), the more dramatic those fluctuations become. When the universe was young, it could *not* have been smooth. There must have been dramatic variations ("ripples") in the density and temperature of the universe from place to place.

If the universe had expanded slowly, those ripples would have smoothed themselves out. But the universe expanded much too rapidly for this. Different parts of the universe could not have "communicated" with each other (telling each other to smooth out the ripples) rapidly enough to

> The CBR is smoother than the early universe should have been.

smooth these ripples out. There just wasn't enough time after the Big Bang for a signal to travel from one region to the other. So when we look at the universe today, we should see the fingerprint of those early ripples imprinted on the cosmic background radiation—but we do not! The fact that the CBR is so smooth is referred to as the **horizon problem** in cosmology: different parts of the universe are too much like other parts of the universe that should have been "over their horizon" and beyond the reach of any signals that might have smoothed out the early quantum fluctuations. Basically, the horizon problem is this: how can different parts of the universe that underwent different fluctuations and were never able to communicate with one another still show the very same temperature in the cosmic background radiation?

Inflation Solves the Problems

In the early 1980s, American physicist Alan Guth (1947–) offered a solution to the flatness and horizon problems of cosmology. Guth suggested that for a brief time the young universe had undergone a period of **inflation** during which the universe itself had expanded at a rate *far* in excess of the speed of light. Like so many things we have seen on our journey, the numbers used to describe inflation are far beyond our ability to grasp intuitively. Between the ages of about 10^{-35} and 10^{-33} second, the scale factor R_U of the universe increased by a factor of at least 10^{30} and perhaps very much more. In that incomprehensibly brief instant, the size of the observable universe grew from 10-trillionths the size of the nucleus of an atom to a region about 3 meters across. That is like a grain of very fine sand growing to the size of today's *entire universe*—all in a billionth the time it takes light to cross the nucleus of an atom!

At this point you may well be saying to yourself, "Wait a minute! What happened to all that business about nothing

> Inflation was a period of intense expansion of the universe.

traveling faster than the speed of light?" Inflation does not violate the rule that no signal can travel *through* space at greater than the speed of light. During inflation, space *itself* expanded so rapidly that the distances between *points in space* increased faster than the speed of light.

To understand how inflation solves the flatness and horizon problems of cosmology, imagine that you are an ant living in the *two-dimensional* universe defined by the surface of a golf ball, as shown in **Figure 21.10**. Your universe would have two very apparent characteristics: First, it would be obviously curved. If you were to walk around the circumference of a circle in your two-dimensional universe and then measure the radius of the circle, you would find the circumference to be less than 2π times the radius. If you were to draw a triangle in your universe, the sum of its angles would be greater than 180°. The second obvious characteristic would be the dimples, approximately a half millimeter deep, on the surface of the golf ball.

Now imagine how this situation would change if your golf ball universe suddenly grew to the size of Earth. (Granted, this change is nowhere near comparable to the inflation experienced by the real universe, but you get the

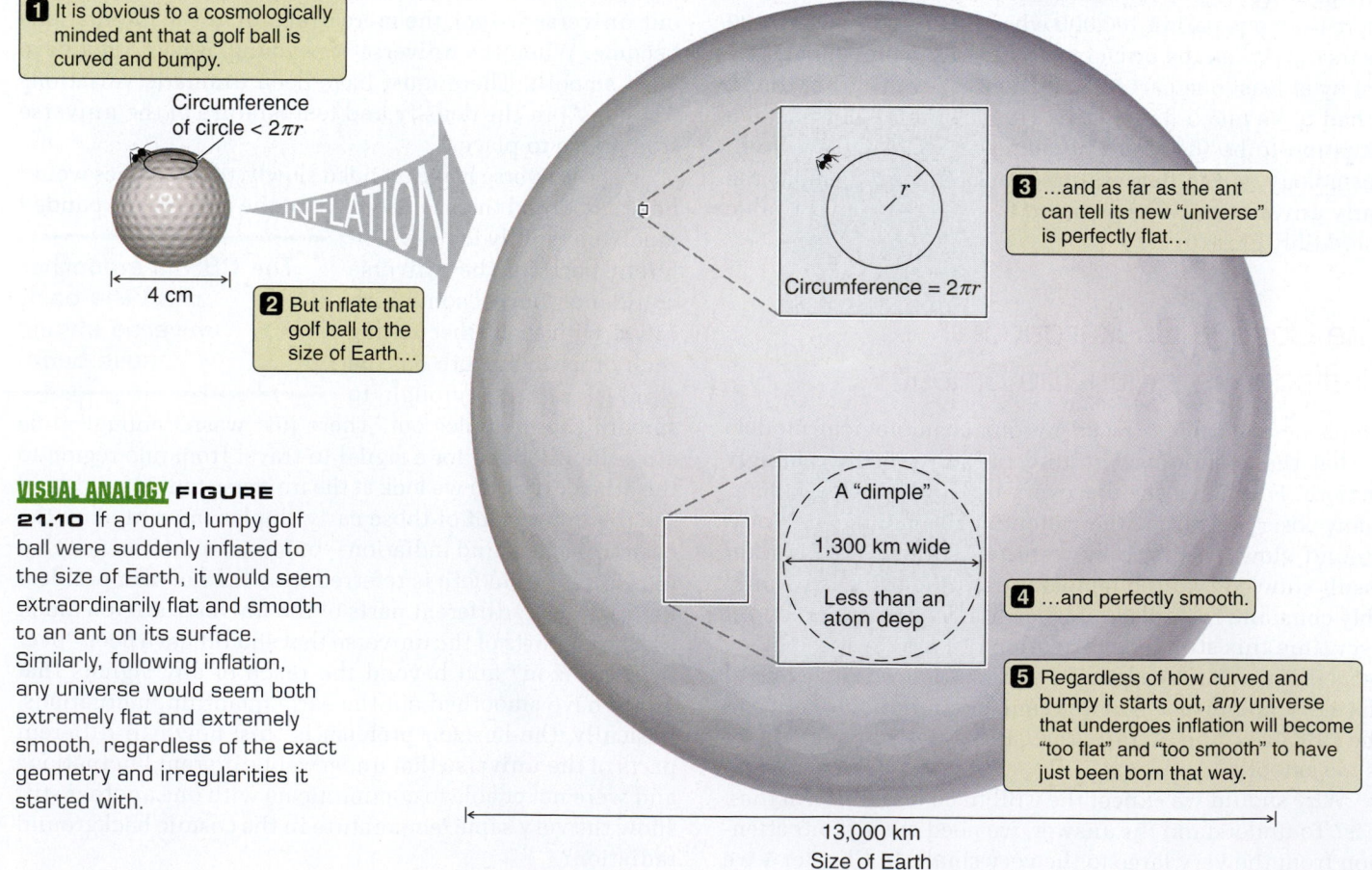

1 It is obvious to a cosmologically minded ant that a golf ball is curved and bumpy.

Circumference of circle < $2\pi r$

4 cm

2 But inflate that golf ball to the size of Earth…

3 …and as far as the ant can tell its new "universe" is perfectly flat…

Circumference = $2\pi r$

A "dimple"

1,300 km wide

Less than an atom deep

4 …and perfectly smooth.

5 Regardless of how curved and bumpy it starts out, *any* universe that undergoes inflation will become "too flat" and "too smooth" to have just been born that way.

13,000 km
Size of Earth

VISUAL ANALOGY FIGURE **21.10** If a round, lumpy golf ball were suddenly inflated to the size of Earth, it would seem extraordinarily flat and smooth to an ant on its surface. Similarly, following inflation, any universe would seem both extremely flat and extremely smooth, regardless of the exact geometry and irregularities it started with.

idea.) First, the curvature of your universe would no longer be apparent. An ant walking along the surface of Earth would be hard-pressed to tell that Earth is not flat. The circumference of a circle would be 2π times its radius, and there would be 180° in a triangle. In fact, it took us most of our history as a species to realize that Earth really is a sphere. In the case of inflationary cosmology, the universe after inflation would be extraordinarily flat (that is, with $\Omega_{mass} + \Omega_\Lambda$ extraordinarily close to 1) *regardless* of what the geometry of the universe was before inflation. Because the universe was inflated by a factor of at least 10^{30}, $\Omega_{mass} + \Omega_\Lambda$ immediately after inflation must have been one to within one part in 10^{60}, which is flat enough for $\Omega_{mass} + \Omega_\Lambda$ to remain close to 1 today. Today's universe is not flat by chance. It is flat because *any* universe that underwent inflation would emerge with a value for $\Omega_{mass} + \Omega_\Lambda$ that was within a gnat's eyelash of 1.

Inflating the size of a dimpled sphere makes it seem flatter.

So much for the flatness problem. What about the horizon problem? When our golf ball universe inflated to the size of Earth, the dimples that covered the surface of the golf ball were stretched out as well. Instead of being a half millimeter or so deep and a few millimeters across, these dimples now are only an atom deep but are hundreds of kilometers across. Again, our ant would be hard-pressed to detect any dimples at all. In the case of the real universe, inflation took the large fluctuations in conditions caused by quantum uncertainty in the preinflationary universe and stretched them out so much that they are unmeasurable in today's postinflationary universe. The slight irregularities that we do see in the CBR are the faint ghosts of quantum fluctuations that occurred as the universe inflated.

Huge expansion also smooths out inhomogeneities.

An early era of inflation in the history of the universe offers a handy way of solving the horizon and flatness problems, but why would the real universe have the audacity to do such a thing? It seems quite remarkable that the universe should undergo a period during which it expanded at such an "astronomical" rate. But there are many remarkable things about the universe. It is remarkable that spacetime has a shape. It is remarkable that light is both a wave and a particle. It is remarkable that galaxies and stars and planets exist at all. Still, it is fair to ask what caused the universe to undergo inflation in the first place, and the answer surely lies in the fundamental physics that governs the behavior of matter and energy at the earliest moments of the universe.

21.3 The Earliest Moments

To understand the universe requires that we understand the *forces* that govern the behavior of all matter and energy in the universe. There are four fundamental forces in nature, and everything in the universe is a result of their action (**Table 21.1**). Chemistry and light are products of the **electromagnetic force** acting between protons and electrons in atoms and molecules. The energy produced in fusion reactions in the heart of the Sun comes from the **strong nuclear force** that binds together the

There are four fundamental forces in nature.

TABLE 21.1

The Four Fundamental Forces of Nature

Force	Relative Strength	Range of Force	Particles That Can Carry the Force	Example of What the Force Does
Strong nuclear	1	10^{-15} m	Gluons	Holds protons and neutrons together in atomic nuclei.
Electromagnetic	10^{-2}	Infinite	Photons	Binds the electrons in an atom to the nucleus.
Weak nuclear	10^{-4}	10^{-16} m	W^+, W^-, and Z^0	Responsible for beta decay.
Gravitational	10^{-38}	Infinite	Gravitons	Holds you to Earth; binds both the Solar System and the Milky Way together.

protons and neutrons in the nuclei of atoms. Beta decay of nuclei, in which a neutron decays into a proton, an electron, and an antineutrino, is governed by the **weak nuclear force**. Finally, there is our old friend **gravity**, which has played such a major role at every point along our journey. We summarize these four forces in Table 21.1. How these forces—these physical laws—came into being is part of the history of the universe as well.

May the Forces Be with You

In Chapter 4 we spoke of electromagnetism using electric and magnetic fields, but we also spoke of the quantum mechanical description of light as a stream of particles called photons. Because there is only one reality, both of these descriptions of electromagnetism have to coexist. The branch of physics that deals with this reconciliation is called **quantum electrodynamics**, or **QED**.

QED treats charged particles almost as if they were baseball players engaged in an endless game of catch. As baseball players throw and catch baseballs, they experience forces. Similarly, in QED, charged particles "throw" and "catch" an endless stream of "virtual photons," as illustrated in **Figure 21.11**. Earlier we grappled with the idea that quantum mechanics is a science of probabilities rather than certainties. The QED

> In QED, photons carry the electromagnetic force.

description of the electromagnetic interaction between two charged particles is an average of all the possible ways that the particles could throw photons back and forth. The resulting force acts, over large scales, like the classical electric and magnetic fields described by Maxwell's equations. Physicists speak of the electromagnetic force being "mediated by the exchange of photons." As is always the case with quantum mechanics, the world described by QED is hard to picture. Even so, QED is one of the most accurate, well tested, and precise branches of physics. As of this writing, not even the tiniest measurable difference between the predictions of the theory and the outcome of an actual experiment has been found.

The central idea of QED—forces mediated by the exchange of carrier particles—provides a template for understanding two of the other three fundamental forces in nature. The electromagnetic and weak nuclear forces have been combined into a single theory called **electroweak theory**. This theory predicts the existence of three particles—labeled W^+, W^-, and Z^0—that mediate the weak nuclear force. In the 1980s, physicists identified these particles in laboratory experiments, thus confirming the essential predictions of electroweak theory.

> The weak nuclear and electromagnetic forces combine in electroweak theory.

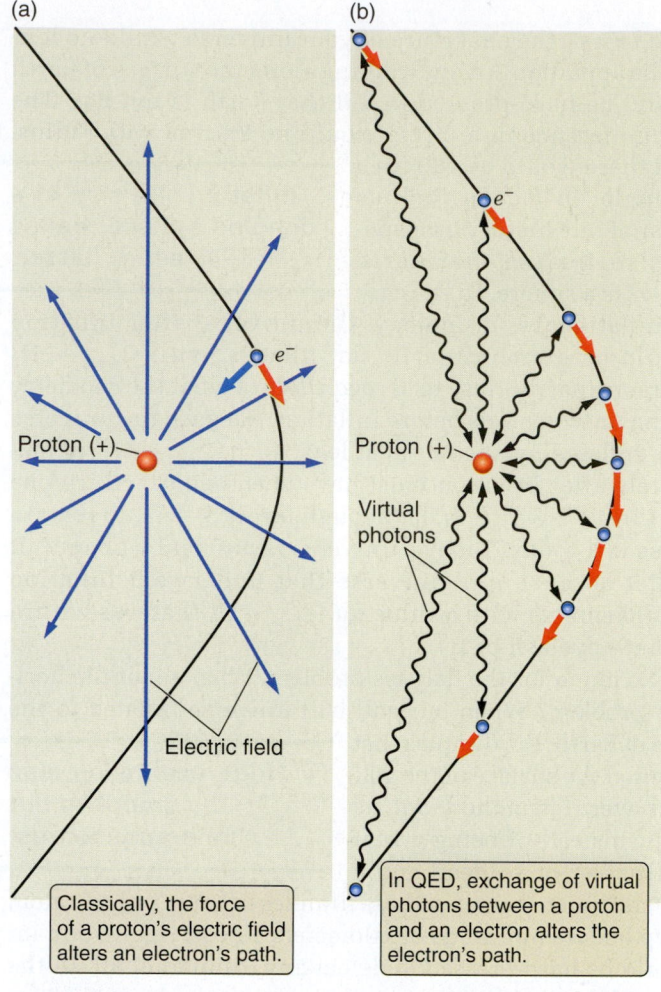

(a)

Proton (+)

e^-

Electric field

(b)

Proton (+)

Virtual photons

e^-

Classically, the force of a proton's electric field alters an electron's path.

In QED, exchange of virtual photons between a proton and an electron alters the electron's path.

FIGURE 21.11 (a) The classical view of an electron being deflected from its course by the electric field from a proton. (b) According to quantum electrodynamics, the interaction is properly viewed as an ongoing exchange of virtual photons between the two particles.

The strong nuclear force is described by a third theory, called **quantum chromodynamics**, or **QCD**. In this theory, particles such as protons and neutrons are composed of more fundamental building blocks, called **quarks**, that are bound together by the exchange of another type of carrier particle, dubbed **gluons**. Together, electroweak theory and QCD are referred to as the **standard model** of particle physics. A deeper investigation of the standard model must await another journey. Here we leave the discussion by pointing out that, excluding gravity, the standard model is able to explain all the currently observed interactions of matter and has made many predictions that were subsequently confirmed by laboratory experiments. However, the standard model leaves many questions unan-

> Electroweak theory + QCD = The standard model of particle physics.

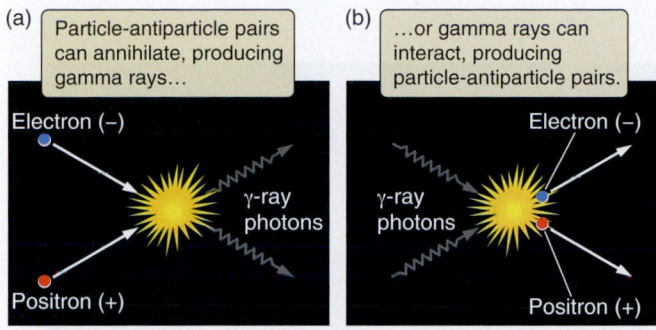

(a) Particle-antiparticle pairs can annihilate, producing gamma rays...

Electron (−)

γ-ray photons

Positron (+)

(b) ...or gamma rays can interact, producing particle-antiparticle pairs.

Electron (−)

γ-ray photons

Positron (+)

FIGURE 21.12 (a) An electron and a positron annihilate, creating two gamma-ray photons that carry away the energy of the particles. (b) In the reverse process, pair production, two gamma-ray photons collide to create an electron-positron pair.

swered, such as whether neutrinos have mass, or why strong interactions are *so* much stronger than weak interactions.

A Universe of Particles and Antiparticles

For modern theories of particle physics to make any sense, every type of particle in nature must also have an alter ego—an **antiparticle**—that is the opposite of that particle in every way described by quantum mechanics. For the electron there is the antielectron, otherwise known as the positron. For the proton there is the antiproton, for the neutron the antineutron, and so on down the list. One fascinating property of these particle-antiparticle pairs is that if you bring such a pair together, the two particles will annihilate each other.

When a particle-antiparticle pair annihilates, the mass of the two particles is converted into energy in accord with Einstein's special theory of relativity ($E = mc^2$). For example, in **Figure 21.12a** an electron and a positron annihilate each other, and the energy is carried away by a pair of photons. (This is the idea behind *Star Trek*'s "antimatter" engines.) When we run time backward (as we are allowed to do in particle physics), we see two high-energy photons colliding with each other, as in **Figure 21.12b**, creating in their place an electron-positron pair. This is an example of how an energetic event can create a particle and its corresponding antiparticle—a process called **pair production**.

In principle, *any* type of particle and its antiparticle can be created in this way. The only limitation comes when there is not enough energy available to supply the rest mass of the particles being created. A specific example helps to show how this works. An electron (or positron) has a mass of $m_e = 9.11 \times 10^{-31}$ kilograms (kg), which corresponds to an energy ($E = m_e c^2$) of 8.20×10^{-14} joules (J). If two gamma-ray photons with a combined energy greater than 16.40×10^{-14} J

($E = 2 \times m_e c^2$) collide, then the two photons may disappear and leave an electron-positron pair behind in their place. If the photons have more than the necessary energy, then the extra energy goes into the kinetic energy of the two newly formed particles.

Now we apply this idea to a hot universe awash in a bath of Planck radiation. Using Wien's law from Chapter 4, which is stated as $\lambda_{peak} = (2{,}900\ \mu m\ K)/T$, and the expression for photon energy ($E = hc/\lambda$), we can show that when the universe was less than about 100 seconds old and had a temperature greater than a billion kelvins, it was filled with photons that had enough energy to create electron-positron pairs. Under these conditions, photons were constantly colliding, creating electron-positron pairs; and electron-positron pairs were constantly annihilating each other, creating pairs of gamma-ray photons. The whole process reached an equilibrium, determined strictly by temperature, in which pair creation and pair annihilation exactly balanced each other. Rather than being filled only with a swarm of photons, at this time the universe was filled with a swarm of photons, electrons, and positrons, as illustrated in **Figure 21.13a**. Similarly, when the universe was even hotter, photons would have produced a swarm of protons and antiprotons.

> The early universe was filled with photons, electrons, and positrons.

The Frontiers of Physics

Physics has given us the tools to understand all the structures we have seen so far on our journey. There are still many gaps in what we know, but we are confident that everything from planets to stars to galaxies can be understood if we apply our current knowledge of the four fundamental forces. In the first decade of the 21st century, most of astronomy has been engaged in a struggle with the complexity of the universe, rather than a shortcoming of our understanding of the fundamental laws governing matter, energy, and spacetime. As we push further back toward the Big Bang itself, however, the nature of the game changes. We need new physical theories.

In Foundations 10.1 we explored the power of symmetry—the idea that we can learn a lot about nature just by thinking about the way one part of something matches up with another. There we were talking about the gravitational forces that hold planets together, but other kinds of symmetry exist as well. In the process of pair creation there is a symmetry between matter and **antimatter**: for every particle created, its antiparticle is created as well.

As the universe cooled, there was no longer enough energy to support the creation of particle pairs, so the swarm of particles and antiparticles that filled the early universe annihilated each other and were not replaced. When this

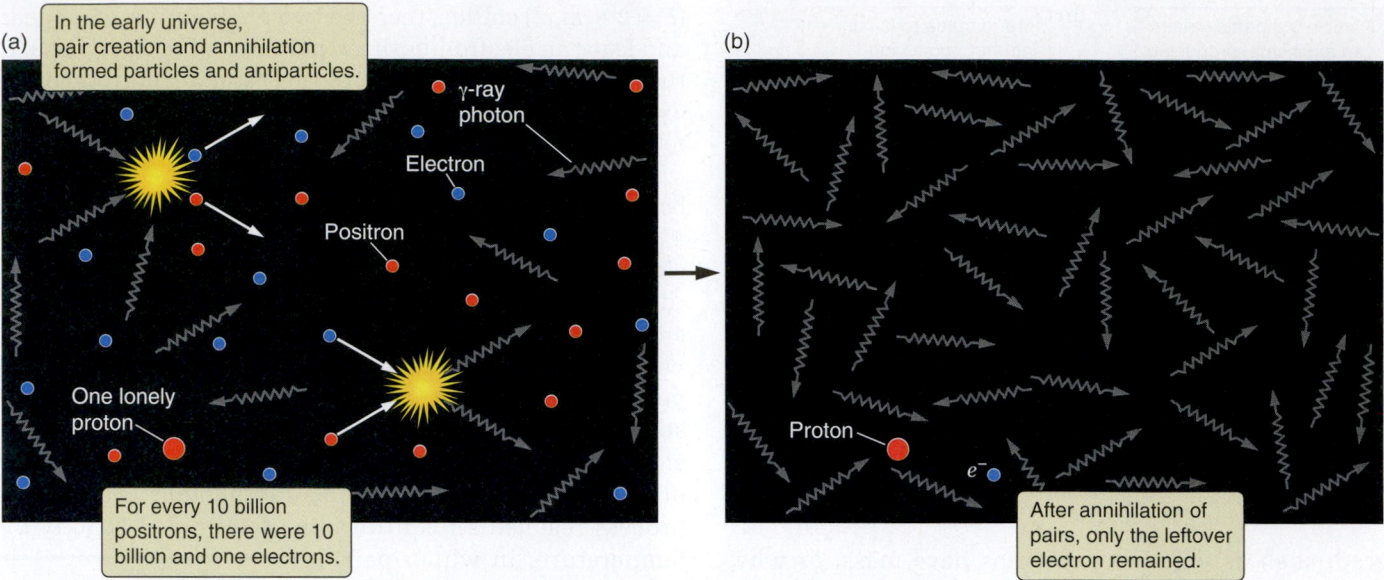

(a)

In the early universe, pair creation and annihilation formed particles and antiparticles.

γ-ray photon

Electron

Positron

One lonely proton

For every 10 billion positrons, there were 10 billion and one electrons.

(b)

Proton

e^-

After annihilation of pairs, only the leftover electron remained.

FIGURE 21.13 (a) A swarm of electrons, positrons, and photons in the very early universe. For every 10 billion positrons, there were 10 billion and one electrons. (b) After these particles annihilated, only the one electron was left.

happened, every electron should have been annihilated by a positron; and every proton should have been annihilated by an antiproton. This was almost the case, but not quite. For every electron in the universe today, there were 10 billion and one electrons in the early universe, but only 10 billion positrons. This one-part-in-10-billion excess of electrons over positrons meant that when electron-positron pairs finished annihilating each other, some electrons were left over—enough to account for all the electrons in all the atoms in the universe today (**Figure 21.13b**). Similarly, there was an excess of protons over antiprotons in the early universe, and the protons we see today are all that is left from the annihilation of proton-antiproton pairs.

> For every 10 billion positrons there were 10 billion and one electrons.

If the standard model of particle physics were a complete description of nature, then the one-part-in-10-billion imbalance between matter and antimatter would not have been there in the early universe. The symmetry between matter and antimatter would have been complete. No matter at all would have survived into today's universe, and we would not exist. The fact that you are reading this page demonstrates that something more needs to be added to the model.

According to current ideas, the symmetry between matter and antimatter may be broken in a theory that joins the electroweak and strong nuclear forces together in much the same way that electroweak theory unified our understanding of electromagnetism and the weak nuclear force. Such a theory, which combines three of the four fundamental forces into a single grand, unified force, is referred to as a **grand unified theory**, or **GUT**.

Many possible grand unified theories exist, and they make many predictions about the universe. Unfortunately, most of those predictions are impossible to test with even the largest of today's particle colliders. The problem is that the particles that carry the force or mediate GUTs are so massive that it takes enormous amounts of energy to bring them into existence—roughly a trillion times as much energy as can be achieved in today's particle accelerators! Even so, some predictions of GUTs are testable with current technology. For example, GUTs predict that protons should be unstable particles that, given enough time, will decay into other types of elementary particles. This is a *very* slow process. Over the course of your life, GUTs predict that there may be as much as a 1 percent chance that *one* of the 10^{28} or so protons in your body will decay. As of this writing, proton decay has yet to be observed, and only the very simplest of GUTs have been ruled out. As we speak, however, large arrays of detectors, such as those at Super-Kamiokande in Japan (pictured in **Figure 21.14**) are peering into huge tanks of water, waiting to see the signature of such a proton decay event. Perhaps soon we will see a news story heralding the confirmation of this central prediction of GUTs.

> GUTs predict that even the proton will decay.

Just as grand unified theories predict that the proton will decay, they can also explain why the universe is composed of matter rather than antimatter. Recall that in the early universe, for every 10 billion positrons there were 10

(a)

FIGURE 21.14 Water tank and detectors used to search for proton decay at the Super-Kamiokande neutrino observatory in Japan. (a) The empty tank. (b) The tank filled with water.

(b)

billion plus one electrons. GUTs provide a possible explanation for this excess of particles over antiparticles. To predict the *exact* amount of this excess, however, would require us to know the *correct* grand unified theory—which as yet we don't. Still, this class of models can at least explain *why* there is an excess of matter over antimatter.

The particles that mediate GUTs may be beyond the reach of today's high-energy physics labs. But when the universe was *very* young (younger than about 10^{-35} second) and *very* hot (hotter than about 10^{27} K), there was enough

energy available for these particles to be freely created. During this time, the distinction among the electromagnetic, weak nuclear, and strong nuclear forces had not yet come into being. There was only the one grand, unified force. Welcome to the era of GUTs.

During this era of GUTs, the apparent size of the entire observable universe was less than a trillionth the size of a single proton. Such an infinitesimally small size may seem virtually incomprehensible, yet the basic ideas needed to understand this universe are already in place. As we move backward in time, however, there is one threshold we have yet to cross. The story of advances in our understanding of physical law has been a story of unification—of the electromagnetic and weak forces into the electroweak theory, and then of these and the strong nuclear force into a GUT. But this program is incomplete. How does gravity fit into this scheme?

General relativity provides a beautifully successful description of gravity that correctly predicts the orbits of planets, describes the ultimate collapse of stars, and even enables us to calculate the structure of the universe.

> Gravity does not fit into the GUT picture.

Yet general relativity's description of gravity "looks" very different from our theories of the other three forces. Rather than talking about the exchange of photons or gluons or other carrier particles, general relativity talks about the smooth, continuous canvas of spacetime upon which events are painted.

We might be tempted to say, "Oh well. Gravity works one way and the other forces work another way, and that is how the universe happens to be." In practice, that is exactly what we do when we call on quantum mechanics to tell us about the properties of atoms and then use relativity to describe the passage of time or the expansion of the universe. Even the era of GUTs is described perfectly if we treat gravity as a separate force. As we push back even closer to the moment of the Big Bang, however, this happy coexistence between relativity and quantum mechanics turns instead into a brutal confrontation.

Toward a Theory of Everything

When the universe was younger than about 10^{-42} second old, its density was incomprehensibly high. The observable universe was so small that 10^{60} universes would have fit into the volume of a single proton! Under these extreme conditions, quantum mechanical fluctuations in the matter and radiation making up the universe involved immense amounts of mass—so much mass that quantum fluctuations made mincemeat of spacetime. Rather than a smooth sheet, spacetime was a quantum mechanical froth. The failure

> In the Planck era, the whole universe was a quantum mechanical froth.

of general relativity to describe this early universe is reminiscent of the failure of Newtonian mechanics to describe the structure of atoms. An electron in an atom must be thought of in terms of probabilities rather than certainties. Similarly, there is no unique history for the earliest moments after the Big Bang. This era in the history of the universe is referred to as the **Planck era**, signifying that we can understand the structure of the universe itself during this period only by using the ideas of quantum mechanics.

The conflict between the continuous and the discrete—between general relativity and quantum mechanics—brings us to the current limits of human knowledge. The physics that we know can take us back to a time when the universe was a millionth of a trillionth of a trillionth of a trillionth of a second old; but to push back any further, we need something new. We need a theory that combines general relativity and quantum mechanics into a single theoretical framework unifying all four of the fundamental forces. Here we have reached the holy grail of modern physics. To understand the earliest moments of the universe, we need a **theory of everything** (**TOE**).

A successful theory of everything would do more than unify general relativity with quantum mechanics. It would tell us which of the possible GUTs is correct and would provide an answer for the nature of dark matter. A successful theory of every-

> Superstring theory is a possible theory of everything.

thing would also necessarily answer several outstanding issues in cosmology, including the how, when, and why of inflation, and the underlying physics of the dark energy that is accelerating the expansion of the universe. Physicists are currently grappling with what a TOE might look like. Currently, the most promising contender for the title is **superstring theory**. Here, elementary particles are viewed not as points but as tiny loops called "strings." A guitar string vibrates in one way to play an F, another way to play a G, and yet another way to play an A. According to superstring theory, different types of elementary particles are like different "notes" played by vibrating loops of string.

In principle, superstring theory provides a way to reconcile general relativity and quantum mechanics, but there is a price to pay for this success. To make superstring theory work, we have to imagine that these tiny loops of string are vibrating in a universe with 10 spatial dimensions! (Adding time to the list would make our universe 11-dimensional.) How can that be, when we clearly experience only 3 spatial dimensions? Whereas the three spatial dimensions that we know spread out across the vastness of our universe, the other seven dimensions predicted by superstring theory wrap tightly around them-

> Superstring theory predicts 10 spatial dimensions, but we experience only 3 of them.

selves (**Figure 21.15**), extending no further today than they did a brief instant after the Big Bang.

To better appreciate how this bizarre notion works, imagine what it would be like to live in a three-dimensional universe (like the one we experience) in which one of those dimensions extended for only a tiny distance. Living in such a universe would be like living within a thin sheet of paper that extended billions of light-years in two directions but was far smaller than an atom in the third. In such

FIGURE 21.15 It is virtually impossible to visualize what seven spatial dimensions wrapped up into structures far smaller than the nucleus of an atom would be like. Here such geometries are projected onto the two-dimensional plane of the paper.

Order "Froze Out" of the Cooling Universe

To understand the very earliest moments in the history of the universe, we have been looking backward to higher and higher energies and correspondingly to earlier and earlier times. We have seen how the four forces become unified in stages—from four separate forces to electroweak theory to GUTs and then to a TOE. We find that our universe started out with one theory of everything, and as the universe expanded and cooled the various forces emerged to take on lives of their own. **Figure 21.16** illustrates how the four fundamental forces emerged in the evolving universe. In the first 10^{-43} second after the Big Bang, as described by the TOE, the physics of elementary particles and the physics of spacetime were one and the same. As the universe expanded and cooled, gravity parted ways with the forces described by the GUT. Spacetime took on the properties described by general relativity. Inflation may also have been taking place at this time.

As the universe continued to expand and its temperature fell further, less and less energy was available for the creation of particle-antiparticle pairs. When the particles responsible for GUT interactions could no longer form, the strong force split off from the electroweak force. One might speak of this transition as the strong force "freezing out" of the GUT because this and other similar transitions are reminiscent of the phase change that occurs as water changes to ice and molecules become more constrained in their motions. Somewhere along the line, as the unity of the original TOE was lost, the symmetry between matter and antimatter was broken. As a result, the universe ended up with more matter than antimatter.

> As the universe cooled, atomic nuclei formed first, followed by atoms.

The next big change took place when the particles responsible for unifying the electromagnetic and weak nuclear forces froze out, leaving these two forces independent of each other. The four fundamental forces of nature that govern today's universe had now come into their own. The temperature of the universe had fallen to a chilly 10^{16} K, and a 10-trillionth of a second had ticked off the cosmic clock. It was a full minute or two before the universe cooled to the billion-kelvin mark, below which not even pairs of electrons and positrons could form.

Although the universe was now too cool to form additional particles and their antiparticles, it was still hot enough for the thermal motions of protons to overcome the electric barriers between them, allowing nuclear reactions to take place. These reactions formed the least massive elements, including helium, lithium, beryllium, and boron, but could not create more massive elements.

Big Bang nucleosynthesis had come to an end by the time the universe was 5 minutes old, and the tempera-

a universe we would easily be aware of length and width—we could move in those directions at will. In contrast, we would have no freedom to move in the third dimension at all, and we might not even recognize that the third dimension existed. Perhaps our only inkling of the true nature of space would come from the fact that in order to explain the results of particle physics experiments, we would have to assume that particles extended into a third, unseen dimension. In like fashion, if superstring theory is correct, we see three spatial dimensions extending on possibly forever, but we are unaware of the fact that each point in our three-dimensional space actually has a tiny but finite extent in seven other dimensions at the same time!

Superstring theory is only a pale shadow of the sort of well-tested theories that we have made use of throughout this book. In some respects, superstring theory is no more than a promising idea providing direction to theorists searching for a TOE. It is worth noting that we will probably never be able to build particle accelerators that enable us to directly search for the most fundamental particles predicted by a TOE. The energies required are simply too high. Fortunately, however, nature has provided us with the ultimate particle accelerator: the Big Bang itself! The structure of the universe that we see around us today is the observable result of that grand experiment.

One of the most intriguing questions that a successful TOE may answer is whether the universe could have been different. Do we live in only one of many possible universes, each with different physical laws? Or are the physical laws that govern our universe the *only* consistent set of physical laws that could exist? We do not know the answer yet, but possibly within your lifetime we will.

> Is ours the only possible universe?

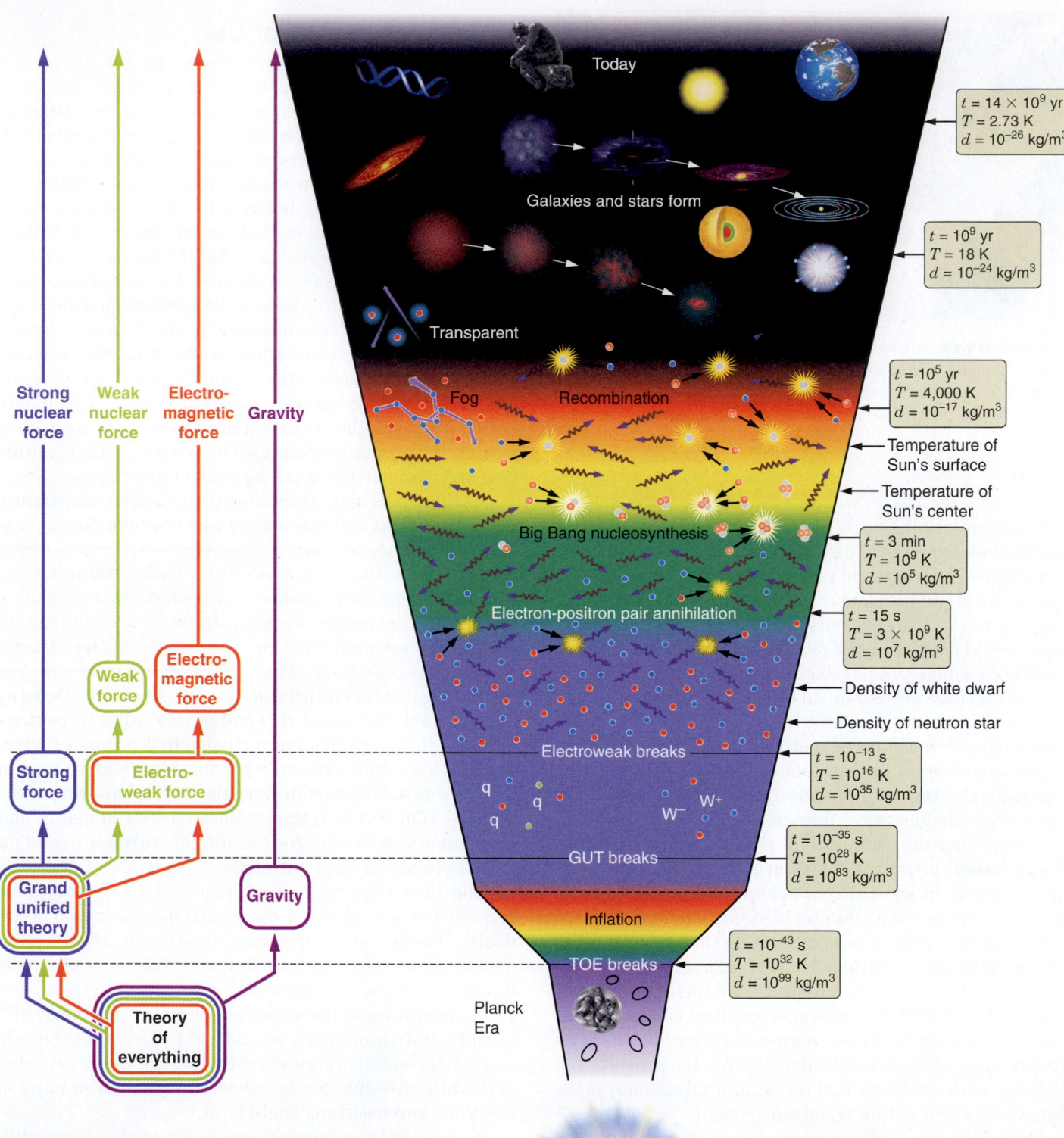

Strong nuclear force

Weak nuclear force

Electro-magnetic force

Gravity

Weak force

Electro-magnetic force

Strong force

Electro-weak force

Grand unified theory

Gravity

Theory of everything

Today
$t = 14 \times 10^9$ yr
$T = 2.73$ K
$d = 10^{-26}$ kg/m^3

Galaxies and stars form
$t = 10^9$ yr
$T = 18$ K
$d = 10^{-24}$ kg/m^3

Transparent

Fog   Recombination
$t = 10^5$ yr
$T = 4{,}000$ K
$d = 10^{-17}$ kg/m^3

Temperature of Sun's surface

Temperature of Sun's center

Big Bang nucleosynthesis
$t = 3$ min
$T = 10^9$ K
$d = 10^5$ kg/m^3

Electron-positron pair annihilation
$t = 15$ s
$T = 3 \times 10^9$ K
$d = 10^7$ kg/m^3

Density of white dwarf

Density of neutron star

Electroweak breaks
$t = 10^{-13}$ s
$T = 10^{16}$ K
$d = 10^{35}$ kg/m^3

q   q W^- W^+
q

GUT breaks
$t = 10^{-35}$ s
$T = 10^{28}$ K
$d = 10^{83}$ kg/m^3

Inflation

TOE breaks
$t = 10^{-43}$ s
$T = 10^{32}$ K
$d = 10^{99}$ kg/m^3

Planck Era

Big Bang

FIGURE 21.16 Eras in the evolution of the universe. As the universe expanded and cooled following the Big Bang, it went through a series of phases determined by what types of particles could be created freely at that temperature. Later the structure of the universe was set by the gravitational collapse of material to form galaxies and stars, and by the chemistry made possible by elements formed in stars.

ture of the universe had dropped below about 800 million K (see Section 18.5). The density of the universe at this point had fallen to only about a tenth that of water. Normal matter in the universe now consisted of atomic nuclei and electrons, awash in a bath of radiation. So the universe remained for the next several hundred thousand years, until finally the temperature dropped so far that electrons were able to combine with atomic nuclei to form neutral atoms. We have encountered this event before. This was the era of recombination, which we see directly when we look at the cosmic background radiation. At this stage the radiation background could no longer dominate over matter, and gravity began playing its role in forming the vast structure of the universe that we now see. Understanding this structure will be the subject of the next chapter.

Seeing the Forest for the Trees

The journey that we have taken in *21st Century Astronomy* has involved many amazing discoveries. We have peered into the heart of the Sun, watched as planetary systems formed, and witnessed the violent death throes of massive stars. We have seen our ideas of cause and effect shattered by the quantum mechanical world of atoms, and we have come to accept that space and time are joined in a four-dimensional fabric of spacetime that is bent and even torn by the presence of mass. Yet even when witnessed within the context of such a remarkable journey of discovery, what we see when we look at the structure of the universe itself is truly mind-boggling.

We live in a universe in which the spacetime described by general relativity has been expanding outward for about 14 billion years—an expansion that can be traced back to a single moment when spacetime and all that it contains came into existence. The Big Bang is a theory, but one of the lessons we have learned along the way is that a well-tested and well-corroborated scientific theory is the closest that humans can ever come to certain knowledge. The predictions of the Big Bang have been borne out time and again, in everything from observations of the cosmic background radiation, to the predictions the theory makes about the abundances of chemical elements, to the fact of the expansion of the universe itself. Although the jury is still out, the best models of the universe early in the 21st century tell of a future in which the expansion of the universe continues forever at an ever-accelerating pace.

Seldom have novelists, theologians, or philosophers dared dream of anything so extraordinary as the universe of modern cosmology. Yet we arrive at our conclusion not by fantasy or conjecture, but by following a path laid down by hard-won observation and well-tested physical law. In this chapter we have discovered the shape of the canvas on which our existence is painted. Yet our world is not defined by spacetime alone. The trees and planets and stars are still there for us to see. We are part of a universe that is filled with structure. How did that structure arise? What are the acts in the grand play that continues on the cosmic stage? In the next chapter we will run the clock forward, watching as the structure and complexity of today's universe emerge.

We have described the canvas of the universe in this chapter. It is now time to move on to the painting itself. Hold on to your hats!

Summary

- Both gravity and the cosmological constant (or dark energy) determine the fate of the universe. Observations suggest that rather than slowing down, the expansion of the universe is accelerating.

- The very early universe may have gone through a brief but dramatic period of exceptionally rapid expansion, called inflation. If true, inflation would explain both the flatness and the homogeneity of the universe we see today.

- Understanding the very earliest moments in the universe requires that we also understand how the four fundamental forces of nature were all unified into one basic phenomenon at early times.

SmartWork, Norton's online homework system, includes algorithmically generated versions of these questions, plus additional conceptual exercises. If your instructor assigns questions in SmartWork, log in at **smartwork.wwnorton.com.**

Student Questions

THINKING ABOUT THE CONCEPTS

1. As applied to the universe, what is the meaning of *critical density*?

2. What set of circumstances would cause an expanding universe to reverse its expansion and end up in a "Big Crunch"?

3. What is the principal difference between normal matter and dark matter?

*4. Describe the observational evidence suggesting that Einstein's cosmological constant (a repulsive force) may be needed to explain the historical expansion of the universe.

5. Describe what astronomers mean by *dark energy*.

6. If the universe is being forced apart by dark energy, why isn't our galaxy, Solar System, or planet being torn apart?

7. Describe the cause and consequences of a "Big Rip."

8. In Chapter 18, we said we could estimate the age of the universe with Hubble time ($1/H_0$). Why does that not give us the best answer?

9. If you could accurately measure a triangle and a circle drawn on the surface of Earth, you would find that the sum of the triangle's interior angles is more than 180° and the circle's circumference is less than $2\pi r$. These results are contrary to what you probably learned about triangles and circles in your introductory geometry class. Explain why we get these results and how they relate to a spherical universe.

10. Explain the geometry of an open universe.

11. Do astronomers believe we live in a closed, flat, or open universe? Explain your answer.

*12. What is the flatness problem and why has it been a problem for cosmologists?

13. During the period of inflation, the universe may have briefly expanded at 10^{30} (a million trillion trillion) or more times the speed of light. Why did this ultra-rapid expansion not violate Einstein's special theory of relativity, which says that neither matter nor communication can travel faster than the speed of light?

14. Why is high-energy physics important to our understanding of the early universe?

15. Name the four fundamental forces in nature.

**16. The fundamental forces of the universe are generally assumed not to change.

a. How would the fate of the universe be affected if Newton's gravitational constant changed with time?

b. What if, instead, the electric force between charged particles changed with time?

17. Of the four fundamental forces in nature, which one depends on electric charge?

18. The standard model cannot explain why neutrinos have mass, or why electron-positron asymmetry existed in the early universe. Do these failings make it an incomplete theory? Should all of its predictions be ignored until the theory can solve these remaining issues?

19. Explain how quantum electrodynamics describes the electromagnetic interaction between two charged particles.

20. What are quarks?

*21. Explain how quantum chromodynamics describes the structure and stability of protons and neutrons.

22. What is an antiparticle, and how do its properties compare with a normal particle?

23. Explain what happens when you bring a particle and an antiparticle together.

24. Why are there so few antiparticles in the universe?

25. Explain the process of pair creation.

26. Describe the Planck era.

27. Is it possible that some planets, stars, or even galaxies are made entirely of antimatter? If so, how might we distinguish these from identical objects made of regular matter?

28. What are the basic differences between a grand unified theory (GUT) and the theory of everything (TOE)?

29. In this chapter we have spoken about GUTs and the TOE as if they exist and we have only to specify and test them. Is this a safe assumption?

30. As the sensitivity of our instrumentation increases, we are able to look ever farther into space and, therefore, ever further back in time. When we reach the era of recombination, however, we run into a wall and can see no further back in time. Explain why.

APPLYING THE CONCEPTS

31. Currently, the Hubble constant has an uncertainty of about 5 percent. What are the corresponding maximum and minimum ages allowed for the universe?

32. How many hydrogen atoms need to be in 1 cubic meter (m^3) of space to equal the critical density of the universe?

**33. Dark energy seems to have strong effects only in the regions of space between groups of galaxies. Estimate the density of mass below which dark energy operates.

34. The universe today has an average density $\rho_0 = 3 \times 10^{-28}$ kg/m^3. Assuming that the average density depends on the scale factor, as $\rho = \rho_0/R_U^{\,3}$, what was the scale factor of the universe when its average density was about the same as Earth's atmosphere at sea level ($\rho = 1.23$ kg/m^3)?

35. The proton and antiproton each have the same mass, $m_p = 1.67 \times 10^{-27}$ kg. What is the energy (in joules) of each of the two gamma rays created in a proton-antiproton annihilation?

**36. There are about 500 million CBR photons in the universe for every hydrogen atom. What is the equivalent mass of these photons? Is it large enough to factor into the overall density of the universe?

37. Suppose you brought together a gram of ordinary-matter hydrogen atoms (each composed of a proton and an electron) and a gram of antimatter hydrogen atoms (each composed of an antiproton and a positron). Keeping in mind that 2 grams is less than the mass of a dime,

a. Calculate how much energy (in joules) would be released as the ordinary matter and antimatter hydrogen atoms annihilated one another.

b. Compare this amount of energy with the energy released by a 1-megaton hydrogen bomb (1.6 × 10^{14} J).

**38. One GUT theory predicts that a proton will decay in about 10^{31} years, which means if you have 10^{31} protons, you should see one decay per year. The Super-Kamiokande observatory in Japan holds about 20 million kg of water in its main detector, and it did not see any decays in 5 years of continual operation. What limit does this observation place on proton decay and on the GUT theory described here?

listen; there's a hell of a good universe next door: let's go.

E. E. CUMMINGS (1894–1962)

An example of galactic cannibalism, the NGC 1532/1531 galaxy pair shows how a large galaxy grows by accreting and subsuming a small galaxy that strays too close.

CHAPTER 22

The Origin of Structure

22.1 Whence Structure?

Throughout the ages there have been as many different ideas about the origin of the universe as there have been cultural traditions. Thoughts about what the universe was like once upon a time have been part of the mythologies and traditions of all great civilizations. We live at a remarkable moment in history, when the nature of the early universe has moved from philosophical speculation to the realm of scientific fact. All we have to do to see the early universe is point our microwave telescopes at the sky. The glow of the cosmic background radiation is there for anyone with the appropriate technology to see.

The early universe was an extraordinary place—an expanding fireball that was far more uniform than the blue of the bluest sky on the clearest day. How different that universe was from the universe we see about us today! Today's universe is one of stars and galaxies, viewed from the surface of a planet with oceans and mountains and uncountable species of living things. The contrast between these two realities cries out for explanation. How did we get from there to here? Thus we come to one of the most philosophically intriguing questions in the history of human thought: What is the origin of structure?

KEY CONCEPTS

The universe that emerged from the Big Bang was incredibly uniform, wholly unlike today's universe of galaxies, stars, and planets. As our journey nears its end, we tackle face on the question that has been with us all along: Where does structure in the universe come from? Here we find that complex structure is a natural, unavoidable consequence of the action of physical law in an evolving universe.

- Just as stars cluster together in galaxies, galaxies gather together to form large and small clusters of galaxies, which themselves cluster together into giant superclusters.

- Galaxies formed as slight ripples in the dark matter emerging from the Big Bang, which then collapsed under the force of gravity, pulling in normal matter as well.

- Larger galaxies form from the merging together of small protogalaxies.

- Galaxies were drawn together by gravity to form the large groupings of galaxies we see today.

- There are two types of dark matter: cold and hot.

- Our own observable universe may be but one among an infinite collection of universes.

- Our current understanding of physics suggests that over the long course of time, our universe will grow very cold and structure itself may slowly disappear.

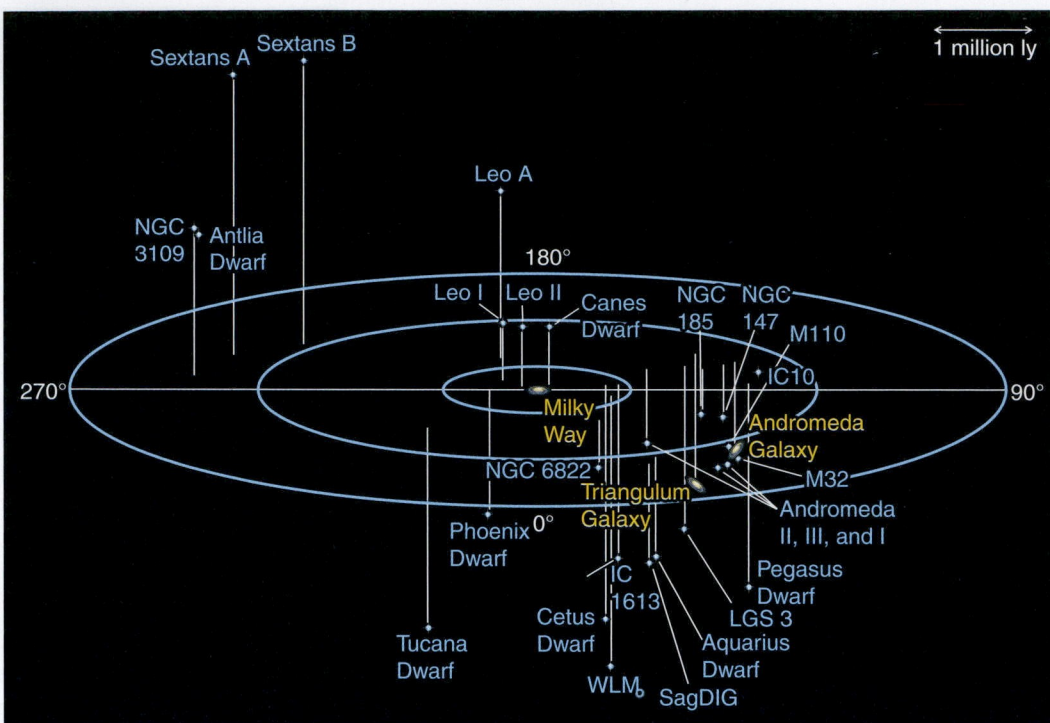

FIGURE 22.1 A graphical map of the galaxies in our own Local Group of galaxies. Most are dwarf galaxies. Spiral galaxies are shown in yellow.

22.2 Galaxies Form Groups, Clusters, and Larger Structures

Just as stars and clouds of glowing gas show us the structure of our galaxy, the distribution of galaxies shows us the structure of our universe. And just as it is gravity that holds galaxies together, giving them their shape, it is gravity that shapes the universe itself. No galaxy exists in utter isolation. The vast majority of galaxies are parts of gravitationally bound collections of galaxies. The smallest and most common of these are called **galaxy groups**. A galaxy group is a structure containing as many as several dozen galaxies, most of them dwarf galaxies. Our Milky Way is a member of the **Local Group**, which consists of two giant spirals (the Milky Way and the Andromeda Galaxy), along with at least 30 smaller dwarf galaxies in a volume of space about 4 million light-years (Mly) in diameter (**Figure 22.1**). Almost 98 percent of all the galaxy mass in the Local Group resides in just its two giant galaxies.

Larger systems of galaxies, called **galaxy clusters**, may consist of hundreds of galaxies, often with a more regular structure than is found in galaxy groups. Galaxy clusters are larger than groups, typically occupying a volume of space 10–15 Mly across. In many ways, our own Local Group can be considered a small cluster of galaxies, and we show its location relative to two well-known clusters (the Virgo Cluster and the Coma Cluster) in **Figure 22.2**. Like groups,

galaxy clusters contain far more dwarf galaxies than giant galaxies. Nevertheless, most of the *mass* in galaxy clusters resides in the giant galaxies. In addition, although spiral galaxies are common in most systems, elliptical galaxies are prevalent in only about one-fourth of galaxy clusters. The Virgo Cluster (**Figure 22.3a**), located 53 Mly from the Local Group (see Figure 22.2), is an example of a cluster containing mostly spiral galaxies. The more distant Coma Cluster (**Figure 22.3b**) is dominated by giant elliptical and S0 galaxies.

Clusters and groups of galaxies themselves bunch together to form enormous **superclusters**, which contain tens of thousands or even hundreds of thousands of galaxies and span regions of space typically more than 100 Mly in size. Our Local Group is part of the Virgo Supercluster, which also includes the Virgo Cluster.

Hubble's law is a powerful tool for mapping the distribution of galaxies, groups, clusters, and superclusters in space. Using Hubble's law, all we need to determine the distance to a galaxy is a single spectrum from which we can measure the galaxy's redshift. Although it is easy in principle to measure the redshift of a galaxy, in practice this can be a time-consuming process. Astronomers measured the first redshifts from spectra recorded on photographic plates. These plates required exposures of several hours to capture the feeble signal, and results rolled in at the breakneck pace of one or two redshifts per night of

> Galaxy redshift surveys measure distances to large numbers of galaxies.

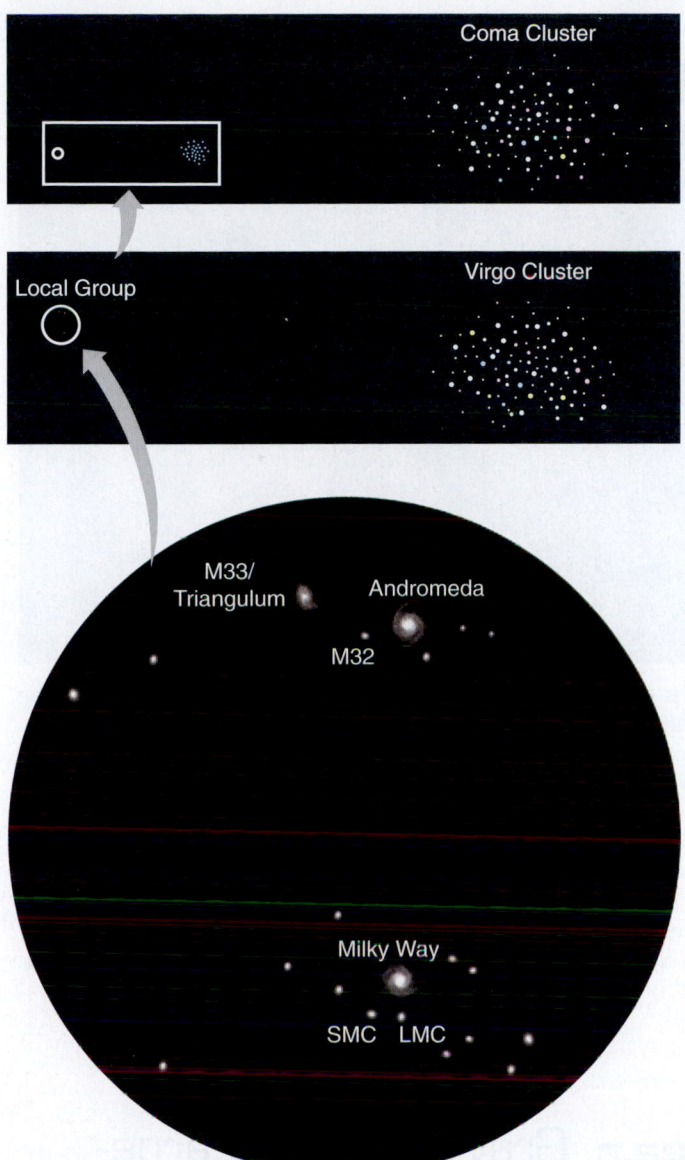

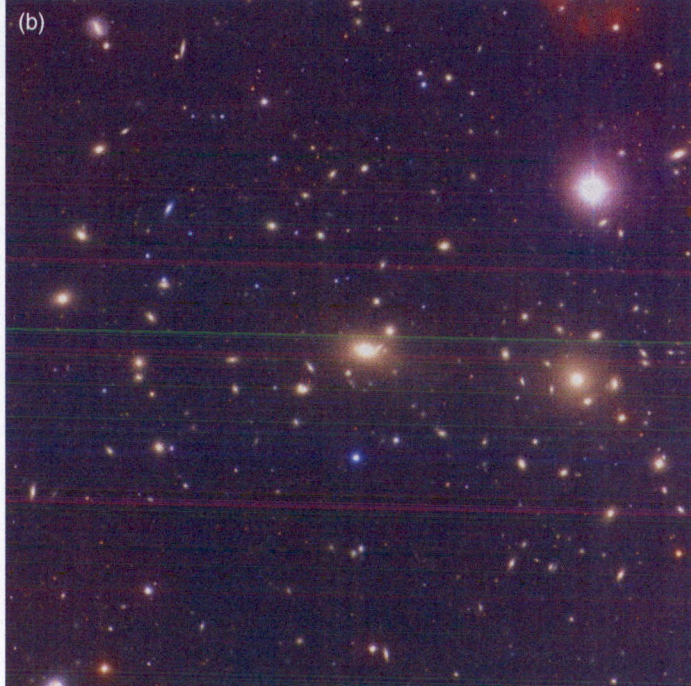

FIGURE 22.2 This figure shows a close-up view of the Local Group (bottom) and its position relative to the Virgo Cluster, which is the nearest cluster, and to the more distant Coma Cluster.

FIGURE 22.3 (a) The Virgo Cluster contains a large fraction of spiral galaxies. (b) The larger, richer Coma Cluster is dominated by elliptical galaxies.

painstaking observation. By 1975, astronomers had documented redshifts for only about a thousand of the hundred billion or so galaxies we can see. Since that time, there has been a happy marriage of larger telescopes, newer instruments (such as CCD detectors and spectrographs capable of observing many galaxies at once), and more powerful computers that process larger amounts of data in rapid, automated fashion. By 1990, astronomers had measured the redshifts of over 10,000 galaxies, and that number is now well over 1 million.

The Harvard-Smithsonian Center for Astrophysics conducted the first large redshift survey and, in 1986, presented the astronomical community with a "slice of the universe," as displayed in **Figure 22.4a**. The observations show that the clusters and superclusters, rather than being scattered randomly through space, are linked together in an intricate network of "filaments" and "walls." The concentrations of galaxies, in turn, surround large **voids**, regions of space

Large-scale structure fills the universe.

that are largely empty of galaxies. These voids represent some of the largest "structures" seen in the universe. Though the voids may seem like "a whole lot of nothing," we do not know that they are empty of *matter*—only that they are largely empty of observable galaxies. Clusters and super-

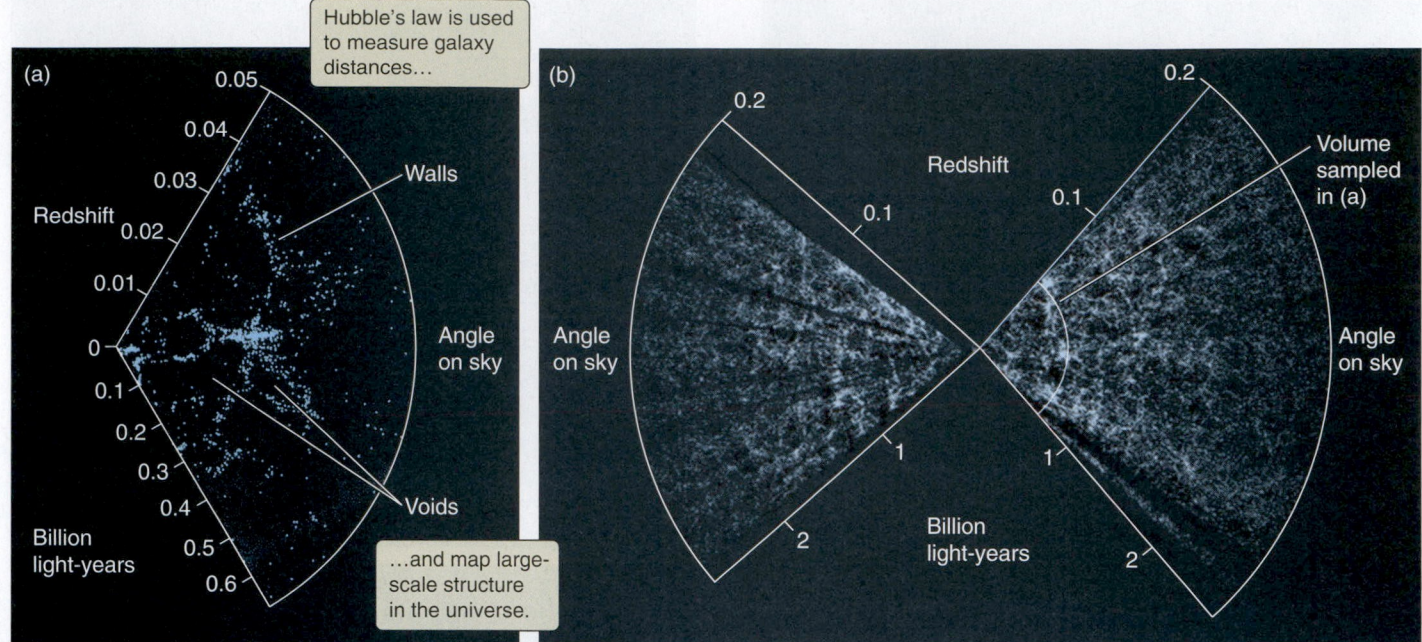

Hubble's law is used to measure galaxy distances…

Walls

Voids

…and map large-scale structure in the universe.

Volume sampled in (a)

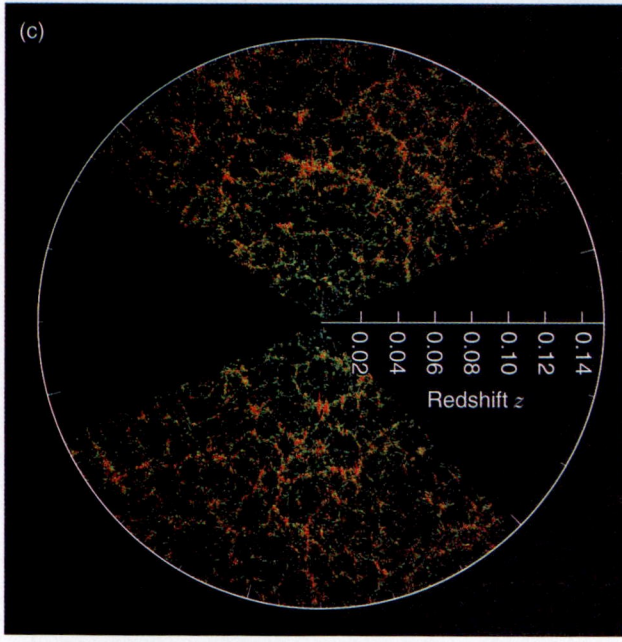

FIGURE 22.4 Redshift surveys use Hubble's law to map the universe. (a) In 1986, the Harvard-Smithsonian Center for Astrophysics redshift survey, called "A Slice of the Universe," was the first to show that clusters and superclusters of galaxies are part of even larger-scale structures. (b) The 2dF Galaxy Redshift Survey, completed in 2003, shows similar structures at even larger distances. (c) The 2008 Sloan Digital Sky Survey map of the universe extends outward to a distance of about 2 billion light-years. Shown here is a sample of 67,000 galaxies colored according to the ages of their stars, with the redder, more strongly clustered points showing galaxies that are made of older stars.

clusters are located within the walls and filaments. This structure, however, is not peculiar to the "nearby" universe. Subsequent surveys have looked at much larger volumes of space. **Figure 22.4b** shows the results of one such survey conducted with the Anglo-Australian Telescope at the Siding Spring Observatory in Australia. Results from a more recent survey, the Sloan Digital Sky Survey (SDSS), are shown in **Figure 22.4c**. For as far out as our observations can currently measure, the universe has a porous structure reminiscent of a sponge. Together, galaxies and the larger groupings in which they are found are referred to as **large-scale structure**.

22.3 Gravity Forms Large-Scale Structure

Large-scale structure cries out for explanation. Clearly it is the consequence of well-defined physical processes, but what processes? Cosmologists have proposed a number of ideas. For example, early on it was suggested that voids were the result of huge expanding blast waves from tremendous explosions that might have occurred in the early universe. The correct answer has turned out to be less fanciful, but far more satisfying: large-scale structure is the fingerprint of our old friend gravity.

In our discussion of star formation in Chapter 15, we learned about gravitational instabilities. As illustrated in Figure 15.14, this concept shows that star formation can begin with a molecular cloud with clumps inside it. Gravity then causes those clumps to collapse faster than their surroundings. If conditions are right, what began as relatively minor variations in the density of a cloud are turned by gravity into stars. The same gravitational instability is responsible for galaxy formation. If you replace the words *molecular cloud* with *universe* and *star* with *galaxy* in the sentence above referring to Figure 15.14, you will have a good starting point for understanding the way galaxies formed from slight variations in the distribution of matter following recombination.

Recall from the previous chapter that the slight ripples in the glow of the cosmic background radiation (CBR) probably result from quantum mechanical variations that imprinted structure on the early universe at the time of inflation. These variations provided the seeds, or "clumps," from which galaxies and collections of galaxies grew. In many ways, it is truly amazing that quantum mechanics, the physics that governs atomic nuclei, atoms, and molecules but that is almost undetectable in our own life experience, is responsible for seeding the very largest structures that we can see in our universe.

> **Ripples in the early universe were the seeds of galaxies and large-scale structure.**

It is one thing to say that galaxies and larger structures formed from gravitational instabilities that began with slight irregularities in the early universe. It is quite another to turn this statement into a real scientific theory with testable predictions. To accomplish that, we return to the same basic technique that we have used to look into the centers of planets and stars and to answer many other questions about things we cannot observe directly. We combine the ideas we want to test with the laws of physics, construct a model, and then compare the predictions of that model with observations of the universe.

> **Models show how gravity turns ripples into large-scale structure.**

To build a model of the formation of large-scale structure, we have to begin with three key pieces of information. First we have to decide what universe we are going to model. That is, what values of Ω_{mass} and Ω_Λ (see Chapter 21) are we going to assume? These values are important in part because they determine how rapidly the universe expands and therefore how difficult it is for gravity to overcome this expansion in a particular region. The more rapidly a universe is expanding, or the less mass it contains, the more difficult it will be for gravity to pull material together into galaxies and larger-scale structures.

The second thing we need to know to construct our model is what the early bumps in the density of the universe looked like. How large and how concentrated were these early bumps? There are several ways to approach this question. One such method uses observations of variations in the CBR. The data from both COBE and WMAP (see Figure 18.20 and the opening image for Chapter 18) provide a good picture of structure in the CBR, from which we can infer what the early inhomogeneities in the universe must have looked like. Future observations, such as those from the European Space Agency's Planck mission, will do even better.

A different way to approach this question is to look at models of inflation, which make predictions about the struc-ture that will emerge following this episode of rapid expansion. These are among the predictions of the inflation model that will be tested in the years to come. These predictions are especially important to test because they tie together the large-scale structure of today's universe with our most basic ideas about what the universe was like in the briefest instant after the Big Bang. While we await final answers about the details of structure, we know enough to say that the early universe was more irregular on smaller (galaxy-sized) spatial scales than it was on the scales of the clusters, superclusters, filaments, and voids that are shown in Figure 22.4. An immediate consequence of this fact is that smaller structures (such as galaxies and even subgalactic clumps) formed first, whereas larger structures took more time to form. The idea of small structure forming first and larger structure forming later is referred to as **hierarchical clustering**. Hierarchical clustering has become one of the most important themes in our growing understanding of how structure in the universe formed.

> **Smaller structures form first. Larger structures form later.**

The third thing we need if we are to model the formation of large-scale structure is a complete list of the types and amounts of ingredients that existed in the early universe. Specifically, we need to know the balance between radiation, normal matter, and dark matter. We also need to make some choices about the nature of the dark matter that we use in our model. We'll discuss this further in a bit.

Once we have these three pieces of information—the shape of the universe we are modeling, the way matter inhomogeneities developed, and the nature and mixture of different forms of matter we are using—the rest is physics and calculations; we simply need to "turn the crank," as the expression goes. As we discuss later, there are some real difficulties in carrying out this strategy. For example, our uncertainties about how gas turns into stars make it difficult to compare our models with the real universe. Despite technical difficulties, we are aiming to answer this key question: What choices lead to a model universe that most resembles the real universe in which we live?

Galaxies Formed Because of Dark Matter

When we actually observe the cosmic background radiation using such satellites as COBE and WMAP, we find variations in the background radiation of about one part in 100,000. The theoretical models clearly show that such tiny variations at the time of recombination (when the universe was about half a million years old) are far too slight to explain the structure we see in today's universe. Gravity is simply not strong enough to grow galaxies and clusters of galaxies from such poor "seeds." These models indicate that for ripples in the density of the universe to have collapsed to form today's galaxies, the density of those ripples must have been at least 0.2 percent greater than the average density of the universe at the time of recombination.

If normal luminous matter in the early universe had been clumped at this level, the variations in the CBR today would be at least 30 times larger than what is observed. At first glance this discrepancy might seem to be an insurmountable problem with our understanding of the origin of structure in the universe, but it has turned out instead to be a crucial result that ties together a number of pieces of the puzzle. The theme of this story is "dark matter," which turns out to be an essential ingredient enabling us to understand how the kind of structure we see in the universe originated.

Dark matter first appeared on our journey back in Chapter 19 as an ad hoc construction—an annoyance, really—used by astronomers to explain the oddly flat rotation curves of spiral galaxies. When we turn our attention to clusters of galaxies, we find that dark matter dominates normal matter on those scales as well. Now we find that irregularities in normal matter in the early universe were too slight to form galaxies. So where will we turn for an answer to the question of how structure in the universe *did* form? You guessed it: dark matter.

In Chapters 19 and 21 we learned that there is much more dark matter in the universe than there is normal matter—consisting of neutrons, protons, and electrons—which we find in stars and gas throughout the universe. Yet our discussion of Big Bang nucleosynthesis (in Chapter 18) revealed that the amount of normal matter we see in the universe predicts just the right abundances of light ele-

ments. Therefore, the much greater amount of dark matter cannot be made of normal matter consisting of neutrons, protons, and electrons. If it were, it would have affected the formation of chemical ele-

> Dark matter provides the seeds for observed structure.

ments in the early universe. The abundances of several isotopes of the least massive elements would be quite different from what we find in nature. Dark matter must be something else—something that has no electric charge (so it does not interact with electromagnetic radiation) and that interacts only feebly with normal matter. Clumps of such dark matter in the early universe would not have interacted with radiation or normal matter, so we would not see them directly when looking at the CBR. This unseen dark matter solves the problems of modeling the formation of galaxies and clusters of galaxies.

What is so different about the behavior of dark matter and normal matter in the early universe? First, pressure waves and radiation did a very good job of smoothing out ripples in the distribution of normal matter in the early universe. Feebly interacting dark matter would be immune to these processes, so clumps of dark matter survived long after bumps in the normal matter had been smoothed out, as illustrated in **Figure 22.5**. Second, the dark matter in these clumps does not glow, so we do not see it directly when we look at the cosmic background radiation. Although these clumps of dark matter do cause slight gravitational redshifts in the light coming from normal matter, the resulting varia-

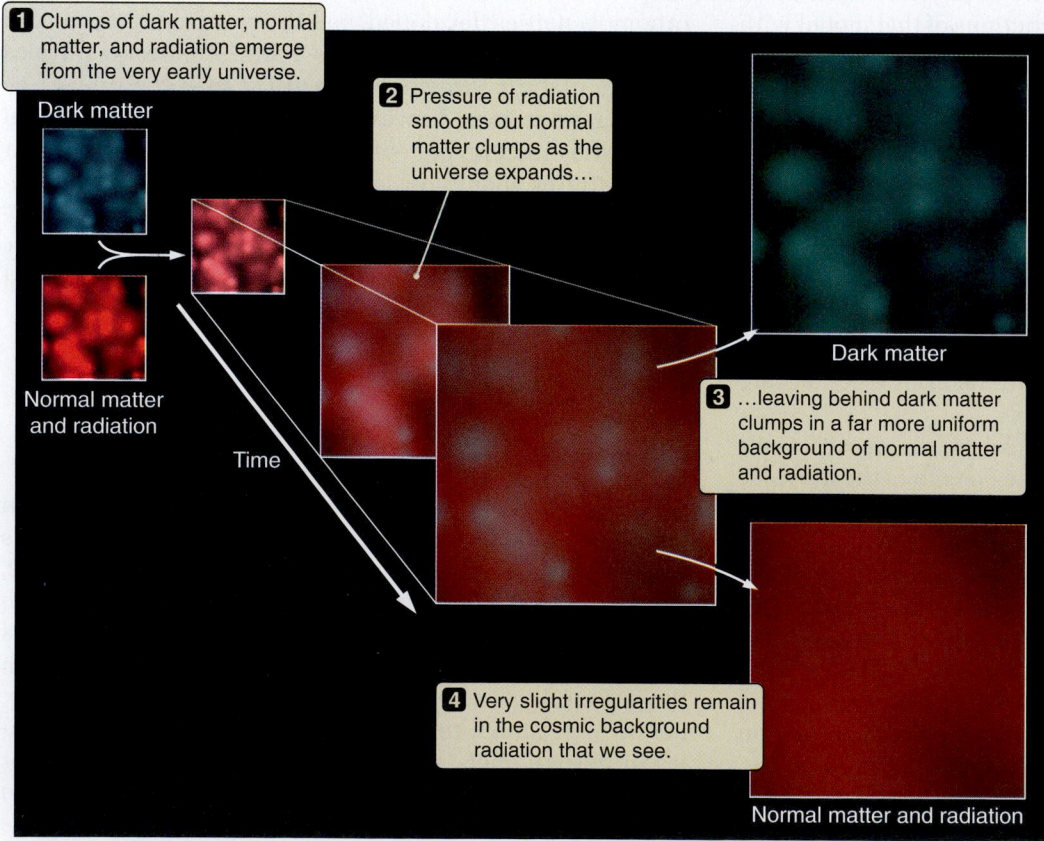

FIGURE 22.5 Radiation pressure and other processes in the early universe smoothed out variations in normal matter, but irregularities in the dark matter survived to become the seeds of galaxy formation.

tions fit well with current observations of the CBR. Thus, clumps of dark matter could have (and, according to our theories of the very early universe, *should* have) existed in the early universe, even though we are unable to see evidence of them directly.

However, just because we cannot see clumps of dark matter in the early universe does not mean they had no effect. In the same way that dark matter dominates the gravitational fields of today's galaxies and clusters, the mass of dark matter controlled the growth of gravitational instabilities in the early universe.

There Are Two Classes of Dark Matter

Here is the story of galaxy formation in a nutshell. Dark matter in the early universe was much more strongly clumped than normal matter. Within a few million years after recombination, these dark matter clumps pulled in the surrounding normal matter. Later, gravitational instabilities caused these clumps to collapse. The normal matter in the clumps went on to form visible galaxies. This story seems plausible enough, but the details of how this happened depend greatly on the properties of dark matter itself. Even though we do not yet know exactly what dark matter in the universe is made of, we can talk about two broad classes of dark matter that are based on how it behaves.

One possibility is that dark matter consists of feebly interacting particles that are moving about relatively slowly, like the slowly moving atoms and molecules in a cold gas. For obvious reasons, this type of dark matter is called **cold dark matter**. There are several candidates for cold dark matter. It is possible that cold dark matter consists of tiny black holes that might have been produced in the early universe. Few physicists and cosmologists favor this idea, however. Most think instead that cold dark matter consists of an unknown **elementary particle**. One candidate is the **axion**, a hypothetical particle first proposed to explain some observed properties of neutrons. Even though axions would have very low mass, they would have been produced in great abundance in the Big Bang. Another candidate is the **photino**, an elementary particle related to the photon. Some theories of particle physics predict that the photino exists and has a mass about 10,000 times that of the proton. Our state of knowledge about the particles that make up cold dark matter could soon change; photinos might be detected in current particle accelerators, and experiments are under way to search for axions and photinos that are trapped in the dark matter halo of our galaxy.

Since the first class of dark matter is called cold dark matter, the other class of dark matter is quite logically known as—are you ready?—**hot dark matter**. Hot dark matter consists of particles that are moving very rapidly. Neutrinos are one example of hot dark matter that we know

> Cold dark matter consists of relatively massive, slowly moving particles.

exist. We have seen that neutrinos interact with matter so feebly that they are able to flow freely outward from the center of the Sun, passing through the dense overlying layers of matter as if they were not there. There is no question that the universe is filled with neutrinos. Calculations indicate that about 300 million of these cosmic relics of the Big Bang fill each cubic meter of space. Measurements of the neutrino mass obtained at the Super-Kamiokande detector in Japan, at the Sudbury Neutrino Observatory in Canada, and by the Main Injector Neutrino Oscillation Search (MINOS) experiment in the United States suggest that neutrinos might account for as much as 5 percent of the mass of the universe. Although this percentage is not enough to account for all of the dark matter in the universe, it may be enough to have had a noticeable effect on the formation of structure.

> Hot dark matter consists of less massive, rapidly moving particles.

The different effects of cold and hot dark matter on structure formation have to do with differences in the way the two types of dark matter cluster in a gravitational field. The faster an object is moving, the harder it is for gravity to hold on to it. This is the same basic idea that we used to explain why hydrogen and helium are able to escape Earth's atmosphere but molecules such as oxygen and nitrogen are not. It is also the same idea we used to explain why the hot gas in elliptical galaxies extends far beyond their visible boundaries. Slow-moving particles are more easily corralled by gravity than are fast-moving particles, so particles of cold dark matter clump together more easily into galaxy-sized structures than do particles of hot dark matter. As a result, models show that on the largest scales of massive superclusters, both hot dark matter and cold dark matter can form the kinds of structures we see; but on much smaller scales, only in models of cold dark matter can the dark matter clump enough to produce structures like the galaxies we see filling the universe. The result of the calculations of galaxy formation is clear: to account for the formation of today's galaxies, we need cold dark matter.

> Cold dark matter forms galaxies.

A Galaxy Forms within a Collapsing Clump of Dark Matter

We can best see how models of galaxy formation work by following the events predicted by the models step by step. Let's consider the case of a universe made up of 90 percent cold dark matter and 10 percent normal matter, clumped together in a manner consistent with observations of the cosmic background radiation. On the scale of an individual galaxy, the effect of the cosmological constant is so small that it can be ignored.

Figure 22.6a shows the state of affairs of one clump of dark matter at the time of recombination. The dark matter is less uniformly distributed than normal matter; but over-all, the distribution of matter is still remarkably uniform. By a few million years after recombination (**Figure 22.6b**), the universe of our model calculation has expanded sever-alfold. Spacetime is expanding, so the clump of dark matter is also expanding. However, the clump of dark matter is not expanding as rapidly as its surroundings are, because its self-gravity has slowed down its expansion. The clump now stands out more with respect to its surroundings. Another change that has taken place is that the gravity of the dark matter clump has begun to pull in normal matter as well. By the stage shown in **Figure 22.6c**, normal matter is clumped in much the same way as dark matter.

A ball thrown in the air will slow, stop, and then fall back to Earth. In like fashion, the clump of dark matter will stop expanding when its own self-gravity slows and then stops its initial expansion. By the time the universe is about a billion years old, the clump

> **Clumps of dark matter can collapse only so far.**

of dark matter has reached its maximum size and is begin-ning to collapse (see Figure 22.6c). The collapse of the dark matter clump stops when the clump is about half its maxi-mum size, however, because the particles making up the cold dark matter are moving too rapidly to be pulled in any

closer (**Figure 22.6d**). The clump of cold dark matter is now given its shape by the orbits of its particles, in the same way that an elliptical galaxy is given its shape by the orbits of the stars it contains.

Unlike dark matter (which cannot emit radiation), the normal matter in the clump is able to radiate away energy and cool. As the gas cools, it loses pressure; and as it loses pressure, it collapses. Models show that small concentra-tions of normal matter within the dark matter clump collapse

> **Normal matter cools and falls toward the center of the dark matter halo.**

under their own gravity to form clumps of normal matter that range from the size of globular clusters to the size of dwarf galaxies. These clumps of normal matter then fall inward toward the center of the dark matter clump, as shown in **Figure 22.6e**. We discussed a similar chain of events in Chapter 15: the collapse of a molecular-cloud core on its way to becom-ing a star. It is important to note that, according to models, gas in our universe can cool quickly enough to fall in toward the center of the dark matter clump only if the clump has a mass of 10^8–10^{12} solar masses ($M_\odot$). This is just the range of masses of observed galaxies! This agreement between theory and observation is an important success of the theory that galaxies form from cold dark matter.

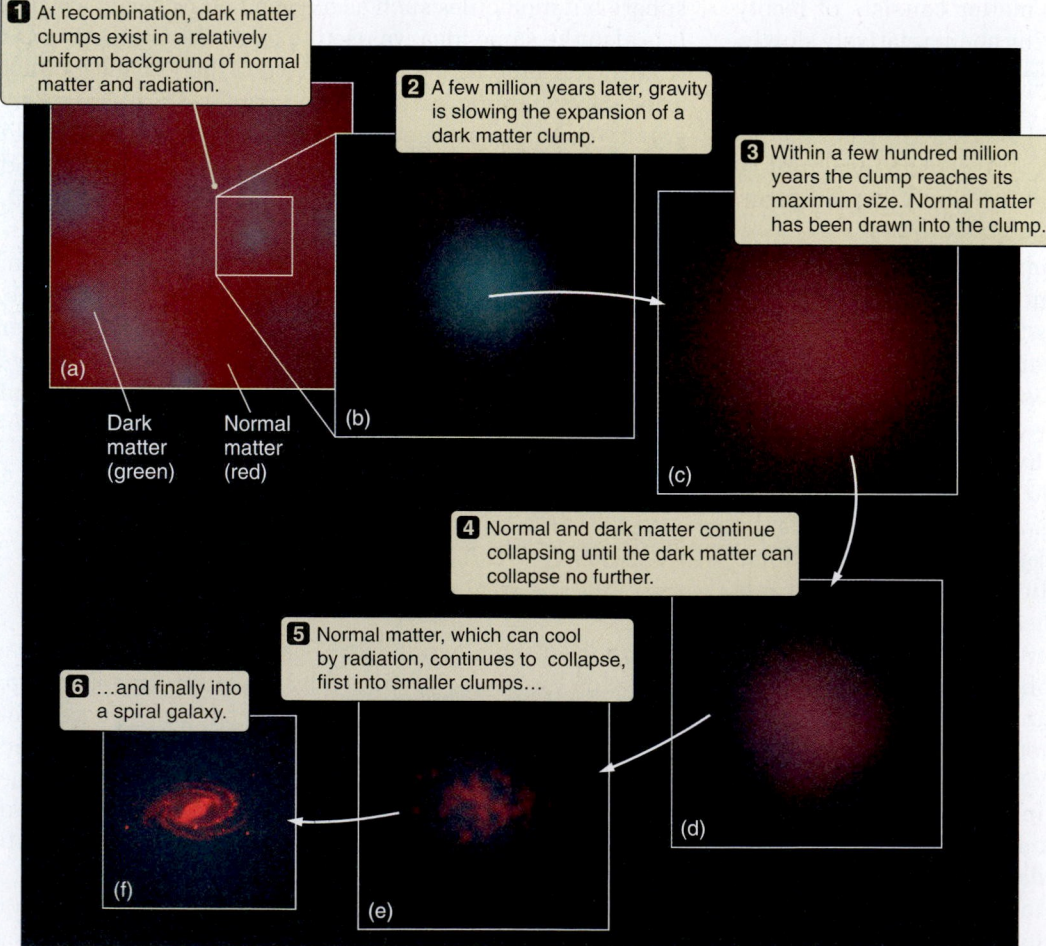

1 At recombination, dark matter clumps exist in a relatively uniform background of normal matter and radiation.

2 A few million years later, gravity is slowing the expansion of a dark matter clump.

3 Within a few hundred million years the clump reaches its maximum size. Normal matter has been drawn into the clump.

4 Normal and dark matter continue collapsing until the dark matter can collapse no further.

5 Normal matter, which can cool by radiation, continues to collapse, first into smaller clumps…

6 …and finally into a spiral galaxy.

Dark matter (green)   Normal matter (red)

(a) (b) (c) (d) (e) (f)

FIGURE 22.6 Stages in the formation of a spiral galaxy from the collapse of a clump of cold dark matter.

As discussed in **Connections 22.1**, there are many similarities between the collapse of a molecular cloud to form stars and the collapse of a clump in the early universe to form galaxies. Another process that we saw at work in star formation also comes into play at this point. The clumps of matter collapsing to form galaxies do not exist in isolation. They have been tugged on by the gravity of neighboring clumps and have been pushed around by the pressure waves that ran through the young universe, smoothing out its structure. As a result, each protogalactic clump has a little bit of rotation when it begins its collapse. As normal matter falls inward toward the center of the dark matter clump, this rotation forces much of the gas to settle into a rotating disk (**Figure 22.6f**), just as the collapsing cloud that was to become our Sun settled first into an accretion disk.

And just as the accretion disk around the Sun provided the raw materials for the planets of our Solar System, the disk formed by the collapse of each protogalactic clump becomes the disk of a spiral galaxy.

Searching for Signs of Dark Matter

Dark matter has come to play a very important role in our understanding of the universe. We might reasonably be uncomfortable having our models of galaxy formation rely so heavily on something we cannot see directly. Fortunately, dark matter shows itself in a variety of ways. The flat rotation curves of spiral galaxies, for example, are compelling evidence for the existence of extended halos made up of dark

---

CONNECTIONS 22.1

Parallels between Galaxy and Star Formation

As you read about galaxy formation, you might find it enlightening to think back to our discussion of star formation in Chapter 15. Both processes involve the gravitational collapse of vast clouds to form denser, more concentrated structures. To help you with the comparison, we list here a few of the similarities and differences between the two.

Gravitational instability. In both star and galaxy formation, the collapse begins with a gravitational instability. Regions only slightly denser than their surroundings are pulled together by their own self-gravity. As the matter in these regions becomes more compact, gravity becomes stronger, and the collapse process snowballs. One key difference is that for a galaxy to form, the dark matter clump must collapse rapidly enough to counteract the overall expansion of the universe itself.

Fragmentation. In both cases the original cloud separates into smaller pieces as a result of the gravitational instability. However, the order of fragmentation differs between star and galaxy formation. In molecular clouds, first large regions begin to collapse, and then they fragment further to form individual stars. In contrast to this "top-down" process, galaxy formation is "bottom-up": smaller structures collapse first and then merge to form galaxies and eventually assemblages of galaxies.

Compression, heating, and thermal support. As an interstellar molecular cloud collapses, its temperature climbs and the pressure in the cloud increases. The higher pressure would eventually be enough to prevent further collapse, except that the cloud core is able to radiate away thermal energy. That is the bright infrared radiation that enables us to see star-forming cores. Compare this process with galaxy formation: As a dark matter clump collapses, its temperature also climbs, and it, too, reaches a point where there is a balance between gravity and the thermal motions of the dark matter. However, dark matter is *not* able to radiate away energy, so once this balance is reached, the collapse of the dark matter is over. Only the normal matter within the cloud of dark matter is able to radiate away thermal energy and continue collapsing. That is why normal matter collapses to form galaxies, while the dark matter remains in much larger dark matter halos.

As galaxies form, dark matter remains in extended halos. Dark matter is too hot to settle into galaxy disks. It is far too hot to become concentrated into even smaller structures such as molecular clouds or to take part in the formation of stars. Dark matter may be the dominant form of matter in the universe, and it may determine the structure of galaxies; but dark matter can never collapse enough to play a role in the processes that shape stars, planets, or the interstellar medium.

Angular momentum and the formation of disks. Conservation of angular momentum is responsible for the formation of disk galaxies, just as it is responsible for the formation of the accretion disks around young stars. The Milky Way and the Solar System are both flat for the same reasons. The origin of the angular momentum is different, though. Whereas turbulent motions within star-forming molecular clouds produce the net angular momentum for stellar disks, gravitational interactions with nearby clumps are responsible for the angular momentum of the Milky Way.

The end product. Once a stellar accretion disk forms, most of the matter moves inward and is collected into a star. In contrast, much of the matter in a spiral galaxy remains in the disk.

(a) For gravity to hold onto this hot X-ray emitting gas…

(b) …this galaxy cluster must be more massive than the galaxies we see.

FIGURE 22.7 (a) An X-ray image of the massive cluster Abell 1689, about 2.3 billion light-years away, shows that the cluster is full of hot, tenuous gas. (b) An image of the same field of view taken in visible light. (c) This combined image of Abell 1689 shows both the galaxies detected by the Hubble Space Telescope and the X-ray image of the 100-million-K gas taken with the Chandra X-ray Observatory.

(c) Galaxy clusters are mostly dark matter.

again ask, How strong must gravity be to hold this cluster together? And again, the answer is that the total mass of the clusters, including dark matter, must be about 10 times greater than the normal matter they contain.

A third way to look for dark matter relies on the predictions of Einstein's general theory of relativity, which states that mass distorts the geometry of spacetime, causing even light to bend near a massive object. In particular, light from a distant object is bent by the gravity of a galaxy or cluster of galaxies, so that images of the distant object can be seen magnified on either side of the intervening galaxy or cluster. The result is a **gravitational lens** (**Figure 22.8a**).

We have already encountered gravitational lensing, in Chapter 20's discussion of MACHOs, where we found that lenses can make background objects appear brighter. When considering the complex three-dimensional geometry of gravitational lenses, we find that lenses can show multiple images of background objects, and that these magnified images are often drawn out into an arclike appearance. The greater the gravitational lensing, the greater the mass that must be in the cluster. **Figure 22.8b** shows an image of a galaxy cluster that is acting as a gravitational lens for a number of background galaxies. By analyzing such images, we can determine the mass of the lensing cluster. **Figure 22.8c** shows six examples of more distant gravitational lenses from the Cosmological Evolution Survey (COSMOS), which is using most of the major space-based telescopes and several ground-based telescopes to catalog these lenses.

> Gravitational lenses are one way to measure the masses of galaxy clusters.

Regardless of how we measure the masses of galaxy clusters—by looking at the motions of their galaxies, by measuring their hot gas, or by using them as gravitational lenses—the results are the same. By mass, galaxy clusters, like the galaxies they are made of, are dominated by the dark matter they contain.

> Dark matter dominates the mass of galaxy groups and clusters.

matter. Similarly, the ability of elliptical galaxies to hold on to hot gas convincingly demonstrates that they contain far more mass than can be accounted for by their stars alone.

We can use a similar approach to look for evidence of dark matter on larger scales. Galaxy clusters are filled with extremely hot gas (10 million to 100 million kelvins [K]), which occupies the space between galaxies (**Figure 22.7**). Even though this gas is *extremely* tenuous, the volume of space that it occupies is enormous: the mass of this hot gas can be up to *five times* the mass of all the stars in that cluster. X-ray spectra show that this gas contains significant amounts of massive elements that must have formed in stars. This chemically enriched gas has been either blown out of galaxies in winds driven by the energy of massive stars, or stripped from galaxies during encounters with neighboring galaxies. Were it not for the gravity of the dark matter filling the volume of the cluster, this hot gas would have dispersed long ago.

We can also look at the motions of the cluster's galaxies (which behave like very large "atoms" in this context) and

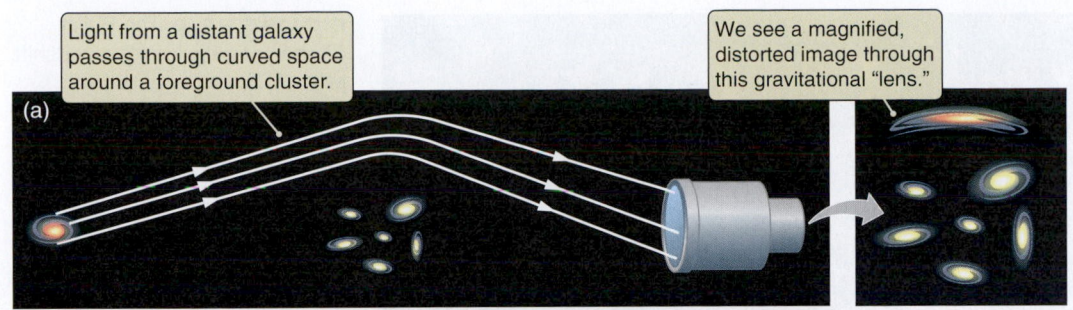

Light from a distant galaxy passes through curved space around a foreground cluster.

We see a magnified, distorted image through this gravitational "lens."

(a)

FIGURE 22.8 (a) The geometry of a gravitational lens. A mass can gravitationally focus the light from a distant object, thereby magnifying and distorting the image. (b) A Hubble Space Telescope image of the cluster Abell 2218, showing many gravitationally lensed galaxies, seen as arcs. (c) Gravitational lenses from several distant individual galaxies.

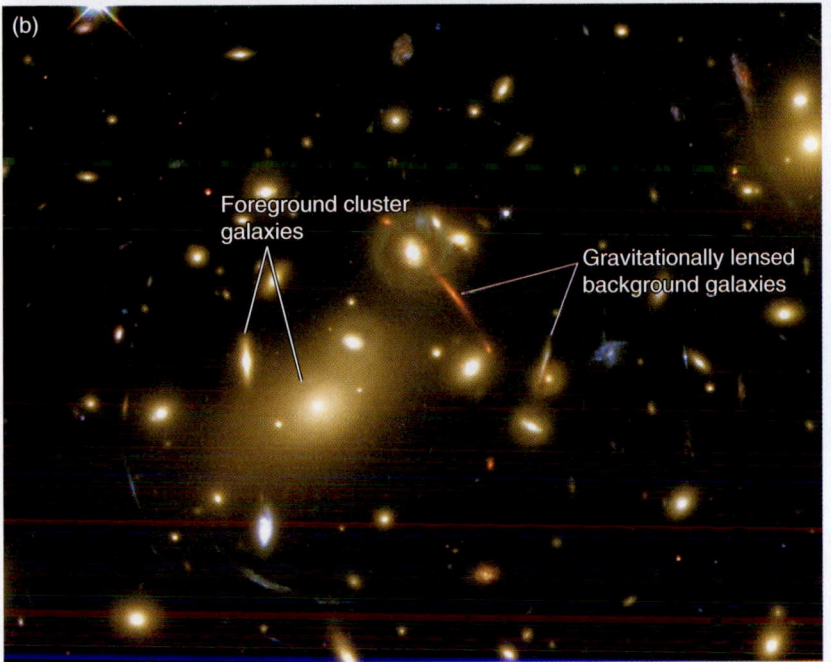

(b)

Foreground cluster galaxies

Gravitationally lensed background galaxies

(c)

0038+4133

0211+1139

5921+0638

0018+3845

0013+2249

0047+5023

22.4 Observations Help Fill Gaps in the Models

The picture of galaxy formation given in the previous sections is much cleaner and more idealized than what happened in reality. In our hierarchical clustering picture, smaller structures collapsed first. When an original clump of galaxy-sized dark matter contained a substructure, that substructure would itself have collapsed into subgalaxy-sized objects before the galaxy as a whole finished forming. This hierarchical model gives us a way to understand the existence of such objects as the globular star clusters and dwarf companion galaxies near the Milky Way and other galaxies. These objects formed inside the larger clump of cold dark matter when the luminous matter in the clump was still settling toward what would become the disk of the galaxy.

Mergers Play a Large Role in Galaxy Formation

As we saw in star formation, galaxies do not always form in isolation. It is likely that the same larger clump often produced more than one galaxy, and that these galaxies later interacted or even merged. The merging of two spiral (or protospiral) galaxies made all sorts of commotion likely. The tidal interactions between the galaxies and the collisions between gas clouds in the galaxies probably triggered many regions of star formation throughout the combined system. Recall our discussion of galaxy mergers at the end of Section 19.4. The collision and merging of two giant clusters is nicely illustrated in **Figure 22.9**, which shows the Bullet Cluster of galaxies at a redshift of $z = 0.3$. The stars contained within the galaxies of this system have less total mass than the X-ray–emitting gas, which is itself much less

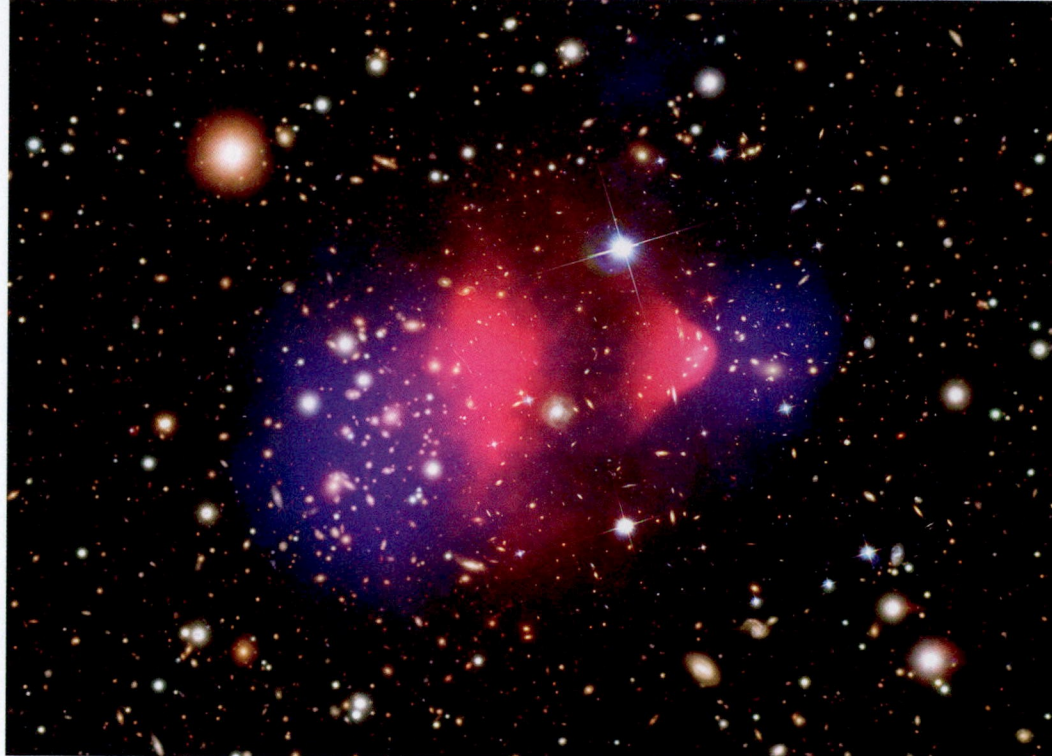

FIGURE 22.9 The Bullet Cluster of galaxies, at a redshift of $z = 0.3$, represents the collision of two giant clusters of galaxies in the process of merging. The galaxies in this system constitute less mass than the X-ray–emitting gas (shown in red), which is itself much less massive than the dark matter (shown in blue) measured from the gravitational lensing by this cluster.

massive than the dark matter, whose mass we deduce from the gravitational lensing produced by the cluster. The fact that the cluster's galaxies, hot gas, and dark matter are so displaced from one another is testament to the violence of the collision that formed this cluster.

The merging of galaxies also answers another puzzle. We have seen how spiral galaxies could form in the early universe, but what about the formation of elliptical galaxies? Ellipticals are now thought to result from the merger of two or more spiral galaxies. **Figure 22.10** shows a computer simulation of such a merger. If the merging galaxies were not originally spinning in the same direction, then the resulting merged galaxy loses its disklike character. The dark matter halos of the galaxies merge, and the stars eventually settle down into the bloblike shape of an elliptical galaxy. As this picture suggests, elliptical galaxies are known to be more common in dense clusters where mergers are likely to have been more frequent.

> Ellipticals form from the mergers of spirals.

Such events also played a major role in feeding supermassive black holes, which themselves must have formed very early in the collapse of galaxies and protogalaxies. The same mergers and interactions that formed the giant galaxies we see today also provided the fuel to power the quasars and other AGNs that were common in the early universe.

With galaxies crashing together, supermassive black holes forming, quasars flooding the young universe with intense radiation and powerful jets, and star formation running amok, galaxy formation in the young universe must have been a violent, messy business. This conclusion has clear consequences for what we should expect to see when looking back to a time in the history of the universe when galaxies were actively forming. Rather than the well-formed spirals and ellipticals that dominate today's universe, the early universe should have contained many clumpy, irregular objects. As expected, images of the young universe such as the Hubble Space Telescope image shown in Figure 18.1 are filled with lumpy, distorted objects that are still in the process of settling in. We show some of these high-redshift objects as insets in **Figure 22.11**.

When we look at the bright, active galaxies that populate the early universe, it is important to remember that what we actually see is only the tip of the iceberg. The luminous galaxies are small in comparison with the dark matter clumps that surround them. These clumps are still found in today's universe as the invisible dark matter halos responsible for the flat rotation curves of spiral galaxies.

Star Formation Leaves a Gap in the Models

We would like to be able to turn our models of collapsing clumps of cold dark matter into detailed models of what galaxies in the young universe should look like, but we have not yet reached such a level of sophistication. One major problem is that we lack a good understanding of how stars form in young galaxies. We know that most of the normal matter in a galaxy winds up as stars; but when and where do those stars form, and how long does it take? Although we can say a great deal about how individual stars form

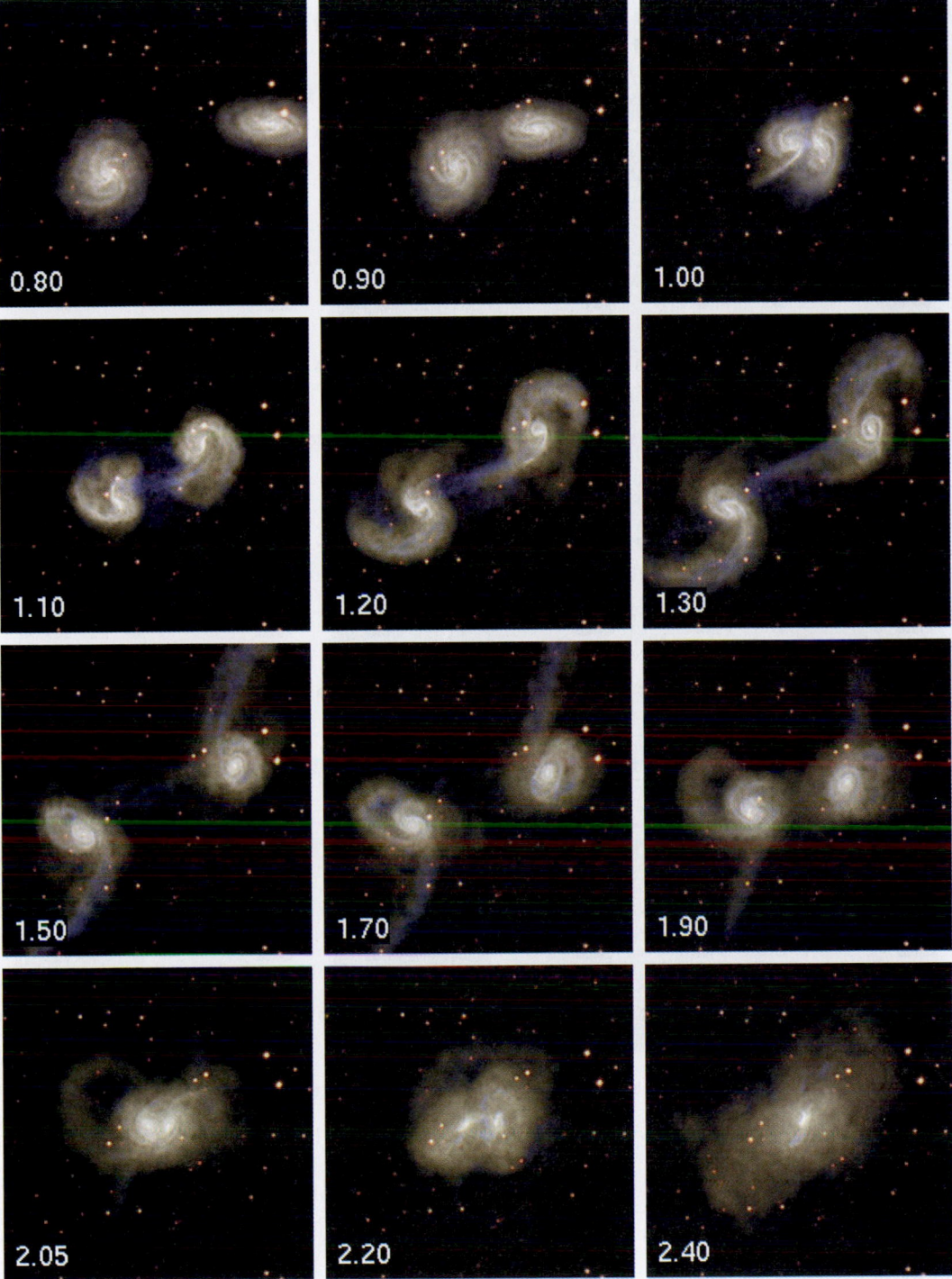

FIGURE 22.10 In this computer simulation, two spiral galaxies merge to form an elliptical galaxy. Numbers indicate time since the interaction began, where 1.00 is the time it took for the galaxies to fall together.

in our galaxy, as we did in Chapter 15, we do not yet have theories or models that predict such important pieces of the puzzle as what fractions of stars should have what masses. Nor do we clearly understand the differences between star formation in the early universe and the star formation going on around us today. In a universe devoid of massive elements, there were no dense, dusty molecular clouds from which stars might form, and no spiral disks in which such clouds might congregate. Instead, the first stars must have formed from the collapse of clouds of hydrogen and helium gas within the overall clump of matter of which they were a part.

Although a detailed understanding of star formation in a forming galaxy is in the future, we can still make a number of factual statements about when and where stars form in a collapsing galaxy. These clues come from the study of our own galaxy. The fact that the atmospheres of even the oldest halo stars in the Milky

Star formation began very early in collapsing galaxies.

FIGURE 22.11 Another picture of the Hubble Ultra Deep Field (HUDF)—the farthest glimpse into the universe's distant past. The insets show high-redshift objects that are examples of the youngest galaxies in the process of forming and merging.

Way contain some amount of massive elements tells us that a significant amount of star formation must have taken place *very* early during the collapse of the Milky Way. If we could closely inspect a collapsing protogalactic clump in the stages depicted in Figures 22.6d and e, we would expect to see stars already forming and supernovae exploding from place to place in the collapsing halo. The halo stars that we see today must have formed while the Milky Way was still collapsing, before the gas from which they formed coalesced into the disk. In contrast, disk stars (all of which have relatively high abundances of massive elements) formed from gas that was enriched by early generations of halo stars before it settled into the galaxy's disk.

Thus, although we cannot say exactly how stars formed in the early universe, we can say that star formation must have been going on throughout the process of galaxy formation. The pattern of stellar ages and massive-element abundances evident in our galaxy today fits naturally with our current models, in which galaxies formed from collapsing clumps of cold dark matter.

Uncertainties in our understanding of star formation complicate quantitative comparisons between models of galaxy formation and observations of the early universe. Fortunately, though, when we talk about structure on scales much larger than galaxies, star formation becomes less of an issue. To understand the formation of clusters and larger structures, we need worry only about the way that galaxy-sized clumps of matter themselves fall together under the force of gravity.

Figure 22.12 shows the results of one computer simulation that follows the motions of millions of clumps of dark matter as they fall through space under their mutual gravitational attraction. The scale of this simulation is so large that we cannot follow

Clusters, filaments, and voids form after galaxies form.

the outcome of individual galaxies but instead are looking at the overall distribution of dark matter. The simulation shows that the smaller-scale structures develop first. During the first few billion years, dark matter falls together into structures comparable in size to today's clusters of galaxies. Only later do the spongelike filaments, walls, and voids become well defined.

The similarities between the results of the models and observations of large-scale structure, such as those shown in Figure 22.4, are quite remarkable. Not just any model will do, however. Only model universes with certain combinations of shape, mass, nature of ripples, type of dark matter, and values for the cosmological constant will produce structure similar to what we actually see. This is a very important result. Our models contain assumptions consistent with our knowledge of the early universe, and they predict the formation of large-scale structure similar to what we actually see in today's universe. **Figure 22.13** provides a succinct summary of the galaxy formation process, in which smaller objects form first and merge into ever-larger structure, leading ultimately to the Hubble Ultra Deep Field picture shown in Figures 22.11 and 18.1.

Peculiar Velocities Trace the Distribution of Mass

In our comparison of models and observations, there is an important caveat: whereas the models show where the *mass* is, our observations show where the *light* is. We have already noted that our understanding of star formation in the early universe is incomplete at best. What if a clump

> The distribution of light may not be the same as the distribution of mass.

of dark matter formed and collapsed, but for some reason the normal matter associated with that clump did not produce stars? Such clumps could still contain a significant amount of mass and could even affect the geometry of the universe, but they would themselves not be seen in images of the sky.[1]

There is a way to get at the question of the overall distribution of dark matter more directly. If the clumping of galaxies is a result of the action of gravity, we should be able to look at how galaxies are moving and see them falling together. In fact, measurements of the way galaxies fall together should enable us to infer the gravitational field they are experiencing, and hence the overall distribution of dark matter.

---

[1]Clumps of dark matter that are devoid of stars might be revealed indirectly by gravitational lensing.

FIGURE 22.12 A computer simulation of the formation of very large-scale structure in a universe filled with cold dark matter.

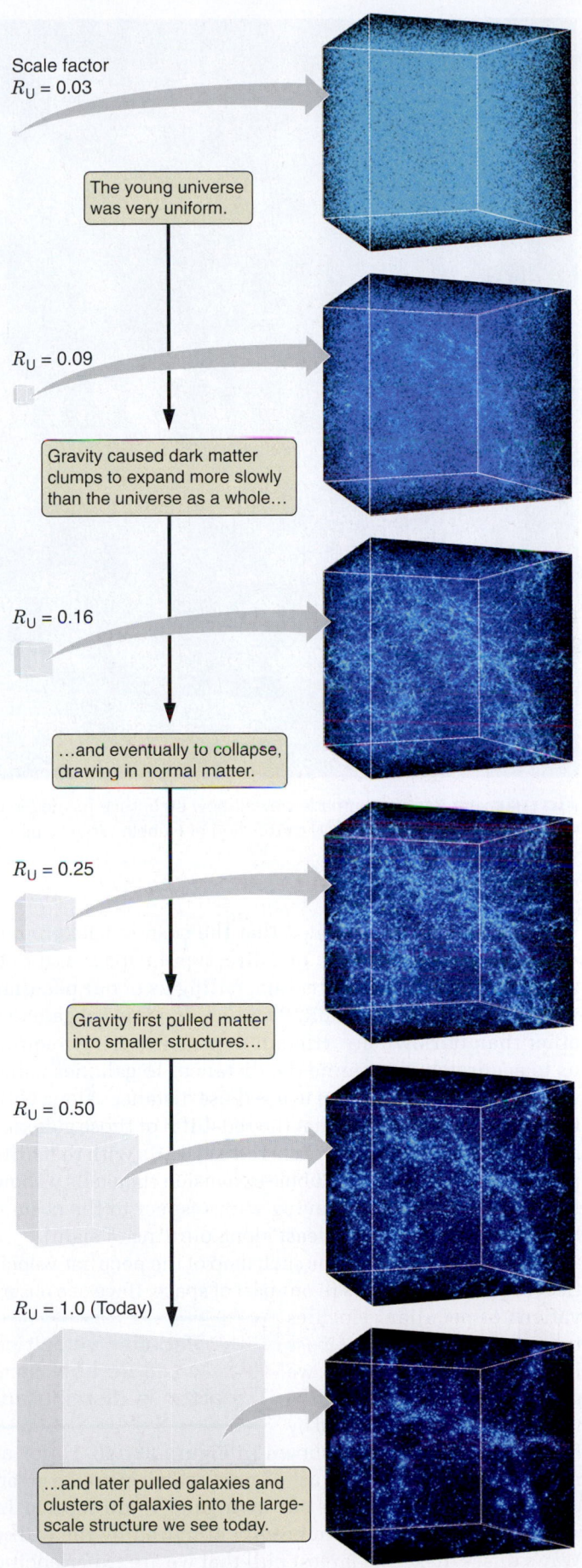

Scale factor
$R_U = 0.03$

The young universe was very uniform.

$R_U = 0.09$

Gravity caused dark matter clumps to expand more slowly than the universe as a whole…

$R_U = 0.16$

…and eventually to collapse, drawing in normal matter.

$R_U = 0.25$

Gravity first pulled matter into smaller structures…

$R_U = 0.50$

$R_U = 1.0$ (Today)

…and later pulled galaxies and clusters of galaxies into the large-scale structure we see today.

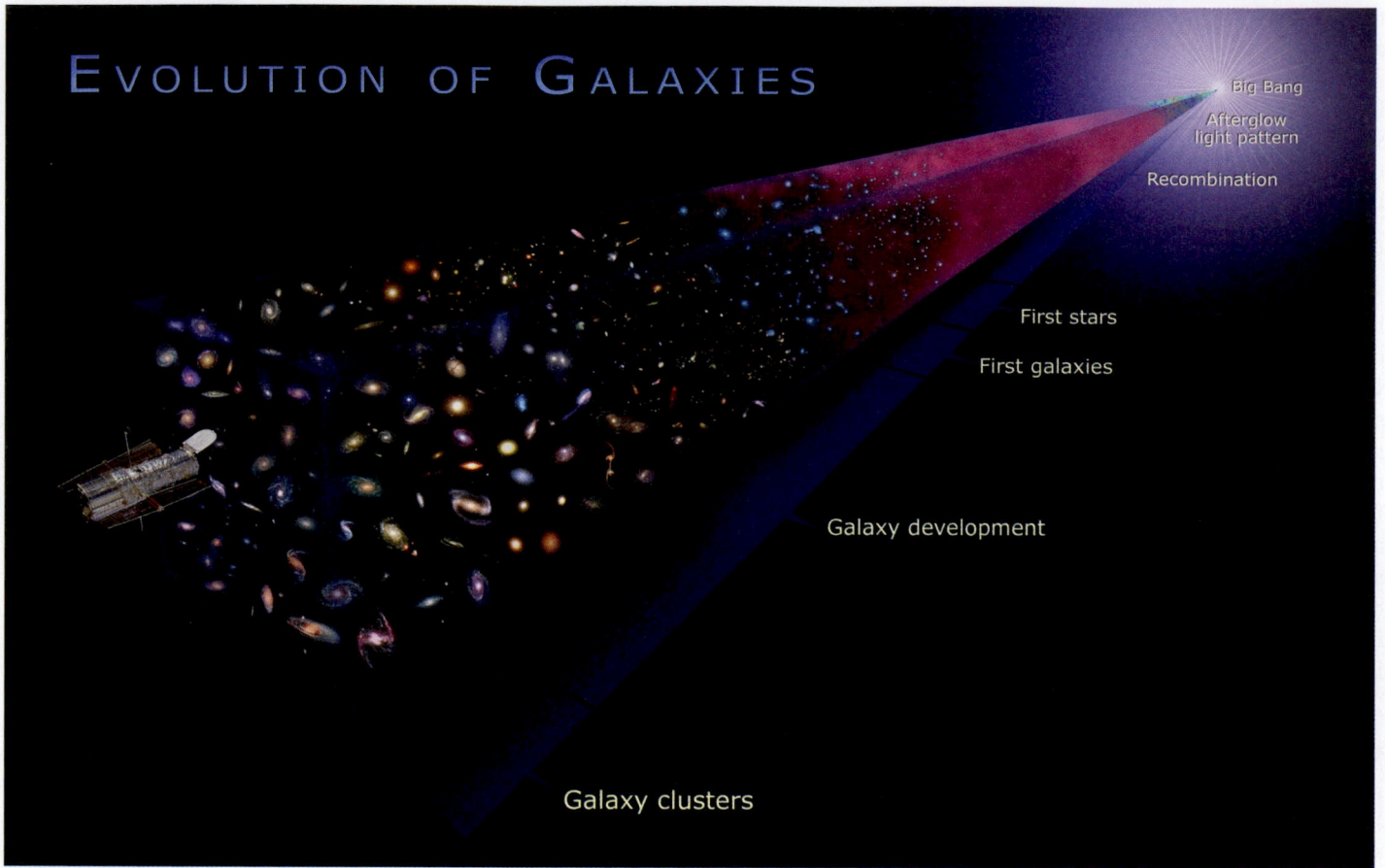

EVOLUTION OF GALAXIES

Big Bang

Afterglow
light pattern

Recombination

First stars

First galaxies

Galaxy development

Galaxy clusters

FIGURE 22.13 A schematic view of how structure formed in the universe, from smaller systems to larger ones, leading ultimately to the kind of Hubble Ultra Deep Field picture shown in Figure 22.11.

In Chapter 18 we learned that the cosmic background radiation is blueshifted in one direction in space and redshifted in the opposite direction, telling us of our **peculiar velocity** relative to the CBR. Peculiar velocities of galaxies other than our own are difficult to measure. They require us to accurately determine the distances to galaxies using standard candles, and then to use those distances along with Hubble's law to predict what the redshifts of those galaxies should be. Comparison of observed redshifts with redshifts predicted on the basis of Hubble expansion then tells us how fast these galaxies are moving with respect to the cosmic background radiation (at least along our line of sight).

Figure 22.14a shows one such map of the peculiar velocities inferred for galaxies in our part of space. If we use observations of peculiar velocities to map concentrations of mass in this part of the universe, we get a picture that looks like two mountains surrounded by foothills and valleys, as shown in **Figure 22.14b**. Unfortunately, these two "mountains" of mass, which exert the greatest gravitational pulls on our own galaxy, are located in regions of the sky that are heavily obscured by the Milky Way's dusty disk. The largest pull that we are experiencing

> Peculiar velocities
> tell us how dark
> matter is distributed.

comes from a region dubbed the Great Attractor, so called because of its strong gravitational tug.

The picture painted in this chapter of the formation of large-scale structure is almost certainly correct in its broad outlines. Even so, uncertainties in star formation, our understanding of the exact nature of dark matter, and our knowledge of the shape of the universe limit the amount of detail in any comparisons we can make between models and observations. In the first decade of the 21st century, these areas were a focus of very active research. Several projects—such as the Sloan Digital Sky Survey (currently collecting large amounts of data on the distribution and redshifts of galaxies), the WMAP satellite, and ESA's Planck mission—can dramatically improve our knowledge of the structure in the CBR. Larger and more sophisticated models are being built all the time, while studies of star-forming regions in our galaxy and in galaxies with lower massive-element abundances are continuing to improve our understanding of this important process. There is also hope that new theories and laboratory experiments will identify the particles that make up dark matter. A more complete understanding of the formation of galaxies and large-scale structure in the universe will not likely come in a single "eureka" moment, but will instead arrive in bits

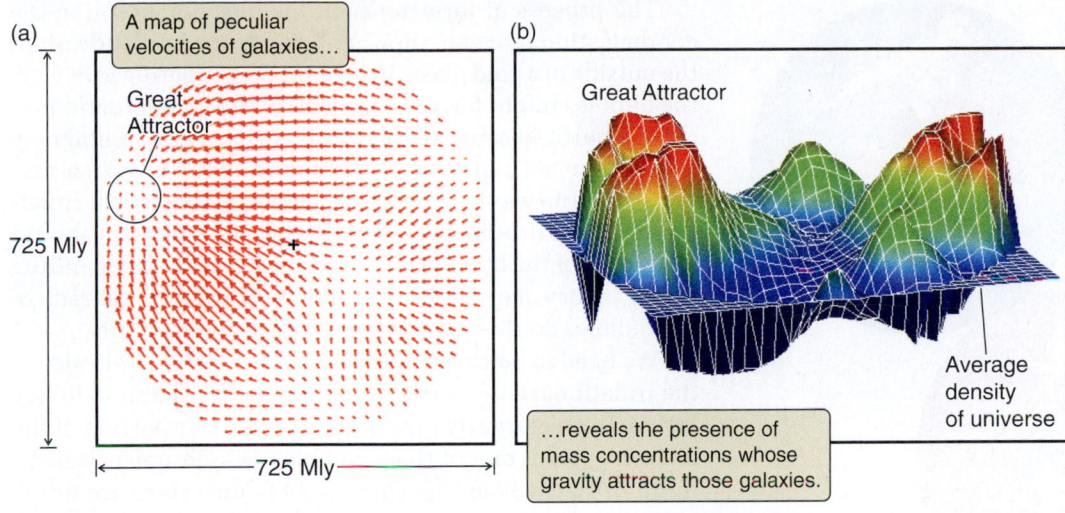

FIGURE 22.14 (a) A map of peculiar velocities of galaxies in our neighborhood. Arrows show the velocities of material lying in the plane of our own local supercluster. Motions seem to be converging onto the position of the Great Attractor. The black cross marks the location of the Milky Way Galaxy. (b) A map of the mass distribution inferred from observations of peculiar velocities.

and pieces throughout the first part of the 21st century as a result of advances in all of these different areas.

22.5 Multiple Multiverses

We have described in detail the origin of structure in our universe from the scale of asteroids and planets all the way to the largest clusters and superclusters. We'll now turn to ideas that are more speculative. Have you ever wondered if there is even more structure on an even larger scale—perhaps parallel universes to our own either separated in space or even occupying the exact same space as ours? These hypothetical universes seem like material for science fiction rather than for a serious science textbook, but it might not be as strange as you may think. In fact, many cosmologists think seriously about the idea of **multiverses**, or collections of parallel universes.

Let's begin with the simplest example of such parallel universes, illustrated in **Figure 22.15**. We know that the age of the universe—that is, the amount of time that has passed since the Big Bang—is 13.7 billion years. That means that light reaching us today can have traveled a distance of only about 13.7 billion light-years (Gly). (Actually, the real distance is a bit larger than that because of the past expansion of the universe, but we'll ignore that difference.) Therefore, everything that we can possibly observe today—that is, our observable universe—must be within a sphere of radius 13.7 Gly. We cannot possibly see anything outside our observable universe. Yet if the geometry of space is flat, as our cosmological parameters already suggest, then our universe is truly infinite in size and must therefore contain an infinite number of disjoint observable universes like our own. These disjoint universes are simply too far away for us ever to be able to observe them.

It is true, of course, that as our universe ages and expands gradually, the size of our observable universe expands as well, and we can see more and more material. However, when the universe becomes completely dominated by dark energy (see Chapter 21), it will expand so fast that the separate observable universes will never be observable from each other. Our own observable part of the universe will become truly isolated from all the other parallel observable universes.

> Our own observable part of the universe will become isolated from all others.

What are these other parallel universes like? We may never know for sure in detail, but we can say several things. First, if the cosmological principle holds, then on large scales, each of these observable universes should look pretty much like our own. But the details could be very different among these parallel universes. Still, in an infinite universe there must be an infinite number of observable universes exactly like ours, with an exact genetic and experiential copy of you reading an identical version of *21st Century Astronomy*. We know this because if our own observable universe is cooler everywhere than about 10^8 K, then there can be no more than 10^{118} particles in the observable universe, and only so many ways that those particles can be distributed. If you then ask how far you must go before you are sure to find an observable universe just like—and we mean *identical to*—our own, the answer is about $10^{10^{118}}$ Mly. Yes, that's 10 raised to the power 10^{118}, and for numbers that large, it doesn't matter if you measure the distance in megalight-years or in nanometers. So, in an infinite universe—as enormous as it might be—we still know how far away an identical copy of you must be!

> There are parallel universes identical to our own.

The parallel observable universes we just described represent the first of four types of multiverses about which cosmologists have theorized (**Figure 22.16a**). The inflationary universe model forms the basis of the second type of multi-

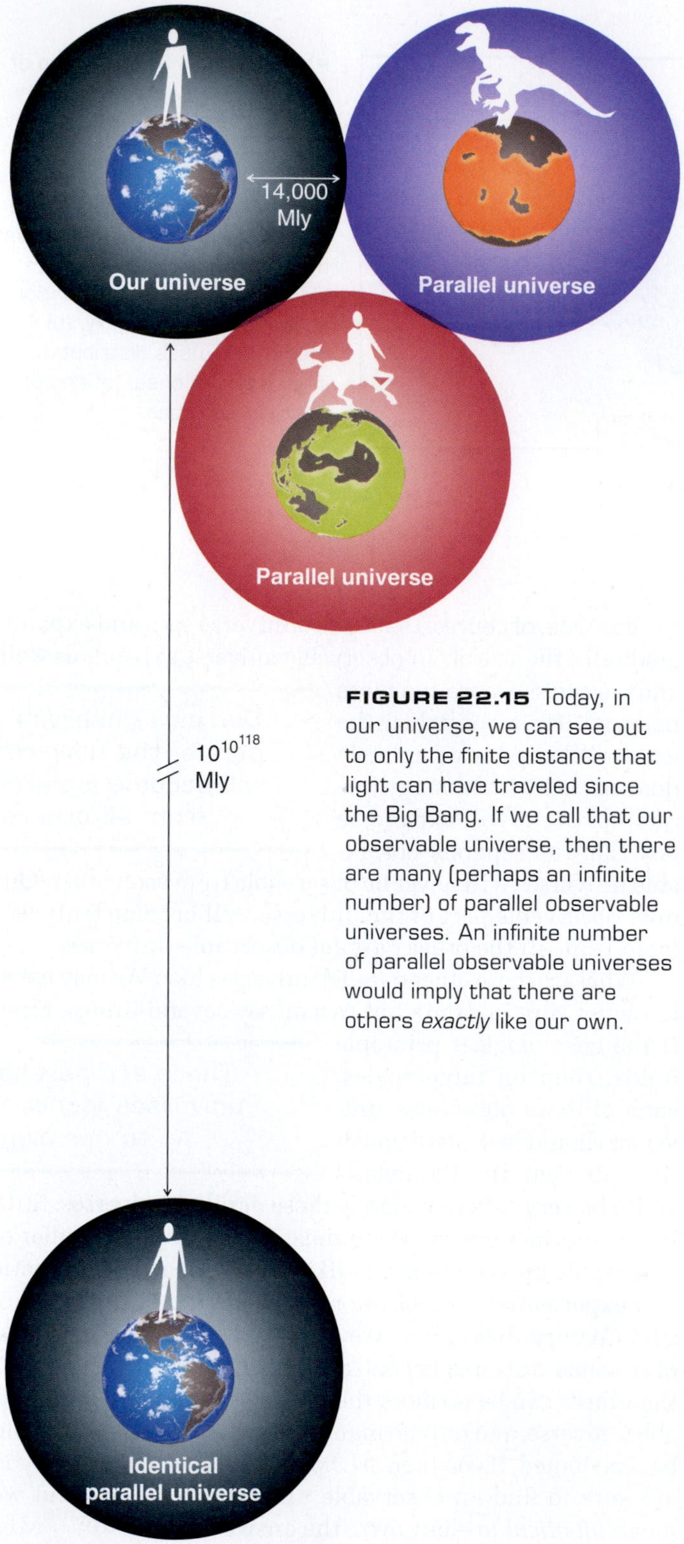

14,000 Mly

Our universe

Parallel universe

Parallel universe

$10^{10^{118}}$ Mly

Identical parallel universe

FIGURE 22.15 Today, in our universe, we can see out to only the finite distance that light can have traveled since the Big Bang. If we call that our observable universe, then there are many (perhaps an infinite number) of parallel observable universes. An infinite number of parallel observable universes would imply that there are others *exactly* like our own.

The process of forming such bubbles is much like the condensation of water vapor to form drops in clouds or on the outside of a cold glass. In an eternally inflating universe, the bubbles might form because of quantum fluctuations—occasionally, because of quantum variations, a region may slow down just a little bit from the rest of the universe, and as a result the whole region may form a bubble whose inflating phase will soon end. Exactly the same physics should govern all of the bubbles in such a universe, and depending on the theory of everything (TOE), it is possible that each of the bubbles could have different physical constants.

We need to better understand the underlying physics of the inflationary universe to know whether that possibility is real. In an eternally inflating universe, our portion of the universe is but one of these bubbles, which must itself be finite in size. Given that these bubble universes are finite, could we be living near the edge of such a bubble? This scenario is possible, but the multiverses may be so huge that it is very unlikely we are anywhere near the edge. One nice feature of eternal inflation is that it neatly answers the question of what there was before the Big Bang. Since the universe has inflated and will inflate forever, there was no beginning or end. Our own bubble or parallel universe separated from the rest of the universe at a time we call the Big Bang, but other bubbles are constantly separating and becoming their own parallel-universe big bangs.

> New big bangs are forever beginning in an eternally inflating universe.

The third type of multiverse has its origins purely in quantum mechanics (**Figure 22.16c**). As we described in Chapter 4, the quantum description of the world is a probabilistic description. There is a certain probability that a radioactive element will decay within a day or that an atom in an excited state will decay in a second. One interpretation of this probability is that the atomic state either decays or it doesn't decay in that second, with the relative likelihoods determined by quantum physics. A second interpretation, however, is that the atom splits into two parallel universes: one in which it decays and one in which it does not decay. If we are observing that atomic state, watching for it to decay, then we, too, split into two parallel universes: one in which we see the decay and one in which we do not. With this interpretation of quantum mechanics, new universes are forming all the time and occupying the same space, even though they cannot communicate with one another.

The fourth type of multiverse is characterized by parallel universes that have different mathematical structure to describe the different physics within these universes (**Figure 22.16d**). This type is the all-encompassing case in which almost any behavior for the universe is possible.

It is fair to ask whether the idea of parallel universes, or multiverses, is really science. Throughout this book we have emphasized over and over again that any legitimate scientific theory must be testable and ultimately falsifiable. Are there tests capable of proving these multiverse ideas to be wrong? Possibly. For example, we do test the Type I

verse. Imagine a universe that undergoes inflation forever, with no beginning or end to the inflation. This idea, called **eternal inflation**, was hypothesized by Russian-American physicist Andrei Linde (1948–), who realized that if such a universe exists, then individual regions could occasionally slow their expansion relative to the rest of the universe and form an expanding bubble within a surrounding inflationary universe (**Figure 22.16b**).

(a) Type I: Regions beyond our cosmic horizon

(b) Type II: Other postinflation bubbles

(c) Type III: The many worlds of quantum physics

(d) Type IV: Other mathematical structures

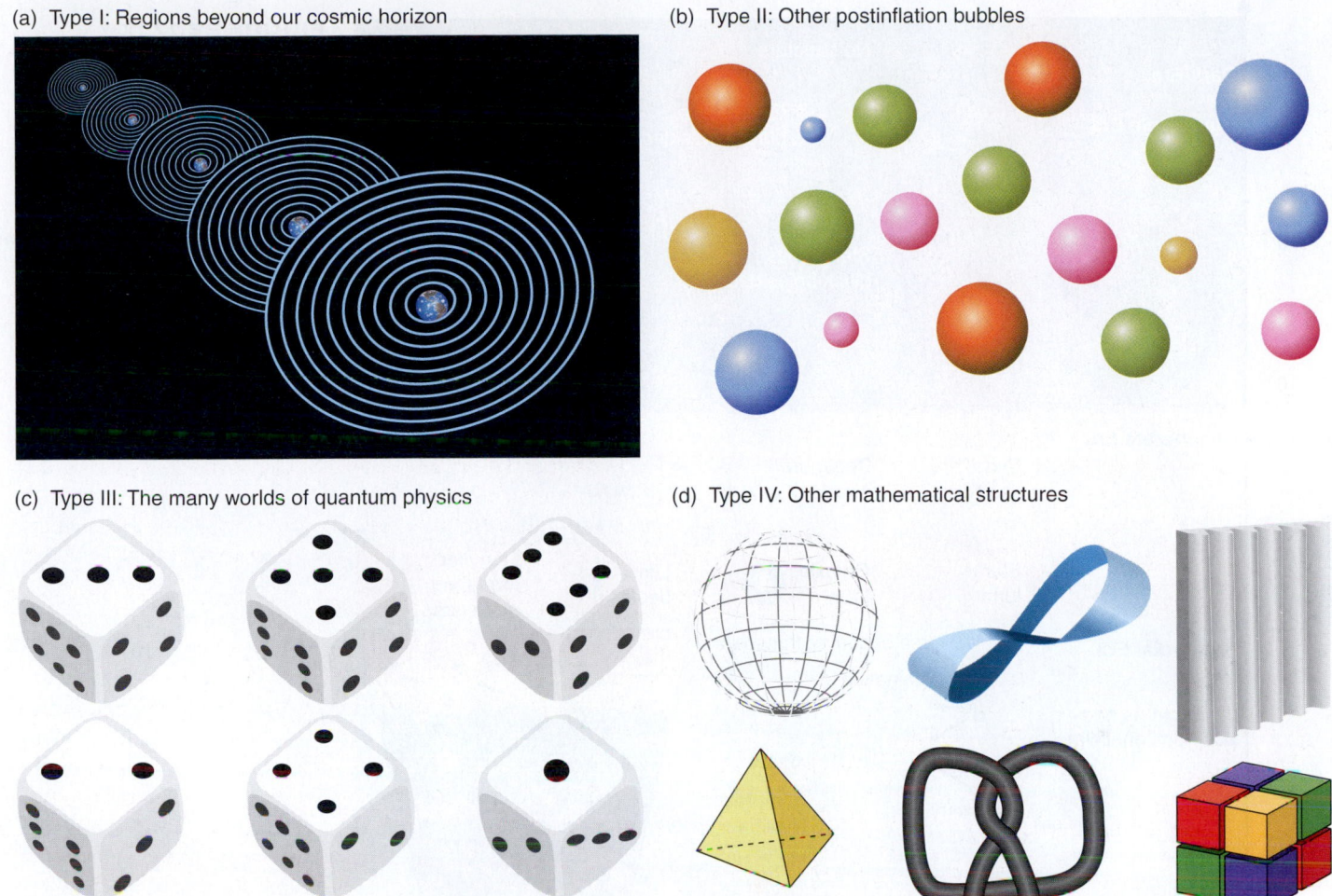

FIGURE 22.16 Cosmologists have proposed four types of multiverses: (a) parallel observable universes; (b) the inflationary model; (c) the quantum mechanical model; (d) parallel universes with different physics.

multiverse, involving distinct observable bubbles, when we measure the isotropy of the CBR or the large-scale distribution of galaxies. Since this type follows inevitably from our standard cosmological models, it makes predictions that can be tested readily. The eternal inflation model is harder to test, because we will never directly observe parallel universes. But if we obtain a theory of everything that predicts eternal inflation, and if that theory of everything is itself falsifiable, then there is a connection between eternal inflation and observation. In addition, if the bubbles in eternal inflation have very different properties from one another, we can also test whether such bubbles can expand into a parallel universe in which human beings can evolve. If such parallel universes evolve too fast or don't produce stars and planets, then our very existence would be unlikely and such models would be highly suspect. Types III and IV might also be modeled with quantum computers and with an ultimate TOE.

There is considerable debate within the scientific community as to whether the multiverse hypotheses, except for the Type I case, can be truly falsified. Even if there are some

tests, the concern is that the tests are not truly meaningful. For example, you might want to test Newton's theory of gravity by releasing an apple and watching whether it falls upward or downward. But this test is not very discerning—it could not always distinguish two sensible theories from one another. Similarly, for the multiverse hypothesis it is still debatable whether the tests that seem possible would be meaningful tests of the theory.

22.6 The Deep Future

Let's turn back now to our own observable universe and ask what the future holds in store for our universe and the structure it has formed. In Chapter 21 we discussed how the universe evolved to its current state, and we found that if the mass of the universe is large enough and the cosmological constant is small enough, gravity will win in the end. Hubble expansion will eventually reverse, and the universe

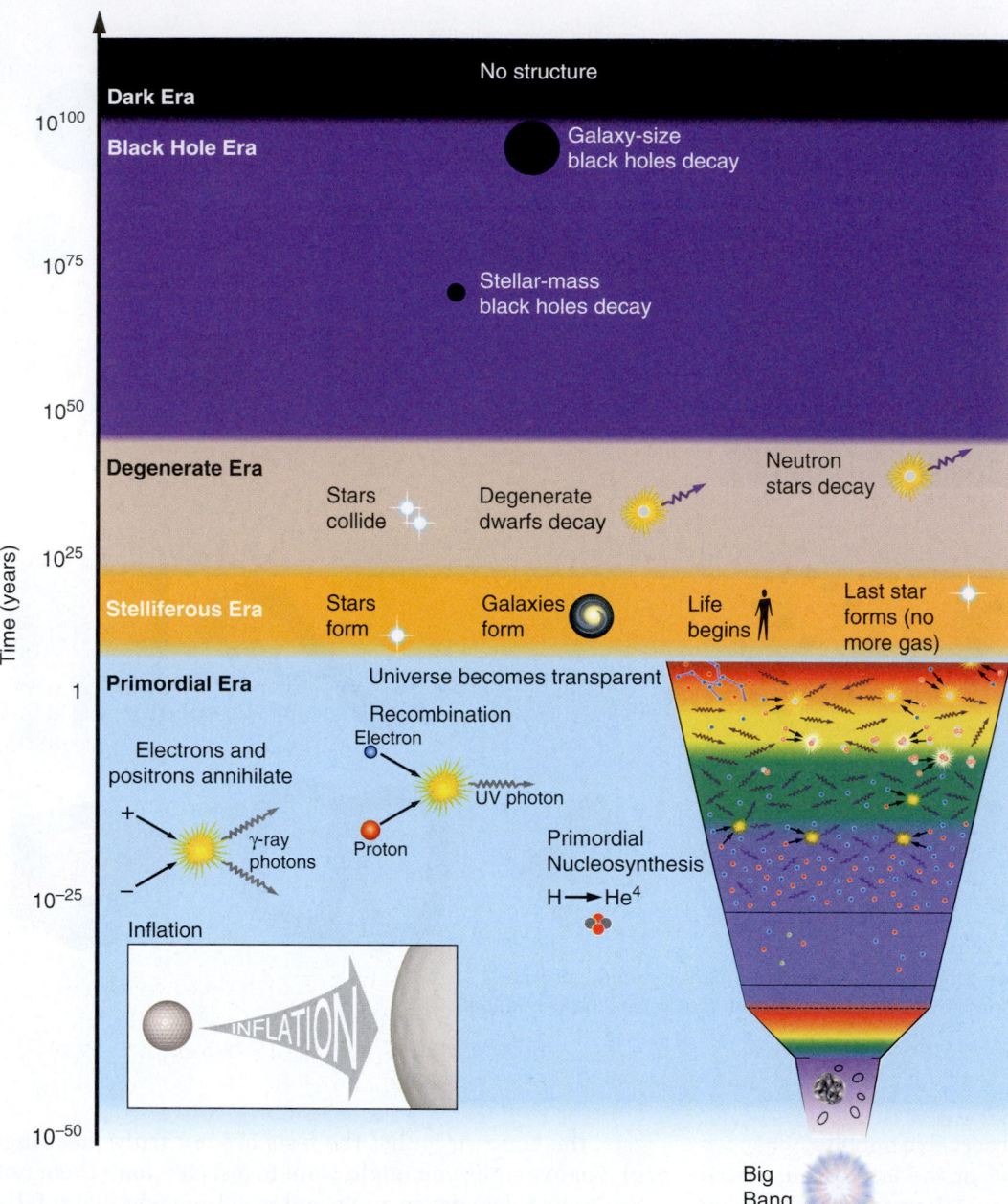

FIGURE 22.17 How structure in the universe evolves—from the Primordial Era to the present, and then to the future Dark Era—as postulated by Fred Adams and Gregory Laughlin.

will collapse back into a state resembling that of the young, hot universe. Galaxies, stars, planets, molecules, atoms, and subatomic particles—all might cease to exist as matter is replaced by pure energy. Perhaps from such a state, a new universe would emerge. Early in the 21st century, however, few cosmologists see such a "Big Crunch" in the future of the universe. It appears that our universe is expanding at an ever-accelerating rate. As we discussed in Chapter 21, if the dark energy's effects are increasing with time, the universe may expand so fast that it becomes infinite in size in a finite time—the "Big Rip" that would tear all structure apart. However, the cosmological parameters we measure suggest that the universe will expand forever, perhaps even at an ever-accelerating pace. Does this mean the universe will go on expanding without end? Yes, but its constituents will appear very different from what we see today.

Using well-established physics, we might speculate how the existing structures in the universe will evolve over a very long time. In 1997, American astrophysicists Fred Adams and Gregory Laughlin published their calculations of the great eras, past and future, in the history of the universe. We illustrate these eras in **Figure 22.17**. During the first era, the "Primordial Era"—the first 500,000 years after the Big Bang and before recombination—the universe was a swarm of radiation and elementary particles. Today we live during the second era, the "Stelliferous Era" ("Era of Stars"); but this era, too, will end. Some 100 trillion (10^{14}) years from now the last molecular cloud will collapse to form stars, and a mere 10 trillion years later the least massive of these stars will evolve to form white dwarfs.

Following the Stelliferous Era, most of the normal matter in the universe will be locked up in degenerate stellar objects: brown dwarfs, white dwarfs, and neutron stars. During this "Degenerate Era," the occasional star will still flare up as ancient substellar brown dwarfs collide, merging to form low-mass stars that burn out in a trillion years or so. However, the main source of energy during this era will come from the decay of protons and neutrons and the annihilation of particles of dark matter. Even these processes will eventually run out of fuel. In 10^{39} years, white dwarfs will have been destroyed by proton decay (see Chapter 21), and neutron stars will have been destroyed by the beta decay of neutrons (see Chapter 14.)

As the Degenerate Era comes to an end, the only significant concentrations of mass left will be black holes. These will range from black holes with the masses of single stars to greedy monsters that grew during the Degenerate Era to have masses as large as those of galaxy clusters. During the period that follows—the "Black Hole Era"—these black holes will slowly evaporate into elementary particles through the emission of Hawking radiation (see Chapter 17). A black hole with a mass of a few solar masses will evaporate into elementary particles in 10^{65} years, and galaxy-sized black holes will evaporate in about 10^{98} years. By the time the universe reaches an age of 10^{100}

> Ultimately, all structure in the universe will decay.

years, even the largest of the black holes will be gone. A universe vastly larger than ours will contain little but photons with colossal wavelengths, neutrinos, electrons, positrons, and other waste products of black hole evaporation. The "Dark Era" will have arrived as the universe continues to expand forever—into the long, cold, dark night of eternity. This may be the final victory of **entropy**—the **heat death** of the universe.

If you think about it, the evolution of structure in our universe, from the earliest moments to the distant future, is a journey of highs and lows. Any chaotic structure present in the Big Bang is soon smoothed out by the enormous expansion of the universe during inflation, the very time when the seeds of later structure arise from quantum fluctuations. Then, until recombination, the radiation in the universe resists the pull of gravity to form structure, after which the dark matter begins to form galaxies, pulling the normal matter along with it. Then stars form and recycle their gas into newer stars, until all the gas is used up. Stellar evolution eventually ceases, leaving behind white dwarfs, neutron stars, and black holes, which in turn ultimately must decay through quantum effects. This leaves our universe a very dull and lifeless place with no structure at all—unless there are parallel universes still undergoing structure formation.

Seeing the Forest for the Trees

As our journey nears its end, we have reached high ground, a vantage point from which we can look back and survey the terrain we have covered. From this perspective it becomes clear that from the beginning, our journey has been guided by a single underlying quest: to understand how the structures we see in the universe came to be. Whether we are talking about the formation of galaxies and clusters, the origin and evolution of stars, the activity of black holes, the history of our Solar System, the geology of our own world, the changes that shaped the universe itself during its earliest moments, or the eras of the far future, 21st century astronomy (the science, as well as the book) is organized around the desire to better understand the origin of structure in the universe.

During the past century we have made remarkable strides toward this goal, and along the way a pattern has emerged. Look back at the quotation opening Chapter 16. What statement remains true in all circumstances? "And this, too, shall pass away." Structure is ephemeral. The universe is not about a final destination; it is a place of process and change. The things that "matter" are the things that happen along the way!

When you see your breath fog up on a cold winter day, think about elementary particles freezing out of the early universe, giving rise not only to matter but to the fundamental forces of nature. As you watch the clouds build before a thunderstorm, think of galaxies coalescing within halos of dark matter. Glance at a crystalline snowflake as it lands on the sleeve of your coat, and in your mind's eye see a star that condensed out of clouds of interstellar gas and dust. The origin of structure is everywhere around us and is no less remarkable for its familiarity.

One of the lessons of our journey is that structure in the universe does not revolve around us. Just as Copernicus showed that we do not occupy a special place at the center of the Solar System, we have now seen that the material of Earth and of ourselves—the protons, neutrons, and electrons that comprise normal matter—is not the major constituent of the universe. In fact the dark matter in our universe is essential for galaxies like our Milky Way to even form. Our own constituents could never coalesce to form the galaxies we see. And of course, without galaxies, there could be no stellar nucleosynthesis of heavy elements and therefore no Earth. We have even seen that the mysterious dark matter itself is not the major constituent of our universe, which is dominated by the even more mysterious dark energy. Indeed, we have learned that our own universe may not even be special—just one among an infinite number of parallel universes.

Another lesson we have learned is that the physics of the smallest particles is really crucial on the very largest scales. Dark energy and dark matter, described by particle theory, dominate the universe on its largest scales. And the quantum physics that is virtually irrelevant in our macroscopic world turns out to be crucial on large scales. Quantum fluctuations may be responsible for parallel universes being formed. In addition, it is quantum fluctuations that produce the ripples in the early universe that in turn seed the formation of galaxies and clusters of galaxies, the very largest structures we can see.

In virtually all cases—from tiny planets to stars, black holes, galaxies, and superclusters of galaxies—it is our old friend gravity that forms structure in our universe and holds it together. Without gravity holding structure together, our universe would be a dull place indeed. But as we complete our journey in the next chapter, we will study one fascinating type of structure that does not form through the action of gravity: the structure we call life.

Summary

- Galaxies reside in groups, clusters, and larger structures, which generally formed after galaxies did.

- On very large scales, the structures in the universe reside mainly on filaments and walls surrounding large regions devoid of galaxies.

- Structure formed in the universe through the gravitational collapse of inhomogeneities in cold dark matter that arise from the early universe.

- Observed galaxies come from complex mergers, with the visible gas in galaxies cooling and falling inward to form the visible stars, which are surrounded by a dark matter halo.

- In the standard Big Bang model, our observable universe is but one of an infinite number of simultaneously existing observable universes, so there must be universes identical to our own.

- If our universe is eternally inflating and forming new bubbles, as some scientists have speculated, then it had no beginning and has been expanding forever.

- Structure will continue to form in our universe for at least another 10^{15} years, but our current understanding of quantum effects suggests that they ultimately lead to the decay of all structure, from planets to black holes.

SmartWork, Norton's online homework system, includes algorithmically generated versions of these questions, plus additional conceptual exercises. If your instructor assigns questions in SmartWork, log in at **smartwork.wwnorton.com.**

Student Questions

THINKING ABOUT THE CONCEPTS

1. Suppose you could view the early universe at a time when galaxies were first forming. How would it be different from the universe we see today?

*2. How do we know that the early universe was "far more uniform than the blue of the bluest sky on the clearest day," as stated in Section 22.1?

3. As clumps containing cold dark matter and normal matter collapse, they heat up. When a clump collapses to about half its maximum size, the increased thermal motion of particles tends to inhibit further collapse. Whereas normal matter can overcome this effect and continue to collapse, dark matter cannot. Explain the reason for this difference.

4. What are the main differences between galaxy groups and galaxy clusters?

5. Why is it reasonable to call our Local Group of galaxies a "group" rather than a cluster? Why is it safe to call our Local Group a "cluster"?

6. What is the difference between a galaxy cluster and a supercluster? Is our galaxy part of either? How do we know this?

7. How do astronomers use the following to measure the amount of dark matter contained in a cluster of galaxies?
   a. Motions of individual members of the cluster.
   b. Extremely hot gas that fills the intergalactic space within the cluster.
   c. Gravitational lensing by the cluster.

8. Can a galaxy be located inside a void? Explain.

9. Is it likely that voids are filled with dark matter? Why or why not?

10. Imagine that there are galaxies in the universe composed mostly of dark matter with relatively few stars or other luminous normal matter. If this were true, how might we learn of the existence of such galaxies?

11. How are the processes of star formation and galaxy formation similar? How do they differ?

12. What is the origin of large-scale structure?

13. Why is dark matter so essential to the galaxy formation process?

14. Which do we believe is the correct evolutionary sequence: (a) small star clusters formed first, which were bound

together into galaxies, which were later bound together in clusters and superclusters; or (b) supercluster-sized regions collapsed to form clusters, which then later collapsed to form galaxies, which formed small clusters of stars? Justify your answer.

15. Why does the current model of large-scale structure require that we include the effects of dark matter?

16. What do we think dark matter is made of, or at least *isn't* made of?

17. Why do we think that some hot dark matter exists?

18. Why do we ignore the cosmological constant (dark energy) when thinking about the formation of a galaxy?

19. Describe the stages of galaxy formation, for both the dark matter and gas components.

20. Why can gas consisting of hydrogen and helium collapse to a much smaller size than can nonbaryonic dark matter?

21. How does a roughly spherical cloud of gas collapse to form a disklike, rotating spiral galaxy?

22. What are some of the observational signs that dark matter exists?

23. How do astronomers believe elliptical galaxies formed? Does this formation process explain the observed differences between spirals and ellipticals?

24. On the basis of our understanding of galaxy formation, describe how galaxies should appear as you look further back in time. Are the features you described observed?

25. How do we know that the dark matter in the universe must be composed mostly of cold, rather than hot, dark matter?

26. Using broad strokes, describe the process of structure formation in the universe, starting at recombination (half a billion years after the Big Bang) and ending today.

*27. How can we be certain that gravity, and not the other fundamental forces, is responsible for large-scale structure?

28. We have never observed a star with zero heavy metals in its atmosphere. What does this fact imply about the history of star formation in the early universe?

APPLYING THE CONCEPTS

*29. Our theory of cosmology assumes that on large scales, the structure in the universe is uniform no matter where you look. Maps of structure, like that in Figure 22.4c, support this assumption. Does the presence of large masses like the Great Attractor violate this principle?

30. In previous chapters we painted a fairly comprehensive picture of how and why stars form. Why, then, is it difficult to model the star formation history of a young galaxy? Is this difficulty a failure of our theories?

**31. If 300 million neutrinos fill each cubic meter of space, and if neutrinos account for only 5 percent of the mass density (including dark energy) of the universe, estimate the mass of a neutrino.

32. What is the approximate mass of
   a. An average group of galaxies?
   b. An average cluster?
   c. An average supercluster?

33. The lifetime of a black hole varies in direct proportion to the cube of the black hole's mass. How much longer does it take a supermassive black hole of 10^{30} solar masses to decay compared to a stellar black hole of 3 solar masses?

34. Knowing what elliptical galaxies are made of, estimate how old they must be. Knowing that ellipticals form via mergers of spirals, and knowing when we believe galaxies first formed, estimate how long it took to complete the merging events that formed the elliptical galaxies we see today.

35. Currently, the Milky Way and Andromeda galaxies (M31) are separated by about 2.3 million light-years and moving toward each other at about 120 kilometers per second (km/s). Estimate how long it may take for the two to collide. Why do you think this may or may not be a good estimate of how long it will take these galaxies to fully merge?

36. Compare the timescales you found in the previous two questions. Is our impending merger with the Andromeda Galaxy typical of those that happened in the past? Can this timescale place any constraints on the distances or ages of galaxies that merged to form the elliptical galaxies seen today?

*37. The Great Attractor is about 200 Mly from the Milky Way, and by some estimates the Milky Way is moving toward it at a speed of about 1,000 km/s. Will we collide with it? Why or why not?

38. It is likely that the initial fluctuations leading to large-scale structure arose from quantum fluctuations in the early universe. How would the universe look different today if those fluctuations were ten times bigger? What about ten times smaller?

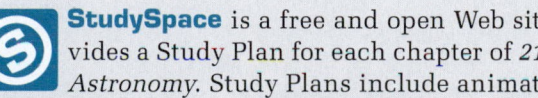

StudySpace is a free and open Web site that provides a Study Plan for each chapter of *21st Century Astronomy*. Study Plans include animations, reading outlines, vocabulary flashcards, and multiple-choice quizzes, plus links to premium content in SmartWork and the ebook. Visit **www.wwnorton.com/studyspace**.

If one could conclude as to the nature of the Creator from a study of his creation it would appear that God has a special fondness for stars and beetles.

J. B. S. HALDANE (1892–1964)

The search for signs of intelligent life in other stellar systems has been an ongoing pursuit of modern science.

CHAPTER 23

Life

23.1 Life Is a Form of Structure

As we near the end of our journey, let's briefly appreciate the distance we have covered. We have followed the origin of structure in the universe from the earliest instants after the Big Bang, when the fundamental forces of nature came to be, to formation of the galaxies and other large-scale structure that is visible today. Earlier on our journey, we watched as stars (including our Sun) formed from clouds of gas and dust within these galaxies, and planets (including our Earth) formed around those stars. We learned of the geological processes that shaped early Earth into the planet we see around us today. In short, we have traced the origin of structure from the instant the universe came into existence up until the modern day. To help put these events into perspective, **Excursions 23.1** shows the relative timing of the major events in the history of the universe if scaled down to fit into a single 24-hour day that began with the Big Bang and ended today. But there is one piece of the story that we have left out. Although the focus of our journey is *astronomy*, no discussion of how structure evolved in the universe would be complete without some consideration of the origin of the particular type of structure we refer to as **life**.

KEY CONCEPTS

Our journey has brought us to a new understanding of the universe and all it contains. Still, one very important and very personal component remains: you, the reader of this book. What combination of events—some probable, others much less likely—has led to your existence as a sentient being, living on a small rocky planet, orbiting a typical middle-aged star? Are you and your fellow humans unique, or are there others like you inhabiting planets scattered throughout the vast universe? Here we will learn that

- Like planets, stars, and galaxies, life is a form of structure that evolved through the action of the physical processes that shape the universe—the universe may teem with life.

- Life is a form of complex chemistry.

- Terrestrial life likely originated in Earth's oceans.

- Evolution is a natural process.

- The conditions needed for life to form and evolve are commonplace throughout the universe.

- Although the search goes on, there is not yet any evidence for extraterrestrial life.

- All life on Earth must eventually come to an end.

23.2 Life on Earth

Before delving into the questions of how life first appeared on Earth and how it has since evolved, we should first ask, How do we define life? The answer might at first seem obvious, but it is not immediately self-evident. Many scientists will tell us that there is no universal definition of life. Although we may have a feeling for what we mean when speaking about terrestrial life, a universal definition would necessarily have to encompass very different forms of life that may occur elsewhere in the universe. To arrive

---

EXCURSIONS 23.1

Forever in a Day

We have seen that the events that ultimately led to our existence here as inhabitants of planet Earth stretched out over many billions of years—intervals that even astronomers have difficulty visualizing. We can sometimes get a better grasp of such enormous spans of time by compressing them into much shorter intervals with which we have day-to-day experience. Try, then, to imagine the age of the universe and those important events associated with our origins as if they were all taking place within a single day. Our cosmic day begins at the stroke of midnight:

12:00:00 A.M. The embryonic universe is a mixed broth—minute specks of matter suspended in a vast soup of radiation. The entire universe exists only as an extraordinarily hot bath of photons and a zoo of elementary particles.

12:00:02 A.M. It is now just 2 seconds after midnight. All of the early eras in the history of the universe have passed. The fundamental forces have frozen out, Big Bang nucleosynthesis has formed the universe's original complement of atomic nuclei, and things have cooled down enough for atomic nuclei to combine with electrons to produce neutral atoms. Both normal and dark matter are now available to create galaxies and stars, but that process will take some time.

12:20 A.M. The first stars, quasars, and galaxies appear. At some point—we are not sure exactly when—our own galaxy becomes visible as star formation begins. Throughout the cosmic day, stars will continue to form. The more massive stars each go through their brief life cycles in only 5–10 seconds of our imaginary 24-hour clock. Each massive star shines briefly, creates its heavy elements, and then disperses this material throughout interstellar space as it dies in a violent supernova explosion. Stars similar to our Sun go through less dramatic life cycles, each lasting about 16 cosmic hours. Stars with mass less than 0.8 solar mass ($M_\odot$) will last several dozen cosmic hours and will survive beyond the end of our cosmic day.

4:00 P.M. Our Solar System forms out of a giant cloud of gas and dust. Collapse of the cloud's protostellar core and then the appearance of the Sun and the planets—including Earth—all take place within the span of a single cosmic minute.

4:05 P.M. A Mars-sized planetesimal crashes into Earth, forming the Moon.

5:20 P.M. The first primitive life appears on Earth. It evolves into the simplest life-forms: unicellular organisms such as bacteria, cyanobacteria, and archaebacteria.

8:40 P.M. More complex single-celled organisms appear, making it possible for multicellular life to develop.

9:30 P.M. The first multicellular organisms (fungi) appear on dry land.

11:00 P.M. Multicellular organisms become abundant. Their evolution paves the way for still larger and more complex life-forms.

11:20 P.M. The first animals (fish) make the transition from ocean to dry land—a major step toward our existence.

11:35 P.M. The first dinosaurs make their appearance. Various small animals appear as well, but they remain dominated by larger and more prevalent life-forms.

11:53:10 P.M. A large comet or asteroid crashes into Mexico's Yucatán Peninsula. Seventy percent of all species on Earth (including the dinosaurs) suddenly vanish. In the minutes that follow, the mammals, being more adaptable to the changed environment, gain prominence.

11:59:25 P.M. Our earliest human ancestors finally appear on the plains of Africa just 35 seconds before the end of our cosmic day.

11:59:59.8 P.M. Modern humans arrive with only a fifth of a second to spare—a fraction of a heartbeat before the day's end.

12:00:00 A.M. Just as our cosmic day draws to a close, will a worldwide catastrophe occur? Will 21st century humans permanently scar the face of the planet, driving many of its life-forms into oblivion by polluting the atmosphere and poisoning the land, the rivers, and the oceans? Will our progeny finally break free of their gravitational bondage to their planetary home and begin to claim the rest of the Solar System as their own? In comparison to all that came before, what happens in the next century will occur in a blur—less than a blink of the eye.

at a complete definition, we would have to take into account not only life-forms of which we are completely unaware, but also those that exceed the limits of our imagination. We will speculate about alien life later in the chapter.

A Definition of Life

For now, we'll turn to a definition of life that meets our terrestrial experience. Life is first of all a chemical process. Complex biochemical processes enable living organisms to grow and sustain themselves by drawing energy from their environment. With the assistance of special molecules such as ribonucleic acid (RNA) and deoxyribonucleic acid (DNA), organisms are able to reproduce and evolve. All terrestrial life involves carbon-based chemistry and employs liquid water as its biochemical "solvent." (A more detailed look at the chemistry of life appears in Section 23.3.)

> Life is a chemical process.

With at least a basic idea of what we mean by *life*, we turn to the greater question of how it got its start here on Earth. As we learned in earlier chapters, Earth's secondary atmosphere was formed in part by carbon dioxide and water vapor that poured forth as a product of volcanism; a heavy bombardment of comets likely added large quantities of methane and ammonia to the mix. Note that these are all simple molecules, incapable of carrying out life's complex chemistry. However, early Earth had abundant sources of energy, such as lightning and ultraviolet solar radiation, that could tear these relatively simple molecules apart, thereby creating fragments that could reassemble into molecules of greater mass and complexity. As rain carried them out of the atmosphere, these heavier organic molecules ended up in Earth's oceans, forming a "primordial soup."

In 1952, American chemists Harold Urey (1893–1981) and Stanley Miller (1930–2007) attempted to create conditions similar to what they believed existed on early Earth. To a laboratory jar containing liquid water as an "ocean," they added methane, ammonia, and hydrogen as a primitive atmosphere; electric sparks simulated lightning as a source of energy (**Figure 23.1**). Within a week, the Urey-Miller experiment yielded 11 of the 20 basic amino acids that link together to form proteins, the structural molecules of life. Other organic molecules that are components of nucleic acids, the precursors of RNA and DNA, also appeared in the mix. We now know that Earth's early secondary atmosphere contained carbon dioxide and nitrogen, rather than hydrogen, but recent experiments with more realistic atmospheric compositions have produced results similar to those of Urey and Miller. From laboratory experiments such as these, scientists have developed various models in which life got its start in Earth's oceans, rich in prebiotic organic molecules.

The details of where and how these prebiotic molecules evolved into the molecules of life is not so clear. Some biologists believe life began in the ocean depths, where volcanic vents provided the hydrothermal energy needed to create the highly organized molecules responsible for biochemistry (**Figure 23.2**). Others believe that life originated in tide pools, where lightning and ultraviolet radiation supplied the energy (**Figure 23.3**). In either case, short strands of self-replicating molecules may have formed first, later evolving into RNA, and finally into DNA, the huge molecule that serves as the biological "blueprint" for self-replicating organisms.

> Terrestrial life probably began in Earth's oceans.

Finally, we should briefly mention the suggestion by a few scientists that life on Earth may have been "seeded" from space in the form of microbes brought here by meteoroids or comets. Although this hypothesis might tell us how life began on Earth, it does not explain its origin elsewhere in the Solar System or beyond. We note that there is no scientific evidence at this time to support the "seeding" hypothesis.

FIGURE 23.1 Chemist Stanley Miller, with his experiment simulating an early-Earth atmosphere.

The First Life

If life did indeed get its start in Earth's oceans, when did it happen? In Chapter 6 we learned that young Earth suffered severe bombardment by Solar System debris for several

(a)

(b)

FIGURE 23.2 (a) Life on Earth may have arisen near ocean geothermal vents like this one. Similar environments might exist elsewhere in the Solar System. (b) Living organisms around geothermal vents, such as the giant tube worms shown here, rely on geothermal rather than solar energy for their survival.

hundred million years following its formation roughly 5 billion years ago. These were hardly the conditions under which life could form and gain a foothold. However, once the bombardment had abated and Earth's oceans had appeared, the opportunities for living organisms to evolve were greatly improved. The earliest *indirect* evidence for terrestrial life is carbonized material found in Greenland rocks dating back to 3.85 billion years ago. Stronger and more direct evidence for early life appears in the form of fossilized masses of simple microbes called "stromatolites," which go back about 3.6

> Life on Earth may have begun more than 3.6 billion years ago.

billion to 3.8 billion years. Fossilized stromatolites have been found in western Australia and southern Africa, and living examples still exist to this day (**Figure 23.4**). We may never know the exact date when life first appeared on Earth, but if the material found in the Greenland rocks is indeed of biological origin, then terrestrial life quickly took advantage of the favorable environment that followed cessation of Earth's pummeling by errant planetesimal fragments. Thus, the current evidence suggests that the earliest life-forms appeared within a billion years after the formation of the Solar System, and within 500 million years of the end of young Earth's catastrophic bombardment by leftover planetesimals.

FIGURE 23.3 Life may have begun in tide pools.

FIGURE 23.4 Modern-day stromatolites growing in colonies along an Australian shore.

The earliest organisms are examples of what biologists call "extremophiles." These life-forms not only survive, but thrive, under extreme environmental conditions. Extremophiles include organisms living in subfreezing environments or in water temperatures as high as 120°C, which occur in the vicinity of deep-ocean hydrothermal vents. Other extremophiles are found under the severe conditions of extraordinary salinity, pressure, dryness, acidity, or alkalinity. Among these earlier life-forms was an ancestral form of "cyanobacteria," single-celled organisms otherwise known as "blue-green algae." Many scientists have found supporting evidence that cyanobacteria were responsible for creating oxygen in Earth's atmosphere by photosynthesizing carbon dioxide and releasing oxygen as a waste product. Oxygen, however, is a highly reactive gas, and the newly released oxygen was quickly removed from Earth's atmosphere by the oxidation of surface minerals. Only when the exposed minerals could no longer absorb more oxygen did atmospheric levels of the gas begin to rise. Oxygenation of Earth's atmosphere and oceans began about 2 billion years ago, and the current level was reached only about 250 million years ago (see Chapter 8). Without cyanobacteria and other photosynthesizing organisms, Earth's atmosphere would be as oxygen-free as the atmospheres of Venus and Mars.

Biologists comparing genetic DNA sequences find that terrestrial life is divided into three major domains: Bacteria, Archaea, and Eukarya—known informally as bacteria, archaeans, and eukaryotes. Bacteria and archaeans are organisms known as "prokaryotes" and consist of free-floating DNA inside a cell wall. (Archaeans are similar in appearance to bacteria but have a metabolic chemistry closer to that of eukaryotes.) Prokaryotes are simple organisms, lacking both cell structure and a nucleus (**Figure 23.5a**). Even today, the largest component of Earth's total biomass is in the form of simple microbes. Eighty percent of the history of terrestrial life has been microbial. "Eukaryotes" have a more complex form of DNA contained within the cell's membrane-enclosed nucleus (**Figure 23.5b**). The first eukaryote fossils go back about 2 billion years, coincident with the rise of free oxygen in the oceans and atmosphere, although the first multicellular eukaryotes did not appear until a billion years later.

Life Becomes More Complex

All life, whether prokaryotic or eukaryotic, shares a similar genetic code, which may have originated from a common ancestor. DNA sequencing enables biologists to trace backward to the time when different types of life first appeared on Earth, and to identify the species they evolved from. Scientists have used DNA sequencing to establish the so-called evolutionary tree of life, which describes the interconnectivity of all species. The tree has revealed some interesting relationships; for example, it places animals (including us) as most similar to fungi, which branched off the evolution-

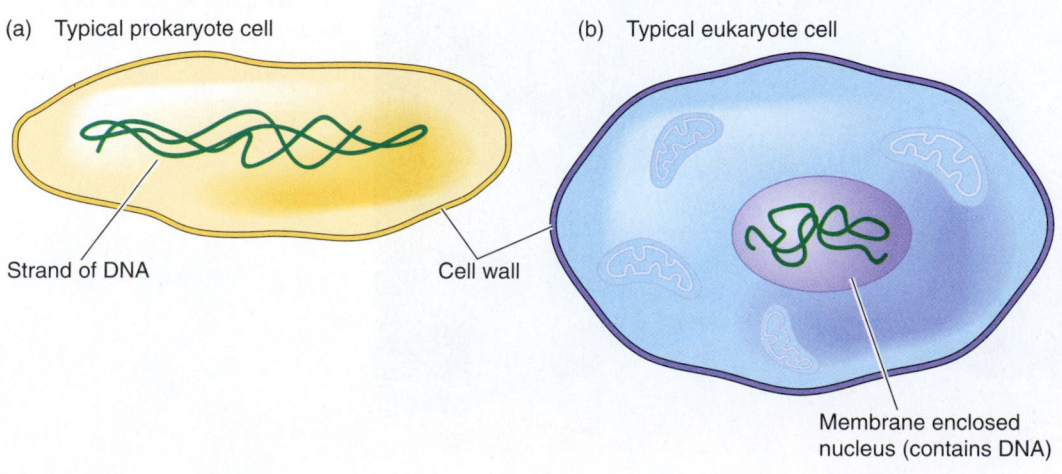

(a) Typical prokaryote cell

(b) Typical eukaryote cell

Strand of DNA   Cell wall

Membrane enclosed
nucleus (contains DNA)

FIGURE 23.5 (a) A simple prokaryotic cell contains little more than the cell's genetic material. (b) A eukaryotic cell contains several membrane-enclosed structures, including a nucleus that houses the cell's genetic material.

ary tree after slime molds and plants! The earliest primates branched off from other mammals about 70 million years ago, and the great apes (gorillas, chimpanzees, bonobos, and orangutans) split off from the lesser apes about 20 million years ago. DNA tests show that humans and chimpanzees share about 98 percent of their DNA; the two groups are believed to have evolved from a common ancestor about 6 million years ago.

The pace of evolution proceeded very slowly over the eons that followed the first appearance of terrestrial life. Living creatures in Earth's oceans remained much the same—a mixture of single-celled and relatively primitive multicellular organisms, for more than 3 billion years. Then, between 540 million and 500 million years ago, something extraordinary happened. The number and diversity of biological species increased spectacularly. Biologists call this event the "Cambrian explosion." The cause of this sudden surge in biodiversity remains unknown, but possibilities include rising oxygen levels, an increase in genetic complexity, major climate change, or a combination thereof. The "Snowball Earth" hypothesis suggests that prior to the Cambrian explosion, Earth was in a period of extreme cold between about 750 million and 550 million years ago and was covered almost entirely by ice. Astrophysicists speculate that the Sun's output at that time was 5–10 percent weaker than it is at present. During this period, predatory animals may have died out, thus making it easier for new species to adapt and thrive. Another possibility is that a marked increase in atmospheric oxygen (see Figure 8.5) would have been accompanied by a corresponding increase in stratospheric ozone, which, as we learned in Chapter 8,

> **The number of new species exploded about a half-billion years ago.**

shields us from deadly solar ultraviolet radiation. With a protective ozone layer in place, life was free to leave the oceans and move to land (**Figure 23.6**).

The first plants appeared on land about 475 million years ago, and large forests and insects go back 360 million years. The age of dinosaurs began 230 million years ago and ended abruptly 65 million years ago, when a small asteroid or comet collided with Earth and exterminated more than 70 percent of all existing plant and animal species (see Chapter 7). Animals were the big winners in the aftermath, creating an evolutionary pathway that has led to us as human beings. Our earliest human ancestors appeared a few million years ago, and the first civilizations occurred a mere 10,000 years ago. Our industrial society, barely more than two centuries old, is but a footnote in the history of life on our planet.

At the beginning of this chapter we mentioned that you, the student, are here today because of a series of events that have occurred throughout the history of the universe. Some of these events seem inevitable, such as the creation of heavy elements by earlier generations of stars and the formation of life-supporting planets, including Earth. Other events may have been less likely, such as the creation of self-replicating molecules that led to Earth's earliest life. A few events stand out as random, such as the life-destroying impact by a piece of space debris 65 million years ago—an incident that eliminated the dominant species of that era. This event—fortuitous for us, but not for the dinosaurs—made possible the evolution of advanced forms of mammalian life and, ultimately, our existence as human beings. We

> **You exist because of a series of events— some inevitable and others more random.**

FIGURE 23.6 A fossil recently found in the Canadian Arctic shows *Tiktaalik*, a fish with limblike fins and ribs—an animal in a midevolutionary step of leaving the water for dry land.

might then ask the question, Would you be here today reading this text if Earth hadn't experienced that fateful encounter? The answer is, probably not! With that sobering thought in mind, let's now return to the question of how prebiotic molecules became the molecules of life.

Evolution Is a Means of Change and Advancement

Imagine that just once during the first few hundred million years after the formation of Earth, a single molecule formed by chance somewhere in Earth's oceans. That molecule had a very special property: Chemical reactions between that molecule and other molecules in the surrounding water resulted in that molecule's making a copy of itself. Now there were two such molecules. Chemical reactions would produce copies of each of these molecules as well, making four. Four became 8, 8 became 16, 16 became 32, and so on. By the time the original molecule had copied itself just 100 times, over a *million trillion trillion* (10^{30}) of these molecules would have existed. That is 100 million times more of these molecules than there are stars in the observable universe! Such unconstrained replication would be highly

> To create life, only one self-replicating molecule needed to form by chance.

unlikely for a number of reasons, such as the limited availability of raw materials needed for reproduction and competition from similar molecules. The point we would like to make here is this: the likelihood that a copying error will occur while a molecule is replicating increases significantly with the number of copies being made.

Chemical reactions are never perfect. Sometimes when a copy is attempted, it is not an exact duplicate of its predecessor. The imperfection in the attempted copy is called a **mutation**. Most of the time such an error is devastating, leading to a molecule that can no longer reproduce at all. But occasionally a mutation is actually helpful, leading to a molecule that is *more* successful in duplicating itself than the original was. Even if imperfections in the copying process crop up only once every 100,000 times that a molecule reproduces itself, and even if only one out of 100,000 of these errors turns out to be beneficial, after only 100 generations there will still be a hundred million trillion (10^{20}) errors that, by blind luck, might improve on the original design. Copies of each of these improved molecules will inherit the change. These molecules will have happened upon a form of **heredity**—the ability of one generation of structure to pass on its characteristics to future generations.

As molecules of the early Earth continued to interact with their surroundings and make copies of themselves, they split into many different varieties. Eventually, these descendants of our original molecule became so numerous that the building blocks they needed in order to reproduce became scarce. In the face of this scarcity of resources, vari-

eties of molecules that were more successful than others in reproducing themselves became more numerous. Varieties that could break down other varieties of self-replicating molecules and use them as raw material were especially successful in this world of limited resources. Competi-

> Success breeds success.

tion, predation, and even cooperation had entered the picture. After a few generations, certain species of molecules came to dominate the mix, while less successful varieties became less and less common. This process, in which better-adapted molecules thrive and less well-adapted molecules die out, is referred to as **natural selection**.

Four billion years is a long time—enough time for the combined effects of heredity and natural selection to shape the descendants of that early self-copying molecule into a huge variety of complex, competitive, successful structures. Geological processes on Earth, such as sedimentation, have preserved a fossil record of the history of these structures (**Figure 23.7**). Among these descendants are "structures" capable of thinking about their own existence and unraveling the mysteries of the stars.

The molecules of DNA (**Figure 23.8**) that make up the chromosomes in the nuclei of the cells throughout your body are direct descendants of those early self-duplicating molecules that flourished in

> Evolution is inevitable.

the oceans of a young Earth. Although the game played by the molecules of DNA in your body is far more elaborate than the game played by those early molecules in Earth's oceans, the fundamental rules remain the same. We now realize that this process is inevitable: any system that combines the elements of heredity, mutation, and natural selection *must* and *will* evolve.

In *The Selfish Gene* (1976), Richard Dawkins points out that, when viewed from a purely utilitarian perspective, a human being is a machine whose "purpose" is to produce copies of its genetic material. The remarkable thing about humanity is that our intelligence (itself a very powerful tool for survival in a competitive world) has also enabled us to develop more noble pursuits (for example, astronomy) than those dictated by our basic biochemical imperative.

23.3 Life beyond Earth

The story of the formation and evolution of life cannot be separated from the narrative of astronomy in the 21st century. We know that we live in a universe full of stars and that systems of planets orbit many—probably most—of those stars. Many biologists also believe that there is nothing mysterious or unique about the origin of life or the processes that cause it to evolve. In fact, as discussed in **Connections 23.1**, the evolution of life on Earth is but

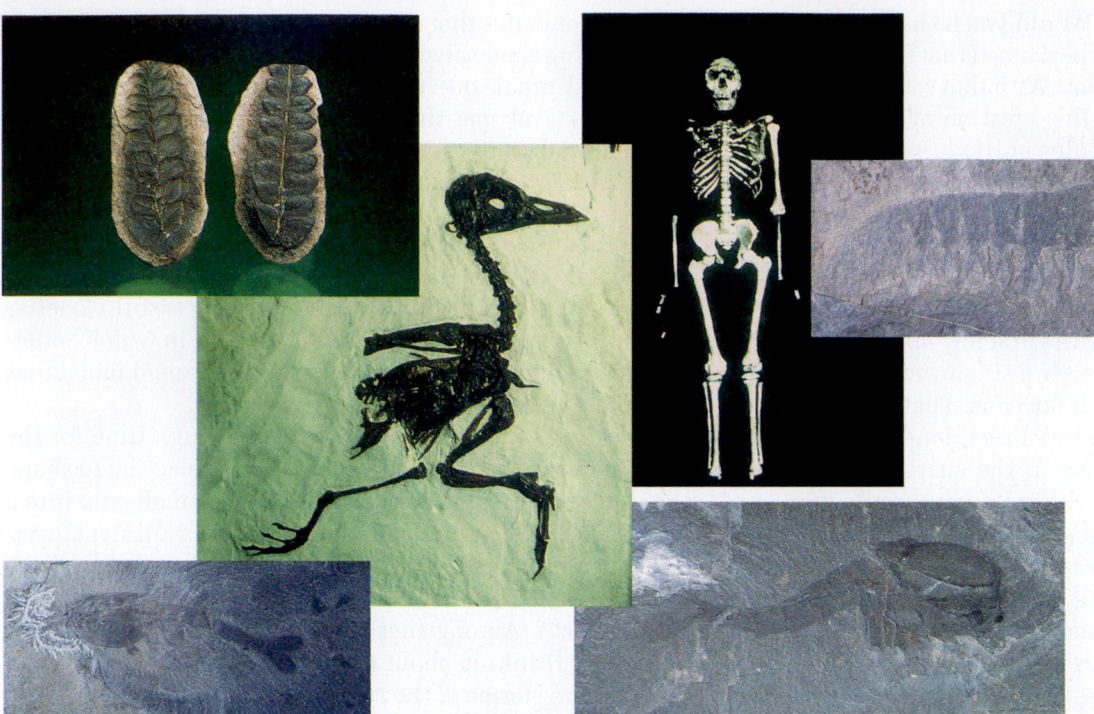

FIGURE 23.7 Fossils record the history of the evolution of life on Earth.

one of the many examples that we have encountered of the emergence of structure in an evolving universe. This point leads naturally to one of the more profound questions that we ask about the universe: Has life arisen elsewhere? To explore this question, we need to take a closer look at the chemistry of life on Earth.

The Chemistry of Life

When we speak about the chemistry of life, what we really mean is life itself. All *known* organisms are composed of a more or less common suite of chemicals, and very complex ones at that. We begin by looking at ourselves. Approximately two-thirds of the atoms in our bodies are hydrogen (H), about one-fourth are oxygen (O), a tenth are carbon (C), and a few hundredths are nitrogen (N). The remaining atomic elements, and there are several dozen of them, make up only 0.2 percent of our total inventory. It turns out that all living creatures—at least those that we know of—are an assemblage of molecules composed almost entirely of these four atomic elements (some-times called CHON), along with small amounts of phosphorus and sulfur. Some of these molecules are enormous. Consider DNA, which is responsible for our genetic code. DNA is made entirely from only *five* atomic elements: CHON and phosphorus. But the DNA in each cell of our bodies is composed of combinations of *tens of billions* of atoms of these same five elements. Then there are proteins, the huge

> All known life is composed primarily of only six elements.

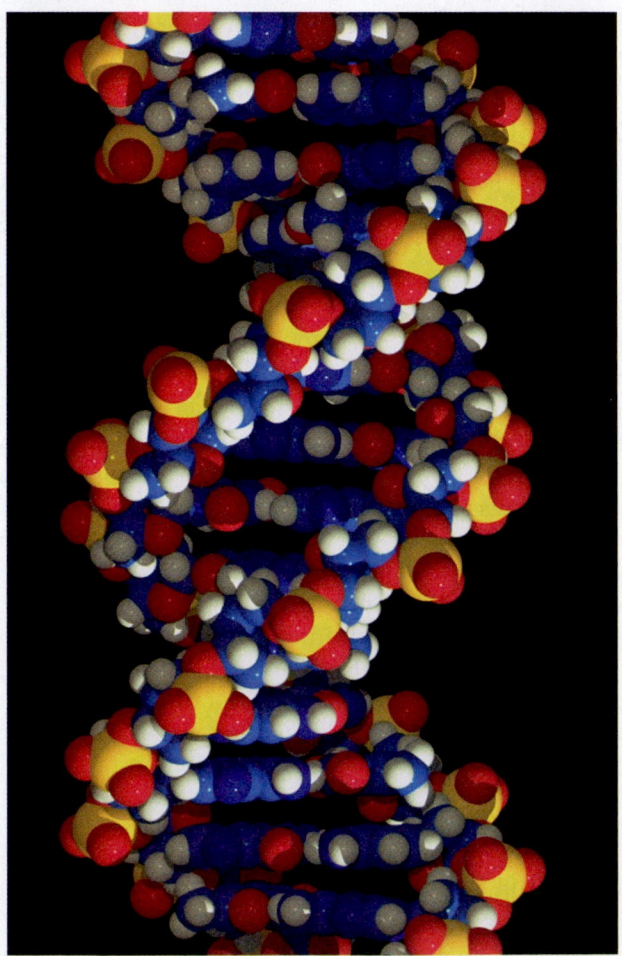

FIGURE 23.8 DNA is the blueprint for life.

CONNECTIONS 23.1

Life, the Universe, and Everything

Popular discussions of the origin of life frequently grapple with the question of how complex, highly ordered structures such as living things could have emerged from simpler, less ordered antecedents. Physicists discuss the degree of order of a system using the concept of **entropy**, which is a measure of the number of different ways a system could be rearranged and still appear the same. For example, place a drop of ink in a glass of water and watch what happens. The ink spreads out, diffusing through the water until the only sign that the ink is there is the fact that the water is a different color. The order represented by the discrete drop of ink naturally fades away as the ink spreads out through the water. Yet no matter how long we watch, we will never see that drop of ink spontaneously reassemble itself.

Initially, the drop of ink and the glass of clear water form a neatly ordered system with low entropy. By contrast, the glass of inky water, which can be stirred or turned and still look the same, is a disordered system with higher entropy. The **second law of thermodynamics** says that, left on its own, an isolated system will always move toward higher entropy—that is, from order toward disorder. This is common sense. As time goes by, we are more likely to find a system in a state that is, well, more likely.

In light of the inescapable march toward disorder that is dictated by the second law, how can ordered structure emerge spontaneously? Creationist or intelligent-design claims that the origin of life flies in the face of the second law of thermodynamics are about as common as reports of Elvis sightings in supermarket tabloids. Yet such claims make a crucial mistake: they focus attention on one player while ignoring the rest of the game.

We have all seen what happens when we set a glass of ice water out on a hot, humid day (**Figure 23.9**). Water vapor from the surrounding air condenses into drops of liquid water on the surface of the cold glass. This is amazing: We have just watched as ordered structure (drops of water) spontaneously emerged from disorder (water vapor in the air). It is almost as if we saw the drop of ink reassemble itself. This simple, everyday event *appears* to violate the second law of thermodynamics—but it does not.

To understand why, we need to step back and look at more than just the drops of water on the glass. When the drops condensed, they released a small amount of thermal energy that slightly heated both the glass and the surrounding air. Heating

FIGURE 23.9 On a humid day, water condenses into droplets on the surface of a cold glass. The second law of thermodynamics does not mean that ordered structure cannot spontaneously emerge.

something increases its entropy. The decrease in entropy due to the formation of the drops is more than made up for by the increase in the entropy of their surroundings. Ordered structure spontaneously emerged, but *overall*, disorder increased.

It is important to note that energy can be used to reduce entropy at one location while increasing entropy somewhere else. An air conditioner provides a good example. Electric energy produced at a power plant is used to pump thermal energy from inside a home and dump that thermal energy outside. When an air conditioner is run in reverse to provide heating, it is referred to as a "heat pump," but "entropy pump" might be a more accurate description. As you sit in your armchair on a summer day, all you immediately notice is that when the air conditioner comes on, the temperature drops. Entropy decreases inside your house. If you look at the system as a whole, on the other hand—including the heating of air by the outside coils and the entropy produced by the burning of coal or natural gas during the production of electricity—you see that turning on the air-conditioning causes an overall *increase* in entropy.

This idea is especially important when we consider the origin of life. A living thing, whether an amoeba or a human being, represents a huge local increase in order. However, no violation of the second law of thermodynamics is involved. In our everyday lives, the food we eat gives us the energy we need to stave off the relentless advance of entropy. Viewed even more broadly, the evolution of life on Earth was powered primarily by energy striking Earth in the form of sunlight. A local increase in order on Earth (such as you) is "paid for" in the end by the much greater decrease in order that accompanies thermonuclear fusion in the heart of the Sun. Order emerges in localized regions within a system, but the second law of thermodynamics is obeyed overall.

The unifying theme of these last two chapters and, in many ways, this entire book has been understanding the origin of structure. The answer we have found is clear. Whether we are discussing the freezing out of matter and the fundamental forces in the young universe; the gravitational collapse of stars, planets, and galaxies; the evolution of life; or water beading up on the outside of a cold glass—ordered structure *does* emerge spontaneously as an unavoidable consequence of the action of physical law.

molecules responsible for the structure and function of living organisms. Proteins are long chains of smaller molecules called amino acids. Terrestrial life employs 20 specific amino acids, which also contain no more than five atomic elements—in this case CHON plus sulfur instead of phosphorus.

Can this be all there is to our body chemistry? No; the chemistry of life is far too complex to get by with a mere half-dozen atomic species. Many others are present in smaller amounts but are essential to the complicated chemical processes that make living organisms tick. They include sodium, chlorine, potassium, calcium, magnesium, iron, manganese, and iodine. Finally, there are the so-called trace elements, such as copper, zinc, selenium, and cobalt. Trace elements also play a crucial role in life chemistry but are needed in only tiny amounts.

We know that the infant universe was composed basically of hydrogen and helium and very little else. But fast-forward to a later time—some 9 billion years later. By this time, all the heavier chemical elements essential to life were present and available in the molecular cloud that gave birth to our Solar System. As we learned in Chapters 16 and 17, those heavy elements—up to and including iron—were forged in the nuclear furnaces of earlier generations of low- and high-mass stars and were then dispersed into space. At times, this dispersal was passive. For example, low-mass stars, such as dying red giants, lose their gravitational grip on their overly extended atmospheres. Along with hydrogen and helium, the newly created heavy elements are blown off into space, eventually finding their way into molecular clouds. Other dispersals were more violent. Most of the trace elements essential to biology are more massive than iron, so they are not produced in the interiors of main-sequence stars. They are instead created within a matter of minutes during the violent supernova explosions that mark the death of high-mass stars. They, too, are thrown into the chemical mix found in molecular clouds.

Heavy elements produced by generations of stars provide the raw materials for life—but what kind of life? The only form that we have discussed so far is terrestrial, because, after all, it is the only one we know of. As we have already pointed out, terrestrial life is organic, or carbon-based. What, then, is so special about carbon? Carbon is the lightest among the "tetravalent" atoms. (*Tetra* means "four," and *valence* refers to an atom's ability to attach to or "bond" with other atoms or molecules.) This chemical property allows as many as four other atoms or molecules to bind themselves to each carbon atom. Now, consider what can happen if the attached molecules also contain carbon. The result can be an enormous variety of long-chain molecules. This great versatility enables carbon to form the complex molecules that provide the basis for terrestrial life's chemistry.

Does this mean that all carbon-based life should be like us? Not at all. There could be forms of extraterrestrial life that are also carbon-based but have chemistries quite different from our own. For example, consider amino acids, the building blocks of protein. There are countless varieties of amino acids, but only 20 among them are employed by terrestrial life. Furthermore, molecules other than RNA and DNA may be capable of self-replication.

We have pointed out that carbon is the lightest of the tetravalent atoms, but silicon is also tetravalent. Does this mean that somewhere in the universe there might exist a form of life that is silicon-based? We see no reason why not. As a potential life-enabling atom, silicon has both advantages and disadvantages when compared to carbon. An important advantage is that silicon-based molecules remain stable at much higher temperatures than carbon-based molecules, enabling a possible silicon-based life to thrive in high-temperature environments, perhaps on exoplanets that orbit close to their parent star. But silicon has a serious disadvantage. Silicon is a larger and more massive atom than carbon, and it lacks the ability to form molecules as complex as those based on carbon. Any silicon-based life would necessarily be simpler than what we have here on Earth. Even so, we cannot rule out the possibility of silicon-based life existing in high-temperature niches somewhere within this almost limitless universe.

Finally, although carbon's unique properties make it readily adaptable to the chemistry of life on Earth, we don't know what other chemistries life might adopt. As we learned from examples of terrestrial extremophiles here and in Chapter 11, life is highly adaptable and tenacious; when it comes to the form that extraterrestrial life might take, nothing should surprise us.

Life within Our Solar System

In our quest to find evidence of extraterrestrial life, the logical place to start would be right here in our own Solar System. As humans we have long been curious about the possibility of life beyond Earth. Some early conjectures seem ridiculous, considering what we now know about the Solar System. Two centuries ago, the eminent astronomer Sir William Herschel, discoverer of Uranus, proclaimed, "We need not hesitate to admit that the Sun is richly stored with inhabitants." In 1877, Italian astronomer Giovanni Schiaparelli (1835–1910) observed what appeared to be linear features on Mars and dubbed them *canali*, meaning "channels" in Italian. In one of astronomy's great ironies, the famous American observer of Mars, Percival Lowell (1855–1916), misinterpreted Schiaparelli's *canali* as "canals," suggesting that they were constructed by intelligent beings. Initial public fascination with Martians turned to hysteria in 1938 when Orson Welles aired H. G. Wells's fictitious *War of the Worlds*, as a "live" radio news coverage of militant Martians

> **Life-forms other than carbon-based are possible.**

> **Terrestrial life is based on carbon.**

invading Earth. Panic ensued when many listeners believed that the "invasion" was actually happening.

The next several decades saw only modest progress in the search for life in the Solar System. During the mid–20th century, groundbased telescopes discovered that Mars possesses an atmosphere and water,[1] both considered essential for any terrestrial-type life to get its start and evolve. During the 1960s, the United States and the Soviet Union sent reconnaissance spacecraft to the Moon, Venus, and Mars, but the instrumentation carried aboard these spacecraft was more suited to learning about the physical and geological properties of these bodies than to searching for life. Serious efforts to look for signs of life—past or present—would have to await advanced spacecraft with specialized bioinstrumentation.

In the meantime, astronomers and biologists alike were discussing where to look and what to look for; and thus was born the science of **astrobiology**, the study of the origin, evolution, distribution, and future of life in the universe. The question of where to look was a relatively easy one. Astrobiologists knew that Mercury and the Moon lacked atmospheres, ruling them out as possibilities. The giant planets and their moons were thought to be too remote and too cold to sustain life. Venus, they knew, was far too hot. As Goldilocks might have concluded, Mars was just right. In the mid-1970s, two American *Viking* spacecraft were sent to

Mars is a planet that may harbor extraterrestrial life in our own Solar System.

Mars with detachable landers containing a suite of instruments designed to find evidence of a terrestrial type of life. (Some scientists were critical of the specific sites chosen, claiming that higher-latitude locations, where water ice might exist, would have been preferable.) When the *Viking* landers failed to find any evidence of life on Mars, hopes faded for finding life on any other body orbiting our Sun.

Since that time, however, optimism has been renewed. A better understanding of the history of Mars indicates that at one time the planet was wetter and warmer, leading many scientists to believe that fossil life or even living microbes might yet be buried under the planet's surface. The first decade of the 21st century saw a return to Mars and a continuation in the search for evidence of current or preexisting life (see Chapter 7). In 2008, NASA's *Phoenix* spacecraft landed at a far-northern latitude, inside the planet's arctic circle, where specialized instruments dug into and analyzed the martian water-ice permafrost (see Figure 7.37). *Phoenix* found that the martian arctic soil has a chemistry similar to the Antarctic dry valleys on Earth, where life exists deep below the surface at the ice-soil boundary. Aqueous minerals, such as calcium carbonate, reveal the existence of ancient oceans on the planet. However, *Phoenix* did not

find direct evidence of life, so the long-standing question of whether there is life on Mars remains unanswered.

The 1980s brought NASA's instrumented robots to the outer Solar System, and what they found surprised many astrobiologists. Although the outer planets themselves did not appear to be habitats for life, some of their moons became objects of special interest. Jupiter's moon Europa is covered with a layer of water ice that appears to overlie a great ocean of liquid water. Impacts by comet nuclei may have added a mix of organic material, another essential ingredient for life. Once thought to be a

Life might exist on icy moons in the outer Solar System.

frozen, inhospitable world, Europa is now a candidate for biological exploration. Saturn's moon Titan appears to be rich in organic chemicals, many of which are thought to be precursor molecules of a type that existed on prebiotic Earth. That 1980s message from Titan was clear: the chemistry necessary for life does indeed exist elsewhere in the Solar System.

Two decades later, the *Cassini* mission found additional evidence for a variety of prebiotic molecules in Titan's atmosphere, and it identified another potential site for life: Saturn's tiny moon Enceladus. In one of the bigger surprises from *Cassini*, the spacecraft detected water-ice crystals spouting from cryovolcanoes near the moon's south pole. Liquid water must lie beneath its icy surface, and Enceladus therefore joins Europa as a possible habitat of extremophile life, perhaps similar to that found near geothermal vents deep within Earth's oceans.

Our efforts to find evidence of life on other bodies within our own Solar System have so far been unsuccessful, but the quest continues. The discovery of life on even one Solar System body beyond Earth would be exciting: if life arose independently *twice* in the same planetary system, then that might increase the likelihood that life exists throughout the universe.

Habitable Zones around Parent Stars

Searching for life among the countless exoplanets within our own Milky Way Galaxy can be daunting. Nevertheless, astronomers are narrowing the search by concentrating on planets that provide environments conducive to the formation and evolution of life (at least, life as we understand it), while eliminating planets that are clearly unsuitable for supporting life. As we learned in Chapter 6, most of the exoplanets discovered so far have environments that are much too harsh to provide a haven for any life-forms that we know of.

To begin their quest, astronomers can optimize their search by focusing on planetary systems that are stable. Planets in stable systems remain in well-behaved, nearly circular orbits that preserve relatively uniform climatological and oceanic environments. Planets in noncircular

[1] The martian polar caps are made up of frozen water ice and carbon dioxide ice.

orbits can experience wild temperature swings that could be detrimental to the survival of life. A stable temperature that maintains the existence of water in a liquid state might be important. We know that liquid water was essential for the formation and evolution of life on Earth. Of course, we don't know if liquid water is an absolute requirement for life elsewhere, but it's a good starting point when we think about where to look. On planets that are too close to their parent star, water would exist only as a vapor—if at all. On planets that are too far from their star, water would be permanently frozen as ice. Still another consideration is planet size. Large planets such as Jupiter retain most of their light gases—hydrogen and helium—during formation and so become gas giants without a surface. Planets that are very small may have insufficient surface gravity to retain their atmospheric gases and so end up like our Moon.

We know that the location of a planet relative to its parent star is critical. A candidate planet should orbit at a distance from its star that provides a range of temperature suitable for life. Astronomers refer to this as a **habitable zone**. Let's think about the habitable zone in our own Solar System. We can see that Venus, which orbits at 0.7 times Earth's distance from the Sun, has become an inferno because of its runaway greenhouse effect (see Chapter 8). Any liquid water that might once have existed on Venus has long since evaporated and been lost to space. Mars orbits about 1.5 times farther from the Sun than Earth, and all of the water that we see on Mars today is frozen. But the orbit of Mars is more elliptical and variable than Earth's, giving the planet a greater variety of climate, including long-term cycles that might occasionally permit liquid water to exist. Most astrobiologists put the habitable zone of our Solar System at about 0.9–1.4 astronomical units (AU), which includes Earth but just misses Venus and Mars.

> Exoplanets harboring life would most likely orbit within a star's habitable zone.

Yet this limit may be too narrow, because it does not account for possible extremophiles that may be thriving in liquid water beneath the surfaces of some icy moons of Jupiter and Saturn.

Astronomers must also think about the type of star they are observing in their search for life-supporting planets (**Figure 23.10**). Stars that are less massive than the Sun and thus cooler will have narrower habitable zones, minimizing the chance that a habitable planet will just happen to form within that slender zone. Stars that are more massive than the Sun are hotter and will have a larger habitable zone. As we learned in Chapter 16, however, a star's main-sequence lifetime depends on its mass. For example, a star of $2\,M_\odot$ would enjoy relative stability on the main sequence for only about a billion years before the helium flash incinerated everything around it! On Earth, a billion years was long enough for bacterial life to form and cover the planet, but insufficient for anything more advanced to evolve. Of course, we don't know if evolution might happen at a different pace elsewhere; we have only our one terrestrial case as an example! Still, main-sequence lifetime is a sufficiently strong consideration that most astronomers prefer to focus their efforts on stars with longer lifetimes—specifically, spectral types F, G, and K.

Finally, some astronomers think about a "galactic habitable zone," referring to a star's location within the Milky Way Galaxy. Stars that are situated too far from the galactic center may have protoplanetary disks with insufficient quantities of heavy elements—such as oxygen and silicon (silicates), iron, and nickel—that make up terrestrial-type planets like Earth. Stars that are too close to the galactic center also face problems. These regions experience less star formation, and therefore less recycling of heavy elements. Perhaps more serious is the high-energy radiation environment near the galactic center (X-ray and gamma-ray), which is damaging to RNA and DNA. Even so, for many stars the galactic habitable zone may not remain as a permanent

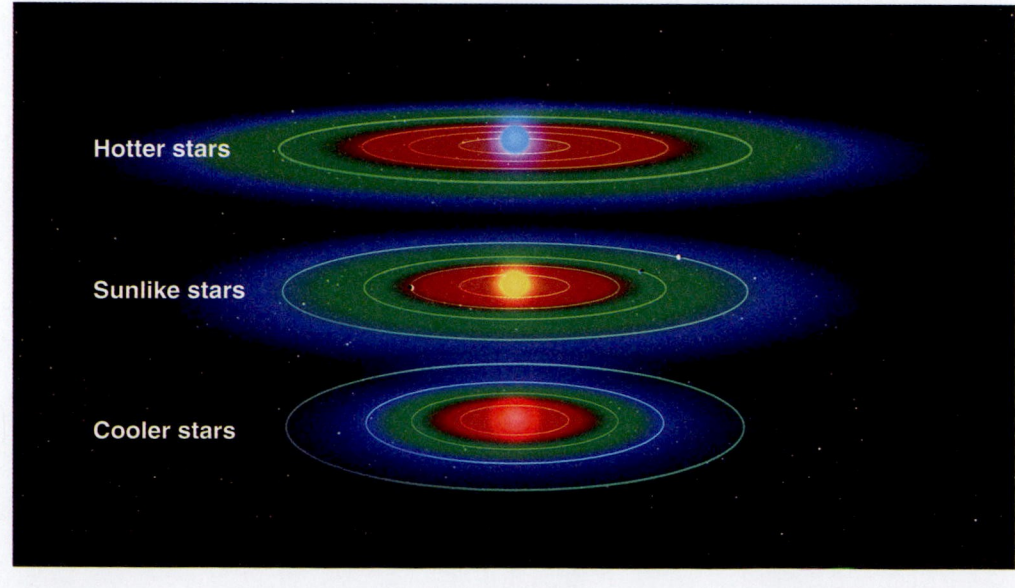

FIGURE 23.10 The distance and extent of a habitable life zone (green) surrounding a star depends on the star's temperature—in other words, its mass. Regions too close to the star are too hot (red), and those too far are too cold (blue.)

Hotter stars

Sunlike stars

Cooler stars

home if they tend to migrate within the galaxy and change their distance from the galactic center over time. In short, astronomers try to narrow down the vast numbers of stars as they conduct their search, but they acknowledge that these types of arguments—based on what worked well for planet Earth—might not be applicable when we are looking at other planetary systems.

The search for Earth-like planets from space is already under way. In 2006, the European Space Agency (ESA) launched CoRoT, the first spacecraft dedicated to extrasolar planet detection. CoRoT monitors nearby stars for changes in brightness that would indicate the presence of a transiting exoplanet (see Chapter 6). The ESA spacecraft has already made several discoveries, including one planet that is only 1.7 times the size of Earth. In 2009, NASA launched *Kepler* into a solar orbit that trails Earth in its own orbit about the Sun. Staying well behind Earth keeps our planet from getting in *Kepler*'s field of view, thereby enabling uninterrupted monitoring of *Kepler*'s target stars. Over the next several years, the spacecraft's photometer will simultaneously and continually monitor the brightness of more than 100,000 stars, looking for transiting exoplanets. *Kepler* can detect planets as small as 0.8–2.2 times the size of Earth.

23.4 The Search for Signs of Intelligent Life

We have come to a point where it seems that, given the right conditions—an Earth-type planet orbiting within a habitable zone around a solar-mass star—life of some kind might invariably arise. It took but a few hundred million years for life to form in Earth's oceans. We know that this earliest life was primitive, and that it took another 3.5 billion years for this early life to climb the evolutionary ladder and arrive at a species with the intelligence and technological capability to contemplate the universe that gave it birth and begin a search through the vastness of space for others like itself. As humans, we have reached that stage at a time when our star is only halfway through its lifetime. Such would not be the case for life arising on planets orbiting stars much more massive than the Sun. As we learned in Chapter 16, a 3-$M_\odot$ star has a main-sequence lifetime of only a few hundred million years. Our terrestrial experience suggests that just as life was getting its start on one of the star's habitable planets, it would quickly be doomed as the star reached the end of its brief period of stability. A star 1½ times as massive as the Sun has a main-sequence lifetime of a couple of billion years, insufficient for any life to reach the equivalent of Earth's "Cambrian explosion." If we are to search for signs of intelligent life, we should look to planets surrounding stars of solar mass or smaller. And now that we have some ideas about where to search for intelligent life, how do we establish communication?

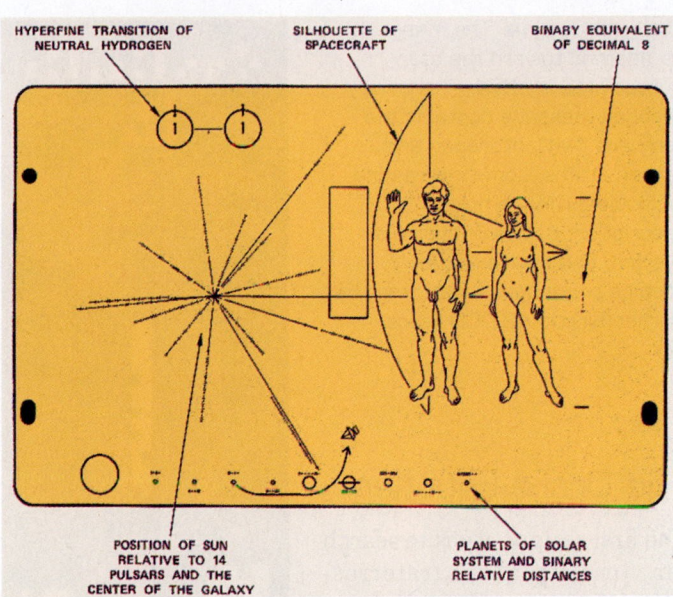

FIGURE 23.11 The plaque included with the *Pioneer 11* probe, which was launched in 1973 and will eventually leave the Solar System to travel through the millennia in interstellar space.

During the 1970s, humans made preliminary efforts to "reach out and touch someone." The *Pioneer 11* spacecraft, which will probably spend eternity drifting through interstellar space, carries the plaque shown in **Figure 23.11**. It describes ourselves and our location to any future interstellar traveler who might happen to find it. Another message to the cosmos accompanied the two *Voyager* spacecraft on identical phonograph records that contain greetings from planet Earth in 60 languages, samples of music, animal sounds, and a message from then-President Jimmy Carter. Interestingly, these messages created concern among some politicians and nonscientists who felt that scientists were dangerously advertising our location in the galaxy, even though radio signals had been broadcast into space for nearly 80 years at the time. The messages were also criticized by some philosophers, who claimed that we were making ridiculous anthropomorphic assumptions about the aliens being sufficiently like us to decode these messages. (Ironically, most Earthlings could not play those phonograph records today!)

Sending messages on spacecraft may not be the most efficient way to make contact with the universe, but it was a significant gesture nonetheless. A somewhat more practical effort was made in 1974, when astronomers used the 300-meter-wide dish of the Arecibo radio telescope to beam a message in binary code[2] (**Figure 23.12**) toward the star cluster M13. (If someone from M13 answers, we will know in 48,000 years!)

---

[2]Binary code consists only of zeros and ones, with zero equivalent to "off" and one equivalent to "on." It may be a universal means of communication that does not depend on any specific language or numerical system.

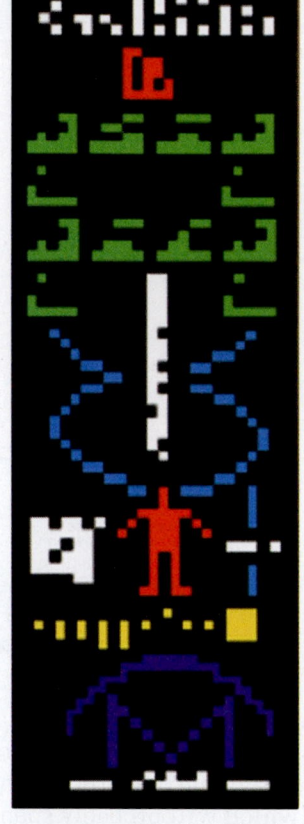

The Drake Equation

The first serious effort to search for intelligent extraterrestrial life was made by astronomer Frank Drake (1930–) in 1960. Drake used what was then astronomy's most powerful radio telescope to listen for signals from two nearby stars. Although his search revealed nothing unusual, it prompted him to develop an equation that still bears his name. The **Drake equation** estimates the likely number (N) of intelligent civilizations that may exist within the Milky Way Galaxy:

$$N = Ff_p\, n_{pm}\, f_l f_c\, L$$

The six factors on the right side of the equation relate to the conditions that Drake thought must be met for a civilization to exist:

The Drake equation estimates the number of technologically advanced civilizations in the galaxy.

1. F is the total number of stars in our galaxy. We know this to be several hundred billion stars.

2. f_p is the fraction of stars that form planetary systems. Questions remain, but we do know that planets form as a natural by-product of star formation and that many—perhaps most—other stars have planets. For this calculation we will assume that f_p is between 0.5 and 1.

3. n_{pm} is the average number of planets and moons in each planetary system capable of supporting life. If such planets are like Earth, they will have to be of the terrestrial type and orbit within a "habitable zone," where temperatures are neither too hot nor too cold. If our own Solar System is a good example, there would be one or two such planets in each planetary system. (In addition to Earth, Mars appears capable of supporting life, whether or not life presently exists there.) Although we now know that many planetary systems do not resemble our own, it is possible that moons orbiting giant planets near stars could also harbor life. For purposes of calculation, we put the value of n_{pm} between 0.1 and 2.

4. f_l is the fraction of planets and moons capable of supporting life on which life actually arises. Remember that just a single self-replicating molecule may be enough to get the ball rolling. Many biochemists now believe that if the right chemical and environmental conditions are present, then life *will* develop. If they are correct, then f_l is close to 1. For f_l we will use a range of 0.01 to 1.

5. f_c is the fraction of those planets harboring life that eventually develop technologically advanced civilizations. With only one example of a technological civilization to work with, f_c is hard to estimate. Intelligence is certainly the kind of survival trait that might often be strongly favored by natural selection. On the other hand, on Earth it took 4 billion years—half the expected lifetime of our star—to evolve tool-building intelligence. The correct value for f_c might be close to 1, or it might be closer to one in 1,000. The truth is, we just don't know.

6. L is the likelihood that any such civilization exists *today*. This factor is certainly the most difficult of all to estimate because it depends on the long-term stability of advanced civilizations. We have had a technological civilization on Earth for about 100 years, and during that time we have developed, deployed, and used weapons with the potential to eradicate our civilization and render Earth hostile to life for many years to come. We have also so degraded our planet's ecosystem that many respectable biologists and climatologists wonder if we are nearing the brink. Do all technological civilizations destroy themselves within a thousand years? A thousand years is only one 10-millionth (10^{-7}) of the lifetime of our star. Conversely, if most technological civilizations learn to use their technology for survival rather than self-destruction, might they instead survive for a 100 million years, or about a hundredth (10^{-2}) of the Sun's lifetime? For L in our calculation, we will use a range of between 10^{-7} (for civilizations that live 1,000 years) and 10^{-2} (for civilizations that live 100 million years).

As illustrated in **Figure 23.13**, the conclusions we draw using the Drake equation depend a great deal on the assumptions we make. For the most pessimistic of our estimates, the Drake equation sets the likelihood of finding a technological civilization in our galaxy at 1 percent, or one chance in 100. If this is correct, then in all likelihood we are the *only* technological civilization in the Milky Way. In fact, this result would suggest that only one in 100 galaxies would contain a technological civilization *at all*. Such a universe would still be full of intelligent life. With a hundred billion galaxies in the observable universe, even these pessi-

The nearest advanced civilization may be as far away as 30 Mly . . .

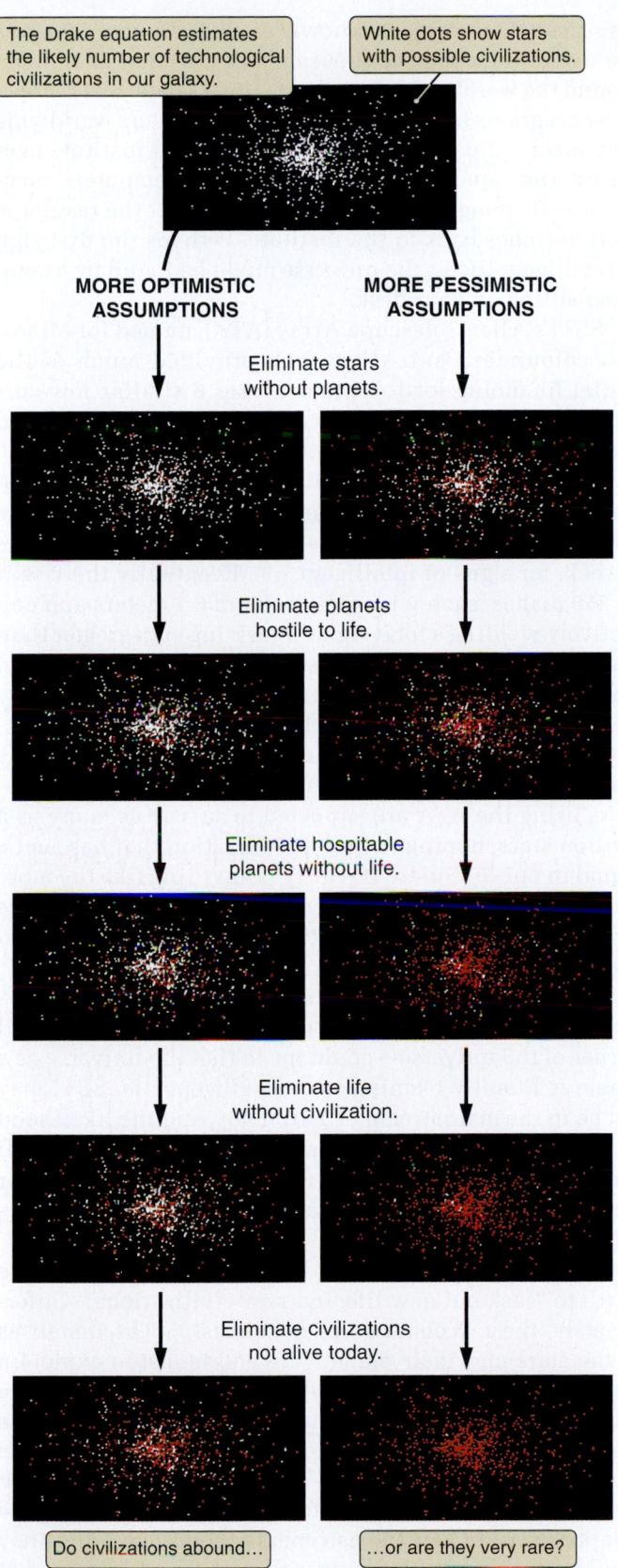

The Drake equation estimates the likely number of technological civilizations in our galaxy.

White dots show stars with possible civilizations.

MORE OPTIMISTIC ASSUMPTIONS   **MORE PESSIMISTIC ASSUMPTIONS**

Eliminate stars without planets.

Eliminate planets hostile to life.

Eliminate hospitable planets without life.

Eliminate life without civilization.

Eliminate civilizations not alive today.

Do civilizations abound...   ...or are they very rare?

FIGURE 23.13 Estimates of the existence of intelligent civilizations in our galaxy based on the Drake equation. Notice how widely these estimates vary, given optimistic and pessimistic assumptions about the six factors (see the text) affecting the prevalence of intelligent extraterrestrial life.

mistic assumptions would mean that there are a billion technological civilizations out there somewhere. On the other hand, we would have to go a *very* long way (30 million light-years [Mly] or so) to find our nearest neighbors.

At the other extreme, what if we take the most optimistic numbers, assuming that intelligent life arises and survives everywhere it gets the chance? The Drake equation then says that there should be 40 million technological civilizations in our galaxy alone! In this case, our nearest neighbors may be "only" 40 or 50 light-years away. If scientists in that civilization are listening to the universe with their own radio telescopes, hoping to answer the question of other life in the universe for themselves, then as you read this page they may be puzzling over an episode of the original *Star Trek* television series.

> . . . or as near as 40 light-years.

If we did run across another technologically advanced civilization, what would it be like? Looking back at the Drake equation, we can see it is highly unlikely that we have neighbors nearby, unless civilizations typically live for many thousands or even millions of years. In this case, any civilization that we encountered would almost certainly have been around for much longer than we have. Having survived that long, would its members have learned the value of peace, or would they have developed strategies for controlling their pesky neighbors? Movie theaters and the science fiction shelves of libraries and bookstores are filled with amusing and thoughtful stories that explore what life in the universe might be like (**Figure 23.14**). For the moment, though, let's set aside such speculation to look at the question as scientists. How might we go about finding a real answer?

> Any civilization we discover will almost certainly be advanced.

Technologically Advanced Civilizations

During lunch with colleagues, the famous physicist Enrico Fermi (1901–1954), a firm believer in extraterrestrial beings, is reported to have asked, "If the universe is teeming with aliens . . . where is everybody?" To which Hungarian-born physicist Leo Szilard (1898–1964) is said to have replied, "They are here among us, but they call themselves Hungarians." We can put Szilard's tongue-in-cheek reply aside, but Fermi's question—first posed in 1950 and sometimes called the "Fermi paradox"—remains unanswered. Consider also the following closely related question: If intelligent life-forms are common but interstellar travel is difficult or impossible, why don't we detect their radio transmissions? The truth is that we have failed so far to detect any alien radio signals—but it is not for lack of trying.

Drake's original project of listening for radio signals from intelligent life around two nearby stars has grown over the years into a much more elaborate program that is referred to as the Search for Extraterrestrial Intelligence,

FIGURE 23.14 The classic 1951 film *The Day the Earth Stood Still* portrayed intelligent extraterrestrials who threatened to destroy Earth if we carried our violent ways into space.

or **SETI**. Scientists from around the world have thought carefully about what strategies might be useful for finding life in the universe. Most of these have focused on the idea of using radio telescopes to listen for signals from space that bear an unambiguous signature of an intelligent source. Some have listened intently at certain key frequencies, such as the frequency of the interstellar 21-centimeter (cm) line from hydrogen gas. The assumption behind this approach is that if a civilization wanted to be heard, its denizens would tune their broadcasts to a channel that astronomers throughout the galaxy should be listening to. More recent searches have made use of advances in technology to record as broad a range of radio signals from space as possible. Analysts then use computers to search these databases for types of regularity in the signals that might suggest they are intelligent in origin.

Unlike much astronomical research, SETI receives its funding from private rather than government sources, and SETI researchers have found ingenious ways to continue the search for extraterrestrial civilizations with limited

> SETI listens for radio signals from other civilizations.

resources. One project, known as SETI@home, involves the use of hundreds of thousands of personal computers around the world to analyze the institute's data. SETI screen saver programs installed on personal computers worldwide download radio observations from the SETI Institute over the Internet, analyze these data while the computers' owners are off living their lives, and then report the results of their searches back to the institute. Perhaps the first sign of intelligent life in the universe might be found by a computer sitting on your desk.

SETI's Allen Telescope Array (ATA), named for Microsoft cofounder Paul Allen, who provided much of the initial financing for the project, uses a similar low-cost approach. A joint venture between the SETI Institute and the University of California, the ATA consists of a "farm" of small, inexpensive radio dishes like those used to capture signals from orbiting communication satellites (**Figure 23.15**). The ATA searches the sky 24 hours a day, 7 days a week, for signs of intelligent life. Eventually there will be 350 dishes, each with a diameter of 6.1 meters and collectively yielding a total signal-receiving area greater than that of a 100-meter radio telescope. Just as your brain can sort out sounds coming from different directions, this array of radio telescopes will be able to determine the direction a signal is coming from, allowing it to listen to many stars at the same time. Over several years' time, astronomers using the ATA are expected to survey as many as a million stars, hoping to find a civilization that has sent a signal in our direction. If reality is anything like the more optimistic of the assumptions we used in evaluating the Drake equation (see Figure 23.13), this project will stand a good chance of success.

Finding even *one* nearby civilization in our galaxy—that is, a *second* technological civilization in our own small corner of the universe—could mean that the universe as a whole is literally teeming with intelligent life. SETI may not be in the mainstream of astronomy, and the likelihood of its success is difficult to predict, but its potential payoff is enormous. Few discoveries would do more to change our understanding of ourselves than certain knowledge that we are not alone.

Science fiction is filled with tales of humans who leave Earth to "seek out new life and new civilizations." Unfortunately, these scenarios are not realistic. The distances to the stars and their planets are enormous; to explore a significant sample of stars would require extending the physical search over tens or hundreds of light-years. As we learned in Chapter 17, special relativity limits how fast we can travel. The speed of light is the limit, and even at that rate it would take over 4 years to get to the *nearest* star. Time dilation would favor the astronauts themselves, and they would return to Earth younger than if they had stayed at home. But suppose they visited a star 15 light-years distant. Even if they traveled at speeds close to that of light, by the time they returned to Earth, 30 years would have passed! Some science fiction enthusiasts get around this problem

FIGURE 23.15 When complete, the Allen Telescope Array will listen for evidence of intelligent life from as many as a million stellar systems.

by invoking "warp speed" or "hyperdrive" to travel faster than the speed of light, or by using wormholes as shortcuts across the galaxy. In truth, there is absolutely no evidence that any of these shortcuts are possible.

Some people claim that the aliens have already visited us: tabloid newspapers, books, and websites are filled with tales of UFO sightings, government conspiracies and cover-ups, alleged alien abductions, and UFO religious cults. However, none of these reports meet the basic standards of science, and we must conclude that there is no scientific evidence for any alien visitations.

23.5 The Future

We have used our understanding of physics and cosmology to look back through time and watch as structure formed throughout the universe. We have peered into the future and seen the ultimate fate of the universe, from its enormous clusters of galaxies down to its tiniest components. Now, as our journey nears its end, we examine our own destiny and contemplate the fate that awaits Earth, humanity, and the star on which all life depends.

Eventually, the Sun Must Die

As we have learned, about 5 billion years from now the Sun will end its long period of relative stability. Shedding its identity as the passive, benevolent star that has nurtured life on Earth for nearly 4 billion years, the Sun will expand to become a red giant and later an asymptotic giant branch (AGB) star, swelling to hundreds of times its present size. The giant planets, orbiting outside the extended red giant atmosphere, may survive the Sun's cranky old age in some form. Even so, they will suffer the blistering radiation from a Sun grown thousands of times more luminous than it is today.

The terrestrial planets will not fare as well. Some—perhaps all—of the worlds of the inner Solar System will be engulfed by the expanding Sun. Just as an artificial satellite is slowed by drag in Earth's tenuous outer atmosphere and eventually falls to the ground in a dazzling streak of white-hot light, so, too, will a terrestrial planet caught in the Sun's atmosphere be consumed by the burgeoning star. If this is Earth's fate, our home world will leave no trace other than a slight increase in the amount of massive elements in the Sun's atmosphere. As the Sun loses more and more of its atmosphere in an AGB wind, our atoms may be expelled back into the reaches of interstellar space from which they came, perhaps to become incorporated into new generations of stars, planets, and even life itself.

Another planetary fate is possible, however. In this scenario, as the red giant Sun loses mass in a powerful wind, its gravitational grasp on the planets will weaken, and the orbits of both the inner and outer planets will spiral outward. If Earth moves out far enough, it may survive as a seared cinder, orbiting the white dwarf that the Sun will become. Barely larger than Earth and with its nuclear fuel exhausted, the white dwarf Sun will slowly cool, eventually becoming a cold, inert sphere of degenerate carbon, orbited by what remains of its retinue of planets. Thus, the

> Far-future Earth will be consumed by the Sun or left as an icy cinder.

ultimate outcome for our Earth—consumed in the heart of the Sun or left behind as a frigid, burned rock orbiting a long-dead white dwarf—is not yet certain. In either case, Earth's status as a garden spot will be at an end.

The Fate of Life on Earth

Life on our planet has even less of a future, for it will not survive even long enough to witness the Sun's departure from the main sequence. Well before that cataclysmic event takes place, the luminosity of our heretofore benevolent star will begin to rise. As solar luminosity increases, so will temperatures on all the planets, including our own. Eventually Earth's temperatures will climb so high that all animal and plant life will perish. Even the extremophiles that inhabit the oceanic depths will die, as the oceans themselves boil away. Models of the Sun's evolution are still too imprecise to predict with certainty when that fatal event will occur, but the end of all terrestrial life may be only 1 billion or 2 billion years away. That is, of course, a comfortably distant time from now, but it is well short of the Sun's departure from the main sequence.

It is far from certain, however, that the descendants of today's humanity will even be around a billion years from now. Some of the threats that await us come from beyond Earth. For the remainder of the Sun's life, the terrestrial planets, including Earth, will continue to be bombarded by asteroids and comets. Perhaps a hundred or more of these impacts will involve kilometer-sized objects, capable of causing the kind of devastation that eradicated the dinosaurs and most other species 65 million years ago. Although these events may create new surface scars, they will have little effect on the integrity of Earth itself. Earth's geological record is filled with such events, and each time they happen, life manages to recover and reorganize.

It seems likely, then, that some form of life will survive to see the Sun begin its march toward instability. However, individual species do not necessarily fare so well when faced with cosmic cataclysm. If the descendants of humankind survive, it will be because we chose to become players in the game by changing the odds of such planetwide biological upheavals. We are rapidly developing technology that could enable us to detect most threatening asteroids and modify their orbits well before they could strike Earth. Comets are more difficult to guard against because long-period comets appear from the outer Solar System with little warning. To offer protection, various means of defending ourselves would have to be in place, ready to be used on very short notice. So far, we have been slow to take such threats seriously. Although impacts from kilometer-sized objects are infrequent, objects only a few dozen meters in size, carrying the punch of a several-megaton bomb, strike Earth about once every 100 years.

> To survive, humanity must learn to manage the threat of impacts.

Perhaps an explosion like the 1908 Tunguska blast in Russia (see Chapter 12), occurring over New York or Paris, would be enough to convince us that such precautions are worthwhile.

We might protect ourselves from the fate of the dinosaurs, but in the long run the descendants of humanity will either leave this world or die out. Planetary systems surround other stars, and all that we know tells us that many other Earth-like planets should exist throughout our galaxy. Colonizing other planets is currently the stuff of science fiction, but if our descendants are ultimately to survive the death of our home planet, off-Earth colonization must become science fact at some point in the future.

Although humankind may soon be capable of protecting Earth from life-threatening comet and asteroid impacts, in other ways we are our own worst enemy. We are poisoning the atmosphere, the water, and the land that form the habitat for all terrestrial life.

> Humanity is the worst threat to its own survival.

As our population grows unchecked, we are occupying more and more of Earth's land and consuming more and more of its resources, while sending thousands of species of plants and animals to their extinction each year. At the same time, human activities are dramatically affecting the balances of atmospheric gases. The climate and ecosystem of Earth constitute a finely balanced, complex system capable of exhibiting chaotic behavior. The fossil record shows that Earth has undergone sudden and dramatic climatic changes in response to even minor perturbations. Such drastic changes in the overall balance of nature would certainly have consequences for our own survival. When politics is added to the mix, even more immediate dangers await. For the first time in human history, we possess the means to unleash nuclear or biological disasters that could threaten the very survival of our species. In the end, the fate of humanity will depend more than anything on whether we accept stewardship of ourselves and of our planet.

Life in a Future Universe

The universe of the far distant future seems likely to be an extremely dull and lifeless place—a universe filled only with decaying black holes, magnetic fields, plasmas, and ever-cooling radiation. But will that necessarily be so?

> We cannot predict what structure might arise in the far future.

Imagine for a moment that intelligent life somehow evolved amid the swarm of exotic particles, such as free quarks and gluons, that filled the universe shortly after the Big Bang. Such organisms would have been far smaller than today's atoms and perhaps would have lived out their lives in 10^{-40} second or so. To such organisms the universe—all 3 meters of it—would have seemed incomprehensibly vast. These creatures and their entire civilization would have had to evolve, live, and die out in a millionth of a trillionth of a trillionth of the time that it takes for a single synapse in our brains to fire. If such creatures ever calculated the conditions in *today's* universe, they would have recoiled in horror. They would have imagined a frozen time when the temperature of the universe was only a thousandth of a trillionth of a trillionth of what they knew—a time when most matter had ceased to exist altogether and the tiny fraction that remained was spread out over a volume of space 10^{75} times greater than that of their universe. In short, such creatures would have looked forward and seen *our* universe as the frozen, desolate future. It seems doubtful they could have foreseen the existence of stars, galaxies, planets, and intelligent creatures for whom a single thought took longer than a billion trillion times the entire history of the universe they knew.

Now turn the tables, and think *forward* to a time when the universe is 10^{50} times older than it is today. Who can say that there will not be life then as well? Perhaps the life-forms of this distant future will be organisms of magnetic fields and tenuous electron plasmas, spread across countless trillions of light-years of space, whose lives unfold over untold eons of time. For such organisms, if they ever exist, it will be *we* who are the impossible creatures, alive for the briefest of instants, still immersed in the momentary fireball of the Big Bang.

Seeing the Forest for the Trees

As it is for galaxies and stars and planets, so it is for ourselves. A recurring theme on our journey of discovery has been the systematic dismantling of conceptual walls that in our minds separated us from the larger universe. Humanity does not stand outside the processes that shape the universe. We are instead one more variety of the structure to which the universe has given birth—a way station on a long road of evolving structure stretching back 14 billion years. Few single words are capable of eliciting as much emotional reaction from some people as *evolution*. Yet the public controversy surrounding evolution cannot change the scientific standing of this theory as one of the best-tested and most successful theories in all of science. As stated by Stephen Jay Gould, "The theory of evolution is in about as much trouble as the theory that Earth revolves around the Sun." By any reasonable standards of scientific knowledge, evolution is a fact.

Opponents of evolutionary theory often act as if evolution were a tiny piece of science that could simply be cast aside without doing violence to the rest. Having traveled the journey of *21st Century Astronomy*, you should see clearly how absurd such a claim is. Modern astronomy would simply cease to be, were it stripped of our understanding of the origin and evolution of galaxies, stars, planets, and every other component of the universe that we observe. Everything we see says that the universe formed 14 billion years ago, not 6,000. Cosmology is the ultimate evolutionary science—the science of the origin and evolution of the universe itself. During the 20th century, geology became the science of the evolution of the surface of Earth, while planetary science applied our understanding of terrestrial geology to understanding the evolution of other planets and their moons. The most fundamental questions in physics concern the origin and evolution of physical laws. So, too, is the case in modern biology, which simply makes no sense until it is organized around the theme of evolution by natural selection.

Evolution is anything but a scientific appendix that can be harmlessly removed. It is an observed fact. It is the backbone and central nervous system of modern science. In *Darwin's Dangerous Idea* (1995), Daniel Dennett speaks of the concept of evolution as "universal acid: it eats through just about every traditional concept, and leaves in its wake a revolutionized worldview." As the 21st century enters its second decade, evolution of structure is *the* unifying theme that ties the breadth of modern science together into a beautiful, powerful, comprehensive whole. If we tried to pull the thread of evolution out of this tapestry, the whole cloth would unravel before our eyes.

Summary

- Life is basically a form of complex carbon-based chemistry, made possible by special molecules capable of reproducing themselves.

- Life likely formed in Earth's oceans, evolving chemically from an organic "soup" of prebiotic molecules into self-replicating organisms.

- All terrestrial life is composed mostly of only six elements: carbon, hydrogen, oxygen, nitrogen, sulfur, and phosphorus.

- Life-forms that are very different from ours, including those based on silicon chemistry, cannot be ruled out.

- Space-based instruments have begun the search for Earth-like planets.

- Astronomers prefer to focus their search for extraterrestrial life on Earth-like planets orbiting in habitable zones surrounding solar-type stars.

- Although the Milky Way Galaxy may be teeming with extraterrestrial life, none has yet been detected.

- Long before the Sun ends its period of stability on the main sequence, all terrestrial life will have perished.

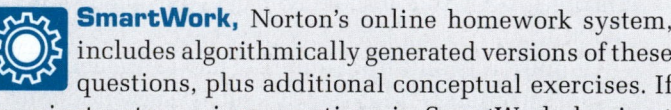 **SmartWork,** Norton's online homework system, includes algorithmically generated versions of these questions, plus additional conceptual exercises. If your instructor assigns questions in SmartWork, log in at **smartwork.wwnorton.com.**

Student Questions

THINKING ABOUT THE CONCEPTS

1. Why do we generally talk about molecules such as DNA or RNA when discussing life on Earth?

**2. Provide a definition for *life* that does not rely on RNA or DNA. Is this definition sufficient to account for new forms that may be discovered in the future, either on Earth or elsewhere?

3. How do we suspect that the building blocks of DNA first formed on Earth?

4. If processes on Earth were energetic enough to form DNA, shouldn't they have been able to destroy it as well? Why, then, does life exist today?

5. When do we believe life first appeared on Earth? What evidence supports this belief?

6. Today, most known life enjoys moderate climates and temperatures. Compare this environment to some of the conditions in which early life developed.

7. Discuss some extremophiles living on Earth today.

8. How was Earth's carbon dioxide atmosphere changed into today's oxygen-rich atmosphere. How long did that transformation take?

9. What are the similarities and differences between prokaryotes and eukaryotes?

10. Tracing back on the evolutionary tree, what kinds of life do we find humans are similar to, and when did the evolutionary steps leading from those kinds of life to humans happen?

11. What was the Cambrian explosion and what might have caused it?

12. Why do you suppose plants and forests appeared in large numbers before large animals did?

13. Why was evolution inevitable on Earth?

14. What were the general conditions needed on our planet for life to arise?

15. Is biological evolution underway on Earth today? If so, how might humans continue to evolve?

16. Where did all the atoms in your body come from?

17. The *Viking* spacecraft did not find evidence of life on Mars when it visited that planet in the late 1970s, nor did the *Phoenix* lander when it examined the martian soil in 2009. Does this imply that life never existed on the planet? Why or why not?

18. A few scientists believe we may be the only advanced life in the galaxy today. If this were indeed the case, which factors in the Drake equation would have to be extremely small?

19. The second law of thermodynamics says that the entropy (a measure of disorder) of the universe is always increasing. Yet living organisms exist as a result of order being created from disorder. Why does this not violate the second law of thermodynamics?

20. What is a habitable zone? What defines its boundaries?

21. In searching for intelligent life elsewhere, why is listening with radio telescopes currently our best method?

22. Why is it likely that life on Earth as we know it will end long before the Sun runs out of nuclear fuel?

APPLYING THE CONCEPTS

23. Excursions 23.1 ("Forever in a Day") takes events spread out over enormous intervals of time and compresses

them into the more comprehensible interval of a single 24-hour day. Make your own "Life in a Day" by compressing all the important events of your lifetime into a single day, starting with your birth at the stroke of midnight and continuing to the present at the end of the day.

24. As noted in Section 23.2, some scientists suspect the early Earth was "seeded" with primitive life stored in comets and meteoroids. Knowing when and how our Solar System, galaxy, and universe formed, what time line is required for such seeding to be possible?

25. Most life on Earth today relies directly or indirectly on the Sun. For example, the food chain of life near the ocean's surface relies on algae, which use photosynthesis. Organisms living on ocean floors generally feed from surface material that sank, so these bottom dwellers rely indirectly on the Sun. Are there any forms of life today that do not rely in any way on solar energy for life? If so, what does this imply about the prospects of life in remote places in our Solar System?

26. If the chance that a given molecule will make a copying error is one in 100,000, how many generations are needed before, on average, at least one mutation has occurred?

27. Consider an organism Beta that, because of a genetic mutation, has a 5 percent greater probability of survival than its nonmutated form, Alpha. Alpha has only a 95 percent probability (p_r) of reproducing itself compared to Beta. After n generations, Alpha's population within the species would be $S_p = (p_r)^n$ compared to Beta's. Calculate Alpha's relative population after 100 generations. (Note: You may need a scientific calculator or help from your instructor to evaluate the quantity 0.95^{100}.)

28. To fully appreciate the power of heredity, mutation, and natural selection, consider Alpha's relative population (from question 27) after 5,000 generations if Beta has a mere 0.1 percent survivability advantage over Alpha.

29. Suppose the early Earth had oceans of ammonia (NH_3) rather than water (H_2O). Could life have evolved in these conditions? Why or why not? What if Earth had been totally dry—that is, with no liquid of any kind? Explain how this may have affected the formation of terrestrial life.

30. Knowing the history and composition of Mars, speculate on whether life evolved there before the planet became geologically dead. If so, how complex did the life become?

31. Speculate on whether life could have evolved on Jupiter's moon Europa. If so, how complex do you predict it to be?

32. Speculate on whether life could have evolved on Saturn's moon Titan. If so, how complex do you predict it to be?

33. Describe the properties of a planet, and its star, that would make it more probable for life to have evolved there.

34. Using your textbook or the Internet, suggest a few nearby stars that could be potential candidates for life-bearing planets. Explain why you selected each one.

35. Why do you think we sent a coded radio signal to the globular cluster M13 in 1974—rather than, say, to a nearby star?

StudySpace is a free and open Web site that provides a Study Plan for each chapter of *21st Century Astronomy*. Study Plans include animations, reading outlines, vocabulary flashcards, and multiple-choice quizzes, plus links to premium content in SmartWork and the ebook. Visit **www.wwnorton.com/studyspace**.

Sometimes the light's all shining on me
Other times I can barely see
Lately it occurs to me
What a long strange trip it's been.

ROBERT HUNTER (1941–)

Epilogue: The Universe and You

The Long and Winding Road

Go out at night and look at the stars, and feel the same sense of wonder and awe that our kind has always felt at the sight. Take it in, be amazed at the majestic canopy overhead, and let your imagination roam, just as our ancestors have done for thousands of years. As you do, drift back down the road we have followed in this book. Reflect on all we have come to know about the universe, and on how much more magnificent the heavens are than our ancestors ever could have imagined. Our journey has been more than a description of the universe and what it contains. It has been a travelogue of the struggle and triumph of the human mind and spirit. Ever since humans first recognized that the patterns shaping our lives are echoed in the sky, we have searched for the threads connecting us to the cosmos and have sought to understand our place in it. We have the privilege of being among the first generations of humans to find those threads and to learn real answers to those age-old questions.

We have no way to count how many different stories have been told about the heavens, but we do know that for thousands of years most of those stories shared a common foundation. One cornerstone was the belief that Earth occupies a special place in the scheme of things. In our minds we were at the center of the universe, fixed and immovable, and all that we saw was present only to give meaning to our existence. A second cornerstone of this traditional worldview was the belief that the heavens are fundamentally different from Earth. To our ancestors, our world was the realm of the ordinary and mundane—a terrestrial existence built from Aristotle's earth, wind, fire, and water. In contrast, the heavens had their own separate reality. There

we saw a realm of gods and angels, of mysticism and magic, of the perfect and unchanging fifth element.

So it remained for thousands of years until, at the dawn of the Renaissance, a Polish monk dared to challenge the wisdom of the ages and to think the unthinkable. Reviving a notion that had been discarded long before by the Greeks, Nicolaus Copernicus allowed himself to imagine that perhaps it was the motion of Earth, rather than the motions of the Sun and stars, that shaped the passage of the days and the years. In so doing, he not only conceptually dislodged Earth from its moorings, but he also broke the shackles that had for so long constrained the human mind. What began as a crack in the foundation of our preconceptions would, in the end, turn that old view of the world to rubble. In its place we would construct an edifice of knowledge that has given us dominion over our world and carried our thoughts to the farthest reaches of the universe.

Copernicus was one of a succession of people with great minds who could not rest without first picking at the loose threads of the ideas they had been raised to believe in. Water runs into the cracks in a slab of granite and freezes, expanding and pushing the cracks open, exposing the flaws in the rock. As the seasons come and go, the imposing boulder stands no chance in the face of this persistent onslaught. In like fashion, the persistent questioning and probing by great minds would eventually shatter the reign of enforced ignorance and entrenched authority. We have met a few of these great minds on our journey; there were many others. They were of different nationalities, different upbringings, and different dispositions and beliefs. But their work shared a common theme: the answers to questions about the world come not from the pronouncements of authority or the prejudices taught in childhood, but from observing nature itself and thinking carefully, honestly, and openly about what we see.

In Isaac Newton's famous thought experiment, a ball fired from an imaginary cannon moves rapidly enough that it falls around the world in a circle. Such a "cannon" could not have been built with the technology available in Newton's day. That achievement would have to await the launch of Sputnik hundreds of years later. Yet although Newton could not make his cannon a reality, he did not have to. He had only to look at the sky and watch as the Moon traced out its monthly path. Newton realized that the force holding the Moon in its orbit about Earth is the same force that gives us weight and guides the path of a ball thrown into the air. In fact, all of us ride Newton's cannonball as the force of the Sun's gravity holds Earth in its yearly orbit.

This thought experiment was an important step in Newton's work. Eventually it led him to invent calculus, which he used to calculate the motions of the planets, making predictions that were confirmed by Johannes Kepler's empirical laws. The philosophical and scientific significance of Newton's thought experiment goes far deeper, however: it signifies the final collapse of the barriers that humankind had placed between Earth and the heavens. With Newton's insight we came to see the heavens as part of the world around us—made of the same substance and shaped by the same physical laws. It is ironic that for knowledge to progress, we had to turn the early cornerstones of our thinking upside down. The modern foundation of our understanding of the universe—the cosmological principle—is the literal negation of those early beliefs: Not only is Earth *not* the center of the universe, but Earth occupies no special place in the universe *at all*. The heavens are *not* "the other," but are instead "the same." We can know the heavens by going into terrestrial laboratories and learning about the nature of matter and energy and radiation, and then applying this knowledge to careful observations of a universe of stars, planets, and galaxies that is governed by physical law.

In our journey we have followed the trail of discovery that grew from this profound change in our understanding. We have watched as stars and planets formed, as stars lived out their lives and died, and as galaxies coalesced out of the primordial fireball of the Big Bang. We have followed our physics back to the very briefest of instants after the event that brought space and time into existence, and we have seen the hints of theories that may in our lifetimes carry us the final step. There is no doubting the wonder of what we have seen. Yet for you, there is another aspect to our journey that, in a practical sense, should be even more significant. While learning *about* the universe, you have also come to better appreciate *how we know* those things—and in so doing you have found a powerful, workable definition for what it means "to know."

The three most common standards that people apply to knowledge, even today, are (1) "It is true because I believe it," (2) "It is true because we believe it," and (3) "It is true because I want to believe it." Of course, none of these has anything to do with what really *is* true. To learn about the universe and our world, we have had to set aside these notions, which blur the line between reality and fantasy, and replace them with a tough, unforgiving, and very different standard: "It is *provisionally* true because we have worked very hard to show that it is false but so far have failed." This standard alone places reality itself in front of our parochial ideas and beliefs. It puts what *is* true ahead of what we would *like* to be true. Only by testing the falsifiable predictions of our theories about the world have we learned to push aside the comfortable notions that for so long prevented us from truly seeing our world and our universe.

We Are Stardust in Human Form

As evidenced by the obelisks of Stonehenge or the ruins of a Mayan pyramid, humans have always built temples to the stars. We still build temples to the stars today. They are seen as an array of radio telescopes spread across the high desert of New Mexico, or a city of domes atop the summit of a dormant Hawaiian volcano, or a satellite telescope carried into orbit and subsequently repaired by space shuttle astronauts, or tiny rovers crawling across the surface of Mars. These modern temples are the legacy of insights by Copernicus, Kepler, Galileo, Newton, Einstein, Hubble, and countless others.

The discoveries that pour forth from these modern-day temples stretch the mind and stir the imagination. We have walked on the Moon and have come to see the planets not as points of light in the sky, but as worlds as rich and complex as our own. We have looked at the remnants of stars that exploded long ago and have peered into eerie columns of glowing interstellar gas within which new stars are being born. We have gazed back in time at galaxies forming when the universe was young, and we have even learned to recognize the birth of the universe itself in the faint glow of the cosmic background radiation. As we contemplate those wonders, the words from act I, scene V, of Shakespeare's *Hamlet* seem almost frighteningly appropriate. As Hamlet faces the challenges and revelations brought by the ghost of his father, Horatio cries out,

O day and night, but this is wondrous strange!

To this comes Hamlet's immortal reply:

And therefore as a stranger give it welcome.
There are more things in heaven and earth, Horatio,
Than are dreamt of in your philosophy.

Here is a message to shout back through the ages. At the start of our journey we asked the most basic question about what we see in the sky—"How big is it?"—and the answers were enough to expose the comedy of humanity's ancient conceits. Traveling at the speed of a modern jetliner, it would take us over 5 million years to cross the distance to even the nearest star beyond our Sun. Even so, we live in a galaxy containing hundreds of billions of such stars, which are themselves outnumbered by other galaxies filling a universe that may stretch on forever. Using the speed of light as our yardstick, we have come to realize that Earth—the world of our birth and the stage on which all of human history has been played—is to the expanse of the observable universe what a single snap of our fingers is to the aeons that have transpired since time itself came into existence, roughly 14 billion years ago.

As we stare at the images that have come to symbolize modern astronomy and consider what they show, it is easy to understand why some people recoil from these insights. "Wouldn't it be nice," they say, "if we could just go back to imagining that Earth is only 6,000 years old, and that humanity occupies a special place at the pinnacle of creation?" Indeed, if our story ended here—with the fact of our seeming insignificance in the universe—we might *all* long for an excuse to retreat into ignorance. Fortunately, this is not where our story ends. Rather, this is where our true journey of discovery begins. Although modern science may have shattered our egotistical views about our exalted place in the scheme of things, it has also offered us a wonderful new appreciation and understanding of ourselves to fill that void.

When we look at distant galaxies, the light we see is produced by stars like our Sun. Each of those stars formed when a cloud of interstellar gas and dust collapsed under the same force of gravity that guided Newton's cannonball. As each of those clouds collapsed, it spun faster and faster, obeying the same laws of motion that accelerate the spin of an Olympic skater as she pulls her arms and leg ever more tightly to her body. This spin prevented those clouds from collapsing directly into stars, forcing them instead to settle into flat, rotating disks. We see such disks today when we look at the newest generation of stars. Inside those disks, grains of dust stick together to make larger grains, which stick together to make still larger grains—this is the beginning of a bottom-up process that culminates with planetesimals crashing together to make planetary worlds. Our Earth is one such world. We have come to view our Sun, Earth, and Solar System as products of natural processes still going on around us today. As we watch new generations of stars form and we search for the planets that surround them, we are witnessing a replay of the birth of our own world, 4.6 billion years ago. We have found our roots in the stars.

Go out at sunset on an evening when a crescent Moon hangs low above the western horizon and several planets stretch out across the darkening sky. The plane of the ecliptic is there in front of you, and your mind's eye might even envision the flat, rotating accretion disk from which our Solar System formed. Once you realize what you are looking at, the cradle of our world and ourselves hangs there in the night sky for all to see.

Viewed in this way, the sunset takes on a whole new significance. It will never be the same. Yet even the majesty of the planets spread across the sky fails to capture the intimacy of our connection with the universe. In the most basic sense, the question "What are we?" is easily answered. Humans and all other terrestrial life are an organized assemblage of various organic molecules, most of which are very complex. Counting the numbers of atoms in our bodies, we are approximately 60 percent hydrogen, 26 percent oxygen, 11 percent carbon, and 2 percent nitrogen, with a small fraction of metals and other heavy elements mixed in. Over the course of our journey we have witnessed the history of those atoms.

A very long time ago—roughly 14 billion years—something wonderful happened. The universe came into being, and time began. From an infinitesimally small volume of concentrated energy, the universe expanded, growing ever larger in size. Within the first few minutes, particles of solid matter, including protons and electrons, condensed out of this dense, primordial ball of energy. Nuclear reactions caused some of the protons to fuse into other light nuclei. Several hundred thousand years later, when the universe had cooled to a temperature of a few thousand kelvins, those nuclei combined with electrons to form atoms. Of those atoms, roughly 90 percent were hydrogen atoms and 10 percent were helium atoms. There were traces of lithium, beryllium, and boron as well, but that was all that existed in the way of normal luminous matter as the universe expanded past the threshold of recombination.

The hydrogen atoms in our bodies date back to this early time in the history of the universe, but what of the rest? Having taken our journey of discovery, you know the answers. As the universe emerged from the Big Bang, clumps of dark matter began to collapse under the force of gravity, pulling normal matter along with it. Within these collapsing protogalaxies the first generations of stars formed—nuclear furnaces powered by the fusion of those original hydrogen atoms into increasingly massive elements. Carbon, oxygen, silicon, sulfur—elements all the way up to iron and nickel—were formed in those stellar infernos. As those first generations of stars ended their lives, they blasted this nuclear ash back into the reaches of interstellar space.

Nucleosynthesis did not end with fusion, however. In the extreme environments of supernovae, free neutrons were captured by the products of fusion, building more massive elements still. Atoms of copper, zinc, tin, silver, and gold—all the way up to the most massive naturally occurring element, uranium—were formed and expelled into space. Here were the chemical elements to fill the periodic table and to build the compounds of life. As early protogalaxies merged to form early galaxies, more generations of stars continued to enrich the universe with the fruits of their alchemy. As early galaxies settled into the well-ordered ellipticals and spirals of today's universe, still more generations of stars came and went, adding to the chemical richness of the universe. When our Solar System appeared on the scene 4.6 billion years ago, it formed from interstellar material that carried the chemical building blocks of planets and of life—atoms produced both in the Big Bang and in the hearts of generations of stars that had lived and died during the 9 billion years that had transpired since the universe began. "What are we, and how did we get here?" We are stardust in human form.

Even as we contemplate our own place in the universe, however, we have come to realize that ordinary matter—the atoms and molecules that constitute our bodies, our planet, our Sun, and the stars and galaxies that surround us—is not the main constituent of our universe. Most of the universe consists of *dark* matter and *dark* energy, unseen components that we do not yet understand. In summary, we have now learned a key lesson about the importance of humankind in the universe. First, Copernicus taught us that our home, our Earth, is not the center of the universe. Then, Darwin reminded us that humans are but a stage in the inevitable evolution of life on Earth. And now, modern cosmology tells us that we and all that we can see about us are but a tiny component of all that is. It is true we are stardust—but we aren't as important in the overall scheme of the universe as we might like to think.

The Future Arrives Every Day

While traveling the highways and back roads of 21st century astronomy, we have come to see our world and ourselves in a very different light. Even so, nothing we have seen has changed the most basic circumstances of our day-to-day existence. Earth remains our world, our home. The hopes and dreams we humans feel are no less real today than they were a thousand years ago. In 1968, Stanley Kubrick and Arthur C. Clarke collaborated to make the film *2001: A Space Odyssey*. This provocative piece of speculative fiction captured the imagination of a generation, and the year 2001 came to signify the future. That future is now here: yet little of our modern-day life is recognizable in those cinematic prognostications dating from more than four decades ago. It is ironic that while we can forecast the future of the Sun with great accuracy and can even calculate the fate of the universe itself, our vision of our own future is so much less certain.

There is no denying the fact that we live in an evolving universe—a place of ongoing and unending change that continues to shape humanity as certainly as it shapes the cosmos itself. The grim prospects mentioned in the closing sections of Chapter 23 are real. Saying that they do not exist

or choosing to push them from our minds will not make them go away. Yet at the same time modern medicine, food production, transportation, and a thousand other technologies that push back the ancient scourges of humanity are equally real. Differences among people remain, but modern communication offers the hope of spreading understanding. Meanwhile, a picture of Earthrise above the lunar horizon taken by the *Apollo 8* astronauts more than four decades ago (see Figure 7.1) remains forever a part of the human experience and perspective. It shows us, as nothing else could, that Earth is a tiny fragile island to be cherished. Whether we like it or not, humanity is a single, interrelated, interdependent global village that will face the future together—or not at all.

At the beginning of the 21st century, science has shown us the wonders of the universe and at the same time has given us the knowledge and power to shape our world and choose our future. The future of humanity may depend entirely on how well we treat Earth and ourselves over the few decades and centuries ahead. If 21st century astronomy does nothing else, it forces us to change our perspective. As we study the laws that govern the workings of atoms, we learn something about the conditions of our own lives: the future is not yet written. As we use our telescopes to stare at galaxies 10 billion light-years away, collecting light from stars that died billions of years before our Sun was even born, manifest destiny seems a pretty silly concept. There are no guarantees that things will work out in the end for one tiny world or for the species to which it gave birth. If we choose to destroy our world, through either direct action or simple neglect, so be it. The universe as a whole will carry on in sublime indifference to our fate.

This is not, however, a message of despair. Instead, it is a message of hope, responsibility, and maturity. We have the power to make our Earth a paradise or to leave our children's children to cope with a world choking from our shortsighted excess. The choices are ours, whether we want them or not. To paraphrase from the book of Genesis, we have truly "tasted of the tree of knowledge." As we stand at the second decade of the 21st century, we face a future filled with choices; but one choice we are not allowed is refusing to acknowledge responsibility for our own destiny.

With that thought, we come to the end of our journey, with the hope that it has helped open your eyes to the wonders of the world and the universe around you. Even more, we hope this journey has given you pause to reflect on who we are as humans and on our place in the larger reality in which we find ourselves. Finally, we hope this journey has changed the shape of the way you think—not only about the sights you see in the night sky, but also about the events of your daily life. If any or all of these hopes are fulfilled, then the journey will have been worth taking—worthy of your time and thought and of ours.

Mathematical Tools

Working with Proportionalities

Most of the mathematics in *21st Century Astronomy* involves proportionalities—statements about the way that one physical quantity changes when another quantity changes. Here we offer a practical guide to working with proportionalities.

To use a statement of proportionality to compare two objects, begin by turning the proportionality into a ratio. For example, the price of a bag of apples is **proportional** to the weight of the bag:

$$\text{Price} \propto \text{Weight}.$$

Here the symbol "$\propto$" is pronounced "is proportional to." What this means is that the ratio of the prices of two bags of apples is equal to the ratio of the weights of the two bags:

$$\text{Price} \propto \text{Weight} \quad means \quad \frac{\text{Price of A}}{\text{Price of B}} = \frac{\text{Weight of A}}{\text{Weight of B}}.$$

Let's work a specific example. Suppose bag A weighs 2 pounds, and bag B weighs 1 pound. That means, of course, that bag A will cost twice as much as bag B. We can turn our proportionality into this equation::

$$\frac{\text{Price of A}}{\text{Price of B}} = \frac{\text{Weight of A}}{\text{Weight of B}} = \frac{2 \text{ lb}}{1 \text{ lb}} = 2.$$

In other words, the price of bag A is two times the price of bag B.

Now let's work another, more complicated example. In Chapter 13 we discuss how the luminosity, brightness, and distance of stars are related. The luminosity of a star—the total energy that the star radiates each second—is proportional to the star's brightness multiplied by the square of its distance:

$$\text{Luminosity} \propto \text{Brightness} \times \text{Distance}^2.$$

What this proportionality means is that if we have two stars—call them A and B—then

$$\frac{\text{Luminosity of A}}{\text{Luminosity of B}} = \frac{\text{Brightness of A}}{\text{Brightness of B}} \times \left(\frac{\text{Distance of A}}{\text{Distance of B}}\right)^2.$$

If we use the symbols L, b, and d to represent luminosity, brightness, and distance, respectively, this equation becomes

$$\frac{L_A}{L_B} = \frac{b_A}{b_B} \times \left(\frac{d_A}{d_B}\right)^2.$$

As an example, suppose that star A appears twice as bright in the sky as star B, but star A is located 10 times as far away as star B. Compare the luminosities of the two stars. Because we know that

$$\text{Luminosity} \propto \text{Brightness} \times \text{Distance}^2,$$

we write

$$\frac{\text{Luminosity of A}}{\text{Luminosity of A}} = \frac{\text{Brightness of A}}{\text{Brightness of B}} \times \left(\frac{\text{Distance of A}}{\text{Distance of B}}\right)^2$$

$$= \frac{2}{1} \times \left(\frac{10}{1}\right)^2 = 200.$$

In other words, star A is 200 times as luminous as star B.

A final note: In our original example we said that the price of a bag of apples is proportional to the weight of the bag, which allowed us to say that a 2-pound bag of apples costs twice as much as a 1-pound bag of apples. This is a statement about the way the world *works*. A 2-pound bag of apples costs *more* than a 1-pound bag, not *less* than a 1-pound bag. To figure out how much a bag of apples actually will cost, we need another piece of information: the price per pound. The price per pound is an example of a **constant of proportionality**. Wrapped up in the price per pound is all sorts of information about the cost of growing apples, the cost of transporting them, what the market is like at the moment, the profit the grocer needs to make, and so forth. In other words, this constant of proportionality is a statement about the way the world *is*.

Proportionalities help us understand how the world works and let us compare one object to another. Constants of proportionality enable us to calculate real values for things. In *21st Century Astronomy*, it is usually the "understanding"— the proportionality itself—that we care about.

Scientific Notation

Astronomy is a science of both the very large and the very small. The mass of an electron, for example, is

0.00000000000000000000000000000009109 kilograms (kg),

whereas the distance to a galaxy far, far away might be about

100,000,000,000,000,000,000,000,000 meters.

All it takes is a quick glance at these two numbers to see why astronomers, like most other scientists, make heavy use of **scientific notation** to express numbers.

Powers of 10

Our number system is based on powers of 10. Going to the left of the decimal place,

$$10 = 10 \times 1,$$

$$100 = 10 \times 10 \times 1,$$

$$1{,}000 = 10 \times 10 \times 10 \times 1,$$

and so on. Going to the right of the decimal place.

$$0.1 = \frac{1}{10} \times 1,$$

$$0.01 = \frac{1}{10} \times \frac{1}{10} \times 1,$$

$$0.001 = \frac{1}{10} \times \frac{1}{10} \times \frac{1}{10} \times 1,$$

and so on, for as long as we care to continue. In other words, each place to the right or left of the decimal place in a number represents a power of 10. For example, 1 million can be written

$$\begin{aligned} 1 \text{ million} &= 1{,}000{,}000 \\ &= 1 \times 10 \times 10 \times 10 \times 10 \times 10 \times 10. \end{aligned}$$

That is, 1 million is "1 multiplied by six factors of 10." Scientific notation combines these factors of 10 in convenient shorthand. Rather than writing out all six factors of 10, instead we combine them in easy shorthand using an exponent:

$$1 \text{ million} = 1 \times 10^6$$

which also means "1 multiplied by six factors of 10."

Moving to the right of the decimal place, each step we take *removes* a power of 10 from the number. One-millionth can be written

$$1 \text{ millionth} = 1 \times \frac{1}{10} \times \frac{1}{10} \times \frac{1}{10} \times \frac{1}{10} \times \frac{1}{10} \times \frac{1}{10}.$$

We are removing powers of 10, so we express this number using a negative exponent. We write

$$\frac{1}{10} = 10^{-1}$$

and

$$1 \text{ millionth} = 1 \times 10^{-1} \times 10^{-1} \times 10^{-1} \times 10^{-1} \times 10^{-1} \times 10^{-1}$$

$$= 1 \times 10^{-6}$$

Returning to our earlier examples, the mass of an electron is 9.109×10^{-31} kg, and the distant galaxy is located 1×10^{26} meters away. These are *much* more convenient ways of writing this information. Notice that *the exponent in scientific notation gives you a feel for the size of a number at a glance*. The exponent of 10 in the electron mass is −31, which tells us instantly that it is a very small number. The exponent of 10 in the distance to a remote galaxy, +26, tells us immediately that it is a very large number. This exponent is often called the "order of magnitude" of a number. When you see a number written in scientific notation while reading *21st Century Astronomy* (or elsewhere), just remember to look at the exponent to better understand what the number is telling you.

Scientific notation is also convenient because it makes multiplying and dividing numbers easier. Again, let's look at an example. Two billion multiplied by eight-thousandths can be written

$$2{,}000{,}000{,}000 \times 0.008,$$

but it is more convenient to write these two numbers using scientific notation, as

$$(2 \times 10^9) \times (8 \times 10^{-3}).$$

We can regroup these expressions in the following form:

$$(2 \times 8) \times (10^9 \times 10^{-3})$$

The first part of the problem is just $2 \times 8 = 16$. The more interesting part of the problem is the multiplication in the right-hand parentheses. The first number, 10^9, is just shorthand for $10 \times 10 \times 10 \dots$ nine times. That is, it represents nine factors of 10. The second number stands for three factors of 1/10—or removing three factors of 10 if you prefer to think of it that way. Altogether, that makes $9 - 3 = 6$ factors of 10. In other words,

$$10^9 \times 10^{-3} = 10^{9-3} = 10^6.$$

Putting the problem together, we get

$$(2 \times 10^9) \times (8 \times 10^{-3}) = (2 \times 8) \times (10^9 \times 10^{-3})$$
$$= 16 \times 10^6.$$

By convention, when a number is written in scientific notation, only one digit is placed to the left of the decimal point. In this case, there are two. However, 16 is 1.6×10, so we can add this additional factor of 10 to the exponent at right, making the final answer

$$1.6 \times 10^7.$$

Dividing is just the inverse of multiplication. Dividing by 10^3 means removing three factors of 10 from a number. Using the previous number.

$$(1.6 \times 10^7) \div (2 \times 10^3) = (1.6 \div 2) \times (10^7 \div 10^3)$$
$$= 0.8 \times 10^{7-3}$$
$$= 0.8 \times 10^4.$$

This time we have only a zero to the left of the decimal point. To get the number into proper form, we can substitute 8×10^{-1} for 0.8, giving us

$$0.8 \times 10^4 = (8 \times 10^{-1}) \times 10^4 = 8 \times 10^3.$$

Adding and subtracting numbers in scientific notation is somewhat more difficult, because all numbers must be written as values multiplied by the *same* power of 10 before they can be added or subtracted. However, almost all modern calculators (and even some cell phones) have scientific notation built in. They keep up with the powers of 10 for you. If you do not have such a calculator, you may want to buy one and learn to use it before tackling the mathematical problems in this book. There are more examples on our website that you can use to better learn how to work with scientific notation.

Significant Figures

In the previous example, we actually broke some rules in the interest of explaining how powers of 10 are treated in scientific notation. The rules we broke involve the *precision* of the numbers we are expressing. In everyday speech we might say, "The store is a kilometer away," by which we probably mean that the store is *roughly* a kilometer away. If it turned out to be 0.8 or 1.2 kilometers (km), it is unlikely that the recipient of our directions would quibble. But when expressing quantities in science, it is extremely important to know not only the value of a number, but also how precise that value is.

The most complete way to keep track of the precision of numbers is to actually write down the uncertainty in the

number. For example, if we know that the distance to the store (call it d) is between 0.8 and 1.2 km, we can write

$$d = 1.0 \pm 0.2 \text{ km,}$$

where the symbol ± is pronounced "plus or minus." In this example, d is between $1.0 - 0.2 = 0.8$ km and $1.0 + 0.2 = 1.2$ km. This is an unambiguous statement about the limitations on our knowledge of the value of d, but carrying along the formal errors with every number we write would be cumbersome at best. Instead, we keep track of the approximate precision of a number by using "significant figures."

The convention for significant figures is this: We assume the number we write has been rounded from a number that had one additional digit to the right of the decimal point. If we say that a quantity d, which might represent the distance to the store, is "1.", what we mean is that d is close to 1. It is likely not as small as "0.", and it is likely not as large as "2.". If we say instead

$$d = 1.0,$$

then we mean that d is likely not 0.9 and is likely not 1.1. It is roughly 1.0 to the nearest tenth. The greater the number of significant figures, the more precisely the number is being specified. For example, 1.00000 is *not* the same number as 1.00. The first number, 1.00000, represents a value that is probably not as small as 0.99999 and probably not as large as 1.00001. The second number, 1.00, represents a value that is probably not as small as 0.99 nor as large as 1.01. The number 1.00000 is much more precise than the number 1.00.

When we carry out mathematical operations, significant figures are important. For example, $2.0 \times 1.6 = 3.2$. It does *not* equal 3.20000000000. *The product of two numbers cannot be known to any greater accuracy than the numbers themselves!* As a general rule, when we multiply and divide, the answer should have the same number of significant figures as the less precise of the numbers being multiplied or divided. In other words, $2.0 \times 1.602583475 = 3.2$. Because all we know is that the first factor is probably closer to 2.0 than to 1.9 or 2.1, all we know about the product is that it is between about 3.0 and 3.4. It is 3.2. It is not 3.205166950 (*even if that is the answer your calculator gives!*). The rest of the digits to the right of 3.2 just do not mean anything.

When we add and subtract, the rules are a bit different. If one number has a significant figure with a particular place value but another number does not, their sum or difference cannot have a significant figure in that place value. For example,

$$1,045.$$
$$+1.34567$$
$$\overline{1,046.}$$

The answer is "1,046.", *not* "1,046.34567". Again, the extra digits to the right of the decimal place have no meaning because "1,045." is not known to that accuracy.

There is always a fly in the ointment, and this is no exception. What is the precision of the number 1,000,000? As it is written, the answer is unclear. Are all those zeros really significant, or are they placeholders? If we write the number in scientific notation, on the other hand, there is never a question. Instead of 1,000,000, we write 1.0×10^6 for a number that is known to the nearest hundred thousand or so; or we write 1.00000×10^6 for a number that is known to the nearest 10.

So our earlier example would have been more correct had we said

$$(2.0 \times 10^9) \times (8.0 \times 10^{-3}) = 1.6 \times 10^7.$$

Algebra

There are many branches of mathematics. The branch that tells us about the relationships between quantities is called **algebra**. If you are reading this book, you have almost certainly taken an algebra class, but you are not alone if you feel a little review is in order. Basically, algebra begins by using symbols to represent quantities. For example, we might write the distance you travel in a day as d. As it stands, d has no value. It might be 10,000 miles. It might be 30 feet. It does, however, have **units**—in this case, the units of distance.

The average speed at which you travel is equal to the distance you travel divided by the time you take. If we use the symbol v to represent your average speed and the symbol t to represent the time you take, then instead of writing out, "Your average speed is equal to the distance you travel divided by the time you take," we can write

$$v = \frac{d}{t}.$$

The meaning of this algebraic expression is exactly the same as the sentence quoted before it, but it is much more concise. As it stands, v, d, and t still have no specific values. There are no numbers assigned to them yet. However, this expression tells us what the relationship between those numbers will be when we *do* look at a specific example. For example, if you go 500 km ($d = 500$ km) in 10 hours ($t = 10$ hours), this expression tells you that your average speed is

$$v = \frac{d}{t} = \frac{500 \text{ km}}{10 \text{ hours}} = 50 \text{ km/h}.$$

Notice that the units in this expression act exactly like the numerical values. They are just multiplicative factors. When we say "500 km," what we really mean is "500 × kilometers." Likewise, 10 hours means "10 × hours." When we divide the two, we find that the units of v are kilometers divided by hours, or km/h (pronounced "kilometers per hour").

We introduced algebra as shorthand for expressing relations between quantities, but it is far more powerful than

that. Algebra provides rules for manipulating the symbols used to represent quantities. We begin with a bit of notation for "powers" and "roots." When we talk about raising a quantity to a power, we mean multiplying the quantity by itself some number of times. For example, if S is a symbol for something (anything), then S^2 (pronounced "S squared" or "S to the second power") means $S \times S$, and S^3 (pronounced "S cubed" or "S to the third power") means $S \times S \times S$. Suppose S represents the length of the side of a square. The area of the square is given by

$$\text{Area} = S \times S = S^2.$$

If $S = 3$ meters (m), then the area of the square is

$$S^2 = 3 \text{ m} \times 3 \text{ m} = 9 \text{ m}^2$$

(pronounced "9 square meters"). It should be obvious why raising a quantity to the second power is called "squaring" the quantity. We could have done the same thing for the sides of a cube and found that the volume of the cube is

$$\text{Volume} = S \times S \times S = S^3.$$

If $S = 3$ meters, then the volume of the cube is

$$S^3 = 3 \text{ m} \times 3 \text{ m} \times 3 \text{ m} = 27 \text{ m}^3$$

(pronounced "27 cubic meters"). Again, it is clear why raising a quantity to the third power is called "cubing" the quantity.

Roots are the reverse of this process. The square root of a quantity is the value that, when squared, gives the original quantity. The square root of 4 is 2, which means that 2 × 2 = 4. The square root of 9 is 3, which means that 3 × 3 = 9. Similarly, the cube root of a quantity is the value that, when cubed, gives the original quantity. The cube root of 8 is 2, which means that 2 × 2 × 2 = 8. Roots are written with the symbol $\sqrt{\ }$. For example, we write

$$\sqrt{9} = 3$$

for the square root of 9, and

$$\sqrt[3]{8} = 2$$

for the cube root of 8. If the volume of a cube is $V = S^3$, we can also write

$$S = \sqrt[3]{V} = \sqrt[3]{S^3}.$$

Roots can also be written as powers. Powers and roots behave exactly like the exponents of 10 in our discussion of scientific notation. (They had better—the exponents used in scientific notation are just powers of 10.) For example, if a, n, and m are all algebraic quantities, then

$$a^n \times a^m = a^{n+m}, \quad \text{and} \quad \frac{a^n}{a^m} = a^{n-m}.$$

(To see if you understand all this, explain why the square root of a can also be written $a^{\frac{1}{2}}$ and the cube root of a can be written $a^{\frac{1}{3}}$.)

Some of the rules of algebra are listed next. These are really no more than the rules of arithmetic applied to the symbolic quantities of algebra. The important thing is this: as long as we apply the rules of algebra properly, then the relationships among symbols we arrive at through our algebraic manipulations remain true for the physical quantities that those symbols represent.

Here we summarize a few algebraic rules and relationships. In this summary, a, b, c, m, n, r, x, and y are all algebraic quantities:

Associative rule:

$$a \times b \times c = (a \times b) \times c = a \times (b \times c)$$

Commutative rule:

$$a \times b = b \times a$$

Distributive rule:

$$a \times (b + c) = (a \times b) + (a \times c)$$

Cross multiplication:

$$\text{If } \frac{a}{b} = \frac{c}{d}, \text{ then } ad = bc.$$

Working with exponents:

$$\frac{1}{a^n} = a^{-n} \qquad a^n a^m = a^{n+m}$$

$$\frac{a^n}{a^m} = a^{n-m} \qquad (a^n)^m = a^{n \times m} \qquad \left(\frac{a}{b}\right)^n = \frac{a^n}{b^n}$$

Equation of a line with slope m and y-intercept b:

$$y = mx + b$$

Equation of a circle with radius r centered at $x = 0$, $y = 0$:

$$x^2 + y^2 = r^2$$

Angles and Distances

The farther away something is, the smaller it appears. This is common sense and everyday experience. In astronomy, where we seldom get to walk up to the object we are studying and measure it with a meterstick, our knowledge about the sizes of things usually depends on knowing how the size of an object, its distance, and the angle it covers in the sky are related.

The natural way to measure angles is to use a unit called the **radian**. As shown in **Figure A1.1a**, the size of an angle in radians is just the length of the arc subtending the angle, divided by the radius of the circle. In the figure, the angle $x = S/r$ radians.

Because the circumference of a circle is 2π multiplied by the radius, $C = 2\pi r$, a complete circle has an angular measure of $(2\pi r)/r = 2\pi$ radians. In more conventional angular measure, a complete circle is 360°, so we can say that

$$360° = 2\pi \text{ radians}$$

or that

$$1 \text{ radian} = \frac{360°}{2\pi} = 57.2958°.$$

When talking about stars and galaxies, we often use seconds of arc (**arcseconds**) to measure angles. A degree is broken into 60 minutes of arc (**arcminutes**), each of which is broken into 60 seconds of arc—so there are 3,600 seconds of arc in a degree. Therefore,

$$3{,}600 \; \frac{\text{arcseconds}}{\text{degree}} \times 57.2958 \; \frac{\text{degrees}}{\text{radian}} = 206{,}265 \; \frac{\text{arcseconds}}{\text{radian}}.$$

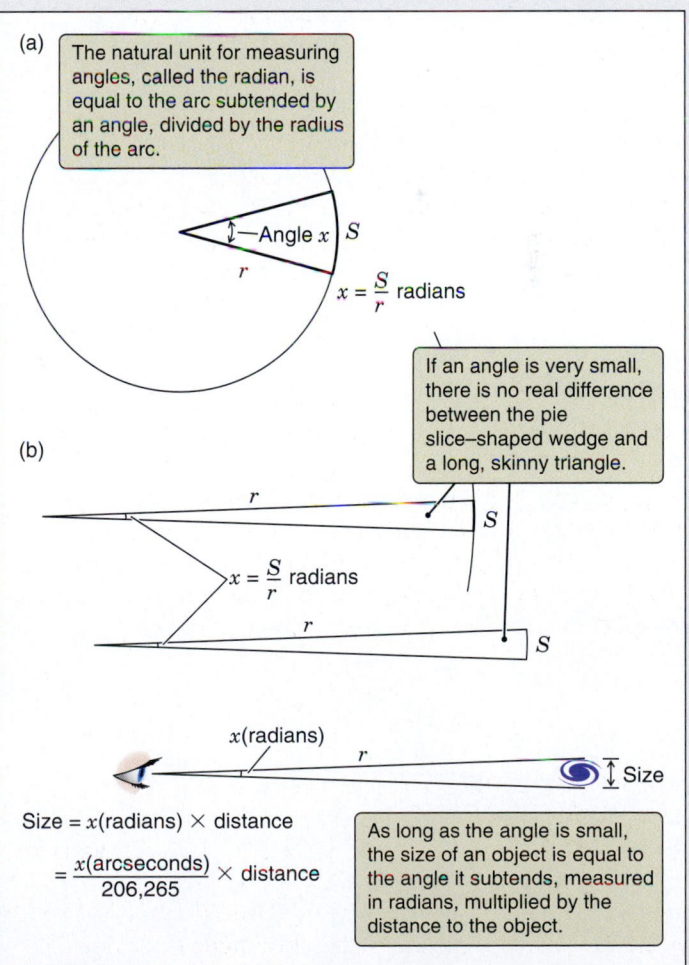

(a) The natural unit for measuring angles, called the radian, is equal to the arc subtended by an angle, divided by the radius of the arc.

—Angle x S

r

$x = \dfrac{S}{r}$ radians

If an angle is very small, there is no real difference between the pie slice–shaped wedge and a long, skinny triangle.

(b)

r

S

$x = \dfrac{S}{r}$ radians

r

S

x(radians)

r

Size

Size $= x$(radians) $\times$ distance

$= \dfrac{x(\text{arcseconds})}{206{,}265} \times$ distance

As long as the angle is small, the size of an object is equal to the angle it subtends, measured in radians, multiplied by the distance to the object.

FIGURE A1.1 Measuring angles.

If the angle is small enough (which it usually is in astronomy), there is precious little difference between the pie slice just described and a long skinny triangle with a short side of length S, as **Figure A1.1b** illustrates. So, if we know the distance d to an object and we can measure the angular size x of the object, then the size of the object is just

$$S = x \text{ (in radians)} \times d = \frac{x \text{ (in degrees)}}{57.2958 \text{ degrees/radian}} \times d$$

$$= \frac{x \text{ (in arcseconds)}}{206,265 \text{ arcseconds/radian}} \times d,$$

which is all we need to turn our knowledge of the angular size and the distance to an object into a measurement of the object's physical size.

Circles and Spheres

To round out our mathematical tools, here are a few useful formulas for circles and spheres. The circle or sphere in each case has a radius r.

$$\text{Circumference}_{\text{circle}} = 2\pi r$$

$$\text{Area}_{\text{circle}} = \pi r^2$$

$$\text{Surface area}_{\text{sphere}} = 4\pi r^2$$

$$\text{Volume}_{\text{sphere}} = \frac{4}{3}\pi r^3$$

Physical Constants and Units

Fundamental Physical Constants

Constant	Symbol	Value
Speed of light in a vacuum	c	2.99792×10^8 m/s
Universal gravitational constant	G	6.673×10^{-11} N m^2/kg^2
Planck constant	h	6.62607×10^{-34} J-s
Electric charge of electron or proton	e	1.60218×10^{-19} C
Boltzmann constant	k	1.38065×10^{-23} J/K
Stefan-Boltzmann constant	σ	5.67040×10^{-8} W/(m^2 K^4)
Mass of electron	m_e	9.10938×10^{-31} kg
Mass of proton	m_p	1.67262×10^{-27} kg

Source: Data from the National Institute of Standards and Technology (http://physics.nist.gov).

Unit Prefixes

Prefix*	Name	Factor[†]
n	nano-	10^{-9}
μ	micro-	10^{-6}
m	milli-	10^{-3}
k	kilo-	10^{3}
M	mega-	10^{6}
G	giga-	10^{9}
T	tera-	10^{12}

These prefixes (*), when appended to a unit, change the size of the unit by the factor ([†]) given. For example, 1 km (kilometer) is 10^3 meters (m).

Units and Values

Quantity	Fundamental Unit	Values
Length	meters (m)	radius of Sun $(R_\odot) = 6.96265 \times 10^8$ m astronomical unit (AU) $= 1.49598 \times 10^{11}$ m 1 AU $= 149{,}598{,}000$ km Light-year (ly) $= 9.4605 \times 10^{15}$ m 1 ly $= 6.324 \times 10^4$ AU 1 parsec (pc) $= 3.261$ ly $= 3.0857 \times 10^{16}$ m 1 m $= 3.281$ feet
Volume	meters^3 (m^3)	1 m^3 $= 1{,}000$ liters $= 264.2$ gallons
Mass	kilograms (kg)	1 kg $= 1{,}000$ grams mass of Earth $(M_\oplus) = 5.9736 \times 10^{24}$ kg mass of Sun $(M_\odot) = 1.9891 \times 10^{30}$ kg
Time	seconds (s)	1 hour (h) $= 60$ minutes (min) $= 3{,}600$ s solar day (noon to noon) $= 86{,}400$ s sidereal day (Earth rotation period) $= 86{,}164.1$ s tropical year (equinox to equinox) $= 365.24219$ days $= 3.15569 \times 10^7$ s sidereal year (Earth orbital period) $= 365.25636$ days $= 3.15581 \times 10^7$ s
Speed	meters/second (m/s)	1 m/s $= 2.236$ miles/h 1 km/s $= 1{,}000$ m/s $= 3{,}600$ km/h $c = 3.00 \times 10^8$ m/s $= 300{,}000$ km/s
Acceleration	meters/second^2 (m/s^2)	gravitational acceleration on Earth $(g) = 9.81$ m/s^2
Energy	joules (J)	1 J $= 1$ kg m^2/s^2 1 megaton $= 4.19 \times 10^{15}$ J
Power	watts (W)	1 W $= 1$ J/s Solar luminosity $(L_\odot) = 3.827 \times 10^{26}$ W
Force	newtons (N)	1 N $= 1$ kg m/s^2 1 pound (lb) $= 4.448$ N 1 N $= 0.22481$ lb
Pressure	newtons/meter^2 (N/m^2)	atmospheric pressure at sea level $= 1.013 \times 10^5$ N/m^2 $= 1.013$ bar
Temperature	kelvins (K)	absolute zero $= 0$ K $= -273.15°$C $= -459.67°$F

Sources: Data from the National Space Science Data Center (2002); *Observer's Handbook 2002* (Royal Astronomical Society of Canada, 2001); and National Institute of Standards and Technology (2002).

Periodic Table of the Elements

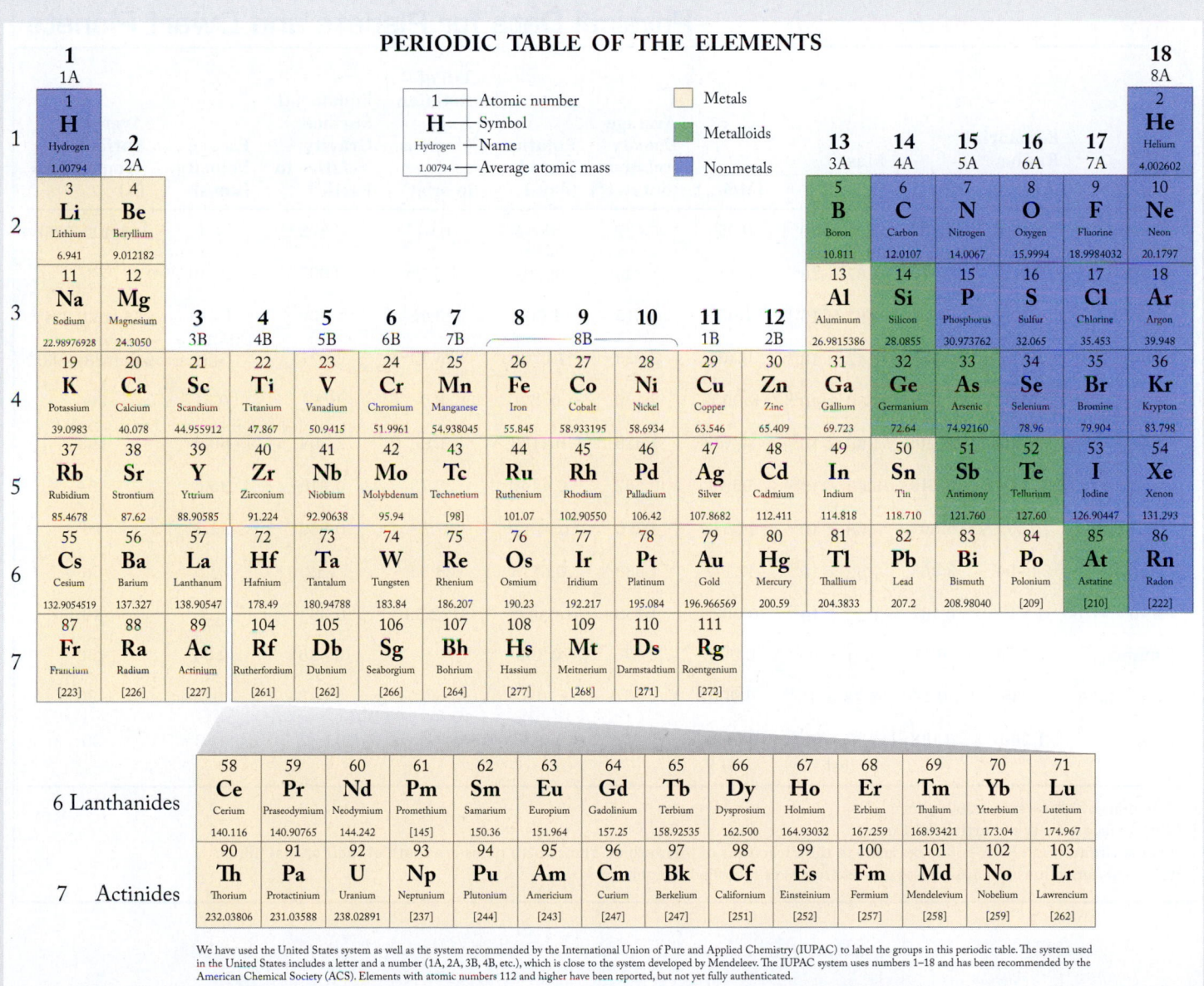

PERIODIC TABLE OF THE ELEMENTS

1 1A																	18 8A
1 **H** Hydrogen 1.00794	2 2A											13 3A	14 4A	15 5A	16 6A	17 7A	2 **He** Helium 4.002602
3 **Li** Lithium 6.941	4 **Be** Beryllium 9.012182											5 **B** Boron 10.811	6 **C** Carbon 12.0107	7 **N** Nitrogen 14.0067	8 **O** Oxygen 15.9994	9 **F** Fluorine 18.9984032	10 **Ne** Neon 20.1797
11 **Na** Sodium 22.98976928	12 **Mg** Magnesium 24.3050	3 3B	4 4B	5 5B	6 6B	7 7B	8	9 8B	10	11 1B	12 2B	13 **Al** Aluminum 26.9815386	14 **Si** Silicon 28.0855	15 **P** Phosphorus 30.973762	16 **S** Sulfur 32.065	17 **Cl** Chlorine 35.453	18 **Ar** Argon 39.948
19 **K** Potassium 39.0983	20 **Ca** Calcium 40.078	21 **Sc** Scandium 44.955912	22 **Ti** Titanium 47.867	23 **V** Vanadium 50.9415	24 **Cr** Chromium 51.9961	25 **Mn** Manganese 54.938045	26 **Fe** Iron 55.845	27 **Co** Cobalt 58.933195	28 **Ni** Nickel 58.6934	29 **Cu** Copper 63.546	30 **Zn** Zinc 65.409	31 **Ga** Gallium 69.723	32 **Ge** Germanium 72.64	33 **As** Arsenic 74.92160	34 **Se** Selenium 78.96	35 **Br** Bromine 79.904	36 **Kr** Krypton 83.798
37 **Rb** Rubidium 85.4678	38 **Sr** Strontium 87.62	39 **Y** Yttrium 88.90585	40 **Zr** Zirconium 91.224	41 **Nb** Niobium 92.90638	42 **Mo** Molybdenum 95.94	43 **Tc** Technetium [98]	44 **Ru** Ruthenium 101.07	45 **Rh** Rhodium 102.90550	46 **Pd** Palladium 106.42	47 **Ag** Silver 107.8682	48 **Cd** Cadmium 112.411	49 **In** Indium 114.818	50 **Sn** Tin 118.710	51 **Sb** Antimony 121.760	52 **Te** Tellurium 127.60	53 **I** Iodine 126.90447	54 **Xe** Xenon 131.293
55 **Cs** Cesium 132.9054519	56 **Ba** Barium 137.327	57 **La** Lanthanum 138.90547	72 **Hf** Hafnium 178.49	73 **Ta** Tantalum 180.94788	74 **W** Tungsten 183.84	75 **Re** Rhenium 186.207	76 **Os** Osmium 190.23	77 **Ir** Iridium 192.217	78 **Pt** Platinum 195.084	79 **Au** Gold 196.966569	80 **Hg** Mercury 200.59	81 **Tl** Thallium 204.3833	82 **Pb** Lead 207.2	83 **Bi** Bismuth 208.98040	84 **Po** Polonium [209]	85 **At** Astatine [210]	86 **Rn** Radon [222]
87 **Fr** Francium [223]	88 **Ra** Radium [226]	89 **Ac** Actinium [227]	104 **Rf** Rutherfordium [261]	105 **Db** Dubnium [262]	106 **Sg** Seaborgium [266]	107 **Bh** Bohrium [264]	108 **Hs** Hassium [277]	109 **Mt** Meitnerium [268]	110 **Ds** Darmstadtium [271]	111 **Rg** Roentgenium [272]							

Legend:
- 1 — Atomic number
- H — Symbol
- Hydrogen — Name
- 1.00794 — Average atomic mass
- Metals
- Metalloids
- Nonmetals

6 Lanthanides

58 **Ce** Cerium 140.116	59 **Pr** Praseodymium 140.90765	60 **Nd** Neodymium 144.242	61 **Pm** Promethium [145]	62 **Sm** Samarium 150.36	63 **Eu** Europium 151.964	64 **Gd** Gadolinium 157.25	65 **Tb** Terbium 158.92535	66 **Dy** Dysprosium 162.500	67 **Ho** Holmium 164.93032	68 **Er** Erbium 167.259	69 **Tm** Thulium 168.93421	70 **Yb** Ytterbium 173.04	71 **Lu** Lutetium 174.967

7 Actinides

90 **Th** Thorium 232.03806	91 **Pa** Protactinium 231.03588	92 **U** Uranium 238.02891	93 **Np** Neptunium [237]	94 **Pu** Plutonium [244]	95 **Am** Americium [243]	96 **Cm** Curium [247]	97 **Bk** Berkelium [247]	98 **Cf** Californium [251]	99 **Es** Einsteinium [252]	100 **Fm** Fermium [257]	101 **Md** Mendelevium [258]	102 **No** Nobelium [259]	103 **Lr** Lawrencium [262]

We have used the United States system as well as the system recommended by the International Union of Pure and Applied Chemistry (IUPAC) to label the groups in this periodic table. The system used in the United States includes a letter and a number (1A, 2A, 3B, 4B, etc.), which is close to the system developed by Mendeleev. The IUPAC system uses numbers 1–18 and has been recommended by the American Chemical Society (ACS). Elements with atomic numbers 112 and higher have been reported, but not yet fully authenticated.

Properties of Planets, Dwarf Planets, and Moons

Physical Data for Planets and Dwarf Planets

Planet	Equatorial Radius (km)	(R/R⊕)	Mass (kg)	(M/M⊕)	Average Density (relative to water*)	Rotation Period (days)	Tilt of Rotation Axis (degrees, relative to orbit)	Equatorial Surface Gravity (relative to Earth†)	Escape Velocity (km/s)	Average Surface Temperature (K)
Mercury	2,440	0.383	3.30×10^{23}	0.055	5.427	58.64	0.01	0.378	4.3	340 (100, 725)[§]
Venus	6,052	0.949	4.87×10^{24}	0.815	5.243	243.02[‡]	177.36	0.907	10.36	737
Earth	6,378	1.000	5.97×10^{24}	1.000	5.515	1.000	23.45	1.000	11.19	288 (183, 331)[§]
Mars	3,397	0.533	6.42×10^{23}	0.107	3.933	1.0260	25.19	0.377	5.03	210 (133, 293)[§]
Ceres	475	0.075	9.60×10^{20}	0.0002	2.100	0.378	3.0	0.27	0.51	200
Jupiter	71,492	11.209	1.90×10^{27}	317.83	1.326	0.4136	3.13	2.364	59.5	165
Saturn	60,268	9.449	5.68×10^{26}	95.16	0.687	0.4440	26.73	0.916	35.5	134
Uranus	25,559	4.007	8.68×10^{25}	14.537	1.270	0.7183[‡]	97.77	0.889	21.3	76
Neptune	24,764	3.883	1.02×10^{26}	17.147	1.638	0.6713	28.32	1.12	23.5	58
Pluto	1,195	0.187	1.25×10^{22}	0.0021	1.750	6.387[‡]	122.53	0.083	1.3	40
Haumea	~700	0.11	4.0×10^{21}	0.0007	~3	0.163	?	0.045	0.84	<50
Makemake	750	0.12	4.18×10^{21}	0.0007	~2	0.32	?	0.048	0.8	~30
Eris	1,200	0.188	1.5×10^{22} (est.)	0.0025 (est.)	~2	>0.3?	?	0.082	~1.3	30

*The density of water is 1,000 kg/m³.
†The surface gravity of Earth is 9.81 m/s².
‡Venus, Uranus, and Pluto rotate opposite to the directions of their orbits. Their north poles are south of their orbital planes.
§Where given, values in parentheses give extremes of recorded temperatures.

Orbital Data for Planets and Dwarf Planets

Planet	Mean Distance from Sun (A*)		Orbital Period (P) (sidereal years)	Eccentricity	Inclination (degrees, relative to ecliptic)	Average Speed (km/s)
	$(10^6$ km)	(AU)				
Mercury	57.9	0.387	0.2408	0.2056	7.005	47.87
Venus	108.2	0.723	0.6152	0.0067	3.395	35.02
Earth	149.6	1.000	1.000	0.0167	0.000	29.78
Mars	227.9	1.524	1.8809	0.0935	1.850	24.13
Ceres	413.9	2.767	4.6027	0.097	9.73	17.88
Jupiter	778.6	5.204	11.8618	0.0489	1.304	13.07
Saturn	1,433.5	9.582	29.4566	0.0565	2.485	9.69
Uranus	2,872.5	19.201	84.0106	0.0457	0.772	6.81
Neptune	4,495.1	30.047	164.7856	0.0113	1.769	5.43
Pluto	5,906.38	39.48	247.6753	0.2488	17.16	4.72
Haumea	6,447.8	43.1	283.3	0.195	28.22	4.48
Makemake	6,836.7	45.7	309.9	0.159	28.96	4.42
Eris	10,123	67.668	557.	0.4418	44.187	3.44

*A is the semimajor axis of the planet's elliptical orbit.

Properties of Selected Moons[*]

Planet	Moon	Orbital Properties		Physical Properties		
		P (days)	A $(10^3$ km)	R (km)	M $(10^{20}$ kg)	Relative Density[†] (water = 1.00)
Earth (1 moon)	Moon	27.32	384.4	1,737.4	735	3.34
Mars (2 moons)	Phobos	0.32	9.38	13.5 × 10.8 × 9.4	0.0001	2.0
	Deimos	1.26	23.46	7.5 × 6.1 × 5.5	0.00002	1.5
Jupiter (63 known moons)	Metis	0.29	127.97	20	0.00036	0.9
	Amalthea	0.50	181.30	131 × 73 × 67	0.0208	0.9
	Io	1.77	421.60	1,815	894	3.55
	Europa	3.55	670.90	1,569	480	3.01
	Ganymede	7.16	1,070	2,631	1,480	1.94
	Callisto	16.69	1,883	2,403	1,080	1.86
	Himalia	250.57	11,480	93	0.042	2.8
	Pasiphae	735[‡]	23,500	25	0.0019	2.9
	Callirrhoe	759[‡]	24,100	4.3	0.00001	2.6

(continued)

Properties of Selected Moons*

		Orbital Properties		Physical Properties		
(continued)						
Planet	Moon	P (days)	A (10^3 km)	R (km)	M (10^{20} kg)	Relative Density[†] (water = 1.00)
Saturn (61 known moons)	Pan	0.58	133.58	14	0.00005	—
	Prometheus	0.61	139.35	72.5 × 42.5 × 32.5	0.0016	0.5
	Pandora	0.63	141.70	57 × 42 × 31	0.0014	0.5
	Mimas	0.94	185.52	196	0.38	1.17
	Enceladus	1.37	238.02	250	1.08	1.6
	Tethys	1.89	294.66	530	6.17	0.97
	Dione	2.74	377.40	560	10.5	1.43
	Rhea	4.52	527.04	765	24.9	1.23
	Titan	15.95	1,222	2,575	1,350	1.88
	Hyperion	21.28	1,481	205 × 130 × 110	0.0558	0.57
	Iapetus	79.33	3,561	735	18.8	1.21
	Phoebe	550.48[‡]	12,952	107	0.08	1.6
	Paaliaq	686.9	15.200	11	0.0001	2.3
Uranus (27 known moons)	Cordelia	0.34	49.75	20	0.0004	1.3
	Miranda	1.41	129.78	236	0.64	1.15
	Ariel	2.52	191.24	579	13.6	1.66
	Umbriel	4.14	265.97	585	11.7	1.39
	Titania	8.71	435.84	789	34.9	1.70
	Oberon	13.46	582.60	761	30.3	1.64
	Setebos	2,225[‡]	17,418	24	0.0008	1.5
Neptune (13 known moons)	Naiad	0.29	48.0	48 × 30 × 26	0.002	1.3
	Larissa	0.55	73.6	108 × 102 × 84	0.05	1.3
	Proteus	1.12	117.6	210 × 208 × 202	0.5	1.3
	Triton	5.88[‡]	354.8	1,353	214	2.07
	Nereid	360.14	5,513.40	170	0.3	1.5
Pluto (3 moons)	Charon	6.39	19.60	593	16.2	1.85
Haumea (2 moons)	Namaka	18	25.66	85	0.018	?
	Hi'iaka	49	49.88	155	0.179	?
Eris	Dysnomia	15.8	37.4	50–125?	?	?

*Innermost, outermost, largest, and/or a few other moons for each planet.
[†]The density of water is 1,000 kg/m^3.
[‡]Irregular moon (has retrograde orbit).

Nearest and Brightest Stars

Stars within 12 Light-Years of Earth

Name*	Distance (ly)	Spectral Type[+]	Relative Visual Luminosity[‡] (Sun = 1.000)	Apparent Magnitude	Absolute Magnitude
Sun	1.58×10^{-5}	G2V	1.000	−26.74	4.83
Alpha Centauri C (Proxima Centauri)	4.24	M5.5V	0.000052	11.09	15.53
Alpha Centauri A	4.36	G2V	1.5	0.01	4.38
Alpha Centauri B	4.36	K0V	0.44	1.34	5.71
Barnard's star	5.96	M4Ve	0.00044	9.53	13.22
CN Leonis	7.78	M5.5	0.000020	13.44	16.55
BD +36-2147	8.29	M2.0V	0.0056	7.47	10.44
Sirius A	8.58	A1V	22	−1.43	1.47
Sirius B	8.58	DA2	0.0025	8.44	11.34
BL Ceti	8.73	M5.5V	0.000059	12.54	15.40
UV Ceti	8.73	M6.0	0.000039	12.99	15.85
V1216 Sagittarii	9.68	M3.5V	0.00050	10.43	13.07
HH Andromedae	10.32	M5.5V	0.00010	12.29	14.79
Epsilon Eridani	10.52	K2V	0.28	3.73	6.19
Lacaille 9352	10.74	M1.5V	0.011	7.34	9.75
FI Virginis	10.92	M4.0V	0.00033	11.13	13.51
EZ Aquarii A	11.26	M5.0V	0.000047	13.33	15.64
EZ Aquarii B	11.26	M5e	0.000050	13.27	15.58
EZ Aquarii C	11.26	—	0.000025	14.03	16.34
Procyon A	11.40	F5IV-V	7.31	0.38	2.66

(continued)

Stars within 12 Light-Years of Earth

(continued)

Name*	Distance (ly)	Spectral Type†	Relative Visual Luminosity‡ (Sun = 1.000)	Apparent Magnitude	Absolute Magnitude
Procyon B	11.40	DA	0.00054	10.70	12.98
61 Cygni A	11.40	K5.0V	0.086	5.21	7.49
61 Cygni B	11.40	K7.0V	0.040	6.03	8.31
Gliese 725 A	11.52	M3.0V	0.0029	8.90	11.16
Gliese 725 B	11.52	M3.5V	0.0014	9.69	11.95
Groombridge 34 A	11.62	M1.5V	0.0063	8.08	10.32
Groombridge 34 B	11.62	M3.5V	0.00041	11.06	13.30
Epsilon Indi A	11.82	K5Ve	0.15	4.69	6.89
Epsilon Indi B (brown dwarf)	11.82	T1.0	—	—	—
Epsilon Indi C (brown dwarf)	11.82	T6.0	—	—	—
DX Cancri	11.82	M6.5V	0.000014	14.78	16.98
Tau Ceti	11.88	G8V	0.45	3.49	5.68
Gliese 1061	11.99	M5.5V	0.000067	13.09	15.26

*Stars may carry many names, including common names (such as Sirius), names based on their prominence within a constellation (such as Alpha Canis Majoris, another name for Sirius), or names based on their inclusion in a catalog (such as BD +36-2147). Addition of letters A, B, and so on, or superscripts indicates membership in a multiple-star system.

†Spectral types such as M3 are discussed in Chapter 13. Other letters or numbers provide additional information. For example, *V* after the spectral type indicates a main-sequence star, and III indicates a giant star. Stars of spectral type T are brown dwarfs.

‡*Luminosity* in this table refers only to radiation in "visual" light.

The 25 Brightest Stars in the Sky

Name	Common Name	Distance (ly)	Spectral Type	Relative Visual Luminosity* (Sun = 1.000)	Apparent Visual Magnitude	Absolute Visual Magnitude
Sun	Sun	1.58×10^{-5}	G2V	1.000	−26.8	4.82
Alpha Canis Majoris	Sirius	8.60	A1V	22.9	−1.47	1.42
Alpha Carinae	Canopus	313	F0II	13,800	−0.72	−5.53
Alpha Bootis	Arcturus	36.7	K1.5IIIFe-0.5	111	−0.04	−0.29
Alpha¹ Centauri	Rigel Kentaurus	4.39	G2V	1.50	−0.01	4.38
Alpha Lyrae	Vega	25.3	A0Va	49.7	0.03	0.58
Alpha Aurigae	Capella	42.2	G5IIIe+G0III	132	0.08	−0.48
Beta Orionis	Rigel	770	B8Ia:	40,200	0.12	−6.69
Alpha Canis Minoris	Procyon	11.4	F5IV-V	7.38	0.34	2.65
Alpha Eridani	Achernar	144	B3Vpe	1,090	0.50	−2.77
Alpha Orionis	Betelgeuse	427	M1-2Ia-Iab	9,600	0.58	−5.14
Beta Centauri	Hadar	350	B1III	8,700	0.60	−5.03
Alpha Aquilae	Altair	16.8	A7V	11.1	0.77	2.21
Alpha Tauri	Aldebaran	65.1	K5+III	151	0.85	0.63
Alpha Virginis	Spica	262	B1III-IV+B2V	2,230	1.04	−3.55
Alpha Scorpii	Antares	604	M1.5Iab-Ib+B4Ve	11,000	1.09	−5.28
Beta Geminorum	Pollux	33.7	K0IIIb	31.0	1.15	1.09
Alpha Piscis	Fomalhaut	25.1	A3V	17.2	1.16	1.73
Alpha Cygni	Deneb	3,200	A2Ia	260,000	1.25	−8.73
Beta Crucis	Becrux	353	B0.5III	3,130	1.30	−3.92
Alpha² Centauri	Alpha Centauri B	4.39	K1V	0.44	1.33	5.71
Alpha Leonis	Regulus	77.5	B7V	137	1.35	−0.52
Alpha Crucis	Acrux	321	B0.5IV	4,000	1.40	−4.19
Epsilon Canis Majoris	Adara	431	B2II	3,700	1.51	−4.11
Gamma Crucis	Gacrux	88.0	M3.5III	142	1.59	−0.56

Sources: Data from *The Hipparcos and Tycho Catalogues*, 1997, European Space Agency SP-1200; SIMBAD Astronomical Database (http://simbad.u-strasbg.fr/simbad); and Research Consortium on Nearby Stars (www.chara.gsu.edu/RECONS).
*Luminosity in this table refers only to radiation in "visual" light

Observing the Sky

The purpose of this appendix is to provide enough information so that you can make sense of a star chart or list of astronomical objects, as well as find a few objects in the sky.

Celestial Coordinates

In Chapter 2 we discuss the **celestial sphere**—the imaginary sphere with Earth at its center upon which celestial objects appear to lie. A number of different coordinate systems are used to specify the positions of objects on the celestial sphere. The simplest of these is the "altitude-azimuth coordinate system." The altitude-azimuth coordinate system is based on the "map" direction to an object (the object's azimuth, with north = 0°, east = 90°, south = 180°, and west = 270°) combined with how high the object is above the horizon (the object's altitude, with the horizon at 0° and the zenith at 90°). For example, an object that is 10° above the eastern horizon has an altitude of 10° and an azimuth of 90°. An object that is 45° above the horizon in the southwest is at altitude 45°, azimuth 225°.

The altitude-azimuth coordinate system is the simplest way to tell someone where in the sky to look at the moment, but it is not a good coordinate system for cataloging the positions of objects. The altitude and azimuth of an object are different for each observer, depending on the observer's position on Earth, and they are constantly changing as Earth rotates on its axis. If we need to specify the direction to an object in a way that is the same for everyone, we need a coordinate system that is fixed relative to the celestial sphere. The most common such coordinates are called "celestial coordinates."

Celestial coordinates are illustrated in **Figure A6.1**. Celestial coordinates are much like the traditional system of latitude and longitude used on the surface of Earth. On Earth, latitude specifies how far you are from Earth's equator, as discussed in Chapter 2. If you are on Earth's equator, your latitude is 0°. If you are at Earth's North Pole, your latitude is 90° north. If you are at Earth's South Pole, your latitude is 90° south.

The latitude-like coordinate on the celestial sphere is called "declination," often signified with the Greek letter δ (delta). The celestial equator has δ = 0°. The north celestial pole has δ = +90°. The south celestial pole has δ = −90°. (See Chapter 2 if you need to refresh your memory about the celestial equator or celestial poles.) Declination is usually expressed in degrees, minutes of arc, and seconds of arc. For example, Sirius, the brightest star in the sky, has δ = −16° 42′ 58″, meaning that it is located not quite 17° south of the celestial equator.

On Earth, east–west position is specified by longitude. Lines of constant longitude run north–south from one pole to the other. Unlike latitude, for which the equator provides a natural place to call "zero," there is no natural starting point for longitude, so we just have to invent one. By arbitrary convention, the Royal Observatory in Greenwich, England, is defined to lie at a longitude of 0°. On the celestial sphere the longitude-like coordinate is called "right ascension," often signified with the Greek letter α (alpha). Unlike the case with longitude, there *is* a natural point on the celestial sphere to use as the starting point for right ascension: the vernal equinox, or the point at which the ecliptic crosses the celestial equator with the Sun moving from the southern sky into the northern sky. The vernal equinox defines the line of right ascension at which α = 0°. The autumnal equinox, located on the opposite side of the sky, is at α = 180°.

Normally, right ascension is measured in units of time rather than degrees. It takes Earth 24 hours (of sidereal time) to rotate on its axis, so the celestial sphere is divided into 24 hours of right ascension, with each hour of right ascension corresponding to 15°. Hours of right ascension are then subdivided into minutes and seconds of time. Right ascension increases going to the east. The right ascension of Sirius, for example, is α = 06^h 45^m 08.9^s, meaning that Sirius is about 101° (that is, 06^h 45^m) east of the vernal equinox. Time is a natural unit for measuring right ascension because time naturally tracks the motion of objects due to Earth's rotation on its axis. If stars on the meridian at a certain time have α = 06^h, then an hour later the stars on the meridian will have 07^h, and an hour after that they will have 08^h. The "local sidereal time," or "star time," at

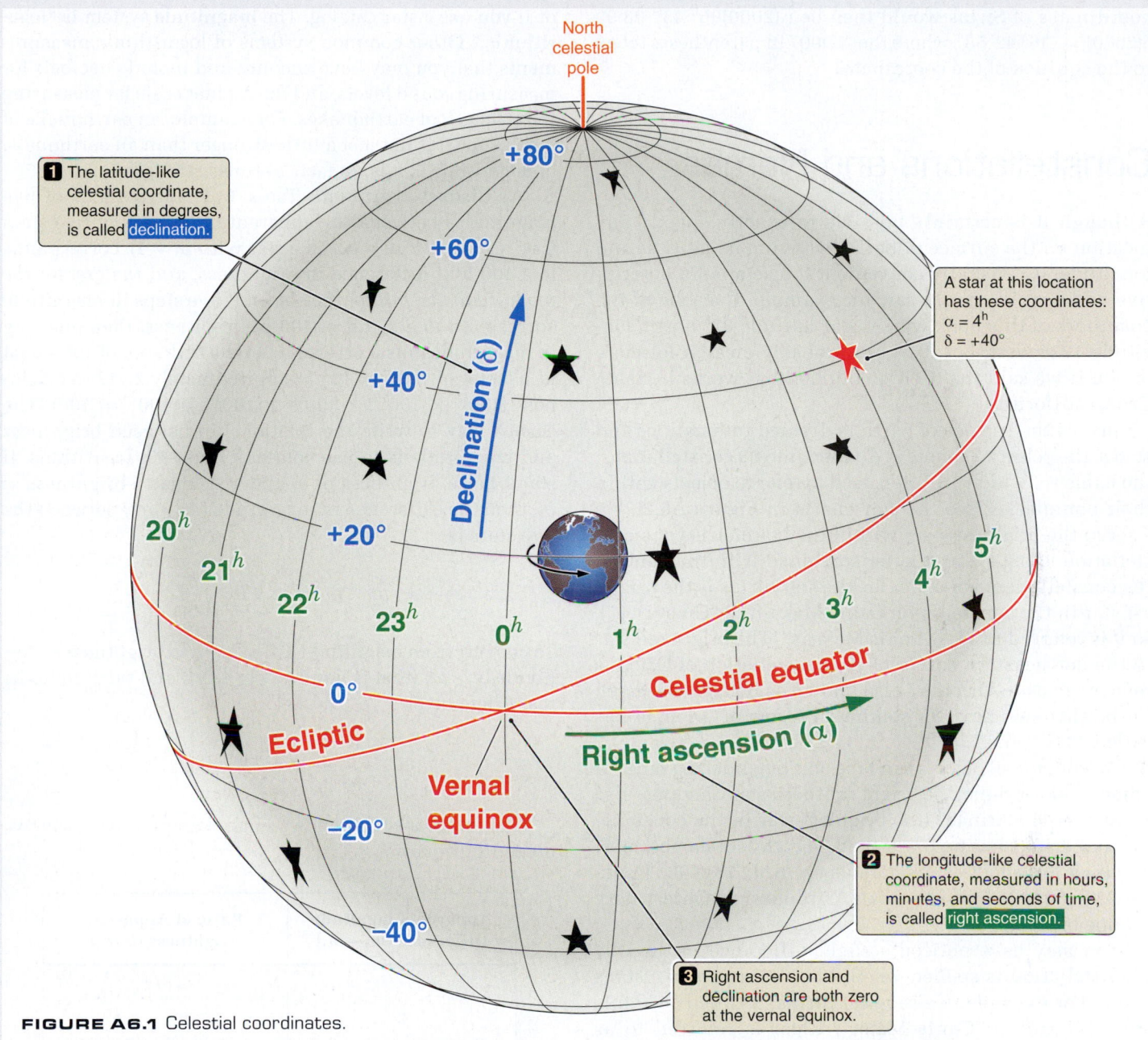

1 The latitude-like celestial coordinate, measured in degrees, is called declination.

North celestial pole

+80°

+60°

A star at this location has these coordinates:
$\alpha = 4^h$
$\delta = +40°$

+40°

Declination (δ)

20^h

21^h

+20°

22^h

23^h

0^h

1^h

2^h

3^h

4^h

5^h

0°

Celestial equator

Ecliptic

Right ascension (α)

Vernal equinox

−20°

2 The longitude-like celestial coordinate, measured in hours, minutes, and seconds of time, is called right ascension.

−40°

3 Right ascension and declination are both zero at the vernal equinox.

FIGURE A6.1 Celestial coordinates.

your location right now is equal to the right ascension of the stars that are on your meridian at the moment. Because of Earth's motion around the Sun, a sidereal day is about 4 minutes shorter than a solar day, so local sidereal time constantly gains on solar time. At midnight on September 23, the local sidereal time is 0^h. By midnight on December 22, local sidereal time has advanced to 06^h. On March 21, local sidereal time at midnight is 12^h. And at midnight on June 21, local sidereal time is 18^h.

Putting this all together, right ascension and declination provide a convenient way to specify the location of any object on the celestial sphere. Sirius is located at $\alpha = 06^h 45^m 08.9^s$, or $-16° 42' 58''$, which means that at midnight on December 22 (local sidereal time = 06^h) you will find

Sirius about 45^m east of the meridian, not quite 17° south of the celestial equator.

There is just one final caveat. As we discussed in Chapter 2, the directions of the celestial equator, celestial poles, and vernal equinox are constantly changing as Earth's axis wobbles like the axis of a spinning top. In Chapter 2 we called this 26,000-year wobble the **precession of the equinoxes**, meaning that the location of the equinoxes is slowly advancing along the ecliptic. So when we specify the celestial coordinates of an object, we need to specify the date at which the positions of the vernal equinox and celestial poles were measured. By convention, coordinates are usually referred to with the position of the vernal equinox on January 1, 2000. A complete, formal specification of the coordinates of Sirius would then be $\alpha(2000)06^h\ 45^m\ 08.9^s$, $\delta(2000) = -16°\ 42'\ 58''$, where the "2000" in parentheses refers to the equinox of the coordinates.

Constellations and Names

Although it is certainly possible to exactly specify any location on the surface of Earth by giving its latitude and longitude, it is usually convenient to use a more descriptive address. We might say, for example, that one of the coauthors of this book works near latitude 37° north, longitude 122° west; but it would probably mean a lot more to you if we said that George Blumenthal works in Santa Cruz, California.

Just as the surface of Earth is divided into nations and states, the celestial sphere is divided into 88 **constellations**, the names of which are often used to refer to objects within their boundaries (see the star charts in **Figure A6.2**). We refer to the brightest stars within the boundaries of a constellation using a Greek letter combined with the name of the constellation. For example, the star Sirius is the brightest star in the constellation Canis Major (the "Great Dog"), so it is referred to as "α Canis Majoris." The bright red star in the northeastern corner of the constellation of Orion is referred to as "α Orionis," also known as Betelgeuse. Rigel, the bright blue star in the southwest corner of Orion, is also called "β Orionis."

Astronomical objects can take on a bewildering range of names. For example, the bright southern star Canopus, also known as "α Carinae" (the brightest star in the constellation of Carina), has no fewer than 34 different names, most of which are about as memorable as "SAO 234480" (number 234,480 in the Smithsonian Astrophysical Observatory catalog of stars).

You may have noticed a slight difference in the way a constellation is spelled when it becomes part of a star's name. For example, we see that Sirius is called "α Canis Majoris," not "α Canis Major"; Rigel is referred to as "β Orionis," not "β Orion"; and Canopus becomes "α Carinae," not "α Carina." This is because we use the Latin genitive or possessive case with star names; for example, *Orionis* means "of Orion."

Astronomical Magnitudes

Apparent Magnitudes

We first introduced magnitudes in Math Tools 13.1; here we provide some additional information. Throughout the text we refer to the brightness of objects; but when discussing the appearance of an object in the sky, astronomers normally speak instead of the object's **apparent magnitude**. You may come across this system if you take a lab course in astronomy

or if you use a star catalog. The **magnitude** system is "logarithmic." Other common systems of logarithmic measurements that you may have encountered include decibels for measuring sound levels, and the Richter scale for measuring the strength of earthquakes. For example, an earthquake of magnitude 6 is not just a little stronger than an earthquake of magnitude 5; it is, in fact, 10 times stronger.

As discussed in Math Tools 13.1, a difference of five magnitudes between the apparent brightness of two stars (say, a star with $m = 6$ and a star with $m = 1$), corresponds to a 100-fold difference in brightness, and *the greater the magnitude, the fainter the object*. If five steps in magnitude correspond to a factor of 100 in brightness, then one step in magnitude must correspond to the fifth root of 100—that is, a factor of $100^{1/5} = 10^{2/5} =$ approximately 2.512 in brightness ($100^{1/5} \times 100^{1/5} \times 100^{1/5} \times 100^{1/5} \times 100^{1/5} = 100$). The easiest way to write the relationship between brightness and magnitude is to use common (base-10) logarithms. If star 1 has a brightness of b_1 and star 2 has a brightness of b_2, then the difference in magnitude $(m_2 - m_1)$ between the two stars is

$$m_2 - m_1 = -2.5\ log_{10}\ \frac{b_2}{b_1}.$$

To convert from magnitude differences to brightness ratios, divide by −2.5 (that is, multiply by −0.4) and raise 10 to the resulting power:

$$\frac{b_2}{b_1} = 10^{-0.4 \times (m_2 - m_1)}.$$

The following table shows some examples using the two preceding equations.

Apparent Magnitude Difference $(m_2 - m_1)$	Ratio of Apparent Brightness (b_1/b_2)
1	2.512
2	$2.512^2 = 6.3$
3	$2.512^3 = 15.8$
4	$2.512^4 = 39.8$
5	$2.512^5 = 100$
10	$2.512^{10} = 100^2 = 10,000$
15	$2.512^{15} = 100^3 = 1,000,000$
20	$2.512^{20} = 100^4 = 10^8$
25	$2.512^{25} = 100^5 = 10^{10}$

Absolute Magnitudes

Recall that stars differ in their brightness for two reasons: the amount of light they are actually emitting, and their distance from Earth. The magnitude system is also used for luminosity, as well as for brightness with the same scale: a difference of five magnitudes corresponds to a 100-fold difference in luminosity. We call these **absolute magnitudes** (M_{abs}), and the idea is to imagine how bright the star would be if it were at a distance of 10 parsecs (pc). Absolute magnitudes allow us to compare how luminous two stars really are, without the factor of distance. Our Sun is very bright because we are so close (apparent visual magnitude = −27); but if the Sun were at a distance of 10 pc, its magnitude would be only about 5.[1] Thus we say the absolute magnitude of the Sun is $M_{abs} = 5$. Recall that we usually express the luminosity of stars by comparing it with the luminosity of the Sun. Thus, a star that is 10,000 times more luminous than the Sun will be 10 absolute magnitudes brighter, or $M_{abs} = -5$, and a star that is 100 times fainter than the Sun will be 5 absolute magnitudes fainter, or $M_{abs} = 10$.

The following is a table of luminosity (where $L_{\odot} = 1$) versus absolute magnitude.

L_{star}/L_{Sun}	M_{abs}
1,000,000	−10
10,000	−5
100	0
1	5
1/100	10
1/10,000	15

Distance Modulus

The difference between the apparent magnitude and the absolute magnitude depends on the star's distance. A star at a distance of exactly 10 pc will have an apparent magnitude equal to its absolute magnitude. We can always measure the brightness of a star and thus its apparent magnitude, and we can estimate the luminosity of a star and thus its absolute magnitude using the H-R diagram. This is the way the distances to most stars are found.

Using the preceding equations and the definition of absolute magnitude, we can get to this relatively simple expression:

$$m - M_{abs} = 5 \log_{10} d - 5,$$

where distance d is in parsecs.

We can rewrite this equation as

$$d = 10^{\left(\frac{m - M_{abs} + 5}{5}\right)}.$$

[1] The apparent and absolute magnitudes of the Sun are −26.74 and +4.83, respectively. We use +5 for the Sun's absolute magnitude as an approximation.

The following table shows how the difference between an object's apparent and absolute magnitudes solves for its distance in parsecs.

$m - M_{abs}$	Distance (pc)
−3	2.5
−2	4.0
−1	6.3
0	10
1	16
2	25
3	40
4	63
5	100
10	1,000
15	10,000
20	100,000

Although the system of astronomical magnitudes is convenient in many ways—which is why astronomers continue to use it—it can also be confusing to new students. Just remember three things and you will probably get by:

1. The greater the magnitude, the fainter the object.

2. One magnitude smaller means about two and a half times brighter.

3. The brightest stars in the sky have magnitudes of less than 1, and the faintest stars that can be seen with the naked eye on a dark night have magnitudes of about 6.

A final note: In Chapters 13 and onward, we used "colors" based on the ratio of the brightness of a star as seen in two different parts of the spectrum. The "b_B/b_V color," for example, was just the ratio of the brightness of a star seen through a blue filter, divided by the brightness of a star seen through a yellow-green (visual) filter. Normally, astronomers instead discuss the "B-V color" of a star, which is equal to the difference between a star's blue magnitude and its visual magnitude. We can use the previous expression for a magnitude difference to write

$$\text{B-V color} = m_B - m_V = -2.5 \log_{10} (b_B/b_V).$$

Thus, a star with a b_B/b_V color of 1.0 has a B-V color of 0.0, and a star with a b_B/b_V color of 1.4 has a B-V color of −0.37. Notice that, as with magnitudes, B-V colors are "backward." The bluer a star, the greater its b_B/b_V color but the less its B-V color.

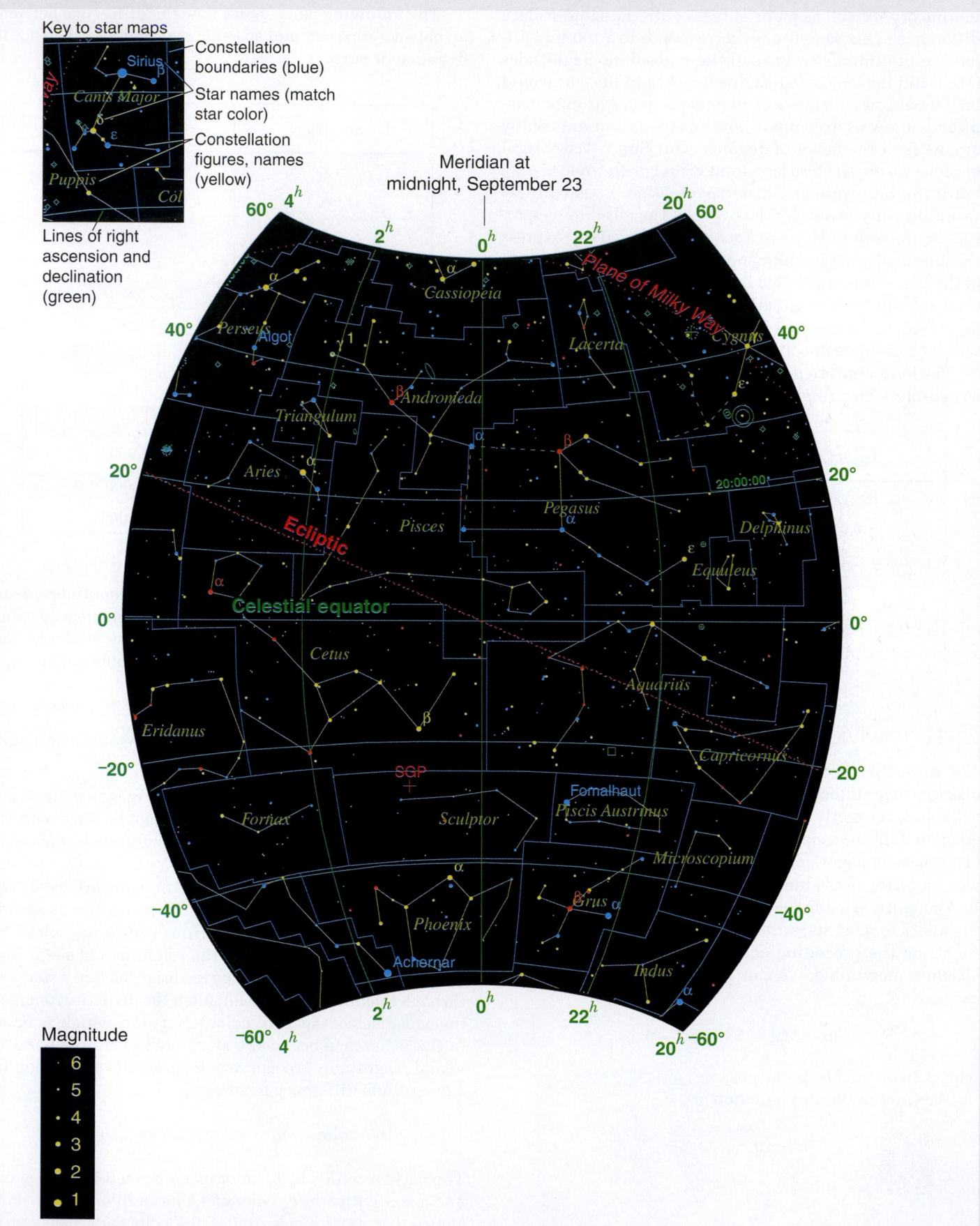

Key to star maps

Constellation boundaries (blue)

Star names (match star color)

Constellation figures, names (yellow)

Lines of right ascension and declination (green)

Sirius

Canis Major

Puppis

Col

Meridian at midnight, September 23

Cassiopeia

Perseus Algol

γ1

Triangulum

β *Andromeda*

α

Aries α

Pisces

Plane of Milky Way

Lacerta

Cygnus

ε

β

20:00:00

Pegasus α

Delphinus

ε

Equuleus

Ecliptic

Celestial equator

Cetus

Eridanus

β

Aquarius

Capricornus

SGP

Fornax

Sculptor

Fomalhaut

Piscis Austrinus

Microscopium

α

Grus α

Phoenix

Achernar

Indus

α

Magnitude

6
5
4
3
2
1

FIGURE A6.2A The sky from right ascension 20ʰ to 04ʰ and declination −60° to +60°.

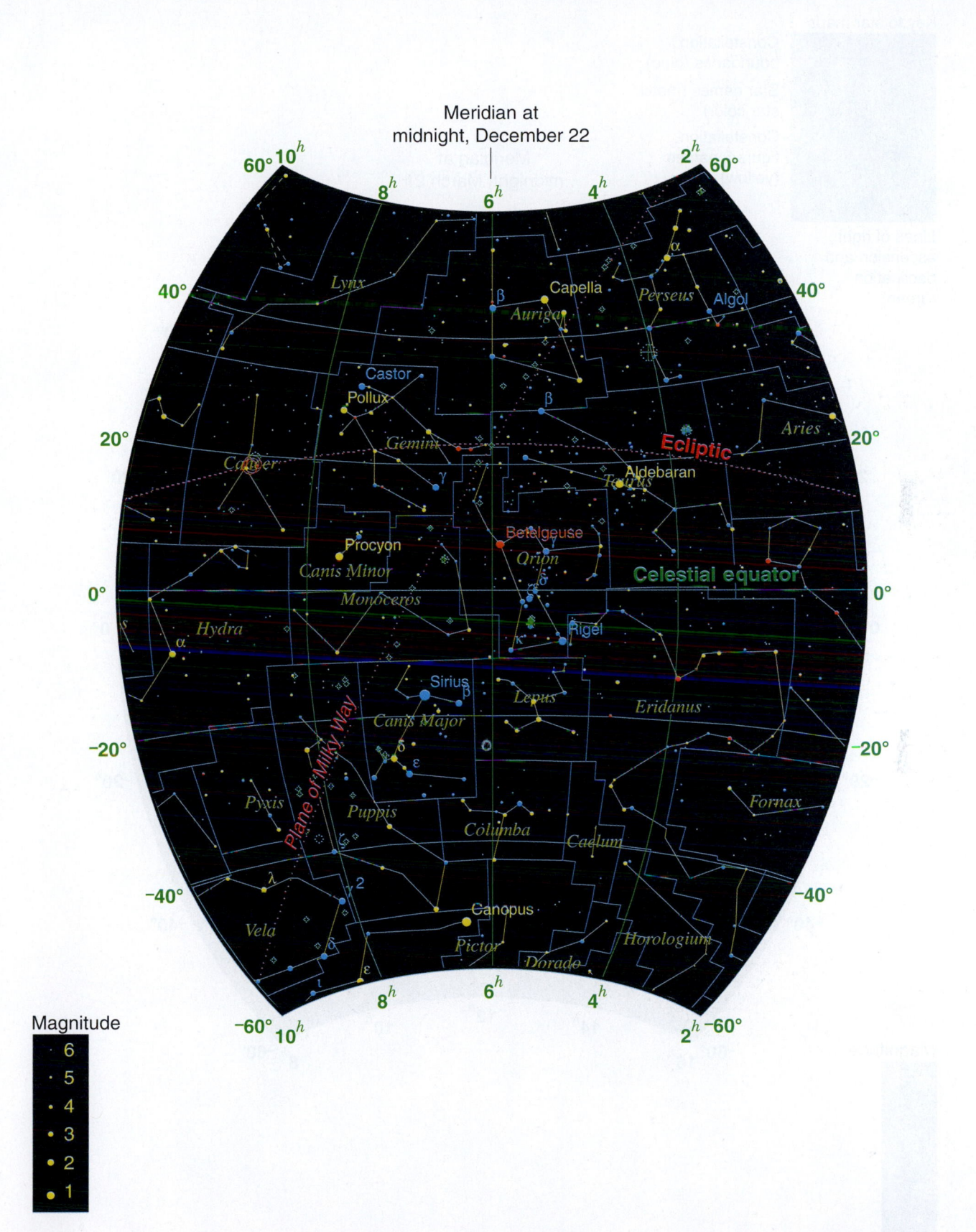

FIGURE A6.2B The sky from right ascension 02ʰ to 10ʰ and declination −60° to +60°.

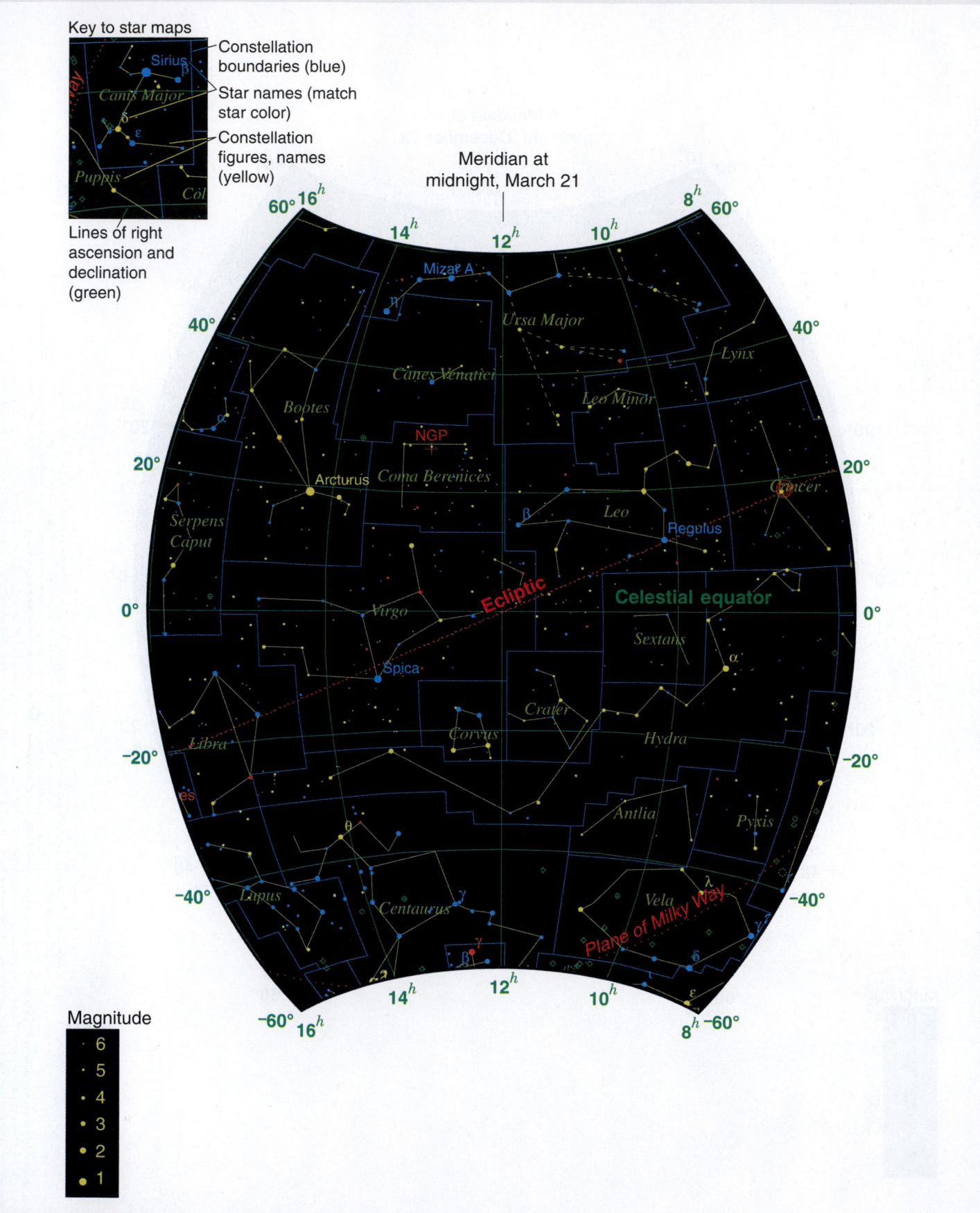

FIGURE A6.2C The sky from right ascension 08ʰ to 16ʰ and declination −60° to +60°.

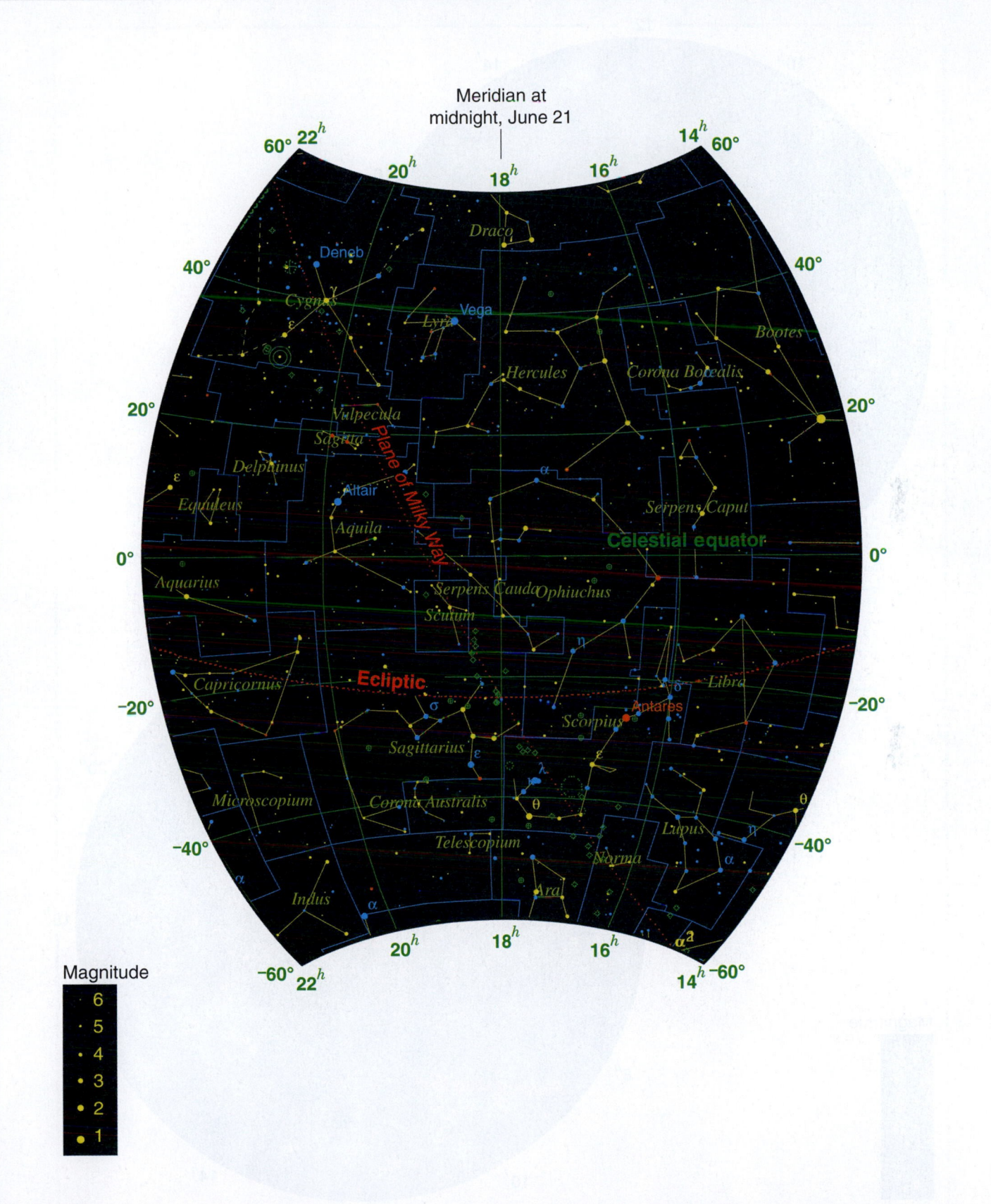

FIGURE A6.2D The sky from right ascension 14ʰ to 22ʰ and declination −60° to +60°.

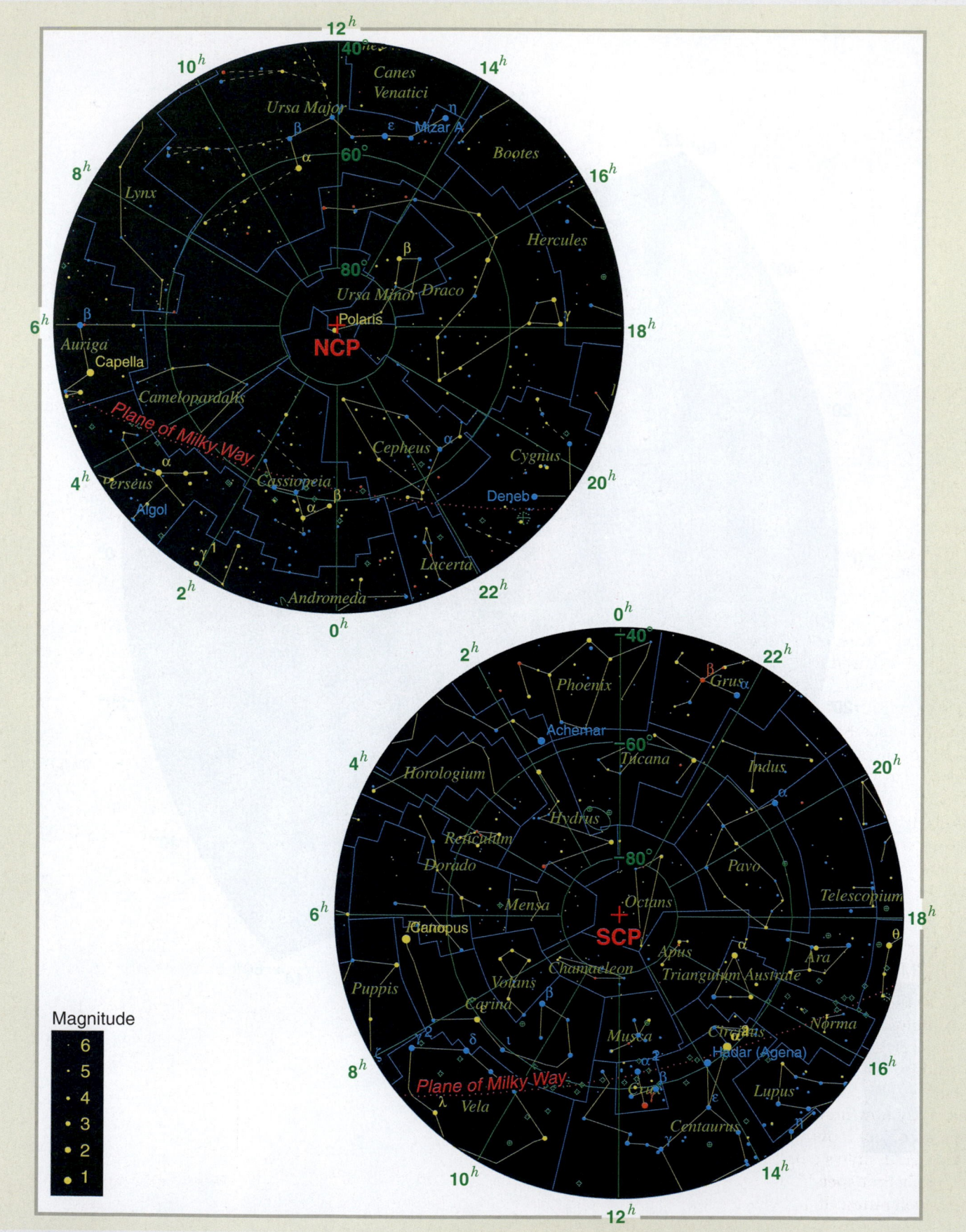

FIGURE A6.2E The regions of the sky north of declination +40° and south of declination −40°.
NCP, north celestial pole; SCP, south celestial pole.

Uniform Circular Motion and Circular Orbits

Uniform Circular Motion

In Chapter 3 (see Section 3.5 and Figure 3.24) we discuss the motion of an object moving in a circle at a constant speed. This motion, called **uniform circular motion**, is the result of the fact that centripetal force always acts toward the center of the circle. The key question when thinking about uniform circular motion is, How hard do I have to pull to keep the object moving in a circle? Part of the answer to this question is pretty obvious: the more massive an object is, the harder it will be to keep it moving on its circular path. According to Newton's second law of motion, $F = ma$; or in this case, the centripetal force equals the mass multiplied by the centripetal acceleration. The larger the mass, the greater the force required to keep it moving in its circle.

The centripetal force needed to keep an object moving in constant circular motion also depends on two other quantities: the speed of the object and the size of the circle. The faster an object is moving, the more rapidly it has to change direction to stay on a circle of a given size. The second quantity that influences the needed acceleration is the radius of the circle. The smaller the circle, the greater the pull needed to keep it on track. You can understand this by looking at the motion. A small circle requires a continuous "hard" turn, whereas a larger circle requires a more gentle change in direction. It takes more force to keep an object moving faster in a smaller circle than it does to keep the same object moving more slowly in a larger circle. (To get a better feel for how this works, think about the difference between riding in a car that is taking a tight curve at high speed and a car that is moving slowly around a gentle curve.)

To arrive at the circular velocity and other results discussed in Chapter 3, we need to turn these intuitive ideas about uniform circular motion into a quantitative expression of exactly how much centripetal acceleration is needed to keep an object moving in a circle with radius r at speed v. **Figure A7.1** shows a ball moving around a circle of radius r at a constant speed v at two different times. The centripetal acceleration that is keeping the ball on the circle is a.

Remember that the acceleration is always directed toward the center of the circle, whereas the velocity of the ball is always perpendicular to the acceleration. The ball's velocity and its acceleration are always at right angles to each other. As the object moves around the circle, the direction of motion and the direction of the acceleration change together in lockstep.

Figure A7.1 contains two triangles. Triangle 1 shows the velocity (speed and direction) at each of the two times. The arrow labeled "Δv" connecting the heads of the two velocity arrows shows how much the velocity changed between time 1 (t_1) and time 2 (t_2). This change is the effect of the centripetal acceleration. If we imagine that points 1 and 2 are very close together—so close that the direction of the centripetal acceleration does not change by much between the two—then we can say that the centripetal acceleration equals the change in the velocity divided by the time between the two, $\Delta t = t_2 - t_1$. So, we have $\Delta v = a\Delta t$.

Triangle 2 shows something similar. Here the arrow labeled "Δr" indicates the change in the position of the ball between time 1 and time 2. Again, if we imagine that the time between the two points is very short, we can say that Δr is equal to the velocity multiplied by the time, or $\Delta r = v\Delta t$.

The line between the center of the circle and the ball is always perpendicular to the velocity of the ball. So if the direction of the ball's velocity changes by an angle α, then the direction of the line between the ball and the center of the circle must also change by the same angle α. In other words, triangles 1 and 2 are "similar triangles." They have the same *shape*. If the triangles are the same shape, the ratio of two sides of triangle 1 must equal the ratio of the two corresponding sides of triangle 2. Using this fact, we can write

$$\frac{a\Delta t}{v} = \frac{v\Delta t}{r}.$$

If we divide by Δt on both sides of the equation and then cross-multiply, we obtain

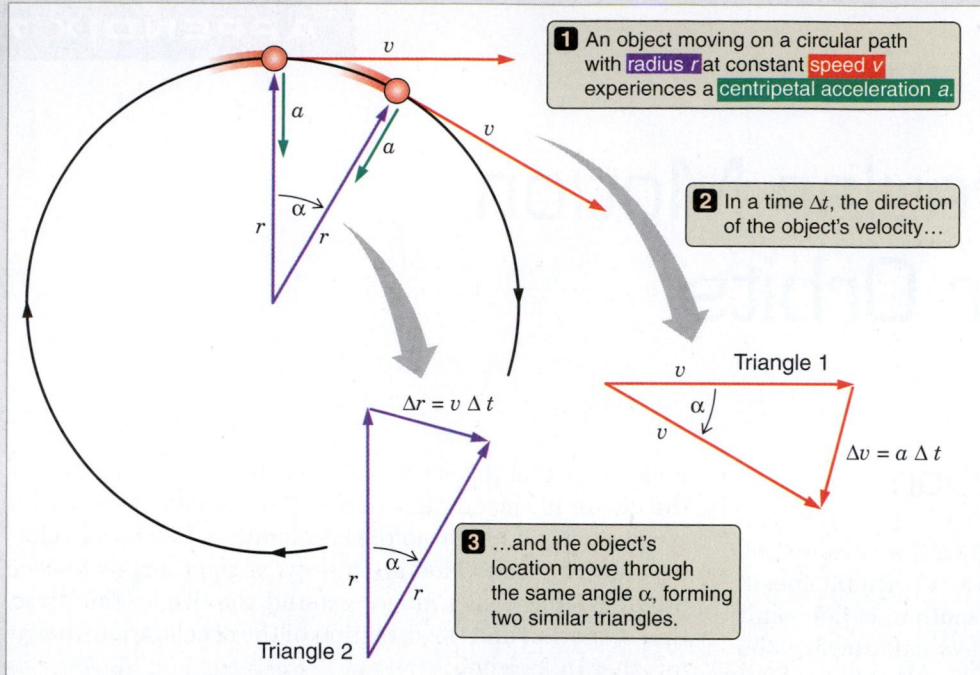

FIGURE A7.1 Similar triangles are used to find the centripetal force that is needed to keep an object moving at a constant speed on a circular path.

$$ar = v^2,$$

which, after we divide both sides of the equation by r, becomes

$$a_{\text{centripetal}} = \frac{v^2}{r}.$$

We have added the subscript "centripetal" to a to signify that this is the centripetal acceleration needed to keep the object moving in a circle of radius r at speed v. The centripetal force required to keep an object of mass m moving on such a circle is then

$$F_{\text{centripetal}} = m a_{\text{centripetal}} = \frac{m v^2}{r}.$$

Circular Orbits

In the case of an object moving in a circular orbit, there is no string to hold the ball on its circular path. Instead, this force is provided by **gravity**.

Think about an object with mass m in orbit about a much larger object with mass M. The orbit is circular, and the distance between the two objects is given by r. The force needed to keep the smaller object moving at speed v in a circle with radius r is given by the previous expression for $F_{\text{centripetal}}$. The force actually provided by gravity (see Chapter 3) is

$$F_{\text{gravity}} = G \frac{M m}{r^2}.$$

If gravity is responsible for holding the mass in its circular motion, then it should be true that $F_{\text{gravity}} = F_{\text{centripetal}}$. That is, if mass m is moving in a circle under the force of gravity, the force provided by gravity *must* equal the centripetal force needed to explain that circular motion. Setting our two expressions for $F_{\text{centripetal}}$ and F_{gravity} equal to each other gives

$$\frac{m v^2}{r} = G \frac{M m}{r^2}.$$

All that remains is a bit of algebra. Dividing by m on both sides of the equation and multiplying both sides by r gives

$$v^2 = G \frac{M}{r}.$$

Taking the square root of both sides then brings us to the desired result:

$$v_{\text{circular}} = \sqrt{\frac{GM}{r}}.$$

This is the **circular velocity** we presented in Chapter 3. It is the velocity at which an object in a circular orbit *must* be moving. If the object were not moving at this velocity, then gravity would not be providing the needed centripetal force, and the object would not move in a circle.

IAU 2006 Resolutions: Definition of a Planet in the Solar System, and Pluto

August 24, 2006, Prague

Resolutions

Resolution 5 is the principal definition for the IAU usage of "planet" and related terms.

Resolution 6 creates for IAU usage a new class of objects, for which Pluto is the prototype. The IAU will set up a process to name these objects.

Resolution 5

Definition of a Planet in the Solar System

Contemporary observations are changing our understanding of planetary systems, and it is important that our nomenclature for objects reflect our current understanding. This applies, in particular, to the designation "planets." The word "planet" originally described "wanderers" that were known only as moving lights in the sky. Recent discoveries lead us to create a new definition, which we can make using currently available scientific information.

The IAU therefore resolves that planets and other bodies, except satellites, in our Solar System be defined into three distinct categories in the following way:

*Source: International Astronomical Union.

(1) A "planet"[1] is a celestial body that (a) is in orbit around the Sun, (b) has sufficient mass for its self-gravity to overcome rigid body forces so that it assumes a hydrostatic equilibrium (nearly round) shape, and (c) has cleared the neighborhood around its orbit.

(2) A "dwarf planet" is a celestial body that (a) is in orbit around the Sun, (b) has sufficient mass for its self-gravity to overcome rigid body forces so that it assumes a hydrostatic equilibrium (nearly round) shape,[2] (c) has not cleared the neighborhood around its orbit, and (d) is not a satellite.

(3) All other objects,[3] except satellites, orbiting the Sun shall be referred to collectively as "Small Solar-System Bodies."

Resolution 6

Pluto

The IAU further resolves:

Pluto is a "dwarf planet" by the above definition and is recognized as the prototype of a new category of Trans-Neptunian Objects.[4]

---

[1] The eight planets are: Mercury, Venus, Earth, Mars, Jupiter, Saturn, Uranus, and Neptune.

[2] An IAU process will be established to assign borderline objects into either dwarf planet and other categories.

[3] These currently include most of the Solar System asteroids, most Trans-Neptunian Objects (TNOs), comets, and other small bodies.

[4] An IAU process will be established to select a name for this category.

Glossary

A

aberration of starlight The apparent displacement in the position of a star that is due to the finite speed of light and Earth's orbital motion around the Sun.

absolute magnitude A measure of the intrinsic brightness of a celestial object, generally a star. Specifically, the apparent brightness of an object, such as a star, if it were located at a standard distance of 10 parsecs (pc). Compare *apparent magnitude*.

absolute zero The temperature at which thermal motions cease. The lowest possible temperature. Zero on the Kelvin temperature scale.

absorption The capture of electromagnetic radiation by matter. Compare *emission*.

absorption line An intensity minimum in a spectrum that is due to the absorption of electromagnetic radiation at a specific wavelength determined by the energy levels of an atom or molecule. Compare *emission line*.

acceleration The rate at which the speed and/or direction of an object's motion is changing.

accretion disk A flat, rotating disk of gas and dust surrounding an object, such as a young stellar object, a forming planet, a collapsed star in a binary system, or a black hole.

achondrite A stony meteorite that does not contain chondrules. Compare *chondrite*.

active comet A comet nucleus that approaches close enough to the Sun to show signs of activity, such as the production of a coma and tail.

active galactic nucleus (AGN) A highly luminous, compact galactic nucleus whose luminosity may exceed that of the rest of the galaxy.

adaptive optics Electro-optical systems that largely compensate for image distortion caused by Earth's atmosphere.

AGB See *asymptotic giant branch*.

AGN See *active galactic nucleus*.

albedo The fraction of electromagnetic radiation incident on a surface that is reflected by the surface.

algebra A branch of mathematics in which numeric variables are represented by letters.

alpha particle A ^{4}He nucleus, consisting of two protons and two neutrons. Alpha particles get their name from the fact that they are given off in the type of radioactive decay referred to as "alpha decay."

Amors A group of asteroids whose orbits cross the orbit of Mars but not the orbit of Earth. Compare *Apollos* and *Atens*.

amplitude In a wave, the maximum deviation from its undisturbed or relaxed position. For example, in a water wave the amplitude is the vertical distance from crests to the undisturbed water level.

angular momentum A conserved property of a rotating or revolving system whose value depends on the velocity and distribution of the system's mass.

annular solar eclipse The type of solar eclipse that occurs when the apparent diameter of the Moon is less than that of the Sun, leaving a visible ring of light ("annulus") surrounding the dark disk of the Moon. Compare *partial solar eclipse* and *total solar eclipse*.

Antarctic Circle The circle on Earth with latitude 66.5° south, marking the northern limit where at least 1 day per year is in 24-hour daylight. Compare *Arctic Circle*.

anticyclonic motion The rotation of a weather system resulting from the Coriolis effect as air moves outward from a region of high atmospheric pressure. Compare *cyclonic motion*.

antimatter Matter made from antiparticles.

antiparticle An elementary particle of antimatter identical in mass but opposite in charge and all other properties to its corresponding ordinary matter particle.

aperture The clear diameter of a telescope's objective lens or primary mirror.

aphelion (pl. **aphelia**) The point in a solar orbit that is farthest from the Sun. Compare *perihelion*.

Apollos A group of asteroids whose orbits cross the orbits of both Earth and Mars. Compare *Amors* and *Atens*.

apparent magnitude A measure of the apparent brightness of a celestial object, generally a star. Compare *absolute magnitude*.

arcminute (arcmin) A minute of arc ('), a unit used for measuring angles. An arcminute is 1/60 of a degree of arc.

arcsecond (arcsec) A second of arc ("), a unit used for measuring very small angles. An arcsecond is 1/60 of an arcminute, or 1/3,600 of a degree of arc.

Arctic Circle The circle on Earth with latitude 66.5 north, marking the southern limit where at least 1 day per year is in 24-hour daylight. Compare *Antarctic Circle*.

asteroid Also called *minor planet*. A primitive rocky or metallic body (planetesimal) that has survived planetary accretion. Asteroids are parent bodies of meteoroids.

asteroid belt Also called *main asteroid belt*. The region between the orbits of Mars and Jupiter that contains most of the asteroids in our Solar System.

astrobiology An interdisciplinary science combining astronomy, biology, chemistry, geology, and physics to study life in the cosmos.

astrology The belief that the positions and aspects of stars and planets influence human affairs and characteristics, as well as terrestrial events.

astronomical seeing A measurement of the degree to which Earth's atmosphere degrades the resolution of a telescope's view of astronomical objects.

astronomical unit (AU) The average distance from the Sun to Earth: approximately 150 million kilometers (km).

astronomy The scientific study of planets, stars, galaxies, and the universe as a whole.

astrophysics The application of physical laws to the understanding of planets, stars, galaxies, and the universe as a whole.

asymptotic giant branch (AGB) The path on the H-R diagram that goes from the horizontal branch toward higher luminosities and lower temperatures, asymptotically approaching and then rising above the red giant branch.

Atens A group of asteroids whose orbits cross the orbit of Earth but not the orbit of Mars. Compare *Amors* and *Apollos*.

atmosphere The gravitationally bound, outer gaseous envelope surrounding a planet, moon, or star.

atmospheric greenhouse effect A warming of planetary surfaces produced by atmospheric gases that transmit optical solar radiation but partially trap infrared radiation. Compare *greenhouse effect*.

atmospheric probe An instrumented package designed to provide *in situ* measurements of the chemical and/or physical properties of a planetary atmosphere.

atmospheric window A region of the electromagnetic spectrum in which radiation is able to penetrate a planet's atmosphere.

atom The smallest piece of a chemical element that retains the properties of that element. Each atom is composed of a nucleus (neutrons and protons) surrounded by a cloud of electrons.

AU See *astronomical unit*.

aurora Emission in the upper atmosphere of a planet from atoms that have been excited by collisions with energetic particles from the planet's magnetosphere.

autumnal equinox 1. One of two points where the Sun crosses the celestial equator. 2. The day on which the Sun appears at this location, marking the first day of autumn (about September 23 in the Northern Hemisphere and March 21 in the Southern Hemisphere). Compare *vernal equinox*.

axion A hypothetical elementary particle first proposed to explain certain properties of the neutron and now considered a candidate for cold dark matter.

B

backlighting Illumination from behind a subject as seen by an observer. Fine material such as human hair and dust in planetary rings stands out best when viewed under backlighting conditions.

bar A unit of pressure. One bar is equivalent to 10^5 newtons per square meter—approximately equal to Earth's atmospheric pressure at sea level.

barred spiral A spiral galaxy with a bulge having an elongated, barlike shape.

basalt Gray to black volcanic rock, rich in iron and magnesium.

beta decay 1. The decay of a neutron into a proton by emission of an electron (beta ray) and an antineutrino. 2. The decay of a proton into a neutron by emission of a positron and a neutrino.

Big Bang The event that occurred 13.7 billion years ago that marks the beginning of time and the universe.

Big Bang nucleosynthesis The formation of low-mass nuclei (H, He, Li, Be, B) during the first few minutes following the Big Bang.

Big Rip A hypothetical cosmic event in which all matter in the universe, from stars to subatomic particles, is progressively torn apart by expansion of the universe.

binary star A system in which two stars are in gravitationally bound orbits about their common center of mass.

binding energy The minimum energy required to separate an atomic nucleus into its component protons and neutrons.

biosphere The global sum of all living organisms on Earth (or any planet or moon.) Compare *hydrosphere* and *lithosphere*.

bipolar outflow Material streaming away in opposite directions from either side of the accretion disk of a young star.

black hole An object so dense that its escape velocity exceeds the speed of light; a singularity in spacetime.

blackbody An object that absorbs and can reemit all electromagnetic energy it receives.

blackbody spectrum See *Planck spectrum*.

blueshift The Doppler shift toward shorter wavelengths of light from an approaching object. Compare *redshift*.

Bohr model A model of the atom, proposed by Niels Bohr in 1913, in which a small positively charged nucleus is surrounded by orbiting electrons, similar to a miniature solar system.

bolide A very bright, exploding meteor.

bound orbit A closed orbit in which the velocity is less than the escape velocity. Compare *unbound orbit*.

bow shock 1. The boundary at which the speed of the solar wind abruptly drops from supersonic to subsonic in its approach to a planet's magnetosphere; the boundary between the region dominated by the solar wind and the region dominated by a planet's magnetosphere. 2. The interface between strong collimated gas and dust outflow from a star and the interstellar medium.

brightness The apparent intensity of light from a luminous object. Brightness depends on both the *luminosity* of a source and its distance. Units at the detector: watts per square meter (W/m^2).

brown dwarf A "failed" star that is not massive enough to cause hydrogen fusion in its core. An object whose mass is intermediate between that of the least massive stars and that of supermassive planets.

bulge The central region of a spiral galaxy that is similar in appearance to a small elliptical galaxy.

C

C See *Celsius*.

C-type asteroid An asteroid made of material that has largely been unmodified since the formation of the Solar System; the most primitive type of asteroid. Compare *M-type asteroid* and *S-type asteroid*.

carbon-nitrogen-oxygen (CNO) cycle One of the ways in which hydrogen is converted to helium (hydrogen burning) in the interiors of main-sequence stars. See also *triple-alpha process* and *proton-proton chain*.

carbon star A cool red giant or asymptotic giant branch star that has an excess of carbon in its atmosphere.

carbonaceous chondrite A primitive stony meteorite that contains chondrules and is rich in carbon and volatile materials.

Cassini Division The largest gap in Saturn's rings, discovered by Jean-Dominique Cassini in 1675.

catalyst An atomic and molecular structure that permits or encourages chemical and nuclear reactions but does not change its own chemical or nuclear properties.

CBR See *cosmic background radiation*.

CCD See *charge-coupled device*.

celestial equator The imaginary great circle that is the projection of Earth's equator onto the celestial sphere.

celestial sphere An imaginary sphere with celestial objects on its inner surface and Earth at its center. The celestial sphere has no physical existence but is a convenient tool for picturing the directions in which celestial objects are seen from the surface of Earth.

Celsius (C) Also called *centigrade scale*. The arbitrary temperature scale—defined by Anders Celsius (1701–1744)—that defines 0°C as the freezing point of water and 100°C as the boiling point of water at sea level. Unit: °C. Compare *Fahrenheit* and *Kelvin scale*.

center of mass The location associated with an object system at which we may regard the entire mass of the system as being concentrated. The point in any isolated system that moves according to Newton's first law of motion.

centigrade scale See *Celsius*.

centripetal force A force directed toward the center of curvature of an object's curved path.

Cepheid variable An evolved high-mass star with an atmosphere that is pulsating, leading to variability in the star's luminosity and color.

Chandrasekhar limit The upper limit on the mass of an object supported by electron degeneracy pressure; approximately 1.4 solar mass ($M_\odot$).

chaos Behavior in complex, interrelated systems in which tiny differences in the initial configuration of a system result in dramatic differences in the system's later evolution.

charge-coupled device (CCD) A common type of solid-state detector of electromagnetic radiation that transforms the intensity of light directly into electric signals.

chemistry The study of the composition, structure, and properties of substances.

chondrite A stony meteorite containing chondrules. Compare *achondrite*.

chondrule A small, crystallized, spherical inclusion of rapidly cooled molten droplets found inside some meteorites.

chromatic aberration A detrimental property of a lens in which rays of different wavelengths are brought to different focal distances from the lens.

chromosphere The region in the Sun's atmosphere located between the photosphere and the corona.

circular velocity The orbital velocity needed to keep an object moving in a circular orbit.

circumpolar Referring to the part of the sky, near either celestial pole, that can always be seen above the horizon from a specific location on Earth.

circumstellar disk See *protoplanetary disk*.

classical mechanics The science of applying Newton's laws to the motion of objects.

classical planets The eight major planets of the Solar System: Mercury, Venus, Earth, Mars, Jupiter, Saturn, Uranus, and Neptune.

climate The state of an atmosphere averaged over an extended time.

closed universe A finite universe with a curved spatial structure such that the sum of the angles of a triangle always exceeds 180°. Compare *flat universe* and *open universe*.

CNO cycle See *carbon-nitrogen-oxygen cycle*.

cold dark matter Particles of dark matter that move slowly enough to be gravitationally bound even in the smallest galaxies. Compare *hot dark matter*.

color index The color of a celestial object, generally a star, based on the ratio of its brightness in blue light (b_B) to its brightness in "visual" (or yellow-green) light (b_V). The difference between an object's blue (B) magnitude and visual (V) magnitude, B-V.

coma (pl. comae) The nearly spherical cloud of gas and dust surrounding the nucleus of an active comet.

comet A complex object consisting of a small, solid, icy nucleus; an atmospheric halo; and a tail of gas and dust.

comet nucleus A primitive planetesimal, composed of ices and refractory materials, that has survived planetary accretion. The "heart" of a comet, containing nearly the entire mass of the comet. A "dirty snowball."

comparative planetology The study of planets through comparison of their chemical and physical properties.

complex system An interrelated system capable of exhibiting chaotic behavior. See also *chaos*.

composite volcano A large, cone-shaped volcano formed by viscous, pasty lava flows alternating with pyroclastic (explosively generated) rock deposits. Compare *shield volcano*.

compound lens A lens made up of two or more elements of differing refractive index, the purpose of which is to minimize chromatic aberration.

conservation law A physical law stating that the amount of a particular physical quantity (such as energy or angular momentum) of an isolated system does not change over time.

conservation of angular momentum The physical law stating that the amount of angular momentum of an isolated system does not change over time.

conservation of energy The physical law stating that the amount of energy of an isolated, closed system does not change over time.

constant of proportionality The multiplicative factor by which one quantity is related to another.

constellation An imaginary image formed by patterns of stars; any of 88 defined areas on the celestial sphere used by astronomers to locate celestial objects.

constructive interference A state in which the amplitudes of two intersecting waves reinforce one another. Compare *destructive interference*.

continental drift The slow motion (centimeters per year) of Earth's continents relative to each other and to Earth's mantle. See also *plate tectonics*.

continuous radiation Electromagnetic radiation with intensity that varies smoothly over a wide range of wavelengths.

convection The transport of thermal energy from the lower (hotter) to the higher (cooler) layers of a fluid by motions within the fluid driven by variations in buoyancy.

convective zone A region within a star in which energy is transported outward by convection.

core 1. The innermost region of a planetary interior. 2. The innermost part of a star.

core accretion A process for forming giant planets, whereby large quantities of surrounding hydrogen and helium are gravitationally captured onto a massive rocky core.

Coriolis effect The apparent displacement of objects in a direction perpendicular to their true motion as viewed from a rotating frame of reference. On a rotating planet, different latitudes rotating at different speeds cause this effect.

corona The hot, outermost part of the Sun's atmosphere.

coronal hole A low-density region in the solar corona containing "open" magnetic field lines along which coronal material is free to stream into interplanetary space.

coronal mass ejection An eruption on the Sun that ejects hot gas and energetic particles at much higher speeds than are typical in the solar wind.

cosmic background radiation (CBR) Isotropic microwave radiation from every direction in the sky having a 2.73-kelvin (K) Planck spectrum. The CBR is residual radiation from the Big Bang.

cosmic ray A very fast-moving particle (usually an atomic nucleus); cosmic rays fill the disk of our galaxy.

cosmological constant A constant, introduced into general relativity by Einstein, that characterizes an extra, repulsive force in the universe due to the vacuum of space itself.

cosmological principle   The (testable) assumption that the same physical laws that apply here and now also apply everywhere and at all times, and that there are no special locations or directions in the universe.

cosmological redshift (z)   The redshift that results from the expansion of the universe rather than from the motions of galaxies or gravity (see *gravitational redshift*).

cosmology   The study of the large-scale structure and evolution of the universe as a whole.

Crab Nebula   The remnant of the Type II supernova explosion witnessed by Chinese astronomers in 1054 C.E.

crescent   Any phase of the Moon, Mercury, or Venus in which the object appears less than half illuminated by the Sun. Compare *gibbous*.

Cretaceous-Tertiary (K-T) boundary   The boundary between the Cretaceous and Tertiary periods in Earth's history. This boundary corresponds to the time of the impact of an asteroid or comet and the extinction of the dinosaurs.

critical density   The value of the mass density of the universe that, ignoring any cosmological constant, is just barely capable of halting expansion of the universe.

crust   The relatively thin, outermost, hard layer of a planet, which is chemically distinct from the interior.

cryovolcanism   Low-temperature volcanism in which the magmas are composed of molten ices rather than rocky material.

cyclone   See *hurricane*.

cyclonic motion   The rotation of a weather system resulting from the Coriolis effect as air moves toward a region of low atmospheric pressure. Compare *anticyclonic motion*.

Cygnus X-1   A binary X-ray source and probable black hole.

D

dark energy   A form of energy that permeates all of space (including the vacuum) producing a repulsive force that accelerates the expansion of the universe.

dark matter   Matter in galaxies that does not emit or absorb electromagnetic radiation. Dark matter is thought to comprise most of the mass in the universe. Compare *luminous matter*.

dark matter halo   The centrally condensed, greatly extended dark matter component of a galaxy that contains up to 95 percent of the galaxy's mass.

daughter product   An element resulting from radioactive decay of a more massive *parent element*.

decay   1. The process of a radioactive nucleus changing into its daughter product. 2. The process of an atom or molecule dropping from a higher-energy state to a lower-energy state.

density   The measure of an object's mass per unit of volume. Units: kilograms per cubic meter (kg/m^3).

destructive interference   A state in which the amplitudes of two intersecting waves cancel one another. Compare *constructive interference*.

differential rotation   Rotation of different parts of a system at different rates.

differentiation   The process by which materials of higher density sink toward the center of a molten or fluid planetary interior.

diffraction   The spreading of a wave after it passes through an opening or past the edge of an object.

diffraction limit   The limit of a telescope's angular resolution caused by diffraction.

diffuse ring   A sparsely populated planetary ring spread out both horizontally and vertically.

dispersion   The separation of rays of light into their component wavelengths.

distance ladder   A sequence of techniques for measuring cosmic distances; each method is calibrated using the results from other methods that have been applied to closer objects.

Doppler effect   The change in wavelength of sound or light that is due to the relative motion of the source toward or away from the observer.

Doppler redshift   See *redshift*.

Doppler shift   The amount by which the wavelength of light is shifted by the Doppler effect.

Drake equation   A prescription for estimating the number of intelligent civilizations existing elsewhere.

dust devil   A small tornado-like column of air containing dust or sand.

dust tail   A type of comet tail consisting of dust particles that are pushed away from the comet's head by radiation pressure from the Sun. Compare *ion tail*.

dwarf galaxy   A small galaxy with a luminosity ranging from 1 million to 1 billion solar luminosities ($L_\odot$). Compare *giant galaxy*.

dwarf planet   A body with characteristics similar to those of a classical planet except that it has not cleared smaller bodies from the neighboring regions around its orbit. Compare *planet* (definition 2).

dynamic equilibrium   A state in which a system is constantly changing but its configuration remains the same because one source of change is exactly balanced by another source of change. Compare *static equilibrium*.

dynamo   A device that converts mechanical energy into electric energy in the form of electric currents and magnetic fields. The "dynamo effect" is thought to create magnetic fields in planets and stars by electrically charged currents of material flowing within their cores.

E

eccentricity (e)   A measure of the departure of an ellipse from circularity; the ratio of the distance between the two foci of an ellipse to its major axis.

eclipse   1. The total or partial obscuration of one celestial body by another. 2. The total or partial obscuration of light from one celestial body as it passes through the shadow of another celestial body.

eclipse season   Any time during the year when the Moon's line of nodes is sufficiently close to the Sun for eclipses to occur.

eclipsing binary   A binary system in which the orbital plane is oriented such that the two stars appear to pass in front of one another as seen from Earth. Compare *visual binary* and *spectroscopic binary*.

ecliptic   1. The apparent annual path of the Sun against the background of stars. 2. The projection of Earth's orbital plane onto the celestial sphere.

ecliptic plane   The plane of Earth's orbit around the Sun. The ecliptic is the projection of this plane on the celestial sphere.

effective temperature   The temperature at which a black body, such as a star, appears to radiate.

ejecta   1. Material thrown outward by the impact of an asteroid or comet on a planetary surface, leaving a crater behind. 2. Material thrown outward by a stellar explosion.

electric field   A field that is able to exert a force on a charged object, whether at rest or moving. Compare *magnetic field*.

electric force   The force exerted on a

charged particle by an electric field. Compare *magnetic force*.

electromagnetic force The force, including both electric and magnetic forces, that acts on electrically charged particles. One of four fundamental forces of nature. The force mediated by photons.

electromagnetic radiation A traveling disturbance in the electric and magnetic fields caused by accelerating electric charges. In quantum mechanics, a stream of photons. Light.

electromagnetic spectrum The spectrum made up of all possible frequencies or wavelengths of electromagnetic radiation, ranging from gamma rays through radio waves and including the portion our eyes can use.

electromagnetic wave A wave consisting of oscillations in the electric-field strength and the magnetic-field strength.

electron (e^-) A subatomic particle having a negative charge of 1.6×10^{-19} coulomb (C), a rest mass of 9.1×10^{-31} kilograms (kg), and mass-equivalent energy of 8×10^{-14} joules (J). The antiparticle of the *positron*. Compare *proton* and *neutron*.

electron-degenerate Describing the state of material compressed to the point at which electron density reaches the limit imposed by the rules of quantum mechanics.

electroweak theory The quantum theory that combines descriptions of both the electromagnetic force and the weak nuclear force.

element One of 92 naturally occurring substances (such as hydrogen, oxygen, and uranium) and more than 20 human-made ones (such as plutonium). Each element is chemically defined by the specific number of protons in the nuclei of its atoms.

elementary particle One of the basic building blocks of nature that is not known to have substructure, such as the *electron* and the *quark*.

ellipse A conic section produced by the intersection of a plane with a cone when the plane is passed through the cone at an angle to the axis other than 0° or 90°.

elliptical galaxy A galaxy of Hubble type "E" class, with a circular to elliptical outline on the sky containing almost no disk and a population of old stars. Compare *spiral galaxy*, *S0 galaxy*, and *irregular galaxy*.

emission The release of electromagnetic energy when an atom, molecule, or particle drops from a higher-energy state to a lower-energy state. Compare *absorption*.

emission line An intensity peak in a spectrum that is due to sharply defined emission of electromagnetic radiation in a narrow range of wavelengths. Compare *absorption line*.

empirical science Scientific investigation that is based primarily on observations and experimental data. It is descriptive rather than based on theoretical inference.

energy The conserved quantity that gives objects and systems the ability to do work. Units: joules (J).

energy transport The transfer of energy from one location to another. In stars, energy transport is carried out mainly by radiation or convection.

entropy A measure of the disorder of a system related to the number of ways a system can be rearranged without its appearance being affected.

equator The imaginary great circle on the surface of a body midway between its poles that divides the body into northern and southern hemispheres. The equatorial plane passes through the center of the body and is perpendicular to its rotation axis. Compare *meridian*.

equilibrium The state of an object in which physical processes balance each other so that its properties or conditions remain constant.

equinox Literally, "equal night." 1. One of two positions on the ecliptic where it intersects the celestial equator. 2. Either of the two times of year (the *autumnal equinox* and *vernal equinox*) when the Sun is at one of these two positions. At this time, night and day are of the same length everywhere on Earth. Compare *solstice*.

equivalence principle The principle stating that there is no difference between a frame of reference that is freely floating through space and one that is freely falling within a gravitational field.

erosion The degradation of a planet's surface topography by the mechanical action of wind and/or water.

escape velocity The minimum velocity needed for an object to achieve a parabolic trajectory and thus permanently leave the gravitational grasp of another mass.

eternal inflation The idea that a universe might inflate forever. In such a universe, quantum effects could randomly cause regions to slow their expansion, eventually stop inflating, and experience an explosion resembling our Big Bang.

event A particular location in spacetime.

event horizon The effective "surface" of a black hole. Nothing inside this surface—not even light—can escape from a black hole.

evolutionary track The path that a star follows across the H-R diagram as it evolves through its lifetime.

excited state An energy level of a particular atom, molecule, or particle that is higher than its ground state. Compare *ground state*.

exoplanet See *extrasolar planet*.

extrasolar planet Also called *exoplanet*. A planet orbiting a star other than the Sun.

F

F See *Fahrenheit*.

Fahrenheit (F) The arbitrary temperature scale—defined by Daniel Gabriel Fahrenheit (1686–1736)—that defines 32°F as the melting point of water and 212°F as the boiling point of water at sea level. Unit: °F. Compare *Celsius* and *Kelvin scale*.

fault A fracture in the crust of a planet or moon along which blocks of material can slide.

filter An instrument element that transmits a limited wavelength range of electromagnetic radiation. For the optical range, such elements are typically made of different kinds of glass and take on the hue of the light they transmit.

first quarter Moon The phase of the Moon in which only the western half of the Moon, as viewed from Earth, is illuminated by the Sun. It occurs about a week after a new Moon. Compare *third quarter Moon*.

fissure A fracture in the planetary lithosphere from which magma emerges.

flat rotation curve A rotation curve of a spiral galaxy in which rotation rates do not decline in the outer part of the galaxy, but remain relatively constant to the outermost points.

flat universe An infinite universe whose spatial structure obeys Euclidean geometry, such that the sum of the angles of a triangle always equals 180°. Compare *closed universe* and *open universe*.

flatness problem The surprising result that the sum of Ω_{mass} plus Ω_Λ is extremely close to unity in the present-day universe; equivalent to saying that it is surprising the universe is so close to being exactly flat.

flux The total amount of energy passing through each square meter of a surface each second. Units: watts per square meter (W/m²).

flux tube A strong magnetic field contained

within a tubelike structure. Flux tubes are found in the solar atmosphere and connecting the space between Jupiter and its moon Io.

flyby A spacecraft that first approaches and then continues flying past a planet or moon. Flybys can visit multiple objects, but they remain in the vicinity of their targets only briefly. Compare *orbiter*.

focal length The optical distance between a telescope's objective lens or primary mirror and the plane (called the focal plane) on which the light from a distant object is focused.

focal plane The plane, perpendicular to the optical axis of a lens or mirror, in which an image is formed.

focus (pl. **foci**) 1. One of two points that define an ellipse. 2. A point in the focal plane of a telescope.

frame of reference A coordinate system within which an observer measures positions and motions.

free fall The motion of an object when the only force acting on it is gravity.

frequency The number of times per second that a periodic process occurs. Unit: hertz (Hz), 1/s.

full Moon The phase of the Moon in which the near side of the Moon, as viewed from Earth, is fully illuminated by the Sun. It occurs about two weeks after a *new Moon*.

G

galaxy A gravitationally bound system that consists of stars and star clusters, gas, dust, and dark matter; typically greater than 1,000 light-years across and recognizable as a discrete, single object.

galaxy cluster A large, gravitationally bound collection of galaxies containing hundreds to thousands of members; typically 10–15 mega-light-years (Mly) across. Compare *galaxy group* and *supercluster*.

galaxy group A small, gravitationally bound collection of galaxies containing from several to a hundred members; typically 4–6 mega-light-years (Mly) across. Compare *galaxy cluster* and *supercluster*.

gamma ray Electromagnetic radiation with higher frequency, higher photon energy, and shorter wavelength than all other types of electromagnetic radiation.

gas giant A giant planet formed mostly of hydrogen and helium. In our Solar System, Jupiter and Saturn are the gas giants. Compare *ice giant*.

gauss A basic unit of magnetic flux density.

general relativistic time dilation The verified prediction that time passes more slowly in a gravitational field than in the absence of a gravitational field. Compare *time dilation*.

general relativity See *general theory of relativity*.

general theory of relativity Sometimes referred to as simply *general relativity*. Einstein's theory explaining gravity as the distortion of spacetime by massive objects, such that particles travel on the shortest path between two events in spacetime. This theory deals with all types of motion. Compare *special theory of relativity*.

geocentric A coordinate system having the center of Earth as its center. Compare *heliocentric*.

geodesic The path an object will follow through spacetime in the absence of external forces.

geometry A branch of mathematics that deals with points, lines, angles, and shapes.

giant galaxy A galaxy with luminosity greater than about 1 billion solar luminosities ($L_\odot$). Compare *dwarf galaxy*.

giant molecular cloud An interstellar cloud composed primarily of molecular gas and dust, having hundreds of thousands of solar masses.

giant planet One of the largest planets in the Solar System (Saturn, Jupiter, Uranus, or Neptune), typically 10 times the size and many times the mass of any *terrestrial planet* and lacking a solid surface.

gibbous Any phase of the Moon, Mercury, or Venus in which the object appears more than half illuminated by the Sun. Compare *crescent*.

global circulation The overall, planet-wide circulation pattern of a planet's atmosphere.

globular cluster A spherically symmetric, highly condensed group of stars, containing tens of thousands to a million members. Compare *open cluster*.

gluon The particle that carries (or, equivalently, mediates) interactions due to the strong nuclear force.

grand unified theory (GUT) A unified quantum theory that combines the strong nuclear, weak nuclear, and electromagnetic forces but does not include gravity.

granite Rock that is cooled from magma and is relatively rich in silicon and oxygen.

grating An optical surface containing many narrow, closely and equally spaced parallel grooves or slits that spectrally disperse reflected or transmitted light.

gravitational lens A massive object that gravitationally focuses the light of a more distant object to produce multiple brighter, magnified, possibly distorted images.

gravitational lensing The bending of light by gravity.

gravitational potential energy The stored energy in an object that is due solely to its position within a gravitational field.

gravitational redshift The shifting to longer wavelengths of radiation from an object deep within a gravitational well.

gravitational wave A wave in the fabric of spacetime emitted by accelerating masses.

gravity 1. The mutually attractive force between massive objects. 2. An effect arising from the bending of spacetime by massive objects. 3. One of four fundamental forces of nature.

great circle Any circle on a sphere that has as its center the center of the sphere. The celestial equator, the meridian, and the ecliptic are all great circles on the sphere of the sky, as is any circle drawn through the zenith.

Great Red Spot The giant, oval, brick-red anticyclone seen in Jupiter's southern hemisphere.

greenhouse effect The solar heating of air in an enclosed space, such as a closed building or car, resulting primarily from the inability of the hot air to escape. Compare *atmospheric greenhouse effect*.

greenhouse molecule One of a group of atmospheric molecules such as carbon dioxide that are transparent to visible radiation but absorb infrared radiation.

Gregorian calendar The modern calendar. A modification of the Julian calendar decreed by Pope Gregory XIII in 1582. By this time the less accurate Julian calendar had developed an error of 10 days over the 13 centuries since its inception.

ground state The lowest possible energy state for a system or part of a system, such as an atom, molecule, or particle. Compare *excited state*.

GUT See *grand unified theory*.

H

H II region A region of interstellar gas that has been ionized by UV radiation from nearby hot massive stars.

H-R diagram The Hertzsprung-Russell diagram, which is a plot of the luminosities versus the surface temperatures of stars. The evolving properties of stars are plotted as tracks across the H-R diagram.

habitable zone The distance from its star at which a planet must be located in order to have a temperature suitable for life; often assumed to be temperatures at which water exists in a liquid state.

Hadley circulation A simplified, and therefore uncommon, atmospheric global circulation that carries thermal energy directly from the equator to the polar regions of a planet.

half-life The time it takes half a sample of a particular radioactive parent element to decay to a daughter product.

halo The spherically symmetric, low-density distribution of stars and dark matter that defines the outermost regions of a galaxy.

harmonic law See *Kepler's third law*.

Hawking radiation Radiation from a black hole.

Hayashi track The path that a protostar follows on the H-R diagram as it contracts toward the main sequence.

head The part of a comet that includes both the nucleus and the inner part of the coma.

heat death The possible eventual fate of an open universe, in which entropy has triumphed and all energy- and structure-producing processes have come to an end.

heavy element Also called *massive element*. 1. In astronomy, any element more massive than helium. 2. In other sciences (and sometimes also in astronomy), any of the most massive elements in the periodic table, such as uranium and plutonium.

Heisenberg uncertainty principle The physical limitation that the product of the position and the momentum of a particle cannot be smaller than a well-defined value, Planck's constant (h).

heliocentric A coordinate system having the center of the Sun as its center. Compare *geocentric*.

helioseismology The use of solar oscillations to study the interior of the Sun.

helium flash The runaway explosive burning of helium in the degenerate helium core of a red giant star.

Hayashi track The path that a protostar follows on the H-R diagram as it contracts toward the main sequence.

Herbig-Haro (HH) object A glowing, rapidly moving knot of gas and dust that is excited by bipolar outflows in very young stars.

heredity The process by which one generation passes on its characteristics to future generations.

hertz (Hz) A unit of frequency equivalent to cycles per second.

Hertzsprung-Russell diagram See *H-R diagram*.

HH object See *Herbig-Haro object*.

hierarchical clustering The "bottom-up" process of forming large-scale structure. Small-scale structure first produces groups of galaxies, which in turn form clusters, which then form superclusters.

high-mass star A star with a main-sequence mass greater than about 9 solar masses ($M_\odot$). Compare *intermediate-mass star* and *low-mass star*.

high-velocity star A star belonging to the halo found near the Sun, distinguished from disk stars by moving far faster and often in the direction opposite to the rotation of the disk and its stars.

homogeneous In cosmology, describing a universe in which observers in any location would observe the same properties.

horizon The boundary that separates the sky from the ground.

horizon problem The puzzling observation that the cosmic background radiation is so uniform in all directions, despite the fact that widely separated regions should have been "over the horizon" from each other in the early universe.

horizontal branch A region on the H-R diagram defined by stars burning helium to carbon in a stable core.

hot dark matter Particles of dark matter that move so fast that gravity cannot confine them to the volume occupied by a galaxy's normal luminous matter. Compare *cold dark matter*.

hot Jupiter A large, Jovian-type extrasolar planet located very close to its parent star.

hot spot A place where hot plumes of mantle material rise near the surface of a planet.

Hubble constant (H_0) The constant of proportionality relating the recession velocities of galaxies to their distances. See also *Hubble time*.

Hubble time An estimate of the age of the universe from the inverse of the Hubble constant, $1/H_0$.

Hubble's law The law stating that the speed at which a galaxy is moving away from us is proportional to the distance of that galaxy.

hurricane Also called *cyclone* or *typhoon*. A large tropical cyclonic system circulating counterclockwise in the Northern Hemisphere and clockwise in the Southern Hemisphere. Hurricanes can extend outward from their center to more than 600 kilometers (km) and generate winds in excess of 300 kilometers per hour (km/h).

hydrogen burning The release of energy from the nuclear fusion of four hydrogen atoms into a single helium atom.

hydrogen shell burning The fusion of hydrogen in a shell surrounding a stellar core that may be either degenerate or fusing more massive elements.

hydrosphere The portion of Earth that is largely liquid water. Compare *biosphere* and *lithosphere*.

hydrostatic equilibrium The condition in which the weight bearing down at a particular point within an object is balanced by the pressure within the object.

hypothesis A well-thought-out idea, based on scientific principles and knowledge, that leads to testable predictions. Compare *theory*.

Hz See *hertz*.

I

ice The solid form of a volatile material; sometimes the *volatile material* itself, regardless of its physical form.

ice giant A giant planet formed mostly of the liquid form of volatile substances (ices.) In our Solar System, Uranus and Neptune are the ice giants. Compare *gas giant*.

ideal gas law The relationship of pressure (P) to number density of particles (n) and temperature (T) expressed as $P = nkT$, where k is Boltzmann's constant.

igneous activity The formation and action of molten rock (magma).

impact crater The scar of the impact left on a solid planetary or moon surface by collision with another object. Compare *secondary crater*.

impact cratering A process involving collisions between solid planetary objects.

index of refraction (n) The ratio of the speed of light in a vacuum (c) to the speed of light in an optical medium (v).

inert gas A gaseous element that combines with other elements only under conditions of extreme temperature and pressure. Examples include helium, neon, and argon.

inertia The tendency for objects to retain their state of motion.

inertial frame of reference 1. A frame of reference that is not accelerating. 2. In general relativity, a frame of reference that is falling freely in a gravitational field.

inflation An extremely brief phase of ultra-rapid expansion of the very early universe. Following inflation, the standard Big Bang models of expansion apply.

infrared (IR) radiation Electromagnetic radiation with frequencies and photon energies occurring in the spectral region between those of visible light and microwaves. Compare *ultraviolet radiation*.

instability strip A region of the H-R diagram containing stars that pulsate with a periodic variation in luminosity.

integration time The time interval during which photons are collected and added up in a detecting device.

intensity (of light) The amount of radiant energy emitted per second per unit area. Units for electromagnetic radiation: watts per square meter (W/m^2).

intercloud gas A low-density region of the interstellar medium that fills the space between interstellar clouds.

interference The interaction of two sets of waves producing high and low intensity, depending on whether their amplitudes reinforce (*constructive interference*) or cancel (*destructive interference*).

interferometer Also called *interferometric array*. A group or array of separate but linked optical or radio telescopes whose overall separation determines the angular resolution of the system.

interferometric array See *interferometer*.

intermediate-mass star A star with a main-sequence mass between about 3 and 9 solar masses ($M_\odot$). Compare *high-mass star* and *low-mass star*.

interstellar cloud A discrete, high-density region of the interstellar medium made up mostly of atomic or molecular hydrogen and dust.

interstellar dust Small particles or grains (0.01–10 micrometers [μm]) of matter, primarily carbon and silicates, distributed throughout interstellar space.

interstellar extinction The dimming of visible and ultraviolet light by interstellar dust.

interstellar medium The gas and dust that fill the space between the stars within a galaxy.

inverse square law The rule that a quantity or effect diminishes with the square of the distance from the source.

ion An atom or molecule that has lost or gained one or more electrons.

ion tail A type of comet tail consisting of ionized gas. Particles in the ion tail are pushed directly away from the comet's head in the antisolar direction at high speeds by the solar wind. Compare *dust tail*.

ionize The process by which electrons are stripped free from an atom or molecule, resulting in free electrons and a positively charged atom or molecule.

ionosphere A layer high in Earth's atmosphere in which most of the atoms are ionized by solar radiation.

IR Infrared.

iron meteorite A metallic meteorite composed mostly of iron-nickel alloys. Compare *stony-iron meteorite* and *stony meteorite*.

irregular galaxy A galaxy without regular or symmetric appearance. Compare *elliptical galaxy*, *spiral galaxy*, and *S0 galaxy*.

irregular moon A moon that has been captured by a planet. Some irregular moons revolve in the opposite direction from the rotation of the planet, and many are in distant, unstable orbits. Compare *regular moon*.

isotopes Forms of the same element with differing numbers of neutrons.

isotropic In cosmology, describing a universe whose properties observers find to be the same in all directions.

J

J See *joule*.

jansky (Jy) The basic unit of flux density. Units: watts per square meter per hertz ($W/m^2/Hz$).

jet 1. A stream of gas and dust ejected from a comet nucleus by solar heating. 2. A collimated linear feature of bright emission extending from a protostar or active galactic nucleus.

joule (J) A unit of energy or work. 1 J = 1 newton meter.

Jy See *jansky*.

K

K See *kelvin*.

K-T boundary See *Cretaceous-Tertiary boundary*.

KBO See *Kuiper Belt object*.

kelvin (K) The basic unit of the Kelvin scale of temperature.

Kelvin scale The temperature scale—defined by William Thomson, better known as Lord Kelvin (1824–1907)—that uses Celsius-sized degrees, but defines 0 K as absolute zero instead of as the melting point of water. Compare *Celsius* and *Fahrenheit*.

Kepler's first law A rule of planetary motion, inferred by Johannes Kepler, stating that planets move in orbits of elliptical shapes with the Sun at one focus.

Kepler's laws The three rules of planetary motion inferred by Johannes Kepler from the data acquired by Tycho Brahe.

Kepler's second law Also called *law of equal areas*. A rule of planetary motion, inferred by Johannes Kepler, stating that a line drawn from the Sun to a planet sweeps out equal areas in equal times as the planet orbits the Sun.

Kepler's third law Also called *harmonic law*. A rule of planetary motion inferred by Johannes Kepler that describes the relationship between the period of a planet's orbit and its distance from the Sun. The law states that the square of the period of a planet's orbit, measured in years, is equal to the cube of the semimajor axis of the planet's orbit, measured in astronomical units: $(P_{years})^2 = (A_{AU})^3$.

kinetic energy (E_K) The energy of an object due to its motions. $E_K = \frac{1}{2}mv^2$. Units: joules (J).

Kirkwood gap A gap in the main asteroid belt related to orbital resonances with Jupiter.

Kuiper Belt A disk-shaped population of comet nuclei extending from Neptune's orbit to perhaps several thousand astronomical units (AU) from the Sun. The highly populated innermost part of the Kuiper Belt has an outer edge approximately 50 AU from the Sun.

Kuiper Belt object (KBO) Also called *trans-Neptunian object*. An icy planetesimal (comet nucleus) that orbits within the Kuiper Belt beyond the orbit of Neptune.

L

Lagrangian equilibrium point One of five points of equilibrium in a system consisting of two massive objects in nearly circular orbit around a common center of mass. Only two Lagrangian points (L_4 and L_5) represent stable equilibrium. A third smaller body located at one of the five points will move in lockstep with the center of mass of the larger bodies.

lander An instrumented spacecraft designed to land on a planet or moon. Compare *rover*.

large-scale structure Observable aggregates on the largest scales in the universe, including galaxy groups, clusters, and superclusters.

latitude The angular distance north (+) or south (−) from the equatorial plane of a nearly spherical body.

law of equal areas See *Kepler's second law*.

law of gravitation See *universal law of gravitation*.

leap year A year that contains 366 days. Leap years occur every 4 years when the year is divisible by 4, correcting for the accumulated excess time in a normal year, which is approximately 365¼ days long.

Leonids A November meteor shower associated with the dust debris left by comet Tempel-Tuttle. Compare *Perseids*.

libration The apparent wobble of an orbiting body that is tidally locked to its companion (such as Earth's Moon) resulting from the fact that its orbit is elliptical rather than circular.

life A biochemical process in which living organisms can reproduce, evolve, and sustain themselves by drawing energy from their environment. All terrestrial life involves carbon-based chemistry, assisted by the self-replicating molecules ribonucleic acid (RNA) and deoxyribonucleic acid (DNA).

light All electromagnetic radiation, which comprises the entire electromagnetic spectrum.

light-year (ly) The distance that light travels in 1 year—about 9 trillion kilometers (km).

limb darkening The darker appearance caused by increased atmospheric absorption near the limb of a planet or star.

line of nodes 1. A line defined by the intersection of two orbital planes. 2. The line defined by the intersection of Earth's equatorial plane and the plane of the ecliptic.

lithosphere The solid, brittle part of Earth (or any planet or moon), including the crust and the upper part of the mantle. Compare *biosphere* and *hydrosphere*.

lithospheric plate A separate piece of Earth's lithosphere capable of moving independently. See also *continental drift* and *plate tectonics*.

Local Group The small group of galaxies of which the Milky Way and the Andromeda galaxies are members.

long-period comet A comet with an orbital period of greater than 200 years. Compare *short-period comet*.

longitudinal wave A wave that oscillates parallel to the direction of the wave's propagation. Compare *transverse wave*.

look-back time The time that it has taken the light from an astronomical object to reach Earth.

low-mass star A star with a main-sequence mass of less than about 3 solar masses ($M_\odot$). Compare *high-mass star* and *intermediate-mass star*.

luminosity The total flux emitted by an object. Unit: watts (W). See also *brightness*.

luminosity class A spectral classification based on stellar size, from the largest supergiants to the smallest white dwarfs.

luminosity-temperature-radius relationship A relationship among these three properties of stars indicating that if any two are known, the third can be calculated.

luminous matter Also called *normal matter*. Matter in galaxies—including stars, gas, and dust—that emits electromagnetic radiation. Compare *dark matter*.

lunar eclipse An eclipse that occurs when the Moon is partially or entirely in Earth's shadow. Compare *solar eclipse*.

lunar tide A tide on Earth that is due to the differential gravitational pull of the Moon. Compare *solar tide*. See also *tide* (definition 2).

ly See *light-year*.

M

M-type asteroid An asteroid that was once part of the metallic core of a larger, differentiated body that has since been broken into pieces; made mostly of iron and nickel. Compare *C-type asteroid* and *S-type asteroid*.

MACHO Literally, "massive compact halo object." MACHO's include brown dwarfs, white dwarfs, and black holes, which are candidates for being considered dark matter. Compare *WIMP*.

magma Molten rock, often containing dissolved gases and solid minerals.

magnetic field A field that is able to exert a force on a moving electric charge. Compare *electric field*.

magnetic force A force associated with, or caused by, the relative motion of charges. Compare *electric force*. See also *electromagnetic force*.

magnetosphere The region surrounding a planet that is filled with relatively intense magnetic fields and plasmas.

magnitude A system used by astronomers to describe the brightness or luminosity of stars. The brighter the star, the smaller its magnitude.

main asteroid belt See *asteroid belt*.

main sequence The strip on the H-R diagram where most stars are found. Main-sequence stars are fusing hydrogen to helium in their cores.

main-sequence lifetime The amount of time a star spends on the main sequence, fusing hydrogen into helium in its core.

main-sequence turnoff The location on the H-R diagram of a single-aged stellar population (such as a star cluster) where stars have just evolved off the main sequence. The position of the main-sequence turnoff is determined by the age of the stellar population.

mantle The solid portion of a rocky planet that lies between the crust and the core.

mare (pl. maria) A dark region on the Moon, composed of basaltic lava flows.

mass 1. Inertial mass: the property of matter that resists changes in motion. 2. Gravitational mass: the property of matter defined by its attractive force on other objects. According to general relativity, the two are equivalent.

mass-luminosity relationship An empirical relationship between the luminosity (L) and mass (M) of main-sequence stars expressed as a power law—for example, $L \propto M^{3.5}$.

mass transfer The transfer of mass from one member of a binary star system to its companion. Mass transfer occurs when one of the stars evolves to the point that it overfills its Roche lobe, so that its outer layers are pulled toward its binary companion.

massive element See *heavy element*.

matter 1. Objects made of particles that have mass, such as protons, neutrons, and electrons. 2. Anything that occupies space and has mass.

Maunder Minimum The period from 1645 to 1715, when very few sunspots were observed.

megabar A unit of pressure equal to 1 million bars.

mega-light-year (Mly) A unit of distance equal to 1 million light-years.

meridian The imaginary arc in the sky running from the horizon at due north through the zenith to the horizon at due south. The meridian divides the observer's sky into eastern and western halves. Compare *equator*.

meteor The incandescent trail produced by a small piece of interplanetary debris as it travels through the atmosphere at very high speeds. Compare *meteoroid* and *meteorite*.

meteor shower A larger-than-normal display of meteors, occurring when Earth passes through the orbit of a disintegrating comet, sweeping up its debris.

meteorite A *meteoroid* that survives to reach a planet's surface. Compare *meteor* and *meteoroid*.

meteoroid A small cometary or asteroidal fragment ranging in size from 100 micrometers (μm) to 100 meters. When entering a planetary atmosphere, the meteoroid creates a *meteor*, which is an atmospheric phenomenon. Compare *meteor* and *meteorite*; also *planetesimal* and *zodiacal dust*.

micrometer (μm) Also called *micron*. 10^{-6} meter; a unit of length used for the wavelength of electromagnetic radiation. Compare *nanometer*.

micron See *micrometer*.

microwave radiation Electromagnetic radiation with frequencies and photon energies occurring in the spectral region between those of infrared radiation and radio waves.

Milky Way Galaxy The galaxy in which our Sun and Solar System reside.

minor planet See *asteroid*.

minute of arc See *arcminute*.

Mly See *mega-light-year*.

μm See *micrometer*.

modern physics Usually, the physical principles, including relativity and quantum mechanics, that have been developed since James Maxwell's equations were published.

molecular cloud An interstellar cloud composed primarily of molecular hydrogen.

molecular-cloud core A dense clump within a molecular cloud that forms as the cloud collapses and fragments. Protostars form from molecular-cloud cores.

molecule Generally, the smallest particle of a substance that retains its chemical properties and is composed of two or more atoms. A very few types of molecules, such as helium, are composed of single atoms.

momentum The product of the mass and velocity of a particle. Units: kilograms times meters per second (kg m/s).

moon A less massive satellite orbiting a more massive object. Moons are found around planets, dwarf planets, asteroids, and Kuiper Belt objects.

multiverse A collection of parallel universes that together comprise all that is.

mutation In biology, an imperfect reproduction of self-replicating material.

N

N See *newton*.

nadir The point on the celestial sphere located directly below an observer, opposite the *zenith*.

nanometer (nm) One billionth (10^{-9}) of a meter; a unit of length used for the wavelength of light. Compare *micrometer*.

natural selection The process by which forms of structure, ranging from molecules to whole organisms, that are best adapted to their environment become more common than less well-adapted forms.

NCP See *north celestial pole*.

neap tide An especially weak tide that occurs around the time of the first or third quarter Moon when the gravitational forces of the Moon and the Sun on Earth are at right angles to each other, thus producing the least pronounced tides. Compare *spring tide*. See also *tide* (definition 2).

near-Earth asteroid An asteroid whose orbit brings it close to the orbit of Earth. See also *near-Earth object*.

near-Earth object (NEO) An asteroid, comet, or large meteoroid whose orbit intersects Earth's orbit.

nebula (pl. **nebulae**) A cloud of interstellar gas and dust, either illuminated by stars (bright nebula) or seen in silhouette against a brighter background (dark nebula).

nebular hypothesis The first plausible theory of the formation of the Solar System, proposed by Immanuel Kant in 1734. Kant hypothesized that the Solar System formed from the collapse of an interstellar cloud of rotating gas.

NEO See *near-Earth object*.

neutrino A very low-mass, electrically neutral particle emitted during beta decay. Neutrinos interact with matter only very feebly and so can penetrate through great quantities of matter.

neutrino cooling The process in which thermal energy is carried out of the center of a star by neutrinos rather than by electromagnetic radiation or convection.

neutron A subatomic particle having no net electric charge, and a rest mass and rest energy nearly equal to that of the proton. Compare *electron* and *proton*.

neutron star The neutron-degenerate remnant left behind by a Type II supernova.

new Moon The phase of the Moon in which the Moon is between Earth and the Sun, and from Earth we see only the side of the Moon not being illuminated by the Sun. Compare *full Moon*.

newton (N) The force required to accelerate a 1-kilogram (kg) mass at a rate of 1 meter per second per second (m/s²). Units: kilograms times meters per second squared (kg m/s²).

Newton's first law of motion The law, formulated by Isaac Newton, stating that an object will remain at rest or will continue moving along a straight line at a constant speed until an unbalanced force acts on it.

Newton's laws See *Newton's first law of motion*, *Newton's second law of motion*, and *Newton's third law of motion*.

Newton's second law of motion The law, formulated by Isaac Newton, stating that if an unbalanced force acts on a body, the body will have an acceleration proportional to the unbalanced force and inversely proportional to the object's mass: $a = F/m$. The acceleration will be in the direction of the unbalanced force.

Newton's third law of motion The law, formulated by Isaac Newton, stating that for every force there is an equal and opposite force.

nm See *nanometer*.

normal matter See *luminous matter*.

north celestial pole (NCP) The northward projection of Earth's rotation axis onto the celestial sphere. Compare *south celestial pole*.

North Pole The location in the Northern Hemisphere where Earth's rotation axis intersects the surface of Earth. Compare *South Pole*.

nova (pl. **novae**) A stellar explosion that results from runaway nuclear fusion in a layer of material on the surface of a white dwarf in a binary system.

nuclear burning Release of energy by fusion of low-mass elements.

nuclear fusion The combination of two less massive atomic nuclei into a single more massive atomic nucleus.

nucleosynthesis The formation of more massive atomic nuclei from less massive nuclei, either in the Big Bang (Big Bang nucleosynthesis) or in the interiors of stars (stellar nucleosynthesis).

nucleus (pl. **nuclei**) 1. The dense, central part of an atom. 2. The central core of a galaxy, comet, or other diffuse object.

O

objective lens The primary optical element in a telescope or camera that produces an image of an object.

oblateness The flattening of an otherwise spherical planet or star caused by its rapid rotation.

obliquity The inclination of a celestial body's equator to its orbital plane.

observational uncertainty The fact that real measurements are never perfect; all observations are uncertain by some amount.

Occam's razor The principle that the simplest hypothesis is the most likely; named after William of Occam (circa 1285–1349), the medieval English cleric to whom the idea is attributed.

Oort Cloud A spherical distribution of comet nuclei stretching from beyond the Kuiper Belt to more than 50,000 astronomical units (AU) from the Sun.

opacity A measure of how effectively a material blocks the radiation going through it.

open cluster A loosely bound group of a few dozen to a few thousand stars that formed together in the disk of a spiral galaxy. Compare *globular cluster*.

open universe An infinite universe with a negatively curved spatial structure (much like the surface of a saddle) such that the sum of the angles of a triangle is always less than 180°. Compare *closed universe* and *flat universe*.

orbit The path taken by one object moving around another object under the influence of their mutual gravitational or electric attraction.

orbital resonance A situation in which the orbital periods of two objects are related by a ratio of small integers.

orbiter A spacecraft that is placed in orbit around a planet or moon. Compare *flyby*.

organic Describing a substance, not necessarily of biological origin, that contains the element carbon.

P

P wave See *primary wave*.

pair production The creation of a particle-antiparticle pair from a source of electromagnetic energy.

paleomagnetism The record of Earth's magnetic field as preserved in rocks.

parallax 1. The apparent shift in the position of one object relative to another object, caused by the changing perspective of the observer. 2. In astronomy, the displacement in the apparent position of a nearby star caused by the changing location of Earth in its orbit.

parent element A radioactive element that decays to form more stable *daughter products*.

parsec (pc) The distance to a star with a parallax of 1 arcsecond using a base of 1 astronomical unit (AU). One parsec is approximately 3.26 light-years.

partial solar eclipse The type of eclipse that occurs when Earth passes through the penumbra of the Moon's shadow, so that the Moon blocks only a portion of the Sun's disk. Compare *annular solar eclipse* and *total solar eclipse*.

pc See *parsec*.

peculiar velocity The motion of a galaxy relative to the overall expansion of the universe.

penumbra (pl. penumbrae) 1. The outer part of a shadow, where the source of light is only partially blocked. 2. The region surrounding the *umbra* of a sunspot. The penumbra is cooler and darker than the surrounding surface of the Sun but not as cool or dark as the umbra of the sunspot.

penumbral lunar eclipse A lunar eclipse in which the Moon passes through the penumbra of Earth's shadow. Compare *total lunar eclipse*.

perihelion (pl. perihelia) The point in a solar orbit that is closest to the Sun. Compare *aphelion*.

period The time it takes for a regularly repetitive process to complete one cycle.

period-luminosity relationship The relationship between the period of variability of a pulsating variable star, such as a Cepheid or RR Lyrae variable, and the luminosity of the star. Longer-period Cepheid or RR Lyrae variables are more luminous than their shorter-period cousins.

Perseids A prominent August meteor shower associated with the dust debris left by comet Swift-Tuttle. Compare *Leonids*.

phase One of the various appearances of the sunlit surface of the Moon or a planet caused by the change in viewing location of Earth relative to both the Sun and the object. Examples include crescent phase and gibbous phase.

photino An elementary particle related to the photon. One of the leading candidates for cold dark matter.

photochemical reaction A chemical reaction driven by the absorption of electromagnetic radiation.

photodissociation The breaking apart of molecules into smaller fragments or individual atoms by the action of photons.

photoelectric effect An effect whereby electrons are emitted from a substance illuminated by photons above a certain critical frequency.

photometry The process of measuring the brightness of a source of light, generally over a specific range of wavelength.

photon Also called *quantum of light*. A discrete unit or particle of electromagnetic radiation. The energy of a photon is equal to Planck's constant (*h*) multiplied by the frequency (*f*) of its electromagnetic radiation: $E_{photon} = h \times f$. The photon is the particle that mediates the electromagnetic force.

photosphere The apparent surface of the Sun as seen in visible light.

physical law A broad statement that predicts a particular aspect of how the physical universe behaves and that is supported by many empirical tests. See also *theory*.

pixel The smallest picture element in a digital image array.

Planck era The early time, just after the Big Bang, when the universe as a whole must be described with quantum mechanics.

Planck spectrum Also called *blackbody spectrum*. The spectrum of electromagnetic energy emitted by a blackbody per unit area per second, which is determined only by the temperature of the object.

Planck's constant (*h*) The constant of proportionality between the energy of a photon and the frequency of the photon. This constant defines how much energy a single photon of a given frequency or wavelength has. Value: $h = 6.63 \times 10^{-34}$ joule-second.

planet 1. A large body that orbits the Sun or other star that shines only by light reflected from the Sun or star. 2. In the Solar System, a body that orbits the Sun, has sufficient mass for self-gravity to overcome rigid body forces so that it assumes a spherical shape, and has cleared smaller bodies from the neighborhood around its orbit. Compare *dwarf planet*.

planet migration The theory that a planet can move from its formation distance around its parent star to a different distance though gravitational interactions with other bodies or loss of orbital energy from interaction with gas in the protoplanetary disk.

planetary nebula The expanding shell of material ejected by a dying asymptotic giant branch star. A planetary nebula glows from fluorescence caused by intense ultraviolet light coming from the hot, stellar remnant at its center.

planetary system A system of planets and other smaller objects in orbit around a star.

planetesimal A primitive body of rock and ice, 100 meters or more in diameter, that combines with others to form a planet. Compare *meteoroid* and *zodiacal dust*.

plasma A gas that is composed largely of charged particles but also may include some neutral atoms.

plate tectonics The geological theory concerning the motions of lithospheric plates, which in turn provides the theoretical basis for continental drift.

positron A positively charged subatomic particle; the antiparticle of the *electron*.

power The rate at which work is done or at which energy is delivered. Unit: watts (W) or joules per second (J/s).

precession of the equinoxes The slow change in orientation between the ecliptic plane and the celestial equator caused by the wobbling of Earth's axis.

pressure Force per unit area. Units: newtons per square meter (N/m^2) or bars.

primary atmosphere An atmosphere, composed mostly of hydrogen and helium, that forms at the same time as its host planet. Compare *secondary atmosphere*.

primary mirror The principal optical mirror in a reflecting telescope. The primary mirror determines the telescope's light-gathering power and resolution. Compare *secondary mirror*.

primary wave Also called *P wave*. A longitudinal seismic wave, in which the oscillations involve compression and decompression parallel to the direction of travel (that is, a pressure wave). Compare *secondary wave*.

principle A general idea or sense about how the universe is that guides us in constructing new scientific theories. Principles can be testable theories.

prograde motion 1. Rotational or orbital motion of a moon that is in the same sense as the planet it orbits. 2. The counterclockwise orbital motion of Solar System objects as seen from above Earth's orbital plane. Compare *retrograde motion*.

prominence An archlike projection above the solar photosphere often associated with a sunspot.

proportional Describing two things whose ratio is a constant.

proton (*p* or *p*⁺) A subatomic particle having a positive electric charge of 1.6×10^{-19} coulomb (C), a mass of 1.67×10^{-27} kilograms (kg), and a rest energy of 1.5×10^{-10} joules (J). Compare *electron* and *neutron*.

proton-proton chain One of the ways in which hydrogen burning can take place. This is the most important path for hydrogen burning in low-mass stars such as the Sun. See also *carbon-nitrogen-oxygen cycle* and *triple-alpha process*.

protoplanetary disk Also called *circumstellar disk*. The remains of the accretion disk around a young star from which a planetary system may form.

protostar A young stellar object that derives its luminosity from the conversion of gravitational energy to thermal energy, rather than from nuclear reactions in its core.

pulsar A rapidly rotating neutron star that beams radiation into space in two searchlight-like beams. To a distant observer, the star appears to flash on and off, earning its name.

pulsating variable star A variable star that undergoes periodic radial pulsations.

Q

QCD See *quantum chromodynamics*.

QED See *quantum electrodynamics*.

quantized Describing a quantity that exists as discrete, irreducible units.

quantum chromodynamics (QCD) The quantum mechanical theory describing the strong nuclear force and its mediation by gluons. Compare *quantum electrodynamics*.

quantum efficiency The fraction of photons falling on a detector that actually produces a response in the detector.

quantum electrodynamics (QED) The quantum theory describing the electromagnetic force and its mediation by photons. Compare *quantum chromodynamics*.

quantum mechanics The branch of physics that deals with the quantized and probabilistic behavior of atoms and subatomic particles.

quantum of light See *photon*.

quark The building block of protons and neutrons.

quasar Short for *quasi-stellar radio* source. The most luminous of the active galactic nuclei, seen only at great distances from our galaxy.

R

radial velocity The component of velocity that is directed toward or away from the observer.

radian The angle at the center of a circle subtended by an arc equal to the length of the circle's radius. Therefore, 2π radians equals 360°, and 1 radian equals approximately 57.3°.

radiant The direction in the sky from which the meteors in a meteor shower seem to come.

radiation belt A toroidal ring of high-energy particles surrounding a planet.

radiative transfer The transport of energy from one location to another by electromagnetic radiation.

radiative zone A region in the interior of a star through which energy is transported outward by radiation.

radio galaxy A type of elliptical galaxy with an active galactic nucleus at its center and having very strong emission (10^{35}–10^{38} watts [W]) in the radio part of the electromagnetic spectrum. Compare *Seyfert galaxy*.

radio telescope An instrument for detecting and measuring radio frequency emissions from celestial sources.

radio wave Electromagnetic radiation in the extreme long-wavelength region of the spectrum, beyond the region of microwaves.

radiometric dating Use of radioactive decay to measure the ages of materials such as minerals.

ratio The relationship in quantity or size between two or more things.

ray 1. A beam of electromagnetic radiation. 2. A bright streak emanating from a young impact crater.

recombination 1. The combining of ions and electrons to form neutral atoms. 2. An event early in the evolution of the universe in which hydrogen and helium nuclei combined with electrons to form neutral atoms. The removal of electrons caused the universe to become transparent to electromagnetic radiation.

red giant A low-mass star that has evolved beyond the main sequence and is now fusing hydrogen in a shell surrounding a degenerate helium core.

red giant branch A region on the H-R diagram defined by low-mass stars evolving from the main sequence toward the horizontal branch.

reddening The effect by which stars and other objects, when viewed through interstellar dust, appear redder than they actually are. Reddening is caused by the fact that blue light is more strongly absorbed and scattered than red light.

redshift Also called *Doppler redshift*. The shift toward longer wavelengths of light by any of several effects, including Doppler shifts, gravitational redshift, or cosmological redshift. Compare *blueshift*.

reflecting telescope A telescope that uses mirrors for collecting and focusing incoming electromagnetic radiation to form an image in their focal planes. The size of a reflecting telescope is defined by the diameter of the primary mirror. Compare *refracting telescope*.

reflection The redirection of a beam of light that is incident on, but does not cross, the surface between two media having different refractive indices. If the surface is flat and smooth, the angle of incidence equals the angle of reflection. Compare *refraction*.

refracting telescope A telescope that uses objective lenses to collect and focus light. Compare *reflecting telescope*.

refraction The redirection or bending of a beam of light when it crosses the boundary between two media having different refractive indices. Compare *reflection*.

refractory material Material that remains solid at high temperatures. Compare *volatile material*.

regular moon A moon that formed together with the planet it orbits. Compare *irregular moon*.

relative humidity The amount of water vapor held by a volume of air at a given temperature compared (stated as a percentage) to the total amount of water that could be held by the same volume of air at the same temperature.

relative motion The difference in motion between two individual frames of reference.

relativistic Describing physical processes that take place in systems traveling at nearly the speed of light or located in the vicinity of very strong gravitational fields.

relativistic beaming The effect created when material moving at nearly the speed of light beams the radiation it emits in the direction of its motion.

remote sensing The use of images, spectra, radar, or other techniques to measure the properties of an object from a distance.

resolution The ability of a telescope to separate two point sources of light. Resolution is determined by the telescope's aperture and the wavelength of light it receives.

rest wavelength The wavelength of light we see coming from an object at rest with respect to the observer.

retrograde motion 1. Rotation or orbital motion of a moon that is in the opposite sense to the rotation of the planet it orbits. 2. The clockwise orbital motion of Solar System objects as seen from above Earth's orbital plane. Compare *prograde motion*.

ring An aggregation of small particles orbiting a planet or star. The rings of the four giant planets of the Solar System are composed variously of silicates, organic materials, and ices.

ring arc A discontinuous, higher-density region within an otherwise continuous, narrow ring.

ringlet A narrowly confined concentration of ring particles.

Roche limit The distance at which a planet's tidal forces exceed the self-gravity of a smaller object, such as a moon, asteroid, or comet, causing the object to break apart.

Roche lobe The hourglass- or figure eight-shaped volume of space surrounding two stars, which constrains material that is gravitationally bound by one or the other.

rotation curve A plot showing how the orbital velocity of stars and gas in a galaxy changes with radial distance from the galaxy's center.

rover A remotely controlled instrumented vehicle designed to traverse and explore the surface of a terrestrial planet or moon. Compare *lander*.

RR Lyrae variable A variable giant star whose regularly timed pulsations are good predictors of its luminosity. RR Lyrae stars are used for distance measurements to globular clusters.

S

S-type asteroid An asteroid made of material that has been modified from its original state, likely as the outer part of a larger, differentiated body that has since broken into pieces. Compare *C-type asteroid* and *M-type asteroid*.

S wave See *secondary wave*.

S0 galaxy A galaxy with a bulge and a disklike spiral, but smooth in appearance like ellipticals. Compare *elliptical galaxy*, *spiral galaxy*, and *irregular galaxy*.

satellite 1. An object in orbit about a more massive body. 2. A moon.

scale factor (R_U) A dimensionless number proportional to the distance between two points in space. The scale factor increases as the universe expands.

scattering The random change in the direction of travel of photons, caused by their interactions with molecules or dust particles.

scientific method The formal procedure—including hypothesis, prediction, and experiment or observation—used to test (attempt to falsify) the validity of scientific hypotheses and theories.

scientific notation The standard expression of numbers with one digit (which can be zero) to the left of the decimal point and multiplied by 10 to the exponent required to give the number its correct value. Example: $2.99 \times 10^8 = 299,000,000$.

SCP See *south celestial pole*.

second law of thermodynamics The law stating that the entropy or disorder of an isolated system always increases as the system evolves.

second of arc See *arcsecond*.

secondary atmosphere An atmosphere that formed—as a result of volcanism, comet impacts, or another process—sometime after its host planet formed. Compare *primary atmosphere*.

secondary crater A crater formed from ejecta thrown from an *impact crater*.

secondary mirror A small mirror placed on the optical axis of a reflecting telescope that returns the beam back through a small hole in the *primary mirror*, thereby shortening the mechanical length of the telescope.

secondary wave Also called *S wave*. A transverse seismic wave, which involves the sideways motion of material. Compare *primary wave*.

seismic wave A vibration due to an earthquake, a large explosion, or an impact on the surface that travels through a planet's interior.

seismometer An instrument that measures the amplitude and frequency of seismic waves.

self-gravity The gravitational attraction among all the parts of the same object.

semimajor axis Half of the longer axis of an ellipse.

SETI The Search for Extraterrestrial Intelligence project, which uses advanced technology combined with radio telescopes to search for evidence of intelligent life elsewhere in the universe.

Seyfert galaxy A type of spiral galaxy with an active galactic nucleus at its center; first discovered in 1943 by Carl Seyfert. Compare *radio galaxy*.

shepherd moon A moon that orbits close to rings and gravitationally confines the orbits of the ring particles.

shield volcano A volcano formed by very fluid lava flowing from a single source and spreading out from that source. Compare *composite volcano*.

short-period comet A comet with an orbital period of less than 200 years. Compare *long-period comet*.

sidereal period An object's orbital or rotational period measured with respect to the stars. Compare *synodic period*.

silicate One of the family of minerals composed of silicon and oxygen in combination with other elements.

singularity The point where a mathematical expression or equation becomes meaningless, such as the denominator of a fraction approaching zero. See also *black hole*.

solar abundance The relative amount of an element detected in the atmosphere of the Sun, expressed as the ratio of the number of atoms of that element to the number of hydrogen atoms.

solar day The 24-hour period of Earth's axial rotation that brings the Sun back to the same local meridian where the rotation started.

solar eclipse An eclipse that occurs when the Sun is partially or entirely blocked by the Moon. Compare *lunar eclipse*.

solar flare Explosive events on the Sun's surface associated with complex sunspot groups and strong magnetic fields.

solar maximum (pl. **maxima**) The time, occurring about every 11 years, when the Sun is at its peak activity, meaning that sunspot activity and related phenomena (such as prominences, flares, and coronal mass ejections) are at their peak.

solar neutrino problem The historical observation that only about a third as many neutrinos as predicted by theory seemed to be coming from the Sun.

Solar System The gravitationally bound system made up of the Sun, planets, dwarf planets, moons, asteroids, comets, and Kuiper Belt objects, along with their associated gas and dust.

solar tide A tide on Earth that is due to the differential gravitational pull of the Sun. Compare *lunar tide*. See also *tide* (definition 2).

solar wind The stream of charged particles emitted by the Sun that flows at high speeds through interplanetary space.

solstice Literally, "sun standing still." 1. One of the two most northerly and southerly points on the ecliptic. 2. Either of the two times of year (the *summer solstice* and *winter solstice*) when the Sun is at one of these two positions. Compare *equinox*.

south celestial pole (SCP) The southward projection of Earth's rotation axis onto the celestial sphere. Compare *north celestial pole*.

South Pole The location in the Southern Hemisphere where Earth's rotation axis intersects the surface of Earth. Compare *North Pole*.

spacetime The four-dimensional continuum in which we live, and which we experience as three spatial dimensions plus time.

special relativity See *special theory of relativity*.

special theory of relativity Sometimes referred to as simply *special relativity*. Einstein's theory explaining how the fact that the speed of light is a constant affects nonaccelerating frames of reference. Compare *general theory of relativity*.

spectral type A classification system for stars based on the presence and relative strength of absorption lines in their spectra. Spectral type is related to the surface temperature of a star.

spectrograph A device that spreads out the light from an object into its component wavelengths. See also *spectrometer*.

spectrometer A *spectrograph* in which the spectrum is generally recorded digitally by electronic means.

spectroscopic binary A binary star pair whose existence and properties are revealed only by the Doppler shift of its spectral lines. Most spectroscopic binaries are close pairs. Compare *eclipsing binary* and *visual binary*.

spectroscopic parallax Use of the spectroscopically determined luminosity and the observed brightness of a star to determine the star's distance.

spectroscopy The study of electromagnetic radiation from an object in terms of its component wavelengths.

spectrum (pl. **spectra**) 1. The intensity of electromagnetic radiation as a function of wavelength. 2. Waves sorted by wavelength.

speed The rate of change of an object's position with time, without regard to the direction of movement. Units: meters per second (m/s) or kilometers per hour (km/h). Compare *velocity*.

spherically symmetric Describing an object whose properties depend only on distance from the object's center, so that the object has the same form viewed from any direction.

spin-orbit resonance A relationship between the orbital and rotation periods of an object such that the ratio of their periods can be expressed by simple integers.

spiral density wave A stable, spiral-shaped change in the local gravity of a galactic disk that can be produced by periodic gravitational kicks from neighboring galaxies or from nonspherical bulges and bars in spiral galaxies.

spiral galaxy A galaxy of Hubble type "S" class, with a discernible disk in which large spiral patterns exist. Compare *elliptical galaxy*, *S0 galaxy*, and *irregular galaxy*.

spoke One of several narrow radial features seen occasionally in Saturn's B ring. Spokes appear dark in backscattered light and bright in forward, scattering light, indicating that they are composed of tiny particles. Their origin is not well understood.

sporadic meteor A meteor that is not associated with a specific meteor shower.

spreading center A zone from which two tectonic plates diverge.

spring tide An especially strong tide that occurs near the time of a new or full Moon, when lunar tides and solar tides reinforce each other. Compare *neap tide*. See also *tide* (definition 2).

stable equilibrium An equilibrium state in which the system returns to its former condition after a small disturbance. Compare *unstable equilibrium*.

standard candle An object whose luminosity either is known or can be predicted in a distance-independent way, so its brightness can be used to determine its distance via the inverse square law of radiation.

standard model The theory of particle physics that combines electroweak theory with quantum chromodynamics to describe the structure of known forms of matter.

star A luminous ball of gas that is held together by gravity. A normal star is powered by nuclear reactions in its interior.

star cluster A group of stars that all formed at the same time and in the same general location.

static equilibrium A state in which the forces within a system are all in balance so that the system does not change. Compare *dynamic equilibrium*.

Stefan-Boltzmann constant (σ) The proportionality constant that relates the flux emitted by an object to the fourth power of its absolute temperature. Value: $5.67 \times 10^{-8} \, \text{W/(m}^2 \, \text{K}^4)$ (W = watts, m = meters, K = kelvin).

Stefan-Boltzmann law The law stating that the amount of electromagnetic energy emitted from the surface of a body, summed over the energies of all photons of all wavelengths emitted, is proportional to the fourth power of the temperature of the body.

stellar mass loss The loss of mass from the outermost parts of a star's atmosphere during the course of its evolution.

stellar occultation An event in which a planet or other Solar System body moves between the observer and a star, eclipsing the light emitted by that star.

stellar population A group of stars with similar ages, chemical compositions, and dynamic properties.

stereoscopic vision The way an animal's brain combines the different information from its two eyes to perceive the distances to objects around it.

stony-iron meteorite A meteorite consisting of a mixture of silicate minerals and iron-nickel alloys. Compare *iron meteorite* and *stony meteorite*.

stony meteorite A meteorite composed primarily of silicate minerals, similar to those found on Earth. Compare *iron meteorite* and *stony-iron meteorite*.

stratosphere The atmospheric layer immediately above the *troposphere*. On Earth it extends upward to an altitude of 50 kilometers (km).

strong nuclear force The attractive short-range force between protons and neutrons that holds atomic nuclei together; one of the four fundamental forces of nature, mediated by the exchange of gluons. Compare *weak nuclear force*.

subduction zone A region where two tectonic plates converge, with one plate sliding under the other and being drawn downward into the interior.

subgiant A giant star smaller and lower in luminosity than normal giant stars of the same spectral type. Subgiants evolve to become giants.

subgiant branch A region of the H-R diagram defined by stars that have left the main sequence but have not yet reached the red giant branch.

sublimation The process in which a solid becomes a gas without first becoming a liquid.

subsonic Moving within a medium at a speed slower than the speed of sound in that medium. Compare *supersonic*.

summer solstice 1. One of two points where the Sun is at its greatest distance from the celestial equator. 2. The day on which the Sun appears at this location, marking the first day of summer (about June 21 in the Northern Hemisphere and December 22 in the Southern Hemisphere). Compare *winter solstice*.

sungrazer A comet whose perihelion is within a few solar diameters of the surface of the Sun.

sunspot A cooler, transitory region on the solar surface produced when loops of magnetic flux break through the surface of the Sun.

sunspot cycle The approximate 11-year cycle during which sunspot activity increases and then decreases. This is one-half of a full 22-year cycle, in which the magnetic polarity of the Sun first reverses and then returns to its original configuration.

supercluster A large conglomeration of galaxy clusters and galaxy groups; typically, more than 100 million light-years (Mly) in size and containing tens of thousands to hundreds of thousands of galaxies. Compare *galaxy cluster* and *galaxy group*.

superluminal motion The appearance (though not the reality) that a jet is moving faster than the speed of light.

supermassive black hole A black hole of 1,000 solar masses ($M_\odot$) or more that resides in the center of a galaxy, and whose gravity powers active galactic nuclei.

supernova (pl. **supernovae**) A stellar explosion resulting in the release of tremendous amounts of energy, including the high-speed ejection of matter into the interstellar medium. See also *Type I supernova* and *Type II supernova*.

supersonic Moving within a medium at a speed faster than the speed of sound in that medium. Compare *subsonic*.

superstring theory The theory that conceives of particles as strings in 11 dimensions of space and time; the current contender for a theory of everything.

surface brightness The amount of electromagnetic radiation emitted or reflected per unit area.

surface wave A seismic wave that travels on the surface of a planet or moon.

symmetry 1. The property that an object has if the object is unchanged by rotation or reflection about a particular point, line, or plane. 2. In theoretical physics, the correspondence of different aspects of physical laws or systems, such as the symmetry between matter and antimatter.

synchronous rotation The case in which the period of rotation of a body on its axis equals the period of revolution in its orbit around another body. A special type of spin-orbit resonance.

synchrotron radiation Radiation from electrons moving at close to the speed of light as they spiral in a strong magnetic field; named because this kind of radiation was first identified on Earth in particle accelerators called "synchrotrons."

synodic period An object's orbital or rotational period measured with respect to the Sun. Compare *sidereal period*.

T

T Tauri star A young stellar object that has dispersed enough of the material surrounding it to be seen in visible light.

tail A stream of gas and dust swept away from the coma of a comet by the solar wind and by radiation pressure from the Sun.

tectonism Deformation of the lithosphere of a planet.

telescope The basic tool of astronomers. Working over the entire range from gamma rays to radio, astronomical telescopes collect and concentrate electromagnetic radiation from celestial objects.

temperature A measure of the average kinetic energy of the atoms or molecules in a gas, solid, or liquid.

terrestrial planet An Earth-like planet, made of rock and metal and having a solid surface. In our Solar System, the terrestrial planets are Mercury, Venus, Earth, and Mars. Compare *giant planet*.

theoretical model A detailed description of the properties of a particular object or system in terms of known physical laws or theories. Often, a computer calculation of predicted properties based on such a description.

theory A well-developed idea or group of ideas that are tied solidly to known physical laws and make testable predictions about the world. A very well-tested theory may be called a *physical law*, or simply a fact. Compare *hypothesis*.

theory of everything (TOE) A theory that unifies all four fundamental forces of nature: strong nuclear, weak nuclear, electromagnetic, and gravitational forces.

thermal conduction The transfer of energy in which the thermal energy of particles is transferred to adjacent particles by collisions or other interactions. Conduction is the most important way that thermal energy is transported in solid matter.

thermal energy The energy that resides in the random motion of atoms, molecules, and particles, by which we measure their temperature.

thermal equilibrium The state in which the rate of thermal-energy emission by an object is equal to the rate of thermal-energy absorption.

thermal motion The random motion of atoms, molecules, and particles that gives rise to thermal radiation.

thermal radiation Electromagnetic radiation resulting from the random motion of the charged particles in every substance.

third quarter Moon The phase of the Moon in which only the eastern half of the Moon, as viewed from Earth, is illuminated by the Sun. It occurs about one week after the full Moon. Compare *first quarter Moon*.

tidal bulge A distortion of a body resulting from tidal stresses.

tidal locking Synchronous rotation of an object caused by internal friction as the object rotates through its tidal bulge.

tidal stress Stress due to differences in the gravitational force of one mass on different parts of another mass.

tide 1. The deformation of a mass due to differential gravitational effects of one mass on another because of the extended size of the masses. 2. On Earth, the rise and fall of the oceans as Earth rotates through a tidal bulge caused by the Moon and Sun.

time dilation The relativistic "stretching" of time. Compare *general relativistic time dilation*.

TNO See *trans-Neptunian object*.

TOE See *theory of everything*.

topographic relief The differences in elevation from point to point on a planetary surface.

tornado A violent rotating column of air, typically 75 meters across with 200-kilometer-per-hour (km/h) winds. Some tornadoes can be more than 3 km across, and winds up to 500 km/h have been observed.

total lunar eclipse A lunar eclipse in which the Moon passes through the umbra of Earth's shadow. Compare *penumbral lunar eclipse*.

total solar eclipse The type of eclipse that occurs when Earth passes through the umbra of the Moon's shadow, so that the Moon completely blocks the disk of the Sun. Compare *annular solar eclipse* and *partial solar eclipse*.

transform fault The actively slipping segment of a fracture zone between lithospheric plates.

trans-Neptunian object (TNO) See *Kuiper Belt object*.

transverse wave A wave that oscillates perpendicular to the direction of the wave's propagation. Compare *longitudinal wave*.

triple-alpha process The nuclear fusion reaction that combines three helium nuclei (alpha particles) together into a single nucleus of carbon. See also *carbon-nitrogen-oxygen cycle* and *proton-proton chain*.

Trojan asteroid One of a group of asteroids orbiting in the L_4 and L_5 Lagrangian points of Jupiter's orbit.

tropical year The time between one crossing of the vernal equinox and the next. Because of the precession of the equinoxes, a tropical year is slightly shorter than the time that it takes for Earth to orbit once about the Sun.

Tropics The region on Earth between latitudes 23.5° south and 23.5° north, and in which the Sun appears directly overhead twice during the year.

tropopause The top of a planet's troposphere.

troposphere The convection-dominated layer of a planet's atmosphere. On Earth, the atmospheric region closest to the ground within which most weather phenomena take place. Compare *stratosphere*.

tuning fork diagram The two-pronged diagram showing Hubble's classification of galaxies into ellipticals, S0s, spirals, barred S0s and spirals, and irregular galaxies.

turbulence The random motion of blobs of gas within a larger cloud of gas.

Type I supernova A supernova explosion in which no trace of hydrogen is seen in the ejected material. Most Type I supernovae are thought to be the result of runaway carbon burning in a white dwarf star onto which material is being deposited by a binary companion.

Type II supernova A supernova explosion in which the degenerate core of an evolved massive star suddenly collapses and rebounds.

typhoon See *hurricane*.

U

ultraviolet (UV) radiation Electromagnetic radiation with frequencies and photon energies greater than those of visible light but less than those of X-rays, and wavelengths shorter than those of visible light but longer than those of X-rays. Compare *infrared radiation*.

umbra (pl. **umbrae**) 1. The darkest part of a shadow, where the source of light is completely blocked. 2. The darkest, innermost part of a sunspot. Compare *penumbra*.

unbalanced force The nonzero net force acting on a body.

unbound orbit An orbit in which the velocity is greater than the escape velocity. Compare *bound orbit*.

uncertainty principle See *Heisenberg uncertainty principle*.

unified model of AGNs A model in which many different types of activity in the nuclei of galaxies are all explained by accretion of matter around a supermassive black hole.

uniform circular motion Motion in a circular path at a constant speed.

unit A fundamental quantity of measurement—for example, metric units or English units.

universal gravitational constant (G) The constant of proportionality in the universal law of gravity. Value: $G = 6.673 \times 10^{-11}$ Newtons times meters squared per kilogram squared (N m²/kg²).

universal law of gravitation The law stating that the gravitational force between any two objects is proportional to the product of their masses and inversely proportional to the square of the distance between them: $F \propto (m_1 m_2/r^2)$.

universe All of space and everything contained therein.

unstable equilibrium An equilibrium state in which a small disturbance will cause a system to move away from equilibrium. Compare *stable equilibrium*.

UV Ultraviolet.

V

vacuum A region of space that contains very little matter. In quantum mechanics and general relativity, however, even a perfect vacuum has physical properties.

variable star A star with varying luminosity. Many periodic variables are found within the instability strip on the H-R diagram.

velocity The rate and direction of change of an object's position with time. Units: meters per second (m/s) or kilometers per hour (km/h). Compare *speed*.

vernal equinox 1. One of two points where the Sun crosses the celestial equator. 2. The day on which the Sun appears at this location, marking the first day of spring (about March 21 in the Northern Hemisphere and September 23 in the Southern Hemisphere). Compare *autumnal equinox*.

virtual particle A particle that, according to quantum mechanics, comes into existence only momentarily. According to theory, fundamental forces are mediated by the exchange of virtual particles.

visual binary A binary system in which the two stars can be seen individually from Earth. Compare *eclipsing binary* and *spectroscopic binary*.

void A region in space containing little or no matter. Examples include regions in cosmological space that are largely empty of galaxies.

volatile material Sometimes called *ice*. Material that remains gaseous at moderate temperature. Compare *refractory material*.

volcanism The occurrence of volcanic activity on a planet or moon.

vortex (pl. vortices) Any circulating fluid system. Specifically, 1. an atmospheric anticyclone or cyclone; 2. a whirlpool or eddy.

W

W See *watt*.

waning The changing phases of the Moon as it becomes less fully illuminated between full Moon and new Moon as seen from Earth. Compare *waxing*.

watt (W) A measure of *power*. Units: joules per second (J/s).

wave A disturbance moving along a surface or passing through a space or a medium.

wavefront The imaginary surface of an electromagnetic wave, either plane or spherical, oriented perpendicular to the direction of travel.

wavelength The distance on a wave between two adjacent points having identical characteristics. The distance a wave travels in one period. Unit: meter.

waxing The changing phases of the Moon as it becomes more fully illuminated between new Moon and full Moon as seen from Earth. Compare *waning*.

weak nuclear force The force underlying some forms of radioactivity and certain interactions between subatomic particles. It is responsible for radioactive beta decay and for the initial proton-proton interactions that lead to nuclear fusion in the Sun and other stars. One of the four fundamental forces of nature, mediated by the exchange of *W* and *Z* particles. Compare *strong nuclear force*.

weight 1. The force equal to the mass of an object multiplied by the local acceleration due to gravity. 2. In general relativity, the force equal to the mass of an object multiplied by the acceleration of the frame of reference in which the object is observed.

white dwarf The stellar remnant left at the end of the evolution of a low-mass star. A typical white dwarf has a mass of 0.6 solar mass ($M_\odot$) and a size about equal to that of Earth; it is made of nonburning, electron-degenerate carbon.

Wien's law A relationship describing how the peak wavelength, and therefore the color, of electromagnetic radiation from a glowing blackbody changes with temperature.

WIMP Literally, "weakly interacting massive particle." A hypothetical massive particle that interacts through gravity but not with electromagnetic radiation and is a candidate for dark matter. Compare *MACHO*.

winter solstice 1. One of two points where the Sun is at its greatest distance from the celestial equator. 2. The day on which the Sun appears at this location, marking the first day of winter (about December 22 in the Northern Hemisphere and June 21 in the Southern Hemisphere). Compare *summer solstice*.

X

X-ray Electromagnetic radiation having frequencies and photon energies greater than those of ultraviolet light but less than those of gamma rays, and wavelengths shorter than those of UV light but longer than those of gamma rays.

X-ray binary A binary system in which mass from an evolving star spills over onto a collapsed companion, such as a neutron star or black hole. The material falling in is heated to such high temperatures that it glows brightly in X-rays.

Y

year The time it takes Earth to make one revolution around the Sun. A solar year is measured from equinox to equinox. A sidereal year, Earth's true orbital period, is measured relative to the stars.

Z

zenith The point on the celestial sphere located directly overhead from an observer. Compare *nadir*.

zodiac The 12 constellations lying along the plane of the ecliptic.

zodiacal dust Particles of cometary and asteroidal debris less than 100 micrometers (μm) in size that orbit the inner Solar System close to the plane of the ecliptic. Compare *meteoroid* and *planetesimal*.

zodiacal light A band of light in the night sky caused by sunlight reflected by zodiacal dust.

zonal wind The planetwide circulation of air that moves in directions parallel to the planet's equator.

Credits

These pages contain credits for all 23 chapters in the complete volume of *21st Century Astronomy*, Third Edition. The Solar System Edition does not include Chapters 15–22. The Stars and Galaxies Edition does not include Chapters 6–12.

Chapter 1

Opener NASA, ESA, Hubble Heritage (STScI/AURA); D. Carter (LJMU) et al. and the Coma HST ACS Treasury Team. **1.2a, b** Neil Ryder Hoos; **1.2c** Owen Franken/Corbis; **1.2d** Monkey Business Images/Dreamstime.com; **1.2e** PhotoDisc; **1.2f** American Museum of Natural History; **1.2g** American Museum of Natural History; **1.2h** NASA/JPL/Caltech. **1.3** Bettmann/Corbis. **1.4 left** NASA/STScI/Arizona State University/Hester/Ressmeyer/Corbis. **1.4 right** Ron Watts/Corbis. **1.5a** Michael J. Tuttle (NASM/NASA); **1.5b** NSSDC. **1.6** NRAO/AUI, James J. Condon, John J. Broderick, and George A. Seielstad. **1.7** Fermilab. **1.9 top left** Special Relativity: The M.I.T. Introductory Physics Series by A. P. French. ©1968, 1966 by Massachusetts Institute of Technology. Used by permission of W. W. Norton & Company, Inc; **1.9 center** Seth Joel/Corbis; **1.9 bottom left** Musée d'Orsay, Paris, France. Photo: Réunion des Musées Nationaux/Art Resource, NY. Photograph by Herve Lewandowski; **1.9 bottom right** Robbie Jack/Corbis. **1.10** Bettmann/Corbis. **1.11a** Howard Bond (Space Telescope Science Institute), Robin Ciardullo (Pennsylvania State University) and NASA; **1.11b** NASA and The Hubble Heritage Team (STScI/AURA). **1.12** Four seasons photographs, Jim Schwabel/Index Stock Imagery/PictureQuest. **1.13** NON SEQUITUR by Wiley Miller. Dist. by Universal Press Syndicate. Reprinted with permission. All rights reserved. **1.14** Craig Lovell/Corbis.

Chapter 2

Opener Anthony Ayiomamitis (http://www.perseus.gr). **2.1** British Library. **2.8 left** Pekka Parviainen/Photo Researchers, Inc.; **2.8 right** David Nunuk/Photo Researchers, Inc. **2.14a** Larry Landolfi/Photo Research; **2.14b** Larry Landolfi/Photo Research. **2.16 top left** AFP/Getty Images; **2.16 top right** AP Images; **2.16 bottom left** Lindsay Hebberd/Corbis. **2.16 bottom right** Lynn Goldsmith/Corbis. **2.21** Roger Ressmeyer/Corbis. **2.23** Nick Quinn. **2.24** Dennis L. Mammana; **2.25a** Reuters New Media Inc./Corbis; **2.25b** ©2001 by Fred Espenak, courtesy of www.MrEclipse.com. **2.27** Fred Espenak NASA/GSFC. **2.28a** Johannes Schedler; **2.28b** Anthony Ayiomamitis (TWAN).

Chapter 3

Opener NASA; **3.1** Nicolaus Copernicus (1473–1543) (oil on canvas), Pomeranian School, (16th century)/Nicolaus Copernicus Museum, Frombork, Poland/Lauros/Giraudon/The Bridgeman Art Library. **3.2** SSPL/The Image Works. **3.3a** Science Photo Library. **3.4** Bettmann/Corbis. **3.5** Tunç Tezel. **3.6** The Granger Collection, NY. **3.12** Lebrecht Music and Arts Photo Library/Alamy. **3.13** Portrait of Galileo Galilei (1564–1642) (oil on canvas), Sustermans, Justus (1597–1681) (school of)/Galleria Palatina, Palazzo Pitti, Florence, Italy/The Bridgeman Art Library. **3.18** Courtesy of Alan Bean.

Chapter 4

Opener David Malin. **4.7** Eyewire Collection/Getty Images. **4.14b** Tom Pantages. **4.24** NASA/NSSDC/GSFC.

Chapter 5

Opener NASA/JPL. **5.1** Junenoire Photography. **5.2** Gianni Tortoli/Photo Researchers, Inc. **5.3** Jim Sugar/Corbis. **5.4a** Yerkes Observatory photographs: Richard Dreiser; **5.4b** W. M. Keck Observatory/CARA. **5.5** ESO. **5.8b** ASU Physics Instructional Resource Team. **5.9b** ASU Physics Instructional Resource Team. **5.11** ASU Physics Instructional Resource Team. **5.13c** Steve Percival/Photo Researchers, Inc. **5.14** ASU Physics Instructional Resource Team. **5.17a, b top** Keck Observatory; **5.17a, b bottom** Gemini Observatory, US National Science Foundation, and University of Hawaii Institute for Astronomy. **5.18 left to right** NASA; NASA/CXC/SAO; Graphic courtesy Orbital Sciences Corp; NOAO/AURA/NSF; NASA/JPL/Caltech; Joint Astronomy Center in Hilo, Hawaii; NRAO VLA Image Gallery; The National Radio Astronomy Observatory, Green Bank; Dave Finley, courtesy National Radio Astronomy Observatory and Associated Universities, Inc. **5.19a** Science Photo Library; **5.19b** Hulton-Deutsch Collection/Corbis. **5.20** NASA/IPAC/NED/JPL/Caltech/SSDS. **5.21b** Jean-Charles Cuillandre (CFHT); **5.21c** Junenoire Photography. **5.22a** Courtesy of Ed Grafton; **5.22b** Steve Larson/University of Arizona. **5.24a** Roger Ressmeyer/Corbis; **5.24b** David Parker/Photo Researchers, Inc. **5.25** Photo by Dave Finley, courtesy National Radio Astronomy Observatory and Associated Universities, Inc. **5.26** ESO. **5.27** NASA/NSSDC. **5.28 a, b** NASA/JPL. **5.29** U.S. Department of Energy. **5.30** CERN. **5.31** Frank Summers, Space Telescope Science Institute; Chris Mihos, Case Western Reserve University; Lars Hernquist, Harvard University.

Chapter 6

Opener Caltech/NASA/JPL. **6.2a** STScI photo: Karl Stapelfeldt (JPL); **6.2 b–d** D. Padgett (IPAC/Caltech), W. Brandner (IPAC), K. Stapelfeldt (JPL) and NASA. **6.3** Photograph by Pelisson, SaharaMet. **6.5** Reuters/Corbis. **6.12** NASA/Johns Hopkins University Applied Physics Laboratory/Carnegie Institution of Washington. **6.13** NASA, ESA, J. R. Graham and P. Kalas (University of California, Berkeley), and B. Matthews (Hertzberg Institute of

University of California, Irvine, for the Super-Kamiokande Collaboration; **14.9c** Lawrence Berkely National Laboratory/ Photo Researchers, Inc.; **14.9d** Evelyn Malkus/ IceCube Research. **14.10** NOAO/ AURA/NSF. **14.11** SOHO/ESA/NASA. **14.13** Nigel Sharp, NOAO/NSO/Kitt Peak FTS/AURA/NSF. **14.14a** Stefan Seip, www. astromeeting.de; **14.14b, c** 2001 by Fred Espenak, courtesy of www.MrEclipse. com. **14.15** NASA/Photo Researchers, Inc. **14.16** SOHO/ESA/NASA. **14.19** SOHO/ ESA/NASA. **14.20** T. Rimmele (NSO), M. Hanna (NOAO)/AURA/NSF, Göran Scharmer/Institute for Solar Physics of the Royal Swedish Academy of Sciences, Sweden. **14.22d** SOHO/ESA/NASA. **14.24a** SOHO/ESA/NASA; **14.24b** SOHO-EIT Consortium, ESA, NASA. **14.25a** SOHO/ EIT; **14.25b** SOHO/LASCO/ESA and NASA. **14.26** SOHO/ESA and NASA.

Chapter 15

Opener Canada-France-Hawaii Telescope/ J.-C. Cuillandre/Coelum. **15.1a** Image courtesy of Dr. A. Mellinger; **15.1b** NASA/ NSSDC/GSFC Diffuse Infrared Background Experiment (DIRBE) instrument on the Cosmic Background Explorer (COBE). **15.4a** NSSDC/GSFC Infrared Astronomical Satellite (IRAS); **15.4b** JAXA. **15.5** NSSDC/ GSFC Roentgen Satellite (ROSAT). **15.6** Dr. Douglas Finkbeiner. **15.7a** Courtesy of Stefan Seip; **15.7b** Courtesy of Emmanuel Mallart www.astrophoto-mallart.com & www.axisinstruments.com. **15.8** ESO Telescope Systems Division. **15.9** C. Jones, Smithsonian Astrophysical Observatory. **15.10a** Canada-France-Hawaii Telescope & Coelum; **15.10b, c** ESO. **15.15** Jeff Hester and Paul Scowen (Arizona State University), Bradford Smith (University of Hawaii), Roger Thompson (University of Arizona), and NASA. **15.19a inset** Karl Stapelfeldt/JPL/NASA; **15.19b** NASA/ JPL-Caltech/UIUC. **15.20a, b, c, d** Jeff Hester (Arizona State University), WFPC2 Team, NASA. **15.21a** Anglo-Australian Observatory/Royal Observatory, Edinburgh, photograph from UK Schmidt, plates by David Malin; **15.21b** NASA/JPL-Caltech/J. Stauffer (SSC/Caltech).

Chapter 16

Opener NASA, ESA and H. E. Bond (STScI). **16.11a** Minnesota Astronomical Society; **16.11b** NASA and The Hubble Heritage Team (STScI/AURA). **16.12 top left** Dr. Raghvendra Sahai (JPL) and Dr. Arsen R. Hajian (USNO), NASA and The Hubble

Heritage Team (STScI/AURA); **16.12 center left** A. Hajian (USNO) et al., Hubble Heritage Team (STScI/AURA), NASA; **16.12 bottom left** J. P. Harrington and K. J. Borkowski (University of Maryland), HST, NASA; **16.12 top right** NASA, A. Fruchter and the ERO team (STScI); **16.12 center right** Anglo-Australian Telescope, photograph by David Malin; **16.12 bottom right** Bruce Balick (University of Washington), Vincent Icke (Leiden University, The Netherlands), Garrelt Mellema (Stockholm University), and NASA.

Chapter 17

Opener NASA/HST/ASU/J. Hester. **17.7** N. Smith, J. A. Morse (U. Colorado) et al., NASA. **17.10a, b** Anglo-Australian Observatory, photographs by David Malin. **17.11a** Joachim Trumper, Max-Planck-Institut für extraterrestrische Physik; **17.11b** Courtesy Nancy Levenson and colleagues, American Astronomical Society; **17.11c** Jeff Hester, Arizona State University, and NASA. **17.15a** European Southern Observatory, ESO PR Photo 40f199; **17.15b, c** Jeff Hester, Arizona State University, and NASA. **17.31** ESA/Hubble.

Chapter 18

Opener NASA/WMAP Science Team. **18.1** NASA, ESA, S. Beckwith (STScI) and the HUDF Team. **18.2** Jason T. Ware/ Photo Researchers, Inc. **18.4** TIMEPIX/ Getty Images. **18.5** Courtesy of Jeff Hester. **18.5** NASA William C. Keel (University of Alabama, Tuscaloosa). **18.8** WIYN/NOAO/ NSF, WIYN Consortium, Inc. **18.17** Lucent Technologies' Bell Labs. **18.18** Ann and Rob Simpson. **18.20 a, b, c** NASA COBE Science Team.

Chapter 19

Opener NASA/ESA/STScI/Photo Researchers, Inc. **19.2 top left** NASA; **19.2 center left** NASA; **19.2 bottom left** NASA. **19.2 top right** Canada-France-Hawaii Telescope/J.-C. Cuillandre/Coelum. **19.2 center right** NASA; **19.2 bottom right** NASA. **19.3** The Hubble Heritage Team (AURA/ STScI/NASA). **19.4 top left** NASA, ESA, and the Hubble Heritage Team (STScI/AURA); **19.4 top right** NASA, ESA, A. Aloisi (STScI/ ESA), and the Hubble Heritage (STScI/ AURA)-ESA/Hubble Collaboration; **19.4 bottom left** NASA, ESA, and the Hubble Heritage (STScI/AURA)-ESA/Hubble Collaboration; **19.4 bottom right** NASA, ESA, and the Hubble Heritage Team (STScI/

AURA)-ESA/Hubble Collaboration. **19.5** NOAO/AURA/NSF. **19.6** David Malin. **19.7** NASA and the Hubble Heritage Team (STScI/AURA). **19.9a, b** European Southern Observatory (ESO). **19.10a, b** Johannes Schedler/Panther Observatory. **19.11a** NASA/JPL-Caltech; **19.11b** John Gleason, Celestial Images. **19.12a** Todd Boroson/NOAO/AURA/NSF; **19.12b** Richard Rand, University of New Mexico. **19.14** image100/SuperStock. **19.15** David Malin. **19.17** X-ray: NASA/CXC/Penn State/G. Garmire; Optical: NASA/ESA/STScI/M. West. **19.18 top left to right** Courtesy of Dr. Michael Brotherton; **19.18 bottom left and center left** NASA, ESA, ESO, Frédéric Courbin (Ecole Polytechnique Federale de Lausanne, Switzerland) & Pierre Magain (Universite de Liege, Belgium); **19.18 bottom right and center right** HST. **19.19** NRAO/ Alan Bridle et al. **19.22b** Courtesy Bill Keel, University of Alabama, and NASA; **19.22c** Holland Ford, STScI/Johns Hopkins University, and NASA; **19.22d** R. P. van der Marel, STScI, F. C. van den Bosch, University of Washington, and NASA. **19.23** F. Owen, NRAO, with J. Biretta, STScI, and J. Eilek, NMIMT. **19.24** Holland Ford, STScI/Johns Hopkins University; Richard Harms, Applied Research Corp.; Zlatan Tsvetanov, Arthur Davidsen, and Gerard Kriss at Johns Hopkins University; Ralph Bohlin and George Hartig at STScI; Linda Dressel and Ajay K. Kochhar at Applied Research Corp. in Landover, MD; and Bruce Margon from the University of Washington, Seattle/NASA. **19.25** John Biretta, STScI. **19.26** NASA, ESA, the Hubble Heritage (STScI/AURA)-ESA/Hubble Collaboration, and A. Evans (University of Virginia, Charlottesville/NRAO/Stony Brook University).

Chapter 20

Opener Tony Rowell/Astrophotostore. com. **20.1a** Axel Mellinger; **20.1b** C. Howk and B. Savage (University of Wisconsin); N. Sharp (NOAO)/WIYN, Inc. **20.2a** US Patent Office Library; **20.2b** NASA/ JPL-Caltech. **20.3** Hubble Heritage Team (AURA/STScI/NASA). **20.4a** NOAO/ AURA/NSF. **20.5** Courtesy of Jeff Hester. **20.8** Anglo-Australian Observatory. **20.11** The Electronic Universe Project. **20.16a** NASA/CXC/MIT/F. K. Baganoff; **20.16b** Hubble, NASA, ESA, & D. Q. Wang (U. Mass, Amherst); Spitzer: NASA, JPL, & S. Stolovy (SSC/Caltech). **20.17** This image was created by Prof. Andrea Ghez and her research team at UCLA and is from data sets obtained with the W. M. Keck

Telescopes. Image creators include Andrea Ghez, Angelle Tanner, Seth Hornstein, and Jessica Lu. **20.18** NRAO/AUI. **20.19** NASA/MIT/F.Baganoff. **20.20** Anglo-Austrailian Observatory, David Malin.

Chapter 21

Opener NASA/JPL/Origins. **21.4 top** NASA, ESA, The Hubble Key Project Team, and The High-Z Supernova Search Team; **21.4 bottom** Smithsonian Astrophysical Observatory, 1993; **21.4 center** NASA/WMAP Science Team. **21.5** Robert Kirshner, Harvard University. **21.14** Kamioka Observatory, ICRR (Institute for Cosmic Ray Research), The University of Tokyo. **21.15** From *The Elegant Universe* by Brian Greene. ©1999 by Brian R. Greene. Used by permission of W. W. Norton & Company, Inc

Chapter 22

Opener Robert Gendler, Jan-Erik Ovaldsen, Allan Hornstrup, IDA Image data: ESO/Danish 1.5m telescope at La Silla, Chile–2008. **22.3a** Anglo-Australian Observatory; **22.3b** Omar Lopez-Cruz and Ian Shelton/NOAO/AURA/NSF. **22.4a, b** Max Tegmark/SDSS Collaboration; **22.4c** Jordan Raddick/SDSS. **22.7a** NASA/CXC/MIT/E.-H Peng et al.; Optical: NASA/STScI; **22.7b** NASA; ESA; L. Bradley (Johns Hopkins University); R. Bouwens (University of California, Santa Cruz); H. Ford (Johns Hopkins University); and G. Illingworth (University of California, Santa Cruz); **22.7c** NASA/CXC/MIT/E.-H Peng et al.; NASA/STScI. **22.8b** NASA, A. Fruchter and the ERO team (STScI, ST-ECF), NASA, ESA; **22.8c** Faure (Zentrum Für Astronomie, University of Heidelberg) and J. P. Kneib (Laboratoire d'Astrophysique de Marseille). **22.9** NASA/CXC/CfA/M. Markevitch et al.; and NASA/STScI; Magellan/U. Arizona/D. Clowe et al.; Lensing Map: NASA/STScI; ESO WFI; Magellan/U. Arizona/D. Clowe et al. **22.10** Images produced by Donna Cox Smithsonian Institution and Motorola Corporation. **22.11** NASA, ESA, and N. Pirzkal (STScI/ESA). **22.13** NASA, ESA, and A. Feild (STScI).

Chapter 23

Opener SETI/NASA. **23.1** Jim Sugar/Corbis. **23.2a** Dr. Michael Perfit, University of Florida, Robert Embley/NOAA; **23.2b** Woods Hole Oceanographic Institution, Deep Submergence Operations Group, Dan Fornari. **23.3** Gaertner/Alamy. **23.4** Chris Boydell/Australian Picture Library/Corbis. **23.6** Courtesy of National Science Foundation. **23.7 top left** Mark A. Schneider/Visuals Unlimited; **23.7 center middle left** K. Sandved/Visuals Unlimited; **23.7 bottom left** Ken Lucas/Visuals Unlimited; **23.7 top right** Science VU/National Museum of Kenya/Visuals Unlimited; **23.7 middle right** Ken Lucas/Visuals Unlimited; **23.7 bottom right** Ken Lucas/Visuals Unlimited. **23.8** Kenneth E. Ward/Biografx/Photo Researchers, Inc. **23.9** Jeff J. Daly/Visuals Unlimited. **23.10** Kepler Mission NASA. **23.11** NASA. **23.12** NASA. **23.13** F. Drake (UCSC) et al., Arecibo Observatory (Cornell, NAIC). **23.14** Twentieth Century Fox Film Corporation/Photofest. **23.15** Seti/ NASA.

Index

This index contains page references for all 23 chapters in the complete volume of *21st Century Astronomy*, Third Edition. The Solar System Edition does not include Chapters 15–22 (pp. 416–629). The Stars and Galaxies Edition does not include Chapters 6–12 (pp. 154–359).

Page numbers in *italics* refer to illustrations. An italic *n* after a number refers to a footnote.